AF544921

MATHEMATICS *in* LIFE, SOCIETY, *and the* WORLD

HAROLD PARKS
GARY MUSSER
ROBERT BURTON
WILLIAM SIEBLER

OREGON STATE UNIVERSITY

PRENTICE HALL
UPPER SADDLE RIVER, NJ 07458

Library of Congress Cataloging-in-Publication Data
Mathematics in life, society, and the world / Gary Musser ... [et al.].
p. cm.
Includes index.
ISBN 0-02-385460-X (hard cover : alk. paper)
1. Mathematics. I. Musser, Gary L.
QA39.2.M3845 1997
510—DC20 96-26528
CIP

Acquisition Editor: *Sally Denlow*
Supplements Editor: *Audra Walsh*
Editorial Assistant: *Joanne Wendelken*
Editorial Director: *Tim Bozik*
Assistant Vice President of Production and Manufacturing: *David W. Riccardi*
Editorial/Production Supervision: *Jack Casteel*
Managing Editor: *Linda Mihatov Behrens*
Executive Managing Editor: *Kathleen Schiaparelli*
Manufacturing Buyer: *Alan Fischer*
Manufacturing Manager: *Trudi Pisciotti*
Marketing Manager: *Evan Girard*
Marketing Assistant: *Jennifer Pan*
Creative Director: *Paula Maylahn*
Art Director: *Amy Rosen*
Assistant to Art Director: *Rod Hernandez*
Art Manager: *Gus Vibal*
Copy Editing: *Phyllis Coyne, et al.*
Proofreading: *Laura Finney*
Illustration: *Academy Artworks, Inc.*
Photo Research: *Melinda Alexandria*
Cover Designer: *Tom Nery*
Cover Photographs: *Lori Adanski-Peek, John Ortner, Tony Stone Images;* Courtesy of *Tony Stone Images*
Interior Designer: *Circa 86*

© 1997 by **PRENTICE-HALL, INC.**
Simon & Schuster/A Viacom Company
Upper Saddle River, NJ 07458

All rights reserved. No part of this book may be reproduced, in any form or by any means, without permission in writing from the publisher.

Printed in the United States of America
10 9 8 7 6 5 4 3 2 1

ISBN 0-02-385460-X

Prentice-Hall International (UK) Limited, *London*
Prentice-Hall of Australia Pty. Limited, *Sydney*
Prentice-Hall Canada, Inc., *Toronto*
Prentice-Hall Hispanoamericana, S.A., *Mexico*
Prentice-Hall of India Private Limited, *New Delhi*
Prentice-Hall of Japan, Inc., *Tokyo*
Simon & Schuster Asia Pte. Ltd., *Singapore*
Editora Prentice-Hall do Brasil, Ltda., *Rio de Janeiro*

Dedications

To:
Susan, Katie, Paul, and David.

HRP

To:
Irene, my wonderful wife of 35 years; Greg, my great son, for valuing education; Marge, my mother, the best one could hope for, G.L., my father who passed away before I could thank him enough.

GLM

To:
Sharon, my wife and to my children Taber, Maya, and Sarah, all at the center of my life.

RMB

To:
Jane, my wife and the best partner I could hope to have; our sons, Lance, Brooke, and Sean; and the students and colleagues who have enriched my life.

WAS

CONTENTS

PREFACE

Traditionally, mathematics departments throughout the country have devoted much of their undergraduate teaching effort to teaching calculus and calculus preparation. However, in recent years, mathematicians have a renewed commitment to students not majoring in science and engineering. This book is designed to be the textbook in such a course. In writing this book, we have kept before us certain goals.

Relevance. The title of the book ***"Mathematics in Life, Society, and the World"*** sums it up. We cover topics that play an important role in everyday life (for example Chapter 6-*Consumer Mathematics*), in civic life (for example Chapter 9 - *Voting and Apportionment*), or in their general appreciation of the world (for example Chapter 11 - *Growth and Scaling*). In class testing the material in this book, we have noticed that students pick up on the relevance of the material and the question "What is this good for?" is seldom heard.

Accessibility. In writing the text we have sought to make the material accessible by developing topics in the most logical and compelling manner. This involves isolating the truly important points and presenting them without needless technical complication. We have avoided writing any impressively complicated derivations, and have tried to make it possible to solve everything with an inexpensive calculator (that can compute x^y), some graph paper, and working knowledge of high school algebra.

Pedagogy. All the material in this book has been class tested many times. The book contains more exercises and problems than most texts of its kind. The exercises and problems, including the applied problems, have a wide range of difficulty so instructors can tailor the assessments to their classroom needs.

OUTLINE

The text is organized into three Parts (12 Chapters). The material in Chapter 2, *"Numbers and Numeracy"* is used throughout later chapters of the book, but the other chapters are relatively independent. Review topics are included near the end of the book for students who need a brief refresher.

Part I–Mathematics in Life. These chapters consist of the mathematics a student will encounter on a regular basis. Chapter 1, *Critical Thinking, Logical Reasoning, and Problem Solving,* sets the tone for this part by showing how mathematics appears in the media (often incorrectly). Then, reasoning patterns are introduced so that students can learn to analyze and construct sound arguments and to recognize incorrect reasoning. Problem solving strategies are presented to aid students in solving problems. Chapter 2, *Numbers and Numeracy,* develops the number systems that are interwoven throughout the book. Our approach to this topic is new in that numeracy is integrated throughout, especially in the coverage of mental math and estimation. Chapters 3-5 provide rich contexts in which statistics and probability are developed. These chapters contain many everyday uses of mathematics. Chapter 6, *Consumer Mathematics,* contains most of the mathematics that students will typically use in their personal financial dealings.

Part II–Mathematics in Society. Composed of three modern topics whose mathematics underlies much of the social structure and interactions around us. Chapter 7, *Game Theory,* shows how games of a social nature can be analyzed and how optimal winning strategies can be developed. Chapter 8, *Management Mathematics,* provides many interesting applications where the techniques involving linear programming and networks can be used to solve problems, especially in the business world. Chapter 9, *Voting and Apportionment,* contains an analysis of a variety of strategies that may be used by politicians to bring "fairness' into a democratic system.

Part III–Mathematics in the World. These three chapters are geometrical in nature. Since much of the world, most notably in the large-scale structure of objects and living beings, is governed and described by geometry in one way or another. Chapter 10, *Geometry,* develops the mathematics behind a variety of patterns in the world, including the analysis of tilings. Conic sections are also studied due to their underlying importance in naturally occurring phenomena. Chapter 11, *Growth and Scaling,* provides many real world applications which illustrate the importance that growth and decay have in Chapter 12, *Recursion and Fractals,* provides insight into these modern topics by showing how they occur in the world.

Throughout the book, we have sought to be faithful to recommendations of our professional organizations such as the MAA, the AMATYC, and NCTM.

ACKNOWLEDGEMENTS

A leading force in the development of the type of course for which this textbook is intended has been the Consortium for Mathematics and Its Applications, and we acknowledge their valuable pioneering efforts. We have benefitted from student feedback as we have tested these materials, and we thank those students for their help and advice. We also thank the Oregon State University Department of Mathematics and its Chair, Professor Francis Flaherty, who have been very supportive of our efforts.

The following individuals reviewed various stages of the manuscript and shared many helpful insights with us:

David Jabon, Eastern Washington University
Henry Kepner, University of Milwaukee
Edwin Kingham, University of Arizona
Eric Matsuoka, University of Hawaii
Leticia Oropesa, University of Miami
Matthew Pickard, University of Puget Sound
Victor Sung, Central Connecticut State University
Lyndon Weberg, University of Wisconsin

We thank Jack Casteel who expertly guided the manuscript through the production process. We also thank Bob Pirtle, our Macmillan editor before that company was integrated into Prentice Hall, for encouraging us to write this book. Finally, we have great praise for our editor, Sally Denlow, for her many creative ideas and persistence in guiding the manuscript from our first draft form to the book you are holding in your hands.

TEXT PREVIEW

CHAPTER 12

RECURSION AND FRACTALS

MARINE BIOLOGIST COMPLETES 3000 MILE TREK ON OREGON COAST

In what may be the most ambitious study of the Oregon coast ever undertaken, Hal Walker, Ph.D., has just completed the first inventory of all accessible tide pools on the Oregon coast. Dr. Walker covered the entire coastline of Oregon by foot—from the California state line, near Brookings, to Astoria at the mouth of the Columbia River. In following the coastline, the scientist tried to stay within one yard of the water's edge except for impassable stretches and dangerous rocks. Bridges were used only when rivers and bays were too deep to be waded. To measure the distance covered, Dr. Walker used a pedometer, recording 2,928.5 miles by the trip's end. For the return trip from Astoria to Brookings, he used a small plane which hugged the coastline at low level. The scientist said the 317 mile air trip was the most beautiful he had ever taken. The support team from the university used a van to return to Brookings on Highway 101, a distance of 339 miles.

Page 586

Chapter opening anecdotes offer realistic media stories to illustrate how mathematics can be used to solve real problems.

Page 537

Chapter Goals: Summarize the main ideas students should be learning within the chapter.

CHAPTER GOALS

1. Solve problems involving similar triangles.
2. Transform geometric objects into larger or smaller objects of similar shape through scaling.
3. Compute quantities such as perimeter, area, volume, and weight of scaled objects.
4. Model population growth, radioactive decay, and other quantities that change in proportion to their size.
5. Model physical objects and compute some related quantities.

Initial Problem: Each Section within a chapter begins with a problem which draws upon the skills a student will learn within the section.

Page 473

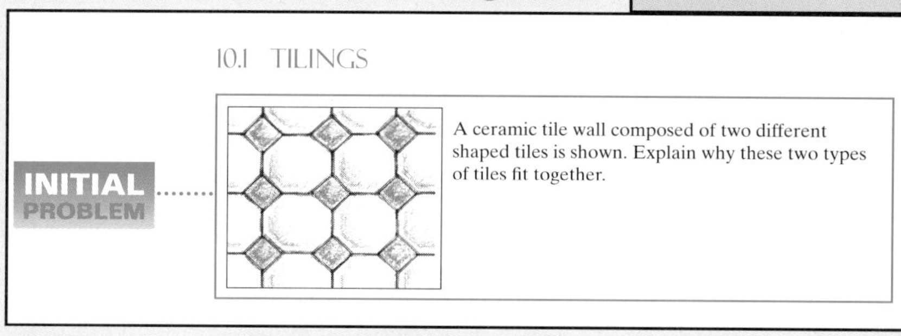

INITIAL PROBLEM SOLUTION

A ceramic tile wall composed of two different shaped tiles is shown. Explain why these two types of tiles fit together.

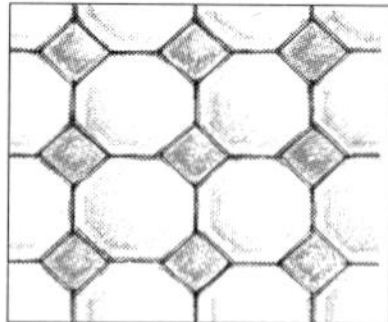

SOLUTION The angle measure in a square is 90° and in a regular octagon is 135° (Table 10.1). Each vertex contains a square and two regular octagons, and 90° + 2(135°) = 360°. Thus the three tiles will fit at each vertex and, as long as their sides are the same length, will tile a wall. This is one of the eight semiregular tilings.

Page 486

Initial Problem – Solution: The initial problem is repeated and its solution is fully worked for the student at the end of each section.

Page 267

HISTORY: Intriguing commentary of how mathematics was used and/or developed throughout the ages.

HISTORY

The first maritime insurance companies were established in Italy and Holland in the 14th century. These companies carried out calculations of chances since larger risks made for larger insurance premiums. For shipping by sea, premiums amounted to about 12% to 15% of the cost of the goods.

Page 566

TIDBIT: These marginal notes call out interesting facts and information that relate to many various aspects of life.

TIDBIT

The body's metabolism of drugs works like radioactive decay. Among the data available to prescribing physicians is the half-life of particular drugs in the average person's body. There is a lot of variability from person to person; for example, the half-life of caffeine varies from three to seven hours.

THE HUMAN SIDE OF MATHEMATICS

One of the oldest breweries in the world is the Guinness Brewing Company in Dublin. Guinness began as a family business in 1759. Its markets grew worldwide. However, the brewing process was overseen by master brewers using arcane methods handed down from master to apprentice. The Guinness corporation was interested in making this process scientific and constructing an exact recipe that could be used worldwide. This was a novel idea that required new techniques.

W. S. Gossett

W. S. Gossett, born in 1876, studied chemistry at the university in Dublin. He was hired by Guinness as a brewmaster in 1899 to work on the problem of making brewing a science. One questions found in the data could be accidental or simply due to natural variation. On the other hand, they could be the result of differences in treatment or process and thus lead to better methods of brewing. There was no way to tell which was which. Gossett went to work at the laboratory of the biometrician, Karl Pearson, to study statistics. During this time, Gossett solved the problem of data variation and developed new techniques. Then, in 1907 he returned to Guinness to be brewer-in-charge. Because of his connection with the Guinness company it was decided that he would not publish his ideas under his own name, but rather use the pseudonym, "Student." His

Page 188

The Human Side of Mathematics: Mathematics was developed by real people – these short biographical sketches are of people whose work was relevant to each chapter.

PROBLEM SET 1.2

In problems 1 through 4, state the premises (or hypotheses) and conclusion in each of the arguments. Are there any premises that aren't explicitly stated?

1. **(a)** If the room is warm, then I'll be uncomfortable. The room is warm, so I'll be uncomfortable.
   (b) If the weather is bad, I'll go to the movies. It's raining heavily, so I'll go to the movies.
2. **(a)** If the weather is good, I'll want to play golf. If I decide to play golf, I'll see if Marty can go. It's warm and clear outside; I guess I'll call Marty.
   (b) If Dr. Goldberg teaches the course, I'll register for her class. If you take a class from Dr. Goldberg, you can count on having to do a term paper. Dr. Goldberg is going to teach the course, so I'll have to write a term paper.
3. **(a)** If the weather is good, Barry will paint the house. Barry didn't paint the house, so the weather wasn't good.
   (b) If you average at least 90% on the tests, you'll get an A for the term. You didn't get an A for the term, so you didn't average 90% on the tests.
4. **(a)** Jenna is going to go swimming or play tennis. Jenna didn't go swimming, therefore she must have played tennis.
   (b) We're either going to the play or the movie. The tickets to the play are all sold out, so we're going to the movie.

In problems 5 through 18, translate the argument into symbolic form. Determine whether the argument is valid or invalid. You may compare the form of the argument to one of the four standard forms or use a truth table.

5. If the movie is good, the people will go.
   The people will go to the movie.
   Therefore the movie is good
6. If the sun is shining, I'll wear a hat.
   The sun isn't shining.
   Therefore I won't wear a hat.
7. If it's cold outside, my hands will be cold.
   It's freezing outside.
   Therefore my hands will be cold.
8. These shoes are not expensive
   If shoes are expensive, I won't buy them.
   Therefore I won't buy these shoes
9. If we miss the bus, we'll have to walk.
   If we walk, then we'll be late.
   Therefore if we miss the bus, we'll be late.
10. If you do the work, you'll get paid.
    If you get paid, you can go to the show.
    Therefore if you do the work, you can go to the show.
11. If Spike Lee is the director, then the movie should be good. Spike Lee didn't direct the movie, so it probably isn't good.
12. The night is cold and dark. The night is not dark or it is cold. Therefore the night is not cold.
13. If Carlos passes his entrance exam, he will attend college. Carlos will not be attending college; therefore he did not pass his entrance exam.
14. If you arrive on time, you'll get a good seat. If you get a good seat, you'll enjoy the play. Therefore if you didn't enjoy the play, you didn't arrive on time.
15. If I can't go to the movie, then I'll go to the park. I can go to the movie. Therefore, I will not go to the park.
16. If you score at least 90%, then you'll earn an A. If you earn an A, then your parents will be proud. You have proud parents. Therefore, you scored at least 90%.
17. If you work hard, then you will succeed. You do not work hard. Therefore you will not succeed.
18. If it doesn't rain, then the street won't be wet. The street is wet. Therefore it rained.

In problems 19 through 26, identify which form of argument (Modus Ponens, the chain rule, Modus Tollens, or disjunctive syllogism) is being used.

19. If Joe is a professor, then he is well educated. If you are well educated, then you went to college. Joe is a professor, so he went to college.
20. If you have children, then you are an adult. Paul is not an adult, so he has no children.
21. Whenever the weather is bad, I stay inside and work. It's raining heavily outside. I'll stay inside and work.
22. Either I get a raise or I'm going to look for a new job. I didn't get a raise, so I'm going to look for a new job.
23. If I don't have enough money to buy gas, I will ride the bus. When I ride the bus I am always late. I can't afford to buy gas, so I'm going to be late.
24. If you get to the store on time, you can pick up a carton of ice cream. You didn't pick up any ice cream, so you didn't get to the store on time.
25. Kim is going to have her old car painted or buy a new car. Kim didn't buy a new car, so she had her old one painted.
26. If I don't eat breakfast, I'll be hungry by 10:00. If I'm hungry before noon, I always snack before lunch. I skipped breakfast, so I will have a snack before lunch.
27. The *Initial Problem* from section 1.1 was to find a symbolic form for an exercise in logic written by Lewis Carroll. The final form of the argument was:

$$\begin{aligned} \sim p &\Rightarrow \sim q \\ \sim r &\Rightarrow \sim s \\ \sim q &\Rightarrow \sim t \\ u &\Rightarrow \sim p \\ \sim t &\Rightarrow \sim r \end{aligned}$$

Use the antecedent: u (a kitten is green-eyed) and provide a logical conclusion for the argument. State the conclusion in terms of the original statements and show that the argument is valid.

Exercises: Each section has an extensive set of exercises and problems covering all the topics developed in the section. Odd and even numbered problems are matched – with the even problems being slightly more difficult.

Page 23

EXTENDED PROBLEMS

43. Investigate the interest rates available on bank credit cards. Is there any reason a person would choose a card with a high interest rate?
44. Compare the terms at a rent-to-own store with the available retail price for an appliance of interest to you. What is the approximate annual percentage rate if the rent-to-own arrangement is treated like a loan with add-on interest?
45. Research the history of consumer credit. Consider such questions as: Is it a recent phenomenon? When did it become institutionalized and regulated? What legal and/or social issues have received special attention?
46. Consult an encyclopedia, almanac, or other suitable reference source for the total amount of consumer debt during the last 25 years. Prepare a bar graph to display the data. Are there any trends or special features to the graph?
47. Research the effect of inflation and deflation on those who owe money. Has this ever had important political ramifications?
48. What are usury laws? Are such laws in force in your state? Were they ever?

Page 293

EXTENDED PROBLEMS: Each section has a selection of extended problems. Some of these extended problems introduce additional topics that an instructor may wish to cover, while others offer an opportunity to do open-ended projects.

Chapter Six Problem

You are buying a car for $10,000 from a dealer who will give you a loan for 5 years at 10% interest. As an extra incentive he offers either $1200 cash back on the 10% loan or a 6% interest rate (which is a rate that you could expect to get if you deposited funds in a bank or money market account). Which deal should you choose?

SOLUTION

Strategy: Use a Table

One approach to making a rational decision is to compare the total amount you would pay under each option. That means you have to first find the amount of the monthly payments for each choice. From this information, you can compute the total for *all* payments.

Since the interest rate is a whole percent, we can use a table rather than computing the monthly payment from a formula. Table 6.5 can be used for this purpose. The part of the table we are interested in is the following:

Chapter Problem: Each chapter ends with a solved problem illustrating the usefulness of the material covered in the chapter and emphasizing one of the important topics.

Page 315

Section 9.1

Plurality Method 424
Borda Count Method 424
Plurality with Elimination Method 426
Run-off Election 426
Preference Table 426
Pairwise Comparison Method 428

Section 9.2

Majority Criterion 435
Head-to-Head Criterion 437
Monotonicity Criterion 438
Irrelevant Alternatives Criterion 441
Arrow Impossibility Theorem 442

Chapter Review—Vocabulary/Notation: To help organize studying, all key vocabulary words and notation are listed with the sections and page numbers where they can be found.

Page 466

Chapter Seven Review Problems

1. Consider the following game matrix.

$$\begin{bmatrix} 3 & -1 & -2 & 4 \\ 1 & 4 & -3 & 1 \\ -3 & 2 & 5 & -1 \\ -1 & 3 & 2 & -4 \end{bmatrix}$$

What will the payoff be if both players use their third strategy?

2. In the matrix in problem 1, what are the most aggressive strategies? What will the payoff be if both players use their most aggressive strategy?
3. In the matrix in problem 1, what are the most conservative strategies? What will the payoff be if both players use their most conservative strategies?
4. Is the game in problem 1 determined or not? Why? If it is determined what is the value of the game?
5. Construct a game matrix for the following game. Alice is on attack and can either pitch or kick. Bob is on defense and can either play tight or loose. If Alice pitches and Bob plays tight, then Alice loses one yard. If Alice pitches and Bob plays loose, then Alice gains 4 yards. If Alice kicks and Bob plays tight, then Alice gains 5 yards. If Alice kicks and Bob plays loose, then Alice loses 10 yards.
6. What is the most aggressive strategy for Alice and Bob in the game in problem 5? Suppose that Alice knows that Bob will choose his most aggressive strategy. What strategy should Alice employ? How many yards will she gain or lose?
7. Consider the following game matrix. Use the box and circle method to decide if the game is determined and, if so, the value of the game.

$$\begin{bmatrix} 0 & 1 & 2 & -1 \\ -3 & 1 & 3 & -2 \\ 1 & 2 & -3 & -1 \\ -1 & 3 & 2 & -4 \end{bmatrix}$$

8. Consider the following game matrix.

$$\begin{bmatrix} -3 & 3 \\ 4 & -2 \end{bmatrix}$$

Suppose that the first player uses a mixed strategy with frequencies $(\frac{1}{2}, \frac{1}{2})$ and the second player uses a mixed strategy with frequencies $(\frac{1}{3}, \frac{2}{3})$. What is the average payoff to the first player of this game?

9. Suppose you are the first player in the game in problem 8 and learn that the second player has a mixed strategy with frequencies $(\frac{1}{3}, \frac{2}{3})$ What is your optimal strategy in this case, and what is the payoff?
10. What is the optimal mixed strategy for each player in problem 8?

Chapter Review—Problems: Each chapter ends with a set of problems which cover the most important topics in the chapter.

Page 359

SUPPLEMENTS

FOR THE STUDENT

Student Study Guide and Solutions Manual
written by Bill Siebler
Provides chapter summaries, hints, study skills and worked out solutions to all odd—numbered exercises.
(ISBN: 0–13-259417-X)

Life on the Internet—Mathematics (A Student's Guide)
written by Andrew T. Stull and Edward Wright
Guides students and instructors through the complexity of the Internet, offering navigation strategies, practice exercises and lists of resources.
(ISBN: 0–13-268616-3)

New York Times Supplement
A contemporary view of selected topics covering various math topics from the pages of the New York Times Newspaper.
(ISBN: 0–13-261637-8)

FOR THE INSTRUCTOR

Instructors Resource Manual
Contains solutions to all even-numbered exercises, as well as a test bank of 1080 multiple choice, true/false and free response questions.
(ISBN: 0–13-259367-X)

Prentice Hall Custom Test
questions written by Bill Siebler
A fully editable test generator which features algorithmic capabilities, an instructor's gradebook, and on-line testing.
Windows (ISBN: 0–13-259375-0)
MAC (ISBN: 0–13-259409-9)

ABC News Videos and Accompanying Projects
edited and written by Kim Query
Segments from Nightline, World News Tonight and This Week with David Brinkley covering various math topics. Also includes discussion questions and group projects.
(ISBN: 0–13-382755-0)

MATHEMATICS *in* LIFE, SOCIETY, *and the* WORLD

PART 1
MATHEMATICS *in* LIFE

CHAPTER 1

CRITICAL THINKING, LOGICAL REASONING, AND PROBLEM SOLVING

BIG SPENDER OR RESPONSIBLE LEGISLATOR?

The following 1994 radio advertisement was run by Representative Buck who was a candidate for governor running against former state Senator Doe.

Narrator: "To learn what people think about some of the issues in the governor's race, we called a few. We asked how much they know about Doe's record. Did you know Doe voted for $2.7 billion in higher taxes when he was a state senator?"

Male voice: "I had no idea that he voted for that amount in higher taxes."

Female voice: "It's a typical tax and spend attitude."

Male voice: "This state needs less taxes, not more."

Female voice: "The liberals always have their hands in my pocket."

Although the advertisement places Senator Doe in a bad light, is he really a big spender?

Chapter Goals

1. Learn about the logical connectives *and, or, if then, and if and only if* and how they are used in reasoning.
2. Learn about the various forms of "if . . ., then" statements and how they are used in constructing valid arguments.
3. Learn to make valid arguments and to recognize invalid argument forms.
4. Learn a problem-solving framework as well as strategies that can be used throughout the book and in life to solve a variety of problems.

Daily we are inundated with statements and images that are designed to influence us. In the advertisement above, Representative Buck was trying to undermine his opponent in a state where reducing taxes was in vogue. A reasonable person would expect the tax increase figure of $2.7 billion to refer to one year's tax increase, unless there was a statement to the contrary. However, a statewide newspaper published the following analysis of the ad: "the figure of $2.7 billion in tax increases includes estimates of how much a tax would have been raised over varying lengths of time. For example, in one case, the total includes the 10-year impact of a gas-tax increase. In some cases, it is four years. Also, some of the bills cited are not even tax increases. For example, two of the bills removed the exemption from income taxes for public employee retirement pensions in exchange for higher benefits." Thus, we see that in his advertisement Representative Buck was misusing information to gain votes.

The media is replete with information that helps form our opinions and guide our actions. The purpose of this chapter is to help you develop the skills needed to think critically about a variety of problems and issues, especially those involving quantitative information and relationships. In the chapter, you will increase your sensitivity regarding the information and misinformation you face on a daily basis. In addition to developing logical reasoning patterns that will help you analyze arguments better, you will also learn to organize your own arguments (which we hope will be logical and truthful). Finally, you will be introduced to a framework and strategies for solving problems.

THE HUMAN SIDE OF MATHEMATICS

George Boole (1815–1864) was the son of an Irish cobbler. Boole was self-taught in higher mathematics and never received a degree. His father, an amateur mathematician, began teaching mathematics to Boole at an early age. Unfortunately, the father had no money for his son's formal education. Nonetheless, Boole continued his studies and was able to write many high quality mathematical papers. He received enough recognition for his work to be appointed professor of mathematics at Queen's College in Cork, Ireland.

George Boole

Boole became one of the most influential mathematicians of his time, and his influence carried forward into modern times. Bertrand Russell, British mathematician and philosopher, credits Boole with discovering pure mathematics, referring to work published in a book called *The Laws of Thought,* in 1854. Traditional logic was incomplete and failed to account for many principles of inference employed in even elementary mathematical reasoning. Boole's primary concern was to develop a nonnumerical algebra of logic that would provide precise methods for handling more general and varied types of deductions, or proofs, than were covered by traditional logic. His work laid the foundations for modern algebra and the study of symbolic logic.

George Boole was a man of goodwill and reputation. Among the poor people of the neighborhood, he was regarded as an innocent who should not be cheated; among the higher classes he was admired as something of a saint, although a bit odd. In 1864, he was caught in the rain one day while hurrying to class. Concerned for his students because he was late, Boole gave his entire lecture while still dripping wet. Boole had suffered from poor health most of his life and came down with a cold that turned to pneumonia. He died shortly thereafter.

Lewis Carroll, one of the most revered names in children's literature, was the pen name of Charles Dodgson (1832–1898), an English mathematician and logician. Born to an upper-class clergyman's family, Dodgson was very religious and was expected to serve God by joining the ministry. Instead, after some agonizing, he thought God might be served as well by a teacher and a writer. After earning his university degree, he took a job as a lecturer in mathematics at Oxford University under the condition that he never marry. He held the job all his life. A stutterer and very hard of hearing, he had difficulty in social situations but did very well with children. He told them stories and then put the stories on paper, eventually having them published. His most famous contributions to children's literature were *Alice's Adventures in Wonderland* and *Through the Looking Glass.*

Lewis Carroll

Lewis Carroll lived quietly, free from the distractions of life, worldly want, and family responsibilities. He produced some of the most enduring works in children's literature as well as a lasting legacy in mathematics. He made significant contributions in Euclidean geometry and in logic, where he invented a method to test whether logical arguments were valid or not. Queen Victoria was said to be so taken by his children's books that she requested copies of every book he had ever written. Imagine her surprise when she received a pile of mathematics books! Lewis Carroll was interested in logical reasoning, especially as it applied to games, puzzles, and pure mathematics. He considered his texts on logic as his most important works. An interesting aspect of his children's books is that they can also be appreciated from a mathematical standpoint. For example, *Through the Looking Glass* is based on a game of chess and *Alice's Adventures in Wonderland* involves size and proportion.

1.1 STATEMENTS AND LOGICAL CONNECTIVES

INITIAL PROBLEM

The following exercise in logic comes from a textbook on the subject by Lewis Carroll, author of *Alice's Adventures in Wonderland.*

No kitten that loves fish is unteachable.
No kitten without a tail will play with a gorilla.
Kittens with whiskers always love fish.
No teachable kitten has green eyes.
No kittens have tails unless they have whiskers.

Is there a way we can make sense of these statements and draw a conclusion?

HISTORY

The Chinese philosopher Mo Ti founded logic in China during the 400s B.C. In the A.D. 300s the Buddhist philosopher Acarya Dignaga invented symbolic logic. His influence in Buddhist Asia was comparable to that of Aristotle in the West. This system spread to China and then Japan in the A.D. 600s.

Although his lasting contributions were to children's literature, the focus of Lewis Carroll's intellectual life was the study and teaching of logical reasoning. As is true with most of mathematics, the study of logic is made easier by the substitution of symbols for words. Yet in the process of logic, the basic tool with which we test all ideas and also solve most of our everyday problems, we are still at the mercy of the inadequacies and clumsiness of words. Some collections of words are meaningful in a mathematical context, and some are not. We begin our study of logic by defining what we mean by statements. Then we will see how statements are combined, modified, and organized to form meaningful arguments. In the second section, we will see how arguments are analyzed for validity and how conclusions are drawn.

STATEMENTS

Our written and spoken language is organized into units called sentences. For the study of logic, the declarative sentence is the most important type since such a sentence makes an assertion, and thus is either true or false. Sentences that can be classified as true or false are called **statements.** Examples of statements are:

1. Based on area, Alaska is the largest state of the United States. (True)
2. Based on population, Texas is the largest state of the United States. (False)
3. $2 + 3 = 5$. (True)
4. $3 < 0$. (False)

It may be impossible to determine if some sentences are true or false. For example, an interrogative sentence or an exclamation is typically neither true nor false. The paradoxical statement "This sentence is false" can be neither true nor false. The following are not statements as defined in logic.

1. My home state is the best state. (Subjective)
2. Help! (An exclamation)
3. Where were you? (A question)

4. The rain in Spain. (Not a sentence)
5. This sentence is false. (Neither true nor false!)

Statements are represented symbolically by lowercase letters (e.g., p,q,r, and s). New statements can be created from existing statements in several ways. For example, if p represents the statement "The sun is shining," then the **negation** of p, written $\sim p$ and read "not p," is the statement "The sun is not shining."

EXAMPLE 1.1 For each of the following statements, decide whether the statement is true or false, construct the negation of each statement, and decide whether the negation is true or false.

(a) The sun sets in the west.
(b) $2 + 3 = 5$.
(c) Gasoline is nonflammable.

SOLUTION

(a) The statement "The sun sets in the west" is true. The negation is the statement "The sun does not set in the west," which is false.
(b) The statement "$2 + 3 = 5$" is true. The negation is the statement "$2 + 3 \neq 5$," which is false.
(c) The statement "Gasoline is nonflammable" is false. The negation is the statement "Gasoline is flammable," which is true. ◆

Notice from the example that when a statement is true, its negation is false; and when a statement is false, its negation is true. That is, a statement and its negation have opposite truth values. We summarize this relationship between a statement and its negation using a **truth table:**

p	$\sim p$
T	F
F	T

The statement p can be either true, indicated by T, or false, indicated by F; we call these the possible **truth values** of the statement p. This table shows that when the statement p is true, then $\sim p$ is false and when p is false, $\sim p$ is true.

LOGICAL CONNECTIVES

Two or more statements can be joined, or connected, to form **compound statements.** Next we study the four commonly used **logical connectives** *and, or, if-then* and *if and only if.*

And

If p is the statement "It is raining" and q is the statement "The sun is shining," then the **conjunction** of p and q is the statement "It is raining *and* the sun is shining." The conjunction is represented symbolically as "$p \wedge q$." The conjunction of two statements p and q is true only when both p and q are true. That is, we say that the statement "It is raining and the sun is shining" is true only in the case where it is true that the sun is shining *and* it is raining. Notice that the two statements p and q each have two possible truth values: T and F. Hence there are four possible combinations of T and F to consider for both statements. The next truth table displays the possible truth values of p and q with the corresponding truth values of $p \wedge q$. For instance, the shaded row in the following truth table tells us that if p is true and q is false, then $p \wedge q$ is false.

p	q	$p \wedge q$
T	T	T
T	F	F
F	T	F
F	F	F

Notice that the conjunction of two statements is true only when *both* statements are true and false if at least one of the statements is false.

Or

The **disjunction** of statements p and q is the statement "p or q," which is represented symbolically as "$p \vee q$." In practice, there are two common uses of "or": the exclusive "or" and the inclusive "or." The statement "I will go or I will not go" is an example of the use of the **exclusive "or"** since either "I will go" is true or "I will not go" is true, but both cannot be true simultaneously. The **inclusive "or"** (called "and/or" in everyday language) includes the possibility that both parts are true. For example, the statement "It will rain or the sun will shine" uses the inclusive "or." It is true if (1) it rains, (2) the sun shines, or (3) it rains and the sun shines. That is, the inclusive "or" in $p \vee q$ allows for both p and q to be true. In mathematics, we agree to use the inclusive "or," whose truth values are summarized in the next truth table.

p	q	$p \wedge q$
T	T	T
T	F	T
F	T	T
F	F	F

Notice that the disjunction of two statements is false only when *both* statements are false, and it is true if at least one of the statements is true.

EXAMPLE 1.2 Let p represent the statement "The rain is wet," let q represent the statement "The snow is cold," and let r represent the statement "The sunshine is warm." Translate the following compound statements into symbolic form.

(a) The rain is wet, and the snow is cold.
(b) The snow is cold, but the sunshine is warm.
(c) The rain is wet or the sunshine is warm.

SOLUTION

(a) $p \wedge q$.
(b) Logically *but* means the same as *and* so this statement translates to $q \wedge r$.
(c) $p \vee r$. ◆

EXAMPLE 1.3 Decide if the following statements are true or false, where p represents the statement "Rain is wet" and q represents the statement "Black is white."

(a) $\sim p$ **(b)** $p \wedge q$ **(c)** $(\sim p) \vee q$
(d) $p \wedge (\sim q)$ **(e)** $\sim(p \wedge q)$ **(f)** $\sim[p \vee (\sim q)]$

SOLUTION

(a) p is T, so $\sim p$ is F.
(b) p is T and q is F, so $p \wedge q$ is F.
(c) $\sim p$ is F and q is F, so $(\sim p) \vee q$ is F.
(d) p is T and $\sim q$ is T, so $p \wedge (\sim q)$ is T.
(e) p is T and q is F, so $p \wedge q$ is F and $\sim(p \wedge q)$ is T.
(f) p is T and $\sim q$ is T, so $p \vee (\sim q)$ is T and $\sim[p \vee (\sim q)]$ is F. ◆

If-Then

One of the most important compound statements is the implication. The statement "If p, then q," denoted by "$p \Rightarrow q$," is called an **implication** or **conditional statement.** The statement p is called the **hypothesis** or **antecedent** and q is called the **conclusion** or **consequent.** To construct the truth table for $p \Rightarrow q$, consider the following conditional promise given to a math class: "If you average at least 90% on all tests, then you will earn an A." Let p represent "You average at least 90% on all tests" and q represent "You earn an A." Then there are four possibilities:

Average at Least 90%	Earn an A	Promise Kept
Yes	Yes	Yes
Yes	No	No
No	Yes	Yes
No	No	Yes

Notice that the only way the promise can be broken is in line 2. In lines 3 and 4, the promise is not broken since an average of at least 90% was not attained. (In these cases, a student may still earn an A, so the promise no longer applies.) This example suggests the following truth table for the conditional $p \Rightarrow q$.

p	q	$p \Rightarrow q$
T	T	T
T	F	F
F	T	T
F	F	T

It is important to notice that when the hypothesis is false (lines 3 and 4 in the truth table) the conditional statement is nevertheless true.

EXAMPLE 1.4 Decide whether each conditional statement is true or false.

(a) If the sun rises in the east, then rain is wet.
(b) If $2 + 2 = 5$, then $3 + 3 = 7$.
(c) If rain is wet, then $2 + 2 = 5$.
(d) If $2 + 2 = 5$, then snow is cold.

SOLUTION (Refer to the truth table for $p \Rightarrow q$.)

(a) The hypothesis "The sun rises in the east" and the conclusion "Rain is wet" are both true. We see that the conditional statement $p \Rightarrow q$ is true.
(b) Both the hypothesis and conclusion are false; the conditional is true.
(c) The hypothesis is true, and the conclusion is false; the conditional is false.
(d) The hypothesis is false and the conclusion is true. Whenever the hypothesis is false the conditional is true. ◆

Related Conditionals

The **converse** of the conditional "If p, then q" is the conditional "If q, then p"; that is, the hypothesis and conclusion of the original conditional are interchanged to form its converse. For example, the conditional "If you have broken your leg, then you will feel pain" has the converse "If you feel pain, then you have broken your leg." Although the original conditional is true, its converse is not. For example, you could feel pain from a headache, a broken arm, or a variety of causes. Thus, a conditional may be true while its converse is false. In general, the truth of a converse of a conditional may or may not agree with the truth of the conditional itself. Typically, advertisers like to entice you to buy their products by reasoning from the converse (whether it is true or not). For example, the ad stating "Michael Jordan eats Wheaties" translates to the conditional "If you are Michael Jordan, then you eat Wheaties." However, the maker of Wheaties wants you to think "If I eat Wheaties, then I'll be like Mike (soar on the basketball court and be rich)." Unfortunately, that reasoning from the converse is unlikely to be true.

EXAMPLE 1.5 Identify the hypothesis and conclusion. Then write the converse of the following conditionals and determine if the conditionals and their converse are true or false.

(a) If $x = 2$ and $y = 2$, then $x + y = 4$.
(b) If rain is wet, then pigs can fly.

SOLUTION

(a) Hypothesis: $x = 2$ and $y = 2$.
Conclusion: $x + y = 4$.
The conditional is true. The converse of the conditional is:
If $x + y = 4$, then $x = 2$ and $y = 2$.
The converse is false, because it could be that $x = 1$ and $y = 3$.
(b) Hypothesis: Rain is wet.
Conclusion: Pigs can fly.
The conditional is false because the hypothesis is true, but the conclusion is false. The converse of the conditional is:
If pigs can fly, then rain is wet.
The converse is true since the hypothesis (pigs can fly) is false. ◆

Two other forms derived from the conditional "If p, then q" are its **contrapositive** "If not q, then not p" and its **inverse,** "If not p, then not q."

EXAMPLE 1.6 Identify the hypothesis and conclusion. Then write the contrapositive and inverse of the following conditionals and determine if the conditionals and their contrapositives and inverses are true or false.

(a) If rain is wet, then snow is cold.
(b) If pigs can fly, then the moon is round.

SOLUTION

(a) Hypothesis: Rain is wet.
Conclusion: Snow is cold.
The conditional is true since both the hypothesis and conclusion are true.
The contrapositive is: If snow is not cold, then rain is not wet.
The inverse is: If rain is not wet, then snow is not cold.
Both the contrapositive and the inverse are true because each has a false hypothesis.

(b) Hypothesis: Pigs can fly.
Conclusion: The moon is round.
The conditional is true since its hypothesis is false.
The contrapositive is: If the moon is not round, then pigs cannot fly.
The inverse is: If pigs cannot fly, then the moon is not round.

The contrapositive is true (verify), but the inverse is false (verify). ◆

The following truth table displays the various truth values for the basic conditional "If p, then q" and its contrapositive, converse, and inverse. Remember that the only case in which a conditional is false is when the hypothesis is true and the conclusion is false.

p	q	$\sim p$	$\sim q$	Conditional $p \Rightarrow q$	Contrapositive $\sim q \Rightarrow \sim p$	Converse $q \Rightarrow p$	Inverse $\sim p \Rightarrow \sim q$
T	T	F	F	T	T	T	T
T	F	F	T	F	F	T	T
F	T	T	F	T	T	F	F
F	F	T	T	T	T	T	T

Notice that the columns of truth values under the conditional $p \Rightarrow q$ and its contrapositive are the same. When this is the case, we say that the two statements are logically equivalent. In general, two statements are **logically equivalent** when they have the same truth tables. Similarly, the converse of $p \Rightarrow q$ and the inverse of $p \Rightarrow q$ have the same truth table. Therefore they, too, are logically equivalent. In mathematics, replacing a conditional with a logically equivalent conditional often facilitates the solution of a problem. In fact, many mathematical statements in the form of conditionals are proved by using the contrapositive. This is done by showing that whenever the original conclusion is false, the hypothesis (which is known or assumed to be true) would have to be false. This method of proof, called *reductio ad absurdum,* or indirect proof, will be examined in the next section.

If and Only If

One other common form derived from the conditional "If p, then q" is the **biconditional** "p if and only if q." A biconditional is true when both its hypothesis and conclusion are true or both are false. Suppose the statement "You will receive an A if and only if your average is at least 90%" is true (as it is by some professors' standards). This biconditional tells a student in the clearest terms what one must do to get an A. Notice that the conditional "If you score at least 90%, then you will get an

A" does not say what happens to the student who earns 89%. It allows a kindheart-ed professor to still give that person an A.

EXAMPLE 1.7 Determine if the following biconditionals are true.

(a) Rain is wet if and only if snow is cold.
(b) Rain is wet if and only if pigs can fly.
(c) Pigs can fly if and only if the moon is a balloon.

SOLUTION

(a) Both "Rain is wet" and "Snow is cold" are true; the biconditional is true.
(b) Since "Rain is wet" is true, but "Pigs can fly" is false, the biconditional is false.
(c) Both "Pigs can fly" and "The moon is a balloon" are false, so the biconditional is true. ◆

The connective "p if and only if q," is written $p \Leftrightarrow q$. It is the conjunction of $p \Rightarrow q$ and its converse $q \Rightarrow p$. That is, $p \Leftrightarrow q$ is logically equivalent to $(p \Rightarrow q) \wedge (q \Rightarrow p)$. The truth table of $p \Leftrightarrow q$ follows.

p	q	$p \Rightarrow q$	$q \Rightarrow p$	$(p \Rightarrow q) \wedge (q \Rightarrow p)$	$p \Leftrightarrow q$
T	T	T	T	T	T
T	F	F	T	F	F
F	T	T	F	F	F
F	F	T	T	T	T

Notice that the biconditional $p \Leftrightarrow q$ is true when p and q have the same truth values and false otherwise.

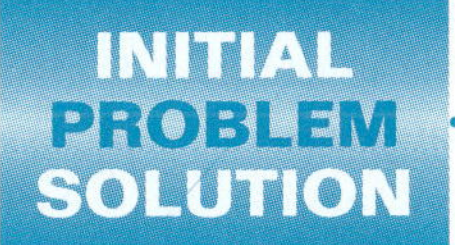

The following exercise in logic comes from a textbook on the subject by Lewis Carroll, author of *Alice's Adventures in Wonderland.*

No kitten that loves fish is unteachable.
No kitten without a tail will play with a gorilla.
Kittens with whiskers always love fish.
No teachable kitten has green eyes.
No kittens have tails unless they have whiskers.

Is there a way we can make sense of these statements and draw a conclu-sion?

SOLUTION

Our first step is to reword and rearrange the statements into conditionals. After some trial and error, one possible set of conditionals is:

1. If a kitten cannot be taught, then it does not love fish.
2. If a kitten has no tail, then it will not play with a gorilla.
3. If a kitten does not love fish, then it has no whiskers.
4. If a kitten is green-eyed, then it cannot be taught.
5. If a kitten has no whiskers, then it has no tail.

Next we substitute letters for statements as follows:
Let

p be "a kitten can be taught,"
q be "a kitten loves fish,"
r be "a kitten has a tail,"
s be "a kitten will play with a gorilla,"
t be "a kitten has whiskers,"
u be "a kitten is green-eyed."

The conditionals in symbolic form are:

1. $\sim p \Rightarrow \sim q$
2. $\sim r \Rightarrow \sim s$
3. $\sim q \Rightarrow \sim t$
4. $u \Rightarrow \sim p$
5. $\sim t \Rightarrow \sim r$

Now we have a set of standard symbolic statements that can be analyzed. As we will see in the next section, the only conclusion from this set of conditionals is $u \Rightarrow \sim s$: "If a kitten is green-eyed, then it will not play with a gorilla."

PROBLEM SET 1.1

1. Determine which of the following are statements.
   (a) What is your name?
   (b) The rain in Spain falls mainly in the plain.
   (c) Happy New Year!
   (d) Five is an odd number.

2. Determine which of the following are statements.
   (a) Is $\triangle ABC$ isosceles?
   (b) Go away!
   (c) The sum of the angle measures.
   (d) The area of a circle is $2\pi r$.

3. Determine which of the following are statements.
   (a) A pentagon has six sides.
   (b) What is a statement?
   (c) The one that is always true.
   (d) Alexander the Great was a student of Aristotle's.

4. Determine which of the following are statements.
   (a) Determine which of the following are not statements.
   (b) The ones with questions marks.
   (c) Statements must be true or false.
   (d) Your last answer was correct.

In problems 5 and 8, write the negation for each of the given statements. If possible, see if you can express the negation in two or more ways.

5. **(a)** A kitten is teachable.
   (b) A kitten has whiskers.

6. **(a)** A kitten has a tail.
   (b) A kitten will play with a gorilla.

7. **(a)** Phil bought a new car.
   (b) The weather is sunny.

8. **(a)** The Rockets won the 1995 NBA title.
   (b) Bill Clinton is president of the United States.

9. Decide the truth value of each statement.
   (a) Alexander Hamilton was once president of the United States.
   (b) The world is flat.
   (c) If dogs are cats, then the sky is blue.

10. Decide the truth value of each statement.
   (a) If Tuesday follows Monday, then the sun is hot.
   (b) If Christmas Day is December 25, then Texas is the largest state in the United States.
   (c) The moon is made out of green cheese if and only if Elvis Presley is alive.

11. Decide the truth value of each statement.
   (a) Alexander the Great was a student of Aristotle's.
   (b) If Bertrand Russell wrote *Hamlet,* then Alexander Hamilton was president of the United States.
   (c) George Bush was president of the United States or Lewis Carroll was prime minister of England.

12. Decide the truth value of each statement.
   (a) A disjunction is false if either of the statements is false.
   (b) If the conditional is false, then the conjunction is false.
   (c) The negation of any statement is false.

13. Let r, s, and t be the following statements:
r: Roses are red.
s: The sky is blue.
t: Turtles are green.
Translate the following statements into English.
(a) $r \wedge s$ **(b)** $r \wedge (s \vee t)$
(c) $s \Rightarrow (r \wedge t)$ **(d)** $(\sim t \wedge \sim s) \Rightarrow \sim r$

14. Let r, s, and t be the following statements:
r: Water is wet.
s: The earth is flat.
t: Western logic began in Greece.
Translate the following statements into English.
(a) $r \vee s$ **(b)** $r \vee (s \wedge t)$
(c) $s \Rightarrow (r \vee t)$ **(d)** $(\sim t \vee t) \Rightarrow \sim r$

15. Write the following in symbolic form using p, q, r, $\sim$, $\Rightarrow$, where p, q, and r represent the following statements:
p: A kitten is green-eyed.
q: A kitten can be taught.
r: A kitten loves fish.
(a) If a kitten is green-eyed, then it cannot be taught.
(b) If a kitten cannot be taught, then it does not love fish.

16. Write the following in symbolic form using p, q, $\sim$, $\Rightarrow$, where p and q represent the following statements:
p: Bill Clinton was president of the United States in 1996.
q: Newt Gingrich is vice president of the United States in 1996.
(a) If Bill Clinton was president of the United States in 1996, then Newt Gingrich was not vice-president of the United States in 1996.
(b) If Newt Gingrich was vice president of the United States in 1996, then Bill Clinton was not president of the United States in 1996.

17. Write the following in symbolic form using p, q, r, $\sim$, $\wedge$, $\vee$, $\Rightarrow$, $\Leftrightarrow$, where p, q, and r represent the following statements:
p: The sun is shining.
q: It is raining.
r: The grass is green.
(a) If it is raining, then the sun is not shining.
(b) It is raining and the grass is green.
(c) The grass is green if and only if it is raining and the sun is shining.
(d) Either the sun is shining or it is raining.

18. Write the following in symbolic form using p, q, r, $\sim$, $\wedge$, $\vee$, $\Rightarrow$, $\Leftrightarrow$, where p, q, and r represent the following statements:
p: A pentagon has six sides.
q: A square is a rectangle.
r: A rhombus is an equilateral quadrilateral.
(a) If a pentagon doesn't have six sides, then a square is a rectangle.
(b) Either a pentagon has six sides, or a square is a rectangle.
(c) A square is a rectangle if and only if either a pentagon has six sides or a rhombus is an equilateral quadrilateral.
(d) If a pentagon has six sides and a square is a rectangle, then a rhombus is an equilateral quadrilateral.

19. Fill in the headings of the following truth table using p, q, $\sim$, $\wedge$, $\vee$, and $\Rightarrow$.

p	q	(a)	(b)
T	T	T	F
T	F	T	F
F	T	T	F
F	F	F	T

20. Fill in the headings of the following truth table using p, q, $\sim$, $\wedge$, $\vee$, and $\Rightarrow$.

p	q	(a)	(b)
T	T	T	F
T	F	F	T
F	T	T	F
F	F	T	F

21. Fill in the headings of the following truth table using p, q, $\wedge$, $\vee$, $\sim$, and $\Rightarrow$.

p	q	(a)	(b)
T	T	T	F
T	F	F	T
F	T	T	T
F	F	T	T

22. Fill in the headings of the following truth table using p, q, $\wedge$, $\vee$, $\sim$, and $\Rightarrow$.

p	q	(a)	(b)
T	T	T	T
T	F	F	T
F	T	F	F
F	F	F	T

23. If p is T, q is F, and r is T, find the truth values for the following:
(a) $p \wedge \sim q$ **(b)** $\sim(p \vee q)$
(c) $(\sim p) \Rightarrow r$ **(d)** $(\sim p \wedge r) \Leftrightarrow q$

24. If p is F, q is T, and r is T, find the truth values for the following:
(a) $(\sim q \wedge p) \vee r$ **(b)** $(r \wedge \sim p) \vee (r \wedge \sim q)$
(c) $p \vee (q \Leftrightarrow r)$ **(d)** $(p \wedge q) \Rightarrow (q \vee \sim r)$

25. If p is T, q is T, and r is F, find the truth values for the following:
(a) $p \wedge \sim q$ **(b)** $\sim(p \Leftrightarrow q)$
(c) $(\sim p) \Rightarrow r$ **(d)** $(\sim p \wedge r) \Leftrightarrow q$

26. If p is F, q is F, and r is T, find the truth values for the following:
(a) $(\sim q \wedge p) \Rightarrow r$ **(b)** $(r \wedge \sim p) \Leftrightarrow (r \wedge \sim q)$
(c) $p \Leftrightarrow (q \Leftrightarrow r)$ **(d)** $(p \wedge q) \Rightarrow (q \Leftrightarrow \sim r)$

27. Suppose that $p \Rightarrow q$ is known to be false. Give the truth values for the following.
(a) $p \vee q$ **(b)** $p \wedge q$
(c) $q \Leftrightarrow p$ **(d)** $\sim q \Leftrightarrow p$

28. Suppose that $p \Leftrightarrow q$ is known to be false. Give the truth values for the following. If more than one truth value is possible, explain why.
(a) $p \vee q$ **(b)** $p \wedge q$
(c) $q \Rightarrow p$ **(d)** $\sim q \Rightarrow p$

29. For each of the following conditionals, identify the antecedent (or hypothesis) and the consequent (or conclusion). Form the converse, inverse, and contrapositive.
(a) If the weather is good, we'll go to the game.
(b) If I don't go to the movie, I'll study my math.
(c) If I get an A on the final, I'll get an A for the course.

30. For each of the following conditionals, identify the antecedent (or hypothesis) and the consequent (or conclusion). Form the converse, inverse, and contrapositive.
(a) If we each do our share, the job will get done.
(b) If I don't go to work, I won't get paid.
(c) If the game is over, we will leave.

31. For each of the following conditionals, identify the antecedent (or hypothesis) and the consequent (or conclusion). Form the converse, inverse, and contrapositive.
(a) Your car won't start if you don't have gasoline in the tank.
(b) I'll graduate if I can pass this class.

32. For each of the following conditionals, identify the antecedent (or hypothesis) and the consequent (or conclusion). Form the converse, inverse, and contrapositive.
(a) My bicycle is not in the rack where I left it, so somebody must have stolen it.
(b) I'd buy a new car if I had the money.

33. Construct one truth table that contains truth values for all of the following statements and determine which are logically equivalent.
(a) $(\sim p) \vee (\sim q)$
(b) $(\sim p) \vee q$
(c) $\sim(p \wedge q)$

34. Construct one truth table that contains truth values for all of the following statements and determine which are logically equivalent.
(a) $p \Rightarrow q$
(b) $\sim(p \wedge q)$
(c) $\sim(p \vee q)$

35. Use a truth table to determine which of the following are always true.
(a) $(p \Rightarrow q) \Rightarrow (q \Rightarrow p)$
(b) $[p \wedge (p \Rightarrow q)] \Rightarrow q$

36. Use a truth table to determine which of the following are always true.
(a) $(p \vee q) \Rightarrow (p \wedge q)$
(b) $(p \wedge q) \Rightarrow p$

37. If possible, determine the truth value of each statement. Assume that a and b are true, p and q are false, and x and y have unknown truth values. If a value can't be determined, write "unknown."
(a) $p \Rightarrow (a \vee b)$ **(b)** $b \Rightarrow (p \vee a)$
(c) $x \Rightarrow p$ **(d)** $a \vee p$
(e) $b \Rightarrow q$ **(f)** $b \Rightarrow x$

38. If possible, determine the truth value of each statement. Assume that a and b are false, p and q are true, and x and y have unknown truth values. If a value can't be determined, write "unknown."
(a) $a \wedge (b \vee x)$ **(b)** $(y \vee x) \Rightarrow a$
(c) $(y \wedge b) \Rightarrow p$ **(d)** $x \Rightarrow a$
(e) $(a \vee x) \Rightarrow (b \wedge q)$ **(f)** $x \vee (\sim x)$

39. Determine if each of the following biconditionals is true or false.
(a) A triangle has equal angles if, and only if, it has sides of equal length.
(b) Fish can swim if, and only if, pigs can fly.
(c) A quadrilaterial is a square if, and only if, it has three right angles.

40. Determine if each of the following biconditionals is true or false.
(a) A quadrilateral has equal angles if, and only if, it has sides of equal length.
(b) Chickens has teeth if, and only if, snakes have legs.
(c) The sum of two numbers is even if, and only if, the numbers are even.

EXTENDED PROBLEMS

DeMorgan's Laws

Two interesting, and often useful, relationships between conjunction, disjunction, and negations are known as DeMorgan's Laws. The first of these states that the negation of a conjunction of two statements is the disjunction of their negations. In symbols, we write

$$\sim(p \wedge q) \Leftrightarrow \sim p \vee \sim q.$$

The second law states that the negation of a disjunction of two statements is the conjunctions of their negations:

$$\sim(p \vee q) \Leftrightarrow \sim p \wedge \sim q.$$

Note: One way to translate the symbolic form $\sim(p \wedge q)$ to words is to read it as "It is not the case that p and q are true."

41. (a) Translate each of DeMorgan's laws into a standard English sentence without using the terms conjunction, disjunction or negation.

(b) Use the following statements for p and q to illustrate the two laws:

p: The moon is dark;

q: The night is cold.

42. Show that DeMorgan's laws are correct by using truth tables.

43. Illustrate the use of DeMorgan's laws with the following statements:

p: Unemployment is low.

q: The stock market is going up.

(a) Write English statements for $\sim(p \wedge q)$ and $\sim(p \vee q)$.

(b) Write English statements for the equivalent forms of DeMorgan's laws, clearly indicating which pairs of statements belong together.

The Double Negation Law

Another law that is often useful in simplifying more complex statements is the double negation law. In symbols, it states

$$\sim(\sim p) \Leftrightarrow p.$$

44. (a) Use a truth table to show the double negation law.

(b) Illustrate the double negation law using the statement p: I like music.

45. Use DeMorgan's laws and the double negation law to simplify the following symbolic statements.

(a) $\sim(p \wedge \sim q)$

(b) $\sim(\sim p \vee \sim q)$

The Distributive Laws

We often combine "or" with "and" to form compound statements in the following way: I'll go to the movie, and I'll eat at home or at the Blue Pine Restaurant." Logically, this is equivalent to the statement: I'll go to the movie and I'll eat at home, or I'll go to the movie and eat at the Blue Pine Restaurant." With symbols: $p \wedge (q \vee r) \Leftrightarrow (p \wedge q) \vee (p \wedge r)$. This is one of two laws known as the distributive laws. The second law replaces $\wedge$ by $\vee$ and $\vee$ by $\wedge$.

46. (a) Verify the first distributive law with an appropriate truth table.

(b) Write the symbolic statement for the second distributive law and verify the law with an appropriate truth table.

47. Illustrate the use of the distributive laws with the following statements:

p: I plan to work next summer.

q: I'll enroll for classes in the fall.

r: I'll travel to Asia in the spring.

(a) Write English statements for $p \wedge (q \vee r) \Leftrightarrow (p \wedge q) \vee (p \wedge r)$.

(b) Write English statements for the second distributive law using the same choices for p, q, and r and the form $p \vee (q \wedge r)$.

1.2 DEDUCTION

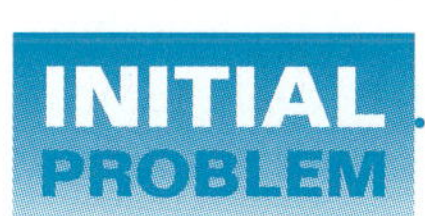

A number of states have provisions that allow initiative measures to be placed before the voters either to pass new laws or to amend the Constitution. These measures qualify for the ballot if supporters collect a sufficient number of nominating signatures. In a recent election in a western state, voters were confronted with an extraordinarily large number of measures. This prompted the second-ranking state official (who is also in charge of elections) to call for more stringent requirements to qualify a ballot measure. What is the reasoning behind the official's decision? Is the call for more stringent requirements justified?

HISTORY

The first teacher of rhetoric was Corax of Syracuse, whose advice was much in demand. The establishment of democracy in Syracuse in 466 B.C. led to massive disputes on claims of property by former exiles. Trials were based on the Athenian system in which each side gave a speech in front of a large jury of at least 201 citizens.

Much of what we believe is based on circumstantial evidence or on assumed relationships. We hear or experience one thing and conclude another: there are a large number of measures, therefore the process must be too easy. How valid is this thinking? In the last section we learned how to assess the truth of statements and conditionals. Now we see how to combine statements to form valid arguments and what can happen when the logical structure breaks down.

LOGICAL ARGUMENTS

When you present a sequence of statements to try to convince someone that some other statement is true, you are presenting an **argument.** The goal of an argument is persuasion, and classically, the art of persuading others by spoken or written language is known as **rhetoric.** In rhetoric there are three recognized means of persuasion available:

1. Appeal to reason *(logos)*.
2. Appeal to emotions *(pathos)*.
3. Appeal by personality and character *(ethos)*. A person well-known to the audience will already have established an impression of his or her personality and character, but even unknown speakers or writers establish such an impression by their speaking or writing.

The first method, appeal to reason, is of interest mathematically. The best argument that appeals to reason uses deductive reasoning, which proceeds step by step from assumed statements, called **premises** or **hypotheses,** to a statement, called the **conclusion.** A typical argument arising from a conditional statement is

PREMISES: If you play your stereo too loud, then you'll blow your ears out.
You play your stereo too loud.
CONCLUSION: Therefore, you will blow your ears out.

Here, the first two sentences are the premises and the third one is the conclusion. A **valid argument** is an argument in which the conclusion must be true *whenever* the premises are true. In ordinary arguments using words, it may be difficult to follow the logic being used (or misused). By introducing notation for the various English sentences involved in an argument, you can translate the argument into symbolic form in order to understand the underlying logical structure of the argument.

In mathematical logic, we reason from absolutely true premises to necessarily true conclusions. This reassuringly solid state of affairs does not hold in real life. In real life, we often must argue from probable premises to a tentative conclusion. Although "too loud" (and for how long) in the conditional above is open to debate, medical science accepts that prolonged exposure to loud noise damages the hearing. Thus, the argument above is valid.

There are three important valid argument forms involving conditional statements that are used repeatedly. The most common one is Modus Ponens.

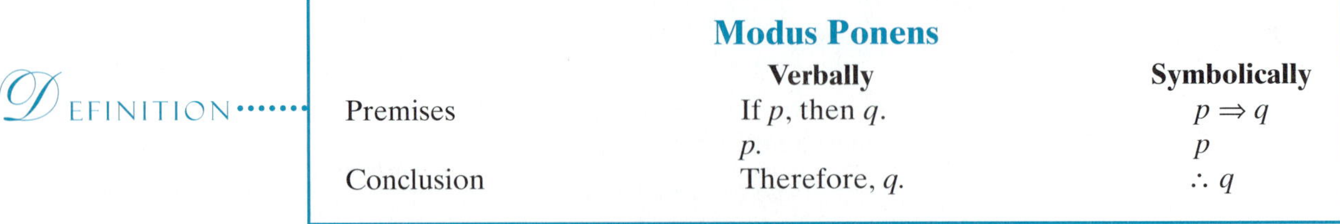

DEFINITION

Modus Ponens

	Verbally	**Symbolically**
Premises	If p, then q.	$p \Rightarrow q$
	p.	p
Conclusion	Therefore, q.	$\therefore q$

TIDBIT

In classical rhetoric an argument from probable premises to tentative conclusion was called an *enthymeme*. This word also refers to a syllogism in which one premise is not explicitly stated.

In words, the **modus ponens** (which is also called the **law of detachment** or **affirming the antecedent**) says that whenever a conditional statement and its hypothesis are true, the conclusion is also true. That is, the conclusion can be detached from the conditional. An example of the use of this law follows.

PREMISES: If a number ends in zero, then it is a multiple of 10.
Forty is a number that ends in zero.
CONCLUSION: Therefore, 40 is a multiple of 10.

Modus ponens is also used in nonmathematical settings.

EXAMPLE 1.8 Identify the premises and conclusion in the following argument as well as the form of the argument.

If you score at least 90% on three midterms, then you are excused from the final.
Your midterm scores are 95%, 91%, and 94%.
Therefore, you are excused from the final.

SOLUTION

PREMISES: If you score at least 90% on three midterms, then you are excused from the final.
Your midterm scores are 95%, 91%, and 94%.
CONCLUSION: Therefore, you are excused from the final.

This argument is an example of modus ponens. ◆

The following truth table shows that modus ponens is a valid argument. The table has columns for each of the basic statements involved in the argument, in this case, p and q. Then we have columns for each of the premises, in this case, $p \Rightarrow q$ and p. Finally, there is a column for the conclusion, in this case, q. In the columns for the basic statements, we insert all possible truth values. Then we locate any rows of the table in which *all* the premises are true. To be a valid argument, the conclusion must also be true in those rows where the premises were all true. There is only one row in the table for which all the premises are true, and we note that the conclusion is also true.

		Premises		Conclusion	
p	q	$p \Rightarrow q$	p	q	
T	T	T	T	T	◀ Premises all true
T	F	F	T	F	
F	T	T	F	T	
F	F	T	F	F	

TIDBIT

The chain rule is also called the hypothetical syllogism, but classical scholars used the term *syllogism* to refer to an argument with two premises and the term *hypothetical syllogism* to refer to *any* syllogism involving a conditional statement.

The following valid argument form is used frequently, often in conjunction with modus ponens.

DEFINITION

Chain Rule

	Verbally	**Symbolically**
Premises	If p, then q.	$p \Rightarrow q$
	If q, then r.	$q \Rightarrow r$
Conclusion	Therefore, if p then r.	$\therefore p \Rightarrow r$

The following argument is an application of this law.

PREMISES: If a number is a multiple of 12, then *it is a multiple of 6.*
If *a number is a multiple of 6,* then it is a multiple of 2.
CONCLUSION: Therefore, if a number is a multiple of 12, it is a multiple of 2.

EXAMPLE 1.9 At lunch on Monday, Sally was chatting with her friend. Sally said, "Ed [Sally's husband] called just a little while ago. He said he has to work late tonight. Whenever he misses Monday night football, he's a real grouch for days. I sure don't look forward to tomorrow." Was there a logical deduction involved here, and if so, was it valid?

SOLUTION There was a deduction involved. One way to analyze the argument is to strip away all of the verbiage by introducing the following notation.

L = Ed works late on Monday. M = Ed misses Monday night football.
G = Ed is a grouch. U = Sally's day is unpleasant.

Thus we have the following statements and conditionals.

PREMISES:	L	(Ed's message)
	$L \Rightarrow M$	(A fact of life not explicitly stated)
	$M \Rightarrow G$	(Ed's personal quirk, well known to Sally)
	$G \Rightarrow U$	(A fact of married life)
CONCLUSION:	Therefore, U.	(Poor Sally)

If the chain rule is applied twice in rows 2–4, the argument reduces to

L
$L \Rightarrow U$
Therefore, U.

Thus, by modus ponens, since Ed had to work late (L), it looks like Sally is in for a bad time. ◆

Notice how the chain rule differs from modus ponens in that the conclusion of the chain rule is a conditional.

We can also check the validity of the chain rule using the following truth table. We observe that on every row where all the *premises* are true, the conclusion is also true.

			Premises		Conclusion	
p	q	r	$p \Rightarrow q$	$q \Rightarrow r$	$p \Rightarrow r$	
T	T	T	T	T	T	◀ Premises all true
T	T	F	T	F	F	
T	F	T	F	T	T	
T	F	F	F	T	F	
F	T	T	T	T	T	◀ Premises all true
F	T	F	T	F	T	
F	F	T	T	T	T	◀ Premises all true
F	F	F	T	T	T	◀ Premises all true

A third valid argument form involving conditional statements is often used in mathematical reasoning.

DEFINITION

Modus Tollens

	Verbally	Symbolically
Premises	If p, then q.	$p \Rightarrow q$
	Not q.	$\sim q$
Conclusion	Therefore, not p.	$\therefore \sim p$

In words, the modus tollens (which is also called the **law of contraposition** or **denying the consequent** says that whenever a conditional statement is true and its conclusion is false, then the hypothesis is also false. Consider the following argument.

PREMISES: If a quadrilateral is a square, then it has four 90° angles.
$ABCD$ does not have four 90° angles.
CONCLUSION: Therefore, $ABCD$ is not a square.

The following truth table shows the validity of modus tollens.

p	q	$p \Rightarrow q$ (Premises)	$\sim q$ (Premises)	$\sim p$ (Conclusion)	
T	T	T	F	F	
T	F	F	T	F	
F	T	T	F	T	
F	F	T	T	T	◀ Premises all true

Notice that in the row where the premises are true, the conclusion is also true. Since a conditional and its contrapositive $\sim q \Rightarrow \sim p$ are logically equivalent, modus tollens and modus ponens are logically equivalent. Thus, you can be doubly certain that modus tollens is a valid argument form.

One of the principles of logic as formulated by Aristotle is that each sentence must be either true or false. Thus if one of the two possibilities "True" or "False" is ruled out, then the other must hold. This argument form is known as the **disjunctive syllogism** or the **Law of the Excluded Middle.**

DEFINITION

Disjunctive Syllogism

	Verbally	Symbolically
Premises	p or q.	$p \vee q$
	Not q.	$\sim q$
Conclusion	Therefore p.	$\therefore p$

p	q	$p \vee q$ (Premises)	$\sim q$ (Premises)	p (Conclusion)	
T	T	T	F	T	
T	F	T	T	T	◀ Premises all true
F	T	T	F	F	
F	F	F	T	F	

The disjunctive syllogism is related to the method of mathematical proof called **argument by contradiction** (also called **indirect reasoning** or **reductio ad absurdum**) in which one shows that if the statement one is trying to prove were false, then a valid argument leads to a contradiction. Thus the statement under consideration cannot be false, and the only other possibility is for the statement to be true.

EXAMPLE 1.10 Use indirect reasoning to show that if a positive whole number has the digit 8 in the ones place, then the square root of that whole number is not a whole number.

SOLUTION Given a whole number, either the square root is a whole number or it is not. Let n be a whole number with an 8 in the ones place. If the square root of n were a whole number, say k, then k must have some digit, d, in its ones place. Now, think about multiplying $k \times k$ by hand. The first step is to multiply together the ones digits of k, that is, $d \times d$, and put the ones digit of the result under the line, and that digit ends up as the ones digit of $k \times k$. For example, consider the computation of $129 \times 129 = 16{,}641$ in Figure 1.1.

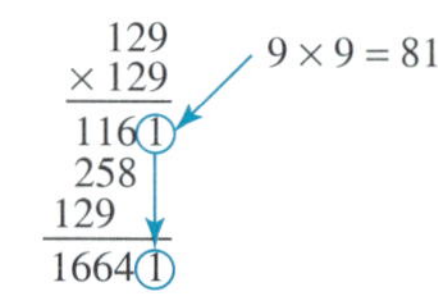

FIGURE 1.1

Thus we see that the ones place of $k \times k$ is filled by the ones place digit of $d \times d$. The square of the digits 0, 1, 2, 3, 4, 5, 6, 7, 8, 9 are 0, 1, 4, 9, 16, 25, 36, 49, 64, 81 of which none has an 8 in its ones place, contradicting the assumption that n does have an 8 in the ones place. We conclude that the square root of n is not a whole number. ◆

The disjunctive syllogism often shows itself in everyday circumstances.

EXAMPLE 1.11 When driving in an unfamiliar area on a cloudy night, you find yourself at an intersection with the North-South road you were looking for. Unfortunately, after all the winding around you have done, you can no longer tell which way is north and which way is south. How does the disjunctive syllogism apply?

SOLUTION When you turn onto the North-South road, you are either going the direction you wish to or you are not. If you were unlucky enough to have chosen the wrong direction, proof of that should appear fairly soon, and you can turn around. ◆

INCORRECT REASONING PATTERNS

People often draw erroneous conclusions by misusing the modus ponens and modus tollens argument forms. These mistakes are referred to as (i) the **fallacy of affirming the consequent** (a misuse of modus ponens) and (ii) the **fallacy of denying the antecedent** (misuse of modus tollens).

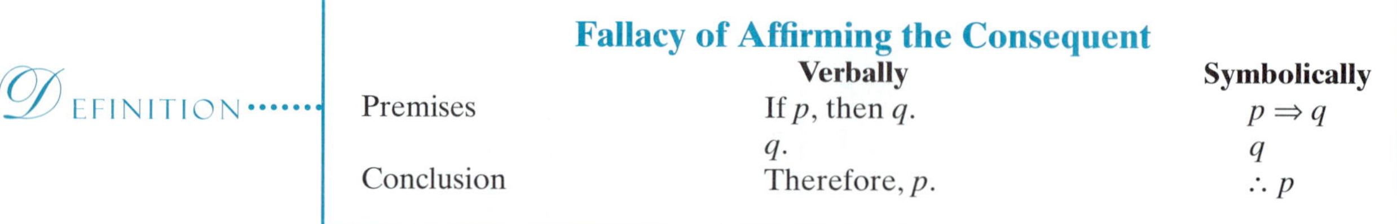
DEFINITION

Fallacy of Affirming the Consequent

	Verbally	**Symbolically**
Premises	If p, then q.	$p \Rightarrow q$
	q.	q
Conclusion	Therefore, p.	$\therefore p$

This fallacy looks like modus ponens, except that the conclusion q is assumed rather than the hypothesis p. The following is an example of this fallacy.

If a person has unhealthy habits, soon he will become ill.
Mr. Frank has had a heart attack.
Therefore, Mr. Frank must have had unhealthy habits.

This invalid argument suggests that people who have heart attacks must have unhealthy habits. However, many people have excellent health habits, yet die from heart attacks. Jim Fixx, a noted running expert in the 1970s and 80s, collapsed and died while running. Had we been given the premises "If a person has unhealthy habits, soon he will become ill" and "Mr. Frank has unhealthy habits," we could validly infer that Mr. Frank will soon become ill.

The following truth table shows why affirming the consequent is a fallacy. We see that there is a row in the truth table on which all the premises are true, but the conclusion is false.

		Premises		Conclusion	
p	q	$p \Rightarrow q$	q	p	
T	T	T	T	T	◀ Premises all true
T	F	F	F	T	
F	T	T	T	F	◀ Premises all true but the conclusion is false! Thus, this argument is invalid.
F	F	T	F	F	

DEFINITION

Fallacy of Denying the Antecedent

	Verbally	Symbolically
Premises	If p, then q,	$p \Rightarrow q$
	not p.	$\sim p$
Conclusion	Therefore, not q.	$\sim q$

This fallacy looks like modus tollens, except that the negation of the hypothesis p is assumed rather than the negation of the conclusion q. Following is an example of this fallacy.

If a person has unhealthy habits, soon he will become ill.
I have no unhealthy habits.
Therefore, I will not become ill.

The premises "If a person has unhealthy habits, soon he will become ill" and "I have never been ill" could validly imply that "I have no unhealthy habits"; this is the contrapositive of the original implication. Unfortunately, healthy living, while prudent, is no guarantee of continued good health.

The following truth table shows why denying the antecedent is a fallacy. Again we see that there is a row in the truth table in which all the premises are true, but the conclusion is false.

		Premises		Conclusion	
p	q	$p \Rightarrow q$	$\sim p$	$\sim q$	
T	T	T	F	F	
T	F	F	F	T	
F	T	T	T	F	◀ Premises all true but the conclusion is false! Thus, this argument is invalid.
F	F	T	T	T	◀ Premises all true

Another way to view this fallacy is to replace the "If p, then q" with its equivalent contrapositive "If $\sim q$, then $\sim p$." Then fallacy of denying the antecedent is equivalent to the fallacy of affirming the consequent.

EXAMPLE 1.12 Translate each of the following arguments into symbolic form. Identify the form of the argument and conclude whether or not it is a valid argument.

(a) If the sun is shining, then Professor Rossini wears a hat.
Professor Rossini is wearing a hat. Therefore, the sun is shining.
(b) If the weather is cold, then my mother wears a sweater.
The weather is cold. Therefore my mother is wearing a sweater.

SOLUTION

(a) Let p be the statement "The sun is shining," and let q be the statement "Professor Rossini wears a hat." The symbolic form of the argument is

PREMISES: $p \Rightarrow q$
q
CONCLUSION: $\therefore p$

This is an example of the fallacy of affirming the consequent; thus it is an invalid argument.

(b) Let r be the statement "The weather is cold," and let s be the statement "My mother wears a sweater." The symbolic form of the argument is

PREMISES: $r \Rightarrow s$
r
CONCLUSION: $\therefore s$

This is an example of the modus ponens; thus it is a valid argument. ◆

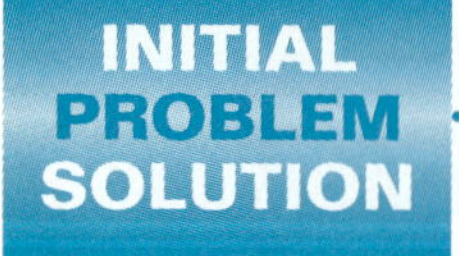

A number of states have provisions that allow initiative measures to be placed before the voters either to pass new laws or to amend the Constitution. These measures qualify for the ballot if supporters collect a sufficient number of nominating signatures. In a recent election in a western state, voters were confronted with an extraordinarily large number of measures. This prompted the second ranking state official (who is also in charge of elections) to call for more stringent requirements to qualify a ballot measure. What is the reasoning behind the official's decision? Is the call for more stringent requirements justified?

SOLUTION The state official might have been reasoning as follows:

If the nominating process is too easy, there will be too many measures.
There were too many measures.
Therefore, the nominating process is too easy.

This is not a valid argument. If we assume that the state official is wise enough not to intend an invalid argument, then we must conclude that she feels there should be a limit on the number of initiative measures, and anything that accomplishes that goal should be considered—for example, making the qualifying process more difficult.

PROBLEM SET 1.2

In problems 1 through 4, state the premises (or hypotheses) and conclusion in each of the arguments. Are there any premises that aren't explicitly stated?

1. **(a)** If the room is warm, then I'll be uncomfortable. The room is warm, so I'll be uncomfortable.

(b) If the weather is bad, I'll go to the movies. It's raining heavily, so I'll go to the movies.

2. **(a)** If the weather is good, I'll want to play golf. If I decide to play golf, I'll see if Marty can go. It's warm and clear outside; I guess I'll call Marty.

(b) If Dr. Goldberg teaches the course, I'll register for her class. If you take a class from Dr. Goldberg, you can count on having to do a term paper. Dr. Goldberg is going to teach the course, so I'll have to write a term paper.

3. **(a)** If the weather is good, Barry will paint the house. Barry didn't paint the house, so the weather wasn't good.

(b) If you average at least 90% on the tests, you'll get an A for the term. You didn't get an A for the term, so you didn't average 90% on the tests.

4. **(a)** Jenna is going to go swimming or play tennis. Jenna didn't go swimming, therefore she must have played tennis.

(b) We're either going to the play or the movie. The tickets to the play are all sold out, so we're going to the movie.

In problems 5 through 18, translate the argument into symbolic form. Determine whether the argument is valid or invalid. You may compare the form of the argument to one of the four standard forms or use a truth table.

5. If the movie is good, the people will go.
The people will go to the movie.
Therefore the movie is good

6. If the sun is shining, I'll wear a hat.
The sun isn't shining.
Therefore I won't wear a hat.

7. If it's cold outside, my hands will be cold.
It's freezing outside.
Therefore my hands will be cold.

8. These shoes are not expensive
If shoes are expensive, I won't buy them.
Therefore I won't buy these shoes

9. If we miss the bus, we'll have to walk.
If we walk, then we'll be late.
Therefore if we miss the bus, we'll be late.

10. If you do the work, you'll get paid.
If you get paid, you can go to the show.
Therefore if you do the work, you can go to the show.

11. If Spike Lee is the director, then the movie should be good. Spike Lee didn't direct the movie, so it probably isn't good.

12. The night is cold and dark. The night is not dark or it is cold. Therefore the night is not cold.

13. If Carlos passes his entrance exam, he will attend college. Carlos will not be attending college; therefore he did not pass his entrance exam.

14. If you arrive on time, you'll get a good seat. If you get a good seat, you'll enjoy the play. Therefore if you didn't enjoy the play, you didn't arrive on time.

15. If I can't go to the movie, then I'll go to the park. I can go to the movie. Therefore, I will not go to the park.

16. If you score at least 90%, then you'll earn an A. If you earn an A, then your parents will be proud. You have proud parents. Therefore, you scored at least 90%.

17. If you work hard, then you will succeed. You do not work hard. Therefore you will not succeed.

18. If it doesn't rain, then the street won't be wet. The street is wet. Therefore it rained.

In problems 19 through 26, identify which form of argument (Modus Ponens, the chain rule, Modus Tollens, or disjunctive syllogism) is being used.

19. If Joe is a professor, then he is well educated. If you are well educated, then you went to college. Joe is a professor, so he went to college.

20. If you have children, then you are an adult. Paul is not an adult, so he has no children.

21. Whenever the weather is bad, I stay inside and work. It's raining heavily outside. I'll stay inside and work.

22. Either I get a raise or I'm going to look for a new job. I didn't get a raise, so I'm going to look for a new job.

23. If I don't have enough money to buy gas, I will ride the bus. When I ride the bus I am always late. I can't afford to buy gas, so I'm going to be late.

24. If you get to the store on time, you can pick up a carton of ice cream. You didn't pick up any ice cream, so you didn't get to the store on time.

25. Kim is going to have her old car painted or buy a new car. Kim didn't buy a new car, so she had her old one painted.

26. If I don't eat breakfast, I'll be hungry by 10:00. If I'm hungry before noon, I always snack before lunch. I skipped breakfast, so I will have a snack before lunch.

27. The *Initial Problem* from section 1.1 was to find a symbolic form for an exercise in logic written by Lewis Carroll. The final form of the argument was:

$$\sim p \Rightarrow \sim q$$
$$\sim r \Rightarrow \sim s$$
$$\sim q \Rightarrow \sim t$$
$$u \Rightarrow \sim p$$
$$\sim t \Rightarrow \sim r$$

Use the antecedent: u (a kitten is green-eyed) and provide a logical conclusion for the argument. State the conclusion in terms of the original statements and show that the argument is valid.
Hint: You may need to rearrange the order of the argument.

In problems 28 through 34, provide a logical conclusion for the argument. Show that the argument is valid with the conclusion you provided by constructing appropriate truth tables or comparing the argument to one of the standard forms of argument.

28. If you study hard, then you will pass the course. You study hard.

29. You will do well in math if you do your assignments. You do not do well in math.

30. If the team wins the rest of their games, they will go to the tournament. The team didn't lose any more games.

31. Michelle finishes her assignment or she goes to the movie. Michelle doesn't go to the movie.

32. If Jon passes algebra and biology he'll have his math and science requirements completed. Jon didn't complete his math and science requirements.

33. If Gunder majors in science or engineering, he'll go to the technical university. Gunder is going to major in engineering and design.

34. If the store is open on the holiday, I'll have to work. If I have to work, I won't get the paper written for history. I didn't have to work.

In problems 35 through 40, show that the argument is invalid and identify the incorrect reasoning patterns involved.

35. If it's snowing, the bus will be late. The bus was late, therefore it was snowing.

36. If it's snowing, the bus will be late. It isn't snowing, so the bus will be on time.

37. If Tonya works hard and has a good attitude she will be promoted. Tonya worked hard but had a bad attitude, therefore she didn't get promoted.

38. I'll vacuum the rug if you'll fix the meal. I vacuumed the rug, so you have to fix the meal.

39. We always have to do term papers when we take a class from Smyth. We didn't take a class from Smyth, so we won't have to do a term paper.

40. We'll have to work on Saturday if the project isn't completed by Friday. We finished the project Friday morning, so we won't have to work on Saturday.

41. Use indirect reasoning to show that if two lines are perpendicular to another line, then they are parallel to each other. Use the following information: the sum of the angles in a triangle is 180°; if two lines are perpendicular, the angle between them is 90°; if two lines are parallel, they don't meet. Assume that the two lines aren't parallel. What are the consequences?

42. Use indirect reasoning to show that the square root of -1 is not a real number. Use the following information: every real number can be classified as positive, negative, or zero. Assume the square root of -1 is a real number. What are the consequences?

EXTENDED PROBLEMS

43. Find examples in advertising for each of the valid forms of argument. Identify the premises, including those that are implied, and the conclusion for each argument.

44. Find examples in the opinion section of the newspaper for each of the valid forms of argument. Identify the premises, including those that are implied, and the conclusion for each argument.

45. Find examples in advertising for each of the fallacies (incorrect reasoning patterns) in the section. Identify the premises, including those that are implied, and the conclusion for each argument.

46. Find examples in the opinion section of the newspaper for each of the fallacies (incorrect reasoning patterns) in this section. Identify the premises, including ones that are implied, and the conclusion for each argument.

47. Truth and validity are two basic concerns when discussing arguments. Is it possible for an argument to be valid if it is not truthful? Discuss these two concepts, using truth tables and/or examples.

For problems 48 and 49, show that the arguments are valid or invalid. Use truth tables or identify the patterns of reasoning involved.

48. $(\sim p \vee \sim q) \Rightarrow (r \wedge s)$
$r \Rightarrow t$
$\sim t$
____
therefore p

49. p
$p \vee q$
$q \Rightarrow (r \Rightarrow s)$
$t \Rightarrow r$
____
Therefore $\sim s \Rightarrow \sim t$.

1.3 CATEGORICAL SYLLOGISMS

The following headline appeared on June 15, 1995:

Senate moves to bar "filth" on the Internet.

The brief summary that introduces the story from the *New York Times*

News Service says, "Few object as lawmakers endorse strict penalties for those who spread pornography on computer networks." The article described what it called "the most aggressive step by Congress to regulate cyberspace," and how the Senate voted. The closing paragraph says, "Many senators viewed Wednesday's debate as something of a sideshow to the broader telecommunications bill, though few were prepared to cast a vote against the measure, which opponents might later attack as support for pornography."

What is meant by that last sentence? Is there a valid concern that a vote against the bill can be interpreted as a vote in support of pornography? What are the implications of this interpretation to the political process?

In the last section we looked at ways to analyse arguments and determine their validity based on the structure of the argument in a symbolic form either by using truth tables or comparing the form of the given argument to the forms of argument that are known to be valid. Now we look at another form of argument, called a syllogism, that can be analyzed by using graphical forms called Euler diagrams.

EULER DIAGRAMS

The argument forms we studied in the preceding section involved reasoning with sentences where the fundamental difficulty is dealing with compound sentences, such as conditionals and disjunctions. An entirely different type of difficulty arises when sentences involve quantifiers—that is, words such as *all, some,* and *every.* The truth of a sentence involving a quantifier depends on whether or not some or all things in one category are also in another category. For example, the sentence "Every president of the United States has been a man" is true (in 1996) because the 41 persons in the category of people who have served as president of the United States are also in the category consisting of men. An **Euler diagram,** which uses circles to represent interrelationships among statements, can be used to illustrate the truth of this sentence by showing the set of people who have served as president of the United States as a subset of the set of men (Figure 1.2). A brief review about sets and set relationships is included in Topic 1.

The sentences we will be considering in this section are of the following six types:

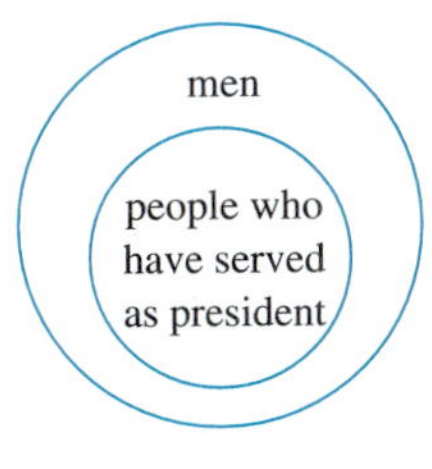

FIGURE 1.2

1. All X are Y. (For example, "All mammals are warm-blooded.")
   This sentence is true when the Euler diagram is like that in Figure 1.3.
2. No X are Y. (For example, "No snakes have legs.")
   This sentence is true when the Euler diagram is like that in Figure 1.4.
3. Some X are Y. (For example, "Some cars are convertibles.")
   This sentence is true when the Euler diagram looks like the one in Figure 1.5. The dot in the intersection of X and Y indicates that there is at least one element in that set.

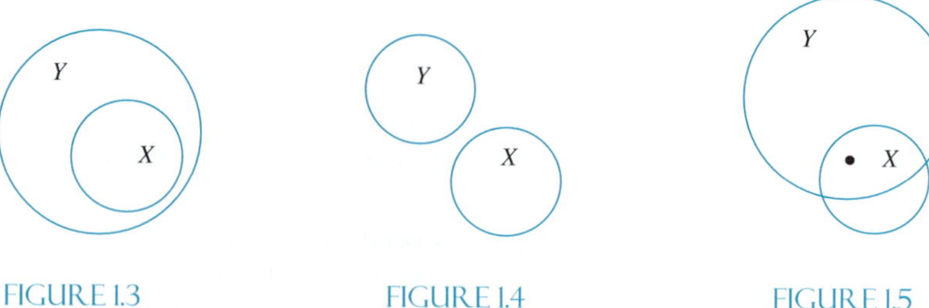

FIGURE 1.3 FIGURE 1.4 FIGURE 1.5

4. Some X are not Y. (For example, "Some cars are not air conditioned.") This sentence is true when the Euler diagram is as in Figure 1.6. The dot in X, but not in Y, indicates that there is at least one element in X that is not in Y.
5. x is a Y, where x is one element. (For example, "My car is a convertible.") This sentence is true when the Euler diagram looks like the one in Figure 1.7. The dot represents the specific person or thing x.
6. x is not a Y. (For example, "My car is not a convertible.") This sentence is true when the Euler diagram looks like the one in Figure 1.8. Again the dot represents the specific person or thing x.

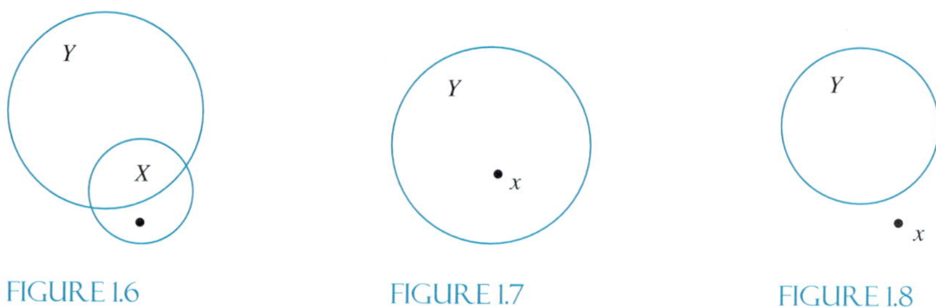

FIGURE 1.6 FIGURE 1.7 FIGURE 1.8

TIDBIT

This most famous of all syllogisms is attributed to Aristotle, who was a student of Plato. Plato was greatly influenced by his friendship with Socrates, though Plato was not actually a student of Socrates.

VALID CATEGORY ARGUMENTS

A **syllogism** is an argument consisting of two statements called **premises** followed by one statement called the **conclusion.** A syllogism that involves the use of statements about categories is called a **categorical syllogism.** A syllogism is **valid** if whenever the two premises are true, the conclusion must also be true. Consider the following syllogism.

PREMISES: All men are mortal.
Socrates is a man.
CONCLUSION: Therefore, Socrates is mortal.

The truth of the first premise requires that the Euler diagram for men and mortal beings be as in Figure 1.9(a). The truth of the second premise requires that the Euler diagram for men and Socrates be as in Figure 1.9(b). If Figures 1.9(a) and (b) are combined into Figure 1.9(c), we see that the conclusion must be true. This syllogism is valid.

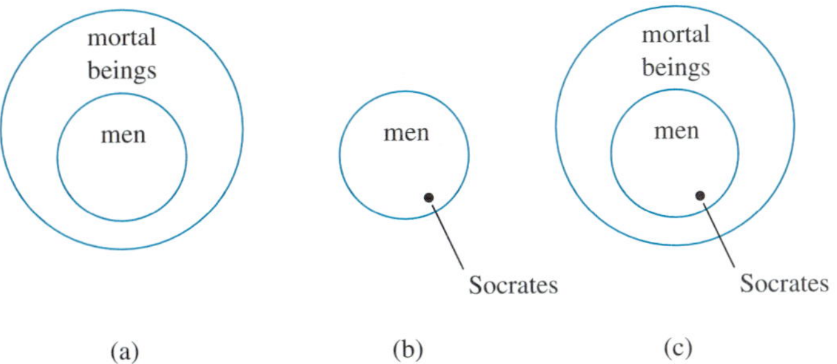

FIGURE 1.9

EXAMPLE 1.13 Identify the premises and conclusion in the following categorical syllogism, and show that it is valid.

All tall men in town shop at Al's Big and Tall Men's Store.
All big men in town shop at Al's Big and Tall Men's Store.
Therefore, all the men in town who are tall or big shop at Al's big and Tall Men's Store.

SOLUTION The premises are:

All tall men in town shop at Al's Big and Tall Men's Store.
All big men in town shop at Al's Big and Tall Men's Store.

The conclusion is

Therefore, all the men in town who are tall or big shop at Al's Big and Tall Men's Store.

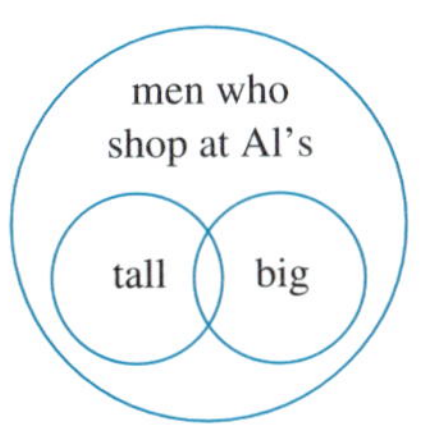

FIGURE 1.10

We have three categories to consider: tall men, big men, men who shop at Al's. We consider the Euler diagram in Figure 1.10. The categories of tall men and big men are both placed inside the category of men who shop at Al's because we always assume the premises are true. The categories of tall men and big men are given their most general position relative to each other because the premises give us no additional information about the relationship of these two categories. You can then see that any man who is tall *or* big shops at Al's. ◆

Sometimes an argument involving categories has more than two premises, and sometimes you may need to draw more than one Euler diagram. An argument is valid if the conclusion must be true whenever the premises are true.

EXAMPLE 1.14 Identify the premises and conclusion, and show with categories that the following is a valid argument.

All tall men in town shop at Al's Big and Tall Men's Store.
All big men in town shop at Al's Big and Tall Men's Store.
Bubba weighs 300 pounds (so he is tall or big).
Therefore, Bubba shops at Al's Big and Tall Men's Store.

SOLUTION The premises are:

All tall men in town shop at Al's Big and Tall Men's Store.
All big men in town shop at Al's Big and Tall Men's Store.
Bubba weighs 300 pounds (so he is tall or big).

The conclusion is

Therefore, Bubba shops at Al's Big and Tall Men's Store.

We now need to put a dot for Bubba somewhere on our Euler diagram from Figure 1.10, but the premises do not tell us whether Bubba is just tall, just big, or both tall and big. Thus we need to consider the three cases in Figure 1.11 depending on whether Bubba is just tall, just big, or both tall and big. We see that in all cases the conclusion is true.

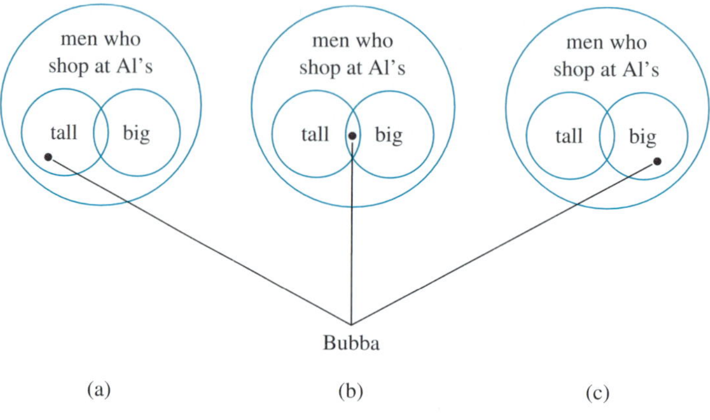

FIGURE 1.11 ◆

EXAMPLE 1.15 Use Euler diagrams to determine if the following argument is valid.

All mammals are warm-blooded animals.
A snake is not warm-blooded.
Therefore, a snake is not a mammal.

SOLUTION The truth of the first premise requires that the Euler diagram for *warm-blooded animals* and *mammals* look like the one in Figure 1.12(a). The truth of the second premise requires that the Euler diagram for a *snake* and the *warm-blooded animals* look like the one in Figure 1.12(b). When the Euler diagrams in Figures 1.12(a) and (b) are combined into Figure 1.12(c), we see that the conclusion must be true. Thus, the argument is valid.

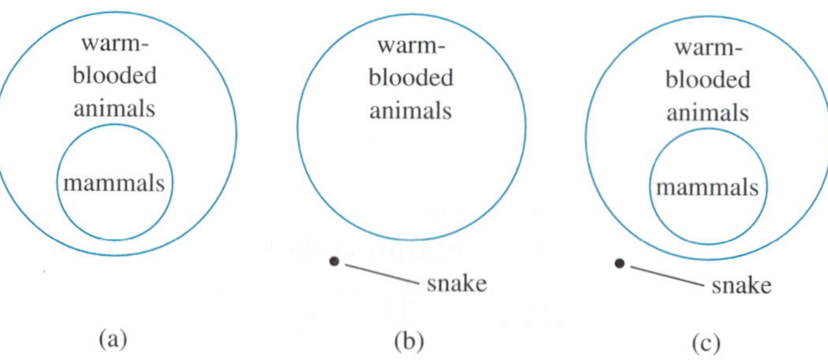

FIGURE 1.12 ◆

INVALID CATEGORY ARGUMENTS

If the conclusion of an argument is not guaranteed by the truth of the premises, then the syllogism is **invalid.** Consider the following:

PREMISES: All presidents have been men.
William Jennings Bryan was a man.
CONCLUSION: Therefore, William Jennings Bryan was president.

The truth of the first premise requires that the Euler diagram for presidents and men look like the one in Figure 1.13(a). The truth of the second premise requires that the Euler diagram for men and William Jennings Bryan look like the one in Figure 1.13(b). We know that most men weren't presidents, and there is not information in the premises to force us to put the dot representing Bryan in a particular location inside the category of men, in particular, not in the category of presidents. When Figures 1.13(a) and (b) are combined into Figure 1.13(c), we see that the conclusion is false. This syllogism is invalid.

TIDBIT

William Jennings Bryan, the most famous orator of his time, was the Democratic Party's candidate for president in the 1896 election. He lost to the Republican nominee William McKinley.

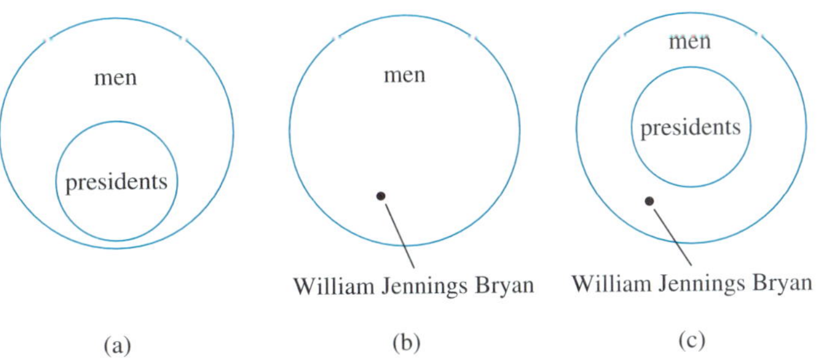

FIGURE 1.13

If the name *William Jennings Bryan* in the syllogism is replaced everywhere by the name *William McKinley,* then the conclusion would be true, but the truth of the conclusion would be unrelated to the truth of the premises. The syllogism is invalid because the truth of the premises does not guarantee the truth of the conclusion. If an argument with categories is invalid, it takes only one Euler diagram to prove it, that is, one in which the premises are true and the conclusion is false. For this syllogism, the case of William Jennings Bryan shows the syllogism is invalid.

EXAMPLE 1.16 Show that the following argument is invalid.

All rock stars have green hair.
No presidents of banks are rock stars.
Therefore, no presidents of banks have green hair.

SOLUTION An Euler diagram that represents this argument is shown in Figure 1.14, Where G represents all people with green hair, R represents all rock stars, and P represents all bank presidents. Note that Figure 1.14 allows for presidents of banks to have green hair, since the circles G and P may have an element in common. Thus the argument, as stated, is invalid since the premises can be true while the conclusion is false. ◆

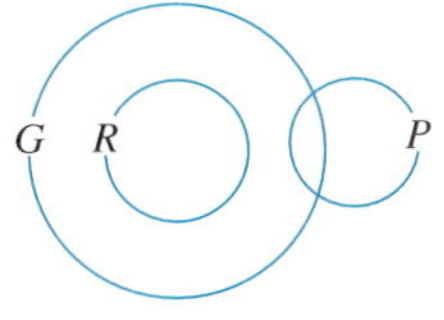

FIGURE 1.14

Any invalid argument involving categories is called a **category fallacy.** When arguments are presented in ordinary English, some of these can slip by a person's logical defenses. If you take time to think carefully or if you sketch an Euler diagram to represent the situation, then you should be able to spot category fallacies. It is very important to consider all possible relationships among the categories that are consistent with the premises before drawing any conclusion. We illustrate this in the next example.

EXAMPLE 1.17 Use Euler diagrams to determine the validity of the following argument.

All professors are teachers.
Some professors are researchers.
Therefore, some researchers are not teachers.

SOLUTION The conclusion seems to be true, but that does not tell us whether or not the argument is valid. The truth of the first premise requires that the Euler diagram for professors and teachers be as in Figure 1.15(a). The truth of the second premise requires that the Euler diagram for professors and researchers be as in Figure 1.15(b) where the dot indicates that there is at least one professor who is a researcher. The two Euler diagrams in Figures 1.15(a) and (b) can be combined as in Figure 1.15(c) *or* as in Figure 1.15(d), since both Figures 1.15(c) and (d) are consistent with the premises. To be a valid argument the truth of the conclusion would need to follow from both Figures 1.15(c) and (d). Since the Euler diagram in Figure 1.15(d) shows that the conclusion can be false, the argument is invalid.

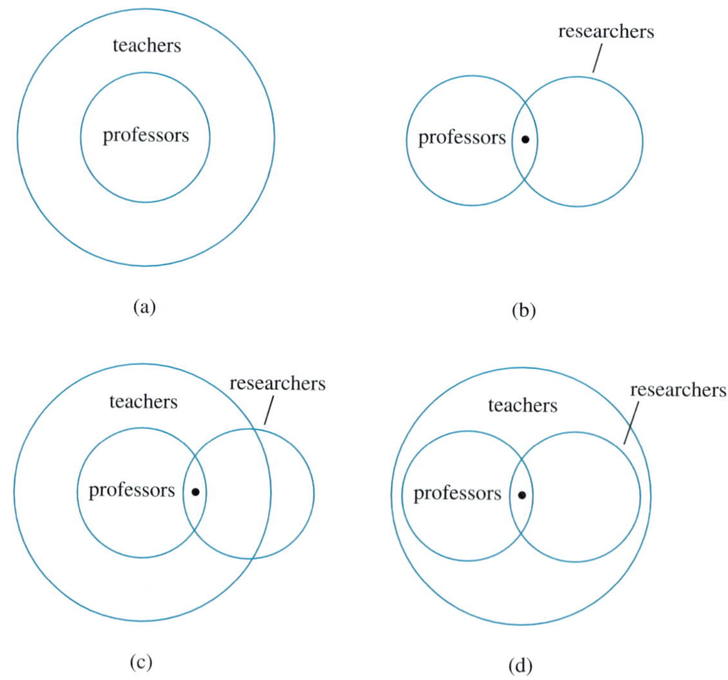

FIGURE 1.15

◆

INITIAL PROBLEM SOLUTION

The following headline appeared on June 15, 1995:

Senate moves to bar "filth" on the Internet.

The brief summary that introduces the story from the *New York Times* News Service says, "Few object as lawmakers endorse strict penalties for those who spread pornography on computer networks." The article described what it called "the most aggressive step by Congress to regulate cyberspace" and how the Senate voted. The closing paragraph says, "Many senators viewed Wednesday's debate as something of a sideshow to the broader telecommunications bill, though few were prepared to cast a vote against the measure, which opponents might later attack as support for pornography."

What is meant in that last sentence? Is there a valid concern that a vote against the bill can be interpreted as a vote in favor of pornography? What are the implications of this interpretation to the political process?

SOLUTION

There is an implied syllogism that goes something like this:

Everyone who supports pornography will vote against the bill.
You voted against the bill.
Therefore, you support pornography.

As we have seen, this form of argument is not valid (and in this case, most likely not true). It is, however, a prevalent form of logic in our lives, perhaps based on a perspective of the world in polarized terms: "It's us or them, and if you're not for us, you're against us." It has a chilling effect on political debate and, several times in our history, serious political consequences.

PROBLEM SET 1.3

In problems 1 through 4, use the given Euler diagram to determine which of the statements are true. Assume there is at least one person or object in every region within the circles.

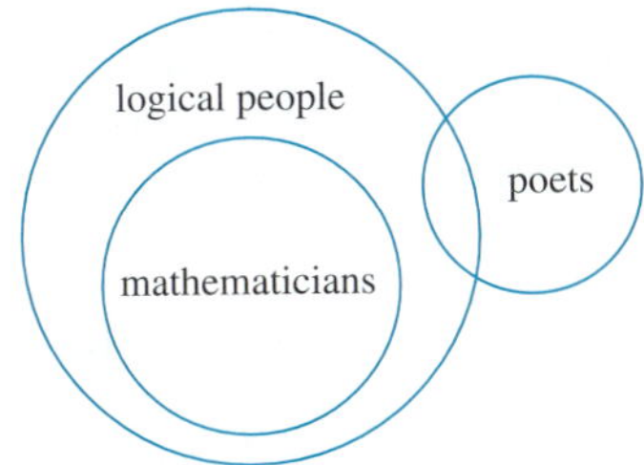

1. **(a)** All mathematicians are logical.
   (b) Lewis Carroll was a logical person.
   (c) Logical people are mathematicians.
   (d) Some poets are mathematicians.
   (e) All poets are logical.

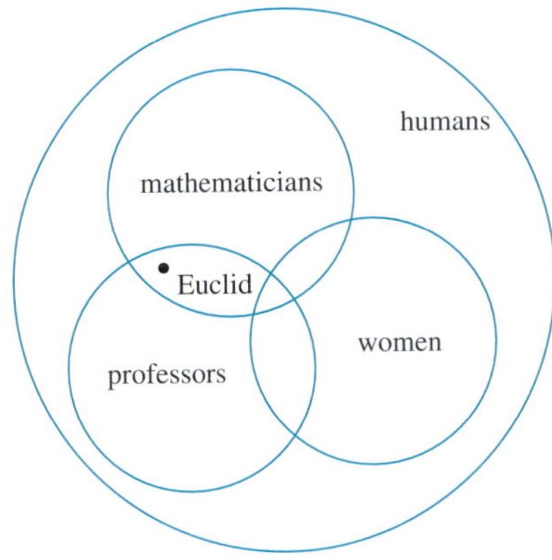

2. **(a)** All women are mathematicians.
   (b) Euclid was a woman.
   (c) All mathematicians are men.
   (d) All professors are humans.
   (e) Some professors are mathematicians.
   (f) Euclid was a mathematician and human.

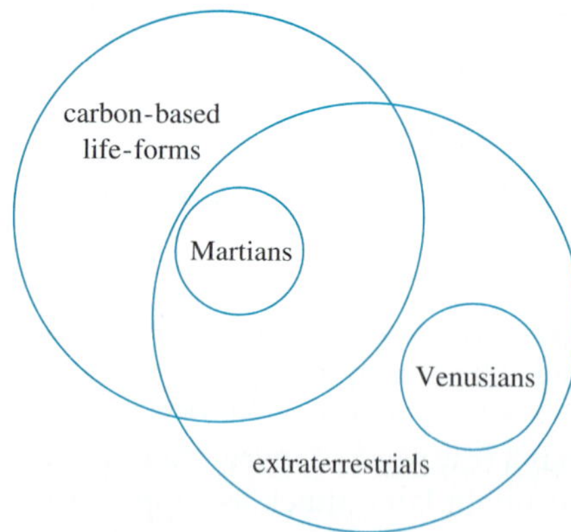

3. **(a)** All Martians are carbon-based life-forms.
   (b) All Venusians are carbon-based life-forms.
   (c) Some carbon-based life-forms are extraterrestrials.
   (d) Some extraterrestrials are not carbon-based life-forms.
   (e) Some Venusians are carbon-based life-forms.
   (f) Some Martians are extraterrestrials.

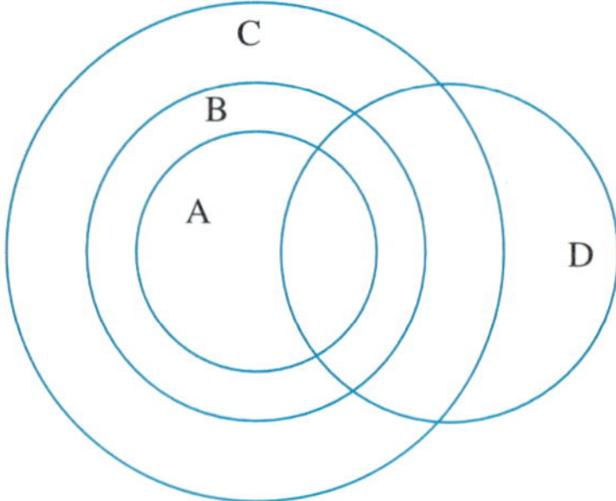

4. **(a)** Every A is a B.
   (b) Every C is a B.
   (c) Some D's are A's.
   (d) Some D's are not C's.
   (e) Some B's are not C's.

In problems 5 through 24, use Euler diagrams to determine the validity of the arguments.

5. All actors are handsome.
   Some actors are tall.
   Therefore, some handsome people are tall.
6. Some arps are bomps.
   All bomps are cirts.
   Therefore, some arps are cirts.
7. All students at Big Time University have to take a World Views class.
   Janet is registered at Big Time University.
   Therefore, Janet has to take a World Views class.
8. No candy is good for you.
   Yogurt is good for you.
   Therefore, yogurt is not candy.
9. No prime numbers are divisible by 2.
   Thirty-five is not divisible by 2.
   Therefore, 35 is not a prime number.
10. All people who vote have to be registered.
    Manuel voted in the election.
    Therefore, Manuel is registered.
11. All equilateral triangles are equiangular.
    All equiangular triangles are isosceles.
    Therefore, all equilateral triangles are isosceles.
12. Some women are teachers.
    All teachers are college graduates.
    Therefore, all women are college graduates.
13. All football players are extroverts.
    Tony is a football player.
    Therefore, Tony is an extrovert.
14. All penguins are elegant swimmers.
    No elegant swimmers can fly.
    Therefore, penguins do not fly.
15. Some men are teachers.
    Sam Jones is a teacher.
    Therefore, Sam Jones is a man.
16. All weight lifters are strong.
    Professor Jones is weak.
    Therefore, Professor Jones is not a weight lifter.
17. All philosophers are intelligent.
    Sigmund Freud was intelligent.
    Therefore, Sigmund Freud was a philosopher.
18. All numbers that are divisible by 6 are divisible by 3.
    My age is divisible by 3.
    Therefore, my age is divisible by 6.
19. Some senators are conservationists.
    No conservationists are pro-development.
    Therefore, some senators are not pro-development.
20. Some houses are made of wood.
    All paper is made of wood.
    Therefore, houses are made of paper.
21. All squares are rectangles.
    All rectangles are quadrilaterals.
    Some quadrilaterals are equilateral.
    Therefore, all squares are equilateral.
22. All chimpanzees are monkeys.
    All monkeys are mammals.
    Some mammals have two legs.
    Therefore, some chimpanzees have two legs.
23. All logicians are mathematicians.
    All mathematicians are philosophers.
    Some philosophers believe in divine beings.
    Therefore, some logicians believe in divine beings.
24. Some people love physics.
    All people who love physics love mathematics.
    Some people who love mathematics love music.
    Therefore, some people love music.

In problems 25 through 40, determine a valid conclusion for each of the given premises. If you can't determine a valid conclusion, show an appropriate Euler diagram.

25. All professional football players are strong.
Joe Montana was a professional football player.

26. All lizards are reptiles.
All reptiles are cold-blooded.

27. No Venusians are carbon-based life-forms.
Some extraterrestrials are Venusians.

28. All Martians are carbon-based life-forms.
Beldar is a Martian.

29. No sane person is an early riser.
My roommate is an early riser.

30. All students have to pay athletic fees.
Joe didn't have to pay athletic fees.

31. All rich people own fancy cars.
Deon owns a fancy car.

32. Some lawyers are dishonest.
Some dishonest people are rich.

33. Some politicians are economists.
No politicians are trustworthy.

34. Some actors are singers.
All singers can read music.

35. All freshmen and sophomore students must take a P.E. class.
All students who take a P.E. class have to buy their own locks.
Kareem is a sophomore.

36. All judges are lawyers.
Some politicians are lawyers.
Margaret isn't a lawyer.

37. All musicians can read music.
All composers can read music.
Jack can't read music.

38. All engineering students have to take physics.
No physics students have to take biology.
Kim-Li is majoring in engineering.

39. All men who are big or tall shop at Al's.
Tony doesn't shop at Al's.
Tony is short.

40. The senator was either a Republican or a Democrat.
None of the Republicans voted for an increase in the capital gains tax.
None of the Democrats voted for a reduction in education funds.
The senator voted for an increase in the capital gains tax.

EXTENDED PROBLEMS

41. In section 1.2, three forms of logical fallacy were introduced. There are many other forms of logical fallacies, including the fallacy of faulty generalization, which is relevant when working with categories. Use an Euler diagram to show that the following argument is not valid.

John has friends who are criminals.
John must be a criminal.

42. Give an example from politics of the fallacy of false generalization.

43. Find two examples from advertising that use the fallacy of false generalization.

44. Analyze the argument form of Modus Ponens with an Euler diagram.

45. Analyze the argument form of Modus Tollens with an Euler diagram.

46. Analyze the argument form of the chain rule with an Euler diagram.

1.4 PROBLEM SOLVING

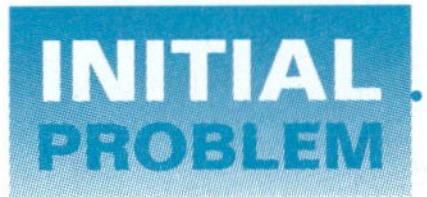

A friend bets you that at least two people in the group of five friends has the same astrological sign. Should you take the bet or not?

Once, at an informal meeting, a social scientist asked a mathematics professor, "What's the main goal of teaching mathematics?" The reply was, "Problem solving." In return, the mathematician asked, "What is the main goal of teaching the social sciences?" Once more the answer was, "Problem solving." All successful engineers, scientists, social scientists, lawyers, accountants, physicians, business managers, and other professionals must be good problem solvers. Although the problems that people encounter may be very diverse, there are common elements and an underlying structure that can help to facilitate problem solving. This section introduces a problem-solving process together with several strategies that will help you solve problems.

PÓLYA'S FOUR STEPS

A famous mathematician, George Pólya, devoted much of his teaching to helping students become better problem solvers. His major contribution is what has become known as **Pólya's four-step process** for solving problems.

STEP 1: Understand the Problem

- Do you understand all the words?
- Can you restate the problem in your own words?
- Can you indentify what information is given?
- Do you know what the goal is?
- Is there enough information?
- Is there extraneous information?
- Is this problem similar to another problem you have solved?

STEP 2: Devise a Plan

Try using one of the following strategies. (A **strategy** can be defined as an artful means to an end.) This list does not include *all* possible strategies, just some of the more useful ones.

1. Guess and Test
2. Look for a Pattern
3. Make a List
4. Use a Variable
5. Look for a Formula
6. Solve an Equation or Inequality
7. Draw a Picture
8. Draw a Graph
9. Draw a Diagram
10. Use a Model
11. Do a Simulation

STEP 3: Carry Out the Plan

- Implement the strategy or strategies that you have chosen until the problem is solved or until a new course of action suggests itself.
- Give yourself a reasonable amount of time in which to solve the problem. Impatience is one of the biggest difficulties in problem solving; even geniuses must take time to think through an unfamiliar problem.
- If you are not successful, seek hints from others or put the problem aside for awhile. (You may have a flash of insight, when you least expect it!)
- Do not be afraid of starting over. A fresh start and a new strategy will often lead to success.

STEP 4: Look Back

- Is your solution correct? Does your answer satisfy the statement of the problem?
- Can you see an easier solution?
- Can you see how you can extend your solution to a more general case?

Usually, a problem is stated in words, either orally or in writing. To solve the problem, translate the words into an equivalent problem using mathematical symbols, solve this equivalent problem, and interpret the answer. This process is summarized in Figure 1.16.

TIDBIT

Pólya's book, *How To Solve It*, has been translated into more than 15 languages. As to why he chose mathematics, he said, "It is . . . not quite wrong to say: I thought I am not good enough for physics and too good for philosophy. Mathematics is in between."

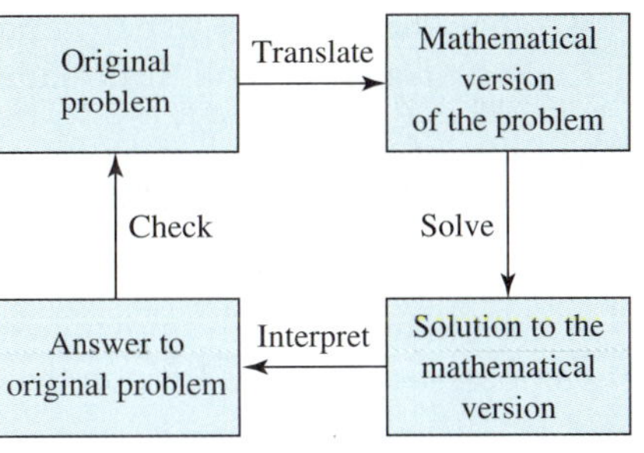

FIGURE 1.16.

Learning to utilize Pólya's four steps and the diagram in Figure 1.16 are the first steps in becoming a good problem solver. In particular, the "Devise a Plan" step is very important. In this section and throughout the book, you will learn how to use the strategies listed under the "Devise a Plan" step. However, selecting an appropriate strategy is critical! As we have worked with students who were successful problem solvers, we asked them to share **clues** that they discovered in the statements of problems that helped them select appropriate strategies. Their clues are listed after each corresponding strategy. Thus, in addition to learning *how* to use the various strategies, these clues can help you decide *when* to select an appropriate strategy or combination of strategies. Problem solving is as much an art as it is a science. You will find that with experience you will develop a feeling or intuition for when to use one strategy over another by recognizing certain clues, perhaps subconsciously. Also, you will find that some problems may be solved in several ways using different strategies.

STRATEGY: GUESS AND TEST

Often good problem solvers begin to solve a problem by 'messing around' using a strategy called Guess and Test or Trial and Error.

EXAMPLE 1.18 Place the digits 1, 2, 3, 4, 5, and 6 in the circles in Figure 1.17 so that the sum of the three numbers on each side of the triangle is 12.

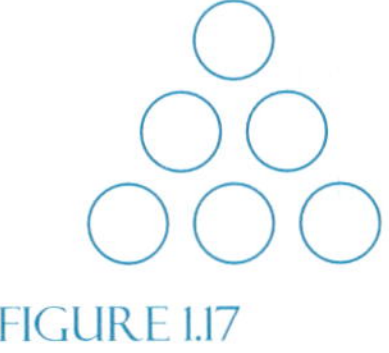

FIGURE 1.17

SOLUTION As its name suggests, to use the Guess and Test strategy, you guess at a solution and test to see if you are correct. If you are incorrect, you refine your Guess and Test. This process is repeated until you obtain a solution.

STEP 1: Understand the Problem

Each number must be used exactly one time when arranging the numbers in the triangle. The sum of the three numbers on each side must be 12.

First Approach: Random Guess and Test

STEP 2: Devise a Plan

Tear off six pieces of paper, mark the numbers 1 through 6 on them, and then try combinations until one works.

STEP 3: Carry Out the Plan

Arrange the pieces of paper in the shape of an equilateral triangle and check sums. Keep rearranging until three sums of 12 are found.

Second Approach: Systematic Guess and Test

STEP 2: Devise a Plan

Rather than randomly moving the numbers around, begin by placing the smallest numbers—namely, 1, 2, 3—in the corners. If that does not work, try 1, 2, 4, and so on.

STEP 3: Carry Out the Plan

With 1, 2, 3 in the corners, the sums of the sides are too small; similarly with 1, 2, 4. Try 1, 2, 5 and 1, 2, 6. The side sums are still too small. Next try 2, 3, 4, then 2, 3, 5, and so on, until a solution is found. We also could begin with 4, 5, 6 in the corners, then try 3, 4, 5, and so on.

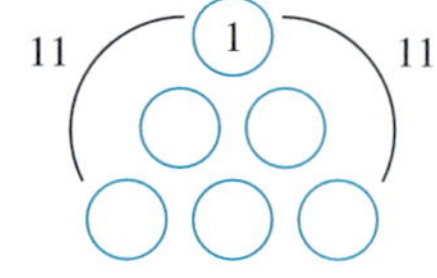

FIGURE 1.18

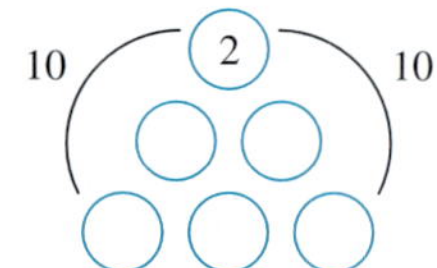

FIGURE 1.19

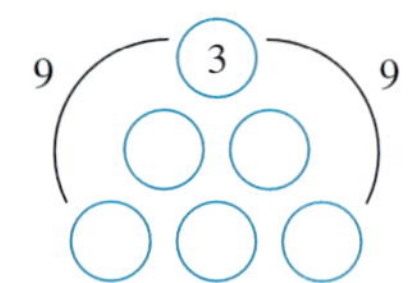

FIGURE 1.20

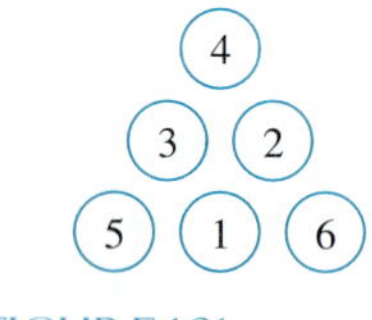

FIGURE 1.21

Third Approach: Inferential Guess and Test

STEP 2: Devise a Plan

Start by assuming that 1 must be in a corner and explore the consequences.

STEP 3: Carry Out the Plan

If 1 is placed in a corner, we must find *two* pairs from the remaining five numbers whose sum is 11 (Figure 1.18). However, out of 2, 3, 4, 5, and 6, only $6 + 5 = 11$. Therefore, we conclude that 1 cannot be in a corner. If 2 is in a corner, there must be two pairs left that add to 10 (Figure 1.19). But only $6 + 4 = 10$. Therefore, 2 cannot be in a corner. Finally, suppose that 3 is in a corner. Then we must satisfy the condition in Figure 1.20. However, only $5 + 4 = 9$ among the remaining numbers. Thus, if there is a solution, 4, 5, and 6 will have to be in the corners (Figure 1.21). By placing 1 between 5 and 6, 2 between 4 and 6, and 3 between 4 and 5, we obtain a solution.

STEP 4: Look Back

Notice how we have solved this problem in three different ways using Guess and Test. Random Guess and Test is often used to get started, but it is easy to lose track of the various trials. Systematic Guess and Test is better because you develop a scheme to ensure that you have tested all possibilities. Generally, Inferential Guess and Test is superior to both of the previous methods because it usually saves time and provides more information regarding possible solutions. ◆

Clues for Guess and Test

The Guess and Test strategy may be appropriate when:

- There is a limited number of possible answers to test.
- You want to gain a better understanding of the problem.
- You have a good idea of what the answer is.
- You can systematically try possible answers.
- Your choices have been narrowed down by the use of other strategies.
- There is no other obvious strategy to try.

STRATEGY: LOOK FOR A PATTERN

When using the Look for a Pattern strategy, one usually lists several specific instances of a problem and then looks to see if a pattern emerges that suggests a solution to the problem. For example, consider the sums produced by adding consecutive odd numbers starting with 1:

$$1$$
$$1 + 3 = 4\ (= 2 \times 2)$$
$$1 + 3 + 5 = 9\ (= 3 \times 3)$$
$$1 + 3 + 5 + 7 = 16\ (= 4 \times 4)$$
$$1 + 3 + 5 + 7 + 9 = 25\ (= 5 \times 5)$$

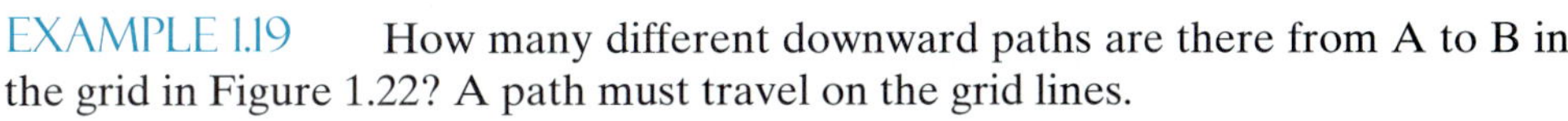

and so on. Based on the pattern generated by these five examples, one might expect that such a sum is always going to be a perfect square.

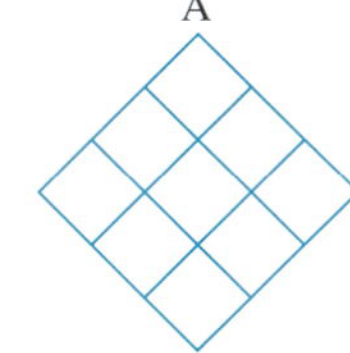

FIGURE 1.22

EXAMPLE 1.19 How many different downward paths are there from A to B in the grid in Figure 1.22? A path must travel on the grid lines.

SOLUTION

STEP 1: Understand the Problem

What do we mean by different and downward? Figure 1.23 illustrates two paths. Notice that each such path will be 6 units long. Different means that they are not exactly the same; that is, some part or parts are different.

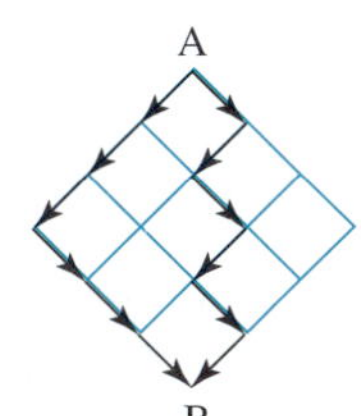

FIGURE 1.23

STEP 2: Devise a Plan

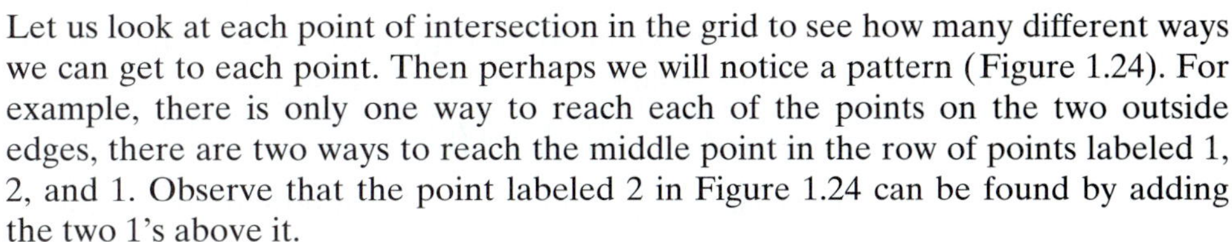

Let us look at each point of intersection in the grid to see how many different ways we can get to each point. Then perhaps we will notice a pattern (Figure 1.24). For example, there is only one way to reach each of the points on the two outside edges, there are two ways to reach the middle point in the row of points labeled 1, 2, and 1. Observe that the point labeled 2 in Figure 1.24 can be found by adding the two 1's above it.

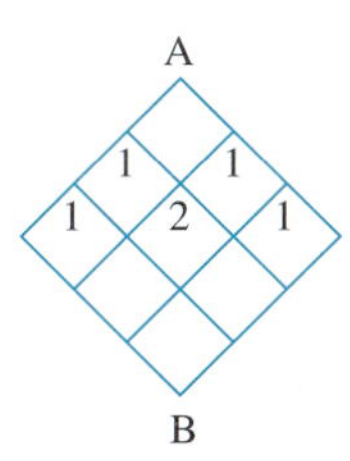

FIGURE 1.24

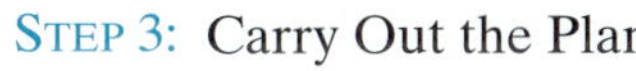

STEP 3: Carry Out the Plan

To see how many paths there are to any point, observe that you need only add the number of paths required to arrive at the point or points immediately above. To reach a point beneath the pair 1 and 2 in Figure 1.24, the paths to 1 and 2 are extended downward resulting in $1 + 2 = 3$ paths to that point. The resulting number pattern is shown in Figure 1.25. Notice, for example, that $4 + 6 = 10$ and $20 + 15 = 35$. The surrounded portion of this pattern applies to the given problem; thus the answer to the problem is 20.

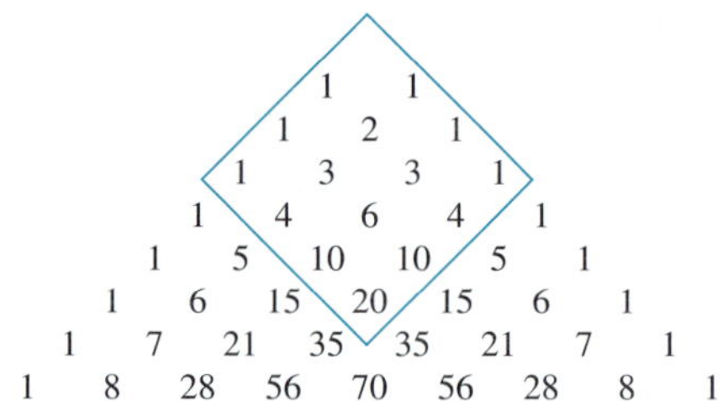

FIGURE 1.25

STEP 4: Look Back

Can you see how to solve a similar problem involving a larger square array—say, a 4 × 4 grid? How about a 10 × 10 grid? How about a rectangular grid? ◆

The Look for a Pattern strategy is often used in a form of reasoning called **inductive reasoning.** Using inductive reasoning you draw a conclusion based on several observations. For example, you likely recognize 376 as an even number simply by observing that 6 is an even number. The fact that a number is even if and only if its ones digit is even can be proven deductively. However, most people travel through life accepting this fact based on their observations.

The strategy **Make a List** is often combined with the Look for a Pattern strategy to suggest a solution to a problem. For example, here is a list of all the squares of the numbers 1 to 20 with their ones digits in boldface.

1, **4**, **9**, 1**6**, 2**5**, 3**6**, 4**9**, 6**4**, 8**1**, 10**0**,
12**1**, 14**4**, 16**9**, 19**6**, 22**5**, 25**6**, 28**9**, 32**4**, 36**1**, 40**0**

The pattern in this list can be used to see that the ones digits of squares must be one of 0, 1, 4, 5, 6, or 9. This suggests that a perfect square can never end in a 2, 3, 7, or 8.

Clues for Look for a Pattern

The Look for a Pattern strategy may be appropriate when:

- A list of data is given.
- A sequence of numbers is involved.
- Listing special cases helps you deal with complex problems.
- You are asked to make a prediction or generalization.
- Information can be expressed and viewed in an organized manner, such as in a table.

STRATEGY: USE A VARIABLE

A **variable** is a letter or symbol that represents a number. Variables are used extensively in mathematics to solve word problems by restating them as equivalent problems involving symbolic expressions. Invariably, problems stated in terms of variables are much easier to solve.

EXAMPLE 1.20 What is the greatest number that evenly divides the sum of any three consecutive whole numbers?

SOLUTION By trying several examples, you might guess that 3 is the greatest such number. However, it is necessary to use a variable to account for all possible instances of three consecutive numbers.

STEP 1: Understand the Problem

The **whole numbers** are 0, 1, 2, 3, . . . and consecutive whole numbers differ by 1. An example of three consecutive whole numbers is the triple 3, 4, 5. The sum of three consecutive whole numbers has a factor of 3 if 3 multiplied by another whole number produces the given sum. In the example of 3, 4, and 5, the sum is 12 and 3×4 equals 12. Thus $3 + 4 + 5$ has a factor of 3.

STEP 2: Devise a Plan

If we let x represent any whole number, then every triple of consecutive whole numbers can be expressed as follows: $x, x + 1, x + 2$. Now we can proceed to see if the sum has a factor of 3.

STEP 3: Carry Out the Plan

The sum of x, $x + 1$, and $x + 2$ is

$$x + (x + 1) + (x + 2) = 3x + 3 = 3(x + 1).$$

Thus $x + (x + 1) + (x + 2)$ is three times the whole number $x + 1$.

We have shown that the sum of any three consecutive whole numbers has a factor of 3. The case where $x = 0$ shows that 3 is the *greatest* such number because $0 + 1 + 2 = 3$ is divisible by 3 but has no larger factor.

STEP 4: Look Back

Is it also true that the sum of any four consecutive whole numbers has a factor of 4? Or, will the sum of any n consecutive whole numbers have a factor of n? Can you think of any other generalizations? ◆

Two additional strategies used in this book in which variables play a key role are **Use a Formula** and **Solve an Equation or Inequality.** Formulas are used extensively in Chapter 6 to determine principal and interest in financial settings. Formulas also are used extensively to describe conic sections in Chapter 10 and to describe recursions in Chapter 12. Solving equations is used often throughout the book. Solving inequalities is the central concept in linear programming, which is covered in Chapter 8.

Clues for Use a Variable

The Use a Variable and associated strategies may be appropriate when:

- A phrase similar to *for any number* is present or implied.
- A problem suggests an equation or inequality.
- There is an unknown quantity related to known quantities.
- The words *is*, *is equal to*, or *equals* appear in a problem.
- You are trying to develop a general formula, perhaps based on a pattern.

STRATEGY: DRAW A PICTURE

For problems involving physical situations, drawing a picture can often help you better understand the problem so that you can formulate a plan to solve the problem. As you proceed to solve the following pizza problem, see if you can visualize the solution *without* looking at any pictures first. Then work through the given solution using pictures to see how helpful they can be.

EXAMPLE 1.21 Can you cut a pizza into 11 pieces with four straight cuts?

SOLUTION

STEP 1: Understand the Problem

Do the pieces have to be the same size and shape?

STEP 2: Devise a Plan

An obvious beginning is to draw a picture showing how a pizza is usually cut and count the pieces (Figure 1.26). Unfortunately, we get only eight pieces using four straight cuts in the usual way. Since we need to get 11 pieces, we must try some unusual cuts.

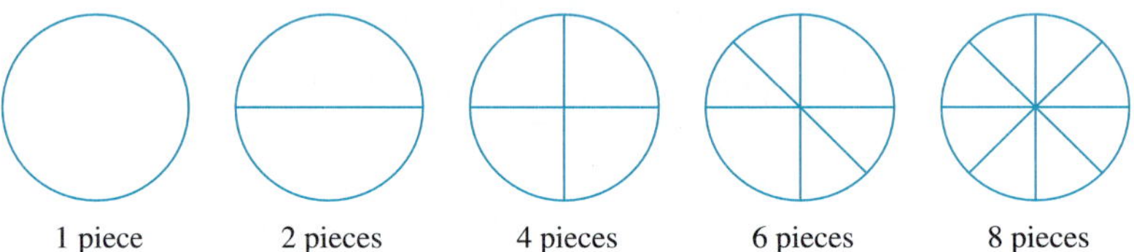

FIGURE 1.26

STEP 3: Carry Out the Plan

Now we experiment with unusual straight cuts to increase the number of pieces as in Figure 1.27.

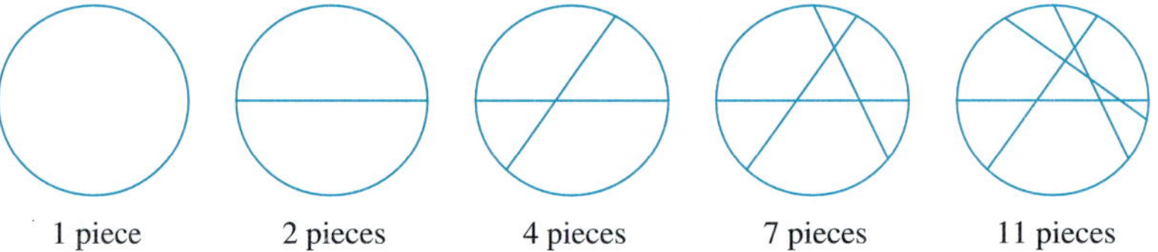

FIGURE 1.27

STEP 4: Look Back

Were you concerned about cutting equal pieces when you started? That is normal. In the context of cutting a pizza, the focus is usually on trying to cut equal pieces rather than on the number of pieces cut. Suppose that circular cuts were allowed? Does it matter if the pizza is circular or square? How many pieces can you get with five straight cuts? With n straight cuts? ◆

Two strategies are closely related to Draw a Picture: **Draw a Graph** and **Draw a Diagram.** Several different kinds of graphs appear in this book. In Chapter 3, bar graphs, line graphs, and circle graphs are used to picture data. In Chapter 4, graphs of distributions are used to interpret data, and scatterplots are graphed to make predictions. In Chapter 8, graphs of inequalities are used to solve problems involving maximizing and minimizing. In Chapter 10, coordinate graphs are used to study conic sections. Probability tree diagrams are used in Chapter 5, and diagrams picturing networks are helpful in Chapter 8.

Clues for Draw a Picture

The Draw a Picture and associated strategies may be appropriate when:

- Geometric figures or measurements are involved.

- A problem can be represented using two variables.
- A visual representation of the problem is possible.
- Finding representations of lines and other geometric figures.

STRATEGY: USE A MODEL

The Use a Model strategy is useful in problems involving geometric figures or their applications. Often, we acquire mathematical insight about a problem by seeing a physical embodiment of it. A model, then, is any physical object that resembles the object in the problem. It may be as simple as a paper, wooden, or plastic shape or as complicated as a carefully constructed replica that an architect or engineer might use.

EXAMPLE 1.22 Which of the shapes in Figure 1.28 can be folded into a closed box?

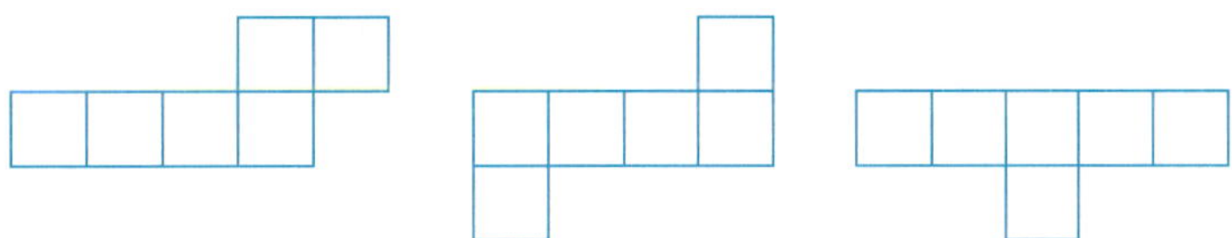

FIGURE 1.28

SOLUTION

STEP 1: Understand the Problem

The closed box will have six faces. Moreover, the six faces must be joined so that when edges are folded appropriately there are no overlapping faces.

STEP 2: Devise a Plan

To make a physical model, draw each shape on a piece of paper, cut it out, and fold along appropriate edges to see if it can form a closed box.

STEP 3: Carry Out the Plan

Make your models. You should observe that only the middle shape forms a closed box.

STEP 4: Look Back

Another related problem is to find how many different ways you can arrange six squares connected on a side such as those in this problem. How many of these shapes can you fold into a closed box? State similar problems for five squares and an open box. ◆

Modeling is often used in the study of probability. In particular, **Do a Simulation** is a common strategy where events are modeled using objects such as coins, dice, cards, and so on.

Clues for Use a Model

The Use a Model strategy may be appropriate when:

- Physical objects can be used to represent the ideas involved.

- A drawing is either too complex or inadequate to provide insight into the problem.
- A problem involves three-dimensional objects.
- A problem involves a complicated probability problem.
- A problem has a repeatable process that can be done experimentally.

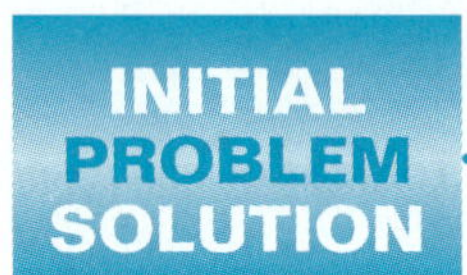

A friend bets you that at least two people in the group of five friends has the same astrological sign. Should you take the bet or not?

SOLUTION One way to solve this problem is to do a simulation. A simple method would be to write each of the names of the 12 astrological signs on a piece of paper, put the 12 pieces of paper in a hat, draw, record the sign, and replace. Do this 5 times to see if there is a match. Repeat this process at least 20 times to determine if the odds are in your favor (the more times you do it, the better your result). A more sophisticated way to do this simulation is to run it on a computer a million times. Surprisingly, there is approximately a 60% chance of a match. Thus, you should *not* take the bet.

PROBLEM SET 1.4

Problems 1 and 2 refer to Example 1.19.

1. If the original grid in Figure 1.22 had four squares on each side instead of three, how many downward paths would there be from point A to point B? What if there were five squares on each side?

2. Is it possible to generalize? Set up a list giving the number of squares to a side and the corresponding number of downward paths from A to B. Do this for grids having from one to five squares on a side. If you find a pattern, test it using Figure 1.25 for a grid having six squares on a side, adding any necessary numbers to those in Figure 1.25.

Problems 3 and 4 refer to Example 1.20.

3. (a) Does the sum of any four consecutive whole numbers have a factor of 4? If yes, explain how you got your answer; what was your strategy? If no, find the largest factor. Explain your strategy and how you determined your answer.

(b) Repeat part (a) for the sum of any five consecutive whole numbers.

4. (a) When does the sum of n consecutive whole numbers have a factor of n? Can you generalize? Make a list and use the Look for a Pattern strategy.

(b) If the sum of n consecutive whole numbers doesn't have a factor of n, can you determine any factors it *would* have? Explain your strategy and conclusions (if any).

Problems 5 and 6 refer to Example 1.21.

5. (a) Draw a picture using four straight cuts to divide a pizza into nine pieces.

(b) Draw a picture using four straight cuts to divide a pizza into ten pieces.

(c) What is the maximum number of pieces possible with four straight cuts? Show a drawing with the number of cuts. Explain why no greater number of pieces is possible.

6. (a) What is the maximum number of pieces possible when dividing a pizza using five straight cuts? Explain your strategy.

(b) Is it possible to generalize? That is, what is the maximum number of pieces possible when dividing a pizza using n straight cuts? Explain your strategy.

Problems 7 and 8 refer to Example 1.22.

7. Find at least six different ways to arrange six squares that are connected on a side as in Figure 1.28 so that the form can be folded into a closed box? If one form can be turned or rotated so that it looks like another, then the forms are not different. Explain your strategy.
8. Find at least six different ways to arrange five squares that are connected on a side as in Figure 1.28 so that the form can be folded into an open box? Show all the forms that are unique. If one form can be turned or rotated so that it looks like another, then the forms are not different. Explain your strategy.

In the problems that follow, you should practice the four-step process and clearly show each step. Point out the information or clues that you use. Practicing the four-step process with simple problems will make you more competent (and confident) when using it with the more difficult ones.

9. If you have a square with the diagonals drawn in, how many triangles of all sizes can be formed?
Hint: There are more than 4.
10. A multiple of eleven I be,
    not odd, but even you see.
    My digits, a pair,
        when multiplied there,
        make a cube and a square
        out of me.
    Who am I?
11. Using the symbols +, −, ×, and ÷, fill in the following three blanks to make a true equation. (A symbol may be used more than once.)

    6____6____6____6 = 13
12. Using three of the symbols +, −, ×, and ÷ *once each*, fill in the following three blanks to make a true equation. (Parentheses are allowed.)

    6____6____6____6 = 66
13. In the following figure (called an **arithmogon**), the number that appears in a square is the sum of the numbers in the circles on each side of it. Determine what numbers belong in the circles.

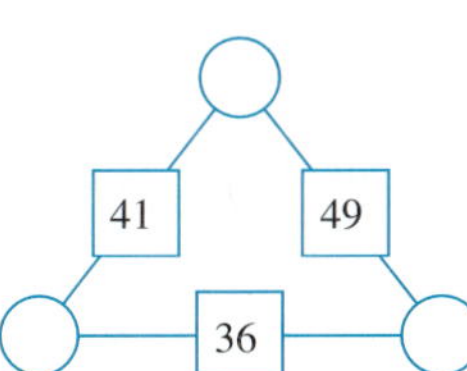

14. In the following arithmogon, the number that appears in a square is the product of the numbers in the circles on each side of it. Determine what numbers belong in the circles.

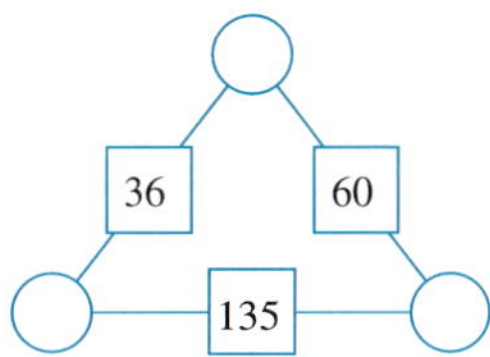

15. Talia walks to school at point B from her house at point A, a distance of six blocks. For variety, she likes to try different routes each day. How many different paths can she take if she always moves closer to B? One route is shown.

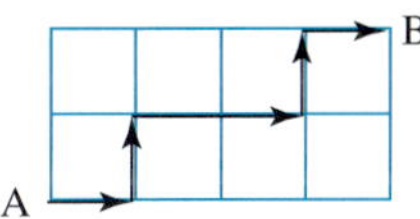

16. Using the numbers 1 through 8, place them in the following eight squares so that no two consecutive numbers are in touching squares (touching includes entire sides or simply one point.)

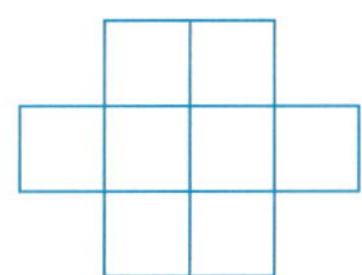

17. Place ten stools along four walls of a room so that each of the four walls has the same number of stools.
18. Scott and Greg were asked to add two whole numbers. Instead, Scott subtracted the two numbers and got 10, and Greg multiplied them and got 651. What was the correct sum?
19. Five friends were sitting on one side of a table. Gary sat next to Bill. Mike sat next to Tom. Howard sat in the third seat from Bill. Gary sat in the third seat from Mike. Who sat on the other side of Tom?
20. Five women participated in a 10-kilometer (10 K) Volkswalk, but started at different times. At a certain time in the walk, the following descriptions were true.
    1. Rose was at the halfway point (5 K).
    2. Kelly was 2 K ahead of Cathy.
    3. Janet was 3 K ahead of Ann.
    4. Rose was 1 K behind Cathy.
    5. Ann was 3.5 K behind Kelly.

    (a) Determine the order of the women at that point in time. That is, who was nearest the finish line, who was second closest, and so on?

    (b) How far from the finish line was Janet at that time?

21. Arrange the numbers 1, 2, . . . , 9 in the following triangle so that each side sums to 23.

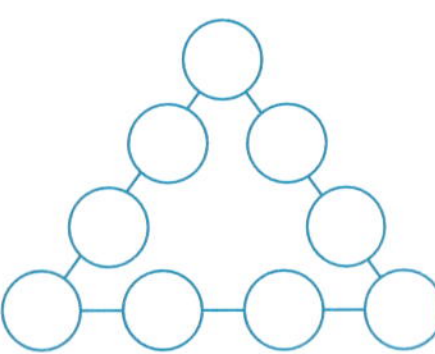

22. Fill in the circles using the numbers 1 through 9 once each and have the sum along each of the five rows total 17.

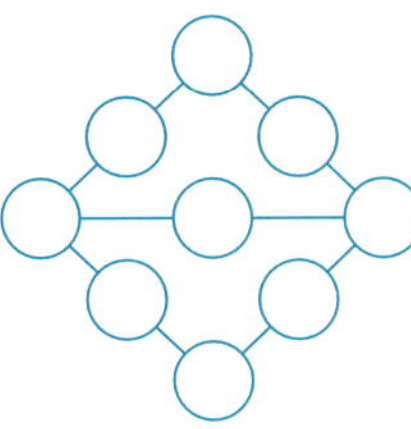

23. Place the digits 1 through 9 so that you can count from 1 to 9 by following the arrows in the diagram.

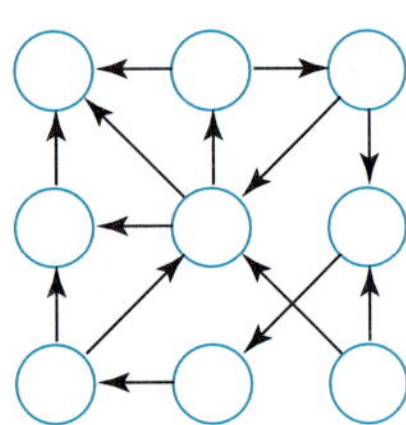

24. Using the numbers 9, 8, 7, 6, 5, and 4 once each, find the following.

(a) The largest possible sum:

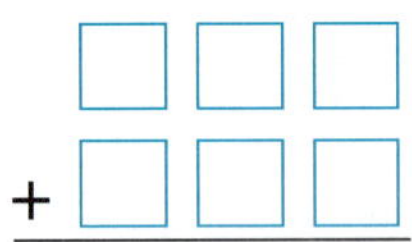

(b) The smallest possible (positive) difference.

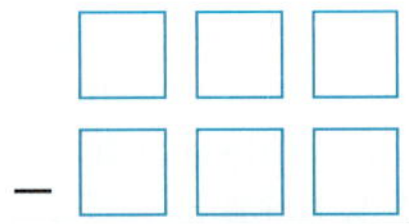

25. Suppose the classified employees went on strike for 22 working days. One of the employees, Juanita, made \$9.74 per hour before the strike. Under the old contract, she worked 240 six-hour days per year. If the new contract is for the same number of days per year, what increase in her hourly wage must Juanita receive to make up for the wages she lost during the strike in one year?

26. Two friends shopping together encounter a special shoe sale in which if two pairs of shoes are purchased at the regular price, a third pair (of lower or equal value) is free. Neither friend wants three pairs of shoes, but Pat would like to buy a \$56 and a \$39 pair while Chris is interested in a \$45 pair. If they buy the shoes together to take advantage of the sale, what is the fairest share for each to pay?

27. Solve this **cryptarithm** where each letter represents a digit and no letter represents two different digits.

$$\begin{array}{r} \text{USSR} \\ +\ \ \text{USA} \\ \hline \text{PEACE} \end{array}$$

28. Find digits A, B, C, and D that solve the following cryptarithm.

$$\begin{array}{r} \text{ABCD} \\ \times\ \ \ \ 4 \\ \hline \text{DCBA} \end{array}$$

29. Susan has 10 pockets and 44 dollar bills. She wants to arrange the money so that there is a different number of dollars in each pocket. Can she do it? Explain.

30. Mike said that when he opened his book, the product of the page numbers of the two facing pages were 7007. Performing only mental calculations, prove that he was wrong.

31. A man's age at death was $\frac{1}{29}$ of the year of his birth. How old was he in 1949?

32. Place numbers 1 through 19 into the 19 circles so that any three numbers in a line through the center will give the same sum.

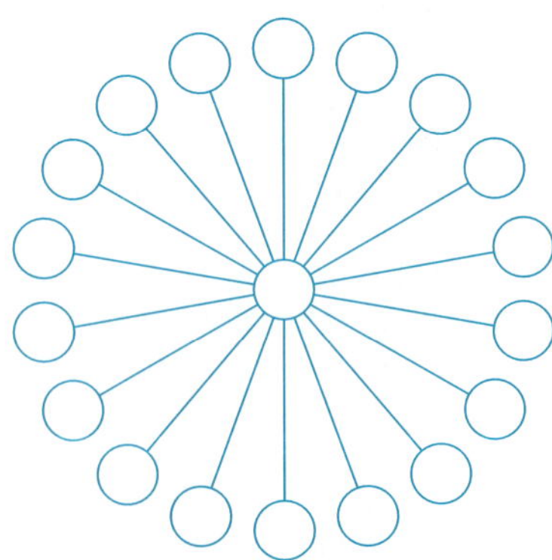

33. In the following square array, the corner numbers were given and the boldface numbers were found by adding the adjacent corner numbers. Following the same rules, what corner numbers are needed for the other square array?

6	**19**	13	___	**10**	___
8		**14**	**15**		**11**
2	**3**	1	___	**16**	___

34. An additive magic square has the same sum in each row, column, and diagonal. Find the error in this magic square and correct it.

47	56	34	22	83	7
24	67	44	26	13	75
29	52	3	99	18	48
17	49	89	4	53	37
97	6	3	11	74	28
35	19	46	87	8	54

35. Three people on the first floor of a building wish to take the elevator up to the top floor. The maximum weight that the elevator can carry is 300 pounds. Also, the elevator is very old and one of the three people must be in the elevator to operate it. If the people weigh 130, 160, and 210 pounds, how can they get to the top floor?

36. You have five identical coins and a balance scale. One of these coins is counterfeit and either heavier or lighter than the other four. Explain how the counterfeit can be identified and whether it is lighter or heavier than the others with only three weighings on the balance scale. (Hint: Solve a simpler problem—Given just three coins, can you find the counterfeit in two weighings?)

37. Consider the following products. Use your calculator to verify that the statements are true.

$1 \times (1) = 1^2$

$121 \times (1 + 2 + 1) = 22^2$

$12321 \times (1 + 2 + 3 + 2 + 1) = 333^2$

Predict the next line in the sequence of products. Use your calculator to check your answer.

38. Consider the following differences. Use your calculator to verify that the statements are true.

$6^2 - 5^2 = 11$

$56^2 - 45^2 = 1111$

$556^2 - 445^2 = 111{,}111$

(a) Predict the next line in the sequence of differences. Use your calculator to check your answer.

(b) What will the eighth line be?

39. Find the missing term in each pattern.

(a) 10 17 ___ 37 50 65

(b) 1 $\frac{3}{2}$ ___ $\frac{7}{8}$ $\frac{9}{16}$

(c) 243 324 432 ___ 768

(d) 234 ___ 23,481 234,819 2,348,200

40. Find the missing term in each pattern.

(a) 256 128 64 ___ 16 8

(b) 1 $\frac{1}{3}$ $\frac{1}{9}$ ___ $\frac{1}{81}$

(c) 7 9 12 16 ___

(d) 127,863 12,789 ___ 135 18

41. Place the numbers 1 through 8 in the circles on the vertices of the following cube so that the difference of any two connected circles is greater than 1.

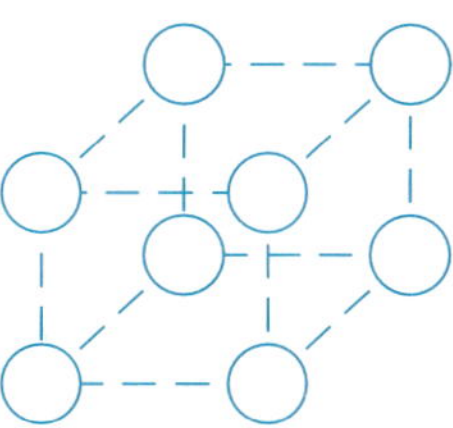

42. Three nickels, one penny, and one dime are placed as shown. You may move only one coin at a time to an adjacent empty square. Move the coins so that the penny and the dime have exchanged places and the lower middle square is empty. Try to find the minimum number of such moves.

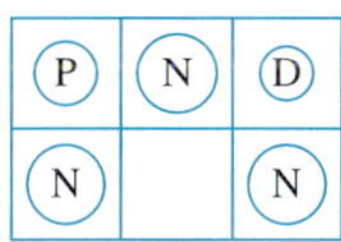

43. A gumball machine contains gumballs in eight different colors. Assume that there are a large number of gumballs equally divided among the eight colors.

(a) Estimate how many pennies you will have to use to get one of each color.

(b) Cut out eight identical pieces of paper and mark them with the digits 1–8. Put the pieces of paper in a container. Without looking, draw one piece and record its number. Replace the piece, mix the pieces up, and draw again. Repeat this process until all digits have appeared. Record how many draws it took. Repeat this experiment a total of 10 times and average the number of draws needed.

44. You are among 20 people called for jury duty. If there are to be two cases tried in succession and a jury consists of 12 people, what are your chances of serving on the jury for at least one trial? Assume that all potential jurors have the same chance of being called for each trial. (Do a simulation using 20 numbered slips of paper and repeating at least 100 times)

45. How many equilateral triangles of all sizes are there in the $3 \times 3 \times 3$ equilateral triangle shown?

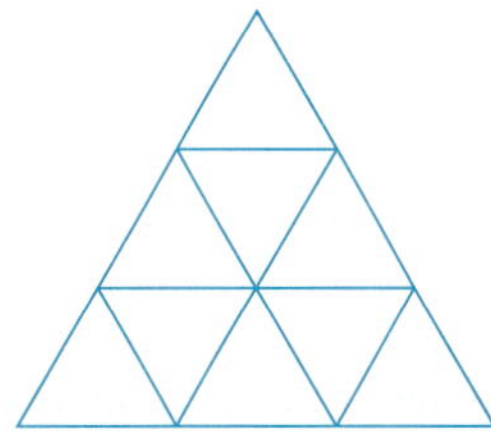

46. How many triangles are in the picture?

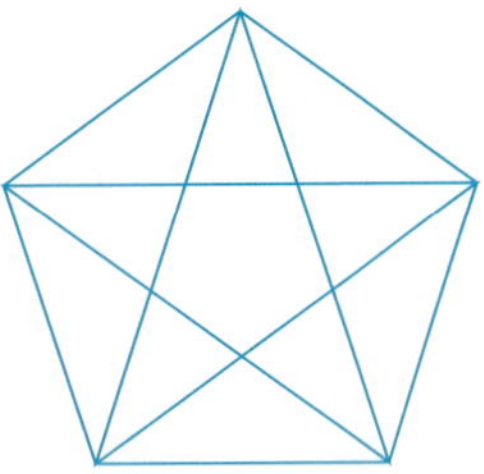

47. Sketch a figure that is next in each sequence.

(a)

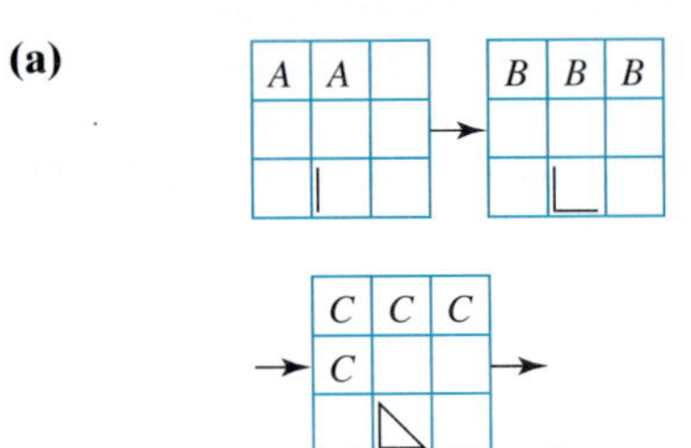

(b)

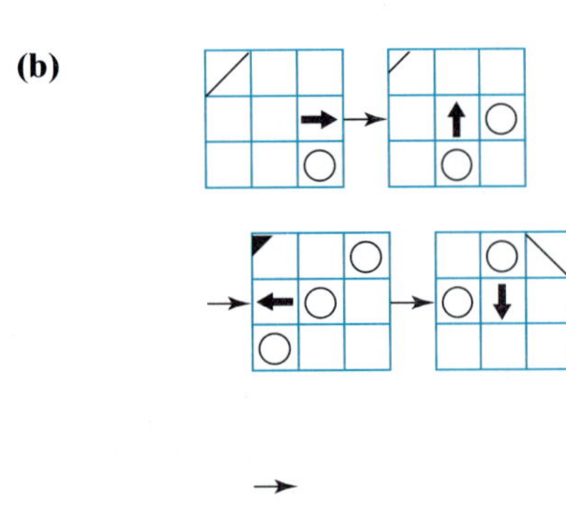

48. Sketch a figure that is next in each sequence.

(a)

(b)

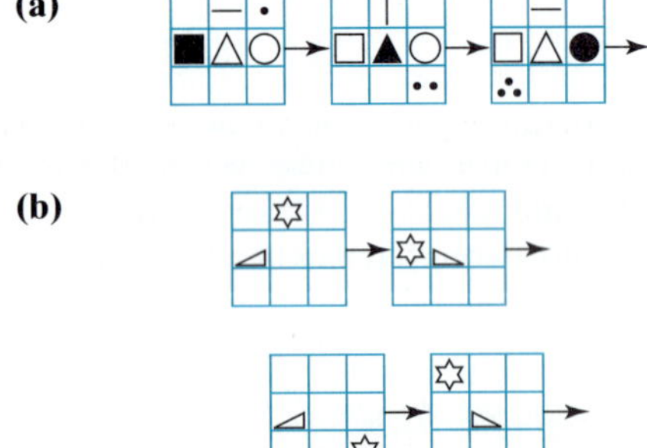

49. A candy bar company is having a contest. The letters N, U, and T are printed in the ratio 3:2:1 on the inside of each package. To determine the number of packages you should buy to spell NUT, perform the following simulation.

1. Using a die, let 1, 2, 3 represent N, let 4, 5 represent U and let 6 represent T.
2. Roll the die and record the corresponding letter. Repeat rolling the die until each letter has been obtained.
3. Repeat step 2 twenty times.

Average the number of rolls of the die required.

50. Since you forgot to study for your math test, a 10-question true-false test, you decide to guess on each question. To determine your chances of getting a score of 70% or better, perform the following simulation.

1. Use a coin where H = true and T = false.
2. Toss the coin 10 times, recording the corresponding answers.
3. Repeat step 2 twenty times.
4. Repeat step 2 one more time. This is the answer key of correct answers. Correct each of the 20 tests.

(a) How many times was the score 70% or better?
(b) What is the probability of a score of 70% or better?

51. How many cubes are in the 100th collection of cubes in this sequence?

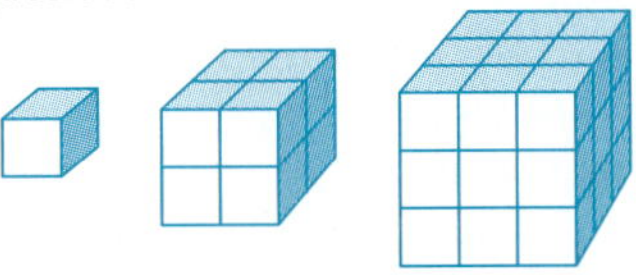

52. How many cubes are in the 10th collection of cubes in this sequence?

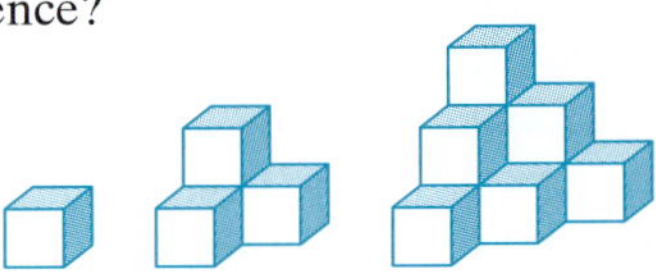

53. Consider the following sequence of shapes. The sequence starts with one square. Then, at each step, squares are attached around the outside of the figure, one square per exposed edge in the figure.

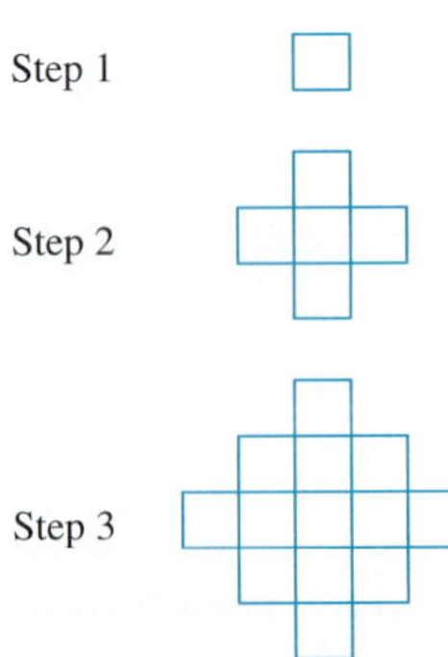

(a) Draw the next two figures in the sequence.
(b) Make a table listing the number of unit squares in the figure at each step. Look for a pattern in the number of unit squares. (Hint: Consider the number of squares attached at each step.)

(c) Based on the pattern you observed, predict the number of squares in the figure at step 6. Draw the figure to check your answer.

(d) How many squares would there be in the 10th figure? in the 20th figure? in the 50th figure?

54. Consider the following sequence of shapes. The sequence starts with one triangle. Then, at each step, triangles are attached around the outside of the figure, one triangle per exposed edge in the figure.

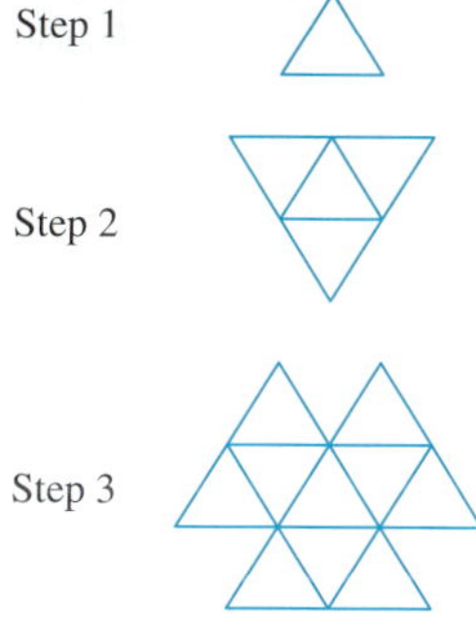

(a) Draw the next two figures in the sequence.

(b) Make a table listing the number of unit triangles in the figure at each step. Look for a pattern in the number of unit triangles. (Hint: Consider the number of triangles added at each step.)

(c) Based on the pattern you observed, predict the number of squares in the figure at step 6. Draw the figure to check your answer.

(d) How many triangles would there be in the 10th figure? in the 20th figure? in the 50th figure?

EXTENDED PROBLEMS

55. The **Fibonacci sequence** is 1, 1, 2, 3, 5, 8, 13, 21, . . . , where each successive number is the sum of the preceding two. For example, $13 = 5 + 8$, $21 = 8 + 13$, and so on. Observe the following pattern.

$1^2 + 1^2 = 1 \times 2$
$1^2 + 1^2 + 2^2 = 2 \times 3$
$1^2 + 1^2 + 2^2 + 3^2 = 3 \times 5$

Write out six more terms of the Fibonacci sequence and use the sequence to predict what $1^2 + 1^2 + 2^2 + 3^2 + \ldots + 144^2$ is without computing the sum. Then use your calculator to check your prediction.

56. Write out 16 terms of the Fibonacci sequence and observe the following pattern.

$1 + 2 = 3$
$1 + 2 + 5 = 8$
$1 + 2 + 5 + 13 = 21$

Use the pattern you observed to predict the sum

$1 + 2 + 5 + 13 + \ldots + 610$

without actually computing the sum. Then use your calculator to check your result.

57. Observe the following pattern based on the Fibonacci sequence.

$1 + 1 = 3 - 1$
$1 + 1 + 2 = 5 - 1$
$1 + 1 + 2 + 3 = 8 - 1$
$1 + 1 + 2 + 3 + 5 = 13 - 1$

Write out six more terms of the Fibonacci sequence and use the sequence to predict the answer to

$1 + 1 + 2 + 3 + 5 + \ldots + 144$

without actually computing the sum. Then use your calculator to check your result.

58. Write out the first 16 terms of the Fibonacci sequence.

(a) Notice that the fourth term in the sequence (called F_4 is odd: $F_4 = 3$. The sixth term in the sequence (F_6) is even: $F_6 = 8$. Look for a pattern in the terms of the sequence and describe which terms are even and which are odd.

(b) Which of the following terms of the Fibonacci sequence are even and which are odd: F_{38}, F_{51}, F_{150}, F_{200}, F_{300}?

(c) Look for a pattern in the terms of the sequence and describe which terms are divisible by 3.

(d) Which of the following terms of the Fibonacci sequence are multiples of 3: F_{48}, F_{75}, F_{196}, F_{379}, F_{1000}?

59. Write out the first 16 terms of the Fibonacci sequence and observe the following pattern.

$1 + 3 = 5 - 1$
$1 + 3 + 8 = 13 - 1$
$1 + 3 + 8 + 21 = 34 - 1$

Use the pattern you observed to predict the answer to

$1 + 3 + 8 + 21 + \ldots + 377$

without actually computing the sum. Then use your calculator to check your result.

60. **Pascal's triangle,** shown next, is where each entry other than a 1 is obtained by adding the two entries in the row immediately above it.

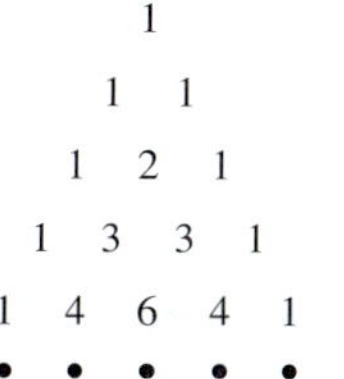

(a) Find the sums of the numbers on the diagonals in Pascal's triangle as indicated in the next figure.

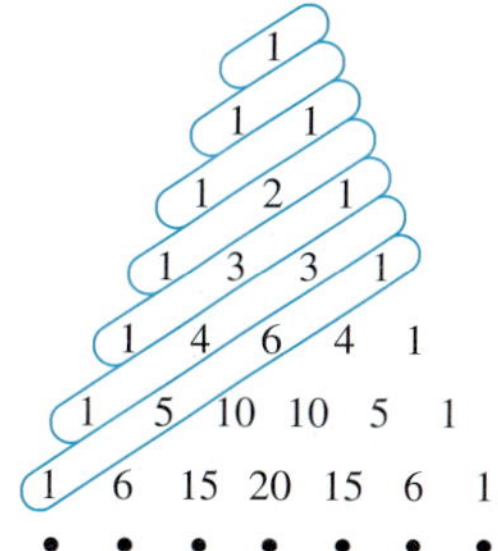

(b) Predict the sums along the next three diagonals in Pascal's triangle without actually adding the entries. Check your answers by adding the appropriate entries.

61. A square can be divided into four identical smaller copies of itself as follows.

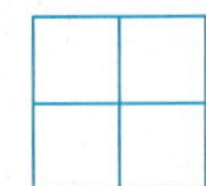

Any subdivision of a shape into identically shaped smaller copies of itself is called a **reptile dissection.** Each of the following shapes can also be divided into four smaller, identical copies of themselves. Show how this can be done in each case.

(a)

(b)

62. Find a reptile dissection of the following shape.

63. If the following four figures are referred to as stars, the first one is a three-pointed star and the second one is a six-pointed star. (Note: If this pattern of constructing a new equilateral triangle on each line segment that is part of the existing figure is continued indefinitely, the resulting figure is called the **Koch curve,** or **Koch snowflake,** which is shown last.)

(a) How many points are there in the third star?
(b) How many points are there in the fourth star?
(c) If the pattern is continued, how many points are there in the nth figure?

64. If the following pattern is continued indefinitely, the resulting figure is called the *Sierpinski triangle* or the *Sierpinski gasket.*

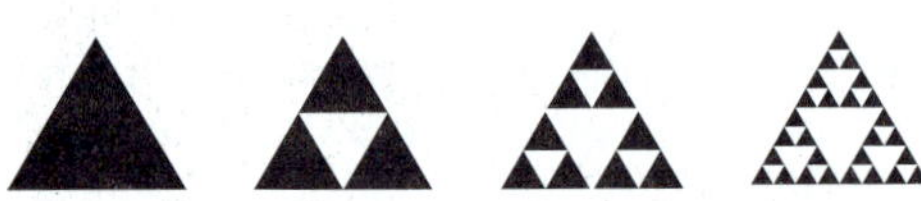

(a) How many black triangles are there in the fourth figure?
(b) How many white triangles are there in the fourth figure?
(c) If the pattern is continued, how many black triangles are there in the nth figure?
(d) If the pattern is continued, how many white triangles are there in the nth figure?

Chapter One Problem

There are four married couples in a room. The men include Tony, Dany, Carlos, and Willie, and the women are Melissa, Laura, Shelly, and Juanita. Use the following clues to determine who is married to whom.

(1) Shelly is Dan's sister.
(2) Melissa is married to Carlos.
(3) Juanita's husband is an only child.
(4) Willie is not married to Shelly.

Solution

Strategy: Make a Table

We make a table to organize the information. Each row will correspond to a woman, and each column corresponds to a man. The intersection of a row and column represents the marriage of the pair. We will put NO in any blank if we know that the couple is not married and YES in any blank if they are. Clue (1)

	Tony	Dan	Carlos	Willie
Melissa	_____	_____	YES	_____
Laura	_____	_____	_____	_____
Shelly	_____	NO	_____	_____
Juanita	_____	_____	_____	_____

tells us that Shelly is Dan's sister. Assuming the social behavior of our time and culture, we know that

Shelly and Dan are not married so we put NO in the entry that has Shelly as a row and Dan as a column. Similarly, Clue (2) tells us that Melissa is married to Carlos so we put a YES below Carlos and in the third space to the right of Melissa.

Since Melissa and Carlos are married to each other we know that they are not married to anyone else. We may put NOs in the rest of Melissa's row and in the rest of Carlos's column.

	Tony	Dan	Carlos	Willie
Melissa	NO	NO	YES	NO
Laura	_____	_____	NO	_____
Shelly	_____	NO	NO	_____
Juanita	_____	_____	NO	_____

Clue (3) says that Juanita's husband is an only child.This means that he has no siblings. In particular, he cannot be Dan since Shelly is Dan's sister. In short, Juanita is not married to Dan. Put a NO at the bottom of Dan's column.

	Tony	Dan	Carlos	Willie
Melissa	NO	NO	YES	NO
Laura	_____	_____	NO	_____
Shelly	_____	NO	NO	_____
Juanita	_____	NO	NO	_____

We know Dan is married. The only possibility that is not ruled out is Laura. Put a YES in the Laura–Dan intersection. Since Laura is married to Dan, she is not married to anyone else; put NOs in the remaining entries of Laura's row.

	Tony	Dan	Carlos	Willie
Melissa	NO	NO	YES	NO
Laura	NO	YES	NO	NO
Shelly	_____	NO	NO	_____
Juanita	_____	NO	NO	_____

Clue (4) says that Willie is not married to Shelly. Put a NO in the Shelly–Willie entry.

	Tony	Dan	Carlos	Willie
Melissa	NO	NO	YES	NO
Laura	NO	YES	NO	NO
Shelly	_____	NO	NO	_____
Juanita	_____	NO	NO	

The solution is nearly at hand. The only possibility for Willie is that he is married to Juanita. Put a YES in the Juanita–Willie entry. Since Juanita cannot be married to Tony, put a NO in the Juanita–Tony entry.

	Tony	Dan	Carlos	Willie
Melissa	NO	NO	YES	NO
Laura	NO	YES	NO	NO
Shelly	_____	NO	NO	NO
Juanita	NO	NO	NO	YES

The only remaining possibility for Shelly and Tony is that they be married to each other; put a YES in their position. The chart is complete.

	Tony	Dan	Carlos	Willie
Melissa	NO	NO	YES	NO
Laura	NO	YES	NO	NO
Shelly	YES	NO	NO	NO
Juanita	NO	NO	NO	YES

Reading from the chart, the married couples are Melissa–Carlos, Laura–Dan, Shelly–Tony, and Juanita–Willie. Notice that the chart gave us a way to systematically use all the information from the clues. Also we needed to use all the information we had about marriage that was not included in the problem, but assumed as common knowledge.

Chapter One Review

Key Ideas and Questions

The following questions review the main ideas of this chapter. Write your answers to the questions and then refer to the pages listed by number to make certain that you have mastered these ideas.

1. Describe how statements may be combined using logical connectives to make new statements and how the truth of these combined statements may be determined using truth tables. 6
2. What are the three common variants of a conditional? Give an example of a conditional statement together with examples of each of its three variants. 9
3. What are the four main argument forms? What are two common forms of argument that are fallacies? Give examples of these. 16
4. What are Euler diagrams used for? Show how they work in practice. 25
5. Describe Pólya's four-step process of problem solving, list at least six problem-solving strategies, and illustrate how these are used to solve a problem. 34

Vocabulary/Notation

Following is a list of key vocabulary, notation, and ideas for this chapter. Mentally review each of these items, write down the meaning of each term, and use it in a sentence. Then refer to the page numbers and restudy any material you are unsure of before solving the Chapter One Review Problems.

Section 1.1

Statement 5
Negation 6
Truth table 6
Truth values 6
Compound statements 6
Logical connectives 6
Conjunction (*and*) 6
Disjunction (*or*) 7
Exclusive "or" 7
Inclusive "or" 7
Implication 8
Conditional 8
Hypothesis 8
Antecedent 8
Conclusion 8
Consequent 8
Converse 9
Contrapositive 9
Inverse 9
Logically equivalent 10
Biconditional 10

Section 1.2

Argument 16
Rhetoric 16
Premise 16
Hypothesis 16
Conclusion 16
Valid deductive argument 16
Modus ponens 16
Law of detachment 16
Affirming the antecedent 16
Chain rule 17
Hypothetical syllogism 17
Modus tollens 19
Law of contraposition 19
Denying the consequent 19
Disjunctive syllogism 19
Law of the excluded middle 19
Argument by contradiction 20
Indirect reasoning 20
Reducio ad absurdum 20
Fallacy of affirming the consequent 20
Fallacy of denying the antecedent 21

Section 1.3

Euler diagram 25
Syllogism 26
Premise 26
Conclusion 26
Categorical syllogism 26
Valid 26
Invalid 29
Category fallacy 30

Section 1.4

Pólya's four-step process 34
Strategy 34
Clues 35
Guess and test 35
Look for a pattern 37
Inductive reasoning 38
Make a list 38
Use a variable 38
Use a formula 39
Solve an equation or inequality 39
Draw a picture 39
Draw a graph 40
Draw a diagram 40
Use a model 41
Do a simulation 41

Chapter One Review Problems

1. Suppose that p represents the statement "Snow is cold" and q represents the statement "Pigs can fly." Which of the following statements are true?
   (a) $\sim p$ **(b)** $p \wedge q$ **(c)** $p \vee q$
   (d) $\sim(\sim p \vee \sim q)$ **(e)** $p \Rightarrow q$ **(f)** $q \Rightarrow p$
2. Identify the hypothesis and the conclusion of the following conditional statement: "If there is a will, then there is a way."
3. Write the converse, inverse, and contrapositive of the following conditional statement: "If it rains the lawn will get wet." Which of these four statements are true?
4. Is the following biconditional statement true? "Fire is hot if, and only if, either pigs can fly or horses can run."
5. Decide if the following arguments are valid and identify the type of argument (even if a fallacy) by name.
   (a) If you live in Omaha, then you live in Nebraska. You do not live in Omaha. Therefore you do not live in Nebraska.
   (b) State senators are either Democrats or Republicans. Your state senator is not a Republican. Therefore your state senator is a Democrat.
   (c) If it could rain cats, then most birds would not fly. It cannot rain cats. Therefore most birds can fly.
   (d) If he has a midterm tomorrow, then he will not go to the party tonight. He has a midterm tomorrow. Therefore he will not go to the party tonight.
   (e) If she has a paper due, then she will use the computer lab. If she uses the computer lab, then she will have to bring her student ID. Therefore, if she has a paper due, she will have to bring her student ID.
6. Determine the validity of the following argument: Drinking lots of beer makes people think clearer. These people are drinking lots of beer. Therefore they will think clearer. Is this argument valid? Is it true?
7. Use truth tables to show that $p \Rightarrow q$ is equivalent to $\sim p \vee q$.
8. Use truth tables to determine if $\sim(p \wedge (p \Rightarrow q))$ is equivalent to $\sim q$.
9. Make an Euler diagram to represent each of the following statements.
   (a) All horses have four legs. Some horses are brown.
   (b) All mammals have fur. Dogs and cats are mammals. No dogs are cats.
   (c) Some pets are dogs. My pet is not a dog.
   (d) There is a quog that is not a quig.
10. Consider the following statement: "If a person is late, then he has to do the dishes. You are late, so you have to do the dishes." Is this a syllogism? Can it be restated as a syllogism? What are the premises and conclusion?
11. Use the following Euler diagram to determine which of the following statements are true. Assume that all regions contain some members.
    (a) All R are B.
    (b) Some Q are R.
    (c) All B are A.
    (d) Some Q are not A.

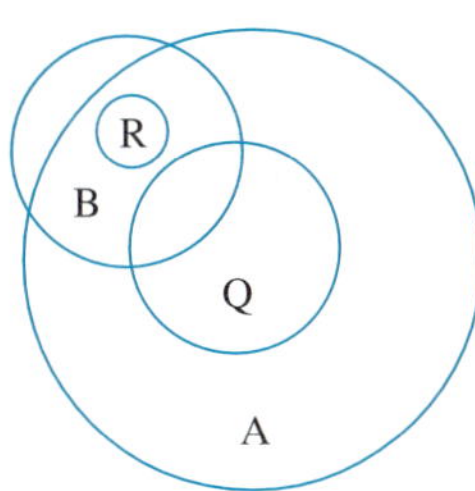

12. Make an Euler diagram to represent the following statements:
    Emily Dickinson was a poet.
    All poets are writers.
    All novelists are writers.
    Emily Dickinson was not a novelist.
    Is the following conclusion valid?
    Therefore some writers are not novelists.
13. Use an Euler diagram to determine the validity of each of the following arguments:
    (a) Blue whales swim north in the spring.
    Blue whales are not extinct.
    Therefore, some whales swim north in the spring.
    (b) Some people are runners.
    All runners are athletes.
    Some runners are teachers.
    Therefore, all teachers are athletes.
14. State and prove a fact about the product of any two consecutive whole numbers. (Hint: First look for a pattern, then make a clear statement of the fact, then give an argument showing why it is true.)
15. Solve the following problem: You are older than your brother. The difference of your ages is 8 and $\frac{2}{3}$ of your age is 2 years more than your brother's age. How old is your brother? What strategy did you use?
16. The electric company, gas company, and TV cable company are on the same street. They need to make underground connections to two houses on the other side of the street. Is it possible for them to do this in such a way that no two connections cross each other or overlie each other? What if there are three houses across the street? (Hint: Draw a picture.)
17. Place the numbers 1, 2, 3, 4, 5, 6, 7, 8, 9 on a 3×3 square so that every row, every column, and every diagonal adds up to 15. Describe all the ways there are to do this. What strategy did you use?

18. How many ways is it possible to unfold a regular tetrahedron (a three-dimensional shape having exactly four faces, each in the shape of an equilateral triangle)? What strategy did you use?

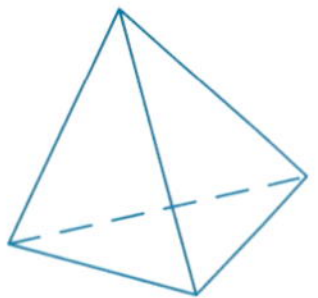

19. How many pieces can a pizza be cut into using circular cuts if you are allowed two cuts? three circular cuts? What strategy did you use?

20. Is the chance that a thumbtack will land point downward more or less than a half? What strategy did you use?

CHAPTER 2

NUMBERS AND NUMERACY

DON'T TRUST THOSE LYING LIE DETECTORS

Many powerful arguments have been made against the use of the lie detector. Here is another compelling objection.

Lie detectors make mistakes with both liars and truth-tellers. Let us assume that a lie detector has an accuracy rate of 80% (a generous estimate) and that no subject knows how to beat the lie detector (also questionable). Let us further assume that 900 out of 1,000 people are telling the truth. Of the 900 truth-tellers, the lie detector will identify 720 (80% of 900) as truth-tellers but will incorrectly identify 180 (20% of 900) as liars. Of the 100 liars, the lie detector will identify 80 as liars (80% of 100) but will incorrectly identify 20 (20% of 100) as truth-tellers.

Thus, of the 260 people (180 + 80) the lie detector identifies as "liars", 180 are telling the truth. In other words, the lie detector is incorrect 70% of the time when it brands someone a liar.

CHAPTER GOALS

1. Learn about earlier systems of numeration and understand the properties of the Hindu-Arabic numeration system we use today.
2. Calculate using "mental math".
3. Estimate results before and after doing a calculation.
4. Use the properties of addition, multiplication, exponentiation, and percentages to solve problems.
5. Solve problems involving exponents and roots.

The selection on the previous page is a synopsis of a letter that appeared in the New York Times. The writer is presenting an argument against the practice of an increasing number of companies that are considering using lie detectors to make hiring decisions among new applicants. On the other hand, when lie detectors are used in the criminal justice system the percentage of liars will be much higher than 10%. If this percentage of liars is as high as 90%, then the argument in the article turns around to show that the percentage of people incorrectly labeled truth-tellers is 70%. Thus the fact that a defendant passes a lie detector test does not necessarily exonerate him. It also appears that there are people who do have the ability to fool a lie detector as well as people that, although honest, are unable to pass a lie detector test.

The letter uses percentages and basic arithmetic to describe how lie detectors, although widely used and believed, are very unreliable. This type of reasoning is possible because our number system has properties that make computation easy. Facility with numbers, estimation of quantities, and percentages is as important as literacy in modern society. This facility is given the name numeracy.

This chapter will show you some of the different systems of numeration (methods of writing numbers) that led to our own. It will also show you how the properties of numbers may be exploited to do mental estimation of arithmetic problems and the use of arithmetic operations to solve problems.

THE HUMAN SIDE OF MATHEMATICS

Around the year A.D. 775, the Arab traveler and mathematician Al-Khwarismi brought manuscripts containing the knowledge of Hindu mathematics and numerals from India to Baghdad. Along the way, his caravan was attacked by bandits who succeeded in taking everything from the travelers except their lives and the manuscripts. The pages that remained included those showing the new numeration system. The knowledge and numerals that Al-Khwarismi brought to Baghdad survived to spread throughout the Arab world. At this time in history, most of Europe was in the Dark Ages; but the knowledge of science, mathematics, and art were being preserved and advanced by Arab scholars.

The Hindu-Arabic system was introduced into Europe when Moslem control spread over much of the Mediterranean region. The intellectual and cultural influence was quite significant. Many of the Greek manuscripts that had been saved from the destruction or the Great Library at Alexandria or survived other events in ancient times had been preserved by Arab scholars. This knowledge was also brought to Europe, together with the Hindu-Arabic mathematics and the beginnings of algebra, with which Al-Khwarismi is also credited. Another lasting legacy of the Moslem period in Spain are the beautiful buildings, gardens, and tile works such as those of The Alhambra, the palace of the Moorish kings at Granada.

Gerbert of Aurillac

Gerbert of Aurillac (A.D. 945–1003) was a French scholar and mathematician in an age that saw little mathematics in Europe and much in the Moslem world. Roman numerals were the system of numeration in use in Europe at this time and it was very difficult to make computations with them. Until the end of the Middle Ages in Europe, computation was accomplished with a Roman abacus, which, itself, was somewhat complicated. Having studied in the Moslem schools of Spain, Gerbert became acquainted with the science and mathematics of the Arabs. Later, he directed the education of theological scholars at Rheims, and his intellectual influence spread throughout Europe. Gerbert is generally credited with the introduction and spread of Hindu-Arabic numerals to Europe. This occurred not through written documents, but through the practice of teaching people to count with a simple abacus called a Roman counting board. The tradition of written numerals and computations overtook all other methods by the end of the 12th century.

During Gerbert's lifetime, the ability to do computations in arithmetic problems was so rare it was considered by many as evidence of witchcraft or other mysterious powers. (It would certainly take superhuman ability to do a long division problem such as MCMXIV divided by XIX.) Gerbert was even accused of selling his soul to the devil for the ability to compute. Ironically, Gerbert had served the Catholic Church all his adult life. He was able to overcome these accusations and held several other influential positions in the Church. He rose through the hierarchy of the Catholic Church and eventually became Pope Sylvester II in A.D. 999.

2.1 NUMERATION SYSTEMS

You are on a trip to England and get into conversation with someone in a pub. In explaining some of the problems the United States is facing, you mention the $4 trillion national debt we have run up. Your acquaintance is horrified well beyond your expectations. What is it about the number 4 trillion that is so astounding?

Mathematics is sometimes described as a language of quantity, patterns, and relationships. However, as we will see, we do not yet have a universal language. As civilizations developed, their changing needs gave rise to different mathematical systems, including different systems of numeration (symbols that represent numbers). In some cases, the choice of a numeration system even contributed significantly to a civilization's success. In this section, we will consider some of the more important historical numeration systems and the development of the system of numerals used in most of the world today.

NUMBERS

A **number** is an idea, or an abstraction, that represents a characteristic of an object, or group of objects. The symbols that we see, write, or touch when representing numbers are called **numerals.** There are three common uses of numbers. The most common use of *whole* numbers is to describe how many objects are in some particular collection or category. Mathematicians often refer to a collection of any sort as a **set,** and the things in the collection are called the **elements** of the set. A number used to describe how many elements are in a set is called a **cardinal number.** When asked "How many stars are on the American flag?," the answer, 50, is a cardinal number. A second use of numbers is concerned with order. For example, Mary may be second in line, or your team may be fourth in the standings. Numbers used in this way are called **ordinal numbers.** Finally, **identification numbers** are used to give unique names to things. Examples are telephone numbers, bank account numbers, and social security numbers. In this case, the numbers are used as symbols; their values are not important.

EXAMPLE 2.1 Classify each of the following as cardinal, ordinal, or identification numbers.

(a) Scott is the third child in their family.
(b) My social security number is 123-45-6789.
(c) I earned $4800 last summer.

SOLUTION

(a) This is the ordinal use of number since it describes Scott's position in birth order.
(b) A social security number is used to identify a person, thus is an identification number.

(c) The earnings from last summer describes how much (or how many), thus is the cardinal use of number. ◆

To make numbers more useful, systems of symbols, or numerals, have been developed to represent numbers. In fact, throughout history, many different numeration systems have evolved. Since you have been using the Hindu-Arabic system from an early age, it may be somewhat easy to take the system for granted and not realize the strengths that it has for representing numbers and performing calculations.

THE TALLY NUMERATION SYSTEM

We will briefly discuss three ancient systems, and we will note the attributes of our Hindu-Arabic system that were found in these systems.

The **tally numeration system** is composed of single strokes, one for each object being counted (Figure 2.1).

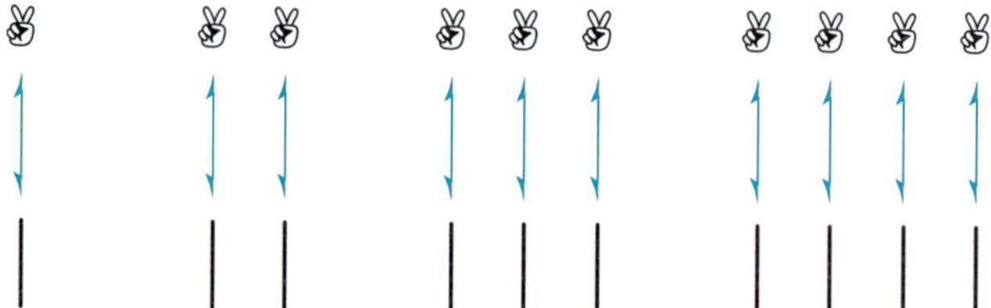

FIGURE 2.1

The next six tally numerals are

An advantage of this system is its simplicity; however, two disadvantages are that (1) large numbers require many individual symbols, and (2) numerals representing large numbers are difficult to read. For example, what number do these tally marks represent?

||||||||||||||||||||||||||||||||||||

The tally system was improved by the introduction of **grouping,** namely, collecting symbols in groups of five. In this case, the fifth tally mark is placed across every four to make a group of five. Thus the last numeral above can be written as follows:

Grouping makes it easier to recognize the number being represented; in this case, there are 37 tally marks.

THE EGYPTIAN NUMERATION SYSTEM

The **Egyptian numeration system,** which developed around 3400 B.C., uses grouping by ten. In addition, this system introduced new symbols for powers of ten (Figure 2.2).

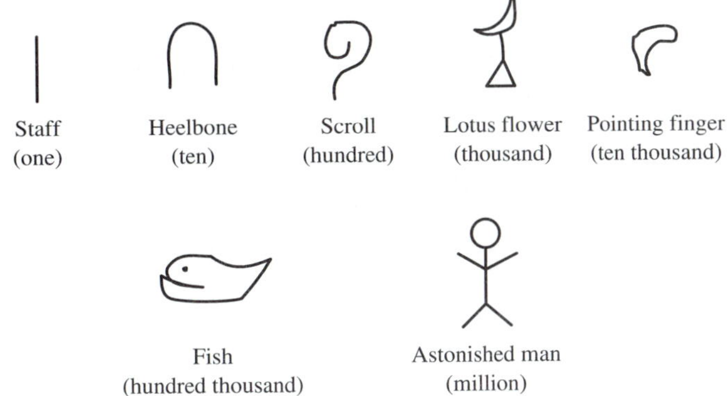

FIGURE 2.2

An example of an Egyptian numeral is shown in Figure 2.3.

FIGURE 2.3

Notice how this system requires far fewer symbols than the tally system once numbers greater than ten were represented. This system is also an example of an **additive system,** since the values for the symbols are added together. The order in which the symbols are written is not important, but arranging the symbols left-to-right in decreasing order of magnitude makes them easier to interpret.

EXAMPLE 2.2 Express the Egyptian numeral in part (a) as a Hindu-Arabic numeral and the numeral in part (b) as an Egyptian numeral.

(a) 

(b) 102,043

SOLUTION

(a) 237

(b) 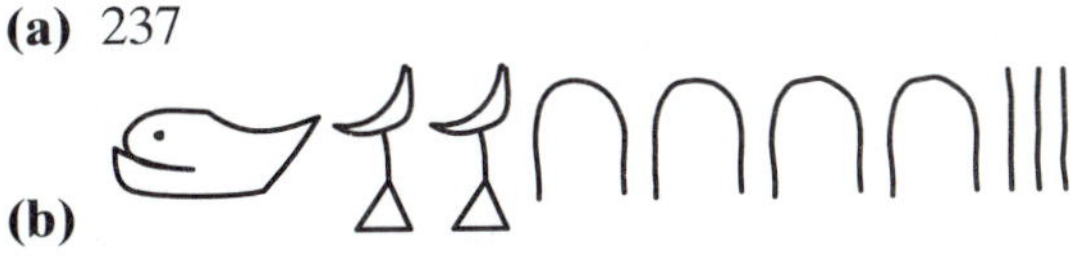 ◆

A major disadvantage of this system is that elementary calculations are cumbersome. Figure 2.4 shows, in Egyptian numerals, the addition problem that we write as 764 + 598 = 1362.

Egyptian Addition

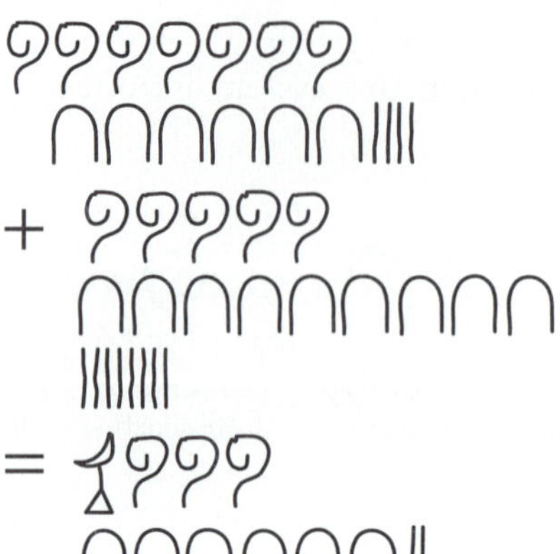

Hindu-Arabic Addition

$$\begin{array}{r} 764 \\ +\,598 \\ \hline 1362 \end{array}$$

FIGURE 2.4

Here 51 individual Egyptian number symbols are needed to express this addition problem, whereas our system requires only 10 digits!

THE ROMAN NUMERATION SYSTEM

The **Roman numeration system,** which developed between 500 B.C. and A.D. 100, also uses grouping, additivity, and a set of several basic symbols that are used to form the numerals. The basic Roman numerals are listed below.

TABLE 2.1

Roman Numeral	Value
I	1
V	5
X	10
L	50
C	100
D	500
M	1000

Roman numerals are made up of combinations of these basic numerals, such as

CCLXXXI (equals 281) and MCVIII (equals 1108).

To find the values of these Roman numerals, add the values of the basic symbols that comprise the numeral. For example,

MCVIII means 1000 + 100 + 5 + 1 + 1 + 1, or 1108.

That is, the Roman system is essentially an additive system.

HISTORY

Despite its awkwardness we still see the Roman numeration system in limited use today, usually to date movies and to name major events such as Super Bowl XXIV. Its survival is due to the wide area controlled by the Romans at the zenith of their empire. Also, the Roman system is simple and for most people with limited needs only the meanings of I, V, X, L, and C were memorized.

Two new attributes that were introduced by the Roman system were a subtractive principle and a multiplicative principle. Both of these principles allow the system to use fewer symbols to represent numbers. The ancient Roman system of numeration used (I) for 1000; repeating the parentheses would imply a multiplication by ten, so ((I)) represented 10,000 and (((I))) represented 100,000. Thus the Roman system was also a **multiplicative system.**

In the Middle Ages, a horizontal bar above a numeral was introduced to represent 1000 times the number. For example, $\overline{\text{V}}$ meant 5 times 1000, or 5000; $\overline{\text{XI}}$ meant 11,000; and so on. The **subtractive principle,** which was not used by the ancient Romans but was introduced in the Middle Ages, permits simplifications using combinations of basic Roman numerals. For example, IV (I to the left of V means five minus one) for 4 rather than using IIII, IX (ten minus one) for 9 instead of VIIII, XL for 40, XC for 90, CD for 400, and CM for 900. Reading from left to right, if the values of any two consecutive symbols increase, group the pair together. The value of this pair, then, is the value of the larger numeral minus the value of the smaller. To evaluate a complex Roman numeral, look to see if any of these subtractive pairs are present, group them together mentally, and then add values from left to right. For example,

in MCMXLIV,

think M CM XL IV, which is $1000 + 900 + 40 + 4 = 1944$.

Without the subtractive principle 14 individual Roman numerals would be required to represent 1944 instead of the seven numerals used in MCMXLIV. Also, because of the subtractive principle, the Roman system is a **positional system.** The position of a numeral can affect the value of the number. For example, VI is six, whereas IV is four.

Although expressing numbers using the Roman numeration system requires fewer symbols than the Egyptian system, it still requires many more symbols than our current system and is cumbersome for doing arithmetic. In fact, the Romans used an abacus to perform calculations instead of paper and pencil methods.

EXAMPLE 2.3 Express the Roman numeral in part (a) as a Hindu-Arabic numeral and the numeral in part (b) as a Roman numeral.

(a) CMLXXIV **(b)** 2793

SOLUTION

(a) CM = 900, LXX = 70, and IV = 4, thus CMLXXIV = 974.
(b) MMDCCXCIII ◆

THE HINDU-ARABIC NUMERATION SYSTEM

The **Hindu-Arabic system** that we use today was developed around the year A.D. 800. The following list features the basic numerals and various attributes of this system.

1. *Digits:* 0, 1, 2, 3, 4, 5, 6, 7, 8, 9. These 10 symbols, or **digits,** can be used in combination to represent all nonnegative whole numbers.

2. *Grouping by tens (decimal system).* Grouping into sets of 10 is a basic prin-

HISTORY

The Hindu-Arabic system of numerals was never used extensively by the Arabs; the Arabs were primarily the translators. The Arabs had, and still have, their own numeration system.

ciple of this system, probably because we have 10 "digits" on our two hands. In fact, the word *digit* also means finger or toe. Ten ones are replaced by one ten, ten tens are replaced by one hundred, ten hundreds are replaced by one thousand, and so on. Figure 2.5 shows how grouping is helpful when representing a collection of objects. The number of objects grouped together is called the **base** of the system; thus our Hindu-Arabic system is a base ten system.

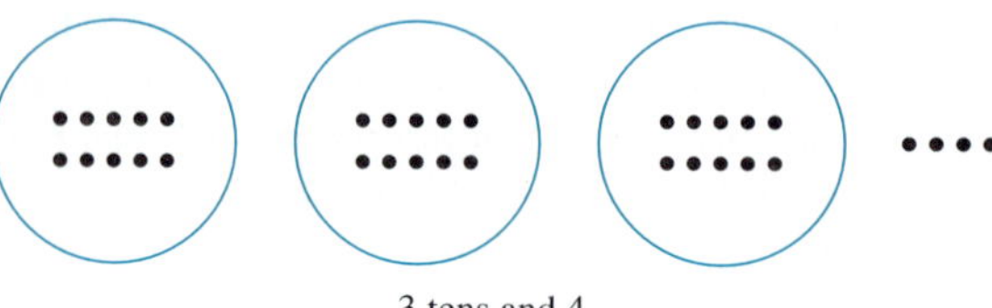

FIGURE 2.5

3. *Place value (hence positional).* The position of each digit shows the size of the groups represented by the digit. That size is the **place value.** The digit itself shows how many groups of that size are represented. For example, each of the places in the numeral 6523 has its own value.

HISTORY

Another ancient numeration system used by the Babylonians used place value and, after 300 B.C., had a place holder. The Babylonians used groupings of 60, and one residue of the Babylonian system which still exists in our modern world is the division of the circle into 360 degrees.

PlaceValue	thousand	hundred	ten	one
Digit	6	5	2	3

The 6 in the thousand place represents 6 thousands, the 5 in the hundred place represents 5 hundreds, the 2 in the ten place represents 2 tens, and the 3 represents 3 ones. The digit 0 is important as a **place holder,** so the position of all the nonzero digits can be determined. For example, in 307, the 0 forces the 3 to be in the hundred place. Thus, 307 is three hundred seven whereas 37 is thirty-seven.

4. *Additive and multiplicative.* The value of a Hindu-Arabic numeral is found by *multiplying* each place value by its corresponding digit and then by *adding* all the resulting products.

Place Value:	thousand		hundred		ten		one
Digits:	6		5		2		3
Numeral value:	6×1000	$+$	5×100	$+$	2×10	$+$	3×1
Numeral:	6523						

Expressing a numeral as the sum of its digits times their respective place values is called the numeral's **expanded form** or **expanded notation.** The expanded form of 83,507 is

$$8 \times 10{,}000 + 3 \times 1000 + 5 \times 100 + 0 \times 10 + 7 \times 1.$$

Because $7 \times 1 = 7$, we can simply write 7 in place of 7×1 when expressing 83,507 in expanded form.

EXAMPLE 2.4 Express the following numbers in expanded form.

(a) 437 **(b)** 3001

SOLUTION

(a) $437 = 4 \times 100 + 3 \times 10 + 7$
(b) $3001 = 3 \times 1000 + 0 \times 100 + 0 \times 10 + 1$ ◆

HISTORY

Acceptance of the Hindu-Arabic system in Europe was gradual, and there was some direct opposition. In 1299 a law was passed in Florence forbidding its use. The concern was apparently over the possibility of fraud since Hindu-Arabic numerals are more easily altered.

One or more of the four features of the Hindu-Arabic system already existed in each ancient numeration system. There must then be some reason (or reasons) why we are not using one of those ancient systems. In particular, there must be a reason that about a thousand years ago the Roman system was gradually replaced by the Hindu-Arabic. We can identify two advantages of the Hindu-Arabic system. It typically requires fewer symbols to represent each number, and it is notationally efficient. In addition, the Hindu-Arabic system is far superior for performing calculations using paper and pencil methods. The second advantage is believed to have been important historically, but it matters less now that calculators are widely used. Even so, the compactness of the Hindu-Arabic numeration system should keep it popular for another millennium.

Associated with each Hindu-Arabic numeral is a word name. Here are a few observations about the naming procedure.

1. The numbers 0, 1, . . . , 12 all have unique names.

2. The numbers 13, 14, . . . , 19 are the "teens," and are composed of a combination of earlier names, with the ones place named first. For example, "thirteen" for "three ten," which means "ten plus three," and so on.

3. The numbers 20, . . . , 99 are combinations of earlier names but *reversed* from the teens in that the tens place is named first. For example, 57 is "fifty-seven," which means "five tens plus seven," and so on. The method of naming the numbers from 20 to 99 is better than the way we name the teens, due to the left-to-right agreement with the way the numerals are written.

4. The numbers 100, . . . , 999 are combinations of hundreds and previous names. For example, 538 is read "five hundred thirty-eight," and so on.

5. In numerals containing more than three digits, groups of three digits are usually set off by commas. When the numeral is written, the groups of digits are used to form the name of the numeral. For example, the number

123,	456,	789,	987,	654,	321
quadrillion	trillion	billion	million	thousand	

is read "one hundred twenty-three quadrillion four hundred fifty-six trillion seven hundred eighty-nine billion nine hundred eighty-seven million six hundred fifty-four thousand three hundred twenty-one." Notice that the word "and" does not appear in any of these names.

The discussion above applies to what is done in the United States. In the United Kingdom "one billion" refers to 1,000,000,000,000 instead of 1,000,000,000, which it means in America. The British call 1,000,000,000 "one milliard."

Above their billion, the increasing British names correspond to six additional zeros (for us it is three zeros), so in the United Kingdom "one trillion" refers to 1,000,000,000,000,000,000, a quantity we call one quintillion.

INITIAL PROBLEM SOLUTION

You are on a trip to England and get into conversation with someone in a pub. In explaining some of the problems the United States is facing, you mention the $4 trillion national debt we have run up. Your acquaintance is horrified well beyond your expectations. What is it about the number 4 trillion that is so astounding?

SOLUTION Your acquaintance in the pub thinks you have just told him that the United States' national debt is

$4,000,000,000,000,000,000

a factor of a million times greater than it actually is. That is, the friend has translated *your* "trillion" into *his* "trillion." The interest alone on such a debt would exceed the gross domestic product of any nation on Earth.

PROBLEM SET 2.1

In problems 1 through 4, classify each number used as cardinal, ordinal, or identification numbers and give a short explanation for your choice. If you think a number can be interpreted two ways, write a short justification for each.

1. **(a)** I am the third child in my family.
(b) My car's license plate reads NUM 123.
(c) The Beavers won 6 games this season.
(d) There are 12 months in a year.

2. **(a)** My address is 4321 Happy Street.
(b) I live in the second house on the right.
(c) There are seven houses on the block.
(d) Her credit card number is 12345678.

3. **(a)** You are now working on problem number 3.
(b) There are 3 other parts to this problem.
(c) This is part 3 of the problem.
(d) You should get at least 3 of the parts of the problem correct.

4. **(a)** I was born on September 12.
(b) There are four children in my family.
(c) OSU finished in fifth place in the conference.
(d) Our taxicab was number 215.

In problems 5 through 8, write short (but complete) sentences that show the given number used in each of the following ways.

(a) As a cardinal number.
(b) As an ordinal number.
(c) As an identification number.

If you are unable to use the given number in one of the ways, briefly explain why.

5. 45
6. 1
7. 0
8. 12
9. Is the Roman numeral XXVII a cardinal, ordinal, or identification number? Explain your answer. Hint: This is a trick question.

10. Suppose ancient Roman and Egyptian archeologists discovered an artifact from an earlier civilization that bore these marks:

|||||||||||||||||||||||||||||

If this was found in a cave that had been used for storage, how might they interpret this, and how would each one write down the information for his records?

11. Change to Hindu-Arabic numerals.

(a) ∩∩∩∩||

(b) 𓍢𓍢𓍢𓍢||||

12. Change to Hindu-Arabic numerals.

(a)  **(b)** 

13. Change to Egyptian numerals.
(a) 9 (b) 23 (c) 1231

14. Change to Egyptian numerals.
(a) 453 (b) 2222 (c) 10,352

15. Change to Hindu-Arabic numerals.
(a) MCMXCI (b) CMLXXVI

16. Change to Hindu-Arabic numerals.
(a) MMMCCXLV (b) MCCXLVII

17. Change to Roman numerals.
(a) 76 (b) 434 (c) 1999

18. Change to Roman numerals.
(a) 396 (b) 697 (c) 2001

19. Change to Egyptian numerals.
(a) MMXCVII (b) MCDLXIV

20. Change 1997 to each of the following:
(a) Egyptian numerals.
(b) Roman numerals.

21. A newspaper advertisement introduced a new car as follows:
"IV Cams, XXXII Valves, CCLXXX Horsepower, coming December XXVI—the new 1993 Lincoln Mark VIII."
(a) Change 1993 to Roman numerals.
(b) Explain why a car dealer might be reluctant to have 1993 changed to Roman numerals, but not the others.

22. After the credits for a film roll by, the Roman numeral MCMLXXXIX appears, which represents the year in which the film was made. Express the year in the Hindu-Arabic numeration system.

23. Express each of the following numbers in expanded form.
(a) 437 (b) 5603

24. Express each of the following numbers in expanded form.
(a) 840 (b) 35,072

25. Write the following as standard numerals.
(a) One hundred five thousand eight hundred forty-two.
(b) Twenty-two million sixty thousand three hundred.

26. Write the following as standard numerals.
(a) Seventy-five thousand two hundred thirty-six.
(b) Three trillion four hundred fifty-seven billion, eighty million.

27. Express each of the following numerals with words.
(a) 345678 (b) 102620057

28. Express each of the following numerals with words.
(a) 3065183 (b) 8409350000000

EXTENDED PROBLEMS

The following Chinese numerals are part of one of the oldest numeration systems known.

一	1	十	10
二	2	百	100
三	3	千	1000
四	4		
五	5		
六	6		
七	7		
八	8		
九	9		

The numerals are written vertically. Some examples follow:

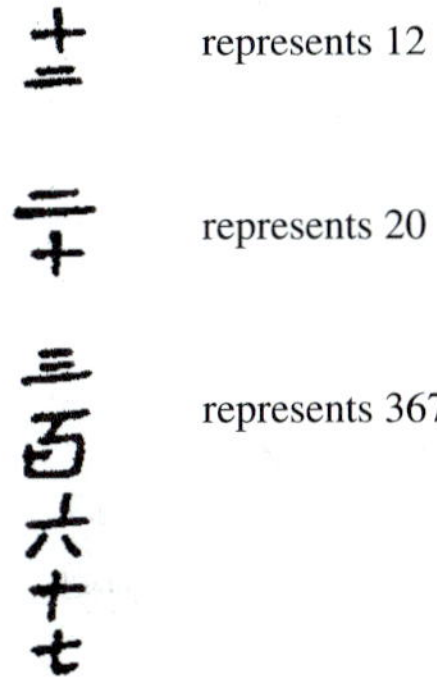

29. Express each of the following numerals in the Hindu-Arabic numeration system.
(a) 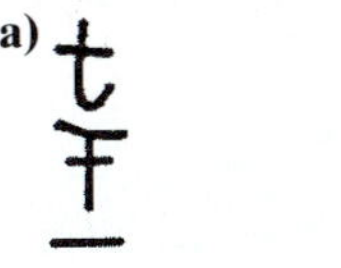 (b) 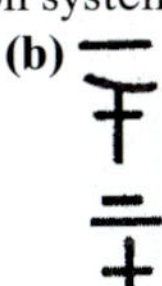

30. Express each of the following numerals in this Chinese numeration system.
(a) 19 (b) 400 (c) 6031

31. Express each of the following numerals in this Chinese numeration system.
(a) 63 (b) 580 (c) 2546

32. One system of numeration used in Greece in about 300 B.C., called the *Ionian system,* was based on the letters of the Greek alphabet. The different symbols used for numbers less than 1000 are as follows:

α	β	γ	δ	ε	ζ	ξ	η	θ	ι	κ	λ	μ	ν
1	2	3	4	5	6	7	8	9	10	20	30	40	50

Ξ	ο	π	φ	ρ	σ	τ	υ	ϕ	χ
60	70	80	90	100	200	300	400	500	600

ψ	ω	Ψ
700	800	900

To represent multiples of 1000, an accent mark was used. For example, 'ε was used to represent 5000. The accent mark might be omitted if the size of the number being represented was clear without it.

Express the following Ionian numerals in our numeration system.

(a) μβ **(b)** σλγ **(c)** 'ηχπδ

33. Express the following Ionian numerals as Hindu-Arabic numerals.

(a) ϕμδ **(b)** 'δσξ **(c)** Ψφβ

34. Express the following numerals in the Ionian numeration system.

(a) 85 **(b)** 247 **(c)** 1997

Braille numerals are formed using dots in a two-dot by three-dot Braille cell. Numerals are preceded by a backwards "L" dot symbol. The following shows the basic elements for Braille numerals and two examples.

0 1 2 3 4 5 6 7 8 9 ,

One billion, four hundred sixty-seven million, seventy thousand, two hundred seventy-nine

Eight hundred four million, six hundred forty-seven thousand, seven hundred

35. Express these Braille numerals in the Hindu-Arabic numeration system.

(a)

(b)

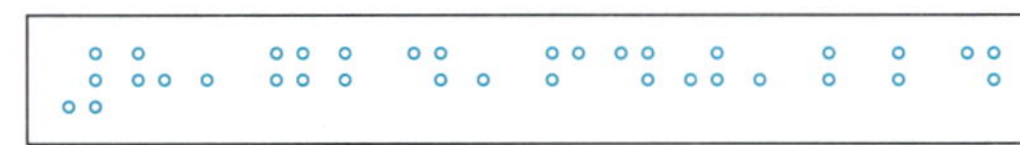

36. Express the following numerals in Braille numerals.

(a) 87 **(b)** 1349 **(c)** 1,234,567

37. Express the following numerals in Braille numerals.

(a) 156 **(b)** 2450 **(c)** 586,507

2.2 MENTAL MATH AND ESTIMATION

You are considering buying a new car. The salesperson wants you to look at a fairly expensive model with a sticker price of \$15,350. You point out that you are a recent graduate without any big down payment and only a modest salary. The salesperson says, "Don't worry, try a test drive; the payments will probably only run about \$160 a month." Use mental math to decide if the salesperson's estimate is reasonable.

The availability and widespread use of calculators and computers have permanently changed the way we work with numbers. Most of the time, but not always, calculators perform their computations flawlessly and computers correctly execute the commands given them. Such is not the case for the person pushing the buttons on the calculator or writing the computer program; each of us is quite likely to occasionally make mistakes. To find and correct errors before damage is done to something important (such as your grade), we need to develop estimating skills to check the reasonableness of results obtained electronically. Since computational estimation involves simple calculations, a good set of basic skills for performing mental math is very helpful. This section begins with several techniques for doing calculations mentally.

MENTAL MATH

Properties

Commutativity, associativity, and distributivity play an important role in simplifying calculations so that they can be performed mentally. These properties (also called rules or laws) are summarized below.

Properties of Arithmetic Operations

$a + b = b + a$ **Commutative Property of Addition**
$a \times b = b \times a$ **Commutative Property of Multiplication**
$a + (b + c) = (a + b) + c$ **Associative Property of Addition**
$a \times (b \times c) = (a \times b) \times c$ **Associative Property of Multiplication**
$a + 0 = 0 + a = a$ **Identity for Addition**
$a \times 1 = 1 \times a = a$ **Identity for Multiplication**
$a \times (b + c) = (a \times b) + (a \times c)$ **Distributivity of Multiplication over Addition**
$a \times (b - c) = (a \times b) - (a \times c)$ **Distributivity of Multiplication over Subtraction**
$(a + b) + c = (a + c) + (b + c)$ **Right Distributivity of Division over Addition**

When you think of arithmetic operations in concrete terms related to counting objects, the various properties are almost self-evident. For example, $2 + 3 = 3 + 2$ is an example of the commutative property of addition. The point is not the creation of fancy names for obvious facts, but rather to bring these facts to our attention so we can freely use them without the necessity of rediscovering them over and over. The following example shows how the properties can be applied in mental calculations.

EXAMPLE 2.5 Calculate the following mentally.

(a) $15 + (27 + 25)$ **(b)** $21 \times 17 - 13 \times 21$
(c) $(8 \times 7) \times 25$ **(d)** $98 + 59$
(e) $87 + 29$ **(f)** $168 \div 3$

SOLUTION

(a) $15 + (27 + 25) = (27 + 25) + 15 = 27 + (25 + 15) = 27 + 40 = 67$. Since our goal is to simplify the calculation, we observed that it was easy to add 15 and 25 mentally. Thus, we used commutativity in the first equality to get the 25 next to the 15 and associativity in the second equality to regroup the numbers so that we could find $15 + 25$ first.

(b) $21 \times 17 - 13 \times 21 = 21 \times 17 - 21 \times 13 = 21 \times (17 - 13) = 21 \times 4 = 84$. Here, 21 was common to 21×17 and 13×21. Thus we might use distributivity. However, commutativity was used first so that distributivity could be applied as it is stated.

(c) $(8 \times 7) \times 25 = (7 \times 8) \times 25 = 7 \times (8 \times 25) = 7 \times 200 = 1400$. Here commutativity is used first, then associativity is used to group the 8 and 25 since their product is 200.

(d) $98 + 59 = 98 + (2 + 57) = (98 + 2) + 57 = 157$. Since $98 + 2 = 100$, 59 was rewritten as $2 + 57$ and then associativity is used.

(e) $87 + 29 = 80 + 20 + 7 + 9 = 100 + 16 = 116$ using associativity and commutativity.

(f) $168 \div 3 = (150 \div 3) + (18 \div 3) = 50 + 6 = 56$. This calculation uses right distributivity of division with respect to addition. Note that 168 was separated into $150 + 18$ since each of those numbers is easily recognized as being divisible evenly by 3. ◆

Compatible Numbers

Compatible numbers are pairs of numbers whose sums, differences, products, or quotients are easy to calculate mentally. Examples of compatible numbers are 86 and 14 under addition (since $86 + 14 = 100$), 25 and 8 under multiplication (since $25 \times 8 = 200$), and 600 and 30 under division since $600 \div 30 = 20$). In part (a) of Example 2.5, adding 15 to 25 produces a number (namely 40) that is easy to add to 27. Notice that numbers are compatible with *respect to an operation.* For example, 86 and 14 are compatible with respect to addition but not with respect to multiplication.

EXAMPLE 2.6 Calculate the following mentally, using properties and/or compatible numbers.

(a) $(4 \times 13) \times 25$ **(b)** $1710 \div 9$ **(c)** 86×15

SOLUTION

(a) Since $4 \times 25 = 100$, we use properties to rearrange the numbers to take advantage of this fact: $(4 \times 13) \times 25 = 13 \times (4 \times 25) = 1300$.

(b) Here we try to rewrite 1710 into two parts that are easily divisible by 9: $1710 \div 9 = (1800 - 90) \div 9 = (1800 \div 9) - (90 \div 9) = 200 - 10 = 190$.

(c) $86 \times 15 = (86 \times 10) + (86 \times 5) = 860 + 430 = 1290$. (Notice that 86×5 is half of 86×10.) ◆

There may be several correct ways to calculate the same result mentally. For instance, in Example 2.6(c), one could also find 86×15 as follows:

$$86 \times 15 = (80 + 6) \times 15 = 80 \times 15 + 6 \times 15 = 1200 + 90 = 1290.$$

Compensation

The sum $43 + (38 + 17)$ can be viewed as $60 + 38 = 98$ using commutativity, associativity, and the fact that 43 and 17 are compatible numbers.

$$43 + (38 + 17) = 43 + (17 + 38) = (43 + 17) + 38 = 60 + 38 = 98.$$

Finding the answer to $43 + (36 + 19)$ is not as easy. However, by thinking of the sum $36 + 19$ as $37 + 18$, we obtain the sum $(43 + 37) + 18 = 80 + 18 = 98$.

Reformulating a sum, difference, product, or quotient as a simpler calculation is called compensation. Some specific techniques using compensation are introduced next.

In the computations of Example 2.5(d), 98 was increased by 2 to 100 and then 59 was decreased by 2 to 57 (a compensation was made) to maintain the same sum. This technique, **additive compensation,** is an application of associativity. Similarly, additive compensation is used when $98 + 59$ is rewritten as either $97 + 60$ or $100 + 57$. The problem $47 - 29$ can be thought of as $48 - 30$ ($= 18$). This use of compensation in subtraction is called the **equal additions method** since the same number (here 1) is added to both 47 and 29 to maintain the same difference. This compensation is performed to make the subtraction easier by subtracting 30 from 48. As another example, the product 48×5 can be found using **multiplicative compensation** as follows: $48 \times 5 = 24 \times 10 = 240$. Here one number is divided by a constant, while the other is multiplied by the same constant.

$$48 \times 5 = (48 \div 2) \times (5 \times 2) = 24 \times 10 = 240.$$

EXAMPLE 2.7 Calculate the following mentally using compensation.

(a) $16 + 61 + 141$ **(b)** $151 - 97$ **(c)** 25×28

SOLUTION

(a) $16 + 61 + 141 = 18 + (60 + 140) = 18 + 200 = 218$. Additive compensation was used when 60 and 140 were each decreased by 1 and consequently 16 was increased by 2.

(b) $151 - 97 = 154 - 100 = 54$. The equal additions method was used when 151 and 97 were increased by 3.

(c) $25 \times 28 = 100 \times 7 = 700$. Multiplicative compensation was used when 25 was multiplied by 4 to obtain 100 and consequently 28 was divided by 4 to obtain 7. ◆

Left-to-Right Methods

To add 342 to 136, one can first add the hundreds ($300 + 100$), then the tens ($40 + 30$), and then the ones ($2 + 6$), to obtain 478. To add 158 and 279, one can think as follows: $100 + 200 = 300$, $300 + 50 + 70 = 420$, $420 + 8 + 9 = 437$. Alternatively, $158 + 279$ can be found as follows: $158 + 200 = 358$, $358 + 70 = 428$, $428 + 9 = 437$. Subtraction from left to right can be done in a similar manner. Research has found that people who are excellent mental calculators utilize this left-to-right method to reduce memory load, instead of mentally picturing the usual right-to-left written method. The multiplication problem 3×123 can be thought of mentally as $3 \times 100 + 3 \times 20 + 3 \times 3$ using distributivity. Similarly, 4×253 can be thought of mentally as $800 + 200 + 12 = 1012$ or as $4 \times 250 + 4 \times 3 = 1000 + 12 = 1012$.

EXAMPLE 2.8 Calculate the following mentally using left-to-right methods.

(a) $123 + 235$ **(b)** $689 - 243$ **(c)** 3×321

SOLUTION

(a) $123 + 235 = (100 + 200) + (20 + 30) + (3 + 5) = 300 + 50 + 8 = 358$

(b) $689 - 243 = (600 - 200) + (80 - 40) + (9 - 3) = 400 + 40 + 6 = 446$

(c) $3 \times 321 = 3 \times 300 + 3 \times 20 + 3 \times 1 = 900 + 60 + 3 = 963$ ◆

COMPUTATIONAL ESTIMATION

The process of estimation takes various forms. The number of beans in a jar may be estimated using no mathematics, simply a "guesstimate." Also, you may estimate how long a trip will be, based simply on experience. **Computational estimation** is the process of finding an approximate answer (an estimate) to a computation, often using mental math. Almost everyone uses calculators; and while the calculator may do perfect calculations, the person using it may push the wrong buttons. Computational estimation is a valuable skill that allows you to judge the reasonableness of your results. Next we consider various types of computational estimation.

Front-End Estimation

We can estimate the sum 478 + 264 using the **front-end method** as follows: To estimate 478 + 264, think 400 + 200 = 600 (the estimate). The front-end method always provides low estimates in addition problems as well as in multiplication problems. In the case of 376 + 53 + 417, the front-end estimate is 300 + 400 = 700 since there are no hundreds in 53. The front-end estimate of 37 × 43 is 30 × 40 = 1200.

A disadvantage of this method is that the answer is always under the exact answer. The next technique addresses this deficiency.

Range Estimation

Often it is sufficient to know an interval or **range,** that is, a low value and a high value, that will contain an answer. The following example shows how to find range estimates.

EXAMPLE 2.9 Find a range for answers to these computations using only the leading digits.

(a) 257 + 576 **(b)** 294 × 53

SOLUTION

(a) Sum 257 + 576
Low estimate 200 + 500 = 700
High estimate 300 + 600 = 900
Thus a range for the answer is from 700 to 900. Notice that you have to look at only the digits having the largest place values (2 + 5 = 7, or 700) to arrive at the low estimate and these digits each increased by one (3 + 6 = 9, or 900) to find the high estimate.

(b) Product 294 × 53
Low estimate 200 × 50 = 10,000
High estimate 300 × 60 = 18,000
Due to the nature of multiplication, this method gives a wide range, here 10,000 to 18,000. Even so, this method will catch many errors. ◆

Keep in mind that estimates are made to obtain a rough answer, so both forms of front-end estimation above belong in a person's estimation repertoire.

EXAMPLE 2.10 Make an estimate using the indicated method.

(a) 503 × 813 using front-end estimation
(b) 1200 × 35 using range estimation

SOLUTION

(a) To estimate 503×813 using the front-end method, think "$500 \times 800 = 400{,}000$." Using words, think "5 hundred times 8 hundred is 4 hundred thousand."

(b) To estimate a range for 1200×35, think $1200 \times 30 = 36{,}000$ and $1200 \times 40 = 48{,}000$. Thus a range for the answer is from 36,000 to 48,000. You could also use $1000 \times 30 = 30{,}000$ and $2000 \times 40 = 80{,}000$. However, this yields a wider range, and most people are comfortable multiplying by 12. ◆

Rounding

Rounding is perhaps the best known estimation technique. The purpose of rounding is to replace complicated numbers with simpler numbers. Here, again, to obtain an estimate, any of several rounding techniques may be used. Some may be more appropriate than others, depending on the problem. For example, if you are estimating how much money to take on a trip, you would round up to be sure that you had enough. When calculating the cost of gas needed for a trip, you would round down the miles per gallon estimate and round up the price per gallon estimate to ensure that there would be enough money for gas. Unlike the previous estimation techniques, rounding is often applied to an answer as well as to the individual numbers before a computation is performed.

Several different methods of rounding are illustrated next. What is common, however, is that each method rounds to a particular place. You are asked to formulate rules for the following methods in the problem set.

FIGURE 2.6

Round Up (Down) The number 473 **rounded up** to the nearest tens place is 480 since 473 is between 470 and 480 and 480 is *above* 473 (Figure 2.6). The number 473 **rounded down** to the nearest tens place is 470. Rounding down is also called **truncating** (truncate means to cut off). The number 1276 truncated to the hundreds place is 1200.

Round a 5 Up The most common rounding technique is the **round a 5 up** method. This method can be motivated using a number line. Suppose that we wish to round 475 to the nearest ten (Figure 2.7).

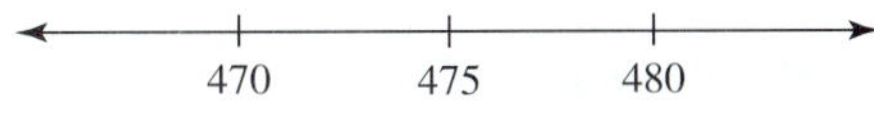

FIGURE 2.7

Since 475 is midway between 470 and 480, we have to decide whether we round 475 to 470 or to 480. The round a 5 up method, also referred to as **rounding off,** always rounds numbers ending in five up, so 475 rounds to 480. Since the numbers 471 to 474 are nearer 470 than 480, they are rounded to 470 when rounding to the nearest ten. The numbers 476 to 479 are rounded to 480.

One disadvantage of this method is that estimates obtained when several 5's are involved tend to be on the high side. For example, the "round a 5 up" to the nearest ten estimate applied to the numbers of the sum $35 + 45 + 55 + 65$ yields $40 + 50 + 60 + 70 = 220$, which is 20 more than the exact sum 200.

EXAMPLE 2.11 Make estimations using the indicated method.

(a) Estimate $2173 + 4359$ by rounding down to the nearest hundreds place.

(b) Estimate $2173 + 4359$ by rounding up to the nearest hundreds place.

(c) Estimate $575 - 398$ by rounding a 5 up to the nearest tens place.

SOLUTION (The symbol $\approx$ means "is approximately.")

(a) $2173 + 4359 \approx 2100 + 4300 = 6400$
(b) $2173 + 4359 \approx 2200 + 4400 = 6600$
(c) $575 - 398 \approx 580 - 400 = 180$ ◆

Round to Compatible Numbers Another rounding technique can be applied to estimate products such as 26×37. A reasonable estimate of 26×37 is $25 \times 40 = 1000$. The numbers 25 and 40 were selected since they are estimates of 26 and 37, respectively, and are compatible with respect to multiplication. (Notice that the rounding up technique would have yielded the considerably higher estimate of $30 \times 40 = 1200$, whereas the exact answer is 962.) **Rounding to compatible numbers** rounds numbers either up or down to compatible numbers to simplify calculation, rather than rounding to specified places. For example, a reasonable estimate of 57×98 is 57×100 (= 5700). Here, only the 98 needed to be rounded to obtain an estimate mentally. The division problem $2716 \div 75$ can be estimated mentally by considering $2800 \div 70$ (= 40). Here 2716 was rounded up to 2800, and 75 was rounded down to 70 because 2800 and 70 are compatible numbers with respect to division.

EXAMPLE 2.12 Make estimates by rounding to compatible numbers in two different ways.

(a) 43×21 **(b)** $256 \div 33$

SOLUTION

(a) $43 \times 21 \approx 40 \times 25 = 1000$
$43 \times 21 \approx 45 \times 20 = 900$ (The exact answer is 903.)
(b) $256 \div 33 \approx 240 \div 30 = 8$
$256 \div 33 \approx 280 \div 40 = 7$ (The exact answer is 7 with remainder 25.) ◆

Rounding is a most useful and flexible technique. Rounding is important because (i) it simplifies calculations while obtaining reasonable answers, and (ii) it gives numerical results that can be easily understood. Any of the methods illustrated here may be used to estimate.

SUMMARY

The following suggestions should help develop your number sense.

1. Master the basic single digit facts for addition and multiplication (e.g., $3 + 4 = 7$, $5 \times 7 = 35$, and so on).
2. Master the concept of place value.
3. Master the basic addition, multiplication, and division properties of real numbers.
4. Develop a habit of using the front-end and left-to-right methods.
5. Practice mental calculations in everyday life; there are many opportunities. For example, if you keep track of the miles traveled between gasoline purchases you can estimate your car's miles-per-gallon efficiency each time you buy gas.
6. Accept approximate answers when exact answers are not needed.
7. Estimate prior to doing exact computations.
8. Be flexible and use a variety of mental math and estimation techniques.

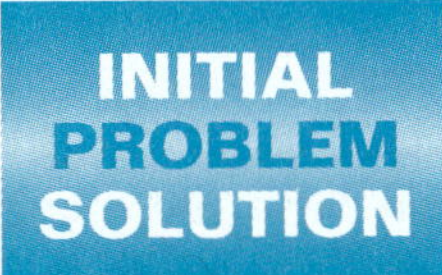

You are considering buying a new car. The salesperson wants you to look at a fairly expensive model with a sticker price of $15,350. You point out that you are a recent graduate without any big down payment and only a modest salary. The salesperson says, "Don't worry, try a test drive; the payments will probably only run about $160 a month." Use mental math to decide if the salesperson's estimate is reasonable.

SOLUTION We round the $160 payment to $150, and we round the sticker price of the car to $15,000, which are compatible numbers. You can see instantly that 100 payments will be needed to pay for the car even if *no interest is charged.* That is over eight years! The salesperson's estimate is not reasonable—it is far too low.

PROBLEM SET 2.2

1. Identify the property that justifies the equation.
   (a) $(23 + 18) + 12 = 23 + (18 + 12)$
   (b) $4 \times 27 = (4 \times 25) + (4 \times 2)$
   (c) $(7 \times 4) \times 25 = 7 \times (4 \times 25)$
   (d) $(17 \div 5) + (18 \div 5) = 35 \div 5$
2. Identify the property that justifies the equation.
   (a) $(7 \times 12) + (7 \times 13) = 7 \times (12 + 13)$
   (b) $14 + (16 + 7) = (14 + 16) + 7$
   (c) $37 \times 100 - 37 \times 1 = 37 \times (100 - 1)$
   (d) $5 \times (2 \times 9) = (5 \times 2) \times 9$
3. Calculate mentally using properties.
   (a) $(37 + 25) + 43$
   (b) $47 \times 15 + 47 \times 85$
   (c) $(4 \times 13) \times 25$
   (d) $26 \times 24 - 21 \times 24$
4. Calculate mentally using properties.
   (a) $(73 + 18) + 27$
   (b) $(5 \times 7) \times 6$
   (c) $(13 \times 5) - (3 \times 5)$
   (d) $(37 \div 5) - (17 \div 5)$
5. Give an example of a number compatible with 93 under addition. Explain your reasoning.
6. Give an example of a number compatible with 25 under multiplication. Explain your reasoning.
7. Calculate the following mentally using properties and/or compatible numbers.
   (a) 38×5
   (b) $483 \div 7$
   (c) $(43 \times 6) + (7 \times 6)$
8. Calculate the following mentally using properties and/or compatible numbers.
   (a) $(73 + 18) + 27$
   (b) $(5 \times 7) \times 6$
   (c) $(13 \times 5) - (3 \times 5)$
9. Find these sums mentally using additive compensation. Write out the steps you thought through.
   (a) $288 + 102$ **(b)** $301 + 199$
10. Find these sums mentally using additive compensation. Write out the steps you thought through.
    (a) $103 + 397$ **(b)** $126 + 234$
11. Find each of these differences mentally using equal additions. Write out the steps you thought through.
    (a) $43 - 17$ **(b)** $132 - 96$
12. Find each of these differences mentally using equal additions. Write out the steps you thought through.
    (a) $62 - 39$ **(b)** $250 - 167$
13. Find these products mentally using multiplicative compensation. Write out the steps you thought through.
    (a) 18×5 **(b)** 222×5
14. Find these products mentally using multiplicative compensation. Write out the steps you thought through.
    (a) 60×25 **(b)** 444×25
15. Calculate mentally from left to right.
    (a) $123 + 456$ **(b)** $467 - 134$
16. Calculate mentally from left to right.
    (a) $587 - 372$ **(b)** $342 + 561$
17. Use front-end estimation to estimate.
    (a) $469 + 555$ **(b)** $137 + 293$
18. Use front-end estimation to estimate.
    (a) $4395 + 7286$ **(b)** $1394 + 4965$
19. Use front-end estimation to estimate.
    (a) 369×657 **(b)** 26×298
20. Use front-end estimation to estimate.
    (a) 3459×8276 **(b)** 1349×5694
21. Use range estimation to estimate.
    (a) $287 + 426$ **(b)** $692 + 128$

22. Use range estimation to estimate.
(a) $7347 + 6428$ **(b)** $4569 + 1358$

23. Use range estimation to estimate.
(a) 231×37 **(b)** 297×384

24. Use range estimation to estimate.
(a) 2395×7682 **(b)** 583×6854

25. Round down to the nearest tens place.
(a) 1235 **(b)** 689
(c) 23,795 **(d)** 16

26. Round down to the nearest hundreds place.
(a) 2657 **(b)** 16,942
(c) 235 **(d)** 63,457

27. Round up to the nearest tens place.
(a) 769 **(b)** 542
(c) 10,111 **(d)** 667

28. Round up to the nearest thousands place.
(a) 23,655 **(b)** 114,879
(c) 2,563,123 **(d)** 283

29. Round off to the nearest hundreds place (recall that round off means round a 5 up).
(a) 4279 **(b)** 6626
(c) 2753 **(d)** 25,750

30. Round off to the nearest thousands place.
(a) 5500 **(b)** 5,499
(c) 13,255 **(d)** 123,123

31. Estimate by rounding to compatible numbers.
(a) $4776 \div 103$ **(b)** 46×202

32. Estimate by rounding to compatible numbers.
(a) 256×39 **(b)** $281 \div 69$

33. Here are 4 ways to estimate 26×12:

$26 \times 10 = 260$
$25 \times 12 = 300$
$30 \times 12 = 360$
$30 \times 10 = 300$

Estimate the following in four ways.
(a) 31×23 **(b)** 35×46

34. Estimate the following in four ways as in problem 33.
(a) 48×27 **(b)** 76×12

35. Estimate the following values and check with a calculator.
(a) 656×74 is between ____000 and ____000.
(b) 143×143 is between ____0000 and ____0000.

36. Estimate the following values and check with a calculator.
(a) 491×3172 is between ____00000 and ____00000.
(b) 183×183 is between ____00 and ____00.

EXTENDED PROBLEMS

37. Cluster Estimation is used to estimate sums and products when there are several numbers that cluster near a single number. For example, the addends in $789 + 810 + 792$ cluster around 800. Therefore, $3 \times 800 = 2400$ is a good estimate of the sum. Estimate the following using cluster estimation.
(a) $347 + 362 + 354 + 336$
(b) $61 \times 62 \times 58$

38. Estimate the following using cluster estimation.
(a) $489 \times 475 \times 523 \times 498$
(b) $782 + 791 + 834 + 812 + 777$

39. The halving and doubling method can be used to multiply two numbers when one factor is a power of 2. For example, to find 8×17, find 4×34 or $2 \times 68 = 176$. Find the following products using this method.
(a) 16×21 **(b)** 4×72

40. Find the following products using the halving and doubling method.
(a) 8×123 **(b)** 16×211

41. Before granting an operating license, a scientist has to estimate the amount of pollutants that should be allowed to be discharged from an industrial chimney. Should she overestimate or underestimate? Why?

42. A scientist has to estimate the distance that lava from an erupting volcano will flow to determine an evacuation zone. Should he overestimate or underestimate? Why?

43. Before approving irrigation allotments for the coming year, a water resources board has to estimate the volume and levels of stream flow for the coming summer. Should they overestimate or underestimate? Why?

44. A biologist wishes to estimate the number of fish that can be harvested without endangering the population of a lake. Should she overestimate or underestimate? Why?

45. Place a multiplication sign or signs so that the products in each problem is correct; for example, in the expression 1 2 3 4 5 6 = 41,472, the multiplication sign should be placed between the 2 and the 3 since $12 \times 3456 = 41{,}472$.
(a) 1 3 5 7 9 0 = 122,310
(b) 6 6 6 6 6 6 = 439,956
(c) 7 8 9 3 4 5 6 = 4,736,070
(d) 1 2 3 4 5 6 7 = 827,115

46. Some mental calculators use the following fact: $(a + b)(a - b) = a \times a - b \times b$. For example, $43 \times 37 = (40 + 3)(40 - 3) = 40 \times 40 - 3 \times 3 = 1600 - 9 = 1591$.
Apply this technique to find the following products mentally.
(a) 54×46
(b) 81×79
(c) 122×118
(d) 1210×1190

2.3 INTEGERS AND RATIONAL NUMBERS

Five-eighths of the students at Moo U live off campus. If one-sixth of these students need parking permits and there are 14,400 students enrolled, how many parking permits will be needed?

While the early development of number systems is directly related to the need for counting and tabulation, it wasn't until the 14th or 15th century that there was a need to represent a number that was "opposite," in some sense, to some other number. Negative numbers came into being and slowly gained acceptance. Fractions, on the other hand, date back to the Egyptians; however, the forms used for fractions and the ways of working with them have gone through many changes. Although working with fractions is a challenge for many, fractions are indispensable in many applications. Fractions may be expressible in a variety of equivalent forms to suit different situations.

THE INTEGERS

The set of natural numbers is the set {1, 2, 3, . . .}. This set together with zero, namely {0, 1, 2, 3, . . .), is called the set of **whole numbers.** If you think of numbers as credits (money you have or are owed) and debits (money you owe), the whole numbers can be used to represent credits, but new numbers are needed to represent debits. The set of **integers,** {. . . , −3, −2, −1, 0, 1, 2, 3, . . .}, can be used to represent both positive and negative numbers. The integers can be represented pictorially using the **integer number line** (Figure 2.8).

HISTORY

Negative numbers were used in India as early as the 6th century and were adopted by the Arabs 200 years later. In Europe, negative numbers were introduced in the 13th century, but not fully accepted until the 16th century.

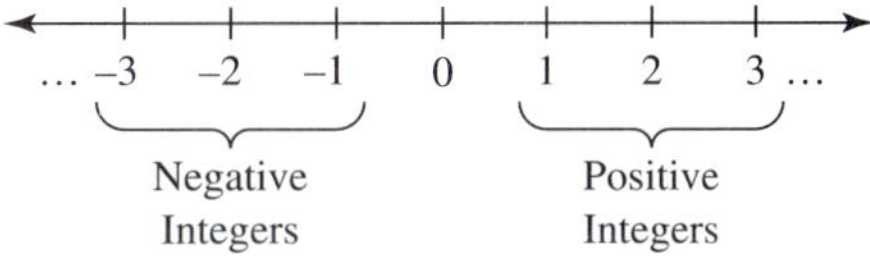

FIGURE 2.8

Since positive integers can be thought of as credits and negative integers as debits, they are opposites of each other. More precisely, on the integer number line, two numbers are said to be **opposites** if they are the same distance from zero, but on the opposite sides of zero. For example, 3 and −3 are opposites. The symbol $-a$ is used to represent the opposite of a. Notice that $-(-3) = 3$; that is, the opposite of the opposite of 3 is 3. In general, we have the following: $-(-a) = a$.

Integers: Addition and Subtraction

Integer arithmetic is an extension of whole number arithmetic. Operations such as addition and subtraction of whole numbers *as integers* are exactly as they are in the set of whole numbers. The credit/debit model can be used to see how other integers are added.

In integer **addition**, $3 + (-4)$ can be thought of as a credit of 3 together with a debit of 4, resulting in a debit of 1. Thus, $3 + (-4) = -1$. Similarly, $5 + (-2) = 3$, $(-3) + (-1) = -4$, and so on. Using this model for addition, notice that a number plus its opposite is zero. Symbolically, $a + (-a) = 0$. The number $-a$ is also called the **additive inverse** of a since it adds to a to produce 0. The number zero is called the **additive identity** since for any integer a, $a + 0 = a$.

Next we consider the subtraction of integers. In the set of whole numbers, $5 - 3 = 2$. In the set of integers, $5 - 3 = 2$ *and* $5 + (-3) = 2$ also. Considering many similar examples will suggest the following definition.

Subtraction of Integers

$$a - b = a + (-b)$$

Subtracting an integer is the same as adding its opposite (additive inverse).

EXAMPLE 2.13 Calculate.

(a) $13 + (-7)$ **(b)** $17 - (-3)$ **(c)** $(-7) - (-11)$

SOLUTION

(a) $13 + (-7) = 6$
(b) $17 - (-3) = 17 + -(-3) = 17 + 3 = 20$
(c) $(-7) - (-11) = (-7) + -(-11) = (-7) + 11 = 4$ ◆

Integers: Multiplication and Division

In a similar manner, integer multiplication and division are extensions of whole number multiplication and division. For **multiplication** consider the following pattern.

$3 \times 2 = 6$
  ↓ 3 less
$3 \times 1 = 3$
  ↓ 3 less
$3 \times 0 = 0$
  ↓ 3 less
$3 \times (-1) = ?$
  ↓ 3 less
$3 \times (-2) = ?$
etc.

When this pattern is continued, we have $3 \times (-1) = -3$, $3 \times (-2) = -6$, and so on. Similar patterns suggest that $5 \times (-7) = -35$, $4 \times (-12) = -48$, and so on. In general, *the product of a positive integer and a negative integer is negative.*

Now consider the next pattern.

$$(-3) \times 2 = (-6)$$
3 more
$$(-3) \times 1 = (-3)$$
3 more
$$(-3) \times 0 = 0$$
3 more
$$(-3) \times (-1) = ?$$
3 more
$$(-3) \times (-2) = ?$$
etc.

TIDBIT

A poet once wrote, "Minus times minus is plus. The reasons why, we need not discuss."

When this pattern is continued, we have $(-3) \times (-1) = 3$, $(-3) \times (-2) = 6$, and so on. Similar patterns suggest that $(-5) \times (-7) = 35$, $(-4) \times (-12) = 48$, and so on. In general, *a negative integer times a negative integer is positive.* Finally, zero times any integer is zero.

Integer **division** is derived from integer multiplication by extending the divi sion of whole numbers. For example, $8 \div 2 = n$ if and only if $8 = n \times 2$. Since $4 \times 2 = 8$, n must be 4. Similarly, $(-6) \div 2 = -3$, and $(-12) \div (-6) = 2$. In general, *the quotient of two positives (or two negatives) is positive* and *the quotient of a positive and a negative is negative.*

EXAMPLE 2.14 Calculate.

(a) $4 \times (-7)$ **(b)** $(-6) \times (-11)$ **(c)** $27 \div (-3)$ **(d)** $(-56) \div (-8)$

SOLUTION

(a) $4 \times (-7) = -28$ **(b)** $(-6) \times (-11) = 66$
(c) $27 \div (-3) = -9$ **(d)** $(-56) \div (-8) = 7$ ◆

The following properties hold for integer addition and multiplication. (Note:

Properties of Integer Operations

$a + b = b + a$ **Commutative Property of Addition**
$a \times b = b \times a$ **Commutative Property of Multiplication**
$a + (b + c) = (a + b) + c$ **Associative Property of Addition**
$a \times (b \times c) = (a \times b) \times c$ **Associative Property of Multiplication**
$a \times (b \pm c) = (a \times b) \pm (a \times c)$ **Distributivity of Multiplication over Addition (and Subtraction)**
$(a \pm b) + c = (a \div c) \pm (b \div c)$ **Right Distributivity of Division over Addition (and Subtraction)**
$a + 0 = 0 + a = a$ **Identity for Addition Property (Zero)**
$a \times 1 = 1 \times a = a$ **Identity for Multiplication Property (One)**
$a + (-a) = (-a) + a = 0$ **Additive Inverse Property**

The symbol "±" below, read "plus or minus," is a convenient notation used to represent two statements at once, one with + and one with −.)

The additive inverse property is a property of integers, but not of whole numbers since the set of whole numbers does not include negative numbers.

RATIONAL NUMBERS

The set of integers extends the whole numbers by providing negative numbers. However, the system of integers does not allow all divisions. For example, $3 \div 4$ is not an integer since there is no integer n such that $3 = n \times 4$; similarly, $-2 \div 5$ does not have an integer answer. As you may recall, fractions are used to represent parts of a whole. The set of rational numbers, which we will study next, extends the integers *and* the fractions.

DEFINITION

The Rational Numbers

The set of **rational numbers** is the set of all numbers that can be represented in the form $\frac{a}{b}$, where a and b are integers and $b \neq 0$.

HISTORY

In ancient times there was no fraction notation such as $\frac{2}{43}$. Instead, all fractions were written as a sum or difference of unit fractions, namely fractions whose numerator is one. As an example, the Rhind Papyrus (circa 1650 B.C.) expresses the ratio of 2 to 43 as $\frac{1}{42} + \frac{1}{86} + \frac{1}{129} + \frac{1}{301}$.

The a in $\frac{a}{b}$ is called its **numerator** and b is called its **denominator.**

The following are examples of rational numbers: $\frac{2}{3}, \frac{-4}{7}, \frac{5}{-6}$, and $\frac{-8}{-12}$. *Notice that the denominator of a rational number can never be zero, since $a \div 0 = c$ would require $a = c \times 0$.* Rational numbers whose numerators and denominators are both positive or both negative are called the **positive rational numbers;** the rational numbers where one of the numerator or denominator is positive and the other is negative are the **negative rational numbers. Fractions** are the positive rational numbers together with zero.

One interesting feature of the rational numbers is that each rational number has infinitely many representations. For example,

$$\frac{1}{3} = \frac{2}{6} = \frac{3}{9} = \frac{4}{12} = \dots .$$

This series of equalities can be visualized by the shaded regions in the following rectangles. The subdivisions are different, but the shaded areas are the same.

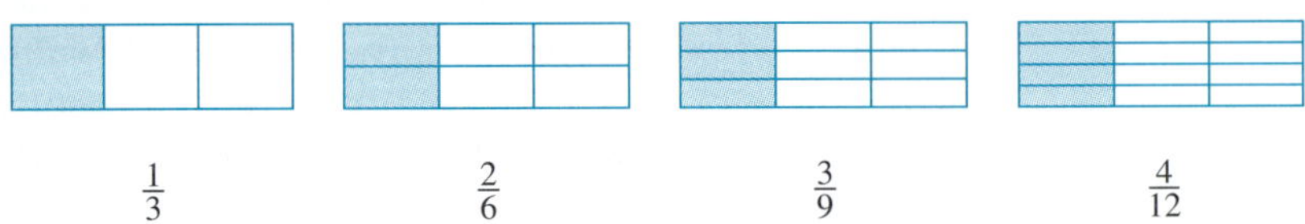

$\frac{1}{3}$ $\frac{2}{6}$ $\frac{3}{9}$ $\frac{4}{12}$

FIGURE 2.9

This idea can be used to simplify or reduce fractions. For example,

$$\frac{12}{18} = \frac{6 \times 2}{6 \times 3} = \frac{2}{3}.$$

Rational Numbers: Simplest Form

Notice that 2 and 3 have only 1 or -1 as a common factor. A rational number is said to be written in **simplest form** if the numerator and denominator have only 1 or -1 as common factors *and* the denominator is positive. (Note: Since every rational number can be written with a positive or negative denominator, the positive one is chosen to ensure that we have a unique simplest form.) The following is a generalization of this idea.

Rational Number Simplification

$\frac{na}{nb} = \frac{a}{b}$ for integers a, b, and n where b and n are nonzero.

In addition to simplifying fractions, this property can be used to determine if two rational numbers are equal. For example, $\frac{12}{18} = \frac{2}{3}$ and $\frac{10}{15} = \frac{2}{3}$. Since both $\frac{12}{18}$ and $\frac{10}{15}$ equal $\frac{2}{3}$, they are equal to each other. Another way to check for equality is to get common denominators, namely denominators that are the same. For example, to see if $\frac{6}{10}$ and $\frac{9}{15}$ are equal, we note that $\frac{6 \times 15}{10 \times 15} = \frac{90}{150}$ and $\frac{9 \times 10}{15 \times 10} = \frac{90}{150}$. Since both numerators and denominators are equal, the two rational numbers must be equal. Notice that the numerators were formed by finding the product of the numerator of one rational number times the denominator of the other. This leads to the following generalization.

Rational Number Equality

$\frac{a}{b} = \frac{c}{d}$ if and only if $a \times d = b \times c$.

EXAMPLE 2.15 Determine if these rational numbers are equal (i) by Rational Number Simplification, (ii) by finding a common denominator, and (iii) by using Rational Number Equality.

(a) $\frac{2}{3}$ and $\frac{12}{15}$ **(b)** $\frac{8}{12}$ and $\frac{12}{18}$

SOLUTION

(a) (i) $\frac{12}{15} = \frac{3 \times 4}{3 \times 5} = \frac{4}{5}$. Since $\frac{2}{3} \neq \frac{4}{5}$, then $\frac{2}{3} \neq \frac{12}{15}$.

(ii) A common denominator is 15.

$\frac{2}{3} = \frac{2 \times 5}{3 \times 5} = \frac{10}{15}$, which is clearly not equal to $\frac{12}{15}$.

(iii) Since $2 \times 15 \neq 3 \times 12$, the fractions are not equal.

(b) (i) $\frac{8}{12} = \frac{2 \times 4}{3 \times 4} = \frac{2}{3}$ and $\frac{12}{18} = \frac{2 \times 6}{3 \times 6} = \frac{2}{3}$. Since both numbers are equal to $\frac{2}{3}$, they are equal.

(ii) A common denominator is 36.

$\frac{8}{12} = \frac{8 \times 3}{12 \times 3} = \frac{24}{36}$ and $\frac{12}{18} = \frac{12 \times 2}{18 \times 2} = \frac{24}{36}$. Since both numbers are equal to $\frac{24}{36}$, they are equal.

(iii) Since $8 \times 18 = 144$ and $12 \times 12 = 144$, the numbers are equal. ◆

Rational Numbers: Addition and Subtraction

Models can also be used to suggest how we add (and subtract) rational numbers (Figure 2.10).

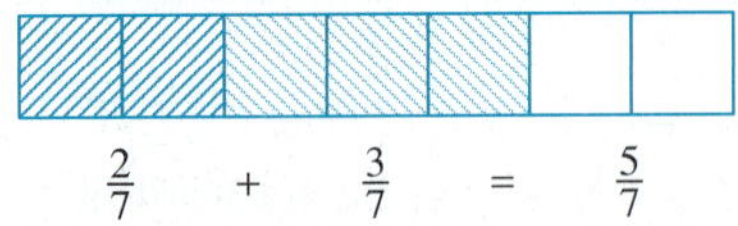

FIGURE 2.10

The model in Figure 2.10 shows that $\frac{2}{7} + \frac{3}{7} = \frac{5}{7}$; that is, the sum of two rational numbers having a common denominator is the sum of their numerators over their common denominator.

The model in Figure 2.10 can also be used to show $\frac{5}{7} - \frac{3}{7} = \frac{2}{7}$ or $\frac{5}{7} - \frac{3}{7} = \frac{5-3}{7}$. In general, addition and subtraction are defined as follows.

DEFINITION

Rational Number Addition and Subtraction

$$\frac{a}{c} + \frac{b}{c} = \frac{a+b}{c} \quad \text{and} \quad \frac{a}{c} - \frac{b}{c} = \frac{a-b}{c}.$$

Notice that

$$\frac{a}{b} + \frac{(-a)}{b} = \frac{a + (-a)}{b} = \frac{0}{b} = 0.$$

Thus, $\frac{-a}{b}$ is the additive inverse of $\frac{a}{b}$. But $-\frac{a}{b}$ is also the additive inverse of $\frac{a}{b}$. Thus, we can conclude the following:

$$-\frac{a}{b} = \frac{-a}{b} = \frac{a}{-b}.$$

Since rational number subtraction extends integer subtraction, an alternate way to define subtraction is

$$\frac{a}{c} - \frac{b}{c} = \frac{a}{c} + \left(-\frac{b}{c}\right).$$

If the denominators of two rational numbers are different, find a common denominator before adding or subtracting.

$$\frac{a}{b} + \frac{c}{d} = \frac{ad}{bd} + \frac{bc}{bd} = \frac{ad+bc}{bd} \quad \text{and} \quad \frac{a}{b} - \frac{c}{d} = \frac{ad}{bd} - \frac{bc}{bd} = \frac{ad-bc}{bd}$$

EXAMPLE 2.16 Calculate and express your answer in simplest form.

(a) $\frac{2}{5} + \frac{4}{35}$ **(b)** $\frac{7}{8} - \frac{5}{12}$

SOLUTION

(a) $\frac{2}{5} + \frac{4}{35} = \frac{2 \times 7}{5 \times 7} + \frac{4}{35} = \frac{14}{35} + \frac{4}{35} = \frac{18}{35}$. Notice that 7 was used as a common factor in the first step to get the common denominator 35.

(b) Here we can use 24 as a common denominator since both 8 and 12 divide into 24 evenly.

$$\frac{7}{8} - \frac{5}{12} = \frac{7 \times 3}{8 \times 3} - \frac{5 \times 2}{12 \times 2} = \frac{21 - 10}{24} = \frac{11}{24}$$ ◆

Rational Numbers: Multiplication and Division

Models also can be used to suggest how rational numbers are multiplied. Figure 2.11 illustrates how to find the product $\frac{2}{3} \times \frac{4}{5}$.

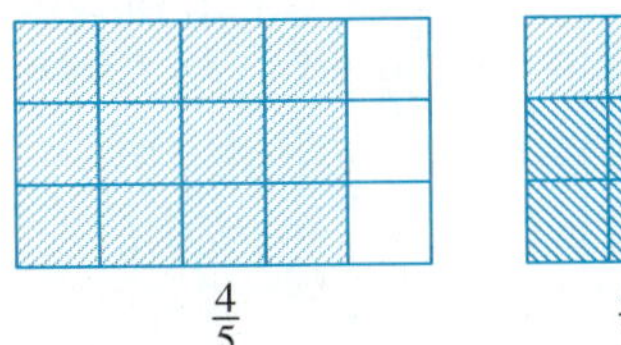

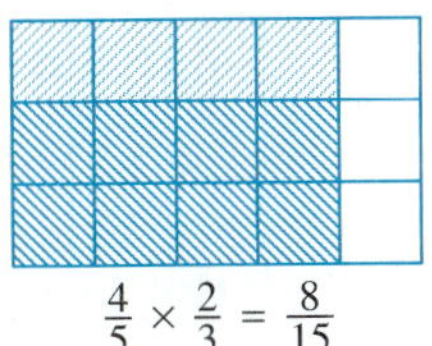

FIGURE 2.11

Since $\frac{2}{3} \times \frac{4}{5}$ means $\frac{2}{3}$ of $\frac{4}{5}$, $\frac{4}{5}$ of the rectangle is shaded on the left, then $\frac{2}{3}$ of the shaded region is designated on the right. Notice that this shows 8 darker shaded squares out of 15, so $\frac{2}{3} \times \frac{4}{5} = \frac{8}{15}$, or $\frac{2}{3} \times \frac{4}{5} = \frac{2 \times 4}{3 \times 5}$.

Multiplication in the rational numbers has both inverse and identity elements. That is, if $b \neq 0$, then $\frac{a}{b} \times \frac{b}{a} = 1$. The number $\frac{b}{a}$ is called the **reciprocal** or **multiplicative inverse** of $\frac{a}{b}$. In words, a rational number times its multiplicative inverse is 1. Also, since $\frac{a}{b} \times 1 = \frac{a}{b}$, 1 is called the **multiplicative identity.**

Division of rational numbers extends integer division. Recall that with integers and whole numbers, $a \div b = c$ if and only if $a = c \times b$. For example, $12 \div 4 = 3$ since $12 = 3 \times 4$. Next, we extend this idea to determine how to divide rational numbers. Consider the following to find $\frac{3}{4} \div \frac{5}{7}$.

$$\textbf{(i)}\quad \frac{3}{4} \div \frac{5}{7} = \frac{x}{y} \text{ if and only if } \frac{3}{4} = \frac{x}{y} \times \frac{5}{7}.$$

By multiplying both sides of the right-hand equation by $\frac{7}{5}$ we have

$$\textbf{(ii)}\quad \frac{3}{4} \times \frac{7}{5} = \frac{x}{y} \times \frac{5}{7} \times \frac{7}{5} \text{ or } \frac{3}{4} \times \frac{7}{5} = \frac{x}{y}.$$

Since $\frac{3}{4} \div \frac{5}{7} = \frac{x}{y}$ in (i) and $\frac{3}{4} \times \frac{7}{5} = \frac{x}{y}$ in (ii), we must have $\frac{3}{4} \div \frac{5}{7} = \frac{3}{4} \times \frac{7}{5}$. The following is a generalization of these ideas.

DEFINITION

Rational Number Multiplication and Division

$$\frac{a}{b} \times \frac{c}{d} = \frac{ac}{bd} \quad \text{and} \quad \frac{a}{b} \div \frac{c}{d} = \frac{a}{b} \times \frac{d}{c}, \text{ when } c \neq 0.$$

The last generalization can be stated in words as follows: The product of two rational numbers is the rational number whose numerator is the product of their numerators and whose denominator is the product of their denominators. The quotient of one rational number divided by a second is the product of the first number and the reciprocal of the second number.

EXAMPLE 2.17 Calculate and express your answer in simplest form.

(a) $\frac{6}{7} \times \frac{-21}{16}$ **(b)** $\frac{-8}{11} \div \frac{10}{-33}$

SOLUTION

(a) $\frac{6}{7} \times \frac{-21}{16} = \frac{6 \times (-21)}{7 \times 16} = \frac{-126}{112} = \frac{-9}{8}$.

An alternate way of obtaining the simplest form of the answer is to simplify along the way. For example,

$$\frac{6 \times (-21)}{7 \times 16} = \frac{3 \times (-21)}{7 \times 8} = \frac{3 \times (-3)}{1 \times 8} = \frac{-9}{8}.$$

(b) $\frac{-8}{11} \div \frac{10}{-33} = \frac{-8}{11} \times \frac{-33}{10} = \frac{-8}{1} \times \frac{-3}{10} = \frac{-4}{1} \times \frac{-3}{5} = \frac{12}{5}$ (or $2\frac{2}{5}$) ◆

The following properties hold for rational number addition and multiplication.

Properties of Rational Number Operations

$\frac{a}{b} + \frac{c}{d} = \frac{c}{d} + \frac{a}{b}$ **Commutative Property of Addition**

$\frac{a}{b} \times \frac{c}{d} = \frac{c}{d} \times \frac{a}{b}$ **Commutative Property of Multiplication**

$\frac{a}{b} + \left(\frac{c}{d} + \frac{e}{f}\right) = \left(\frac{a}{b} + \frac{c}{d}\right) + \frac{e}{f}$ **Associative Property of Addition**

$\frac{a}{b} \times \left(\frac{c}{d} \times \frac{e}{f}\right) = \left(\frac{a}{b} \times \frac{c}{d}\right) \times \frac{e}{f}$ **Associative Property of Multiplication**

$\frac{a}{b} \times \left(\frac{c}{d} \pm \frac{e}{f}\right) = \left(\frac{a}{b} \times \frac{c}{d}\right) \pm \left(\frac{a}{b} \times \frac{e}{f}\right)$ **Distributivity of Multiplication over Addition (and Subtraction)**

$\left(\frac{a}{b} \pm \frac{c}{d}\right) \div \frac{e}{f} = \left(\frac{a}{b} \div \frac{e}{f}\right) \pm \left(\frac{c}{d} \div \frac{e}{f}\right)$ **Right Distributivity of Division over Addition (and Subtraction)**

$\frac{a}{b} + 0 = 0 + \frac{a}{b} = \frac{a}{b}$ **Identity for Addition Property (Zero)**

$\frac{a}{b} \times 1 = 1 \times \frac{a}{b} = \frac{a}{b}$ **Identity for Multiplication Property (One)**

$\frac{a}{b} + \left(-\frac{a}{b}\right) = \left(-\frac{a}{b}\right) + \frac{a}{b} = 0$ **Additive Inverse Property**

$\frac{a}{b} \times \frac{b}{a} = 1$, for $a \neq 0$ **Multiplicative Inverse Property**

The multiplicative inverse property is a property of rational numbers, but not of integers since the set of integers only includes reciprocals for 1 and −1.

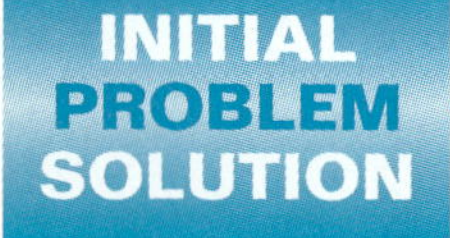

Five-eighths of the students at Moo U live off campus. If one-sixth of these students need parking permits and there are 14,400 students enrolled, how many parking permits will be needed?

SOLUTION Since we want fractional parts of numbers, we can multiply. $\frac{5}{8} \times \frac{1}{6} = \frac{5}{48}$ and $\frac{5}{48} \times 14{,}400 = 1500$; so 1500 students will need permits.

PROBLEM SET 2.3

1. Simplify the following.
   (a) $(-21) + 7$ (b) $(17) + (-12)$
   (c) $(-12) + (-9)$ (d) $(-17) - 6$
   (e) $(14) - (-7)$ (f) $(-8) - (-15)$
2. Simplify the following.
   (a) $14 + (-25)$ (b) $(-12) + 30$
   (c) $(-8) - 18$ (d) $12 - (-4)$
   (e) $(-12) - (-3)$ (f) $(-8) + (-15)$
3. Simplify the following.
   (a) $(-21) \times 7$ (b) $(7) \times (-12)$
   (c) $(-10) \times (-9)$ (d) $(-45) \div (-5)$
   (e) $(63) \div (-7)$ (f) $(-36) \div 4$
4. Simplify the following.
   (a) $(-12) \times 30$ (b) $4 \times (-25)$
   (c) $(-7) \times (-5)$ (d) $(-30) \div (5)$
   (e) $(42) \div (-6)$ (f) $(-18) \div (-3)$
5. Name the property of integer operations demonstrated.
   (a) $6 \times (3 + 4) = 6 \times 3 + 6 \times 4$
   (b) $1 + (2 + 3) = (1 + 2) + 3$
   (c) $5 \times 10 = 10 \times 5$
   (d) $17 + 4 = 4 + 17$
6. Name the property of integer operations demonstrated.
   (a) $(5 + 3) \times 4 = 4 \times (5 + 3)$
   (b) $(-6) \times (4 \times 3) = (-6 \times 4) \times 3$
   (c) $(-5) \times (2 - 7) = (-5 \times 2) - ((-5) \times 7)$
   (d) $(-12) + (12 + 3) = ((-12) + 12) + 3$
7. Reduce the following fractions to simplest form.
   (a) $\frac{24}{30}$ (b) $\frac{15}{18}$
   (c) $\frac{28}{84}$ (d) $\frac{72}{48}$
   (e) $\frac{105}{45}$ (f) $\frac{144}{176}$
8. Reduce the following fractions to simplest form.
   (a) $\frac{48}{72}$ (b) $\frac{18}{21}$
   (c) $\frac{14}{63}$ (d) $\frac{36}{64}$
   (e) $\frac{105}{135}$ (f) $\frac{125}{225}$
9. Determine whether or not the following fractions are equivalent.
   (a) $\frac{6}{9}$ and $\frac{25}{45}$ (b) $\frac{8}{10}$ and $\frac{12}{15}$
   (c) $\frac{33}{121}$ and $\frac{45}{143}$
10. Determine whether or not the following fractions are equivalent.
    (a) $\frac{5}{9}$ and $\frac{45}{81}$ (b) $\frac{18}{30}$ and $\frac{8}{15}$
    (c) $\frac{55}{143}$ and $\frac{65}{165}$
11. For each fraction below, state two equivalent fractions.
    (a) $\frac{9}{12}$ (b) $\frac{1}{5}$
12. For each fraction below, state two equivalent fractions.
    (a) $\frac{6}{18}$ (b) $\frac{15}{24}$
13. Complete the following additions and subtractions, and express your answers in simplest form. If there is a negative sign in the answer, use it with the numerator; that is, use $\frac{-2}{3}$ rather than $\frac{2}{-3}$ or $-\frac{2}{3}$.
    (a) $\frac{24}{30} + \frac{9}{30}$ (b) $\frac{15}{18} + \frac{2}{18}$
    (c) $\frac{28}{84} - \frac{12}{84}$ (d) $\frac{4}{5} + \frac{2}{15}$
    (e) $\frac{5}{6} - \frac{4}{21}$ (f) $\frac{13}{24} - \frac{13}{28}$
14. Complete the following additions and subtractions, and express your answers in simplest form. If there is a negative sign in the answer, use it with the numerator.
    (a) $\frac{-5}{12} + \frac{2}{3}$ (b) $\frac{-2}{3} - \frac{4}{21}$
    (c) $\frac{5}{6} - \frac{-5}{12}$ (d) $\frac{-5}{12} + \frac{-2}{3}$
    (e) $\frac{4}{21} - \frac{5}{6}$ (f) $\frac{-8}{15} - \frac{-3}{4}$
15. Calculate, and express your answers in lowest terms. If there is a negative sign in the answer, use it with the numerator.
    (a) $\frac{-4}{9} \times \frac{15}{8}$ (b) $\frac{-6}{15} \times \frac{10}{-21}$
    (c) $\frac{4}{21} \div \frac{-2}{3}$ (d) $\frac{-2}{3} \div \frac{-5}{12}$
    (e) $\frac{8}{15} \times \frac{-25}{16}$ (f) $\frac{-8}{25} \div \frac{-12}{35}$
16. Calculate, and express your answers in lowest terms. If there is a negative sign in the answer, use it with the numerator.
    (a) $\frac{-5}{12} \times \frac{2}{3}$ (b) $\frac{-2}{3} \div \frac{4}{21}$
    (c) $\frac{5}{6} \div \frac{-5}{12}$ (d) $\frac{-5}{12} \div \frac{-2}{3}$
    (e) $\frac{4}{21} \times \frac{5}{6}$ (f) $\frac{-8}{15} \times \frac{-3}{4}$
17. Name the property of rational number operations demonstrated.
    (a) $\frac{2}{3} \times 1 = 1 \times \frac{2}{3} = \frac{2}{3}$
    (b) $\left(6 + \frac{2}{3}\right) \div \frac{1}{3} = \left(6 \div \frac{1}{3}\right) + \left(\frac{2}{3} \div \frac{1}{3}\right)$
    (c) $\frac{2}{7} \times 4 = 4 \times \frac{2}{7}$
    (d) $\frac{3}{5} \times \frac{5}{3} = 1$
18. Name the property of rational number operations demonstrated.
    (a) $\frac{4}{5} + 0 = 0 + \frac{4}{5}$
    (b) $\frac{2}{3}\left(\frac{2}{5} - \frac{3}{4}\right) = \left(\frac{2}{3}\right)\left(\frac{2}{5}\right) - \left(\frac{2}{3}\right)\left(\frac{3}{4}\right)$
    (c) $\frac{3}{7} \times \frac{-2}{5} = \frac{-2}{5} \times \frac{3}{7}$
    (d) $\frac{-9}{4} \times \frac{4}{-9} = 1$
19. For the number $\frac{m}{n}$, state the following.
    (a) The additive identity (b) The opposite
20. For the number $\frac{m}{n}$, state the following.
    (a) The additive inverse (b) The reciprocal
21. On one winter's day, the low temperatures at four cities were:

    Baltimore, MD: −7°F Fargo, ND: −23°F
    San Diego, CA: 56°F Tulsa, OK: 27°F

What were the temperature differences between
(a) San Diego and Baltimore?
(b) Tulsa and Fargo?
(c) San Diego and Tulsa?
(d) Fargo and Baltimore?

22. On one winter's day, the low temperatures at four cities were

Baltimore, MD: −22°C
Fargo, ND: −31°C
Tulsa, OK: −3°C
San Diego, CA: 13°C

What were the temperature differences between
(a) San Diego and Baltimore?
(b) Tulsa and Fargo?
(c) San Diego and Tulsa?
(d) Fargo and Baltimore?

23. The temperature of air drops about 3°F for each 1000-foot increase in altitude. What would be the approximate temperature for an airplane that is flying at 20,000 feet off the Atlantic coast if the temperature at sea level was 50°F?

24. The temperature of air drops about 1°C for each 600-foot increase in altitude. What would be the approximate temperature for an airplane that is flying at 15,000 feet off the Alaskan coast if the temperature at sea level was 10°C?

25. Dixie had a balance of $115 in her checking account at the beginning of the month. She deposited $384 in the account and then wrote checks for $153, $86, $196, $34, and $79. Then she made a deposit of $123. If at any time during the month the account is overdrawn, a $10 service charge is deducted. At the end of the month, what was Dixie's balance?

26. Paul had a balance of $90 in his checking account at the beginning of the month. He wrote checks for $40 and $35. He also had an automatic payment of $25 that was taken from the account. After those checks had been paid, Paul made a deposit of $150 and wrote additional checks for $60 and $75. If at any time during the month the account is overdrawn, a $10 service charge is deducted. At the end of the month, what was Paul's balance?

27. Find the outcome of each of the following.
(a) In a series of downs, a football team gained 7 yards, lost 4 yards, lost 2 yards, and gained 8 yards. What was the total gain or loss?
(b) In a week, a given stock gained 5 points, dropped 12 points, dropped 3 points, gained 18 points, and dropped 10 points. What was the net change in the stock's worth?
(c) A visitor in an Atlantic City casino won $300, lost $250, and then won $150. Find the gambler's overall gain or loss.

28. Find the outcome of each of the following.
(a) A visitor in Las Vegas won $100 at blackjack, lost $120 at roulette, lost $70 at poker, and won $35 at the slot machines. What was the gambler's overall gain or loss?
(b) In a series of downs, a football team gained 5 yards, lost 3 yards, was penalized 10 yards for holding, and then gained 4 yards. What was the total gained or lost?
(c) Over a five-week period, a certain stock gained three points, lost two points, lost five points, gained three points, and then lost six points. What was the net change in the stock's worth?

29. As of 1991 the state of Illinois was generating approximately 13,100,000 tons of waste per year, and of that amount, 786,000 tons were being recycled. In 1991 the state of Texas was generating approximately 18,000,000 tons of waste, and of that amount, 1,440,000 tons were being recycled. Which state had the higher recycling rate?

30. Mrs. Wills and Mr. Roberts gave the same test to their fourth-grade classes. In Mrs. Wills's class, 28 out of 36 students passed the test. In Mr. Roberts's class, 26 out of 32 students passed the test. Which class had the higher passing rate?

31. A recipe that makes 3 dozen peanut butter cookies calls for $1\frac{1}{4}$ cups of flour.
(a) How much flour would you need if you doubled the recipe?
(b) How much flour would you need for half the recipe?
(c) How much flour would you need to make 5 dozen cookies?

32. In a cost-saving measure, Chuck's company reduced all salaries by $\frac{1}{8}$ of their present salaries. If Chuck's monthly salary was $2400, what will he now receive? If his new salary is $2800, what was his old salary?

EXTENDED PROBLEMS

33. Fill in each empty square so that the number in a square will be the sum of the pair of numbers beneath the square.

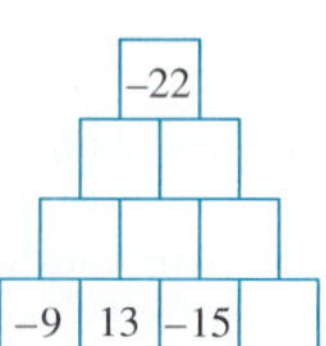

34. An additive magic square has the same sum for each row, column, and diagonal. Complete the magic square using the following integers:

$$10, 7, 4, 1, -5, -8, -11, -14$$

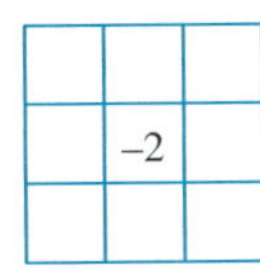

35. **(a)** Make a magic square with the numbers 1, 2, 3, 4, 5, 6, 7, 8, 9.

(b) Show that every magic square using these numbers must have a 5 in the center square.

(c) Show that there are exactly 8 possible magic squares with these numbers.

36. Given two rational numbers find a test that will tell which number is larger.

37. In the following system of arithmetic, there are two numbers: 0 and 1. There are two operations, + and *, defined by the following:

$0 + 0 = 0$	$0 * 0 = 0$
$0 + 1 = 1$	$0 * 1 = 0$
$1 + 0 = 1$	$1 * 0 = 0$
$1 + 1 = 0$	$1 * 1 = 1$

Which of the rules of common arithmetic holds for this system?

38. Operations on sets include union and intersection. Union is roughly analogous to addition of numbers, and intersection is roughly analogous to multiplication. If A and B are sets, then $A \cup B$ is the set of objects that belong to one or both of A and B and $A \cap B$ is the set of objects that belong to both A and B. What set corresponds to the additive identity? Suppose that we are only considering subsets of the integers between 1 and 10. What set corresponds to the multiplicative identity? Which of the laws of arithmetic hold for these operations? Are there some laws that hold for set operations but do not hold for arithmetic?

39. Consider a system of arithmetic based on the clock, having the whole numbers 1 through 12 as the only numbers and addition as the only operation. The operation of addition is based on the way time is "added" on a clock. Five hours after nine o'clock, it will be two o'clock; therefore, $5 + 9 = 2$.

(a) Which rules for common arithmetic hold for this system?

(b) Is there an additive identity and additive inverses? If so, what are they?

2.4 DECIMALS AND REAL NUMBERS

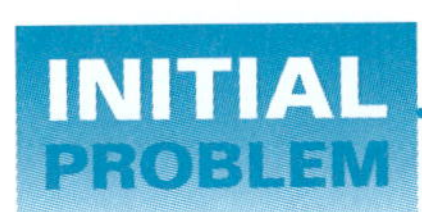

You are planning a trip to Europe, and you would like to pay for lodging ahead of time. The hotel where you'll be staying in Brussels sends you a bill for one week's lodging as follows:

Please remit 28.000 BF (Belgian francs).

Is this a tremendous bargain or is the Belgian franc a very valuable currency?

The common form for fractions has the advantage of being intuitively understandable when we want to talk about parts of the whole. However, operations involving even a few fractions can be complicated, such as adding a set of four fractions with different denominators. The set of decimals is an extension of the set of integers and is a convenient numeration system for fractions. In this section, we also discuss rational and irrational numbers, which comprise the set of real numbers.

DECIMALS

Generally, much effort can be saved by performing arithmetic operations with a calculator. Some calculators permit you to operate directly with fractions, while other calculators require that all numbers be entered in decimal form or show all results in decimal form. In any case, it is important to be able to convert a fraction to a decimal and a decimal to a fraction.

A **decimal fraction** is a fraction that has a power of 10, such as 10, 100, or 1000, for its denominator; for example, $\frac{37}{100}$ and $\frac{291}{1000}$ are decimal fractions. Decimal fractions can be expressed in decimal notation which is an extension of the Hindu-Arabic system of place values. A **decimal point** is a period that separates the whole number on the left from the decimal fraction on the right. The first place to the right of the decimal point has the place value of $\frac{1}{10}$. The place value of each subsequent place to the right is one tenth that of the place to its immediate left. For example, 427.3689 represents

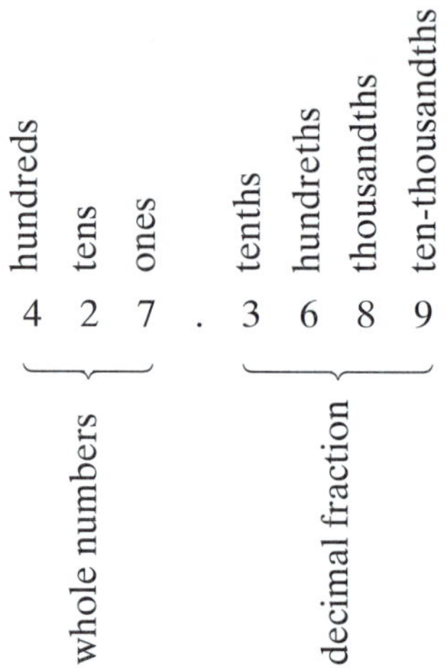

hundreds	tens	ones		tenths	hundredths	thousandths	ten-thousandths
4	2	7	.	3	6	8	9
whole numbers				decimal fraction			

The number 427.3689 is read as "four hundred twenty-seven and three thousand six hundred eighty-nine ten-thousandths." Notice that the word "and" is only used to indicate where the decimal point is located. The number 427.3689 is composed of the whole number 427 *and* the decimal fraction $\frac{3689}{10{,}000}$.

The term **decimal** is often used to refer to any number written using a decimal point and the preceding place value notation. Thus we consider 13.0 and 0.0 to be decimals.

Decimal notation has evolved over the years without universal agreement. Consider the list of decimal expressions for the fraction $\frac{3142}{1000}$ in Table 2.2.

Table 2.2

Notation	Date Introduced
3 142	1522, Adam Riese (German)
3|142 ; 3,142	1579, François Vieta (French)
$\begin{array}{c\|c\|c\|c} 0 & 1 & 2 & 3 \\ \hline 3 & 1 & 4 & 2 \end{array}$	1585, Simon Stevin (Dutch)
3 · 142	1614, John Napier (Scottish)

Today, Americans use a version of Napier's "decimal point" notation (3.142, where the point is on the baseline), but the English retain Napier's original version (3 · 142, where the point is in the middle of the line). The French and Germans retain Vieta's "decimal comma" notation (3,142). The issue of establishing a universal decimal notation remains unresolved to this day.

A fraction is converted to a decimal by performing the indicated division, either by hand or with a calculator.

EXAMPLE 2.18 Convert $\frac{1}{8}$ to a decimal using a calculator.

SOLUTION Using a calculator we have

$$1 \boxed{\div} 8 \boxed{=} \boxed{0.125}.$$

So $\frac{1}{8} = 0.125$. ◆

Trouble arises almost immediately, as the next example shows.

EXAMPLE 2.19 Convert $\frac{1}{3}$ to a decimal.

SOLUTION Using a calculator we have

$$1 \boxed{\div} 3 \boxed{=} \boxed{0.3333333333}.$$

We have $\frac{1}{3} = 0.3333\ldots$, an infinite repeating decimal, although the calculator display truncates the value at ten digits. ◆

A decimal such as 0.125 is called a **terminating decimal** because there are only a finite number of nonzero digits to the right of the decimal point, in this case 3. (Note: It can be shown that the only terminating decimals are those whose fraction representations in simplest form have only 1, 2, or 5 as factors of the denominator.) The decimal 0.3333 . . . has infinitely many nonzero digits, but they occur with a very simple pattern, namely, three is repeated over and over. A decimal that repeats indefinitely is called a **repeating decimal.** Every rational number can be represented by either a terminating decimal or a repeating decimal. For a repeating decimal, a bar is put over the set of digits that repeat. Thus we write

$$\frac{1}{3} = 0.\overline{3}.$$

If you use your calculator to convert a fraction to a decimal and the result is a repeating decimal, then your calculator's answer will not be exactly correct. The calculator's internal representation of the number also will be slightly in error. Of course, the result is almost always close enough for practical purposes.

When you convert a fraction to a decimal, it is natural to ask how you will know if the repeating pattern has been found. In fact, the pattern is completely exposed when the long division has come back to a step previously performed.

EXAMPLE 2.20 Convert $\frac{2}{7}$ to its decimal form.

SOLUTION We compute by long division.

```
      0.285714
    7)2.000000
      0
     ↗20
      14
       60
       56
        40
        35
         50
         49
          10
           7
           30
           28
          ↗20
```

Notice that the number at the second arrow is the same as the number at the first arrow. So the pattern repeats. We conclude that

$$\frac{2}{7} = 0.\overline{285714}.$$ ◆

Note: Terminating decimals have alternate representations as nonterminating repeating decimals. For example, $1.0 = 0.\overline{9}$, $0.5 = 0.4\overline{9}$, and $0.125 = 0.124\overline{9}$. It is preferable to use the terminating form. The infinite sequence of 9's is the only nonterminating repeating decimal that is equivalent to a terminating decimal. If a calculator is used to find the repeating decimal representation of a fraction, the calculator may not have enough places to display a complete pattern. In the case of very long patterns, long division or a series of calculator operations must be used.

ORDERING DECIMALS

When you compare two decimals, that is, order decimals, first be sure that a number that can be written in a terminating form is written in that way. Then compare digits place by place, starting at the highest place value. When a place is reached where the digits disagree, the number with the larger digit is the larger number.

EXAMPLE 2.21 For each pair of decimals, determine which is larger.

(a) 635.48976312 and 635.48986312 **(b)** 27.356 and 126.78
(c) $0.0023\overline{123}$ and $0.00\overline{23}$

SOLUTION

(a) Comparing place by place as follows

```
6 3 5 . 4 8 9 7 6 3 1 2
‖ ‖ ‖   ‖ ‖ ‖ ∧
6 3 5 . 4 8 9 8 6 3 1 2
```

we conclude that $635.48986312 > 635.48976312$ since $8 > 7$ in the thousandths place and all prior places are the same.

(b) We need to introduce some 0's which do not change the value of the numbers represented. Then we compare place by place as follows

$$\begin{array}{l} 0\,2\,7.3\,5\,6 \\ \wedge \\ 1\,2\,6.7\,8\,0 \end{array}$$

to conclude that $126.78 > 27.356$.

(c) We first write down more of the digits that are indicated by the bar over the pattern. Then we compare place by place as follows

$$\begin{array}{ccccccccc} 0 & . & 0 & 0 & 2 & 3 & 1 & 2 & 3\ldots \\ \| & & \| & \| & \| & \| & \wedge & & \\ 0 & . & 0 & 0 & 2 & 3 & 2 & 3 & \ldots \end{array}$$

to conclude that $0.0023\overline{123} < 0.00\overline{23}$. ◆

MULTIPLYING AND DIVIDING DECIMALS BY POWERS OF TEN

There is a shortcut that can be used to multiply or divide a decimal by 10. For example, $10 \times 37.43 = 374.3$. That is, to multiply by 10 move the decimal point one digit to the right. To divide by 10 move the decimal place one digit to the left. Because $100 = 10 \times 10$, to multiply by 100 we just multiply by 10 twice, which amounts to moving the decimal point two digits to the right. Likewise, to divide by 100 we move the decimal point two digits to the left.

Multiplication and Division by Powers of Ten

To multiply a decimal by the number represented by 1 followed by n zeros, move the decimal point n digits to the right.

To divide a decimal by the number represented by 1 followed by n zeros, move the decimal point n digits to the left.

EXAMPLE 2.22 Calculate the following.

(a) 1000×12.28 **(b)** $37.987 \div 10{,}000$

SOLUTION

(a) There are three 0's in 1000, so we move the decimal point three digits to the right. First we must introduce some extra 0's into the decimal representation for $12.28 = 12.2800$.

$$12\,280.0$$

We find $1000 \times 12.28 = 12{,}280$.

(b) There are four 0's in 10,000, so we move the decimal point four digits to the left. Once again some extra 0's must be introduced into the decimal

TIDBIT

When you travel to a foreign country, it is reassuring to already have some of the local currency when you arrive so you can pay for small expenses while you find your way around. Most larger banks can obtain foreign currency with a few days' notice.

representation of 37.987 = 00037.987. Of course, 00037.987 is a nonstandard way to write the number.

0.0037 987

We find 37.987 ÷ 10,000 = 0.0037987. ◆

The next example illustrates the usefulness of this technique.

EXAMPLE 2.23 Using the exchange rate in Table 2.3 from 1994, calculate the number of French francs you will be able to purchase with $100.

TABLE 2.3

Currency	Value in Dollars	Units per Dollar
Argentina, peso	1.01	0.990
Belgium, franc	0.0305	32.8
Brazil, cruzeiro	0.00037	2700
Canada, dollar	0.725	1.38
France, franc	0.183	5.46
Germany, mark	0.629	1.59
Italy, lira	0.000633	1580
Switzerland, franc	0.758	1.32

SOLUTION We locate the row for French francs

France, franc 0.183 5.46

The second number, 5.46, is the number of francs per dollar. To multiply by 100, we move the decimal point two places to the right to obtain 546 francs. ◆

CONVERTING A REPEATING DECIMAL TO A FRACTION

The procedure for converting a repeating decimal to a fraction is algebraic and is included for completeness. We illustrate with an example.

EXAMPLE 2.24 Convert $0.\overline{142857}$ to a fraction.

SOLUTION Let n stand for the number; that is,

$$n = 0.\overline{142857} = 0.142857142857\ldots$$

There are six digits in the repeated pattern, so we multiply n by the number represented by 1 followed by six zeros, that is, by 1,000,000. We find

$$1{,}000{,}000 \times n = 142857.142857\ldots \quad \text{or} \quad 142857.142857142857\ldots$$

because of the repeating digits. We next subtract n from $1{,}000{,}000 \times n$ in two ways. On one hand,

$$1{,}000{,}000 \times n - n = 999{,}999 \times n.$$

On the other hand, since $n = 0.142857142857\ldots$, $(100 \times n) - n$ equals

$$\begin{array}{r} 142857.142857142857\ldots \\ -0.142857142857\ldots \\ \hline 142857.000000000000\ldots \end{array}$$

Since all digits to the right of the decimal point are identical, they must cancel out under subtraction. Since we have computed $1{,}000{,}000 \times n - n$ in two ways, the two values must be equal, so we have

$$999{,}999 \times n = 142{,}857.$$

Dividing both sides by 999,999 we conclude that

$$n = \frac{142857}{999999}.$$

Of course this should be reduced to simplest form. In fact 142,857 evenly divides 999,999; we have $999{,}999 \div 142{,}857 = 7$, so we finally find

$$0.\overline{142857} = \frac{142857}{999999} = \frac{1}{7}.$$ ◆

HISTORY

It is believed that the first example of an irrational number to be recognized was $\sqrt{2}$. This discovery, made by Pythagoras or his followers circa 500 B.C., is said to have been celebrated with the sacrifice of 100 oxen.

REAL NUMBERS

Perhaps you can imagine a nonterminating decimal that does not repeat. For example, we could start with 0.1 followed by one 0, then put another 1 followed by two 0's, then another 1 followed by three 0's, and so on.

$$0.1010010001000010000010000001000000010000000010000000001\ldots$$

Any nonterminating, nonrepeating decimal represents a number, but a number that is not rational. Such numbers are called **irrational numbers.** The number $\sqrt{2}$ is an irrational number. The set of numbers consisting of both the rational numbers and the irrational numbers is called the **real numbers.** Since no irrational number can be faithfully represented by your calculator, you will be using good approximations to real numbers more often than you use exact real numbers.

A common way to visualize the real numbers is with the aid of the **real number line.** Choose two points on a straight line (assume the line extends infinitely both to the left and to the right). The point on the left is labeled 0, and the point on the right is labeled 1. The distance between the two chosen points is used as a unit of measurement, and any real number r corresponds to a point on the line at a distance r from the point labeled 0, with the positive numbers corresponding to points on the right of 0, and the negative numbers corresponding to points on the left. Likewise any point on the line corresponds to some real number (Figure 2.12).

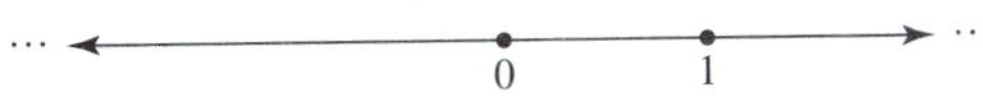

FIGURE 2.12

As the Greeks discovered thousands of years ago, a few simple geometric constructions show that irrational lengths, hence irrational numbers, exist. For example, the circumference of a circle with diameter 1 is the irrational length π. The earliest example of a number that was proved to be irrational is the length of the

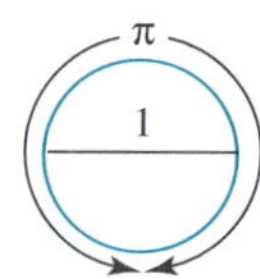

hypotenuse of an isosceles right triangle. If the length of each side is 1, the hypotenuse is $\sqrt{2}$ (Figure 2.13). The arithmetic of the real numbers is an extension of the rational numbers. Since irrational numbers cannot be expressed in a fraction form, decimals are usually used to represent real numbers.

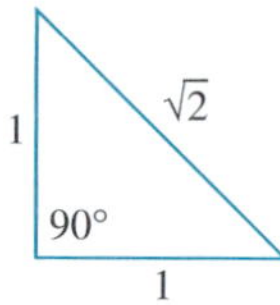

FIGURE 2.13

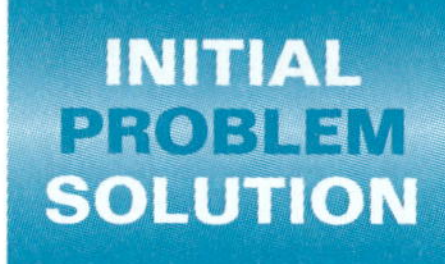

You are planning a trip to Europe, and you would like to pay for lodging ahead of time. The hotel where you'll be staying in Brussels sends you a bill for one week's lodging as follows:

Please remit 28.000 BF (Belgian francs).

Is this a tremendous bargain, or is the Belgian franc a very valuable currency?

SOLUTION Neither of these possibilities turns out to be true. Checking the foreign exchange rates in Table 2.3, we see the Belgian franc is not a particularly valuable currency; it is worth about 3 cents. Nor is the room going to have the bargain price of 28×3 cents, as you might hope. Instead, the Belgians use the decimal for grouping digits, rather than the comma. Therefore the bill is for $28{,}000 \times \$0.0305 = \$854$.

PROBLEM SET 2.4

In problems 1 through 4, write the given numbers in words.

1. **(a)** 3.047 **(b)** 300.47 **(c)** 0.347
2. **(a)** 2.0304 **(b)** 507.823 **(c)** 0.0000255
3. **(a)** 0.075 **(b)** 150.25 **(c)** 5.0175
4. **(a)** 458.276 **(b)** 0.0005 **(c)** 82.479

In problems 5 through 8, change the given words to numerals.

5. **(a)** Three and one hundred five ten-thousandths.
(b) Three hundred twenty-seven and twenty-seven hundredths.
(c) Eighty-five millionths.

6. **(a)** Four thousand three hundred twenty-five ten-thousandths.
(b) Two thousand fifty and twelve thousandths.
(c) Five hundred thirty-five millionths

7. **(a)** Three hundred five ten-thousandths.
(b) Four thousand thirty-eight millionths
(c) Twenty-five and five thousand two hundred eighty-seven hundred-thousandths.

8. **(a)** Eight thousand two hundred twenty and eighty-seven hundredths
(b) Five hundred fifty and ninety-five thousandths.
(c) One thousand seventy-five ten-thousandths.

9. Round each of the following as indicated.
(a) 413.46437 to the nearest hundredth
(b) 413.4999 to the nearest whole number

10. Round each of the following as indicated.
(a) 25.51163 to the nearest ten-thousandth
(b) $456.3\overline{56}$ to the nearest hundred-thousandth

11. Round each of the following as indicated.
(a) 28.87494 to the nearest thousandth
(b) 87.698 to the nearest hundredth

12. Round each of the following as indicated.
(a) 75.95035 to the nearest tenth
(b) $456.3\overline{56}$ to the nearest hundred-thousandth

In problems 13 through 16, fill in the missing entries. If the decimal form of a fraction does not terminate, round your result to the nearest ten-thousandth place.

13.

Fraction	Decimal
$\frac{17}{25}$	**(a)** ___
(b) ___	0.375
$\frac{5}{6}$	**(c)** ___

14.

Fraction	Decimal
$\frac{23}{80}$	**(a)** ___
(b) ___	1.75
$3\frac{4}{9}$	**(c)** ___

15.

Fraction	Decimal
(a) ___	2.375
$\frac{4}{7}$	**(b)** ___
(c) ___	0.84

16.

Fraction	Decimal
(a) ___	0.0625
$\frac{13}{125}$	**(b)** ___
(c) ___	3.456

17. Which of the following is greater?
   (a) 23.345625 or 23.346517
   (b) 6.34277 or 6.342589
   (c) $412.3\overline{45}$ or $412.34\overline{45}$

18. Which of the following is greater?
   (a) $27.\overline{427}$ or $27.42\overline{7}$
   (b) $7.6\overline{36}$ or $7.\overline{63}$
   (c) 143.1279 or 143.128

19. Find the following products and quotients mentally.
   (a) 375×100
   (b) 5.6529×1000
   (c) $0.0045 \div 10{,}000$

20. Find the following products and quotients mentally.
   (a) $0.084 \times 10{,}000$
   (b) $6.5 \div 100$
   (c) $0.0000037 \times 100{,}000$

21. Convert each of the following repeating decimals to a rational number.
   (a) $0.3\overline{27}$ **(b)** $32.\overline{7}$

22. Convert each of the following repeating decimals to a rational number.
   (a) $0.\overline{34}$ **(b)** $0.7\overline{315}$

23. Gary cashed a check from Joan for $29.35. Then he bought two magazines for $1.95 each, a book for $5.95, and a tape for $5.98. He had $21.45 left. How much money did he have before cashing the check?

24. Students in a chemistry lab weighed out the following amounts of potassium for an experiment: 8.5 grams, 4.25 grams, 2.45 grams, and 5.1 grams. Since the supply was getting low, the lab assistant added 25.75 grams to the supply, which brought the total back up to 30.5 grams. How many grams of potassium were in the supply before the lab session started?

25. The weight in grams, to the nearest hundredth, of a particular sample of toxic waste was 28.67 grams.
   (a) What is the minimum amount the sample could have weighed? (Write out your answer to the ten-thousandths place.)
   (b) What is the maximum amount? (Write out your answer to the ten-thousandths place.)

26. The winning height in the pole vault event of a track meet was 5.48 meters, to the nearest hundredth.
   (a) What is the minimum height of the winning vault? (Write your answer to the thousandths place.)
   (b) What is the maximum height? (Write out your answer to the thousandths place.)

27. How many of each of the following currencies could you get for $10,000 based on the exchange rate table in Example 2.23?
   (a) Belgium, franc   **(b)** Italy, lira
   (c) Brazil, cruzeiro
   Note: The exchange rates listed are no longer valid, as most exchange rates fluctuate regularly.

28. How many dollars could you get for each of the following currencies based on the exchange rate table in Example 2.23?
   (a) 15,000 German marks   **(b)** 1,000,000 Italian lira
   (c) 2,500 Swiss francs

29. One of the most popular road running events is the 10-kilometer run. A kilometer is approximately 0.62137 mile. What is the length of the 10-kilometer run to the nearest hundredth of a mile?

30. At a height of 8.488 kilometers, the highest mountain in the world is Mt. Everest in the Himalayas. The deepest part of the oceans is the Marianas Trench in the Pacific Ocean, with a depth of 11.034 kilometers. What is the vertical difference between the highest mountain in the world and the deepest part of the oceans?

31. A kilometer is approximately 0.62137 of a mile. Change each of the following to miles. Round your answer to the nearest tenth of a mile.
   (a) 8.488 kilometers
   (b) 11.034 kilometers

32. The acceleration (the change in speed) of falling objects that is caused by the force of gravity is 32 feet per second per second. This means that for every second, the speed increases by 32 feet per second. Express the acceleration due to gravity in terms of meters per second per second to the nearest hundredth of a meter. A meter is approximately 39.37 inches.

EXTENDED PROBLEMS

33. Determine if the following is an additive magic square. If not, change one entry so that your resulting square is magic.

0.438	0.073	0.584
0.511	0.365	0.219
0.146	0.647	0.292

34. The numbers shown next can be used to form an additive magic square: 10.48, 15.72, 20.96, 26.2, 31.44, 36.68, 41.92, 47.16, 52.4. Determine where to place the numbers in the nine cells of the magic square.

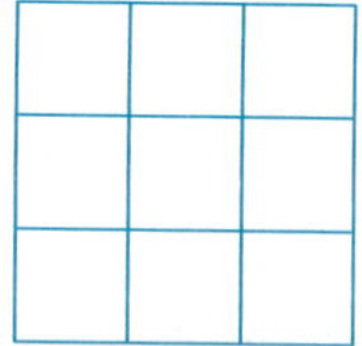

35. Write a brief report on the history and development of the decimal system. When did decimal fractions come into common use? Was there any resistance to their acceptance?

36. In section 2.1 different numeration systems were discussed. How were fractions represented in these systems? Write a brief report.

37. Binary numbers are based on powers of 2, rather than powers of 10, and use only two digits, 0 and 1. Binary numbers and other binary forms played an important role in the development of computers. Write a report on the use of binary numbers, including an explanation of how whole numbers are written in binary form.

38. How are fractions represented in the binary number system? Are there special forms or characteristics such as terminating, repeating, or nonterminating fractions? Write a brief report.

2.5 SOLVING PERCENT PROBLEMS

Your brother-in-law recounts the following story: "I bought some stock and it lost 90% of its value. Luckily, I didn't sell it, because it then increased to 1000% of its depressed value. I made a bundle. Too bad you don't have my financial ability." Did your brother-in-law make a big profit?

TIDBIT

As an example of a proportion per hundred, medical statistics tell us that 4% of the population has an aneurysm (a weak spot) on one of the blood vessels supplying the brain. This is interpreted to mean that an average of 4 out of any group of 100 persons will have such a condition (usually without knowing it or suffering ill effect).

Percents provide another common way to represent fractions; they are particularly suited for easy comparisons and for use in business applications. Percents also provide a convenient way to measure change. To an extent, operations with percents are easier than those with common fractions. However, a percent represents a comparison to a certain reference number or amount, and confusion may result if the reference changes, as in the initial problem.

PERCENT

The term **percent** comes from the Latin phrase *per centum,* which means per hundred or for each hundred. At one time "percent" was written as "per cent" simply abbreviating the Latin. The definition of N percent is $\frac{N}{100}$, abbreviated as N%. Because the denominator is 100, it is very convenient to use decimals to represent a percent. This makes a calculator the perfect tool to calculate percents. For example, $6\% = \frac{6}{100} = 0.06$. The fraction $\frac{6}{100}$ is not in simplest form, so one might prefer to write $6\% = \frac{3}{50}$, but if you are going to be using a calculator, the decimal 0.06 will be more convenient than the fraction $\frac{3}{50}$. To convert a fraction to a percent, first convert the fraction to a decimal (usually with a calculator) and then move the decimal point two places to the right.

Although everyday percents such as 50%, 25%, and $33\frac{1}{3}\%$ are numbers between 0 and 1, *any* number can be expressed as a percent. For example, $2000\% = \frac{2000}{100} = 20$ and $0.07\% = \frac{0.07}{100} = 0.0007$. Figure 2.14 shows that every number can be expressed as a fraction, a decimal, or a percent. Converting among the three forms can be summarized as follows:

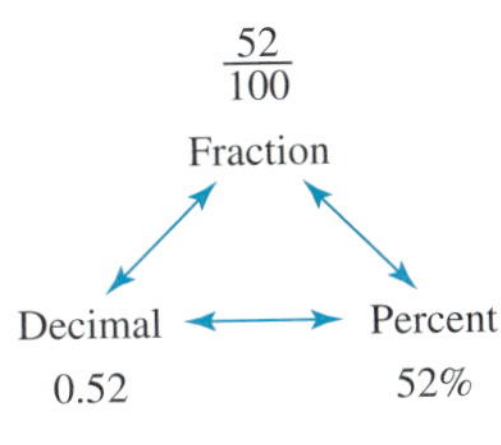

FIGURE 2.14

Decimal-Percent-Fraction Conversions

1. To convert between a decimal and a percent, move the decimal point two places (to the right when converting from decimal to percent and to the left when converting from percent to decimal).
2. To change a fraction to a decimal, divide the numerator by the denominator. To change a fraction to a percent, change it to a decimal and use step 1.
3. To change a percent to a fraction, express the percent in its decimal equivalent, then change the decimal to its fraction form.

EXAMPLE 2.25 Express each of the following in the other two equivalent forms among fractions, decimals, and percents.

(a) $\frac{3}{5}$ **(b)** 225% **(c)** 0.005 **(d)** 3

SOLUTION

(a) $\frac{3}{5} = \frac{6}{10} = 0.6 = 60\%$ **(b)** $225\% = \frac{225}{100} = 2.25$ or $2\frac{1}{4}$

(c) $0.005 = \frac{5}{1000} = 0.5\%$ **(d)** $3 = \frac{300}{100} = 300\%$ ◆

PERCENT OF A NUMBER

A **percentage** is a rate or proportion per hundred, as in the percentage of a group falling into a category. A dictionary will indicate that *percent* and *percentage* are synonyms, so use the word that sounds right in the circumstance.

Since percents are just another expression for fractions (or decimals), you can do all the usual operations of arithmetic with them. However, multiplication turns out to be the most important operation. The most common use of percentage is finding the percent of a number. For example, suppose that you invest $7000 at 5.6% interest per year. To find the amount your money would earn in a year, you multiply $7000 by 5.6% or 0.056, to obtain $392. Some calculators have a percentage key that can be used as follows:

7000 [×] 5.6 [%] [392].

Note: Some calculators require pressing the [=] key after the [%] key, or perhaps some other keystrokes.

HISTORY

The name of Professor Draco used in the text was inspired by a lawyer in Athens in the 7th century B.C., who was one of the first to put the law into a written form. His system of laws was later considered to be unduly harsh, since it called for the death penalty for many trivial crimes. The word *draconian* is derived from Draco's legal code.

A second type of computation is finding what percentage of a group fall into some category. For example, the chair of the mathematics department might be interested in the percentage of students who fail Professor Draco's calculus class. Suppose that 27 of Professor Draco's 31 students received the grade of "F." The department chair would almost surely use his calculator to compute

$$\frac{27}{31} = 0.8709677 = \frac{87.09677}{100}.$$

So 87% of Professor Draco's class failed. (Since there are only 2 digits in the count of students, it is appropriate to round off to 2 digits in the percentage.) Presumably, Draco would fail 87 out of every 100 students, or 87 *per centum.* However, with that kind of failure rate, Professor Draco will probably be having some interesting discussions with the department chair.

A third type of percent problem gives the percent and asks for the original total. A car advertisement states that over 3,000,000, or 96%, of their cars sold in the past 10 years are still on the road. If s represents the total number of cars sold, then we can use the equation $96\% \times s = 3{,}000{,}000$ to find s.

3000000 [÷] 96 [%] [3125000].

Note: Some calculators require an equal sign after the [%] key. If a percent key is not present, use p [÷] 100 in the place of p [%].

The above three types of percent problems can be summarized as follows.

Three Types of Percent Problems

1. To find $p\%$ of n, multiply n by $\frac{p}{100}$. (Find n [×] p [÷] 100 [=] or n [×] p [%] on a calculator.)
2. To find what percent m is of n, express $\frac{m}{n}$ as a percent. (Find m [÷] n [×] 100 [=] on a calculator.)
3. To find $p\%$ of what number is n, calculate $n \div p\%$. (Find n [÷] p [×] 100 [=] or n [÷] p [%] on a calculator.)

EXAMPLE 2.26 Solve.

(a) Find 65% of 240.
(b) Find the percent 30 is of 75.
(c) 40% of what number is 75?

SOLUTION

(a) 240 [×] 65 [%] [156]

(b) 30 [÷] 75 [×] 100 [=] [40]

(c) 75 [÷] 40 [%] = [187.5] (Note: We could also divide 75 by 0.4.) ◆

Often it is useful to estimate the answers to percent problems. For instance, in Example 2.26 (a), 65% is approximately $66\frac{2}{3}\%$, which is $\frac{2}{3}$. Since $\frac{1}{3}$ of 240 is 80, then $\frac{2}{3}$ of 240 is 160. Thus, the answer of 65% of 240 is about 160. This technique of estimating is referred to as using **fraction equivalents.** The next table shows some of the most common fraction equivalents.

TABLE 2.4

Percent	75%	$66\frac{2}{3}\%$	50%	$33\frac{1}{3}\%$	25%	20%	10%
Fraction equivalent	$\frac{3}{4}$	$\frac{2}{3}$	$\frac{1}{2}$	$\frac{1}{3}$	$\frac{1}{4}$	$\frac{1}{5}$	$\frac{1}{10}$

EXAMPLE 2.27 Estimate using approximate fraction equivalents.

(a) 35% of 92 **(b)** 120% of 52 **(c)** 61% of 960

SOLUTION

(a) $35\% \approx 33\frac{1}{3}\% = \frac{1}{3}$. Thus 35% of 92 $\approx \frac{1}{3}$ of 93, or 31.

(b) 120% is 100% + 20%, and 20% $= \frac{1}{5}$. Thus, 120% $= \frac{6}{5}$ and $\frac{6}{5}$ of 52 is approximately 6 times $\frac{1}{5}$ of 50, or 60.

(c) $61\% \approx 60\%$ and 60% can be viewed as 50% + 10%. Thus, 61% of 960 is approximately 480 (50% of 960) plus 96 (10% of 960), or approximately $480 + 100 = 580$. ◆

PERCENT CHANGE

Often one speaks of the **percent change** of some quantity. The procedure is always to

(i) subtract the original value of the quantity from the final value of the quantity,

(ii) divide the difference from (i) by the original value,

(iii) change the quotient from (ii) to its percent form.

Unless the numbers are "nice" or you enjoy long division, it is best to do the division with a calculator and then move the decimal point to the right two places to convert to percent. Symbolically, the percent of change is found as follows:

Percent Change

$$\frac{\text{Final Value} - \text{Original Value}}{\text{Original Value}} \times 100\%$$

If the present value is greater than the original value, the percent change is often called the **percent increase.**

EXAMPLE 2.28 Tuition, which was $1200 per quarter last year, was raised to $1350 this year. What is the percent change in tuition?

SOLUTION

Percent change (increase here) is $\frac{1350 - 1200}{1200} \times 100\% = 12.5\%$. ◆

If the original value is greater than the present value, then the percent change is called the **percent decrease.**

EXAMPLE 2.29 A car dealer was offering cars with a sticker price of $25,000 for sale at $23,000. What was the percent change in price?

SOLUTION

Percent change (decrease) is $\frac{23000 - 25000}{25000} \times 100\% = -8\%$. ◆

TIDBIT

It has been estimated by the Bureau of Labor Statistics that more than half the population of this country (including dependents) has income affected in some way by the CPI. In 1985, income tax brackets began to be adjusted with respect to the CPI.

Almost everyone worries about changes in the prices they pay for all kinds of things, since prices tend to increase rather than decrease. The most widely used measure to keep track of price level changes is the Consumer Price Index published monthly by the U.S. Department of Labor's Bureau of Labor Statistics. The Consumer Price Index (CPI) is a monthly index based on the composite cost of selected goods and services such as housing, food, transportation, and utilities used by working-class households. The costs in the period 1982–83 are used as a basis for comparison (1983 = 100). Although the CPI has come under attack in recent years, it is still considered the best measure of inflation we have in this country. The CPI is often referred to as the *cost-of-living* index. Most newspapers do not report the actual value of the CPI, but only the percent of change. The following table shows how the CPI has changed over time.

TABLE 2.5 CPI (1983 = 100)

YEAR	INDEX	YEAR	INDEX	YEAR	INDEX
1955	28	1975	54	1990	131
1960	30	1980	83	1991	136
1965	32	**1983**	**100**	1992	140
1970	39	1985	108	1993	145

EXAMPLE 2.30 What net monthly salary in 1985 would have the same purchasing power as a net monthly salary of $1500 in 1980?

SOLUTION To have the same purchasing power, the fraction of a salary in 1985 compared to a salary in 1980 would have to be the same as the fraction of the CPI in 1985 to the CPI in 1980. Therefore, if x stands for the net monthly salary in 1985, we have the following equation:

$$\frac{\text{salary in 1985}}{\text{salary in 1980}} = \frac{\text{CPI in 1985}}{\text{CPI in 1980}} = \frac{x}{1500} = \frac{108}{83}.$$

We solve this as $x = (108/83)1500 = \$1952$ (rounded to the nearest dollar). ◆

The next example combines the CPI and the percent of increase.

EXAMPLE 2.31 What is the overall percent increase in the cost-of-living from 1970 to 1985 as measured by the CPI?

SOLUTION We solve this problem by calculating the percent increase in the CPI over this period of time.

$$\frac{\text{CPI in 1985} - \text{CPI in 1970}}{\text{CPI in 1970}} = \frac{108 - 39}{39} = \frac{69}{39} = 1.769 = 176.9\%.$$

We interpret this by saying consumer prices (or the cost-of-living) has increased by 177% between 1970 and 1985. ◆

TIDBIT

The higher a merchant's markup, the greater the profit on selling an item. However, sometimes a high markup lowers profits. A college bookstore offered four 35mm single-lens-reflex cameras for sale at manufacturer's suggested retail price, which included a hefty markup. Since cameras are widely discounted, after 2 years no cameras had been sold. However, 2 had been shoplifted.

Merchants define their **markup** to be the selling price minus the dealer's cost divided by the selling price, then converted to a percent. Many retailers would like to sell at a 50% markup. You might have expected to divide by the retailer's cost, but that would produce a larger number, and retailers find their way more useful. A **markdown** is computed on the original selling price or on the manufacturer's suggested retail price (often abbreviated MSRP). Often, the consumer laws in a state will require that before a price can be declared a sale price, the item must be offered at the higher price for a significant length of time.

Calculating markup and markdown can be summarized as follows.

Percent Markup and Markdown

$$\text{Percent Markup} = \frac{\text{Selling Price} - \text{Cost}}{\text{Selling Price}} \cdot 100\%$$

$$\text{Percent Markdown} = \frac{\text{Selling Price} - \text{Sale Price}}{\text{Selling Price}} \cdot 100\%$$

EXAMPLE 2.32 If a sweater costs a shopkeeper \$30, her markup is 40%, and it is being offered at 20% off,

(a) what was the retail price?
(b) what is the sale price?

SOLUTION

(a) Let r be the retail price. Since the markup is 40%, we have

$$\frac{r - 30}{r} = 40\% = \frac{40}{100}. \text{ Thus,}$$

$$\frac{r - 30}{r} \times 100 = 40, \text{ or } 100r - 3000 = 40r.$$

Rearranging terms in this equation yields

$$60r = 3000, \text{ or } r = \$50.$$

(Notice that the markup is taken on the selling or retail price. If the markup was calculated on the *cost,* it would be $\frac{20}{30}$ or $66\frac{2}{3}\%$ markup.)

(b) The sale price is $50 - 50(20\%) = 50 - 10 = \$40$. ◆

A simpler way to do the above price reduction computation is to first subtract the percent reduction from 100% to find the percentage you must pay. After all, if 20% is "off," there is still 80% "on." We find the purchase price to be 80% of \$50 or $0.80 \times \$50 = \$40$.

Large stores with computerized inventory systems usually do the computation automatically at the cash register. Rounding is often required, which may or may not be correctly programmed.

INITIAL PROBLEM SOLUTION

Your brother-in-law recounts the following story: "I bought some stock and it lost 90% of its value. Luckily, I didn't sell it, because it then increased to 1000% of its depressed value. I made a bundle. Too bad you don't have my financial ability." Did your brother-in-law make a big profit?

SOLUTION Suppose \$1000 was spent on the purchase of the stock. Then 90% of its value is the same as $90/100 = 0.90$ of its value, which is $\$1000 \times 0.90 = \$900$. That was how much value was lost, so the stock was then worth $\$1000 - \$900 = \$100$. When the stock subsequently increased to 1000% of its value, that means its new value was \$100 multiplied by 1000% ($1000\% = 1000/100 = 10$). So the value at which he sold the stock was $\$100 \times 10 = \$1000$, exactly what was paid for it. No profit was made. What is surprising is that the net effect of a 90% loss followed by a 1000% appreciation is that there is no net change. (Indeed in the real world there would be a loss equal to the commissions charged for buying and selling the stock.)

PROBLEM SET 2.5

1. Express each of the following in the other two equivalent forms.
(a) 25% **(b)** 4.25 **(c)** $\frac{2}{5}$

2. Express each of the following in the other two equivalent forms.
(a) 0.75 **(b)** $\frac{1}{8}$ **(c)** $33\frac{1}{3}\%$

3. Express each of the following in the other two equivalent forms.
(a) $\frac{3}{8}$ **(b)** 112% **(c)** 0.875

4. Express each of the following in the other two equivalent forms.
(a) 45% **(b)** 0.1525 **(c)** $\frac{6}{5}$

5. Using your own words, explain how to convert a fraction to a percent.

6. Using your own words, explain how to convert a percent to a fraction.

7. Solve.
(a) Find 40% of 300.
(b) 30% of what number is 80?
(c) 200% of what number is 128?

8. Solve.
(a) 25% of n is 18. What is n?
(b) Find $33\frac{1}{3}\%$ of 297.
(c) 17 is what percent of 85?

9. Estimate mentally.
(a) Find 49% of 201.
(b) 32% of what number is 51?
(c) 21 is about what percent of 79?

10. Estimate mentally.
(a) Find 124% of 84.
(b) 52 is about what percent of 7?
(c) 16% of what number is 3.1?

11. Find the percent of change mentally.
(a) 50 to 70
(b) 75 to 50
(c) 30 to 90

12. Find the percent of change mentally.
(a) 50 to 40

(b) 75 to 90
(c) 90 to 75

13. Find each of the missing values.
   (a) original value = 130, percent of decrease = 20%, final value = ____
   (b) original value = 240, final value = 300, percent of increase = ____
   (c) percent of increase = 30%, final value = 390, original value = ____
14. Find each of the missing values.
   (a) original value = 120, percent of decrease = 15%, final value = ____
   (b) original value = 40, final value = 65, percent of change = ____
   (c) percent of decrease = 25%, final value = 60, original value = ____
15. Find each of the missing values.
   (a) original value = 120, percent of increase = 30%, final value = ____
   (b) original value = 210, final value = 180, percent of change = ____
   (c) percent of increase = 25%, final value = 70, original value = ____
16. Find each of the missing values.
   (a) original value = 80, final value = 200, percentage of change = ____
   (b) original value = 185, percent of change = 42%, final value = ____
   (c) percent of decrease = 24%, final value = 60, original value = ____

   For problems 17 through 20, refer to Table 2.5.

17. What net monthly salary in 1990 would have the same purchasing power as a net monthly salary of $1850 in 1985?
18. What net monthly salary in 1990 would have the same purchasing power as a net monthly salary of $1275 in 1975?
19. What is the percent change in the Consumer Price Index from 1970 to 1990?
20. What is the percentage change in the Consumer Price Index from 1960 to 1990?
21. A woman making $2000 per month has her salary reduced by 10% due to sluggish company sales. One year later, after a dramatic improvement in sales, she is given a 20% raise over her reduced salary.
   (a) What is her salary after the raise?
   (b) What percent change is this from the $2000 per month?
22. An antique dealer sold two pieces of furniture for $480 each. For one of them, this represented a 20% loss; for the other, a 20% profit (based on dealer cost). How much did she make or lose on the total transaction?
23. A shopowner buys an item at wholesale for $36 and marks it for sale at $60. What percent of markup is the shopowner using?
24. A certain furniture store that uses a markup of 40% on its goods buys a sofa that has a wholesale price of $270. At what price will the furniture store mark the sofa for sale?
25. The total price of a new stereo (including 6% sales tax) is $418.70. How much of this total is tax?
26. A bicycle costs a wholesaler $56. What will the retailer sell it for if the wholesaler's markup is 20% of the wholesale selling price and the retailer's markup is 30% of the retailer's selling price?
27. An art dealer has a 50% markup and gives selected customers a "patron's" discount of 10%. If the dealer pays $1275 to an artist for a bronze sculpture, how much will she charge when she sells it to one of her selected customers?
28. If the art dealer in problem 27 sells a pair of blown glass goblets to one of her selected customers for $810, how much did the dealer pay when she bought the goblets from the artist?
29. Suppose a furniture store that uses a markup of 40% on its goods sells a table at a 20% discount. If the total price (including 6% sales tax) is $296.80, what was the wholesale price the store paid for the table?
30. If the furniture store in problem 29 buys a sofa that has a wholesale price of $375 and subsequently puts the sofa on sale at a 25% discount, what will be the final price, including 6% sales tax?

EXTENDED PROBLEMS

The Consumer Price Index (CPI) is one of the most closely watched, utilized, and criticized numbers in our economic system.

31. Write a brief report on the history and development of the Consumer Price Index, with emphasis on the philosophy behind the index and the goods and services upon which it was based.
32. Write a report on the economic impact of the CPI as it relates to labor negotiations.
33. Beginning in the early 1980s, the federal deficit and national debt increased at historically unprecedented rates. One of the chief concerns has been the effect of entitlement programs on the federal budget. Write a brief report on the impact of the CPI on the federal budget.
34. Many critics contend that the CPI overstates the amount of inflation in the consumer economy. What alternatives have been proposed for replacing the Consumer Price Index? What advantages and disadvantages do these methods have?
35. What influence does the CPI have with respect to budgets of the city, county, and state in which you live and the college or university you attend?

2.6 INTEGER EXPONENTS

Despite advice to the contrary, you show up for an exam with an unfamiliar calculator borrowed at the last minute. The displayed answer to a calculation is

1.687 3

The space between the 7 and the 3 looks unusual to you, and 1.6873 seems awfully small to be the correct answer. What should you put down on your exam paper?

In this section, whole number exponents are first introduced as a shorthand form for repeated multiplication. Next, zero and negative exponents are defined in such a way that they are consistent with the properties developed for the whole number exponents. Finally, an adaptation is made that allows us to use exponents for numbers that are either very large or very small. In this way, exponents are used in many important applications.

WHOLE NUMBER EXPONENTS

The first mathematical operation a child learns is counting or, to put it more technically, *incrementing-by-one.* The operation of addition is a series of increment-by-one operations performed all at once. Of course, children often break simple addition problems into their subparts by counting the results on their fingers.

Addition of Whole Numbers: Repeated Increment Approach

$$a + m = a \underbrace{+ 1 + 1 + \ldots + 1}_{m \text{ increments}}$$

The operation of multiplication similarly is a series of addition operations performed all at once.

DEFINITION

Multiplication by Whole Numbers: Repeated Addition Approach

$$a \times m = \underbrace{a + a + \ldots + a}_{m \text{ addends}}$$

Not surprisingly there is a further stage in this development: The operation that performs a series of multiplication operations all at once is called **exponentiation.**

Whole Number Exponent

Let a be a real number and let m be a nonzero whole number. Then

$$a^m = \underbrace{a \times a \times \ldots \times a}_{m \text{ factors}}$$

The number m is called the **power** applied to a, or the exponent of a, and a is called the **base.** The number a^m is read "a to the power m" or "a to the mth power," or simply as "a to the mth." When the exponent has the particular values 2 and 3, the terms "squared" and "cubed" are used. So, for example, $5^2 = 5 \times 5 = 25$ is usually read "5 squared" rather than "5 to the second power." As another example, 2^3 can be read "2 to the third power" or, more often, "2 cubed" and equals $2 \times 2 \times 2 = 8$.

The notation a^m puts a premium on careful reading. The m is a superscript on the a, which means that the letter is placed a small amount higher than the a, is (usually) a little smaller in size than an ordinary m, and there is no extra space between the a and the m. Since the convention in algebra is that two letters side by side are multiplied (so that $a \times m$ can be written am), you could easily confuse "a to the power m" and "a times m." The problem is even more acute when expressions are handwritten. The notation is too accepted for us to even dare to suggest a change, so all we can do is warn you to be careful.

RULES FOR WHOLE NUMBER EXPONENTS

There are several properties of exponents that permit us to represent numbers and do many calculations quickly.

EXAMPLE 2.33 Rewrite each of the following expressions using a single exponent.

(a) $2^3 \times 2^4$ **(b)** $3^2 \times 7^2 \times 11^2$ **(c)** $(5^3)^2$
(d) $5^7 + 5^3$ **(e)** $\frac{2^3}{5^3}$

SOLUTION

(a) $2^3 \times 2^4 = (2 \times 2 \times 2) \times (2 \times 2 \times 2 \times 2) = 2^7 = 2^{3+4}$
(b) $3^2 \times 7^2 \times 11^2 = (3 \times 3) \times (7 \times 7) \times (11 \times 11)$
$= (3 \times 7 \times 11) \times (3 \times 7 \times 11) = (3 \times 7 \times 11)^2$
(c) $(5^3)^2 = 5^3 \times 5^3 = 5^{3+3} = (5 \times 5 \times 5) \times (5 \times 5 \times 5) = 5^6 = 5^{3\times 2}$
(d) $5^7 \div 5^3 = (5 \times 5 \times 5 \times 5 \times 5 \times 5 \times 5) \div (5 \times 5 \times 5) = 5 \times 5 \times 5 \times 5 = 5^4 = 5^{7-3}$
(e) $\frac{2^3}{5^3} = \frac{2 \times 2 \times 2}{5 \times 5 \times 5} = \frac{2}{5} \times \frac{2}{5} \times \frac{2}{5} = \left(\frac{2}{5}\right)^3$ ◆

HISTORY

The notation a^m was introduced by René Descartes in his work *La Géométrie* published in 1637.

The important features in the example are as follows: In part (a) the bases in the factors are the same, in part (b) the exponents in the factors are all the same, in part (c) an exponent is applied to an exponent, in part (d) the bases are again the same, and in part (e) the exponents are again the same. These are instances of **five general rules of exponents** which follow from the definition.

Product of Numbers Having the Same Base

Let a be a real number and let m and n be any nonzero whole numbers. Then

$$a^m \times a^n = a^{m+n}.$$

Product of Numbers Having the Same Exponent

Let a and b be real numbers and let m be any nonzero whole number. Then

$$a^m \times b^m = (a \times b)^m.$$

Exponent Applied to an Exponent

Let a be a real number and let m and n be any nonzero whole numbers. Then

$$(a^m)^n = a^{m \times n}.$$

Quotient of Numbers Having the Same Base

Let a be a nonzero real number and let m and n be any nonzero whole numbers where $m > n$. Then

$$a^m \div a^n = a^{m-n}.$$

Quotient of Numbers Having the Same Exponent

Let a and b be real numbers, with $b \neq 0$, and let n be any nonzero whole number. Then

$$\frac{a^n}{b^n} = \left(\frac{a}{b}\right)^n.$$

CALCULATOR EXPONENT KEYS

There are three common types of exponent keys: $\boxed{10^n}$, $\boxed{x^2}$, and $\boxed{y^x}$.

The $\boxed{10^n}$ key finds powers of 10 in one of two ways, depending on the make of calculator:

$$\boxed{10^n}\ 3\ \boxed{=}\ \boxed{\quad 1000\quad}\ \text{or } 3\ \boxed{10^n}\ \text{results in}\ \boxed{\quad 1000\quad}.$$

The $\boxed{x^2}$ key is used to find squares as follows: 3 $\boxed{x^2}$ $\boxed{\quad 9\quad}$.

The $\boxed{y^x}$ key is used to find more general powers. For example, 7^3 may be found as follows:

$$7\ \boxed{y^x}\ 3\ \boxed{=}\ \boxed{\quad 343\quad}.$$

If you forget which way the $\boxed{y^x}$ key works, try out the possibilities for finding $2^3 = 8$ and $3^2 = 9$.

Note: There may also be a key on your calculator that "undoes" the action of the $\boxed{10^n}$ key. That is the $\boxed{\log}$ key. (The function is called the **common logarithm.**) Scientific calculators also have an exponent key of the form $\boxed{e^x}$. The number $e \approx 2.718$ is the **base of natural logarithms.** The action of this key is undone by the **natural logarithm** key $\boxed{\ln}$.

NEGATIVE AND ZERO EXPONENTS

Notice that we have not yet defined a^0 or a^{-m}. Consider the following pattern:

$$\begin{aligned}
a^3 &= a \times a \times a \\
&\quad \div a \\
a^2 &= a \times a \\
&\quad \div a \\
a^1 &= a \\
&\quad \div a \\
a^0 &= 1 \\
&\quad \div a \\
a^{-1} &= \frac{1}{a} \\
&\quad \div a \\
a^{-2} &= \frac{1}{a^2} \\
&\quad \div a \\
a^{-3} &= \frac{1}{a^3} \\
&\text{etc.}
\end{aligned}$$

As the numbers are divided by a each time, the exponents decrease by 1.

Extending this pattern, we see that the following definition is appropriate.

Negative and Zero Exponents

Let a be a nonnegative real number and let m be any nonzero whole number. Then

$$a^0 = 1 \text{ and } a^{-m} = \frac{1}{a^m}.$$

Notice that 0^0 is not defined. The reason for this is to ensure consistency in our system. Consider the following two patterns:

$$\begin{array}{ll} 3^0 = 1 & 0^3 = 0 \\ 2^0 = 1 & 0^2 = 0 \\ 1^0 = 1 & 0^1 = 0 \\ 0^0 = ? & 0^0 = ? \end{array}$$

The pattern on the left suggests that 0^0 should be 1, whereas the pattern on the right suggests that 0^0 should be 0. Since $1 \neq 0$, we say that 0^0 is undefined.

Applying the definition, we can consider the graph of the values of $2^{-3} = 1/8$, $2^{-2} = 1/4$, $2^{-1} = 1/2$, $2^0 = 1$, $2^1 = 2$, $2^2 = 4$, $2^3 = 8$. For some it is hard to accept that a^0 must be equal to 1, but by looking at the graph of these values in Figure 2.15, you can see that any other value for a^0, other than 1, would be out of place.

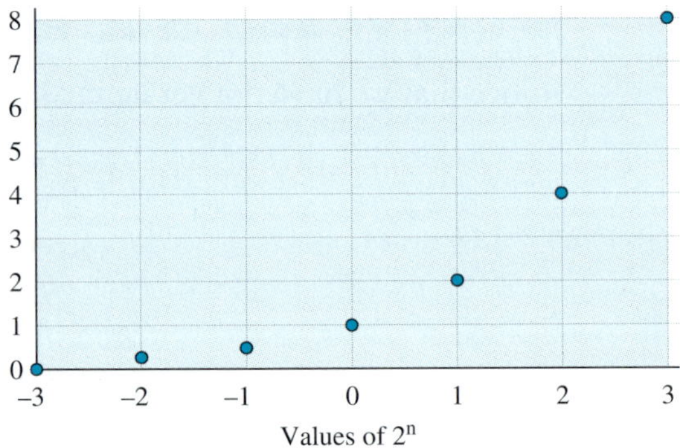

FIGURE 2.15

The rules for exponents given earlier for positive whole number exponents remain true when zero and negative whole number exponents are allowed.

EXAMPLE 2.34 Rewrite each of the following expressions using a single exponent.

(a) $5^7 \times 5^{-4}$ **(b)** $7^{-3} \times 11^{-3}$ **(c)** $(7^3)^{-6}$

(d) $3^{-7} \div 3^{-9}$ **(e)** $\dfrac{4^{-5}}{7^{-5}}$

SOLUTION

(a) $5^7 \times 5^{-4} = 5^{7+(-4)} = 5^3$
(b) $7^{-3} \times 11^{-3} = (7 \times 11)^{-3} = 77^{-3}$
(c) $(7^3)^{-6} = 7^{3(-6)} = 7^{-18}$
(d) $3^{-7} \div 3^{-9} = 3^{(-7)-(-9)} = 3^2$
(e) $\dfrac{4^{-5}}{7^{-5}} = \left(\dfrac{4}{7}\right)^{-5}$ ◆

SCIENTIFIC NOTATION

Scientific notation uses integer exponents to represent very large and very small numbers. For example, a certain wavelength whose value in decimal form is 0.000000573 can be expressed as 5.73×10^{-7}. A number is written in **scientific notation** if it is written in the form $a \times 10^n$ where $1 \le a < 10$ and n is an integer. Numbers written in their scientific notation are easy to compare. For example, 8.73×10^7 is less than 4.31×10^9 since $7 < 9$. Similarly, $6.7 \times 10^{-5} < 4.9 \times 10^{-2}$ since $-5 < -2$. Properties of exponents are useful when calculating products and quotients of large or small numbers using scientific notation. The next example illustrates the process.

EXAMPLE 2.35 Calculate and express your answers in scientific notation.

(a) $(3.9 \times 10^{-7}) \times (2.3 \times 10^{-9})$ **(b)** $(7.1 \times 10^3) \times (4.3 \times 10^5)$

SOLUTION

(a) $(3.9 \times 10^{-7}) \times (2.3 \times 10^{-9}) = 3.9 \times 2.3 \times 10^{-7} \times 10^{-9} = 8.97 \times 10^{-16}$
(b) $(7.1 \times 10^3) \times (4.3 \times 10^5) = 7.1 \times 4.3 \times 10^3 \times 10^5 = 30.53 \times 10^8$

Since scientific notation requires the form $a \times 10^n$ where $1 \le a < 10$, we change as follows: $30.53 \times 10^8 = 3.053 \times 10^1 \times 10^8 = 3.053 \times 10^9$. ◆

Scientific notation is expressed in various ways on calculators. For example, on the 10-digit display of one type of calculator, the product of 123,456,789 and 987 is displayed as [1.218518507^{11}], which means $1.21851850 \times 10^{11}$. When calculating products and quotients of large or small numbers using a calculator, the answer may be expressed in scientific notation even though the numbers were not entered that way. Your calculator's user's manual will show how your calculator represents numbers in scientific notation.

INITIAL PROBLEM SOLUTION

Despite advice to the contrary, you show up for an exam with an unfamiliar calculator borrowed at the last minute. The displayed answer to a calculation is

[1.687 3].

The space between the 7 and the 3 looks unusual to you, and 1.6873 seems awfully small to be the correct answer. What should you put down on your exam paper?

SOLUTION The space between the 7 and the 3 is your clue that the calculator is using scientific notation.

1.687 3

is the calculator's way of representing

$$1.687 \times 10^3 = 1.687 \times 1000 = 1687.$$

Put down 1687 on the exam paper.

PROBLEM SET 2.6

In problems 1 through 10, rewrite each expression using a single exponent.

1. **(a)** $2^3 \times 2^7$
   (b) $5^2 \times 5^4$
   (c) $b^2 \times b^4 \times b^3$
2. **(a)** $7^3 \times 7^4$
   (b) $4^2 \times 4^3$
   (c) $c^3 \times c^2 \times c^3$
3. **(a)** $2^4 \times 4^4$
   (b) $4^3 \times 2^3$
   (c) $2^2 \times 3^2 \times 5^2$
4. **(a)** $5^3 \times 3^3$
   (b) $2^5 \times 3^5$
   (c) $3^2 \times 5^2 \times 4^2$
5. **(a)** $(5^2)^3$
   (b) $(3^2)^4$
   (c) $((b^3)^2)^3$
6. **(a)** $(4^3)^3$
   (b) $(2^5)^2$
   (c) $((c^2)^3)^2$
7. **(a)** $3^7 \div 3^2$
   (b) $5^6 \div 5^2$
   (c) $(b^3)^2 \div b^2$
8. **(a)** $5^5 \div 5^2$
   (b) $4^7 \div 4^2$
   (c) $c^8 \div (c^3)^2$
9. **(a)** $5^3 \div 8^3$
   (b) $4^2 \div 7^2$
   (c) $x^5 \div y^5$
10. **(a)** $8^3 \div 5^3$
    (b) $6^4 \div 5^4$
    (c) $(2x)^5 \div (3y)^5$
11. Use the $\boxed{x^2}$ key on your calculator to find each of the following.
    (a) $(1.55)^2$ **(b)** $(17.2)^2$ **(c)** $(0.031)^2$
12. Use the $\boxed{x^2}$ key on your calculator to find each of the following.
    (a) $(25.7)^2$ **(b)** $(459.2)^2$ **(c)** $(0.23)^2$
13. The Richter Scale is an index used to represent the relative strength of earthquakes, but it does so in an indirect manner as follows: If an earthquake has a reading of x on the Richter Scale, then the relative strength of the earthquake is proportional to 10^x. That is, a reading of 3 on the Richter Scale indicates a relative strength of 10^3, or 1000. Find the relative strength of earthquakes that have the following readings on the Richter Scale (use the $\boxed{10^x}$ key on your calculator).
    (a) 5.3
    (b) 5.8
    (c) 7.8
14. Referring to problem 13, find the relative strength of earthquakes that have the following readings on the Richter Scale (use the $\boxed{10^x}$ key on your calculator).
    (a) 5.4
    (b) 5.5
    (c) 7.3
15. Use the $\boxed{y^x}$ key on your calculator to find each of the following.
    (a) $(5.2)^5$
    (b) $(45.2)^{2.6}$
    (c) $(0.23)^{-1.5}$
16. Use the $\boxed{y^x}$ key on your calculator to find each of the following.
    (a) $(12)^{1.73}$
    (b) $(2.7183)^{2.3}$
    (c) $(6.7)^{-2.74}$

In problems 17 through 22, rewrite the expression in its simplest form as an integer or fraction.

17. **(a)** 2^{-3} **(b)** 3×5^{-2} **(c)** $\dfrac{4}{2^{-2}}$
18. **(a)** 3^{-2} **(b)** $\dfrac{5}{3 \times 4^{-1}}$ **(c)** $\dfrac{2^{-3}}{3^{-2}}$
19. $3^0 \times 2^{-2} \times \dfrac{1}{5^{-1}}$
20. $4 \times 3^{-2} \times 6$
21. $\dfrac{4^{-1}}{3} \times \dfrac{9}{2^{-2}} \times \dfrac{-1}{3^{-1}}$

22. $\dfrac{3^{-1} \times 2}{4^0 \times 2^{-2}} \times \dfrac{5 \times 2^{-1}}{3^{-2} \times 2^3}$

23. Fill in the missing entries in the following table.

Standard Notation	*Scientific Notation*
825,000,000	**(a)**
(b)	2.37×10^5
0.0253	**(c)**
805,000	**(d)**

24. Fill in the missing entries in the following table.

Standard Notation	*Scientific Notation*
(a)	4.05×10^{-2}
(b)	2.19×10^5
0.00000075	**(c)**
(d)	5.6×10^{-5}

25. Which is larger, 2^{20} or 1000^2?
26. Which is larger, 3^{20} or 1000^3?
27. Which is larger, 5^{-4} or 10^{-4}?
28. Which is larger, 6^{-4} or 6^{-5}?

In problems 29 through 32, use the properties of exponents to solve for x.

29. $4^x \times 4^2 = 4^{12}$
30. $5^4 \times 5^x = 5^{11}$
31. $(b^x)^2 = b^{18}$
32. $b^x \div b^3 = b^5$

In problems 33 through 38, rewrite each number in scientific notation.

33. The mean (average) distance from Jupiter to the sun is approximately 484,000,000 miles.
34. The farthest planet from the sun is Pluto. The mean distance from Pluto to the sun is approximately 1,666,000,000 miles.
35. The amount of gold in the earth's crust is estimated at 120,000,000,000,000 kilograms.
36. The volume of a typical neuron is about 0.00000003 cubic centimeters.
37. A joule is about 0.00000028 kilowatt-hours.
38. The coefficient of thermal expansion for aluminum is 0.00001835.

EXTENDED PROBLEMS

39. *Kepler's Third Law* In 1619, Johannes Kepler, a German astronomer, discovered that the period, T (in years), of each planet in our solar system is related to the planet's mean distance, R (in astronomical units), from the sun by the equation

$$T^2/R^3 = k.$$

Test Kepler's equation for the nine planets in our solar system, using the given table. (Astronomical units relate the other planet's period and mean distance to those of the Earth.) Do you get approximately the same value for k in each case?

Planet	**T**	**R**
Mercury	0.241	0.387
Venus	0.615	0.723
Earth	1.000	1.000
Mars	1.881	1.523
Jupiter	11.861	5.203
Saturn	29.457	9.541
Uranus	84.008	19.190
Neptune	164.784	30.086
Pluto	248.350	39.507

40. Asteroids are minor planetary bodies that are typically found between the planetary orbits of Mars and Jupiter (there are approximately 1500 known asteroids). If an asteroid has a mean distance from the sun of 2.873 astronomical units, what would be the period of this asteroid (assume $k = 1$)?
41. Suppose an astronomer discovers a new planet whose mean distance from the sun is 48.125 astronomical units. Use the information in problem 39 to find the period of this planet.
42. Suppose the astronomer in problem 41 discovers another new planet whose period is calculated to be 415 years. What would be the mean distance of this planet from the sun (in astronomical units)?
43. The surface area of a sphere is given by the equation $S = 4 \times \pi \times r^2$. When the radius of a sphere is doubled, how much is the surface area increased? Justify your answer algebraically.
44. The volume of a sphere is given by $V = \frac{4}{3} \times \pi \times r^3$. When the radius of a sphere is doubled, how much is the volume increased? Justify your answer algebraically.
45. A tetrahedron is a pyramid with a triangular base. If all the edges of a tetrahedron are the same length, the tetrahedron is called a *regular* tetrahedron and its volume is given by $V \approx 0.11785s^3$, where s is the length of the edge. What is the approximate length of the edge of a regular tetrahedron that has a volume of 100 cubic inches?
46. The surface area of a regular tetrahedron is given by $S \approx 1.7321s^2$, where s is the length of the edge. What is the approximate volume of a regular tetrahedron having a surface area of 80 square inches? (See problem 45.)

2.7 GENERAL EXPONENTS

Some gamblers use a system in which if the first bet, say \$10, is lost, then the second bet is \$20; that is, double the first. If the second bet is also lost, then the third bet is \$40, and so on until finally the gambler wins back all the losses plus a profit of \$10 equal to the first bet. (We are assuming a game in which the gambler winnings are equal to the amount bet—this is called an *even money* wager.) Why is this doubling system, called a *martingale,* dangerous, if not foolish?

HISTORY

Old-timers (such as the authors) learned in school how to find square roots using a laborious paper and pencil method similar to long division. The "antique" slide rule included a square root scale for finding square roots to two or three significant digits. Tables of values were also used to find estimates.

Many relationships between numbers (but not all) can be reversed in the sense that instead of comparing a to b, we can compare b to a. For example, instead of saying 15 is three times as large as 5, we can say that 5 is one-third as large as 15. In this section, we consider the conditions under which we can reverse the relationship between a number and a power of that number. This leads to the idea of a root, which we express first with a radical and then with a rational numbers exponent. Finally, we see that with a suitable interpretation any real number can be used as an exponent.

ROOTS

Remember that we refer to b^2 as "b squared." Often it is convenient to think of **squaring** as an operation that when applied to the number b produces the number b^2. Your calculator may have a key that does this. Some operations can be undone, like taking your hat off undoes putting it on. The **square root** is the operation that undoes squaring. We write the square root of a number by putting the radical sign ($\sqrt{\ }$) over it. For example, we have $\sqrt{9} = \sqrt{3^2} = 3$. Notice that since both $(-3)^2$ and 3^2 equal 9, there are actually two choices for the square root of 9, either -3 or 3. The symbol $\sqrt{9}$ represents the *nonnegative* choice 3. (Note: $-\sqrt{9} = -3$.) The number $\sqrt{a}$, called the **principal square root of *a*,** is the nonnegative number whose square is a, *if such a real number exists.* Your calculator should have a key that performs the square root operation. If you use your calculator to take the square root of a negative number, then you will probably get an error message. Since b^2 is nonnegative for any real number b, a negative number cannot have a real number square root.

DEFINITION

Square Root

Let a be a nonnegative real number. Then the principal square root of a, written $\sqrt{a}$, is defined by

$$\sqrt{a} = b \text{ where } b^2 = a \text{ and } b \geq 0.$$

EXAMPLE 2.36 Calculate the following square roots.

(a) $\sqrt{4}$ **(b)** $-\sqrt{25}$ **(c)** $\sqrt{19}$

HISTORY

An algebraic method for finding cube roots exists and can be found in various math references. However, the advent of the digital calculator made such algebraic methods obsolete.

SOLUTION

(a) $\sqrt{4} = b$ if and only if $b^2 = 4$ and $b \geq 0$. Since $2^2 = 4$, we have $\sqrt{4} = 2$. Alternatively, we enter 4 into our calculator and press the $\boxed{\sqrt{\ }}$ key to find $\sqrt{4} = 2$.

(b) $\sqrt{25} = b$ if and only if $b^2 = 25$ and $b \geq 0$. Since $5^2 = 25$, we have $\sqrt{25} = 5$. Alternatively, we enter 25 in our calculator and press the $\boxed{\sqrt{\ }}$ key to find $\sqrt{25} = 5$. So $-\sqrt{25} = -5$.

(c) $\sqrt{19} = b$ if and only if $b^2 = 19$ and $b \geq 0$. Since $4^2 = 16$ and $5^2 = 25$, we see that $\sqrt{19}$ is somewhere between 4 and 5, but it is not a whole number. To obtain the answer, we enter 19 into our calculator and press the $\boxed{\sqrt{\ }}$ key to find $\sqrt{19} = 4.358898$. If your calculator has one more decimal place of precision you would get $\sqrt{19} = 4.3588989$. If you have one less decimal place of precision you would get $\sqrt{19} = 4.35889$. None of the calculators is exactly right, because the square is actually an infinitely long nonrepeating decimal. The most honest thing to write is $\sqrt{19} \approx 4.35889$, namely, $\sqrt{19}$ is *approximately* 4.35889 or 4.36 to two decimal places. ◆

Next we generalize the definition of square root to more general types of roots. For example, since $(-2)^3 = -8$, -2 is called the **cube root** of -8. Because of the negative numbers, the definition must be stated in two parts.

DEFINITION

*n*th Root

Let a be a real number and let n be a positive integer.

1. If $a \geq 0$, then $\sqrt[n]{a} = b$ if and only if $b^n = a$ and $b \geq 0$.
2. If $a < 0$, then $\sqrt[n]{a} = b$ if and only if $b^n = a$.

The number a in $\sqrt[n]{a}$ is called the **radicand** and n is called the **index.** The symbol $\sqrt[n]{a}$ is read the **nth root of *a*** and is called a **radical.** The expression $\sqrt[n]{a}$ has not been defined for the case when n is even and a is negative. This is because $b^n \geq 0$ for any real number b when n is an even positive integer. For example, there is no real number b such that $b = \sqrt{-1}$ (or $b^2 = -1$).

EXAMPLE 2.37 Where possible, write the following values in simplest form by applying the previous definition.

(a) $\sqrt[4]{81}$ **(b)** $\sqrt[5]{-32}$ **(c)** $\sqrt[6]{-64}$ **(d)** $\sqrt[3]{29}$

SOLUTION

(a) $\sqrt[4]{81} = b$ if and only if $b^4 = 81$ and $b \geq 0$. Since $3^4 = 81$, we have $\sqrt[4]{81} = 3$.

(b) $\sqrt[5]{-32} = b$ if and only if $b^5 = -32$. Since $(-2)^5 = -32$, we have $\sqrt[5]{-32} = -2$.

(c) We begin as before in applying the definition and write $\sqrt[6]{-64} = b$ if and only if $b^6 = -64$. However, since b^6 must always be positive or zero, there is no real number b such that $\sqrt[6]{-64} = b$.

(d) $\sqrt[3]{29}$ is between $\sqrt[3]{27} = 3$ and $\sqrt[3]{64} = 4$, thus is not a whole number. Your calculator may have a $\boxed{\sqrt[x]{y}}$ key; in that case you can calculate

$$29 \boxed{\sqrt[x]{y}}\, 3 \boxed{=} \boxed{3.072316826} \text{ or } \sqrt[3]{29} \approx 3.07.$$ ◆

RATIONAL AND REAL EXPONENTS

Using the concept of radicals, we can now proceed to define rational exponents. As an example, what would be a good definition of $3^{1/2}$?

If the usual addition property of exponents is to hold, then

$$3^{1/2} \times 3^{1/2} = 3^{1/2+1/2} = 3^1 = 3.$$

But $\sqrt{3} \times \sqrt{3} = 3$. Therefore, $3^{1/2}$ should represent $\sqrt{3}$.

Similarly, $5^{1/3} = \sqrt[3]{5}$, $2^{1/7} = \sqrt[7]{2}$, and so on. We summarize this idea in the next definition.

DEFINITION

Unit Fraction Exponent

Let a be a real number and let n be any positive integer. Then

$$a^{1/n} = \sqrt[n]{a}$$

where

1. n may be any positive integer when $a \geq 0$, and
2. n must be an odd positive integer when $a < 0$.

For example, $(-8)^{1/3} = \sqrt[3]{-8} = -2$, and $81^{1/4} = \sqrt[4]{81} = 3$. To find $37^{1/5}$, use a calculator to find $\sqrt[5]{37}$. Notice that $(-81)^{1/4}$ is not defined.

The combination of the last definition with the definitions for integer exponents leads us to the definition for any rational exponent. For example, taking into account the previous definition and our earlier work with exponents, a natural way to think of $27^{2/3}$ would be $(27^{1/3})^2$. For the sake of simplicity, we restrict our definition to rational exponents of nonnegative real numbers.

DEFINITION

Rational Exponent

Let a be a nonnegative real number, and let $\frac{m}{n}$ be a rational number. Then

$$a^{m/n} = (a^{1/n})^m.$$

Note: $a^{m/n}$ also equals $(a^m)^{1/n}$, provided that $(a^m)^{1/n}$ is defined.

EXAMPLE 2.38 Rewrite the following rational number exponents as radicals and radicals as rational number exponents.

(a) $9^{3/2}$ **(b)** $16^{-5/4}$ **(c)** $\sqrt[3]{125^4}$ **(d)** $\sqrt{\dfrac{1}{237^5}}$

SOLUTION

(a) $9^{3/2} = \sqrt{9^3}$ **(b)** $16^{-5/4} = \sqrt[4]{16^{-5}} = \sqrt[4]{\dfrac{1}{16^5}}$

(c) $\sqrt[3]{125^4} = 125^{4/3}$ **(d)** $\sqrt{\dfrac{1}{237^5}} = \sqrt{237^{-5}} = 237^{-5/2}$ ◆

EXAMPLE 2.39 Express the following values without exponents. Where possible, write the following values in simplest form by applying the previous definition.

(a) $9^{3/2}$ **(b)** $16^{-5/4}$ **(c)** $125^{4/3}$ **(d)** $237^{-5/2}$

SOLUTION

(a) $9^{3/2} = (9^{1/2})^3 = 3^3 = 27$

(b) $16^{-5/4} = (16^{1/4})^{-5} = 2^{-5} = \dfrac{1}{32} = 0.03125$

(c) $125^{4/3} = (125^{1/3})^4 = 5^4 = 625$

(d) To find $237^{-5/2}$ using a calculator, we have to use parentheses as follows:

237 [y^x] ([(−)] 5 [÷] 2) [=] [0.000001156].

Since $-5/2 = -2.5$, we also have

237 [y^x] −2.5 [=] [0.000001156]. ◆

Real-number exponents, such as $b^{\sqrt{2}}$, are defined using more advanced mathematics. Exponentials involving real-number bases and/or real-number exponents can be approximated well by approximating the real numbers by rational numbers. That is, to approximate $5^{\sqrt{2}}$, use $5^{1.414}$ since $\sqrt{2} \approx 1.414$. To find 2^{π}, use either $\frac{22}{7}$ or 3.14 as an approximation to π.

Of course, $2^{22/7}$ is not easy to calculate either. The only practical way to do the computation of either $2^{22/7}$ or 2^{π} is to use a calculator. Just remember that the calculator approximates the real numbers by rational numbers (and truncates any infinite repeating decimals), so the result is just a very good approximation. If your calculator has a [π] key, we find 2^{π} as follows:

2 [y^x] π [=] [8.824977827], or 2^{π} is approximately 8.825.

The following properties hold for real exponents (which includes the case of rational exponents).

Properties of Real Exponents

Let a, b represent positive real numbers, and m, n any real exponents. Then

$$a^m \times a^n = a^{m+n}$$
$$a^m \times b^m = (a \times b)^m$$
$$(a^m)^n = a^{m \times n}$$
$$a^m \div a^n = a^{m-n}$$
$$\frac{a^n}{b^n} = \left(\frac{a}{b}\right)^n$$

TIDBIT

If you like to see things that grow really fast, then n^n should strike your fancy. The digits needed to write out $300{,}000^{300{,}000}$ would fill a high density floppy disk (over 1.6 megabytes would be needed).

GROWTH OF EXPONENTIALS

When considering the behavior of an exponential expression, we fix either the base or the exponent and allow the other quantity to vary. **Exponential growth** occurs when the exponent varies. The difference between the behaviors that results in those two cases is enormous. Just consider the two examples n^2 (a power expression) and 2^n (an exponential expression).

TABLE 2.6

n	n^2	2^n
1	1	2
2	4	4
3	9	8
4	16	16
5	25	32
6	36	64
7	49	128
8	64	256
9	81	512
10	100	1024
⋮	⋮	⋮
15	225	32,678
⋮	⋮	⋮
20	400	1,048,576
⋮	⋮	⋮
25	625	33,554,432
⋮	⋮	⋮
30	900	1,073,741,824

While the growth of 2^n is modest at first, the values soon become astronomical. Table 2.6 shows a comparison of how exponential expressions grow much more rapidly than do power expressions. This fact is summarized next.

Comparison of the Growth of Power Expressions and Exponential Expressions

Let a be any real number greater than 1, and let m be any positive integer. Then for all sufficiently large n

$$a^n \text{ is larger than } n^m.$$

Note: a and m are constants, n is a variable.

INITIAL PROBLEM SOLUTION

Some gamblers use a system in which if the first bet, say \$10, is lost, then the second bet is \$20; that is, double the first. If the second bet is also lost, then the third bet is \$40, and so on until finally the gambler wins and gets back all the losses plus a profit of \$10 equal to the first bet. (We are assuming a game in which if the gambler wins, he or she receives a gain equal to the bet—this is called an *even money wager.*) Why is this doubling system, called a *martingale,* dangerous, if not foolish?

SOLUTION The martingale is dangerous because of the rapid growth of exponentials documented above. After only a few consecutive losses, the required bet becomes very large. If the method is pushed to its logical conclusion, the point will be reached where either the gambling establishment will not accept so large a bet or the gambler will not have enough money left to make the required bet. If the first bet is \$10, then a series of 17 losses requires the gambler to bet more than a million dollars. In practice, the likely outcome is an uncomfortably large loss ending the gaming.

PROBLEM SET 2.7

In problems 1 through 4, find the indicated square roots. Round answers to three decimal places.

1. **(a)** $\sqrt{36}$ **(b)** $\sqrt{64}$ **(c)** $-\sqrt{25}$

2. **(a)** $\sqrt{169}$ **(b)** $-\sqrt{64}$ **(c)** $\sqrt{10000}$

3. **(a)** $\sqrt{61}$ **(b)** $\sqrt{200}$ **(c)** $-\sqrt{50}$

4. **(a)** $\sqrt{45}$ **(b)** $\sqrt{160}$ **(c)** $-\sqrt{14}$

For problems 5 through 8 use the definition to find the following values in simplest form, if they exist.

5. **(a)** $\sqrt[4]{16}$ **(b)** $\sqrt[3]{-8}$ **(c)** $\sqrt[4]{-81}$

6. **(a)** $\sqrt[3]{27}$ **(b)** $\sqrt[3]{-27}$ **(c)** $\sqrt[5]{32}$

7. **(a)** $\sqrt[3]{8}$ **(b)** $\sqrt[4]{81}$ **(c)** $\sqrt[5]{-32}$

8. **(a)** $\sqrt[3]{64}$ **(b)** $\sqrt[4]{-64}$ **(c)** $\sqrt[3]{-1000}$

In problems 9 through 12, rewrite the expression using rational exponents.

9. **(a)** $\sqrt{27}$ **(b)** $\sqrt[3]{15}$ **(c)** $\sqrt[4]{x^3}$

10. **(a)** $\sqrt[3]{32}$ **(b)** $\sqrt[3]{-32}$ **(c)** $\sqrt[5]{25}$

11. **(a)** $\sqrt{45}$ **(b)** $-\sqrt[4]{64}$ **(c)** $\sqrt{x^3}$

12. **(a)** $-\sqrt[4]{100}$ **(b)** $\sqrt[3]{25}$ **(c)** $\sqrt[4]{y^3}$

In problems 13 through 16, rewrite the expressions using radicals. Simplify, if possible.

13. **(a)** $25^{1/3}$ **(b)** $10^{2/3}$ **(c)** $8^{2/3}$

14. **(a)** $4^{3/2}$ **(b)** $6^{2/3}$ **(c)** $16^{3/4}$

15. **(a)** $2x^{1/2}$ **(b)** $(25y)^{1/2}$ **(c)** $(x)^{3/2}$

16. **(a)** $4x^{1/3}$ **(b)** $6y^{2/3}$ **(c)** $(16z)^{3/4}$

For problems 17 and 18 find the following values without exponents and without using a calculator.

17. **(a)** $27^{2/3}$ **(b)** $16^{3/2}$ **(c)** $16^{-5/4}$

18. **(a)** $25^{3/2}$ **(b)** $27^{-4/3}$ **(c)** $125^{-2/3}$

In problems 19 through 22, find the value to three decimal places using a calculator.

19. **(a)** $3^{1/3}$ **(b)** $25^{2/3}$ **(c)** $8^{2.5}$

20. **(a)** $10^{3/2}$ **(b)** $5^{1.7}$ **(c)** $6^{2/3}$

21. **(a)** $\sqrt{10}$ **(b)** $\sqrt{5}^{\sqrt{2}}$

22. **(a)** $\sqrt{3}$ **(b)** $\sqrt{2}^{\sqrt{5}}$

In problems 23 and 24, use the properties of exponents to simplify the expressions.

23. **(a)** $b^{1/2} \times b^{1/6}$ **(b)** $\left(\frac{4x}{9}\right)^{1/2}$

(c) $12x^{25} \div 4x^{15}$ **(d)** $(9y^2)^{3/2}$

24. **(a)** $(5^{2/3})^6$ **(b)** $8x^2 \div 2x^{-1}$

(c) $16^{4/3} \div 16^{5/6}$ **(d)** $\left(\frac{2x^{2/3}}{3}\right)^3$

In problems 25 and 26, use the properties of exponents to find the value of the variable, *k*.

25. **(a)** $4^{2/3} \times 4^k = 4^2$

(b) $5^{3/2} \div 5^k = 25$

(c) $2^k \times 3^k = 36$

26. **(a)** $9^{3/2} \times 9^k = 81$

(b) $(2^k)^2 = 8$

(c) $25^{2/3} \div 25^k = 5$

The surface area and volume of a sphere of radius r are given by the following formulas.

Surface Area: $S = 4\pi r^2$

Volume: $V = \frac{4}{3}\pi r^3$.

27. What is the radius of a sphere that has a surface area of 100 square inches?

28. What is the radius of a sphere that has a volume of 200 cubic inches?

29. The surface area of the Earth is approximately 200,000,000 square miles. Find the approximate volume of the Earth. Hint: You must first find the radius.
30. The volume of the moon is approximately 5,300,000,000 cubic miles. Find the approximate surface area of the moon. Hint: You must first find the radius.

EXTENDED PROBLEMS

Exponential expressions are used as models for both growth (increase) and decay (decrease) of many quantities such as population growth, radioactive decay, and compound interest; subjects covered in more detail in later chapters.

31. An account earning compound interest will have a formula like the following:

$$A = 2000(1.09)^t,$$

where 2000 was the amount originally invested and t is the number of years.
   (a) Make a table of values for A with t going from 0 to 20.
   (b) When does the amount become greater than 4000? When does it become greater than 8000?
32. The amount of a radioactive material will have an equation like the following:

$$A = 100(0.93)^t,$$

where 100 is the amount originally present, and t is the time in years.
   (a) Make a table of values for A with t going from 0 to 20.
   (b) When will the amount of material present be half of the original? When will it be one-fourth?

Before there were calculators and computers, mathematicians and students used a special interpretation of exponents to solve many complex problems.

33. Write a short report on the "invention" and use of logarithms. Why is it possible to multiply, divide, and take powers of numbers with logarithms, but they cannot be used to add or subtract two numbers?
34. What properties or rules govern the use of logarithms? How do these relate to the rules for exponents? Why can't you take the logarithm of a negative number?
35. The creation of logarithms is credited to John Napier in 1614. Napier is also credited with introducing present-day notation by using the decimal point in writing numerals. Write a report on John Napier.
36. Earthquakes are a common and deadly experience in many parts of the world. The intensity of earthquakes is generally measured and reported on the Richter Scale, which is actually designed in terms of logarithms. Write a report on the Richter Scale. Are there any other methods that are used or proposed to study the intensity of earthquakes?

In problems 37 through 41, make a table like Table 2.6 for comparing the growth of power expressions and exponential expressions.

37. **(a)** Compare n^3 and 3^n.
   (b) What is the minimum value of n for which 3^n exceeds n^3?
38. **(a)** Compare n^4 and 4^n.
   (b) What is the minimum value of n for which 4^n exceeds n^4?
39. **(a)** Compare $n^{1.5}$ and $(1.5)^n$.
   (b) What is the minimum value of n for which 1.5^n exceeds $n^{1.5}$?
40. **(a)** Compare $n^{1.4}$ and $(1.4)^n$.
   (b) What is the minimum value of n for which 1.4^n exceeds $n^{1.4}$?
41. For any positive real number real a, compare the growth of a^n to n^n. What is the minimum value of n for which n^n is greater than a^n?

Chapter Two Problem

In an early legend, a wise man Sissa Ben Dasir was allowed by his king to name his own reward for inventing the game of chess. His answer seemed to ask very little, and the King thought him foolish.

A chessboard has a square grid of 64 squares, eight to a side. The wise man asked for one grain of rice for the first square, two grains for the second square, four grains for the third square, eight grains for the fourth square, and so on, doubling the number of grains of rice each time. What is the mathematical expression for the exact number of grains of rice asked for by Sissa Ben Dasir? Is this a lot of rice?

SOLUTION

Strategies: Make a table and look for a pattern.

First we make a table listing the number of grains on each of the first ten squares and the running total.

Square	Number on this square	Total
1	$2^0 = 1$	$1 = 2^1 - 1$
2	$2^1 = 2$	$3 = 2^2 - 1$
3	$2^2 = 4$	$7 = 2^3 - 1$
4	$2^3 = 8$	$15 = 2^4 - 1$
5	$2^4 = 16$	$31 = 2^5 - 1$
6	$2^5 = 32$	$63 = 2^6 - 1$
7	$2^6 = 64$	$127 = 2^7 - 1$
8	$2^7 = 128$	$255 = 2^8 - 1$
9	$2^8 = 256$	$511 = 2^9 - 1$
10	$2^9 = 512$	$1023 = 2^{10} - 1$

Observing the pattern, it appears that there will be 2^{63} grains on the 64th square and a total of $2^{64} - 1$ grains altogether. Using a calculator, we find that $2^{64} - 1 \approx 1.844 \times 10^{19}$. To understand the magnitude of this number consider the following. Assume that 200 grains of rice weigh about a gram. Then 200,000 grains weighs 1000 grams or a kilogram. A metric ton is 1000 kilograms, thus there are about 200,000,000 grains in a metric ton. The world production of rice was about 500 million metric tons in 1985 and has never been as high as a billion metric tons. So the world production of rice is less than 200,000,000,000,000,000 or 2×10^{17} grains in scientific notation. Since $2^{64} - 1$ is approximately 2×10^{19}, the amount Sissa Ben Dasir would have received is over one hundred times the high estimate for world rice production in any single year or more than all the rice produced in the world during the entire twentieth century—a large amount of rice, indeed!

Chapter Two Review

Key Ideas and Questions

The following questions review the main ideas of this chapter. Write your answers to the questions and refer to the pages listed to make certain that you have mastered these ideas.

1. Which attributes of the Hindu-Arabic numeration system make it a good system in comparison with other numeration systems discussed in the chapter? 63
2. What estimation methods give a range and which give an estimate that may be too high or too low? 70
3. What properties do whole-number addition and multiplication share? 67
   Explain how 1 plays the same role for multiplication that 0 plays for addition. 67
   Explain why there is not always a multiplicative inverse with the integers, and why we need the rational numbers. 78
4. Explain why decimals are needed to represent all numbers. 91
   Why is every rational number either a terminating decimal or a repeating decimal? 87
5. Explain how to convert among decimals, percents, and fractions. 95
6. Describe how markdowns and markups are related to percents of change. 99
7. What are the main properties of exponentials? 104 and 113
8. Discuss the similarities and differences between the properties of exponents and properties of multiplication. 102

Vocabulary/Notation

Following is a list of key vocabulary, notation, and ideas for this chapter. Mentally review each of these items, write down the meaning of each term, and use it in a sentence. Then refer to the pages listed by number and restudy any material you are unsure of before solving the Chapter Two Review Problems.

Section 2.1

Number 57
Numeral 57
Set 57
Element of a Set 57
Cardinal Number 57
Ordinal Number 57
Identification Number 57
Tally Numeration System 58
Grouping 58
Egyptian Numeration System 59
Additive System 59
Roman Numeration System 60
Multiplicative System 61
Subtractive Principle 61
Positional System 61
Hindu-Arabic System 61
Digits 61
Base 62
Place Value 62
Place Holder 62
Expanded Form 62
Expanded Notation 62

Section 2.2

Commutative Property 67
Associative Property 67
Identity for Addition 67
Distributivity 67
Compatible Numbers 68
Additive Compensation 69
Equal Additions Method 69
Multiplicative Compensation 69
Computational Estimation 70
Front-end Estimation 70
Range Estimation 70
Rounded Up 71
Rounded Down 71
Truncating 71
Round a 5 Up 71
Rounding Off 71
Round to Compatible Numbers 72

Section 2.3

Whole Numbers 75
Integers 75
Integer Number Line 75
Addition 75
Additive Identity 76
Additive Inverses 76
Subtraction 76
Multiplication and Division 76
Properties of Integer Operations 77
Rational Number 78
Numerator 78
Denominator 78
Positive Rational Numbers 78
Negative Rational Numbers 78
Fractions 78
Simplest Form of a Rational Number 78
Rational Number Simplification 79
Rational Number Equality 79
Rational Number Addition and Subtraction 80
Multiplicative Inverse, Reciprocal 81
Multiplicative Identity 81
Rational Number Multiplication and Division 81
Properties of Rational Number Operations 82

Section 2.4

Decimal Fraction 86
Decimal Point 86
Decimal 86
Terminating Decimal 87
Repeating Decimal 87
Ordering Decimals 88
Irrational Number 91
Real Number 91
Real Number Line 91

Section 2.5

Percent 94
Percentage 95
Fraction Equivalents 97
Percent Change 97
Percent Increase 98
Percent Decrease 98
Percent Markup 99
Percent Markdown 99

Section 2.6

Exponentiation 102
Exponents 103
Power 103
Base 103
Five General Rules of Exponents 104
Common Logarithm 105
Base of Natural Logarithms 105
Natural Logarithm 105
Negative and Zero Exponents 106
Scientific Notation 107

Section 2.7

Squaring 110	Radicand 111	Unit Fraction Exponent 112
Square Root 110	Index 111	Rational Exponent 112
Principal Square Root 110	*n*th Root 111	Real Number Exponents 113
Cube Root 111	Radical 111	Exponential Growth 114

Chapter Two Review Problems

1. How are the following numbers used—as cardinals, as ordinals, or as identification?
   (a) The Ionian runner came in third place in the marathon.
   (b) The Ionian runner had a time of 2 hours, 31 minutes.
   (c) The Ionian runner wore jersey number 17 in the race.
2. Which of the ancient numeration systems are additive systems?
3. Which of the ancient numeration systems are positional systems?
4. Convert the Roman numeral XIX to Hindu-Arabic.
5. Convert 1914 to a Roman numeral.
6. Convert MMVII to an Egyptian numeral.
7. Write out 9876543210 in words and then express it in expanded form.
8. Give a quick mental calculation of the following expressions using the properties of arithmetic. Which properties are you using? When are you using compatible numbers?
   (a) $(117 + 49) - 17$ **(b)** $(2 \times 791) \times 50$
   (c) $14 \times 17 + 17 \times 6$
9. Find the following in your head using compensation.
   (a) 88×25 **(b)** $1998 + 5114$
10. Describe the following calculation using left to right computation: $458 + 293 + 116 = ?$
11. Give an estimate of the sum in problem 10 using: range estimation, round off to the nearest 100, round a 5 up to the nearest ten.
12. Do the following calculations and reduce your answers to simplest form.
    (a) $\frac{3}{9} + \frac{2}{12}$ **(b)** $\frac{-7}{12} + \frac{4}{9}$
    (c) $\frac{4}{15} \times \frac{5}{8}$ **(d)** $\frac{10}{18} \div \frac{5}{3}$
13. Which properties of rational number operations are used below?
    (a) $\frac{7}{2} \times \frac{2}{7} = 1$ **(b)** $\frac{4}{3} \times \frac{2}{9} = \frac{4 \times 2}{3 \times 9}$
    (c) $2 \times \left(\frac{7}{2} - \frac{4}{2}\right) = 7 - 4.$
14. Convert $\frac{3}{7}$ to a decimal.
15. Convert 1.12121212 . . . to a fraction.
16. Round.
    (a) 137.26511 to the nearest hundredth.
    (b) 0.00123456 to the nearest ten-thousandth.
17. Express the following in the other two equivalent forms.
    (a) 40% **(b)** 2.95 **(c)** $\frac{3}{8}$
18. Determine if the two numbers shown are equal and, if not, which is the larger.
    (a) 9/10 or 10/11 **(b)** 3.91540006 or 3.9149996
19. Solve.
    (a) Find 128% of 41.5.
    (b) What percent is 12 of 7?
    (c) 63% of what number is 330?
20. Estimate, using fraction equivalents.
    (a) 149% of 41 **(b)** 65% of 119
21. A car dealership has a markup of 15% on the cars it offers for sale. What will be the sticker price on a car that it purchases for \$9718? Suppose that during a sale the dealership has a 20% markdown. What will be the sale price?
22. Simplify.
    (a) $3^3 \times 2^3 \times 5^3$ **(b)** $(5^3)^2$
    (c) $(2^3)^5$ **(d)** $5^3 \times 5^2$
23. Express in scientific notation.
    (a) 0.000026744 **(b)** 186,360,000
24. Simplify without using a calculator.
    (a) $\left(2^{\sqrt{2}}\right)^{\sqrt{2}}$ **(b)** $(1/81)^{(-3/4)}$
25. Find the values to three decimal places on a calculator.
    (a) $21^{\pi/4}$ **(b)** $2^{\sqrt{2}}$

CHAPTER 3

DESCRIPTIVE STATISTICS—DATA AND PATTERNS

CRIME FIGHTING ACTION PROPOSED

During his reelection campaign Senator Sternmeister presented data to show that increasing amounts of crime are committed by immigrants from the country of Incanda. The chart shows the number of crimes committed by immigrants over the last 7 years. Many immigrants have come from Incanda due to political instability and military rule beginning late in 1985. The senator believes quotas for immigrants should be sharply reduced and the rules in accepting political refugees should be tightened.

Chapter Goals

1. Graph data using various types of graphs to display and compare data.
2. Graph two or more sets of data using comparison graphs.
3. Identify graphs that have been distorted and explain how to correct the distortions.
4. Compute numerical summaries of data such as means and medians.

The statement by the senator has the probable effect, perhaps by design, of causing people to fear immigrants from Incanda and elsewhere. It takes advantage of existing concerns regarding immigration and crime. In the middle of a campaign, the senator also hopes it will cause voters to choose a senator who will protect both their homes and the shores of the United States. In order to fairly understand this situation more information is needed than is given by the graph. The graph shows that the total number of crimes committed by immigrants has gone up, but what the graph does not show is that the increase in the number of crimes is primarily due to the increase in the number of immigrants. A closer study would reveal that the percentage of immigrants committing violent crimes is actually less than that of the population at large. The number of crimes committed by immigrants from Incanda is also a very small fraction of the total number of crimes committed. Clearly the crime problem cannot be solved simply by attacking this small group. This illustrates the use of a graphical display to create a false impression, even though the data presented is true.

In this chapter you will learn to use graphs to display and compare data, and you will become sensitive to the ways those graphs can be misused. You will also learn about the most common numerical expressions used to summarize data.

THE HUMAN SIDE OF MATHEMATICS

John Playfair (1759–1823) was a writer with no mathematical background who was interested in swaying the minds of the people, essentially by producing propaganda. Playfair lived during a time of great revolutions in ideas and governments, both in Europe and America. He saw the value of persuading the common people instead of merely the aristocracy. Rather than using long intricate arguments full of calculations and tables, he discovered that a chart with good visual design could provide more information and much more impact—"a picture is worth a 1000 words." Playfair also found ways to graph data in such a way that he could exaggerate the point that he was trying to make. His work was so revolutionary that it soon permeated the ways people communicate.

John Playfair

Visual displays have become one of our primary methods for communicating quantitative information. In books, newspapers, magazines, and television, we are regularly presented with a wide variety of graphs, charts, and other visuals intended to inform, impress, or persuade the viewer. Playfair invented the standard graph forms: bar charts, line graphs, and circle graphs. Few of us realize that so many of these graphs are the result of one person's genius.

John Tukey (1915–) was a child prodigy who was home-schooled until he entered college. He earned a doctorate in mathematics at Princeton on a topic so abstract that it was not considered to have any practical value. One early theorem of his, called the *Ham-Sandwich Theorem,* proved that any three loaves of bread fixed in space may be simultaneously sliced so that a piece of ham will separate equal amounts of bread on either side. During World War II he changed the direction of his work to graphical display and statistics. He was extremely prolific and made important contributions in a variety of fields ranging from astrophysics to global pollution.

John Tukey

Most of the new graphic forms developed during this century, such as stem and leaf plots and box and whisker plots, were due to Tukey. However, these graphs are only a small part of Tukey's work. He is the leading statistician of the modern age and is credited with liberating statistics from the straitjacket of abstract mathematics. Tukey also invented the term *bit* for the unit of storage in computers.

Nearly all of the graphs and charts that we commonly use to communicate and analyze information are the invention of two people, one born in the 18th century with no background in mathematics and one in the 20th century with a doctorate in one of the most abstract areas of mathematics.

3.1 ORGANIZING AND PICTURING DATA

INITIAL PROBLEM

You have to give a sales report that shows the sales figures of each of three districts having markets that are roughly equal in size. In 1994, district A had \$135,000 in sales, district B had \$85,000 in sales, and district C had \$115,000 in sales. How should you present this data so the comparison of each district to the others is clearly shown?

Statistics is the science (and art) of making sense out of data. Our world is filled with information, especially numerical information. This chapter describes how to organize sets of numbers, called **data sets,** into sensible visual patterns and charts (data sets may include repeated elements). It also shows how the eye may be misled by such charts. Other ways to summarize data are also discussed in this chapter.

TIDBIT

It has been estimated that each year there are between 1 and 3 trillion charts printed; this is on the order of 400 for each human on Earth.

DOT PLOTS

Consider the following set of numbers:

{80, 74, 87, 62, 96, 87, 71, 93, 32, 76, 26, 81, 84, 54, 70, 87, 89, 71, 95, 67}

What can we say about it? Very little on the surface. There is no context to give meaning to these numbers and there is no clear trend or pattern to these numbers. Actually, the most noticeable aspect of this set of numbers may be that they are boring. To put things into a context, suppose that these numbers are the scores of a test taken by an economics class. We can see that there are 20 numbers between 100 and 0 and that the highest number is 96, while the lowest number is 26. Since these numbers represent test scores, we may also make the assumption that the person who got 96 did well, while the person who got 26 did not.

As a first step, this data can be organized by putting the numbers in order. Usually we arrange numbers from lowest to highest as follows:

{26, 32, 54, 62, 67, 70, 71, 71, 74, 76, 80, 81, 84, 87, 87, 87, 89, 93, 95, 96}

Another way to represent this data is to draw a graph. The following **dot plot** can be used to get an initial graphical view of the data (Figure 3.1).

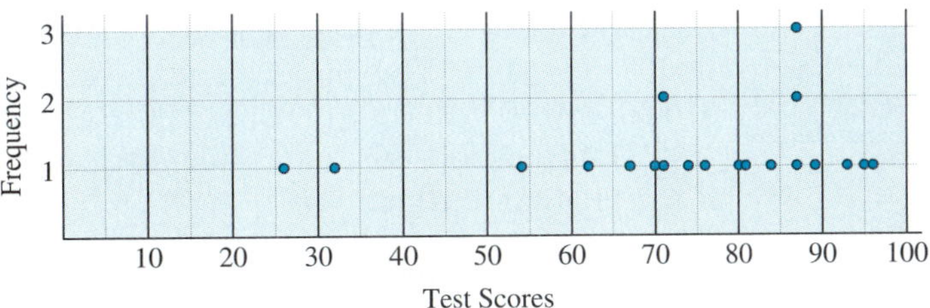

FIGURE 3.1

TIDBIT

Stem and leaf plots were invented by the father of modern data analysis, John Tukey. He wrote, "If we are going to make a mark, it may as well be a meaningful one. The simplest—and most useful—meaningful mark is a digit."

Notice that one dot is placed above 26 since there is one 26 in the data set, whereas there are three dots above 87 to represent the three 87's in the data set. Observe how a dot plot can be used to visually identify numbers that occur most (the tallest column of dots) as well as gaps and numbers that are widely separated from others.

STEM AND LEAF PLOTS

Another popular method of arranging data is to use a stem and leaf plot. To make a stem and leaf plot of the data from the economics class, first arrange the numbers by 10's (Table 3.1).

TABLE 3.1

90's	93, 95, 96
80's	80, 81, 84, 87, 87, 87, 89
70's	70, 71, 71, 74, 76
60's	62, 67
50's	54
40's	
30's	32
20's	26
10's	
0's	

TABLE 3.2

9	3 5 6
8	0 1 4 7 7 7 9
7	0 1 1 4 6
6	2 7
5	4
4	
3	2
2	6
1	
0	

Notice that there are redundancies in Table 3.1 in that the tens digits are repeated in each row. Also, there are blank rows because there are no scores in the 0's, 10's, or 40's. This list can be presented more efficiently by using the **stem and leaf plot** at the left (Table 3.2). Notice how the tens digits are placed down the vertical "stem" of the plot and that the ones digits are arranged in order as "leaves." This visual way of arranging the data allows us to see at a glance the relative sizes of each category. For example, there are more scores in the 80's than in any other category. Also, most of the data is grouped in a general **cluster** between 54 and 96. There is a large **gap** between 54 and the scores 32 and 26. Scores that are separated from the others by gaps, such as 32 and 26, are called **outliers.** Outliers may be the largest or smallest numbers in a set and are usually indicative of data points with different or atypical conditions. In a stem and leaf plot, cluster, gap, and outlier are imprecise terms that might be interpreted differently by different people. However, they can often reveal useful information such as in the following example.

EXAMPLE 3.1 A pizza delivery person has delivered 10 pizzas in the first two hours of a shift. The prices of pizza delivered in increasing order were \$9.20, \$10.50, \$10.70, \$10.80, \$10.80, \$12.00, \$12.10, \$12.20, \$12.20, \$12.30. Make a stem and leaf plot of this data. Identify any clusters and gaps and suggest an interpretation for them.

TABLE 3.3

12	0 1 2 2 3
11	
10	5 7 8 8
9	2

SOLUTION In the stem and leaf plot use the dollar amounts as the stem and tens of cents as the leaves. For example, in the stem and leaf (Table 3.3) 10|5 represents \$10.50. Notice that there are two clusters separated by a gap. One might conjecture that the small pizzas cost about $9 or $10, while the large pizzas cost about $12. In this interpretation there are roughly the same number of small pizzas ordered as large pizzas. ◆

Most graphs are designed to show only the general pattern of the data. One advantage of the stem and leaf plot is that all the data is still contained in the graph and is displayed in a way that makes it easy to see.

HISTOGRAMS

Histograms group data into intervals called **bins.** For example, the data from the stem and leaf plot in Table 3.2 is grouped into bins of length ten, such as 80 to 89 (or length eleven, in the case of 90–100). The number of data points in each bin is called the **frequency** of the interval. The information is collected into a table called a **frequency table** (Table 3.4). Using the data in a frequency table, a **histogram** is plotted by placing a bar with height equal to the frequency of a bin above the interval for each bin in the graph (Figure 3.2).

Interval	Frequency
90–100	3
80–89	7
70–79	5
60–69	2
50–59	1
40–49	0
30–39	1
20–29	1
10–19	0
0–9	0

FIGURE 3.2 Economics test histogram with bins of length 10

TIDBIT

Not every graph was invented by Playfair and Tukey. Florence Nightingale made a radial histogram to show death rates in the Crimean War. Her graphs led to an improvement of medical care.

Notice that if you were given the histogram in Figure 3.2, you could reconstruct the frequency table. Although the horizontal grid lines make the graph easier to read, they may make it more cluttered. Their use is optional.

Histograms show the general picture and suppress specifics of the data. For example, the outliers in Figure 3.2 between 20 and 39 are notable because they are separated from most of the data. However, we cannot tell what their precise values are. Notice that a dot plot can be viewed as a histogram where the bins are intervals of length one.

When working with histograms, it often takes some judgment in determining which intervals are best to use. If we take the data from the economics class and group into 5's, using the intervals of 30–34, 35–39, 40–44, and so on, we get the following histogram (Figure 3.3).

The intervals are a little too small to tell the general pattern easily, and we also have a large number of bins. We can also group the data by 20's, using the intervals 20–39, 40–59, 60–79, 80–100 (Figure 3.4).

These intervals are so large that they obscure much of the information about the class. The histograms in Figures 3.2, 3.3, and 3.4 represent the same data, yet give very different views of the data. Thus, choosing an appropriate-sized interval is critical. It is somewhat like focusing a pair of binoculars.

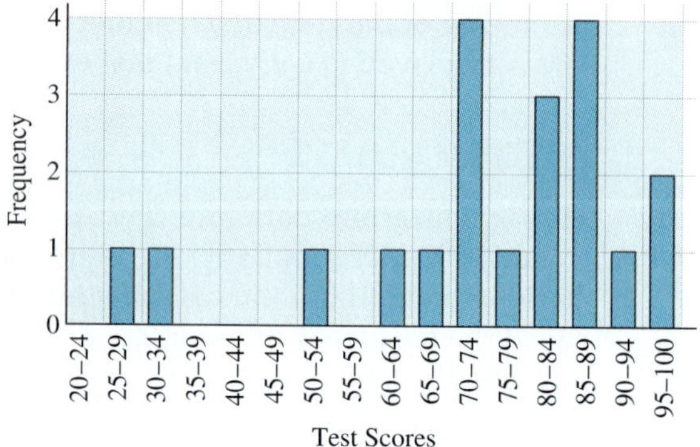

FIGURE 3.3 Economics test histogram with bins of length 5

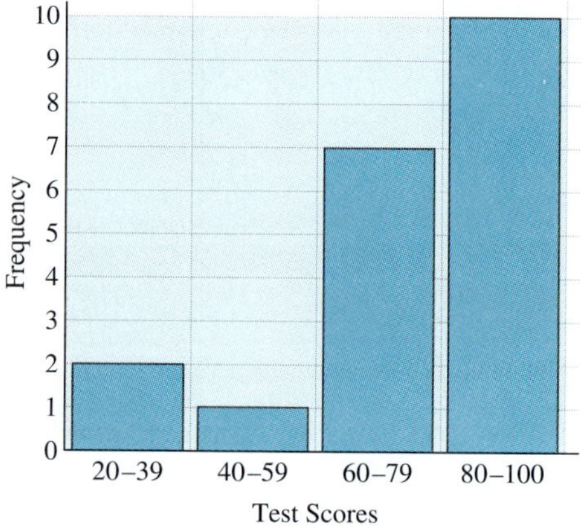

FIGURE 3.4 Economics test histogram with bins of length 20

EXAMPLE 3.2 Make a histogram of the pizza data from Example 3.1: \$9.20, \$10.50, \$10.70, \$10.80, \$10.80, \$12.00, \$12.10, \$12.20, \$12.20, \$12.30.

SOLUTION Since the prices range between \$9 and \$13, we group the prices into bins whose length is one dollar, so that pizzas costing \$9.00 to \$9.99 go into the first bin, \$10.00 to \$10.99 go into the second bin, etc. (Figure 3.5). Then we count the number of data points in each bin.

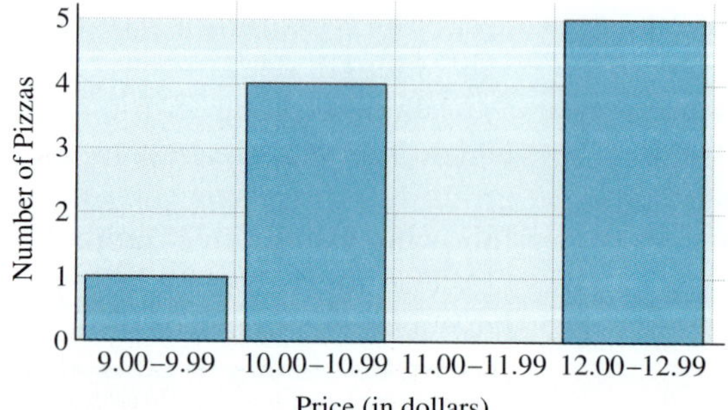

FIGURE 3.5 Pizza prices

If we want to get a finer picture of this data we could group into smaller bins, say bins whose length is 50 cents (Figure 3.6).

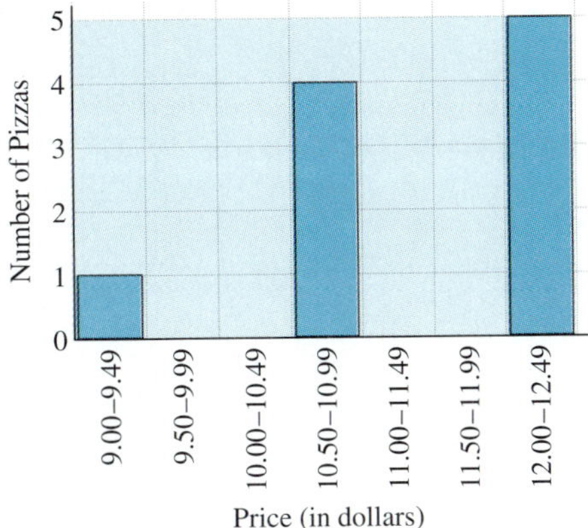

FIGURE 3.6 Pizza prices

In Figure 3.6 an outlier becomes apparent. It could represent a choice of small pizza with no toppings that was very cheap (but not so popular). ◆

BAR GRAPHS

The bars in a histogram represent the number, or frequency, of data in the given bins. However, other information can also be represented using bars. Thus, histograms are a special case of a large class of graphs called **bar graphs** or **bar charts.** A bar graph is any graph in which the length of bars is used to represent frequencies or quantities.

The following table suggests financial rewards of an education (Table 3.5). A bar graph of the data in Table 3.5 is displayed in Figure 3.7.

TABLE 3.5

Educational Attainment of Householder	1991 Median Income in Thousands of Dollars
Less than 9th grade	13
More than 9th grade, no degree	17
High School Graduate	28
Associate Degree	40
Bachelor's Degree	49
Professional Degree	78
Master's Degree	55
Doctorate	70

Notice how the heights of the bars provide a quick visual summary of the data.

When we make a bar graph, the vertical scale should be just a bit larger than the largest data point. In the last case the vertical scale goes from 0 to $80,000 because the largest data point is $77,949. This gives a fair comparison of the amounts. The widths of the intervals and the bars are chosen so that the graph fits in the available space and so that it "looks right."

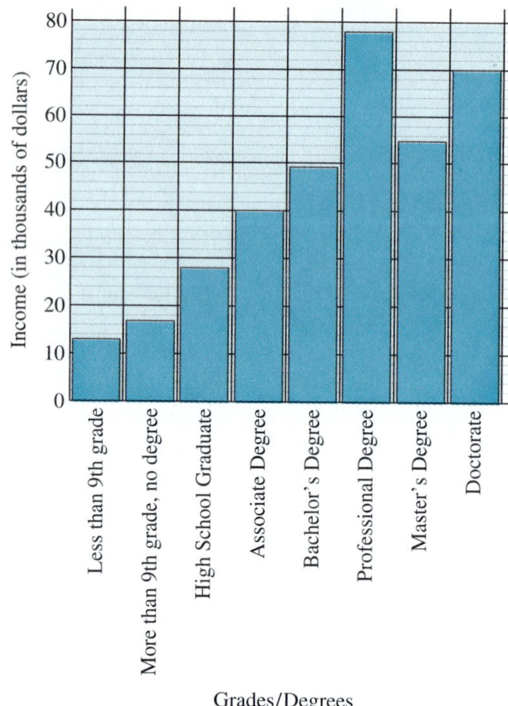

FIGURE 3.7 Income based on educational level

Notice that this information does not prove that more education will increase your income, although it is highly suggestive. It is possible that this merely reflects the possibility that people who have greater opportunities for education also have greater opportunities to make a lot of money.

EXAMPLE 3.3 Some people believe that a person's birthdate influences the chance of playing in a professional sports league. Table 3.6 shows the number of players in the FA Premier Soccer Leagues in England who have birthdays in each season. This data shows the association between professional soccer status and birthdates. Make a bar graph to more vividly illustrate this relationship.

TABLE 3.6

Birthdate Quarter	Number of Players
September–November	288
December–February	190
March–May	147
June–August	136

SOLUTION We make a bar graph whose vertical axis goes from 0 to 300 since the category with the largest number is 288. The chart shows that players whose birthdates are in the September–November range are most likely to play in these leagues (Figure 3.8).

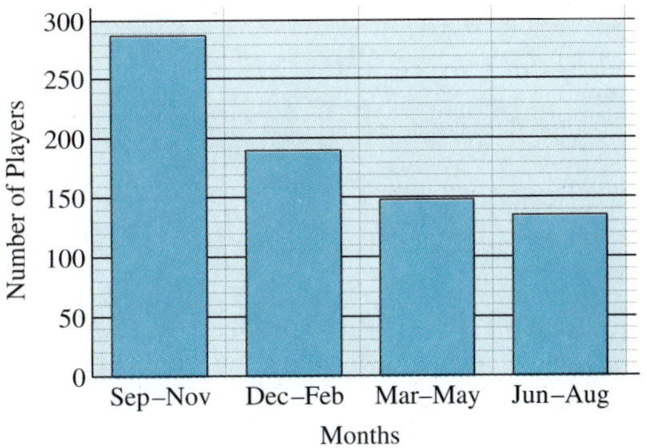

FIGURE 3.8 Birth months of professional soccer players

One fact that may be relevant is that in England children who play soccer are grouped according to age on the first day of the soccer season, which is September 1. ◆

Notice how the bar graph in Figure 3.8 provides a visual picture of the data in Table 3.6, making it easier to see relationships among the data.

Bar graphs are often used to show trends in time. In Figure 3.9, funding for a university is growing over the period from 1988 to 1992.

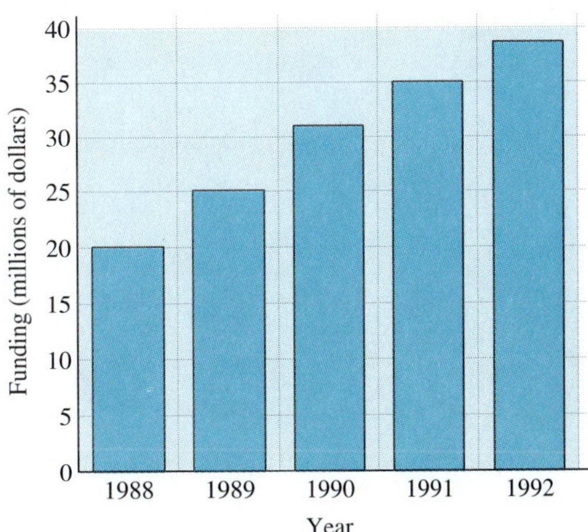

FIGURE 3.9 University funding

EXAMPLE 3.4 Use the bar graph in Figure 3.9 to determine the level of university funding for each year from 1988 through 1992. Which year showed the greatest increase over the previous year?

SOLUTION Reading across the horizontal lines and estimating whenever necessary, we have 1988, 20; 1989, 25; 1990, 31; 1991, 35; 1992, 38. The year that showed the greatest increase over the previous year was 1990 with an increase of approximately $6,000,000 over 1989. ◆

LINE GRAPHS

A line graph is often used to plot data over time. This type of graph shows trends and variation. For example, the same data for the university funding levels shown in Figure 3.9 may be plotted as a **line graph** (Figure 3.10).

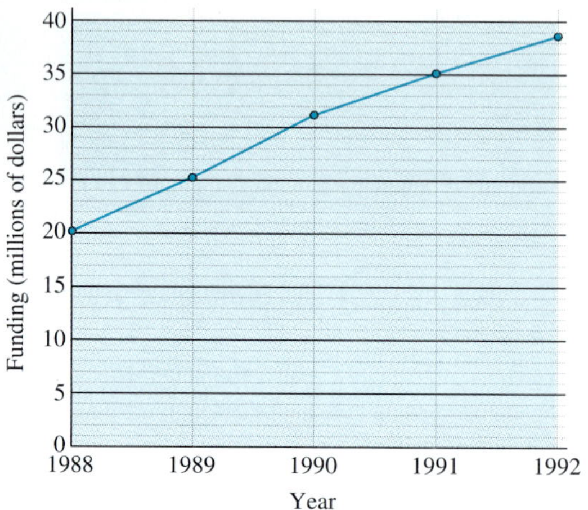

FIGURE 3.10 University funding

It is clear from the bar graph in Figure 3.9 that the trend in funding is up. However, the line graph in Figure 3.10 shows that the rate of increase in the 90s is less than it was in the 80s because the line segments are not as steep.

EXAMPLE 3.5 Draw a line graph of the federal debt over the years from 1980 to 1989 (Table 3.7).

TABLE 3.7

Year	Federal Debt (in billions of dollars)
1980	914
1981	1004
1982	1147
1983	1382
1984	1577
1985	1827
1986	2130
1987	2345
1988	2615
1989	2881

TIDBIT

By 1993, the federal debt had grown to more than 4.5 trillion dollars, or more than \$16,800 per capita.

SOLUTION We use a line graph and plot the data (Figure 3.11).

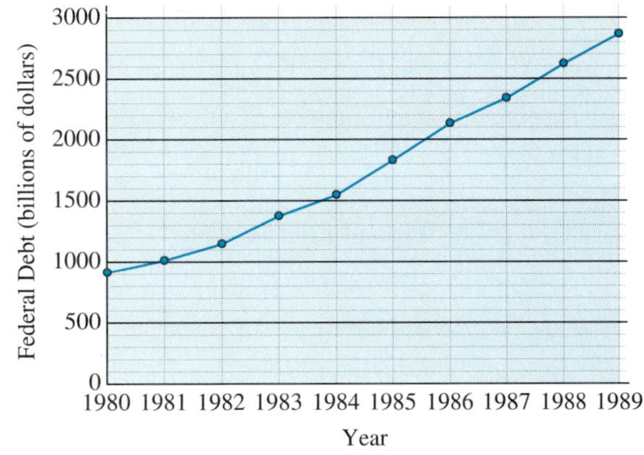

FIGURE 3.11 ◆

The line graph in Figure 3.11 clearly shows the steady upward climb of the national debt.

TIDBIT

The following graph was used to display sales in a bakery at a board of directors meeting.

PIE CHARTS

Pie charts or **circle graphs** are often used to show relative proportions of quantities. The following pie chart shows the average amount of sleep people get during a night (Figure 3.12).

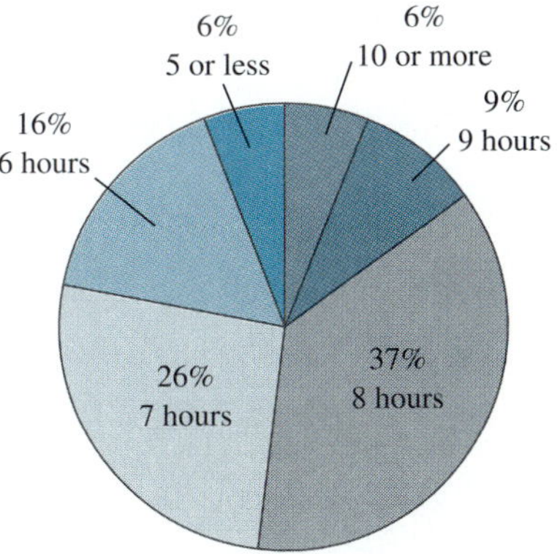

FIGURE 3.12 SLEEP TIMES

It is clear from the pie chart that most people sleep 7 or 8 hours a night. Also, a surprisingly high percentage of people (6%, or more than one out of 17 people) need 5 hours or less of sleep each night.

A pie chart is constructed by finding what portions of a circle each part should represent. In the chart in Figure 3.12, the percentages 6%, 6%, 9%, 16%, 26%, and 37% are represented. Since there are 360° in a circle, the angles of the respective pieces are $6\% \times 360° \approx 22°$, $6\% \times 360° \approx 22°$, $9\% \times 360° \approx 32°$, $16\% \times 360° \approx 58°$, $26\% \times 360° \approx 94°$, and $37\% \times 360° \approx 133°$. (Note: Due to rounding, the sum of these six angles is over 360°, but close enough for measuring angles with a protractor.)

Pie charts are especially useful for information such as budget expenditures because it gives a way to present proportions that are nearly free of distortion.

We summarize the graphs presented in this section in Table 3.8 and give the uses for which they are best suited.

Type of Chart/Graph	Use
Histogram	Displays grouped data
Bar Chart	Displays trends and amounts as lengths
Line Graph	Displays trends and variation
Pie Chart	Displays proportions

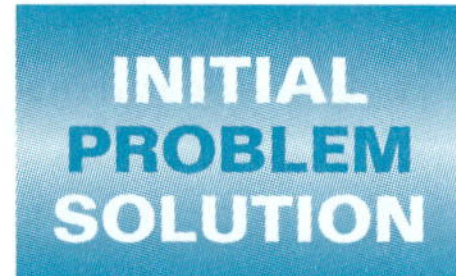

You have to give a sales report that shows the sales figures of each of three districts having markets that are roughly equal in size. In 1994, district A had \$135,000 in sales, district B had \$85,000 in sales, and district C had \$115,000 in sales. How should you present this data so the comparison of each district to the others is clearly shown?

SOLUTION Although the sales data could be presented as a bar chart, it is better to use a pie chart to show the comparisons because it will show the proportion of sales for each of the districts. To do this, we first compute the total sales: \$135,000 + \$85,000 + \$115,000 = \$335,000. Then we must find the relative percentage for each district. Finally, we find what portion of a circle (360°) each of the district sales represents.

$$135{,}000/335{,}000 \approx 0.402985 \approx 40.3\% \text{ and } 40.3\% \times 360^\circ \approx 145^\circ$$
$$85{,}000/335{,}000 \approx 0.253731 \approx 25.4\% \text{ and } 25.4\% \times 360^\circ \approx 91^\circ$$
$$115{,}000/335{,}000 \approx 0.343283 \approx 34.3\% \text{ and } 34.3\% \times 360^\circ \approx 123^\circ$$

Sketching this gives the pie chart in Figure 3.13.

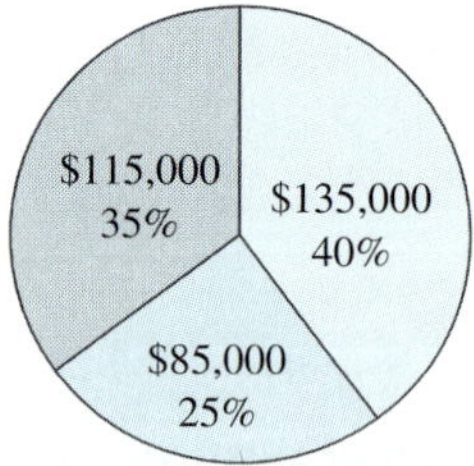

FIGURE 3.13

PROBLEM SET 3.1

1. The scores on a chemistry midterm were as follows:

64	87	76	68	92	88	82	75	51
90	84	83	77	82	70	75	73	92
81	84	74	80	84	97	57	75	86

(a) Make a dot plot of the set of scores.

(b) Make a stem and leaf plot of the scores.

2. Students in a health and wellness class took their resting pulse rates with the following results:

62	74	56	68	48	57	64	58	76
65	82	74	68	63	66	74	62	57
84	72	60	74	78	66	55	64	

(a) Make a dot plot of the set of scores.
(b) Make a stem and leaf plot of the scores.

3. A sample of starting salaries for recent graduates of a university's accounting program were as follows:

\$23,500	\$25,450	\$22,800	\$26,750
\$25,100	\$27,000	\$24,245	\$25,600
\$24,800	\$25,380	\$25,400	\$23,820
\$24,750	\$24,180	\$26,300	\$25,200

(a) Make a dot plot of the salaries after rounding each to the nearest 100.
(b) Make a stem and leaf plot of the scores. Show the leaves in terms of 100's.

4. In 1798 the English scientist Henry Cavendish measured the density of the earth in an experiment with a torsion balance. He made 29 repeated measurements with the same instrument and obtained the data given below. Note: Theoretically, the value would be the same each time; however, small differences are introduced due to measurement error.

5.50	5.61	4.88	5.07
5.26	5.55	5.36	5.29
5.58	5.65	5.57	5.53
5.62	5.29	5.44	5.34
5.79	5.10	5.27	5.39
5.42	5.47	5.63	5.34
5.46	5.30	5.75	5.68
5.85			

SOURCE: Annals of Statistics, 5:1055–1078, 1977.

(a) Make a dot plot of Cavendish's experimental data.
(b) Make a stem and leaf plot of Cavendish's experimental data.

5. Make a stem and leaf plot of the following averages. Is there a general pattern or shape?

American League Batting Champions (1970–1994)

1970	Johnson	.329
1971	Oliva	.337
1972	Carew	.318
1973	Carew	.350
1974	Carew	.364
1975	Carew	.359
1976	Brett	.333
1977	Carew	.388
1978	Carew	.333
1979	Lynn	.333
1980	Brett	.390
1981	Lansford	.336
1982	Wilson	.332
1983	Boggs	.361
1984	Mattingly	.343
1985	Boggs	.368
1986	Boggs	.357
1987	Boggs	.363
1988	Boggs	.366
1989	Puckett	.339
1990	Brett	.328
1991	Franco	.341
1992	Martinez	.343
1993	Olerud	.363
1994*	O'Neill	.359

* shortened season (baseball strike)
SOURCE: 1995 Information Please Almanac.

6. Make a stem and leaf plot of the following averages. Is there a general pattern or shape?

National League Batting Champions (1970–1994)

1970	Carty	.366
1971	Torre	.363
1972	Williams	.333
1973	Rose	.338
1974	Garr	.353
1975	Madlock	.354
1976	Madlock	.339
1977	Parker	.338
1978	Parker	.334
1979	Hernandez	.344
1980	Buckner	.324
1981	Madlock	.341
1982	Oliver	.331
1983	Madlock	.323
1984	Gwynn	.351
1985	McGee	.353
1986	Raines	.334
1987	Gwynn	.370
1988	Gwynn	.313
1989	Gwynn	.336
1990	McGee	.335
1991	Pendleton	.319
1992	Sheffield	.330
1993	Galarraga	.370
1994*	Gwynn	.394

* shortened season (baseball strike)
SOURCE: 1995 Information Please Almanac.

7. Make a histogram for the data in problem 1. Use bins of size 10, beginning with 50–59.

8. Make a histogram for the data in problem 2. Use bins of length 10 beginning with 40–49.

9. Make a histogram for the data in problem 3. Use bins of size 1000, beginning with 22,000–22,999.

10. Make a histogram for the data in problem 4. Use bins of length 0.10 beginning with 4.80–4.89.

The following information and data refer to problems 10 and 11.

What's the state of your state's health? Some states are healthier to live in than others. A study used data from government agencies and health organizations to rate the states on 17 statistical measures of health. Included were such things as smoking, traffic death rates, violent-crime rates, motor-vehicle death rates per 100,000 miles driven, incidence of major infectious diseases, life expectancy at birth and access to health care. The higher scores represent "healthier" states.

State	Score	State	Score
Minnesota	120	Oklahoma	100
Utah	120	Wyoming	100
New Hampshire	119	Delaware	100
Hawaii	118	Missouri	100
Nebraska	116	Washington	99
Connecticut	116	Texas	99
Massachusetts	114	North Carolina	98
Wisconsin	114	Idaho	97
Iowa	114	Georgia	96
Kansas	113	Tennessee	96
Colorado	113	New York	96
Vermont	110	Illinois	96
North Dakota	110	Kentucky	94
Maine	110	Alabama	93
Virginia	109	Arkansas	93
New Jersey	108	Arizona	93
Rhode Island	107	South Carolina	93
Montana	106	Oregon	91
Ohio	105	Florida	90
Pennsylvania	104	New Mexico	88
Indiana	104	Louisiana	88
California	103	Nevada	87
Michigan	102	Mississippi	87
South Dakota	101	West Virginia	85
Maryland	101	Alaska	84

SOURCE: Northwestern National Life Insurance Company.

11. (a) Make a stem and leaf plot of the data.

(b) Make a histogram using bins of length 10, starting with 80–89.

12. (a) Make a histogram using bins of length 10 starting with 81–90.

(b) Make a histogram using bins of length 5 starting with 80–84.

13. The following histogram shows the frequencies of various scores received by students on a 10-point quiz in Psychology 121.

(a) Make a frequency table for the scores.

(b) How many students took the quiz?

14. The following histogram shows the number of days of frost (minimum temperature below freezing) in Greenwich, England, in the month of April over an extended period.

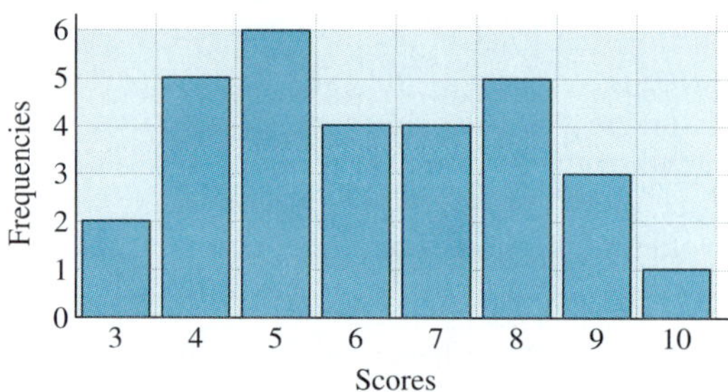

(a) Make a frequency table for the scores.

(b) How many years were included in the data?

15. A 1989 survey showed that American kids received the following weekly allowances based on age:

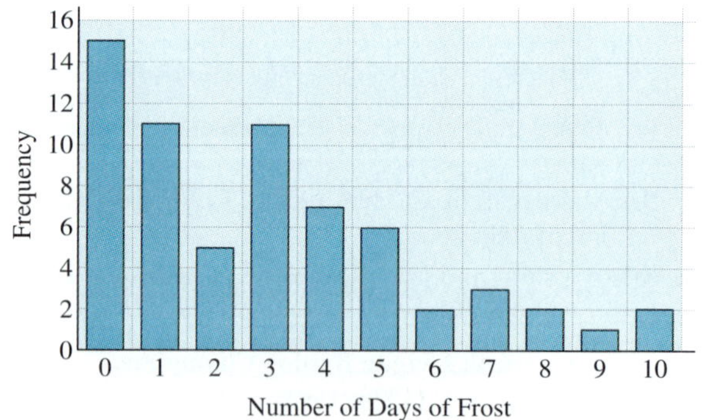

Ages	Amount
6 to 7	$ 1.98
8 to 9	4.15
10 to 11	7.82
12 to 13	17.44
14 to 15	32.44

SOURCE: Knight Rider News Service, 1989.

Draw a bar graph for this data.

16. In 1990, the four most populous nations in the world were

China	1.1 billion
India	833 million
Soviet Union	289 million
United States	250 million

SOURCE: U.S. Census Bureau.

(a) Draw a bar graph for this information.
(b) Explain why you would not use a pie chart.

17. According to the *Statistical Abstract of the United States*, in 1987, more people went to the opera (17.7 million) and the symphony (23.3 million) than went to a National Basketball Association game (14 million), a National Football League game (17 million), or a National Hockey League game (12.4 million). Draw a bar graph for this information.

18. What do Americans spend on gifts? A survey by the Gallup poll for the gift and stationery industry found that the average Christmas gift for a close family member or friend cost $55.50. Wedding gifts came in second, costing $47.90, followed by anniversary gifts at $44.10, and birthday gifts in fourth place at $30.70. Draw a bar graph for the data.

19. Drugs in the workplace have been reported to cause the following percentage of problems:

Absenteeism	54%
Accidents	30%
Increase in medical expenses	30%
Insubordination	30%
Thefts	36%
Product or service problems	33%

Draw a bar graph for the data.

20. The most common things we are allergic to that we eat are:

Dairy	40%
Seafood	21%
Vegetables	20%
Fruits	20%
Chocolate	11%

SOURCE: NPD/Home Testing Institute.

Draw a bar graph for this information.

21. The percentage of participants in low-impact fitness activities who are women.

Fitness walking	65%
Stationary biking	59%
Ski machines	57%
Treadmills	55%

SOURCE: American Sports Data Incorporated.

Draw a bar graph for this data.

22. Average weekly grocery cost per person:

Size of household:	
1 person	$45
2 people	$35
3 to 4 people	$28
5 or more	$22

SOURCE: Food Marketing Institute.

Draw a bar graph for weekly grocery costs.

23. Satisfied Americans

Percentage of people, in 1988 who were satisfied with the following:

Family Life	94%
Health	88%
Free Time	87%
Housing	87%
Standard of Living	85%
Job	76%
Household Income	69%

SOURCE: The Gallup Organization.

Draw a bar graph for this information.

24. **Major Causes—Accidental Death in 1991:**

Motor vehicle	43,500
Falls	12,200
Drowning	4,600
Fire burns	4,200
Ingestion of food or object	2,900
Firearms	1,400
Poisoning	6,400

SOURCE: National Safety Council.

Draw a bar graph for this data.

25. The bar graph shows how the population of the United States changed from 1790 to 1990.

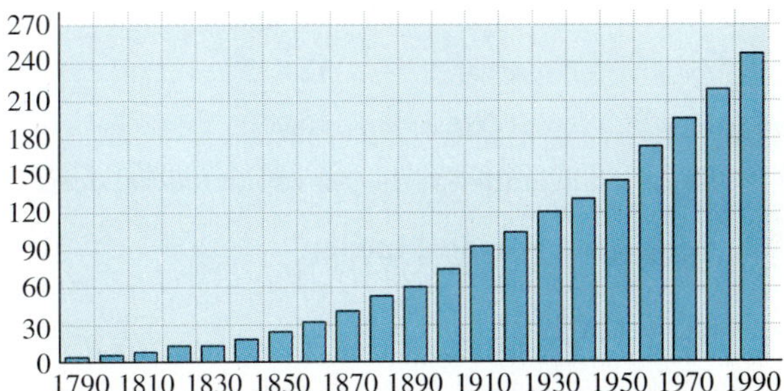

Source: U.S. Department of Commerce, Bureau of the Census, *Statistical Abstract of the United States*, 1991.

(a) Estimate the population of the U.S. in 1790, 1890, and 1990
(b) What was the change in population from 1790 to 1890?
(c) What was the change in population from 1890 to 1990?
(d) What was the percentage change in population from 1790 to 1890?
(e) What was the percentage change in population from 1890 to 1990?

26. The bar graph shows the energy consumed (in Btu's) to make a 12-ounce beverage container.

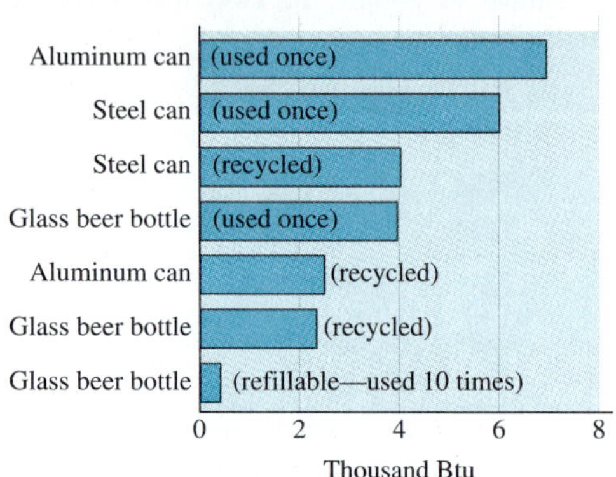

(a) Approximately how many times more energy is used for each aluminum can that is used only once compared to a recycled aluminum can?
(b) Approximately how many times more energy is used for each glass beer bottle that is used only once compared to a recycled glass beer bottle?

27. The federal debt on a per capita basis is given.

1950	$1,688
1955	1,650
1960	1,572
1965	1,612
1970	1,807
1975	2,497
1980	3,970
1985	7,598
1990	12,823
1992	15,650

Draw a line graph for the per capita federal debt.

28. Per Capita Personal Income

1950	$1,501
1955	1,881
1960	2,219
1965	2,773
1970	3,893
1975	5,851
1980	9,910
1985	13,896
1990	18,625
1992	19,841

SOURCE: *Monthly Labor Review,* May 1993.

Draw a line graph for per capita personal income.

29. Mothers with Children under 18 Participating in the Labor Force

1955	27%
1965	35%
1970	not avail.
1975	47%
1980	57%
1985	62%
1990	67%

SOURCE: U.S. Dept. of Labor, Bureau of Labor Statistics.

(a) Draw a line graph for the data.
(b) What assumption(s) do you make in connecting the line between 1965 and 1975?

30. College Graduates (nearest thousand)

1950	432,000
1960	392,000
1970	827,000
1975	979,000
1980	1,000,000
1985	979,000
1990	1,050,000
1992	1,105,000

SOURCE: Department of Education, Center for Education Statistics.

Draw a line graph for the number of college graduates.

31. World Record Times for the Mile Run.

1950	4:01.4 (4 min 1.4 sec.)
1955	3:58.0
1960	3:54.5
1965	3:53.6
1970	3:51.1
1975	3:49.4
1980	3:48.8
1985	3:46.3
1990	3:46.3
1993	3:44.4

SOURCE: USA Track & Field.

Draw a line graph for this data.

32. Harness Racing Records for the Mile

Trotters		Pacers	
1921	1:57.8	1904	1:56
1922	1:57	1938	1:55
1922	1:56.8	1955	1:54.8
1937	1:56.6	1960	1:54.6
1937	1:56	1966	1:54
1938	1:55.2	1966	1:53.6
1969	1:54.8	1971	1:52
1980	1:54.6	1980	1:49.2
1982	1:54	1989	1:48.4
1987	1:52.2	1993	1:46.2

SOURCE: 1995 Information Please Almanac.

(a) Draw a line graph for Trotters.
(b) Draw a line graph for Pacers.

In problems 31 through 44, draw a pie chart for the data.

33. Of the 52.4 million Americans who are 55 or over

51% are retired or prefer not to work
27% are employed
12% prefer to work, but can't work
10% prefer to work, but can't find jobs
SOURCE: *Fortune* July, 1995.

34. A 1989 survey of corporate executives showed the following plans regarding retirement.

Retire before age 60	31%
Retire between ages 60–65	50%
Retire between ages 65–70	9%
Work as long as possible	10%

SOURCE: Korn/Ferry International.

35. In a 1990 nationwide survey of 2000 high school students conducted on behalf of Chrysler Motors, it was found that 70% of those surveyed drank alcohol. Those surveyed reported they drank:

Every day	2%
Few times a week	12%
Once a week	31%
Once a month	29%
Almost never	26%

36. A Lou Harris poll conducted for the Girl Scouts of the United States asked 2000 kids, grades 7–12, who their hero figures were. The results showed:

Actors	38%
Musicians	19%
Athletes	11%
Comedians	11%
Politicians	6%
Others	15%

37. Money for the Arts: 12 percent of the money businesses give to nonprofit organizations goes to support the arts.

Percent of money given to the arts:

Museums	16%
Symphony orchestras	16%
Theater	12%
Dance	8%
Opera	8%
Public radio/TV	8%
All Others	32%

SOURCE: Business Committee for the Arts.

38. United States Resident Population by Race and ethnic Origin, 1993

White	74%
Black	12%
Hispanic Origin	10%
Native American	1%
Asian and Others	3%

SOURCE: U.S. Bureau of the Census.

39. NBA Champions 1980–1993

Team	Championships
Los Angeles Lakers	5
Boston Celtics	3
Chicago Bulls	3
Detroit Pistons	2
Philadelphia 76ers	1

SOURCE: National Basketball Association.

40. NBA Champions 1960–1993

Team	Championships
Boston Celtics	14
Los Angeles Lakers	6
Chicago Bulls	3
New York Knicks	2
Philadelphia 76ers	2
Detroit Pistons	2
Others	5

SOURCE: National Basketball Association.

41. The percentage of American with blood types

O positive	36%
O negative	6%
A positive	38%
A negative	6%
B positive	8%
B negative	2%
AB positive	3%
AB negative	1%

SOURCE: American Red Cross.

(a) Draw a pie chart of the given data.
(b) Draw a pie chart in which types O, A, B, and AB are the only classifications.

42. We drive for the following reasons:

To work	34.3%
Family/personal business	30.4%
Social activities, recreation	30.0%
Civic, education, religious	4.1%
Other	1.2%

SOURCE: The *1990 Highway Fact Book*.

43. The junk bond market represents over $200 billion in face value outstanding. Who owns junk bonds?

Insurance companies	30%
Mutual funds, money managers	30%
Pension funds	15%
Foreign investors	9%
Savings and loans	7%
Individuals	5%
Corporations	3%
Securities dealers	1%

SOURCE: Drexel Burnham Lambert.

44. Reasons for Being Fired:

Incompetence	39%
Inability to get along with others	17%
Dishonesty or lying	12%
Negative attitude	10%
Lack of motivation	7%
Failure to follow instructions	7%
Other reasons	8%

SOURCE: Robert Half International, Inc.

EXTENDED PROBLEMS

45. Histograms were described with bins that had a given length but a variable number of data points. It is also possible and often desirable to have the bins of variable length, but with the same number of data points in each bin. In this kind of histogram, each rectangle may have differing widths, but the heights are adjusted so that each rectangle has the same area. Make such a histogram for the data used in the dot plot in Figure 3.1. Each rectangle should cover 4 data points and should have end points halfway between adjacent data points in neighboring bins. What are the strengths and weaknesses of this type of graph?

46. Repeat the directions from problem 45 with the data from the chemistry midterm in problem 1 of this problem set.

47. Write a brief report on John Playfair and his graphical innovations.

48. Write a brief report on the graphical methods used by Florence Nightingale.

3.2 COMPARISONS

You are the manager of a small stand near the beach that sells hot chocolate, ice cream, and hot dogs. You have to present monthly sales figures to the owner showing how the shop has done over the past year. This information is contained in a table. How should you present this data so as to clearly show the sales trends of each item and to compare the three?

Graphs and charts can help us understand the details and relationships within a set of data, and they can also help us see the nature of changes in quantities that are studied over a period of time. In addition, we can use charts and graphs to make comparisons between different, but related, sets of data. An effective visual presentation can not only show that there are similarities or differences between sets of data, but can also sometimes provide help in understanding why they exist. In the case of the initial problem above, our informational needs are quite high; we want to provide a picture of the operation as a whole while being able to visualize the trends of the business and understand the relationships, if any, among the items we sell. The effective presentation of data is an important component in communication and decision making for companies and organizations.

DOUBLE STEM AND LEAF PLOTS

Stem and leaf plots and histograms may be used to compare two different data sets. Suppose that two classes of an economics course took a test and their scores were (in order):

Class 1—26, 32, 54, 62, 67, 70, 71, 71, 74, 76, 80, 81, 84, 87, 87, 87, 89, 93, 95, 96

Class 2—34, 45, 52, 57, 63, 65, 68, 70, 71, 72, 74, 76, 76, 78, 83, 85, 85, 87, 92, 99.

Which class did better? The answer is not obvious from looking at the data sets.

These two data sets may be combined into the same plot called a **double stem and leaf plot** (Table 3.9).

TABLE 3.9

Class 1		Class 2
6 5 3	9	2 9
9 7 7 7 4 1 0	8	3 5 5 7
6 4 1 1 0	7	0 1 2 4 6 6 8
7 2	6	3 5 8
4	5	2 7
	4	5
2	3	4
6	2	
	1	
	0	

Note that the stem is placed in the middle and the two sets of leaves are placed to either side of the stem like branches on a tree trunk. It appears that Class 1 did somewhat better than Class 2 since there are more leaves near the top left side of the stem than on the right.

COMPARISON HISTOGRAMS

The test data for the two economics classes may be put into a **comparison histogram,** which will make the choice of the better section yet easier (Figure 3.14).

In Figure 3.14 we see that the histogram for Class 1 peaks in the 80–89 bin while Class 2 has a peak in the 70–79 bin. Later, we will develop quantitative methods to compare these classes.

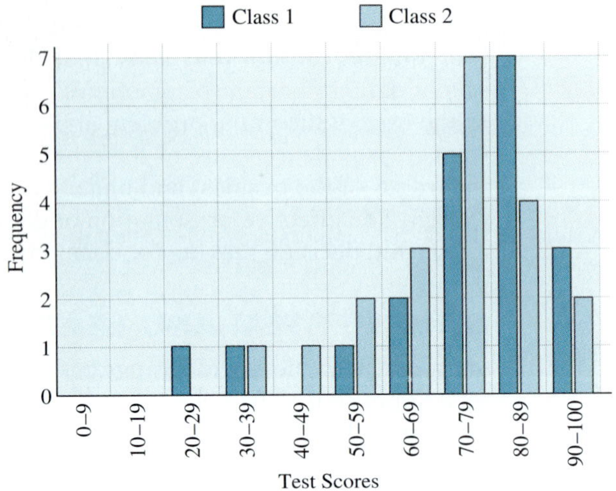

FIGURE 3.14 Test scores for two economics classes

EXAMPLE 3.6 Construct a comparison histogram for the data in Table 3.10.

TABLE 3.10 Number of Doctors Practicing in Selected Medical Specialties in 1992

	Dermatology	Family Practice	Ob/Gyn	Pediatric
Males	6000	41,300	23,500	23,800
Females	1900	9700	8100	16,600

SOLUTION

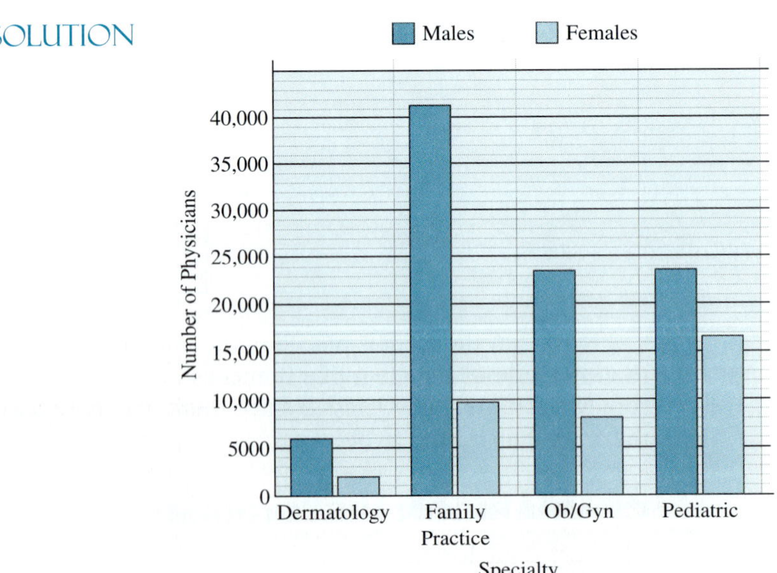

◆

FIGURE 3.15

MULTIPLE BAR GRAPHS

Bar graphs may also be used to show relative strengths. A few years ago a major success in behavior modification resulted from a marketing campaign. For generations people scooped powder laundry detergents from giant boxes into washing machines. More recently, there has been a switch to using superconcentrates. How this change occurred was described in the *Wall Street Journal,* January 5, 1993. The article states that profits for detergent companies had been declining due to changing demographic factors (e.g., sex, age, marital status, educational level, family size, etc.). The number of washer loads per household was down from 10 per week several years before to about 6 per week in 1993. It was believed there were fewer (but larger) washer loads due to more single head of household families and fewer

TABLE 3.11 Comparison of Sales of Each Type of Detergent in 1991 and 1992

Soaps	1991 Sales (in millions)	1992 Sales (in millions)
Tide (super)	151.7	205.0
Tide (regular)	85.6	16.7
Cheer (super)	49.0	66.5
Cheer (regular)	3.4	0.2
Wisk (super)*	33.7	47.5
Surf (super)	7.5	46.3
Surf (regular)	36.3	3.7
Arm & Hammer (super)	9.1	32.1
Arm & Hammer (regular)	25.5	1.3

Note: Wisk never had a regular detergent.

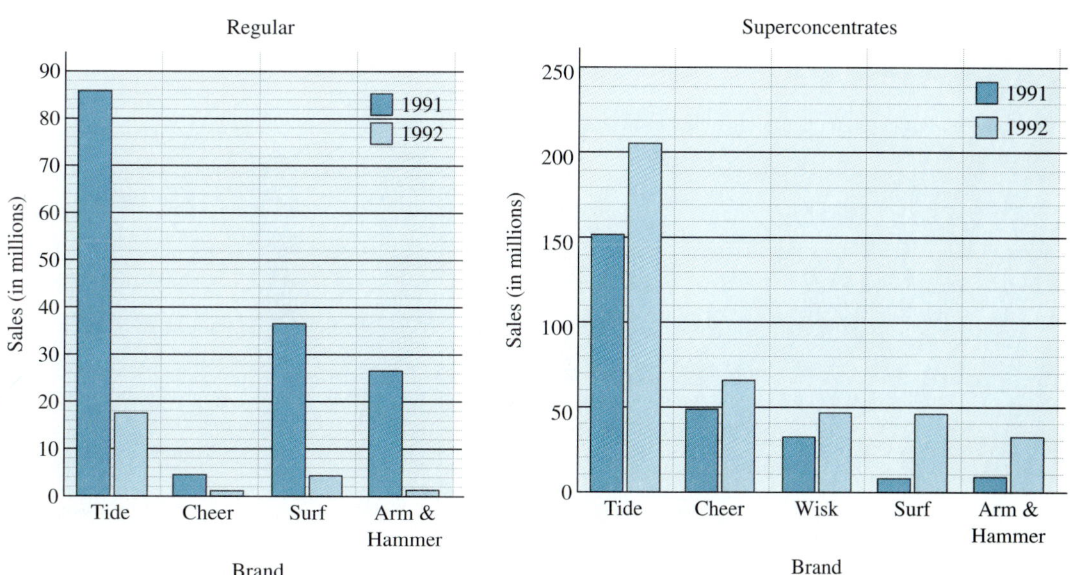

FIGURE 3.16

couples having one partner who stayed home to care for the household. To increase profits, detergent companies introduced superconcentrates that are even more concentrated. They were marketed as more ecologically friendly because the packages would take less space and there would be less waste. This campaign was highly successful as Table 3.11 shows.

To compare what happened to superconcentrate detergent sales during this period as compared with regular detergent, we make a comparison bar graph for each type of detergent that gives 1991 and 1992 sales. These graphs are called **double bar graphs** (Figure 3.16).

It should be clear that the marketing campaign to switch from regular to super was a smashing success since the 1992 bars for the superconcentrates are much taller than the bars for the regular—so much so that regular detergent is nearly unavailable today and its use may be a distant memory to many readers.

EXAMPLE 3.7 The following United States Census Bureau data compares the percentage of homeowners and renters who own certain appliances (Table 3.12). Give a clear graphical presentation of this information.

TABLE 3.12

	Homeowner	**Renter**
Washer	93%	42%
Dryer	84%	33%
Dishwasher	55%	31%

SOLUTION Make a double bar chart having a vertical axis whose units are percents from 0 to 100 (Figure 3.17).

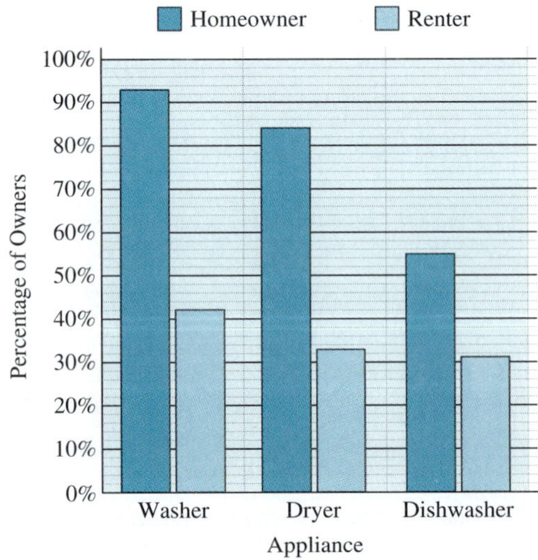

FIGURE 3.17 Percentage of appliances owned by homeowners and renters ◆

MULTIPLE LINE GRAPHS

Like bar graphs, line graphs may be used to show comparisons together with trends. The number of new business incorporations and the number of business failures from 1980 to 1986 are plotted in Figure 3.18 as a **double line graph.**

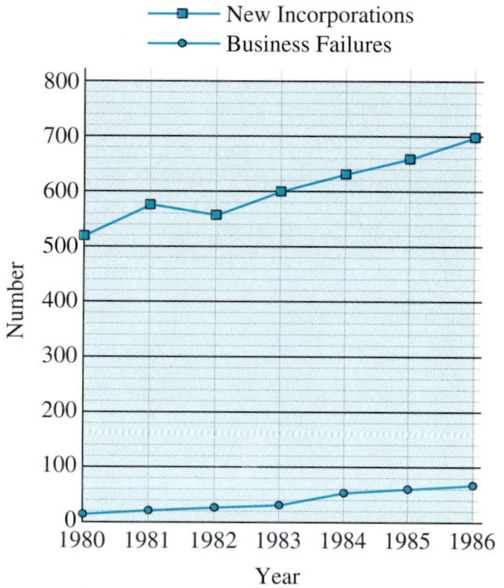

FIGURE 3.18 Numbers of new incorporations and business failures

The overall trend for new incorporations is increasing, although there is a slight dip in 1982, a recession year. The number of failures is relatively small compared to the number of new incorporations and also has an increasing trend.

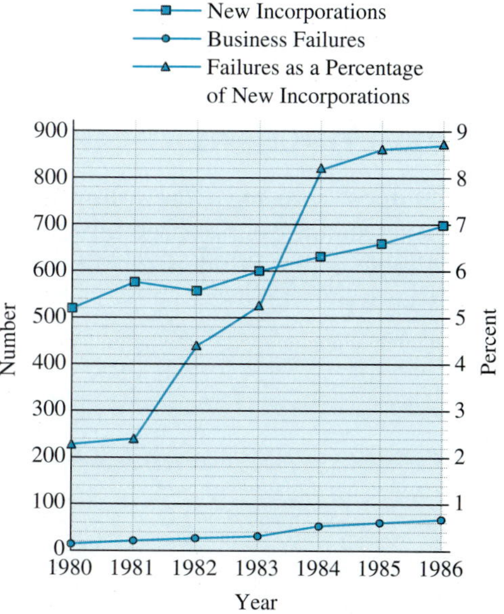

FIGURE 3.19 A comparison of the number of new incorporations and business failures

We can analyze the data further by adding the percentage of failures for new incorporations. This shows an increasing trend that greatly accelerated between 1981 and 1984 before leveling off between 8.5% and 9.0% (Figure 3.19).

Notice how this graph has two vertical scales, one labeled "Number" on the left and one labeled "Percent" on the right. This technique is often used in graphs to provide additional information.

EXAMPLE 3.8 Compare the two economics class test scores using a double line graph.

SOLUTION The double line graph in Figure 3.20 shows the two classes are clearly separated. The graphs of the classes appear similar, but the graph for class 1 is roughly ten points higher than the graph for class 2.

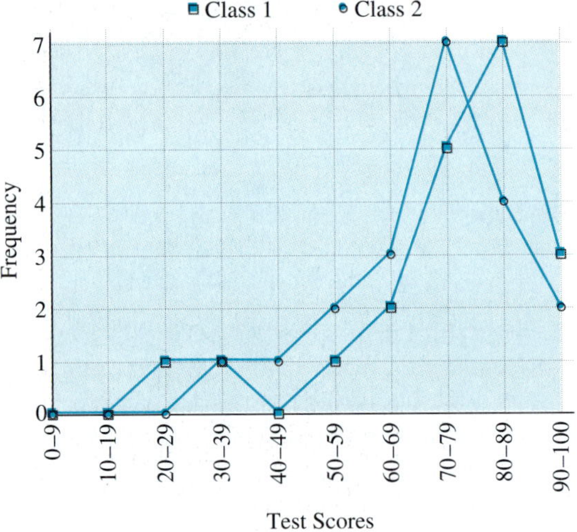

FIGURE 3.20 Test scores for two economics classes ◆

MULTIPLE PIE CHARTS

Suppose you have the following information giving details of the percentage of taxes taken by the three levels of government for the years 1950, 1970, and 1991 (Table 3.13). What is the best way to illustrate these proportions and show the trends over time?

TABLE 3.13

	1950	1970	1991
Local	15%	17%	18%
State	16%	21%	27%
Federal	69%	62%	55%

One way to present these trends is to construct **multiple pie charts.** Figure 3.21 gives pie charts for 1950, 1970, and 1991.

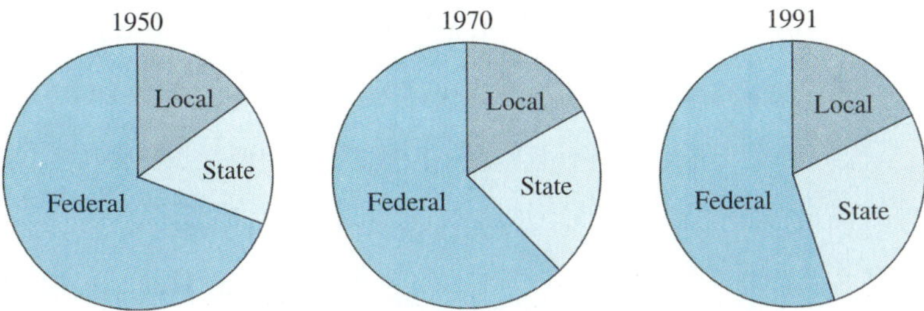

FIGURE 3.21 Percentage of taxes taken by the three main levels of government for the years 1950, 1970, and 1991

Notice how this family of charts displays relative amounts for each year while indicating a trend over the three years. Specifically, a trend toward relatively more state taxes and less federal taxes is indicated.

PROPORTIONAL BAR GRAPHS

Bar graphs can also be used to show relative amounts and trends simultaneously. A **proportional bar graph** can be used for this purpose. Each bar is the same height and corresponds to 100% of the total for the given year. Each bar is divided into pieces whose lengths correspond to the appropriate percentages for each level of government for that year (Figure 3.22).

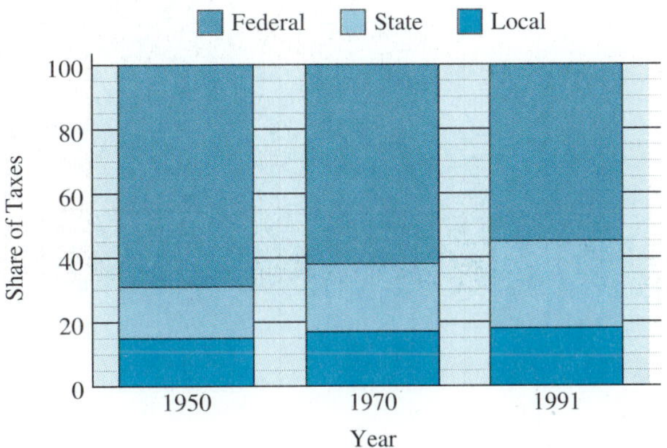

FIGURE 3.22

Figure 3.22 shows that the federal government is still the largest collector of taxes, but its share has been declining over the last 40 years. Further, the state government share has been increasing far more than the local governments.

You are the manager of a small stand near the beach that sells hot chocolate, ice cream, and hot dogs. You have to present monthly sales figures to the owner showing how the shop has done over the past year. This information is contained in a table. How should you present this data so as to clearly show the sales trends of each item and to compare the three?

INITIAL PROBLEM SOLUTION

SOLUTION Listed are the sales figures for the three products for the past year.

	Hot Chocolate	Ice Cream	Hot Dogs
Oct	400	330	220
Nov	470	240	200
Dec	630	200	270
Jan	600	110	190
Feb	670	90	180
Mar	570	120	210
Apr	490	220	250
May	280	370	270
Jun	130	460	310
Jul	70	620	330
Aug	80	660	340
Sep	240	450	260

We display this data using a multiple line graph (Figure 3.23).

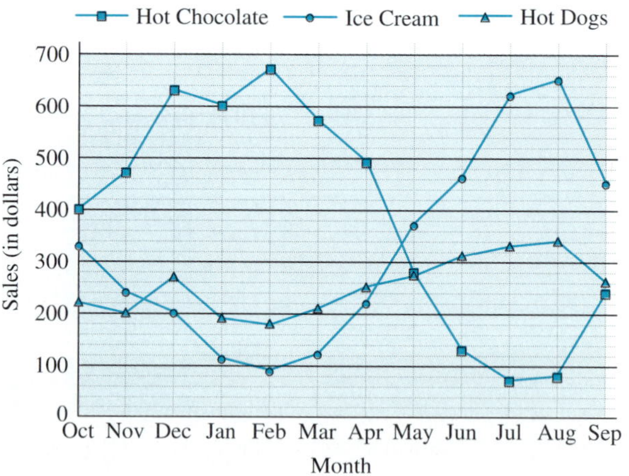

FIGURE 3.23 Sales figures for your refreshment stand

Notice that some interpretations are suggested by the graph. Hot chocolate seems to sell better during the colder months while ice cream has increased sales during the warmer months. Hot dogs do not show as much variation.

PROBLEM SET 3.2

1. Two sociology classes taught by the same professor are scheduled together for a joint midterm. Scores for the two classes were as follows:

Class 1: 85, 73, 84, 76, 73, 92, 64, 86, 84, 95, 66, 87, 63, 74, 84, 92, 76, 80, 86, 77, 91, 74, 76, 85

Class 2: 66, 74, 86, 84, 54, 82, 70, 86, 94, 88, 96, 83, 73, 78, 75, 83, 80, 74, 77, 82, 85, 73, 85, 80, 84, 76, 88

Make double stem and leaf plots for the test scores from the two classes. Use the same stem for both classes, but put the leaves for class 1 on one side and the leaves for class 2 on the other. Are there any significant features to the plots? Are there any significant differences between the groups? Are there any outliers or other features?

2. Suppose that two fifth-grade classes take a reading test, yielding the following scores. (Scores are given in year-month equivalent form. For example, a score of 5.3 means that the student is reading at the fifth-year, third-month level, where "year" means year in school.)

Class 1: 5.3, 4.9, 5.2, 5.4, 5.6, 5.1, 5.8, 5.3, 4.9, 6.1, 6.2, 5.7, 5.4, 6.9, 4.3, 5.2, 5.6, 5.9, 5.3, 5.8

Class 2: 4.7, 5.0, 5.5, 4.1, 6.8, 5.0, 4.7, 5.6, 4.9, 6.3, 7.2, 3.6, 8.1, 5.4, 4.7, 4.4, 5.6, 3.7, 6.2, 7.5

Make double stem and leaf plots of this data. Do there appear to be any significant differences between the groups? Are there any outliers or other features?

3. Babe Ruth was one of the greatest baseball players of all time. Among his many accomplishments were his lifetime and seasonal records for home runs (both since broken). Here are the number of home runs that Babe Ruth hit in each of his 15 years as a New York Yankee:

54, 59, 35, 41, 46, 25, 47, 60, 54, 46, 49, 46, 41, 34, 22

Next to Babe Ruth, the most productive home run hitter to wear a New York Yankee uniform was Mickey Mantle, who died of cancer in 1995. In his 18 years as a Yankee, Mantle had the following home run totals:

13, 23, 21, 27, 37, 52, 34, 42, 31, 40, 54, 30, 15, 35, 19, 23, 22, 18

Make a double stem and leaf plot of these data. How do Ruth and Mantle compare?

4. The 1992 World Series Champion Toronto Blue Jays were the highest paid team in baseball, with opening day salaries averaging $1,707,963. Toronto repeated as winners of the 1993 World Series against the Philadelphia Phillies. Opening day salaries for both teams are as follows:

Toronto	Philadelphia
$5,500,000	$3,500,000
5,425,000	2,625,000
4,833,333	2,600,000
4,250,000	2,466,667
3,583,333	2,450,000
3,500,000	2,146,667
3,250,000	2,000,000
2,500,000	1,375,000
2,325,000	1,050,000
2,133,333	1,000,000
2,100,000	1,000,000
1,487,500	700,000
800,000	600,000
700,000	600,000
625,000	500,000
500,000	315,000
500,000	300,000
500,000	275,000
290,000	250,000
262,000	200,000
215,000	185,000
182,500	150,000
160,000	125,000
157,500	122,500
115,000	122,500
112,500	109,000
109,000	

Draw two double stem and leaf plots for the data. First, use millions as stems and hundred thousands as leaves (i.e., a salary of $4,250,000 has a stem of 4 and a leaf of 2). Then make a second stem and leaf plot splitting each stem in two (for leaves 0 through 4 and leaves 5 through 9).

(a) Which display do you prefer? Why?

(b) How do the teams compare? Are there any significant characteristics or differences? Are there any outliers or other special features?

5. Make a comparison histogram for the data in problem 1.

6. Make a comparison histogram for the data in problem 2.

7. Make a comparison histogram for the data in problem 3.

8. Make a comparison histogram for the data in problem 4.

9. In Example 3.7, U.S. Census Bureau data compared the percentage of homeowners and renters with certain appliances. The same document included the following data:

	Owners	Renters
Dining room	48%	25%
Porch, deck, patio, or balcony	83%	61%
Garage or carport	72%	29%

Make a multiple bar chart for the data.

10. The following table shows the dollar value of recordings shipped by manufacturers during the period from 1989 to 1992.

Manufacturer's Shipments of Recordings (in millions of dollars)

	1989	1990	1991	1992
Compact disc	2,587	3,451	4,337	5,326
Audiocassette	3,345	3,472	3,019	3,116

SOURCE: Recording Industry Association of America.

Draw a multiple bar chart for this data.

11. During the period 1971–1989, the following data show changes that occurred in meat-eating habits in the United States. Draw a multiple bar chart for this data.

Meat Consumption per Person

	1971	1989
Red meat	75%	59%
Poultry	19%	32%
Seafood	6%	9%

SOURCE: U.S. Dept. of Agriculture, 1990.

12. As people grow older, their television watching habits gradually change. A Nielsen Media Research study provided these figures on our new national pastime.

Weekly TV Viewing by Age (in hours and minutes)

Male teens	22 hrs 39 min
Female teens	21 hrs 00 min
Men 18–24	23 hrs 31 min
Women 18–24	23 hrs 54 min
Men 25–54	28 hrs 44 min
Women 25–54	31 hrs 05 min
Men 55 and over	38 hrs 28 min
Women 55 and over	44 hrs 11 min

SOURCE: Nielsen Media Research, 1993.

Draw a multiple bar chart for this data. Use either vertical or horizontal bars.

13. Although the country's economy has had its ups and downs during the last two decades, the size of the labor force has consistently grown. In the following table, the first line is the total number of people employed (either part-time or full-time), and the second line is the number who were employed full-time. Find the number who were employed part-time and compare them to the number who were employed full-time. Show this data with a double bar graph.

Employed Workers: 1970 to 1990 (in thousands)

	1970	1975	1980	1985	1990
Employed	78,700	85,800	99,300	107,200	116,900
Employed Full-time	66,800	71,600	82,600	88,600	98,000

SOURCE: U.S. Dept. of Labor, Bureau of Labor Statistics.

14. Las Vegas has been one of the nation's top tourist attractions for many years. The number of rooms available and the average number occupied (both in thousands) are given in the following table.

	Available	Occupied
1991	76.9	61.6
1992	76.5	64.2
1993	86.1	75.4
1994	88.6	78.9

SOURCE: Las Vegas Convention and Visitors Authority.

Compare the number of rooms available and the average number occupied with a multiple line graph.

15. In the discussion on Welfare Reform, much attention is given to the combined effects of taxes and transfer programs (such as social security). The following table gives the percentage of household private income for the poorest fifth of U.S. families through the richest fifth of U.S. families in 1992.

	Before Taxes and Transfers	After
Poorest 20%	0.9%	4.9%
Next 20%	7.4%	11.0%
Middle 20%	15.4%	16.7%
Next 20%	25.3%	24.0%
Richest 20%	51.0%	43.3%

Compare the percentages of household private income for U.S. families before and after taxes and transfers with a double line graph.

16. As of Jan. 1, 1994, the Monthly Basic Pay Rates, based on years of service, for the top grades of the three classifications for U.S. military personnel were:

Years of Service	2	10	20	26
Commissioned Officers	7040	7311	8822	9371
Warrant Officers	2303	2678	3663	4076
Enlisted Members	1578	2497	2783	3214

SOURCE: 1995 Information Please Almanac.

Use a multiple line graph to compare the amounts and relationships among the three pay scales.

17. Sometimes what appears to be a trend over a short period of time proves otherwise when examined over a longer period. Present the following data graphically with a double line graph. Do this in two ways: First, show the values in the proper scale; then modify the vertical axis so that the base line for the graph is the age of 19, but keep the overall height of the graph as before.

Median Age at First Marriage

	Males	Females
1930	24.3	21.3
1940	24.3	21.5
1950	22.8	20.3
1960	22.8	20.3
1970	23.2	20.8
1980	24.7	22.0
1990	26.1	23.9
1992	26.5	24.4

SOURCE: Department of Commerce, Census Bureau.

Which graph do you prefer? Why?

18. Present the following data with a double line graph.

College Graduates in the United States (in thousands)

	Men	Women
1970	484	343
1975	534	425
1980	526	473
1985	483	497
1990	492	560
1993	529	616

SOURCE: U.S. Department of Education.

19. The following table gives the average cost of tuition and fees at four-year colleges and universities.

U.S. College Tuition & Fees

	Public	Private
1983	1031	4639
1984	1148	5093
1985	1228	5556
1986	1318	6121
1987	1414	6658
1988	1537	7116
1989	1646	7722
1990	1780	8396
1991	1888	9083
1992	2134	9841

SOURCE: The College Board.

Draw a double line graph for the information in the table.

20. The following table shows the average salaries in higher education in the United States.

Average Salaries in Higher Education (in thousands)

	1970	1975	1980	1985	1990
Public (all ranks)	13.1	16.6	22.1	31.2	41.6
Private (all ranks)	13.1	16.6	22.1	33.0	45.1

SOURCE: American Association of University Professors.

(a) Present the data with a double line graph.
(b) Present the data with a double bar graph.
(c) Which graph do you prefer? Why?

21. The following double bar chart compares the use of fertilizer in different areas of the world during the periods 1975–77 and 1982–84.

Consumption of Fertilizer by Continent (tons of active ingredients per acre)

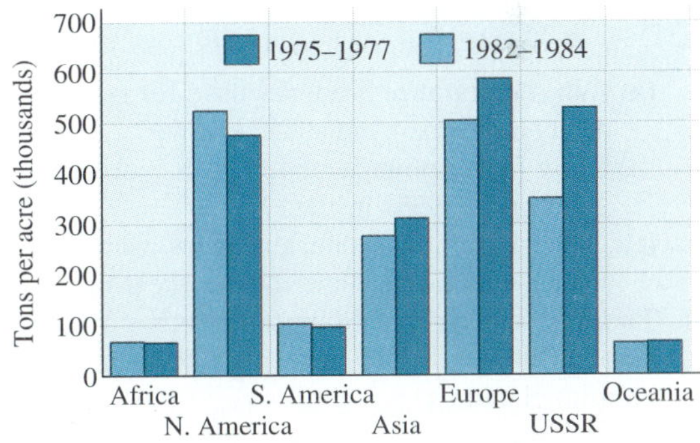

Source: United Nations Environment Programme, Environmental Data Report, 1991/92.

(a) What trends or changes in use of fertilizers can you determine from the graphs?
(b) Can you think of reasons why different regions of the world show different levels of use and changes in use of fertilizers?

22. Modern communication technology has evolved rapidly during the last two decades. The following graph shows the impact of these changes on a regional and world-wide basis.

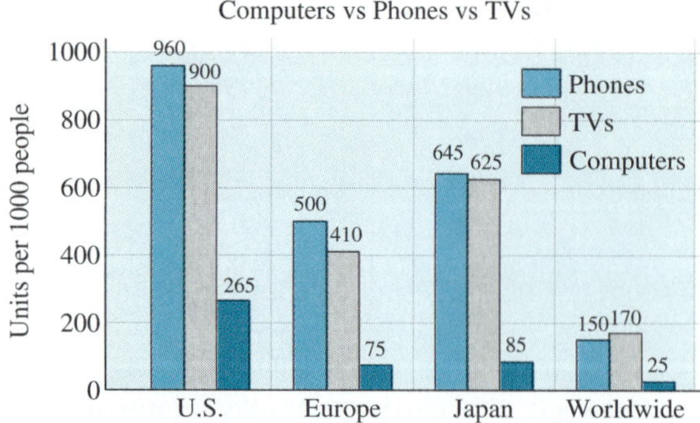

(a) What general conclusions can you draw about the worldwide use of telecommunications technology?

(b) In which of the categories of technology is the U.S. use greatest when compared to other parts of the world?

23. From 1950 to 1985, the population of the world increased dramatically. The following graph shows the changes during that time in the area for grain production and pounds of fertilizer used on a per person basis.

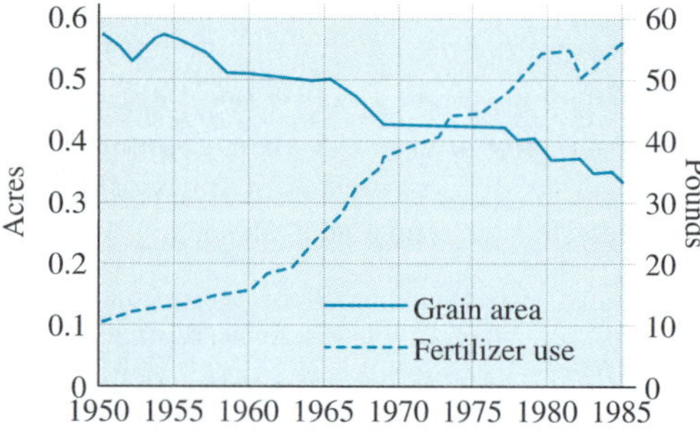

(a) What amount of land was used for grain production (per person) in 1955? in 1985?

(b) How many pounds of fertilizer were used (per person) in 1955? in 1985?

(c) Are there any generalizations or conclusions you can draw from the data in the graph?

24. The following graph shows the number of U.S. civilians (millions) and the percent voting in recent presidential elections.

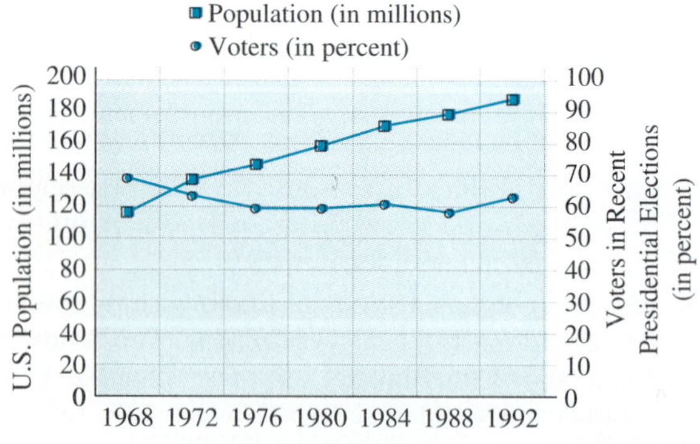

(a) What was the voting age population in 1970?

(b) What percentage voted in 1984?

(c) Are there any generalizations or conclusions you can draw from the data in the graph?

25. The following information on computers shows both the percentages of computers in use and computing power (based on number of operations that can be performed) in different regions of the world in 1992.

	Computers in Use	Computing Power
U.S.	45.3%	51.4%
Europe	24.9%	23.6%
Japan	7.1%	6.8%
Others	22.7%	18.2%

SOURCE: 1995 Information Please Almanac.

Display this information in the form of two pie charts.

26. Present the following data using pie charts.

Victim-Offender Relationship

Relationship	Homicide	Robbery	Assault
Stranger	18%	75%	51%
Acquaintance	39%	17%	35%
Relative	18%	4%	10%
Unknown	25%	4%	4%

SOURCE: Report to the Nation on Crime and Justice.

27. In problem 25, information was presented on the percentage of computers in use and the related computing power in selected regions of the world. Display this same information in the form of two proportional bar graphs.

28. Display the information on Victim-Offender Relationship from problem 26 in the form of a proportional bar graph.

Victim-Offender Relationship

Relationship	Homicide	Robbery	Assault
Stranger	18%	75%	51%
Acquaintance	39%	17%	35%
Relative	18%	4%	10%
Unknown	25%	4%	4%

SOURCE: Report to the Nation on Crime and Justice.

29. Between 1979 and 1989, the amount of petroleum used in the United States decreased by more than 1.25 million barrels per day (a little more than 6%). The table shows how it was used.

Petroleum Use in the United States (million barrels per day)

	1979	1989
Residential/ commercial	1.73	1.40
Industrial	5.34	4.26
Transportation	10.01	10.85
Electric utilities	1.44	0.74

SOURCE: Energy Information Administration Annual Energy Review, 1989.

(a) Present this information with multiple pie charts.
(b) Present this information with a multiple bar graph.
(c) Which graphing approach do you prefer? Why?

30. Here is the data showing the percentage of people in certain age categories for the different ethnic groups as provided by the U.S. Census Bureau. Present this data in two different ways: First use pie charts, one for each ethnic group; then use four proportional bar charts. You do not need to include median age in the graph.

Age According to Race

	Median Age	Percent Under 35	Percent 35–64	Percent 65 and Older
Hispanics	26.1	68	27	5
Blacks	27.7	63	28	8
Native Americans and Asians	29.0	61	32	7
Whites	33.6	53	34	13

SOURCE: U.S. Census Bureau, 1989 estimate.

Which graph do you prefer? Why?

31. The type of visual display chosen to present information can have a significant effect on how the information is perceived. Consider the following information.

People Living Alone (in thousands)

	1970	1980	1992
Males	3532	6966	9613
Females	7319	11,330	14,361
Total	10,851	18,296	23,974

SOURCE: Department of Commerce, Census Bureau.

(a) Present this information with three pie charts.
(b) Present this information with a multiple line graph.
(c) Which graphing approach do you prefer? Why?

32. Problem 10 contained the data on musical recordings produced in different formats. Use the same data and produce the following graphs.
(a) A bar chart using bars whose heights are proportional to total sales.
(b) A proportional bar chart.
(c) How do the two graphs compare? Which do you prefer? Why?

33. Table 3.11 showed the changes in use of laundry detergents. Figure 3.16 showed one way to display the information. Use the information regarding Tide and Surf to produce two different graphs.
(a) Combine the data for regular and super concentrates in a multiple bar chart comparing the two brands in 1991 and 1992. The height of each bar should reflect the total sales volume for each brand.
(b) Use the combined information from part (a) to produce a proportional bar chart to compare the two brands in 1991 and 1992.
(c) How do the two graphs compare? Does either have an advantage over the other?

34. Since 1900, the death rate in the United States has fallen from approximately 1600 per 100,000 population to under 900 per 100,000 by 1992. Much of the decline can be attributed to the control or elimination of many diseases such as tuberculosis, as well as reductions in accidental deaths.

Death Rates for Selected Causes (death rate per 100,000)

	1950	1980	1992
Cancer	139.8	182.5	204.1
Cardiovascular disease	510.8	434.5	357.6
All other causes	309.5	266.4	317.6
Total	960.1	883.4	879.3

SOURCE: National Center for Health Statistics.

(a) Display this information with three pie charts.
(b) Display this information with a multiple line graph.
(c) Which graphing approach do you prefer? Why?

EXTENDED PROBLEMS

35. Gather information about the difference in pay for men and women in the United States over the last few decades. The *Statistical Abstract of the United States* is a good source of information and can be found in nearly all college libraries. Present this information in graphical form with one or more displays.

36. Gather current information about the differences in pay for men and women in selected fields of employment. State or local employment offices should provide this information. Most college placement offices have this information as well. Display this information in the most effective manner. Why did you choose the method you did?

37. The Gross Domestic Product (GDP) is a measure of the total goods and services produced by the United States. Gather the data on the GDP for the most recent ten-year period available. Prepare two bar graphs for the data. One graph should show the actual value of the GDP, and the other should show the percentage change from one year to the next. How are the increases and decreases of the values in one graph related to those of the other?

38. During 1995, the Dow Jones Industrial Average (DJIA) went through significant changes. Gather the data on the end-of-the-month value of the DJIA for 1995. Repeat the instructions for problem 37.

39. Visit a local bakery that provides several types of baked goods, say donuts, cinnamon rolls, and bagels. Find out how many of each type were baked over a two-week period. Display this information graphically. Give an explanation for any unusual features of the graph.

3.3 ENHANCEMENT, DISTRACTION, AND DISTORTION

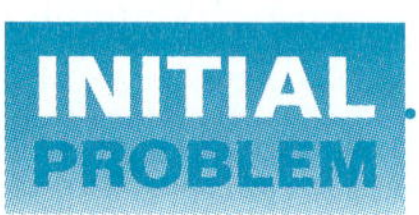

You are on a debating team and know you will argue the question, "Resolved, the most important economic issue facing the country is the federal debt." You do not know which side you will have to argue. You decide to make two graphs, each of which displays the federal debt over time. One graph will show the debt in the most threatening light possible and the other in the most benign light. Make two such graphs using federal debt data from 1952 to 1989.

Correctly communicating your ideas, either verbally or visually, is not always an easy task. When we present quantitative information in a graphical form, we have to consider which type of graph to use, what to emphasize from the data, and how to construct the actual graphs. If some aspect of the graph is distorted, a misleading impression can easily result. Some forms of distortion are considered common practice, and those who use the graphs are aware of what is being done. An example of this is the reporting of stock market information such as the Dow Jones Industrial Average. Only the upper values are shown in the graphs; this emphasizes vertical change over shorter periods of time. In this case, it's what the readers expect and want to see. Distortion may be benign or unintentional, but at other times it is intentional with the purpose to deceive or misdirect the reader. In this section, we will look at ways in which the elements of a graph can be manipulated to create different impressions of the data.

First we consider variations on the basic kinds of graphs. We particularly wish to consider ways that the graph may mislead in subtle ways so that you can determine when you are being misled and how to understand honest ways to put your viewpoint in the most favorable light. Then we consider graphs that have been enhanced by making them more pictorial. These graphs are more interesting and can reinforce your message, but they too can be misleading.

SCALING AND AXIS MANIPULATION

If someone wants the differences among the bars of a histogram or bar chart to look more dramatic, a chart is often displayed with part of the vertical axis missing. Beary Beary Sticks, a children's cereal with a relatively high sugar content level (9 grams per serving), is advertised as wholesome since it has less sugar than other children's cereals. The high sugar content cereals chosen had the following grams of sugar per serving: 15, 14, 13, 11. The bar chart in Figure 3.24 appears on the box.

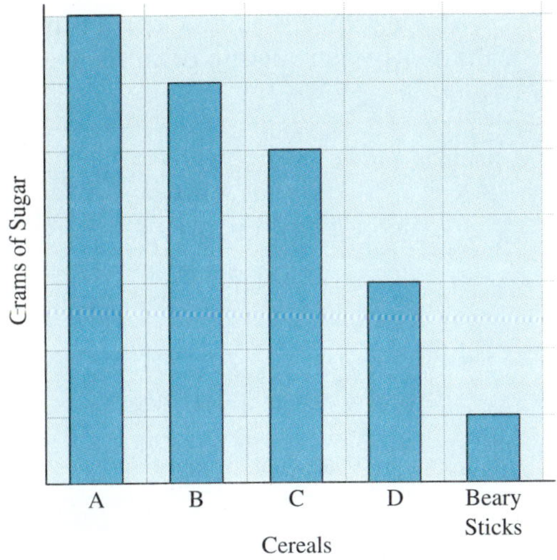

FIGURE 3.24

The scale of the vertical axis is intentionally not shown, and indeed begins at 8 instead of 0. A less misleading graph would look like the one in Figure 3.25.

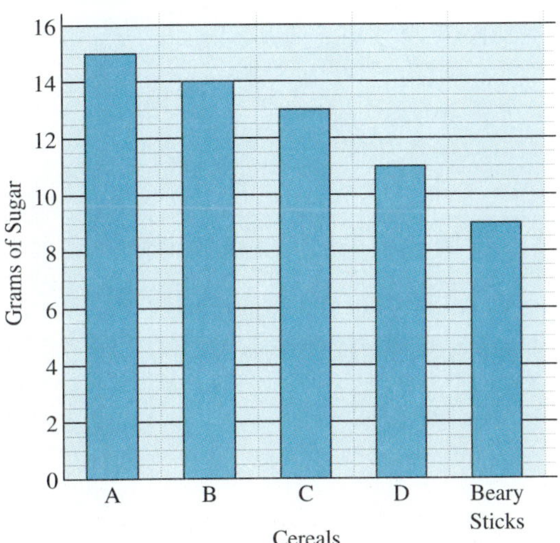

FIGURE 3.25

Notice that the Beary Beary Sticks company did not choose to compare the sugar content of their cereal with either Corn Flakes (2 grams per serving) or Shredded Wheat (0 grams per serving).

EXAMPLE 3.9 The prices of three brands of baked beans are as follows:

Brand X–79¢ Brand Y–89¢ Brand Z–99¢

Draw a bar graph of the data so that Brand X looks like a much better buy than the other two brands.

SOLUTION Brand X can be made to look much cheaper than the other two brands by starting the price scale at 75¢ as shown in Figure 3.26.

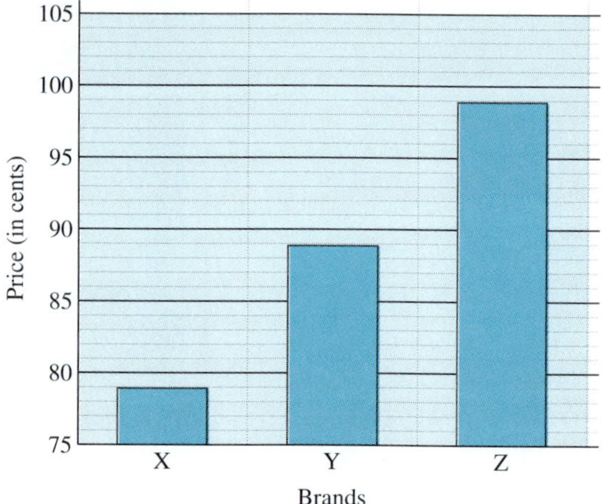

FIGURE 3.26 ◆

Another technique to distort the nature of some data is to reverse the axes or reverse the orientation of one of the axes. Figure 3.27 is a bar chart that shows declining profits of a company.

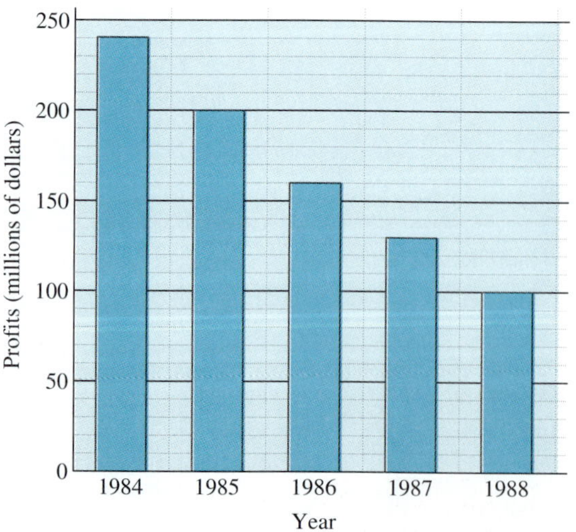

FIGURE 3.27

In Figure 3.28, the same data is displayed in a horizontal bar chart in which the years are in the reverse order.

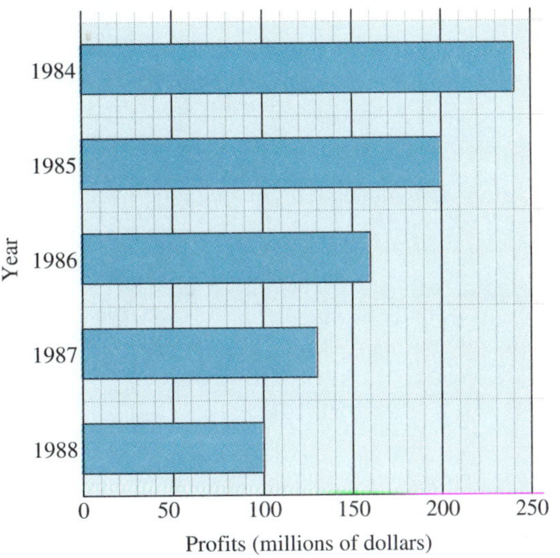

FIGURE 3.28

The chart in Figure 3.28 displays the same information, but has less of a negative connotation because it does not have the "feel" of a decreasing trend.

EXAMPLE 3.10 Some data from a crime-ridden island is given in the following table.

Year	Crimes per 1000 people
1988	25
1989	30
1990	32
1991	34
1992	38

Present this data in a graph that will mislead the citizens into thinking things are getting better rather than worse.

SOLUTION The graph in Figure 3.29 can be used to lead citizens to think that the community is getting safer by (i) starting the graph at 20, (ii) showing the years in decreasing order, and (iii) making the graph narrow so that the apparent declining trend appears to be more significant than it really is.

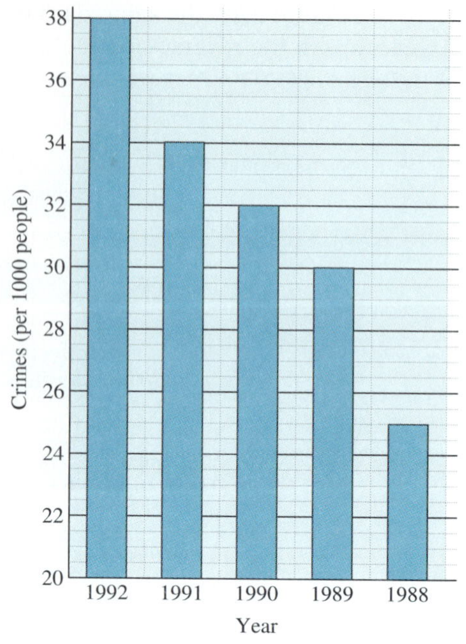

FIGURE 3.29

◆

LINE GRAPHS AND CROPPING

What we have seen regarding bar graphs also applies to line graphs. Figure 3.30 shows the company's profits from Figure 3.27 displayed as a line graph.

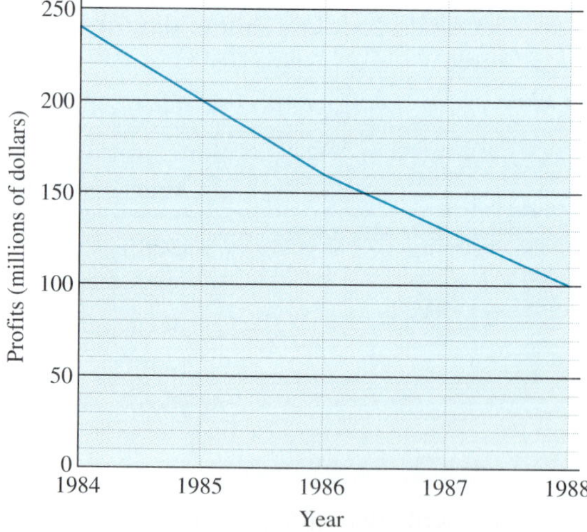

FIGURE 3.30

This decline can be made to appear less dramatic by extending the scale of the vertical axis and using smaller increments as in Figure 3.31.

This kind of scale manipulation is part of a larger phenomenon called cropping. **Cropping** refers to the choice of the window used to view the data. Suppose

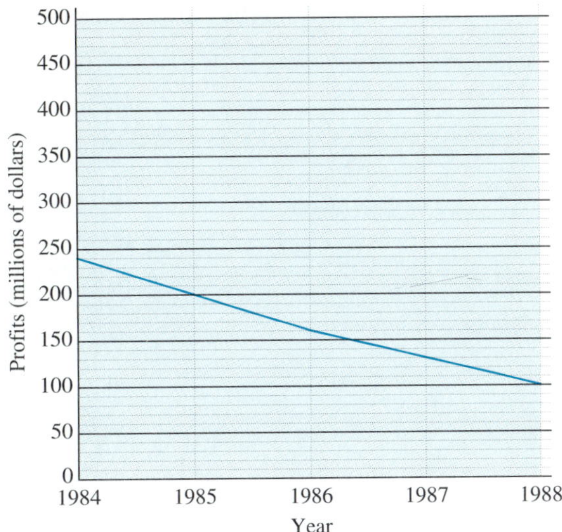

FIGURE 3.31

we wish to present the price of a certain company's stock. We may choose which time period and vertical axis to display. In other words, when we show a picture we have to choose a window to frame it. This is much the same as when you look at a television image on the news, you do not see what is outside the view screen. An image of a violent street demonstration may look much less significant if the camera pans and we see that there really are not very many people demonstrating.

EXAMPLE 3.11 Draw two line graphs of the crime data from Example 3.10 that give different impressions of the situation.

SOLUTION

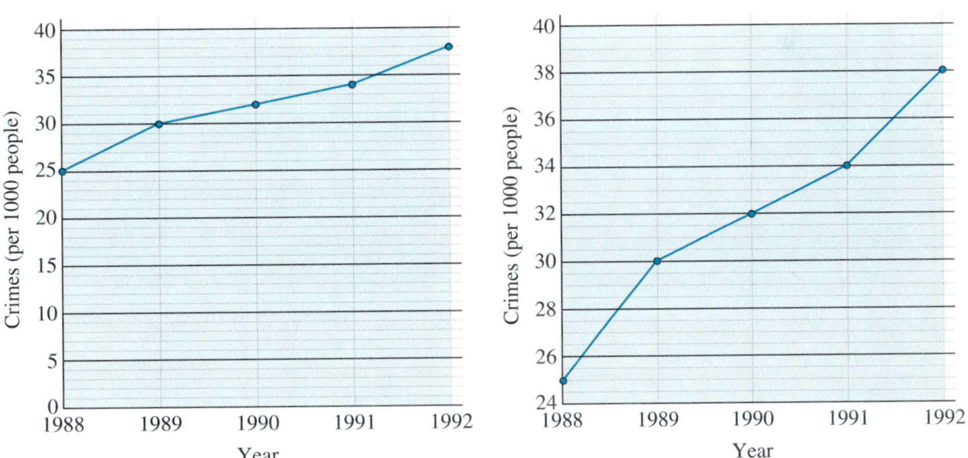

FIGURE 3.32

The graph on the left in Figure 3.32 suggests that the rate of crime is growing slowly, whereas the graph on the right gives the impression that crime is rising more rapidly. ◆

Figure 3.33 shows the value of a stock from January 11 through January 20.

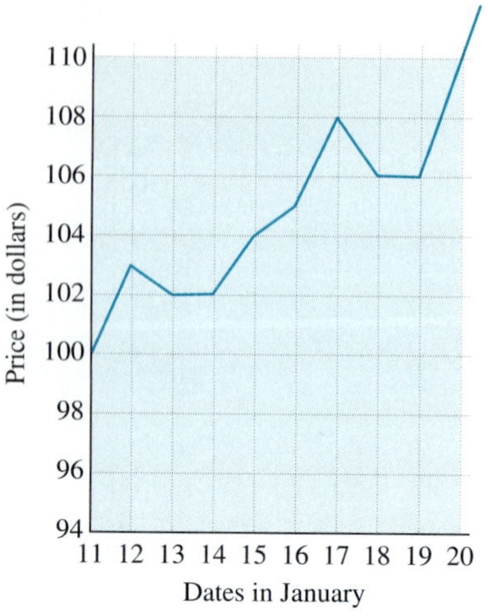

FIGURE 3.33

The stock appears to be a good buy because the price is on an upward trend. Notice that the graph is rising above the edge of the vertical scale. Graphs that do this, or even go to the edge of the scale, make the trend appear more dramatic.

Figure 3.34 shows the value of the stock over the previous five months; the stock price is plotted every 10 days.

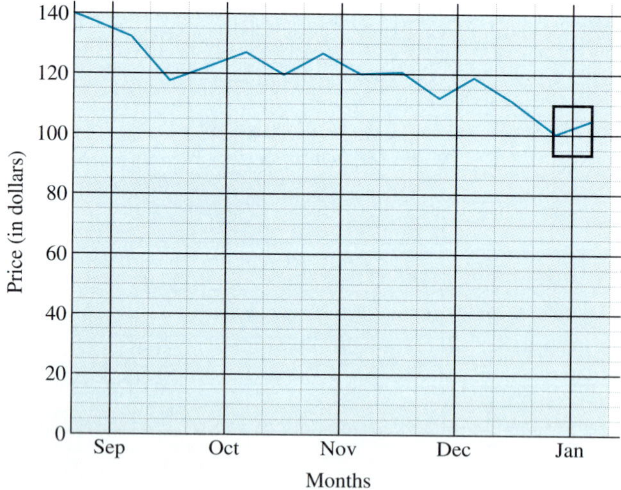

FIGURE 3.34

The data from Figure 3.33 is now contained in the box of Figure 3.34. Thus, this graph gives a very different perception regarding the value of the stock. This different perception is caused by the change in scales.

The downward trend in Figure 3.34 would be even more apparent if we choose the vertical scale to be between 100 and 140. The data from Figure 3.34 is shown in Figure 3.35.

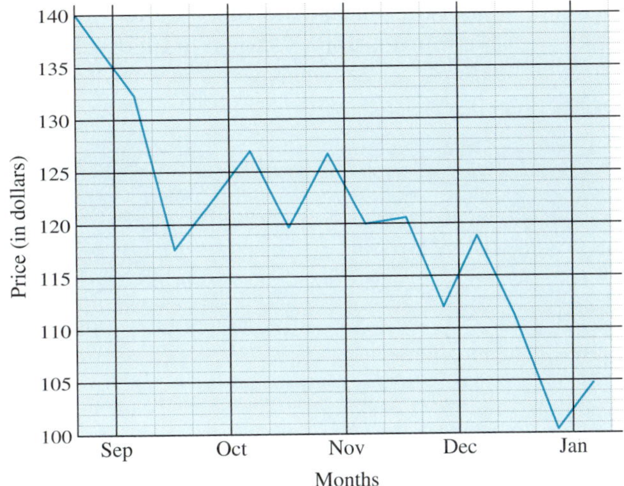

FIGURE 3.35

Notice how by changing the vertical axis, we get a very different impression of the price trend of the company's stock.

THREE-DIMENSIONAL EFFECTS

Three-dimensional effects, which are often found in newspapers and magazines, make a graph more attractive but can also obscure the true picture of the data. These graphs are difficult to draw unless you have computer graphing software.

The data for the profits of a company shown in Figure 3.27 are shown using a bar graph with three-dimensional effects in Figure 3.36.

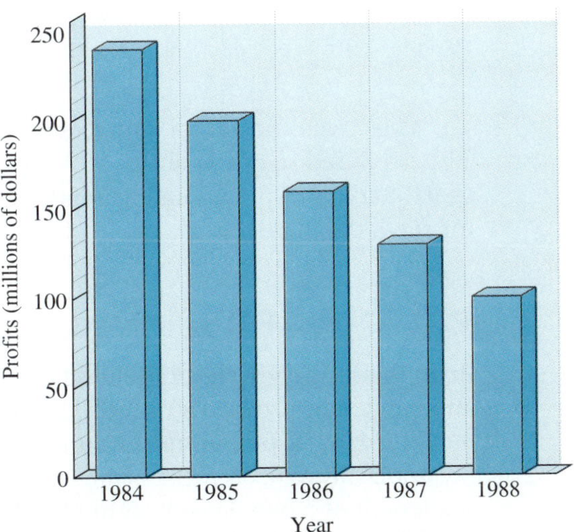

FIGURE 3.36

The perspective of the graph makes it difficult to see exact values. For example, the profits in 1988 were $100,000, but to glance at the graph it could be estimated to be as much as $120,000.

Line charts with three-dimensional effects may also reduce the amount of visible information as shown in Figure 3.37.

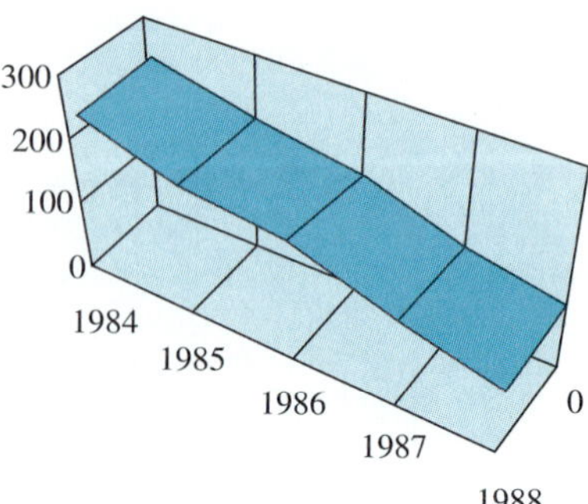

FIGURE 3.37

The downward trend is still apparent but the exact values are very difficult to read. This is a graph of the same data shown in Figure 3.36.

Pie charts can also be manipulated to reinforce a particular message or even to blatantly mislead. It is very common to take a sector of the "pie" and "explode" it, that is, move it slightly away from the center (Figure 3.38).

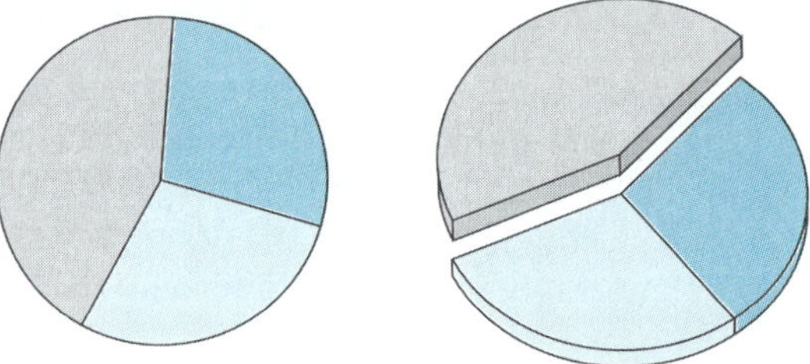

FIGURE 3.38

This gives the sector more emphasis and may make it seem larger than it is. The exploded sector is about 50% larger than each of the others. Rotating the pie chart, making it three-dimensional, and exploding the sector makes the larger sector seem much larger.

PICTOGRAPHS

Pictographs are graphs containing embellishments, which make them more visually appealing and provide a different kind of emphasis. In earlier times the most common form of a **pictograph** was a horizontal or vertical bar chart in which icons (symbols) were used for specific amounts rather than have the total represented by the length of a bar. The chart in Figure 3.39 is a pictograph.

This graph displays the projected population of the world's 10 largest cities in the year 2000. Each person icon represents one million people. In this graph, populations were rounded to the nearest million. However, fractions of millions could

TIDBIT

Archeologists believe that this type of pictogram is even older than numbers. Early accountants kept track of grain and agricultural supplies by making a symbol in clay that was a picture representing a unit of grain. Eventually, the accountants realized that it was more economical to make one picture for the grain and another symbol to tell how many units there were. So numbers were invented.

City	Projected Population in 2000
Mexico City	
San Paulo	
Tokyo	
Shanghai	
New York	
Calcutta	
Bombay	
Beijing	
Los Angeles	
Jakarta	
	≈ 1 Million people

FIGURE 3.39

have been represented by portions of an icon. In 1950, 7 out of the 10 most populated cities were in developed nations. The graph shows that this proportion will be reversed by the year 2000.

Graphs may be embellished with pictures in a variety of ways to make them more interesting. The left-hand part of Figure 3.40 shows a pictorial embellishment of a basic pictograph, while the right-hand part has a pictorial embellishment of a bar chart.

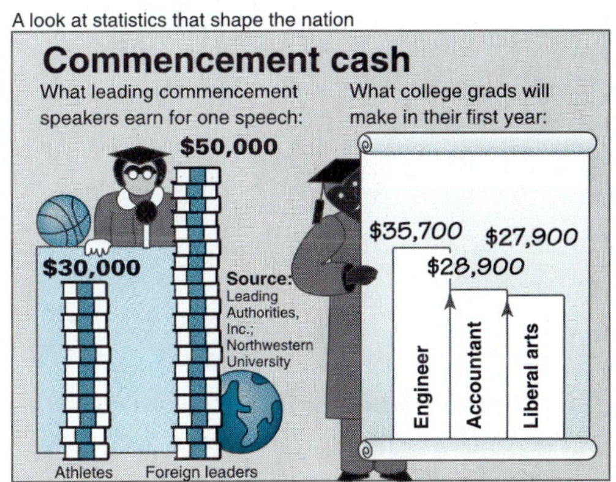

FIGURE 3.40

In Figure 3.40 each bundle of money in the stacks represents about $2700. This is a perfectly valid way to represent the data and get the point across.

Pictorial embellishment can lead to confusion, however, and sometimes the pictographs are downright deceptive. Figure 3.41 displays a bar chart, embedded in a gasoline pump nozzle, which compares the price of gas in Tokyo, Japan; Caracas, Venezuela; and the average price in the U.S.

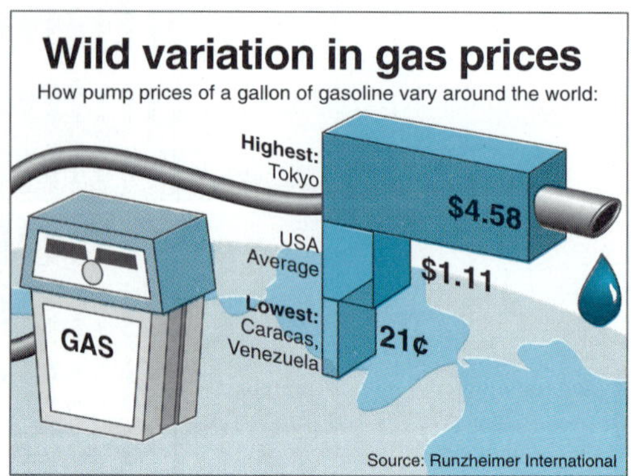

FIGURE 3.41

The chart has visual appeal, but is drawn in a misleading way. The length of the bar corresponding to Tokyo is 1 inch, and represents a price of \$4.58 per gallon. Thus 1 inch of bar represents \$4.58. The length of the bar for the U.S. is $\frac{1}{4}$ inch, so that an inch represents only $\$1.11 \times 4 = \$4.44$. The length for Caracas is $\frac{1}{16}$ inch, giving a scale of $\$0.21 \times 16 = \$3.36$ per inch. These discrepancies, while slight, make the differences appear more pronounced.

Figure 3.42 gives a variation on a bar chart by curving the bars.

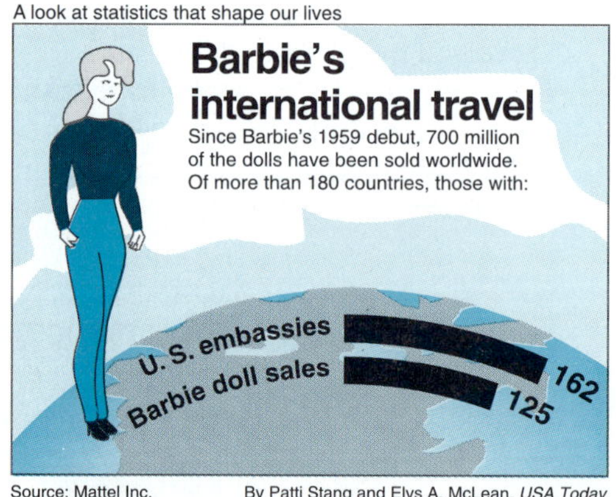

FIGURE 3.42

The point of the graphic seems to be that Barbie dolls may be considered to be ambassadors of the United States almost as much as official representatives of the government. Curving the bars also makes them appear to be closer to the same length, because the lower edge of the "U.S. embassies" bar is compared to the upper edge of the "Barbie doll sales" bar.

Objects, either two-dimensional or three-dimensional, are used to represent quantities. Consider the pictograph of milk cartons showing the increased sales in milk from 1985 to 1991 (Figure 3.43).

The amount of milk sold in 1991 was about twice that sold in 1985. At first

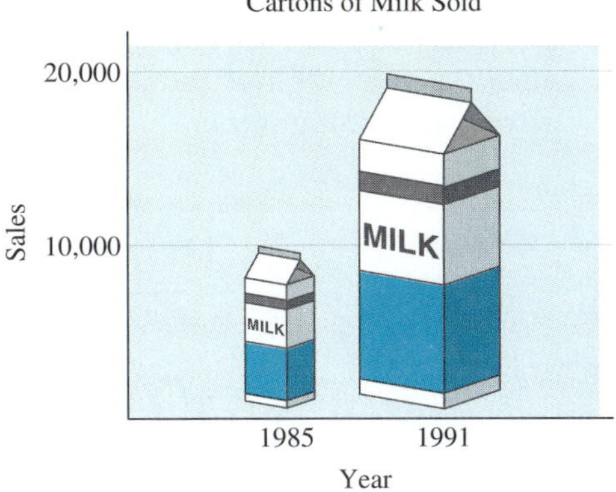

FIGURE 3.43

glance, it might seem appropriate to make the second carton twice as tall as the other. However, looking at the pictures of the two cartons, we get the impression that the taller one has much more than twice the volume of the other. In addition to making the height of the larger twice the height of the smaller, the larger carton's width and depth have also been doubled. Thus, the carton on the right represents a volume that is $2 \times 2 \times 2 = 8$ times as large as the one on the left.

EXAMPLE 3.12 The pictograph in Figure 3.44 compares the average size of a city in the National Football Conference with the average size of a city in the American Football Conference in 1994. What is misleading about it?

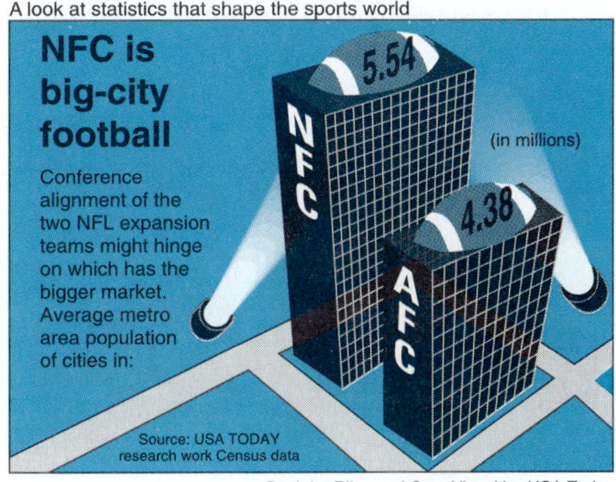

FIGURE 3.44

SOLUTION The population sizes are labeled at the top of the skyscrapers and the skyscrapers appear as bars in a bar chart. However, they are not drawn to scale. Since the NFC skyscraper is $1\frac{5}{16}$ inches tall, a vertical inch represents 4.22 million people ($5.54 \div 1\frac{5}{16} = 4.22$). However, a vertical inch on the AFC skyscraper represents 5.19 million people since it is $\frac{27}{32}$ inches tall ($4.38 \div \frac{27}{32} = 5.19$). There is more deception afoot in this pictograph. The NFC building is wider than the AFC building and the perspective gives the larger building a more imposing presence. ◆

Any graph may be embedded in a picture to make it more eye-catching and provide emphasis so that you interpret the graph in a desired way. Figure 3.45 shows a line graph of the number of babies delivered by midwives. This shows a strong, increasing trend.

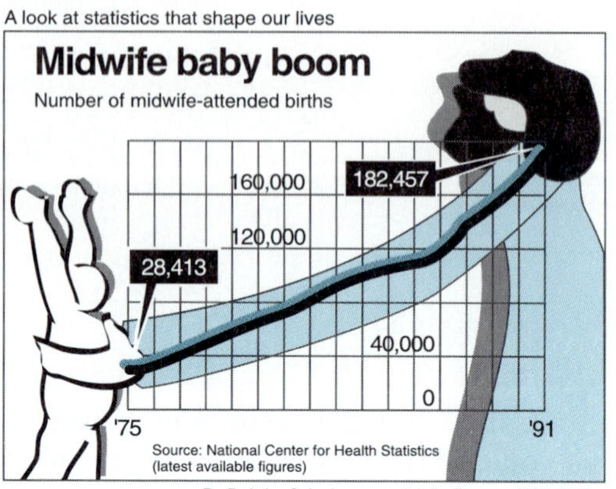

FIGURE 3.45

By making the line of the graph be the arm of the midwife, the eye is directed upward from the infant in the center of the graph along the arm to the midwife. This exaggerates the increasing nature of the graph.

Pie charts may be used and misused in various ways. Distortions may be caused by not labeling the percentages, having percentages that do not add to 100%, or overemphasizing one sector. Exploding a sector of a pie chart is an example that we have already seen. Figure 3.46 displays a pie chart embedded in a picture.

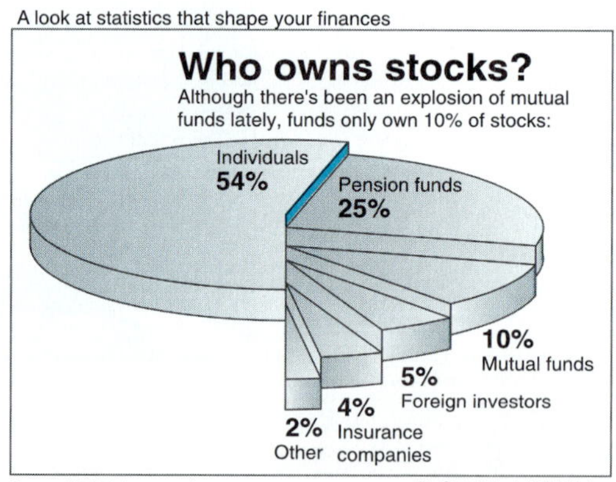

FIGURE 3.46

The graph is not especially misleading, although there is a dominant effect given to the larger sector representing the share of stocks owned by individuals.

EXAMPLE 3.13 Figure 3.47 has what looks like a pie chart embedded in a picture of a hamburger. It conceals a misleading piece of distortion. Can you spot it?

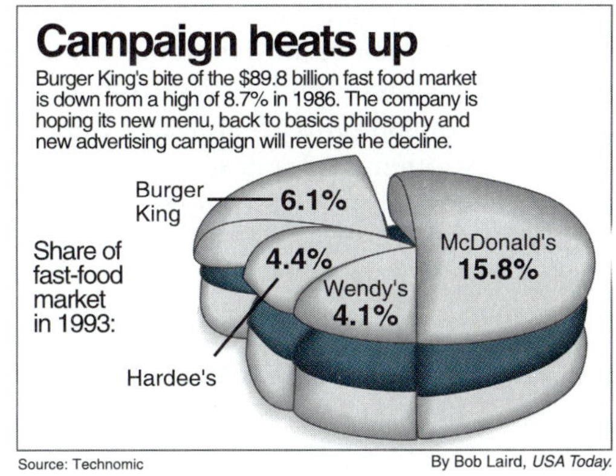

FIGURE 3.47

SOLUTION The percentages do not add up to 100%. There are only a total of 30.4%. The impression is given that McDonald's and the other chains have a much larger share of the market than they actually do. ◆

GRAPHICAL MAPS

Maps can be used to summarize information or show patterns related to national or world concerns. Figure 3.48 shows a **graphical map** indicating which states voted for Nixon or Kennedy in the presidential elections of 1960. There was also a third party candidate, Senator Harry Byrd, who received 15 electoral votes.

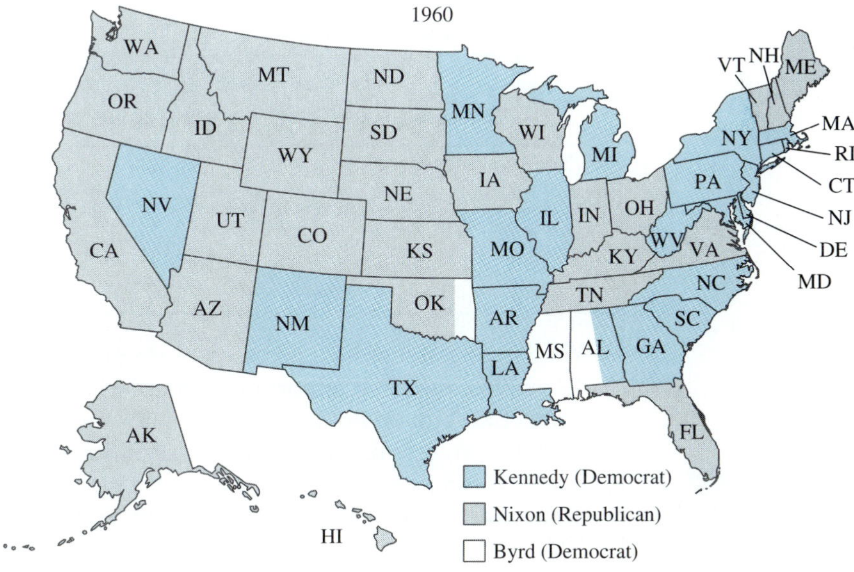

FIGURE 3.48

Although the shaded regions are not proportional to the number of votes received, much information is conveyed by the map. On the grounds of land area alone, Nixon had an advantage. Nixon was also strong in the western, mountain, and plains states, and much of the midwest. Kennedy's strengths were in the northeast, old south, and a strip from New Mexico to Michigan. Kennedy won the electoral vote 303-219-15 although the popular vote was exceedingly close.

Another map you see nearly every day is a national weather map (Figure 3.49). At a glance, you can tell the expected weather in any part of the country. A table giving such data would be less informative.

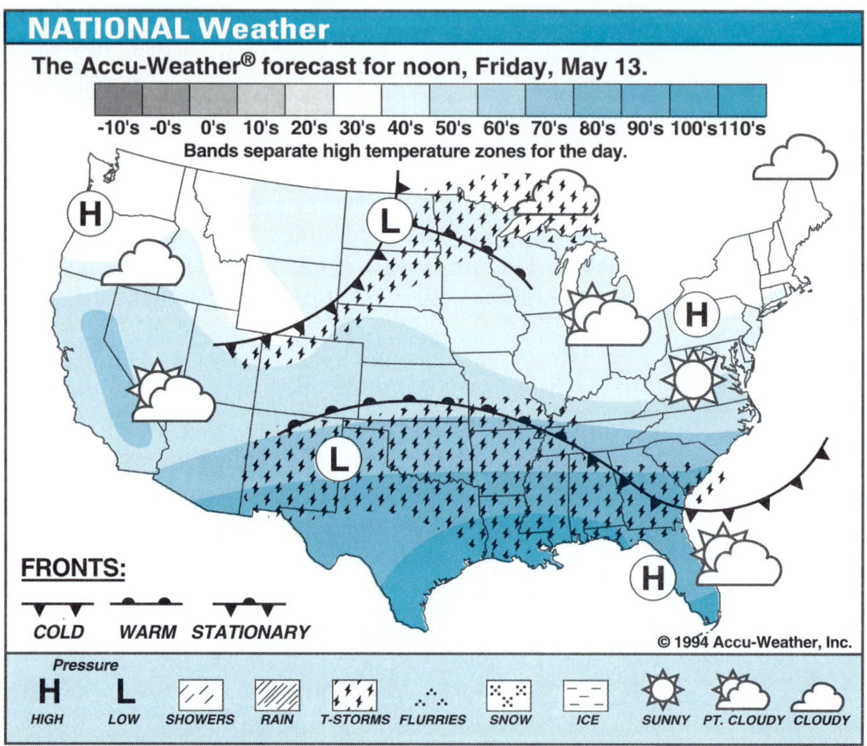

The Associated Press

FIGURE 3.49

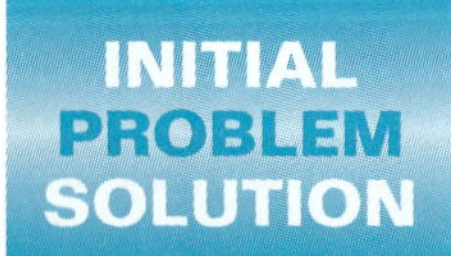

You are on a debating team and know you will argue the question, "Resolved, the most important economic issue facing the country is the federal debt." You do not know which side you will have to argue. You decide to make two graphs, each of which displays the federal debt over time. One graph will show the debt in the most threatening light possible and the other in the most benign light. Make two such graphs using federal debt data from 1952 to 1989.

SOLUTION To make the national debt appear as serious as possible, we should plot many years, showing its recent upward trend in contrast to the past

(Figure 3.50). Also, we can use horizontal and vertical scales that result in a tall thin rectangle, or even make the top of the curve go over the top of the scale. Finally, we could add some pictorial elements such as a rocket to show the debt literally shooting through the roof.

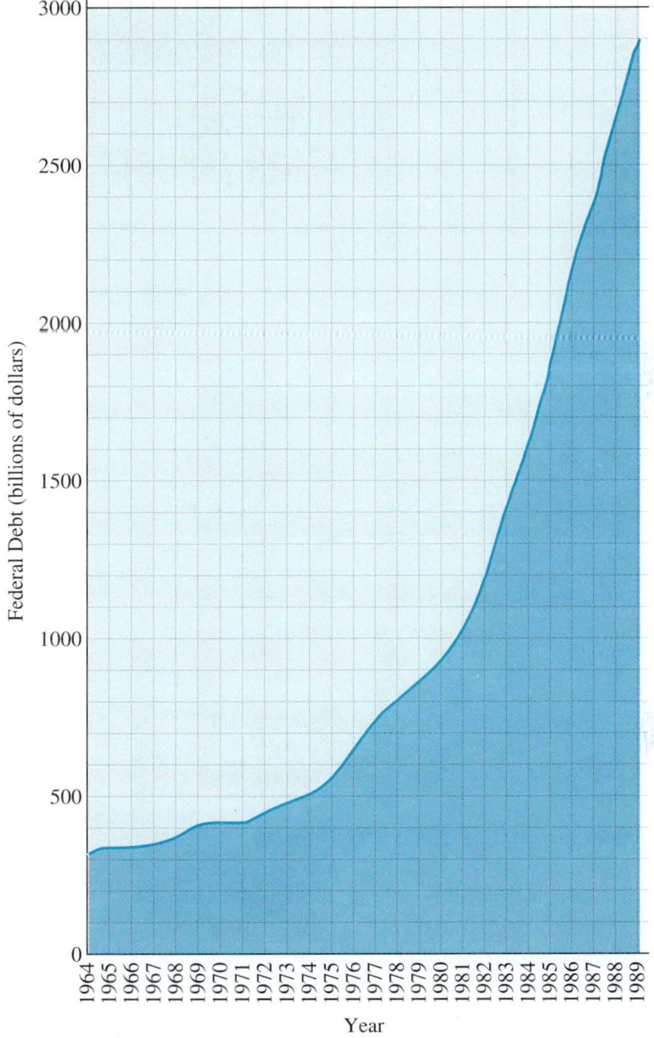

FIGURE 3.50

To make the national debt appear not as serious we can reduce the distance between increments on the vertical scale to compress the variation in the curve. We might also add horizontal scale lines to reinforce the stable, level appearance (Figure 3.51).

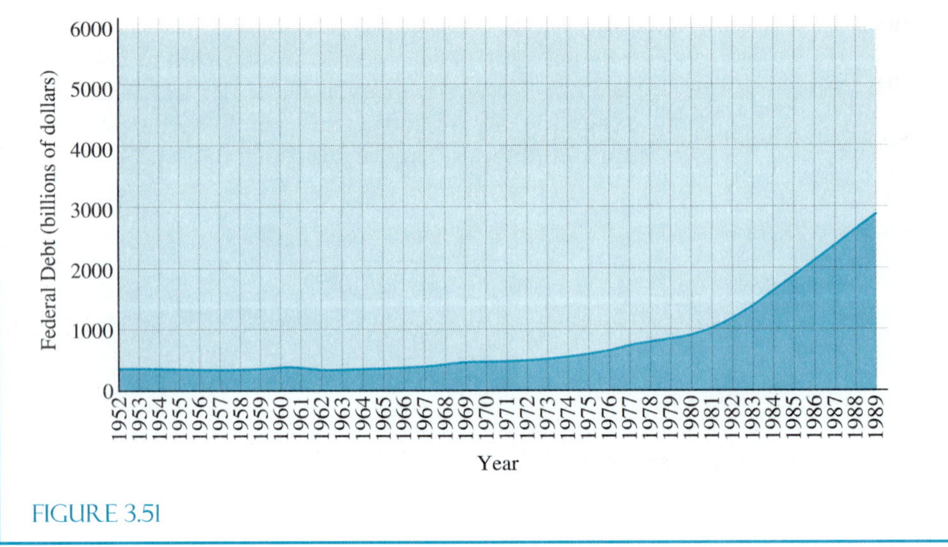

FIGURE 3.51

PROBLEM SET 3.3

1. Suppose the economics professor in section 3.2 had chosen to use the following four bins in the histogram for test scores in his classes (see Table 3.4):

Interval
90–100
80–89
70–79
below 70

(a) Find the frequency table for the distribution of scores.

(b) Draw the histogram.

(c) Are there any features that are significantly changed?

2. A statistics professor gives an 80 point test to his class, with the following scores:

35, 44, 48, 55, 56, 57, 60, 61, 62, 62, 63, 64, 67, 70, 71, 71, 75

As a common practice (and example for his students), the professor generally carries out a data analysis of all test scores, including a frequency table and histogram. He is considering two options for the bins:

(1) Grouping the data into bins of length 10, beginning with 70–80, 60–69, etc., or

(2) Grouping the data into bins of length 8, beginning with 73–80, 65–72, etc.

(a) Find the frequency table for each option.

(b) Draw the histogram for each option.

(c) Give a reason justifying the use of each histogram. Why might he use the first one? Why might he use the second?

3. In section 3.1, the history of the world record time for the mile run was given:

1950	4:01.4	(4 min 1.4 sec)
1955	3:58.0	
1960	3:54.5	
1965	3:53.6	
1970	3:51.1	
1975	3:49.4	
1980	3:48.8	
1985	3:46.3	
1990	3:46.3	
1993	3:44.4	

Source: USA Track and Field.

(a) Draw a line graph of this data using 3:30.0 as the base line for the graph.

(b) What effect does having 3:30.0 as the base line have on the impression made by the graph?

4. Harness Racing Records for the Mile

Trotters		Pacers	
1921	1:57.8	1904	1:56
1922	1:57	1938	1:55
1922	1:56.8	1955	1:54.8
1937	1:56.6	1960	1:54.6
1937	1:56	1966	1:54
1938	1:55.2	1966	1:53.6
1969	1:54.8	1971	1:52
1980	1:54.6	1980	1:49.2
1982	1:54	1989	1:48.4
1987	1:52.2	1993	1:46.2

Source: 1995 Information Please Almanac

(a) Draw a line graph of the data on Trotters using 1:40.0 as the base line for the graph.
(b) What effect does having 1:40.0 as the base line have on the impression made by the graph?

5. Since 1900, the death rate related to certain causes in the United States has fallen, while rising in several others. For major cardiovascular disease, including the heart, the death rate per 100,000 population was as follows:

1950	1980	1990	1992
510.8	434.5	368.3	357.6

Source: National Center for Health Statistics.

(a) Draw a bar graph for this data using the same distance between each of the bars.
(b) Draw a line graph for the data having the years as the baseline with the usual spacing.
(c) Which graphing approach do you prefer? Why?

6. Re-draw the bar graph from Figure 3.27 with horizontal bars, but this time reverse the order of the bars from how they appear in Figure 3.28.
(a) What is the visual impression regarding profits in this graph?
(b) Which graph would you use? Why?

7. Draw a horizontal bar graph for the data in problem 5. Draw the bar graph in such a way that it can give the impression that the death rate is increasing.

Use the following data and graph for problems 8, 9 and 11.

The Federal Tax Burden per Capita, Fiscal Year 1990–1995

1990	1991	1992	1993	1994	1995
$4,026	$4,064	$4,153	$4,382	$4,701	$5,049(est)

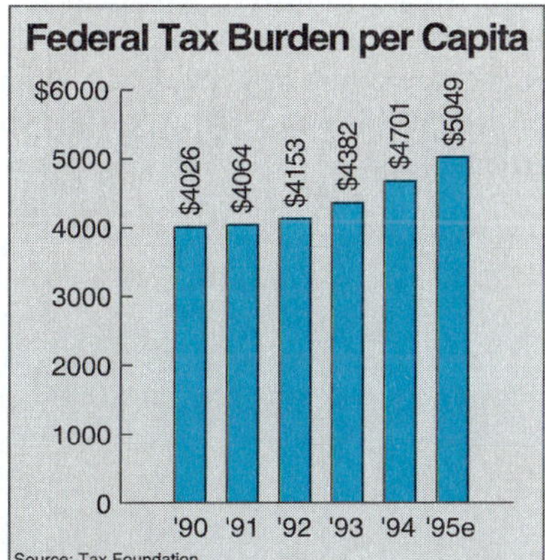

8. Prepare a vertical bar chart for the data on Federal Tax Burden per Capita in such a way that the amount actually appears to be decreasing.

9. Re-draw the graph on the increases in the Federal Tax Burden per Capita, 1990–1995, to emphasize the changes. Manipulate the vertical axis so that the increases appear more dramatic.

Use the following graph for problems 10 and 12.

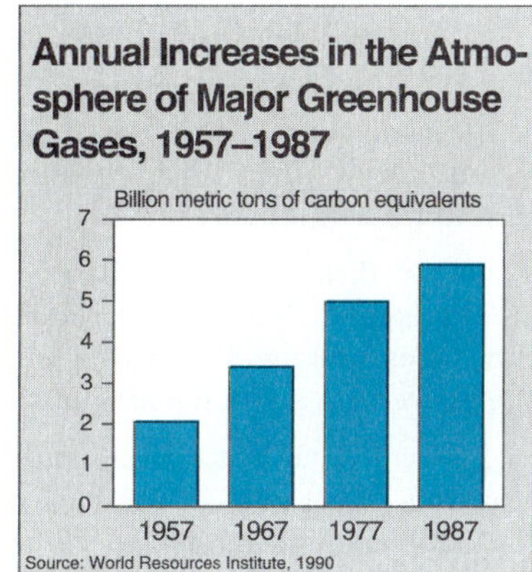

10. Re-draw the graph on Increases in Major Greenhouse Gases to emphasize the changes. Manipulate the horizontal axis so that the increases appear more dramatic.

11. Re-draw the graph on the increases in the Federal Tax Burden per Capita, 1990–1995, to de-emphasize the changes. Manipulate the horizontal and vertical axes so that the increases appear less dramatic.

12. Re-draw the graph on Increases in Major Greenhouse Gases to de-emphasize the changes. Manipulate the horizontal axis so that the increases appear less dramatic.

Use the following for problems 13 through 16.

The pictograph below was taken from the May 17, 1993, issue of Fortune magazine.

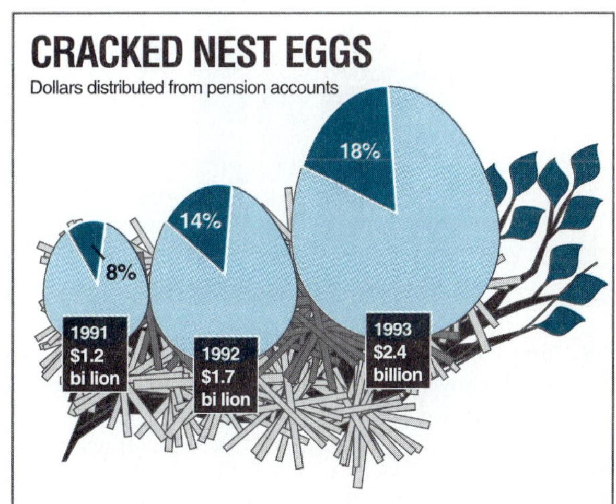

In it, the ovals that represent the "nest-eggs" have lengths that are in proportion to the total amounts in the pension accounts. This tends to exaggerate the amounts they represent. That is, the area of the third oval is actually *four* times the area of the first oval although the amount it represents is only *two* times as great.

13. Create a proportional bar chart based on the data from the pictograph. In a proportional bar chart, all bars are the same height. How does making the bars all the same height affect the impression about the amount of funds distributed?

14. Create a set of three pie charts based on the data from the pictograph. Make all the circles the same size. How does making the circles the same size affect the impression about the amounts involved?

15. Create a *segmented* bar chart based on the data from the pictograph. Make each of the bars proportional in height to the amounts in the pension accounts and then divide each bar in proportion to the amounts distributed.

16. Create a set of three pie charts based on the data in the pictograph. Make the area of each circle proportional to the amount in the pension fund. That is, the area of the circle for 1993 should be twice the area of the circle for 1991.

17. One indicator of how well the economy is doing is the number of "Help Wanted" ads that appear in the newspapers. Redo the following graph so that the increase in 1994 appears even more dramatic than it is.

18. Gun Control has been a major political issue. The graph shows the number of crimes committed with handguns for 1987 to 1992.

(a) Redo the graph so the increase appears even greater.

(b) Redo the graph so the increase is not so obvious.

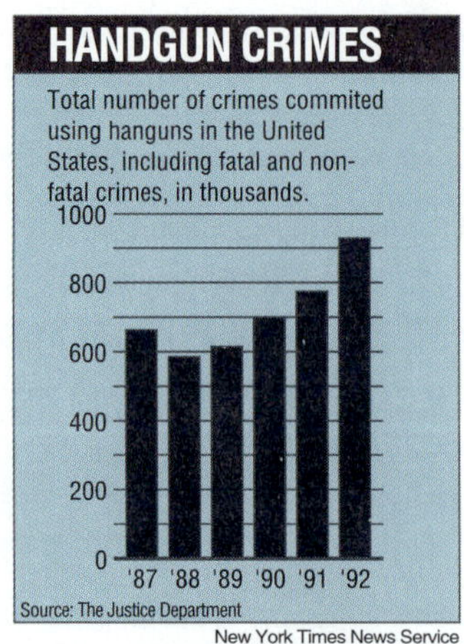

19. Health care costs became a major issue in the last decade for both employers and employees. The following graphs show changes that occurred during this period.

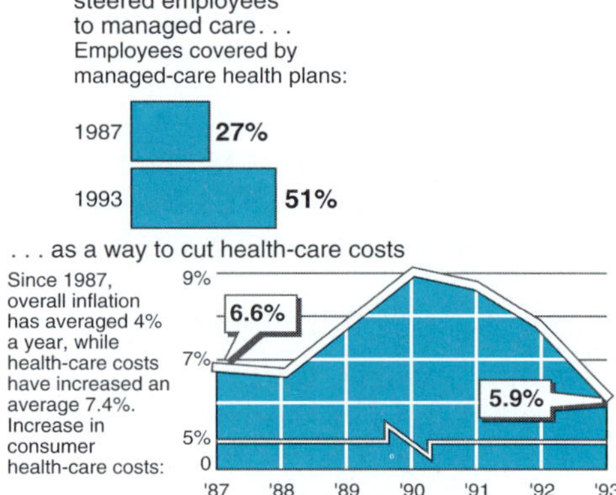

Redo the graph showing percentage of change without shortening the vertical scale.

20. The Dow Jones Industrial Average is one of the most closely followed statistics in the economy. The following chart shows how the DJIA changed during March through May of 1994. Redo the chart using a full vertical scale.

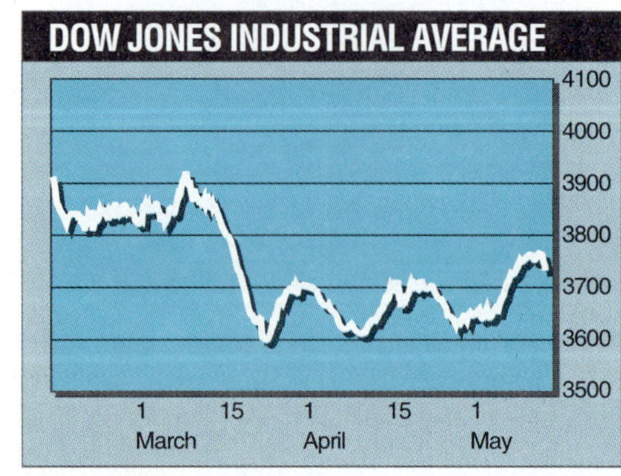

21. From April 1993 to April 1994 the average weekly wages in manufacturing in Oregon went through many changes, as shown in the following graph. Redo the graph with a full vertical scale.

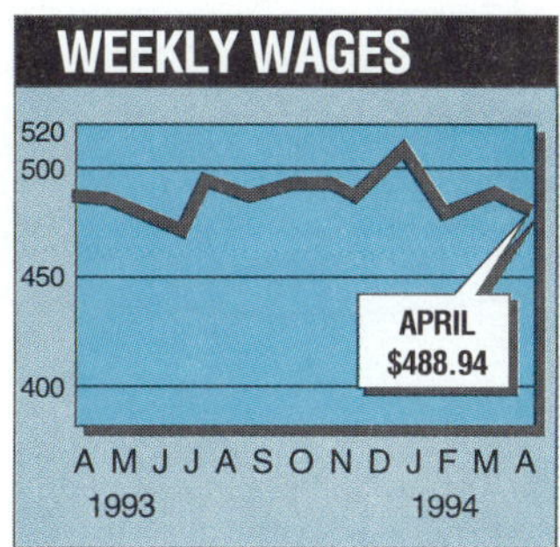

22. Health Care Reform has become a major political issue. The following graph shows health care spending as a percentage of the Gross Domestic Product (GDP). The GDP is the value of all goods and services produced in the national economy.

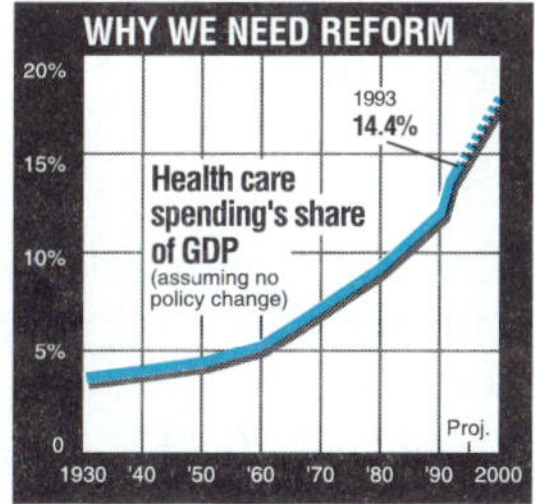

(a) Redo the graph so the increase appears even greater.

(b) Redo the graph so the increase is not so dramatic.

23. Pictograms are often drawn incorrectly even if there is no intent to distort the data. Suppose we want to show that the number of women in the workforce today is twice what it was at some time in the past. One way this could be done is to have two pictures of women representing the number of women in the workforce and draw the one for today twice as tall as the one for the past, similar to what was done with the milk cartons in Figure 3.43. The problem is that most people tend to respond to graphics by comparing areas; we are also used to interpreting depth and perspective in drawings depicting three-dimensional objects.

Suppose we want to compare the revenue of two companies. Suppose company A had revenues of $5,000,000 last year and company B had $10,000,000.

(a) If we want to use the area of circles to represent the revenues of the companies, what should be the radius of the circle for company B if the radius of the circle for company A is 1 inch? Explain.

(b) If we want to use the volume of spheres to represent the revenues of the companies, what should be the radius of the sphere for company B if the radius of the sphere for company A is 1 inch? Explain.

24. Repeat problem 23 with company A having revenues of $8,000,000 and company B having revenues of $36,000,000. Use 1 inch as the radius of the circle and sphere representing company A.

25. Redo the following graph so that Agriculture Prices from May 1993 to May 1994 don't appear to change so much.

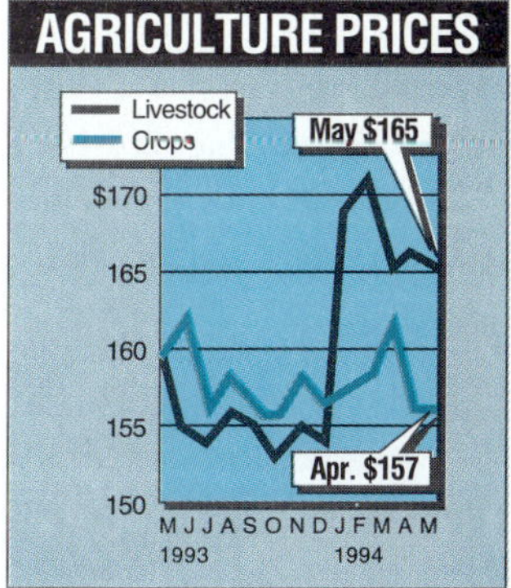

26. During the 80s and early 90s, many changes occurred with respect to the workforce, including downsizing and temporary employees. As a result, job security became a significant concern. The following graph shows the changes in attitude among workers.

(a) Redo the graph so that the downward trend is obvious

(b) Redo the graph so that the trend is apparently even worse than it is.

27. Create a 3-D bar chart for the following data.

Passenger Car Retail Sales (new)

1980	8.98 million
1985	11.04 million
1990	9.30 million

28. Create a 3-D line chart for the following data on the projected number of land fills in the United States.

1985	6000
1990	3300
1995	2600
2000	1500
2005	1100

Use the following pie chart for Meat Consumption per Person, 1989, for problems 29 and 30.

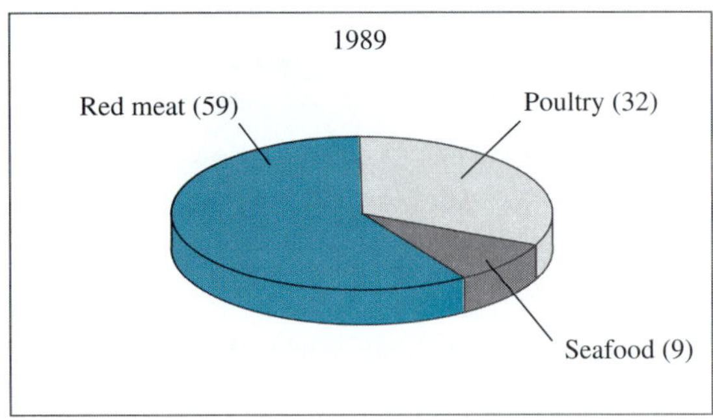

29. Create an "exploded" 3-D pie chart (Figure 3.38) to emphasize the amount of red meat consumed per person.

30. Create an "exploded" 3-D pie chart to emphasize the amount of poultry consumed per person. Rotate the pie chart to further emphasize the poultry.

The following advertisement touts the merits of a new golf ball.

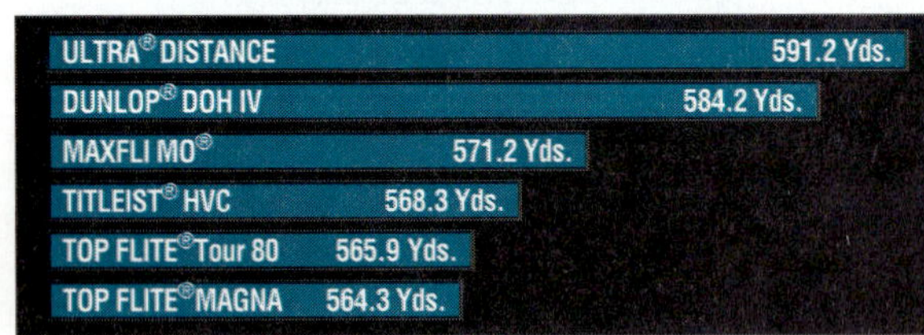

Combined yardage with a driver, #5 iron and #9 iron

31. Use the data in the golf ball advertisement to produce a new bar graph in which the length of each bar is proportional to the combined distances it represents.

32. Using perspective with pie charts can be deceiving.

52 weeks ending Dec 11, 1993 in millions of units

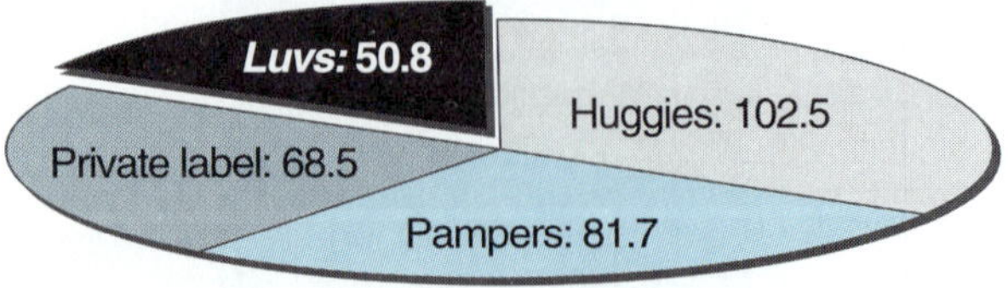

52 weeks ending Jun 13, 1992 in millions of units

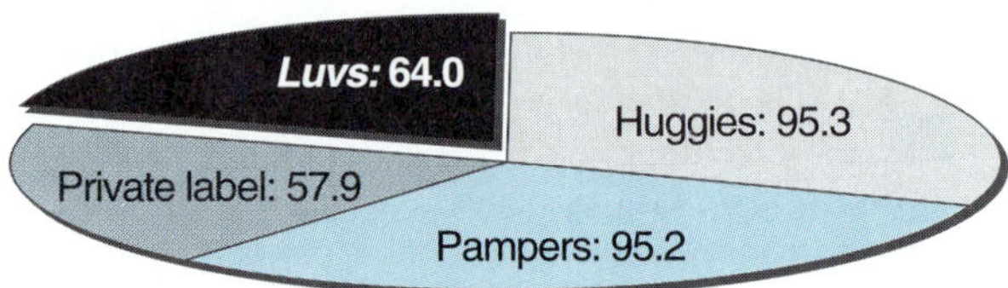

Source: Company reports, Nielson Marketing Research, Investors Business Daily

(a) Use the data from these two pie charts to draw two new pie charts in the usual manner.

(b) How do the pie charts you drew compare to the original ones?

(c) Do the comparative pieces seem the same as before?

33. The following graphs appeared together in an environmental publication. Estimate values from each graph, combine them into a single set of numbers, and produce a single bar graph.

Smog Levels Above Standards, Selected U.S. Cities
(average number of days)

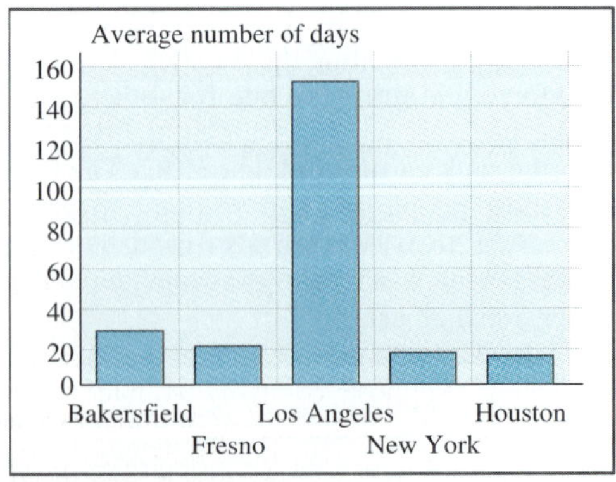

Source: U.S. Environmental Protection Agency.

Smog Levels Above Standards, Selected Canadian Cities
(average number of days)

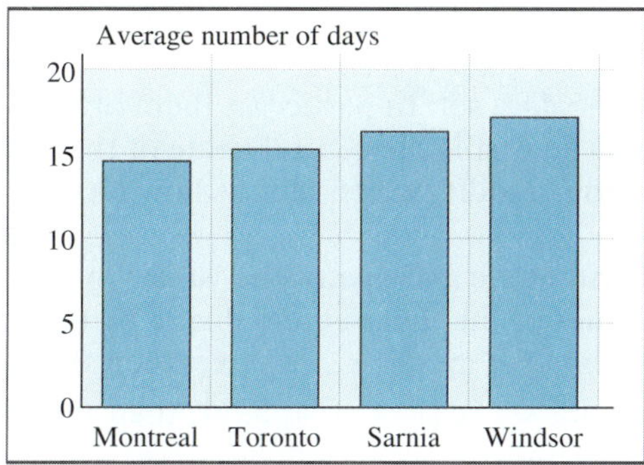

Source: Environment Canada.

EXTENDED PROBLEMS

34. In 1861, a French engineer, Charles Minard, created a graphical presentation of Napoleon's Russian campaign of 1812. This display is considered by some to be the greatest statistical graphic ever created before the advent of computer graphics. Write a report on the graph and explain its features. One source of information on this and other graphs is Edward Tufte's book, *The Visual Display of Quantitative Information.*

35. Write a report on the different types of graphical representations of data that are used in major fields of study. What types of graphics are found in textbooks, reference books, or journals? Are there any specialized types of particular interest?

36. Compare the treatment of data regarding a major national or world event in different publications such as *USA Today,* the *Wall Street Journal, Time Magazine,* or other diverse sources. How does the target audience of a publication influence the choice of the graphics? Write a report including several examples of good graphics use. Include examples of bad use if you find them.

37. Write a report on the use of statistics and graphics in the publications of the college or university you attend. Include examples of good graphics and bad. Are any of them misleading or deceptive?

38. Contact the public relations or advertising department of a large firm in your local area. What types of graphics are commonly used? Does the firm have a policy or guidelines on the use of graphics and statistics in its internal and external publications? Write a brief report.

3.4 MEANS, MEDIANS, AND PERCENTILES

INITIAL PROBLEM

A math class has 15 people. All but Howard take a test. The test scores are 56, 59, 67, 71, 73, 74, 78, 80, 81, 84, 86, 94, 95, 97. The top 5 people in the class are to be given a ticket to a concert. Howard was out of town competing for the university golf team. As a makeup the instructor had Howard take another version of the test with another class. Howard's score was 82 and the others in that class had scores of 47, 60, 66, 71, 72, 76, 80, 85, 86, and 89. Should Howard and the top 4 people from his class be given tickets to the concert, or should Howard stay home and the tickets go to the top 5 from his class? Students in the second class do not receive tickets.

Previously, we considered various ways to describe or display data visually. Often, people who use quantitative information want to summarize or characterize large sets of data with a few calculations or values that can be used to represent

TIDBIT

Means are useful for comparison. The wettest place in the world by far is Mawsynram, in Meghalaya State, India, with an average rainfull of 467.5 inches per year. The driest place is on the Pacific coast of Chile between Arica and Antofagasta with an average rainfall of less than 0.004 inches.

and communicate key features of the entire set. We will look at the most common of these measures and ways in which they can be combined with visual displays to create a more complete description and presentation of the data.

MEANS

It is often useful to give a synopsis of the data set. In particular there are two properties of a data set that are important. The first is the central value of the data. The other is the spread, or deviation from this middle, which tells us how far away from the center the bulk of the points are.

The **mean** of a set of data is the arithmetic average of the values in the set. To find the mean, we add together the values of the data set and divide by the number of data points. If the numbers in our data set are $x_1, x_2, \ldots, x_N$, then the mean is $\frac{x_1 + x_2 + \ldots + x_N}{N}$. For example, if our data set is {1, 2, 3, 4, 5, 6}, then the mean is $\frac{1 + 2 + 3 + 4 + 5 + 6}{6} = \frac{21}{6} = 3.5$ (Figure 3.52).

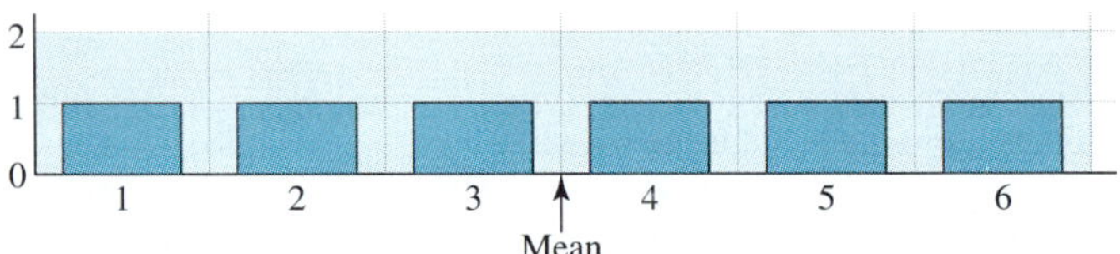

FIGURE 3.52

EXAMPLE 3.14 Find the mean of the following data sets.

(a) {1, 1, 2, 2, 3}
(b) {1, 1, 2, 2, 11}
(c) {1, 1, 2, 2, 47}

SOLUTION

(a) The mean is $\frac{1 + 1 + 2 + 2 + 3}{5} = \frac{9}{5} = 1\frac{4}{5}$.

(b) The mean is $\frac{1 + 1 + 2 + 2 + 11}{5} = \frac{17}{5} = 3\frac{2}{5}$.

(c) The mean is $\frac{1 + 1 + 2 + 2 + 47}{5} = \frac{53}{5} = 10\frac{3}{5}$. ◆

A small number of data points may have a large effect on the mean if they are far away from the rest of the data points. For example, in the data sets in Example 3.14, the mean was affected by simply increasing the largest number.

MEDIANS

The other measure of central tendency is the median. The **median** is the value closest to the middle of the data set. To find this measure, first place the data set in increasing order. If there are an odd number of data points, there is one in the exact middle. This data point is the median, *m,* of the data set. If there is an even number of data points, then there are two data points in the middle. In this case the median is the average of these two points.

Consider the data set {4, 5, 5, 6, 6}. The data is in order and the middle value is 5. The median is 5. The data set {4, 5, 5, 6, 6, 10} is also in order. There is no middle value because there is an even number of data points. We take the median to be the average of the *two* middle data points; that is, the median is the average of 5 and 6 which is $\frac{5+6}{2} = 5.5$.

EXAMPLE 3.15 Find the mean and the median of the following data sets.

(a) $\{0, 2, 4\}$
(b) $\{0, 2, 4, 10\}$
(c) $\{0, 2, 4, 10, 1000\}$

SOLUTION

(a) The median is the middle data point: 2. The mean is

$$\frac{0+2+4}{3} = \frac{6}{3} = 2.$$

(b) Since there is an even number of data points, the median is the average of the two data points nearest the middle: $\frac{2+4}{2} = 3$. The mean is

$$\frac{0+2+4+10}{4} = \frac{16}{4} = 4.$$

(c) Since we are back to having an odd number of data points, the median is the middle data point: 4. The mean is

$$\frac{0+2+4+10+1000}{5} = \frac{1016}{5} = 203.2.$$ ◆

The example shows that a particular number in a data set may have little effect on the median, but may have a large effect on the mean. This is the seesaw effect: Points far away from the center have a larger "weight."

QUARTILES

We also want a measure of dispersion or spread for data sets. We define the **first quartile,** q_1, to be the median of the lower half of the points. If N, the number of data points, is odd, then q_1 is the median of the lower half of the points, not including the middle data point. If N is even, then q_1 is the median of the lower half of the points. The **third quartile,** q_3, is the median of the upper half of the points. (Notice that the second quartile is the median.) The **interquartile range (IQR)** is $q_3 - q_1$. This measures the amount of dispersion or spread in the data.

Let us consider some examples. We saw that the median of $\{4, 5, 5, 6, 6\}$ is 5. The lower half of the data, not including the middle data point, is $\{4, 5\}$, which has a median of 4.5. This is the first quartile. The upper half of the data is $\{6, 6\}$ whose median is 6, which is the third quartile. Thus, the IQR is $6 - 4.5 = 1.5$. In the data set $\{4, 5, 5, 6, 6, 10\}$ the median is 5.5. The lower half of the data set is $\{4, 5, 5\}$, which has a median of 5; and the upper half is $\{6, 6, 10\}$, which has a median of 6. In this case, $q_1 = 5$, $m = 5.5$, $q_3 = 6$, and the IQR is $6 - 5 = 1$.

EXAMPLE 3.16 Consider the economics class test results from section 3.1. Recall that the ranked scores were

26, 32, 54, 62, 67, 70, 71, 71, 74, 76, 80, 81, 84, 87, 87, 87, 89, 93, 95, 96.

Find the median, the first and third quartiles, and the interquartile range.

SOLUTION There are 20 numbers in this data set, so there is no data point that is exactly in the middle. Counting over, we see that the 10th point is 76 and that the 11th is 80, so the median is $m = \frac{76 + 80}{2} = 78$.

Next we find the first quartile, q_1. This is the median of the first half of the data set, which is

$$26, 32, 54, 62, 67, 70, 71, 71, 74, 76.$$

There are also an even number of data points here, so the median is $\frac{67 + 70}{2} = 68.5$. Likewise, the third quartile is the median of the second half of the data set, or $\frac{87 + 87}{2} = 87$. The interquartile range is $q_3 - q_1 = 87 - 68.5 = 18.5$.

Another, much cruder measure of spread is the difference between the largest data point and the smallest data point. The **range** of the data is the difference, namely, $x_N - x_1$, where x_N is the largest and x_1 is the smallest. For the data in Example 3.16 the range is $96 - 26 = 70$. A large or small outlier will affect the range much more dramatically than it will affect the interquartile range.

BOX AND WHISKER PLOTS

Suppose we have a data set and we know that s is the smallest data point, L is the largest data point, m is the median, and q_1 and q_3 are the first and third quartiles. Then the set $\{s, q_1, m, q_3, L\}$ is called the **five-number summary** of the data. These numbers may be graphed in a **box and whisker plot** to give a picture of the data while omitting the details. For example, the five-number summary of the data from the economics class of Example 3.16, namely, {26, 68.5, 78, 87, 96}, is shown in the box and whisker plot in Figure 3.53.

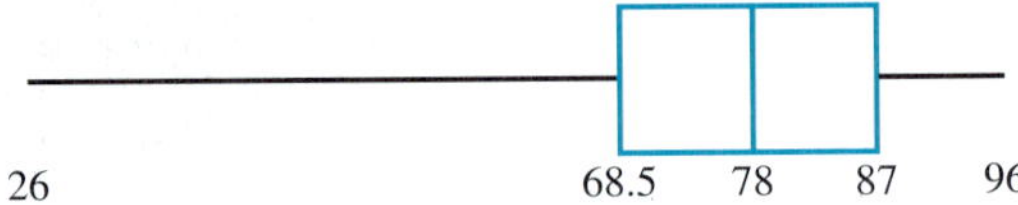

FIGURE 3.53

Box and whisker plots give a quick and easy way to compare two data sets. Suppose the ranked scores for a second economics class were:

$$34, 45, 57, 63, 67, 68, 70, 71, 72, 74, 76, 76, 78, 81, 83, 85, 85, 87, 92, 99.$$

There are still 20 data points. The median is the average of the 10th and 11th values, or $\frac{74 + 76}{2} = 75$. The first quartile is $\frac{67 + 68}{2} = 67.5$ and the third quartile is $\frac{83 + 85}{2} = 84$. Comparing the box and whisker plots for both economics tests we see that most of the class 1 scores are slightly higher than those of class 2 (Figure 3.54).

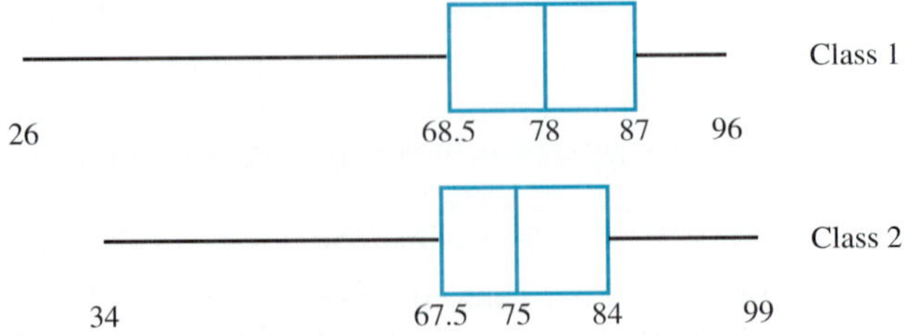

FIGURE 3.54

EXAMPLE 3.17 Monthly rainfall data for two cities is given in the table below. The first number is January's mean rainfall, the second is February's mean rainfall, and so on (Table 3.14).

TABLE 3.14 Monthly Rainfall Data

	Jan	Feb	Mar	Apr	May	June	July	Aug	Sept	Oct	Nov	Dec
St. Louis, MO	2.21	2.31	3.26	3.74	4.12	4.10	3.29	2.96	3.20	2.64	2.64	2.23
Portland, OR	0.46	1.13	1.47	1.61	2.08	2.31	3.05	3.61	3.93	5.17	6.14	6.16

Make box and whisker plots to compare the rainfall in these communities. Which community would you prefer if you wanted to live where there was the least rain? Which community would you prefer if you wanted the least variation in rainfall?

SOLUTION We compute the five-number summaries for each of the cities.

St. Louis, MO

2.21, 2.23, 2.31, 2.64, 2.64, 2.96, 3.20, 3.26, 3.29, 3.74, 4.10, 4.12
median = 3.08 $q_1 = 2.475$ $q_3 = 3.515$
five-number summary: {2.21, 2.475, 3.08, 3.515, 4.12}
interquartile range: 1.04

Portland, OR

0.46, 1.13, 1.47, 1.61, 2.08, 2.31, 3.05, 3.61, 3.93, 5.17, 6.14, 6.16
median = 2.68 $q_1 = 1.54$ $q_3 = 4.55$
five-number summary: {0.46, 1.54, 2.68, 4.55, 6.16}
interquartile range: 3.01

Box and whisker plots may now be drawn side by side for comparison (Figure 3.55).

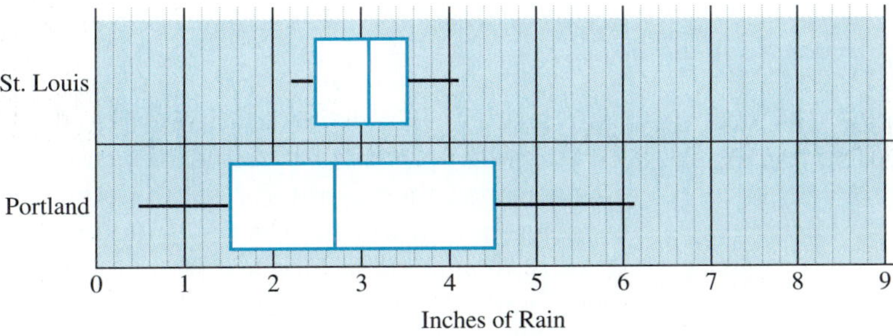

FIGURE 3.55

Several conclusions can be drawn from these two plots. Since the spread of St. Louis is much narrower than that of Portland, the amount of rain or snow is much more even in St. Louis on a month-to-month basis. The wider spread of the plot for Portland indicates that Portland has some months that are drier than St. Louis' driest months and some that are far wetter than St. Louis' wettest months. ◆

PERCENTILES

Every point, p, in a set of data has a **percentile,** which is usually defined to be the percentage of data points having equal or smaller value than p. If there are nine data points, then the seventh largest is the $\frac{7}{9} \approx 0.78 = 78\%$ or 78th percentile. In many cases, the median is the 50th percentile, but not always. If there are nine values in a data set, then the fifth value in rank order is the median. Since there are five values at or below the median, the percentile for the median in this case would be $\frac{5}{9} \approx 0.56 = 56\%$ or the 56th percentile.

Percentiles give a way to compare data points in different sets, as in the next example.

EXAMPLE 3.18 Consider the data sets

(i) $\{-2, 0, 3, 4, 10\}$ and
(ii) $\{26, 28, 33, 37, 41, 59, 77, 101, 102, 110\}$.

Find the percentile of 4 in the first data set and 101 in the second data set.

SOLUTION In the first data set, 4 is the fourth of five numbers $\left(\frac{4}{5} = 80\%\right)$, so it is the 80th percentile. In the second data set, 101 is the eighth of ten numbers $\left(\frac{8}{10} = 80\%\right)$; thus, it is also the 80th percentile. ◆

A math class has 15 people. All but Howard take a test.
The test scores are 56, 59, 67, 71, 73, 74, 78, 80, 81, 84, 86, 94, 95, 97.
The top 5 people in the class are to be given a ticket to a concert. Howard was out of town competing for the university golf team. As a makeup the instructor had Howard take another version of the test with another class. Howard's score was 82 and the others in that class had scores of 47, 60, 66, 71, 72, 76, 80, 85, 86, and 89. Should Howard and the top 4 people from his class be given tickets to the concert, or should Howard stay home and the tickets go to the top 5 from his class? Students in the second class do not receive tickets.

INITIAL PROBLEM SOLUTION

SOLUTION The test that Howard took was more difficult for that class than was the test taken by Howard's class. Not only is the mean of that test (74) less than the mean (78.2) of the other, but the lowest score and highest score are both lower than in the first class. It might not be fair to just use the score of 82 and claim Howard should not get a ticket, because 82 is a lower score than 84, the fifth highest score in his class. To fairly compare, we use percentiles. The score of 84 has rank 10 and 14 people took this test. The percentile of this score is $\frac{10}{14} \times 100 \approx 71\%$. On the other hand, Howard's score has rank 8 out of 11 test-takers. His percentile is $\frac{8}{11} \times 100 \approx 73\%$. Since Howard's percentile is higher, it makes sense to let Howard have the fifth ticket to the concert. Notice that we have rounded the percentiles to the nearest percent.

PROBLEM SET 3.4

Find the mean for each set of data in problems 1 through 8.

1. {3, 7, 12, 9, 10, 15}
2. {5, 5, 7, 10, 20, 25}
3. {2, 4, 7, 10, 11, 12, 15, 21}
4. {−4, 6, −3, −5, 12, −2, 3}
5. {2, 2, 2, 5, 7, 30}
6. {1, 4, 20, 20, 22, 23, 25, 26}
7. {2, 4, 6, 8, 10, 12, 38, 45}
8. {3, 3, 4, 4, 4, 8, 8, 8, 9, 9}
9. In problems 5 through 7, what is the effect on the mean caused by one or two values that are very different from the rest of the values in the set?
10. In problem 8, how can the idea of symmetry be used to find the mean without adding up the values?

Use the following information to answer problems 11 through 14.

During the 1980s, the Rose Bowl produced the following scores:

1980	USC 17, Ohio St. 16
1981	Michigan 23, Washington 6
1982	Washington 28, Iowa 0
1983	UCLA 24, Michigan 14
1984	UCLA 45, Illinois 9
1985	USC 20, Ohio St. 17
1986	UCLA 45, Iowa 28
1987	Arizona St. 22, Michigan 15
1988	Michigan St. 20, USC 17
1989	Michigan 22, USC 14

SOURCE: 1994 Information Please Almanac.

11. What was the mean winning score in the Rose Bowl during the 1980s?
12. What was the mean losing score in the Rose Bowl during the 1980s?
13. What was the mean number of points scored in the Rose Bowl during the 1980s?
14. What was the mean margin of victory in the Rose Bowl during the 1980s?
15. How could the results from problems 11 and 12 be used to answer problem 13?
16. How could the results from problems 11 and 12 be used to answer problem 14?
17. A set of 10 scores from a test in psychology has a mean of 80.7. When the professor goes back to check the grades at a later date, she finds that one of the scores is missing. The remaining scores are {66, 72, 75, 76, 81, 86, 88, 90, 94}. What is the missing score?
18. Twelve tests in a history class are recorded. The average is calculated as 76.5. It is later discovered that a score of 86 was incorrectly recorded as 68. What should the correct average be?

For problems 19 through 26, do the following:

(i) Arrange the data in increasing order.
(ii) Find the median.
(iii) Find the first and third quartiles.
(iv) Find the interquartile range.

19. {10, 8, 9, 3, 12, 15, 4, 6, 1, 5, 11}
20. {2, 5, 10, 20, 6, 4, 12, 15, 9, 8, 16}
21. {10, 21, 13, 6, 12, 24, 14, 26, 9, 18}
22. {7, 3, 5, 13, 20, 6, 4, 12, 15, 10, 9, 16}
23. {2, 5, 10, 10, 8, 4, 12, 15, 9, 8, 6}
24. {3, 7, 4, 6, 8, 4, 12, 9, 8, 10, 4, 7}
25. {4, 1, 2, 5, 8, 2, 6, 9, 4, 3, 1, 5, 10, 4}
26. {22, 31, 38, 30, 25, 29, 31, 26, 40, 34, 26, 29}
27. In the final round of tournament play, the 12 members of the university's golf team recorded the following scores: 78, 81, 77, 76, 84, 81, 73, 95, 78, 86, 80, 79.
(a) Find the five-number summary.
(b) Draw a box and whisker plot for the scores.
28. Students in a literature class received these final exam scores: 82, 80, 93, 88, 98, 85, 82, 77, 90, 78, 83, 75, 86, 66, 91, 85, 93.
(a) Find the five-number summary.
(b) Draw a box and whisker plot for the scores.
29. Although known as a member of the New York Yankees, for whom he played from 1920 until 1934, Babe Ruth started his major league career as a pitcher with the Boston Red Sox from 1915 to 1919. He began playing in the outfield in 1918, and then played there almost exclusively when he was traded to the Yankees. Ruth also played sparingly for the Boston Braves for one year. For the years when Babe Ruth was an outfielder, his home run totals were: 11, 29, 54, 59, 35, 41, 46, 25, 47, 60, 54, 46, 49, 41, 34, 22, 6.
(a) Find the mean and median for Ruth's home run totals as an outfielder.
(b) Draw a box and whisker plot for his home run totals as an outfielder.
30. During his major league career, from 1914 to 1935, Babe Ruth hit 714 home runs, a record that many thought would never be broken. The record lasted nearly 40 years when it was broken by Hank Aaron, who eventually hit a total of 755 home runs. In his 23 years in the majors, Aaron's home run totals were: 13, 27, 26, 44, 30, 39, 40, 34, 45, 44, 24, 32, 44, 39, 29, 44, 38, 47, 34, 40, 20, 12, 10.
(a) Find the mean and median for Aaron's home run totals.

(b) Draw a box and whiskers plot for Aaron's home run totals.

31. In a study of surgical procedures performed by Swiss doctors, the number of hysterectomies performed by a sample of 15 male doctors were: 27, 50, 33, 25, 86, 25, 85, 31, 37, 44, 20, 36, 59, 34, 28. A sample of 10 Swiss female doctors was studied at the same time and the number of hysterectomies recorded by these doctors were: 10, 7, 19, 33, 5, 14, 31, 29, 18, 25. Draw side-by-side box and whisker plots for the male and female doctors.

32. Draw side-by-side box and whisker plots for Babe Ruth's and Hank Aaron's home run totals in problems 29 and 30.

33. Exam scores for a 100-point economics test were as follows:

67	92	75	94	88
73	87	78	84	90
91	68	70	83	82
80	71	76	84	93
74	82	66	81	

(a) What is the percentile rank for a score of 90?
(b) What is the percentile rank for a score of 70?

34. Scores for a 50-point sociology quiz were as follows:

41	37	27	48	40
45	42	36	43	40
38	36	38	41	35
28	37	34		

(a) What is the percentile rank for a score of 45?
(b) What is the percentile rank for a score of 35?

35. A student was enrolled in both the economics class (problem 33) and the sociology class (problem 34). Her score on the economics test was 88 and she had a 43 on the sociology quiz. In comparison to her classmates, in which class did she do better?
(a) Provide a rationale for saying that she did better in the economics class.
(b) Provide a rationale for saying that she did better in the sociology class.

36. In terms of the other exam and quiz scores from problems 33 and 34, which performance would be better: 80% correct on the economics exam or 80% correct on the sociology quiz? Justify your answer.

Use the following information for problems 37 through 40.

In 1993, the United States made the following direct investment in European Economic Community (EEC) countries:

Country	Investment (billions)
Belgium	$11.6
Denmark	1.8
France	23.6
Germany	37.5
Greece	0.4
Ireland	9.6
Italy	13.9
Luxembourg	2.3
Netherlands	19.9
Portugal	1.2
Spain	6.4
United Kingdom	96.4

SOURCE: Survey of Current Business.

37. Find the mean, median, and five-number summary for the data on direct investment in EEC countries.

38. Find the mean, median, and five-number summary when the United Kingdom is removed from the list of countries.

39. Prepare side-by-side box and whisker plots comparing direct investments to EEC countries,
(a) including the United Kingdom.
(b) excluding the United Kingdom.

40. **(a)** What is the effect on the mean when the United Kingdom is excluded?
(b) What is the effect on the median?

Use the following information for problems 41 and 42.

1992 Marriage Rates and Birth Rates for Selected Countries (per 1000 population)

Country	Marriage Rate	Birth Rate
Belgium	5.8	12.4
Denmark	6.2	13.1
France	4.7	12.9
Germany	5.7	10.0
Greece	4.7	10.1
Ireland	4.5	14.5
Italy	6.4	9.9
Netherlands	6.4	13.0
Poland	5.7	13.4
Sweden	4.3	14.1
United Kingdom	6.1	13.5
United States	9.2	15.7

SOURCE: United Nations, Monthly Bulletin of Statistics.

41. **(a)** Find the mean, median, and five-number summary for the marriage rates of the selected countries.
(b) Draw a box and whisker plot for the data.

42. **(a)** Find the mean, median, and five-number summary for the birth rates of the selected countries.
(b) Draw a box and whisker plot for the data.

EXTENDED PROBLEMS

An outlier is a data point that appears to be not typical of the data as a whole. In general, this is imprecise and subject to interpretation. In the case of data sets such as we are studying and in the context of ranked data, many statisticians have agreed to call any data point an outlier if it is more than 1.5 times the interquartile range, 1.5(IQR) below the first quartile, or 1.5(IQR) above the third quartile. In box and whisker plots, outliers are usually denoted by an asterisk and the highest and lowest non-outliers are plotted as the ends of whiskers.

There are two outliers in the economics class test scores considered in Example 3.16. The interquartile range is IQR = 18.5 and 1.5(IQR) = 1.5(18.5) = 27.75. Thus any scores below 68.5 − 27.75 = 40.75 or above 87 + 27.75 = 114.75 are outliers. There are two scores that fit this bill, namely, 26 and 32. We could redo our box and whisker plot to show these scores as outliers:

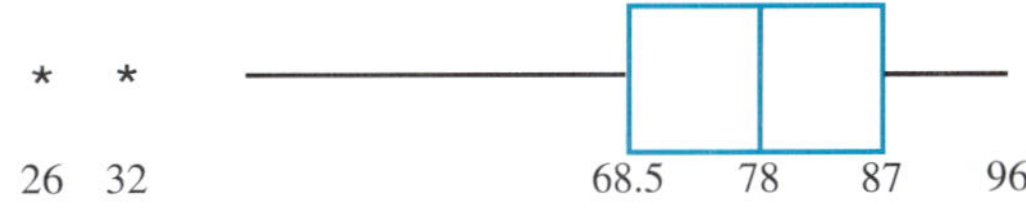

43. For the golf scores in problem 27, identify any outliers and draw a modified box plot.

44. Determine if there are any outliers in the literature exam scores from problem 28. Draw a modified box plot for the data.

45. Are there any outliers in the home run totals of Babe Ruth? Draw a modified box plot.

46. Are there any outliers in the home run totals of Hank Aaron? Draw a modified box plot.

47. **(a)** Show that the United Kingdom is an outlier in problem 37.

(b) Are there any other outliers?

(c) Draw a modified box plot.

Chapter Three Problem

As a member of a watchdog committee, you read that the mayor says the anti-crime program is working because the rate of increase of crime has been decreasing. The following chart is given to show evidence of this. How do you analyze the situation?

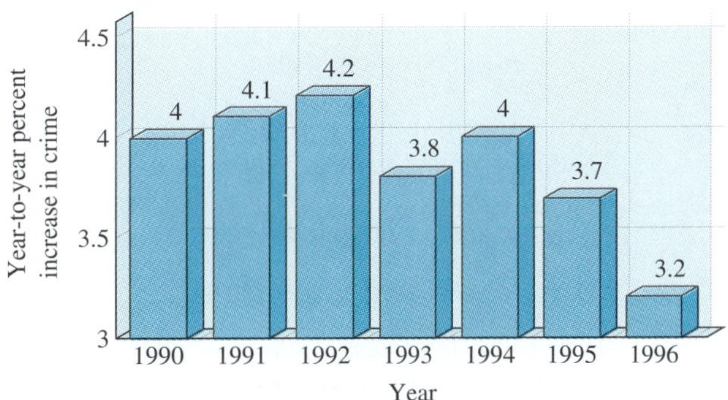

Solution

Strategies: Use a variable, make a table, and make a chart.

First we want to understand the level of crime during this time period, and the mayor provides the year-to-year *rate* of increase, which is decreasing. Let C = the amount of crime in 1989. Since the growth rate in 1990 was 4% of the amount of crime in 1989 (C), the amount of crime in 1990 was C + (0.04)C = 1.04C. The growth rate in 1991 was 4.1%, so the amount of crime is (1.041)(1.04)C ≈ 1.083C. The growth rate in 1992 was 4.2%, so the amount of crime is 1.042 multiplied by last year's amount of crime or (1.042)(1.083)C ≈ 1.128C. Continuing in this way, we make a table of the amount of crime relative to 1989.

Year	Amount of Crime
1989	C
1990	1.04C
1991	1.083C
1992	1.128C
1993	1.171C
1994	1.218C
1995	1.263C
1996	1.303C

We see that although there may have been a decline in the percent of increase of crime, there are still increasing amounts of crime every year. It would seem premature to declare victory in the war on crime until that amount is reduced. To show what was happening to crime, we make a chart displaying the data in the table as follows.

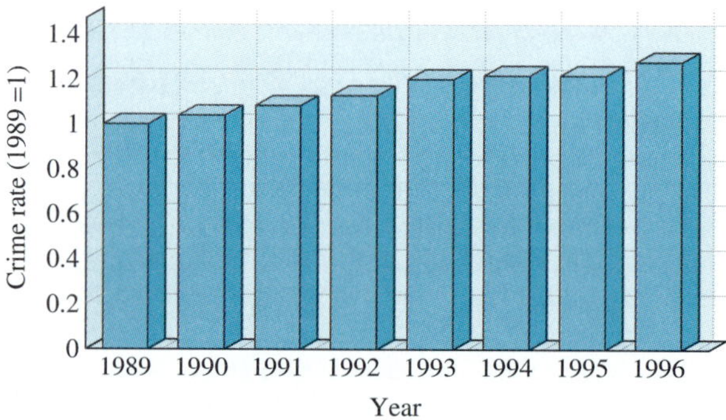

This graph shows a steady, though slowing, *increase* in the amount of crime over the last six years with 1989 having a base of 1.

Chapter Three Review

Key Ideas and Questions

The following questions review the main ideas of this chapter. Write your answers to the questions and then refer to the pages listed by number to make certain that you have mastered these ideas.

1. What are the six main types of graphs in this chapter, and what are the strengths and weaknesses of each of these types? How can these graphs be used to compare different, but related, data sets? **123–145**
2. Describe four ways that graphs may be altered to influence perception of the viewer. **153–160**
3. What are two measures of central tendency for a data set? How can we express the amount of spread in a data set? Describe how a box and whisker plot can graphically compare two data sets. **174–176**

Vocabulary/Notation

Following is a list of key vocabulary, notation, and ideas for this chapter. Mentally review each of these items, write down the meaning of each term, and use it in a sentence. Then refer to the pages listed by number and restudy any material you are unsure of before solving the Chapter Three Review Problems.

Section 3.1

Section 3.2

Section 3.3

Section 3.4

Chapter Three Review Problems

1. A study considered the starting salaries of a group of college graduates that were hired in the communication industry. These salaries were:

\$28,518	\$26,121	\$27,089	\$24,890
\$26,856	\$28,220	\$27,660	\$25,812
\$27,818	\$27,500	\$26,549	\$25,900
\$26,500	\$28,120	\$26,700	\$27,160

(a) Round this data set to the nearest \$100.

(b) Make a dot plot of these rounded salaries.

(c) Make a stem and leaf plot of these rounded salaries.

2. Make a histogram for the salaries in problem 1 rounded to the nearest \$1000.

Use the following data set for problems 3 and 4. These represent grades given to craft projects in a class.
{0.7, 09, 1.1, 1.2, 1.5, 1.7, 1.9, 2.0, 2.0, 2.0, 2.1, 2.4, 2.5, 3.2}.

3. Make a histogram of the data set by grouping into bins of width 0.5 so that 0.6 through 1.0 are in the same bin, 1.1 through 1.5 are in the same bin, and so on.

4. What is the frequency of the bin with the highest frequency? Are there any outliers?

Use the following chart for problems 5 through 8.

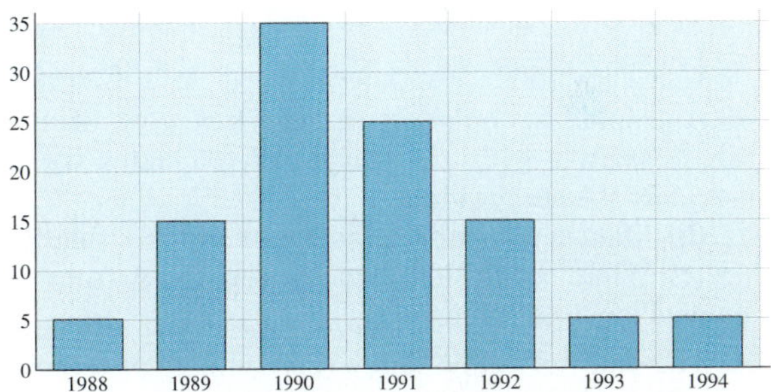

5. The chart gives sales figures of Carnivore Caveman action figures in units of \$100,000 from 1988 to 1994. Convert the graph to a table of values. Which year had the largest sales?

6. Make a line graph of the Carnivore Caveman action figure sales emphasizing the change in sales.

7. What is the percentage change from 1989 to 1990? From 1990 to 1991?

8. What year had the greatest percentage decrease from the previous year?

9. What is the main advantage of a stem and leaf plot as opposed to either a dot plot or a histogram?

10. Suppose the proportion of teenage consumer spending in relation to other age groups has been increasing over the last few years. Which type of graph would be easiest to display this trend?

11. As part of a report you need to show amounts of sales of refrigerators, stoves, dishwashers, and hot water heaters from the years 1950 through 1990. Which type of graph would be clearest to display this information?

12. A survey is done of an ice cream parlor. The parlor serves five kinds of ice cream: vanilla, chocolate, rocky road, cookies and cream, and strawberry. The week's sales of each kind were:

vanilla	120 pounds
chocolate	86 pounds
rocky road	68 pounds
cookies and cream	90 pounds
strawberry	50 pounds

Make a pie chart showing the proportions that were sold. Label each sector with kind of ice cream and percentage.

13. Suppose that the study in problem 1 considers a new group of college graduates who also have MBA degrees (Masters of Business Administration). These starting salaries after rounding to the nearest $100, were:

$29,300	$27,000	$27,900	$25,500
$27,600	$28,100	$28,000	$26,400
$28,100	$27,300	$27,300	$25,800
$27,800	$29,500	$26,500	$26,700

(a) Make a double stem and leaf plot of these salaries with the salaries of graduates without MBAs in problem 1.
(b) Make a comparison histogram of these salaries.
(c) What conclusions would you draw from this data?

14. The data set below shows the numbers of rentals in millions by several automobile rental companies:

	Hurry-Up	Airus	Up&Coming
1988	6.7	4.1	3.2
1989	6.8	4.3	3.4
1990	6.9	4.4	3.7
1991	5.5	4.2	4.3
1992	5.1	4.6	5.6
1993	4.8	4.7	6.9
1994	4.9	4.6	8.1

(a) Make a multiple bar chart to show this data.
(b) Make a multiple line chart to show this data. Do you prefer the multiple bar chart or the multiple line chart? Why?
(c) Which company has had the most growth since 1988?

15. **(a)** Suppose that the CEO of Up&Coming wishes to show the data in problem 14 to his advantage. Make a multiple line graph with new axes to stress the advantage that Up&Coming has. Besides cropping the vertical axis, you might also consider cropping the horizontal axis.
(b) Suppose the CEO of Hurry-Up wishes to graph the data so as to de-emphasize the advantage that Up&Coming has gained. Make a chart that he would like to use.

16. (Continuing problems 14 and 15) Market share is defined as the percentage of the market that a company has. Suppose that all other companies have the following numbers of rentals in millions:

	All Other Rental Companies
1988	10.1
1989	10.2
1990	10.6
1991	10.1
1992	11.0
1993	12.1
1994	12.4

Combine the data from problems 14 and 16. Make proportional bar charts to show the trends in market share.

17. Make a sequence of pie charts showing all car rentals from 1991 to 1994 with the sector corresponding to Up&Coming "exploded" to make the rise in Up&Coming more dramatic.

18. Criticize the following graphs.
(a) The following pie chart was put out by a library to stress how many residents use it.

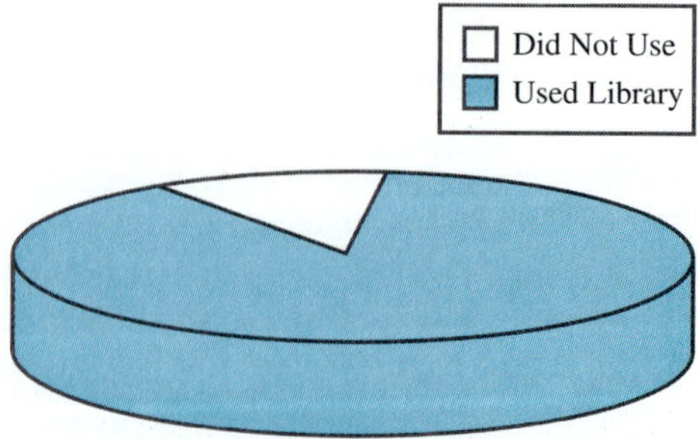

19. Consider the data set {2, 10, 8, 3, 7, 5, 4, 7, 14, 6, 8, 9, 12}. What are the mean and the median of this data set? What is the five-number summary? What is the percentile of the data point 6?

CHAPTER 4

INFERENTIAL STATISTICS—DISTRIBUTIONS, SURVEYS, SAMPLES, AND CONFIDENT PREDICTION

LITERARY DIGEST POLL SHOWS LANDSLIDE VICTORY FOR LANDON OVER ROOSEVELT

The Literary Digest magazine surveyed an astonishing 2,400,000 Americans in 1936. The results showed that 57% were planning to vote for Alf Landon and 43% were planning to vote for President Franklin D. Roosevelt, thus a landslide Landon victory was expected with a return to Republican dominance of the presidency. The election was won by Roosevelt with 62% of the popular vote!

Chapter Goals

1. Identify biased and unbiased samples.
2. Compute percentages from a "normal population".
3. Choose a random sample from a population.
4. Compute the mean and standard deviation of a sample proportion.
5. Compute confidence intervals.

Public opinion polls are a regular feature of our daily lives and have been fairly common for more than a century. Even as early as the presidential election of 1824, the Harrisburg *Pennsylvania* reported a "straw vote taken without discrimination of parties" which indicated Andrew Jackson was the popular choice for president over John Quincy Adams.

In the 1936 election, Roosevelt received 62% of the popular vote; his victory in the Electoral College was even more decisive, winning by a margin of 432 to 8. The Literary Digest poll had a huge number of participants from all over the United States, and should have been accurate *if* the sample had been chosen properly. The problem was that the sample was not representative of the entire American electorate; instead the Digest's sample of voters was drawn from its subscription lists and lists of automobile and telephone owners. In contrast, the Gallup and Roper polls, which used new scientific sampling methods, correctly predicted Roosevelt's victory.

Statistics are used to inform us and to influence our behavior: customer satisfaction statistics induce us to buy a certain car, statistics about drunk drivers teach us to not drink and drive, statistics connect sunspots and stock market prices, and so on. We often encounter statistics that are badly misused or are so counter to our experience that we automatically mistrust them. Unfortunately, we are not always able to tell the difference.

This chapter will describe how to interpret statistics and judge them critically. In particular, it will show how a small sample can provide us with meaningful information about the entire population and how an incorrectly chosen sample can give us incorrect or misleading information.

THE HUMAN SIDE OF MATHEMATICS

W. S. Gossett

One of the oldest breweries in the world is the Guinness Brewing Company in Dublin. Guinness began as a family business in 1759. Its markets grew worldwide. However, the brewing process was overseen by master brewers using arcane methods handed down from master to apprentice. The Guinness corporation was interested in making this process scientific and constructing an exact recipe that could be used worldwide. This was a novel idea that required new techniques.

W. S. Gossett, born in 1876, studied chemistry at the university in Dublin. He was hired by Guinness as a brewmaster in 1899 to work on the problem of making brewing a science. One question concerned finding the best kind of barley to use. Gossett gathered agricultural data and other information about barley. He realized that differences found in the data could be accidental or simply due to natural variation. On the other hand, they could be the result of differences in treatment or process and thus lead to better methods of brewing. There was no way to tell which was which. Gossett went to work at the laboratory of the biometrician, Karl Pearson, to study statistics. During this time, Gossett solved the problem of data variation and developed new techniques. Then, in 1907 he returned to Guinness to be brewer-in-charge. Because of his connection with the Guinness company it was decided that he would not publish his ideas under his own name, but rather use the pseudonym, "Student." His work created the modern field of statistical inference, and the primary method for working with small samples is known as *Student's t-test*.

Ronald A. Fisher

In 1919, Ronald A. Fisher (1890–1962) became statistician at the Rothamsted Experimental Station, the oldest agricultural research station in Great Britain. He conducted field studies and worked on genetics; in the process, he pioneered the use of randomization in experimental design and invented formal statistical methods for analysis of experimental data. Fisher's research produced 55 groundbreaking papers that extended and clarified the revolutionary work of Gossett, Pearson, and others. This research formed the basis for the theory of making inferences from samples, which we use today.

The new scientific method, employing randomization and probability theory, spread quickly through the biological sciences and then to the medical sciences. Ironically, Fisher also played a significant role in what is perhaps the greatest statistical debate in history: Does tobacco cause lung cancer? An experimental study is needed to properly answer the question. However, this requires finding a group of nonsmokers, separating them into two groups, having the members of one group smoke over a long period of time, and then comparing their medical conditions. Such a study has not been done, for ethical reasons. Various other approaches were tried; these included comparing medical histories for groups of smokers and nonsmokers or following groups of smokers and nonsmokers for many years and then comparing rates of lung cancer. By the late 1950s all the studies that had been done came to the same conclusion: There was a significantly higher rate of lung cancer among smokers.

Fisher had always been concerned about how any such nonexperimental studies were interpreted. This, together with his work in genetics, caused him to argue that if a person had a hereditary predilection for smoking and also a hereditary predisposition for disease, the results would be exactly like those that were found in the studies. Fisher repeatedly pointed out that the evidence against tobacco as a health hazard was only circumstantial.

4.1 NORMAL DISTRIBUTIONS

Since the fall of the Berlin Wall in 1990, many companies from the United States and western Europe have either provided assistance to eastern European countries or have established business operations and enterprises of their own. Suppose that the company you work for is establishing offices and manufacturing facilities in the country of Midrovia. In order to design workplace areas you have to estimate the percentage of women in the country who are at least 5′2″ tall. You have found some information on this, namely, that the mean height of Midrovian women is 5′4″ and the standard deviation is 2″. You can find no other information. What should you do?

In this chapter we look at the process of making inferences (educated guesses) about a population based on information from a sample. Although the characteristics in a sample of the population, such as the mean or proportion, may vary from those of the population as a whole, the variations are systematic and can be understood in terms of the variation within the population itself. In this section, we introduce two important statistical tools: standard deviation and the normal distribution.

DEVIATION

The data set {1, 2, 3, 5, 9} has a mean of $\frac{1 + 2 + 3 + 5 + 9}{5} = \frac{20}{5} = 4$. In an effort to determine how the data is spread out, it is useful to consider the difference between a data point and the mean. This is called the **deviation from the mean,** or simply **deviation,** of the point.

EXAMPLE 4.1 Make a table of the data set {1, 2, 3, 5, 9} together with the deviations of each point.

SOLUTION As seen above, the mean of the data set is 4.

Data Point	Deviation
1	$1 - 4 = -3$
2	$2 - 4 = -2$
3	$3 - 4 = -1$
5	$5 - 4 = 1$
9	$9 - 4 = \underline{5}$
	Total = 0

◆

Notice that the sum of all of the deviations from the mean in Example 4.1 is zero. In fact, this can be shown to be true for all data sets. Since the average of the deviations for any data set is the sum of the deviations, namely zero, divided by the number of data points, the average deviation must also always be zero. Thus, the average of the deviations from the mean does not give us any information about the spread of the data.

To obtain some meaningful information about the spread of a data set, we consider the **variance,** which is the average of the squares of the deviations.

EXAMPLE 4.2 Compute the variance of the data set {1, 2, 3, 5, 9}.

SOLUTION We expand the table in Example 4.1 by finding the squares of the deviations. Then the variance is their average.

Data Point	Deviation	Deviation²	
1	$1 - 4 = -3$	$(-3)^2 = 9$	
2	$2 - 4 = -2$	$(-2)^2 = 4$	
3	$3 - 4 = -1$	$(-1)^2 = 1$	
5	$5 - 4 = 1$	$1^2 = 1$	
9	$9 - 4 = \underline{5}$	$5^2 = \underline{25}$	
	Total = 0	Total = 40	Variance $= \frac{40}{5} = 8$

◆

TIDBIT

The standard deviation has been used for well over a century, but sometimes with different names. The mathematician Gauss used the term "mean error" in the early 19th century.

The square root of the variance is called the **standard deviation** of a data set. In Example 4.2, the standard deviation of the data set is $\sqrt{8} \approx 2.83$.

EXAMPLE 4.3 Find the mean, variance, and standard deviation of the data sets {0, 3, 6}, and {2, 3, 4}.

SOLUTION The mean of each set is 3. However, the data are bunched closer together in the set {2, 3, 4}, so the variance should reflect this information.

Data Point	Deviation	Deviation²
0	$0 - 3 = -3$	$(-3)^2 = 9$
3	$3 - 3 = 0$	$0^2 = 0$
6	$6 - 3 = 3$	$3^2 = \underline{9}$
		Total = 18

The variance of {0, 3, 6} is $\frac{18}{3} = 6$ and the standard deviation is $\sqrt{6} \approx 2.45$.

Data Point	Deviation	Deviation²
2	$2 - 3 = -1$	$(-1)^2 = 1$
3	$3 - 3 = 0$	$0^2 = 0$
4	$4 - 3 = 1$	$1^2 = \underline{1}$
		Total = 2

The variance of {2, 3, 4} is $\frac{2}{3}$ and the standard deviation is $\sqrt{2/3} \approx 0.82$. ◆

As shown in Example 4.3, the data set that is more spread out has the larger variance and standard deviation. This is the case in general.

Many scientific calculators have built-in statistical functions that can be used to find the mean, variance, and standard deviation. When using these functions on a calculator, it is important to enter the values from the data set by entering any repeated number as often as it occurs since the calculator uses the *number* of times data is entered as the divisor when finding the mean. The following keystrokes show how one such calculator would be used to find the mean and standard deviation for the data set

$$\{5, 6, 6, 7, 7, 7, 7, 7, 8, 8, 8, 8, 8, 8, 8, 9, 9, 9\}$$

once the calculator is put into its STAT mode.

5 [Σ+] 6 [Σ+] 6 [Σ+] 7 [Σ+] 7 [Σ+] 7 [Σ+] 7 [Σ+]
7 [Σ+] 8 [Σ+] 8 [Σ+] 8 [Σ+] 8 [Σ+] 8 [Σ+] 8 [Σ+]
8 [Σ+] 9 [Σ+] 9 [Σ+] 9 [Σ+] [18]

(Note: The Greek letter sigma, Σ, represents summation in mathematics.) As the data points are entered, the screen keeps a tally. After the last entry, the screen shows that there are 18 data points. The following keystrokes show how to find the (i) mean, $\bar{x}$, and (ii) standard deviation, σ_n.

(i) [2nd function] [$\bar{x}$] [7.5]

(ii) [2nd function] [σ_n] [1.07]

The variance can be found by squaring the standard deviation. That is, variance $= 1.07^2 \approx 1.14$.

Z-SCORES

The mean and the standard deviation can also be used to compare data points from different data sets. To do this, each data point of a data set has a ***z*-score,** which measures the number of standard deviations between the data point and the mean. That is, if a data point is one standard deviation greater than the mean, then it has a z-score of 1. If a data point has a z-score of -2, then it is two standard deviations less than the mean. To compute the z-score of a data point, we use the following formula:

$$z\text{-score} = \frac{\text{data point} - \text{mean}}{\text{standard deviation}}.$$

EXAMPLE 4.4 Compute the z-scores of each data point of $\{1, 2, 3, 5, 9\}$.

SOLUTION From Example 4.2, the mean of the data set is 4 and its standard deviation is $\sqrt{8} \approx 2.83$. Thus the z-scores are as follows:

Data Point	z-score
1	$\frac{1-4}{2.83} \approx -1.06$
2	$\frac{2-4}{2.83} \approx -0.71$
3	$\frac{3-4}{2.83} \approx -0.35$
5	$\frac{5-4}{2.83} \approx 0.35$
9	$\frac{9-4}{2.83} \approx 1.77$

◆

Notice that the data points that are below the mean have negative z-scores and the ones above the mean have positive z-scores. If the mean is a data point in a set, then its z-score is zero. Also, the sum of the z-scores is zero.

Associated with every data set is the set of z-scores for each of its data points. This new set of data has mean 0 and standard deviation 1. Using this idea, we can picture a data set using two different horizontal scales: one that shows the numbers in the data set and one that shows the z-scores (Figure 4.1).

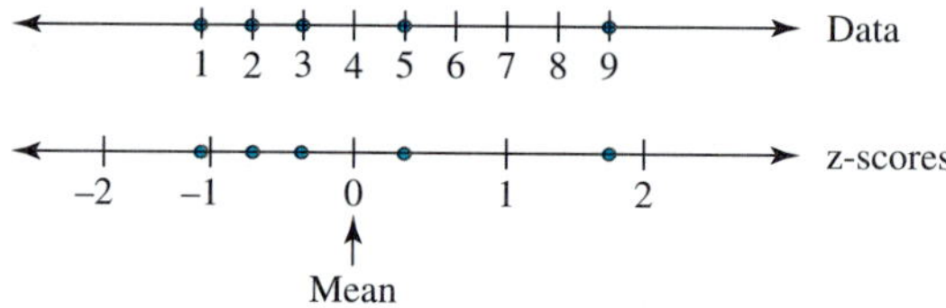

FIGURE 4.1

STANDARD NORMAL DISTRIBUTION

When you have a large collection of data that includes many different values (for example, heights of a large group of people), the histogram that represents the data will often approximate a symmetric bell-shaped curve such as the one illustrated in Figure 4.2.

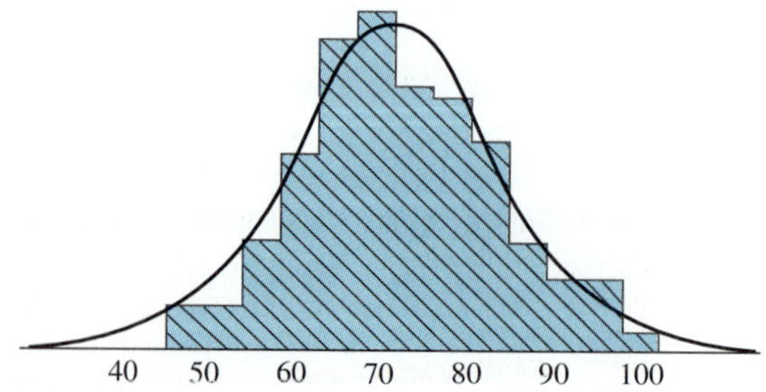

FIGURE 4.2

A data set that is represented by this type of ideal bell-shaped curve is said to be normally distributed and is called a **normal distribution.** If the bell-shaped curve represents a data set that has mean 0 and standard deviation 1, it is called a **standard normal distribution** (Figure 4.3).

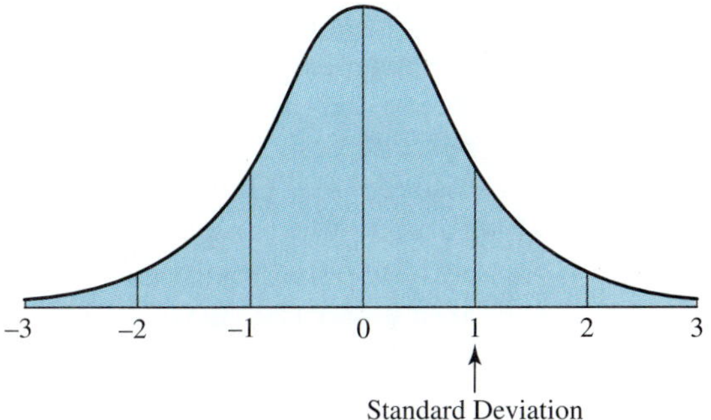

FIGURE 4.3

The z-scores for a large collection of data often approximate the shape of a standard normal distribution. This is one of the reasons that z-scores are so important as a tool for analyzing data.

COMPUTING WITH THE STANDARD NORMAL DISTRIBUTION

Several methods can be used to find the area that lies between any two vertical lines (corresponding to z-scores on the standard normal distribution) as a percentage of the entire area under the curve of the standard normal distribution. Some of these percentages are given in Figure 4.4.

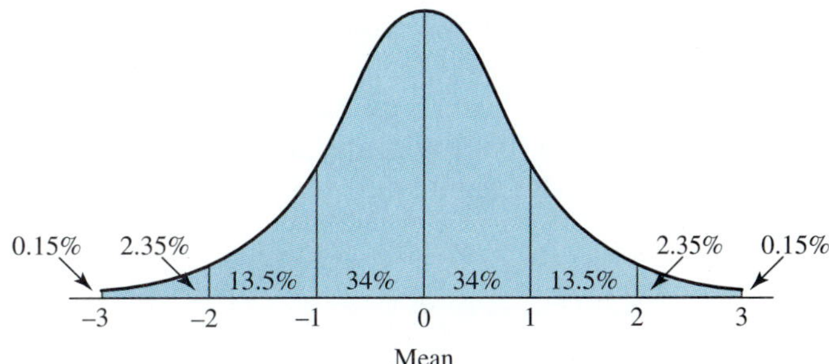

FIGURE 4.4

Note that the area under the entire standard normal curve is 100% of the data and 50% of the data is above (and 50% is below) the mean.

The information about the percentage of area under part of a standard normal curve is used to approximate the percentage of numbers having z-scores in that range. Thus, if we are told that a set of numbers is normally distributed, then the percentage of numbers having a z-score between -1 and $+1$ is obtained by adding the percentages for each included slice as shown in Figure 4.4; that is, 34% + 34% or 68%.

EXAMPLE 4.5 Suppose a population has a standard normal distribution.

(a) Find the percentage of the population that has a value between −2 and 1.
(b) Find the percentage of the population whose value does not lie between 1 and 2.

SOLUTION

(a) Referring to Figure 4.4, there are three regions between −2 and 1. The percentages associated with these regions are 13.5%, 34%, and 34%, respectively. The total percentage is 13.5% + 34% + 34% = 81.5%.
(b) To compute the percentage that does not lie between 1 and 2, we could add the percentages associated with the 7 regions. However, since the total area under the curve is 100% and the area of the region between 1 and 2 is 13.5%, the area outside this region is 100% −13.5% = 86.5%. ◆

EXAMPLE 4.6 A professor in charge of a class of 430 students says that she will be grading on a curve using z-scores where anyone with a z-score of 1 or greater will receive an "A." About how many A's will she give?

SOLUTION Referring to Figure 4.4, the sum of the percentages of the regions exceeding a z-score of 1 is 13.5% + 2.35% + 0.15% = 16%. Since 16% × 430 = 68.8, about 69 students should get A's. ◆

COMPUTING WITH NORMAL DISTRIBUTIONS

In practice, many populations have a normal distribution, but not a standard normal distribution. Most of these populations have a mean that is different from 0 or a standard deviation that is different from 1. For example, IQ scores are normally distributed with a mean of 100 and a standard deviation of 15. Weights of rabbits, crop yields, and lengths of manufactured parts also tend to be normally distributed. The normal distribution is a very common distribution.

Suppose we wish to compute the percentage of people who have an IQ between 85 and 115. To compute these percentages, we add another horizontal axis below the horizontal axis of the standard normal distribution. We call the original axis with 0 at the center and 1 at the standard deviation the ***z*-score axis,** or simply the ***z*-axis** (Figure 4.5).

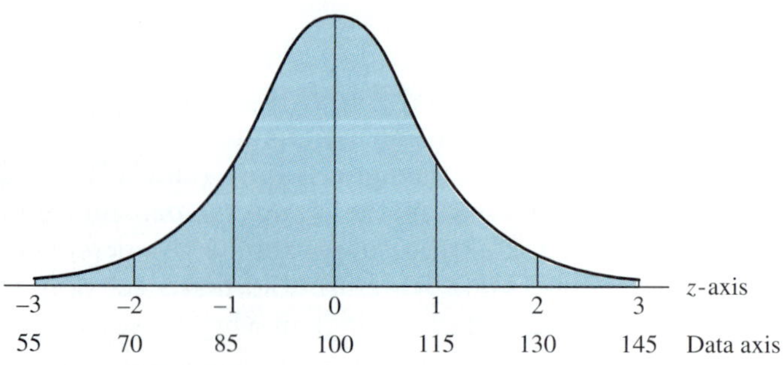

FIGURE 4.5

The axis we add below this is called the **data axis.** We label the data axis so that the mean of our distribution, 100, lies below 0 on the z-axis. We then label the data axis so that below 1 on the z-axis we put 115, which is the mean plus one standard deviation. Beneath -1 we put $85 = 100 - 15$; beneath -2 we put $70 = 100 - 2(15)$ and beneath -3 we put $55 = 100 - 3(15)$. Similarly, we put 130 beneath 2 and 145 beneath 3. Notice that the z-axis gives the z-score of the corresponding point on the data axis.

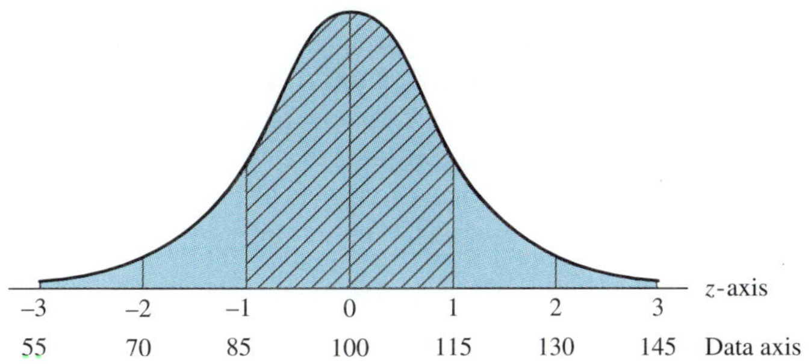

FIGURE 4.6

Now we can compute as before. The percentage of people with scores between 85 and 115 is the sum of the percentages of the two central slices, or $34\% + 34\% = 68\%$ (Figure 4.6).

Similarly, the percentage of people with an IQ score larger than 145 is 0.15%, the percentage associated with the last slice of the normal curve. This means that only 15 people in 10,000 have IQs this large (although perhaps a larger percentage *think* they do).

EXAMPLE 4.7 The mean weight of rabbits is 4.5 pounds and the standard deviation is 0.5 pound. What percentage of rabbits have weights between 4.0 and 5.5 pounds? (You may assume that the weights of rabbits have a normal distribution.)

SOLUTION We set the data axis according to the mean and standard deviation of the population. Below 0 on the z-axis we put 4.5. Below 1 on the z-axis we put $4.5 + 0.5 = 5$, and so on (Figure 4.7).

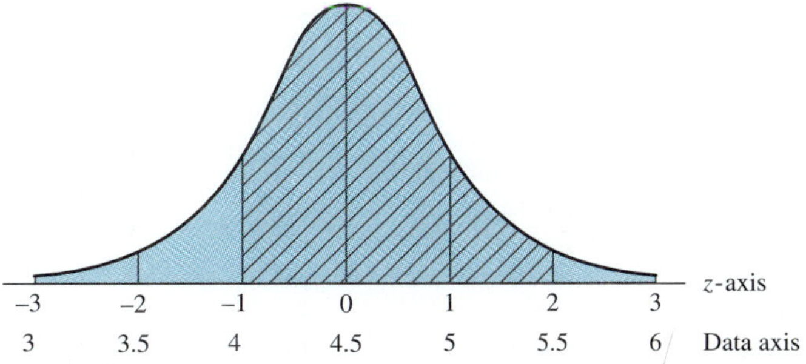

FIGURE 4.7

Looking at the distribution, we see that the percentage of rabbits having a weight between 4.0 and 5.5 pounds is the same as the area under a standard normal curve between −1 and 2. Adding the percentages of these slices yields 34% + 34% + 13.5% = 81.5%. Slightly more than 8 out of 10 rabbits have weights in this range. ◆

Since the fall of the Berlin Wall in 1990, many companies from the United States and western Europe have either provided assistance to eastern European countries or have established business operations and enterprises of their own. Suppose that the company you work for is establishing offices and manufacturing facilities in the country of Midrovia. In order to design workplace areas you have to estimate the percentage of women in the country who are at least 5'2" tall. You have found some information on this, namely, that the mean height of Midrovian women is 5'4" and the standard deviation is 2". You can find no other information. What should you do?

SOLUTION We decide to assume that the heights of women in Midrovia have a normal distribution. This is an assumption that seems to hold for the heights of people in other populations. Putting the mean and standard deviation of the population on the data axis we get the picture in Figure 4.8.

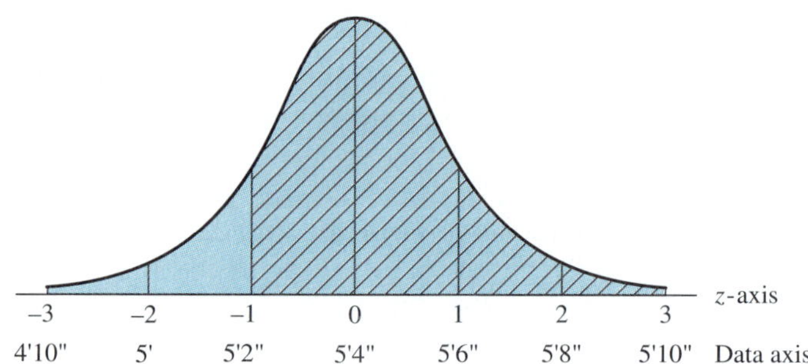

FIGURE 4.8

We see that the percentage of women in Midrovia with heights of at least 5'2" is the same as the area under a standard normal curve with values greater than −1. We use symmetry to note that the percentage for the right half of a standard normal distribution is 50%, while the percentage of the slice between −1 and 0 is 34%. Therefore, 84% (34% + 50%) of the women are at least 5'2" tall.

PROBLEM SET 4.1

1. For the data set {3, 7, 12, 9, 4} the mean is 7. Find the deviation from the mean for each value in the set.
2. For the data set {4, 10, 7, 1, 5} the mean is 5.4. Find the deviation from the mean for each value in the set.
3. Find the variance and standard deviation for the data set in problem 1.
4. Find the variance and standard deviation for the data set in problem 2.
5. Find the mean, variance, and standard deviation for each of the following data sets:
   (a) {4, 6, 7, 10, 13}
   (b) {−2, −2, 1, 2, 4, 12}
   (c) {3, 4, 4, 4, 5, 5, 5, 6}
6. Find the mean, variance, and standard deviation for each of the following data sets:
   (a) {−3, 0, 4, 5, 14}
   (b) {1, 2, 3, 10, 12, 17}
   (c) {5, 5, 5, 6, 10, 11, 11, 11}

7. Find the mean, variance, and standard deviation for the following data set:

$$\{2.72, 3.84, 4.07, 4.80, 5.61, 6.78\}$$

Round all answers to two decimal places.

8. Find the mean, variance, and standard deviation for the following data set:

$$\{11.24, 13.45, 13.82, 14.39, 16.55, 19.71\}$$

Round all answers to two decimal places.

9. For the data set {3, 10, 9, 7, 15}, show that the mean is 8.8 and the standard deviation is 3.92. Modify the data set by adding 5 to each data point. Find the mean and standard deviation for the modified data set and compare them to those of the original data set. What do you notice?

10. For the data set {6, 7, 9, 12, 15}, show that the mean is 9.8 and the standard deviation is 3.31. Modify the data set by subtracting 3 from each data point. Find the mean and standard deviation for the modified data set and compare them to those of the original data set. What do you notice?

11. For the data set {9, 7, 3, 10, 15}, the mean is 8.8 and the standard deviation is 3.92. Modify the given data set by multiplying each data point by 3. Find the mean and standard deviation for the modified data set and compare them to those of the original data set. What do you notice?

12. For the data set {12, 9, 7, 15, 6}, the mean is 9.8 and the standard deviation is 3.31. Modify the data set by dividing each data point by 10. Find the mean and standard deviation for the modified data set and compare them to those of the original data set. What do you notice?

13. Find the z-score for each data point in {3, 7, 11, 14, 15}. Round values to two decimal places.

14. Find the z-score for each data point in {4, 9, 12, 13, 15, 16}. Round values to two decimal places.

15. Convert the data set {13, 9, 7, 11, 10} to a set of z-scores, rounding values to two decimal places. Find the mean and standard deviation for the set of z-scores. Make a dot plot of the data on both a numerical scale and a z-score scale.

16. Convert the data set {6, 15, 4, 10, 12} to a set of z-scores, rounding values to two decimal places. Find the mean and standard deviation for the set of z-scores. Make a dot plot of the data on both a numerical scale and a z-score scale.

17. The number of credits taken by students enrolled in a statistics class were {15, 14, 15, 16, 14, 15, 16, 15, 13, 15, 18, 14, 16, 16, 15, 14, 16, 18, 16, 15}.
    (a) Find the mean and standard deviation for the scores.
    (b) Make a frequency histogram of the data using both a numerical axis and a z-score axis. Round answers to two decimal places.

18. Twenty students were enrolled in a statistics class. During the first four weeks, attendance in class was {18, 20, 18, 17, 19, 15, 20, 18, 17, 17, 16, 18, 17, 18, 20, 16}.
    (a) Find the mean and standard deviation for the scores.
    (b) Make a frequency histogram of the data using both a numerical axis and a z-score axis. Round answers to two decimal places.

19. **(a)** Find the percentage of a standard normal population that has a value between 1 and 3.
    (b) Find the percentage that has a value larger than 2.
    (c) Find the percentage that is not between -1 and 1.

20. **(a)** Find the percentage of a standard normal population that has a value between -2 and 3.
    (b) Find the percentage that has a value less than 1.
    (c) Find the percentage that is not between 0 and 1.

21. **(a)** Find the percentage of a standard normal population that has a value between 2 and 3.
    (b) Find the percentage that has a value less than 2.
    (c) Find the percentage that is not between -2 and 2.

22. **(a)** Find the percentage of a standard normal population that has a value between -3 and 1.
    (b) Find the percentage that has a value less than -2.
    (c) Find the percentage that is not between 1 and 3.

23. Suppose that dressed turkeys from a certain ranch have a weight that is normally distributed and have a mean of 12 pounds with a standard deviation of 2 pounds. What percentage of turkeys have a weight less than 10 pounds? What percentage of turkeys weigh between 10 and 14 pounds? How many turkeys of a shipment of 1000 would you expect to weigh less than 10 pounds?

24. Suppose that there are 100 franchises of Betty's Boutique in 100 similar shopping malls across middle America. The gross sales of these boutiques on a Saturday is normal with a mean of \$4610 and a standard deviation of \$370. What percentage of the Betty's Boutique franchises will have gross sales less than \$3870? What percentage will have gross sales between \$4240 and \$4980? How many stores would you expect to have sales less than \$3870? Round your answer to the nearest whole number.

25. Suppose that a certain brand of tires is good for a mean of 40,000 miles with a standard deviation of 5000 miles. Also suppose that the population of this brand of tires is normal. What is the percentage of tires that last less than 40,000 miles? What is the percentage of tires that last at least 35,000 miles?

26. Suppose that a certain brand of lightbulb lasts a mean of 5000 hours with a standard deviation of 350 hours. Also suppose that these lightbulb lifetimes form a normal population. What is the percentage of lightbulbs that last less than 4300 hours?

What is the percentage that last between 4650 and 5700 hours?

27. In problem 25, suppose that the tire company guarantees tires to last at least 30,000 miles and will replace any tire that does not last this long (on a properly aligned automobile). What percentage of tires will have to be replaced? If the cost of such a replacement is $86, how much will the company expect to pay on each lot of 1000 tires?

28. In problem 26, suppose that the lightbulb company agrees to reimburse anyone who purchases a lightbulb that lasts less than 4300 hours. If lightbulbs cost 50 cents apiece and there are 100,000 lightbulbs sold in a year, what is the expected cost of such a program?

EXTENDED PROBLEMS

When data is collected, it is often grouped by intervals, such as 0–9.9, 10.0–19.9, 20.0–29.9, and so on. Because the actual data is no longer available, only estimates of the mean, variance and standard deviation are possible. To estimate these values for data grouped by intervals, the *midpoint* of each interval is used as the data point for that group of data. For example, if the interval is 10.0–19.9, then 15 is used as the data point (15 is the midpoint between 10.0 and the start of the next interval, 20.0). If the frequency of the interval 10.0–19.9 is 4 (that is, there are 4 values between 10.0 and 19.9), then the estimate for the total of those values is 60. When working with grouped data, the relevant calculation is made for the midpoint of the interval and then multiplied by the frequency of the interval.

29. Find the mean, variance, and standard deviation for the following data: Round answers to two decimal places.

Frequency	Interval
4	0.0–9.9
6	10.0–19.9
4	20.0–29.9
2	30.0–39.9

30. Find the mean, variance, and standard deviation for the following data: Round answers to two decimal places.

Frequency	Interval
2	50.0–59.9
1	60.0–69.9
5	70.0–79.9
8	80.0–89.9
6	90.0–99.9

31. Find the mean, variance, and standard deviation for the following data. Round answers to two decimal places.

Frequency	Interval
2	30.0–39.9
5	40.0–49.9
7	50.0–59.9
3	60.0–69.9
1	70.0–79.9

4.2 SAMPLES AND BIAS

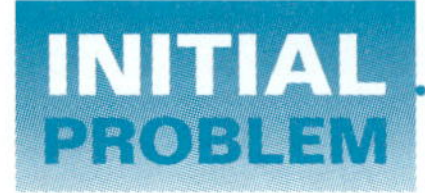

As office manager for a radio station you have three tickets for a vacation in Tahiti. There are 29 workers, all equally deserving of extra recognition. How can you choose 3 of these 29 workers in a way that is fair to all of them?

To find the mean for some characteristic of a population, or look at the pattern of how the values are distributed, we must first get data on individuals in the population. However, this may be difficult to do for the entire population. Instead, we

TIDBIT

The name *statistics* was first applied to collections of data relating to matters important to the State, such as the numbers of the population, the yield of taxation, and so on. An important early example is the *Doomsday Book*, the record of William the Conqueror's survey of England in the latter part of the 11th century.

often study a smaller group taken from the population and assume, with some reservations, that our results apply to the population as a whole.

POPULATIONS AND SAMPLES

One of the most common uses of statistics is gathering and analyzing information about specific groups of people or objects. For example, an insurance company may need to know the average height or weight of 50-year-old males, political advisors may need to know the percentage of people who support the president's foreign policies, a manufacturer may need to know the average "lifetime" of manufactured parts or the percentage of defective parts produced in the manufacturing process. It is often impractical (if not impossible), however, to check every member of the group we are interested in. For example, an insurance company probably doesn't have the time or money needed to weigh and measure *every* 50-year-old male.

Instead of dealing with the entire group in question, called the **population,** we will usually select a portion, or **sample,** from the population and analyze it instead. If a sample is carefully chosen, we may assume it is representative of the population and shares the main characteristics. The results we obtain from the sample, such as means or percentages, can then be used as estimates for the values we would find in the population. However, a great deal of care must be taken in selecting the sample.

EXAMPLE 4.8 Suppose you wish to determine voter opinion regarding the ballot measure to fund the proposed new library. To determine this, you survey potential voters among the pedestrians on Main Street during the lunch hour. What is the population and what is the sample?

SOLUTION The population consists of people who are going to vote in the upcoming election. The sample consists of those interviewed on the street who say they will be voting in the election. ◆

BIAS

If a sample is not representative of the population, we will draw an erroneous conclusion. A **bias** is a flaw in the sampling procedure that makes it more likely the sample will not be representative of the population. As an example, suppose a late night news program wished to have a call-in telephone poll on a gun control issue with a 50 cent cost for participation. Such a telephone poll has many sources of bias. An important source is the fact that it takes an effort and some expense to participate. This means that people who have strong opinions about gun control and are willing to part with 50 cents are more likely to participate. Other sources of bias include the fact that there is nothing to prevent nonresidents from participating or to prevent people from voting more than once. There are other forms of bias that can also affect the results, such as the way questions are worded. In this section, we will discuss how to analyze surveys and polls and how to choose samples that are free of bias.

EXAMPLE 4.9 Suppose you wish to determine voter opinion regarding the elimination of the capital gains tax (a profit made on an investment is called a capital gain). To determine this, you survey potential voters near Wall Street in New York City. Identify a source of bias in this poll.

SOLUTION One source of bias in choosing this sample is that many people involved in trading stocks work on Wall Street and their incomes could be enhanced by the elimination of the capital gains tax. The percentage of people in this sample who favor elimination is likely to be much higher than that of the population as a whole. ◆

The population and sample need not always consist of people, as we see in the next example.

EXAMPLE 4.10 To test the reliability of a lot (a unit of production) of automobile components produced at a certain factory, the first 30 components from a lot of 1000 are tested for defects. Describe the population, the sample, and any potential sources of bias.

SOLUTION The population is the lot of 1000 automobile components that are produced at the factory. The sample is the set of the first 30 components from the lot. Bias results from the fact that the first 30 are chosen. It is possible that these 30 were made with special care or that they were made at the start-up of the process when defects are more likely. ◆

SIMPLE RANDOM SAMPLES

Given a population and a desired sample size, we say a **simple random sample** is any sample that is chosen in such a way that all samples of the same size have the same chance of being chosen. A simple random sample is the only sample of a fixed size that has no bias. If our population had only two members and we wished to choose a simple random sample of size 1, then we could toss a coin to choose the sample. If you are choosing a sample one element at a time, then all unselected elements should have the same chance of being chosen at any step in the process.

Suppose you are trying to decide which two of your four favorite CDs you are going to take to the beach. You need to choose two from Alice in Chains (A), Bon Jovi (B), the Carpenters (C), and Devo (D). To choose a simple random sample of size two, we could list all the possible subsets of size two:

1. {A, B} 2. {A, C} 3. {A, D} 4. {B, C} 5. {B, D} 6. {C, D}

There are 6 possible samples of size 2. To choose a simple random sample, we could roll a die (one of a pair of dice) and let the outcome determine the sample. For example, if we roll the die and a 4 comes up, then we would choose the sample number 4 consisting of Bon Jovi and the Carpenters.

While the method just described works, it is not always practical. If you wanted to choose five cards from a deck of cards, you would not write down all the possibilities (there are 2,598,960 of them) and pick one; instead you would likely shuffle and deal. We need a method for choosing a simple random sample that is more like shuffling and dealing. One such method uses a random number generator or a table of random numbers.

A **random number generator** is a computer or calculator program designed to produce numbers that are as random as possible; that is, there is no apparent pat-

TIDBIT

Random number generators are built into CD players to randomly shuffle selections on a CD or group of CDs.

TABLE 4.1

101	03918	77195	47772	21870	87122	99445
102	10041	31795	63857	64569	34893	20429
103	43537	25368	95237	17707	34280	04755
104	64301	66836	12201	60638	85624	33306
105	43857	49021	49026	93608	51382	49238
106	91823	38333	37006	78545	23827	39103
107	34017	00983	48659	39445	90910	29087
108	49105	95041	94232	50784	59181	44253
109	72479	24246	35932	33358	34853	77573
110	84281	57601	78425	36246	79348	41681
111	61589	93355	41310	17068	65700	54464
112	25318	28496	80120	31632	06746	90642
113	40113	91130	74270	27914	80511	70243
114	58420	96471	28464	72438	37667	16233
115	18075	32457	50011	42175	41029	07733
116	52754	43382	02151	46182	40557	94157
117	05255	73603	15957	99738	62835	62959
118	76032	69846	63316	48201	11580	45699
119	97050	48883	17828	98601	74821	06605
120	29030	55519	63362	55720	15296	78787
121	45609	12114	36541	53609	09322	28694
122	07608	55455	49299	90355	35334	29000
123	94901	06633	04618	82809	76952	21697
124	50581	84325	17532	57302	81752	25570
125	22265	14648	32967	10792	81713	68326
126	59294	06043	86457	78791	44380	62238
127	45473	93910	79160	19436	00813	75916
128	40239	02596	12487	99703	08901	49759
129	30241	44100	59953	83094	05261	46901
130	43837	77175	96514	61955	75287	24839
131	25050	80925	64073	70415	39896	69297
132	01445	23629	74556	24642	01672	92860
133	85236	77764	06026	33455	17737	08377
134	05946	75867	30147	53490	50415	24093
135	61189	32931	99257	50892	66516	45434
136	91267	07544	22194	04212	20015	15407
137	17039	95693	69650	40076	57722	38787
138	58541	34646	17657	30584	94546	09286
139	85563	13994	46354	93939	12491	41648
140	48576	89126	32012	39665	43906	76405
141	00543	87408	87066	74781	13065	35705
142	27954	32772	58815	88341	28322	05945
143	89156	74789	42290	03617	10054	13262
144	62334	04229	42057	10099	35791	10708
145	76172	20142	30526	88296	61844	89118

tern to the numbers. A **random number table** is a table produced with a (very good) random number generator (Table 4.1).

In order to produce a simple random sample using a random number table, we will pick an arbitrary place on the table to begin and then move across the table in a systematic way. Suppose we wish to choose a simple random sample of size 5 from a group of 10 people. First, we give each of the 10 people in the population a one-digit label to identify them: 0, 1, 2, 3, 4, 5, 6, 7, 8, 9. Then, we pick a place on the table to begin; for simplicity, we will begin at the top of the first column of Table 4.1, which begins 03918. We will use only the first digit in each row of this column. We move down the column and take the first five digits, ignoring any duplication. The digits in this (single digit) column are 0, 1, 4, 6, 4, 9, and so on. After crossing out the duplicate 4, we obtain our simple random sample—in this case, 0, 1, 4, 6, 9. Note: In choosing our set of random numbers, we also could have picked numbers from any other column or row and begun in the middle of the table.

EXAMPLE 4.11 Choose a simple random sample of size 4 from the following 12 semifinalists of a contest: Astoria, Beatrix, Charles, Delila, Elsie, Frank, Gaston, Heidi, Ian, Jose, Kirsten, and Lex.

SOLUTION First assign numerical labels to the contestants. They must be two-digit numbers since there are more than 10 in the population. We label as follows: 00 = Astoria, 01 = Beatrix, 02 = Charles, 03 = Delila, 04 = Elsie, 05 = Frank, 06 = Gaston, 07 = Heidi, 08 = Ian, 09 = Jose, 10 = Kirsten, 11 = Lex. We start at the top of the fourth column of random numbers in Table 4.1, which begins 21870. Looking only at the first two digits in each row of the column, we read down the entire column: 21, 64, 17, 60, 93, 78, 39, 50, ... We are only interested in numbers from 00 through 11. Eliminating all numbers in the column larger than 11 gives the numbers 10, 04, 03, 10. Removing the duplicated 10 gives us the numbers 10, 04, 03. We need a sample of size 4, so we need another number. Covering up the first two digits of this column we get new random territory. The first two digits of this new column begin 87, 56, 70, ... Scanning down for the first digits from 00 through 11, we find 06. So the numbers of our simple random sample are 10, 04, 03, 06, or, in order, 03, 04, 06, 10. Looking up the names of the corresponding contestants gives us Delila, Elsie, Gaston, and Kirsten. We used columns because they are easier to read. However, we could have scanned across rows also. The main requirement is that we systematically look at new random numbers, never looking at numbers in the table more than once. ◆

Note: We customarily begin at a random position in the table and then select the numbers in a systematic way. For the sake of clarity in the examples, we will usually start our selection of random numbers in a very convenient place.

EXAMPLE 4.12 Choose a simple random sample of size 8 from the states of the United States.

SOLUTION The first step is to assign numerical labels to the states: We used the first numbered list of the states that we came across that ranks the states by area—Alaska first, Rhode Island fiftieth. To make the choices, we pick a method for going through the random number table. We will start at the top row, left column, and go left to right using the last two digits from each entry:

18, 95, 72, 70, 22, 45, 41, 95, 57, 69, 93, 29, 37, 68, 37, 07, 80, 55, 01 . . .

Of course, we only want numbers from 01 to 50, and we do not use repetitions.

18, 95, 72, 70, **22, 45, 41,** 95, 57, 69, 93, **29, 37,** 68, 37, **07,** 80, 55, **01**

That thins the list to the following: 18, 22, 45, 41, 29, 37, 07, 01, . . .

We have eight numbers, so we look up the states to which they correspond:

Oklahoma	Florida	Massachusetts	West Virginia
Alabama	Kentucky	Nevada	Alaska

◆

INDEPENDENT SAMPLING

TIDBIT

The Internal Revenue Service picks an independent sample of tax returns for audit. Customs agents also frequently pick an independent sample of people entering the country to carefully check for contraband.

A simple random sample gives a sample of a *fixed size* from a fixed population. There is another method of sampling that yields a sample with a random size. **Independent sampling** occurs when each member of the population has a fixed chance of being selected for the sample regardless of whether other members of the population were selected or not. For example, if we wish to sample 50% of the customers coming into a store for a survey, we could toss a coin for each customer and interview those customers when the coin came up heads.

EXAMPLE 4.13 We wish to sample 10% of the Viper automobiles that are produced by a certain factory for quality control testing. Suppose that one day 100 cars are produced. Label these cars 1, 2, . . . , 100, and choose a 10% independent sample using the first column of Table 4.1.

SOLUTION The digits 0, 1, 2, . . . , 9 all have the same chance of occurring in a random number table. Thus the digit 0 occurs 10% of the time. We will only consider the numbers in the five-digit-wide first column and ignore the rest of the table. We will read down the column going from the left to the right, row by row. We look at the first digit; it is a 0, so we choose car 1 as part of the sample. The next digit is a 3, so we do not choose car 2 as part of the sample. Continuing, we see that of the first 100 digits in the column, 0's occur at places 1, 7, 8, 19, 33, 39, 62, 70, 73, 81, 88, 93, 95, 98, 100. ◆

Note that a 10% independent sample may contain more than 10% of the population, less than 10%, or exactly 10%. The sample in Example 4.13 had 15% of the population. If we had wished to find a 20% independent sample, we could have looked for either the digits 0 or 1. To find a 30% independent sample, we could have looked for either the digits 0, 1, or 2, and so on.

EXAMPLE 4.14 Find a 50% independent sample of the 12 semifinalists from Example 4.11: Astoria, Beatrix, Charles, Delila, Elsie, Frank, Gaston, Heidi, Ian, Jose, Kirsten, and Lex.

SOLUTION In a random number table, the five digits 0, 1, 2, 3, 4 have a 50% chance of occurring. We will use column 6. The first 12 digits of the column are 99445 20429 04, where each digit will determine if the contestant in that position is chosen. If the digit is 0, 1, 2, 3, or 4, the contestant is chosen. Crossing out all but the numbers 0, 1, 2, 3, 4, we have **44* 2042* 04. The digits 4 or less occur at places 3, 4, 6, 7, 8, 9, 11, 12, and the corresponding contestants are Charles, Delila, Frank, Gaston, Heidi, Ian, Kirsten, and Lex. ◆

Independent sampling is often done in a process known as continuous quality control. One example of a quality control procedure using independent sampling is Military Standard 781C. Here items produced are independently sampled and tested for defects. If the percentage of defective items exceeds a certain threshold, then the adjustments are made to the process and the proportion of items sampled is increased. When the percentage of defective items returns to a low state, the proportion of items sampled goes back to the lower level.

Selecting a simple random sample is often very difficult, if not impossible. If we wanted to select a simple random sample of 1000 people residing in the United States, we would have to assign every resident a number (perhaps from 1 to 260,000,000) and then use a computer to choose 1000 of these numbers at random. To avoid this exercise in futility, the U.S. Census Bureau uses a nested sampling procedure. The country is divided into regions called Primary Sampling Units (PSUs). A Primary Sampling Unit is usually a collection of neighboring counties. Each PSU is further divided into smaller areas of about 500 people each, called Census Enumeration Districts (CEDs). To sample the residents of the country, the U.S. Census Bureau first chooses some PSUs using a simple random sample. In each PSU, the Bureau uses another simple random sample to choose some CEDs. Now a collection of CEDs has been chosen. From each CED, people are chosen using a new simple random sample. This process contains fairly little bias, although it has somewhat more variability than a simple random sample of the entire population. The advantage of this method lies in the fact that it is relatively quick and inexpensive.

Just because a simple random sample is unbiased does not mean that it is necessarily representative of the population. The point to understand is: Every subset of the right size should have the same chance of being selected as any other subset—even including those subsets that are very atypical. If we choose a simple random sample of 400 voters and ask them if they support Senator Smith, it is possible that we might just happen to choose 400 supporters of Senator Smith, and it is also possible to choose 400 nonsupporters. Anything might happen. Of course, it is much more likely to choose a sample in which the proportion of supporters is fairly representative of the population.

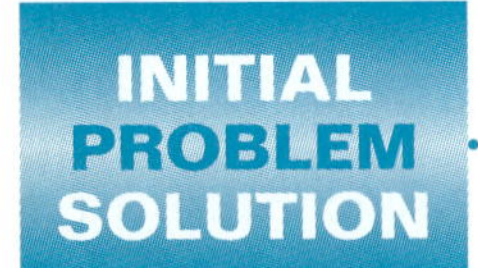

As office manager for a radio station you have three tickets for a vacation in Tahiti. There are 29 workers, all equally deserving of extra recognition. How can you choose 3 of these 29 workers in a way that is fair to all of them?

SOLUTION Choose a simple random sample. Assign each of the 29 workers the numbers 00, 01, . . . , 28 in order. Looking at the first two digits down the last column of Table 4.1, we see 99, 20, 04, 33, 49, 39, 29, 44, 77, 41, 54, 90, 70, 16, 07, . . . The first three numbers that are 28 or less are 20, 04, 16. The workers that have been assigned the numbers 04, 16, and 20 go to Tahiti.

PROBLEM SET 4.2

In problems 1 through 4, identify the population being studied and the sample that is actually observed.

1. A lightbulb company says its bulbs last 2000 hours. To test this, a package of 8 bulbs is purchased and the bulbs are kept lit until they burn out. Five of the bulbs burn out before 2000 hours.
2. A chest of 1000 gold coins is to be presented to the king. The royal minter believes the king will not notice if only one of the coins is counterfeit. The king is suspicious and has 20 coins taken from the top of the chest and tested to see if they are pure gold.
3. The registrar's office is interested in the percentage of full-time students who commute on a regular basis. One hundred full-time students are randomly selected and briefly interviewed; 75 of these students commute on a regular basis.
4. The mathematics department is concerned about the amount of time students regularly set aside for studying. A questionnaire is distributed in three classes having a total of 82 students.

In problems 5 through 8, identify and discuss any sources of bias in the sampling method.

5. A Minnesota-based toothpaste company claims that 90% of dentists prefer the formula in its toothpaste to any other. To prove this, they conduct a study. They send questionnaires to 100 dentists in the Minneapolis–St. Paul area asking if they prefer the company's formula to others.
6. A magazine devoted to exercise, vitamins, and healthy living is interested in the habits of older adults regarding exercise and nutritional supplements. The current issue includes an article on the subject and a questionnaire it asks readers to fill out and mail in.
7. A soft drink company produces a lemon-lime drink that it says people prefer by a margin of two-to-one over its main competitor, a cola. To prove this claim, it sets up a booth in a large shopping mall where customers are allowed to try both drinks. The customers are filmed for a possible television commercial. They are asked which drink they prefer.
8. A sociologist working for a large school system is interested in demographic information on the families having children in the schools served by the system. Two hundred students are randomly selected from the school system's database and a questionnaire is sent to the home address in care of the parents or guardian.

In problems 9 through 12, identify the population being studied, the sample actually observed, and discuss any sources of bias.

9. A biologist wants to estimate the number of fish in a lake. As part of the study, 250 fish are caught, tagged, and released back into the lake. Later, 500 fish are caught and examined; 18 of these fish are found to be tagged and the rest are untagged.
10. A college professor is up for promotion. Teaching performance, as judged through student evaluations, is a significant factor in the decision. The professor is asked to choose one of his classes for student evaluations. The day of the evaluations he passes out questionnaires and then remains in the room to answer any questions about the form and filling it out.
11. A drug company wishes to claim that 9 out of 10 doctors recommend the active ingredients in their product. They commission a study of 20 doctors. If at least 18 doctors say they recommend the active ingredients in the product, the company will feel free to make this claim. If not, the company will commission another study.
12. There are two candidates for student body president of a college. Candidate Johnson believes that the student body resources should be used to enhance the social atmosphere of the college and that the number one priority should be dances, concerts, and other social events. Candidate Jackson believes that sports should be the number one priority and wants to subsidize student sporting events and enlarge the recreation facility. A poll is taken by the student newspaper. One interviewer goes to a coffee house near the college one evening and asks students which candidate they prefer. Another interviewer goes to the gym and ask students which candidate they prefer.
13. Five students are to be randomly selected from a class of 36. The students are labeled with two-digit numbers from 00 to 35. Use Table 4.1 to select the students by taking the last two digits of the third column. Begin with row 115 and proceed down the column. If necessary, continue with the last two digits of the fourth column beginning with the first row of that column.
14. An automobile distributor has received 80 new cars for the sales region. Ten cars are to be randomly selected for detailed inspections before the shipment is finally accepted. The cars are numbered 00 to 79. Use Table 4.1 to select the cars by taking the second and third digits of the fourth column. Begin with row 110 and proceed down the column. If necessary, continue with the second and third digits of the fifth column beginning with the first row of that column.
15. There are 250 graduate students in the university's science departments. Five of the students are going to be selected for interviews regarding financial aid, program requirements, and other matters. The students are numbered 000 to 249. Use Table 4.1 to select the students by taking the first three digits of the second column. Begin with row 110 and proceed down the column. If necessary, continue with the first three digits of the third column beginning with the first row of that column.
16. Repeat problem 15, but this time use the middle three digits of each column. Begin with row 115 of the third column and proceed down the column, continuing with the fourth column if necessary.
17. The 15 members of a university's flying club decide that a committee should be formed to rewrite the club's bylaws. Rather than ask for volunteers, it is determined that four members will be selected at random to serve on the committee. The members of the club are:

Allen	Mary
Fred	John
Patty	Chris
Margaret	Matt
Bill	Dan
Tom	Jamie
Amy	Tyler
Jane	

Beginning with Allen and going down the columns, ending with Tyler, label the members with two digits starting with 00. Use Table 4.1 to select the members of the committee. Take the first two digits of each column for the sequence of random numbers. Begin in row 118 of the second column and read down the column. Continue with the third column if necessary.

18. Repeat problem 17, but this time take the last two digits of each column. Begin selecting the sample on row 125 of the first column.

19. Choose a simple random sample of 6 letters from the 26 letters of the alphabet. Begin the labeling of the letters by using 00 for "a." Use Table 4.1 and take the third and fourth digits of each column. Begin with line 105 of the fourth column and read down the columns.

20. Five members of a college basketball team will be selected for a special study on exercise and conditioning being conducted by a graduate student in sports physiology. The team members are:

Amy	Michelle
Debra	Patti
Ellen	Rebecca
Gina	Sandi
Kari	Shannon
Maria	Teddie

Use the second and third digit of each column of Table 4.1 to select a simple random sample of the team members. Begin on line 110 of the second column.

21. Refer to problem 17.

(a) Select four simple random samples of 5 students from the flying club. In Table 4.1, use the first and second digits of the first column for the first sample; the second and third digits of the second column for the second sample; the third and fourth digits of the third column for the third sample; and the fourth and fifth digits of the fourth column for the fourth sample. In each case, begin on line 105 and continue to the next column if necessary.

(b) Since there are only 15 students and a total of 20 selections are made for the four samples, there had to be some duplication. Were there any students who weren't selected for at least one sample?

22. Refer to problem 20.

(a) Select four simple random samples of 5 players from the basketball team. In Table 4.1, use the first and second digits of the first column for the first sample; the second and third digits of the second column for the second sample; the third and fourth digits of the third column for the third sample; and the fourth and fifth digits of the fourth column for the fourth sample. In each case, begin on line 110 and continue to the next column if necessary.

(b) Since there are only 12 players and a total of 20 selections were made for the four samples, there had to be some duplication. Were there any players who weren't selected for at least one sample?

23. There are 24 people working in an office:

Arnold	Molly
*Bob	Natalie
*Chris	Oliver
Demi	Polly
Esther	Quinten
Freya	*Raul
*Glenda	Sandra
Holly	Teresa
Ingrid	Ursula
Jason	Victor
Kelly	Wesley
Lester	Xia

(a) Choose four simple random samples of 5 people to test for high blood pressure. In Table 4.1, use the first and second digits of the second column for the first sample; the second and third digits of the third column for the second sample; the third and fourth digits of the fourth column for the third sample; and the fourth and fifth digits of the fifth column for the fourth sample. In each case, begin on line 105 and continue to the next column if necessary.

(b) The names with an asterisk indicate the people who actually have high blood pressure. What percentage of people in each sample have high blood pressure?

24. Repeat problem 23, but this time take simple random samples of 10 people.

25. In Example 4.13, the first column of digits was used to find a 10% independent sample from a set of 100 cars. Repeat the example using 0 as the identifying digit, but this time find a 10% independent sample using the third column of digits from Table 4.1.

26. Find a 20% independent sample of the letters of the alphabet (A = 1) using the third column of digits from Table 4.1 in the same way as in Example 4.14. Use 0 and 1 as the identifying digits. If the first digit is a 0 or 1, then A is selected; if the second digit is 0 or 1, then B is selected, and so on.

27. Refer to problem 23. Find a 20% independent sample from the people in the office. Use the second column of Table 4.1, beginning with line 130.

28. Find 50% independent samples from the people in the office in problem 23.

(a) Use the fifth column of Table 4.1, starting on line 110. Use the digits 0, 1, 2, 3, 4 to determine if the person belongs in the sample.

(b) Use the fifth column of Table 4.1, starting on line 110. Use the digits 0, 2, 4, 6, 8 to determine if the person belongs in the sample.

For problems 29 through 35 refer to the following list of the states:

	Number of Representatives	Per Capita Personal Income	Automobile Registration (1000's)	English Language Sunday Papers
Alabama	7	$17,234	2,224	19
Alaska	1	22,846	317	4
Arizona	6	18,121	1,990	15
Arkansas	4	16,143	990	16
California	52	21,821	17,357	68
Colorado	6	21,564	2,153	11
Connecticut	6	28,110	2,421	11
Delaware	1	21,481	427	2
Florida	23	20,857	8,353	35
Georgia	11	19,278	4,251	18
Hawaii	2	23,354	666	5
Idaho	2	17,646	597	8
Illinois	20	22,582	6,615	28
Indiana	10	19,203	3,366	22
Iowa	5	18,315	1,959	10
Kansas	4	20,139	1,273	16
Kentucky	6	17,173	1,951	12
Louisiana	7	16,667	2,012	20
Maine	2	18,895	754	2
Maryland	8	24,044	3,143	7
Massachusetts	10	24,563	3,109	14
Michigan	16	20,453	5,726	27
Minnesota	8	21,063	2,790	14
Mississippi	5	14,894	1,527	15
Missouri	9	19,463	2,885	21
Montana	1	17,322	552	7
Nebraska	3	19,726	905	7
Nevada	2	22,729	633	4
New Hampshire	2	22,659	678	6
New Jersey	13	26,967	5,211	17
New Mexico	3	16,297	851	13
New York	31	24,623	8,442	42
North Carolina	12	18,702	3,854	37
North Dakota	1	17,488	400	7
Ohio	19	19,688	7,401	36
Oklahoma	6	17,020	1,787	41
Oregon	5	19,443	2,013	10
Pennsylvania	21	21,351	6,628	37
Rhode Island	2	21,096	512	3
South Carolina	6	16,923	1,922	14
South Dakota	1	17,666	433	4
Tennessee	9	18,434	3,813	16
Texas	30	19,189	8,746	86
Utah	3	16,180	820	6
Vermont	1	19,467	349	3
Virginia	11	21,634	4,056	15
Washington	9	21,887	3,155	16
West Virginia	3	16,209	776	11
Wisconsin	9	19,811	2,481	20
Wyoming	1	19,539	255	4
United States		**20,817**	**145,740**	**884**

29. Use Table 4.1 to find a 10% independent sample of the states. Record both the names of the states and the per capita personal income. Use column 6.

30. Use Table 4.1 to find a 20% independent sample of the states. Record both the names of the states and the number of English language Sunday papers.

EXTENDED PROBLEMS

31. A 25% independent sample can be obtained using Table 4.1 in the following way. If we disregard the digits 8 and 9, then the digits 0 and 1 occur 25% of the time for the remaining digits. To see if the first item in a list is selected for the sample, we look at the first digit in the sequence of digits we've chosen to use. If the digit is a 0 or 1, the item is selected for the sample; if the digit is a 2, 3, 4, 5, 6, or 7, the item is not selected; if the digit is 8 or 9, we go to the next digit to see if the item was selected. Use Table 4.1 to find a 25% sample of the states. Use the third column of the table, beginning with the first digit on line 120 (you should end with the first digit in line 132). Record both the names of the states and the number of members in the House of Representatives.

32. Find a $33\frac{1}{3}\%$ independent sample of the states. State clearly what digits you will use and how you will use them. Use the first column of Table 4.1 and begin on line 110.

33. Use Appendix A (Table of 10,000 Random Digits) to find ten samples of size 5 from the list of states. Record both the names of the states and the number of representatives. Use column 1 for the first sample, column 2 for the second sample, and so on. Use the first 2 digits in each column and continue to the next page if needed.
  (a) Calculate the mean of the number of representatives for the states in each sample.
  (b) Make a frequency distribution and histogram for the mean number of representatives in each sample.

34. Use Appendix A to find ten samples of size 5 from the list of states. Record both the names of the states and the number of representatives. Use column 1 for the first sample, column 2 for the second sample, and so on. Use the last 2 digits in each column and continue to the next page if needed.
  (a) Calculate the mean of the number of representatives for the states in each sample.
  (b) Make a frequency distribution and histogram for the mean number of representatives in each sample.

35. Use Appendix A to find ten samples of size 5 from the list of states. Record both the names of the states and the number of representatives. Use column 1 for the first sample, column 2 for the second sample, and so on. Use the second and third digits in each column and continue to the next page if needed.
  (a) Calculate the mean of the number of representatives for the states in each sample.
  (b) Make a frequency distribution and histogram for the mean number of representatives for the states in each sample.

4.3 CONFIDENCE INTERVALS AND RELIABLE ESTIMATION

A candy bar company has a promotion in which some of the wrappers have letters printed on the inside that entitle a person to a prize. You buy 400 of these candy bars, and you and your friends pig out. You find letters on 25 of the wrappers. What percentage of wrappers did the company put letters on? You are thinking of buying another 1000 candy bars. How many wrappers would you expect to have letters printed on the inside?

In many statistical applications, the statement that a population is normally distributed really means that the distribution is "almost" normal. The advantage of being able to treat a distribution as being normally distributed is that we can use the standard deviation and mean together with the standard normal distribution to analyze the population in terms of percentages (or the chance of occurrence) of intervals containing data points. In this section we investigate two distributions that are known to be almost normal: the populations of sample means and sample proportions. Knowing how the means or proportions of samples are distributed, together with information about their standard deviations, will be the basis for using information from samples to draw conclusions about populations and assign a level of confidence to our results.

SAMPLE PROPORTION

Suppose you read the results of a poll that tell you that 48% of the American people who are registered to vote support the budget the president submitted to Congress. The poll in question is based on 413 interviews and the margin of error of the results is 5%. How can you interpret these statements? In particular, how can interviews with only 413 people out of 130,000,000 registered voters give us reliable information?

Suppose that in actual fact 50% of the American people support the budget that the president recently submitted to Congress. Since $\frac{65{,}000{,}000}{130{,}000{,}000} = 50\%$, this means that roughly 65,000,000 of 130,000,000 registered voters support the budget. The proportion $\frac{65{,}000{,}000}{130{,}000{,}000} = 50\%$ is called a **population proportion** since it represents a certain fraction of an entire population under consideration. A population proportion is represented by the letter "p."

In the sample of 413 people mentioned above, the pollster divided the number that support the budget, 198, by the total number sampled, 413, to arrive at $\frac{198}{413}$ ($\approx 48\%$). Such a number is called a **sample proportion** since it compares a portion of a sample with the entire sample. A sample proportion is represented by the symbol $\hat{p}$, called "p hat." For a sample of size N, we compute the sample proportion as follows.

DEFINITION

Sample Proportions

If a sample of size N is selected from a population, then the sample proportion of a particular group in the sample is given by

$$\hat{p} = \frac{\text{number sampled that belong to the group}}{N}.$$

EXAMPLE 4.15 Suppose there are 3520 freshmen attending Friendly State College and 1070 of those freshmen have consumed an alcoholic beverage within the previous thirty days. The instructor in a freshman health class of 25 asks his class to fill in an anonymous survey on which one of the questions is "Have you consumed an alcoholic beverage within the last thirty days?" Four of the students in the class respond Yes and the other 21 respond No. What is the population, what is the sample, what is the population proportion, and what is the sample proportion?

SOLUTION The population is freshmen attending Friendly State College. The sample is the set of freshmen in the particular health class.

The population proportion is $\frac{1070}{3520} \approx 0.30 = 30\%$.
The sample proportion is $\frac{4}{25} = 0.16 = 16\%$. ◆

DISTRIBUTION OF SAMPLE PROPORTIONS

Notice that in Example 4.15 the sample proportion differs significantly from the population proportion. You might guess that students in the health class are more health conscious, and consequently are less likely to consume alcohol than the typical freshman; that is, the sample is probably biased. Even if the sample has been chosen in a way that avoids bias, it is most likely for the sample proportion to differ from the population proportion, and it may even differ greatly. Different samples will probably have different proportions. However, by using an unbiased sample of large enough size, we can guarantee that it is very likely the sample proportion will be close to the population proportion.

EXAMPLE 4.16 Consider the population consisting of four men: Al, Bob, Chuck, and Dave. Suppose Al and Bob are left-handed. Draw a histogram for the sample proportion of left-handers in a sample of size 2.

SOLUTION First, we list all samples of size 2, then the number of left-handers in each sample, and the sample proportion. The data is plotted in Figure 4.9.

Sample	Left-handers	Sample Proportion
Al, Bob	2	100%
Al, Chuck	1	50%
Al, Dave	1	50%
Bob, Chuck	1	50%
Bob, Dave	1	50%
Chuck, Dave	0	0%

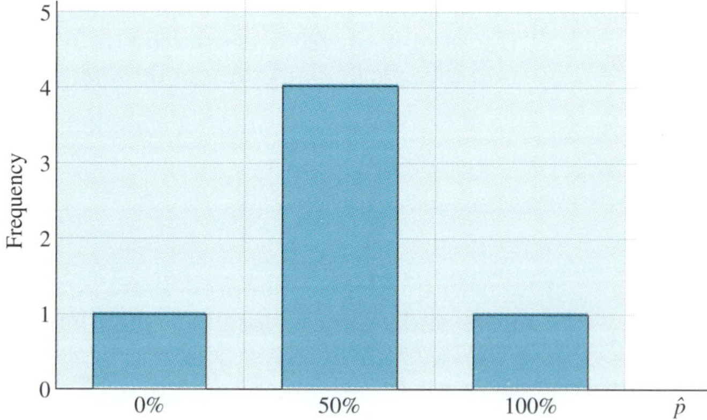

FIGURE 4.9 ◆

Notice that the histogram in Example 4.16 is centered at the population proportion and is also concentrated at the population proportion.

Carrying out the calculations as we did in the last example is generally prohibitive for even moderate-sized populations and samples. The number of possible samples of a given size that can be taken from a population can be surprisingly large, even if the population itself is fairly small. For example, there are 56 samples of size 3 that can be taken from a population of size 8. For a population with 20 members, there are 1140 different samples of size 3. Considering Example 4.15, the number of samples of 25 freshmen from a population of 3520 is enormous ($\approx 2.73 \times 10^{63}$). Fortunately, there are formulas available that quickly and simply give us the most important information about the distribution of sample proportions. The two main facts are summarized next.

Distribution of Sample Proportions

If samples of size N are taken from a population having a population proportion p, then the set of all sample proportions has mean p and standard deviation

$$\sqrt{\frac{p(1-p)}{N}}.$$

EXAMPLE 4.17 Suppose that the population proportion of a group is 0.4, and we choose a simple random sample of size 30. Find the mean and standard deviation of the sample proportion.

SOLUTION Here $p = 0.4$ and $N = 30$. Substituting into the preceding formula, the set of all sample proportions has a mean of 0.4 and standard deviation

$$\sqrt{\frac{(0.4)(0.6)}{30}} = 0.09.$$ ◆

The reason it is so useful to know the mean and standard deviation for the sample proportion is that once the sample contains more than a couple of dozen members, the histogram is nearly a bell-shaped curve. For example, Figure 4.10 shows the histogram for the sample proportions in Example 4.17.

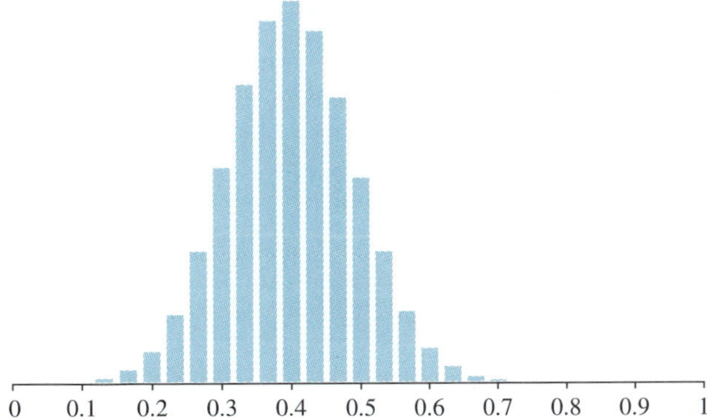

FIGURE 4.10

When we know that a histogram is approximated by a bell-shaped curve, we can tell what fraction of the histogram lies under any part of the curve. In our first example of this section concerning the president's budget proposal, the population is the set of registered American voters, and the group is the set of supporters of the president's budget. The sample size is $N = 413$ and the population proportion is $p = 0.5$. Therefore, the mean of $\hat{p}$ is 0.50, and the standard deviation is $\sqrt{\frac{(0.5)(1-0.5)}{413}} \approx 0.025$. We record this information on the data axis of the standard normal distribution curve in Figure 4.11.

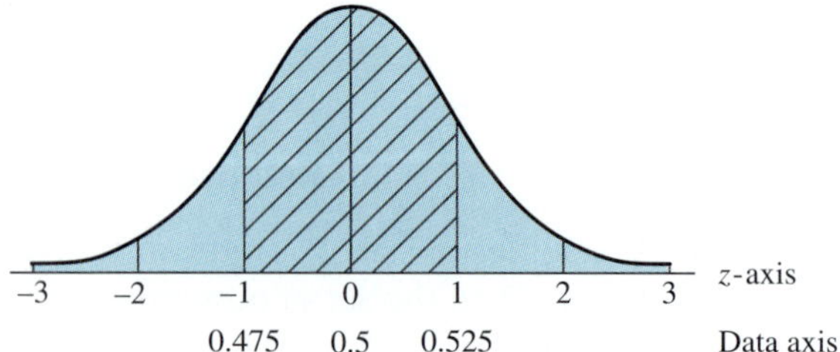

FIGURE 4.11

Figure 4.11 shows that the percentage of samples for which the sample proportion is between 47.5% and 52.5% is the same as the area between -1 and 1 under a standard normal curve. This is 68%, or about $\frac{2}{3}$. Thus it is not surprising that we saw a sample mean of 48%.

EXAMPLE 4.18 A pollster samples 600 people and asks them if they support their congressional representative for reelection. Suppose that 60% of all Americans support their congressional representative. What is the percentage of samples for which the proportion of people in the sample who voice this support is between 58% and 62%?

SOLUTION In this problem the population is the set of American residents and the group is the set of people who support their congressional representative for reelection. Since $p = 0.6$, the sample proportion has a mean of 0.6. The sample size is $N = 600$, so the standard deviation of the sample proportion is

$$\sqrt{\frac{p(1-p)}{N}} = \sqrt{\frac{(0.6)(0.4)}{600}} = \sqrt{0.0004} = 0.02.$$

Putting 0.6 for the mean in the data axis of the normal plot and intervals of 0.02 for the standard deviation as in Figure 4.12 shows us that the percentages of samples between 58% (0.58) and 62% (0.62) is the same as the percentage of a standard normal curve between -1 and 1.

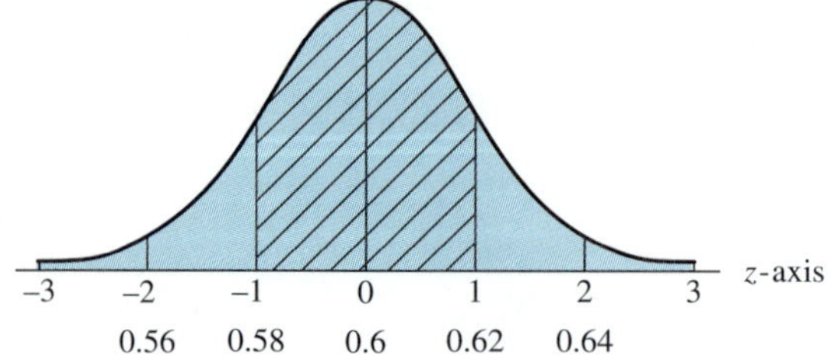

FIGURE 4.12

This percentage is 34% + 34% = 68%. ◆

THE STANDARD ERROR

If we already knew the population proportion, we would not need to bother taking samples and computing the sample proportion. In most situations, we have the reverse problem: We only know the sample proportion, and we need an estimate of how close it is to the true population proportion. When the poll was taken to determine support for the president's budget, the population proportion, p, was unknown. Indeed, the poll was taken to find out what p is. After a sample had been chosen, the sample proportion, $\hat{p}$, was computed to be 48%. This was obtained by dividing the number of people in the sample that supported the budget by 413, the sample size. Clearly, 48% should be our best guess for the population proportion. The question is, how good a guess is it?

The following general formula can be used to answer this question.

DEFINITION

Standard Error

Given a sample of size N with sample proportion $\hat{p}$, the standard deviation of the set of all sample proportions is approximately

$$\hat{s} = \sqrt{\frac{\hat{p}(1-\hat{p})}{N}}$$

which is known as the **standard error** of the sample.

This formula is the standard deviation formula with p replaced by $\hat{p}$ and it is a very good approximation to the true standard deviation of the sample.

EXAMPLE 4.19 What is the standard error in a sample of size 400 if the sample proportion is 35%?

SOLUTION We use the formula above to compute the standard error

$$\hat{s} = \sqrt{\frac{\hat{p}(1-\hat{p})}{N}} = \sqrt{\frac{(0.35)(0.65)}{400}} = \sqrt{0.00056875} \approx 0.024.$$ ◆

CONFIDENCE INTERVALS

For any normal distribution, 95% of the data must be within two standard deviations of the mean (see Figure 4.4). That tells us that 19 out of 20 times, the sample proportion will be within two standard deviations of the population proportion. If the standard deviation is not known, we use the standard error (which can be calculated) and conclude that in 95% of the cases the population proportion is within two standard errors of the sample proportion. Thus, a **95% confidence interval** is the interval of numbers from $\hat{p} - 2\hat{s}$ to $\hat{p} + 2\hat{s}$ (Figure 4.13).

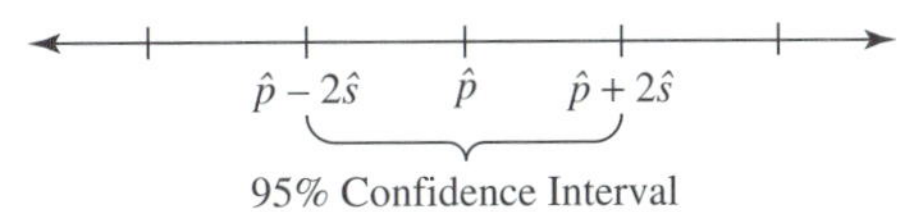

FIGURE 4.13

Any value in this interval is a reasonable estimate for the population proportion, p. It is called a 95% confidence interval because, for 95% of the samples, the interval computed in this way will contain p [Figure 4.14(a) and (b)]. Any guess of a proportion p that is not in the 95% confidence interval is not a good guess [Figure 4.14(c)]. Such a guess is unlikely because less than 5% of the samples taken would give an interval that did not contain p.

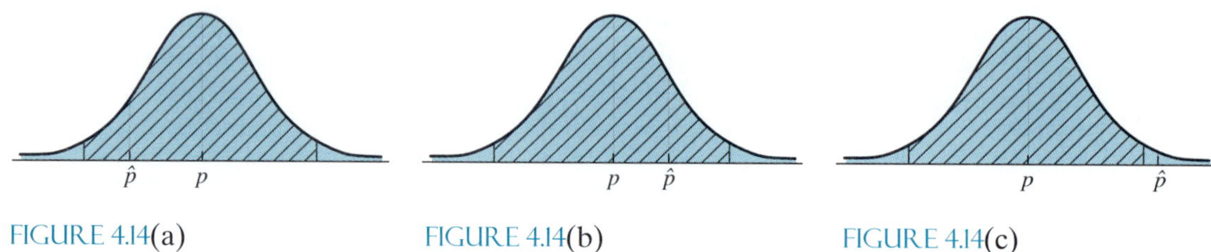

FIGURE 4.14(a) FIGURE 4.14(b) FIGURE 4.14(c)

Returning to the example of the president's budget, the size of our sample is $N = 413$. The sample proportion is $\hat{p} = 0.48$. The standard error is

$$\hat{s} = \sqrt{\frac{\hat{p}(1 - \hat{p})}{N}} = \sqrt{\frac{(0.48)(0.52)}{413}} \approx 0.025.$$

Notice that this is very close to the estimate we obtained in the paragraph following Example 4.17 using p in the formula rather than $\hat{p}$ as we do now. We calculate our interval as $\hat{p} - 2\hat{s} = 0.43$ and $\hat{p} + 2\hat{s} = 0.53$. This means that we have a 95% confidence that the population proportion is between 0.43 and 0.53. To say this another way, 19 times out of 20, a simple random sample taken from this population of registered voters will give an interval with $\hat{p} - 2\hat{s} \leq .50 \leq \hat{p} + 2\hat{s}$.

EXAMPLE 4.20 What is the 95% confidence interval for a sample of size 400 with a sample proportion of 35%?

SOLUTION In Example 4.19 we compute the standard error of this poll to be 0.024 = 2.4%. The confidence interval goes from 35% minus two standard errors to 35% plus two standard errors. That is, the 95% confidence interval is from 35% − 4.8% = 30.2% to 35% + 4.8% = 39.8%; or 30% to 40%. ◆

Note that if the values are rounded to full percentages, they are rounded "outward" so that at least 95% of the sample proportions are still included. That is, in a 95% confidence interval of 41.6% to 46.4%, the percentages would be rounded to 41% and 47%.

EXAMPLE 4.21 Suppose that we sample 600 people and ask them if they have an American-built car as their primary source of transportation. In this sample, 362 people say that they do. Compute a 95% confidence interval for the proportion of the population that has an American-built car.

SOLUTION The sample proportion is $\hat{p} = \frac{362}{600} = 0.603$. The standard error is computed as

$$\hat{s} = \sqrt{\frac{\hat{p}(1-\hat{p})}{N}} = \sqrt{\frac{(0.603)(0.397)}{600}} \approx 0.020.$$

The standard error is $\hat{s} = 0.020$ (we keep the last digit). A 95% confidence interval is the interval of numbers within two standard errors of $\hat{p}$. In this case the confidence interval is the interval between $0.603 - 2(0.020) = 0.563$ and $0.603 + 2(0.020) = 0.643$ (Figure 4.15).

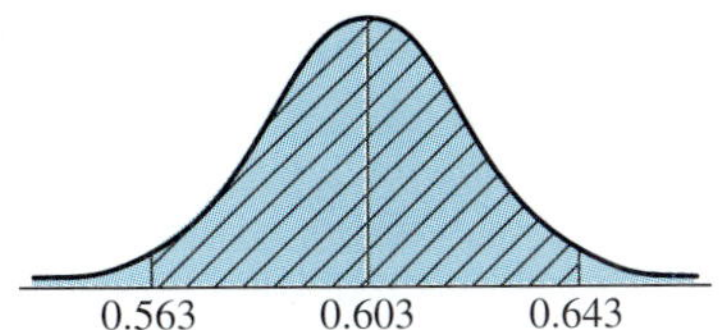

FIGURE 4.15

With a confidence level of 95%, we conclude that the population proportion is between 56.3% and 64.3%. ◆

This type of reasoning also applies to quality control problems. If we take a sample of items from a single day's production and check each one for defects, the proportion of defects in the sample gives the best estimate for the proportion of defects in the entire day's production. This is no different from taking a survey of people.

EXAMPLE 4.22

(a) Suppose that 1000 computer chips are tested for defects and 216 of them test defective. Find the 95% confidence interval for the population proportion of defective chips.

(b) Suppose that instead 10,000 computer chips had been chosen as a sample and that 2160 were found to be defective. What is the 95% confidence interval in this case? Does choosing a larger sample give significantly better results?

SOLUTION

(a) In the first case, $N = 1000$ and $\hat{p} = \frac{216}{1000} = 0.216$. The standard error is $\hat{s} = \sqrt{\frac{(0.216)(1-0.216)}{1000}} \approx 0.013$. Thus, we have $0.216 - 2(0.013) = 0.190$ and $0.216 + 2(0.013) = 0.242$. A 95% confidence interval is the interval between 0.190 and 0.242. Our estimate for the proportion of defective computer chips is between 19.0% and 24.2%.

(b) Here, $N = 10{,}000$ and $\hat{p} = \frac{2160}{10{,}000} = 0.216$, as before. However, the standard error is now $\hat{s} = \sqrt{\frac{(0.216)(1-0.216)}{10{,}000}} \approx 0.004$. Thus, we have $0.216 - 2(0.004) =$

0.208 and $0.216 + 2(0.004) = 0.224$. A 95% confidence interval is the interval of proportions between 20.8% and 22.4%, which is a significantly better estimate than an interval between 19.0% and 24.2%. ◆

INITIAL PROBLEM SOLUTION

A candy bar company has a promotion in which some of the wrappers have letters printed on the inside that entitle a person to a prize. You buy 400 of these candy bars, and you and your friends pig out. You find letters on 25 of the wrappers. What percentage of wrappers did the company put letters on? You are thinking of buying another 10,000 candy bars. How many wrappers would you expect to have letters printed on the inside?

SOLUTION Consider the first 400 bars that were unwrapped. Here, $N = 400$. The sample proportion is $\hat{p} = \frac{25}{400} = 0.0625$. The estimate of mean number of the next 1000 wrappers with letters on the inside is therefore $0.0625(1000) = 62.5$. The standard error here is $\hat{s} = \sqrt{\frac{0.0625(1 - 0.0625)}{400}} \approx 0.0121$.

We calculate our interval as

$$\hat{p} - 2\hat{s} = 0.0625 - 2(0.0121) = 0.0383, \text{ and}$$
$$\hat{p} + 2\hat{s} = 0.0625 + 2(0.0121) = 0.0867.$$

A 95% confidence interval is the interval of numbers between 0.0383 and 0.0867. Thus, out of 1000 candy bars, you should expect between 38 and 87 to have a letter on the wrapper.

PROBLEM SET 4.3

1. There are 6000 cars produced in a factory in a certain week, and 300 of them have significant problems needing correction. Sixty of the cars are selected for a detailed inspection that reveals that 5 have a problem needing correction. What is the population proportion of cars having problems? What is the sample proportion of cars having problems?
2. An assortment of candies is made by mixing 500 caramels with 1000 chocolate-covered nuts. These are then put into half-pound packages. A particular package is opened and found to have 12 caramels and 18 chocolate-covered nuts. What is the population proportion of caramels? What is the sample proportion of caramels in the package?
3. There are 7140 registered voters in a certain city, of which 3460 are Democrats, 3250 are Republicans, and 430 are Independents. A preelection canvassing in a given neighborhood reveals the following numbers of registered voters: 185 Democrats, 210 Republicans, and 25 Independents. What is the population proportion of registered Republicans? What is the sample proportion of registered Republicans in the canvassed neighborhood?
4. There are 7,123,000 people in the country of Leftvia, and 688,000 are left-handed. During a national assessment of physical characteristics that sampled 2400 Leftvians from the population, it was found that 232 of the people in the sample were left-handed. What is the population proportion of left-handed people? What is the sample proportion of left-handed people?
5. Suppose a student has five classes left as requirements in science and humanities and decides to take three of them next term. Since he registers early and there are multiple sections for each course, he feels free to choose any three of the five. The classes are {Science A, Science B, Humanities A, Humanities B, and Humanities C}. His possible selections are:

{SA, SB, HA}	{SA, SB, HB}
{SA, SB, HC}	{SA, HA, HB}
{SA, HA, HC}	{SA, HB, HC}
{SB, HA, HB}	{SB, HA, HC}
{SB, HB, HC}	{HA, HB, HC}

(a) Find the population proportion of required humanities courses he can take.
(b) Find the sample proportion of required humanities courses for each selection of three classes.
(c) Make a histogram for the sample proportions of humanities courses.
(d) Find the mean of the sample proportions.

6. Five students tie for top honors in a graduating class. Since the top two students traditionally give speeches during the graduation ceremonies, school officials decide to pick two of the students at random. The five students are Tom, Maria, Ann, Paul, and Betty. The possible pairs of students are:

{Tom, Maria}	{Tom, Ann}
{Tom, Paul}	{Tom, Betty}
{Maria, Ann}	{Maria, Paul}
{Maria, Betty}	{Ann, Paul}
{Ann, Betty}	{Paul, Betty}

(a) Find the population proportion of females among the top five students.
(b) Find the sample proportion of females in each of the pairs.
(c) Make a histogram of the sample proportions in the possible pairs.
(d) Find the mean of the sample proportions.

7. There are 6000 cars produced in a factory in a certain week and 300 of them have significant problems needing correction. A random sample of 60 cars is selected for a detailed inspection. What is the population proportion of cars having problems? What are the mean and standard deviation of the sample proportion of cars having problems?

8. An assortment of candies is made by mixing 500 caramels with 1000 chocolate-covered nuts. These are then mixed and put into packages with 30 candies each. What is the population proportion of caramels? What are the mean and standard deviation of the sample proportion of caramels in each package?

9. There are 7140 registered voters in a certain city, of which 3460 are Democrats, 3250 are Republicans, and 430 are Independents. Prior to an election, a special interest group picks a random sample of 200 registered voters to call regarding a measure on the ballot. What is the population proportion of registered Republicans? What are the mean and standard deviation of the proportion of registered Republicans in random samples of 200 taken from the voter lists?

10. There are 7,123,000 people in the country of Leftvia, and 688,000 are left-handed. In a national assessment of physical characteristics, 2400 Leftvians were randomly selected from the population. What is the population proportion for left-handed people? What are the mean and standard deviation of the sample proportion for left-handed people when samples of size 2400 are taken from the population?

11. What is the standard error in a sample of size 600 if the sample proportion is 45%?

12. What is the standard error in a sample of size 100 if the sample proportion is 70%?

13. Find the standard error for a sample of size 640 if the sample proportion is 65%.

14. Find the standard error for a sample of size 1620 if the sample proportion is 25%.

15. Five hundred students are randomly selected for a student services survey. Of those selected, 265 are females. Find the standard error for the proportion of females.

16. Sixty cars are randomly selected from the weekly production of cars at a plant. Five of these cars have problems requiring corrections. Find the standard error for the proportion of cars requiring corrections.

17. A special interest group randomly selects people from voter registration lists and asks their opinions on a piece of proposed legislation. Of the 220 people who are contacted, 128 say they are inclined to vote for the legislation in the next election. Find the standard error for the proportion of registered voters who are inclined to vote for the legislation.

18. A general biology class is doing a project on physical characteristics. The students in the class were randomly assigned to the section after preregistration, so the instructor considers them as a random sample of the student body. There are 42 students in the class, and 7 are left-handed. Find the standard error for the proportion of students who are left-handed.

19. A survey is conducted to find out how many people intend to vote for a proposition to limit property taxes. A sample of 431 likely voters chosen at random finds that 209 are planning to vote for the proposition. The sample proportion is 48.5% and the standard error is 2.4%.
(a) Verify the sample proportion and standard error.
(b) Find the 95% confidence interval for the percentage of likely voters who intend to vote for the proposition.

20. A company wishes to determine the level of customer satisfaction with its portable cassette player. It surveys 620 customers chosen randomly and finds that 579 are either satisfied or very satisfied with the cassette player they bought. The percentage of those who are either satisfied or very satisfied in this customer sample is 93.4%, and the standard error is 1.0%.
(a) Verify the sample proportion and standard error.

(b) Find the 95% confidence interval for the percentage of customers who are either satisfied or very satisfied with the cassette player.

21. The student services office of a university is concerned about student acceptance of new registration procedures. A random sample of students is selected and contacted. They are asked whether or not they find the new procedures satisfactory. Of the 280 students who respond, 172 are satisfied with the new procedures. The percentage of students who say they are satisfied with the new procedures is 61.4%, and the standard error is 2.9%.

(a) Verify the sample proportion and standard error.

(b) Find the 95% confidence interval for the percentage of students who say they are satisfied with the new registration procedures.

22. A company that produces flashlight batteries wishes to know what percentage of its batteries will last longer than 30 hours. A random sample of 1000 batteries is selected and tested. Of these batteries, 917 last 30 hours or more. The percentage of batteries that last 30 hours or more is 91.7%, and the standard error is 0.9%.

(a) Verify the sample proportion and standard error.

(b) Find the 95% confidence interval for the percentage of batteries that last 30 hours or more.

23. In a highway safety study, 442 of the 535 truck drivers interviewed said they would not use retreaded tires. Find a 95% confidence interval for the percentage of truck drivers who would not use retreaded tires.

24. A mail order company is studying its processes for filling and shipping customer orders. The company standards are that orders are to be shipped within three working days of the time they are received. A random sample of 120 of the orders received the previous month is selected and examined. Of these, 106 orders were filled and shipped on time. Find a 95% confidence interval for the percentage of orders that were shipped on time for the last month.

25. Administrators at a college are interested in the number of students who are working 10 or more hours per week while taking full-time class loads. A random sample of 240 full-time students reveals that 105 of the students are working 10 or more hours per week. Find a 95% confidence interval for the percentage of full-time students who are working 10 or more hours per week.

26. A random survey of 500 pregnant women conducted in a large northeastern city indicated that 145 of them preferred a female obstetrician to a male obstetrician. Find a 95% confidence interval for the percentage of pregnant women in the city who would prefer a female obstetrician.

EXTENDED PROBLEMS

In previous problems, we have been concerned with finding a 95% confidence interval for a population proportion. While a 95% confidence interval is one that is most commonly used (public opinion polls use it almost exclusively), other confidence intervals can be easily defined and calculated. A 95% confidence interval contains those values that are within 2 standard errors of the sample proportion. This is because the distribution of sample proportions is normal, and 95% of the sample proportions are within two standard errors (standard deviations) of the population proportion. Similarly, we can define a 99.7% confidence interval, which would be based on 3 standard errors, since 99.7% of all sample proportions are within 3 standard errors of the population proportion. Other commonly used confidence intervals are a 99% confidence interval based on 2.58 standard errors and a 90% confidence interval based on 1.65 standard errors.

27. Referring to problem 21,

(a) find a 90% confidence interval for the proportion of students who are satisfied with the new registration procedures.

(b) find a 99.7% confidence interval for the proportion of students.

28. Referring to problem 22,

(a) find a 90% confidence interval for the proportion of batteries that last 30 hours or more.

(b) find a 99.7% confidence interval for the proportion of batteries that last 30 hours or more.

29. Referring to problem 23,

(a) find a 90% confidence interval for the proportion of truck drivers who would not use retreaded tires.

(b) find a 99% confidence interval for the proportion of truck drivers who would not use retreaded tires.

30. A company decides to offer a "double your money back" guarantee on its product, a gigax. The gigax costs $15, and the company promises to refund $30 to any customer who purchases a defective gigax. To determine how much they might expect to pay out on this guarantee, the company tests 800 gigaxes from a random sample. Of these, 28 are defective, and the rest are of high quality. Find a 99% confidence interval for the proportion of defective gigaxes that are manufactured.

31. Referring to problem 30, the company expects to produce 100,000 gigaxes in the next year. Find the 99.7% confidence interval for the amount of money the company can expect to pay out if all defective gigaxes are returned under the conditions of the guarantee.

4.4 SCATTERPLOTS

INITIAL PROBLEM

The following data gives a list of student midterm scores and their corresponding final exam scores.

(Midterm, Final Exam): (124, 250), (120, 176), (60, 148), (153, 283), (79, 240), (135, 241), (170, 255), (145, 281), (114, 210), (120, 272), (210, 299), (94, 220), (126, 233), (116, 249), (128, 285), (137, 272), (84, 207), (68, 202), (38, 209), (156, 213), (77, 270), (138, 275), (200, 275), (166, 266), (123, 260), (172, 263), (205, 292)

Suppose that a student has a midterm score of 180 points. What is our best guess for this student's final exam score? How sure are we that this is a good prediction?

Up to this point, our work has been focused on a single variable, or characteristic, from a population. Now we look at relationships between two variables. We begin with visual displays and will emphasize those that show a linear relationship between the two variables. The question of cause and effect will also be discussed.

SCATTERPLOTS

Sometimes data are grouped into pairs of numbers that may or may not have a relation to each other. For example, data points might be records of sales and temperature, selling price of a house and its appraised value, employment and interest rates, or education and income. Such pairs of numbers can be plotted as points on a portion of the (x, y)-plane forming what is called a **scatterplot.** For example, Table 4.2 lists data on significant earthquakes of the 1960s.

TABLE 4.2 Significant Earthquakes of the 1960s

Date	Place	Deaths	Magnitude
Feb. 29, 1960	Morocco	12,000	5.8
May 21–30, 1960	Chile	5000	8.3
Sept. 1, 1962	Iran	12,230	7.1
July 26, 1963	Yugoslavia	1100	6.0
Mar. 27, 1964	Alaska	131	8.4
Aug. 19, 1966	Turkey	2520	6.9
Aug. 31, 1968	Iran	12,000	7.4

To investigate the possible relationship between the magnitude of an earthquake and the number of deaths that result, we make a scatterplot of the data in the table.

The magnitude scale is placed along the horizontal axis, and the number of deaths scale is placed along the vertical axis. For each earthquake we place a dot at the intersection of the appropriate horizontal and vertical lines. For instance, the dot representing the July 1963 earthquake in Yugoslavia is on the vertical line for magnitude 6 and is on an imagined horizontal line for 1100 deaths, that is, just a little below the horizontal line for 1200 deaths (Figure 4.16).

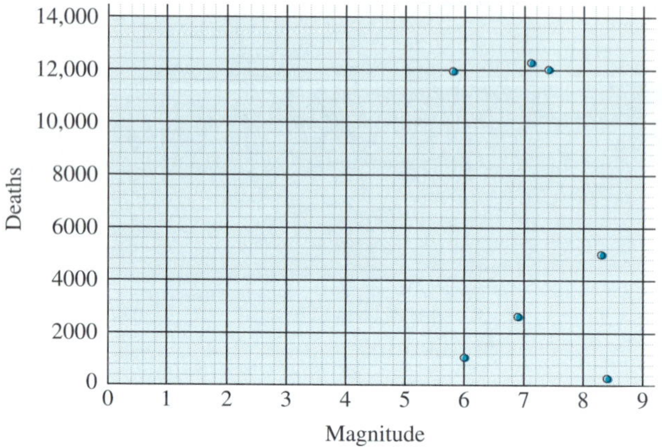

FIGURE 4.16

When we look at Figure 4.16, the scatterplot of the earthquake data, we do not see any particular pattern other than that the magnitude of all the earthquakes is above 5. (Can you explain why there does not appear to be a relationship between the magnitude of the earthquake and the number of deaths it causes?) With other data, however, it often happens that you can see a pattern; and in many of the cases, the data points will appear to lie approximately on a line, as in the next example.

TIDBIT

Census data from 1992 indicated the average annual earnings of a high school graduate to be $18,700 versus $32,600 for a college graduate.

EXAMPLE 4.23 Suppose that ten people are interviewed and asked about their income level and educational attainments (Table 4.3).

TABLE 4.3 Educational Level vs. Income

Person	Years of Education	Yearly Income (1000s)	Data Points
1	12	22	(12, 22)
2	16	63	(16, 63)
3	18	48	(18, 48)
4	10	14	(10, 14)
5	14	2	(14, 2)
6	14	34	(14, 34)
7	13	31	(13, 31)
8	11	97	(11, 97)
9	21	96	(21, 96)
10	16	44	(16, 44)

TIDBIT

Ecologists predict the amount of salmon in a run by observations of the conditions. Using regression analysis they can predict quite accurately the number of salmon that will be in the run in the following year. Sometimes the underlying conditions change without warning. At this point the ecologist has no choice but to wait for enough new data points to be collected so that a new scatterplot and new regression line may be drawn. Worst of all is that there is seldom any warning as to when this will happen!

Plot this information in a scatterplot and draw a line that this data seems to approximate, or "fit."

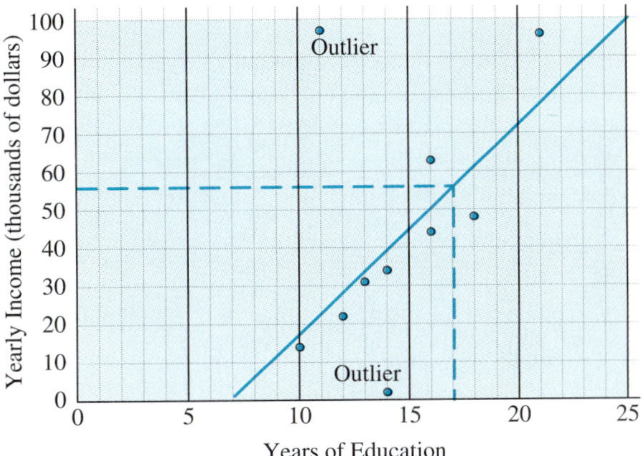

FIGURE 4.17

SOLUTION To visualize this information, we plot it on a graph with Years of Education on the horizontal axis and Yearly Income on the vertical axis (Figure 4.17).

There are two exceptional points in this data that we again call **outliers.** One outlier is a person with an 11th grade education who nonetheless makes $97,000. The interview revealed that this person owned his own successful tulip bulb import business. The other outlier was a person with 2 years of college (14 years of education) who made only $2,000. This unfortunate individual was an unemployed homeless person. Ignoring the outliers, we notice that these points lie roughly on the straight line that has been sketched in. There appears to be a relationship between educational level and yearly income in which higher income levels correspond to higher educational levels. We call such a mutual relationship a *correlation.* This does not imply that one is the cause of the other. ◆

There is a specific line that best fits the data; this line is called the **regression line.** The method for computing the regression line is treated in the problem set. In many problems, you can use a straightedge and eyeball a best-fitting line as we did in the example. A regression line can be very useful. If you know the value of one of the variables, say the educational level, then you can use the regression line to estimate a likely value for the other variable, the income level. For example, if we were to interview another person whose educational level was 17 years (1 year of graduate school), then we could give an educated guess as to what this person's income level might be using the regression line. To make this estimate, you trace a vertical line from 17 on the horizontal axis up to the regression line; then you trace a horizontal line left until it intersects the income axis. The process is shown by the dotted lines in Figure 4.17. In this case, we use the regression line to project that this person's income level is likely to be close to $58,000.

EXAMPLE 4.24 Figure 4.18 shows the scatterplot of per capita gross domestic product and infant mortality per 1000 live births for a selection of countries.

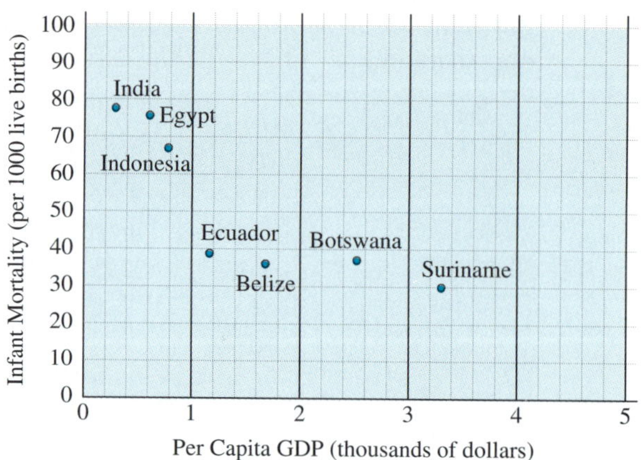

FIGURE 4.18

Draw an approximate regression line and use it to estimate a per capita gross domestic product, which would result in the infant mortality rate being as low as possible with present medical practices.

SOLUTION We draw a line to fit the data points, as shown in Figure 4.19.

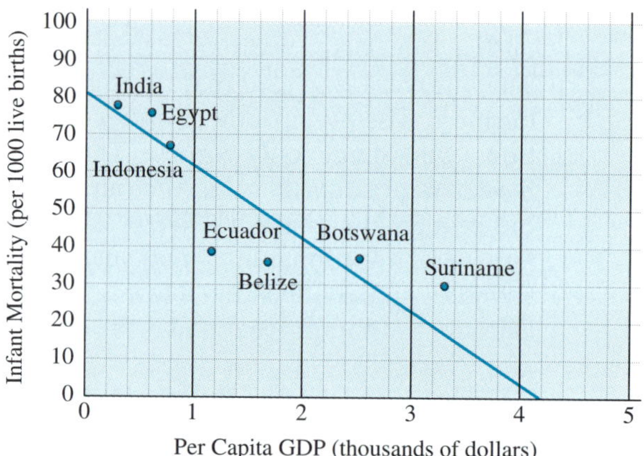

FIGURE 4.19

The approximate regression line intersects the horizontal axis at \$4200 per capita GDP. We estimate that if a country can bring its per capita GDP up to \$4200, then the country's infant mortality will be correspondingly reduced to the lowest possible level (a 0 level of infant mortality is unattainable). ◆

VARIATION FROM A TREND

Once the regression line has been determined, the data may or may not fit the line well. Technically, the fit of the data to the regression line is measured by the **correlation coefficient.** The correlation coefficient is a number between $+1$ and -1, which is computed by more advanced formulas given in the extended problems. If the correlation coefficient is $+1$, then the points fit the line exactly. As one variable increases, the other also increases, and we may predict with perfect accuracy one variable from the other. If the correlation coefficient is -1, then there is still a perfect fit between the points and the line, and we may still predict one variable from the other with perfect accuracy. In this case, however, as one variable increases, the other decreases. A correlation coefficient of 0 means that the two variables are essentially unrelated to each other.

Instead of the correlation coefficient, we will use a more intuitive description of the fit of the data to the regression line. Figures 4.20–4.24 are various scatterplots with regression lines showing the relationship between a student's grade in a class and other possible predictive factors. We will use the terms **strong positive correlation** (Figure 4.20), **weak positive correlation** (Figure 4.21), **no correlation** (Figure 4.22), **weak negative correlation** (Figure 4.23), and **strong negative correlation** (Figure 4.24) to describe the fit of the data to the regression line as illustrated in those figures.

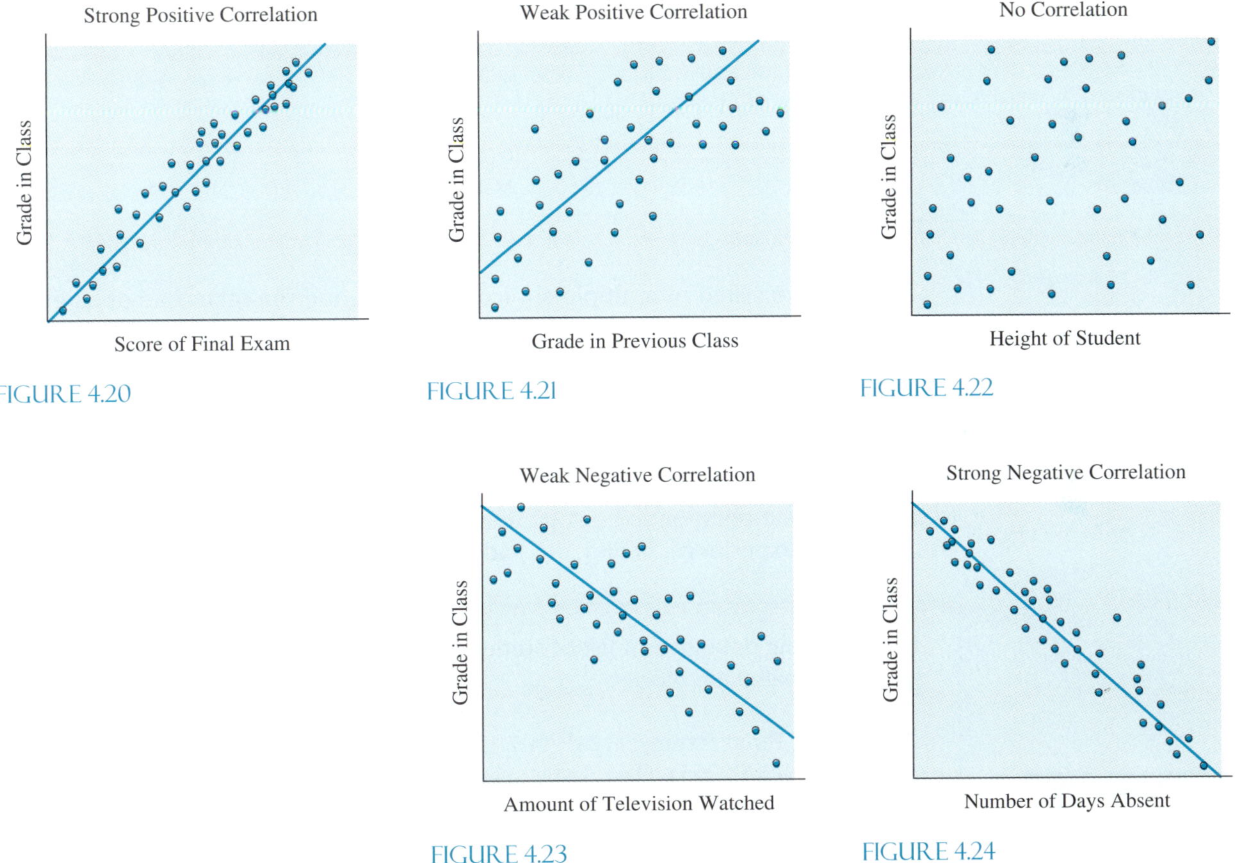

FIGURE 4.20

FIGURE 4.21

FIGURE 4.22

FIGURE 4.23

FIGURE 4.24

CORRELATION AND CAUSATION

"Correlation is not causation" is a well-known proverb in statistics. It means that two quantities may be highly correlated without one quantity being the cause of the other. This is the point often argued by tobacco companies. The fact that lung disease, heart disease, and other health problems are correlated with smoking does not prove that smoking causes these illnesses. Rather, it is possible that some factor, perhaps genetic or environmental, leads a person to enjoy smoking and also causes disease. It is not possible to prove this possibility wrong, although it may seem implausible given that ingredients in tobacco smoke have been shown to cause disease in animals.

For another example, consider the scatterplot where students were given two kinds of tests over the same material (Figure 4.25).

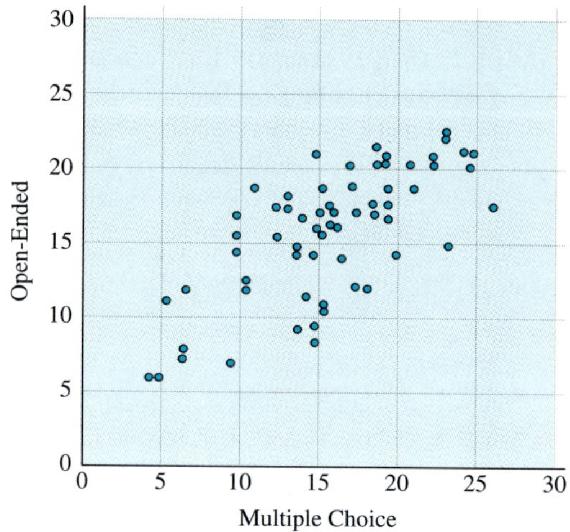

FIGURE 4.25

One test consisted of multiple-choice questions, and the other had open-ended questions. There is a positive correlation between the two variables, although not a strong one. This means that the scores on the multiple-choice test may be used as predictors of scores on the open-ended test. However, it is not a perfect predictor. Roughly the same information is gathered by each test. However, there are some students who do better on open-ended tests and others who do better on multiple-choice tests. Notice that doing well on the multiple-choice test does not cause one to do well on the open-ended test. Rather, the same combination of circumstances (study, talent, experience, health, etc.) allow a student to do well on either.

The following data gives a list of student midterm scores and their corresponding final exam scores.

(Midterm, Final Exam): (124, 250), (120, 176), (60, 148), (153, 283), (79, 240), (135, 241), (170, 255), (145, 281), (114, 210), (120, 272), (210, 299), (94, 220), (126, 233), (116, 249), (128, 285), (137, 272), (84, 207), (68, 202), (38, 209), (156, 213), (77, 270), (138, 275), (200, 275), (166, 266), (123, 260), (172, 263), (205, 292)

Suppose that a student has a midterm score of 180 points. What is our best guess for this student's final exam score? How sure are we that this is a good prediction?

SOLUTION Consider the scatterplot of this data (Figure 4.26).

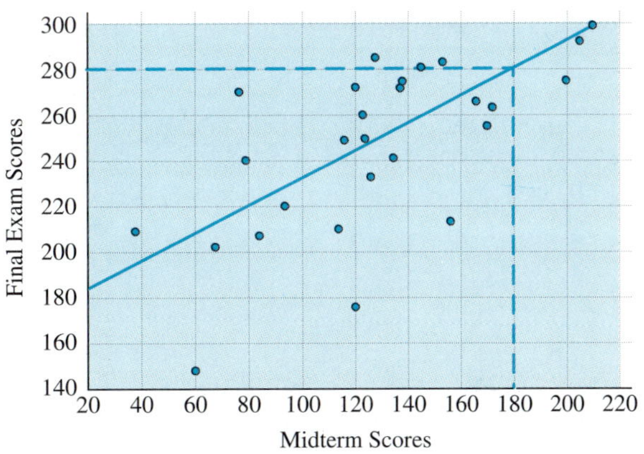

FIGURE 4.26 Scatterplot of Midterm and Final Exam Scores

There is a weak positive correlation. Just viewing the graph, we can tell that there is a positive correlation, due to the lower left to upper right slope of the regression line, but also lots of scatter. The actual correlation coefficient is 0.66. Thus, the midterm score is a good, but not great, predictor of the final exam score. Drawing an approximate regression line and looking at the y-value corresponding to an x-value of 180 gives a predicted final exam score of about 280.

PROBLEM SET 4.4

Complete the following for problems 1 through 4.

(i) Make a scatterplot for the data.
(ii) What kind of correlation is indicated by the scatterplot?
(a) Is it positive or negative?
(b) Is it strong or weak?
(iii) Are there any outliers in the scatterplot?

1. The college admissions office uses high school grade point average (GPA) as one of its selection criteria for admitting new students. At the end of the year, ten students are selected at random from the freshman class and a comparison is made between their high school grade point averages and their grade point averages at the end of their freshman year in college.

High School GPA	Freshman GPA
2.8	2.5
3.2	2.6
3.4	3.1
3.7	3.2
3.5	3.3
3.8	3.3
3.9	3.6
4.0	3.8
3.6	3.9
3.8	4.0

2. A female student thinks that people of similar heights tend to date each other. She measures herself, her roommates, and several others in the dormitory. Then she has them find out the heights of the last man each of the women dated. The heights are given in inches.

Female	Male
64	70
62	71
66	73
65	68
64	72
70	71
61	66
66	69

3. Students taking a speed-reading course produced the following gains in their reading speeds:

Weeks in Program	Speed Gain (words per min.)
2	50
4	100
4	140
5	130
6	170
6	140
7	180
8	230

4. A high school career counselor does a ten-year follow-up study of graduates. Among the data she collects is a list of the number of years of education beyond high school and incomes earned by the graduates. The following list shows the data for ten randomly selected graduates:

Years of Education Beyond High School	Income (1000s)
2	27
5	33
0	22
2	25
7	48
4	35
0	28
6	32
4	22
5	30

Complete the following for problems 5 through 8.

(i) Make a scatterplot for the data.

(ii) Sketch in the regression line. As a line that best fits this data, the line should have a balance of data points that are above it and below it.

5. A golf course professional collected the following data on the average scores for eight golfers and their average weekly practice time.

Practice Time (hours)	Average Score
6	79
3	83
4	92
6	78
3	84
2	94
5	80
6	82

6. An Alaskan naturalist made aerial surveys of a certain wooded area on ten different days, noting the wind velocity and the number of black bears sighted.

Wind Velocity (mph)	Black Bears Sighted
2.1	93
16.7	60
21.1	30
15.9	63
4.9	82
11.8	76
23.6	43
4.0	89
21.5	49
24.4	36

7. A company that assembles electronic parts uses several methods for screening potential new employees. One of these is an aptitude test requiring good eye-hand coordination. The personnel director selects eight employees at random and compares their test results with their average weekly output.

Aptitude Test Results	Weekly Output (dozens of units)
6	30
9	49
5	32
8	42
7	39
5	28
8	41
10	52

8. A high school math teacher has students maintain records on their study time and then compares their average nightly study time to the scores received on an exam. A random sample of the students showed these comparisons:

Study Time (nearest 5 min)	Exam Score
15	58
25	72
50	85
20	75
25	68
30	88
40	80
15	74

(contd.)

25	78
30	70
45	94
35	75

9. In Example 4.24 we looked at the scatterplot of per capita gross domestic product and infant mortality per 1000 live births for a selection of countries. Use the scatterplot and regression line to predict the infant mortality rate in a country that has a per capita gross domestic product of $2000.

10. A study of cognitive development in young children recorded the age (in months) when they spoke their first word and the results of an aptitude test taken much later. The data is contained in the following scatterplot.

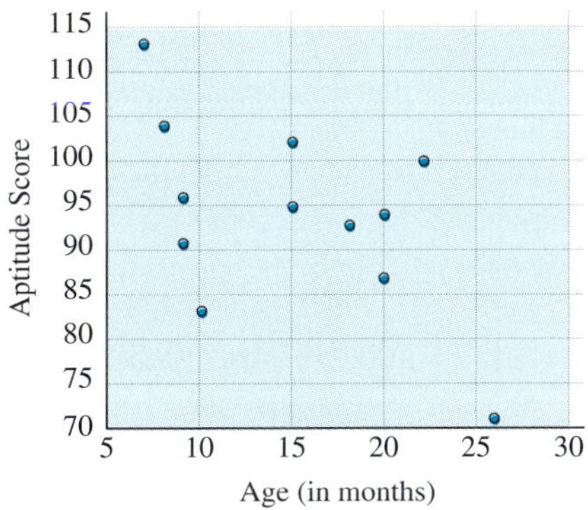

(a) Predict the aptitude test score for a child who spoke his first word at 12 months of age.

(b) Predict the aptitude test score for a child who spoke his first word at 20 months of age.

11. A doctor conducted a study investigating the relationship between weight and diastolic blood pressure of males between 40 and 50 years of age. The scatterplot and regression line indicate the relationship.

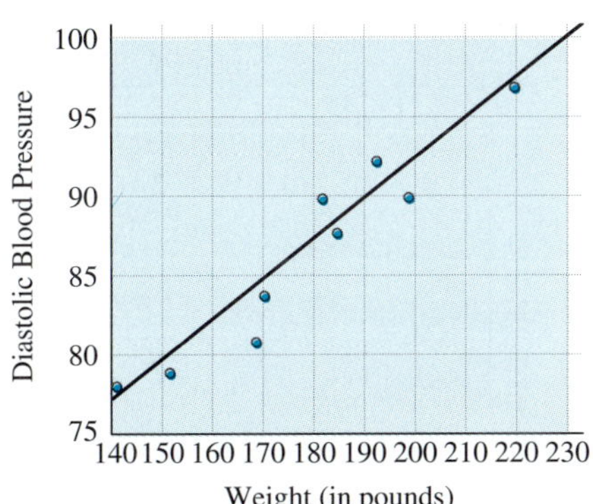

(a) Predict the diastolic blood pressure of a 45-year-old man who weighs 160 pounds.

(b) Predict the diastolic blood pressure of a 42-year-old man who weighs 180 pounds.

12. In a study on obesity involving twelve women, the lean body mass (in kilograms) was compared to the resting metabolic rate. The scatterplot and regression line indicate the data and relationship.

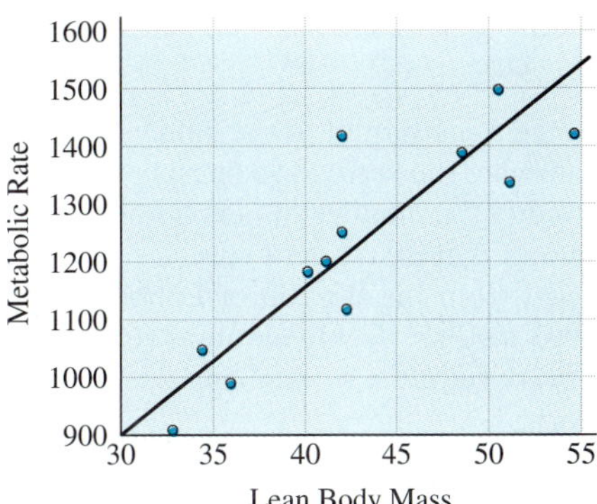

(a) Predict the resting metabolic rate for a woman with a lean body mass of 40 kilograms.

(b) Predict the resting metabolic rate for a woman with a lean body mass of 50 kilograms.

13. A company compared the commuting distance and number of absences for a group of employees, with the following data:

Commuting Distance (mi)	Number of Absences (yr)
8	4
21	5
8	5
8	3
2	2
15	5
17	7
11	4

(a) Make a scatterplot of the data.

(b) Estimate the regression line.

(c) Predict the number of absences (per year) for an employee with a commute of 15 miles.

14. A local bank compared the number of car loans and new home mortgages it processed each month for a year.

Month	Car Loans	Mortgages
Jan	45	6
Feb	36	6
Mar	48	10
Apr	62	14
May	60	15

Jun	72	18
Jul	76	14
Aug	84	15
Sep	67	12
Oct	60	10
Nov	53	9
Dec	68	11

(a) Make a scatterplot of the data.
(b) Estimate the regression line.
(c) Predict the number of new home mortgages in a month that has 50 car loans.

15. A report from the Bureau of Labor Statistics listed the 1993 median weekly earnings (for both men and women) of full-time workers in selected occupational categories.

Median Weekly Earnings

Occupation	Men	Women
Managerial and prof. specialty	791	580
Technical, sales, admin. support	534	376
Service occupations	350	259
Precision production	511	344
Operators, fabricators, laborers	399	288
Transportation	456	358
Handlers, equip. cleaners	319	286
Farming, forestry, fishing	274	242

(a) Make a scatterplot of the data.
(b) Estimate the regression line.
(c) Predict the median weekly salary for a woman if the median weekly salary for a man is $450.

16. During the last two decades corporations have invested in new plants and equipment as corporate profits have continued to increase.

	Corporate Profits (billions)	Expenditures for Plants and Equip. (billions)
1970	69	106
1975	121	163
1980	192	318
1985	223	455
1990	293	592
1993	442	650

SOURCE: 1995 Information Please Almanac.

(a) Make a scatterplot of the data.
(b) Estimate the regression line.
(c) Predict the expenditures for new plants and equipment if corporate profits were $250 billion.

In problems 17 through 20, describe the fit of the data to the regression line as one of the following: strong positive correlation, weak positive correlation, no correlation, weak negative correlation, or strong negative correlation.

17.

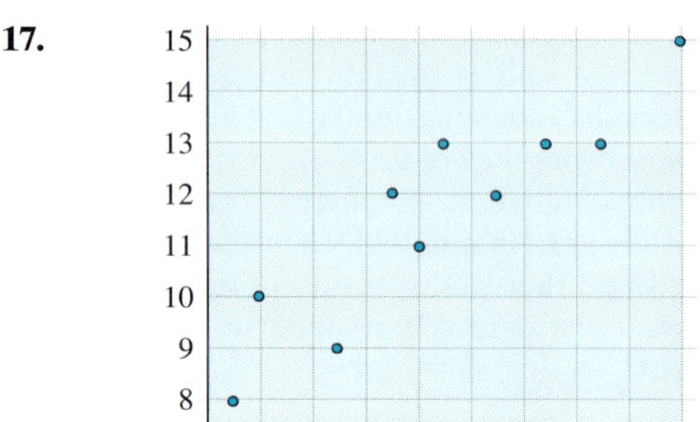

18.

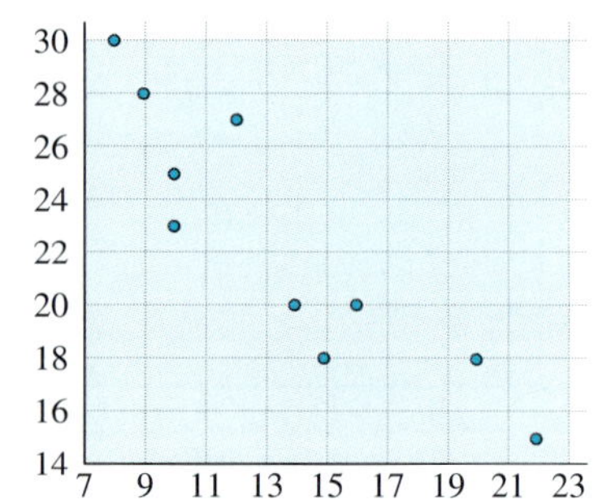

19.

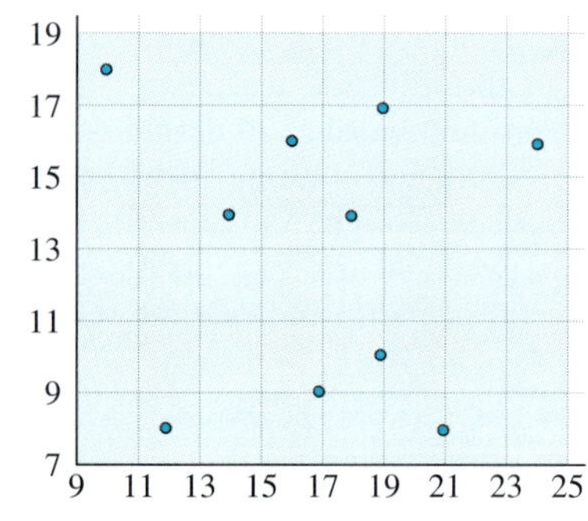

20.

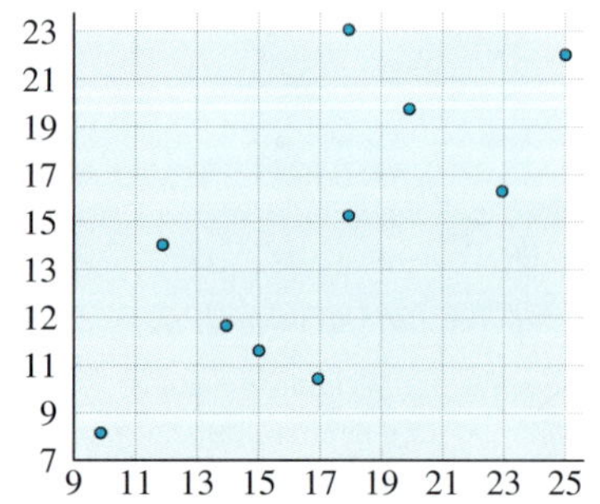

21. In a test of fuel efficiency, a test car ran a race course at varying speeds. The table shows the speed of the car in miles per hour and the accompanying fuel efficiency in miles per gallon.

Speed	Miles per Gallon
30	34
35	31
40	32
45	30
50	29
55	30
60	28
65	27

(a) Make a scatterplot for the data.
(b) Describe the correlation in terms of weak or strong, positive or negative, or no relationship.

22. A large manufacturing company wants to have its employees work overtime rather than add new workers to its labor force. The union claims that accidents increase due to fatigue as workers put in more hours. The records for average hours worked per week and number of accidents for the past eight weeks are given in the following table:

Average Hours Worked	Number of Accidents
38	3
37	3
39	4
41	4
43	8
45	10
49	9
47	7

(a) Make a scatterplot for the data.
(b) Describe the correlation in terms of weak or strong, positive or negative, or no relationship.

23. Fifteen students in a statistics class were asked to record the amount of time they spent studying before they took their statistics exam. Their responses were then matched with their scores on the exam.

Hours	Score
1.00	66
1.25	58
1.50	75
1.50	68
2.00	78
2.25	76
2.75	85
3.00	72
3.00	87
3.00	78
3.50	88
4.00	96
4.00	86
4.50	92
5.00	90

(a) Make a scatterplot for the data.
(b) Describe the correlation in terms of weak or strong, positive or negative, or no relationship.

24. A large retail store compares the records for its monthly expenditures for advertising and its total monthly sales. The figures for last year were as follows:

Advertising ($ thousands)	Sales ($ millions)
21	37
22	38
18	30
19	34
24	40
20	35
21	36
27	47
18	34
23	38
25	42
28	44

(a) Make a scatterplot for the data.
(b) Describe the correlation in terms of weak or strong, positive or negative, or no relationship.

EXTENDED PROBLEMS

Formula for finding the equation of the regression line (line of "best fit")

The line that best fits the data in a scatterplot has two main characteristics:

(i) It goes through the point $(\bar{x}, \bar{y})$ that corresponds to the averages for the two variables. This makes sense, because when we use the line for predictions, we would want the average value for x to predict the average value for y.

(ii) The slope of the line is based on the correlation coefficient and the standard deviations for the two variables. This also makes sense because the correlation coefficient measures the strength of the relationship, and the standard deviations measure the spread of the variables. The equation of the regression line is written $y - \overline{y} = m(x - \overline{x})$, where

$$m = \frac{n\Sigma xy - (\Sigma x)(\Sigma y)}{n\Sigma x^2 - (\Sigma x)^2}$$

This may look difficult at first glance. However, the value can be obtained systematically, as the following example with three pairs of values shows. The symbol Σ means that we are to add up all the respective values;

x	y	x^2	xy
3	5	9	15
5	9	25	45
8	10	64	80

$\Sigma x = 16\ \Sigma y = 24\ \Sigma x^2 = 98\ \Sigma xy = 140$
$\overline{x} = 5.33$ and $\overline{y} = 8$

$$m = \frac{(3)(140) - (16)(24)}{3(98) - (16)^2} = 0.9474 \text{ or } 0.95$$

The equation of the regression line is then $y - 8 = 0.95(x - 5.33)$ or $y = 0.95x + 2.94$.

For problems 25 through 28, use the formula for finding the equation of the regression line.

25. Find the equation for the regression line for the data from problem 1.

26. Find the equation for the regression line for the data from problem 3.

27. Find the equation for the regression line for the data from problem 5.

28. Find the equation for the regression line for the data from problem 7.

Formula for computing the correlation coefficient, *r*

If the x and y are associated with the variables for the horizontal and vertical axes, respectively, then the correlation coefficient, r, can be calculated by the formula

$$r = \frac{n\Sigma xy - (\Sigma x)(\Sigma y)}{\sqrt{(n\Sigma x^2 - (\Sigma x)^2)(n\Sigma y^2 - (\Sigma y)^2)}}.$$

At first, this formula looks very intimidating. However, the value can be obtained systematically, as the following example with three pairs of values shows. The symbol Σ means that we are to add up all the respective values.

x	y	xy	x^2	y^2
3	5	15	9	25
5	9	45	25	81
8	10	80	64	100

$\Sigma x = 16\ \Sigma y = 24\ \Sigma xy = 140\ \Sigma x^2 = 98\ \Sigma y^2 = 206$

$$r = \frac{3(140) - (16)(24)}{\sqrt{(3(98) - (16)^2)(3(206) - (24)^2)}}$$
$$= 0.9011$$

For problems 29 through 32, use the formula for computing the correlation coefficient.

29. Find the correlation coefficient for high school GPA and college GPA in problem 1.

30. Find the correlation coefficient for "weeks in the program" and "speed gain" in problem 3.

31. Find the correlation coefficient for "practice time" and "average scores" in problem 5.

32. Find the correlation coefficient for "aptitude test results" and "weekly output" in problem 7.

Chapter Four Problem

Suppose 121,000 computer motherboards are made in a day. To test the quality, 1000 are chosen at random. These boards are tested for defects, taken apart, and inspected. Of these 1000 boards, 187 boards are found to be defective. How many boards of the remaining 120,000 are likely to be defective? How confident are we of this answer?

Solution

Strategy: Use a variable

We wish to know the number of boards that are likely to be defective. Call this number D. The tools of this chapter do not directly involve D, but rather discuss the population proportion and the sample proportion. The sample proportion may be found from the facts in the problem. Using the notation of the chapter, the sample size is $N = 1000$ and the sample proportion is $\hat{p} = \frac{187}{1000} = 0.187$. The best guess for the population proportion p is the sample proportion itself. This means that 0.187 is our best guess for the population proportion.

$$\text{Recall population proportion} = \frac{\text{\# defectives}}{\text{\# in population}}$$
$$= \frac{D}{120{,}000}, \text{ so our best guess for the ratio is } 0.187.$$

(We have 120,000 instead of 121,000 because the sample of 1000 was already destroyed.)

Setting $\frac{D}{120{,}000} = 0.187$ gives

$$D = 120{,}000 \times 0.187 = 22{,}440.$$

Thus 22,440 boards are likely to be defective. To see how good this guess is, we compute a 95%confidence interval for $\hat{p}$. $\hat{s} = \sqrt{\frac{(0.187)(1 - 0.187)}{1000}} = 0.01233$. The lower bound of a 95% confidence interval is $\hat{p} - 2\hat{s} = 0.16234$. This translates to $120{,}000 \times 0.16234 = 19{,}481$. The upper bound of a 95% confidence interval is $\hat{p} + 2\hat{s} = 0.2166$, which is equivalent to $120{,}000 \times 0.2166 = 25{,}399$. Thus while we may be very unsure that there are exactly 22,440 defective boards, we may be 95% confident that there are between 19,481 and 25,399 defective boards. On the positive side, that means we can expect between 94,601 (120,000 − 25,399) and 100,519 (120,000 − 19,481) good boards. This is another measure of our daily production of boards and would be needed in developing marketing plans.

Chapter Four Review

Key Ideas and Questions

The following questions review the main ideas of this chapter. Write your answers to the questions and refer to the pages listed to make certain that you have mastered these ideas.

1. Describe the characteristics of a normal distribution. **193** How does knowing that a distribution is normal allow you to compute percentages of the distribution that lie between two values? **194**
2. How can you sample so that there is no bias? **200** In what ways might bias be present in other kinds of sampling? **199**
3. What does a 95% confidence interval mean? **213** Why does sampling allow for meaningful results to come from surveys of a relatively small number of people? **210**
4. Is it more important to have a very large sample or to be sure that your sample is as unbiased as possible? **187**
5. How does a scatterplot with a strong correlation allow you to make confident predictions? **222** Does a strong correlation allow you to infer that one variable is the cause of the other? **223**

Vocabulary/Notation

Following is a list of key vocabulary, notation, and ideas for this chapter. Mentally review each of these items, write down the meaning of each term, and use it in a sentence. Then, refer to the pages listed by number and restudy any material you are unsure of before solving the Chapter Four Review Problems.

Section 4.1

Deviation from the Mean 189	z-score 191	z-axis 194
Deviation 189	Normal Distribution 193	Data Axis 195
Variance 190	Standard Normal Distribution 193	
Standard Deviation 190	z-score Axis 194	

Section 4.2

Population 199	Simple Random Sample 200	Independent Sampling 203
Sample 199	Random Number Generator 200	
Bias 199	Random Number Table 201	

Section 4.3

Population Proportion 209	Standard Error 213	
Sample Proportion 209	95% Confidence Interval 213	

Section 4.4

Scatterplot 219	Correlation Coefficient 222	No Correlation 223
Outlier 221	Strong and Weak Positive Correlation 223	Strong and Weak Negative Correlation 223
Regression Line 221		

Chapter Four Review Problems

1. Consider the data set {0, 2, 3, 5, 10}. Compute the mean, variance, and standard deviation of this data set. Compute the z-score of each data point of this set.

2. What percent of a standard normal population has values between -1 and 3? What percent has values greater than 3? What percent is less than -1? Why do these percentages sum to 100%?

3. What percent of a normal population has a z-score greater than 2?

4. The mean size of paperback mysteries in a certain series is 250 pages with a standard deviation of 20 pages. Assume this is a normal population. What percent of mysteries in this series has fewer than 210 pages? What percent has more than 270 pages?

5. We wish to determine the opinion of the voters in a certain town with regard to allowing rollerblading in the town square. A survey is taken of adult passers-by near the local high school one late afternoon. What is the population in this case? What is the sample? What sources of bias might there be in this sampling procedure?

6. The student union wishes to raise student fees so that a new center for bowling and video games can be built. A group opposed to this plan takes a survey of students coming from the library. Are there any sources of bias in this survey? What if the survey is taken in a pool hall downtown?

7. Explain how to choose three people in an unbiased way from a group of five people.

8. A study of heart disease follows the history of 100 people over the course of their lives. For bookkeeping purposes, these people are assigned numbers from 0 to 99. Using a random number table, choose a simple random sample of size 20 from this group.

9. In problem 8, suppose that 20% of the people in the study actually develop heart disease. These people are those numbered 4, 7, 15, 16, 22, 31, 34, 39, 41, 46, 47, 49, 60, 66, 73, 78, 86, 89, 92, 95. How many of these people were chosen in your study? Repeat problem 8 and this problem again and compare the answers. Why is it not surprising if these answers are different?

10. Repeat problem 8 using 20% independent sampling instead of a simple random sample.

11. A sample is taken from a jar of candies that have been well mixed. The sample is size 30 and 9 of these turn out to be chocolate. If there are 200 candies in the jar, what is the best guess for the total number of chocolate candies in the jar?

12. A jar contains 20,000 jelly beans, and 5000 of these are watermelon flavored. Suppose that you take a simple random sample of 48 jelly beans. What is the population proportion of watermelon flavored jelly beans? What are the mean and standard deviation of the sample proportion of watermelon flavored jelly beans for samples of size 48?

13. In problem 12, what is the range of percentages that contains 95% of the sample proportions for samples of size 48?

14. Another jar has a large number of jelly beans. A simple random sample of size 50 is taken and found to contain 12 watermelon flavored jelly beans. What is the sample proportion? What is the standard error? Find a 95% confidence interval for the population proportion of watermelon flavored jelly beans.

15. **(a)** Which of the following plots have strong positive correlations?

(b) Which of the following plots have no correlations?

(c) Which of the following plots have weak negative correlations?

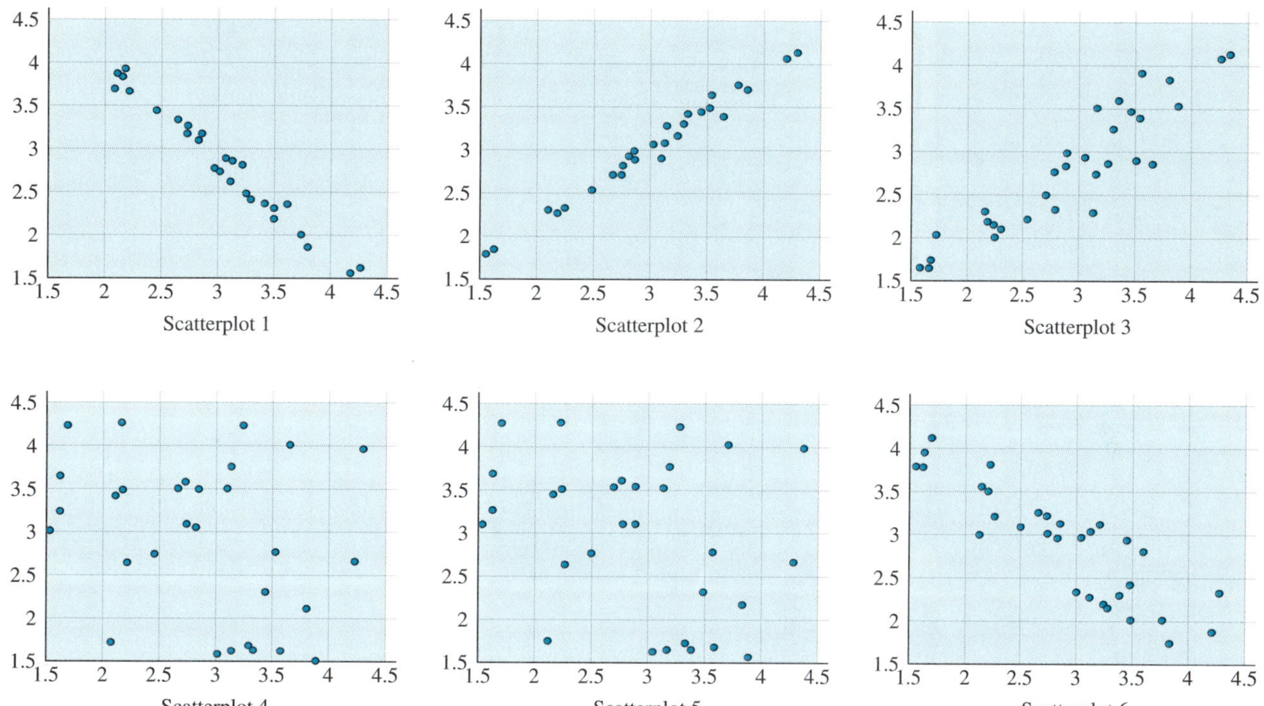

Scatterplot 1 Scatterplot 2 Scatterplot 3

Scatterplot 4 Scatterplot 5 Scatterplot 6

16. The manager of a sporting goods store notes that high levels of rainfall have a negative effect on sales of beach equipment and apparel. The sales in thousands of dollars and the summer rainfall in inches have been measured for various years and are recorded in the table below.

rain (in inches)	**sales (in thousands of dollars)**
10	300
22	120
20	160
2	360
21	180
5	320
18	340

Make a scatterplot of this data. Identify any outliers. Sketch a regression line. Do you think this data is strongly or weakly correlated? Is the correlation positive or negative? If the predicted rainfall for the coming summer is 15 inches, what is the best prediction for sales? If the sales in one year were $260,000 what is the best guess for rainfall that summer?

CHAPTER 5

PROBABILITY

RUN THAT BALL!

The local college football team runs an option offense. In the opening game of the season, the quarterback suffered a broken collarbone, an injury that would cause him to miss a good share of the season. In the option offense, the quarterback is more exposed because he has the option of running the ball himself, handing the ball to a running back, or passing to a receiver. Fans have noticed that quarterback injuries have occurred more frequently since the option offense was installed two years earlier. At the boosters club meeting the following week, the coach was asked if his quarterbacks were going to keep getting hurt. The coach responded, "Four quarterbacks in the country were injured on Saturday. One was an option quarterback, the others were standard drop-back passers. Just because you run the option does not mean that you are going to get hurt. There is risk in any sport."

Chapter Goals

1. Compute probabilities of events in situations where all outcomes have the same chance of occurring.
2. Use tree diagrams to make probability computations.
3. Compute probabilities when there is partial information available.
4. Use expected values to find the true cost of lottery tickets, insurance premiums and similar items.

The coach's reasoning sounds plausible and even somewhat reassuring. His quarterback was injured, but there were injuries to three other quarterbacks as well. The coach's comments suggest that there is no greater risk of injury for option quarterbacks. On closer inspection, however, we see there is a fallacy. There are about 107 NCAA division I-A football teams. Of these, 4 were using the option offense. Thus, on the previous Saturday, 1 out of 4 option teams, or 25%, lost a quarterback. By comparison, only 3 out of 103, or about 3%, of the rest of the teams in the country lost a quarterback. Looked at this way, the option offense seems hazardous to a quarterback's health.

Probability is the branch of mathematics that deals with chance. While most people have an intuitive grasp of probability, even mathematicians are at odds as to what it really means and how it is applied in certain situations. Probability theory is full of apparent paradoxes, and many arguments that appear quite reasonable are nonetheless incorrect. Suppose a person is tested for a rare and fatal disease and learns from his doctor that the test is positive — indicating they have the disease. Suppose further that the test is very reliable, though not perfect. For anyone, this is a real cause for alarm, and some people will believe they have received a death sentence. However, correct reasoning (somewhat similar to our analysis of the option offense) will show that the probability the person actually has the disease may still be quite low, and more testing is needed to make a final determination. This will be discussed in more detail in this chapter.

THE HUMAN SIDE OF MATHEMATICS

David Blackwell

David Blackwell (1919–) entered college at the age of 16 with the ambition to earn a bachelor's degree and become an elementary school teacher. Six years later, in 1942, he earned a doctorate in mathematics and was nominated for a fellowship at the Institute for Advanced Study at Princeton. The position included an honorary membership in the faculty at nearby Princeton University, but the university objected to the appointment of a black man as a faculty member. The director of the institute insisted on appointing Blackwell and eventually won out. From Princeton, Blackwell went on to teach for 10 years at Howard University and then moved to the University of California at Berkeley. He has been a prolific researcher and writer and has made important contributions to probability, statistics, game theory, and set theory.

In addition to being an accomplished research mathematician, David Blackwell was also a dedicated teacher. He expressed his love of teaching when he said, "Why do you want to share something beautiful with someone else? It's because of the pleasure he will get, and in transmitting it you will appreciate its beauty all over again. My high school geometry teacher really got me interested in mathematics."

Marilyn vos Savant

In the September 1990 issue of *Parade* magazine, the following question appeared in the column "Ask Marilyn":

"Suppose you're on a game show, and you're given the choice of three doors: Behind one door is a car; behind the others, goats. You pick a door, say number 1, and the host, who knows what's behind the doors, opens another door, say number 3, which has a goat. He then says to you, 'Do you want to pick door number 2?' Is it to your advantage to switch your choice?" The columnist of "Ask Marilyn," Marilyn vos Savant (who just happened to have the world's highest tested IQ), replied, "Yes, you should switch. The first door has a one-third chance of winning, but the second door has a two-thirds chance."

There was a very strong response to this column. Many professional mathematicians and statisticians (complete with Ph.Ds) informed her, in no uncertain terms, that she was wrong. Several "scolded" her for confusing both the issue and the general public. One wrote: "I'll come straight to the point. In (the question and your answer), you blew it! Let me explain: If one door is shown to be a loser, that information changes the probability of either remaining choice to $\frac{1}{2}$. As a professional mathematician, I'm very concerned with the general public's lack of mathematical skills. Please help by confessing your error and, in the future, being more careful." Another responded: "Your answer to the question is in error. But if it's any consolation, many of my colleagues have also been stumped by this problem." However, vos Savant was not in error with her answer.

Rather than apologize for her response, vos Savant explained the reasoning behind her answer in her next column. The response this time was even more intense than before. Thousands of letters were received and the vast majority insisted that vos Savant was wrong. The letters included one from a deputy director of the Center for Defense Information and another from a research statistician from the National Institutes of Health. Of the letters from the general public, 92% disputed her answer, compared to 65% of the letters from universities. But as one writer, a Ph.D. from the Massachusetts Institute of Technology, put it, "You are indeed correct. My colleagues at work had a ball with this problem, and I dare say that most of them—including me at first—thought you were wrong." While it is quite disconcerting that so many mathematicians and statisticians were in error, many were incorrect because they had preconceived ideas of what the answer should be and did not either read or think clearly enough when analyzing the question.

5.1 COMPUTING PROBABILITIES IN SIMPLE EXPERIMENTS

Following a wedding, the attendants for the groom loaded the wedding gifts into a van and took them to the reception hall. After they had taken all of the presents into the hall, they noticed that three of the presents did not have gift cards from the senders. They returned to the van and found the three cards, but there was no way to tell which card went with which gift. Slightly flustered, they decided to arbitrarily put each card with one of the untagged gifts. What are the chances that at least one of those gifts received the correct card?

Everyone has an intuitive grasp of probability. But when it comes to an explanation of what probability actually *is,* or determining the probability of a complex action, we are sometimes at a loss for the correct answer.

In this section we introduce the language and general concepts for the mathematical discussion of probability and the rules that govern it, beginning with the idea of an "experiment" (an action for which you do not know the results beforehand, although you might know the range of possibilities). For example, if two cards are drawn from a standard deck of 52 cards, we might want to know the probability of getting a pair that match, say, two aces or two sevens.

SIMPLE EXPERIMENTS

Probability is the mathematics of chance, and the terminology used in probability theory occurs many times in daily life. For example, you may hear on the radio "The probability of precipitation today is 80%." This should be interpreted as meaning that on days in the past with atmospheric conditions like those of today, it rained at some time on 80% of those days. You may read in an article about test results in medical science that a patient has a 6 in 10 chance of improving if treated with drug X. We interpret this to mean that in a large group of patients, say 100, who have had the same symptoms as the patient being treated, $\frac{6}{10} \times 100 = 60$ of them improved when administered drug X.

EXAMPLE 5.1 An advertisement for one of the state lottery games says "The chances of winning the lottery game 'Find the Winning Ticket' are 1 in 150,000." How should you interpret this statement?

SOLUTION If 150,000 lottery tickets are printed, only one of the tickets is the winning ticket. If more tickets are printed, the fraction of winning tickets is approximately $\frac{1}{150{,}000}$. ◆

HISTORY

Even though the founding of probability theory as a science is usually attributed to Blaise Pascal in the mid-1600s, the concept of sample space was not formulated until this century (by R. von Mises).

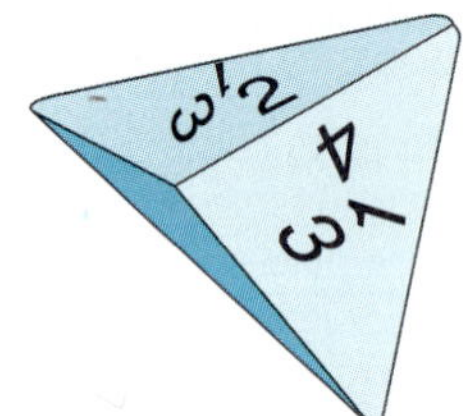

Four-sided die

Eight-sided die

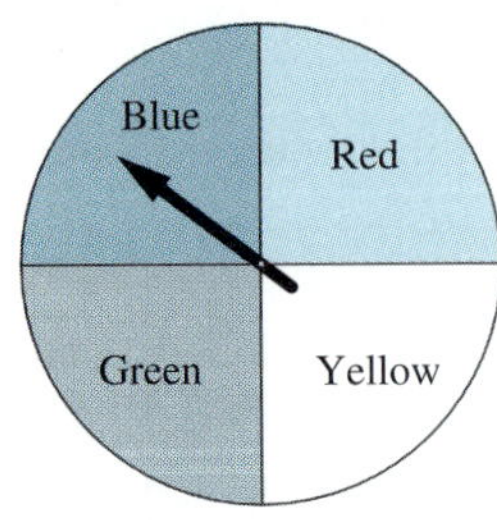

FIGURE 5.2

Probability tells us the relative frequency with which we expect an event to occur. That is, if we repeat an experiment over and over, the fraction of times the event occurs should be the probability of the event. The probability can be reported as a fraction, a decimal, or a percent, but it must always be between zero and one. The greater the probability, the more likely the event is to occur. Conversely, the smaller the probability, the less likely the event is to occur.

To study probability in a mathematically precise way, we need special terminology and notation (the ideas of set theory used in this chapter are reviewed in Topic 1 near the end of the book). Making an observation or taking a measurement of some act, such as flipping a coin, is called an **experiment.** An **outcome** is one of the possible results of an experiment, such as getting a head when flipping a coin. The set of all the possible outcomes is called the **sample space.** Finally, an **event** is any collection of the possible outcomes. That is, an event is a subset of the sample space. These concepts are illustrated in Example 5.2.

EXAMPLE 5.2 List the sample space and an event for each experiment.

(a) ***Experiment:*** Roll a standard six-sided die with 1, 2, 3, 4, 5, 6 dots, respectively, on the six faces (Figure 5.1). Record the number of dots showing on the top face.

FIGURE 5.1

(Note: In the problem set, we will describe experiments with four-sided and eight-sided dice whose faces are all the same size and shape; these dice will be referred to as "regular" dice. A "standard" die will always refer to the one with six faces as shown in Figure 5.1.)

(b) ***Experiment:*** Toss a coin three times and record the results in order.

(c) ***Experiment:*** Spin the spinner in Figure 5.2 twice and record the colors of the regions where it comes to rest.

(d) ***Experiment:*** Roll two dice and record the number of dots on each die.

SOLUTION

(a) ***Sample Space:*** There are six possible outcomes: {1, 2, 3, 4, 5, 6}, where numerals represent the number of dots.
Event: For example, {2, 4, 6} is the event of getting an even number of dots, and {2, 3, 5} is the event of getting a prime number of dots. There are $2^6 = 64$ events in total, each event being a subset of {1, 2, 3, 4, 5, 6}.

(b) ***Sample Space:*** Use three-letter sequences to represent the outcomes. For example, HHH represents tossing three heads. First we list all possible outcomes. Note: HTH represents tossing a head first, a tail second, and a head third.

HHH	3 heads	TTT	3 tails
HHT HTH THH	2 heads, 1 tail	TTH THT HTT	1 head, 2 tails

There are eight possible outcomes in the sample space. With set notation, the sample space is written {HHH, HHT, HTH, THH, TTH, THT, HTT, TTT}. The list of possible outcomes is easier to grasp than the set notation because the list is organized into meaningful categories of outcomes. Such a list is often the best way to display a sample space.

Event: The sample space of eight elements has many subsets (256 subsets, in fact). Any one of its subsets is an event. For example (HTH, HTT, TTH, TTT) is the event of getting a tail on the second coin, since it contains all possible outcomes matching that condition. (Not all events have such simple descriptions, however.)

(c) ***Sample Space:*** Using pairs of letters to represent the outcomes (colors in this experiment), we have:

RR	*YR*	*GR*	*BR*
RY	*YY*	*GY*	*BY*
RG	*YG*	*GG*	*BG*
RB	*YB*	*GB*	*BB*

Event: For example, the event that the colors match is {*RR, YY, GG, BB*}. The event that at least one of the spins is green is {*RG, YG, GG, BG, GR, GY, GB*}.

(d) ***Sample Space:*** We use ordered pairs to represent the outcomes, namely, the number of dots on the faces of the two dice. For example, the ordered pair (1, 3) represents one dot on the first die and three dots on the second.

(1, 1)	(1, 2)	(1, 3)	(1, 4)	(1, 5)	(1, 6)
(2, 1)	(2, 2)	(2, 3)	(2, 4)	(2, 5)	(2, 6)
(3, 1)	(3, 2)	(3, 3)	(3, 4)	(3, 5)	(3, 6)
(4, 1)	(4, 2)	(4, 3)	(4, 4)	(4, 5)	(4, 6)
(5, 1)	(5, 2)	(5, 3)	(5, 4)	(5, 5)	(5, 6)
(6, 1)	(6, 2)	(6, 3)	(6, 4)	(6, 5)	(6, 6)

Event: For example, the event of getting a total of seven dots on the two dice is {(6, 1), (5, 2), (4, 3), (3, 4), (2, 5), (1, 6)}. The event of getting more than nine dots is {(6, 4), (5, 5), (4, 6), (6, 5), (5, 6), (6, 6)}. ◆

HISTORY

Cubical dice marked equivalently to modern dice have been found in Egyptian tombs dated before 2000 B.C. and in Chinese excavations dating to 600 B.C.

EQUALLY LIKELY OUTCOMES

With our intuitive approach, the probability of an event should be the fraction of the time the event occurs in many repetitions of the experiment. One way to find the probability of event E is to make many repetitions of the experiment and simply determine the frequency with which E occurs. The relative frequency of E occuring is called its **experimental probability.** Experimental probability may vary from one set of observations to another.

EXAMPLE 5.3 An experiment consists of tossing two coins 500 times and recording the results. Table 5.1 gives the observed results and experimental probabilities. Let E be the event of getting a head on the first coin. Find the probability of E.

Outcome	Frequency		Experimental Probability
HH	137		$\frac{137}{500}$
HT	115		$\frac{115}{500}$
TH	108		$\frac{108}{500}$
TT	140		$\frac{140}{500}$
Total:	500	Total:	$\frac{500}{500} = 1.00$

The event E is {HH, HT}. From the table, the experimental probability of E is $\frac{137 + 115}{500} = \frac{252}{500}$. Hence, the probability of E is $\frac{252}{500} = 0.504$. ◆

The advantage of finding the experimental probability of an event is that it is determined by making observations. The obvious disadvantage is that it depends on a particular set of repetitions of an experiment and hence may need to be recomputed when more experiments are performed. The number of repetitions in an experiment may influence the probability of events since rare outcomes may not appear.

In many important cases, we can determine what fraction of the time an event is going to occur without actually performing the experiment. For example, we assume that a coin is going to land heads about $\frac{1}{2}$ of the time, and we asume that a die is going to land showing three dots about $\frac{1}{6}$ of the time. The reason we are sure of these fractions for coins and dice is that in each case the sample space has **equally likely outcomes** if the coins and dice are symmetrical. Based on the fact that the outcomes are equally likely, we can compute the probability of each outcome, or the probability of a more complicated event that contains several outcomes, by using the following definition.

DEFINITION

Probability of an Event with Equally Likely Outcomes

Suppose that all of the outcomes in the sample space S are equally likely to occur. Let E be an event. Then the probability of event E, denoted $P(E)$, is

$$P(E) = \frac{\text{number of outcomes in } E}{\text{number of outcomes in } S}.$$

There are two things to notice about this definition. First, if you consider just one outcome, its probability is 1 divided by the number of outcomes in the entire sample space. Thus the sample space for one die has 6 outcomes, so the probability of any particular face showing is $\frac{1}{6}$. Second, the probability of any event is a number from 0 to 1. The event containing no outcomes has probability zero, and the event containing all the outcomes in the sample space has probability 1. Of course, it seems silly to discuss the probability of the event containing no outcomes, but in a complicated problem it might not be immediately obvious that an event actually cannot happen.

HISTORY

The French naturalist Georges-Louis Leclerc de Buffon (1707–1788) tossed a coin 4040 times obtaining 2048 heads (50.69% heads). Around 1900, the English statistician Karl Pearson tossed a coin 24,000 times obtaining 12,012 heads (50.05% heads). While imprisoned by the Germans during World War II, the English mathematician John Kerrich tossed a coin 10,000 times obtaining 5067 heads (50.67% heads).

We cannot be sure that a real world coin or die is perfectly balanced, so we cannot be sure that the outcomes in the sample space are equally likely. Thus, when we apply the definition, we are usually computing **theoretical probabilities.** (In fact, theoretical probabilities work very well for real world problems.) When we wish to make it clear that we are dealing with theoretical probabilities of an ideal coin or an ideal die, we refer to them as a **fair coin** or a **fair die**. Example 5.4 illustrates how to assign theoretical probabilities.

EXAMPLE 5.4 An experiment consists of tossing two fair coins. Find theoretical probabilities for the outcomes and for the event of getting at least one head.

SOLUTION There are four outcomes: HH, HT, TH, TT. If the coins are fair, all outcomes should be equally likely to occur, and each outcome should occur $\frac{1}{4}$ of the time. Hence we make the assignments listed in Table 5.2.

TABLE 5.2

Outcome	Theoretical Probability
HH	$\frac{1}{4} = 0.25$
HT	$\frac{1}{4} = 0.25$
TH	$\frac{1}{4} = 0.25$
TT	$\frac{1}{4} = 0.25$

Let E be the event of getting at least one head; that is, $E = \{\text{HH, HT, TH}\}$. The theoretical probability of E is

$$\frac{\text{number of outcomes in } E}{\text{number of outcomes in } S} = \frac{3}{4}.$$

That is, $P(E) = 0.75$, so we expect to get at least one head approximately 75% of the time when tossing two coins. ◆

EXAMPLE 5.5 We toss two fair dice. Let A be the event of getting a total of 7 dots, B be the event of getting 8 dots, and C be the event of getting at least 4 dots. [Recall that we found in Example 5.2(d) that this sample space has 36 possible outcomes.]

SOLUTION Since the outcomes obtained by tossing two fair dice are equally likely outcomes, we can compute the theoretical probabilities $P(A)$, $P(B)$, and $P(C)$. These are recorded in Table 5.3.

TABLE 5.3

Event	Number of Outcomes	Probability
A	6	$P(A) = \frac{6}{36} = \frac{1}{6}$
B	5	$P(B) = \frac{5}{36}$
C	33	$P(C) = \frac{33}{36} = \frac{11}{12}$

◆

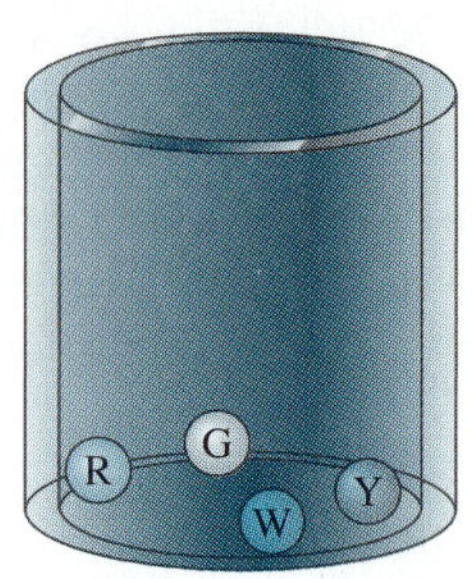

FIGURE 5.3

EXAMPLE 5.6 A jar contains four marbles: one red, one green, one yellow, and one white (Figure 5.3). If we draw two marbles from the jar, one after the other, without replacing the first one drawn, what is the probability of each of the following events?

A: One of the marbles is red.
B: The first marble is red or yellow.
C: The marbles are the same color.
D: The first marble is not white.
E: Neither marble is blue.

SOLUTION The sample space consists of the following outcomes. (RG, for example, means that the first marble is red and the second marble is green.)

RG	GR	YR	WR
RY	GY	YG	WG
RW	GW	YW	WY

The sample space has 12 possible outcomes. Since there is exactly one marble of each color, we assume that all the outcomes are equally likely. Then

$A = \{RG, RY, RW, GR, YR, WR\}$, so $P(A) = \frac{6}{12} = \frac{1}{2}$.

$B = \{RG, RY, RW, YR, YG, YW\}$, so $P(B) = \frac{6}{12} = \frac{1}{2}$.

$C = \varnothing$, the "empty" event because the jar does not contain two marbles having the same color. That is, C is impossible, so $P(C) = \frac{0}{12} = 0$.

$D = \{RG, RY, RW, GR, GY, GW, YR, YG, YW\}$, so $P(D) = \frac{9}{12} = \frac{3}{4}$.

$E =$ the entire sample space, S, because the jar has no blue marbles. So $P(E) = \frac{12}{12} = 1$. ◆

The **union** of two events ($A \cup B$) refers to all outcomes that are in one, or the other, or both events; the **intersection** ($A \cap B$) refers to outcomes that are in both events. Notice that event B in Example 5.6 can be represented as the union of two events corresponding to drawing red on the first marble or yellow on the first marble. That is, if we let $L = \{RG, RY, RW\}$ and $M = \{YR, YG, YW\}$, then $B = L \cup M$. Observe that $L \cap M = \varnothing$. Events such as L and M, which have no outcome in common, are called **mutually exclusive.**

If we compute $P(L \cup M)$, $P(L)$, and $P(M)$, we find $P(L \cup M) = P(B) = \frac{1}{2}$, while $P(L) + P(M) = \frac{3}{12} + \frac{3}{12} = \frac{1}{2}$. Therefore, $P(L \cup M) = P(L) + P(M)$.

Probability of Mutually Exclusive Events

If L and M are mutually exclusive events, then

$$P(L \cup M) = P(L) + P(M).$$

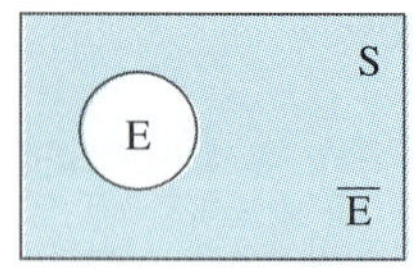

FIGURE 5.4

The set of outcomes in the sample space S, but not in event E, is called the **complement of the event E** (Figure 5.4); this is written $\overline{E}$ (read "not E"). Since $S = E \cup \overline{E}$ and $E \cap \overline{E} = \varnothing$, it follows that $P(E) + P(\overline{E}) = P(S)$. But $P(S) = 1$, so

$P(E) + P(\overline{E}) = 1$; while $P(E) = 1 - P(\overline{E})$ and $P(\overline{E}) = 1 - P(E)$.

The last equation can be used to find the probability of $P(E)$ whenever $P(\bar{E})$ is known. In Example 5.6, D was the event that the first marble is not white; therefore, $P(\bar{D})$ is the probability that the first marble *is* white, namely, $\frac{3}{12}$ or $\frac{1}{4}$. So $P(D) = 1 - P(\bar{D}) = 1 - \frac{1}{4} = \frac{3}{4}$, as we found directly.

EXAMPLE 5.7 Carolan and Mary are playing a number-matching game. Carolan chooses a whole number from 1 to 4 but does not tell Mary, who then guesses a number from 1 to 4. Assume that all numbers are equally likely to be chosen by each player.

(a) What is the probability that the numbers are equal?
(b) What is the probability that the numbers are unequal?

SOLUTION The sample space can be represented as ordered pairs of numbers from 1 to 4, the first being Carolan's number, the second Mary's.

(1, 1)	(1, 2)	(1, 3)	(1, 4)
(2, 1)	**(2, 2)**	(2, 3)	(2, 4)
(3, 1)	(3, 2)	**(3, 3)**	(3, 4)
(4, 1)	(4, 2)	(4, 3)	**(4, 4)**

(a) Let E be the event that the numbers are equal. Outcomes for E are in boldface. Assuming all outcomes are equally likely, $P(E) = \frac{4}{16} = \frac{1}{4}$.
(b) The event that the numbers are unequal is $\bar{E}$. Hence the probability that the numbers are unequal is $1 - \frac{1}{4} = \frac{3}{4}$. [We can verify this directly by counting the 12 outcomes not in E. Hence $P(\bar{E}) = \frac{12}{16} = \frac{3}{4}$.] ◆

EXAMPLE 5.8 Figure 5.5 shows a diagram of a sample space S for an experiment with equally likely outcomes. Events A, B, and C are indicated, with their outcomes represented by points. Find the probability of each of the following events: S, $\varnothing$, A, B, C, $A \cup B$, $A \cap B$, $A \cup C$, $\bar{C}$.

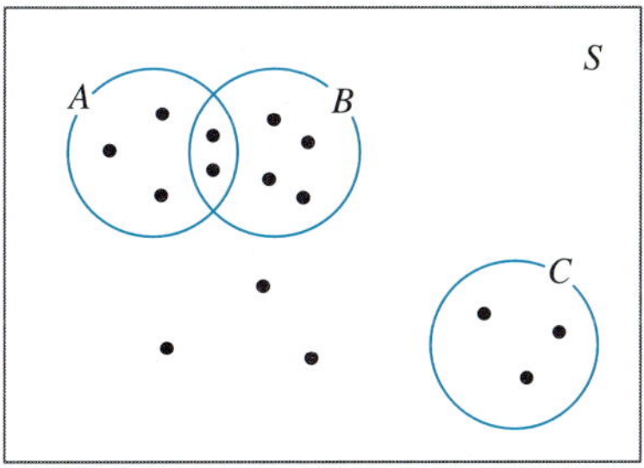

FIGURE 5.5

SOLUTION We can tabulate the number of outcomes in each event and the probabilities (Table 5.4). For example, number of outcomes in A = 5 and number of outcomes in S = 15; $P(A) = \frac{5}{15} = \frac{1}{3}$.

TABLE 5.4

Event E	Number of Outcomes in E	$P(E) = \dfrac{\text{Number of Outcomes in } E}{\text{Number of Outcomes in } S}$
S	15	$\frac{15}{15} = 1$
$\varnothing$	0	$\frac{0}{15} = 0$
A	5	$\frac{5}{15} = \frac{1}{3}$
B	6	$\frac{6}{15} = \frac{2}{5}$
C	3	$\frac{3}{15} = \frac{1}{5}$
$A \cup B$	9	$\frac{9}{15} = \frac{3}{5}$
$A \cap B$	2	$\frac{2}{15}$
$A \cup C$	8	$\frac{8}{15}$
$\overline{C}$	12	$\frac{12}{15} = \frac{4}{5}$

◆

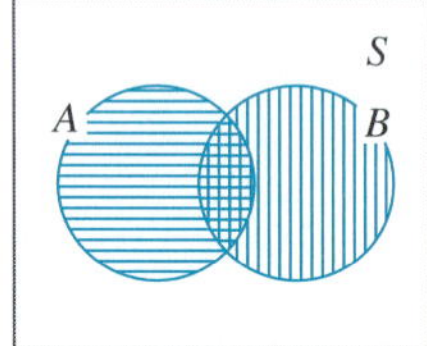

FIGURE 5.6

In Example 5.8, $P(A \cup B) = \frac{9}{15}$, while $P(A) + P(B) - P(A \cap B) = \frac{5}{15} + \frac{6}{15} - \frac{2}{15} = \frac{9}{15} = P(A \cup B)$. Hence $P(A \cup B) = P(A) + P(B) - P(A \cap B)$. This result is true for all events A and B in any sample space. In Figure 5.6, observe how the region $A \cap B$ is shaded twice, once from A and once from B. Thus, for any sets A and B, to find the number of elements in $A \cup B$, we can find the sum of the number of elements in A and B; but we must subtract the number of elements in $A \cap B$ so that these elements are not counted twice. Hence the number of elements in $A \cup B$ equals the number of elements in A plus the number of elements in B minus the number of elements in $A \cap B$. Therefore,

$$P(A \cup B) = P(A) + P(B) - P(A \cap B).$$

We can summarize our observations about the **properties of probability** as follows.

Properties of Probability

1. For any event A, $0 \leq P(A) \leq 1$.
2. $P(\varnothing) = 0$.
3. $P(S) = 1$, where S is the sample space.
4. For mutually exclusive events A and B, $P(A \cup B) = P(A) + P(B)$.
5. For any events A and B, $P(A \cup B) = P(A) + P(B) - P(A \cap B)$.
6. If $\overline{A}$ denotes the complement of event A, then $P(A) = 1 - P(\overline{A})$.

INITIAL PROBLEM SOLUTION

Following a wedding, the attendants for the groom loaded the wedding gifts into a van and took them to the reception hall. After they had taken all the presents into the hall, they noticed that three of the presents did not have gift cards from the senders. They returned to the van and found the three cards, but there was no way to tell which card went with which gift. Slightly flustered, they decided to arbitrarily put each card with one of the untagged gifts. What are the chances that at least one of those gifts received the correct card?

SOLUTION Let E be the event that at least one gift receives the correct card. We will indicate the three gifts by the letters A, B, and C, and their respective cards by a, b, and c. We list all the possible choices for combinations of gifts and cards in the following table. Each line of the table indicates one way in which the gifts can be matched with the cards. For example, the entry (B, c) means that gift B receives the card that belongs with gift C.

(A, a)	(B, b)	(C, c)
(A, a)	(B, c)	(C, b)
(A, b)	(B, a)	(C, c)
(A, b)	(B, c)	(C, a)
(A, c)	(B, a)	(C, b)
(A, c)	(B, b)	(C, a)

There are six outcomes in the sample space and only the fourth and fifth lines correspond to all the gifts receiving the wrong cards. In the other four cases, at least one card is matched with the correct gift. We conclude that $P(E) = \frac{4}{6} = \frac{2}{3}$.

PROBLEM SET 5.1

1. According to the weather report, there is a 20% chance of snow in the county tomorrow. Which of the following statements would be appropriate?
   (i) Out of the next five days, it will snow one of those days.
   (ii) Out of the next 24 hours, snow will fall for 4.8 hours.
   (iii) Of past days when conditions were similar, one out of five had some snow.
   (iv) It will snow on 20% of the area of the county.

2. The doctor says "There is a 40% chance that your problem will get better without surgery." Which of the following statements would be appropriate?
   (i) You can expect to feel 40% better.
   (ii) In the future you will feel better on two out of every five days.
   (iii) Among you and the next four other patients with the same problem, two will get better without surgery.
   (iv) Among patients with symptoms similar to yours who have participated in research studies of nonsurgical treatments, about 40% got better.

3. List the elements of the sample space for each of the following experiments.
   (a) A quarter is tossed.
   (b) A single die with faces labeled A, B, C, D, E, F is rolled.
   (c) A regular tetrahedron die with its four faces labeled 1, 2, 3, 4 is rolled and the number on the bottom face is recorded.

4. List the elements of the sample space for each of the following experiments.
   (a) A \$20 bill is obtained from an automatic teller machine and the right-most digit of the serial number is recorded.
   (b) Some white and black marbles are placed in a jar, mixed, and a marble is chosen without looking.
   (c) The following "Red-Blue-Yellow" spinner is spun once. (All central angles are 120°.)

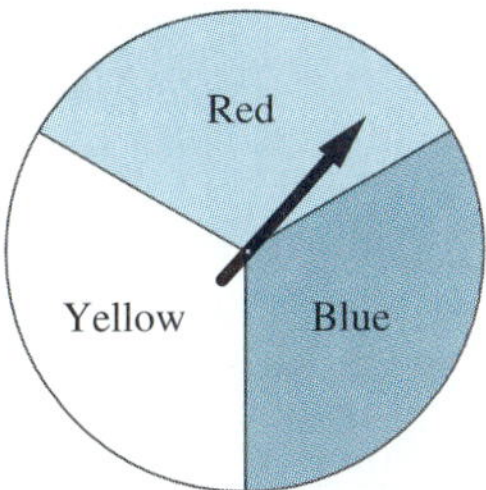

5. An experiment consists of tossing four coins and seeing if each is a head or a tail. List each of the following.
   (a) The sample space
   (b) The event of a head on the first coin
   (c) The event of three heads
   (d) The event of a head or a tail on the fourth coin.
   (e) The event of a head on the second coin and a tail on the third coin

6. An experiment consists of tossing a regular die with 12 faces numbered 1–12. List the following.
   (a) The sample space
   (b) The event of an even number
   (c) The event of a number less than 8
   (d) The event of a number divisible by 2 and 3
   (e) The event of a number greater than 12

One way to find the sample space of an experiment involving two parts is to plot the possible outcomes of one part of the experiment horizontally and of the other part of the experiment vertically, then fill in the pairs of outcomes in the rectangle. For example, an experiment consists of tossing a dime and a quarter.

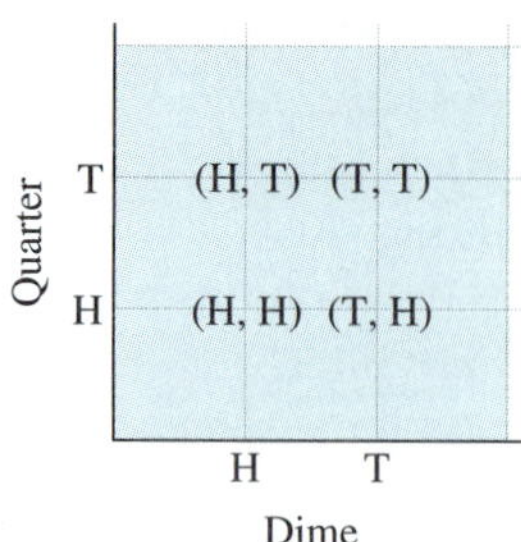

The sample space of the experiment is {(H, H), (H, T), (T, H), (T, T)}.

7. Use the preceding method to construct the sample space of the experiment of tossing a coin and rolling a tetrahedral die (four faces).
8. Use the preceding method to construct the sample space of the experiment of tossing a coin and drawing a marble from a jar containing purple, green, and yellow marbles.
9. A standard die is rolled 60 times with the following results recorded.

Outcome	1	2	3	4	5	6
Frequency	10	9	10	12	8	11

Find the experimental probability of the following events.
   (a) Getting a 4
   (b) Getting an odd number
   (c) Getting a number greater than 3
10. A dropped thumb tack will land with the point up or the point down. The results for tossing a thumb tack 60 times are as follows.

Outcome	Point up	Point down
Frequency	42	18

   (a) What is the experimental probability that the thumbtack lands
      (i) point up?
      (ii) point down?
   (b) If the thumbtack were tossed 100 times, about how many times would you expect it to land
      (i) point up?
      (ii) point down?
11. An experiment consists of tossing three fair coins and counting the number of heads.
   (a) List the outcomes and the theoretical probabilities using a format such as in problems 9 and 10.
   (b) Find the probability for the event of getting at least one head.
12. A jar contains three marbles: one red, one green, and one yellow. An experiment consists of drawing a marble from the jar, noting its color, replacing it in the jar, and drawing a second marble.
   (a) List the outcomes and the theoretical probabilities using a format such as in problems 9 and 10.
   (b) Find the probability for the event of getting at least one red marble.
13. Refer to Example 5.2(d), which gives the sample space for the experiment of rolling two standard dice. Assume the dice are fair, and give the probabilities of the following events.
   (a) A 4 on the second die
   (b) An even number on each die
   (c) At least 7 dots in total
   (d) A total of 15 dots
14. Refer to Example 5.2(d), which gives the sample space for the experiment of rolling two standard dice. Assume the dice are fair, and give the probabilities of the following events.
   (a) A 5 on the first die
   (b) An even number on one die and an odd number on the other die
   (c) No more than 7 dots in total
   (d) A total greater than 1
15. What is the probability of getting yellow on the following spinner?

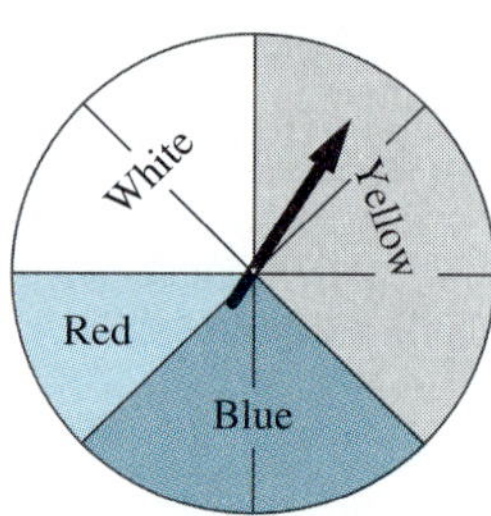

16. What is the probability of getting blue on the preceding spinner?
17. Two regular dice in the shape of tetrahedrons (four sides) have their sides numbered 1 to 4.
   (a) Find the sample space for the experiment of rolling the two dice and adding the numbers.
   (b) Assume the dice are fair and find the probabilities for the following events:
      (i) The total is 5.
      (ii) The total is even.
      (iii) The total is at least 4.
18. Two regular dice are rolled in an experiment. One is a standard die with its sides numbered 1 to 6, and the other is a four-sided die with sides numbered 1 to 4.
   (a) Find the sample space for the experiment of rolling the two dice and adding the numbers.
   (b) Assume the dice are fair and find the probabilities for the following events:

(i) The total is 7.
(ii) The total is odd.
(iii) The total is no greater than 8.

19. A standard die is made that has two faces marked with 2's, three faces marked with 3's, one face marked with a 5. If this die is thrown once, find the probability of:
(a) getting a 2
(b) not getting a 2
(c) getting an odd number
(d) not getting an odd number

20. A regular die with 12 faces has three faces marked with 1's, two faces marked with 2's, three faces marked with 3's, and four faces marked with 4's. If this die is thrown once, find the following probabilities.
(a) Getting a 2
(b) Not getting a 2
(c) Getting an odd number
(d) Not getting an odd number

21. Consider the sample space for Example 5.2(c) and the following events.
A: getting a green on the first spin
B: getting a yellow on the second spin
(a) Find the probability of the event A, of the event B, of the event $A \cap B$, and of the event $A \cup B$.
(b) Verify that the equation $P(A \cup B) = P(A) + P(B) - P(A \cap B)$ holds for the probabilities in part (a).

22. Suppose a jar contains twenty marbles numbered 1 through 20 with each odd numbered marble red, while each even numbered marble is black. A marble is drawn from the jar and its color and number are noted.
(a) Find the sample space.
(b) If A and B are the following events,
A: getting a black marble
B: getting a number divisible by 3,
find the probability of the events
(i) A
(ii) B
(iii) the event $A \cap B$
(iv) the event $A \cup B$.
(c) Verify that the equation
$P(A \cup B) = P(A) + P(B) - P(A \cap B)$
holds for the probabilities you found in part (b).

23. Two regular four-sided dice with their sides numbered 1 to 4 are rolled.
(a) Find the sample space.
(b) If A and B are the following events
A: getting a 3 on the first die.
B: getting an even number on the second die.
Assume the dice are fair and find the probabilities of the events
(i) A
(ii) B
(iii) the event $A \cap B$
(iv) the event $A \cup B$.
(c) Verify that the equation $P(A \cup B) = P(A) + P(B) - P(A \cap B)$ holds for the probabilities you found in part (b).

24. Suppose a jar contains twenty marbles numbered 1 through 20. If the number is divisible by 3, the marble is black; all other marbles are red. A marble is drawn from the jar and its number and color are noted.
(a) List the sample space for this experiment.
Define two events as follows:
A: getting a black marble.
B: getting a number divisible by 6
(b) Find the probabilities of event A, event B, the event $A \cap B$, and the event $A \cup B$.
(c) Verify that the equation $P(A \cup B) = P(A) + P(B) - P(A \cap B)$ holds for the probabilities you found in part (b).

25. In Example 5.6, a jar contains four marbles: one red, one green, one yellow, and one white. Two marbles are drawn from the jar, one after another, without replacing the first one drawn. Define the event A as: The first marble is green or the second marble is white. The complement of A is the event: The first marble is not green *and* the second marble is not white.
(a) Find the probability of the event A and the event $\bar{A}$.
(b) Verify that the equation $P(\bar{A}) = 1 - P(A)$ holds for the probabilities you found in part (a).

26. Two regular four-sided dice have their sides numbered 1 to 4.
(a) Find the sample space for the experiment of rolling the two dice and adding the numbers.
Define the event A as: The sum of the numbers is divisible by 3.
(b) Find the probability of the event A and the event $\bar{A}$.
(c) Verify the equation $P(\bar{A}) = 1 - P(A)$ for the probabilities you found in (b).

EXTENDED PROBLEMS

The probability of a geometric event involving the concept of measure (length, area, volume) of a geometric set of points is determined as follows. Let $m(A)$ and $m(S)$ represent the measure of the event A and of the sample space S, respectively. Then

$$P(A) = \frac{m(A)}{m(S)}.$$

For example, in the first of the following two figures, if the length of S is 12 cm and the length of A is 4 cm, then $P(A) = \frac{4}{12} = \frac{1}{3}$. Similarly, in the second figure, if the area of region B is 10 sq cm and the area of the region S is 60 sq cm, then $P(B) = \frac{10}{60} = \frac{1}{6}$.

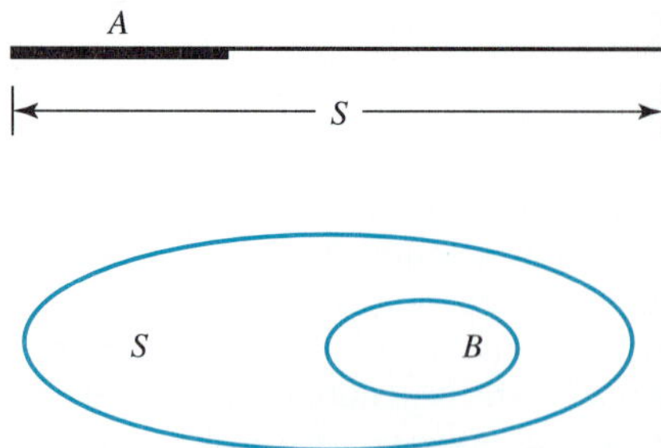

27. A bus travels between Albany and Binghamton, a distance of 100 miles. If the bus has broken down, we want to find the probability that it has broken down within 10 miles of either city.
   (a) The road from Albany to Binghamton is the sample space. What is $m(S)$?
   (b) Event A is that part of the road within 10 miles of either city. What is $m(A)$?
   (c) Find $P(A)$.

28. The dart board illustrated is made up of circles with radii of 1, 2, 3, and 4 units. A dart hits the target randomly. What is the probability that the dart hits the bull's eye? (Hint: The area of a circle with radius r is πr^2.)

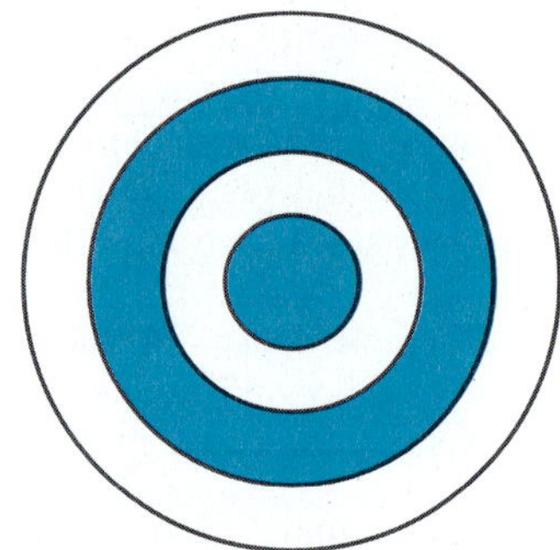

29. Four guests at a party toss their hats on the host's bed. When leaving the party they each pick up a hat at random. What is the probability that at least one guest leaves wearing his own hat?

30. In problem 10, the results of an experiment of dropping a thumbtack and noting whether the point was up or down were described. Carry out the experiment on your own as follows: Drop an ordinary thumbtack from a height of about six feet onto a smooth surface such as a floor or tabletop (try to be fairly consistent with the height).
   (a) Do 60 trials of the experiment and record the results. Let A be the event the point is up; find the experimental probability of event A.
   (b) Repeat part (a) with the same tack. How do the results compare?
   (c) Repeat part (a) with a thumbtack having a longer point but the same size head as the first. Compare the probability that the point is up with the probability you found in part (a). Is there a difference that seems to be due to more than chance alone? What do you imagine would be the result if you tried the experiment again with a thumbtack having an even longer point?

31. A classic problem in geometric probability is known as the *Buffon Needle Problem*. The problem can be stated as follows: If a needle of a certain length (say 2 inches) is dropped at random on a floor made of planks wider than that length (say 3 inches wide), what is the probability the needle will fall across a crack between two planks? To simulate the experiment without a planked floor, get a large piece of paper and draw several parallel lines 3 inches apart.
   (a) Drop a needle onto the "floor" from a consistent height (about 5 feet). Repeat the experiment 60 times, recording whether or not the needle falls across a crack. Compute the experimental probability for the event.
   (b) Repeat part (a) with a longer needle (but still shorter than the distance between the "planks").
   (c) Divide the length of the longer needle by the length of the shorter one (measure the lengths carefully). Next, divide the probability found in part (b) by the probability from part (a). How do the two compare? Is there a relationship or generalization that is suggested?

32. In our examples and problems, we have generally assumed that coins and dice were "fair"; that is, each of the faces was equally likely to appear on top. Complete the following experiment with a die that is possibly not "fair."
   (a) Find a wooden cube that is 1 to 2 inches on each edge (from a set of blocks, perhaps). Number the faces 1 through 6.
   (b) Roll the "die" 100 times, record the results, and compute the experimental probabilities for each of the faces.
   (c) Repeat part (b). Are the results similar to those of part (b)? If not, to what could you attribute any differences?
   (d) Hollow out the center of the face with the "6"; this can be done with a knife or a drill. Make the hollowed region about 1/2 inch deep and covering about half the face (don't cut an edge).
   (e) Roll the hollowed die 100 times, record the results, and compute the experimental probabilities for each of the faces.
   (f) Compare your results with those from part (b). Are the results similar to those of part (b)? If not, to what could you attribute any differences? Are the differences, if any, greater for the face opposite the hollowed face or for those adjacent to it?

5.2 COMPUTING PROBABILITIES IN COMPLEX EXPERIMENTS

A friend who likes to gamble makes you the following wager: He'll bet you two dollars against one dollar that if you toss a coin repeatedly, you'll get a total of two tails before you can get a total of three heads. You reason that two heads are as likely as two tails, and you have a one-out-of-two chance of getting a third head after getting two heads. Should you take the bet?

In the last section, probability was defined in terms of the relative frequency of given events. While this is a simple concept, it is often quite a task to determine how many ways a certain event can occur, as well as keep track of the sequence of activities that may make up an event. We will now introduce two tools to help with these problems; the first of these helps us to visualize an experiment, and the second helps us count in situations where it isn't practical (or even possible) to do a tally.

TREE DIAGRAMS AND COUNTING TECHNIQUES

In some experiments it is difficult to construct a list of all possible outcomes. The list may be too long to write down, or it may not even be clear what pattern to follow in constructing the list. Therefore, we will develop alternative procedures for computing probabilities.

A **tree diagram** is a graphic device that can be used to represent the outcomes of an experiment. The simplest tree diagrams have one stage (when the experiment involves just one action). For example, consider drawing one ball from a box containing three balls: a red (R), a white (W), and a blue (B). The following steps show how to draw the tree diagram in Figure 5.7.

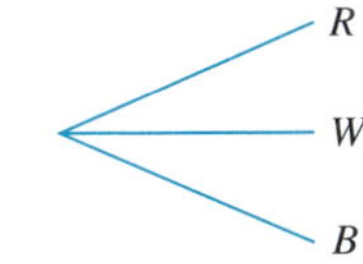

FIGURE 5.7

1. Draw a single dot.
2. Draw one branch for each outcome.
3. Place a label at the end of the branch for each outcome.

Two stage trees are used to represent experiments that consist of a sequence of two experiments. To draw a two stage tree, follow these steps.

1. Draw the one stage tree for the outcomes of the first experiment. The branches in this part of the tree are called **primary branches.**
2. Starting at the end of each branch of the tree in step 1, draw the (one stage) tree for the outcomes of the second experiment. These are called **secondary branches.**

The experiment of drawing two marbles, one at a time from a jar of four marbles without replacing the one drawn, is represented by the two stage tree diagram in Figure 5.8.

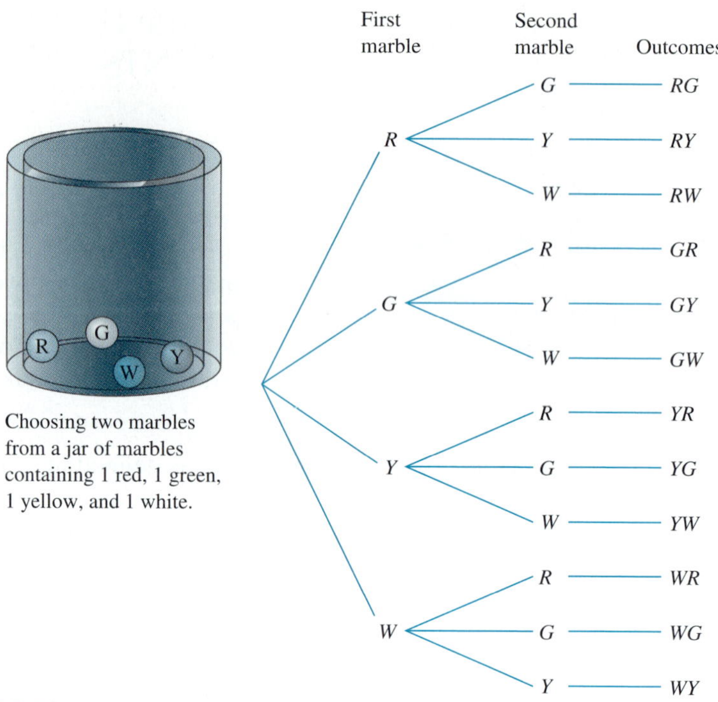

FIGURE 5.8

The diagram in Figure 5.8 shows that there are 12 outcomes in the sample space since there are 12 right-hand endpoints on the tree. Rather than count all the outcomes, we can actually compute the number of outcomes by making a simple observation about the tree diagram. Notice that there are four primary branches corresponding to the color of the first marble, and attached to each primary branch there are three secondary branches. Since there are the same number of secondary branches connected to each primary branch, the total number of outcomes can be found by multiplying the number of primary branches by the number of secondary branches attached to each primary branch; that is, there are $4 \times 3 = 12$ possible outcomes. This counting procedure suggests the following principle.

Fundamental Counting Principle

If an event A can occur in r ways, and (for each of these r ways) an event B can occur in s ways, then the number of ways events A and B can occur, in succession, is $r \times s$.

The next example shows how the **fundamental counting principle** is applied to more than two events.

EXAMPLE 5.9 A one-topping pizza can be ordered in 3 sizes (small, medium, or large), 2 crusts (white or wheat), and 5 toppings (sausage, pepperoni, bacon, onions, or mushrooms). Apply the fundamental counting principle to find the number of possible one-topping pizzas.

SOLUTION Since there are 3 sizes and 2 crusts possible, there are $3 \times 2 = 6$ combinations of the two: (small, white), (medium, white), (large, white), (small, wheat), (medium, wheat), (large, wheat). Each of those 6 combinations can be covered with 5 different toppings to provide $6 \times 5 = 30$ different selections. The number of different possible selections could have been found in one step by finding the product $3 \times 2 \times 5 = 30$. ◆

Next we apply the fundamental counting principle to compute the probability of an event in an experiment.

EXAMPLE 5.10 Find the probability of getting a sum of 11 when tossing a pair of fair dice.

SOLUTION Since each die has six faces and there are two dice, there are $6 \times 6 = 36$ possible outcomes on a pair of dice according to the fundamental counting principle. There are two ways of tossing an 11, namely, (5, 6) and (6, 5). Therefore, the probability of tossing an 11 is $\frac{2}{36}$, or $\frac{1}{18}$. ◆

The next example shows how the fundamental counting principle can be used to find the probability of an event having many outcomes.

EXAMPLE 5.11 Suppose two cards are drawn from a standard deck of 52. What is the probability of getting a pair (Figure 5.9)?

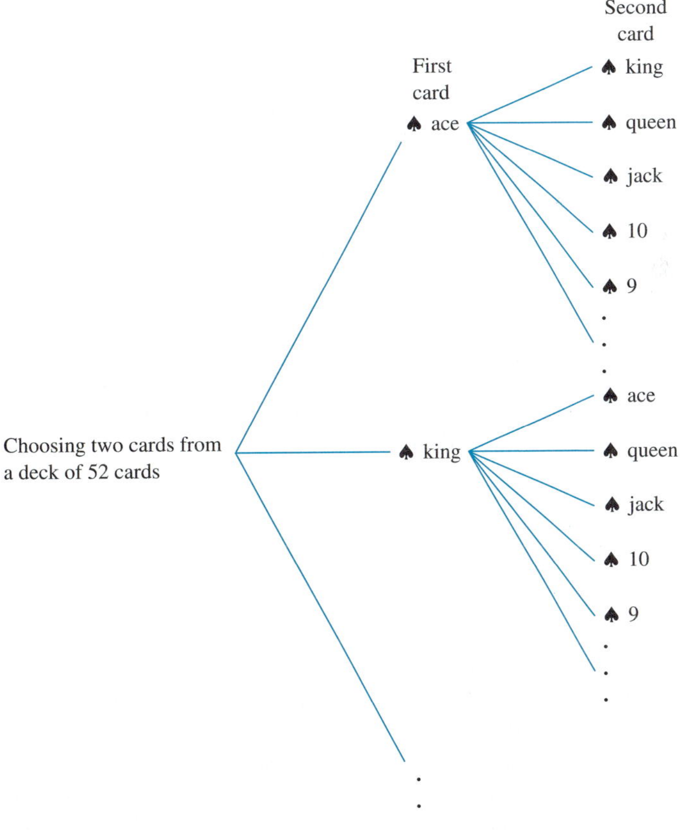

FIGURE 5.9

SOLUTION We consider the tree diagram in Figure 5.9 representing all the ways to draw two cards from the deck, keeping track of which card is drawn first and which card is drawn second.

In Figure 5.9 there are 52 primary branches and attached to each primary branch there are 51 secondary branches. Thus, by the fundamental counting principle, there are $52 \times 51 = 2652$ possible outcomes in the sample space.

In Figure 5.10 we construct another tree containing the outcomes that result in drawing a pair.

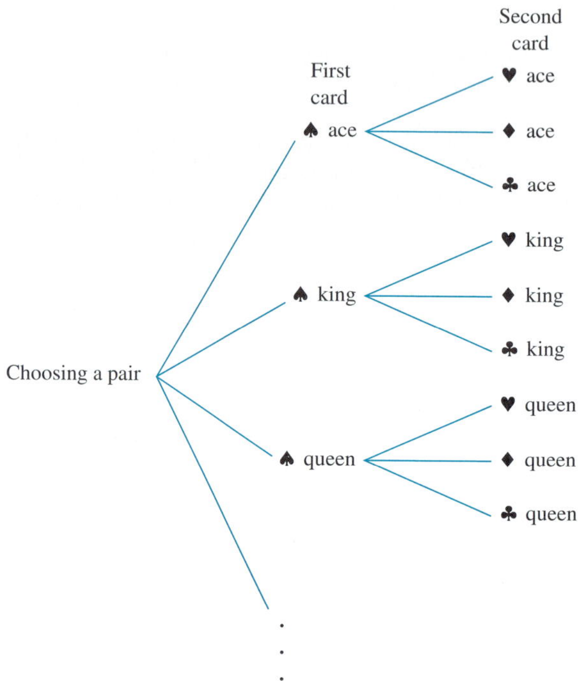

FIGURE 5.10

In Figure 5.10 there are again 52 primary branches, but now there are only 3 secondary branches attached to each primary branch. By the fundamental counting principle, there are 52×3 outcomes in Figure 5.10. If S is the sample space associated with drawing two cards in succession and E is the event that those cards form a pair, then the number of elements in $S = 52 \times 51$ and the number of elements in $E = 52 \times 3$, so

$$P(E) = \frac{\text{number of ways of drawing a pair}}{\text{number of ways of drawing two cards}} = \frac{52 \times 3}{52 \times 51} = \frac{3}{51} = \frac{1}{17}.$$ ◆

PROBABILITY TREE DIAGRAMS

In addition to helping to display and count outcomes, tree diagrams can be used to determine probabilities in complex experiments. Tree diagrams that are labeled using the probabilities of events are called **probability tree diagrams.** Suppose a container has 4 balls: 1 red, 2 white, and 1 blue [Figure 5.11(a)]. Since there are 4 balls, we draw a tree with 4 branches and label them each with probability $\frac{1}{4}$ since they are equally likely to occur [Figure 5.11(b)].

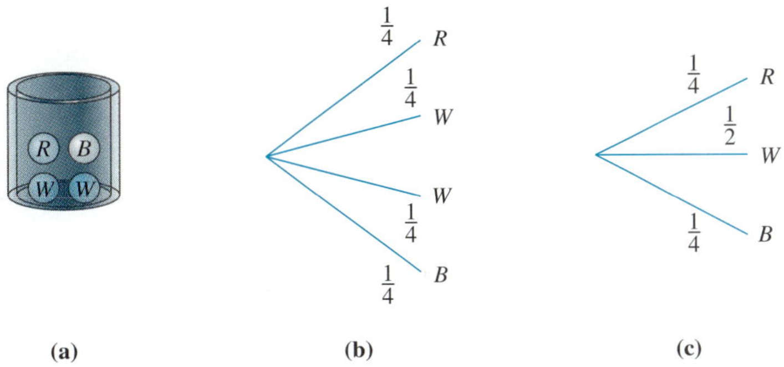

(a) (b) (c)

FIGURE 5.11

To find the probability of drawing a white ball, we add the probabilities on the two branches labeled W; that is, $P(W) = \frac{1}{4} + \frac{1}{4} = \frac{1}{2}$. To simplify the tree, the two branches labeled W are combined into one branch. However, that branch must be labeled $\frac{1}{2}$, the sum of the probabilities on the two branches we combined [Figure 5.11(c)].

EXAMPLE 5.12 Draw a probability tree that represents drawing one ball from a container holding three red balls and two white balls. Combine branches where possible.

SOLUTION See Figure 5.12.

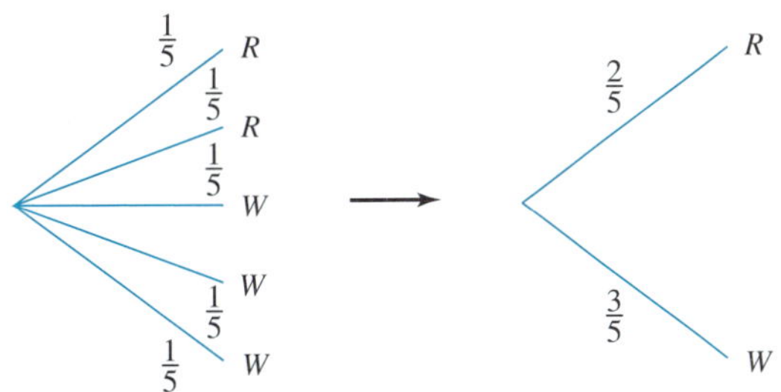

FIGURE 5.12

◆

The concept of adding the probabilities of branches in a probability tree is stated in general terms in the next property.

Additive Property of Probability Tree Diagrams

If a complex event E is the union of simple events $E_1, E_2, \ldots, E_n$, where each pair of the events is mutually exclusive, then

$$P(E) = P(E_1) + P(E_2) + \ldots + P(E_n).$$

The probabilities of the events $E_1, E_2, \ldots, E_n$ can be viewed as those associated with the ends of branches in a probability tree diagram.

[Notice that the additive property of probability tree diagrams is an extension of the property $P(A \cup B) = P(A) + P(B)$, where A and B are mutually exclusive events.]

In Example 5.12, there were two possible outcomes, R and W, but with unequal probabilities, $\frac{2}{5}$ and $\frac{3}{5}$, respectively. In general, the outcomes in a sample space are not equally likely.

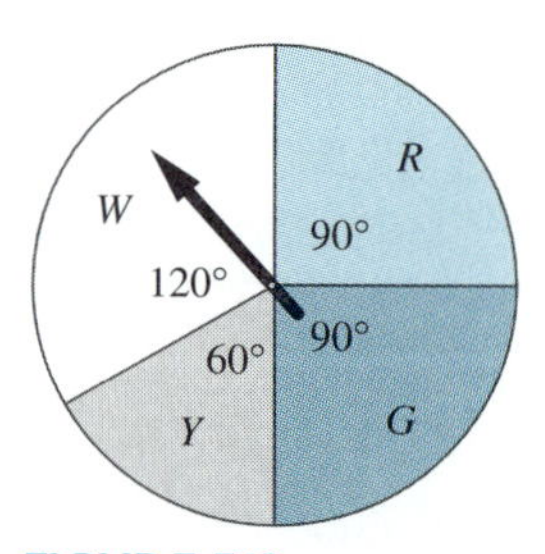

FIGURE 5.13

EXAMPLE 5.13 Draw a probability tree for the spinner in Figure 5.13 and determine the probability of spinning a W or a G.

SOLUTION The spinner has four outcomes: W, R, Y, and G. Using the central angles for each portion of the spinner, we see that $P(W) = \frac{120}{360} = \frac{1}{3}$, $P(R) = \frac{90}{360} = \frac{1}{4}$, $P(Y) = \frac{60}{360} = \frac{1}{6}$, and $P(G) = \frac{90}{360} = \frac{1}{4}$ (Figure 5.14).

Since $P(W) = \frac{1}{3}$ and $P(G) = \frac{1}{4}$, the probability of spinning W or G is $\frac{1}{3} + \frac{1}{4} = \frac{7}{12}$ according to the additive property of probability tree diagrams. ◆

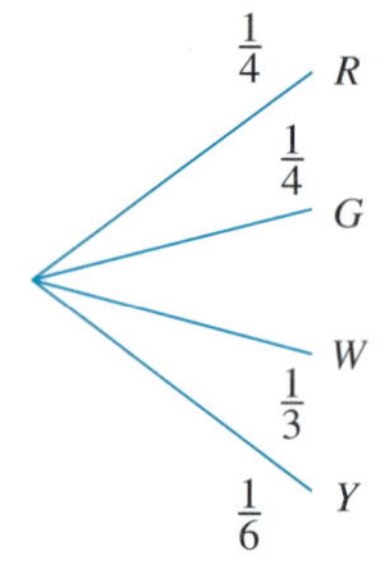

FIGURE 5.14

Notice that even though the probabilities in Example 5.13 are unequal, their sum is one: $\frac{1}{3} + \frac{1}{4} + \frac{1}{4} + \frac{1}{6} = \frac{4}{12} + \frac{3}{12} + \frac{3}{12} + \frac{2}{12} = \frac{12}{12} = 1$. The fact that the sum of the probabilities of all the outcomes of an experiment is one is true whether the outcomes are equally likely or not.

In the next example, a marble is drawn and replaced. This is referred to as **drawing with replacement.** If the marble is not replaced, then the process is called **drawing without replacement.** In Example 5.11, a pair of cards was drawn without replacement.

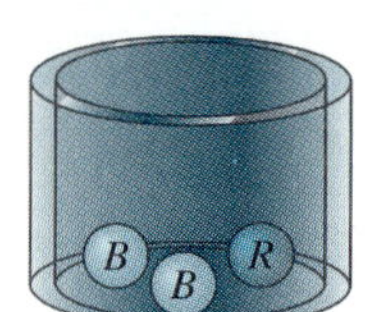

FIGURE 5.15

EXAMPLE 5.14 A jar contains 3 marbles, 2 black and 1 red (Figure 5.15). A marble is drawn, replaced, and a second marble is drawn. What is the probability that both marbles are black? Assume that the marbles are equally likely to be drawn.

SOLUTION Figure 5.16(b) shows how the number of branches in Figure 5.16(a) can be reduced by collapsing branches that represent the same type of outcome and weighting them accordingly.

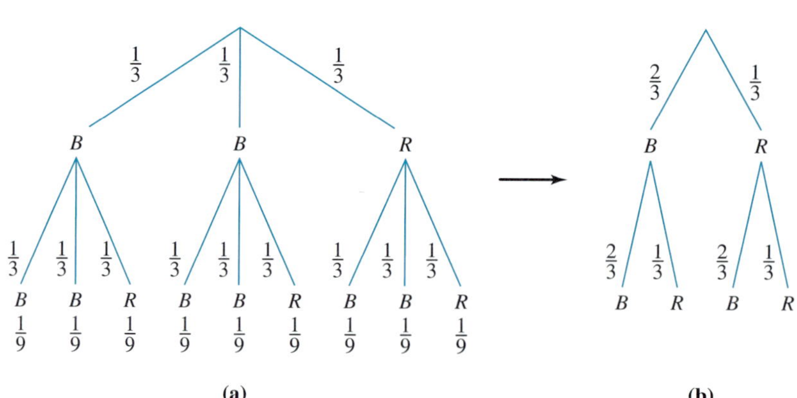

FIGURE 5.16

To determine the weights for the ends of the secondary branches in Figure 5.16(b), consider the left portion of 5.16(a) in Figure 5.17.

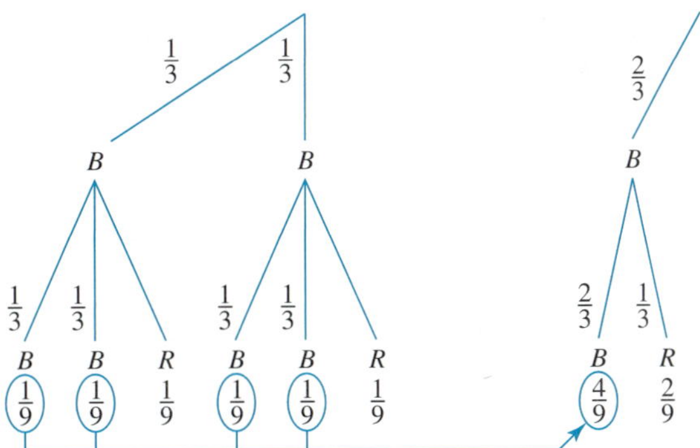

FIGURE 5.17

In Figure 5.17 we see that $P(BB) = \frac{1}{9} + \frac{1}{9} + \frac{1}{9} + \frac{1}{9} = \frac{4}{9}$. Similarly, $P(BR) = \frac{1}{9} + \frac{1}{9} = \frac{2}{9}$, $P(RB) = \frac{1}{9} + \frac{1}{9} = \frac{2}{9}$, and $P(RR) = \frac{1}{9}$ in Figure 5.18(b).

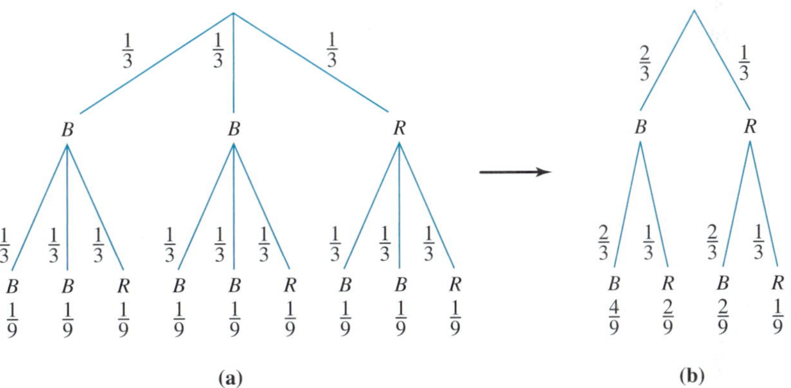

FIGURE 5.18

Notice that the $\frac{4}{9}$ obtained for $P(BB)$ is located at the end of the two branches labeled $\frac{2}{3}$ and $\frac{2}{3}$, and $\frac{2}{3} \times \frac{2}{3} = \frac{4}{9}$. This multiplicative procedure also holds for the remaining branches. ◆

The property of multiplying the probabilities along a series of branches in a probability tree diagram to find the probability at the end of a branch is based on the fundamental counting principle.

Multiplicative Property of Probability Tree Diagrams

Suppose an experiment consists of a sequence of simpler experiments that are represented by branches of a probability tree diagram. Then the probability of any of the simpler experiments is the product of all the probabilities on its branch.

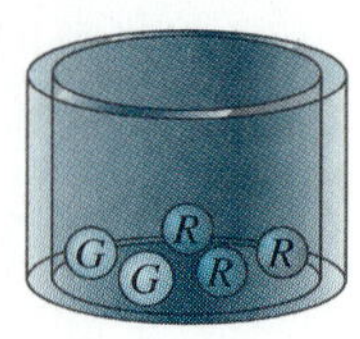

FIGURE 5.19

EXAMPLE 5.15 A jar contains three red balls and two green balls (Figure 5.19). A two-step experiment is performed. First, a coin is tossed. If the coin lands heads, a red ball is added to the jar. If the coin lands tails, a green ball is added to the jar. Second, a ball is chosen from the jar. What is the probability that a red ball is chosen?

SOLUTION The probability tree diagram for the first stage of this experiment, namely, tossing a coin and adding a ball to the jar, is shown in Figure 5.20(a).

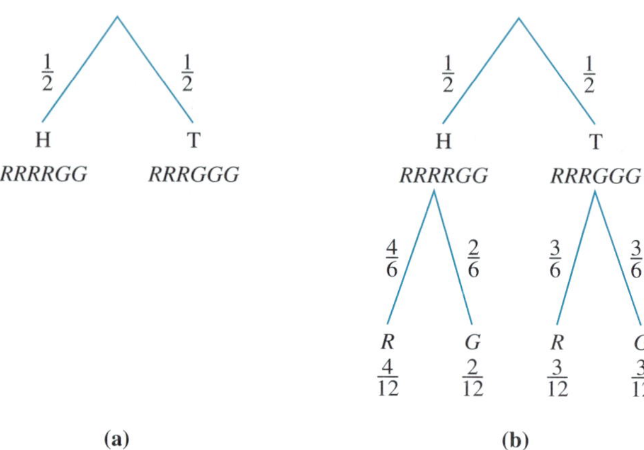

FIGURE 5.20

The second stage of the probability tree diagram, namely, choosing a ball from the jar, is shown in Figure 5.20(b). Notice the probabilities are different on the branches for the second stage. The probability at the end of each branch is then the product of the probabilities along the two branches leading to the end. Finally, the probability that the red ball will be chosen is found by adding the probabilities at the end of the R branches; that is, $P(R) = \frac{4}{12} + \frac{3}{12} = \frac{7}{12}$. ◆

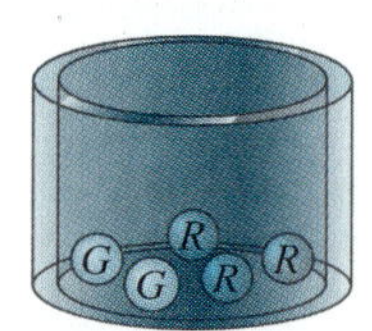

FIGURE 5.21

EXAMPLE 5.16 A jar contains three red gumballs and two green gumballs (Figure 5.21). An experiment consists of drawing gumballs, one at a time from the jar, without replacement, until a red gumball is obtained. Find the probability of the following events:

A: only one draw is needed.
B: exactly two draws are needed.
C: exactly three draws are needed.

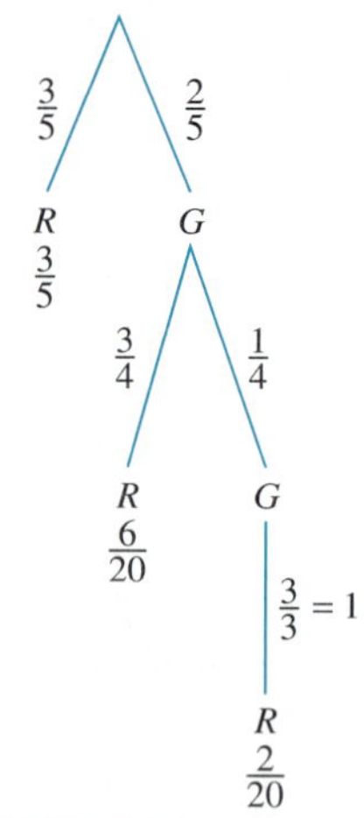

FIGURE 5.22

SOLUTION Make the probability tree diagram (Figure 5.22). Hence $P(A) = \frac{3}{5}$ (a R gumball is drawn), $P(B) = \frac{2}{5} \times \frac{3}{4} = \frac{6}{20} = \frac{3}{10}$ (a G followed by a R are drawn), and $P(C) = \frac{2}{5} \times \frac{1}{4} \times 1 = \frac{2}{20} = \frac{1}{10}$ (two Gs followed by a R are drawn). ◆

In every probability tree diagram, the sum of the probabilities at the end of the branches is 1. This sum is the probability of the entire sample space. For example, in Figure 5.22, $\frac{3}{5} + \frac{6}{20} + \frac{2}{20} = \frac{12}{20} + \frac{6}{20} + \frac{2}{20} = 1$.

Next we illustrate finding the probabilities in a complex experiment using the additive and multiplicative properties of probability tree diagrams.

EXAMPLE 5.17 Both spinners shown in Figure 5.23 are spun. Find the probability that they stop on the same color.

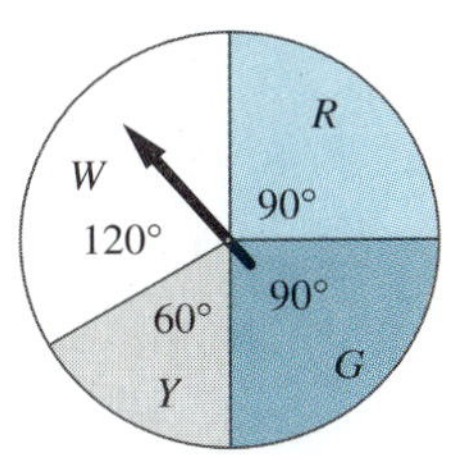

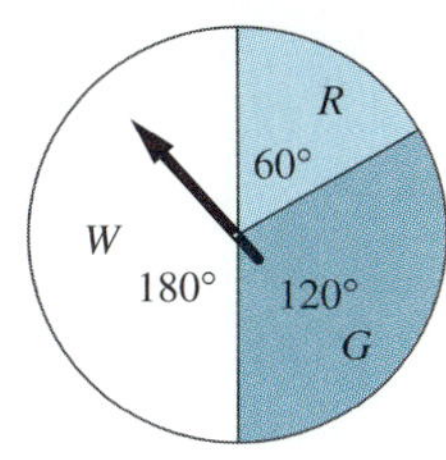

FIGURE 5.23

SOLUTION Draw an appropriate probability tree diagram where the color names are abbreviated by the first letter (Figure 5.24).

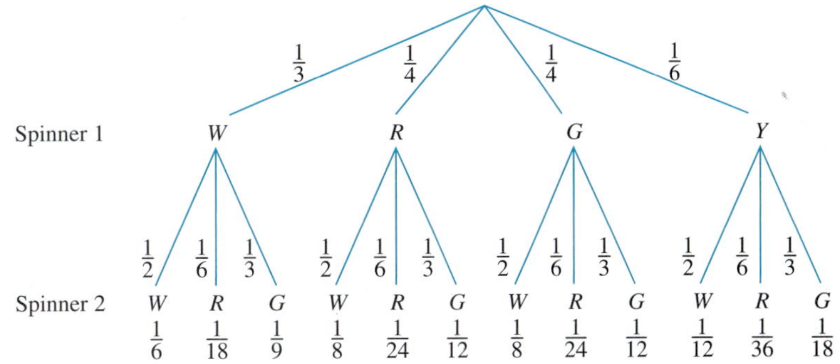

FIGURE 5.24

Notice that the probability of each color is the fraction of 360° occupied by that color. The desired event is {*WW, RR, GG*}. First, we find the probability of getting *WW, RR,* and *GG* separately. By the multiplicative property of probability tree diagrams $P(WW) = \frac{1}{3} \times \frac{1}{2} = \frac{1}{6}$, $P(RR) = \frac{1}{4} \times \frac{1}{6} = \frac{1}{24}$, and $P(GG) = \frac{1}{4} \times \frac{1}{3} = \frac{1}{12}$. Now, we find the probability of getting either *WW, RR,* or *GG*. By the additive property of probability tree diagrams,

$$P(\{WW, RR, GG\}) = \frac{1}{6} + \frac{1}{24} + \frac{1}{12} = \frac{7}{24}.$$ ◆

In summary, the probability of a complex event can be found as follows:

1. Construct the appropriate probability tree diagram.
2. Assign probabilities to each branch in the diagram.
3. Multiply the probabilities along individual branches to find the probability of the outcome at the end of each branch.
4. Add the probabilities of the relevant outcomes, depending on the event.

A friend who likes to gamble makes you the following wager: He'll bet you two dollars against one dollar that if you toss a coin repeatedly, you'll get a total of two tails before you can get a total of three heads. You reason that two heads are as likely as two tails, and you have a one-out-of-two chance of getting a third head after getting two heads. Should you take the bet?

INITIAL PROBLEM SOLUTION

SOLUTION One way to think about the problem is as an experiment with several stages. The experiment can be outlined with a probability tree diagram based on tossing several coins, as many as needed. The branches of the tree we construct will end whenever the particular sequence of outcomes indicates the end of a game (two tails or three heads, whichever comes first). Since we assume the coin is a fair coin, the outcomes at each stage will be equally likely, and we can assign $\frac{1}{2}$ as the probability along each branch. Finally, we calculate the probabilities at the end of each branch to indicate the probability of that particular sequence ending in a win or a loss.

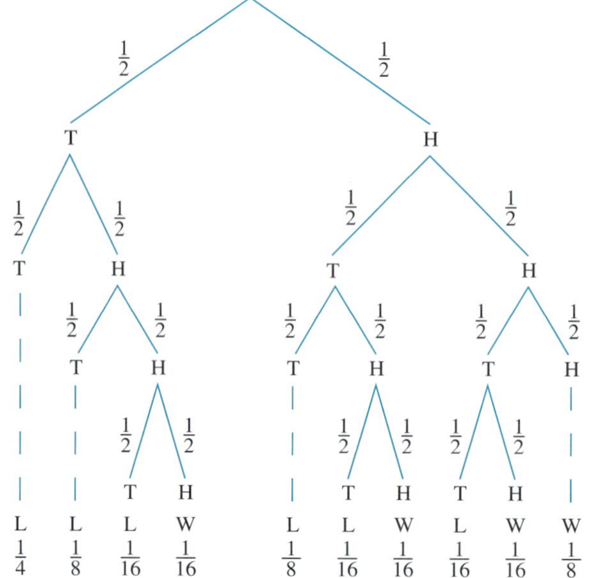

FIGURE 5.25

When we add the probabilities at the ends of each branch, we see that the probability the game ends as a loss is $\frac{1}{4} + \frac{1}{8} + \frac{1}{16} + \frac{1}{8} + \frac{1}{16} + \frac{1}{16} = \frac{11}{16}$, while the probability is only $\frac{5}{16}$ that we will win. Since the chances of losing are more than twice the chance of winning, a payoff of two dollars against our one dollar bet doesn't seem like a good deal. If you only want to play when the game is "fair," you shouldn't take the bet.

PROBLEM SET 5.2

In problems 1 through 6, draw a one stage tree to represent the given experiments.

1. Toss a dime.
2. Draw a marble from a bag containing one red, one green, one black, and one white marble.
3. Pull a dollar bill from your wallet, noting the last digit in the serial number.
4. Roll a regular four-sided die with its faces labeled 1, 2, 3, 4, noting the number on the bottom face.
5. Choose a TV program from among channels 2, 6, 9, 12, and 13.
6. Spin the spinner, where all central angles are 120°.

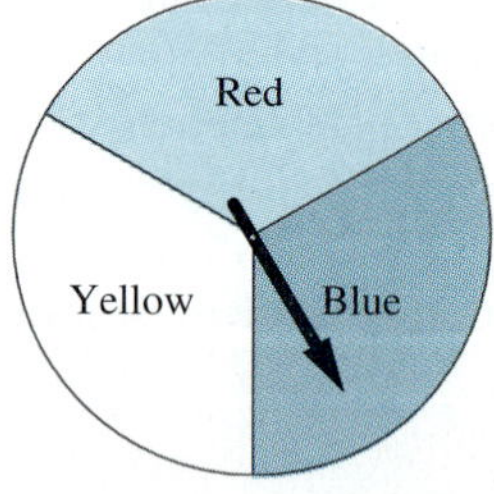

In problems 7 through 12, draw a two stage tree to represent the given experiments.

7. Toss a coin twice.
8. Draw a marble from a box containing one yellow and one green marble, then draw a marble from a box containing one yellow, one red, and one blue marble.
9. Note the gender and birth order of two children in a family.
10. Toss a coin, then roll a die.
11. Roll two regular four-sided dice with faces labeled 1, 2, 3, 4, and note the number on the bottom face of each.
12. Spin the spinner from Problem 6, then roll a regular four-sided die.

In problems 13 through 16, draw trees to represent each of the experiments.

13. Toss a coin three times.
14. Note the genders of four children in a family.
15. Spinning the spinner from Problem 6, then tossing a coin, then rolling a regular four-sided die.
16. Tossing a coin four times.

17. In some cases, what happens at the first stage of the tree affects what can happen at the next stage. For example, one ball is drawn from a box containing one red, one white, and one blue ball, but is not replaced before a second ball is drawn.
   (a) Draw the first stage of the tree.
   (b) If the red ball was selected and not replaced, what outcomes are possible on the second draw? Starting at *R*, draw a branch to represent each of these outcomes.
   (c) If white was drawn first, what outcomes are possible on the second draw? Draw these branches.
   (d) Do likewise for the case that blue was drawn first.
   (e) In total, how many outcomes are possible?

18. Draw the tree to represent the experiment in which one first tosses a coin, then if the coin landed heads, one rolls a regular four-sided die, while if the coin landed tails, one tosses the coin again.

19. Trees need not be symmetrical. For example, from the box containing one red, one white, and one blue ball, we will draw balls (without replacement) until the red ball is chosen. Draw the outcome tree.

20. Trees need not be finite. For example, consider the experiment of tossing a coin until it lands heads. This usually takes only a few tosses (in fact, two on average), but it could take any number of tosses. Draw at least three stages of the tree to show the pattern.

21. An experiment consists of tossing a coin and two dice. How many outcomes are there for the following?
   (a) the coin **(b)** the first die
   (c) the second die **(d)** the experiment

22. For your vacation, you will travel from your home to New York City, then to London. You may travel to New York City by car, train, bus, or plane, and from New York to London by ship or plane.
   (a) Draw a tree diagram to represent possible travel arrangements.
   (b) How many different routes are possible?
   (c) Apply the fundamental counting principle to find the number of possible routes. Does your answer agree with part (b)?

23. Suppose that frozen yogurt can be ordered in 3 sizes (small, medium, large), 2 flavors (vanilla, chocolate), and 4 toppings (plain, sprinkles only, fudge only, sprinkles and fudge).
   (a) Draw a tree diagram to represent possible yogurt selections.
   (b) Apply the fundamental counting principle to find the number of possible yogurt selections.

24. Find the probability of getting a sum of 3 when tossing a pair of regular four-sided dice (each numbered 1 to 4).

25. A pinochle deck contains 48 cards. The cards are arranged in 4 suits. Each suit contains 2 of each of the following: 9, 10, jack, queen, king, ace. Suppose two cards are drawn at random. What is the probability of getting a pair?

26. Use the fundamental counting principle to determine the number of outcomes possible for the following.
   (a) Three coins are tossed.
   (b) How many outcomes are there when 4, 5, and 6 coins are tossed?
   (c) How many outcomes are there when n coins are tossed? (n is a counting number.)

27. A fair coin is flipped 4 times.
   (a) How many outcomes involve exactly two heads?
   (b) Find the probability of getting exactly two heads.

28. A bowl contains three marbles (red, blue, green). A box contains four numbered tickets (1, 2, 3, 4). One marble will be selected at random, and then a ticket will be selected at random.
   (a) Use the fundamental counting principle to find the number of possible outcomes.
   (b) Find the probability you select the red marble and the ticket numbered 3.
   (c) Find the probability you select the green marble.

29. Suppose two cards are drawn from a standard deck of 52 cards.
   (a) How many outcomes are possible for this experiment?
   (b) How many outcomes correspond to the event that the cards are both hearts?
   (c) What is the probability that both cards are hearts?

30. Suppose two cards are drawn (without replacement) from a standard pinochle deck (see problem 25).
   (a) How many outcomes are possible for this experiment?
   (b) How many outcomes correspond to the event that the cards are both face cards?
   (c) What is the probability that both cards are face cards?

31. Draw the probability tree diagram for drawing a ball from the following containers.

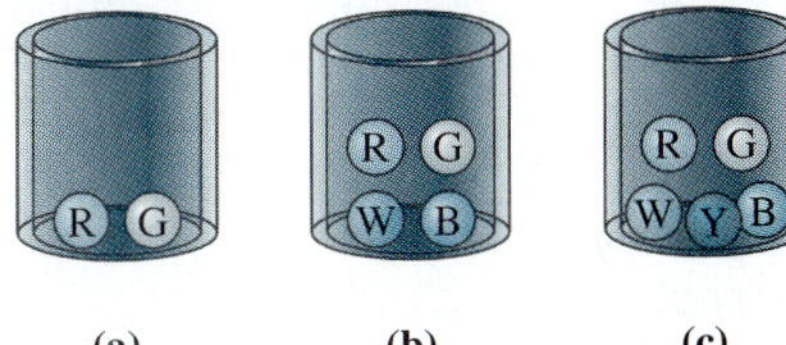

32. A box contains four envelopes each containing a single bill: \$100, \$100, \$5, and \$1. An envelope is drawn from the box and the dollar amount inside noted. It is returned to the box and a second envelope is drawn. Assume the envelopes are equally likely to be drawn.
   (a) Construct a tree combining branches where possible.
   (b) Assign probabilities to each branch of the tree.
   (c) What is the probability that \$100 bills were drawn both times?

33. The following spinner will be spun twice.

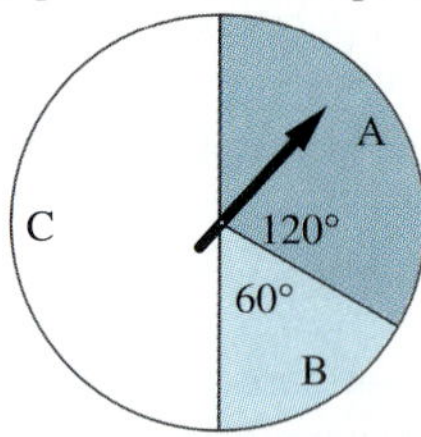

(a) Construct a probability tree diagram for the experiment.
(b) Find the probability the spinner lands on B both times.
(c) Find the probability the spinner lands on C the first time and A the second time.

34. A lightbulb will be selected from box 1 and another from box 2. In box 1, 30% of the bulbs are defective. In box 2, 45% of the bulbs are defective.
(a) Construct a probability tree diagram for this experiment.
(b) Find the probability that both bulbs are defective.
(c) Find the probability that the first bulb is defective and the second is not defective.

35. While still half asleep, you randomly select a black sock from your drawer. Now the drawer contains two white socks and four more black socks. You continue to randomly select one sock at a time from your drawer until another black sock is selected. Construct a probability tree diagram. Find the probability of each of these events.
(a) Exactly one draw is needed.
(b) Exactly two draws are needed.
(c) Exactly three draws are needed.

36. A game at a carnival consists of throwing darts at balloons. There are eight balloons situated in such a way that the player will always pop one of them. Behind two of the balloons a star is hidden. If the player pops a balloon revealing a star, then he wins a prize. A player pays for three attempts. Construct a probability tree diagram and find the probability that the player
(a) wins in one shot.
(b) wins in exactly 2 shots.
(c) wins in exactly three shots.
(d) does not win.

37. You come home on a dark night and find the front porch light burned out. Since you cannot tell which key is which, you randomly try each of the seven keys on your key ring until you find one that opens your apartment door. Two keys on your key ring unlock the door. Find the probability of opening the door with the first or second key.

38. Each individual letter of the word MISSISSIPPI is placed on a piece of paper, and all 11 pieces of paper are placed in a bowl. Two letters are selected at random from the bowl without replacement. Find the probability of:
(a) selecting two P's.
(b) selecting the same letter in two selections.
(c) selecting two consonants.

39. A pair of regular dice is constructed so that each die contains one spot on one side, two spots on two sides, and three spots on three sides. The dice are rolled. Find the probability that
(a) two 3's are rolled.
(b) the same number appears on each die.
(c) two odd numbers are rolled.

40. A pair of regular dice are constructed as in problem 39. The two dice are rolled. What is the probability that:
(a) neither die is a two?
(b) the two dice are different?
(c) one die has an odd number and the other has an even number?

EXTENDED PROBLEMS

The basic definition of probability can be summarized in the following way:

$$P(E) = \frac{\text{number of ways } E \text{ can happen}}{\text{number of possibilities in } S},$$

where E is an event and S is the sample space. The difficulty is in counting the number of ways something can happen.

One of the basic counting methods used in probability is called a combination, or, more completely, "the combination of n things taken k at a time." This means we are going to choose k objects from a collection of n objects. The order in which they are chosen does not matter.

For example, there are 6 ways to select two objects from a set of four objects: Let the objects be called A, B, C, and D. The possible pairs are: $A,B; A,C; A,D; B,C; B,D; C,D$. Most calculators let you find combinations with a few keystrokes. The symbol is usually $C(n,k)$ or something similar.

Check the manual to find out how combinations are found on your calculator.

41. Lotteries have been established in a majority of states, and there are lotteries that are played nationally and around the world. In a typical lottery, a person has to correctly choose 6 numbers out of 44 and match all 6 that are selected by some random method. The order in which the numbers are selected does not matter. The probability of winning can be expressed as:

$$P(W) = \frac{\text{choose 6 out of 6}}{\text{choose 6 out of 44}}.$$

What is the probability of winning this lottery?

42. What is the probability of choosing all the winning numbers in a lottery in which there are 38 numbers and you choose 5?

43. In most lotteries it is possible to be a winner by choosing most of the numbers correctly. This is done to encourage more participation. In the lottery in which 6 numbers are chosen from 44, what is the probability of choosing 5 numbers correctly?

44. Write a report on the lottery in your state, if there is one, or one of the national lotteries. Most lotteries are run by the states, and information is publicly available.

45. The study of probability has its origins in gambling, and it is here that probability can be applied in its purest form. Write a brief report on the history of probability and the major mathematicians who contributed to its development.

5.3 CONDITIONAL PROBABILITY, INDEPENDENCE, AND EXPECTED VALUE

American roulette wheels have 38 slots numbered 00, 0, and 1 through 36 (Figure 5.26). You place a bet on a specific number by putting your wager on the numbered square on the roulette cloth, or layout. Bets may also be placed on more than one number or on combinations of numbers. The wheel is spun in one direction and a ball is rolled in the opposite direction in a surrounding sloped bowl. When the ball slows sufficiently, it drops down into the numbered slots and bounces along until coming to rest on the winning number. If you had bet on the winning number, the croupier (the manager of the table) leaves your bet on the layout and adds to it 35 times as much as you bet. If you chose a wrong number, your wager and the other losing bets are gathered in with a rake. If you bet \$100 on one number, what is your expected gain or loss?

In the solution to the Initial Problem of the previous section, we ended with the idea that a certain game wasn't "fair," although we didn't say how to determine if a game was fair or even define what we meant by a fair game. In this section, we will develop several additional properties of probability that will be used in analyzing complicated events, determining probable origins for certain sequences, and relating values to the outcomes of events.

CONDITIONAL PROBABILITY

At times, conditions are imposed on a sample space that cause us to focus on a portion of the sample space called the **conditional sample space.** That is, certain information is known about the experiment that affects the possible outcomes. Such "given" information is illustrated in the next example.

EXAMPLE 5.18 Describe the sample space for the experiment "the first coin came up heads when tossing three fair coins."

SOLUTION When tossing three coins, there are eight outcomes: HHH, HHT, HTH, THH, HTT, THT, TTH, TTT. In this case, the condition "the first coin came up heads" produces the following conditional sample space: {HHH, HHT, HTH, HTT}. ◆

In Example 5.18, the original sample is reduced to those outcomes having the given condition, namely H, on the first coin. Consider the following example using this conditional sample space.

Let A be the event that exactly two tails appear among the three coins, and B be the event that the first coin among the three coins comes up heads; therefore, $A = \{\text{HTT, THT, TTH}\}$ and $B = \{\text{HHH, HHT, HTH, HTT}\}$. To find the probability of A given B, we mean to find the probability of event A occurring within the conditional sample space B. The notation $P(A \mid B)$ is used to represent "the probability of A given B." In this example, there is only one way for A to occur, HTT, within the set B, which contains four elements. (Note that $A \cap B = \{\text{HTT}\}$.) Thus the probability of A given B is $\frac{1}{4}$. Notice that $P(A \mid B) = \frac{1}{4} = \frac{1/8}{4/8} = \frac{P(A \cap B)}{P(B)}$. That is, $P(A \mid B)$ is the relative frequency of the event A within the conditional sample-space B. This suggests the following.

DEFINITION

Conditional Probability

Suppose A and B are events in a sample space S such that $P(B) \neq 0$. The **conditional probability** that the event A occurs, given that the event B occurs, denoted $P(A \mid B)$, is

$$P(A \mid B) = \frac{P(A \cap B)}{P(B)}.$$

A diagram can be used to illustrate the definition of conditional probability. A sample space S of equally likely outcomes is shown in Figure 5.27(a). The conditional sample space, given that event B occurs, appears in Figure 5.27(b).

TIDBIT

The folk saying "lightning never strikes twice in the same place" asserts that the conditional probability of a lightning strike, given a previous strike, is zero. The folk saying is false. For example, ex-park ranger Roy C. Sullivan was struck by lightning and survived seven times.

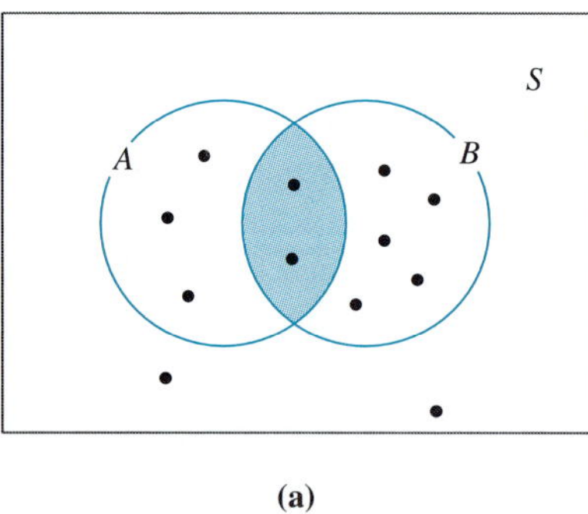

(a)

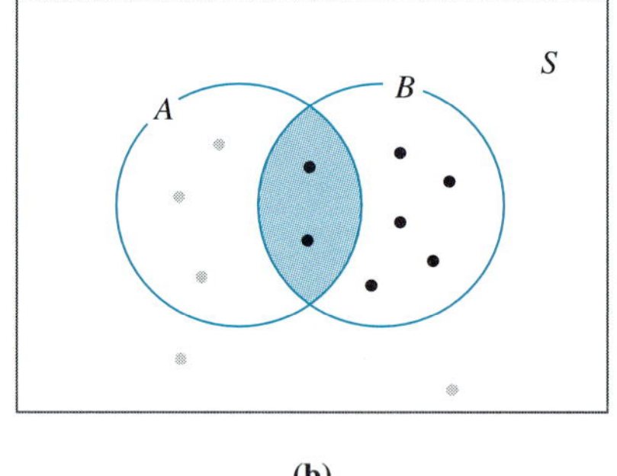

(b)

FIGURE 5.26

Figure 5.26(a) shows $\frac{P(A \cap B)}{P(B)} = \frac{2/12}{7/12} = \frac{2}{7}$. From Figure 5.26(b), we see $P(A \mid B) = \frac{2}{7}$. Thus $P(A \mid B) = \frac{P(A \cap B)}{P(B)}$.

The next example illustrates conditional probability in the case of outcomes that are not equally likely.

EXAMPLE 5.19 Suppose we have two jars of marbles. The first jar contains 2 white marbles and 1 black marble and the second jar contains 1 white marble and 2

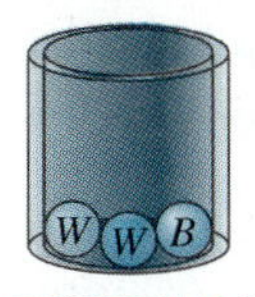

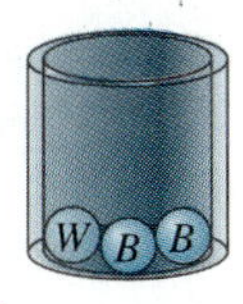

FIGURE 5.27

black marbles (Figure 5.27). A fair coin is tossed. If the coin lands heads, then a marble is drawn from the first jar; but if the coin lands tails, the marble is drawn from the second jar. Find the probability that the coin landed heads, given a black marble was drawn.

SOLUTION First, we construct a probability tree diagram to describe the experiment (Figure 5.28).

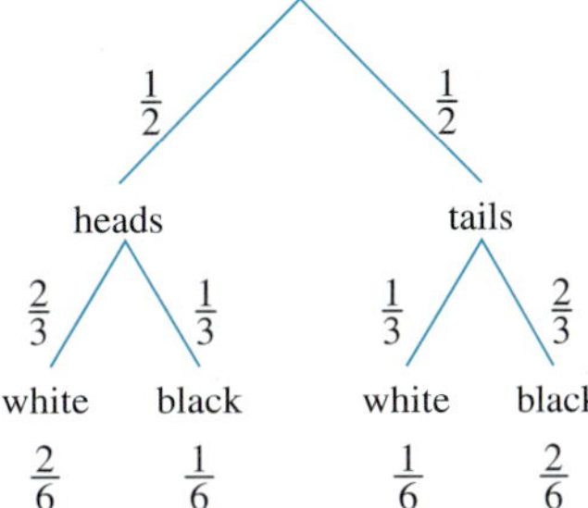

FIGURE 5.28

From the tree diagram we can read off the sample space and probabilities. For example, the left-most path down the tree gives us the element (heads, white) in the sample space and that element has probability $\frac{1}{2} \times \frac{2}{3} = \frac{2}{6} = \frac{1}{3}$. But we do not want to consider the entire sample space. Instead, we want to consider only those paths that end with a black marble being drawn. In Figure 5.29 we have reproduced Figure 5.28, but the paths that end in a black marble have been emphasized. We have also circled the path on which a black marble is drawn *and* on which the coin lands heads.

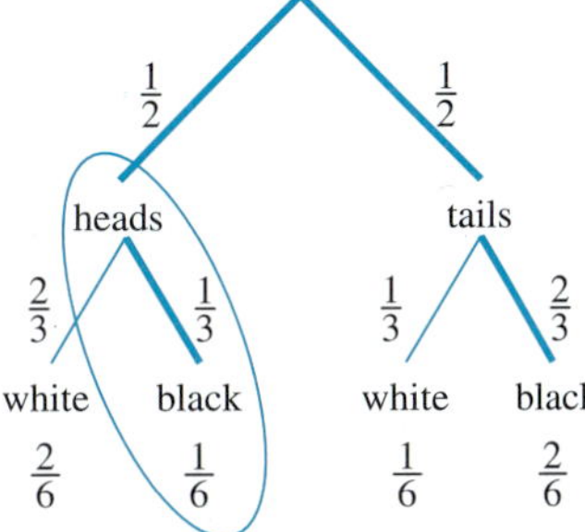

FIGURE 5.29

The sum of the probabilities at the end of the two emphasized paths ending in black is $\frac{1}{6} + \frac{2}{6} = \frac{3}{6}$, and the probability of a head and a black marble, which is circled, is $\frac{1}{6}$. In this example, $P(A \mid B) =$ is interpreted verbally as P(heads given a black marble) $= \frac{P(\text{heads and a black marble})}{P(\text{a black marble})} = \frac{1/6}{3/6} = \frac{1}{3}$. ◆

The conditional probability we obtained in Example 5.19 may seem completely unrelated to anything you will need to consider in the real world. However, the fact is that real world situations often confront us with information that actualy refers to conditional probabilities. When decisions must be made on the basis of such information, it is important to appreciate that the analysis needed may be more subtle than most people realize. We hope the next example will reinforce this point.

EXAMPLE 5.20 Suppose a test is available for detecting the presence of a viral infection, although the test is not 100% accurate. Assume 1/4 of the population is infected, and the other 3/4 is not. Assume that 90% of those infected test positive and 80% of uninfected persons test negative. (Testing positive means the test indicates the presence of the viral infection; testing negative means the test indicates there is no infection.) Given the test is positive, what is the probability that the person is infected?

HISTORY

The now common faith in medical tests only dates back to the first decade of this century when Wassermann, Neisser, and Bruck developed a test to locate the causative agent of syphilis in the blood serum.

SOLUTION This is a two-part experiment similar to that in Example 5.19. In the first stage of the experiment, a person is chosen at random, and that person can be either infected or uninfected. The second stage in the experiment consists of the testing, which can come out either positive or negative. We form the probability tree diagram (Figure 5.30). (We have indicated two paths that correspond to the undesirable situation in which the test results are wrong; that is, false positives and false negatives.)

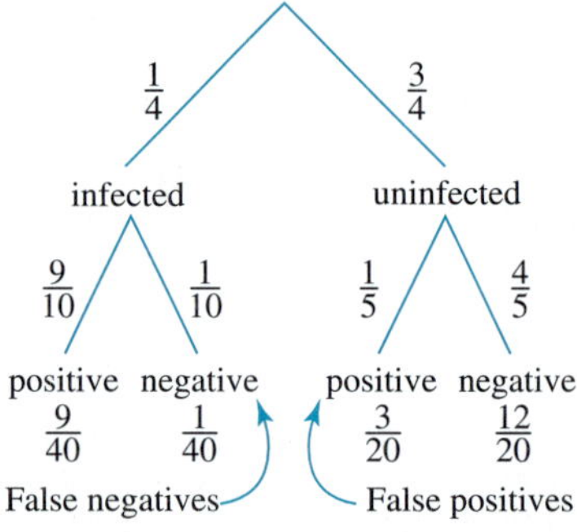

FIGURE 5.30

We want to find P(the person is infected given that the test is positive). To aid in finding the conditional probability, we emphasize the paths corresponding to the event "the test is positive," and we circle the part of the tree that corresponds to "the person is infected and the test is positive," that is, the path through "infected" *and* "positive" (Figure 5.31).

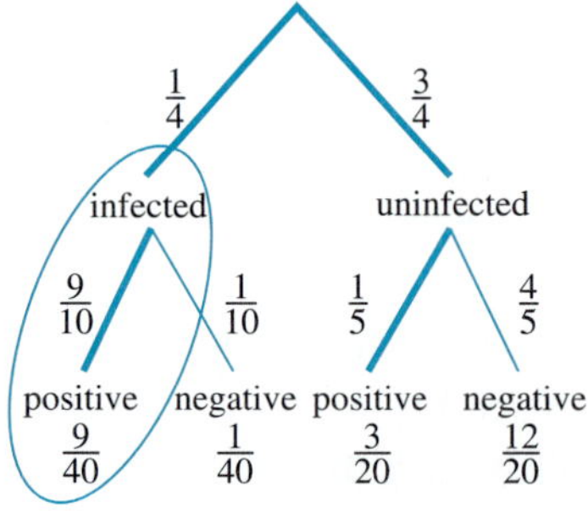

FIGURE 5.31

We see that:

$P(\text{test positive}) = \frac{9}{40} + \frac{3}{20} = \frac{15}{40}$ and $P(\text{infected and test positive}) = \frac{9}{40}$, so

$$P(\text{infected} \mid \text{test positive}) = \frac{9/40}{15/40} = \frac{9}{15} = 0.60.$$

Thus there is a 60% chance that the person is infected. ◆

Notice that even though the test produces correct results more than 80% of the time, the probability of the person being infected given that he or she tested positive is only 60%. This is quite typical. For an ordinary person with no symptoms or reason to think she is at risk, a positive test result for a rare disease is often a false positive, causing a lot of needless worry.

INDEPENDENT EVENTS

If a pair of dice is rolled, the probability of getting a "1" on the first roll is $\frac{1}{6}$, and the probability of getting a "1" on the second roll is also $\frac{1}{6}$. The probability of getting a "1" on the first roll *and* a "1" on the second roll is $\frac{1}{6} \times \frac{1}{6} = \frac{1}{36}$. These two events are called **independent events,** in the sense that one event does not influence the other. Now let us suppose that we have a jar containing 10 white marbles and 2 red marbles. If we mix the marbles and choose one, the probability of getting a red marble is $\frac{2}{12} = \frac{1}{6}$. If we replace the marble and remix, then we are in the same situation as when we started. The event that we obtained a red marble on the first draw has no effect on the chances of getting a red marble on the second draw. The events are independent. The probability of getting red marbles on both the first *and* second draws will be $\frac{1}{6} \times \frac{1}{6} = \frac{1}{36}$. The following statement summarizes the concept of independent events.

Probability of Independent Events

When two events are independent, the probability of both happening equals the product of their probabilities. For independent events A and B,

$$P(A \cap B) = P(A) \times P(B).$$

EXAMPLE 5.21 Consider drawing 2 marbles from the jar containing 10 white marbles and 2 red marbles, but do not replace the first marble before drawing the second. Find the probability that the first marble is red, the probability that the second marble is red, and the probability that both marbles are red.

SOLUTION We construct a probability tree diagram for the experiment (Figure 5.32).

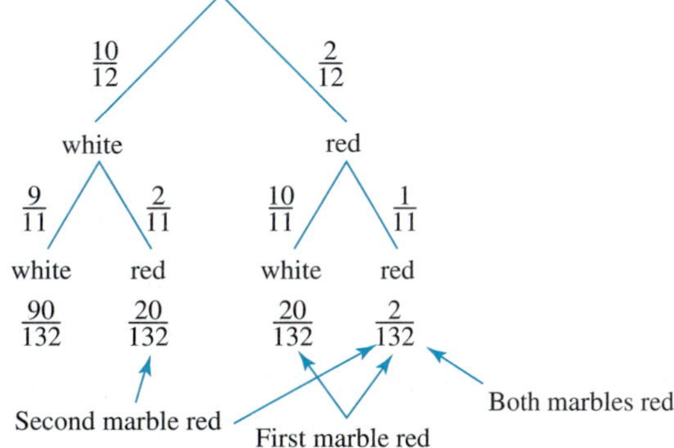

FIGURE 5.32

We see that:

$$P(\text{first marble is red}) = \frac{20}{132} + \frac{2}{132} = \frac{22}{132} = \frac{1}{6},$$

$$P(\text{second marble is red}) = \frac{20}{132} + \frac{2}{132} = \frac{22}{132} = \frac{1}{6},$$

$$P(\text{both marbles are red}) = \frac{2}{132} = \frac{1}{66}.$$

◆

Notice that in Example 5.21 the probability of both events happening is not the product of the probabilities, showing that the two events are not independent; that is, $P(\text{first marble is red}) \times P(\text{second marble is red}) = \frac{1}{6} \times \frac{1}{6} \neq \frac{1}{66} = P$ (both marbles are red).

Recognizing that two events are independent makes the computation of the probability that both happen much easier. It also makes the computation of conditional probabilities easier, as we illustrate. If A and B are independent and if $P(B) > 0$, then the conditional probability of A given B is

$$P(A|B) = \frac{P(A \cap B)}{P(B)} = \frac{P(A) \times P(B)}{P(B)} = P(A).$$

This is consistent with our intuitive understanding. If two events are independent, then the occurrence of one of these events does not affect the probability that the other will occur.

EXPECTED VALUE

Sometimes the possible outcomes of a probability experiment are numbers, such as the number of dots showing on a die. In other cases, the possible outcomes of an experiment may not actually be numbers, but numbers can be associated to the outcomes. For example, Bob wins \$1 from Jennifer if he chooses a higher card from the deck than Jennifer does, otherwise he loses \$1 to Jennifer. We can associate $+1$ with the event that Bob chooses the higher card and associate -1 with the event that Jennifer chooses the higher card. For experiments with numerical outcomes it is useful to know what the average should be for many repetitions of the experiment. This number is called the **expected value**.

Suppose you are asked to play a game where you win \$3 if you toss a three or greater on a standard die and lose \$5 if you toss a one or a two. Let's see if you should play this game. You win \$3 on $\frac{4}{6}\left(=\frac{2}{3}\right)$ of the tosses and lose \$5 on $\frac{2}{6}\left(=\frac{1}{3}\right)$ of the tosses. You should expect to win \$3 an average of two times out of three and lose \$5 the other third of the time. Thus, on average, over the long run, in three tosses you should expect to win \$3 + \$3 = \$6 and lose \$5, yielding a net profit of \$1 in three plays. On a per play basis, the analysis looks like this:

$$\$3 \times \frac{2}{3} + (-\$5) \times \frac{1}{3} = \$2 + \left(\$ - 1\frac{2}{3}\right) = \$\frac{1}{3}, \text{ or about } 33¢.$$

In general, to find the expected value of an experiment with a numerical outcome (or with an associated numerical outcome) multiply each possible numerical outcome by its probability, and add all of the products. More formally, we have the following definition.

Expected Value

Suppose that the outcomes of an experiment are numbers (values) called v_1, $v_2, \ldots, v_n$, and the outcomes have probabilities $p_1, p_2, \ldots, p_n$, respectively. The **expected value,** E, of the experiment is the sum of the products

$$E = (v_1 \times p_1) + (v_2 \times p_2) + \ldots + (v_n \times p_n).$$

We use this definition in the next example.

EXAMPLE 5.22 Compute the expected value of the roll of a fair die.

SOLUTION The possible values are the whole numbers 1 through 6, and each has the probability $\frac{1}{6}$. The computation of the expected value can be organized by putting the necessary information into a table, as in Table 5.5.

TABLE 5.5

Value	1		2		3		4		5		6		
Probability	$\frac{1}{6}$		$\frac{1}{6}$		$\frac{1}{6}$		$\frac{1}{6}$		$\frac{1}{6}$		$\frac{1}{6}$		
Product	$\frac{1}{6}$	+	$\frac{2}{6}$	+	$\frac{3}{6}$	+	$\frac{4}{6}$	+	$\frac{5}{6}$	+	$\frac{6}{6}$	=	$\mathbf{\frac{21}{6}}$
													Expected Value

So we see that the expected value is $E = \frac{21}{6} = \frac{7}{2} = 3.5$, that is, you should expect to average 3.5 dots per toss. ◆

HISTORY

The first maritime insurance companies were established in Italy and Holland in the 14th century. These companies carried out calculations of chances since larger risks made for larger insurance premiums. For shipping by sea, premiums amounted to about 12% to 15% of the cost of the goods.

Common applications of expected values are in determining admissions to games and premiums for insurance. The next example shows a simplified version of how insurance companies use expected values.

EXAMPLE 5.23 Suppose that an insurance company has broken down yearly automobile claims for drivers from ages 16 through 21, as shown in Table 5.6.

TABLE 5.6

Amount of Claim (nearest $2000)	**Probability**
0	0.80
$2000	0.10
4000	0.05
6000	0.03
8000	0.01
10,000	0.01

How much should the company charge as its average premium in order to break even on its costs for claims?

SOLUTION We should think of Table 5.6 as giving us the probabilities for various numerical outcomes of an experiment. Then we compute the expected value of that experiment.

TABLE 5.7

Value	0		2000		4000		6000		8000		10,000		
Probability	0.80		0.10		0.05		0.03		0.01		0.01		
Products	0	+	200	+	200	+	180	+	80	+	100	=	**760**
													Expected Value

Thus the expected value is \$760. Since the average claim value is \$760, the average automobile insurance premium should be set at \$760 per year for the insurance company to break even on its claims costs. ◆

American roulette wheels have 38 slots numbered 00, 0, and 1 through 36 (Figure 5.26). You place a bet on a specific number by putting your wager on the numbered square on the roulette cloth, or layout. Bets may also be placed on more than one number or on combinations of numbers. The wheel is spun in one direction, and a ball is rolled in the opposite direction in a surrounding sloped bowl. When the ball slows sufficiently, it drops down into the numbered slots and bounces along until coming to rest on the winning number. If you had bet on the winning number, the croupier (the manager of the table) leaves your bet on the layout and adds to it 35 times as much as you bet. If you chose a wrong number, your wager and the other losing bets are gathered in with a rake.

If you bet \$100 on one number, what is your expected gain or loss?

SOLUTION This problem describes a probability experiment with numerical outcomes. The probability of winning is $\frac{1}{38}$ since there are 38 equally likely outcomes, namely, 00, 0, 1, 2, 3, . . . , 36; if you win, you win \$3500. The probability of losing is $\frac{37}{38}$; if you lose, you lose \$100. We put this information into Table 5.8 and compute the expected value. The expected value is $\frac{-200}{38} \approx -\$5.26$. Thus, for every \$100 bet, you should expect to lose \$5.26. In other words, on any given bet, you will lose \$100 or win \$3500; but for many bets, you should expect to lose an average of \$5.26.

TABLE 5.8

Value	-100		3500		
Probability	$\frac{37}{38}$		$\frac{1}{38}$		
Products	$\frac{-3700}{38}$	$+$	$\frac{3500}{38}$	$=$	$\mathbf{\frac{-200}{38}}$
					Expected Value

PROBLEM SET 5.3

1. Find the sample space for the experiment "the first die was even when two fair dice were rolled."
2. Find the sample space for the experiment "the second ball was white when two balls were drawn without replacement from a jar with 3 white balls and 2 red balls."
3. Two standard dice are rolled and the numbers added. Find the following:
   (a) the probability that at least one of the dice is a five.
   (b) the probability the sum is eight.
   (c) the probability that one of the dice is a five given the sum is eight.
4. A jar contains five white balls and three green balls. Two balls are drawn, in order, without replacement. Find the following:
   (a) the probability the second ball is green.
   (b) the probability the first ball was white given the second ball was green.
   (c) the probability the first ball was green given the second ball was green.
5. The diagram shows a sample space S of equally likely outcomes and events A and B.

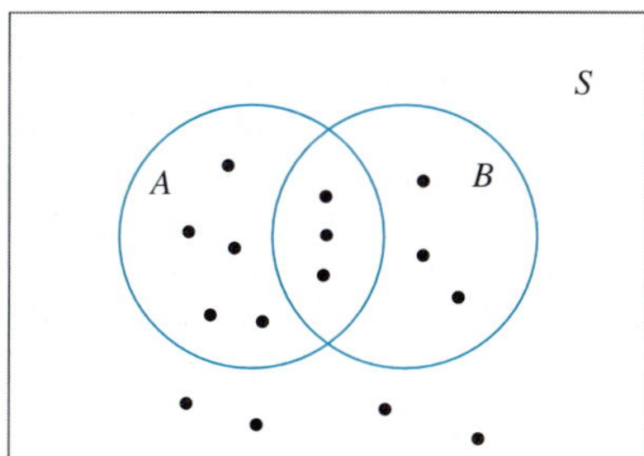

Find the following probabilities.
(a) $P(A)$ **(b)** $P(B)$
(c) $P(A|B)$ **(d)** $P(B|A)$

6. The diagram shows a sample space S of equally likely outcomes and events A and B.

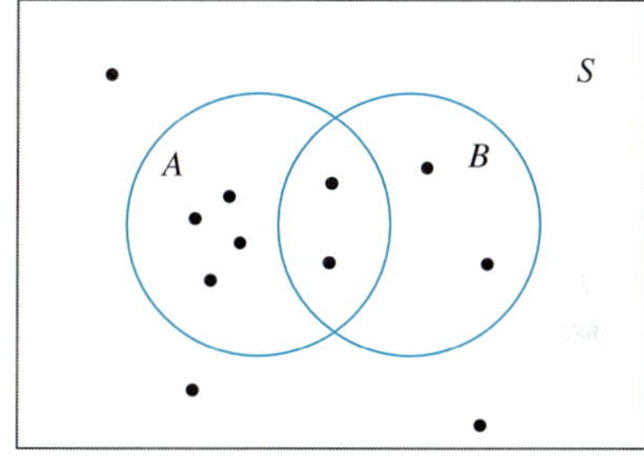

Find the following probabilities:
(a) $P(A \cup B)$ **(b)** $P(A \cap B)$
(c) $P(A|B)$ **(d)** $P(B|A)$

7. If A is the event that a person committed aggravated assault, and D is the event that a person is a drug dealer, then state *in words* what probabilities are expressed by each of the following:
   (a) $P(A|D)$ **(b)** $P(D|A)$
   (c) $P(\bar{A}|D)$ **(d)** $P(\bar{A}|\bar{D})$
8. If H is the event that a student completes her homework each night, and G is the event that a student gets good grades, then state *in words* what probabilities are expressed by each of the following:
   (a) $P(G|H)$ **(b)** $P(\bar{G}|H)$
   (c) $P(G|\bar{H})$ **(d)** $P(\bar{G}|\bar{H})$

9. Given is the probability tree diagram for an experiment.
The sample space $S = \{a, b, c, d\}$. Also, event $A = \{a, b, c\}$ and event $B = \{b, c, d\}$. Find the following probabilities:
(a) $P(A)$ (b) $P(B)$
(c) $P(A \cap B)$ (d) $P(A \cup B)$
(e) $P(A|B)$ (f) $P(B|A)$

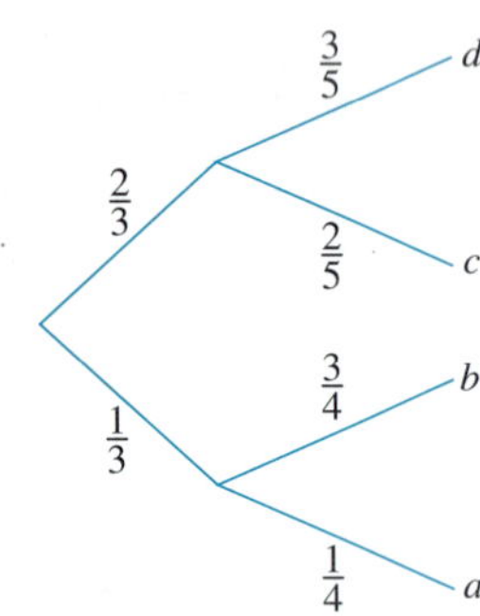

10. Box A contains 7 cards numbered 1 through 7, and box B contains 4 cards numbered 1 through 4. A box is chosen at random and a card is drawn. It is then noted whether or not the card is even. Given is the probability tree diagram for this experiment.

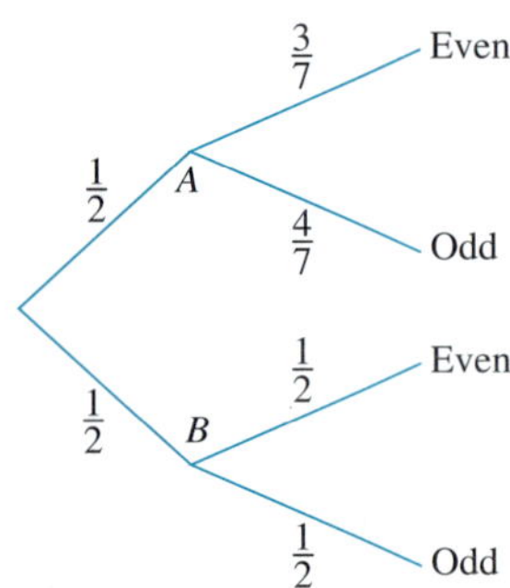

Find the following probabilities:
(a) P(the number is even).
(b) P(the number is odd).
(c) P(the number is even | it came from box A).
(d) P(the number is odd | it came from box B).

11. The spinner is spun once. (All central angles equal 60°.)

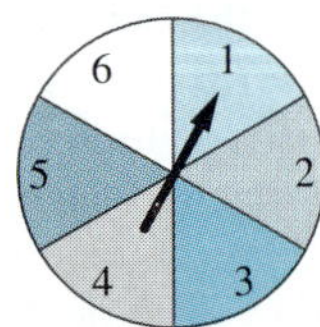

(a) What is the probability that it lands on 4?
(b) If you are told that it landed on an even number, what is the probability that it landed on 4?
(c) If you are told that it landed on an odd number, what is the probability that it landed on 4?

12. In a fuse factory, machines A, B, and C manufacture, respectively, 20, 45, and 35 percent of the total fuses. Of the outputs for each machine, A produces 5% defectives, B produces 2% defectives, and C produces 3% defectives. A fuse is drawn at random and is found to be defective.
(a) What is the probability it came from machine A?
(b) What is the probability it came from machine B?
(c) What is the probability it came from machine C?

13. A random sample of 400 adults are classified according to sex and education level achieved.

Education	Female	Male
Elementary school	64	53
High school	99	87
College	28	39
Graduate school	14	16

If a person is picked at random from this group, find the probability that
(a) the person is female given that the person has a graduate degree.
(b) the person is male given that the person has a high school diploma.
(c) The person does not have a college degree given that the person is female.

14. A study was performed to find out how the number of defective items produced varied between the day, evening, and night shifts.

	Day	Evening	Night
Defective	24	28	47
Not defective	279	224	165

If an item is picked at random, find the probability that:
(a) the item is defective given that it came from the night shift.
(b) the item is not defective given that it came from the day shift.
(c) the evening shift produced the item given that it was not defective.

15. Suppose a screening test for a certain virus is 95% accurate for both infected and uninfected persons. If 10% of the population is infected, find the following:
(a) the probability of a false positive.
(b) the probability of a false negative.
(c) the probability that a person is infected given that the test results are positive.

16. Suppose a screening test for a certain virus is 95% accurate for both infected and uninfected persons. If 2% of the population is infected, find the following:

(a) the probability the person is uninfected given that the test is negative.
(b) the probability the person is infected given that the test is positive.

17. Two classes at a university are studying modern Latin American fiction. Twenty of the 25 students in the first class speak Spanish, and 12 of the 18 students in the second class speak Spanish. If one student is selected at random from each of the classes, what is the probability that both will speak Spanish?

18. Two assembly lines are producing ink cartridges for a desktop printer: 5% of the cartridges produced by the first line are defective, while 10% of those from the other line are defective. If a cartridge is selected randomly from each line, what is the probability that neither one will be defective? Justify the reasoning for your answer.

19. Roll a die twice. Let A be the event a 3 occurs on the first roll, let B be the event that the sum of the two rolls is 7, and let C be the event that the same number is rolled both times. Use the definition of independent events to
(a) determine if events A and B are independent.
(b) determine if events A and C are independent.
(c) determine if events B and C are independent.

20. In a box of computer chips, there are two that are defective and five that are not defective. Two chips are selected, without replacement. Let A be the event that a defective chip is chosen first, let B be the event a nondefective chip is chosen second, and let C be the event that both chips are defective. Use the definition of independent events to
(a) determine if events A and B are independent.
(b) determine if events A and C are independent.

21. Make a table, as in Example 5.22, where the values in the top row are the number of girls possible in a family with four children, and the second row has the probabilities for those numbers. Assume that boys or girls are equally likely and the births are independent.

22. Suppose you play a game with the following rules: two fair dice are rolled; if the dice are different, you lose $2; if the dice are the same, you win an amount equal to $2 plus the sum of the dice. Make a table, as in Example 5.22, that includes the possible payoff values and their probabilities.

23. Referring to problem 21, what is the expected number of girls in a family of four children?

24. Referring to problem 22, what is the expected value of the game?

25. Suppose a standard die is constructed so that the value 6 occurs four times as often as any other value. Compute the expected value for one roll of this "loaded" die.

26. A golf course makes a profit of $900 on fair days, but loses $250 each day of bad weather. The probability for bad weather is 0.35. Find the expected value for the profit or loss on a single day.

27. For any day of the week, the highway department determined that the probabilities of 0, 1, 2, or 3 fatal car accidents are, respectively, 0.52, 0.21, 0.18, and 0.09. Find the expected number of fatal car accidents on a single day.

28. Find the expected number of games that would be played between two teams of equal ability in a best of 3 out of 5 game series.

29. In a lottery there are 50 prizes of $10, 10 prizes of $15, 5 prizes of $30, and 1 prize of $50. Suppose that 1000 tickets are sold.
(a) What is a fair price to pay for one ticket?
(b) What should the price of one ticket be if, on the average, people lose $0.50?
(c) If all 1000 tickets are sold, what can the lottery expect to gain if ticket prices are $2?

30. A church conducts a raffle to raise money for the building fund. One thousand tickets are placed in a box. On each ticket is placed one of the following dollar amounts: $0, $5, $10, $50, or $200.

# of Tickets	Dollars
1	200
4	50
10	10
20	5
965	0

(a) Find the expected value if each ticket costs $3.
(b) If the tickets cost $3 each and all tickets are sold, what will the church make for the building fund?
(c) If you buy a single ticket, what is the probability that you do not win any money?

31. For visiting a resort, you will receive one gift. The probabilities (and manufacturer's suggested retail value) of each gift are: gift A, 1 in 52,000 ($9272.00); gift B, 25,736 in 52,000 ($44.95); gift C, 1 in 52,000 ($2500.00); gift D, 3 in 52,000 ($729.95); gift E, 25,736 in 52,000 ($26.99); gift F, 3 in 52,000 ($1000.00); gift G, 180 in 52,000 ($44.99); gift H, 180 in 52,000 ($63.98); gift I, 160 in 52,000 ($25.00). Find the expected value of your gift.

32. According to a publisher's records, 20% of the books published break even, 30% lose $1000, 25% lose $10,000, and 25% earn $20,000. When a book is published, what is the expected income for the book?

33. Suppose that the InsureAll Insurance Company has broken down the yearly automobile claims for high-risk male and female drivers ages 16 through 25.

Amount of Claim (nearest $2000)	Male Probability	Female Probability
$0	0.43	0.61
$2000	0.23	0.20
$4000	0.16	0.09
$6000	0.09	0.07
$8000	0.06	0.02
$10,000	0.03	0.01

(a) Find the expected claim for male drivers.
(b) Find the expected claim for female drivers.
(c) Should male or female drivers be charged a higher premium? Why?

34. A teacher has a stack of test papers. The paper for any student is equally likely to be anywhere in the stack of papers. If the teacher looks through the papers one at a time starting from the top, how many papers should the teacher expect to look through before finding the paper of a particular student if the stack contains
(a) 5 papers? (Hint: Make a table and compute the expected value.)
(b) 10 papers?
(c) 15 papers?
(d) n papers? [Hint: Can you generalize from parts (a) through (c)?]

EXTENDED PROBLEMS

35. In an effort to fight an apparent growth in the use of illegal drugs, many companies, professional sports teams, and schools have established drug-testing programs. Write a brief report on this topic. How widespread is the use of drug-testing programs? How are they administered? What kind of follow-up programs have been established?

36. What tests are being used for random drug testing programs? How effective are they? What data are available on the incidence of false positives and false negatives? If positive results are obtained on a test, is a second test used for confirmation? Does your college or university use drug testing as a condition of participation in any of its programs?

37. Make a survey of 10 or more of the major employers in your city or community. What percentage of the companies have policies that require drug testing, and what percentage of workers are subject to the policies?

38. What are the legal and constitutional questions related to drug testing? Have any policies been invalidated? If so, why?

39. Screening tests are used for a variety of purposes other than the detection of drugs. Investigate the relationship between false positives and false negatives. Suppose that a screening test is $T\%$ accurate for those that have some characteristic such as a disease and that $P\%$ of the population have this characteristic. What must the effectiveness level of the screening test be with those who don't have the characteristic, to ensure that the number of true positives is greater than the number of false positives?

40. In 1995 the O.J. Simpson trial focused the attention of the nation on DNA testing. How accurate are these tests in general, and what circumstances could lead to a false match? Write a brief report on DNA testing and the probabilities involved.

Chapter Five Problem

You are sitting in the student lounge when an individual offers you two wagers. First, he tells you to take a penny out of your pocket and balance it on the tabletop. If you hit the table with your fist, then the penny will land on a side. If it lands with the tail side up, he will pay you $10, but if it lands with the head side up, then you must pay him $5. Since it is your penny and you will hit the table, there can be no tricks. You should be a winner, on average, for this wager. Should you take the bet?

Second, he says there are about 30 people in the room, and offers to bet you $5 that at least two people have the exact same birthdate. Since there are 365 possible birthdates this seems like a good bet for you. Should you take this bet?

SOLUTION

Strategies: Do a Simulation and Look for a Formula

When anyone offers a bet that appears too good to be true, it probably is. These wagers are no exception. The bet with the penny looks very good. There are two possibilities: either the coin will land on a head or a tail. The coin is an ordinary penny since it came from your pocket. Theoretical probability tells us that half the time we should get heads and half the time we should get tails. Our expected gain is therefore $10 \times \frac{1}{2} + (-5) \times \frac{1}{2} = \$2.50$ per play. This seems quite good, but is there something wrong here? To be certain the game is in your favor, you should excuse yourself and do a simulation. Secretly, take the pennies that you have, balance them on edge, and cause them to fall over by hitting the surface. It will turn out that they land with a head showing nearly every time. You would have lost a lot of money playing this game! Further research would show that pennies are not perfectly cylindrical, but have a cross section that is a trapezoid.

To understand the second wager, we need to know what the probability is that 30 people have the same birthdate. We assume that all 365 birthdates, month and day, have the same probability. First, however, we will do a similar problem with smaller numbers.

If there are three people, what is the probability that at least two of them are born in the same season (Spring, Summer, Fall, Winter)? This is a complicated event to work with. Instead, we calculate the probability of the *complement* of this event; that is, we find the probability that each person was born in a different season. Each person may have a birthdate in any one of the four seasons, so the sample space has $4 \times 4 \times 4$ possible outcomes. To count that number of outcomes in the event that no two people have birthdates in the same season, notice that the first person may have a birthdate in any one of 4 seasons. Then the second person can only have 3 possible seasons for a birthdate (to be different from the first), and the third person can only have 2 possible seasons (to be different from the first two). This event has $4 \times 3 \times 2$ outcomes in it. The probability that no two people have the same season as birthdate is $\frac{4 \times 3 \times 2}{4 \times 4 \times 4} = \frac{3}{8}$. The probability that at least two people have a birthdate in the same season is $1 - \frac{3}{8} = \frac{5}{8}$. This shows us how to find a formula for the more complicated case. If there are 30 people, then each of them could have any one of 365 birthdates. Thus, the sample space has 365^{30} outcomes. The event that no two people have the same birthdate has $365(365 - 1)(365 - 2) \ldots (365 - 29)$ outcomes. The probability that no two people have the same birthdate is the quotient $\frac{365(365 - 1)(365 - 2) \ldots (365 - 29)}{365^{30}} \approx 0.294$. Thus, the probability that at least two do have the same birthdate is approximately $1 - 0.294 = 0.706$. Again this is a bad wager since you would only win about 30% of the time.

Chapter Five Review

Key Ideas and Questions

The following questions review the main ideas of this chapter. Write your answers to the questions, and refer to the pages listed to make certain that you have mastered these ideas.

1. Describe the main features of a probability model with equally likely outcomes, including the sample space, events, outcomes, and how to assign probabilities to events. **238, 240** Illustrate this with the experiment of spinning a roulette wheel with 38 slots.
2. Compare theoretical and experimental probability. **239, 240**
3. How does a probability tree diagram simplify computations in comparison with a tree diagram? **249, 252**
   What are the additive and multiplicative properties of probability tree diagrams? **253, 255** Illustrate this with an example.
4. What is the formula for the conditional probability of an event A given that event B has occurred? **262** Illustrate this with an example.
5. What does the independence of two events mean? Write a description.

What is the formula that independent events must satisfy? **265**
If A and B are independent, then what is the conditional probability of A given B? **266**

6. How do you find the expected value of an experiment?
What is the interpretation of the expected value of an experiment? **267**

Vocabulary/Notation

Following is a list of key vocabulary, notation, and ideas for this chapter. Mentally review each of these items, write down the meaning of each term, and use it in a sentence. Then, refer to the pages listed by number and restudy any material that you are unsure of before solving the Chapter Five Review Problems.

Section 5.1

Experiment 238	Experimental Probability 239	Intersection 242
Outcome 238	Equally Likely Outcomes 240	Mutually Exclusive 242
Sample Space 238	Theoretical Probability 241	Complement of the Event E 242
Event 238	Fair Coin/Fair Die 241	Properties of Probability 244
	Union 242	

Section 5.2

Tree Diagram 249	Fundamental Counting Principle 250	Drawing with Replacement 254
One Stage Tree 249	Probability Tree Diagram 252	Drawing without Replacement 254
Two Stage Tree 249	Additive Property of Probability Tree Diagrams 253	Multiplicative Property of Probability Tree Diagrams 255
Primary Branches 249		
Secondary Branches 249		

Section 5.3

Conditional Sample Space 261	Independent Events 265	Expected Value 267
Conditional Probability 262		

Chapter Five Review Problems

1. Suppose that prizes are put into 100,000 boxes of cereal. One box will have a grand prize worth \$10,000. One hundred boxes will have prizes worth \$20. All the other boxes will have prizes that are nearly worthless. Suppose you buy a box of cereal and check to see if you won a prize. What is the sample space of this experiment? What is the probability you win the grand prize? What is the probability that you win one of the \$20 prizes? What is the probability that you win at least \$20?

2. Describe the sample space for the experiment in which a fair coin is tossed four times. List the outcomes in the event that "more heads appear than tails." What is the (theoretical) probability of this event? List the outcomes in the event that "only tails appear." What is the probability of this event? Without listing or counting the outcomes, find the probability that "at least one head appears." (Hint: Use Property 6, pg. 244).

3. Suppose that a jar has four coins: a penny, a nickel, a dime, and a quarter. You remove two coins at random. Describe the sample space. Describe the following events and compute their probabilities:

(a) The event A that you get less than 12 cents.
(b) The event B that you get the quarter.
(c) The event C that you get the dime.
(d) Which pairs of these events are mutually exclusive?
(e) Compute $P(A \cup B)$ and $P(B \cup C)$.

4. Suppose that you have three pairs of slacks and five shirts. How many possible shirt-slack outfits can be put together from these?

5. What is the probability of getting a sum of 4 when you roll two fair standard dice?

6. What is the probability of drawing two aces in a row from a thoroughly shuffled deck?

7. Suppose that 2 cards are drawn from a deck of 16 cards consisting of the ace, king, queen, and jack of each suit. Find the probability that both cards are the same suit by drawing a probability tree diagram and only keeping track of the suits.

8. A jar contains 20 red marbles, 40 blue marbles, 30 green marbles, and 10 white marbles. Describe the probability model for drawing a marble and noting its color. You are told that either a red or a blue marble has been chosen. What is the conditional probability that this marble is blue?

9. A jar contains 4 marbles: 2 red and 2 blue. Draw marbles one at a time, without replacement, until you have 2 marbles that have the same color. Draw a probability tree diagram to represent this experiment. What is the probability that the first marble drawn is blue? What is the probability that the second marble drawn is blue? What is the probability that three drawings are needed and the final marble drawn is blue? What is the probability that only two drawings are necessary?

10. In problem 9 suppose that it took three drawings to get 2 marbles of the same color and that this color was blue. What is the conditional probability that the first marble drawn was blue?

11. A student has a 0.6 probability of studying for a True/False test. If the student studies, then she has a 0.8 probability of getting an A; if she does not study, then she has a 0.3 probability of getting an A. Make a probability tree diagram for this experiment. What is the probability that she gets an A? If she gets an A, what is the conditional probability that she studied?

12. A jar contains two red jelly beans and two blue jelly beans. Two jelly beans are chosen without replacement from the jar. Let $A =$ the event that both are the same color, $B =$ the event that the first jelly bean chosen was red, and $C =$ the event that at least one jelly bean chosen was blue. Compute the probabilities of these events. Which, if any, pairs of these events are independent?

13. The jar in problem 3 contained a penny, a nickel, a dime, and a quarter. If we take one coin out at random, what is the expected amount of money we get? Suppose we choose two coins at random. What is the expected total amount of money we get?

14. Claims for towing insurance cost the company either \$30 or \$55. The probability of a claim for \$30 in a year is 0.12 and for \$55 in a year is 0.08. The insurance company wishes to make \$10 per year per claim to cover administrative charges. How much should they charge for a premium?

CHAPTER 6

CONSUMER MATHEMATICS—BUYING AND SAVING

ANALYST PREDICTS STOCK MARKET PERFECTLY

Suppose you are considering investing in the stock market. However, you have not developed an investment strategy, nor have you selected a financial adviser. You receive a letter from a stock market consultant, J. J. Herringbone, touting his services. According to the letter, he has astounded the experts with his ability to predict the volatile commodities market. This is a market in which 90% of the investors lose money, yet he has successfully predicted the movements of prices over 80% of the time! As an inducement to subscribe to his services, he provides you with six predictions regarding the market for the next ten days. When the ten days are up, you are surprised to see that Herringbone has been correct on all of his predictions!

The next week, you receive another letter from Herringbone with an offer of his services. Now that the accuracy of his system has been verified, he is offering yearly memberships that will include a newsletter with future predictions. These memberships will cost $5000 per year. The letter ends with a postscript: "Individuals who have followed my earlier tips have already recovered the full cost of their membership and profited substantially. This is a limited time offer! Do not miss out on this once in a lifetime opportunity."

CHAPTER GOALS

1. Calculate simple and compound interest.
2. Become familiar with various types of loans and how to compute finance charges (interest).
3. Learn about financing a house.
4. Find the future value of annuities.

Does this sound like a good offer? If this is a fraud, how was Herringbone able to predict the market 6 times with such accuracy? The probability of doing this by chance is 1 in 64, which makes this very unlikely. Should you run out with the money you have been saving for a down payment on a house? No, you should not.

If something sounds too good to be true, then it probably is. Herringbone could have chosen 64 different cities in which to promote his services and newsletter. For his first prediction, half of the cities are sent the information that stock A will rise, and the other half got the news that stock A will fall. He will be correct in 32 cities. Herringbone could send 16 of these cities the prediction that stock B will rise and the other 16 the prediction that the stock will fall. He will be correct in 8 cities. He continues this process until he is left with one city—yours—in which he has made only correct predictions. Based only on the movement of the stock (rise or fall), Herringbone can develop 64 sets of predictions, one of which is 100% correct, and another six sets of predictions that are correct in five out of six predictions (83%).

This offer is deceptive and may be a crime. The world of stocks, bonds, and other investments is very complex. It is replete with offers suggesting methods you can use to get rich quickly with "little risk," which are just as strange as the above scam, but not necessarily illegal. The purpose of this chapter is to teach you how to use your money carefully, although it will not discuss speculative investments. The chapter shows how interest works, how finance options and strategies can help you, and the advantages and dangers in certain types of borrowing.

THE HUMAN SIDE OF MATHEMATICS

The secret of acquiring great wealth on the stock market is well known: buy low and sell high. The trick is knowing what the market will do. Charles Dow (1851–1902) was a financial journalist who believed that the stock market could be understood using mathematical principles. Dow was a founder of Dow Jones & Company, which became a financial empire. They published an index averaging various stocks as an indicator of how the market is doing as a whole. This publication became the precursor of *The Wall Street Journal*. He invented the "Dow Theory" of buying stocks, which is based entirely on numerical market information. Dow theorists are active in the market to this day. He was a man of great confidence and energy. As a young man he applied for a reporter's job at the *Providence Journal* and was told there was nothing for him to do. He replied that this was all right—he knew what news was and would find things to do. He became their star reporter for the next five years.

Charles Dow

Another person who tried to understand the stock market using mathematical principles was Ralph Elliott (1871–1948). Elliott was an engineer by profession. In the 1930s, he studied the Great Pyramid at Giza and concluded that the design for the structure was based on the Fibonacci numbers, which include 2, 3, 5, 8, 13, 21, 34. He decided that this progression of numbers held the secret to predicting the ups and downs of the stock market. In 1927, Elliott contracted a severe illness and spent many years in convalescence. It was during this time that he developed a theory on price movements now known as the *Elliott Wave Theory*. Before you run out and purchase the secrets of Dow and Elliott you should know that professional practitioners of these methods often come up with different predictions based on the same information. Today, there are thousands of people (called chartists or technicians) with advanced degrees trying to forecast the market using mathematics. In the high-powered world of the stock market, even the slightest advantage can often translate to instant riches.

Ralph Elliot

6.1 INTEREST

You discover that you are the only direct descendant of a man who loaned the Continental Congress $1000 in 1777. However, he was never repaid nor have any of his descendants received repayment. You think it is about time to get the family money back. How much should you demand from the U.S. government? Use an interest rate of 6% and a compounding period of 3 months.

HISTORY

In the past, charging interest on borrowed money was often considered evil, and, in particular, was long prohibited by the Catholic Church. One way people got around the law against paying interest was to borrow in one currency and repay in another, the interest being disguised in the exchange rate.

Not having the money you want when you want it is a too common experience for most people, whether it's having enough money for tuition, buying essentials for daily living, or buying a new car. In this chapter we look at the different ways in which money can be borrowed (or saved), beginning with the familiar concept of simple interest and progressing to fairly complex financial instruments called annuities.

SIMPLE INTEREST

Many people deposit savings in a bank or similar financial institution; likewise, many borrow money for major purchases such as a house or a car. Credit cards are used routinely to borrow smaller amounts of money to make purchases. We expect to receive interest income on our savings, and we also expect to pay interest on the money we borrow. Thus it is important to understand how interest is calculated, so that you can make informed decisions.

Not surprisingly, the simplest type of interest calculation is called **simple interest.** To calculate simple interest, you must know the amount of money on which the interest is being charged, the interest rate, and the time period over which the interest is being charged. Suppose you borrow $1000 at 5% simple interest for two years. The amount of money on which the interest will be charged is called the **principal,** in this case $1000. The **interest rate,** expressed as a percent, is the percentage of the principal that will be paid each year, in this case 5%. A unit of time other than years could be specified, but "per year" is understood if nothing else is said. Since the interest is percent per year, one must know the **time period** in years over which the interest will be charged, in your case 2 years. The formula for calculating interest is given next.

Simple Interest Formula

If I represents interest, P the principal, r the interest rate expressed as a decimal, and t the time, then

$$I = Prt.$$

In words,

$$\text{interest} = \text{principal} \times \text{rate} \times \text{time (in years)}.$$

The (interest) rate in this formula is the annual rate and is normally given as a percent. For the example of borrowing \$1000 at 5% simple interest for two years, you calculate the interest to be

$$\text{interest} = \$1000 \times 0.05 \times 2 = \$100.$$

Note: $r\% = (0.01)r$.

EXAMPLE 6.1 Find the interest on a loan of \$100 at 6% simple interest for 1, 2, and $2\frac{1}{2}$ years.

SOLUTION For one year the interest is $\$100 \times 0.06 \times 1 = \$6$. The interest for 2 years is $\$100 \times 0.06 \times 2 = \$12$. For each year the interest is 6% of the principal of \$100, that is, \$6 per year. Thus for $2\frac{1}{2}$ years the interest would be $\$6 \times 2\frac{1}{2} = \$15$. ◆

The next example shows how a calculator with algebraic logic may be used to do calculations.

EXAMPLE 6.2 What is the simple interest on a \$500 loan at 12% from June 6 until October 12 (in a non-leap year)?

SOLUTION To find the time period as measured in years, we first add up the days: $(30 - 6) + 31 + 31 + 30 + 12 = 128$. Since the problem concerns a non-leap year, there are 365 days in the year, so the time period is $\frac{128}{365}$ of a year. Next compute the interest: 500 [×] 0.12 [×] 128 [÷] 365 [=] [21.041096], or \$21.04. ◆

HISTORY

The length of the "year" has varied considerably over recorded history. The earliest Babylonian year was determined by the occurrence of the lunar eclipse, which happens approximately every six months. This may explain the superhuman ages attained by certain Biblical personages. Around 1800 B.C., the Assyrians had a year of exactly 360 days divided into 12 equal months, just as is used in computing ordinary interest.

The preceding interest calculation may seem complicated. Since all the arithmetic can be done with a calculator, the hardest part is determining the number of days in the time period. To simplify the calculation, a type of simple interest called ordinary interest was created. **Ordinary interest** is based on two accepted conventions: (i) each month is assigned 30 days and (ii) a year is assigned 360 days. If we had used ordinary interest, the time period in the last example would have been 4 months from June 6 to October 6 plus 6 days from October 6 to October 12. That gives us $(4 \times 30) + 6 = 126$ days out of a 360-day year. The interest is then

500 [×] 0.12 [×] 126 [÷] 360 [=] [21], or \$21.

Since ordinary interest makes individual days more expensive, it is not appropriate for computing simple interest on a short time period in days. Because of the widespread use of pocket calculators, the simplification of dividing by 360 instead of 365 or 366 is unimportant. But when the time period is in months as is often the case for short-term borrowing, *it is standard to treat all months as exactly $\frac{1}{12}$ of a year.*

EXAMPLE 6.3 What is the ordinary interest on a \$500 loan at 12% from June 1992 through September 1993?

(Note: our use of "through" will mean that both June and September are included.)

SOLUTION Counting all the months involved as complete months, there are 16 months from June 1992 through September 1993. Hence we have

$$\$500 \times 0.12 \times \frac{16}{12} = \$80.$$ ◆

COMPOUND INTEREST

A problem dating back to the 1200s asks how many rabbits you would have if you started with just 2 and let them reproduce. Assume that a pair of rabbits produces a pair of offspring each month and that each pair of rabbits produces their first offspring at age 2 months. Month by month the number of pairs of rabbits is

Month	1	2	3	4	5	6	7	8	9	10	11	12	13	14	15	16
Pairs of Rabbits	1	2	3	5	8	13	21	34	55	89	144	233	377	610	987	1597

TIDBIT

This problem was posed almost 800 years ago by Leonardo of Pisa who used the pen name Fibonacci. The sequence is known as the *Fibonacci sequence*.

After just 16 months, you have almost 1600 pairs of rabbits!

Like the young rabbits becoming part of the adult breeding stock, reinvesting interest income makes the amount of your money grow faster. This phenomenon, called **compound interest,** is a powerful way to make money grow. To calculate compound interest you need the same information that is used to calculate simple interest, namely, the principal, the interest rate, and the time period. In addition, you need the **compounding period.** The principal is the initial amount deposited in the account. The amount in the account at any time will be called the **balance.** (So the principal is the **initial balance.**) For each compounding period, calculate simple interest on the balance in the account at the start of the compounding period. At the end of the compounding period, add the interest to the balance. This becomes the principal for the next compounding period. Typical compounding periods are 1 month, 3 months, and 6 months; in these cases, ordinary interest computations are used instead of the more laborious simple interest method.

EXAMPLE 6.4 Given \$1000 principal, a 10% interest rate, and a 6-month compounding period, find the balance after 2 years.

SOLUTION Ordinary interest at 10% (= 0.10) for 6 months $\left(\frac{1}{2}\text{ year}\right)$ on \$1000 is

$$\$1000 \times 0.10 \times \tfrac{1}{2} = \$50.$$

At the end of the 6 months this \$50 is added to the balance, so the new balance is \$1050. Ordinary interest for the next 6 months at 10% on \$1050 is

$$\$1050 \times 0.10 \times \tfrac{1}{2} = \$52.50.$$

The new balance at the end of 1 year is \$1050 + \$52.50 = \$1102.50. Ordinary interest for 6 months at 10% on \$1102.50 is

$$\$1102.50 \times 0.10 \times \tfrac{1}{2} = \$55.13.$$

Thus the new balance at the end of $1\frac{1}{2}$ years is \$1157.63. Ordinary interest for 6 months at 10% on \$1157.63 is

$$\$1157.63 \times 0.10 \times \tfrac{1}{2} = \$57.88.$$

The new balance at the end of 2 years is \$1157.63 + \$57.88 = \$1215.51.

Notice that in the preceding example, simple interest at 10% on \$1000 for 2 years is

$$\$1000 \times 0.10 \times 2 = \$200,$$

so compound interest gave \$15.51 more than simple interest. Compound interest is always larger than simple interest at the same rate, and the longer the period, the more striking the difference.

Computing Compound Interest by a Formula

Calculating compound interest is not as complicated as it seems. The important thing to realize is that the balance at the end of a compounding period can be computed by simply multiplying the balance at the start of the compounding period by

$$1 + (\text{interest rate} \times \text{compounding period}),$$

where the interest rate is expressed as a decimal, and the compounding period is expressed in years. For the previous example, the interest rate is 10% and the compounding period is 6 months. Thus you multiply by:

$$1 + \left(0.10 \times \tfrac{1}{2}\right) = 1.05.$$

Since the two years in the example represents 4 compounding periods, we multiply the initial balance by 1.05 four times to get the final balance:

1000 [×] 1.05 [×] 1.05 [×] 1.05 [×] 1.05 [=] [1215.50625]

which rounds to \$1215.51.

Repeated multiplication can be written using power or exponential notation. So

$$1.05 \times 1.05 \times 1.05 \times 1.05 = 1.05^4,$$

and the final balance in the example can be obtained by using the [x^y] (or [y^x]) key of a scientific calculator as follows:

1.05 [x^y] 4 [×] 1000 [=] [1215.50625]

and then rounding-off to the nearest penny.

> **Compound Interest Formula**
>
> If P represents the principal, r the interest rate expressed as a decimal, n the number of equal compounding periods (in a year), and t the time in years, then the final amount, A, is given as
>
> $$A = P \times \left(1 + \tfrac{r}{n}\right)^{nt}.$$

EXAMPLE 6.5 Find the final balance in the following savings accounts having an initial balance of \$2457.

(a) simple interest at 4.5% for 3 years
(b) interest at 4.5% compounded every 4 months for 3 years
(c) interest at 4.5% compounded monthly for 3 years

SOLUTION

(a) 2457 [×] (1 [+] 0.045 [×] 3) [=] [2788.695] (Note: The 0.045 is multiplied by 3, not the 1 + 0.045.)

(b) When compounding every 4 months, $n = 3$.

2457 [×] (1 [+] 0.045 [÷] 3) [x^y] (3 [×] 3) [=] [2809.30917]

(Note: By the Compound Interest Formula, the 0.045 was divided by 3 since there are 3 four-month periods per year.)

(c) When compounding monthly, $n = 12$.

2457 [×] (1 [+] 0.045 [÷] 12) [x^y] (12 × 3) [=] [2811.416924] ◆

There are two important observations that can be made with respect to the solution of Example 6.5.

1. Most scientific calculators use algebraic logic. If yours does not, you will have to make adjustments in your keystrokes. For example, in (b), you might try the following:

0.045 [÷] 3 [+] 1 [=] [x^y] (3 [×] 3) [=] [×] 2457 [=] [2809.30917]

2. The correct use of parentheses is crucial.

ANNUAL YIELD—THE EFFECT OF THE COMPOUNDING PERIOD

To compare different savings plans, you need to have a common basis for making the comparisons. The **annual yield** provides such a basis. The annual yield is the simple interest rate that would have earned the same amount of interest in one year. To find the annual yield, the easiest approach is to compute what happens to $100 over 1 year.

EXAMPLE 6.6 What is the annual yield on a savings account paying 12% with a compounding period of 3 months?

SOLUTION Starting with $100 principal, we use the Compound Interest Formula to find the balance: $\$100 \times (1 + \frac{0.12}{4})^4 = \$112.55$. Since the balance in the account has increased by $12.55 over the year, the annual yield is 12.55%. This compares to the stated rate of 12%. ◆

TIDBIT

The characteristic of being convertible to cash is called *liquidity*. Often, to obtain a high yield you must sacrifice liquidity.

In this example, the dollar increase in the account translates exactly to the annual yield in percent; this is the reason for considering a $100 initial balance. Financial institutions are required to inform the consumer of the annual yield of their various savings options to help consumers make informed decisions. The annual yield should be an important factor in your choice of savings account. Some accounts do not allow you to withdraw your money at will, a provision that should also be an important consideration.

In Example 6.6, $100 was used for convenience in our computation. More formally, the annual yield for each dollar on deposit is given by $y = A - 1$, where A is the balance, with interest, generated by $1.

Table 6.1 Annual Yield Table

	Compounding Periods						
Percent	1 (simple)	2 (semi-ann)	4 (quarterly)	12 (monthly)	365 (daily)	1000	Continuously Compounded
5	5.00000	5.06250	5.09453	5.11619	5.12675	5.12698	5.12711
6	6.00000	6.09000	6.13636	6.16778	6.18313	6.18346	6.18365
7	7.00000	7.12250	7.18590	7.22901	7.25010	7.25056	7.25082
8	8.00000	8.16000	8.24322	8.29995	8.32776	8.32836	8.32871
9	9.00000	9.20250	9.30833	9.38069	9.41621	9.41699	9.41743
10	10.00000	10.25000	10.38129	10.47131	10.51558	10.51654	10.51709
11	11.00000	11.30250	11.46213	11.57188	11.62596	11.62713	11.62781
12	12.00000	12.36000	12.55088	12.68250	12.74746	12.74887	12.74969
13	13.00000	13.42250	13.64759	13.80325	13.88020	13.88188	13.88284
14	14.00000	14.49000	14.75230	14.93420	15.02429	15.02625	15.02738
15	15.00000	15.56250	15.86504	16.07545	16.17984	16.18212	16.18342

HISTORY

In 1777 Jacob DeHaven loaned General George Washington \$50,000 in gold and \$400,000 worth of supplies. When offered Continental money as repayment, DeHaven refused, insisting on gold. He was never repaid. In 1989 members of DeHaven's family sued for repayment. Because of compound interest the amount involved has increased to about \$150 billion.

Substituting $P = 1$ (for \$1), $t = 1$ (for 1 year), and $y = A - 1$ in the Compound Interest Formula, we obtain the following formula. The resulting decimal can be directly translated as a percentage.

Annual Yield Formula

If y represents the annual yield, r the interest rate expressed as a decimal, and n the number of equal compounding periods (in a year), then

$$y = \left(1 + \frac{r}{n}\right)^n - 1.$$

There are two factors that determine the annual yield: the interest rate, r, and the compounding period, n. The annual yield is always at least as large as the interest rate. How much larger it is depends on how often the interest is compounded. Table 6.1 shows the annual yield of various combinations of interest rates and compounding periods.

In the table, the number of compounding periods per year is indicated in the top row, and the interest rates are in the left column. As you read across each row, you can see that the yield keeps getting larger and larger as the number of compounding periods gets larger. However, there is a limit to this growth, and the yield levels out, approaching the well-defined limit in the far right column. For the limiting value, the number of compounding periods in a year is infinite, and the interest is said to be **continuously compounded.** If r is the interest rate expressed as a decimal, the annual yield for continuously compounded interest is

$$e^r - 1,$$

which is easily computed on any calculator with the $\boxed{e^x}$ key.
Note: the value of e is approximately 2.72.

EXAMPLE 6.7 Use (a) the table and (b) a calculator to find the annual yield of an account paying 5% interest compounded monthly.

SOLUTION

(a) In the intersection of the row for 5% interest and the column for monthly compounding, we find 5.11619, so the annual yield is 5.12%.

(b) $y = (1$ [+] 0.05 [÷] 12) [x^y] 12 [−] 1 [=] [0.0511619], or, rounding up, 5.12%. ◆

INITIAL PROBLEM SOLUTION

You discover that you are the only direct descendant of a man who loaned the Continental Congress $1000 in 1777. However, he was never repaid nor have any of his descendants received repayment. You think it is about time to get the family money back. How much should you demand from the U.S. government? Use an interest rate of 6% and a compounding period of 3 months.

SOLUTION Historically, a typical interest rate has been approximately 6% and a compounding period of 3 months has also been typical. To apply the compound interest formula we replace P by $1000, r by 0.06, and n by 4. The time period is from 1777 to 1996, which is 1996 − 1777 = 219 years, so we replace t in the formula by 219 and get

$$\$1000 \times [1 + 0.06 \div 4]^{(219 \times 4)} = \$1000 \times 1.015^{876} = 461{,}586{,}406.$$

You should demand $461,586,406 (but expect to get much less).

PROBLEM SET 6.1

In problems 1 through 12, P is the principal and r is the annual rate as a percent.

1. Find simple interest.
- **(a)** $P = \$600$ $r = 7\%$ $t = 3$ years
- **(b)** $P = \$400$ $r = 12\%$ $t = 5$ years
- **(c)** $P = \$1235$ $r = 7\%$ $t = 10$ years

2. Find simple interest.
- **(a)** $P = \$525$ $r = 5\%$ $t = 2$ years
- **(b)** $P = \$300$ $r = 3\%$ $t = 5$ years
- **(c)** $P = \$7934$ $r = 4.15\%$ $t = 8$ years

3. Find ordinary interest.
- **(a)** $P = \$800$ $r = 6\%$ $t = 36$ months
- **(b)** $P = \$1400$ $r = 12\%$ $t = 30$ months
- **(c)** $P = \$1235$ $r = 7.5\%$ $t = 20$ months

4. Find ordinary interest.
- **(a)** $P = \$525$ $r = 5\%$ $t = 48$ months
- **(b)** $P = \$300$ $r = 3\%$ $t = 40$ months
- **(c)** $P = \$7934$ $r = 4.15\%$ $t = 33$ months

5. Find the simple interest on a $650 loan at 6% from January 10 to November 16 (in a non-leap year).

6. Find the simple interest on a $1200 loan at 10% from February 10 to June 28 (in a leap year).

7. Find the simple interest on a $1600 loan at 7.5% from February 12 to November 6 (in a leap year).

8. Find the simple interest on a $2000 loan at 9% from March 15 to October 8 (in a non-leap year).

9. Find ordinary interest.
- **(a)** $P = \$800$, $r = 6\%$, from 6/92 through 2/94
- **(b)** $P = \$950$, $r = 7.5\%$, from 10/92 through 3/95

10. Find ordinary interest.
- **(a)** $P = \$700$, $r = 5\%$, from 1/92 through 3/95
- **(b)** $P = \$385$, $r = 8.5\%$, from 3/90 through 2/93

11. Find ordinary interest.
- **(a)** $P = \$2350$, $r = 5.6\%$, from 7/92 through 9/95
- **(b)** $P = \$7200$, $r = 5\%$, from 1/92 through 3/95

12. Find ordinary interest.
- **(a)** $P = \$7500$, $r = 6.1\%$, from 5/92 through 11/94
- **(b)** $P = \$1250$, $r = 5\%$, from 12/92 through 3/95

13. Find the balance of a $2000 account that earns 5% simple interest for 3 years.

14. Find the balance of a $3579 account that earns 7.25% simple interest for 5 years.

15. Find the balance for a $3000 deposit that earns 6.5% simple interest for 5 years.

16. Find the balance due on a $4550 loan that earns 7.5% simple interest for 3 years.

17. Find the total owed on a $2575 loan that earns 5.75% ordinary interest for 40 months.

18. Find the balance for a $2400 deposit that earns 6.25% simple interest for 28 months.

19. Given $1500 principal, a 12% interest rate, and a 3-month compounding period, find the balance after one year. Calculate the new balance at the end of each compounding period.

20. Given $2000 principal, an 8% interest rate, and a 6-month compounding period, find the balance after one year. Calculate the new balance at the end of each compounding period.

21. Which is the best deal over 3 years?
(a) investing at 5% compounded annually
(b) investing at 4.95% compounded semiannually
(c) investing at 4.9% compounded monthly

22. Which is the best deal over 5 years?
(a) investing at 8% compounded annually
(b) investing at 7% compounded monthly
(c) investing at 6.8% compounded continuously

23. What is the annual yield to the nearest hundredth of a percent on an account paying 8% compounded every 2 months?

24. What is the annual yield on an account paying 7% compounded every 5 days?

25. John's parents agree to loan him $5000 on the condition that he pays them $6000 at the end of 5 years. What simple rate of interest is John paying to his parents?

26. A loan of $4000 is made with the condition that if it is paid back within a year, a simple interest rate of 8% will be charged.
(a) If the loan was made on June 10, 1995, and repaid on February 17, 1996, what amount of interest should be charged?
(b) What would be the amount charged if ordinary interest is used?

27. An investor bought Signal Microchips stock for $16 a share. If the annual dividend is $1.50 per share and the stock was valued at $21.50 per share at the end of one year, what simple interest rate (including the dividend) was earned on the investment?

28. Marcia loaned $3000 to a friend for 90 days at 12% simple interest. After 30 days, she sold the note to a third party for the original amount of $3000. What interest rate did the third party receive? (Use 360 days in a year.)

29. A student has a savings account earning 6% simple interest. She must pay $1500 for the first semester tuition by September 1 and $1500 for the second semester by January 1. How much must she have on hand at the end of the summer (by September 1) in order to pay the first semester tuition on time and still have the remainder of her funds grow to $1500 by January 1? Use ordinary interest in the calculation.

30. An investor owns several apartment buildings. The taxes on these buildings total $30,000 for the year and are due before April 1. The late fee is 1/2% per month up to 6 months, at which time more severe penalties will be assessed. If the investor has $30,000 available on March 31, will he save money by paying the taxes on time or by investing the money at 8% simple interest and paying the taxes and late fee on September 30?

31. What is the annual yield equivalent to a rate of 8%, compounded monthly?

32. If money is invested at 6.5%, compounded continuously, what is the equivalent annual rate?

In problems 33 through 35, make a guess and check your answer. Then adjust your guess and check again until you have the right answer.

33. How long (to the nearest year) would $1200 have to be invested at 8%, compounded annually, to amount to a total of $1925?

34. How long will it take for $2000 to double at 10% interest compounded annually?

35. How long will it take for $500 to grow to $2500 at 15% compounded semiannually?

36. What amount should be deposited to yield $2000 if it is compounded at 8% annually for 5 years?

37. A child's parents want to have $40,000 in 15 years to pay for her college education. What amount should they deposit if they can earn 8% compounded annually?

38. An IRA (Individual Retirement Account) allows a saver to save tax-deferred (i.e., taxes are paid when the money is withdrawn). If a person deposits $5000 in an IRA earning 5% compounded quarterly, how much will the account be worth in 20 years?

39. How much money would you need to have saved to earn $25,000 interest per year
(a) if you could get 4% interest?
(b) if you could get 6% interest?
(c) if you could get 10% interest?

40. How much money would you need to have saved to earn $100,000 interest per year
(a) if you could get 5% interest?
(b) if you could get 8% interest?
(c) if you could get 12% interest?

If an investment grows in value from P_0 to P_1 over a period of n years, the equivalent annual rate of growth is

$$\left(\frac{P_1}{P_0}\right)^{1/n} - 1,$$

which when converted to a percentage gives the equivalent annual percentage rate.

41. Suppose the value of your mutual fund account increased from \$3000 to \$4500 over five years. Assuming all dividends were reinvested, what was the equivalent annual percentage rate?
42. Suppose you paid \$10,000 for bonds that pay 12% interest. After 3 years you sell the bonds. Since it is prior to maturity and interest rates have changed, you only receive \$9500. What is the equivalent annual percentage rate for this investment?

EXTENDED PROBLEMS

43. Investigate the interest rates available to you from the following sources.
    (a) savings accounts in banks, credit unions, and savings and loans
    (b) certificates of deposit in banks, credit unions, and savings and loans
    (c) United States savings bonds
    (d) money market funds
    (e) Treasury bills and notes
44. Research the relationship between short-term and long-term interest rates. In particular, is one usually higher than the other?
45. Research the historical record on interest rates. Discuss the important factors that seem to influence interest rates.
46. What are "junk bonds"? Why would anyone buy a junk bond?

The compounding of interest is an example of what is known as "exponential growth." This concept has important applications when looking at the growth of human, or other, populations; in this case, our population consists of money. One important way to assess the effects of growth is called the doubling time. Specifically, the doubling time is the length of time needed for an amount to double in value. When an amount is continuously compounded, the compound interest formula may be written as $A = Pe^{rt}$, where $e \approx 2.7183$, r is the rate expressed as a decimal, and t is the time in years. If the amount, A, is to be twice that of P, then $e^{rt} = 2$. Since $e^{0.6932} \approx 2$ (check this on your calculator), we see that the amount doubles when $rt = 0.6932$. Equivalently, the amount doubles when $t = 0.6932/r$. For example, if the rate is 6%, the doubling time is $\frac{0.6932}{0.06} \approx 69/6$ or 11.5 years. To find the doubling time exactly requires the use of logarithms, but the method given in the approximation is satisfactory for most purposes. Since we saw that compounding monthly gives values close to those of compounding continuously, the method also gives good approximations for most periodic compounding situations. To make things even easier, 72 is often used instead of 69 because it has many more integer factors than 69. This method of approximation has a name: the rule of 72. Using the rule of 72, we would say that an amount that is being compounded at 6% will double within 12 years (72/6), while an amount being compounded at 8% will double within 9 years.

47. Find the doubling time for a deposit of \$2500 earning 9% compounded monthly. Check your result by using the compound interest formula.
48. Find the doubling time if an investment earns 7.2% compounded monthly. Check your result by using the compound interest formula.
49. Find the doubling time if an investment earns 4.5% compounded monthly. Check your result by using the compound interest formula.
50. How many years will it take for an investment to quadruple in value if it is being compounded monthly at 6%?

6.2 LOANS

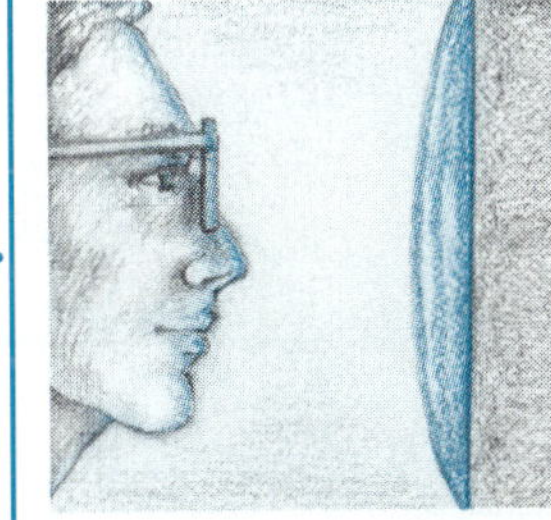

Suppose you can rent a \$500 television for \$30 a month and after 24 months you own it. Is this a good idea, or would it be better to charge it on your credit card and pay off that credit card account at a rate of \$30 a month?

Having discussed the ways in which interest is charged and paid, we will now consider, in more detail, the most common forms of loans. In this section, we look at loans that are based on simple interest.

SIMPLE INTEREST LOANS

TIDBIT

Conditions on most credit cards have become fairly complex. Many national credit cards have a variety of restrictions and conditions reflecting differences in state laws.

The interest on a **simple interest loan** is simple interest on the amount currently owed. Credit card accounts are a common example. Each month the bank or charge card company charges simple interest, called the **finance charge,** based on the balance owed. Many credit cards also have a grace period during which no interest is charged if full payment is received by the payment due date. The most common method for calculating finance charges uses the average daily balance. When the **average daily balance** is used, a cardholder is only charged for the actual number of days each amount owed was carried on the bill. This method converts the annual percentage rate to a daily interest rate.

To calculate the average daily balance, you determine the outstanding balance for each day, and divide the sum of these daily balances by the number of days in the monthly billing period. Any payments or other credits are subtracted from the previous day's balance as they occur. In general, the monthly statement includes the current month's charges, any unpaid balance, and finance charges.

EXAMPLE 6.8 The statement for Bob Chargeit's credit card account shows the following activity:

June 12	auto repair	\$ 45.60
June 18	payment	\$150.00
June 22	gasoline	\$ 20.00
July 3	paint	\$ 78.50

Find the average daily balance, the finance charge, and new balance if the billing period is from June 10 through July 9, inclusive, the previous balance was \$287.84, and the annual percentage rate is 21%.

SOLUTION To find the average daily balance, we use the balance for each day in the billing period and the number of days the balance was in effect, as shown in the table. Each balance is multiplied by the number of days it was in effect; we multiply 6×333.44 rather than add 333.44 for six different days. The values are added, and the total is divided by the number of days in the billing period.

Time Period	Days	Daily Balance
June 10–June 11	2	\$287.84
June 12–June 17	6	\$287.84 + \$45.60 = \$333.44
June 18–June 21	4	\$333.44 − \$150.00 = \$183.44
June 22–July 2	11	\$183.44 + \$20.00 = \$203.44
July 3–July 9	7	\$203.44 + \$78.50 = \$281.94

$$\text{average daily balance} = \frac{2(287.84) + 6(333.44) + 4(183.44) + 11(203.44) + 7(281.94)}{2 + 6 + 4 + 11 + 7}$$

$$= \frac{7521.50}{30} = \$250.72$$

The finance charge is the simple interest on the average daily balance using a daily interest rate. For an annual percentage rate of 21%, the daily percentage rate is 0.057534%, which is $\frac{21\%}{365}$. We then use the simple interest formula $I = Prt$, with the rate as a decimal and time given in terms of days. The calculation could also be done in terms of years, as follows:

$$\text{finance charge} = 250.72 \times 0.21 \times \left(\frac{30}{365}\right) = 4.327496 = \$4.33.$$

The new balance on the account will be the sum of the ending balance and the finance charge:

$$\text{new balance} = \$281.94 + \$4.33 = \$286.27.$$ ◆

The particulars of how the balance on a credit card is determined vary. But what happens if you allow a balance to recur month after month? The answer is: you end up paying compound interest to the credit card company!

ADD-ON INTEREST

Sometimes businesses offer to finance the purchase of furniture, appliances, or automobiles with monthly payments using what is called **add-on interest.** To find the monthly payment for such a purchase, calculate simple interest at the annual interest rate over the length of the loan agreement. Then divide the sum of the purchase price and the interest into equal monthly payments.

EXAMPLE 6.9 A \$1200 motorcycle is financed over a 2-year period with 12% add-on interest. Find the monthly payment.

SOLUTION Simple interest at 12% over 2 years on \$1200 is

$$\$1200 \times 0.12 \times 2 = \$288.$$

Adding the \$288 interest and the \$1200 principal yields \$1488, which must be paid in 24 equal payments. So the monthly payment is $\$1488 \div 24 = \$62.00$. ◆

ANNUAL PERCENTAGE RATE

In 1969, Congress passed the Consumer Credit Protection Act, which is usually known as the Truth-in-Lending Act, requiring lenders to compute and disclose the annual percentage rate of any loan that is not a simple interest loan. The **annual percentage rate (APR)** is the simple interest rate that would require the same payments to pay off the debt. Notice that add-on interest loans charge interest on the *entire* principal over the life of the loan even though you don't have use of all of the money during that period. Computing the annual percentage rate for add-on interest loans is difficult to do from formulas. Instead, tables showing the APRs for different combinations of rates and loan periods are used.

EXAMPLE 6.10 Find the annual percentage rate for the purchase of the motorcycle in the previous example, namely, a 12% loan over 2 years.

SOLUTION The following abbreviated table gives some sample APRs.

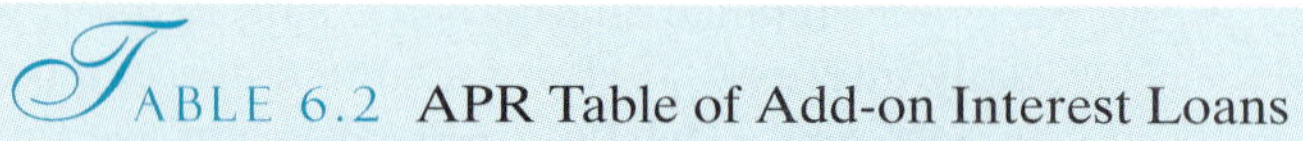

TABLE 6.2 APR Table of Add-on Interest Loans

Nominal Interest Rate	Length of the Loan in Years: 1	2	5
6	10.9	11.1	10.8
8	14.5	14.7	14.1
10	18.0	18.2	17.3
12	21.5	21.6	20.3

Since it was an add-on interest loan with a nominal rate of 12% and length of the loan was 2 years, the APR is 21.6%, the intersection of the "12%" row and the "2 year" column in the APR table. ◆

The APR table can be used to estimate the annual percentage rate for loans. Observe that in the 10% row, the add-on interest rates are 18.0, 18.2, and 17.3. Comparing these rates to 10 shows that the annual percentage rate for an add-on interest loan is approximately *1.8 times* the nominal interest rate! This is a good estimate for most interest rates and time periods. Since there is no simple formula for the APR, you should use the estimate of multiplying by 1.8 when an APR table is unavailable.

EXAMPLE 6.11 Find the approximate annual percentage rate for an add-on interest of 20% for 1 year.

SOLUTION We use the factor of 1.8 for approximations: $20\% \times 1.8 = 36\%$. ◆

RENT-TO-OWN

Since the passage of the Truth-in-Lending Act forced some lenders to reveal the extremely high annual percentage rates they charge, add-on interest loans have become rare. The replacement for add-on interest is the rent-to-own transaction, in which you rent an item you cannot afford to buy outright. After a contracted number of payments, the item becomes yours. Of course, the rental may be for a shorter period of time, and the item is then returned. The effect of a rent-to-own transaction is the same as buying on credit, but technically it is not a credit purchase and is thus not covered by the Truth-in-Lending Act.

For comparison shopping on rates, you can still treat a rent-to-own transaction as a loan and compute the annual percentage rate. Do this as follows: Find the total of all the payments required to buy the item and subtract the best purchase price available at an ordinary retail store; the difference is essentially the add-on interest. Find the simple interest rate that would have to be charged on the retail purchase price to amount to the add-on interest; then multiply by 1.8 to get the approximate annual percentage rate.

EXAMPLE 6.12 Suppose you can rent-to-own a $500 television for 24 monthly payments of $30. Estimate the annual percentage rate you would be paying on rent-to-own.

SOLUTION The payments total $\$30 \times 24 = \$720$. You will be paying $\$720 - 500 = \$220$ as add-on interest over two years, or $110 per year. The simple interest rate needed for that amount of interest is $110/500 = 0.22 = 22\%$. Since we do not have a table that includes the APR for a loan with nominal interest rate of 22% and term of 2 years, we must make an estimate by multiplying the nominal rate by 1.8. So we estimate that the annual percentage rate is approximately $1.8 \times 22\% \approx 40\%$! ◆

The annual percentage rate on rent-to-own transactions will usually turn out to be very high, partly due to the fact that these transactions are, in general, a higher risk for the merchant. If possible, you would be well advised to go elsewhere to borrow the money needed to make the ordinary retail purchase.

INITIAL PROBLEM SOLUTION

Suppose you can rent a $500 television for $30 a month and after 24 months you own it. Is this a good idea, or would it be better to charge it on your credit card and pay off that credit card account at a rate of $30 a month?

SOLUTION The annual interest rate of the rent-to-own option is about 40%, as shown in Example 6.12. Credit card accounts generally have annual percentage rates around 20% or lower. Since even the nominal rate of 22% of the rent-to-own option is higher than 20%, the credit card is clearly the way to go.

PROBLEM SET 6.2

In problems 1 through 4 find the finance charge on a credit card account for the given conditions. When billing periods are given, the use of "through" means the beginning and ending dates are both included.

1. **(a)** Average Daily Balance = $255.00, Annual Percentage Rate = 12.9%, Billing Period = 30 days
(b) Average Daily Balance = $425.80, Annual Percentage Rate = 14.9%, Billing Period = 31 days

2. **(a)** Average Daily Balance = $183.65, Annual Percentage Rate = 16.9%, Billing Period = 31 days
(b) Average Daily Balance = $194.85, Annual Percentage Rate = 16.5%, Billing Period = 30 days

3. **(a)** Average Daily Balance = $315.42, Annual Percentage Rate = 14.9%, Billing Period: May 15 through June 14
(b) Average Daily Balance = $275.65, Annual Percentage Rate = 15.9%, Billing Period: March 11 through April 10

4. **(a)** Average Daily Balance = $583.27, Annual Percentage Rate = 17.9%, Billing Period: April 11 through May 10
(b) Average Daily Balance = $224.85, Annual Percentage Rate = 16.5%, Billing Period: July 15 through August 14

In problems 5 through 10, find the finance charge and new balance for each credit card account.

5. Ending Balance = $320.50, Average Daily Balance = $275.00, Annual Percentage Rate = 18.9%, Billing Period = 30 days

6. Ending Balance = $485.88, Average Daily Balance = $325.80, Annual Percentage Rate = 14.9%, Billing Period = 31 days

7. Ending Balance = $147.85, Average Daily Balance = $155.00, Annual Percentage Rate = 15.9%, Billing Period = 30 days

8. Ending Balance = $227.54, Average Daily Balance = $215.80, Annual Percentage Rate = 18.9%, Billing Period = 31 days

9. Ending Balance = $135.92,
Average Daily Balance = $105.00,
Annual Percentage Rate = 16.9%,
Billing Period = 31 days

10. Ending Balance = $362.00,
Average Daily Balance = $325.80,
Annual Percentage Rate = 14.9%,
Billing Period = 30 days

In problems 11 through 14, the activity in a credit card account is given for one month.

11. Billing period: 10/11 through 11/10;
previous balance: $165.45;
annual interest rate: 12.9%.

October 18	Payment	$100.00
October 25	Restaurant	$ 28.90
November 5	Software	$ 85.64

(a) Find the average daily balance and the finance charge.
(b) What is the new account balance on November 11?

12. Billing period: 9/11 through 10/10; previous balance: $385.56; annual interest rate: 14.9%.

September 15	Payment	$200.00
September 22	Bookstore	$ 42.85
October 2	Clothes	$192.93

(a) Find the average daily balance and the finance charge.
(b) What is the new account balance on October 11?

13. Billing period: 6/11 through 7/10; previous balance: $225.85; annual interest rate: 14.9%.

June 20	Shoes	$ 79.95
June 25	Payment	$125.00
June 28	Books	$ 34.65
July 5	Radio	$ 69.50

(a) Find the average daily balance and the finance charge.
(b) What is the new account balance on July 11?

14. Billing period: 3/11 through 4/10; previous balance: $95.15; annual interest rate: 15.8%.

March 15	Clothes	$113.50
March 20	Hardware	$ 52.93
March 20	CDs (music)	$ 28.67
March 22	Payment	$175.00

(a) Find the average daily balance and the finance charge.
(b) What is the new account balance on April 11?

The formula for the total amount, A, that must be paid for the principal plus interest when add-on interest is used is the same as that for simple interest loans: $A = P + Prt$ or $A = P(1 + rt)$, where r is the rate, as a decimal, and t is the time in years. The difference between the two is that with simple interest, the total is all paid at the end; with add-on interest, you make many payments throughout the life of the loan.

In problems 15 through 20, find the monthly payment on contracts involving add-on interest.

15. A $675 stereo is financed over a two-year period with 15% add-on interest. What is the monthly payment?

16. Louise buys a new bike for $375 and pays for it over two years with 13.5% add-on interest. What is her monthly payment?

17. Sean's new snowboard cost him $425. He financed it over an 18-month period with 15% add-on interest. What is the monthly payment?

18. A used car that sold for $5775 is financed for 30 months with 9.5% add-on interest. What is the monthly payment?

19. Ron and Shannon buy a new refrigerator for $755. The purchase is financed over 30 months with 10.5% add-on interest. What is their monthly payment?

20. What monthly payment is required for a $425 set of skis that are financed for one year with 13.5% add-on interest?

21. What is the original principal for a simple interest loan that had a total value of $4340 after 3 years at 8%?

22. Jerry's grandfather loaned him money to help buy a car. Two years later, when Jerry paid back the loan, his grandfather only charged him 3.5% simple interest on the loan. If Jerry paid his grandfather $2568, how much had he borrowed?

23. What was the original purchase price (to the nearest dollar) of a television purchased with 20 monthly payments of $48.03, which include add-on interest of 12.5%?

24. Lucinda purchased a new watch from a jewelry store and paid for her purchase with 12 monthly payments of $34.53 based on add-on interest of 10.5%. What was the purchase price?

25. Find the APR for each of the following nominal add-on interest rates.
(a) 12% for 5 years
(b) 6% for 2 years
(c) 10% for 1 year
(d) 8% for 2 years

26. Find the APR for each of the following nominal add-on interest rates.
(a) 12% for 1 year
(b) 6% for 5 years
(c) 10% for 2 years
(d) 8% for 5 years

27. If a $600 television is purchased on a rent-to-own agreement at $32 a month for 24 months, what amount is paid above-and-beyond the stated purchase price?

28. A dining room set at a rent-to-own store can be rented for \$65 a month (a deposit and a minimum of three months are required). If the set has a suggested retail value of \$1250 and you own the set after 24 months, how much extra do you pay under this arrangement?

In problems 29 through 32, find the annual add-on interest rate being used.

29. **(a)** Principal = \$500, term = 1 year, monthly payment = \$46.47
(b) Principal = \$750, term = 2 years, monthly payment = \$37.50

30. **(a)** Principal = \$925, term = 2 years, monthly payment = \$48.18
(b) Principal = \$1500, term = 3 years, monthly payment = \$58.17

31. **(a)** Principal = \$600, term = 18 months, monthly payment = \$40.58
(b) Principal = \$450, term= 20 months, monthly payment = \$28.69

32. **(a)** Principal = \$1150, term = 30 months, monthly payments = \$50.31
(b) Principal = \$1800, term = 36 months, monthly payment = \$75.35

33. Estimate the APR for a loan with add-on interest of 12.5%

34. Estimate the APR for a loan with add-on interest of 10.5%

35. If a \$600 television is purchased on a rent-to-own agreement at \$32 a month for 24 months, what nominal rate of interest is being charged if you consider the extra charges as add-on interest? What is the estimated APR?

36. If a dining room set with a suggested retail value of \$1250 is purchased under a rent-to-own agreement of \$65 a month for 24 months, what nominal rate of interest is being charged if you consider the extra charges as add-on interest? What is the estimated APR?

37. If a \$1500 home entertainment center is purchased on a rent-to-own agreement at \$70 a month for 30 months, what nominal rate of interest is being charged if you consider the extra charges as add-on interest? What is the estimated APR?

38. A living room set with a suggested retail price of \$1050 is available at two rent-to-own stores. One store has the set at a rent-to-own rate of \$55.56 a month for 24 months; at the other store, the rent-to-own rate is \$63 a month for 20 months. What are the annual interest rates being charged when you consider the extra charge as add-on interest? What is the estimated APR?

39. Ted buys a used car priced at \$2500 with no down payment and makes payments of \$125 a month for two years. What is the annual interest rate if the finance charges are figured as add-on interest? What is the estimated APR?

40. A student buys a new stereo system that costs \$975 by agreeing to make monthly payments for three years to cover the cost of the set and 12% add-on interest.
(a) What is the amount of the monthly payment?
(b) What is the estimated APR?

41. Jamie decides to get a new table and chairs and finds a suitable set at a rent-to-own dealer. The terms are \$32 a month, with ownership after 30 months. Then, Jamie finds the same basic set for sale at \$629 at a local furniture store. What is the approximate annual percentage rate that Jamie could pay for a conventional loan and still have payments that were no more than \$32 a month?

42. Jorge is either going to buy a new stereo system or get the same system through the rent-to-own dealer. He can buy the set for \$796; at the rent-to-own dealer he would pay \$42.50 for 24 months. What is the approximate annual percentage rate that Jorge could pay for a conventional loan and still have payments that were no more than \$42.50 a month?

EXTENDED PROBLEMS

43. Investigate the interest rates available on bank credit cards. Is there any reason a person would choose a card with a high interest rate?

44. Compare the terms at a rent-to-own store with the available retail price for an appliance of interest to you. What is the approximate annual percentage rate if the rent-to-own arrangement is treated like a loan with add-on interest?

45. Research the history of consumer credit. Consider such questions as: Is it a recent phenomenon? When did it become institutionalized and regulated? What legal and/or social issues have received special attention?

46. Consult an encyclopedia, almanac, or other suitable reference source for the total amount of consumer debt during the last 25 years. Prepare a bar graph to display the data. Are there any trends or special features to the graph?

47. Research the effect of inflation and deflation on those who owe money. Has this ever had important political ramifications?

48. What are usury laws? Are such laws in force in your state? Were they ever?

6.3 AMORTIZED LOANS

You need to borrow \$85,000 to buy a house. Loan rates are at 7%. How much will your monthly payment be for a fifteen-year home loan? What would the payment be for a thirty-year loan?

The majority of loans or debts involve regular payments over a period of several, or many, years. While this was true for add-on interest loans, the stated rate and the APR were quite different. In this section we look at the common time payment loans in which the rates accurately reflect the APR.

TERMINOLOGY

The word *amortize* comes from the Latin *admortiz,* which means "to bring to death." In the context of loans, it is the debt that is brought to death, not the debtor. An **amortized loan** is a simple interest loan with equal monthly payments over the length of the loan. Each payment includes the interest charged since the previous payment and also reduces the balance owed. The size of the equal payments is chosen so that once all the payments are made, the balance is zero, that is, the loan will have been paid-off (due to rounding, the last payment may be more or less than the other monthly payments).

The important factors related to an amortized loan are the principal, the amount borrowed; the interest rate, the annual percentage rate; the length of the loan (also called the **term of loan**); and the **monthly payment.** These four factors are interrelated. If you know any three of them, the fourth can be found. However, it is easier to find some of the factors than it is to find the others.

THE HISTORY OF A LOAN

To understand amortized loans, we will examine one over its entire history. For convenience, we do this with a loan having only a few monthly payments. In reality, though, amortized loans are most often made for a period of a few years (for motor vehicles) and up to 30 years (for homes).

EXAMPLE 6.13 Chart the history of an amortized loan of \$1000 for 3 months at 12% with monthly payments of \$340.

SOLUTION When the first payment is made, $\frac{1}{12}$ of a year has passed; so the interest is $\$1000 \times 0.12 \times \frac{1}{12} = \$10$. The payment goes first toward the interest, then toward reducing the balance; so the payment toward the balance is reduced to $340 − $10 = $330. The $330 is the **net payment** or **payment to principal.** The new balance for the loan after one payment is

$$\$1000 + 10.00 - 340 = \$670 \text{ or } \$1000 - 330 = \$670.$$

We have added the interest to the balance and then subtracted the full payment. The same steps are repeated two more times. When the second payment is made, the interest charged for the time between the first and second payments is

$$\$670 \times 0.12 \times \tfrac{1}{12} = \$6.70.$$

The new balance after the second payment is

$$\$670 + 6.70 - 340.00 = \$336.70.$$

When the third payment is made, the interest charged for the time between the second and third payments is

$$\$336.70 \times 0.12 \times \tfrac{1}{12} = \$3.37.$$

The balance after the third payment is

$$\$336.70 + 3.37 - 340.00 = \$0.07.$$

Thus the third payment must be adjusted upward 7 cents to pay off the loan. The record can be kept neatly in tabular form (Table 6.3).

TABLE 6.3

Payment	Interest	Net Payment	New Balance
340.00	10.00	330.00	670.00
340.00	6.70	333.30	336.70
340.07	3.37	336.70	0.00

◆

Once you know the principal, interest rate, and payment size you can chart the history of any amortized loan. Of course, this procedure will probably be stretched over a period of years. The main pitfall faced is making an error in rounding-off. If you borrow from a financial institution they will do the computation for you, but you may wish to check their work. If you have an amortized loan made between private parties, such as an owner-financed home purchase, it is essential that at least one party knows how to do the calculation or has access to appropriate tables of values.

FINDING THE MONTHLY PAYMENT

Perhaps the most important problem in dealing with amortized loans is finding the payment when you are given the loan amount, the interest rate, and the length of the loan. There are three ways to do this: (a) by using a financial or business calculator, (b) by looking it up in a table, and (c) by apply a formula and using a scientific calculator.

Using a Financial Calculator

Perhaps the easiest way to find a monthly payment is to own a financial calculator. These have built-in functions to calculate the payment required for any amortized loan. Real estate agents routinely use such calculators. The following example shows the keystrokes used on one such calculator.

EXAMPLE 6.14 Using a financial calculator, find the monthly payment on a 5-year loan of $10,000 at 10% interest.

SOLUTION

10000 [PV] (This enters the present value of $10,000.)

0.1 [÷ 12] [i] (This enters the interest rate divided by 12.)

5 [× 12] (This enters 5 × 12 = 60 months.)

[COMP] [PMT] [212.47045] (This computes the payment: $212.47.) ◆

Using an Amortization Table

Another common way to find the payment required for an amortized loan is to consult an **amortization table.** You can look up the payment required for a variety of typical loans. Such amortization tables can be found in most business stationery stores, are relatively inexpensive, and last a lifetime. Typically, each page is devoted to one interest rate. The rows correspond to the amount of the loan and the columns correspond to the length of the loan. The entry in a particular row and column is the monthly payment required for the loan. Table 6.4 shows part of such a table; a more complete table is provided in Appendix 1.

TABLE 6.4 Amortization Table at 10%

Amount	5 years	10 years	15 years	20 years	25 years	30 years	35 years	40 years
100	2.13	1.33	1.08	.97	.91	.88	.86	.85
200	4.25	2.65	2.15	1.94	1.82	1.76	1.72	1.70
500	10.63	6.61	5.38	4.83	4.55	4.39	4.30	4.25
1000	21.25	13.22	10.75	9.66	9.09	8.78	8.60	8.50
2000	42.50	26.44	21.50	19.31	18.18	17.56	17.20	16.99
5000	106.24	66.08	53.74	48.26	45.44	43.88	42.99	42.46
10000	212.48	132.16	107.47	96.51	90.88	87.76	85.97	84.92
20000	424.95	264.31	214.93	193.01	181.75	175.52	171.94	169.83
50000	1062.36	660.76	537.31	482.52	454.36	438.79	429.84	424.58
100000	2124.71	1321.51	1074.61	965.03	908.71	877.58	859.68	849.15

EXAMPLE 6.15 Find the monthly payment on a 5-year loan of $10,000 at 10% interest.

SOLUTION We look for the entry in the row for the amount of 10000 and the column for a term of 5 years. The row and column have been shaded in the table above. We find the monthly payment is $212.48. This compares to the value of $212.47 found in Example 6.14. The difference is due to rounding. ◆

A second type of amortization table assumes a standard loan size such as $1000. The rows then correspond to the interest rate, and the columns correspond to the length of the loan. The entry in the table is the payment per $1000 borrowed. To get the correct payment for a particular loan, multiply the entry by the number of thousands of dollars borrowed. The result must be rounded-up to get the proper payment. If you were to round-down, every payment would be somewhat smaller, and the loan would not be paid-off in the assigned number of payments. When you round-up, the final payment will be less than the regular payment.

TABLE 6.5 Amortization Table for $1000 Loan

Percent	5 years	10 years	15 years	20 years	25 years	30 years
5	18.871234	10.606552	7.907936	6.599557	5.845900	5.368216
6	19.332802	11.102050	8.438568	7.164311	6.443014	5.995505
7	19.801199	11.610848	8.988283	7.752989	7.067792	6.653025
8	20.276394	12.132759	9.556521	8.364401	7.718162	7.337646
9	20.758355	12.667577	10.142666	8.997260	8.391964	8.046226
10	21.247045	13.215074	10.746051	9.650216	9.087007	8.775716
11	21.742423	13.775001	11.365969	10.321884	9.801131	9.523234
12	22.244448	14.347095	12.001681	11.010861	10.532241	10.286126
13	22.753073	14.931074	12.652422	11.715757	11.278353	11.061995
14	23.268251	15.526644	13.317414	12.435208	12.037610	11.848718
15	23.789930	16.133496	13.995871	13.167896	12.808306	12.644440

EXAMPLE 6.16 Find the monthly payment on a 10-year loan of $13,000 at 12% interest.

SOLUTION We look for the entry in the row for a 12% interest rate and the column for a 10-year loan. The row and column have been shaded in Table 6.5. We find

\$14.347095. Since this is the payment per \$1000, we multiply by 13 (for a \$13,000 loan) to get

$$\$14.347095 \times 13 = \$186.512235.$$

The correct payment is \$186.52. ◆

Using a Formula

If one has access to a scientific calculator with an $\boxed{x^y}$ key, the following formula can be used to find a monthly payment.

> **Monthly Payment Formula**
>
> If P is the amount of the loan, r is the annual percentage rate as a decimal, and t is the length of the loan in years, then
>
> $$m = \frac{P \times \frac{r}{12} \times (1 + \frac{r}{12})^{12t}}{[(1 + \frac{r}{12})^{12t} - 1]}$$

Clearly, you do not use this formula without a calculator, and it must be a calculator with an $\boxed{x^y}$ key or the equivalent.

EXAMPLE 6.17 Find the monthly payment for a \$1000 loan for 3 years at 10%.

SOLUTION Here $P = \$1000$, $r = 0.10$, and $t = 3$. By the formula then

$$\begin{aligned} m &= \frac{\$1000 \times \left(\frac{0.10}{12}\right) \times \left(1 + \left(\frac{0.10}{12}\right)\right)^{12\times 3}}{\left[\left(1 + \left(\frac{0.10}{12}\right)\right)^{12\times 3} - 1\right]} \\ &= \frac{\$100 \times 0.008333 \times 1.008333^{36}}{[(1.008333)^{36} - 1]} \\ &\approx \$32.27 \end{aligned}$$

These calculations can be done with one long string of keystrokes on a calculator using parentheses. However, due to the complexity of the formula and the fact that $1 + \left(\frac{0.10}{12}\right)^{12\times 3}$ appears twice, a calculator memory key is useful.

STEP 1: Clear your calculator memory. Then calculate $1 + \left(\frac{0.10}{12}\right)^{12\times 3}$ and store it in memory.

(1 [+] 0.1 [÷] 12) [x^y] 36 [=] [1.348182] [STO]

This stores the value in memory, and you can recall it as you need it.

STEP 2:

1000 [×] 0.1 [÷] 12 [×] [RCL] [÷] ([RCL] [−] 1) [=] [32.267187]

Thus a payment of \$32.27 per month is needed to pay off a \$1000 loan at 10% in 3 years. ◆

You need to borrow \$85,000 to buy a house. Loan rates are at 7%. How much will your monthly payment be for a fifteen-year home loan? What would the payment be for a thirty-year loan?

SOLUTION We are given an interest rate of 7%. Typically, mortgages are for a term of 15 or 30 years. For a 15-year loan at 7%, Table 6.5 gives a payment of \$8.988283 per \$1000. The payment on an \$85,000 loan would be

$$\$8.988283 \times 85 \approx \$764.01.$$

For a 30-year loan at 7%, the table gives a payment of \$6.653025 per \$1000. The payment on an \$85,000 loan would be

$$\$6.653025 \times 85 \approx \$565.51.$$

The choices are \$764.01 per month for 15 years or \$565.51 per month for 30 years. These payments cover the principal and interest only. The typical house payment will also include amounts for taxes and insurance.

PROBLEM SET 6.3

1. A \$2000 loan is made at 10% interest with monthly payments of \$42.50 for 5 years. What is the balance after the first payment is made?
2. A loan of \$5000 is made at 8% interest with monthly payments of \$101.40 for 5 years. What is the balance after the first payment is made?
3. A new car is purchased with a trade-in and a 4.9% loan on the balance of \$13,000. If the payments are \$298.79 a month for 48 months, what is the balance on the loan after the first payment is made?
4. Tom has to have his car painted. He finances the bill of \$1350 with a 12.9% loan for two years from the credit union. If his monthly payments are \$64.12, what is the balance on the loan after the first payment is made?
5. Chart the first three months' history of an amortized loan of \$5000 for 5 years at 12% with monthly payments of \$111.23.
6. Chart the first three months' history of an amortized loan of \$10,500 for 4 years at 12% with monthly payments of \$276.51.
7. Chart the history of a loan of \$600 that is paid back with 5 monthly payments of \$123.78 based on a 12.5% rate. What is the amount of the final payment?
8. Chart the history of a loan of \$900 that is paid back with 4 monthly payments of \$231.25. If the rate is 13.25%, what is the amount of the final payment?
9. Use Table 6.4 to find the monthly payment on a 10-year loan of \$10,000 at 10% interest.
10. Use Table 6.4 to find the monthly payment on a 20-year loan of \$50,000 at 10% interest.
11. Use Table 6.4 to find the monthly payment on a 5-year loan of \$18,000 at 10% interest.
12. Use Table 6.4 to find the monthly payment on a 15-year loan of \$28,000 at 10% interest.
13. Use Table 6.5 to find the monthly payment on a 5-year loan of \$18,000 at 12% interest.
14. Use Table 6.5 to find the monthly payment on a 10-year loan of \$15,000 at 9% interest.
15. Use Table 6.5 to find the monthly payment on a 15-year loan of \$22,500 at 6% interest.
16. Use Table 6.5 to find the monthly payment on a 20-year loan of \$25,750 at 8% interest.
17. Use the formula to find the monthly payment on a 10-year loan of \$20,000 at 10% interest. Verify the value by comparing it to Table 6.4.
18. Use the formula to find the monthly payment on a 5-year loan of \$5000 at 10% interest. Verify the value by comparing it to Table 6.4.
19. Find the monthly payment on a 5-year loan of \$1000 at 9% interest. Use the formula, and compare your result with the value in Table 6.5.
20. Find the monthly payment on a 10-year loan of \$1000 at 12% interest. Use the formula, and compare your result with the value in Table 6.5.
21. Find the monthly payment for a \$6500 loan for 30 months at 9%.

22. Find the monthly payment for a \$4800 loan for 2 years at 9.8%.

23. What is the monthly payment on a loan of \$4850 for 20 months at 7.25%?

24. What is the monthly payment on a loan of \$10,800 for 4 years at 11.75%?

25. What is the monthly payment on a car loan for \$6725 financed for 4 years at 12.8% interest?

26. What is the monthly payment needed to finance a new car purchase of \$12,400 for 40 months at 7.9% interest?

27. What is the monthly payment on a furniture purchase of \$3250 financed for 2 years at 10.5% interest after a 20% down payment is made?

28. A home priced at \$85,000 is sold with a 10% down payment and the balance financed at 8% for 30 years. What is the monthly payment?

29. John bought a new car for \$16,285 by paying 20% down and financing the balance at 11% for 5 years. What will his monthly payment be?

30. After graduation, Wendy took a tour to the major art museums of eastern Europe. The tour was financed with a down payment of 10% and the balance paid in 30 monthly payments at 11.5% interest. The tour package Wendy chose was priced at \$2795. What was her monthly payment?

31. Use the amortization tables to find the size of loan that can be financed at 9% for 20 years with a monthly payment of \$500.

32. Use the amortization tables to find the size of loan that can be financed at 12% for 15 years with a monthly payment of \$300.

33. Use Table 6.5 to find the size of loan that can be financed at 9.5% for 15 years with a monthly payment of \$600. Since 9.5% isn't listed on the table, use an appropriate approximation.

34. As in problem 33, use Table 6.5 to find the size of loan that can be financed at 8.75% for 10 years with a monthly payment of \$250.

35. Tim needs to buy a car. After going over his budget, he decides he can afford \$250 a month for a car payment. If he pays no money down and gets financing for 5 years at 8% interest, how much can he afford to pay for a car?

36. The Romeros decide to get a home improvement loan. After going over their finances, they determine that the most they can budget for a monthly payment is \$300. If they can get a 10-year loan at 9%, what is the maximum loan possible?

37. What is the approximate annual interest rate for a 30-year loan of \$40,000 with monthly payments of \$321.85?

38. What is the approximate annual interest rate for a 25-year loan of \$30,000 with monthly payments of \$231.55?

39. What is the approximate annual interest rate for a 15-year loan of \$25,000 with monthly payments of \$292.68?

40. What is the approximate annual interest rate for a 10-year loan of \$50,000 with monthly payments of \$626.30?

EXTENDED PROBLEMS

41. Compare the terms available through local financial institutions for amortized loans to buy the following items.
- **(a)** used car
- **(b)** new car
- **(c)** manufactured home
- **(d)** conventional home

42. Discuss why the terms you found in problem 41 differ from each other.

43. Typically, interest payments for a loan secured by your home are deductible on your income tax.
- **(a)** If you had the cash to pay for your house, should you do so, or should you finance the house and invest the cash?
- **(b)** How does the answer to (a) depend on interest rates and tax bracket?

The Formula for Finding the Loan Amount

If you are given the interest rate, the length of the loan, and the monthly payment, you can use the formula for finding the monthly payment "in reverse" to find the amount of the loan.

If r = interest rate (as a decimal),
t = length of the loan (in years)
m = the monthly payment, and
P = principal of the loan,

then $$P = m \times \frac{12}{r} \times \left[1 - \frac{1}{\left(1 + \frac{r}{12}\right)^{12t}}\right]$$

44. Use the formula to find the size of loan that can be financed at 9.5% for 15 years with a monthly payment of \$600.

45. Use the formula to find the size of loan that can be financed at 10.75% for 20 years with a monthly payment of \$420.

46. Verify the formula given above. Begin with the formula for finding the monthly payment, and solve for P.

6.4 BUYING A HOUSE

Suppose you have saved \$15,000 towards a down payment on a home, and your total household income is \$35,000 per year. What is the most you could afford to pay for a home? Assume that (1) your insurance costs will be 0.25% of the value of your home, (2) your taxes will be 2% annually, (3) your closing costs will be about \$2000, and (4) you can obtain a fixed rate mortgage for 30 years at 8% interest.

Not only is the purchase of a new home the biggest financial commitment in most people's lives, it is also one of the most complicated. Many factors beyond the price of the home have to be considered, beginning with the ability to pay the initial costs as well as the monthly costs that may last for 30 years. This section will cover the basic mechanics of a home mortgage. We will not cover some topics of interest, such as the tax advantages of home ownership, which are beyond the scope of this book.

HISTORY

Part of the English feudal system involved a duty to the lord on whose land you had tenure—for example, military service. Most of the Colonies adopted this system, imposing the least burdensome duty of "socage," which was usually payment of rent to the Crown. This system was distasteful to the colonists and led to incidents of rebellion as early as 1676.

RULES OF THUMB

A central part of the "American Dream" is owning your own home. This requires money, and most people must borrow the bulk of that money. We have already covered the general topic of loans, but when real estate is involved there are a number of additional conditions that can make such transactions extremely complex.

A natural question might be "Why is real estate so complicated?" One answer is "History!" The laws governing real estate in most of the United States evolved from the laws of England while the states were still colonies. Those English real estate laws were mainly derived from ancient feudal laws and from even earlier common law predating the Norman conquest. Unfortunately, as real estate law evolved, it was not simply a case of the new replacing the old. Instead, the new was added to the old, making a progressively more complex legal structure. One consequence is that there is no single document to prove conclusively that a particular person owns a particular property. In any case, we cannot cover real estate law, nor would we want to. We will focus on the main mathematical issues that affect the ordinary person. If you purchase a home, we trust that you will have professionals help sort out the legal details.

Traditionally, a few general guidelines, "rules of thumb," have been used to estimate how much a buyer could afford to spend on a house. Here are two of the most common ones.

1. The home you purchase should not cost more than 3 times your annual family income.
2. You should limit your monthly housing expenses including mortgage payment, property taxes, and homeowner's insurance to no more than 25% of your monthly gross income (that is, income before deductions).

If your planned house purchase fits under both of these guidelines, then you can almost surely afford it.

EXAMPLE 6.18 If your annual family income is \$30,000, what do the traditional rules of thumb tell you regarding a purchase price and monthly expenses for your potential home?

SOLUTION The houses you consider should not cost more than

$$3 \times \$30{,}000 = \$90{,}000$$

and the monthly expenses for mortgage payment, property taxes, and homeowner's insurance should not exceed

$$0.25 \times \frac{1}{12} \times \$30{,}000 = \$625.$$ ◆

In practice, the most important question is whether or not a bank or other financial institution will approve your loan application. Among the things that will be considered is your other debt. Having car payments and credit card balances may affect your ability to buy a house. As of this writing, many banks are allowing up to 38% of the borrower's monthly income to go for mortgage payment, property taxes, and homeowner's insurance. For the situation in the Example 6.18, assuming there are no other significant debts, the limit on monthly housing expenses may go as high as

$$0.38 \times \frac{1}{12} \times \$30{,}000 = \$950.$$

EXAMPLE 6.19 Suppose Andrew and Barbara both have jobs, each earning \$24,000 per year, and they have no debts. What are the low and high estimates of how much they can afford to pay for monthly housing expenses?

SOLUTION The low estimate for acceptable monthly housing expenses is 25% of gross monthly income, or in the case of Andrew and Barbara

$$0.25 \times \frac{24{,}000 + 24{,}000}{12} = \$1000.$$

The high estimate for acceptable monthly housing expenses is 38% of gross monthly income, or

$$0.38 \times \frac{24{,}000 + 24{,}000}{12} = \$1520$$ ◆

THE MORTGAGE

A **mortgage** is a loan that is guaranteed by real estate. If the borrower fails to make the payments, the lender can take control of the property. Technically only certain loans secured by real estate are called mortgages, but we will use the term in the sense of everyday conversation rather than its full legal meaning.

There are many types of mortgages available. In an era of financial stability, such diversity would not make much sense, but the 1970s and 1980s were decades of high inflation and high interest rates. Those factors forced financial institutions to introduce new types of mortgages to meet the needs of clients and attract otherwise marginal borrowers.

The two main categories of mortgage are *fixed rate* and *adjustable rate.* For a **fixed rate mortgage,** the interest rate is set, once and for all, at the time the loan is made. For an **adjustable rate mortgage** the interest rate can change from year to year. The actual interest rate is usually a specified amount higher than some particular financial index, for example the interest rate paid by Treasury bonds; often there is also a limit (called a **cap**) on how much the interest rate is allowed to rise in a single year.

The second major distinction among mortgages is the **term** of the mortgage. Typically the choices are 15 year or 30 year. The longer term loan usually carries a higher interest rate because the money is used for a longer period of time, and the lender's risk is extended.

Another variable in choosing a mortgage loan is commonly referred to as **points.** One point is one percent of the amount of the loan. Points are generally charged in two ways: (1) a **loan origination fee** for making the loan at all and (2) a **discount charge** for offering a lower interest rate. For a fixed rate mortgage the combined effect of the interest rate, fee, and discount charge can be summed up in the annual percentage rate. For an adjustable rate mortgage the fee and discount charge are typically smaller, and the annual percentage rate cannot be computed because the interest rate will be changing.

EXAMPLE 6.20 If you are going to borrow \$80,000 for a home and there is a 1 point loan origination fee and a 1 point discount charge, what will be your added costs?

SOLUTION Each of these expenses will cost 1% of \$80,000, or \$800. Thus your added costs will be \$1600. ◆

One question is "Do I want a fixed rate mortgage or an adjustable rate mortgage?" Unfortunately, the answer depends on what is going to happen to interest rates in the future. For example, if you knew that interest rates were going to go down significantly over the life of your mortgage, then you would want to have an adjustable rate mortgage so that the interest rate you pay would go down also. On the other hand, if you knew that interest rates were going to go up over the life of your mortgage, then you would want a fixed rate mortgage to lock in the initial lower interest rate. Of course, if you really knew in advance what interest rates were going to do, you could use your psychic powers to make significant amounts of money in the stock market, since changes in interest rates have well-known effects on stocks and bonds.

In the real world, you do not have advance knowledge of the direction of interest rates, although the rates are carefully monitored and analyzed by financial professionals. Interest rates reflect national economic policy and economic conditions, with the Federal Reserve setting the pace. You must make your choice on the basis of what you can afford when you initially borrow the money, your expectations concerning your future income, and the amount of uncertainty you can accept. If rates are changing, timing may be the most important factor.

A sample table of mortgage loan information is included in the table below.

TABLE 6.6 Interest Rate Table

Banks	Term	Rate	F + DC	APR
Bank of America	15	6.625	1.75	6.94
# (4207) 1-800-284-5626	30	7.125	1.75	7.32
Bank of Newport	15	6.5	1.875	6.8277
624-5864	30	7.125	1.5	7.2945
Clackamas Co. Bank	15	6.375	1.775	6.706
# (4208) 669-7205	30	6.875	1.925	7.097
First Interstate	15	6.5	1.75	6.865
# (4209) 225-3922	30	7	1.75	7.23
First Security Bank of Oregon	15	6.5	2	6.9199
643-5934	30	7	2	7.2672
Key Bank	15	6.5	2.125	6.842
292-7360	30	7	2.125	7.2142
Liberty Federal Bank	15	6.625	1	6.818
231-4370	30	7.125	1	7.246
U.S. Bank	15	6.375	2.25	6.777
1-800-392-2412	30	7	1.875	7.215
West One Bank	15	6.5	1.75	6.7892
684-6633	30	7	1.75	7.182

Similar tables are published weekly in most major newspapers. (The columns in the table are mostly self-explanatory, but the "F + DC" column is not. This column gives the points charged.) Scanning through the table you can see that the annual percentage rate on a 30-year fixed mortgage was about 7%.

THE DOWN PAYMENT

A typical **down payment** on a house is 20% of the total value of the house. The published rates in the table assume such a 20% down payment. If the down payment is 20% of the value of the property, then the value of the property you can buy is the down payment divided by 0.20.

EXAMPLE 6.21 If you have \$20,000 for a down payment, what is the highest priced home you can afford if a 20% down payment is required?

SOLUTION $\frac{\$20{,}000}{0.20} = \$100{,}000$. Thus you can shop for a \$100,000 house. ◆

In general

$$\text{total value} = \text{down payment dollars} \div \text{down payment percentage}$$

where the percentage must be converted to a decimal.

A lender prefers a large percentage down payment as protection against the borrower defaulting on the loan. If payments are missed and the lender has to take control of the property, the down payment is a significant part of the value that can be recovered. This lessens the risk to the lender. If a smaller down payment is made, you must expect to pay in some way for the lender's increased risk. The interest rate may be higher, or you may be required to purchase Private Mortgage Insurance—which has a price in points—to insure the lender against a default on the loan. Another possibility for qualified buyers is an FHA (Federal Housing Administration) loan, which not only insures the lender against default, but also requires a lower down payment. Down payments smaller than 20% are becoming more common, in particular in programs aimed at attracting first-time buyers. Returning to Example 6.21, if the \$20,000 is used to make a 5% down payment, then the value of the property being considered could be as high as

$$\frac{\$20{,}000}{0.05} = \$400{,}000.$$

Note that if you bought the \$400,000 house with 5%, or \$20,000, down, that would mean a loan of \$380,000. Even with a low rate of 6%, the interest alone would be \$1900 the first month!

EXAMPLE 6.22 The Garcia family purchases a home for \$150,000. They pay 20% down and pay all the closing costs. If the balance is financed at 7% for 30 years, what will their monthly payment be for principal and interest?

SOLUTION The 20% down payment is $\$150{,}000 \times 0.20 = \$30{,}000$. The balance to finance is $\$150{,}000 - \$30{,}000 = \$120{,}000$. Using Table 6.5 we see that the payment for principal and interest on \$120,000 at 7% over 30 years is

$$120 \times 6.653025 \approx \$798.37.$$ ◆

TIDBIT

Because of the complex way ownership of real estate is determined, it is necessary to purchase insurance against the possibility that someone else actually owns the house you bought and paid for! This is called title insurance.

REMEMBER THE POINTS

The loan origination fee and the discount charge were mentioned earlier. Table 6.6 shows the F + DC factor varies from a low of 1 point to a high of 2.25 points. If the down payment is less than 20% there may well be mortgage insurance points. Additionally, there may be other charges when the transaction is finalized. This finalizing of the purchase is called the **closing**, and the additional expenses are called **closing costs.** If you purchase a home, you should ask your realtor in advance how much this will cost, so that you can have enough in your checking account to cover it. The costs vary from state to state, and whether the buyer or seller pays for a particular item may be negotiable.

INITIAL PROBLEM SOLUTION

Suppose you have saved \$15,000 towards a down payment on a home, and your total household income is \$35,000 per year. What is the most you could afford to pay for a home? Assume that (1) your insurance costs will be 0.25% of the value of your home, (2) your taxes will be 2% annually, (3) your closing costs will be about \$2000, and (4) you can obtain a fixed rate mortgage for 30 years at 8% interest.

SOLUTION First, we estimate the maximum loan you can afford. In addition to the down payment, a limiting factor to the amount of loan you can obtain is the monthly payment you can afford to make. Your monthly obligation includes your payment, home insurance, and any real estate taxes (which will vary from state to state).

Your total income is \$35,000 per year, and you have \$15,000 available to use for the purchase. However, \$2000 will be needed for closing costs, leaving \$13,000 for your down payment. By the first rule of thumb, you might look for a home in the $3 \times \$35{,}000 = \$105{,}000$ range. By the second rule of thumb, you should be able to afford to make monthly payments of 25% of $\frac{1}{12}$ of \$35,000, or about \$730 per month. The following calculations will determine if you can afford a house in this price range.

Cost of the house	\$105,000
Down payment	13,000
Amount needed to finance	92,000
Monthly payment on 30 year 8% loan (use Table 6.5)	675
Cost of insurance and real estate taxes $\left(\frac{1}{12}\text{ of } 2.25\% \text{ of } \$105{,}000\right)$	197
Total monthly payment	\$872

According to the second rule of thumb, you will be able to afford a monthly payment of \$730. Since \$872 exceeds \$730, there are several things you might do to afford the house in this price range:

1. wait for interest rates to fall,
2. increase your income(s),
3. come up with a larger down payment.

With the same down payment, a \$100,000 house with a 7% loan would require a monthly payment of \$767—still a bit too high. If you increased the down payment to \$25,000, then the monthly payment for the 8% loan would be \$738, which is close enough to have a loan approved.

If the bank uses the more liberal guideline of 38%, you could afford a monthly payment of approximately \$1110, and the loan would most likely be approved.

PROBLEM SET 6.4

1. If a family's annual income is \$35,000, what do the two rules of thumb tell them regarding the purchase price and monthly payments for a potential home purchase?
2. If a family's annual income is \$52,800, what do the two rules of thumb tell them regarding the purchase price and monthly payments for a potential home purchase?
3. Martha and Ed have a combined income of \$43,550 a year. Based on the two rules of thumb, what are their considerations concerning price and monthly payments if they decide to buy a house?
4. After he graduates from college, Philip is going to invest in a house rather than rent. He anticipates going to work for a software engineering company for at least \$32,000 per year. If he follows the rules of thumb, what is the maximum he should consider paying for a house? What about monthly housing expenses?
5. Using the two rules of thumb, what is the maximum Paul and Ann should pay for a house if his annual income is \$22,000 and hers is \$31,000? How much should they be willing to pay for monthly housing expenses?
6. Richard and Rose Sheng are considering buying a new home. He has an annual income of \$35,000 and hers is \$24,000. What guidelines should they follow regarding purchase price and monthly housing expenses?
7. If a couple has a combined monthly income of \$3650, what are the low and high estimates of what they can afford to pay for monthly housing expenses?
8. If a couple has a combined monthly income of \$2980, what are the low and high estimates of what they can afford to pay for monthly housing expenses?
9. The Baileys are going to borrow \$95,000 for a new home. If there is a 1 point loan origination fee and a 1 point discount charge, what are their costs for these two items?

10. If there is a 1 point loan origination fee and a 1 point discount charge on a home loan of $78,500, what is the total for these two charges?
11. A home loan of $92,000 is subject to a 1.5 point loan origination fee and a 1 point discount charge. What is the total for these two charges?
12. There is a 1.25 point loan origination fee and a 1 point discount charge on a new home loan for $165,000. What is the total for these two charges?
13. John and Sue have been saving their money to buy a home and have $23,000 available for a down payment. The realtor says that they should expect to need 20% for a down payment. What is the maximum price they will be able to afford based upon the down payment?
14. Referring to problem 13: If John and Sue are able to qualify for a special loan program for new home buyers, the realtor tells them they will only need a down payment of 10%. If this is true, how much will they be able to afford based upon their down payment?
15. If a down payment of 20% is needed and a couple has $18,000 for a down payment, what price home could they buy? What annual income should they have in order to follow the rule of thumb regarding purchase price?
16. If a 10% down payment is needed for a home in a new subdivision, what is the maximum price the Addams can consider if they have $15,000 for a down payment?
17. The Davis family sold some property and are in the market for a new home. They have $18,000 available for the home purchase and initial costs. What is the maximum price they will be able to pay if they need $2500 for closing costs and 10% for a down payment?
18. Referring to problem 17: Suppose the Davis family finds a house they wish to buy priced at $135,000, but the seller requires a 20% down payment. How much additional money do the Davis's need?
19. Use Table 6.5 to find the monthly payment for principal and interest on a loan of $72,000 at 12% interest for 30 years.
20. Use Table 6.5 to find the monthly payment for principal and interest on a $91,800 loan at 10% interest for 15 years.
21. A family decides to buy a home at an agreed upon price of $135,000. After making a down payment of 20% and paying all closing costs, the balance was financed at 8% for 30 years. What is the monthly payment for principal and interest? Use Table 6.5.
22. A home was sold for $92,500 with a down payment of 20%. The remaining balance was financed at 9% for 20 years. What is the monthly payment for principal and interest? Use Table 6.5
23. A home priced at $127,500 sold with a 10% down payment. The remaining balance was then financed at 9.5% for 30 years. Estimate the monthly payment for principal and interest. Use Table 6.5 and average the amounts corresponding to 9% and 10%.
24. A home sold for $115,500 with a down payment of 10%. The remaining balance was financed at 10.5% for 20 years. Estimate the monthly payment for principal and interest. Use Table 6.5 and average the amounts corresponding to 10% and 11%.
25. A home was sold for $95,500 with a down payment of 10%. The remaining balance was financed at 9.6% for 30 years. Find the monthly payment for principal and interest using the monthly payment formula.
26. A loan balance of $74,200 was financed at 8.85% for 20 years. Find the monthly payment for principal and interest using the monthly payment formula.

Home mortgages are often financed through lending agencies that require reserve funds be collected as part of the monthly mortgage payment to cover property taxes and homeowner's insurance. In problems 27 through 30, you'll be given the assessed value of the house, the amount that is financed, the rate, the term, and the information on taxes and insurance. Use Table 6.5 and find the monthly payment for principal, interest, taxes, and insurance.

27. Assessed value = $150,000
Loan amount = $115,000
Rate = 9%
Term = 30 years
Taxes = 2.5% of assessed value
Insurance = $650 per year
28. Assessed value = $120,000
Loan amount = $105,000
Rate = 8%
Term = 20 years
Taxes = 2.5% of assessed value
Insurance = $480 per year
29. Assessed value = $127,700
Loan amount = $75,000
Rate = 10%
Term = 15 years
Taxes = 2.25% of assessed value
Insurance = $740 per year
30. Assessed value = $225,000
Loan amount = $165,000
Rate = 9%
Term = 30 years
Taxes = 1.85% of assessed value
Insurance = $1260 per year

? EXTENDED PROBLEMS

31. Find current data on interest rates for mortgages in your area.
32. Consider the difference between 15-year and 30-

year mortgages on a $100,000 home. Use typical current rates in your area if possible.

(a) How much is the difference in monthly payments?

(b) How much is the difference in the total money paid over the entire loan? Include points if that information is available.

(c) Which seems to be the better deal?

33. The saying "Possession is nine points of the law," is due to the 17th century writer Thomas Fuller. The fact that ownership of real estate is proved by title refutes Fuller for that one form of property. For what other types of property is ownership provable by title?

34. The familiar frame house was a technical innovation developed in the 1830s in Chicago, Illinois, and Rochester, New York. Why was this type of construction invented, and what was its importance?

35. What was the significance of Levittown to widespread home ownership?

36. What programs are available through the federal government that promote home ownership through low down payments or low interest rates?

6.5 ANNUITIES AND SINKING FUNDS

In 18 years you would like to have $50,000 saved for your child's college education. At 6% interest, compounded monthly, what monthly deposit must be made to accomplish this goal?

Many of our important purchases or financial commitments are taken care of after the fact through loans and mortgages. However, this is not a practical strategy for many situations, such as retirement. Many people also prefer to have their funds set aside before obligating the funds, while government entities may legally be required to do so, and businesses may find it necessary or prudent. These situations require thoughtful planning, and the key to success is making appropriate regular payments to carry out the plan.

TERMINOLOGY

HISTORY

Large amounts of cash sitting in a sinking fund have sometimes been an irresistible temptation for governments and companies and have been raided for other purposes than originally intended.

An **annuity** is an account into which a sequence of equal, regular payments are made. For example, an account to which a person contributes one hundred dollars a month until retirement is an annuity. An annuity is really just a saving strategy. It is a useful strategy for two reasons. First, the regular nature of the deposits helps the saver maintain discipline. Second, if the payments are made over a long time period, the interest received contributes significantly to the growth of the annuity, especially the compounding of the interest.

Unfortunately, the financial world also has another meaning for the word "annuity." In the second context, if you purchase an annuity you will receive a sequence of equal, regular payments. To distinguish between the two meanings, we will call the latter type of annuity an **income annuity.**

A **sinking fund** is a type of annuity in which the goal is to have a particular amount of money saved at the end of a given time period. Historically, the term "sinking fund" referred to the method state and local governments used to accumulate enough money to pay off bonds issued for improvements such as roads and bridges. To raise money for such improvements, bonds (a type of promissory note) would be sold at a discount from face value (the value stated on the bond). Bonds are generally issued for periods of 10 to 30 years, and the issuer of the bonds has no obligation to pay until the end of the bonding period, known as the maturity date. When the maturity date arrives, the issuer pays the holders of the bonds the face value of the bonds. This is called "redeeming" the bonds. To insure that the money for redemption was available when the maturity date arrived, regular deposits would be made to a sinking fund during the life of the bonds. The size of the regular payments was chosen so that there would be sufficient money in the fund.

There are two basic questions with annuities. One is to determine how much will accumulate over time when given periodic payments are made, and the other is to determine what periodic payments will be necessary in order to accumulate a specific amount in a given period of time. In a sense, these questions are the reverse of those when amortizing loans.

Suppose you want to pay for something that costs more than you can afford from your paycheck, say a vacation. One strategy is to borrow the money; another is to set aside a smaller amount from each paycheck until you have accumulated enough for what you want to do. If you regularly set aside a given amount of money in an account that pays interest, then you have an annuity.

EXAMPLE 6.23 Mary wants to take a nice vacation trip, so she begins setting aside \$250 per month. If she deposits this money on the first of each month in a savings account that pays 6% interest compounded monthly, how much will she have at the end of 10 months?

SOLUTION Mary's first payment of \$250 will earn interest for ten months. In order to make the calculation, we use the Compound Interest Formula

$$A = P\left(1 + \frac{r}{n}\right)^{nt}, \text{ where } r = 0.06,\ n = 12, \text{ and } t = \frac{10}{12} = \frac{5}{6} \text{ (year).}$$

Ten months later that \$250 will have grown to

$$250 \times \left(1 + \frac{0.06}{12}\right)^{12(5/6)} = 250 \times (1.005)^{10} = 262.79.$$

The amount of \$262.79 is referred to as the future value of the first payment. The future value of the second payment is

$$250 \times (1.005)^9 = 261.48,$$

since that payment earns interest over nine months. The future values of the other payments are

Payment	Future Value	Payment	Future Value
3rd	$250 \times (1.005)^8 = 260.18$	4th	$250 \times (1.005)^7 = 258.88$
5th	$250 \times (1.005)^6 = 257.59$	6th	$250 \times (1.005)^5 = 256.31$
7th	$250 \times (1.005)^4 = 255.04$	8th	$250 \times (1.005)^3 = 253.77$
9th	$250 \times (1.005)^2 = 252.51$	10th	$250 \times (1.005)^1 = 251.25$

Totaling up the future values, we see that Mary will be able to withdraw \$2569.80 for her vacation trip. Since she only deposited \$2500, the additional \$69.80 was interest earned. ◆

Notice that the preceding calculation used the fact that the compounding period for the interest is the same as the length of time between payments, called the **payment period.** An annuity for which the interest compounding period is the same as the payment period is called a **simple annuity.** We will restrict our attention to simple annuities. The term of an annuity is the length of time from the beginning of the first payment period until the end of the last payment period. In Example 6.23, the term of the annuity was 10 months. At the end of the term of the annuity, the annuity is said to have **expired** and the money may be withdrawn.

Payments to an annuity may be due at the beginning of each payment period, as in the example, or at the end of each payment period. An annuity for which payments are due at the beginning of each payment period is called an **annuity due** (Example 6.23 illustrates an annuity due). An annuity for which payments are due at the end of each payment period is called an **ordinary annuity.** The difference between an annuity due and an ordinary annuity is that each payment to an ordinary annuity will be on deposit for one less payment period. Therefore, every payment to an ordinary annuity will earn interest for one less compounding period than an annuity due. The difference can be significant.

FINDING THE FUTURE VALUE OF AN ANNUITY

If an annuity involves many payments, then working out the future value of each payment and adding them together as we did in the example will be too time-consuming. Instead, the following formulas can be used to find the total at the end of the term. The amount of money available for withdrawal when the annuity expires is called the **future value of the annuity** and will be represented by the letter F in the various formulas in this section. We will also use i to represent the **periodic interest rate,** which is the annual interest rate, r, divided by the number of periods.

Future Value of an Annuity

If p is the payment size, r the annual rate (expressed as a decimal), and n the number of payment periods, then the periodic interest rate is $i = \frac{r}{m}$ (where m is the number of payment periods per year), and the future value is

$$F = p \times \frac{(1 + i)^n - 1}{i} \quad \text{for an ordinary annuity,}$$

$$F = p \times (1 + i) \times \frac{(1 + i)^n - 1}{i} \quad \text{for an annuity due.}$$

It is when n, the number of payment periods, is large that the effect of compound interest becomes important.

EXAMPLE 6.24 Find the future value of an ordinary annuity with a term of 25 years, payment period of one month, payment size of \$50, and annual interest rate of 6%.

SOLUTION With monthly payments, we have $n = 12 \times 25 = 300$. The interest rate per payment period is $i = 0.06 \div 12 = 0.005$. The payment size is $p = \$50$. Using a calculator to apply the ordinary annuity formula above, the future value of the annuity is

50 [×] [[ ] [[ ] 1 [×] 0.005 [ ]] [x^y] 300 [−] 1 [ ]] [÷] 0.005 = [34649.69812],

or $F = \$34,649.70$ ◆

Note that in this example the total of 300 \$50 payments is only \$15,000. This means that the interest earned is over \$19,000. In many annuities, particularly retirement accounts, the interest earned will significantly exceed the amount contributed by the saver.

FINDING PAYMENTS FOR SINKING FUNDS

In setting up the formulas for the future value of an annuity, we assumed that the payment size had been chosen in advance and that the future value was to be found. The distinguishing feature of a *sinking fund* is that the future value is chosen in advance, and the payment size has to be found. We will continue to use p for the payment size, i for the interest rate per period, and n for the total number of periods. The formulas for sinking funds are as follows:

Payments for Sinking Funds

If p is the payment size, F the future value, i the interest rate per period as a decimal, and n the number of payment periods, then

$$p = F \times \frac{i}{(1+i)^n - 1} \text{ for an ordinary annuity,}$$

$$p = F \times \frac{i}{(1+i) \times [(1+i)^n - 1]} \text{ for an annuity due.}$$

EXAMPLE 6.25 Suppose you decide to use a sinking fund to save \$10,000 for a car. If you plan to make 60 monthly payments and you receive 1% interest per month, what is the required payment for an ordinary annuity?

SOLUTION We apply the formula for Payments for Sinking Funds with $F = \$10,000$, $n = 60$, and $i = .01$. We find

$$p = 10,000 \times \frac{0.01}{(1+0.01)^{60} - 1} \approx \$122.45.$$ ◆

INCOME ANNUITIES

An income annuity provides a sequence of regular equal payments to an individual, the **annuitant,** or to several individuals, the annuitants. The payments may be made over a fixed period of time, in which case the annuity is called an **annuity certain,** or the payments may continue indefinitely until some contingency event occurs. The most common contingency that ends an annuity is the death of the

TIDBIT

Variable annuities are annuities in which the payment size is based on some economic index. This is done to provide protection against inflation.

annuitant. An annuity that is terminated by a death is called a **life annuity;** it is good for the lifetime of the annuitant. Therefore, the life expectancy of the annuitant is a major factor in determining the cost of the annuity. Retirement benefits are life annuities and are often variable with a cost-of-living adjustment. By contract, the cost of an annuity certain can be readily computed without reference to the age or health status of the annuitant.

EXAMPLE 6.26 The parents of a college student give her a lump sum of $300 with the admonition "This is your pocket money for the term. Make it last." Assuming the term is 10 weeks long, describe how the student can turn this into an income annuity.

SOLUTION The student could ask her parents to keep the money at home, and send her $30 each week. By giving up the lump sum of $300 the student has effectively purchased an annuity certain with a term of 10 weeks, a payment period of 1 week, and a payment size of $30. ◆

The student in Example 6.26 protected herself from the temptation to spend the pocket money too fast, but she did not earn any interest on the money. Usually, one expects an annuity certain to earn interest. The process involved in an annuity certain is the same as putting a specific amount of money into an interest paying account and making periodic equal withdrawals, but no deposits, until the account is empty.

Table 6.7 allows the computation of the purchase of an annuity certain with a payment period of one year and a compounding period of one year. The term of the annuity (that is the number of years for which the payments will be received) determines the row of the table in which to look, and the interest rate determines the column. The figure in the intersection of the row and column is called the **annuity factor.** For each dollar of yearly income desired you must pay that many dollars to purchase the annuity. The first payment to you will be made one year after purchase of the annuity. For example, suppose you would like to purchase an annuity that provides a yearly income of $10,000 for 10 years, and the offered interest rate is 6%. You will need to make a lump sum payment of $10,000 × 7.36 = $73,600, where the number 7.36 is the entry in Table 6.7 in the 10-year row and in the 6% column.

TABLE 6.7

	Return			
Years	**3%**	**6%**	**8%**	**10%**
10	8.53	7.36	6.71	6.14
15	11.94	9.71	8.56	7.61
20	14.88	11.47	9.82	8.51
25	17.41	12.78	10.67	9.08
30	19.60	13.76	11.26	9.43

EXAMPLE 6.27 What is the price to purchase a 20-year income annuity paying $6000 per year if the return on investment is 8%?

SOLUTION Consulting the table, we find the entry in the 20-year row and 8% column is 9.82. The cost of the annuity would be

$$\$9.82 \times 6000 = \$58,920.$$

Over the life of the annuity, the annuitant will receive a total of $120,000. ◆

INITIAL PROBLEM SOLUTION

In 18 years you would like to have $50,000 saved for your child's college education. At 6% interest, compounded monthly, what monthly deposit must be made to accomplish this goal?

SOLUTION We must determine the payment size for a sinking fund with a future value of $F = \$50{,}000$. The term of the sinking fund is 18 years and the payment period is one month, so the number of payments is

$$n = 12 \times 18 = 216.$$

The interest received per period is

$$i = 0.06 \div 12 = 0.005.$$

It is not specified whether payment is to be made at the beginning of each period or at the end of each period, so we will do it both ways. Plugging the given information into the two formulas, we find

$$p = \$128.44$$

if payment is made at the beginning of each month (an annuity due) and

$$p = \$129.08$$

if payment is made at the end of each month (an ordinary annuity). In actual practice you would most likely round up to $130 for convenience.

PROBLEM SET 6.5

1. The Russell family decides to save for a vacation by putting aside $300 at the start of each month for the next six months in an account paying 6% interest compounded monthly. When they withdraw the money and interest six months later how much will they have for the vacation?
2. Theo Amundsen wants to save some money for a big screen television. He puts aside $200 at the start of each month in an account paying 8% interest compounded monthly. After 12 months he will withdraw the money and the interest. How much will he have for the television?
3. Gerry needs to save some money for orthodontia. At the start of each month he puts $150 in an account paying 8% interest compounded monthly. After ten months how much money will he have saved, including interest?
4. Shawna decides to save money for a down payment on a new car. She sets aside $175 at the start of each month in an account paying 5% interest compounded monthly. After 12 months how much money will she have saved, including interest?
5. Find the future value of an ordinary annuity with a term of 25 years, payment period of one month, payment size of $50, and annual interest rate of 6% compounded monthly.
6. Find the future value of an ordinary annuity with a term of 20 years, payment period of one month, payment size of $75, and annual interest rate of 5%.
7. Traci makes a deposit of $60 at the end of each month into an account paying 5% interest. How much will she have in the account at the end of two years?
8. Brian saves for a sailboat by making payments of $100 to an account at the end of each month. If the account pays 6% interest, how much will be in the account at the end of 5 years?
9. Find the future value of an annuity due with a term of 25 years, payment period of one month, payment size of $50, and an annual interest rate of 6%.
10. Find the future value of an ordinary annuity with a

20-year term, payment period of one month, payment size of $80, and an annual interest rate of 7%.

11. If an ordinary annuity has a 5 year term, a payment period of one month, $80 payments, and an annual interest rate of 6%, what is its future value?

12. What is the future value of an annuity due with a payment of $75 a month, a term of 5 years, and an annual interest rate of 5.4% compounded monthly?

13. Find the future value of an annuity due with a term of 25 years, payment period of three months, payment size of $100, and an annual interest rate of 6%.

14. Find the future value of an ordinary annuity with a 20-year term, payment period of three months, payment size of $250, and an annual interest rate of 7%.

15. If an ordinary annuity has a 10-year term, a payment period of six months, $200 payments, and an annual interest rate of 6%, what is its future value?

16. What is the future value of an annuity due that has a payment of $300 every four months, a term of 10 years, and an annual interest rate of 5.4%?

17. What monthly payment is required for an annuity due to accumulate $25,000 in a sinking fund in five years with 12% annual interest?

18. If an ordinary annuity is used, what size monthly payment is required if $6,000 is to be accumulated in a sinking fund in four years with 10% annual interest?

19. Suppose you decide to use a sinking fund to save $10,000 for a car. If you plan on 60 monthly payments and you receive 1% interest per month, what is the required payment for an annuity due?

20. Suppose you use a sinking fun to save $12,000 for a car. If you plan on 60 monthly payments and you receive 0.5% interest per month, what is the required payment for an ordinary annuity?

21. A family decides to use a sinking fund to save $15,000 for the down payment on a house. If they plan on 48 monthly payments and receive 0.8% interest per month, what is the required payment for an ordinary annuity?

22. A small company sets up a sinking fund to save $50,000 for a new phone system. If the company makes 60 monthly payments and receives 0.75% interest per month, what is the required payment for an annuity due?

23. The county government sells a series of bonds that have a face value of $8,000,000 that are to be redeemed in 10 years. The county sets up a sinking fund with 6% APR to cover the bonds. If monthly payments begin with the current month (an annuity due) what minimum monthly payment is required so that the bonds will be covered by the date of redemption?

24. The school district issues $6.5 million in bonds for new high school computer laboratories. The district will issue the bonds on the same date as the beginning of the district's budget year and start a sinking fund to redeem the bonds when they mature. Ten annual payments will be made beginning next year (an ordinary annuity). What size payment is required if the district can get 9% interest?

25. What is the price to purchase a 20-year income annuity paying $6,000 per year if the return on investment is 10%?

26. What is the price to purchase a 20-year income annuity paying $10,000 per year if the return on investment is 6%?

27. What amount is needed to purchase a 30-year income annuity paying $20,000 per year if the return on investment is 8%?

28. Find the purchase price of a 15-year income annuity paying $15,000 per year if the return on investment is 6%.

29. The Jordan's set up a fund for their son's college expenses. They will put $40 a month into the fund beginning on his third birthday and ending on his 18th birthday. How much will be in the fund when he turns 18 if the fund earns 5.4% annually, compounded monthly? (The first payment is on his third birthday and the last is on his 18th birthday.)

30. A couple decides to save money for the down payment on a home by putting $250 into an account at the beginning of each month for the next four years. How much will they have available if the annual interest rate is 6%?

EXTENDED PROBLEMS

31. The Browns establish an account that they will use to pay cash when buying a new car. They put $5000 into the account when they open it and will deposit $200 at the end of each month for the next three years. If their account has an annual interest rate of 5.75%, how much will they have in the fund at the end of three years? (Hint: treat this as *two* problems.)

32. Suppose you want to have $10,000 available to buy a new car in 3 years and decide to use the following strategy: you will make a lump sum deposit into an account now and then make monthly payments at the beginning of each month. Each of the sources of funds should generate $5000 toward the total. If you can get a 6% rate of interest, how much should you deposit as a lump sum, and what should the monthly payment be?

33. On her 35th birthday, Merrie decides that when she is 50 she would like to take two years off from work so that she can travel and see the world. To be able to do this, she will make monthly deposits into a fund that will pay an annual rate of 6%, and then when she turns 50 start making monthly withdrawals. Merrie estimates that she will need $2500 per month for her expenses when she starts to travel.

(a) How much will Merrie need to have available in the account when she is 50?

(b) How much should Merrie deposit at the beginning of each month to achieve her goal?

34. When Ted begins the seventh grade, his mother begins saving for his college living expenses. Her plan is to make monthly deposits into an account earning 6% interest until Ted begins college in six years, then she will make regular withdrawals from the account to cover his living expenses. She estimates that Ted will need $600 a month for four years, including summer. How much should she deposit each month?

35. Research the options open for individuals who want to establish tax-sheltered annuities. What are eligibility requirements? What are the limitations on the amounts that can be deposited? What other restrictions exist?

36. What are the advantages of tax-sheltered annuities to employees? To employers? To the government?

37. Why would a Swiss annuity (bought in Switzerland and paying Swiss francs) be preferable to an annuity purchased in the United States?

38. One of the difficulties with long-term investing is that the entire nature of the economy can change over the course of a lifetime. Early in this century stock in railroad companies was considered by experts to be the soundest of investments. Research the economic fortunes and stock prices of the Pennsylvania Railroad (now Penn Central) during this century.

Chapter Six Problem

You are buying a car for $10,000 from a dealer who will give you a loan for 5 years at 10% interest. As an extra incentive he offers either $1200 cash back on the 10% loan or a 6% interest rate (which is a rate that you could expect to get if you deposited funds in a bank or money market account). Which deal should you choose?

SOLUTION

Strategy: Use a Table

One approach to making a rational decision is to compare the total amount you would pay under each option. That means you have to first find the amount of the monthly payments for each choice. From this information, you can compute the total for *all* payments.

Since the interest rate is a whole percent, we can use a table rather than computing the monthly payment from a formula. Table 6.5 can be used for this purpose. The part of the table we are interested in is the following:

Amount per $1000

Percent	5 years
6	19.332802
10	21.247045

We see that if the interest were 10% on a 5-year loan, then the monthly payments would be $21.247045 for a $1000 loan. Multiplying by 10 gives payments of $212.48 for a $10,000 loan. (Note that we rounded-up.) The total of the 60 monthly payments is $60 \times 212.48 = \$12{,}748.80$.

If the interest rate were 6%, then the monthly payments would be $19.332802 for a $1000 loan, which turns into $193.34 for a $10,000 loan. The total of the 60 monthly payments is $60 \times 193.34 = \$11{,}600.40$. (Note that the difference between the monthly payments is $19.14.)

One way to look at the problem is that you have a choice to either have $1200 now or have an extra $19.14 a month for the next 60 months. The total amount of money that you will keep by going with the lower interest is $\$19.14 \times 60 = \$1148$. Thus you should take the \$1200 rather than the lower interest rate since you will have \$1200 to spend now or to invest during the entire 60 months. If you can invest the \$1200 at 6% compounded monthly for the next 5 years, it would be worth \$1618. The net amount you would pay for the car would be $60 \times 212.48 - 1618 =$ \$11,130.80 compared to the \$11,600.40 you would pay for the car with the lower rate.

Chapter Six Review

Key Ideas and Questions

The following questions review the main ideas of this chapter. Write your answers to the questions, and refer to the pages listed to make certain that you have mastered these ideas.

1. How does simple interest differ from compound interest? 279, 281
Which seems to be the more fair method of paying interest? 279, 281
How can you compare different compound interest programs? For example, how would you compare borrowing money at 8% compounded annually with 7.6% compounded monthly? 283

2. Why is add-on interest less advantageous to the customer than simple interest? 289
How can you compare different loan arrangements such as add-on interest and rent-to-own at different rates? 289

3. Describe the history of an amortized loan and explain why the amount of principal paid increases each month. 294

4. Describe the process of buying a house and the rules of thumb designed to prevent a buyer from becoming overextended. 301

5. How does the purchase of an income annuity work? 312
Why is an annuity like an amortized loan in reverse? 308

Vocabulary/Notation

Following is a list of key vocabulary, notation, and ideas for this chapter. Mentally review each of these items, write down the meaning of each term, and use it in a sentence. Then refer to the pages listed by number, and restudy any material that you are unsure of before solving the Chapter Six Review Problems.

Section 6.1

Simple Interest 279
Principal 279
Interest Rate 279
Time Period 279
Ordinary Interest 280
Compound Interest 281
Compounding Period 281
Balance 281
Initial Balance 281
Annual Yield 283
Continuously Compounded Interest 284

Section 6.2

Simple Interest Loan 288
Finance Charge 288
Average Daily Balance 288
Add-on Interest 289
Annual Percentage Rate (APR) 289

Section 6.3

Amortized Loan 294
Term of Loan 294
Monthly Payment 294
Net Payment 295
Payment to Principal 295
Amortization Table 296

Section 6.4

Mortgage 302
Fixed Rate Mortgage 303
Adjustable Rate Mortgage 303
Cap 303
Term 303
Points 303
Loan Origination Fee 303
Discount Charge 303
Down Payment 304
Closing Costs 305

Section 6.5

Chapter Six Review Problems

1. Suppose that the principal on a loan is \$650, the interest rate is 8%, and the loan is made for a period of 4 years. What is the interest on the loan if it is calculated as simple interest?

2. Suppose that the terms of the loan in problem 1 are changed so that the rate is still 8%, but the loan was made June 1, 1995 and was to be repaid September 1, 1997. How much interest should be charged if it is calculated as ordinary interest?

3. Suppose a loan of \$650 is made and is to be repaid in 4 years with an interest rate of 8% compounded yearly. What total amount will be due in 4 years? How much of this is interest?

4. Suppose that the interest in problem 3 is to be compounded monthly. What total amount will be due in 4 years? How much of this is interest? Compare this with the interest payments in problem 1 and problem 3.

5. What is the annual yield of an investment at 9% if interest is compounded monthly?

6. Suppose that during a 30-day month the average daily balance on a credit card is \$412, and the interest rate is 19%. What is the finance charge for the month if the billing period is 30 days?

7. Suppose that on June 1 the balance on a credit card is \$0. On June 20 the card is used to purchase a television set for \$612. What will be the average daily balance for the month of June if this is the only transaction? What will the finance charge be if the interest rate is 21%?

8. What is the monthly payment if you borrow \$750 with an interest rate of 15% for 36 months, and the interest is computed as add-on interest? What is the approximate APR for this loan?

9. A television set and a VCR are offered for \$46 per month on a rent-to-own agreement. After 24 months, the television and VCR are yours. Suppose that the actual cost of the system is \$819. What is the approximate APR that you have paid?

10. A loan of \$61,000 is made at an interest rate of 9% and is to be paid over a 10-year period. The monthly payments are \$772.72 per month. Chart the first three months' history of this amortized loan.

11. Use a table to compute the monthly payment on a 10-year loan of \$16,000 at 11% interest per year.

12. Using the formula, find the monthly payment on a 3-year loan of \$5125 at 10% interest.

13. Suppose that Pat and Chris are hoping to buy a house. Their combined income is \$51,000 per year. Using the rules of thumb given in the text, what is the maximum purchase price they should pay for a home?

14. In problem 13, what are the high and low estimates of what they can afford to pay for a monthly payment, including mortgage, taxes and insurance?

15. Pat and Chris have found a house they like for \$110,000. They have \$14,000 in savings for the down payment and closing costs. The closing costs are expected to be \$3000. The insurance for a year is \$450, and they can assume taxes will be 3.5% of the value of the home. The bank will give them a 30-year loan at 8%. What will the monthly payments be? Can they afford these payments? Refer to problem 14.

16. Juanita sets up a savings account in which she pays \$40 on the first of each month beginning with February 1. The last payment will be made November 1, and the money will be withdrawn on December 1. How much money will she have if the interest rate is 6%?

17. In order to have \$50,000 in 10 years, how much should be deposited in an account every six months if interest is compounded semiannually with a rate of 6%? Assume the deposit is made at the beginning of each payment period.

18. Find the future value of an ordinary annuity with a term of 20 years, an annual interest rate of 5%, and a monthly payment size of \$60.

19. Using the annuity table find the amount of money needed to purchase a 25-year annuity paying \$10,000 per year if the return on investment is 6%.

PART 2

MATHEMATICS *in* SOCIETY

CHAPTER 7

GAME THEORY

HIGH SCHOOL PITCHER ANNOUNCES UNUSUAL PITCHING STRATEGY

During the last weeks of the regular baseball season, Fernando Valedictorian, the star pitcher for Gamesville High School, had increasing problems getting his curveball and slider in the strike zone. Before the start of the first game of the state playoffs, he told reporters his pitching strategy for the game. "I'm going to throw a curve ball or a slider for my first pitch to each batter, and then I'm only going to throw fastballs. That is my best pitch, it is the hardest to hit, and the batters will be confused since they will be looking for curves and sliders."

Chapter Goals

1. Compute outcomes in two-person games.
2. Determine when an optimal strategy is forced on players.
3. Compute the optimal mixed strategy for a player and the value of a game.
4. Find the average payoff for a player.

Whether you are preparing to play a game, take an exam, or begin a business negotiation, you should have a plan that is designed to increase your chances of success. In any activity where there are two sides with different goals in mind, such as a game or negotiations, any part of the plan that is designed to gain an advantage or fool the opposition is called a *strategy.*

Fernando has a strategy, but no coach would allow him to use it. Even though the fastball is Fernando's best pitch, it will be much less effective if the batter expects it. In baseball, and most other games, one side pays careful attention to what the other side is doing to see if they can discover the strategy being used. It would not take very long for Fernando's strategy to be revealed. In this situation, the correct strategy should be what is called a mixed strategy, one in which there is no evident pattern to the type of pitch being thrown. The pitch with the most likelihood of being chosen should be the fastball, but there should always be some element of surprise. The pitcher and catcher should also keep in mind the strengths and weaknesses of each batter and the game situation.

There are many situations when two people or groups of people have opposing interests. These include rival companies bidding on a project, negotiations between a union and an employer, two people playing tennis, countries negotiating a trade agreement, and so on. Both sides usually choose a strategy and do so in secret. This chapter covers the mathematics of competition and conflict in which there are two sides, and one side's loss is the other side's gain. It also describes how optimal strategies may be found from among the available options for action.

THE HUMAN SIDE OF MATHEMATICS

John von Neumann

John von Neumann (1903–1957) was one of the great mathematicians of the 20th century and a true genius. Even as a child of six he could divide 8-digit numbers in his head. His photographic memory was legendary. He could glance at a telephone directory page and recite from memory the numbers on the page. Von Neumann was born to a Hungarian Jewish family during a time of great persecution in Europe. After relocating from Hungary to Germany, he moved to the United States before the Nazi regime came to power.

Von Neumann found a permanent home at the Institute for Advanced Study at Princeton, New Jersey, as one of its founding members (along with Albert Einstein). Although a brilliant theoretical mathematician, von Neumann is best known for the application of mathematics in other fields. His first major contribution was the mathematical foundation of quantum mechanics: the description of the behavior of matter, in particular on an atomic scale. During World War II, von Neumann was a consultant to the theoretical group at Los Alamos, New Mexico that was charged with solving computational problems in building the first atomic bomb. He became intrigued with the design of machines that were being built to carry out the vast number of calculations needed to compile artillery tables for the army. These machines, the first primitive computers, were "hard-wired," which meant there were actual wires that had to be changed with each problem; this was a complex and time-consuming task. Von Neumann conceived of developing a logical design for the computers that could accept a flexible *stored* program. In doing so, he developed the theoretical basis of a modern computer. In 1952, von Neumann and his colleagues at Princeton created MANIAC (Mathematical Analyzer, Numerical Integrator, and Computer), which would serve as a model for future computers.

As early as the 17th century, mathematicians had considered ways to use mathematical methods to study human interaction and conflict. In 1927, von Neumann made his first important contributions to this effort. In collaboration with the economist Oskar Morgenstern, von Neumann is credited with founding game theory in 1944 with the publication of the book *Theory of Games and Economic Behavior.* When John von Neumann died in 1957 at the age of 54, he was still at the height of his intellectual powers.

Bertrand Russell

Bertrand Russell (1872–1970) was an English mathematician, philosopher, and political activist. He was born to unconventional freethinking parents and was the grandson of a prime minister to Queen Victoria. As a young mathematician he tried to reduce all of mathematics to a few axioms of logic and set theory. This turned out to be impossible. He then turned to philosophy, especially the philosophy of science. He wrote popular books of philosophy that were very liberal in terms of social mores. As a youth he changed from a militaristic proponent of the British empire to a pacifist. His beliefs led him more toward social writing and political action. He organized some of the first peace marches and designed the now familiar peace sign, which was a composite of the semaphore signal for N and D, signifying nuclear disarmament. Surprisingly, during the early days of the cold war, Russell favored a preemptive nuclear attack against the Soviet Union. This reasoning was based on von Neumann's mathematical theory of games and the belief that war with the Soviet Union was inevitable. Of course, his advice was not heeded by the political leaders of the day. Peace was found without resorting to nuclear war.

7.1 THE DESCRIPTION OF A GAME

Suppose two partners, Sid and Mark, have come to a parting of the ways and wish to dissolve their partnership, which has assets of about $250,000. They could come to an amicable agreement between themselves where each gets equal value, or they could settle the matter in court. Suppose legal expenses for the court battle would run about $12,500 for each side, and a fair and independent observer would likely give Sid 60% of the assets. Is it possible to look at their strategies in terms of a game? If so, how?

Conflict is an inevitable part of the human experience, occurring in settings ranging from romance to warfare. Throughout history people have tried to avoid conflict or eliminate it, some have tried to exploit it for gain, and many have also tried to understand it. Game theory is a modern branch of mathematics developed to analyze competitive situations. In this chapter, we introduce the mathematical framework for analyzing such situations. Some of our examples will, in fact, be games—perhaps ones that you've played and enjoyed. In a larger sense, however, what we are dealing with is a contest between two "opponents" who have different goals and how each achieves the maximum possible result.

GAMES

HISTORY

The first suggestions that mathematics could be applied to situations of conflict were attributed to Ernst Zermelo (1912) and Émile Borel (1921).

The following features are typically present when game theory is applied.

1. Conflicting interests of two or more participants. For example, two people playing chess, people playing poker, or several countries involved in trade negotiations. Each player wants to win.
2. Incomplete information. In card games you do not know what cards the other players hold. Incomplete information is not always a feature. In some games, such as chess, there is no hidden information since both players can view the entire board and know what moves are allowed.
3. An interplay of chance and rational choice. You assume the other players in the game are going to do the smart thing, but sometimes they may have the chance to make an unpredictable choice. Bridge players try to play the right card at the right time, but a poker player can bluff.

GAME TABLES

The participants in a game are called **players.** Each player has a selection of possible actions to take or choices to make. The possible choices or actions are called **strategies.** Once the choices have been made, the players receive a **payoff** determined by the choices made.

Many of the applications of game theory include very serious and important conflicts like warfare. Those situations are too complicated to use for a first example, and many others require specialized knowledge in a particular subject.

Instead, we will use simple or familiar situations to show how you go from a real-world game to a mathematical game theory analysis. The terminology and the principles introduced in the simple examples are the same as many of those in the more complex real-world situations.

EXAMPLE 7.1 There is a children's game known as "Rock, Scissors, Paper." This game has two players. On the count of three, each player extends a hand and does one of the following three things (Figure 7.1).

(1) Forms the hand into a fist. The fist resembles a rock.
(2) Holds out the first two fingers, slightly separated so that the two fingers resemble a pair of scissors.
(3) Holds out a flat hand. The flat hand resembles a piece of paper.

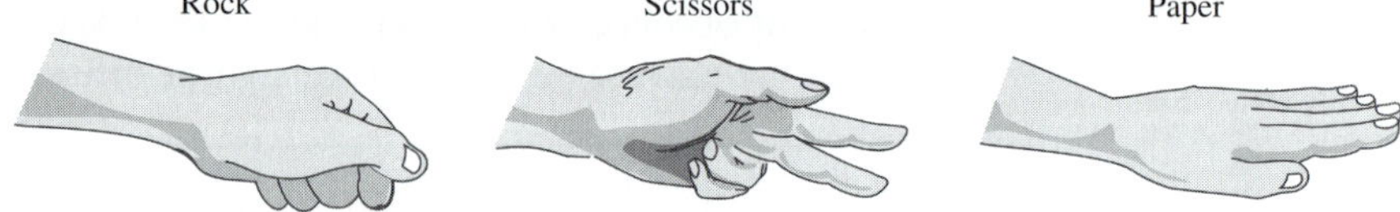

FIGURE 7.1

The two players do this simultaneously, and neither knows ahead of time what the other will do. If the two players make the same choice, that particular play of the game is considered a draw—neither person wins or loses. If the two players make differing choices, then one wins and the other loses. The following rules determine who wins.

- Rock breaks scissors (the player who chooses rock wins, the player who chooses scissors loses).
- Scissors cut paper (the player who chooses scissors wins, the player who chooses paper loses).
- Paper covers rock (the player who chooses paper wins, the player who chooses rock loses).

What are the strategies and payoffs?

SOLUTION Each player has the choice of the same strategies of rock, scissors, and paper. We can denote these by R, S, P, respectively. The possible payoffs are winning, losing, and drawing. We can denote these by W, L, D. We gather all the data about this game in a table as follows:

Rock, Scissors, Paper—Strategies and Payoffs

Player 1 Strategy	**Player 2 Strategy**	**Player 1 Payoff**	**Player 2 Payoff**
R	R	D	D
R	S	W	L
R	P	L	W
S	R	L	W
S	S	D	D
S	P	W	L
P	R	W	L
P	S	L	W
P	P	D	D

◆

In the preceding example, the strategies are the same for both players, but this is not generally required, and it is not always the case. In ordinary usage the word "strategy" usually means that some sort of well developed plan is involved. That is not the meaning here—strategy is just a name for a possible choice.

Each player has various **preferences** among the payoffs. For example, the players of Rock, Scissors, Paper are expected to prefer to win. Before we can analyze a game mathematically, we need to express the payoffs that are part of the game and the preferences of the players as a final numerical payoff. This may not be straightforward if the original payoffs are not quantitative. Even if the payoffs are initially numbers, perhaps battle casualties, the preferences may involve moral judgments or psychological factors that influence how these numbers are weighted. For example, reducing friendly casualties may be considered more important than increasing enemy casualties. Sometimes, final answers may only be given by history with the hope that we can learn from it.

In the Rock, Scissors, Paper game, the payoff was not quantitative. Somehow we must assign numerical values to win, lose, and draw. The choice of numbers is somewhat arbitrary. We will assign +1 to win, −1 to lose, and 0 to draw on the basis that win is preferred to draw and draw is preferred to lose. With this understanding, we rewrite the preceding table describing the game.

Rock, Scissors, Paper—Strategies and Numerical Payoffs

Player 1 Strategy	**Player 2 Strategy**	**Player 1 Payoff**	**Player 2 Payoff**
R	R	0	0
R	S	+1	−1
R	P	−1	+1
S	R	−1	+1
S	S	0	0
S	P	+1	−1
P	R	+1	−1
P	S	−1	+1
P	P	0	0

There is one reason for choosing to assign +1, −1, and 0 as we did. The numerical payoffs we assigned were such that, for any choices the two players made in one play, the sum of the numerical payoffs received by the players is zero. That is, the sum of the numbers across each row of the table is 0. Such a game is called **zero-sum game,** that is, a game where amounts won equal amounts lost. Poker games among friends are usually a zero-sum game, but if some of the money bet is devoted to buying refreshments, more is bet than is returned in winnings.

If we know that the game under consideration is a zero-sum game and if it also involves only two players, called a **two-person game,** then it is not necessary to list the payoff to the second player. The second player's payoff is automatically the opposite of the first player's payoff. We then can simplify the table by omitting the fourth column.

Rock, Scissors, Paper—Strategies and Numerical Payoffs

Player 1 Strategy	Player 2 Strategy	Player 1 Payoff
R	R	0
R	S	+1
R	P	−1
S	R	−1
S	S	0
S	P	+1
P	R	+1
P	S	−1
P	P	0

So far, we have taken a real-world game, Rock, Scissors, Paper, and described it as a two-person zero-sum game. We have shown how all the mathematically important information about this game, or any other two-person zero-sum game, can be given in a table listing the players' possible strategies and the numerical payoff to the first player.

LABELED GAME MATRICES

It is customary to further compress the information that describes a two-person, zero-sum game by putting all the data into a rectangular array called the **game matrix** (mathematicians use the term **matrix** (plural matrices) to refer to any rectangular array of numbers). For the Rock, Scissors, Paper game, we construct the game matrix as follows (Figure 7.2).

ROCK, SCISSORS, PAPER—GAME MATRIX

		Second Player		
		R	S	P
First Player	R	0	1	–1
	S	–1	0	1
	P	1	–1	0

FIGURE 7.2

The strategies R, S, and P along the leftmost column of the matrix always refer to the strategy chosen by the first player, and the R, S, and P along the top row refer to the strategy chosen by the second player, and the entries in the matrix are the payoffs for the first player.

MATRICES

A matrix is usually displayed between a pair of large brackets as in Figure 7.3.

$$M = \begin{bmatrix} 17 & 2.5 & 4 \\ 8 & -2 & -1 \\ 3 & 1 & 43 \\ -6 & 33 & 8 \end{bmatrix}$$

FIGURE 7.3

Each number in the matrix is called an **entry,** or **element,** of the matrix. All the entries in a horizontal line form a **row** of the matrix, and all entries in a vertical line form a **column** of the matrix. The rows of a matrix are numbered from the top to the bottom, and the columns are numbered from left to right. The second row of the matrix, M, has the elements (8, −2, −1), and the third column has the elements (4, −1, 43, 8). These are highlighted in Figure 7.4.

$$\begin{bmatrix} 17 & 2.5 & 4 \\ 8 & -2 & -1 \\ 3 & 1 & 43 \\ -6 & 33 & 8 \end{bmatrix} \quad \begin{bmatrix} 17 & 2.5 & 4 \\ 8 & -2 & -1 \\ 3 & 1 & 43 \\ -6 & 33 & 8 \end{bmatrix}$$

FIGURE 7.4

The **size of a matrix** is given as the number of rows and the number of columns, so M is a 4 by 3 matrix (4 rows and 3 columns). You pick out a particular entry of a matrix by giving its row and column position. For example, −1 is in the second row and third column of M (Figure 7.5).

$$\begin{bmatrix} 17 & 2.5 & 4 \\ 8 & -2 & -1 \\ 3 & 1 & 43 \\ -6 & 33 & 8 \end{bmatrix}$$

FIGURE 7.5

EXAMPLE 7.2 What is the size of the matrix A given below? What number is in the third row and fourth column (Figure 7.6)?

$$A = \begin{bmatrix} 2 & -3 & 0 & 4 & 43 \\ 8 & -2 & -7 & 55 & 8 \\ 3 & 1 & 1 & 29 & 0 \\ 88 & 33 & 5 & 7 & -1 \\ 0 & -6 & 31 & 0 & 9 \\ 19 & -1 & 3 & -1 & 0 \end{bmatrix}$$

FIGURE 7.6

SOLUTION We count six rows and five columns, so A is a 6 by 5 matrix. To find the entry in the third row and fourth column, we highlight that row and column, and notice that 29 is the number in both (Figure 7.7).

$$A = \begin{bmatrix} 2 & -3 & 0 & 4 & 43 \\ 8 & -2 & -7 & 55 & 8 \\ 3 & 1 & 1 & 29 & 0 \\ 88 & 33 & 5 & 7 & -1 \\ 0 & -6 & 31 & 0 & 9 \\ 19 & -1 & 3 & -1 & 0 \end{bmatrix}$$

FIGURE 7.7

ABSTRACT GAME MATRIX

For purposes of mathematical analysis, the details of each strategy are not important. That is, it does not matter that the game Rock, Scissors, Paper involves the possibility of forming a fist, sticking out two fingers, or holding the hand flat. The mathematician only cares that each player has three strategies and that the payoff is zero for both if they choose the same strategy, and so on. The abstract game matrix just uses the row numbers as names for the first player's strategies (e.g. the first player's second strategy is associated with the second row of the matrix) and the column numbers as the names of the second player's strategies (e.g. the second player's third strategy is associated with the third column of the matrix). It is also customary to eliminate the grid and just keep the matrix of payoff numbers. Recall that the payoffs of the second player are the opposites of the numbers in the matrix for a two–person zero–sum game.

The following matrix represents the Rock, Scissors, Paper game, which is a two-person zero-sum game (Figure 7.8).

$$\begin{bmatrix} 0 & 1 & -1 \\ -1 & 0 & 1 \\ 1 & -1 & 0 \end{bmatrix}$$

FIGURE 7.8

The rows represent the first player's three strategies, and the columns represent the second player's three strategies. Saying that the game is zero-sum means that the payoff for the second player is the opposite of the payoff for the first player; it has nothing to do with the numbers in the matrix adding up to 0. That is, the matrix $\begin{bmatrix} 2 & 3 \\ -4 & 2 \end{bmatrix}$ can represent a two-person zero-sum game.

We can use the matrix to determine the payoff for any choice of strategies. For example, suppose the first player chooses her third strategy and the second player chooses her second strategy. The payoff for the first player is then the entry of the matrix in the third row and second column (Figure 7.9).

$$\begin{bmatrix} 0 & 1 & -1 \\ -1 & 0 & 1 \\ 1 & -1 & 0 \end{bmatrix}$$

FIGURE 7.9

The payoff for the first player is −1, and the payoff for the second player is +1.

EXAMPLE 7.3 Suppose a particular two-person zero-sum game has two strategies for the first player and three strategies for the second player. Assuming the payoffs are given as in the following table, construct the abstract game matrix for the game.

Strategies and Payoffs

Player 1 Strategy	**Player 2 Strategy**	**Player 1 Payoff**	**Player 2 Payoff**
I	I	1	−1
I	II	7	−7
I	III	3	−3
II	I	−6	6
II	II	5	−5
II	III	−4	4

SOLUTION Player 1 has two strategies, I and II, and player 2 has three strategies, I, II, and III. Since each strategy of the first player corresponds to a row of the matrix and each strategy of the second player corresponds to a column of the matrix, we are going to be constructing a 2 by 3 matrix (Figure 7.10).

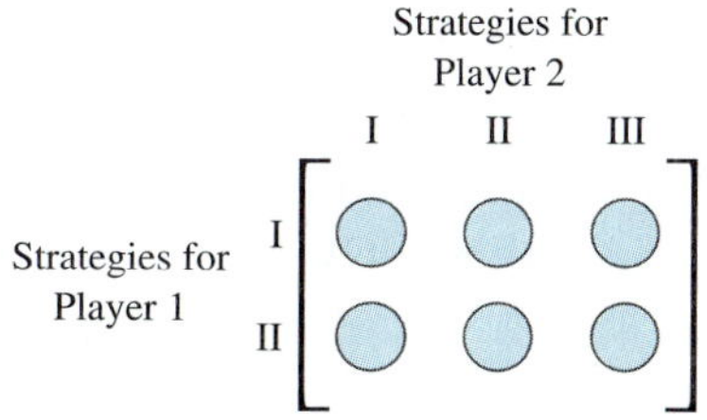

FIGURE 7.10

To fill in the entries of this matrix, we work from the table to arrive at the following matrix (Figure 7.11).

$$\begin{bmatrix} 1 & 7 & 3 \\ -6 & 5 & -4 \end{bmatrix}$$

FIGURE 7.11

All the entries represent the payoffs for player 1. On the other hand, the *opposites* of the numbers are the payoffs for player 2. ◆

EXAMPLE 7.4 Suppose we are given the following game matrix for a two-person zero-sum game (Figure 7.12).

$$\begin{bmatrix} 1 & 2 & 0 & 1 & 4 \\ 3 & 4 & -1 & 0 & 3 \\ -1 & 0 & -4 & -3 & 2 \\ 0 & 1 & -2 & -1 & 1 \end{bmatrix}$$

FIGURE 7.12

(a) How many strategies does the first player have to choose from?
(b) How many strategies does the second player have to choose from?
(c) If the first player chooses her second strategy and the second player chooses his third strategy, what payoff do both players receive?

SOLUTION

(a) There are four rows in the matrix, so the first player must have four strategies to choose from.
(b) There are five columns in the matrix, so the second player must have five strategies to choose from.
(c) The first player's second strategy corresponds to the second row of the matrix (Figure 7.13).

$$\begin{bmatrix} 1 & 2 & 0 & 1 & 4 \\ 3 & 4 & -1 & 0 & 3 \\ -1 & 0 & -4 & -3 & 2 \\ 0 & 1 & -2 & -1 & 1 \end{bmatrix}$$

FIGURE 7.13

The second player's third strategy corresponds to the third column of the matrix (Figure 7.14).

$$\begin{bmatrix} 1 & 2 & 0 & 1 & 4 \\ 3 & 4 & -1 & 0 & 3 \\ -1 & 0 & -4 & -3 & 2 \\ 0 & 1 & -2 & -1 & 1 \end{bmatrix}$$

FIGURE 7.14

The entry in the second row and third column gives the payoff for the first player, in this case -1. The second player's payoff is the opposite, or $+1$. ◆

Note that a two-person game does *not* require the row sum or column sum of the game matrix to be zero. It only means that the second player's payoff is the opposite of the first player's payoff.

HISTORY

American football is a descendent of the English game of rugby. A stone plaque at Rugby school reads "This stone commemorates the exploit of William Webb Ellis who with a fine disregard for the rules of football as played in his time first took the ball in his arms and ran with it, thus originating the distinctive feature of the Rugby game, A.D. 1823."

The American sport of football is particularly useful for illustrating game theory. The action in football is conveniently divided into separate units called "plays," and there are two players, the offensive and defensive teams. The basic choices for the offensive team on a typical football play are to either run or pass. (Of course, there are complicated combinations and variations which we ignore.) This choice is made before the play, when the offensive team huddles. The defensive team also has the opportunity to choose a defensive formation and plan. The defensive team has the basic strategies of using a defense designed to stop the run, a defense designed to stop the pass, or an all-purpose defense. The payoff for the offensive team, which we will consider the first player, is measured by yards gained by the offensive team on that play. In the next example we deal with the beginning of the analysis of a football play.

EXAMPLE 7.5 Suppose the following statistics about NFL teams are known.

- A pass attempted against a pass defense gains an average of -3 yards.
- A pass attempted against a run defense gains an average of 7 yards.
- A pass attempted against an all-purpose defense gains an average of 4 yards.
- A run attempted against a pass defense gains an average of 5 yards.
- A run attempted against a run defense gains an average of 1 yard.
- A run attempted against an all-purpose defense gains an average of 2 yards.

Set up the labeled game matrix.

SOLUTION We consider the first player to be the offensive team, and the second player will be the defensive team. The first piece of information "A pass attempted against a pass defense gains an average of -3 yards" corresponds to putting a -3 in the box that is in both the Pass row and the PD (for Pass Defense) column. The other five pieces of information are given as follows.

- A pass against a run defense results in a 7 in the Pass row and RD column.
- A pass against an all-purpose defense results in a 4 in the Pass row and AD (for All–purpose–Defense) column.
- A run against a pass defense results in a 5 in the Run row and PD column.
- A run against a run defense results in a 1 in the Run row and RD column.
- A run against an all-purpose defense results in a 2 in the Run row and AD column (Figure 7.15).

	PD	RD	AD
Pass	–3	7	4
Run	5	1	2

FIGURE 7.15 ◆

To begin to understand how the game matrix is used to analyze the situation, consider the strategy that offers the possibility of the maximum payoff, without regard to risk; this is called the **most aggressive strategy.** Consider the abstract game matrix associated with the football game in Example 7.5. To find the most aggressive strategy for the first player (the offensive team) we find the largest element in the game matrix and mark it with a star (Figure 7.16).

$$\begin{bmatrix} -3 & 7^* & 4 \\ 5 & 1 & 2 \end{bmatrix}$$

FIGURE 7.16

Because the largest element, 7, is in the first row, the most aggressive strategy for the first player is the one corresponding to the first row—Pass. The offense wants to gain yardage, and passing the football has the potential to gain the most.

Similarly, to find the most aggressive strategy for the second player (the defensive team) we find the smallest element in the game matrix, and mark it with two stars. Since the payoff to the second player is the opposite of the entry in the matrix, smaller is better from the second player's point of view (Figure 7.17).

$$\begin{bmatrix} -3^{**} & 7 & 4 \\ 5 & 1 & 2 \end{bmatrix}$$

FIGURE 7.17

Because the smallest element, -3, is in the first column, the most aggressive strategy for the second player is the one corresponding to the first column—Pass Defense. The defense wants to see the offense gain as few yards as possible or even lose yardage. The greatest potential for this is when the defense is using a pass defense against a passing play—they may be able to tackle the quarterback behind the line of scrimmage as he prepares to throw the football. If each of the players chooses his most aggressive strategy, the resulting payoffs are -3 for the offense and $+3$ for the defense (Figure 7.18).

$$\begin{bmatrix} -3 & 7 & 4 \\ 5 & 1 & 2 \end{bmatrix}$$

FIGURE 7.18

EXAMPLE 7.6 Given the following game matrix, find the most aggressive strategy for each player. What would be the payoff to each player if both players adopt the most aggressive strategy (Figure 7.19)?

$$\begin{bmatrix} -4 & 0 & 1 & 2 \\ 3 & -1 & 0 & 3 \\ 1 & -3 & -1 & 5 \end{bmatrix}$$

FIGURE 7.19

SOLUTION We find the most aggressive strategy for the two players, put a star next to the largest entry in the matrix, 5, and two stars next to the smallest entry in the matrix, -4. The row containing the entry with one star is the most aggressive strategy for the first player since 5 is the maximum possible payoff for the first player. The column containing the entry with two stars is the most aggressive strategy for the second player since the -4 implies that the first player loses 4, hence the second wins 4, the maximum possible payoff for the second player.

To find the payoff if both players adopt the most aggressive strategy, we locate the entry in the third row and first column. That is a 1 as shown in Figure 7.20.

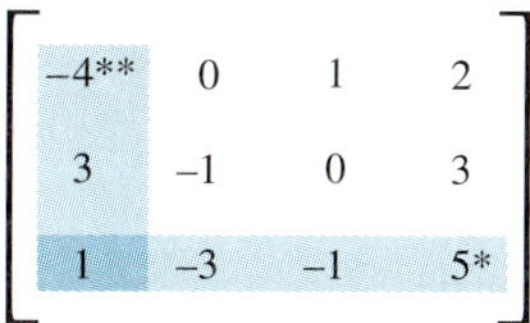

FIGURE 7.20

Thus, if each player chooses his own most aggressive strategy, the payoffs will be 1 for the first player and −1 for the second. ◆

A player's most aggressive strategy offers that player the greatest possible gain, but getting that gain depends on the other player doing just the right thing, and the other player is not likely to be so cooperative. For example, if both players in Example 7.6 use the most aggressive strategy neither will get the gain he was hoping for. Adopting the most aggressive strategy does not guarantee the best payoff.

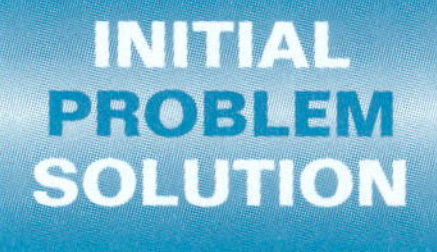

Suppose two partners, Sid and Mark, have come to a parting of the ways and wish to dissolve their partnership, which has assets of about $250,000. They could come to an amicable agreement between themselves where each gets equal value, or they could settle the matter in court. Suppose legal expenses for the court battle would run about $12,500 for each side, and a fair and independent observer would likely give Sid 60% of the assets. Is it possible to look at their strategies in terms of a game? If so, how?

SOLUTION This is a two-person game in which each player has two possible strategies: Settle or Sue. It is not a zero-sum game because if either player chooses the strategy of suing, there will be $25,000 lost to legal expenses. We can put the information into the form of a table.

Strategies and Payoffs

Sid Strategy	Mark Strategy	Sid Payoff	Mark Payoff
Settle	Settle	$125,000	$125,000
Settle	Sue	$137,500	$87,500
Sue	Settle	$137,500	$87,500
Sue	Sue	$137,500	$87,500

Looked at coldly and rationally, Mark should be happy to settle and should not antagonize Sid. In practice, there may be psychological reasons that drive Mark to sue Sid, even though it is going to benefit Sid and harm Mark. If Mark feels this way, then the dollar amounts are not the real payoffs in the game as far as Mark is concerned. That is, Mark views winning in terms of something other than money.

PROBLEM SET 7.1

1. **(a)** What is the size of the matrix below?
   (b) What element is in the first row and the second column?

$$\begin{bmatrix} 4 & -2 & 5 \\ 0 & 3 & -4 \\ 2 & -3 & 0 \end{bmatrix}$$

2. **(a)** What is the size of the matrix below?
   (b) What element is in the second row and the second column?

$$\begin{bmatrix} 2 & 1 & -3 \\ -2 & 0 & 1 \end{bmatrix}$$

3. **(a)** What is the size of the matrix below?
   (b) What element is in the second row and the third column?

$$\begin{bmatrix} -3 & 2 & -1 & -1 \\ -2 & 3 & 1 & -1 \\ 4 & 0 & -2 & 3 \end{bmatrix}$$

4. **(a)** What is the size of the matrix below?
   (b) What element is in the third row and the second column?

$$\begin{bmatrix} -1 & 0 & 4 \\ 1 & 1 & -3 \\ -2 & 2 & -2 \\ -1 & 4 & 1 \end{bmatrix}$$

In problems 5 through 12, the table gives the payoffs to the players in a two-person zero-sum game for all combinations of strategies the two players have available. Construct the game matrix for each table.

5. Each player has two strategies.

Strategies and Payoffs

Player 1 Strategy	Player 2 Strategy	Player 1 Payoff	Player 2 Payoff
I	I	2	−2
I	II	4	−4
II	I	−3	3
II	II	−1	1

6. Each player has two strategies.

Strategies and Payoffs

Player 1 Strategy	Player 2 Strategy	Player 1 Payoff	Player 2 Payoff
I	I	1	−1
I	II	2	−2
II	I	3	−3
II	II	4	−4

7. The first player has two strategies and the second player has three strategies.

Strategies and Payoffs

Player 1 Strategy	Player 2 Strategy	Player 1 Payoff	Player 2 Payoff
I	I	3	−3
I	II	4	−4
I	III	−2	2
II	I	−3	3
II	II	−1	1
II	III	2	−2

8. The first player has three strategies and the second player has two strategies.

Strategies and Payoffs

Player 1 Strategy	Player 2 Strategy	Player 1 Payoff	Player 2 Payoff
I	I	0	0
I	II	4	−4
II	I	−5	5
II	II	−1	1
III	I	0	0
III	II	3	−3

9. Each player has three strategies.

Strategies and Payoffs

Player 1 Strategy	Player 2 Strategy	Player 1 Payoff	Player 2 Payoff
I	I	1	−1
I	II	2	−2
I	III	3	−3
II	I	4	−4
II	II	5	−5
II	III	−4	4
III	I	−3	3
III	II	−2	2
III	III	−1	1

10. Each player has three strategies.

Strategies and Payoffs

Player 1 Strategy	Player 2 Strategy	Player 1 Payoff	Player 2 Payoff
I	I	−4	4
I	II	−3	3
I	III	−2	2
II	I	−1	1
II	II	0	0
II	III	1	−1
III	I	2	−2
III	II	3	−3
III	III	4	−4

11. The first player has three strategies and the second player has four strategies.

Strategies and Payoffs

Player 1 Strategy	Player 2 Strategy	Player 1 Payoff	Player 2 Payoff
I	I	1	−1
I	II	−6	6
I	III	2	−2
I	IV	−5	5
II	I	3	−3
II	II	−4	4
II	III	4	−4
II	IV	−3	3
III	I	5	−5
III	II	−2	2
III	III	6	−6
III	IV	−1	1

12. The first player has four strategies and the second player has three strategies.

Strategies and Payoffs

Player 1 Strategy	Player 2 Strategy	Player 1 Payoff	Player 2 Payoff
I	I	1	−1
I	II	−6	6
I	III	2	−2
II	I	−5	5
II	II	3	−3
II	III	−4	4
III	I	4	−4
III	II	−3	3
III	III	5	−5
IV	I	−2	2
IV	II	6	−6
IV	III	−1	1

In problems 13 through 20, a game situation is described. Find the payoffs and set up the labeled game matrix.

13. Two children, Patti and Patrick, are playing the game of Two-Finger Morra. They play the game by simultaneously holding out either 1 or 2 fingers. If the sum is even Patti wins a dime; if the sum is odd, Patrick wins a dime.

14. Patti and Patrick (problem 13) decide to make the game more interesting. They each take a dime and a quarter and hold the coins behind their backs. On the count of three, each chooses one of the coins and shows it. If the coins match, Patti wins both coins; if they don't match, Patrick wins both coins.

15. Carmen and Rose are playing a card game. Each has an ace and a king. Each chooses a card and places it on the table. The ace beats the king and it wins two dollars. If both play aces, then Carmen wins a dollar. If both play kings, then Rose wins three dollars.

16. Carlos and Rob play a card game. Carlos has two cards, a 3 and a 7; and Rob has three cards, a 2, a 4, and a 5. Each player picks a card and places it on the table. If the sum is even, Carlos pays Rob \$3; if the sum is odd, Rob pays Carlos \$6.

17. A pitcher and a batter face each other in a baseball game. The pitcher can throw a fastball and a curveball. Prepared to hit a fastball, the batter hits .450 against the fastball, but only .200 against the curveball. Prepared for a curveball, the batter hits .400 against the curveball and .240 against the fastball.

18. A pitcher and a batter face each other in a baseball game. The pitcher can throw a fastball, a curveball, or a screwball. Prepared to hit a fastball, the batter hits .400 against a fastball, .200 against a curveball, and .100 against a screwball. Prepared for a curveball, the batter hits .400 against a curveball, .220 against a fastball, and .160 against a screwball. Prepared for a screwball, the batter hits .380 against a screwball, .280 against a curveball, and .120 against a fastball.

19. A businessman is considering cheating on his income tax. If he cheats and isn't audited, he will be \$1500 ahead. If he cheats and is audited, he will pay a fine of \$3000 and the \$1500 that he owes. If he doesn't cheat and is audited, he figures that he loses \$200 for his effort and time lost. If he doesn't cheat and doesn't get audited, he's not sure if he'll feel good or bad, so he will call it even (\$0).

20. Suppose you own a home furniture store and must place orders for your major lines of appliances. If the economy is good and consumer confidence is up, you figure you should place a big order, anticipating a high volume in sales and a profit of \$50,000. If the economy isn't good and consumer confidence is down, you should place a small order, anticipating slow sales and profits of \$30,000. If the economy is good, but you've ordered a small quantity, you anticipate profits of \$20,000. However, if the economy is not good and you've placed a big order, the most you can make is \$5,000.

For problems 21 through 26, construct the table of strategies and payoffs that corresponds to the two-person zero-sum game with the given game matrix. That is, reverse the process shown in Example 7.3.

21. $\begin{bmatrix} 1 & -2 \\ -1 & 2 \end{bmatrix}$

22. $\begin{bmatrix} 1 & 0 \\ 0 & -1 \end{bmatrix}$

23. $\begin{bmatrix} 0 & 2 & -1 \\ 2 & 1 & -2 \\ -1 & -2 & 2 \end{bmatrix}$

24. $\begin{bmatrix} 4 & 1 & -1 \\ -2 & 3 & 0 \\ -4 & -3 & 2 \end{bmatrix}$

25. $\begin{bmatrix} 0 & -2 & 0 & -2 \\ -1 & 1 & -1 & 1 \\ 2 & 0 & 2 & 0 \end{bmatrix}$

26. $\begin{bmatrix} 3 & 0 & 3 & 0 & 3 \\ -1 & 2 & -1 & 2 & 2 \\ 1 & -2 & 1 & 1 & -2 \\ -3 & 0 & 0 & -3 & 0 \end{bmatrix}$

27. Using the game matrix in problem 21, find the most aggressive strategy for
(a) the first player.
(b) the second player.

28. Using the game matrix in problem 22, find the most aggressive strategy for
(a) the first player.
(b) the second player.

29. Using the game matrix in problem 24, find the most aggressive strategy for
(a) the first player.
(b) the second player.

30. Consider the game matrix in problem 26.
(a) Find the most aggressive strategy for the first player.
(b) In finding the most aggressive strategy for the second player, one is led to something different from the other examples. What is it?

EXTENDED PROBLEMS

Another way of modeling games considers the game to be a series of decisions made by the players. This is especially appropriate for games such as chess and checkers where the players respond in turn to a fully known situation. To illustrate, consider a simplified version of tic-tac-toe in which players X and O seek to get just two matching letters side-by-side on a grid of three adjacent squares; X plays first. A sample game is illustrated in the next figure.

X			First move by X
X	O		First move by O blocks X
X	O	X	Only one choice for X's second move. Game ends in a draw.

To describe all possible plays we number the squares 1, 2, and 3 from left to right. Then X-1 will indicate that player X plays in square 1, O-2 will indicate that player O plays in square 2, and so on. The next figure gives all possible ways this game could be played.

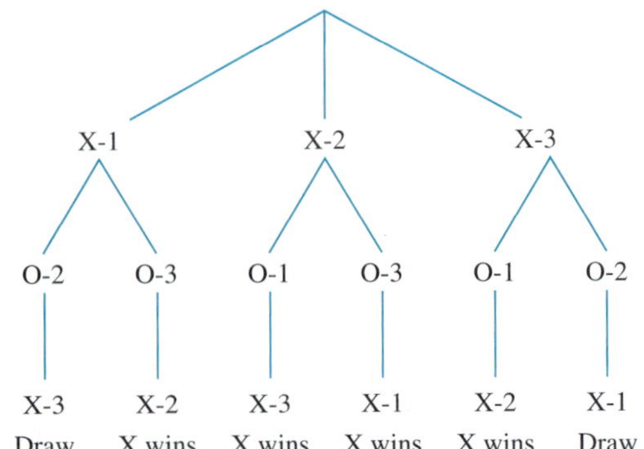

The diagram looks like a tree. From the tree we can see that player O cannot win. Also, player X can find a winning strategy by pruning from the tree any branch on which there is a Draw as shown in the next figure.

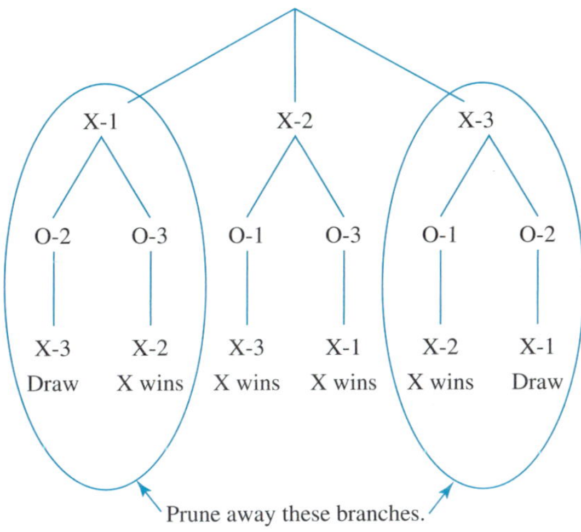

31. Consider the simplified version of tic-tac-toe in which players X and O seek to get just two matching letters side-by-side on a grid of four adjacent squares; X plays first.

(a) Draw the tree diagram for this game.

(b) Prune the tree to find the winning strategies for player X.

(c) Can this process be done for the standard game of tic-tac-toe?

32. What does the tree diagram analysis tell us about checkers and chess?

33. In problem 16, Carlos has two cards, a 3 and a 7; and Rob has three cards, a 2, a 4, and a 5. Each player picks a card and places it on the table. If the sum is even, Carlos pays Rob \$3; if the sum is odd, Rob pays Carlos \$6. Draw the tree diagram for this game. Is there a winning strategy you can identify?

34. Game theory can be applied to military situations. Which generals would you consider to have been likely to choose the most aggressive strategy? Research and justify your conclusions.

35. Research the ways in which games have been used to develop military strategy.

36. Game theory can be applied to economics. Research some particular application, for example, to labor negotiations.

37. In your favorite sport, analyze some aspect in terms of game theory.

7.2 DETERMINED GAMES

Suppose the two theaters in town are across the street from one another. The owners have noticed that if one theater is showing a wholesome family movie and the other is showing a movie with sex and violence in it, the theater with the racier movie gets the majority of the night's business. However, if they both are showing the same type of movie, they draw roughly equal crowds. Represent this situation by a game matrix. Determine if there is a winning strategy for either theater.

In Section 7.1 we focused on the mathematical description of a game in terms of strategies and payoffs. Now, we focus our attention on identifying and selecting the most appropriate strategies to maximize the result of the game to each player.

MOST CONSERVATIVE STRATEGY

The opposite approach to maximizing possible gain is to minimize possible loss; this is known as the **most conservative strategy**. Consider a two-person zero-sum game with the following game matrix (Figure 7.21):

$$\begin{bmatrix} -4 & 0 & 1 & 2 \\ 3 & -1 & 0 & 3 \\ 1 & -3 & -1 & 5 \end{bmatrix}$$

FIGURE 7.21

Let's consider the worst case scenarios for each of the first player's strategies:

- in the first strategy (row 1), the worst payoff is -4 when the second player plays his first strategy (column 1).
- in the second strategy (row 2), player 1's worst possible payoff is -1 (column 2).
- in the third strategy (row 3), the worst possible payoff is -3 (column 2).

The worst case scenario for the first player is a payoff of -4, -1, or -3 when choosing the first, second, or third strategy, respectively. Since -1 is the largest of those possibilities, the most conservative strategy for the first player is the second strategy: the first player's second strategy minimizes possible loss for the first player. To summarize, what we have done is find the minimum entry in each row and then selected the maximum from among the minima (plural for minimum). This is referred to as the **maximin procedure**.

A similar analysis applies to the second player. The details are different because each entry in the matrix represents the opposite of the second player's payoff. If the second player chooses his first strategy, the worst thing that can happen is that the first player chooses his second strategy resulting in the payoff of -3 to the second player, because 3 is the largest entry in the first column of the matrix. Remember that when considering payoffs for the second player, you must think of the *opposites* of the numbers in the matrix. That is, the worst case for the second player is the best case for the first player.

To find the worst outcome for each of the second player's other strategies, we locate the largest element in the other columns. Since the matrix is written from the first player's perspective, it makes sense to analyze it with the first player in mind. From the perspective of the first player, the entries 3, 0, 1, 5 (the maxima, plural for maximum) are the worst cases corresponding to the second player's strategies. Player two's most conservative strategy is selecting the minimum from among the maxima. This is known as the **minimax procedure**.

In the next example, we illustrate a systematic method for finding the most conservative strategy for each player.

EXAMPLE 7.7 Recall the labeled game matrix for a football play from Example 7.5, with the offensive team as the first player and the defensive team as the second player (Figure 7.22).

	PD	RD	AD
Pass	–3	7	4
Run	5	1	2

FIGURE 7.22

(a) What is the most conservative strategy for the offense?
(b) What is the most conservative strategy for the defense?
(c) What is the payoff if both teams use their most conservative strategy?

SOLUTION

(a) To find the most conservative strategy for the offensive team (the first player), circle the smallest number in each row of the abstract game matrix. Looking through the circled numbers, we find that the largest of the circled numbers occurs in the second row. The row containing that largest circled number corresponds to the most conservative strategy for the first player. We have highlighted that row (Figure 7.23).

Most conservative strategy for the first player

$$\begin{bmatrix} \textcircled{-3} & 7 & 4 \\ 5 & \textcircled{1} & 2 \end{bmatrix}$$

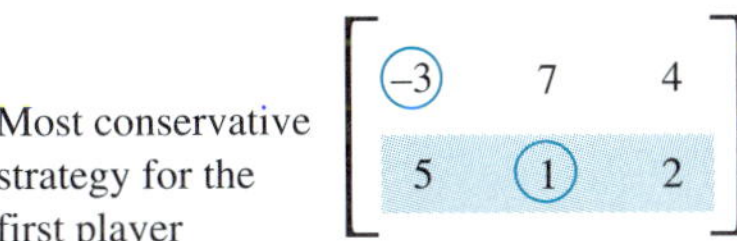

FIGURE 7.23

The first player is using the maximin strategy. (There may be more than one such row, in which case we will consider them all to be the most conservative strategies.) In this example the second strategy for the offensive team is to run the football. Our conclusion is that running the football is the most conservative strategy, that is, the safest strategy.

(b) To find the most conservative strategy for the second player, draw a box around the largest number in each column. (Remember the second player's payoff is the opposite of the number in the matrix.) Looking through the boxed numbers, we find the smallest of the boxed numbers, 4, occurs in the third column. The column containing that smallest boxed number corresponds to the most conservative strategy for the second player. (Again there may be more than one.)(Figure 7.24)

Most conservative strategy for the second player

$$\begin{bmatrix} -3 & \boxed{7} & \boxed{4} \\ \boxed{5} & 1 & 2 \end{bmatrix}$$

FIGURE 7.24

In this example the third strategy for the defensive team is to use an all-purpose defense. The second player is using the minimax procedure. Our conclusion is that the most conservative thing for the defensive team to do is to use the all-purpose defense.

(c) To find the payoff if both teams use their most conservative strategy, we must find the entry of the matrix in the second row and third column (Figure 7.25).

$$\begin{bmatrix} -3 & 7 & 4 \\ 5 & 1 & 2 \end{bmatrix}$$

FIGURE 7.25

The entry in question is 2. If both teams adopt their most conservative strategy, the offense will gain 2 yards on average, and the defense will be giving up 2 yards on average. ◆

The offensive team in football cannot be satisfied with making only 2 yards on each play, so they will not continue to use their most conservative strategy. They will start passing. If the offense starts consistently passing and the defensive team stays with the all-purpose defense, then the payoff will be determined by the entry in the first row and third column of the matrix, that is, 4 (Figure 7.26).

$$\begin{bmatrix} -3 & 7 & 4 \\ 5 & 1 & 2 \end{bmatrix}$$

FIGURE 7.26

We see that the offense will be gaining an average of 4 yards per play. To counter this, the defense will start using a pass defense.

The pass defense corresponds to the first column of the matrix, so the payoff will become the entry in the first row and first column, that is, -3 (Figure 7.27).

$$\begin{bmatrix} -3 & 7 & 4 \\ 5 & 1 & 2 \end{bmatrix}$$

FIGURE 7.27

Constantly faced with a pass defense, the offense will be giving up 3 yards per play if they continue to always try passing the football. Instead, they will try to run the ball. And so it will continue, with each side switching strategies to get an advantage over the other. That is part of what makes the game of football interesting to watch.

DETERMINED GAMES

Not every game requires that the players mix up their strategies to get an advantage. There are some game matrices for which the players can find a pair of satisfactory choices.

EXAMPLE 7.8 Given the following game matrix, find the most conservative strategy for both players (Figure 7.28).

$$\begin{bmatrix} -4 & -2 & 1 & 2 \\ -5 & -2 & -7 & 1 \\ 3 & -1 & 0 & 3 \\ 1 & -3 & -1 & 5 \end{bmatrix}$$

FIGURE 7.28

(a) What would be the payoff to each player if both players adopt the most conservative strategy?

(b) If both players adopt the most conservative strategy and the game is repeated over and over, will either player want to change strategies?

SOLUTION We find the most conservative strategy for the first player by the method of circling the smallest entry in each row and finding the row containing the largest circled number (Figure 7.29).

Most conservative strategy for the first player

$$\begin{bmatrix} -4 & -2 & 1 & 2 \\ -5 & -2 & -7 & 1 \\ 3 & -1 & 0 & 3 \\ 1 & -3 & -1 & 5 \end{bmatrix}$$

FIGURE 7.29

The most conservative strategy for the first player is her third strategy.

We find the most conservative strategy for the second player by the method of boxing the largest entry in each column and finding the column containing the smallest boxed number (Figure 7.30).

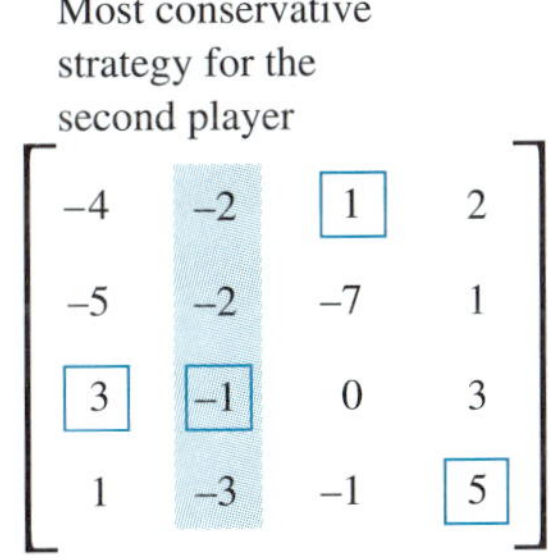

FIGURE 7.30

The most conservative strategy for the second player is her second strategy.

To find the payoff if both players adopt the most conservative strategy, we locate the entry in the third row and second column, which is -1 (Figure 7.31).

$$\begin{bmatrix} -4 & -2 & 1 & 2 \\ -5 & -2 & -7 & 1 \\ 3 & -1 & 0 & 3 \\ 1 & -3 & -1 & 5 \end{bmatrix}$$

FIGURE 7.31

So the payoff to the first player is -1, and the payoff to the second player is 1.

If the game is repeated over and over, the first player will not wish to change strategies (as long as the second player continues to choose her second strategy). A change by the first player will result in a payoff of -2 or -3, which is less desirable than the payoff of -1 associated with the third strategy.

Similarly, if the game is repeated over and over, the second player will also not wish to change strategies, because a change will result in a payoff of 0 or -3 to the second player, which is less desirable than the payoff of 1 to the second player associated with the second strategy. ◆

A two-person zero-sum game for which each player adopting the most conservative strategy makes the other player's most conservative strategy the best choice is called a **determined game**. The two choices of strategies are called **optimal** because neither player can reliably improve his or her payoff.

EXAMPLE 7.9 Show that the following game matrix is the matrix of a determined game (Figure 7.32).

$$\begin{bmatrix} -4 & -1 & 2 & 5 \\ 3 & 0 & 1 & 3 \\ 1 & -3 & -1 & 2 \end{bmatrix}$$

FIGURE 7.32

SOLUTION A useful shortcut is to do the circling and boxing used to find the most conservative strategies on the same copy of the matrix. Recall that the circles indicate the smallest entry in each row, and the boxes indicate the largest entry in that column (Figure 7.33).

$$\begin{bmatrix} \textcircled{-4} & -1 & \boxed{2} & \boxed{5} \\ \boxed{3} & \boxed{\textcircled{0}} & 1 & 3 \\ 1 & \textcircled{-3} & -1 & 2 \end{bmatrix}$$

FIGURE 7.33

When an entry of the matrix winds up with both a box and a circle around it, then the game is determined, so this game is determined. ◆

The entry that is both boxed and circled in a matrix is called the **value of the game**, and the row and column containing it represent the optimal strategies. A game with value 0 is called a **fair game.** The game in the last example is a fair game.

The next example refers to the game of tennis. A singles court is illustrated schematically in Figure 7.34. There are two basic styles of play: *baseline,* in which the player stays back near the baseline, and *serve and volley,* in which the player moves up to the net after the service.

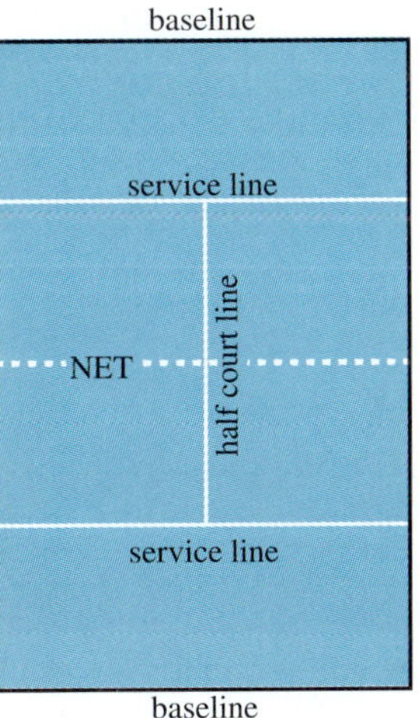

FIGURE 7.34

EXAMPLE 7.10 Suppose statistics have been kept on the performance of two tennis players, Chris and Martina, when they have played each other. These statistics reveal the percentage of the time the first player, Chris, wins when each uses the style of play indicated. Is this a determined game? What is the significance of your answer (Figure 7.35)?

		Martina	
		Baseline	Serve and Volley
Chris	Baseline	54	52
	Serve and Volley	49	47

FIGURE 7.35

SOLUTION We form the abstract game matrix and use the circle and box method to determine the most conservative strategy for both players (Figure 7.36).

$$\begin{bmatrix} 54 & 52 \\ 49 & 47 \end{bmatrix}$$

FIGURE 7.36

We see this is a determined game in which the first player, Chris, should play from the baseline, and the second player, Martina, should play serve and volley. Chris will have a slight advantage, winning 52% of the time. ◆

In some cases the box and circle method can be a bit tricky to apply. The following example shows why.

EXAMPLE 7.11 Show that the following game matrix represents a determined game. Find the optimal strategies and the value of the game (Figure 7.37).

$$\begin{bmatrix} -4 & -1 & -1 & 5 \\ 3 & 0 & 0 & 3 \\ 1 & -3 & -1 & 2 \end{bmatrix}$$

FIGURE 7.37

SOLUTION We apply the box and circle method. Recall that circles indicate the smallest integer in a row, and boxes indicate the largest entry in a column (Figure 7.38).

$$\begin{bmatrix} -4 & -1 & -1 & 5 \\ 3 & 0 & 0 & 3 \\ 1 & -3 & -1 & 2 \end{bmatrix}$$

FIGURE 7.38

In this case, there are *two* entries that are both boxed and circled, so this is a determined game. The second strategy is optimal for the first player. For the second player, both the second and third strategies are optimal. The value of the game is 0. ◆

Further reflection on the game in Example 7.11 shows that the second player's strategy is always preferable to the third strategy. Note that every entry in the second column is less than or equal to the corresponding entry in the third column. In the third row, the -3 is strictly less than -1. We say that the second strategy **dominates** the third. A dominated strategy, the third column here, should never be used since there is another that gives at least as favorable an outcome no matter what the opponent does. In this game the second player's third strategy should never be used because the second strategy always gives as good a result for the second player.

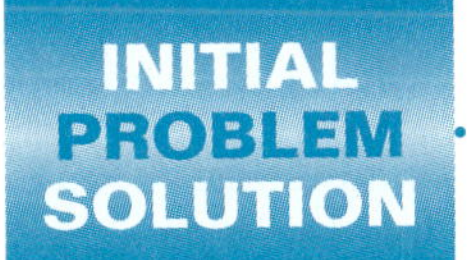

Suppose the two theaters in town are across the street from one another. The owners have noticed that if one theater is showing a wholesome family movie and the other is showing a movie with sex and violence in it, the theater with the racier movie gets the majority of the night's business. However, if they both are showing the same type of movie they draw roughly equal crowds. Represent this situation by a game matrix. Determine if there is a winning strategy for either theater.

SOLUTION We will consider the theater getting a major portion of the night's business the winner and award that theater +1 (since limited information is given, more precise units are inappropriate). The labeled game matrix is the following (Figure 7.35).

	Family Movie	Sex and Violence
Family Movie	0	–1
Sex and Violence	1	0

FIGURE 7.39

From the labeled game matrix we form the abstract matrix and apply the circle and box method (Figure 7.40).

$$\begin{bmatrix} 0 & -1 \\ 1 & 0 \end{bmatrix}$$

FIGURE 7.40

The game is determined. This analysis indicated that both theaters should only show movies with sex and violence if their only concern is getting the largest share of the audience.

Of course this may not be a zero-sum game in reality. Some people feel that the major television networks used this sort of thinking to win the nightly ratings battle, while driving away a significant portion of the audience.

PROBLEM SET 7.2

In problems 1 through 10, a game matrix is given. Do the following.

(a) Determine the most conservative strategy for each player.

(b) State whether or not the game is determined.

(c) If the game is determined, find the value.

1. $\begin{bmatrix} 1 & -2 \\ -1 & 2 \end{bmatrix}$

2. $\begin{bmatrix} 1 & 0 \\ 0 & -1 \end{bmatrix}$

3. $\begin{bmatrix} 1 & 3 \\ 0 & 2 \end{bmatrix}$

4. $\begin{bmatrix} 1 & 3 \\ 2 & 0 \end{bmatrix}$

5. $\begin{bmatrix} 0 & 2 & -1 \\ 2 & 1 & -2 \\ -1 & -2 & -3 \end{bmatrix}$

6. $\begin{bmatrix} 4 & 1 & -1 \\ -2 & 3 & 0 \\ -4 & -3 & 2 \end{bmatrix}$

7. $\begin{bmatrix} 0 & 1 & 2 \\ 3 & 4 & 5 \\ 6 & 7 & 8 \end{bmatrix}$

8. $\begin{bmatrix} 0 & 3 & 6 \\ 7 & 1 & 4 \\ 5 & 8 & 2 \end{bmatrix}$

9. $\begin{bmatrix} 0 & -2 & 0 & -3 \\ -2 & 1 & -1 & 2 \\ 2 & 0 & 2 & 1 \end{bmatrix}$

10. $\begin{bmatrix} 3 & 0 & 3 & 0 & 3 \\ -1 & 2 & -1 & 2 & 2 \\ 1 & -2 & 1 & 1 & -2 \\ -3 & 0 & 0 & -3 & 0 \end{bmatrix}$

In problems 11 through 20, do the following.

(a) Write the game matrix.

(b) Determine the most conservative strategy for each player.

(c) State whether or not the game is determined.

(d) If the game is determined, find the value.

11. Matching Pennies is one of the simplest games of all. Two players each put down a penny, either head or tail up, without letting the other player see. Then they both uncover their coins. If the coins match, both heads or both tails, then the first player wins. If the coins don't match, the second player wins.

12. Two children, Patti and Patrick, are playing the game of Two-Finger Morra. They play the game by simultaneously holding out either 1 or 2 fingers. If the sum is even, Patti wins a dime; if the sum is odd, Patrick wins a dime.

13. A pitcher and a batter face each other in a baseball game. The pitcher can throw a fastball or a curveball. Prepared to hit a fastball, the batter hits .450 against the fastball, but only .200 against the curveball. Prepared for a curveball, the batter hits .400 against the curveball and .240 against the fastball.

14. A pitcher and a batter face each other in a baseball game. The pitcher can throw a fastball, a curveball, or a screwball. Prepared to hit a fastball, the batter hits .400 against a fastball, .200 against a curveball, and .100 against a screwball. Prepared for a curveball, the batter hits .400 against a curveball, .220 against a fastball, and .160 against a screwball. Prepared for a screwball, the batter hits .380 against a screwball, .280 against a curveball, and .120 against a fastball.

15. The sky is overcast and there is a possibility of rain. Deciding whether or not to take an umbrella when you walk is a "guessing" game (nature is the other player). Jane analyzes the choices and consequences as follows and assigns relative values as payoffs:

take the umbrella; it rains $= -2$

take the umbrella; no rain $= -1$

leave the umbrella; it rains $= -5$

leave the umbrella; no rain $= 3$

16. Charlane is thinking about investing some of her savings in bonds, stocks, and money market funds. Her expected returns will depend on whether interest rates rise or fall during the coming year. Her financial advisor tells her the expected return on stocks will be 10% if the rates rise and 12% if the rates fall. For bonds, the expected returns will be 6% (rates rise) and 11% (rates fall); while for money market funds they will be 15% (rates rise) and 10% (rates fall).

17. Each of two players, Alex and Bart, has two cards, a 2 and a 3. Each player selects a card and they show their cards simultaneously. If the cards match, then Alex wins \$1. If Alex's card is greater than Bart's, then Alex wins \$3. If Bart's card is greater than Alex's, then Bart wins \$5.

18. Carlos and Rob play a card game. Carlos has two cards, a 3 and a 7; and Rob has three cards, a 2, a 4, and a 5. Each player picks a card and places it on the table. If the sum is even, Carlos pays Rob \$3; if the sum is odd, Rob pays Carlos \$6.

19. McDonald's and Burger King are both planning to open new franchises in a small city. There are four suitable locations for either business: two in the downtown core area and two out by the new mall. If they locate in the same general area, they will split the business. If McDonald's builds downtown and Burger King at the mall, then Burger King will get 60% of the business; however, if Burger King is downtown and McDonald's is at the mall, McDonald's will get 65% of the business.

20. A businessman is considering cheating on his income tax. If he cheats and isn't audited, he will be \$1500 ahead. If he cheats and is audited, he will pay a fine of \$3000 and the tax that he owes. If he doesn't cheat and is audited, he figures that he loses \$200 for his effort and time lost. If he doesn't cheat and doesn't get audited, he's not sure if he'll feel good or bad, so he will call it even (\$0).

Dominant strategies can be used to reduce the size of a matrix game, making it easier to analyze the game and find optimum strategies. As an example, consider the game matrix from Example 7.11.

$$\begin{bmatrix} -4 & -1 & -1 & 5 \\ 3 & 0 & 0 & 3 \\ 1 & -3 & -1 & 2 \end{bmatrix}$$

Since the second column dominates the third column, we can eliminate the third column.

$$\begin{bmatrix} -4 & -1 & 5 \\ 3 & 0 & 3 \\ 1 & -3 & 2 \end{bmatrix}$$

The second column also dominates the last column. We reduce the matrix again.

$$\begin{bmatrix} -4 & -1 \\ 3 & 0 \\ 1 & -3 \end{bmatrix}$$

We now notice that the resulting second row dominates both the first and third rows; we have

$$[3 \quad 0]$$

Since the second column now dominates the first column, we are left with

$$[0]$$

As we saw in Example 7.11, this is the value of the game. Many game matrices have dominant strategies, although we can't reduce to a single row/column except when the game is a determined game.

In problems 21 through 28, use dominant strategies, if any, to reduce the given game matrix as far as possible. Begin with rows and then switch to columns. Find the value of the game if the matrix reduces far enough.

21. $\begin{bmatrix} 1 & -4 & 0 \\ -1 & -2 & 1 \end{bmatrix}$

22. $\begin{bmatrix} -1 & 4 \\ -2 & -3 \\ 2 & -1 \end{bmatrix}$

23. $\begin{bmatrix} 4 & -2 & -2 \\ 0 & 4 & 1 \\ -1 & 0 & 0 \end{bmatrix}$

24. $\begin{bmatrix} 0 & 3 & 6 \\ 8 & 5 & 6 \\ 9 & 4 & 5 \end{bmatrix}$

25. $\begin{bmatrix} 0 & 2 & 1 & 3 \\ -4 & -1 & 2 & -1 \\ -5 & 3 & 0 & -2 \end{bmatrix}$

26. $\begin{bmatrix} 1 & -2 & -5 & 0 \\ 4 & 0 & -3 & 5 \\ 1 & 2 & 2 & 2 \end{bmatrix}$

27. $\begin{bmatrix} 3 & -2 & -1 \\ -2 & 3 & -2 \\ 1 & 1 & 2 \end{bmatrix}$

28. $\begin{bmatrix} -3 & 1 & -1 & 4 \\ 2 & 1 & 0 & -2 \end{bmatrix}$

29. As a concert promoter, you have an opportunity to book a very popular band. You can hold the concert in the outdoor stadium which seats 20,000 or indoors at the coliseum which seats only 10,000. The outdoor arena will cost $15,000 and the coliseum will cost $10,000. Experience tells you that you can sell tickets for $30 each. There is the possibility of very bad weather canceling an outdoor concert, and moderately bad weather could reduce ticket sales by 30% for an outdoor event. The band will get 50% of gate receipts and is guaranteed $50,000 even if the weather forces cancellation. Rental fees are guaranteed.
   - **(a)** Make a table of the possibilities.
   - **(b)** Convert the table to a game matrix.
   - **(c)** What is the most aggressive strategy?
   - **(d)** What is the most conservative strategy?
30. You run a travel agency in a college town. The local football team has a chance of going to a major bowl game, and if they do and if you have in place appropriate tour packages you can expect to make $40,000 profit. If the team goes to a major game and your agency isn't ready, you can still make a profit of $15,000 through regular bookings. In order to have the tour packages ready to offer you must put one of your employees to work on making arrangements at a cost of $2000.
   - **(a)** Make a table of the possibilities.
   - **(b)** Convert the table to a game matrix.
   - **(c)** What is the most aggressive strategy?
   - **(d)** What is the most conservative strategy?

EXTENDED PROBLEMS

31. Consider a two-person game which is *not* a zero-sum game.
   - **(a)** How should the most aggressive and most conservative strategies be defined and found?
   - **(b)** Is there a situation in which the game should be said to be determined?
32. "The prisoner's dilemma" is a non–zero-sum game that is a classic of game theory. In the game, two men who are suspected of a felony are arrested in the course of committing a misdemeanor. The police claim they have more than enough evidence to convict the pair. The prisoners are separated for interrogation, and each is given the option of confessing to the felony (and being a witness against the partner) or remaining silent. If they both remain silent, they can each expect to spend one year in prison for the misdemeanor. If only one confesses, he will receive a suspended sentence, while the other will spend at least five years in prison. If both confess, they can both expect to spend three years in prison. Analyze the prisoner's dilemma. Is this a determined game?
33. The prisoner's dilemma has been used as a model for other situations such as nuclear disarmament and business competition. The "best" solution for both parties is to abandon their narrow self-interests and cooperate to some degree. Research the prisoner's dilemma and write a brief paper.
34. An entry in a game matrix that ends up surrounded by both a circle and a square when we find the players' most conservative strategies is called a saddle point.
   - **(a)** Plot a surface above the matrix in Example 7.7 where the height is the entry in the matrix.
   - **(b)** Why is the point called a saddle point?
35. The saddle point argument in game theory was discovered by John von Neumann. Research who von Neumann was and what he accomplished.
36. Game theory can be applied to military situations. Which generals would you consider to have been likely to choose the most conservative strategy? Research and justify your conclusions.
37. During World War II a critical battle in the Pacific, the battle of the Bismark Sea, was fought for control of New Guinea. The allied leader, General Kennedy, had intelligence reports that indicated the Japanese would move a troop and supply convoy in the region. The Japanese leader had two choices: one that passed north of the island of New Britain or one that passed south. By either route, the trip would take the Japanese three days.

    General Kennedy had to decide whether to concentrate his reconnaissance aircraft on the northern route or the southern. Once the convoy was spotted, it could be bombed until it reached its port. On the northern route, poor visibility was almost certain, while on the southern route the weather was likely to be clear.

    Kennedy's staff assessed the outcomes as follows. If they concentrated on the southern route and the Japanese took the southern route they would be spotted early, and there would be three days of bombing. However, if the Japanese took the northern route, they would get only one day of bombing.

    If the aircraft were concentrated on the northern route, Kennedy's staff estimated they would get two days of bombing whichever route the Japanese took.
   - **(a)** Analyze this situation as a game.
   - **(b)** Find the value of the game.
   - **(c)** Research the Battle of the Bismark Sea. What really happened?

7.3 MIXED STRATEGIES

There are two likely invasion routes leading into General Latka's remote land: the mountain pass or the beach. The forces at Latka's disposal to defend these routes are not adequate, so he will need to choose one to heavily defend, while the other receives virtually a token defense. Suppose the level of casualties the attackers will suffer is estimated as in the table (these numbers represent estimates of the relative significance of the expected casualties, not specific units) (Figure 7.41). What should Latka do?

	Attack the Beach	Attack the Pass
Defend the Beach	6	1
Defend the Pass	2	4

FIGURE 7.41

In the games we have analyzed thus far, we were able to find optimum strategies and the values for determined games. We also were able to simplify the structure of some games by finding strategies that dominated others. In this section, we will look at more general games and techniques for finding the most effective strategies for the players.

OPTIMAL MIXED STRATEGY

In a determined game each player can use just one strategy all the time, and the opponent can even be told in advance what strategy will be used. But not every game is a determined game. For example, the following abstract game matrix is not the matrix of a determined game. We know this is not a determined game because no entry in the matrix is both circled and boxed (Figure 7.42).

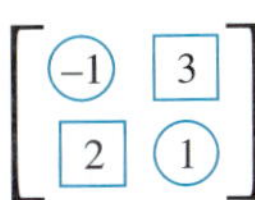

FIGURE 7.42

EXAMPLE 7.12 Consider the matching game in which both players simultaneously hold out either one or two fingers. If the two players show matching numbers the first player wins $1 from the second player, while if the two players show different numbers, the second player wins $1 from the first player. Is this a determined game?

SOLUTION The labeled game matrix for this game is the following (Figure 7.43).

	One finger	Two fingers
One finger	1	–1
Two fingers	–1	1

FIGURE 7.43

From the labeled game matrix we form the abstract matrix and apply the circle and box method (Figure 7.44).

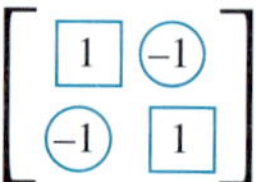

FIGURE 7.44

No entry in the matrix is both circled and boxed, so the game is not determined. ◆

If a game is not determined, it is not possible to select one strategy for constant use without giving the other player an advantage. To optimize his results in a game that is not determined, a player must use a **mixed strategy.** This means two or more strategies are used at random. The trick is to determine what percentage of the time to use each strategy. The matching game described in Example 7.12 may be familiar to you. If not, try it with a friend (but for pennies instead of dollars). Each player in the matching game should try to use each strategy (that is, one or two fingers) equally often and mix them up so the other player cannot anticipate what is coming.

There is a good reason that both players should use their strategies equally often in the matching game: using that system makes each player's average payoff the best it can be without the opponent's cooperation.

DEFINITION

Average Payoff of a Game

The **average payoff of a game** is computed by multiplying each payoff by the fraction of the time the related the strategy is used and adding those products.

If both players in the matching game show one or two fingers equally often at random, half the time there will be a match, and half the time there will not be a match. If both use one finger $\frac{1}{2}$ of the time, then they will match with one finger $\frac{1}{4}$ of the time. Similarly, there will be a match with two fingers $\frac{1}{4}$ of the time. The average payoff then would be

$$\tfrac{1}{2} \times (+1) + \tfrac{1}{2} \times (-1) = 0.$$

In the next example, we see what happens if the players above do not follow the advice to use both strategies equally often.

EXAMPLE 7.13 Use the matching game from Example 7.12 above. Suppose the first player chooses her first strategy $\frac{1}{4}$ of the time and chooses her second strategy $\frac{3}{4}$ of the time. Also, suppose the second player chooses her first strategy $\frac{1}{3}$ of the time and her second strategy $\frac{2}{3}$ of the time. (These fractions are chosen arbitrarily to show what happens if you go against the advice to use one and two fingers equally often.) Assume the choices of strategy are made independently and randomly so neither player can predict what the other is going to do. What is the average payoff for each player?

SOLUTION The first player will be showing one finger $\frac{1}{4}$ of the time. Simultaneously, the second player will be making a choice of strategy so that she shows one finger $\frac{1}{3}$ of the time. Both players will be showing one finger $\frac{1}{4} \times \frac{1}{3} = \frac{1}{12}$ of the time. All the information about the choice of strategies by both players can be put into a table as follows:

First Player		Second Player		Both Players
Strategy	**Fraction**	**Strategy**	**Fraction**	**Fraction of Time**
1	$\frac{1}{4}$	1	$\frac{1}{3}$	$\frac{1}{4} \times \frac{1}{3} = \frac{1}{12}$
1	$\frac{1}{4}$	2	$\frac{2}{3}$	$\frac{1}{4} \times \frac{2}{3} = \frac{2}{12}$
2	$\frac{3}{4}$	1	$\frac{1}{3}$	$\frac{3}{4} \times \frac{1}{3} = \frac{3}{12}$
2	$\frac{3}{4}$	2	$\frac{2}{3}$	$\frac{3}{4} \times \frac{2}{3} = \frac{6}{12}$

When we know what strategy both players have chosen, we can determine the payoff to the first player from the description of the game or the game matrix. This payoff information can be included as another column in the table.

First Player		Second Player		Both Players	
Strategy	**Fraction**	**Strategy**	**Fraction**	**Fraction of Time**	**Payoff**
1	$\frac{1}{4}$	1	$\frac{1}{3}$	$\frac{1}{4} \times \frac{1}{3} = \frac{1}{12}$	+1
1	$\frac{1}{4}$	2	$\frac{2}{3}$	$\frac{1}{4} \times \frac{2}{3} = \frac{2}{12}$	−1
2	$\frac{3}{4}$	1	$\frac{1}{3}$	$\frac{3}{4} \times \frac{1}{3} = \frac{3}{12}$	−1
2	$\frac{3}{4}$	2	$\frac{2}{3}$	$\frac{3}{4} \times \frac{2}{3} = \frac{6}{12}$	+1

The average payoff to the first player is the sum (for all possible pairs of strategies) of the product of the relative frequency (fraction of the time) each pair of strategies is chosen and the payoff to the first player, in this case

$$\frac{1}{12} \times (+1) + \frac{2}{12} \times (-1) + \frac{3}{12} \times (-1) + \frac{6}{12} \times (+1) = \frac{1 - 2 - 3 + 6}{12} = \frac{2}{12} = \frac{1}{6}.$$

The average payoff to the first player is $\frac{1}{6}$. Since this is a zero-sum game, the average payoff to the second player is the opposite of the average payoff to the first player or $-\frac{1}{6}$. ◆

The calculation in Example 7.13 tells us that if the two players mix up their strategies with those relative frequencies, then the first player is going to do better than the second player. Over the course of many games, the first player is going to win an average of about \$0.17 each time they play.

We have already said that the best thing for the players to do in the matching game is to use their two strategies equally often and at random. This advice was based on experience. What should you do if you are confronted with a game in which the best mixing of strategies is not known? Fortunately, there is a formula that can be used for two-person zero-sum games. This formula gives the **optimal mixed strategy;** this tells you the relative frequency with which each player should use the available strategies. With the optimal mixed strategy, neither player can do better without the cooperation of the opponent.

Optimal Mixed Strategy

For a two-person zero-sum game with matrix

$$\begin{bmatrix} A & B \\ C & D \end{bmatrix}$$

if the game is not a determined game, then the optimal mixed strategy is for the first player to choose her first strategy with relative frequency

$$\frac{C - D}{(B + C) - (A + D)}$$

and the second player to choose his first strategy with relative frequency

$$\frac{B - D}{(B + C) - (A + D)}.$$

For the game matrix in the matching game shown in Figure 7.44 we have

$$A = 1, \quad B = -1, \quad C = -1, \quad D = 1.$$

So

$$\frac{C - D}{(B + C) - (A + D)} = \frac{(-1) - 1}{[(-1) + (-1) - (1 + 1)]} = \frac{-2}{-4} = \frac{1}{2}$$

and

$$\frac{B - D}{(B + C) - (A + D)} = \frac{(-1) - 1}{[(-1) + (-1) - (1 + 1)]} = \frac{-2}{-4} = \frac{1}{2},$$

confirming the advice to use the strategies equally often.

EXAMPLE 7.14 Suppose statistics have been kept on the performance of two tennis players, Carlos and Rob, when playing each other. These statistics reveal the percentage of the time the first player, Carlos, wins when each uses the style of play indicated (Figure 7.45).

		Rob	
		Baseline	Serve and Volley
Carlos	Baseline	48	52
	Serve and Volley	54	48

FIGURE 7.45

What strategies should these players use?

SOLUTION We form the abstract game matrix, and use the circle and box method to show that this game is not determined because no entry is both circled and boxed (Figure 7.46).

$$\begin{bmatrix} \textcircled{48} & \boxed{52} \\ \boxed{54} & \textcircled{48} \end{bmatrix}$$

FIGURE 7.46

The players should use mixed strategies. To apply the formula above we note

$$A = 48, \quad B = 52, \quad C = 54, \quad D = 48,$$

so

$$\frac{C - D}{(B + C) - (A + D)} = \frac{54 - 48}{(52 + 54) - (48 + 48)} = \frac{6}{10} = \frac{3}{5}$$

and

$$\frac{B - D}{(B + C) - (A + D)} = \frac{52 - 48}{(52 + 54) - (48 + 48)} = \frac{4}{10} = \frac{2}{5}.$$

We conclude that Carlos should use his first strategy, the baseline style, about $\frac{3}{5}$ of the time, and Rob should use his first strategy, the baseline strategy, about $\frac{2}{5}$ of the time. ◆

EXAMPLE 7.15 What strategies should players use for the following game matrix (Figure 7.47)?

$$\begin{bmatrix} 2 & -3 \\ -1 & 3 \end{bmatrix}$$

FIGURE 7.47

SOLUTION The first player should use her first strategy $\frac{-1-3}{-4-5} = \frac{4}{9}$ of the time. The second player should use his first strategy $\frac{-3-3}{-4-5} = \frac{-6}{-9} = \frac{2}{3}$ of the time. ◆

INITIAL PROBLEM SOLUTION

There are two likely invasion routes leading into General Latka's remote land: the mountain pass or the beach. The forces at Latka's disposal to defend these routes are not adequate, so he will need to choose to defend one route heavily, while the other receives virtually a token defense. Suppose the level of casualties the attackers will suffer is estimated in the table (these numbers represent estimates of the relative significance of the expected casualties, not specific units) Figure 7.48. What should Latka do?

	Attack the Beach	Attack the Pass
Defend the Beach	6	1
Defend the Pass	2	4

FIGURE 7.48

SOLUTION The information can be organized as a game matrix. You should verify that the game is not determined (Figure 7.49).

$$\begin{bmatrix} 6 & 1 \\ 2 & 4 \end{bmatrix}$$

FIGURE 7.49

In the notation from the definition, we have

$$A = 6, \quad B = 1, \quad C = 2, \quad D = 4.$$

The optimal strategy for the first player (General Latka) is to use the first strategy (heavily defend the beach) with probability

$$\frac{C - D}{(B + C) - (A + D)} = \frac{2 - 4}{(1 + 2) - (6 + 4)} = \frac{2}{7}.$$

The second strategy (heavily defend the mountain pass) should occur with probability $1 - \frac{2}{7} = \frac{5}{7}$.

General Latka must decide on one action now, so he may as well go with the strategy with the higher probability by game theory.

HISTORY

In 1944, while the Allied Forces prepared to invade the mainland of Europe to begin the final phase of defeating Nazi Germany, a fundamental question for the German high command was where would the invasion take place. Pas de Calais and Normandy were the two leading candidates. The Germans decided that the Allied landing would occur in Pas de Calais rather than at Normandy (where it actually took place).

PROBLEM SET 7.3

1. Suppose both players choose their first and second strategies half the time at random. Find the average payoff for the first player assuming the following game matrix.

$$\begin{bmatrix} 1 & -2 \\ -1 & 3 \end{bmatrix}$$

2. Suppose both players choose their first and second strategies half the time at random. Find the average payoff for the first player assuming the following game matrix.

$$\begin{bmatrix} 1 & -3 \\ -2 & 2 \end{bmatrix}$$

3. Suppose the first player chooses his first and second strategy half the time at random, but the second player chooses her first strategy $\frac{1}{3}$ of the time and her second strategy $\frac{2}{3}$ of the time at random. Find the average payoff for the first player assuming the following game matrix.

$$\begin{bmatrix} 1 & -2 \\ -1 & 2 \end{bmatrix}$$

4. Suppose the first player chooses her first and second strategy half the time at random, but the second player chooses his first strategy $\frac{1}{3}$ of the time and his second strategy $\frac{2}{3}$ of the time at random. Find the average payoff for the first player assuming the following game matrix.

$$\begin{bmatrix} 1 & -3 \\ -2 & 2 \end{bmatrix}$$

5. Suppose the first and second players both choose their first, second, and third strategies $\frac{1}{3}$ of the time at random. Find the average payoff for the first player assuming the following game matrix.

$$\begin{bmatrix} 0 & 2 & -1 \\ 2 & 1 & -2 \\ -1 & -2 & -3 \end{bmatrix}$$

6. Suppose the first player chooses her first, second, and third strategies $\frac{1}{3}$ of the time at random, but the second player chooses his first strategy $\frac{1}{6}$ of the time, his second strategy $\frac{1}{3}$ of the time, and his third strategy $\frac{1}{2}$ of the time at random. Find the average payoff for the first player assuming the game matrix in problem 5.

In problems 7 through 10 find the optimal mixed strategy for each game matrix.

7. $$\begin{bmatrix} 1 & -2 \\ -1 & 2 \end{bmatrix}$$

8. $$\begin{bmatrix} 1 & 3 \\ 2 & 0 \end{bmatrix}$$

9. $$\begin{bmatrix} 1 & -3 \\ -1 & 3 \end{bmatrix}$$

10. $$\begin{bmatrix} 4 & -2 \\ -3 & 3 \end{bmatrix}$$

Suppose the frequencies that the second player will use are known. If that is certain, then the first player's optimal strategy is pure: one of the available strategies should be used all the time. To decide which, evaluate the average payoff for each possibility and choose that which gives the larger payoff.

It is possible that two or more strategies will give the same payoffs. For example, if in Rock, Scissors, Paper the second player is committed to using the three strategies with equal frequencies, then it does not matter what the first player does; the average payoff will be zero.

For example, suppose the game matrix is

$$\begin{bmatrix} 1 & -2 \\ -2 & 3 \end{bmatrix}$$

and the second player is certain to use the first strategy with frequency $\frac{1}{3}$ and the second strategy with frequency $\frac{2}{3}$. If the first player uses the first strategy with frequency p and the second strategy with frequency $1 - p$, the average payoff is

$$-\frac{7}{3}p + \frac{4}{3}.$$

If we graph the payoff as a function of p over the interval from $p = 0$ to $p = 1$, we obtain a straight line segment. Necessarily that line segment is either horizontal or is higher at one end than the other. The high end corresponds to the optimal strategy for the first player, in this case $p = 0$ with payoff $\frac{4}{3}$. A similar argument applies if the first player has more than just two strategies from which to choose.

For each of the game matrices in problems 11 through 16, find the optimal strategy for the first player and the average payoff of the game given the frequency of choices for the second player.

11. Second player chooses the first strategy $\frac{1}{3}$ of the time, the second strategy $\frac{2}{3}$ of the time.

$$\begin{bmatrix} 1 & -2 \\ -1 & 2 \end{bmatrix}$$

12. Second player chooses the first strategy $\frac{1}{2}$ of the time, the second strategy $\frac{1}{2}$ of the time.

$$\begin{bmatrix} 1 & 3 \\ 2 & 0 \end{bmatrix}$$

13. Second player chooses the first strategy $\frac{1}{4}$ of the time, the second strategy $\frac{1}{4}$ of the time, and the third strategy $\frac{1}{2}$ of the time.

$$\begin{bmatrix} 1 & -3 & 2 \\ -1 & 3 & -2 \end{bmatrix}$$

14. Second player chooses the first strategy $\frac{1}{2}$ of the time, the second strategy $\frac{1}{4}$ of the time, and the third strategy $\frac{1}{4}$ of the time.

$$\begin{bmatrix} 1 & -3 & 2 \\ -1 & 3 & -2 \end{bmatrix}$$

15. Second player chooses the first strategy $\frac{1}{4}$ of the time, the second strategy $\frac{1}{4}$ of the time, and the third strategy $\frac{1}{2}$ of the time.

$$\begin{bmatrix} 1 & -3 & 2 \\ -2 & 4 & -1 \\ 1 & -1 & -1 \end{bmatrix}$$

16. Second player chooses the first strategy $\frac{1}{2}$ of the time, the second strategy $\frac{1}{4}$ of the time, and the third strategy $\frac{1}{4}$ of the time.

$$\begin{bmatrix} 1 & -3 & 2 \\ -2 & 4 & -1 \\ 1 & -1 & -1 \end{bmatrix}$$

In problems 17 through 22, use dominant strategies to reduce the size of the given game matrix, and find the optimal mixed strategies for each player.

17.
$$\begin{bmatrix} 3 & -1 & -1 \\ -2 & 2 & 3 \end{bmatrix}$$

18.
$$\begin{bmatrix} 3 & 0 & 2 \\ -5 & -1 & 1 \end{bmatrix}$$

19.
$$\begin{bmatrix} 4 & 2 & -1 & -1 \\ -2 & -3 & 2 & 3 \end{bmatrix}$$

20.
$$\begin{bmatrix} -2 & 2 \\ 3 & -1 \\ -1 & 3 \end{bmatrix}$$

21.
$$\begin{bmatrix} 1 & -2 & -1 \\ -3 & 3 & 2 \\ 2 & -2 & 3 \end{bmatrix}$$

22.
$$\begin{bmatrix} -1 & -1 & 3 \\ 2 & 3 & -2 \\ -2 & 0 & 2 \end{bmatrix}$$

23. A pitcher and a batter face each other in a baseball game. The pitcher can throw a fastball or a curveball. Prepared to hit a fastball, the batter hits .450 against the fastball, but only .200 against the curveball. Prepared for a curveball, the batter hits .400 against the curveball and .240 against the fastball. What strategies should batter and pitcher use?

24. Carlos and Rob play a card game. Carlos has two cards, a 3 and a 7; and Rob has three cards, a 2, a 4, and a 5. Each player picks a card and places it on the table. If the sum is even, Carlos pays Rob \$3; if the sum is odd, Rob pays Carlos \$6. What strategies should Carlos and Rob use?

25. Charlane is thinking about investing some of her savings in bonds, stocks, and money market funds. Her expected returns will depend on whether interest rates rise or fall during the coming year. Her financial advisor tells her the expected return on stocks will be 10% if the rates rise, 12% if the rates fall. For bonds, the expected returns will be 6% (rates rise) and 11% (rates fall); while for money market funds they will be 15% (rates rise) and 10% (rates fall). What percentage of her savings should she invest in each type of investment?

26. A politician is planning a reelection campaign. He may use one of three types of advertising: newspaper, television, or radio. The campaign committee has raised \$100,000 to be allocated to the media depending on the opponent's campaign. If the opponent decides to attack the incumbent's record, then 70% of the money should be television and the remainder equally divided between the other two. But if the opponent's strategy is to focus on future issues, then only 50% should be spent on television, 30% on newspaper ads, and 20% on radio. Express the possible choices in a game matrix. How should the committee allocate funds to advertising?

Problems 27 and 28 refer to two computer software companies, Action Faction and Mega Sports, that place new games on the market at the same time. Each company uses television and newspaper advertising to market its products. The payoff matrix shows increases in sales (in millions of dollars) for Action Faction depending on the choices each company makes of its advertising campaign.

Action Faction	Mega Sports: Newspaper	Mega Sports: Television
Newspaper	3	−2
Television	−1	2

27. What is the optimum strategy for Action Faction if Mega Sports spends 25% of its advertising budget on newspaper ads and the remainder on television ads? What is the payoff to Action Faction if that strategy is used?

28. What is the optimum strategy for Action Faction if Mega Sports spends 75% of its advertising budget on newspaper ads and the remainder on television ads? What is the payoff to Action Faction if that strategy is used?

Problems 29 and 30 refer to a gubernatorial campaign. The Democratic and Republican candidates can campaign in either the urban or the rural areas. The political advisers have analyzed the possible payoff for campaigning as follows (the numbers are gains or losses in tens of thousands of votes).

		Republican	
		Urban	Rural
Democrat	Urban	−5	3
	Rural	4	2

29. What is the Democrat's optimum strategy if the Republican spends 80% of her campaign effort in urban areas? What is the payoff to the Democratic candidate?
30. What is the Democrat's optimum strategy if the Republican spends 40% of her campaign effort in urban areas? What is the payoff to the Democratic candidate?
31. What are the optimal mixed strategies for Action Faction and Mega Sports in problems 27 and 28? What is the payoff for each player?
32. What are the optimal mixed strategies for the Democrat and Republican in problems 29 and 30? What is the payoff for each player?

EXTENDED PROBLEMS

Suppose in the game of Rock, Scissors, Paper the first player chooses strategies R, S, and P with frequencies p, q, and $1 - p - q$, respectively, while the second player uses frequencies $r, s, 1 - r - s$ respectively. The payoff for the first player is

$$\begin{aligned} &ps - p(1 - r - s) - qr + q(1 - r - s) + \\ &r(1 - p - q) - s(1 - p - q) \\ &= p(s - 1 + r + s - r + s) + \\ &\quad q(-r + 1 - r - s - r + s) + (r - s) \\ &= p(3s - 1) + q(1 - 3r) + (r - s). \end{aligned}$$

By requiring the coefficients of p and q in the preceding formula for the payoff to be zero we can arrange for the payoff to be unchanged even if p and q change. That is, by setting $r = \frac{1}{3}$ and $s = \frac{1}{3}$ the first player cannot improve his or her payoff by changing frequency. Similarly, by setting $p = \frac{1}{3}$ and $q = \frac{1}{3}$ the second player cannot improve his or her payoff by changing frequency. This is the optimal mixed strategy.

To summarize, the optimal mixed strategy is found by writing the payoff in the form

$p \times$ [an expression involving only r, s, and constants]

$+ q \times$ [an expression involving only r, s, and constants]

$+$ [an expression involving only r, s, and constants]

and setting the coefficients of p and q equal to zero. Then the process is carried out again with the roles of p and q interchanged with the roles of r and s.

33. Find the optimal mixed strategies for the game with the following matrix.

$$\begin{bmatrix} 1 & -3 & 2 \\ -2 & 4 & -1 \\ 1 & -1 & -1 \end{bmatrix}$$

34. Find the optimal mixed strategies for the game with the following matrix.

$$\begin{bmatrix} 1 & -3 & -1 \\ -1 & 3 & 2 \\ 1 & -1 & -1 \end{bmatrix}$$

35. A pitcher and a batter face each other in a baseball game. The pitcher can throw a fastball, a screwball, or a curveball. Prepared to hit a fastball, the batter hits .400 against a fastball, .200 against a curveball, and .100 against a screwball. Prepared for a curveball, the batter hits .400 against a curveball, .220 against a fastball, and .160 against a screwball. Prepared for a screwball, the batter hits .380 against a screwball, .280 against a curveball, and .120 against a fastball.
    (a) Find the matrix for the game between the pitcher and batter.
    (b) Find the optimal mixed strategies for the game.
36. In problems 33 through 35, one is finding a "Nash equilibrium." Research who John Nash was and what he did.

Chapter Seven Problem

In a western duel there are three gunslingers, the Good, the Bad, and the Ugly. The Good always hits her target with 100% accuracy, the Bad hits his target 80% of the time, and the Ugly only has 60% accuracy. In this duel there can be only one winner. First Ugly takes a shot, then Good takes a shot (if she is still remaining in the duel), then Bad takes a shot. This continues in rotation until there is only one gunslinger left. What is each player's best strategy, and what is their chance of winning if they employ this strategy?

SOLUTION

Strategy: Draw a diagram.

On the surface Ugly has two options: (i) shoot at Bad or (ii) shoot at Good.

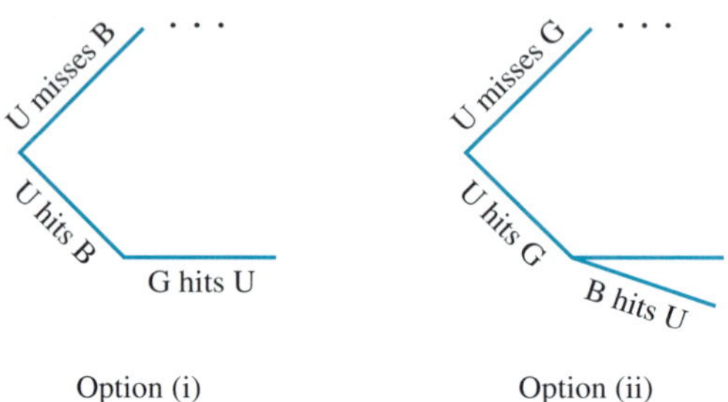

Option (i) Option (ii)

If he shoots at Bad and hits him, then Ugly has no chance of surviving since Good will surely kill him on the next shot. If Ugly hits Good, then Bad will have an 80% chance of killing Ugly, so the chance that Ugly even survives the first round would only be 20%.

Are there any other options? Yes, in option (iii), Ugly could purposely miss.

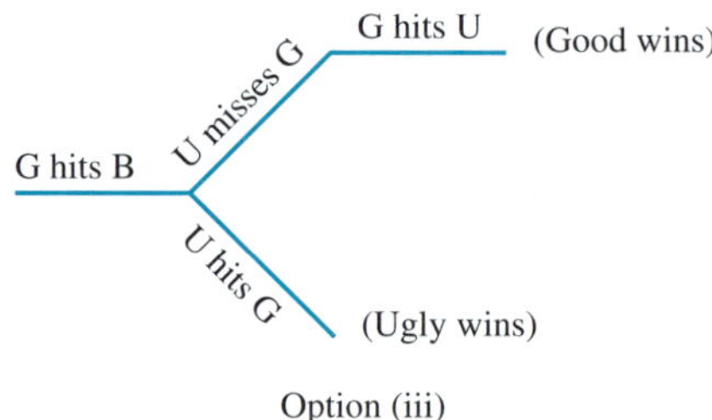

Option (iii)

This strategy is not expected, but it is within the rules of the problem. This may seem paradoxical, but Ugly's other options are not good. If Ugly misses, then Good will shoot at her most dangerous opponent, namely Bad, and hit him. Now there are only Ugly and Good left. Ugly hits Good with 60% chance. If Ugly misses, he will be shot by Good. If each duelist uses his best strategy, then each lives with the following probabilities.

Good 40% Bad 0% Ugly 60%

Chapter Seven Review

Key Ideas and Questions

The following questions review the main ideas of this chapter. Write your answers to the questions, and then refer to the pages listed by number to make certain that you have mastered these ideas.

1. How is a game matrix for a zero-sum game constructed? 326, 328
   Why is the most aggressive strategy typically a poor strategy? 331
2. Describe the most conservative strategy. 338
   Suppose that both players use their most conservative strategies, and in this case the optimal outcome occurs for both. Why is this game determined? 340

3. Describe why a mixed strategy may be beneficial over a strategy in which your opponent knows what you will do. 352

Vocabulary/Notation

Following is a list of key vocabulary, notation, and ideas for this chapter. Mentally review each of these items, write down the meaning of each term, and use it in a sentence. Then refer to the pages listed by number, and restudy any material you are unsure of before solving the Chapter Seven Review Problems.

Section 7.1

Players 323	Zero-sum Game 325	Row 327
Strategies 323	Two–Person Game 325	Column 327
Payoffs 323	Game Matrix 326	Size of a Matrix 327
Preferences 325	Entry/Element 327	Most Aggressive Strategy 331

Section 7.2

Most Conservative Strategy 338	Determined Game 342	Fair Game 343
Maximin Procedure 338	Optimal 342	Dominates 344
Minimax Procedure 338	Value of the Game 343	

Section 7.3

Mixed Strategy 350
Average Payoff of a Game 350
Optimal Mixed Strategy 352

Chapter Seven Review Problems

1. Consider the following game matrix.

$$\begin{bmatrix} 3 & -1 & -2 & 4 \\ 1 & 4 & -3 & 1 \\ -3 & 2 & 5 & -1 \\ -1 & 3 & 2 & -4 \end{bmatrix}$$

What will the payoff be if both players use their third strategy?

2. In the matrix in problem 1, what are the most aggressive strategies? What will the payoff be if both players use their most aggressive strategy?
3. In the matrix in problem 1, what are the most conservative strategies? What will the payoff be if both players use their most conservative strategies?
4. Is the game in problem 1 determined or not? Why? If it is determined what is the value of the game?
5. Construct a game matrix for the following game. Alice is on attack and can either pitch or kick. Bob is on defense and can either play tight or loose. If Alice pitches and Bob plays tight, then Alice loses one yard. If Alice pitches and Bob plays loose, then Alice gains 4 yards. If Alice kicks and Bob plays tight, then Alice gains 5 yards. If Alice kicks and Bob plays loose, then Alice loses 10 yards.
6. What is the most aggressive strategy for Alice and Bob in the game in problem 5? Suppose that Alice knows that Bob will choose his most aggressive strategy. What strategy should Alice employ? How many yards will she gain or lose?
7. Consider the following game matrix. Use the box and circle method to decide if the game is determined and, if so, the value of the game.

$$\begin{bmatrix} 0 & 1 & 2 & -1 \\ -3 & 1 & 3 & -2 \\ 1 & 2 & -3 & -1 \\ -1 & 3 & 2 & -4 \end{bmatrix}$$

8. Consider the following game matrix.

$$\begin{bmatrix} -3 & 3 \\ 4 & -2 \end{bmatrix}$$

Suppose that the first player uses a mixed strategy with frequencies $(\frac{1}{2}, \frac{1}{2})$ and the second player uses a mixed strategy with frequencies $(\frac{1}{3}, \frac{2}{3})$. What is the average payoff to the first player of this game?

9. Suppose you are the first player in the game in problem 8 and learn that the second player has a mixed strategy with frequencies $(\frac{1}{3}, \frac{2}{3})$ What is your optimal strategy in this case, and what is the payoff?
10. What is the optimal mixed strategy for each player in problem 8?

CHAPTER 8

MANAGEMENT MATHEMATICS

MATHEMATICIAN'S BREAKTHROUGH WORTH MILLIONS TO INDUSTRY

In 1984, Narendra Karmarkar, a young mathematician working at AT&T Bell Laboratories, found a way to greatly reduce the amount of time needed to solve linear programming problems, one of the most common problems confronted by modern business. Large industries (such as airlines and oil companies) with complex networks and processes for the manufacture or distribution of a wide range of services and products realized a great economic benefit from this breakthrough.

CHAPTER GOALS

1. Solve linear inequalities to find optimal mixtures of resources under several constraints.
2. Solve routing problems using graph theory.

It is a fairly rare event when an advance in mathematics makes headlines in newspapers around the world, including *The Wall Street Journal.* What was all the fuss about?

The type of problem Karmarkar was trying to solve for AT&T Bell Laboratories dealt with routing millions of telephone calls over a vast network. This meant determining the best way to use resources to maximize the number of calls that could be handled. The economic stakes were very high. The best solution could result in saving the company many millions of dollars.

Even small businesses must choose among options with varying costs and outcomes. Consider the problem of formulating an adequate diet. Certain nutrients must be included in a diet and various food sources have differing costs and differing amounts of nutrients. The problem involved may be to find the least expensive selection of foods that contains, in total, all the necessary nutrients in sufficient amounts. The best possible way of achieving the goal can be found by a process known as linear programming. In this chapter you will learn to formalize problems such as the dietician problem, and solve them using linear programming.

Karmarkar's work involved a linear programming problem with about 800,000 variables and many conditions. Karmarkar's solution was estimated to be more than *50 times* faster than the most efficient other method. Many companies face problems as large in scope. An airline may deal with more than 1000 jets, 100 cities, and 100,000 passengers. It is in such complex situations that Karmarkar's algorithm is useful.

We will also discuss the application of mathematics to the analysis of networks, including ways to find the most efficient routes within a network.

THE HUMAN SIDE OF MATHEMATICS

George Dantzig

George Dantzig (1914–) was the oldest son of a writer and mathematician who hoped his two sons would follow in his footsteps. In 1939, while in graduate school at the University of California at Berkeley, Dantzig arrived late to a statistics class to find two problems written on the board. Assuming they were homework problems, Dantzig copied them down; he solved the problems later and turned them in. After about six weeks, his professor informed Dantzig that the problems he thought were homework were, in fact, two famous, and previously unsolved, problems in statistics. The problems became the basis for Dantzig's Ph.D. thesis in mathematics.

During World War II, George Dantzig served in the Air Force as a mathematician. His task was to find practical and effective ways to distribute men, weapons, and supplies to the war front. Following the war, Dantzig was assigned to the Pentagon where he continued work on the resource allocation planning process. In the aftermath of the war, the question arose as to whether the process could be formulated as a mathematical system, and how computers (also a product of the war effort) could be used to solve such a system. The result of his work is known as linear programming. In 1948, when relations between the Soviet Union and the Allies deteriorated into the Cold War, Dantzig's newly formulated linear programming was used in the Berlin Air Lift to thwart the Soviet effort to gain total control of Berlin.

In the postwar world, linear programming was quickly applied to a wide variety of business, economic, and environmental problems. The first major industry-wide adoption of linear programming was in petroleum refining where it was used for blending gasoline as well as for transportation problems. Planners in a broad spectrum of fields rely heavily on linear programming.

Ronald L. Graham

Ronald L. Graham (1935–) is one of the leading mathematicians in the world and also one of the world's top jugglers. Graham is presently the director of the mathematics research department of AT&T Bell Laboratories. He is actively engaged in his own research and supervises the work of a team of outstanding mathematicians who are working on complex problems related to management and industry. Graham's mathematical research is concerned with combinatorial structures that apply to problems of networks. In simpler terms, he deals with finding the number of ways something can be done. For example, when you make a telephone call across the country there is no direct line between you and the other party. The question that concerns Graham is how the telephone system should route your call among the countless possible paths so that a connection is made as quickly and reliably as possible. Of course, he has to consider the millions of other calls also being placed on the network.

In addition to being a leading mathematical researcher in industry, Graham has served as professor of mathematics at UCLA, Stanford, Princeton, Cal Tech, and Rutgers. He is a past president of the American Mathematical Society and is in much demand as a speaker, having appeared at numerous industrial conferences and college campuses around the country.

Graham also has the honor of actually using the largest number ever reported in a legitimate scientific publication. Ron Graham is one of the brilliant people in mathematics whose influence reaches industry, government, and academia.

8.1 LINEAR CONSTRAINTS

INITIAL PROBLEM

A gardener would like to apply at least five pounds each of nitrogen, phosphorus, and potassium to her garden. The amount of these elements in any fertilizer is usually given by three numbers, a–b–c, which represent the percentage by weight of the content of nitrogen, phosphorus, and potassium. She intends to use a mixture of packaged chicken manure rated 3–2–2, and a general purpose chemical fertilizer rated 5–15–10. Make a graph showing the possible ways the gardener could accomplish her goal by applying these two types of fertilizer.

There are some types of problems that are faced by all managers. They must obtain needed supplies from various sources, decide how much of various products to produce, and generally allocate resources between various competing demands. For example, oil companies obtain crude oil from all over the world, have refineries in numerous locations, produce several different products (gasoline, diesel fuel, motor oil, etc.), and supply customers all across the nation. One of the most fundamental management problems is the allocation of resources among various alternative uses.

Even if you are a student and feel that you have almost no resources, you are in fact managing a small-scale enterprise with at least one valuable resource that you may not have thought of: your time. Let us consider the time-management problem of a typical student. Suppose the student has a part-time job in addition to taking courses. There are only 24 hours in a day, and the student must devote some of that time to studying and attending classes, some to working, and some to other less important incidentals—eating, sleeping, relaxing, socializing, and so on—all of which we will group under the term "living." Suppose that, because of an upcoming exam and a shortage of money, the student has decided to forgo the "living" portion of her life for a day, and to keep the nose to the grindstone by only studying and working, then the possible allocations of time range from spending all 24 hours studying, to spending all 24 hours working. Introducing the variables s for time spent studying, and w for time spent working, we have the equation

$$s + w = 24$$

because the time spent studying and the time spent working must add up to the entire 24 hours in the day. Some allocations of time can be listed as pairs of numbers

$$(24, 0)\ (23, 1)\ (22, 2)\ (18, 6)\ (12, 12)\ (9, 15)\ (13.4, 10.6)\ \ldots$$

where we will agree to write the value of s as the first element (i.e. the element on the left) in the pair, and the value of w as the second element; each pair has the form (s, w). For example, the pair (18, 6) indicates the student chooses to spend 18 hours studying and 6 hours working. If we allow tenths of an hour to be used, it is

TIDBIT

Sleep deprivation experiments show that, in addition to experiencing extreme muscular weariness, subjects become very irritable, their thoughts become disorganized, and they may even hallucinate. Not good conditions for taking an exam.

not practical for us to list all the pairs. A more useful way to visualize the possible time allocations is to graph all the possible pairs of points. As is customary, the first entry in each pair determines the position of the point in the graph in the horizontal direction, and the second entry determines the position in the vertical direction. The graph of the equation $s + w = 24$ is shown in Figure 8.1.

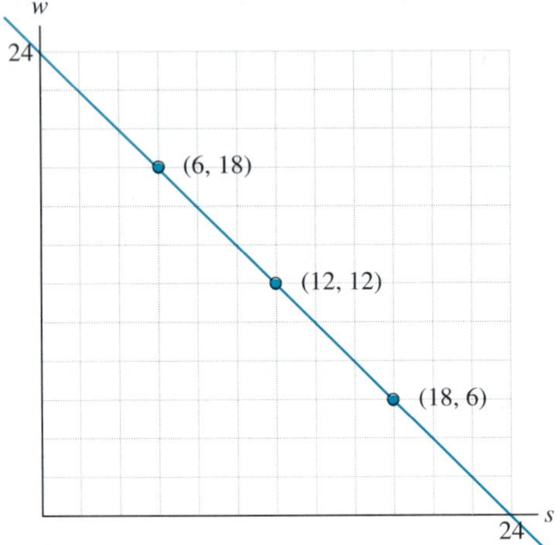

FIGURE 8.1

Here we are only interested in that part of the line where both s and w are nonnegative, because there is no way to spend a negative amount of time studying or a negative amount of time working. The graph of the possible time allocations is the line segment from the point (24, 0), representing the choice to spend all the time studying, to the point (0, 24), representing the choice to spend all the time working. This is shown in Figure 8.2.

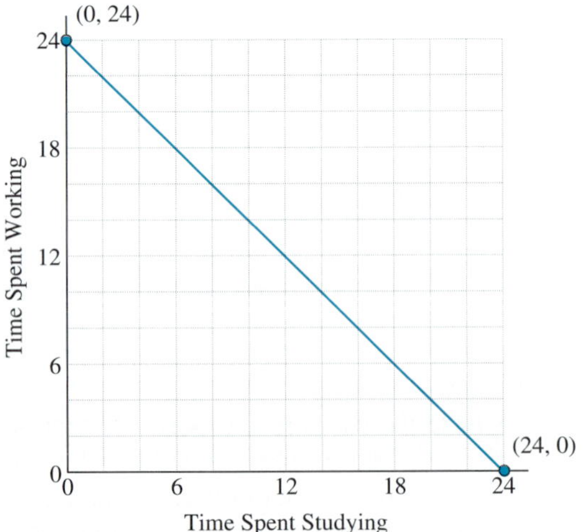

FIGURE 8.2

Most of the time our student is not going to be willing or able to stick to a regimen of only studying and working. Some time will be needed for living, including getting some sleep. For example, the student might devote several hours of the day to living. The remaining time can be divided between studying and working in many different ways, but no longer will the values of s and w add to 24. Instead they add to a number smaller than 24. For example, if the student devotes 6 hours to basic living, then there are at most 18 hours left for studying and working. That is, the following inequality will hold:

$$s + w \leq 24.$$

Note that it is still possible that the student will occasionally spend all 24 hours working and studying.

An inequality of the form $ax + by \leq c$ where a, b, and c are constants and x and y are variables, is called a **linear inequality.** The set of points that satisfy a linear inequality, called its **solution set,** consists of the points on one side or the other of the line determined by replacing the inequality sign with an "=." If the inequality is **strict** (i.e. the sign is $<$ or $>$), then the boundary line is not part of the set satisfying the inequality; if the inequality is **inclusive** (i.e. the sign is $\leq$ or $\geq$), then the boundary line is part of the set satisfying the inequality. For example, Figure 8.3(a) shows the solution set of $x + y < 4$ where the line $x + y = 4$ is dashed to indicate that it is not part of the solution set, and Figure 8.3(b) shows the solution set of $2x + 3y \geq 6$ where the line $2x + 3y = 6$ is solid to indicate that it is part of the solution set.

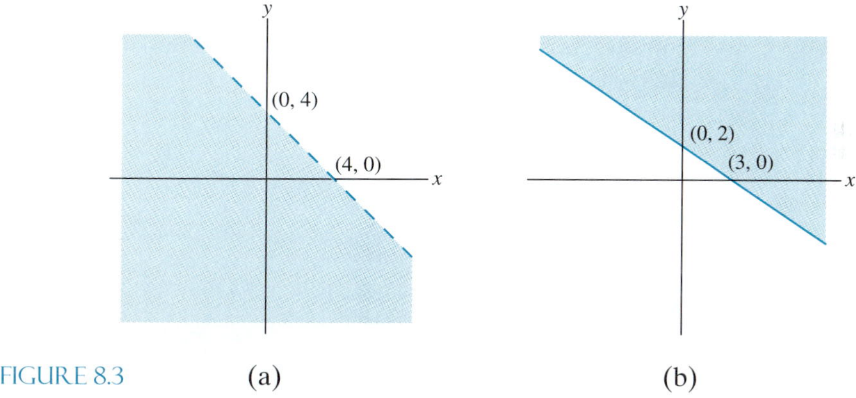

FIGURE 8.3 (a) (b)

Plotting the points satisfying a linear inequality involves two main tasks:

(i) plotting the line of points satisfying the associated equation, and
(ii) determining on which side of the line the points satisfy the inequality.

It is important to remember that, as far as the linear inequality is concerned, all the points on one side of the line are the same; they either all satisfy the inequality, or they all fail to satisfy the inequality. We can use this fact to our advantage. Just find a convenient point not on the line, and determine whether or not it satisfies the inequality. Once you know that, the same is true for all other points on the same side of the line.

EXAMPLE 8.1 Graph the inequality $x + y \leq 24$.

SOLUTION

STEP 1: Plot the line $x + y = 24$: since (24, 0) and (0, 24) both satisfy the equation, they determine the line [Figure 8.4(a)]. The line is solid due to the inclusive inequality $\leq$.

STEP 2: Find which side of the line is determined by the $\leq$: since $0 + 0 \leq 24$, the point (0, 0) is in the solution set. Thus all points on the same side of line $x + y = 24$ as (0, 0) are in the solution set [Figure 8.4(b)].

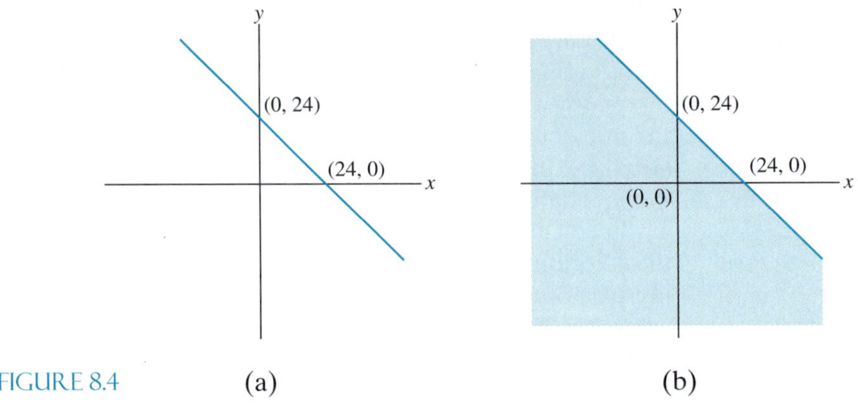

FIGURE 8.4 (a) (b) ◆

In the case of the time-management problem and the inequality $s + w \leq 24$, the studying hours, s, and working hours, w, cannot be negative. Thus the set of points shaded in Figure 8.5 consists of all possible allocations of the student's time between studying and working.

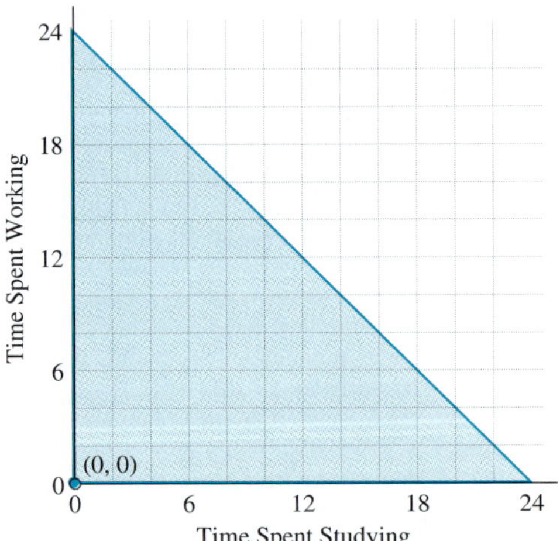

FIGURE 8.5

Whenever there is more than one inequality that must be satisfied, we call it a **system of inequalities.**

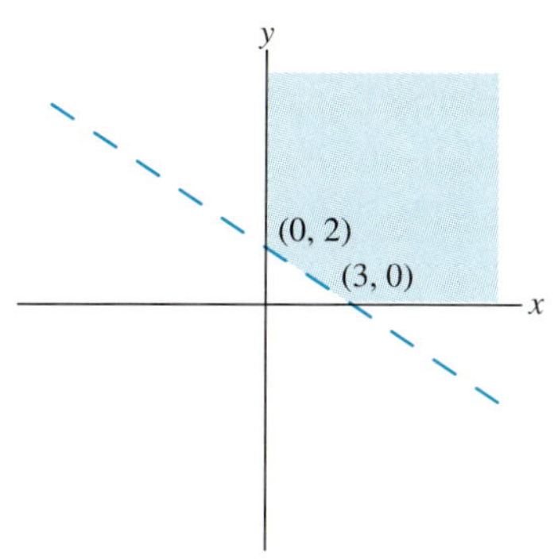

FIGURE 8.6

EXAMPLE 8.2 Graph the solution set of the system of inequalities

$$2x + 3y > 6, \quad x \geq 0, \quad \text{and} \quad y \geq 0.$$

SOLUTION The shaded region in Figure 8.6 is the solution set.

Observe that line $2x + 3y > 6$ is dashed due to the strict inequality $>$. Also, (0, 0) is not in the solution set of $2x + 3 > 6$ so the portion of the plane above the line is shaded. Since $x \geq 0$ and $y \geq 0$, only the portion above the x-axis and to the right of the y-axis is shaded. ◆

Any allocation of the student's time in the time–management problem must satisfy the following three conditions.

$$s + w \leq 24 \qquad s \geq 0 \qquad w \geq 0.$$

In this context, conditions that must be satisfied are called constraints. Since the constraints here are linear inequalities, we call them **linear constraints.** The region consisting of the points that satisfy all the constraints is called the **feasible region** determined by the constraints.

EXAMPLE 8.3 Consider a furniture company that produces both unfinished and finished chairs. It takes 6 person-hours to produce an unfinished chair and 10 person-hours to produce a finished chair. Plot the feasible region for a week's production assuming the company has 480 person-hours of manufacturing labor available each week.

HISTORY

Little furniture was made during the later middle ages (14th and 15th centuries) because the insecurity of the times dictated that no one could own more than he could defend. Possibly because of this, both the French word for furniture, *meuble,* and the German, *Mobel,* have the sense of being movable.

SOLUTION First assign variables.

Let x = the number of unfinished chairs produced during the week

and y = the number of finished chairs produced during the week.

Then

$6x$ = the number of person-hours to produce x unfinished chairs

and $10y$ = the number of person-hours to produce y finished chairs.

Then the linear inequality that must be satisfied is

$$6x + 10y \leq 480.$$

The associated equation is $6x + 10y = 480$. We can find the feasible region of $6x + 10y \leq 480$ by first plotting the line $6x + 10y = 480$.

Since a line is determined by any two of its points, we select two points that are convenient, namely, those where $x = 0$ or $y = 0$. For $6 \times 0 + 10y = 480$, we have

$y = 48$. Thus (0, 48) is one of the points of the line. Similarly, when $6x + 10 \times 0 = 480$, we have $6x = 480$, or $x = 80$. Thus (80, 0) is also on the line. We graph the points satisfying the equation by drawing the line segment from (80, 0) to (0, 48). As before, we have only plotted nonnegative values since a negative number of chairs cannot be produced (Figure 8.7).

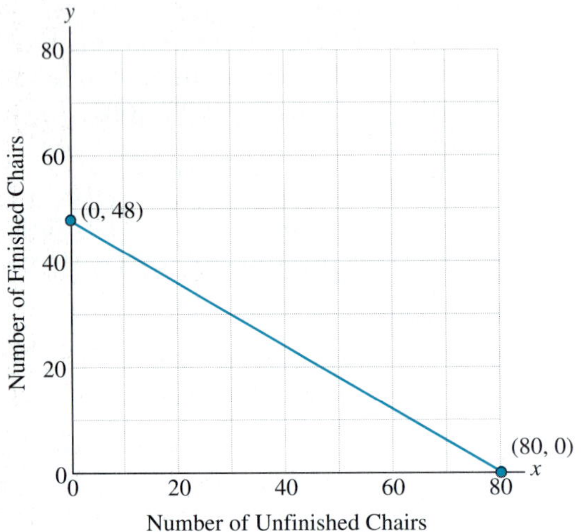

FIGURE 8.7

To plot the feasible region, we choose a convenient point not on the line of points satisfying the equation and determine whether or not that point satisfies the inequality. A convenient point is (0, 0), and it does satisfy the inequality, namely, $6 \times 0 + 10 \times 0 < 480$ [the point (0, 0) means that the company can choose to make no chairs of either type in some week if its warehouse is full]. The feasible region is plotted in Figure 8.8.

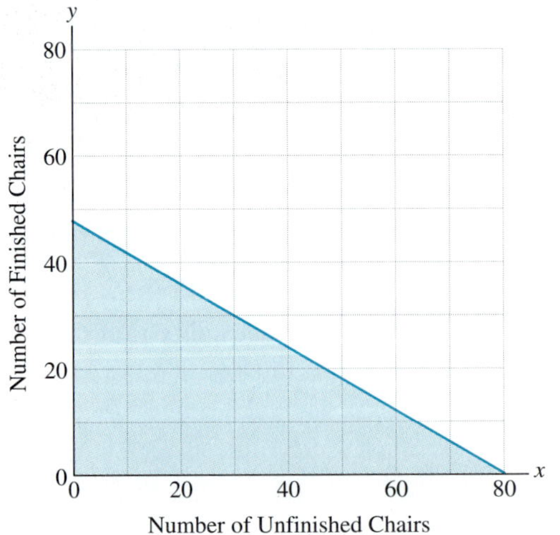

FIGURE 8.8

◆

HISTORY

Marco Polo is supposed to have brought a recipe for "milk ice" to the western world. Later, chefs invented ice cream for the European nobility. Dolley Madison, first lady when James Madison was president, introduced ice cream that was made with strawberries grown in the White House garden to American society.

The next example illustrates the situation when there is more than one linear constraint (in addition to the standard nonnegativity requirements). This will involve solving two inequalities.

EXAMPLE 8.4 A creamery produces two qualities of ice cream. Each gallon of the regular ice cream requires 0.4 gallons of cream and 0.6 gallons of milk, and each gallon of the deluxe ice cream requires 0.5 gallons of cream and 0.5 gallons of milk. If the creamery's suppliers can provide a maximum of 1000 gallons of cream and a maximum of 1200 gallons of milk, plot the feasible region for the creamery's ice cream production.

SOLUTION Let r denote the number of gallons of regular ice cream produced, and let d denote the number of gallons of deluxe ice cream produced. We will plot r along the horizontal axis and d along the vertical axis. The maximum available cream gives rise to one linear constraint.

$$0.4r + 0.5d \leq 1000.$$

(Note: r gallons of regular takes $0.4r$ gallons of cream, and d gallons of deluxe takes $0.5d$ gallons of cream. No more than 1000 gallons are available.)

The maximum available milk gives rise to a second linear constraint

$$0.6r + 0.5d \leq 1200.$$

The associated equations are

$$0.4r + 0.5d = 1000 \qquad \text{and} \qquad 0.6r + 0.5d = 1200.$$

To plot the lines associated with these two equations, find ordered pairs where one of the coordinates is 0. For the first equation, we draw the line segment connecting (2500, 0) and (0, 2000) since these two points satisfy the equation $0.4r + 0.5d = 1000$. For the second equation, we draw the line segment connecting (2000, 0) to (0, 2400) (Figure 8.9).

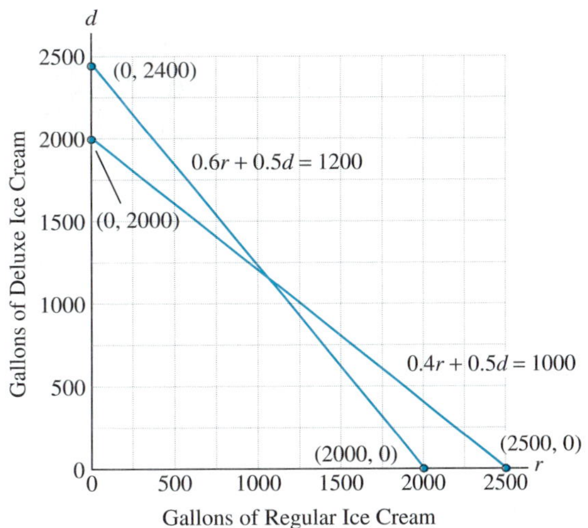

FIGURE 8.9

The region satisfying the two inequalities can be determined by finding one point in the region (or even a point not in the region); the origin is usually the most convenient point to check. Here the origin satisfies both inequalities. In Figure 8.10(a) we shade the points satisfying the first inequality, and in Figure 8.10(b) we shade the points satisfying the second inequality.

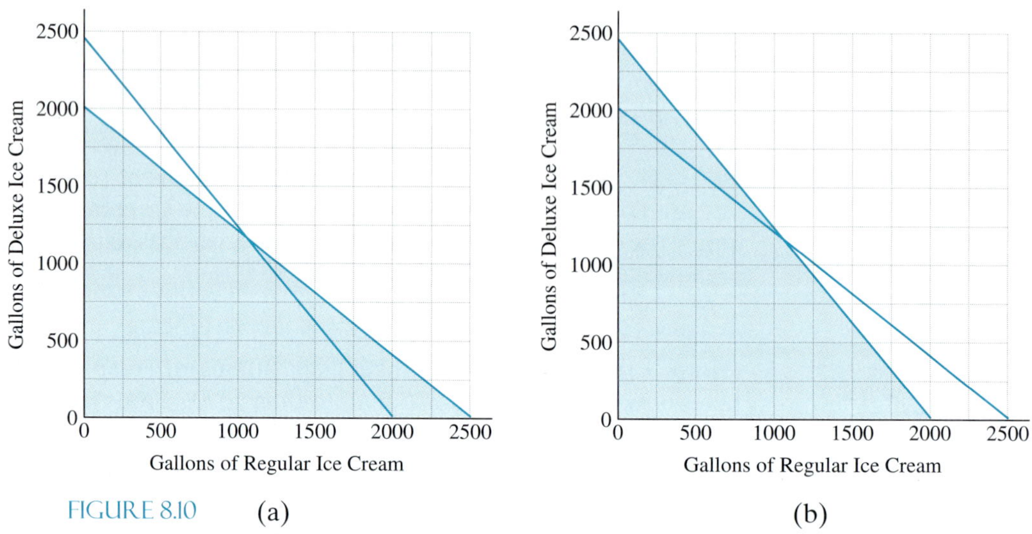

FIGURE 8.10 (a) (b)

In Figure 8.11 we shade the points satisfying both inequalities. The region in Figure 8.11 is the feasible region, namely, the region that contains the pairs of points that are solutions to the two inequalities (whose coordinates provide all solutions to the original problem). Any point in the feasible region represents a plan for ice cream production that can be carried out with the available supply of milk and cream. For example, the point (1000, 500) is in the feasible region, so the creamery could produce 1000 gallons of regular ice cream and 500 gallons of deluxe ice cream. To do so, it would use 650 gallons of cream and 850 gallons of milk.

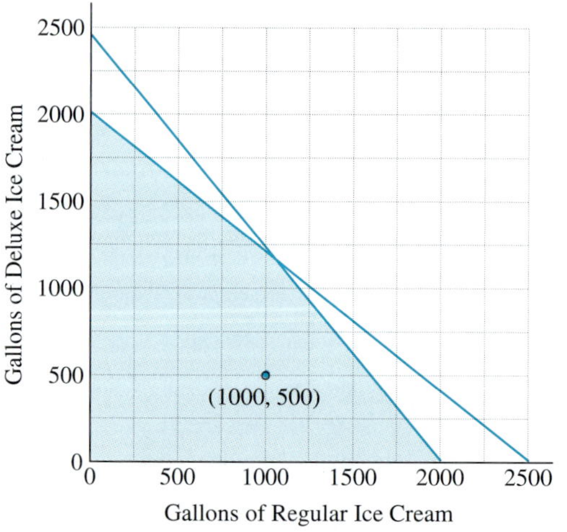

FIGURE 8.11 ◆

The following summarizes how you find the feasible region of a system of linear inequalities.

The Feasible Region of a System of Linear Inequalities

1. Graph the lines determined by the given equations.
2. Determine if the lines are to be solid or dashed based on the inequalities.
3. Determine the feasible region for each inequality.
4. Find the intersection of all feasible regions in 3.

A gardener would like to apply at least five pounds each of nitrogen, phosphorus, and potassium to her garden. The amount of these elements in any fertilizer is usually given by three numbers, a–b–c, which represent the percentage by weight of the content of nitrogen, phosphorus, and potassium. She intends to use a mixture of packaged chicken manure rated 3–2–2, and a general purpose chemical fertilizer rated 5–15–10. Make a graph showing the possible ways the gardener could accomplish her goal by applying these two types of fertilizer.

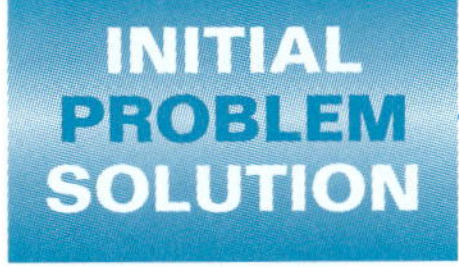

SOLUTION We need to graph the feasible region. Let x denote the amount in pounds of chicken manure that is to be used, and let y denote the amount in pounds of chemical fertilizer that is to be used. The requirement of applying at least 5 pounds of nitrogen leads to the inequality

$$0.03x + 0.05y \geq 5.$$

That is, 3% of x pounds of chicken manure plus 5% of y pounds of chemical fertilizer must equal at least five pounds of nitrogen.

Likewise the requirement of applying at least five pounds of both phosphorus and potassium leads to the inequalities

$$0.02x + 0.15y \geq 5$$
$$0.02x + 0.10y \geq 5.$$

The variables x and y must both be nonnegative. Multiplying all the inequalities by 100 will eliminate the decimals. The complete set of constraints is

$$3x + 5y \geq 500$$
$$2x + 15y \geq 500$$
$$2x + 10y \geq 500$$
$$x \geq 0, y \geq 0.$$

In Figure 8.12, we have plotted the three lines satisfying the associated equations.

HISTORY

The Romans knew enough about soil fertility to recommend rotating crops, adding manure, liming acidic soil, and growing legumes (which fix atmospheric nitrogen). The first scientific observations concerning fertilizer were made by the 17th century German J. R. Glaubner who experimented with the use of potassium nitrate (saltpeter).

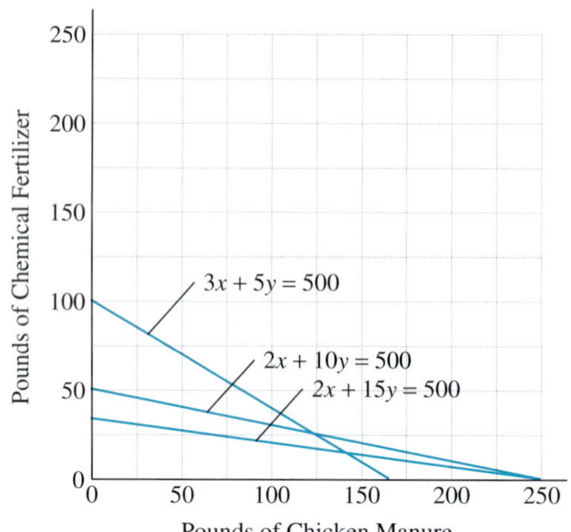

FIGURE 8.12

The origin does not satisfy any of the inequalities, so we know the region satisfying each inequality is on the side not containing the origin. We shade the intersection of these three regions in Figure 8.13.

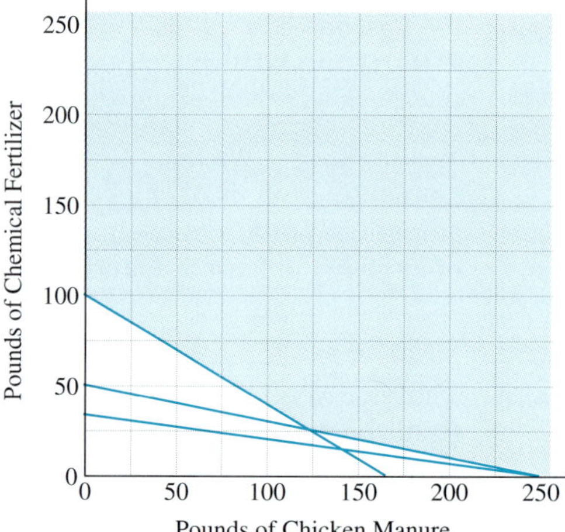

FIGURE 8.13

Notice that the line segment satisfying $2x + 15y = 500$ is entirely below the line segment satisfying $2x + 10y = 500$, while the feasible region lies above the latter line segment. This tells us that the inequality $2x + 15y \geq 500$ has no real bearing on the problem and can be ignored in the solution.

PROBLEM SET 8.1

In problems 1 through 8 determine whether the points satisfy the given inequality.

1. $x + y \geq 5$
  (a) $(0, 0)$ **(b)** $(6, -2)$ **(c)** $(3, 4)$

2. $x - y \leq 2$
  (a) $(0, 0)$ **(b)** $(3, -2)$ **(c)** $(5, 3)$

3. $2x - 3y \leq 5$
  (a) $(0, 0)$ **(b)** $(4, 1)$ **(c)** $(6, 2)$

4. $3x - y \geq 3$
  (a) $(0, 0)$ **(b)** $(-1, -5)$ **(c)** $(1, -1)$

5. $x + y > 7$
  (a) $(0, 0)$ **(b)** $(2, 6)$ **(c)** $(9, -2)$

6. $2x - y < 5$
(a) $(0, 0)$ **(b)** $(3, -1)$ **(c)** $(4, 3)$

7. $x + 3y < 8$
(a) $(0, 0)$ **(b)** $(2, 2)$ **(c)** $(10, -1)$

8. $2x - y < 5$
(a) $(0, 0)$ **(b)** $(3, -1)$ **(c)** $(4, 3)$

In problems 9 through 16, sketch the graph of each linear inequality. Shade those points that satisfy the inequality.

9. $x + y \geq 2$
10. $x + 2y \leq 6$
11. $2x + y \geq 6$
12. $2x - 3y \leq 12$
13. $-x + y \geq 5; x \geq 0, y \geq 0$
14. $5x - 2y \leq 10; x \geq 0, y \geq 0$
15. $2x + 3y \leq 12; x \geq 0, y \geq 0$
16. $3x + 2y \leq 15; x \geq 0, y \geq 0$

In problems 17 through 20, match the solution region of each system of linear inequalities with one of the four regions in the following figure.

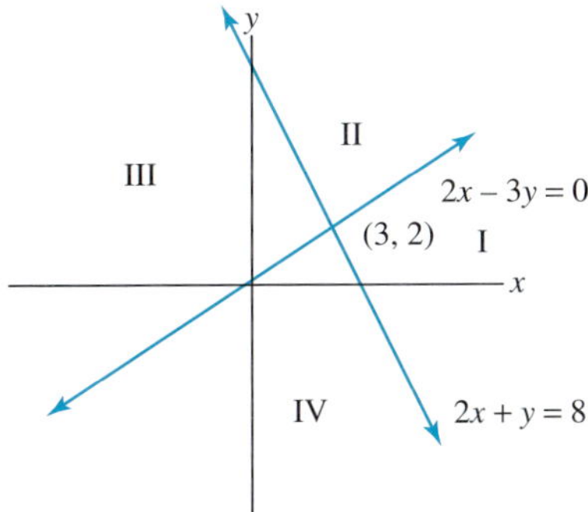

17. $2x - 3y \leq 0; 2x + y \leq 8$
18. $2x - 3y \leq 0; 2x + y \geq 8$
19. $2x - 3y \geq 0; 2x + y \leq 8$
20. $2x - 3y \geq 0; 2x + y \geq 8$

In problems 21 through 24, match the solution region of each system of linear inequalities with one of the four regions in the following figure.

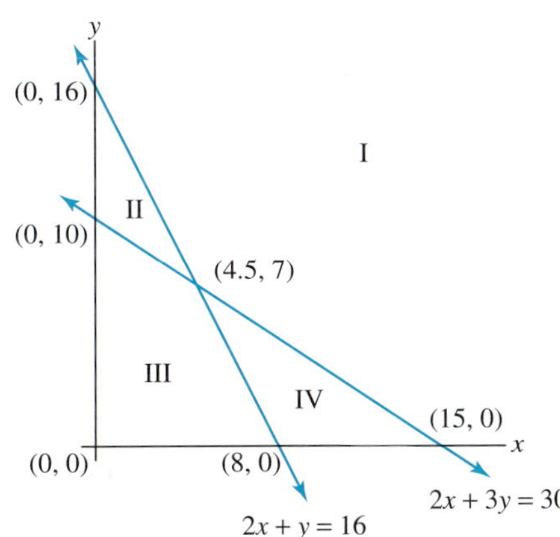

21. $2x + 3y \geq 30$
$2x + y \leq 16$
$x \geq 0, y \geq 0$

22. $2x + 3y \geq 30$
$2x + y \geq 16$
$x \geq 0, y \geq 0$

23. $2x + 3y \leq 30$
$2x + y \geq 16$
$x \geq 0, y \geq 0$

24. $2x + 3y \leq 30$
$2x + y \leq 16$
$x \geq 0, y \geq 0$

In problems 25 through 38, graph the feasible region determined by the given constraints.

25. $x + y \leq 8$
$2x + y \leq 10$
$x \geq 0, y \geq 0$

26. $2x + 3y \leq 12$
$2x + y \leq 8$
$x \geq 0, y \geq 0$

27. $2x + y \geq 4$
$x + 2y \geq 4$
$x \geq 0, y \geq 0$

28. $x + y \geq 5$
$2x + y \geq 8$
$x \geq 0, y \geq 0$

29. $3x + 2y \geq 6$
$x + y \leq 5$
$x \geq 0, y \geq 0$

30. $x - y \geq 2$
$x + y \geq 6$
$x \geq 0, y \geq 0$

31. $x + y \geq 6$
$5x + 3y \leq 30$
$x \geq 0, y \geq 0$

32. $x + y \leq 8$
$x + 2y \geq 6$
$x \geq 0, y \geq 0$

33. $x + y \geq 6$
$5x + 4y \geq 40$
$x \geq 0, y \geq 0$

34. $2x + y \leq 10$
$-x + y \leq 2$
$x \geq 0, y \geq 0$

35. $x + 3y \geq 6$
$x + y \geq 4$
$3x + y \geq 6$
$x \geq 0, y \geq 0$

36. $x + y \leq 8$
$2x + 3y \leq 18$
$3x + 2y \leq 18$
$x \geq 0, y \geq 0$

37. $x + y \le 8$
$x + 3y \le 12$
$-x + 2y \le 2$
$x \ge 0, y \ge 0$

38. $x + y \ge 5$
$2x + y \ge 6$
$x + 2y \ge 6$
$x \ge 0, y \ge 0$

In problems 39 through 44, express the given conditions as linear inequalities.

39. A woodworker produces bookcases and tables in his shop. He uses 40 feet of 12-inch boards to make a bookcase and 68 feet of 12-inch boards to make a table. He has only 800 feet of 12-inch boards available.

40. An animal breeder raises two breeds of dogs, which he feeds special diets. One breed of dog requires 4 oz. per serving of food source A, and the other breed requires 2 oz. per serving of the same source. The breeder can obtain, at most, 40 pounds of food source A on a regular daily basis.

41. A tool company manufactures two types of electric drills, one of which is cordless. The cord-type drill requires 2 labor hours to make, and the cordless drill requires 3 hours. The company has only 600 labor hours available each day.

42. The packaging department of the company in problem 41 can package 250 drills per day at the most.

43. A recreational manufacturer makes two types of tents: a 4-person tent that costs $100 to make and a 2-person tent that costs $60 to make. The manufacturer can budget no more than $9000 to produce the tents.

44. The manufacturer in problem 43 can make no more than 120 tents and must make at least 60 tents. At least 40 of the tents must be 4-person models.

In Problems 45 through 50, express the given conditions as linear inequalities and graph the feasible region.

45. A private fishing resort has bass and trout in its lake. The owner provides two types of food, A and B, for these fish. Each week, the bass require 2 units of food A and 4 units of food B, and the trout require 5 units of food A and 2 units of food B. The owner can obtain up to 800 units of each food on a regular weekly basis.

46. On the final weekend before the election, a candidate wishes to use a combination of radio and television ads in her campaign. Research has shown that during the prime viewing and listening hours each 30-second spot on television can reach 72,000 people, and each 30-second spot on radio can reach 12,000 people. The candidate wants to have the ads reach at least 2.16 million people (including duplications) and use a total of at least 40 minutes of ads.

47. A large travel agency has sold 1200 tour packages, including airfare, to the NCAA "Final Four" National championships. They have two types of airplanes for the charter flights. Type 1 aircraft carry 100 passengers and type 2 aircraft carry 150 passengers. Each flight of a type 1 aircraft will cost $9000 and each flight of a type 2 aircraft will cost $15,000. The association can lease no more than ten planes.

48. A door manufacturer produces doors in two styles, a regular wooden door and a deluxe model with glass panels. The manufacturer spends $175 to make a regular door, which it sells for $270, while each deluxe door costs $250 to make and sells for $380. The daily production capacity is 110 doors, and costs cannot exceed $21,000.

49. A large company is having its annual national sales meeting. Planners must provide accommodations for at least 400 attendees. They have two types of rooms available: motel rooms that sleep three people and hotel rooms that sleep two people. For meals they have budgeted $20 daily per person for those in motel rooms and $40 daily per person for those in hotels, and they must not spend more than $12,000 in meal money per day.

50. The director of the computer center at a large university wants to staff consulting stations with two types of shifts: type A will have 2 senior programmers and 1 student assistant; type B will have a senior programmer and 4 student assistants. Type A shifts will serve for two hours, and type B shifts will serve for three hours. The consulting stations will be open at least 48 hours during the week, and the director wants to use no more than 210 hours of staff time.

EXTENDED PROBLEMS

51. Linear programming problems can often get fairly large. How many variables and how many constraints (both resource and minimum) would there be for a company that uses 8 resources (materials and processes) to produce 5 products? Explain your reasoning.

52. Although there is no Nobel Prize in mathematics, several prizes have been awarded for mathematical applications in other fields, particularly economics. Write a report on the recipients of the 1975 Nobel Prize in economics and their work.

53. During recent decades, many colleges and universities have been facing pressures with respect to increasing enrollments, demands for services, and reduced resources. How does your college or university deal with the challenge? What planning or resource allocation methods are used, for example, for scheduling classes?

54. Contact a major local manufacturer, transportation company, or distributor. What factors must this business take into account besides customer needs and resource constraints? What methods are used for planning, scheduling, or resource allocation? What role do computers play in their work?

8.2 LINEAR PROGRAMMING

A feedlot is fattening lambs for market. The lambs can be fed an expensive high protein food costing \$0.80 per kilogram or a cheaper food costing only \$0.40 per kilogram. Suppose the high protein food supplies 125 grams of protein and 4500 calories per kilogram, while the cheaper food only supplies 50 grams of protein but 7500 calories per kilogram. If each lamb needs to be fed at least 100 grams of protein and 4500 calories per week, what is the least expensive mix of feed to accomplish this goal?

In the preceding section we showed how applied problems can lead to a combination of several conditions, or constraints, that must be satisfied simultaneously. We saw how to represent these constraints as linear inequalities and how, when the number of variables is two, to plot the region containing all possible combinations of the variables that satisfy the problem constraints. In this section we learn how to define the goal, or objective, of our problem-solving efforts and to find the solution that best accomplishes our goal from among the feasible alternatives.

OBJECTIVE FUNCTIONS

The purpose of linear programming is to give a rational process for deciding what should be done in a situation involving a number of restrictions that can be expressed as linear constraints. So far we have only been able to describe what solutions are possible (that is, those in the feasible region), but we have not been able to say which solutions are most appropriate. An additional ingredient is required. An **objective function** provides a numerical assessment of how good any particular course of action is. In a business situation, the objective function is generally profit, which is to be maximized. In another context, the objective function might be fuel usage, which is to be minimized. The type of objective function we will study has the form $F = ax + by + c$ where a, b, and c are constants, and x and y are variables; such a function is called a **linear function** where a, b and c are called **coefficients.** Notice that in a linear function, the variables x and y have an exponent of one.

HISTORY

Linear programming was developed during World War II, when the problem of logistics was investigated by mathematicians. Our use of the word "logistics" can be attributed to the 19th century book *Precis de l'art de la guerre* by Baron Antoine Henri Jomini.

EXAMPLE 8.5 Consider the following functions. Which ones are linear, and which are not? For each function that is not linear tell why. For each function that is linear give the coefficients of the variables.

(a) $Q = 1 + 2x + 3y$ **(b)** $F = \dfrac{1 + x}{1 - y}$

(c) $P = x^3 + 2y$ **(d)** $H = x^y$

SOLUTION

(a) The function Q is linear since its variables are x and y, and they only occur to the first power. The coefficient of x is 2, and the coefficient of y is 3. There is a constant term of 1.

(b) The function F is not linear because the variable y is in the denominator of the fraction.

(c) The function P is not linear because a variable is raised to a power other than 1.

(d) The function H is not linear because a variable occurs as an exponent. ◆

In Section 8.1, we considered an example involving a furniture company that manufactures both unfinished and finished chairs. Suppose the profit on an unfinished chair is \$6, and the profit on a finished chair is \$4. If the company produces x unfinished chairs and y finished chairs during a week, then the expected profit for the week, P, is given by

$$P = 6x + 4y.$$

Here we note that the value of P depends on the values of the two variables x and y, so P is said to be a function of those variables. When we maximize the profit, then P becomes our objective function (that is, our objective is to maximize the value of P). This type of problem is called a **linear programming** problem. Our objective is to maximize the linear function P over the set of all points in the feasible region determined by linear constraints.

LEVEL LINES

Suppose that our objective function is $P = 6x + 4y$. Then our profit associated with values chosen for x and y is $6x + 4y$. For example, if we decide on production of $x = 5$ and $y = 2$, then our profit is $6(5) + 4(2) = 38$. We would like to choose x and y so that the objective function is as large as possible. Here, we might like to choose the values of x and y both extremely large, say equal to a thousand each. However, such large numbers may not be in the feasible region. You might like to draw out a thousand dollars from your savings account and a thousand dollars from your checking account, but this action may not be feasible for you, while it might be feasible for someone else. The problem is to choose values for x and y in the feasible region that make the objective function as large as possible. Two choices of (x, y) that give the same value for the objective function are equally good. To find the best such pairs, we first plot lines for which all points on the lines have the same objective function value. Such lines are called **level lines.**

EXAMPLE 8.6 Plot the level line of the objective function $P = 6x + 4y$ corresponding to the value $P = 120$. Then plot the level lines corresponding to the values $P = 240$, $P = 360$, $P = 480$, $P = 600$, $P = 720$, $P = 840$, and $P = 960$.

SOLUTION The level line corresponding to the value 120 consists of the set of points, (x, y), that satisfy $120 = 6x + 4y$. This is a linear equation that we can plot using the usual methods. We choose a convenient value for x, say $x = 0$. Substituting 0 for x in the equation $120 = 6x + 4y$ yields $y = 30$. So $(0, 30)$ is a point on the line. Setting $y = 0$ yields $x = 20$. Thus the graph of the objective function when $P = 120$ is the line through the points $(0, 30)$ and $(20, 0)$. This line and the other level lines are given in Figure 8.14.

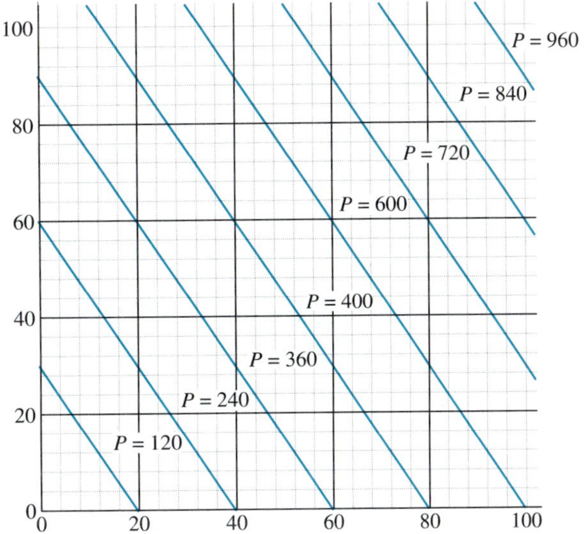

FIGURE 8.14

Notice that all the level lines are parallel. Also the values, P, corresponding to the level lines are increasing as we move away from the origin. Whenever the objective function is linear, the level lines will be parallel lines with the objective value increasing in one direction or the other. ◆

FUNDAMENTAL PRINCIPLE OF LINEAR PROGRAMMING

Next we incorporate the feasible region from Figure 8.8, and consider the relationship of the level lines to the feasible region (Figure 8.15).

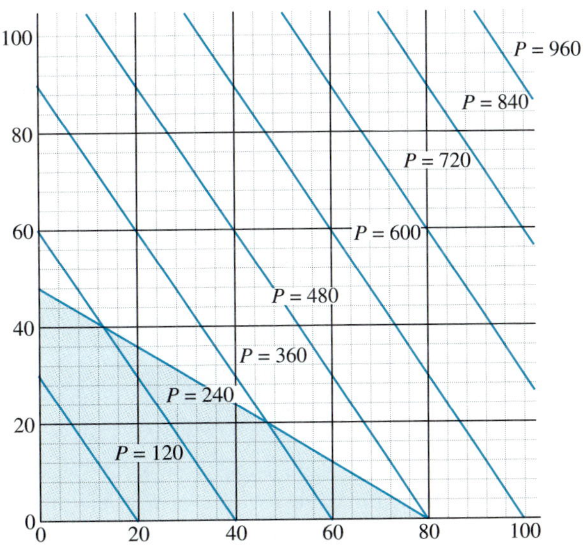

FIGURE 8.15

Notice that the level line representing the largest value of P that intersects the feasible region is $P = 480$, and that this level line intersects the feasible region in the corner of the region corresponding to the choices $x = 80$ and $y = 0$. This is true because the line where $P = 480$ is the line farthest from the origin that still intersects the feasible region. The conclusion from this analysis is that if the furniture company wants to maximize profit it should produce only unfinished chairs.

This example is a special case of the following principle.

> **Fundamental Principle of Linear Programming**
>
> If the feasible region is bounded, then the maximum of a linear objective function is attained at a corner of the feasible region, and likewise the minimum of a linear objective function also is attained at a corner of the feasible region.

EXAMPLE 8.7 Suppose the feasible region is as in Figure 8.16. Find the maximum value and where it is obtained for the objective function $P = 7 + 3x + 5y$.

WARNING

The choice of objective function is decisive. During World War II, the unit in charge of supplying the forces at the front seemed to think tonnage shipped was the objective function. As General Patton observed "It is perfectly useless to get a thousand tons of gasoline when you need five hundred tons of gasoline, two hundred tons of ammunition, and three hundred tons of bridging material."

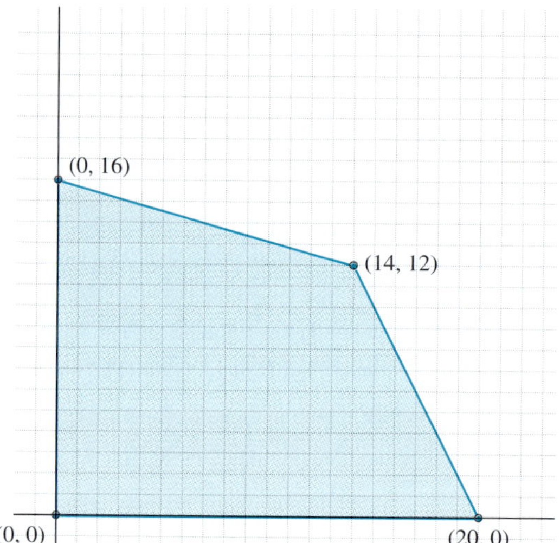

FIGURE 8.16

SOLUTION To find the maximum value of this objective function, evaluate the objective function at the four corners of the feasible region.

Point	Value of P
(0, 0)	$7 + 3 \times 0 + 5 \times 0 = 7$
(20, 0)	$7 + 3 \times 20 + 5 \times 0 = 67$
(14, 12)	$7 + 3 \times 14 + 5 \times 12 = 109$
(0, 16)	$7 + 3 \times 0 + 5 \times 16 = 87$

The maximum value of P is 109 and is attained when $x = 14$ and $y = 12$. ◆

The next example shows how the technique used in Example 8.7 can be used to solve a practical problem.

EXAMPLE 8.8 A sporting equipment company makes baseballs and softballs. It takes 5 minutes to make a baseball and 6 minutes to make a softball. The plant can allocate 100 person-hours to making balls. The shipping department can pack and ship 1100 balls a day. If baseballs are sold for \$3.25 and softballs for \$3.50, how many of each type of ball should be made to maximize revenue?

SOLUTION Let

$$x = \text{number of baseballs to be made and}$$
$$y = \text{number of softballs to be made.}$$

Then

$$5x + 6y = \text{number of minutes devoted to making balls.}$$

Since 100 person-hours = 6000 person-minutes, we have

$$5x + 6y \leq 6000.$$

Since the shipping department can handle at most 1100 balls a day, we have

$$x + y \leq 1100.$$

The amount of revenue (in dollars) produced by x baseballs and y softballs is

$$3.25x + 3.5y.$$

Thus we wish to maximize $F = 3.25x + 3.5y$ subject to the constraints

$$5x + 6y \leq 6000, x + y \leq 1100, x \geq 0, y \geq 0.$$

The feasible region is shown in Figure 8.17.

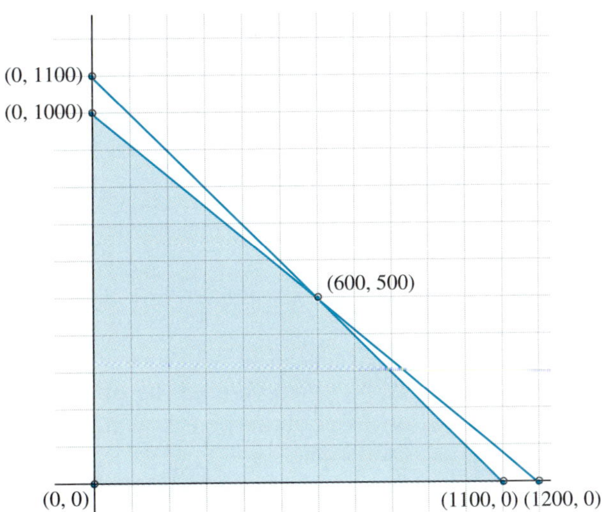

FIGURE 8.17

The point (600, 500) was found by solving the two equations $5x + 6y = 6000$ and $x + y = 1100$ simultaneously (see Topic 4 near the end of the book for a review of this topic). We find the maximum values of $F = 3.25x + 3.5y$ by evaluating the function at the three corners other than (0, 0) [since (0, 0) would represent the situation when no balls are produced].

Point	Value of $F = 3.25x + 3.5y$
(0, 1000)	$3.25 \times 0 + 3.5 \times 1000 = \$3500$
(600, 500)	$3.25 \times 600 + 3.5 \times 500 = \$3700$
(1100, 0)	$3.25 \times 1100 + 4 \times 0 = \$3600$

Thus 600 baseballs and 500 softballs should be made with a maximum revenue of \$3700. ◆

If the feasible region is unbounded, then there may or may not be a maximum/minimum for any particular objective function. However, as long as the minimum or maximum exists, it is still going to occur at a corner of the feasible region.

EXAMPLE 8.9 Suppose the feasible region is the infinite region indicated in Figure 8.18. Find the maximum value and minimum value, and where they are obtained for the objective function $P = 7 + 3x + 5y$.

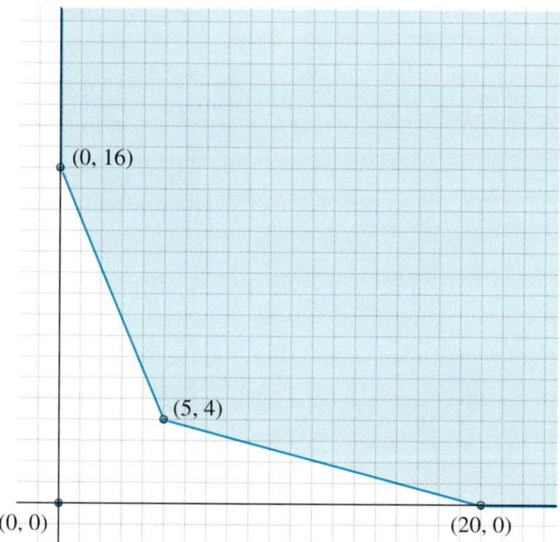

FIGURE 8.18

SOLUTION We evaluate the objective function at the three corners of the feasible region. We obtain the following information.

Point	Function Value
(20, 0)	$7 + 3 \times 20 + 5 \times 0 = 67$
(5, 4)	$7 + 3 \times 5 + 5 \times 4 = 42$
(0, 16)	$7 + 3 \times 0 + 5 \times 16 = 87$

The graphs of level lines associated with the objective function P will be similar to those shown in Figure 8.14, except for their slopes. Thus P can take on values as large as one wants. We conclude that no maximum value is attained, and there is a minimum value of 42 attained when $x = 5$ and $y = 4$. ◆

THE SIMPLEX ALGORITHM

In this section we have explored a method for solving linear programming problems by listing all the corners of the feasible region and the value of the objective function on those corners. From that list the smallest or largest value of the objective function may be selected. Algebraically, each corner can be found as the simultaneous solution of two of the equations associated with the constraint inequalities. When the number of variables is larger than two, graphical methods become difficult or impossible to apply, and we must proceed by using solely algebraic methods. Once programmable digital computers were invented there was a need for a highly organized systematic procedure to do this. It was George Dantzig who, in the late 1940s, developed such a systematic procedure known as the **simplex algorithm.**

The beauty of the simplex algorithm is that it efficiently proceeds from corner to corner along the boundary of the feasible region and recognizes when the minimum or maximum has been found, so the complete list of corners need not be generated. The simplex algorithm is widely used in industry; for example, almost every oil refinery is run using the results of applying the simplex algorithm. Problems nowadays often involve thousands of constraints and a million variables.

A feedlot is fattening lambs for market. The lambs can be fed an expensive high protein food costing \$0.80 per kilogram or a cheaper food costing only \$0.40 per kilogram. Suppose the high protein food supplies 125 grams of protein and 4500 calories per kilogram, while the cheaper food only supplies 50 grams of protein but 7500 calories per kilogram. If each lamb needs to be fed at least 100 grams of protein and 4500 calories per week, what is the least expensive mix of feed to accomplish this goal?

SOLUTION Let x = number of kilograms per week of expensive feed and y = number of kilograms per week of the cheaper feed. Then $125x + 50y$ = number of grams of protein supplied. We must have

$$125x + 50y \geq 100.$$

Also $4500x + 7500y$ = number of calories supplied. We must have

$$4500x + 7500y \geq 4500.$$

The cost of this feed is

$$0.80x + 0.40y.$$

Thus we wish to minimize $F = 0.80x + 0.40y$ subject to the constraints $125x + 50y \geq 100$, $4500x + 7500y \geq 4500$, $x \geq 0$, $y \geq 0$.

The feasible region is shown in Figure 8.19.

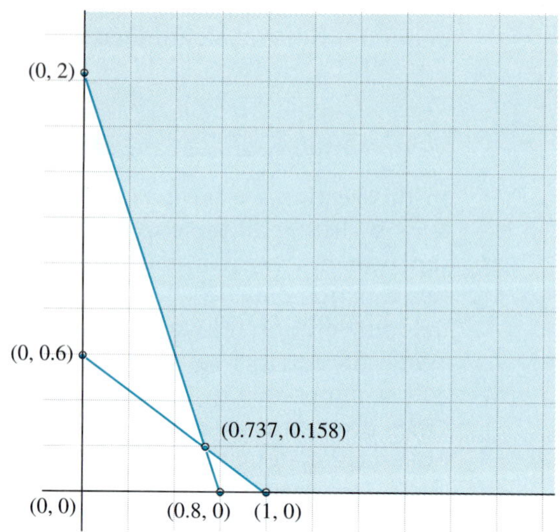

FIGURE 8.19

The point (0.737, 0.158) was found by solving the two equations $125x + 50y = 100$ and $4500x + 7500y = 4500$ simultaneously. The objective function will have a minimum since using larger and larger amounts of feed will cost more and more. We find the minimum value of $F = 0.80x + 0.40y$ by evaluating the function at the three corners.

Point	**Value of $F = 0.80x + 0.40y$**
(0, 2)	$0.80 \times 0 + 0.40 \times 2 = \$0.80$
(0.737, 0.158)	$0.80 \times 0.737 + 0.40 \times 0.158 = \$0.65$
(1, 0)	$0.80 \times 1 + 0.40 \times 0 = \$0.80$

Therefore, we can use 0.737 kilograms of the high protein food and 0.158 kilograms of the cheaper food, and our minimum feeding costs will be \$0.65 per lamb per week.

PROBLEM SET 8.2

For problems 1 through 4, consider the given functions. Which ones are linear, which are not? For each function that is linear, give the coefficients of the variables (first x, then y).

1. **(a)** $F = 3x - 7y$
(b) $G = x^2 + 2y - 2$
(c) $H = \dfrac{x + 2y}{3x - y}$
(d) $J = y + 2x$

2. **(a)** $F = 3x^y + x - y$
(b) $G = y + x$
(c) $H = \sqrt{3x + 4y}$
(d) $K = \dfrac{5x + 12y}{3}$

3. **(a)** $P = xy + 3$
(b) $Q = 4 - x + 2y$
(c) $R = 10x + 6y$
(d) $S = y - x^3$

4. **(a)** $P = 15 + 10x + 8y$
(b) $R = x + 2(5 + y)$
(c) $S = 15y + 25x$
(d) $U = 2x + y^2$

In problems 5 through 12, plot the level lines corresponding to the given values for the objective function.

5. Objective function: $P = 10x + 8y$
$P = \{20, 40, 60, 80, 100\}$

6. Objective function: $P = 5x + 8y$
$P = \{40, 60, 80, 100, 120\}$
7. Objective function: $C = 8x + 10y$
$C = \{20, 40, 60, 80, 100\}$
8. Objective function: $M = x + 2y$
$M = \{5, 10, 15, 20, 25\}$
9. Objective function: $R = 5x + 4y + 20$
$R = \{30, 35, 40, 45, 50, 55\}$
10. Objective function: $S = 10x + 15y$
$S = \{20, 40, 60, 80, 100, 120\}$
11. Objective function: $P = 15x + 10y$
$P = \{30, 60, 90, 120, 150, 180\}$
12. Objective function: $R = 4x + 5y + 10$
$R = \{30, 50, 70, 90, 110, 130\}$

In problems 13 through 16, graph the feasible region, and plot the level lines for the objective function that correspond to the given values. Estimate the maximum or minimum value of the level line that would contact the feasible region without intersecting it.

13. $P = 4x + 6y$; $P = \{12, 18, 24, 30\}$
$x + y \leq 4$
$x + 2y \leq 6$
$x \geq 0, y \geq 0$
14. $M = 2x + 3y$; $M = \{6, 9, 12, 15, 18\}$
$4x + 3y \leq 18$
$2x + y \leq 8$
$x \geq 0, y \geq 0$
15. $C = 2x + 5y$; $C = \{10, 15, 20, 25, 30\}$
$2x + y \geq 6$
$x + 2y \geq 8$
$x \geq 0, y \geq 0$
16. $S = 3x + 2y$; $S = \{12, 18, 24, 30, 36\}$
$x + y \geq 5$
$2x + y \geq 8$
$x \geq 0, y \geq 0$

In problems 17 through 20, the graph of the solution to each system of inequalities is shown below the system. Find the corner points for each feasible region.

17. $x + 2y \leq 8$
$x + y \leq 7$
$x \geq 0, y \geq 0$

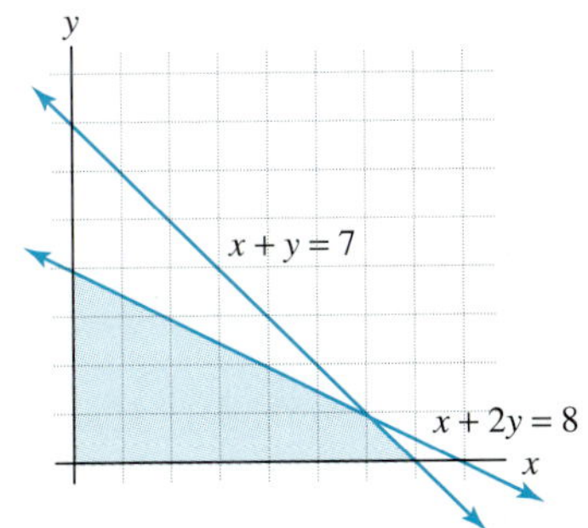

18. $-x + 2y \leq 4$
$x + y \leq 5$
$x \geq 0, y \geq 0$

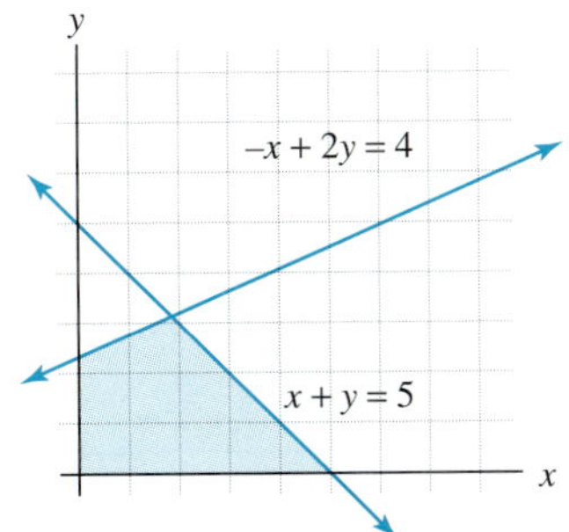

19. $x + 2y \leq 8$
$x + 4y \leq 12$
$x + y \leq 7$
$x \geq 0, y \geq 0$

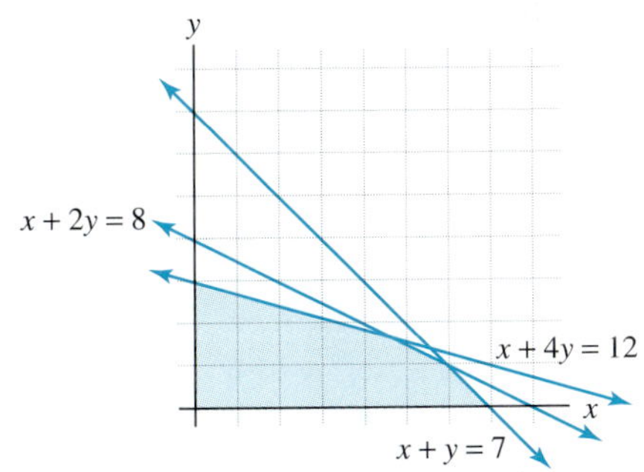

20. $x + y \geq 6$
$x + 5y \geq 10$
$2x + y \geq 8$
$x \geq 0, y \geq 0$

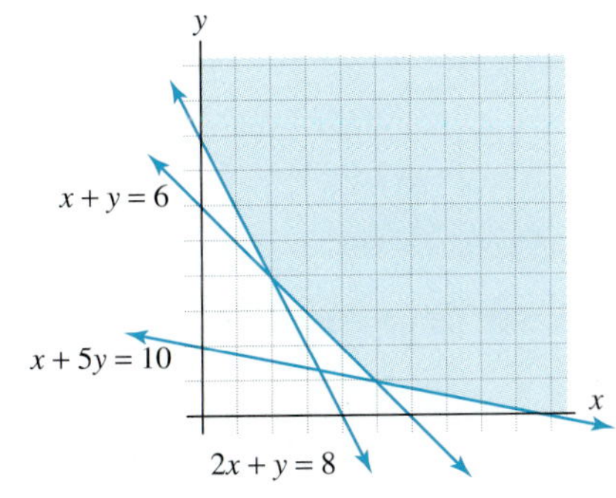

In problems 21 through 24, use the given feasible regions to find the maximum and minimum of the given objective function (if they exist).

21. $P = 3x + 2y$

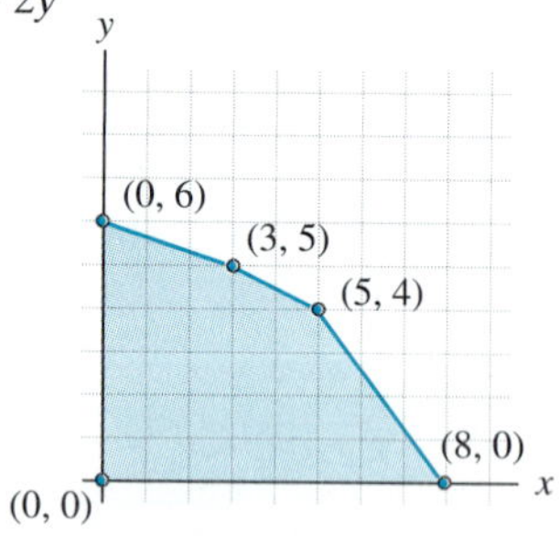

22. $C = 4x + 5y$

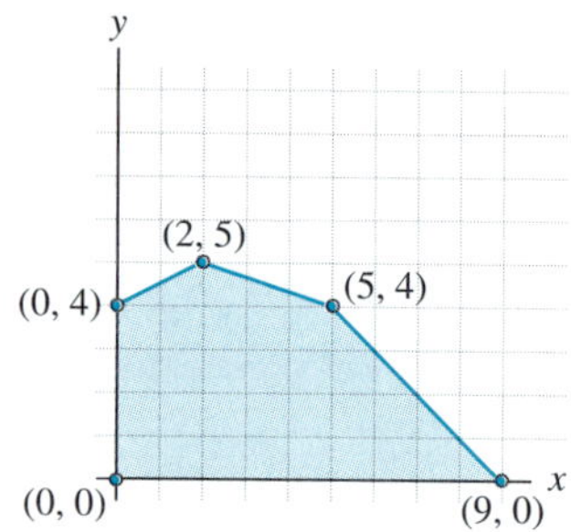

23. $M = 3x + 4y$

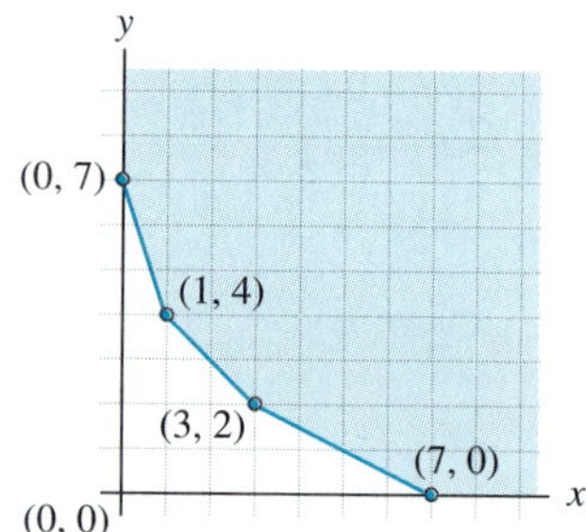

24. $F = 4x + 5y$

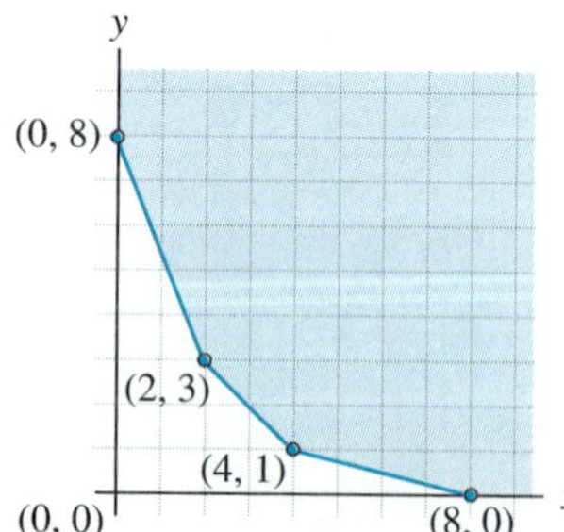

In problems 25 through 30, find the indicated maximum or minimum value of the objective function in the linear programming problem.

25. Maximize $F = 5x + 2y$
subject to $2x + y \leq 8$
$2x + 3y \leq 12$
$x \geq 0, y \geq 0$

26. Maximize $M = 4x + 7y$
subject to $2x + y \leq 8$
$-x + y \leq 2$
$x \geq 0, y \geq 0$

27. Minimize $C = 10x + 6y$
subject to $3x + 2y \geq 12$
$x + y \geq 5$
$x \geq 0, y \geq 0$

28. Minimize $F = 12x + 15y$
subject to $x + y \geq 8$
$x + 3y \geq 12$
$x \geq 0, y \geq 0$

29. Maximize $P = 7x + 5y$
subject to $x + 2y \leq 10$
$x + y \leq 7$
$2x + y \leq 12$
$x \geq 0, y \geq 0$

30. Minimize $N = 8x + 10y$
subject to $x + y \geq 5$
$2x + y \geq 6$
$x + 2y \geq 6$
$x \geq 0, y \geq 0$

In problems 31 through 38, write the mathematical model for the problem, including the objective function, problem constraints, and nonnegativity constraints. Graph the feasible region, find the corner points, and solve the problem. If the nature of the problem requires that the answers be whole numbers adjust your answer to the most suitable values still in the feasible region. For example, you wouldn't schedule a fraction of an ad or lease a fraction of an airplane.

31. A tool company manufactures two types of electric drills, one of which is cordless. The cord-type drill requires 2 labor hours to make, and the cordless drill requires 3 hours. The company has only 600 labor hours available each day, and the packing department can package 250 drills per day at the most. If the cordless drills sell for \$60 and the cord-type drills sell for \$45, how many of each type should the company produce each day to maximize revenue?

32. A recreational manufacturer markets two types of tents: a 4-person model and a 2-person model. The 4-person tent costs \$100 to make and sells for \$160; the 2-person tent costs \$60 dollars to make and sells for \$110. The manufacturer can budget no more than \$9000 to produce the tents. The capacity of the plant is limited to 120 tents, and at least 40 of the tents must be 4-person models. How many of each type of tent should be produced to maximize profit?

33. A private fishing resort has bass and trout in its lake.

The owner provides two types of food, A and B, for these fish. Each week, the bass require 2 units of food A and 4 units of food B, the trout require 5 units of food A and 2 units of food B. The owner can get up to 800 units of each food on a weekly basis. Find the maximum number of fish the lake can support.

34. On the final weekend before the election, a candidate wishes to use a combination of radio and television ads in her campaign. Research has shown that during the prime viewing and listening hours each 30-second spot on television can reach 72,000 people, and each 30-second spot on radio can reach 12,000 people. The candidate wants to have the ads reach at least 2.16 million people (including duplications) and use a total of at least 40 minutes of ads. How many minutes of each medium should be used to minimize costs if radio ads cost $100 per spot and television ads cost $500 per spot?

35. An association of travel agents has been allocated 1200 tickets to the NCAA "Final Four" National championships. The weekend package includes airfare, and they have two types of airplanes for the charter flights. Type 1 aircraft can carry 100 passengers and type 2 aircraft can carry 150 passengers. Each flight of a type 1 aircraft will cost $9000, and each flight of a type 2 aircraft will cost $15,000. The association is allowed to lease no more than ten planes. How many airplanes of each type should be leased to minimize costs? Hint: assume at least 1200 seats are needed.

36. A door manufacturer produces doors in two styles, a regular wooden door and a deluxe model with glass panels. The manufacturer spends $175 to make each regular door, which it sells for $270, while each deluxe door costs $250 to make and sells for $380. The daily production capacity is 110 doors, and daily costs cannot exceed $21,000. What is the average number of doors of each type that should be produced per day to maximize profit? How does your answer change if only complete doors can be scheduled for daily production?

37. A large company is having its annual national sales meeting. Planners must provide accommodations for at least 400 attendees. They have two types of rooms available: motel rooms that sleep three people and hotel rooms that sleep two people. For meals they have budgeted $20 daily per person for those in motel rooms and $40 daily per person for those in hotels, and they must not spend more than $12,000 in meal money per day. If the daily costs for motel rooms are $135 per room and hotel rooms are $120 per room, how many rooms of each type should be reserved to minimize room costs?

38. The director of the computer center at a large university wants to staff consulting stations with two types of shifts: type A will have 2 senior programmers and 1 student assistant; type B will have a senior programmer and 4 student assistants. Type A shifts will serve for two hours and type B shifts will serve for three hours. The consulting stations will be open at least 48 hours during the week, and the director wants to use no more than 210 hours of staff time. Personnel costs are $48 per hour for a type A shift and $52 per hour for a type B shift. How many shifts of each type should be scheduled to minimize personnel costs? Only full shifts can be scheduled.

? EXTENDED PROBLEMS

39. The corner points for the feasible region determined by the linear constraints

$$x + 2y \leq 16$$
$$3x + y \leq 18$$
$$x \geq 0, y \geq 0$$

are $O = (0, 0)$, $A = (0, 8)$, $B = (4, 6)$, $C = (6, 0)$. If $P = ax + by$ $(a > 0, b > 0)$, what conditions on a and b will ensure that the maximum value of P occurs?

(a) only at A **(b)** only at B
(c) only at C **(d)** at both A and B
(e) at both B and C

When there are more than two variables in a linear programming problem, you cannot use the graphical method of solution (unfortunately, most linear programming applications in the real world involve more than two variables). The simplex method is used to solve these general type problems. When the problems get very large, even more sophisticated methods must be used. However, the process for describing the problem with constraints and objective functions remains the same.

In problems 40 and 41, write the constraints based on resources, the minimum value constraints, and the objective function.

40. Suppose the creamery in Example 8.4 decides to expand and add a "lite" ice cream to its product line. Each gallon of regular ice cream requires 0.4 gallons of cream and 0.6 gallons of milk; each gallon of the deluxe ice cream requires 0.5 gallons of cream and 0.5 gallons of milk; and each gallon of "lite" ice cream requires 0.2 gallons of cream and 0.8 gallons of milk. The creamery's suppliers can provide a maximum of 1500 gallons of cream and 2000 gallons of milk. The creamery makes a profit of $0.60 on a gallon of regular ice cream, $0.80 on a gallon of deluxe, and $0.50 on a gallon of "lite." The creamery wants to maximize its profit.

41. A large store has a maximum of $30,000 to spend on television advertising for a sale. All ads will be placed with a single station. A series of 30-second ads will be run. These cost $1200 on daytime TV with 12,000 viewers, $2250 on prime-time TV with 21,000 viewers, and $1600 on late-night TV with

16,000 viewers. The store wants to run at least 10 ads, and the television station will not run more than 15 ads in all three time periods. The store wants to maximize the number of viewers.

Problems 42 through 44 show the application of linear programming to major league baseball scheduling.

42. Prior to 1994, the National League was split into two divisions, with six teams in each division. There were 162 games in a regular season. Suppose a team played each of the other five teams in its division x times and each of the six teams in the other division y times. Then

$$5x + 6y = 162.$$

Since each team should play more games against teams in its own division, we require

$$x > y.$$

Another consideration was that there should be a series of at least two games between each pair of teams in each park, so that $y \geq 4$.

(a) Find the feasible region for the two inequalities.
(b) Find all pairs (x, y) of whole numbers in the feasible region satisfying

$$5x + 6y = 162.$$

(c) Why might the National League choose one solution over the others?

43. Prior to 1994, the American League had 14 teams split into two divisions. If each team played the other teams in its division x times and the teams in the other division y times, then

$$6x + 7y = 162.$$

Other constraints (see problem 42) are

$$x > y \text{ and } y \geq 4.$$

(a) Find the feasible region for the two inequalities.
(b) Find all pairs (x, y) of whole numbers in the feasible region satisfying

$$6x + 7y = 162.$$

(c) Why might the American League choose one solution over the others?

44. For many years each league had only 8 teams (no divisions), and both leagues played 154 game schedules.
(a) With the pre-1994 alignment in the National League (problem 42), how many games would each team play against teams from its own division in a 154 game schedule?
(b) With the pre-1994 alignment in the American League (problem 43), how many games would each team play against teams from its own division in a 154 game schedule?

8.3 ROUTING PROBLEMS

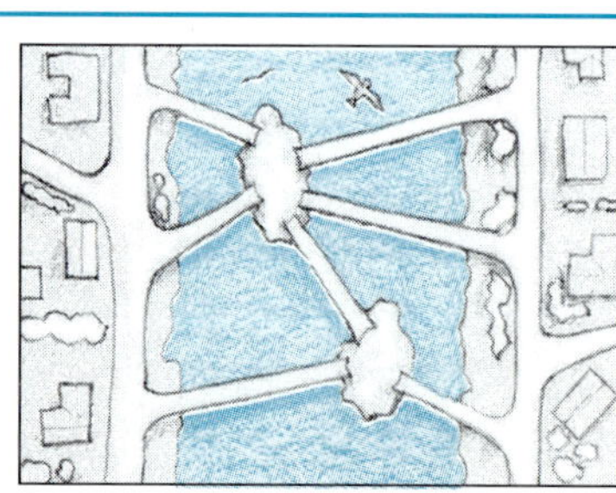

The 18th century town of Königsberg was built on both sides of the Pregel River and on two islands in the river. The people enlivened their Sunday strolls through town by trying to make a circuit that crossed each of the town's seven bridges once and only once (Figure 8.20). Is such a route possible?

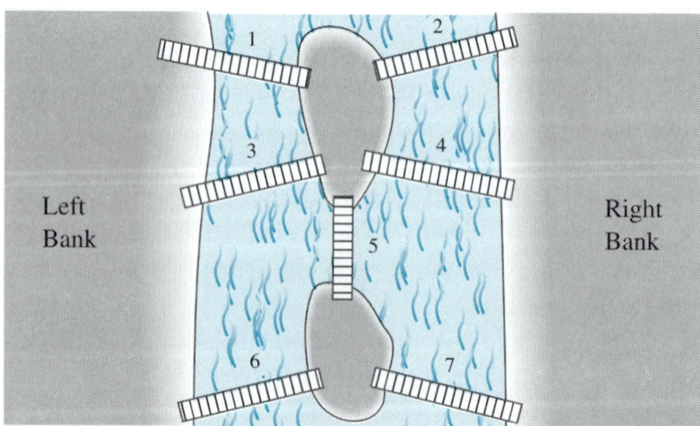

FIGURE 8.20

Another important management problem involves finding efficient ways to route the delivery of goods or services to various destinations. The destinations involved are thought of as part of a network, together with the connecting links. The goal is to use the network efficiently to accomplish a task, such as the delivery of the mail or collection of the garbage. While there are a number of variations on the theme of routing problems, we will concentrate on the type presented in the initial problem, where we seek a route, or circuit, that uses each connection in a network once and only once. In this section we begin with the description and analysis of networks.

The initial problem is known as the Königsberg bridge problem. Since it was originally solved by Leonhard Euler, problems of this sort are called Euler circuit problems. A tool that mathematicians use to deal with Euler circuit problems and other routing problems is graph theory. In this context a graph is not a plot of data such as a bar graph. A **graph** refers to a collection of points, called **vertices** (singular is **vertex**), and the paths connecting them, called **edges,** which may be straight or curved. Typically we draw a sketch of a graph where the vertices are represented by black dots, and the edges are represented by line segments or arcs (Figure 8.21).

FIGURE 8.21

Each edge either has two ends with different vertices at those ends or it is a **loop** connecting a vertex to itself (Figure 8.22).

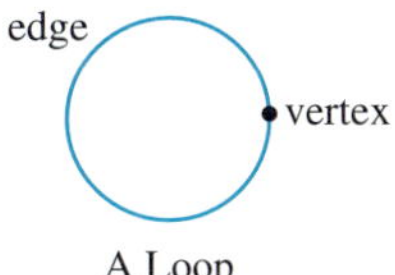

FIGURE 8.22

Just because two of the arcs representing edges appear to cross, it does not mean there is a vertex there! This is similar to the way two roads may appear to cross each other on a highway map without there being a way to drive from one to the other because there is an overpass but no interchange. Finally, two graphs are considered to be the same if they have the same number of vertices connected to each other in the same way even if the edges look different. We will consider Figures 8.23(a) and (b) to be two drawings of the same graph; we have labeled the vertices, to show the connections are the same.

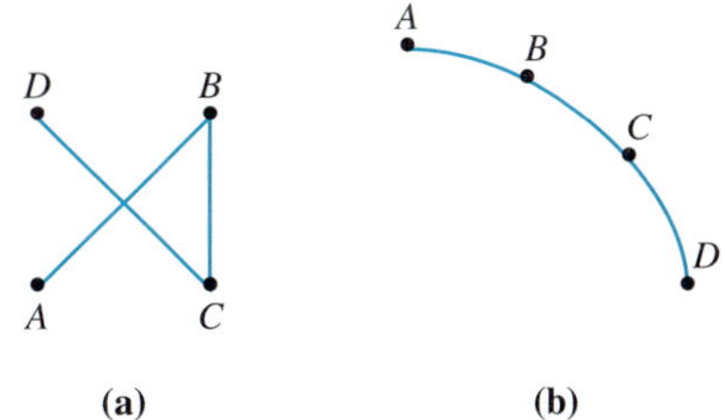

FIGURE 8.23

EXAMPLE 8.10 Consider the map of the interstate highways in the states of Idaho, Oregon, and Washington as shown in Figure 8.24.

FIGURE 8.24

Sketch a graph representing the interstate highway connections between the cities of Boise, Olympia, Portland, Salem, Seattle, and Spokane using I-5, I-84, and I-90.

SOLUTION The cities of Seattle, Olympia, Portland, and Salem lie along I-5 in that order. From Seattle you can take I-90 to Spokane, and from Portland you can take I-84 to Boise. There are no other connections using the designated highways. We put this information into a sketch of a graph in Figure 8.25.

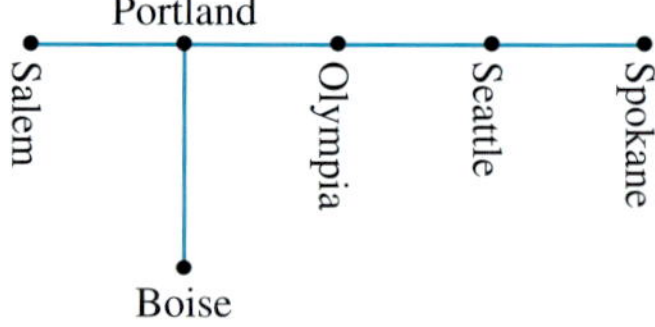

FIGURE 8.25 ◆

Two vertices in a graph are said to be adjacent in a graph if there is an edge connecting them.

EXAMPLE 8.11 For the graph illustrated in Figure 8.26, list the pairs of adjacent vertices.

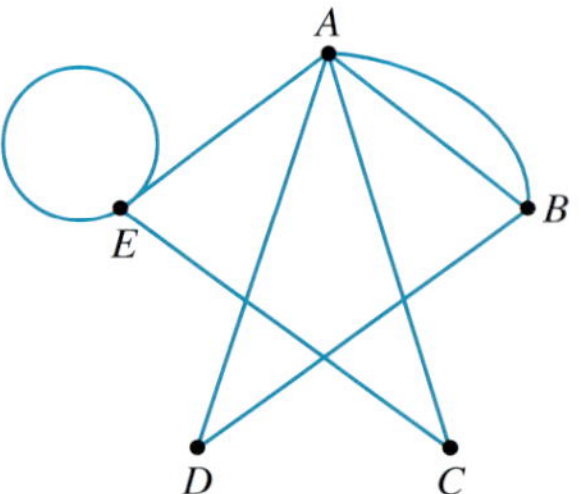

FIGURE 8.26

SOLUTION To find the adjacent pairs systematically, first list all the pairs of adjacent vertices involving vertex A, then all those involving B but not A, then all those involving C but neither A nor B, then all those involving D but not A, B, or C. Lastly, check to see if E is adjacent to itself. (It is.) The adjacent pairs are A and B, A and C, A and D, A and E, B and D, C and E, E and E (the order in which the vertices are mentioned is unimportant). ◆

EXAMPLE 8.12 Draw two different pictures of a graph with four vertices A, B, C, D with the following adjacent pairs of vertices: A and C, B and C, B and D, C and D.

SOLUTION Two possible pictures are shown in Figure 8.27.

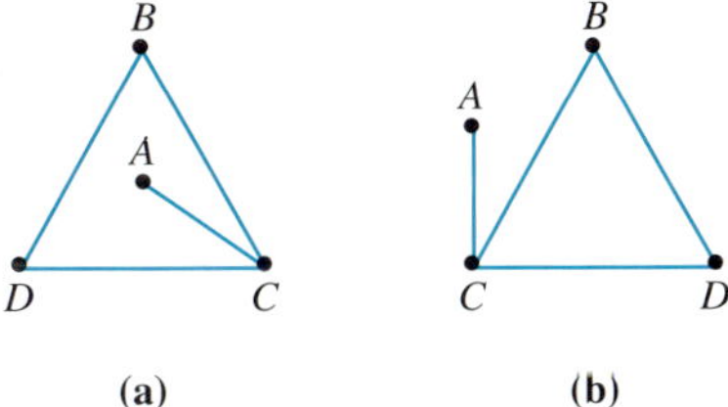

FIGURE 8.27 ◆

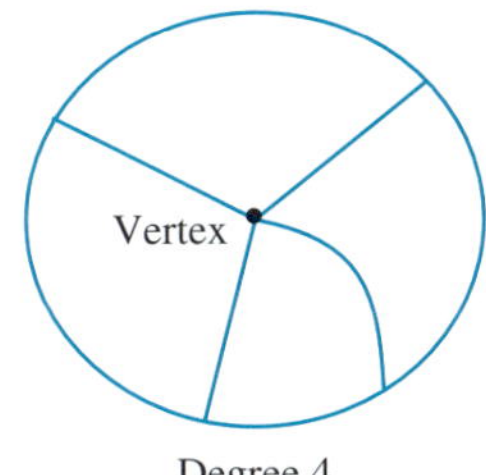

FIGURE 8.28

The **degree of a vertex** in a graph is the total number of edges at that vertex. If a loop connects a vertex to itself, then we will agree that the loop contributes 2 to the degree of the vertex. Visually, the degree of a vertex can be found by counting the number of segments or arcs that are attached at the vertex. Figure 8.28 shows a close-up of one vertex in a graph. Since we see four segments and arcs attached at that vertex, its degree is 4.

EXAMPLE 8.13 Find the degree of each vertex of the graph in Figure 8.26.

SOLUTION Simply count all the ends of edges that attach at each vertex.

Vertex	**Degree**
A	5
B	3
C	2
D	2
E	4

◆

There is a useful but not surprising relationship between the number of edges in a graph and the sum of the degrees of the vertices.

EXAMPLE 8.14 Compute the sum of the degrees of all the vertices of the graph in Figure 8.26, and compare that number to the number of edges in the graph.

SOLUTION In Example 8.13 we have already found the degree at each vertex, so the sum of the degrees over all vertices is $5 + 3 + 2 + 2 + 4 = 16$. Referring to Figure 8.26 we count that there are 8 edges, so the sum of the degrees is twice the number of edges. ◆

The sum of the degrees of a graph is always twice the number of edges. This is because each edge of a graph has two ends with a vertex at each, and the degree of a vertex is the number of ends of edges attached to the vertex. We consider a loop to have two ends which just happen to attach to the same vertex.

> The sum of the degrees of all the vertices in a graph is twice the number of edges in the graph.

A **path** in a graph is a list of vertices in the graph such that each vertex in the list is adjacent to the next vertex in the list, and each edge that connects adjacent vertices is used at most once. One thinks of traveling from vertex to vertex via the edges, with each edge used at most once. However, a vertex may appear more than once. The thicker segments in Figure 8.29 illustrate a path from vertex A to vertex B.

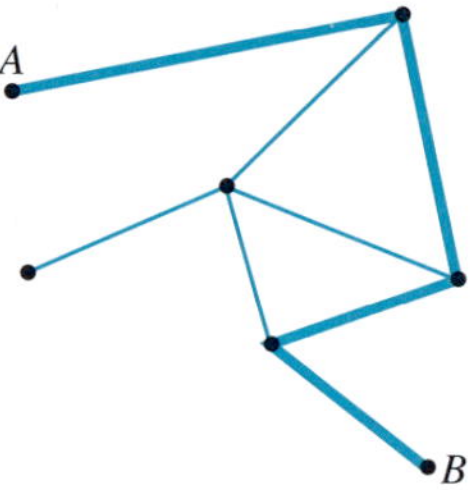

FIGURE 8.29

A path that ends at the same vertex at which it starts is called a **circuit.** The more heavily drawn segments in Figure 8.30 illustrate a circuit.

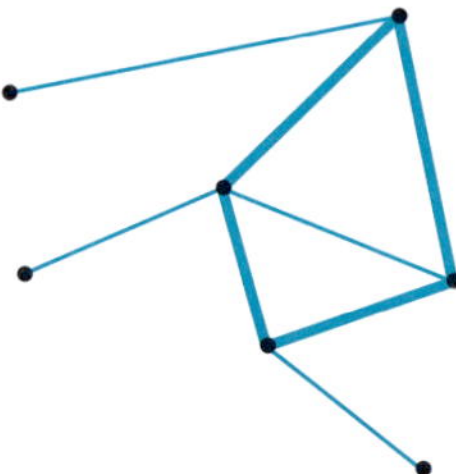

FIGURE 8.30

A path that uses every edge once is called an **Euler path.** Figure 8.31 illustrates an Euler path.

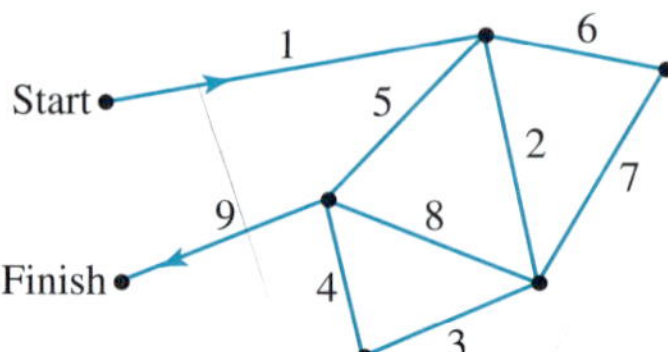

FIGURE 8.31

It is no longer enough to draw each edge in the path with a heavier line because every edge is in the path. Instead we have indicated a starting vertex and an arrow to show which way to go first. All the edges are numbered, so when you arrive at a vertex, just take the next numbered edge.

A circuit that uses every edge once is called an **Euler circuit.** Figure 8.32 illustrates an Euler circuit.

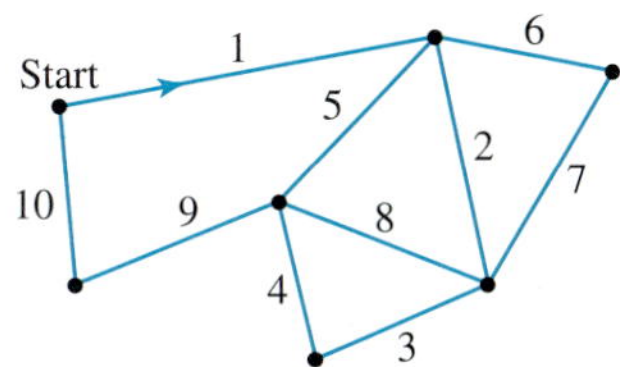

FIGURE 8.32

Again a starting vertex is indicated, and the numbering of the edges tells you how to proceed. It actually does not matter where you start as long as you follow the edges in order (using edge 1 after edge 10, of course).

Another way to describe a path is to name the vertices, and then list them in the order they are visited on the path. For example, the list R, S, T would indicate that you begin at vertex R, then go to S, and finally to T. The edges involved are between R and S and between S and T, in that order. You should number the edges as you go from vertex to vertex. If there are two or more edges between two vertices, you can only use each edge once, but a vertex can be visited more than once.

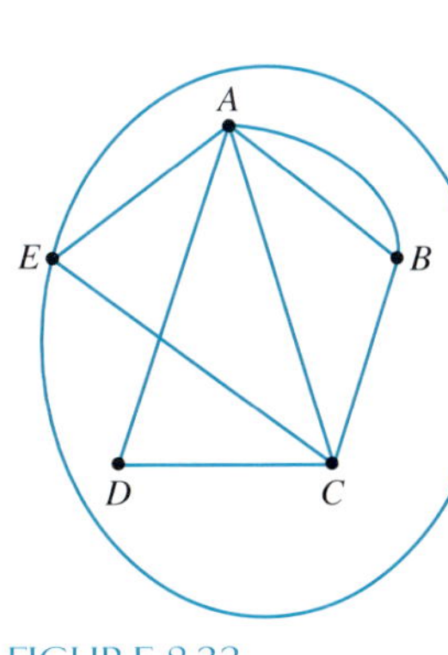

FIGURE 8.33

EXAMPLE 8.15 Which of the following lists of vertices form a path in the graph pictured in Figure 8.33? Is any a circuit, Euler path, or Euler circuit?

(a) A, B, C, D
(b) A, B, C, D, E
(c) A, B, A
(d) A, E, E, A
(e) $A, B, C, D, A, E, E, C, A, B$

SOLUTION

(a) This is a path.
(b) This is not a path because D and E are not adjacent since there is no edge with D and E as its endpoints.
(c) This is a path, and it is also a circuit, if the two different edges are used.
(d) This is not a path because there is only one edge with A and E as its endpoints, and we are not allowed to use it twice in a path. Note that A, E, E is a path because there is a loop at E.
(e) This is an Euler path because each edge is used once and only once. It is not a circuit because it begins at A and ends at B. ◆

A graph is **connected** if, for every pair of vertices, there is a path that contains them. For example, the graph in Figure 8.34 is connected. In fact, all the graphs we have considered so far have been connected.

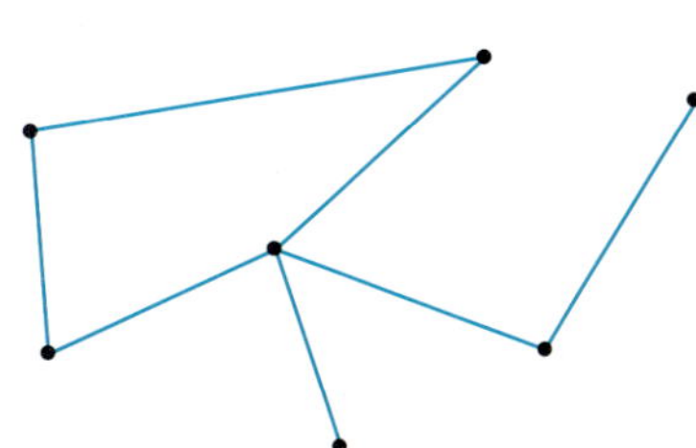

FIGURE 8.34

Whether or not a graph is connected is often very important, especially for communication or economic purposes. Historically, the development of railroad links between cities in the interior of the United States had a tremendous impact on the growth of the nation's economy.

As a convention, we agree that a graph with only one vertex is automatically connected, whether or not there are any edges. If a graph is not connected, then it is said to be **disconnected.** A disconnected graph can always be separated into connected pieces of the largest possible size (maximal connected pieces) called components. To find the **components of a graph** use the following steps.

1. Pick any vertex and highlight it.
2. Highlight all the edges connecting to the highlighted vertex and all the vertices at the ends of those edges.
3. Do step (2) again for all edges connected to any highlighted vertex.
4. When no new vertices get highlighted, you have a connected part of the graph that is as big as it can be (maximal) while still being connected; that is, you have a component of the graph.

This process is illustrated in Figure 8.35.

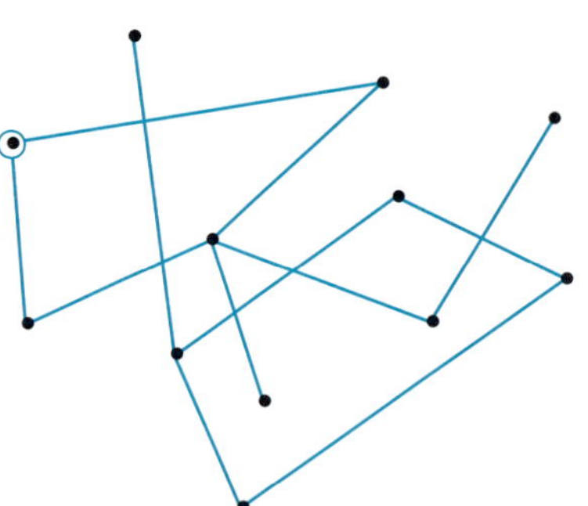

Highlight a vertex.

(a)

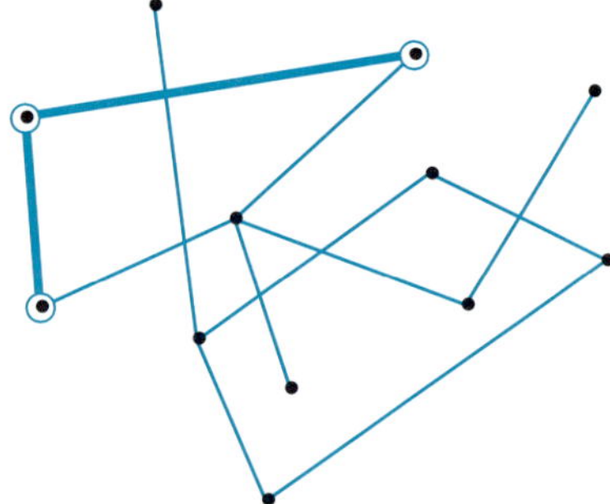

Highlight edges attached to highlighted vertex.
Highlight vertices at ends of edges.

(b)

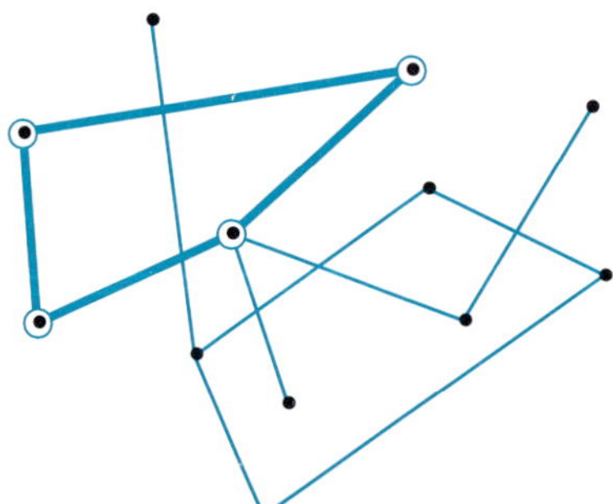

Highlight edges attached to highlighted vertices.
Highlight vertices at ends of edges.

(c)

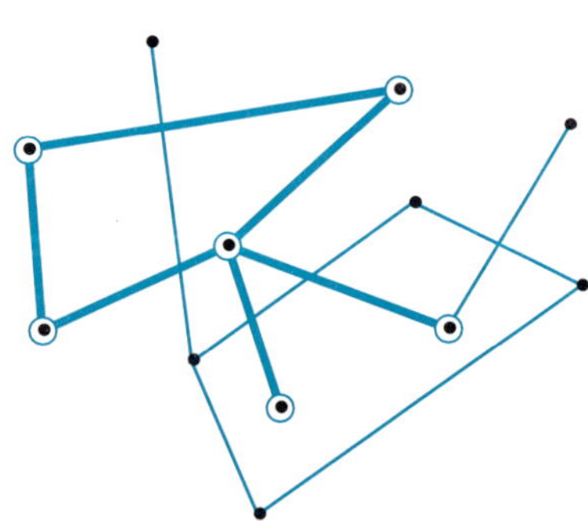

Do it again.

(d)

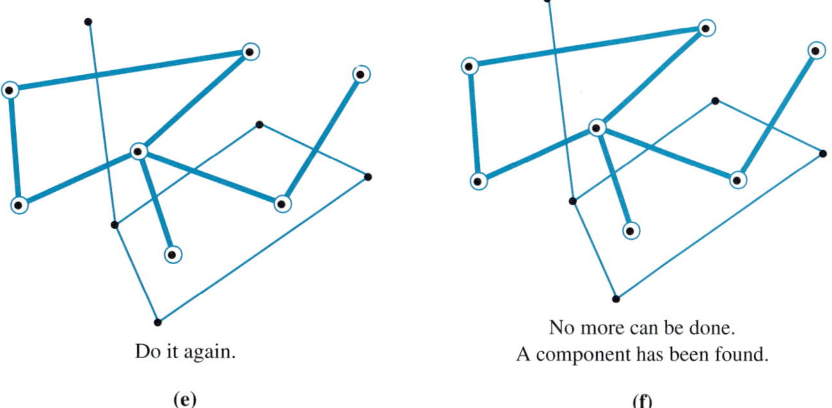

FIGURE 8.35

The five remaining vertices also form a second component for the graph. We have a disconnected graph with two maximal components.

A disconnected graph may or may not be drawn with the components neatly isolated like islands. Figure 8.36 shows a disconnected graph in which the components can be readily seen.

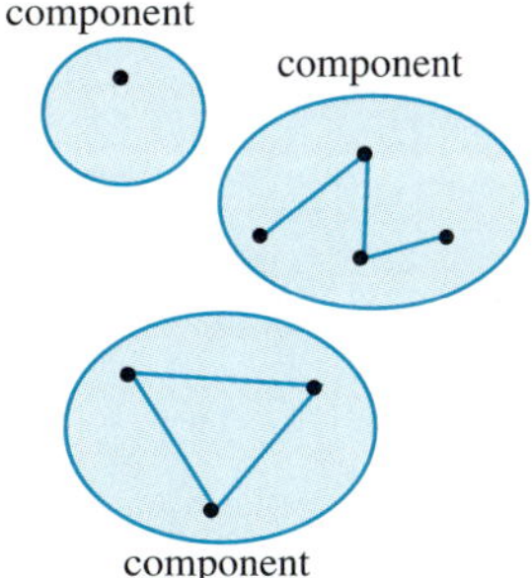

FIGURE 8.36

For simple graphs you may not need to go through the step-by-step process described above for constructing components, but real-life applications such as analyzing the connections in a computer or a highway system require the step-by-step approach.

EXAMPLE 8.16 Find the components of the graph in Figure 8.37.

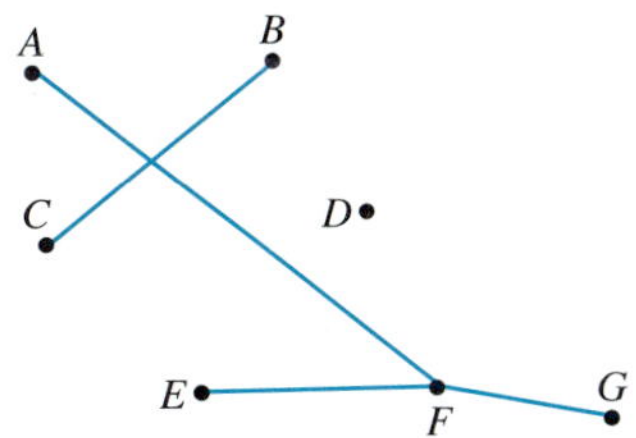

FIGURE 8.37

SOLUTION The vertex A is connected only to F, which in turn is connected to E and G, but E and G are connected to no other vertices. So the vertices A, E, F, and G and all edges involving those vertices form a component. The vertex B is connected only to C, and C is connected to no other vertex. Thus the vertices B and C and the edge connecting them form a component. Finally, the vertex D by itself forms a component. In Figure 8.38 we have shown the same graph as in Figure 8.37, but we have rearranged it to emphasize the components.

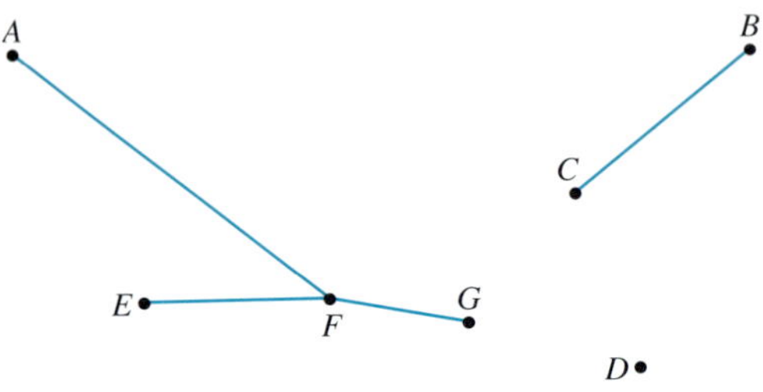

FIGURE 8.38 ◆

An edge in a connected graph is called a **bridge** if its removal from the graph would leave behind a graph that is not connected. In Figure 8.39 the graph is connected, and there is one edge that is a bridge.

FIGURE 8.39

Recall that the sum of the degrees of all the vertices in a graph is twice the number of edges in the graph, so the sum of degrees is always even. Every vertex in a graph is either of even or odd degree. Since the total of an odd number of odd numbers is odd, and the total of an even number of odd numbers is even, we conclude that a graph must have an even number of vertices with odd degree. That may have sounded confusing, but we have the following result.

> Any graph must have an even number of vertices with odd degree.

The question of whether or not a given graph has an Euler circuit was first asked, just as a curiosity, in the Königsberg bridge problem. But now that we have delivery services, garbage pickup, street sweepers, etc., using an Euler circuit can save a service company some significant expense. Over two hundred years ago, Euler found the key to answering the question of whether or not there is an Euler circuit in a given graph: you only need to know the degrees of the vertices in the graph. The precise result is stated next (remember no graph can have exactly one vertex with odd degree, and no graph can have exactly three vertices with odd degree, etc.).

THEOREM

Theorem (L. Euler)

For a connected graph

1. If the graph has no vertices of odd degree, then it has at least one Euler circuit (which is also an Euler path), and if a graph has an Euler circuit, then it has no vertices of odd degree.
2. If the graph has exactly two vertices of odd degree, then it has at least one Euler path, but does not have an Euler circuit. Any Euler path in the graph must start at one of the two vertices with odd degree and end at the other.
3. If the graph has four or more vertices of odd degree, then it does not have an Euler path.

EXAMPLE 8.17 For the graph in Figure 8.40, decide whether or not the graph has an Euler circuit or Euler path. Find the Euler circuit or Euler path if the graph has one.

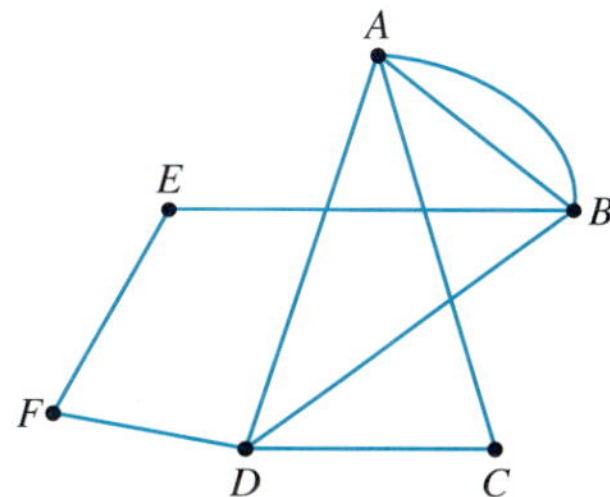

FIGURE 8.40

SOLUTION The first step is to find the degree for each vertex (Table 8.1). (As a check, we count 9 edges, and note that the sum of degrees, 18, does equal twice the number of edges as required.)

Since there are no vertices of odd degree, Euler's theorem tells us that there must be an Euler circuit. Since the graph is small, we can find an Euler circuit by trial and error (but shortly we will describe a systematic method that can be applied when the graph is too large for trial and error). For the given graph, such a circuit goes from F to E to B to A to B to D to C to A to D to F via the edges numbered 1 through 9 in Figure 8.41.

TABLE 8.1

Vertex	Degree
A	4
B	4
C	2
D	4
E	2
F	2

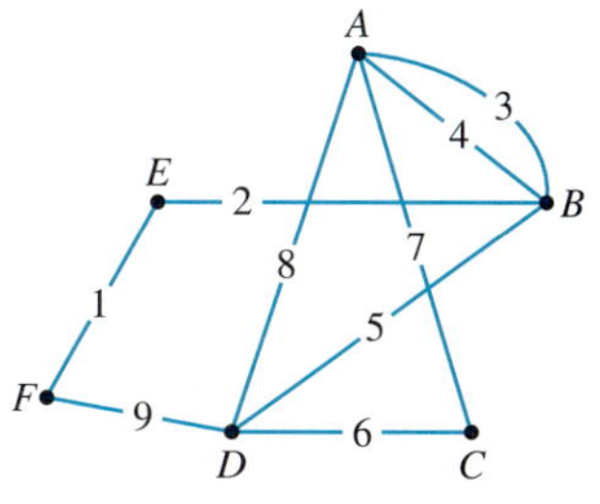

FIGURE 8.41

◆

Euler's theorem only tells us whether or not there is an Euler circuit or Euler path in a graph. The following algorithm tells us how to find one if it exists.

FLEURY'S ALGORITHM

For a large graph that has an Euler circuit or Euler path, we may not be able to find one easily. What we need is a procedure that is guaranteed to work. **Fleury's algorithm** is such a procedure. If Euler's theorem guarantees the existence of an Euler circuit, then we can find at least one Euler circuit by the following steps.

1. Draw a new graph that is just a copy of the vertices of the original graph, but does not include the edges (Figure 8.42).

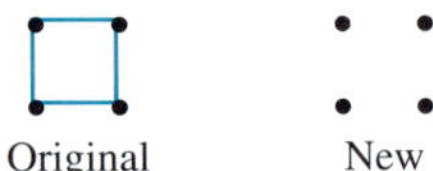

FIGURE 8.42

2. Select any vertex of the original graph and mark it as the present position (Figure 8.43).

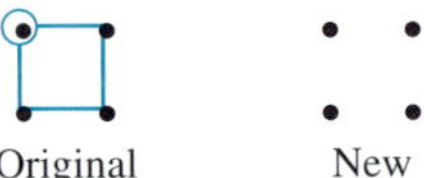

FIGURE 8.43

3. Consider all the edges connected to the present position vertex. Remove one edge, shift it to the new graph, and connect it to copies of the same points. Do not choose an edge that leaves behind a disconnected graph when it is removed (that is, do not remove a bridge), unless the only edge attached to the present position vertex is a bridge. Give the shifted edge a number to keep track of the order in which the path is being constructed (Figure 8.44).

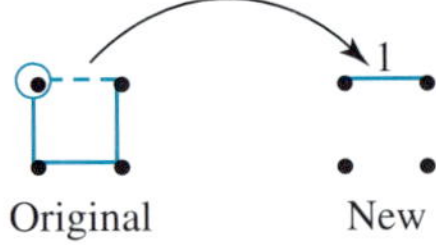

FIGURE 8.44

4. If the edge you removed was the only edge connected to the present position vertex, then remove the present position vertex from the original graph because it will not be needed again.
5. If the edge you removed was not the last edge left, call the edge's other end vertex (on the original graph) the new present position, and go back to step 3 (Figure 8.45).

FIGURE 8.45

6. If the edge you removed was the last edge left in the original graph, then you are done. The edges in the new graph are numbered to give an Euler path.

We will illustrate the use of the algorithm in the next example, even though the graph in the example is small enough that it is easy to construct an Euler circuit by inspection. The algorithm is needed for very large graphs.

EXAMPLE 8.18 The graph in Figure 8.46 has at least one Euler circuit. Find one by Fleury's algorithm.

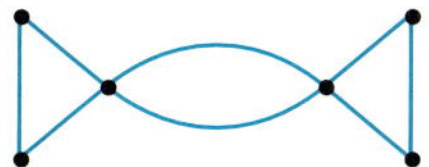

FIGURE 8.46

SOLUTION We start at the top left vertex. The process is illustrated in Figures 8.47(a)–(i). In step 1, we could have chosen either edge attached to the present position vertex. Once the edge has been removed, two other edges in the graph become bridges and have been marked. One of the edges attached at the present position vertex in step 2 is a bridge, so that edge may not be removed. If the only edge attached to the present position vertex is a bridge, then it may be removed because the vertex also will be removed, and what is left behind will not be disconnected after all.

When all the edges have been removed, the new graph has become a copy of the original graph, but with the edges numbered to show an Euler circuit.

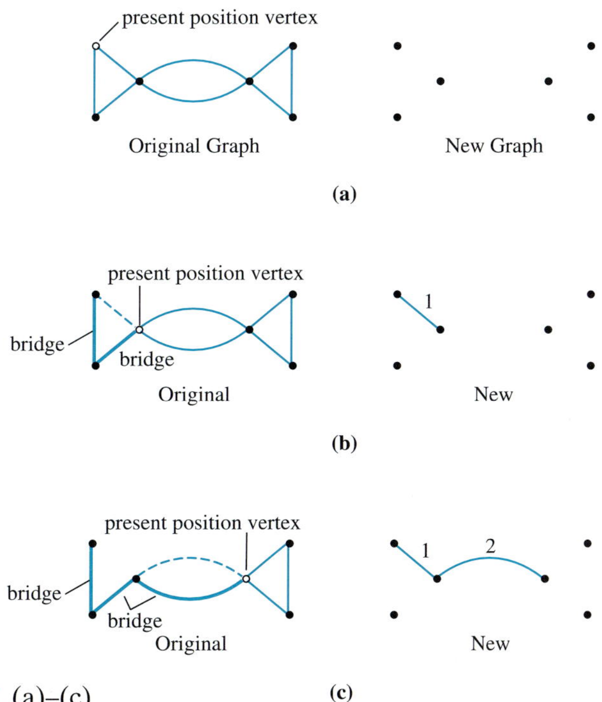

FIGURE 8.47 (a)–(c)

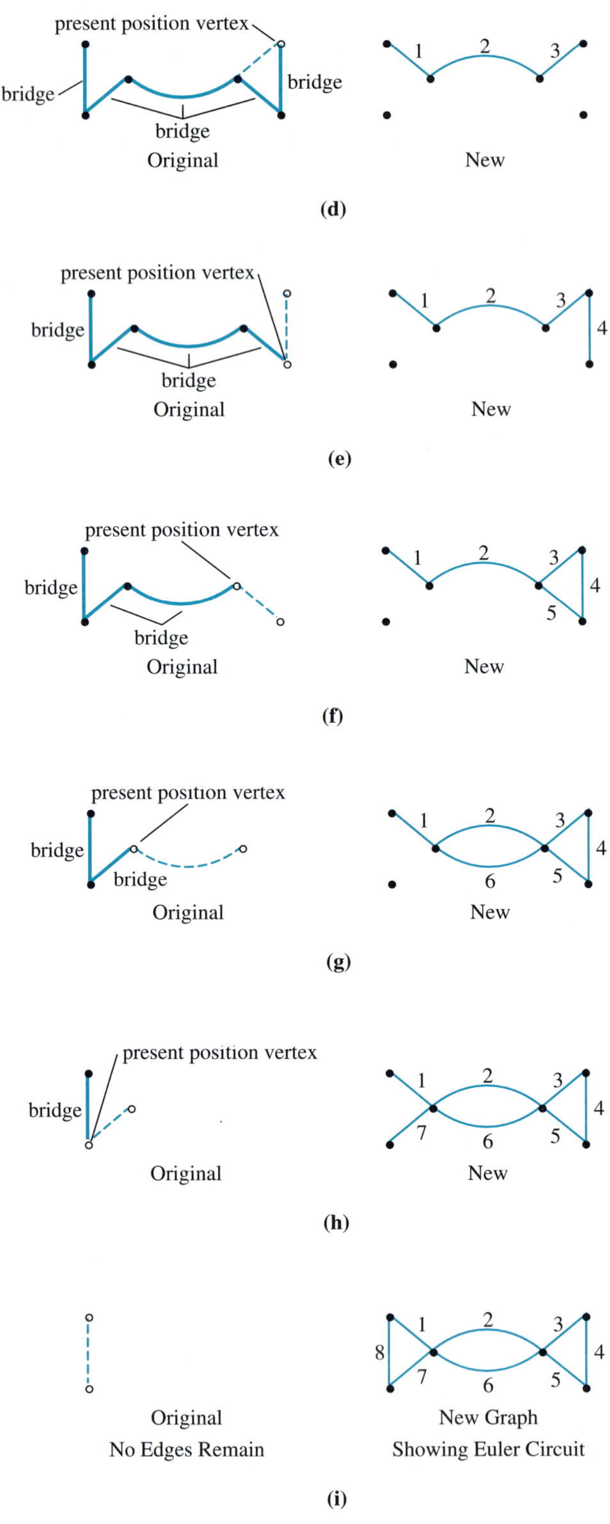

FIGURE 8.47 (d)–(i)

INITIAL PROBLEM SOLUTION

The 18th century town of Königsberg was built on both sides of the Pregel River and on two islands in the river. The people enlivened their Sunday strolls through town by trying to make a circuit that crossed each of the town's seven bridges once and only once (Figure 8.20). Is such a route possible?

SOLUTION In Figure 8.48 we show a graph equivalent to the system of bridges in Königsberg.

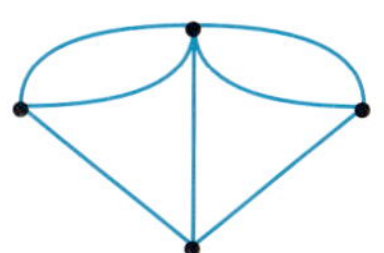

FIGURE 8.48

All four vertices have odd degree, so by Euler's theorem not only is there no Euler circuit, there is no Euler path.

PROBLEM SET 8.3

For problems 1 through 4, refer to Figure 8.24, and sketch a graph representing the interstate highway connections between the given cities using I-5, I-84, and I-90.

1. Bellingham, Boise, Eugene, Olympia, Portland, and Seattle
2. Bellingham, Olympia, Portland, Salem, Seattle, and Spokane
3. Bellingham, Boise, Portland, Salem, Seattle, and Spokane
4. Boise, Eugene, Olympia, Pendleton, Portland, Seattle, and Spokane

In problems 5–8, list (a) the vertices, and (b) the pairs of adjacent vertices.

5.

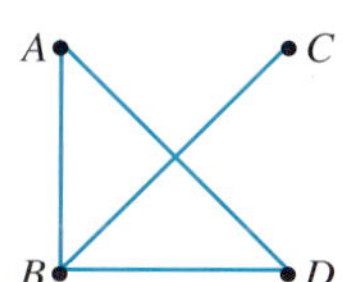

6.

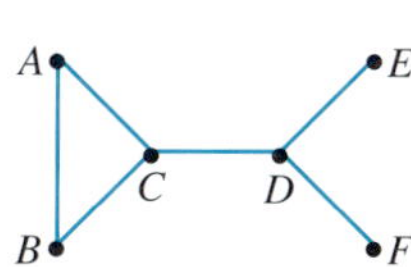

7.

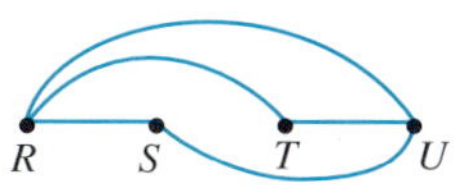

8.

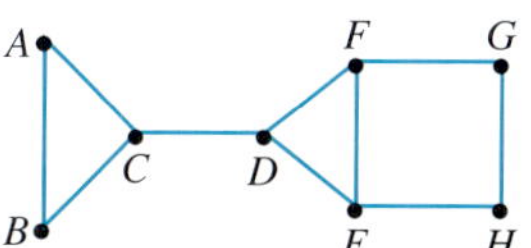

For problems 9 and 10, refer to Figure 8.24. Suppose a new interstate highway is built between Boise and Spokane. Sketch a graph representing the interstate highway connections between the given cities using I-5, I-84, I-90, and the new highway.

9. Boise, Olympia, Portland, Salem, Seattle, and Spokane
10. Bellingham, Boise, Pendleton, Portland, Salem, Seattle, and Spokane

For each of the graphs in problems 11 through 14, list (a) the vertices, and (b) the pairs of adjacent vertices. If there are two or more edges for a pair of vertices, indicate that as part of your answer.

11.

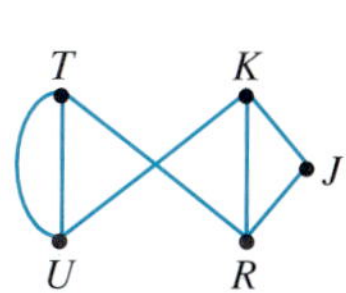

12.

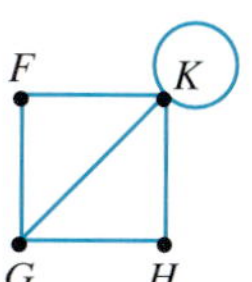

13.

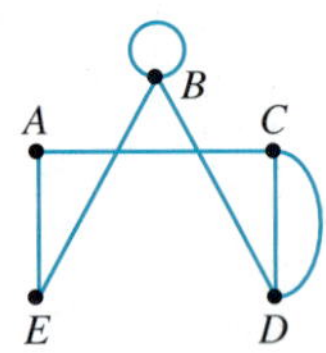

14.

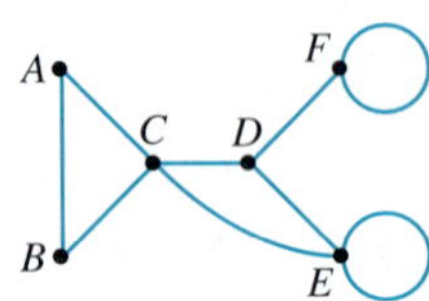

15. In problems 15 and 16, list the vertices and edges. To list the edges, use the adjacent vertices to name an edge; i.e. *AB* is the edge between vertices *A* and *B*.

(a)

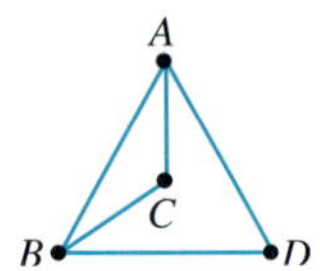

(b)

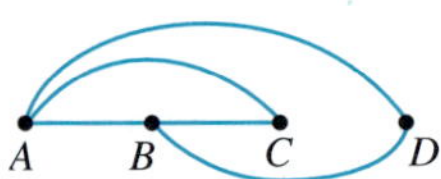

16. (a)

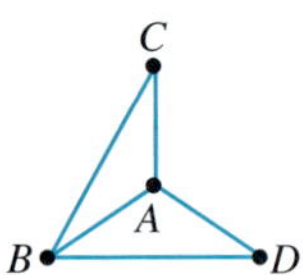

(b)

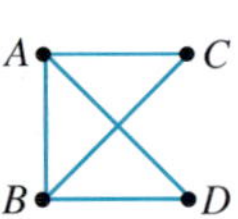

17. In what way are the graphs in problem 15 similar? In what way are they different?

18. In what way are the graphs in problem 16 similar? In what way are they different?

19. Draw two different pictures for each of the following graphs.

(a) Vertices: *A*, *C*, *E*, *G*
Adjacent vertices: *A* and *C*, *A* and *E*, *C* and *E*, *C* and *G*, *E* and *G*.

(b) Vertices: *R*, *S*, *T*, *U*
Edges: *R* is adjacent to *S*, *T*, and *U*; *S* is adjacent to *R* and *U*; *T* is adjacent to *R*; *U* is adjacent to *R*, *S*, and *U*.

20. Draw two different pictures for each of the following graphs.

(a) Vertices: *A*, *B*, *C*, *D*, *E*
Adjacent vertices: *A* and *B*, *A* and *C*, *A* and *E*, *B* and *C*, *B* and *D*, *B* and *E*, and *C* and *D*.

(b) Vertices: *K*, *R*, *U Z*, *T*
Edges: *K* is adjacent to *R* and *U*; *R* is adjacent to *K*, *Z*, and *T*; *U* is adjacent to *K* and *Z*; *Z* is adjacent to *R*, *U*, and *T*; *T* is adjacent to *R* and *Z*.

In problems 21 through 24, find the degree of each vertex in the graph.

21.

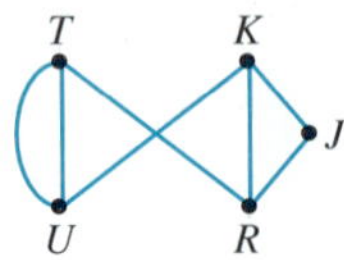

22.

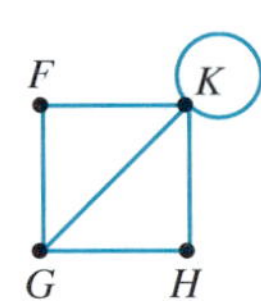

23.

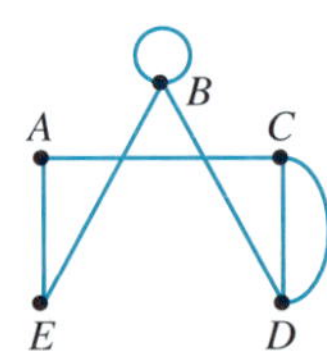

24.

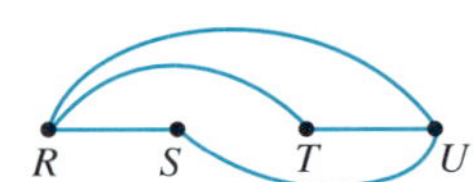

In problems 25 through 28, find the sum of the degrees of the vertices and the number of edges in the graph. How do these compare?

25.

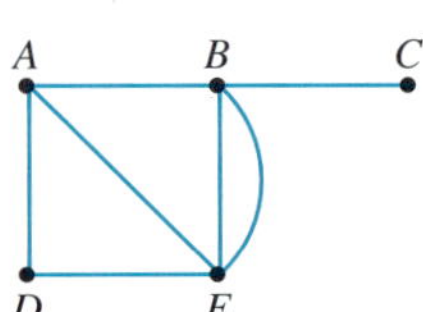

26.

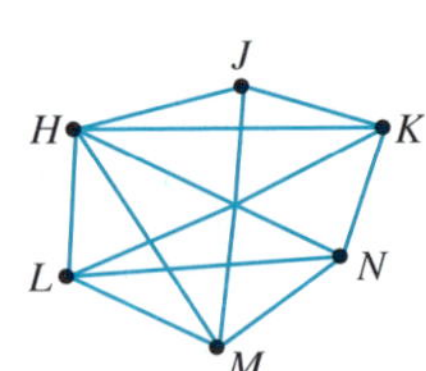

27.

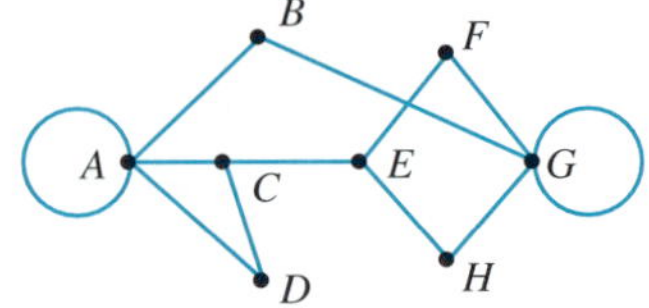

28.

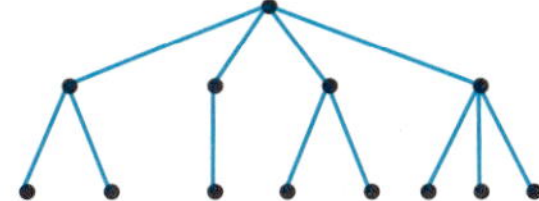

In problems 29 and 30, identify the edges that are bridges, and show the components of the graph if that bridge is removed.

29.

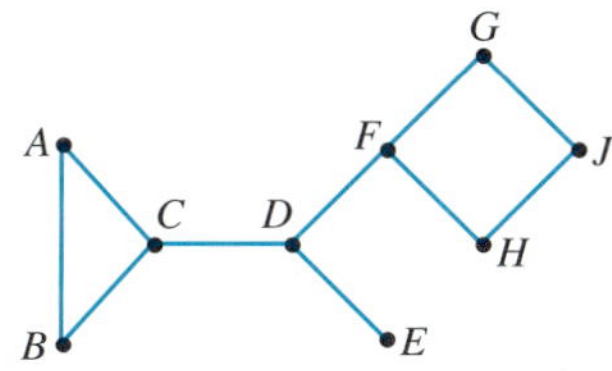

30.

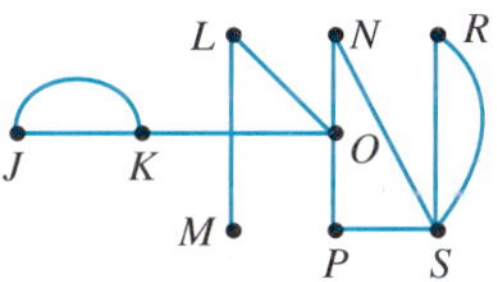

In problems 31 through 34, which of the lists of edges forms a path in the given graph? Are any of the paths a circuit, Euler path, or Euler circuit? See Example 8.15.

31.

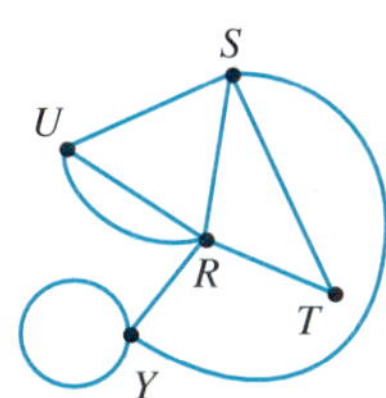

(a) *RU*, *US*, *ST*
(b) *RU*, *US*, *ST*, *TY*
(c) *RU*, *UR*
(d) *RY*, *YY*, *YR*
(e) *RU*, *US*, *ST*, *TR*, *RY*, *YY*, *YS*, *SR*

32.

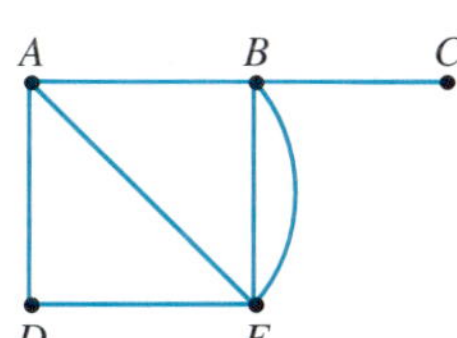

(a) *AE*, *EB*, *BC*
(b) *AD*, *DE*, *EB*
(c) *ED*, *DA*, *AE*, *EB*, *BE*
(d) *AD*, *DE*, *EA*, *AB*, *BE*, *EB*, *BC*
(e) *EB*, *BE*, *EA*, *AD*, *DE*, *EA*, *AB*, *BC*

33.

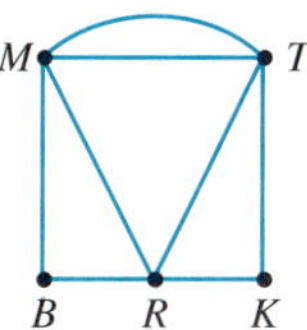

(a) *MT*, *TR*, *RB*
(b) *BR*, *RM*, *MK*
(c) *RM*, *MT*, *TK*, *KR*, *RB*, *BM*, *MT*, *TR*
(d) *KT*, *TR*, *RM*, *MT*, *TM*, *MB*, *BR*, *RK*

34.

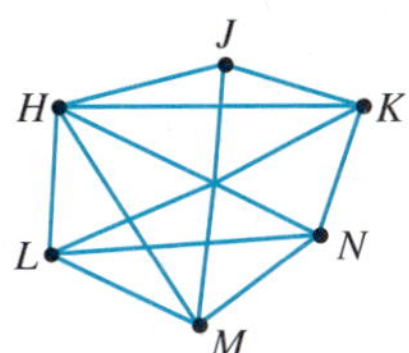

(a) *HN*, *NL*, *LJ*, *JK*
(b) *LK*, *KH*, *HJ*, *JM*, *ML*
(c) *JM*, *MN*, *NL*, *LK*, *KH*, *HJ*, *JK*
(d) *HM*, *ML*, *LH*, *HK*, *KJ*, *JH*, *HN*, *NM*, *MH*

35. Explain why the graph in problem 31 can have an Euler path but can't have an Euler circuit.

36. Explain why the graph in problem 33 must have an Euler circuit.

Problems 37 and 38 refer to the following graph from Example 8.17.

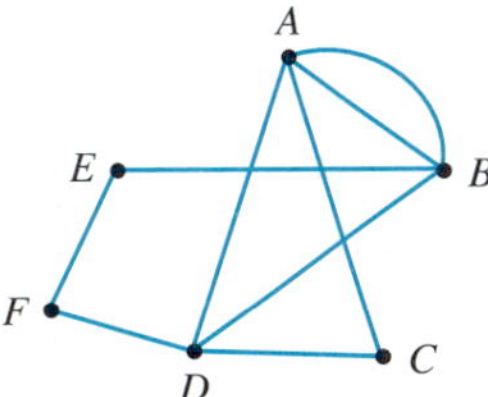

37. The graph was shown to have an Euler circuit of *FE*, *EB*, *BA*, *AB*, *BD*, *DC*, *CA*, *AD*, *DF*. Find another Euler circuit with *FE* being the first edge.

38. Is it possible to find an Euler circuit that begins with *DF* as the first edge? If it is, find such an Euler circuit. If it isn't possible, explain why.

For problems 39 through 44, determine whether or not the given graph has any Euler paths or Euler circuits. If it does, use Fleury's algorithm to find the path and/or circuit.

39.

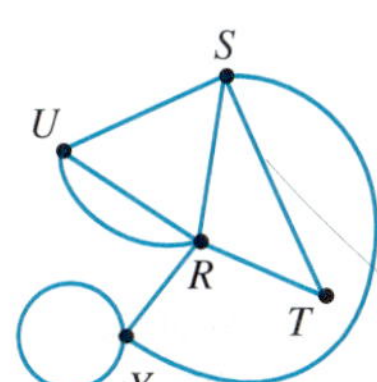

40.

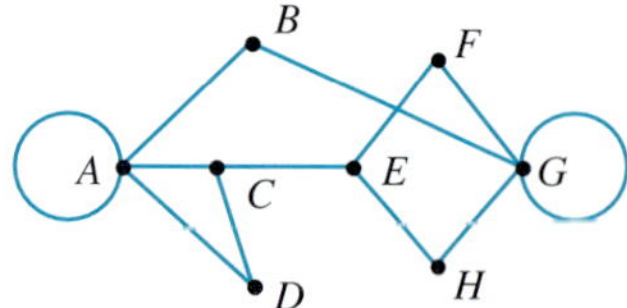

41.

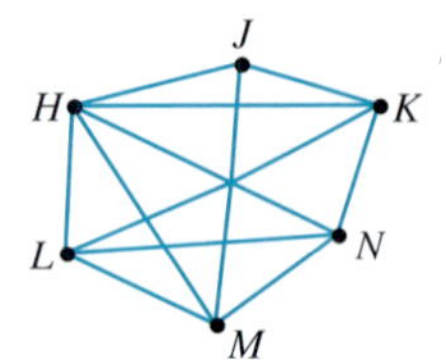

42.

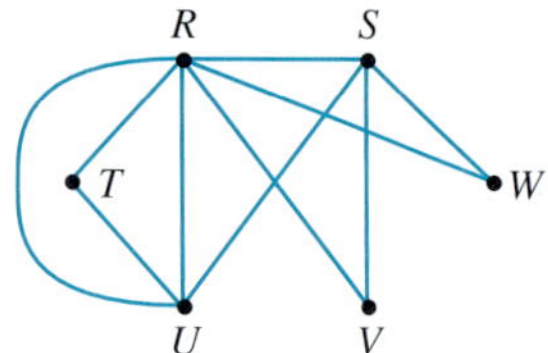

43.

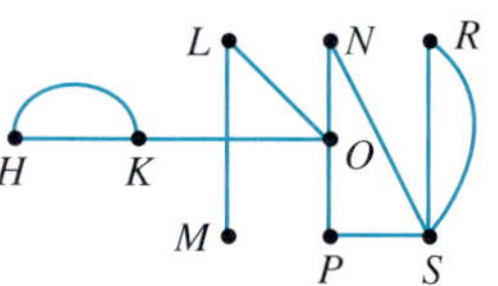

44.

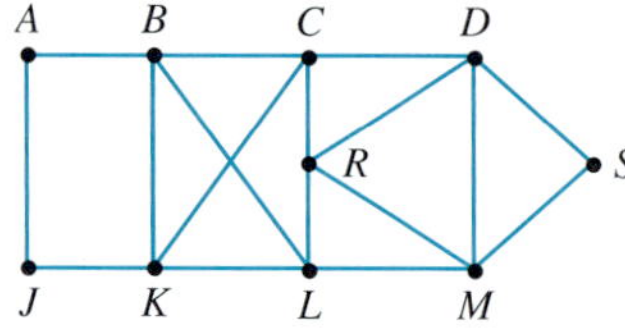

45. The following map contains at least one Euler path. Why?

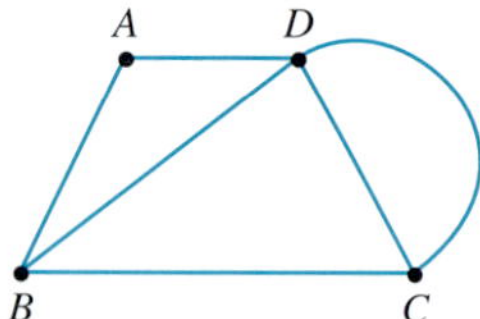

(a) How many Euler paths originate at point A?
(b) How many Euler paths originate at point B?
(c) How many Euler paths originate at point C?
(d) How many Euler paths originate at point D?

46. Referring to problem 45, is there an edge that can be removed to create a new map that has an Euler circuit? If yes, show a new map; if no, explain why it is not possible.

Problems 47 through 51 refer to the following map and description.

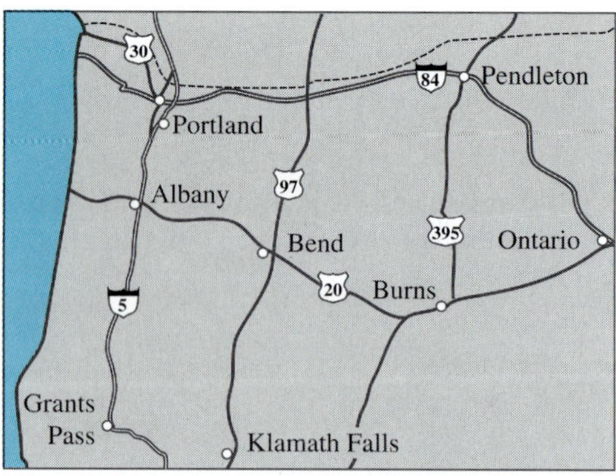

The Trans-Oregon Agricultural Products company (TOAP) has offices in Portland (shipping offices), Albany (grass seed), Pendleton (wheat), Burns (livestock), and Ontario (potatoes). The company began in Pendleton, and head offices are located there. Several years ago, TOAP established a courier service to deliver orders, invoices, and other documents between the various offices. The courier cannot cover all of the territory in a single day and must stay overnight somewhere before completing a visit to all of the cities. TOAP has signed a contract with a motel chain that has a motel in each of the cities, and the driver doesn't mind living on the road. The courier needs to drive a route that covers each road at least once each trip to allow for picking up or leaving documents at other destinations on the route.

47. In order to minimize costs, the courier should cover each road exactly once. Show that this is possible by:
(a) drawing an appropriate route.
(b) basing your reasoning on theory.

48. (a) If the courier begins a trip in Pendleton, in which cities will it be possible to spend the night after making a trip over all of the roads?
(b) If the courier begins a trip in Burns, in which cities will it be possible to spend the night after making a trip over all of the roads?

49. TOAP is considering moving the main offices to Portland because of increased international shipping. What effect will a move to Portland have on the courier service if the courier service begins trips from Portland?

50. Suppose TOAP moves the main offices to Portland but decides that also moving the courier service to Portland is not a good idea. Which cities are possible sites for locating the courier service if it must be in a city from which you can begin or end a trip under the conditions previously established while the main offices were in Pendleton? Leaving the courier service in Pendleton is one option. Are there any others?

51. If TOAP establishes a new office in Bend and includes the section of Highway 97 between Bend and Highway 84 as part of the route, what effects will this have on the operation of the courier service?

Problems 52 and 53 refer to the following map.

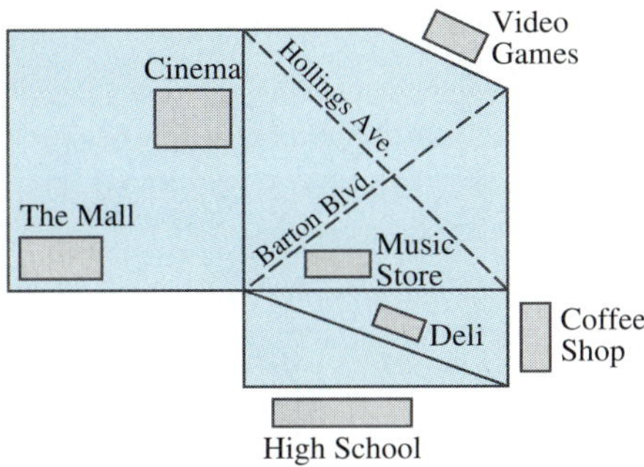

52. Many of the teenagers in Central City like to cruise around town on the weekend using only certain streets (shown as solid lines on the map). Is it possible to choose a route that uses each of these streets exactly once? If yes, show such a route; if no, explain why it is not possible.
53. Suppose a new teenage club is located at the intersection of Hollings Ave. and Barton Blvd. If only one of these streets can be added to the route, which one would you choose, and why?

EXTENDED PROBLEMS

54. Explain why the number of edges in any graph will be half the sum of the degrees of all the vertices.
55. According to Euler's theorem, if there are exactly two vertices of odd degree, there will be at least one Euler path; and any Euler path must start at one of the vertices of odd degree and end at the other. Explain why an Euler path beginning at one of the vertices of odd degree could not possibly be an Euler circuit.
56. Are any of the bridges of Königsberg "bridges" in the sense of graph theory? Explain.
57. Is there any way to connect the city of Königsberg by a system of bridges in such a way that each bank of the river and each island is accessible by an even number of bridges? If so, give an example; if not, explain why.
58. If the citizens of Königsberg had decided to build a new bridge in the city, is there a location for the bridge that would make it possible for a person to make an Euler circuit of the bridges on a single walk? If yes, show an appropriate circuit; if no, explain why it is not possible to make an Euler circuit.
59. Referring to problem 58, would the addition of a single bridge make it possible for a person to make an Euler path of the city on a single walk? If yes, show an appropriate path; if no, explain why it is not possible to make an Euler path.
60. Suppose that a flood destroyed one of the bridges as shown in the map below:

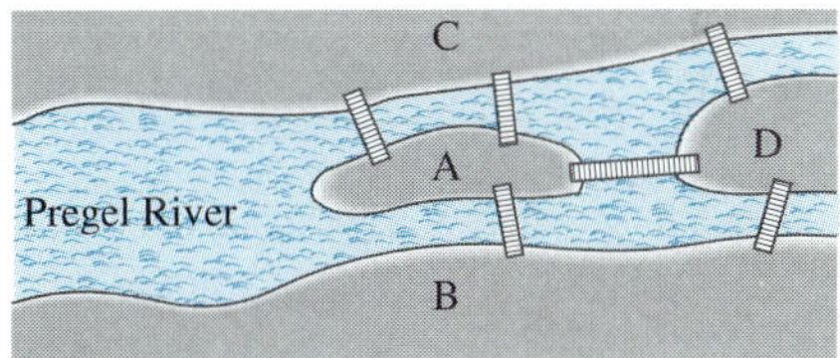

The six bridges of Königsberg

Is it possible for a person to make an Euler path of the city on a single walk? If yes, show an appropriate path; if no, explain why it is not possible to make an Euler path.

61. Referring to problem 60, has the elimination of the bridge helped make an Euler circuit? If yes, show an appropriate circuit; if no, explain why it is not possible to make an Euler circuit.

8.4 NETWORK PROBLEMS

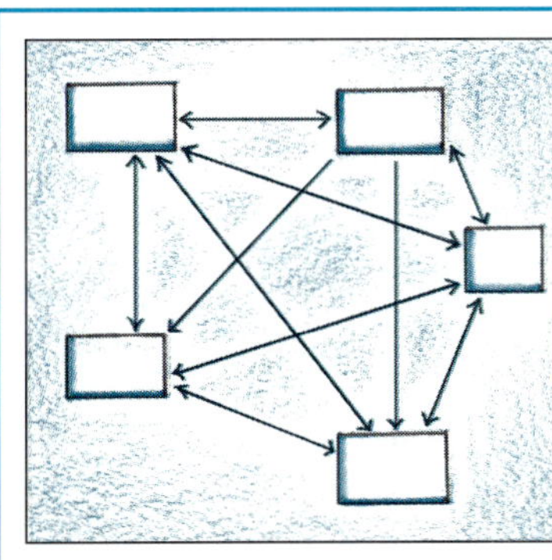

During the first year on its new campus, Cold Region State College had several unfortunate cases of frostbite among students walking to class. To avoid this happening in the future, the administration intends to convert several of the existing walkways to protected walkways so that students can go from any building on campus to another without being exposed to the inclement weather. To minimize expenses and the time of construction, they wish to draw up plans for the minimum total length of walkways that need to be converted. How can this be done? The campus map is shown below.

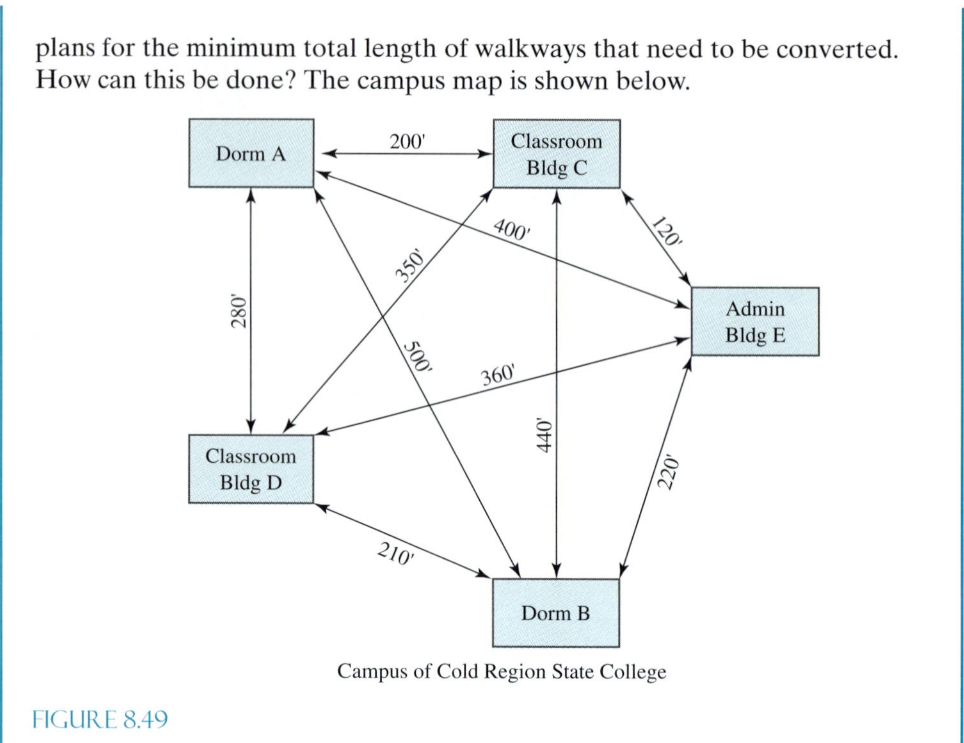

Campus of Cold Region State College

FIGURE 8.49

Section 8.3 focused on the description and analysis of networks. This included discussion of the general forms of graphs and the paths or circuits that could be used to traverse them. In that setting, all edges of a graph were treated equally. But, an edge of a graph that represents a road over the mountains is not the same as an edge that represents a road in the valley. With this in mind, we will consider how graphs can be made more applicable to real-world problems.

WEIGHTED GRAPHS

Another type of problem that arises in business and other contexts is designing efficient networks. For example, you might want to minimize the length of the wire used in a computer network or the mileage delivery vans must cover. Such problems can often be dealt with using graph theory, but we must consider graphs hav-

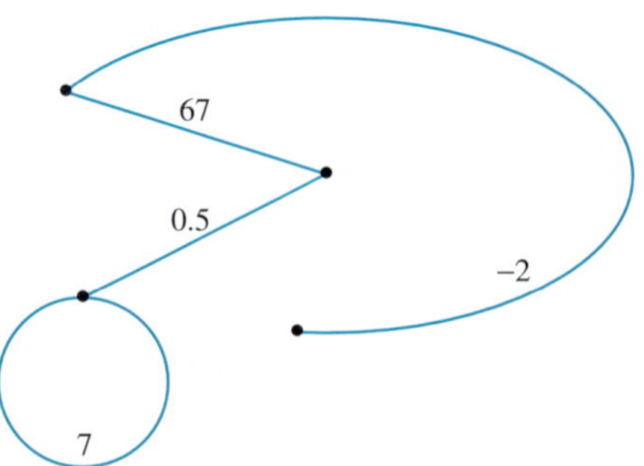

FIGURE 8.50

ing characteristics that measure efficiency. The type of enhanced graph we need is a **weighted graph** whose edges have numbers associated with them. The number associated to the edge is called the **weight** of the edge. An example of a weighted graph is given in Figure 8.50.

If the graph is associated with some practical problem or physical situation, the weights should be assigned in a way relevant to the problem. However, the lengths of the edges drawn for the graph do not need to be proportional to the weights, as is indicated in Figure 8.50.

EXAMPLE 8.19 Suppose it is a 35-minute drive from Ed's home to his workplace, a 15-minute drive from work to his health club, and a 25-minute drive from his health club to his home. Draw a diagram of the associated weighted graph.

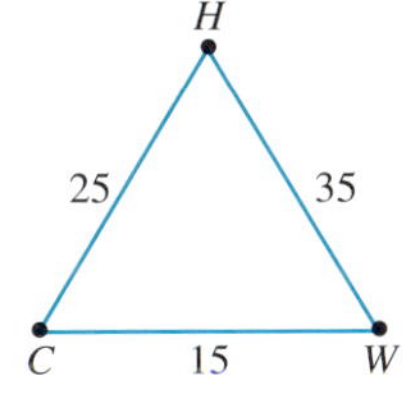

FIGURE 8.51

SOLUTION The vertices of the graph will correspond to Ed's home, H, his workplace, W, and his health club, C. Each pair of vertices will be connected by an edge, and the weight associated with an edge will be the driving time, not the miles, between the two vertices. The result is shown in Figure 8.51. Note that the lengths of the edges do not have to be proportional to the weights. This is a definite advantage when we need to draw complex graphs. ◆

SUBGRAPHS

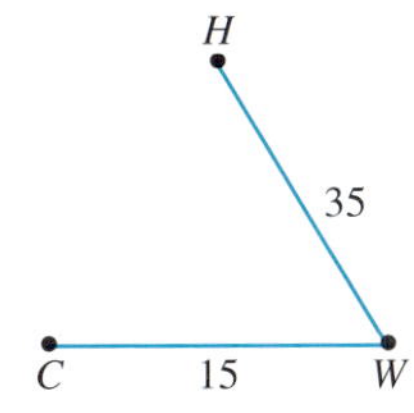

FIGURE 8.52

One way for you to increase the efficiency of a network is to remove redundant connections. In terms of the weighted graphs we have been considering, this corresponds to selecting a smaller set of edges from a graph. Technically, we are interested in a **subgraph,** by which we mean a set of vertices and edges chosen from among those of the original graph. For example, referring to Figure 8.51, if the direct road between Ed's health club and Ed's home is closed for construction work, then the subgraph of his route is illustrated in Figure 8.52.

If Ed's health club is also closed for its yearly maintenance work, he has the even smaller subgraph shown in Figure 8.53.

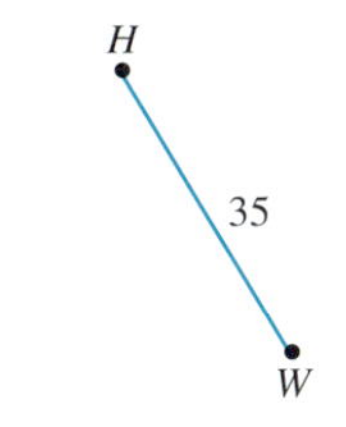

FIGURE 8.53

TREES

In a network, if there is a path that starts somewhere and returns without using any connection twice, then there must be a redundant connection. If you are seeking efficiency, then you want to avoid having such redundancy. A graph that is connected (that is, you can go from any vertex to any other vertex) and has no circuits is called a **tree;** trees then will have no redundant connections. In Figure 8.54 we show some examples of graphs that are trees and a graph that is not a tree. In the graph that is not a tree, the circuit it contains is shown using thicker lines.

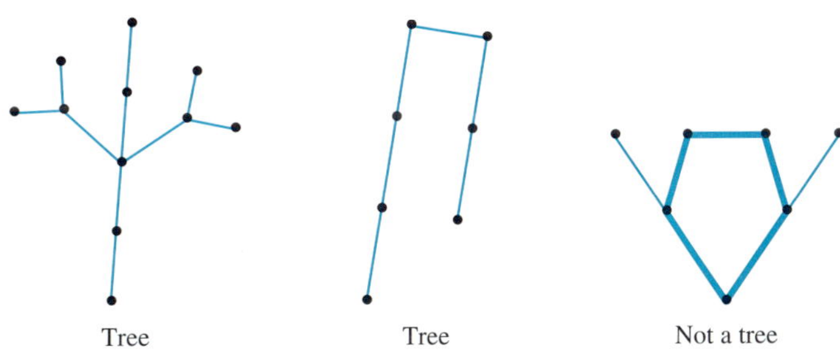

FIGURE 8.54

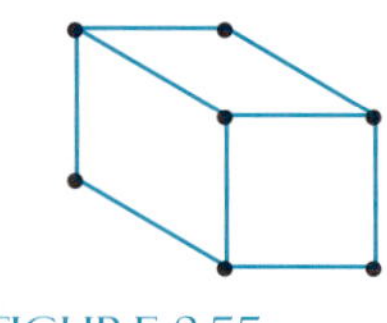

FIGURE 8.55

EXAMPLE 8.20 The graph in Figure 8.55 is not a tree. Darken the edges forming a circuit, then remove edges until obtaining a subgraph that is a tree.

SOLUTION In Figure 8.56(a) we have darkened the edges of one of the circuits in the graph. In Figure 8.56(b) we remove one edge that was in the circuit in (a). In (c) we find there is still a circuit, which we have darkened. In (d) we remove one of the edges from the circuit in (c). In (e) we see there is still a circuit, which we have darkened. Finally in (f) we have removed an edge from the circuit in (e), and we see there is no longer any circuit, so we have a tree.

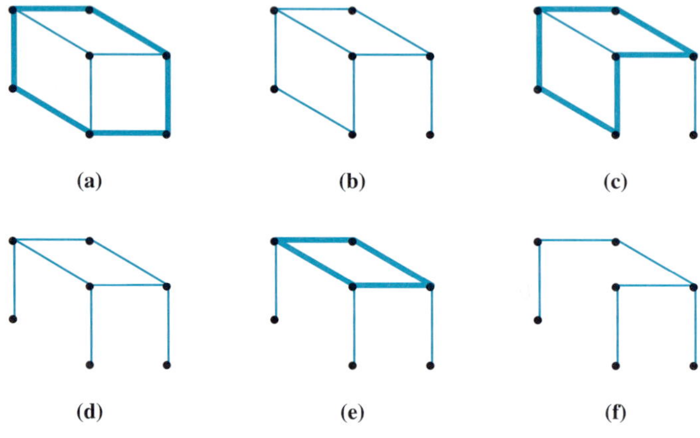

FIGURE 8.56 ◆

A subgraph that contains all the original vertices, is connected, and contains no circuits is called a **spanning tree.** Figure 8.57(b) shows a spanning tree for the graph in Figure 8.57(a). Several other spanning trees are possible.

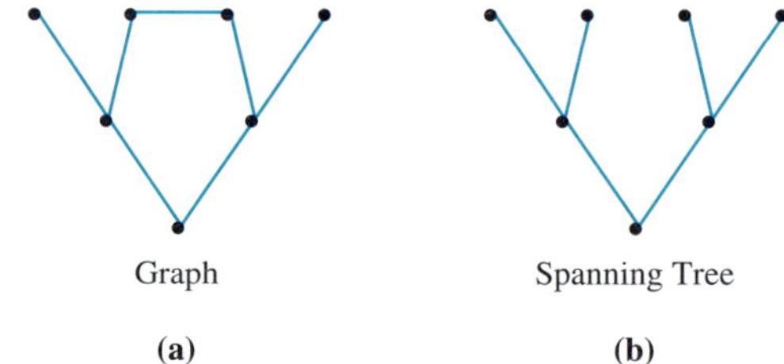

FIGURE 8.57

MINIMAL SPANNING TREES

If you begin with a connected weighted graph representing some physical network and you want to construct the most efficient network, then you should try to find a connected subgraph of smallest total weight that contains all of the vertices. Thus you want to construct a spanning tree with the smallest possible total weight; such a tree is called a **minimal spanning tree.** For example, in Figure 8.58(b) and (c) we have shown two spanning trees for the graph in (a). Notice that the total weight for the graph in Figure 8.58(c) is less than that in (b). In fact Figure 8.58(c) shows a minimal spanning tree for the graph in (a).

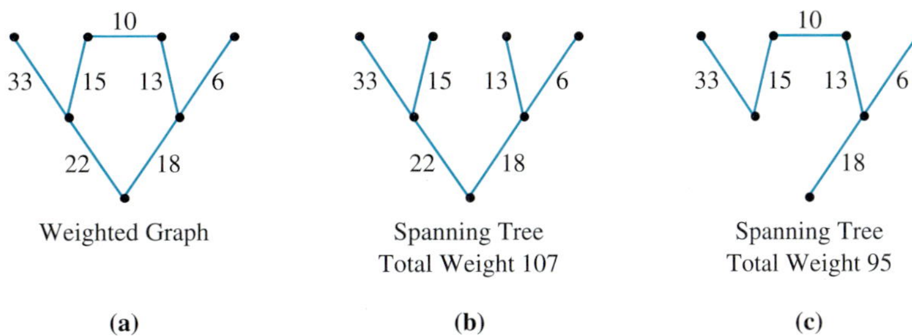

FIGURE 8.58

Here is a procedure for finding a minimal spanning tree in a weighted graph. Instead of removing edges, we will start with only the vertices and add edges until the resulting graph is connected. All you need to do at each stage is look at the list of edges that have not been used and *add the acceptable edge of smallest weight* according to the following rules.

Acceptable

(i) An edge that does not share a vertex with any of the edges already chosen is always acceptable because it certainly cannot complete a circuit (Figure 8.59).

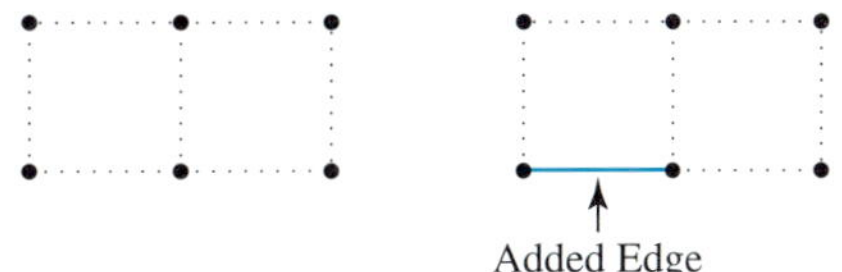

FIGURE 8.59

(ii) An edge that connects together two components of the subgraph is also acceptable (Figure 8.60).

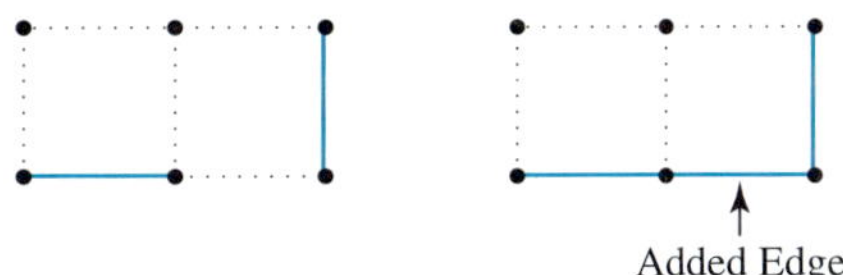

FIGURE 8.60

(iii) An edge that connects to a component of the subgraph and brings a new vertex into the subgraph is also acceptable (Figure 8.61).

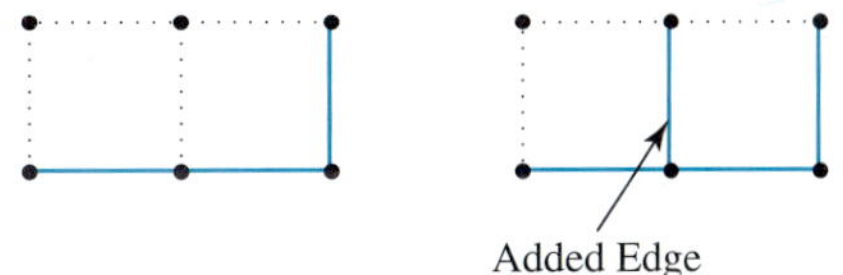

FIGURE 8.61

Unacceptable

(iv) Adding an edge to a component of the subgraph without also adding a vertex (Figure 8.62).

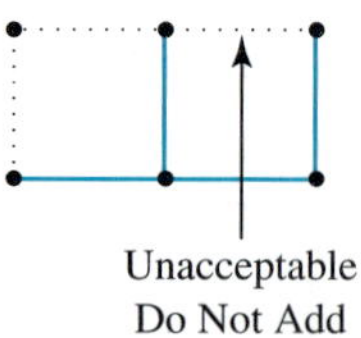

FIGURE 8.62

EXAMPLE 8.21 Construct a minimal spanning tree in the weighted graph shown in Figure 8.63.

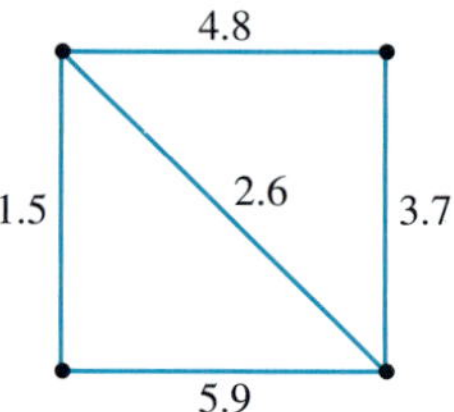

FIGURE 8.63

SOLUTION We start with the four vertices [Figure 8.64(a)]. Then pick the edge with weight 1.5 since that is the smallest [Figure 8.64(b)]. For the next addition, all the edges would be acceptable by the preceding rules. We choose the edge

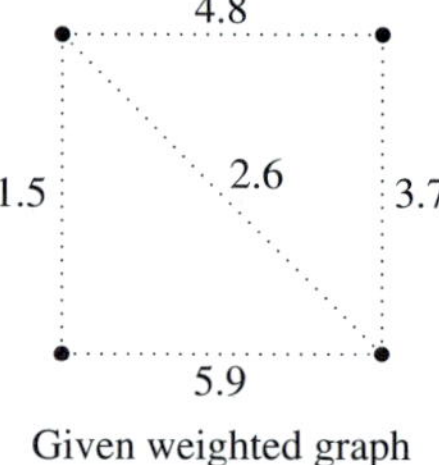

Given weighted graph

(a)

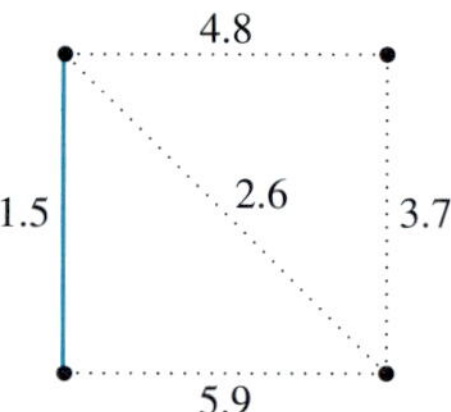

First choice: Choosing the acceptable edge with the smallest weight

(b)

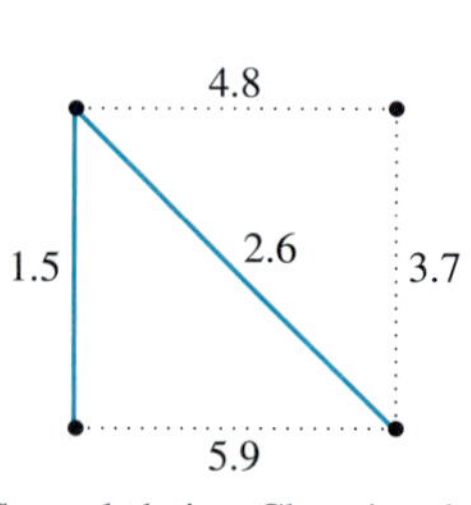

Second choice: Choosing the remaining acceptable edge with the smallest weight

(c)

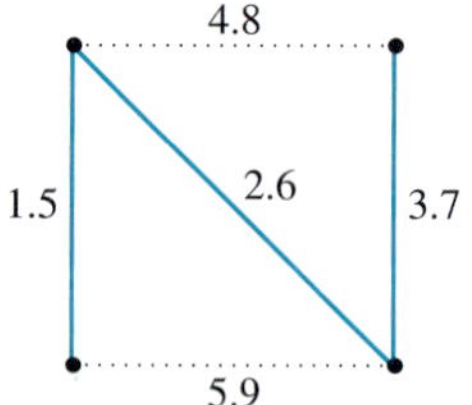

Third choice completes minimal spanning tree

(d)

FIGURE 8.64

◆

with smallest weight: 2.6 [Figure 8.64(c)]. There are two remaining acceptable edges, one with weight 4.8 and one with weight 3.7. Whenever there is a choice, we choose the edge with the smaller weight. We add the edge with weight 3.7 to the subgraph and have constructed a minimal spanning tree [Figure 8.64(d)]. It has total weight $1.5 + 2.6 + 3.7 = 7.8$.

This method for finding the minimal spanning tree for a graph is known as **Kruskal's algorithm;** it was developed by David Kruskal at AT&T Bell Laboratories. ◆

INITIAL PROBLEM SOLUTION

During the first year on its new campus, Cold Region State College had several unfortunate cases of frostbite among students walking to class. To avoid this happening in the future, the administration intends to convert several of the existing walkways to protected walkways so that students can go from any building on campus to another without being exposed to the inclement weather. To minimize expenses and the time of construction, they wish to draw up plans for the minimum total length of walkways that need to be converted. How can this be done? The campus map is shown below.

SOLUTION We assume that all the protected walkways will go from building to building along existing paths rather than constructing new hub buildings from which new walkways could be built. We form the weighted graph with vertices corresponding to buildings, edges corresponding to existing walkways, and with the weight of an edge being the distance between the buildings (Figure 8.65).

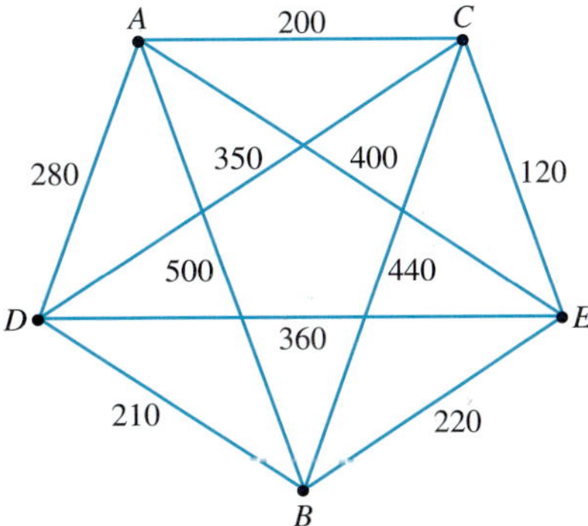

FIGURE 8.65

We then follow the procedure of adding edges without completing circuits, always adding the edge with smallest weight. Edges that are added are indicated by thickened lines. Edges that are unacceptable are indicated by broken lines (Figure 8.66).

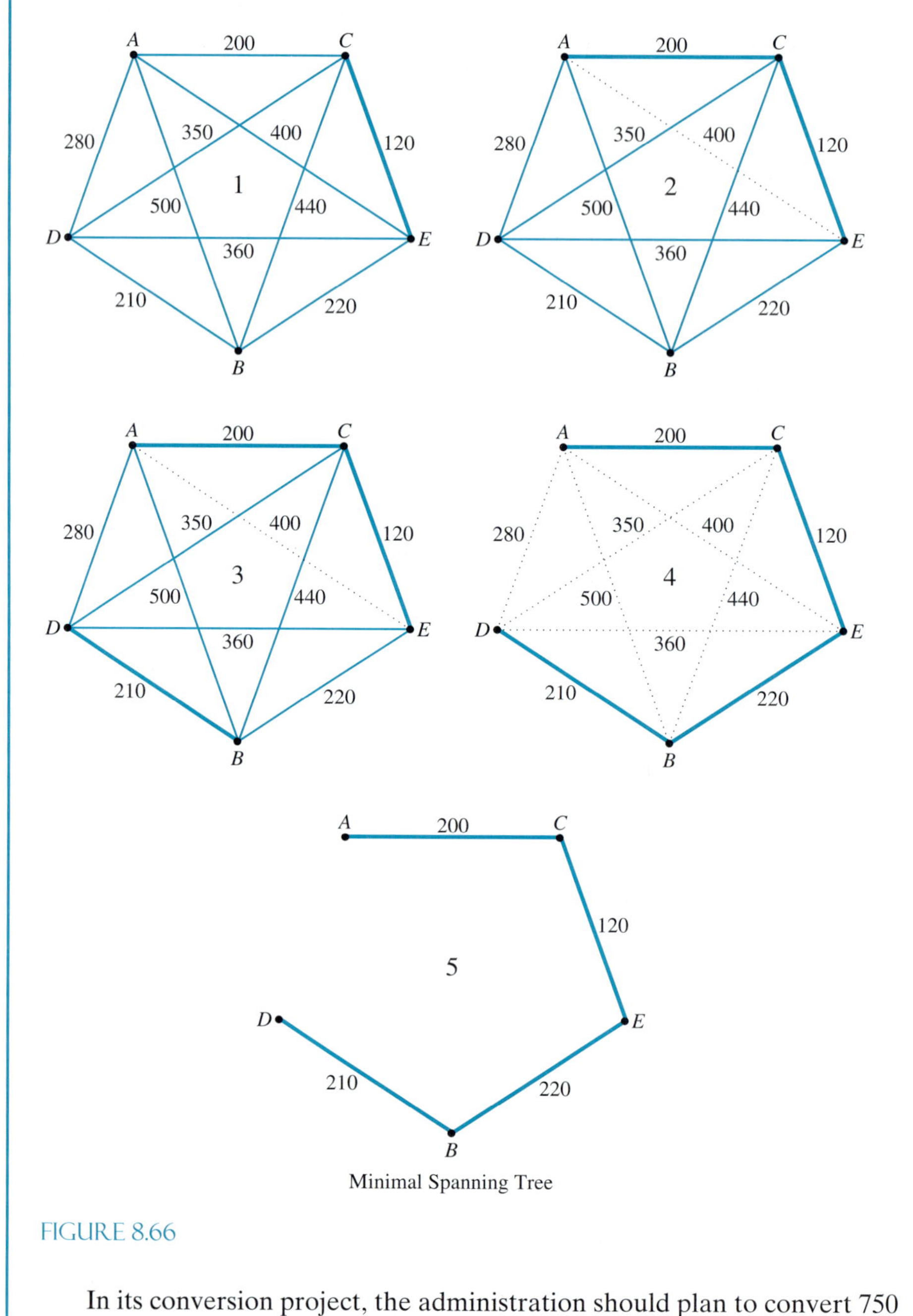

FIGURE 8.66

In its conversion project, the administration should plan to convert 750 feet of existing walkways to protected walkways.

PROBLEM SET 8.4

In problems 1 through 8, draw a weighted graph with the given information. The only essential considerations are the identification of the vertices and the edges connecting them, as in a graph of the bridges of Königsberg.

1. (a) Marsha is attending college and working part-time at a clothing store. It is 10 minutes from Marsha's apartment to the college, 15 minutes from her apartment to the store, and 20 minutes from the college to the store. Also, it is 10 minutes from Marsha's apartment to the rest home where she volunteers on Sundays (Marsha never goes from either the store or the college to the rest home).

(b) If Marsha quits her job at the store, what does the modified graph look like?

2. Suppose that Ed, in Example 8.19, has a business at home and makes daily trips to the post office. The driving times between the post office and the other destinations on Ed's route are:

15 minutes to his home
20 minutes to the health club
25 minutes to his workplace

Modify the graph in Figure 8.50 to include this new information.

3. The Trans-Oregon Agricultural Products company (TOAP) has offices in Albany, Portland, Pendleton, Ontario, and Burns. The distances of the highway connections the company uses for its courier service are:

Albany-Portland	68 mi
Portland-Pendleton	208 mi
Pendleton-Ontario	167 mi
Pendleton-Burns	198 mi
Burns-Ontario	130 mi

4. A central California company has offices in Hayward, Oakland, Sacramento, San Francisco, San Jose, and Stockton. The highway distances the company uses when traveling between offices are:

Hayward-Oakland	12 mi
Hayward-San Jose	32 mi
Hayward-Stockton	65 mi
Oakland-Sacramento	92 mi
Oakland-San Francisco	11 mi
San Francisco-San Jose	47 mi
Sacramento-San Francisco	85 mi

5. A midwestern commuter airline provides services between Cleveland, Chicago, Minneapolis and St. Louis. Mileage between the cities is listed in the given table. Draw a graph representing the airline system.

Cl	—			
Ch	335	—		
M	740	405	—	
StL	530	290	550	—
	Cl	Ch	M	StL

6. Draw a graph representing the airline system in problem 5 if Memphis is added to the regular schedule. Distances from Memphis to the other cities are: Cleveland, 710; Minneapolis, 830; Chicago, 530; and St. Louis, 285.

7. A western equipment company has facilities in San Francisco, Butte, Denver, Salt Lake City, and Los Angeles. The railroad distances between these cities are listed in the given chart. Draw a graph representing the railroad connections between the company's facilities.

LA	—				
SF	470	—			
SLC	780	820	—		
B	1220	1180	430	—	
D	1350	1370	570	890	—
	LA	SF	SLC	B	D

8. Draw the map for the railway connections between facilities of the equipment company in problem 7 if it adds facilities in Albuquerque. The railroad distances between Albuquerque and the other cities are: Butte, 1370; Los Angeles, 890; Salt Lake City, 990; Denver, 480; and San Francisco, 1210.

9. Which of the following graphs are trees?

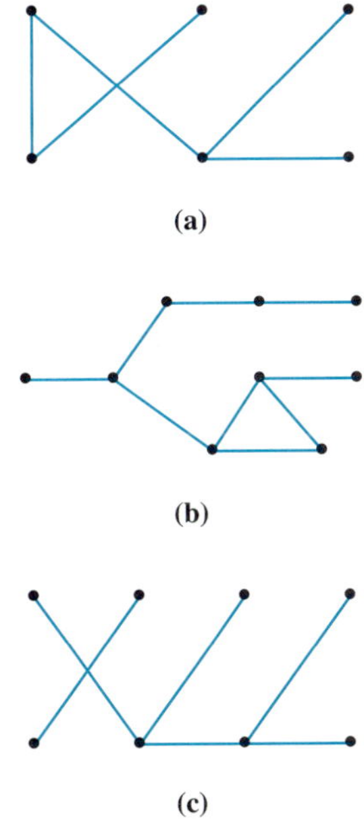

(a)

(b)

(c)

10. Which of the following graphs are trees?

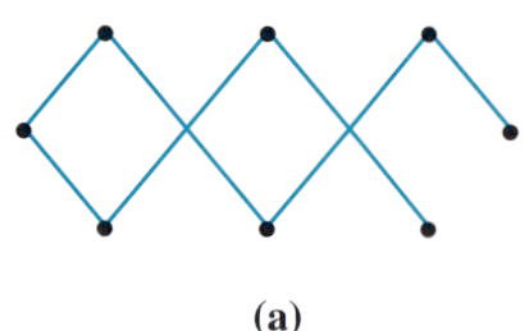

(a)

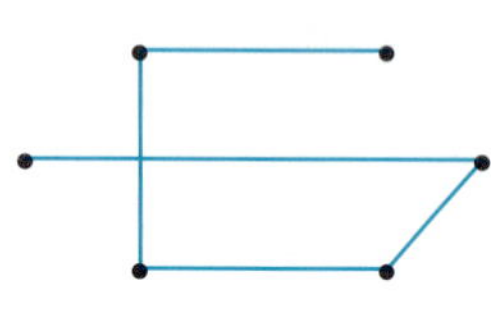

(b)

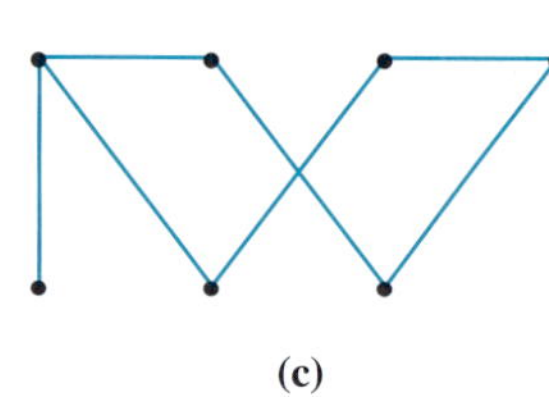

(c)

In problems 11 through 14, the given graphs are not trees because they contain circuits. Find all possible spanning trees for each graph.

11.

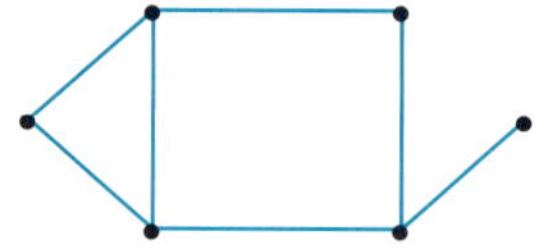

12.

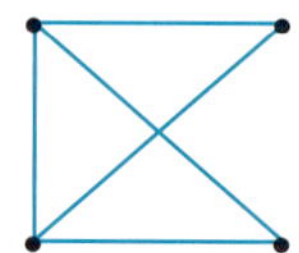

13.

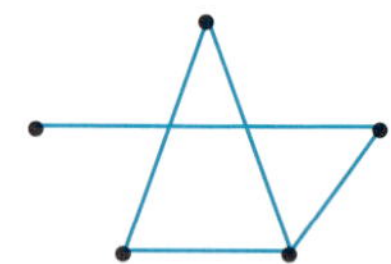

14.

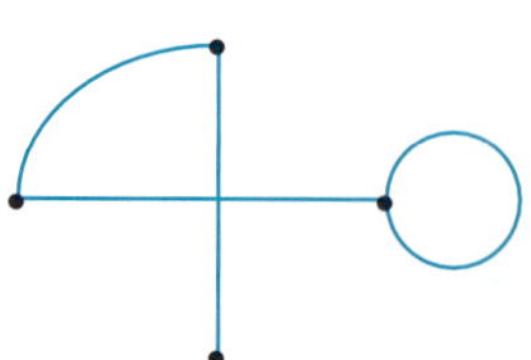

In problems 15 through 18, do the following for each of the given graphs:

(i) Determine the number of edges that must be removed to form a tree.

(ii) Identify which edges cannot be removed when forming a tree from the graph.

(iii) Determining the number of different spanning trees that can be produced from the graph.

15.

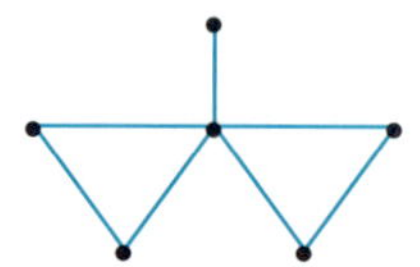

16.

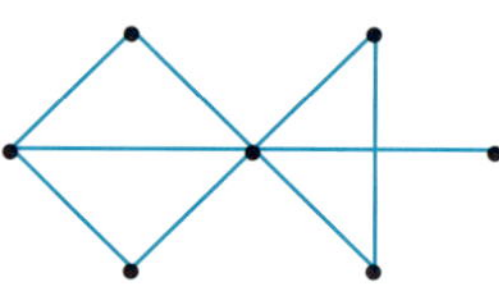

17.

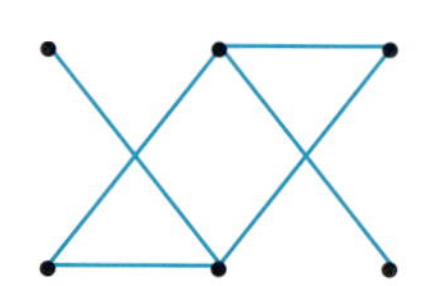

18.

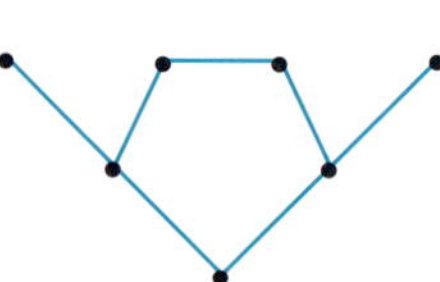

For problems 19 through 22, find all the spanning trees for the given graphs.

19.

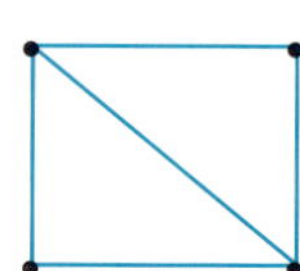

20.

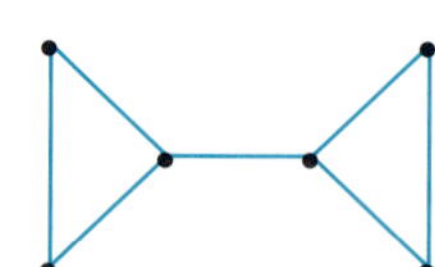

21.

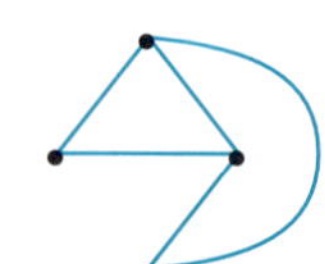

22.

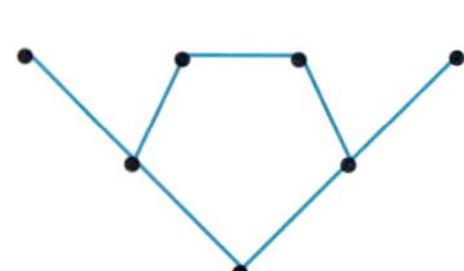

For problems 23 through 26, find at least three spanning trees for the given graphs.

23.

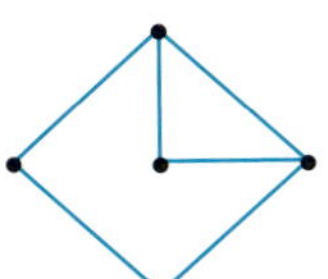

24.

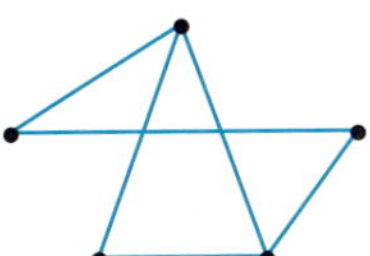

25.

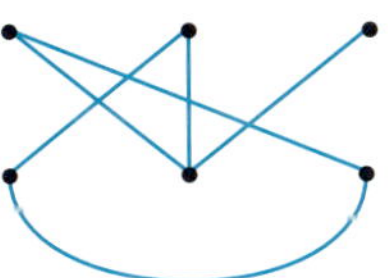

26.

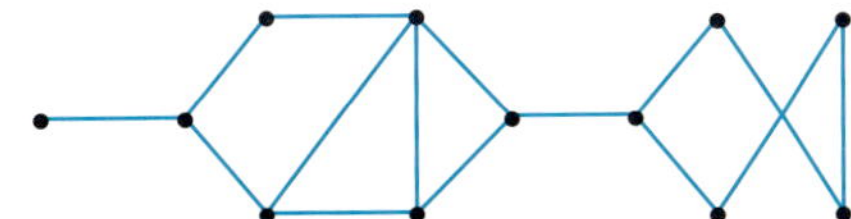

In problems 27 through 30, use Kruskal's algorithm to find minimal cost spanning trees for each of the given graphs. List the edges in the order they are selected.

27.

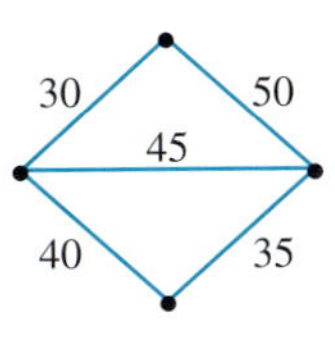

(a)

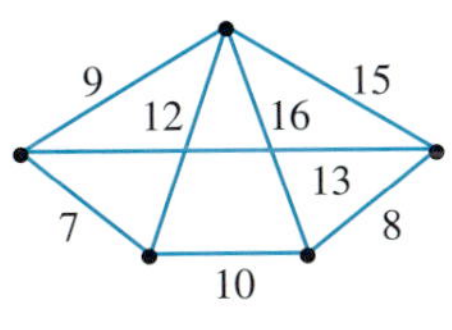

(b)

28.

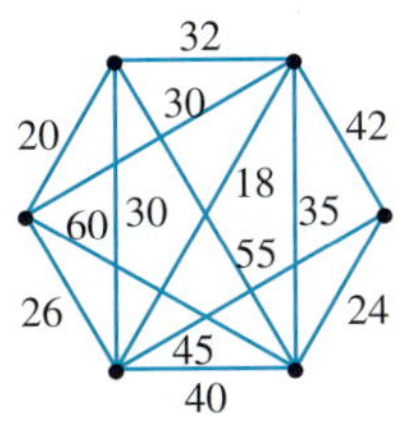

29.

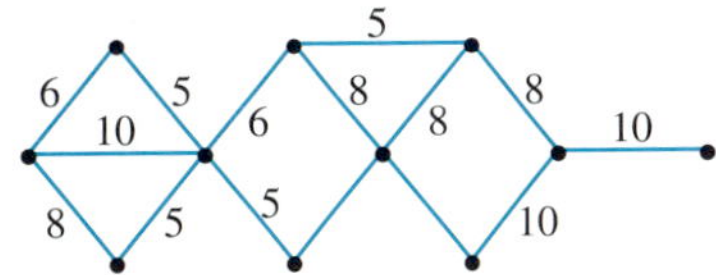

30.

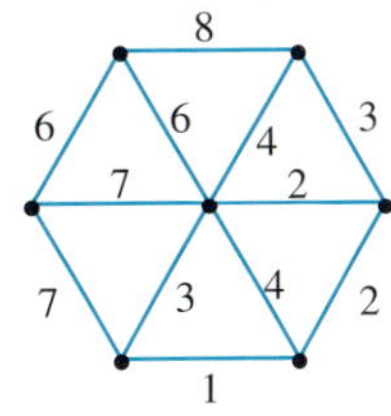

Networks, whether they are highway systems, airline routes, telephone connections, or collection routes, can be represented by graphs. In some applications, it is desirable to use the maximum weights in the graph.

In problems 31 through 34, modify Kruskal's algorithm to find a maximal spanning tree for each of the given graphs.

31.

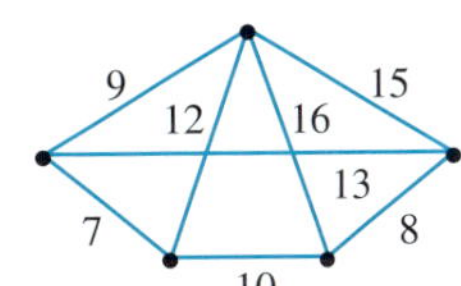

32.

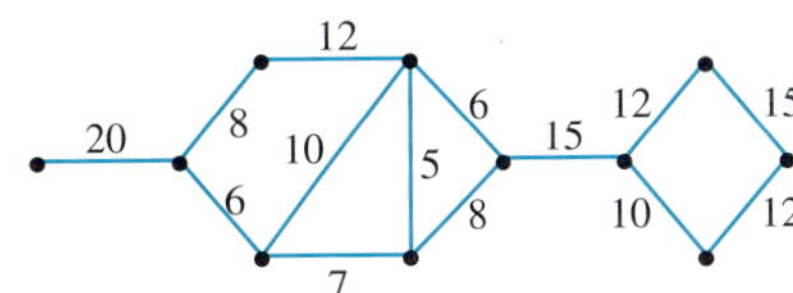

33.

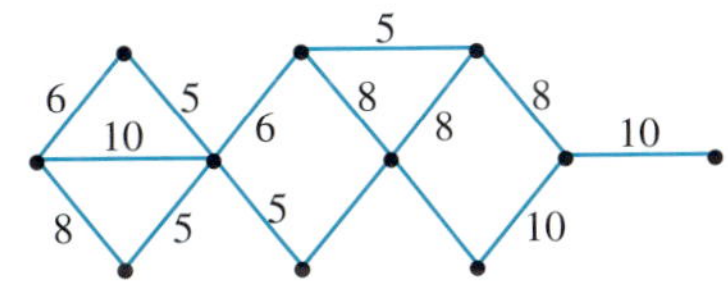

34.

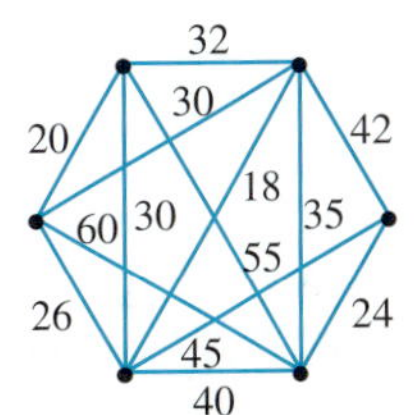

EXTENDED PROBLEMS

Visiting EDGES versus visiting VERTICES.

For some networks, such as a highway system joining a number of cities, it is important that each "edge" be visited. A highway road inspector, for example, would travel each road in the system on an inspection tour. The most efficient tour would have the inspector travel each road exactly once, although a city might be visited more than once; this would be an Euler path. If the inspector wanted to end up at the beginning vertex, then the tour would be an Euler circuit. Another user of the network may simply want to be connected to each "vertex" in the most efficient way. A delivery service or a traveling salesperson would want to visit each city in the network. In this case, the most efficient tour would visit each city only once without traveling on all the roads. This type of path is called a Hamiltonian path. It is called a Hamiltonian circuit if it begins and ends at the same vertex. Finding the minimum-cost Hamiltonian path or circuit is referred to as the Traveling Salesperson Problem (TSP).

As an example, consider a simple highway system with six towns and nine highway sections. This can be visualized in a graph as:

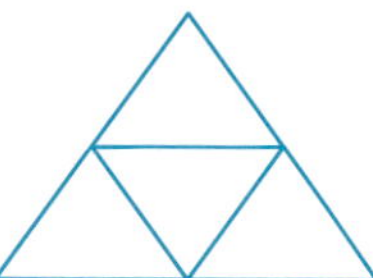

For the highway inspector and salesperson who begin and end their routes at the same town, their routes could be as follows:

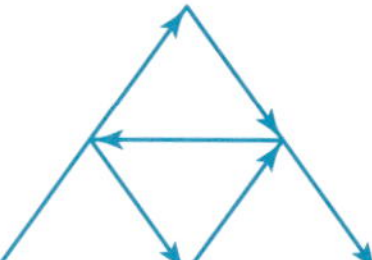

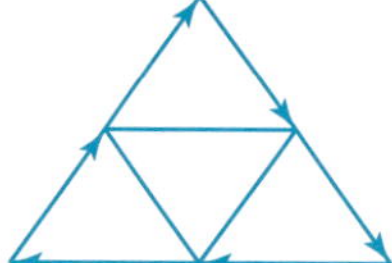

35. It is possible to have an Euler path that is not a Hamiltonian path? If it is, give an example. If it is not, give a brief explanation.

36. Is it possible to tell whether or not a graph has an Euler circuit by determining the degree of the vertices in the system (Euler's theorem)? No similar method has been found to tell if a graph has a Hamiltonian circuit, but certain types of graphs are known NOT to have Hamiltonian circuits. Explain, in your own words, why the following graph cannot have a Hamiltonian circuit.

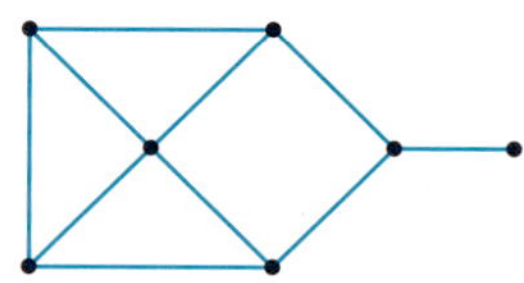

37. Explain why the path ABCDECA is an Euler circuit but not a Hamiltonian circuit.

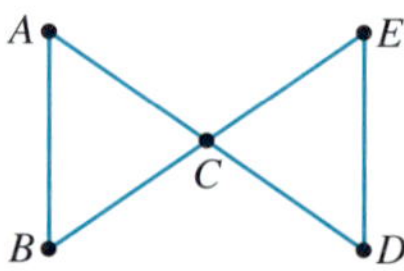

38. Show that you can add one edge to the graph in problem 20 to create a graph with a Hamiltonian circuit; identify the circuit by listing the vertices in order. Does the new graph have an Euler circuit?

39. Find a Hamiltonian circuit for each of the following graphs.

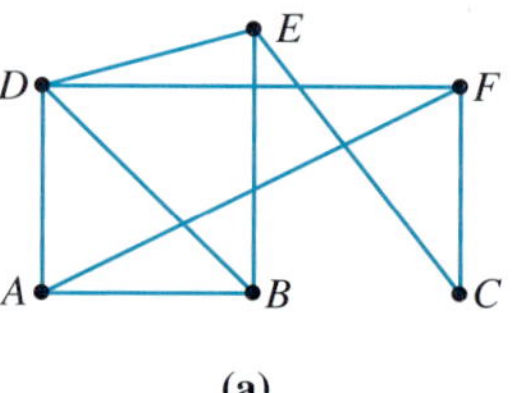

(a)

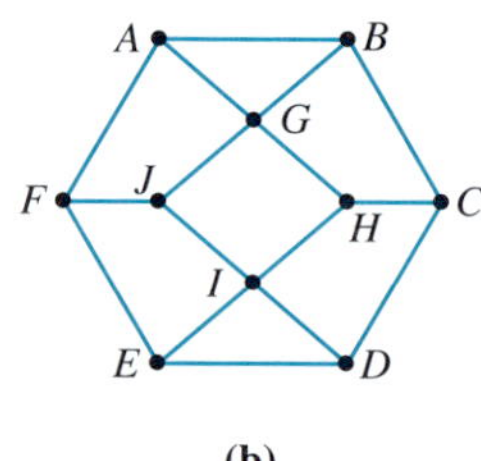

(b)

40. Explain why the following graph cannot have a Hamiltonian circuit.

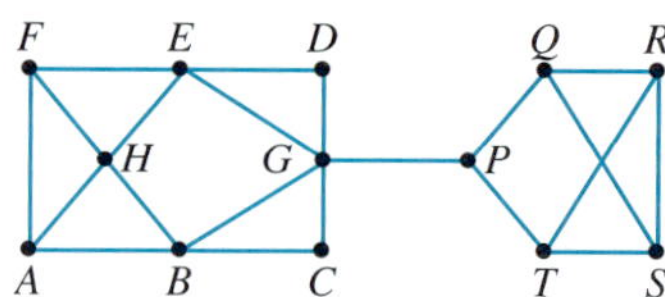

Because the number of Hamiltonian circuits may become very large, finding the minimum length (or cost) circuit may be difficult. However, there are several methods that will generally produce a value that is acceptably close. One of the methods that applies to a complete graph is the Nearest Neighbor algorithm.

With the Nearest Neighbor algorithm, you begin at the initial vertex. From this vertex, or anywhere in the route, you pick the next vertex on the route from among those that have not already been included, and pick the one that has the smallest weight, the one "nearest;" in case of a tie, select at random. When all vertices have been visited, return to the initial vertex. If you have a choice as to the initial vertex, then work out the route from each vertex and keep the best one as the solution.

In problems 41 through 43, use the Nearest Neighbor algorithm to find an approximation to the traveling salesperson problem beginning at each vertex of the given graph.

41.

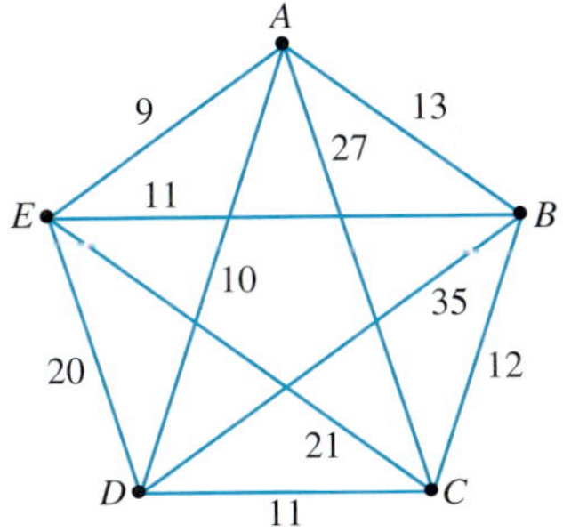

42.

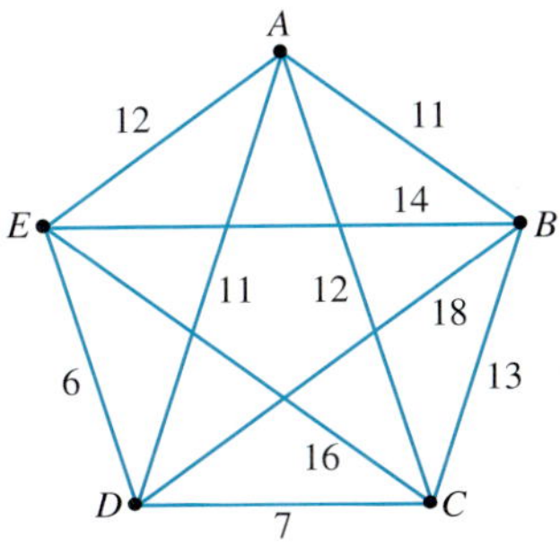

43.

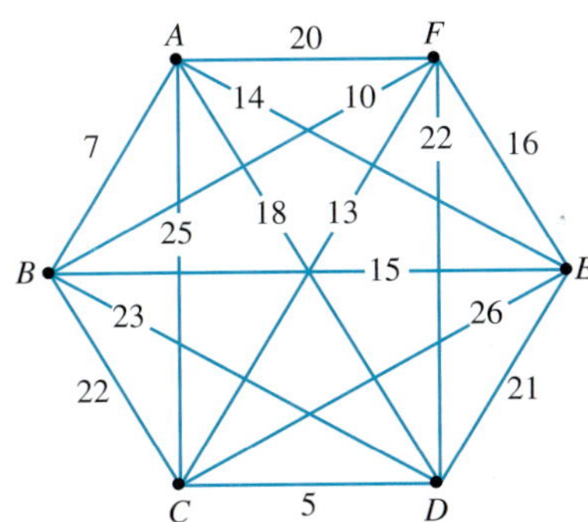

Three dimensional networks can often be represented as a "flat" graph. A graph is appropriate as a model of a network if it shows all the vertices and the connections between them. The method may require some ingenuity, but the following examples for a tetrahedron and a cube illustrate the approach.

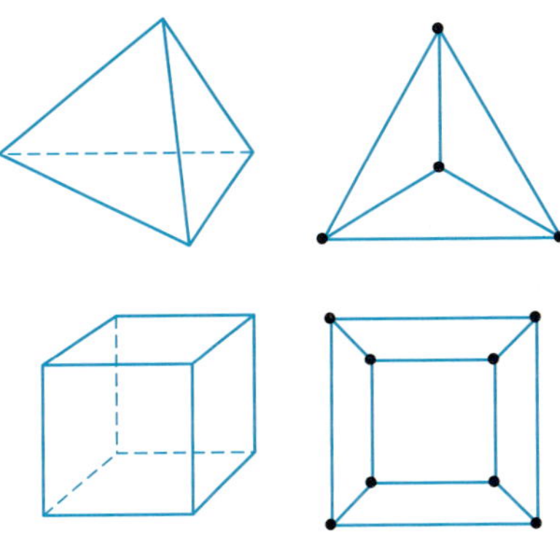

Hamiltonian circuits are named for the Irish mathematician, Sir William Rowan Hamilton. In 1859, Hamilton created and marketed a rather peculiar puzzle (much like Rubik did with his cube). The main part of the puzzle was a regular dodecahedron made of wood. The dodecahedron has regular pentagons for each of its 12 faces, with the edges of three pentagons meeting at each of its 20 vertices. In all, there are 30 edges. Each corner of Hamilton's dodecahedron was marked with the name of an important city: London, Frankfurt, New York, and so on, from around the world. The object of the game was to travel along the edges of the dodecahedron, visiting each city just once. You had to begin and end at the same city, and Hamilton made the puzzle more challenging by requiring that several cities be visited first.

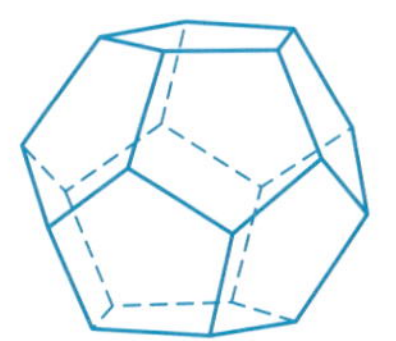

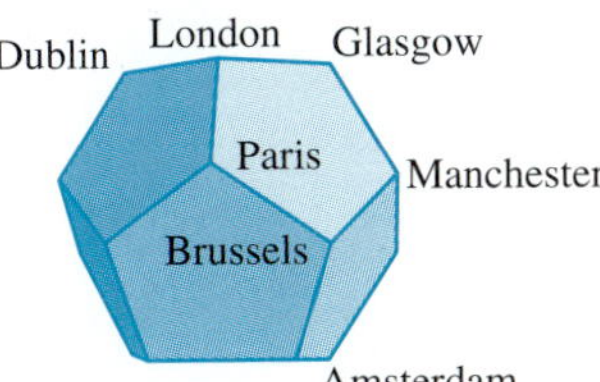

44. Explain why the following graph is an appropriate representation of the dodecahedron.

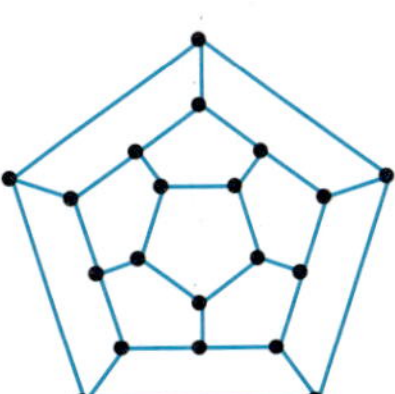

45. Beginning at any vertex, find a Hamiltonian circuit for the dodecahedron.
46. Graphs provide us with tools for organizing and visualizing data in ways that emphasize the characteristics that are important and eliminating those that are not. Hamilton's puzzle gives us a good example of this. It is difficult for most people to trace around the edges of the three-dimensional dodecahedron, but relatively easy to trace a route on the "flat" graph that represents the dodecahedron. A graph contains all the essential characteristics, by points, lines, numbers, or other symbols. As other characteristics become important in a network, graphs can be modified appropriately. In all the discussion of graphs so far, it has been assumed that you could go in either direction between vertices on an edge.
   (a) Can you think of an instance when the direction in a network might be important? How might this be indicated on a graph?
   (b) Can you think of an instance when a link in a network might not be available at all times? How might this be indicated on a graph?

Chapter Eight Problem

Gertrude Olsen and her husband attend a gathering with four other married couples. When they arrive all the other couples are there, and a certain amount of handshaking ensues. However, no one shakes hands with the same person more than once, no one shakes hands with a spouse, and (of course) no one shakes hands with himself (or herself). Gertrude asks each of the other guests (including her husband) how many people they shook hands with. Surprisingly, each of the other guests gives a different answer. How many people did Gertrude's husband shake hands with?

Solution

Strategy: Draw a Diagram and look for a pattern

The key to solving this problem is to visualize the possibilities and keep track of them. Gertrude asked the other guests about the number of handshakes they had. All had shaken hands with a different number of people. We visualize the people at the gathering with a diagram.

• • • • •

• • • • •

We represent the 10 people as points in two rows of 5 (five couples).

Since there are nine other people (and the most number of handshakes any person may have is 8) these nine people must have had (in some order) 0, 1, 2, 3, 4, 5, 6, 7, 8 handshakes. Let the first point represent the person with eight handshakes. Draw connecting edges between this point and all other points except the opposing point in the other row (representing the spouse). Someone shakes hands with all possible eight people. Someone shakes hands with no one. These two people must be married. Otherwise the person who shakes hands with eight people would not be able to shake hands with two people (the spouse and the person who shakes with no one). Draw these connections. Now someone shakes exactly seven persons' hands and someone shakes exactly one person's hand. These two people must be married since the person who shakes seven persons' hands already does not shake the hand of the person who shakes no one's hand (and is married to someone else), thus the second person whose hand is not shaken is the spouse. In a similar fashion, the person who shakes six persons' hands must be married to the person who shakes two persons' hands, and the person who shakes five persons' hands must be married to the person that shakes three persons' hands. This leaves only the person who shakes four persons' hands, who must be married to Gertrude.

Chapter Eight Review

Key Ideas and Questions

The following questions review the main ideas of this chapter. Write your answers to the questions and then refer to the pages listed by number to make certain that you have mastered these ideas.

1. How is a set of linear inequalities graphed in the plane? **365**
2. How do you use regions in the plane as feasible regions for a problem involving linear constraints? **371**
3. What method is used to find the maximum or minimum of a linear objective function over a feasible region? **378**
4. What is the test to determine whether a graph has an Euler path? **395**
   What is the test for an Euler circuit? **395**
5. How does one use Fleury's algorithm to construct Euler circuits? **396**
6. Explain why a circuit is created if an edge is added to a spanning tree. **406**
   Explain why a disconnected graph is created if any edge is removed from a spanning tree. **406**
7. Describe the method for constructing a minimal spanning tree of a weighted graph. **407**

Vocabulary/Notation

Following is a list of key vocabulary, notation, and ideas for this chapter. Mentally review each of these items, write down the meaning of each term, and use it in a sentence. Then refer to the pages listed by number, and restudy any material you are unsure of before solving the Chapter Eight Review Problems.

Section 8.1

Solution Set 365	System of Inequalities 366	Feasible Region 367
Strict 365	Linear Inequality 365	
Inclusive 365	Linear Constraints 367	

Section 8.2

Objective Function 375	Coefficient 375	Level Lines 376
Linear Function 375	Linear Programming 376	Simplex Algorithm 381

Section 8.3

Graph 387	Path 390	Disconnected 392
Vertex, Vertices 387	Circuit 390	Components of a Graph 392
Edge 387	Euler Path 390	Bridge 394
Loop 387	Euler Circuit 391	Fleury's Algorithm 396
Degree of a Vertex 389	Connected 391	

Section 8.4

Chapter Eight Review Problems

1. Graph the inequalities $x + y \geq 6$, $2x + y \geq 8$, $x \geq 0$, $y \geq 0$. Does the point (2, 3) satisfy this set of inequalities? Does the point (6, −1)?
2. Graph the inequalities $x + 2y \leq 4$, $4x + y \leq 4$, $x \geq 0$, $y \geq 0$. Does (1, 1) satisfy this set of inequalities? Does (0.5, 1)?
3. Suppose an objective function associated with the feasible region in problem 2 is $P = 2x - y$. What (x, y) pair in the feasible region maximizes P? What is this maximal P value? What (x, y) pair in the feasible region minimizes P? What is this minimal P value?
4. Suppose an objective function associated with the feasible region in problem 2 is $R = 2x + y$. What (x, y) pair in the feasible region maximizes R? What is this maximal R value? What (x, y) pair in the feasible region minimizes R? What is this minimal R value?
5. A company can make two types of peanut butter: "creamy" and "chunky". Peanuts are the only ingredients used in either type, and a pound of peanuts yields a pound of peanut butter. The first uses 40% grade A peanuts and 60% grade B peanuts. The second uses 60% grade A peanuts and 40% grade B peanuts. The company may purchase up to 2400 pounds of each grade of peanut. Graph the feasible region for this problem and label the corner points.
6. Referring to problem 5, suppose the company is able to sell all the peanut butter it can make and decides it will make as much as possible. What is the objective function? How many of each type of peanuts should be bought, how much of each type of peanut butter should be produced, and how much peanut butter is made in total?
7. Referring to problem 5, suppose the profit is \$0.60 per pound for creamy and \$0.80 per pound for chunky. If the company wants to maximize profit, what is the objective function? How many pounds of each type of peanuts should be bought, how much of each type of peanut butter should be produced, and what is the maximum profit?
8. Is this graph connected? What is the degree of each vertex? List any bridges. Does an Euler path exist? Why or why not?

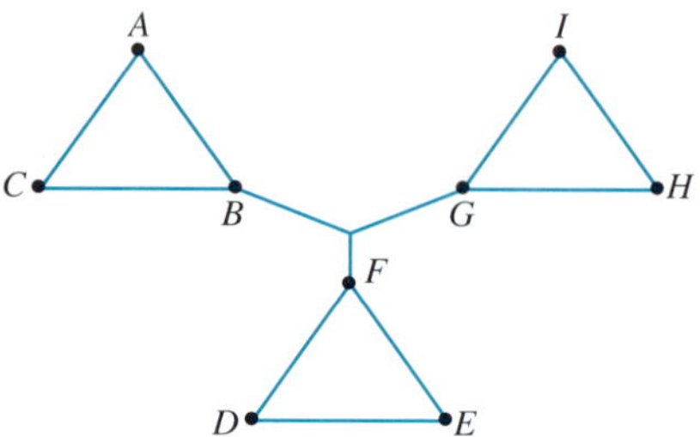

9. Is this graph connected? What is the degree of each vertex? Does an Euler path exist? If so, find one. Does an Euler circuit exist? If so, find one.

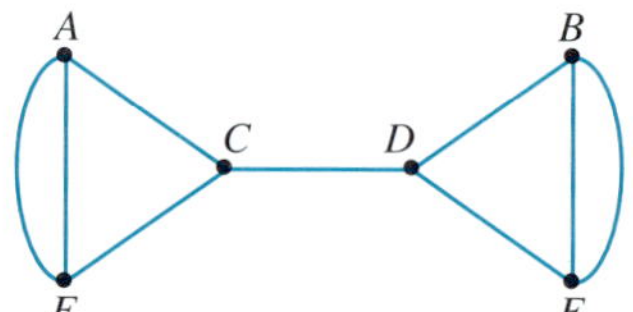

10. Consider the system of five bridges connecting the river banks and two islands. Show there is a path that traverses each bridge exactly once, but there is no path that traverses each bridge exactly once and begins and ends on an island.

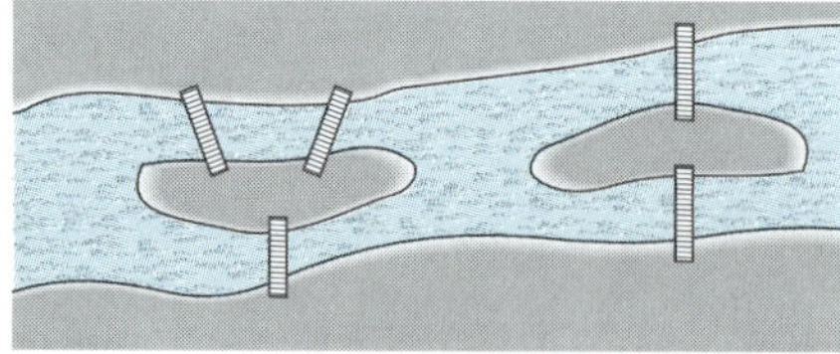

11. In problem 10, show there is a path that begins and ends at the same point and traverses each bridge exactly twice.
12. Explain why for *any* graph there is a circuit that traverses every edge exactly two times.
13. List all the spanning trees of the following graph.

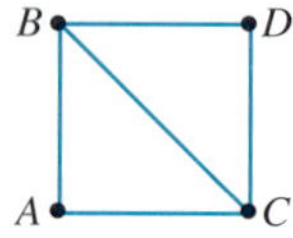

14. Find a minimal spanning tree for the following weighted graph.

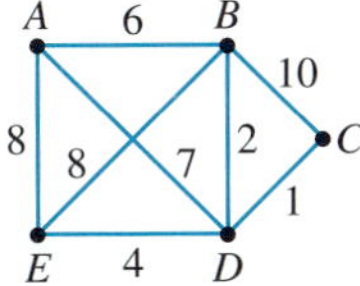

15. Four siblings live in four cities. They wish to form a telephone tree so that important family news may be shared in a cost-effective way. This will be a spanning tree. As soon as one person learns important news, the news is to be shared along the edges of the spanning tree. Choose a spanning tree to minimize the total cost of long distance charges.

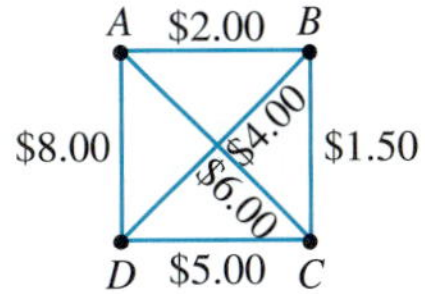

CHAPTER 9

VOTING AND APPORTIONMENT

COACH APOLOGIZES FOR TELLING PLAYERS TO LOSE

In the last game of the regular season, the coach of the Redwings realized that his team was not likely to win. With the game tied and time running out, the other team had adopted a style of play that seemed to guarantee the game would end in a tie. For the other team, a tie meant they would advance to the state playoffs. For the Redwings, a tie was worse than a loss. Because of the league rules, a loss for the Redwings would actually give them a chance to advance to the playoffs, but a tie would not. The coach instructed his team to allow the opponents to score a virtually uncontested goal, take the loss, and hope to advance to the playoffs. After the game, this was realized by officials and other coaches, and the Redwings' coach issued an apology. The commissioner later announced that new playoff rules (not subject to the flaw) would be found prior to the start of the next season.

Chapter Goals

1. Use various voting methods to determine election results.
2. Show how different voting methods satisfy, or fail to satisfy, the fairness properties.
3. Use different apportionment methods to determine fair shares.
4. Identify the various flaws and paradoxes of apportionment methods.

Going into the final game between the Doves and the Redwings, the Doves were a half-point behind the league leading Pigeons, and the Redwings were another half-point behind the Doves. A win is worth one point, and a tie is worth half a point. The league sends two teams to the state playoffs: The league's first place team and the winner of a wildcard game between the second and third place teams. *But* in the event of a two-way tie for first place in the league, those two teams go to state and the third place team is out.

The following table shows the point totals of the top three teams and the result of each possible outcome of the game. Teams guaranteed to go to the state playoffs are marked with an asterisk(*).

	Before the Final Game	Redwings Win	Redwings Lose	Tie Game
Pigeons	21	*21	21	*21
Doves	20.5	20.5	*21.5	*21
Redwings	20	*21	20	20.5

We see that a tie game puts the Redwings out of the playoffs. Winning is best, but a tie is worse than a loss.

The playoff rules were intended to send the two best teams to the state playoffs, but something unfair has happened: Finding the best team is similar to choosing the best candidate in an election. It turns out that if there are more than two candidates, then whatever election rules are used, something unfair can happen.

In this chapter, you will also study apportionment, in which one tries to determine "fair shares." As with voting, you will see that every apportionment system has flaws.

THE HUMAN SIDE OF MATHEMATICS

The Marquis de Condorcet (1743–1794), an aristocrat, was one of the leading mathematicians, sociologists, economists, and political thinkers of France at the time of the American and French Revolutions. He has been called the French Thomas Jefferson because his influence on scientific and political thought was so pervasive. His early intellectual interests were in the applications of integral calculus and probability to science. Later, he turned his attention to the solution of social problems. Condorcet was a member of a liberal group of thinkers known as the encyclopédistes, and his ideas were influential in the events leading to the French Revolution. He believed that humanity was evolving historically on a course toward political and economic enlightenment. He was certain that science could be used for the benefit of the people and that principles of fair government could be discovered mathematically. Thus he analyzed voting methods and soon discovered the dismaying fact that sometimes there is no clear way to choose a winner of an election. Condorcet showed that it was possible to have three candidates, A, B, and C, for whom the electorate would prefer A to B and B to C, but then prefer C to A. One method of choosing the winner of an election is often called the Condorcet method (referred to as the pairwise comparison method in this text). Unfortunately, as with the candidates A, B, and C, this method does not always produce a winner, and it can produce some unexpected results.

The Marquis de Condorcet

At the close of the French Revolution, the Jacobins (a political faction of governmental deputies at Versailles) seized power. Their leader, Robespierre, became virtual dictator of France, establishing the Terror and eliminating his rivals. Condorcet was arrested for his political views and for being a member of the aristocracy. He died in prison soon afterwards, many believed by suicide, some believed by murder.

Kenneth J. Arrow (1921–) is an economist who has spent his life trying to understand how corporations, organizations, and societies make decisions; how such decisions may be made in the best possible way; and how to maintain accountability for these decisions. Arrow's initial interests were in the theory of corporate decision making, particularly where stockholders were concerned. He next worked with the RAND Corporation (a government–industry think tank) on the application of mathematical concepts, in particular game theory, to rational decision making in military and diplomatic affairs. In his work, he constructed a set of properties that a fair and reasonable decision making, or voting, system should have. One such property, for example, was that an alternative that was supported by a majority should be selected. He also considered the property of transitivity: that if A was preferred to B, and B was preferred to C, then it was logical that A would be preferred to C. Like Condorcet before him, he found that this property was easily violated in actual practice. He was able to show that no system of voting can satisfy all the properties he could identify as being fair and reasonable. In other words, any system of voting will seem to be unfair with the right set of circumstances. This is called the Arrow Impossibility Theorem, which he proved in 1951. Although Arrow proved that finding a perfect voting system is not possible, the goal of his work remained in finding the best possible systems for guiding the decision making of governments and corporations. In 1972, Arrow was awarded the Nobel Prize in economics for his contributions to the theory of general economic equilibrium.

Kenneth Arrow

9.1 VOTING SYSTEMS

INITIAL PROBLEM

The city council must select among three locations for building the new sewage treatment plant. Before the election, councilor Jones is able to talk individually to each of the other councilors and finds out that site A, which he prefers, is preferred by a majority of councilors over site B, and is also preferred by a majority of councilors over site C. That is, a majority prefer A over B and A over C. However, when the actual vote is taken, site B is selected. Jones feels betrayed, and believes that either some of the councilors lied to him or changed their minds. Is Jones necessarily correct?

The United States is a constitutional democracy. In the late eighteenth century, forming such a government was an extraordinary undertaking. The challenges were enormous. (1) A constitution needed to be written with enough detail to keep the new nation on the proper course, but with enough flexibility to allow for the tremendous changes that occurred as the nation grew. (2) Once a republic was established, it was necessary to prove that such a government could function. From the outset, the ability of the Federal government to enforce its laws was very much in doubt. (3) The last major challenge was to prove that the nation could hold together despite regional differences. Ultimately the last question was settled by the Civil War.

TIDBIT

While today it is only a footnote, the first instance of the Federal government successfully opposing a challenge to the administering of its laws occurred in the 1790s with the suppression of the Whiskey Rebellion, in western Pennsylvania, against the taxes levied on the indigenous rye whiskey industry. Pennsylvania distilled rye whiskey is still available.

As citizens of a constitutional democracy, we consider our right to vote sacred. We want a voice in the decisions that affect us, but we seldom realize that the method by which we vote is sometimes as decisive for the outcome as the actual votes. The effect is subtle. Faced with the usual contest between the two major parties, we make our choice, and the candidate with the most votes wins. It is clear, clean, and simple. But as soon as there are more than two choices, the issue becomes muddied. In fact, there is no method that is always fair when a choice must be made among three or more alternatives. This is not obvious and was not proved until 1951. This fact is now known as the Arrow Impossibility Theorem.

In this section we will describe the most popular voting systems and how they are implemented. In the next section we will examine the weaknesses of each. The following methods are discussed:

1. plurality method
2. Borda count method
3. plurality with elimination method
4. pairwise comparison method

PLURALITY METHOD

If there are three or more candidates vying in an election, then it often happens that no one candidate receives a majority of the votes. When this happens, there must be provisions for some other way of deciding the election. The plurality method is one way of settling who is the winner in such an election where no candidate has a majority. When the plurality method is used to choose among several candidates, each voter simply votes for her preferred choice, and the candidate receiving the most votes is selected.

DEFINITION

The Plurality Method

Vote for one candidate.
The candidate receving the most votes is selected.

The plurality method offers several advantages. For one, it requires a simple choice by the voter: vote for your favorite. No complicated ranking decisions are needed. A second advantage is that it is easy to determine the winner of the election after the votes are cast.

EXAMPLE 9.1 Four persons are running for student body president: Aaron, Bonnie, Charles, and Dion. They receive the following vote totals.

Aaron 2359 Bonnie 2457 Charles 2554 Dion 2288

Under the plurality method who is elected?

HISTORY

The Borda count method was proposed by Jean-Charles de Borda (1733–1799), a French cavalry officer and naval captain.

SOLUTION The person having the most votes, Charles with 2554, is elected student body president. However, he received less than 27% of the total 9658 votes cast. ◆

BORDA COUNT METHOD

When the Borda count method is applied to choose among several candidates, each voter must rank all the candidates. A voter's last choice is given one point, the next-to-last choice is given two points, and so on until the voter's first choice is given as many points as there are candidates. The points for each candidate are totaled, and the one with the most points wins.

The Borda Count Method

Voters rank the m candidates.
A voter's mth choice gets one point, $(m-1)$st choice gets two points, . . . , second choice gets $m-1$ points, and first choice gets m points. The candidate receiving the most points is selected.

The main advantage of the Borda count method is that it lets the voters provide more information than the plurality method. Variants of the Borda count method are widely used. For example, the recipient of the Heisman trophy, given each year to the best college football player in the country, is chosen using a variant of the Borda count method. The voters (about 900 sportswriters and former Heisman trophy winners) submit ballots listing their first, second, and third choices out of the thousands of potential candidates playing intercollegiate football each year. Players receive three points for each first place vote, two points for each second place vote, and one point for each third place vote.

EXAMPLE 9.2 Four persons are running for student body president: Aaron, Bonnie, Charles, and Dion. Voters are asked to rank the candidates first through fourth. They receive the following vote totals.

	First Place Votes	Second Place Votes	Third Place Votes	Fourth Place Votes
Aaron	2359	1368	2786	3145
Bonnie	2457	3499	2474	1228
Charles	2554	2367	1734	3003
Dion	2288	2424	2664	2282

Under the Borda count method who is elected?

SOLUTION We need to convert the votes to points as in the next table.

	Points from First Place Votes	Points from Second Place Votes	Points from Third Place Votes	Points from Fourth Place Votes
Aaron	$2359 \times 4 = 9436$	$1368 \times 3 = 4104$	$2786 \times 2 = 5572$	3145
Bonnie	$2457 \times 4 = 9828$	$3499 \times 3 = 10{,}497$	$2474 \times 2 = 4948$	1228
Charles	$2554 \times 4 = 10{,}216$	$2367 \times 3 = 7101$	$1734 \times 2 = 3468$	3003
Dion	$2288 \times 4 = 9152$	$2424 \times 3 = 7272$	$2664 \times 2 = 5328$	2282

Then we need to total the points for each candidate.

Aaron $9436 + 4104 + 5572 + 3145 = 22{,}257$
Bonnie $9828 + 10{,}497 + 4948 + 1228 = 26{,}501$
Charles $10{,}216 + 7101 + 3468 + 3003 = 23{,}788$
Dion $9152 + 7272 + 5328 + 2282 = 24{,}034$

Since Bonnie has received the largest point total, she is elected student body president using the Borda count method. ◆

Note that the votes received by each candidate in Example 9.1 are the same as the first place votes received in Example 9.2. Although Charles had more first place votes, Bonnie is elected under the Borda count method because her point total is higher due to the fact that she is the second choice of many voters.

PLURALITY WITH ELIMINATION METHOD

When the plurality with elimination method is used for choosing among several candidates, a series of votes may be required. In the first round:

- Each voter votes for his or her preferred candidate.
- If one candidate wins a majority of votes, then that candidate is selected.
- If no candidate attains a majority, then the candidate receiving the fewest votes is eliminated. If there are ties for this distinction, all are dropped.

Another round of voting is performed under the same rules, so again either one candidate attains a majority or one (or more) candidate is eliminated. Eventually a decision is reached.

The Plurality with Elimination Method

Vote for one candidate.
If a candidate receives a majority of votes, that candidate is selected. If no candidate receives a majority, eliminate the candidate(s) receiving the fewest votes, and do another round of voting.

Plurality with elimination methods sometimes use other rules to decide which candidates are in the second (or later) round of voting; for example, the top two may be the only candidates left for the second round. (If only two candidates are left for the second round of voting, then the second round is often called a **run-off election.**) Including all its variations, the plurality with elimination method is probably the most widely used voting method; for example, the President of France is now chosen using a plurality with elimination method.

Many rounds of voting may be required to carry out the plurality with elimination method we described in the definition because our rules only eliminate one candidate after each round of voting. To consider what happens in the later rounds of voting, we will assume each voter has a ranking of all the candidates, and in each election the voter casts her vote for her highest ranking candidate still in the election. For example, if (before the first election) one particular voter were to rank Garcia, Johnson, and Smith first, second, and third, and then Garcia was eliminated, she would vote for Johnson in the second round. This is a simplifying assumption, since real people do not always behave in such a predictable way.

In the next example, we require the voters to rank all the candidates. These are displayed in what we call a **preference table.** In the first vote, each voter casts her vote for her first choice. If a candidate must be eliminated, the first place votes of the eliminated candidate go instead to the second place choices of those voters. This simulates running another ballot. In an actual election, voters may dislike ranking all the candidates because this forces them to make choices that may turn out to be unnecessary.

EXAMPLE 9.3 Four persons are running for department chairperson: Alice, Bob, Carlos, and Donna. The seventeen voters are asked to rank the candidates first through fourth. Candidates receive the following votes, with each vertical column in the preference table representing one voter's ballot.

Alice	1	1	4	3	4	4	1	2	2	3	3	1	4	3	2	1	1
Bob	4	3	3	4	1	3	4	1	1	4	4	2	1	4	3	3	2
Carlos	3	2	2	1	2	2	3	4	4	1	1	3	2	1	4	2	3
Donna	2	4	1	2	3	1	2	3	3	2	2	4	3	2	1	4	4

Under the plurality with elimination method, who is elected?

SOLUTION First it is necessary to count the first place votes received by each candidate. To find Alice's first place votes, we count the 1's in the first row of the previous table. Similar counts are done for the other candidates.

First Round First Place Votes
Alice 6 Bob 4 Carlos 4 Donna 3

No candidate received a majority, so Donna, the candidate with the fewest first place votes, is eliminated. Consider the voter whose ballot is represented by the first column in the table:

Alice	1
Bob	4
Carlos	3
Donna	2

Donna was that voter's second choice, but Donna has been eliminated. This voter's third choice of Carlos becomes the voter's second choice, and his fourth choice is now his third choice. The preference table must be altered to account for these changes and the similar changes for all voters except those who had Donna as their last choice. The new preference table is the following.

Alice	1	1	3	2	3	3	1	2	2	2	2	1	3	2	1	1	1
Bob	3	3	2	3	1	2	3	1	1	3	3	2	1	3	2	3	2
Carlos	2	2	1	1	2	1	2	3	3	1	1	3	2	1	3	2	3

Second Round First Place Votes:
Alice 7 Bob 4 Carlos 6

Again, no candidate received a majority. This time Bob is eliminated. The choices of the voters whose first or second choice was Bob are retabulated. The new preference table follows.

Alice	1	1	2	2	2	2	1	1	1	2	2	1	2	2	1	1	1
Carlos	2	2	1	1	1	1	2	2	2	1	1	2	1	1	2	2	2

Third Round First Place Votes:
Alice 9 Carlos 8

Alice is elected chairperson. ◆

PAIRWISE COMPARISON METHOD

When a choice among several candidates is made by the pairwise comparison method, each voter must make a choice between every possible pair of candidates. For example, if the candidates are Franklin, Goldstein, and Hernandez, we could ask the voters to vote three times in contests of Franklin versus Goldstein, then Franklin versus Hernandez, and finally Goldstein versus Hernandez. Instead of three separate ballots, we will require the voters to rank all the candidates. So if one particular voter ranks the three candidates Hernandez, Goldstein, and Franklin as first, second, and third choices, then we assume the voter would vote for Hernandez in the Goldstein versus Hernandez contest, would vote for Hernandez in the Franklin versus Hernandez contest, and would vote for Goldstein in the Franklin versus Goldstein contest.

Using the rankings, we go through every possible pairing of candidates and determine which of the two is preferred based on the rankings. Each candidate will be assigned points based on how well they do with respect to the other candidates: one point is awarded to the candidate preferred by the greatest number of voters, and $\frac{1}{2}$ point is awarded to each candidate if they are preferred by the same number of voters.

The Pairwise Comparison Method

Voters rank all the candidates.
For each pair of candidates X and Y, determine how many voters prefer X to Y and vice versa.
If X is preferred to Y, then X receives 1 point.
If Y is preferred to X, then Y receives 1 point.
If the candidates tie, then each receives $\frac{1}{2}$ point.
The candidate receiving the most points is selected.

EXAMPLE 9.4 After Donna becomes tired of the incessant elections being held and withdraws, three persons are left in the running for department chairperson: Alice, Bob and Carlos. The seventeen voters are asked to rank the candidates first through third. When Donna withdraws from the election, the preference schedules are modified the same as they were in Example 9.3 when Donna eliminated at the first stage of the plurality with elimination method.

Alice	1	1	3	2	3	3	1	2	2	2	2	1	3	2	1	1	1
Bob	3	3	2	3	1	2	3	1	1	3	3	2	1	3	2	3	2
Carlos	2	2	1	1	2	1	2	3	3	1	1	3	2	1	3	2	3

Under the pairwise comparison method who is elected?

SOLUTION There are three pairs to consider

Alice vs. Bob   Alice vs. Carlos   Bob vs. Carlos

For each pair we consider just the part of the above preference table that includes the two candidates in question.

Alice vs. Bob

Alice	1	1	3	2	3	3	1	2	2	2	2	1	3	2	1	1	1
Bob	3	3	2	3	1	2	3	1	1	3	3	2	1	3	2	3	2

We see that Alice is preferred to Bob by a margin of 11 to 6. Alice receives one point.

Alice vs. Carlos

Alice	1	1	3	2	3	3	1	2	2	2	2	1	3	2	1	1	1
Carlos	2	2	1	1	2	1	2	3	3	1	1	3	2	1	3	2	3

Alice is preferred to Carlos 9 to 8. Alice receives another point.

Bob vs. Carlos

Bob	3	3	2	3	1	2	3	1	1	3	3	2	1	3	2	3	2
Carlos	2	2	1	1	2	1	2	3	3	1	1	3	2	1	3	2	3

Carlos is preferred to Bob 10 to 7. Carlos receives one point.

Since Alice has received two points, Bob has received no points, and Carlos has received one point, Alice is elected. ◆

HISTORY

During the 1850s, one of the existing parties in the two party system, the Whigs, fell apart and was replaced by the then new Republican party. This was a consequence of the bitter sectional conflict over the extension of slavery into the territories. The same sectional conflict so weakened the Democratic party that a Republican president, Lincoln, was elected in only the party's second presidential campaign. The election of Lincoln was so distasteful to the South that states seceded even before he assumed office.

TIE BREAKING

The four voting systems we have examined can produce different winners even when the voter preferences are the same. Any one of them can also produce a tie between two or more of the alternatives. In some cases the voter preferences are perfectly balanced, and a tie is just in the nature of things, for example when there are two alternatives, an even number of voters, and exactly the same number of supporters for each alternative. The only way to break a tie caused by perfectly balanced support is to either make an arbitrary choice (such as by flipping a coin) or bring in another voter. For example, the Vice President of the United States is President of the U. S. Senate, but only has a vote in the Senate when the rest of the Senate is deadlocked.

Sometimes a tie can be broken in a more rational fashion than by a coin flip. For example, if the Borda count method is used, then it may be possible to break a tie based on which candidate obtained the most first place rankings. Choosing different tie breaking methods can result in different winners, so the proper thing to do is decide the tie breaking method in advance.

INITIAL PROBLEM SOLUTION

The city council must select among three locations for building the new sewage treatment plan. Before the election, councilor Jones is able to talk individually to each of the city councilors and finds out that site A, which he prefers, is preferred by a majority of councilors over site B, and is also preferred by a majority of councilors over site C. That is, a majority prefer A over B and A over C. However, when the actual vote is taken, site B is selected. Jones feels betrayed and believes that either some of the councilors lied to him or changed their minds. Is Jones necessarily correct?

SOLUTION Jones is wrong, although he may be hard to convince. The method of voting used by the council may be the reason for the apparent discrepancy. The city council uses plurality with elimination when there is no majority. Consider the following preferences for the eleven council members.

3 favor A over B over C
4 favor B over A over C
4 favor C over A over B

Site A is eliminated in the first round, and site B is the winner over site C by a vote of 7 to 4.

PROBLEM SET 9.1

1. Four candidates are running for mayor. They receive the following vote totals.

Abrahms 2067 Morrita 2987
Bache 1875 Steiner 2765

Under the plurality method, who is elected?

2. There were three candidates for governor. The vote totals they received were:

Froelich 385,542 Morgan 212,473 Drumm 326,764

Under the plurality method, who is elected?

3. The nine-member city council is choosing from three possible locations for a new fire station. They rank each of the alternatives in their order of preference. The results are summarized below.

Councilor	A	B	C	D	E	F	G	H	I
9th Street	2	2	3	1	2	1	2	2	2
Davis Ave	1	3	2	3	1	3	1	3	1
Beca Blvd	3	1	1	2	3	2	3	1	3

Under the plurality method, which location is selected?

4. The members of the football team select their team captain from three of the seniors: Jorgensen, Petrini, and Rameriz. Each player ranks his choices from first to third, and 88 players turn in completed ballots. There are six ways to rank the choices, and the following table shows the number of players who ranked the candidates in a given order.

No. of players	11	18	20	14	15	10
Jorgensen	1	1	2	2	3	3
Petrini	2	3	1	3	1	2
Rameriz	3	2	3	1	2	1

Under the plurality method, who is selected as team captain?

5. The planning commission is going to select a consultant for a management study. Each of the seven members on the commission ranks the consultants from first to third, with the following results:

Commissioner Consultant	A	B	C	D	E	F	G
Gorman	3	2	2	1	3	3	3
Finster	2	1	1	3	2	2	2
Yamada	1	3	3	2	1	1	1

Under the Borda count method, which consultant is selected?

6. The senior class is selecting a president and vice president from among four candidates. Voters are asked to rank the candidates first through fourth. The ones receiving the two highest point totals will be elected president and vice president, respectively.

	First	Second	Third	Fourth
Aaron	135	223	127	105
Denise	185	164	139	102
Garth	106	168	176	140
Kermit	164	35	148	243

Under the Borda count method, who is selected as president and vice president?

7. In problem 3, what location is selected if the Borda count method is used?

8. In problem 4, who is selected as team captain if the Borda count method is used?

9. In problem 5, which consultant is selected if the plurality method is used?

10. In problem 6, who is selected as president and vice president if the one with the most first place votes is president and the one with the next-most first place votes is vice president?

11. In problem 3, the nine-member city council is choosing from three possible locations for a new fire station. They rank each of the alternatives in their order of preference. The results are summarized below.

Councilor	A	B	C	D	E	F	G	H	I
9th Street	2	2	3	1	2	1	2	2	2
Davis Ave	1	3	2	3	1	3	1	3	1
Beca Blvd	3	1	1	2	3	2	3	1	3

Under the plurality with elimination method, which location is selected?

12. In problem 5, the planning commission is going to select a consultant for a management study. Each of the seven members on the commission ranks the consultants from first to third, with the following results:

Commissioner Consultant	A	B	C	D	E	F	G
Gorman	3	2	2	1	3	3	3
Finster	2	1	1	3	2	2	2
Yamada	1	3	3	2	1	1	1

Under the plurality with elimination method, which consultant is selected?

In problems 13 through 16, each column represents a preference table for a group of voters, with the first choice for all voters listed at the top.

13. A new president is being elected by the ceramics guild, and three members, Ann, Eno, and Pat, have been nominated. The members of the guild are asked to rank the candidates from first to third. The 48 ballots are grouped as follows, with the number at the top indicating how many members voted this way:

12	8	6	10	8	4
Ann	Ann	Eno	Eno	Pat	Pat
Eno	Pat	Ann	Pat	Ann	Eno
Pat	Eno	Pat	Ann	Eno	Ann

Who is elected president using the plurality with elimination method?

14. The board of directors of a large company is choosing a site for a new branch office in the southwest. The cities being considered are Albuquerque (A), Phoenix (P), Santa Fe (S), and Tucson (T). The members of the board are asked to rank the cities from first to fourth. The results are summarized below. Although there are 24 ways to rank the cities, only the following are used, with the number at the top indicating how many board members voted this way.

4	3	3	3	3	2	2	1	1	1	1	1
P	A	A	S	T	S	T	A	P	P	S	S
T	S	T	A	P	A	A	P	T	T	P	T
A	P	S	P	A	T	P	S	A	S	A	A
S	T	P	T	S	P	S	T	S	A	T	P

Which city is chosen using the plurality with elimination method?

15. The Jimenez family is deciding on which national park to visit next summer; the choices are: Yellowstone (Y), the Grand Canyon (G), or Mount St. Helens (M). The family's preferences are:

dad	mom	boy 1	boy 2	boy 3	girl 1	girl 2
Y	G	M	G	Y	Y	M
M	M	G	Y	G	M	Y
G	Y	Y	M	M	G	G

Using the pairwise comparison method, which park is selected for next summer?

16. After the last performance, the cast and crew of the school play are going out for dinner. The choices of restaurants are Chinese (C), Italian (I), or Mexican (M). The choices are summarized below, with the number above indicating the number who had this order of preferences.

5	4	4	5	3	6
C	C	I	I	M	M
I	M	C	M	C	I
M	I	M	C	I	C

Using the pairwise comparison method, which restaurant is selected?

17. In problem 14, which city is selected if the pairwise comparison method is used?

18. If there are four candidates in an election,

(a) how many different ways can they be arranged in order of preference?

(b) how many pairings are needed using the pairwise comparison method?

Use the following information for problems 19 and 20. There are three candidates running for president of the senior class: Peter, Carmen, and Shawna. The voters all mark their ballots to indicate first, second, and third choice among the candidates (only ballots with all three marked are valid). The results are summarized as follows:

Candidate	First Place Votes	Second Place Votes	Third Place Votes
Peter	33	68	34
Carmen	53	28	54
Shawna	49	39	47

19. Who is elected president using the Borda count method?

20. Who is elected president using the plurality method?

Use the following information for problems 21 and 22. Three candidates—Able, Boastful, and Charming—are running for the office of mayor of Tinytown. The voters all mark their ballots to indicate first, second, and third choice among the candidates (only ballots with all three choices marked are valid). The results are summarized as follows:

Candidate	First Place Votes	Second Place Votes	Third Place Votes
Able	33	33	34
Boastful	39	19	42
Charming	28	48	24

21. Who wins the new mayor's spot using the Borda count method?

22. Who is the new mayor of Tinytown using the plurality method?

A modified Borda count method is used in many types of contests other than elections, such as athletic events. A fairly common practice is to emphasize first place finishes, or de-emphasize other finishes, by changing the way in which points are awarded (such as 5 points for first place, 3 points for second place, 2 points for third place, and 1 point for fourth place).

23. Who is elected senior class president (Problem 19) if 4 points are given for first place votes, 2 points for second place, and 1 point for third place using the Borda count method?

24. Who is elected mayor of Tinytown (Problem 21) if 4 points are given for first place votes, 2 points for second place, and 1 point for third place using the Borda count method?

Four teams—the Raiders (R), the Spartans (S), the Titans (T), and the Vikings (V)—are competing for the gymnastics team championship in four events. The results are:

Place	Events			
	Beam	Horse	Bars	Floor
First	R	S	V	S
Second	S	T	T	V
Third	T	V	R	R
Fourth	V	R	V	V

(Note: each team was allowed two competitors in each event.)

25. How do the teams rank in this competition if the finishing positions are scored 4, 3, 2, and 1, respectively, using the Borda count method?

26. How do the teams rank in this competition if the finishing positions are scored 5, 3, 1, and 0, respectively, using a modified Borda count method?

The following information is used in problems 27 through 33. The Board of Commissioners for Baker County must pick a site for a new jail. Three locations have been determined to be suitable, and each commissioner ranks her preferences in order. The preference schedules are:

Voter	1	2	3	4	5	6	7	8	9	10	11	12	13	14	15
	A	A	B	C	B	A	C	B	B	A	C	A	A	B	C
	B	C	C	B	A	C	B	C	C	B	B	C	C	C	B
	C	B	A	A	C	B	A	A	A	C	A	B	B	A	A

Which site is selected by each of the following methods?

27. plurality

28. plurality with elimination

29. Borda count (3, 2, 1)

30. modified Borda count (4, 2, 1)

31. pairwise comparison method

Just as the Borda count method can be modified, there are several ways to eliminate candidates for a run-off election. What jail site would the county commissioners select if they used elimination in the following ways?

32. Plurality with elimination in which the alternative with the most *last* place votes is eliminated.

33. Plurality with elimination having a: run-off election between the alternatives that rank second and third in terms of first place votes, and then the winner of the run-off against the choice that originally has the most first place votes.

The following information is used for problems 34 through 41. Four seniors on the baseball team, Joe Aaron (A), Billy Bonds (B), Mike Griffey (G), and Tim Ruth (R), are being considered for team captain. The 21 other members of the team are asked to rank them in order of preference from first to fourth. The ballots are grouped as follows, with the number at the top of each ballot indicating the number of players voting this way.

4	3	2	2	2	2	2	1	1	1	1
A	B	R	G	G	A	A	B	G	G	R
B	G	B	A	B	B	R	R	R	R	G
G	R	A	R	R	G	G	A	B	B	B
R	A	G	B	A	G	B	A	B	A	A

Who is selected as team captain using the following methods?

34. plurality

35. plurality with elimination

36. Borda count (4, 3, 2, 1)

37. modified Borda count (5, 3, 1, 0)

38. pairwise comparison method

39. Plurality with elimination: run-off election between the alternatives that rank second and third in terms of first place votes, and then the winner of the run-off against the choice that originally has the most first place votes

40. Plurality with elimination: first, using a run-off election between the two candidates with the lowest number of first place votes; then using the winner of that contest and the candidate with the second highest number of first place votes; and finally, using the winner of that contest and the candidate with the most original first place votes. If the number of first place votes is a tie, the order is determined by the number of second place votes.

41. Who would be selected as baseball team captain if plurality with elimination is used, and the candidate with the most *last* place votes is eliminated at each step? This method eliminates candidates who are *least* preferred, which may be an advantage to a person who may be a consensus-builder and get cooperation.

42. Who would be selected as department chairperson in Example 9.3 if plurality with elimination is used, and the candidate with the most *last* place votes is eliminated at each step?

EXTENDED PROBLEMS

In some elections, such as for Commissions or Boards of Directors, there may be several vacancies and voters may select as many candidates as there are vacancies, or vote for fewer candidates (even "none").

43. There are two vacancies on the Executive Committee for the United Way Board of Directors. Four members, Malcolm Adams (A), Jennifer Barrons (B), Jesse Calderone (C), and Angela Darden (D), have indicated their willingness to serve. A ballot is taken, and each member of the Board is directed to vote for a maximum of two candidates. The 25 ballots are marked as follows:

A	x		x		x	x	x			x		x	
B	x	x		x		x		x	x				x
C		x						x		x		x	
D			x		x		x		x		x		x

A		x		x		x		x		x	x	x	
B			x				x	x			x		x
C	x			x	x			x	x	x			x
D	x	x	x		x	x		x	x			x	

Which two candidates win seats on the Executive Committee if the two with the highest vote counts win?

Another method of voting that is gaining favor in elections in which more than one candidate can win is called approval voting. In approval voting, each voter may give one vote to each candidate he likes, with no limit set on the number of candidates a voter can give votes to. Voters don't have to pick a favorite, only those they would be willing to see elected. If they don't approve of a candidate, they can withhold their votes from that individual. The winners are the ones that have the largest number of approval votes.

44. Suppose the United Way Board of Directors had used approval voting to fill the two vacancies on the Executive Committee, and the 25 ballots had been marked as follows:

A	x		x		x	x	x		x	x		x	
B	x	x	x	x	x	x		x	x				x
C		x			x			x		x		x	
D	x		x		x		x		x		x		x

A		x		x		x			x	x	x	x
B	x		x			x	x	x		x		x
C	x			x	x			x	x			x
D	x	x	x		x	x		x			x	

Which two candidates win seats on the Executive Committee?

When there are several candidates or alternatives in an election and there is no clear winner with a majority, run-offs are often used to determine the winner. Sometimes the order of the run-offs is determined by the number of votes received, while at other times it may be determined by some other method (even as simple as drawing straws).

45. Ten members of the city council are voting on three budget options, referred to as A, B, and C. The preferences of the council members are summarized as follows:

Four members prefer A over B, and B over C
Three members prefer B over C, and C over A
Three members prefer C over A, and A over B

(a) Which option is selected if the council first chooses between A and B, and then chooses between the winner and C? (Note: you should assume that if voters prefer A over B, and B over C, then they would prefer A over C when selecting between those two choices. This principle is referred to as transitivity, and underlies all our work with preference schedules.)

(b) Which option is selected if A and C are considered first, with the winner against B?

(c) Which option is selected if B and C are considered first, with the winner against A?

46. The school board is considering three options for a new career counseling program. The preferences of the board members are summarized as follows:

Two members prefer A over B, and B over C
Five members prefer A over C, and C over B
Four members prefer B over A, and A over C
Four members prefer C over B, and B over A

(a) Which option is selected if the council first chooses between A and B, and then chooses between the winner and C?

(b) Which option is selected if A and C are considered first, with the winner against B?

(c) Which option is selected if B and C are considered first, with the winner against A?

9.2 FLAWS OF THE VOTING SYSTEMS

INITIAL PROBLEM

The Compromise of 1850 averted civil war in the United States for 10 years. This compromise began as a group of eight resolutions presented to the Senate on January 29, 1850 by Henry Clay of Kentucky. Six months of speeches, bargaining, and amendments culminated in the defeat of Clay's measure on July 31, 1850. Yet shortly thereafter, essentially the same proposals were shepherded to passage by Stephen Douglas of Illinois. How is this possible?

We began this chapter with a discussion of Kenneth Arrow's Impossibility Theorem (The Human Side of Mathematics). Basically, the theorem says that there is no voting method that will always satisfy all the properties we would believe to be rational and reasonable in a good voting system. In the last section, we examined several common voting methods and saw (among other things) that the choice of the method used can be as decisive as the actual preferences in determining the outcome in an election of any kind. This applies to a very wide range of decision-making situations.

Now we consider some of the properties of a voting system that we would believe to be rational and reasonable. We will see that there are circumstances where the use of each voting method we studied can fail to satisfy those conditions. We will refer to these properties as *criteria* since they will be used as standards in judging whether a voting system is always fair and sensible.

THE MAJORITY CRITERION

If one candidate is the first choice of a majority of the voters, then most people would think that candidate ought to win the election. This property is called the majority criterion. If we follow the majority criterion and there are three candidates, Leroy, Melvin, and Nancy, with 501 out of 1000 voters thinking Leroy is the best choice, then Leroy should be elected no matter what people think about how Melvin and Nancy compare.

Majority Criterion

If a candidate is the first choice of a majority of voters, then that candidate should be selected.

The majority criterion only tells us who should be elected if there is one candidate who is the first choice of a majority of voters.

The Borda count method, in which all the candidates are ranked and assigned points based on their rankings, sometimes fails to satisfy the majority criterion. The next example will illustrate this. In fact, we will show how to create a voting situation in which this happens. Note that there is no set method such as a formula to construct solutions to problems like our next example. Such problems should be treated like puzzles. This will give you a chance to exercise your problem-solving skills.

HISTORY

The writers of the United States Constitution felt the opinion of the majority sometimes needs to be tempered by the wisdom of elected representatives. One example of this tempering of majority rule is the election of the President of the United States by the Electoral College rather than by a direct vote of the people.

EXAMPLE 9.5 Give an example of a preference table for three voters confronted by four candidates for which the Borda count method violates the majority criterion.

SOLUTION We call the candidates A, B, C, and D. We need to create a preference table in which one candidate, say A, is the first choice of a majority of the voters, but for which applying the Borda count rules leads to the election of another candidate, say B. To make candidate A the first choice of a majority we let the two voters, represented by the first two columns, have A as their first choice.

A	1	1	
B			
C			
D			

To try to arrange that B will be elected by the Borda count method we assign B the highest possible rankings we can without changing the first place rankings already assigned to A.

A	1	1	
B	**2**	**2**	**1**
C			
D			

To keep candidate A's Borda count score as low as possible we have the third voter rank A as low as possible.

A	1	1	**4**
B	2	2	1
C			
D			

To complete the preference table we need to fill in rankings for candidates C and D. It is reasonable to expect that the details of how C and D rank will be irrelevant, so we fill those in arbitrarily.

A	1	1	4
B	2	2	1
C	**3**	**3**	**2**
D	**4**	**4**	**3**

Now we check to see if the proposed solution works. Alternative A is the majority winner because two of three voters chose it first. Assigning four points for first place, three points for second place, two points for third place, and one point for last place, the point totals for the Borda count method are

$$\begin{array}{ll} \text{A} & 4 + 4 + 1 = 9 \\ \text{B} & 3 + 3 + 4 = 10 \\ \text{C} & 2 + 2 + 3 = 7 \\ \text{D} & 1 + 1 + 2 = 4, \end{array}$$

which makes B the winner under Borda count rules. ◆

Looking back at the solution of Example 9.5 we see that candidates C and D mainly served to increase the point differences between candidates A and B that resulted from the third voter's rankings. With only two choices, and two of the three voters giving A their first place votes, B would lose by the Borda count method no matter what the third voter did. With three choices, the best B can possibly do is a tie. Only when there are four choices and three voters does B have a chance to win if two of the voters make A their first choice.

THE HEAD-TO-HEAD CRITERION

It seems reasonable that if there is one candidate the voters favor when compared, in turn, to each of the other candidates, then that candidate ought to win the election. For example, if Jesse would beat Colleen in an election, and Jesse would also beat Eric, then in a three-way election, it seems reasonable that Jesse would win. This thinking gives us the head-to-head criterion. The requirement is not that the winner always win in every head-to-head comparison, because there usually is no such alternative.

EFINITION

Head-to-Head Criterion

If a candidate is favored when compared separately with every other candidate, then the favored candidate should be elected.

Notice that if a method fails to satisfy the majority criterion, then it automatically fails to satisfy the head-to-head criterion, since a majority winner is also the winner of every head-to-head contest in which it is involved.

EXAMPLE 9.6 Give an example of a preference table for seven voters confronted by three candidates for which the plurality method violates the head-to-head criterion.

SOLUTION We name the candidates A, B, and C. We need to create a preference table in which one of the candidates, say A, is the winner of the election by the plurality method, but there is another candidate, say B, that is preferred to A and is preferred to C in head-to-head contests.

We assign the first place votes so that A wins a plurality, but just barely.

A	**1**	**1**	**1**				
B				**1**	**1**		
C						**1**	**1**

Since we want B to win when compared head-to-head against A, the four voters who did not have A as their first choice must all prefer B to A. That tells us how to assign the rankings of the last two voters.

A	1	1	1			**3**	**3**
B				1	1	**2**	**2**
C						1	1

TIDBIT

The head-to-head criterion is also known as the Condorcet criterion after the Marquis de Condorcet (1743–1794).

Since we also want candidate B to be preferred to candidate C, we need to have at least two of the first three voters rank B higher than C. This will ensure that B is preferred to C by a majority. In our example, we will have all three of these voters prefer B to C.

A	1	1	1			3	3
B	**2**	**2**	**2**	1	1	2	2
C	**3**	**3**	**3**			1	1

No matter how we fill in the final rankings for the fourth and fifth voters, the head-to-head criterion is going to be violated. As an extra frill, we can have those voters prefer C to A. Then both candidates B and C would win against A in head-to-head elections.

A	1	1	1	**3**	**3**	3	3
B	2	2	2	1	1	2	2
C	3	3	3	**2**	**2**	1	1

Although A wins a plurality 3 to 2 to 2, in head-to-head elections B is preferred to both A and C, and C is preferred to A. ◆

THE MONOTONICITY CRITERION

Typically, an election is preceded by a campaign or at least a discussion. As persons become more informed or have time to reflect, they sometimes change their preferences. If a candidate gains support at the expense of the other candidates, then the chances of this candidate winning the election should be increased. Certainly, if a candidate was already in position to win an election, then increasing support for the candidate should only help. This thinking leads to our third criterion, the monotonicity criterion.

DEFINITION

Monotonicity Criterion

If a candidate is selected in an election and, in a reelection, all the voters who change their preferences make the previous winner their first choice, then this candidate should be selected in the reelection.

EXAMPLE 9.7 Give an example of a pair of voting patterns for 41 voters confronted with three alternatives for which the plurality with elimination method violates the monotonicity criterion.

SOLUTION In this example we are required to deal with so many voters that a preference table will be quite awkward. A better way to keep track of the rankings is to notice that there are only six possible ways to rank three candidates:

Ranking
A first, B second, C third
A first, C second, B third
B first, A second, C third
B first, C second, A third
C first, A second, B third
C first, B second, A third

We will record the number of voters whose ranking matched each of the six possibilities. Some "guess and test" may be needed.

Ranking	Votes
A first, B second, C third	?
A first, C second, B third	?
B first, A second, C third	?
B first, C second, A third	?
C first, A second, B third	?
C first, B second, A third	?

To solve this problem we need to replace the question marks by whole numbers adding up to 41 (the total number of voters) in such a way that one of the candidates, say A, wins using the plurality with elimination method, and then we need to construct a second table in which some of the votes change in a way that moves A up in the rankings, but another candidate, say B, then wins using the plurality with elimination method. The idea is that voters who improve their opinion of A should do so at the expense of C rather than B. This is hard to do, but after a certain amount of guessing and testing we try the following pattern as the first one that we hope leads to victory for A.

Ranking	Votes
A first, B second, C third	0
A first, C second, B third	14
B first, A second, C third	12
B first, C second, A third	0
C first, A second, B third	5
C first, B second, A third	10

Under plurality with elimination, alternative B would be eliminated in the first round, since it had 12 first place votes versus 14 for A and 15 for C. The preferences between A and C would be as listed in the following table in which we keep track of the same groups of voters.

Ranking	Votes
A first, C second	0
A first, C second	14
A first, C second	12
C first, A second	0
C first, A second	5
C first, A second	10

The first three rows can be combined and the last three can be combined to make the following simpler table. Thus A is selected using the plurality with elimination method.

Ranking	Votes
A first, C second	26
C first, A second	15

Now we need to change the voting pattern in such a way that voters are improving their ranking of the winning candidate A. Suppose, for some reason, a reelection has to take place and, prior to the vote, four of the five voters who ranked C first, A second, and B third change their rankings to A first, C second, and B still third. Perhaps since A won the original election, they could reasonably conclude that A will also win the reelection and might want to be on record as favoring the eventual winner. Thus the support for the winning alternative A is increased (at the expense of C). The new preference table is as follows.

Ranking	Votes
A first, B second, C third	0
A first, C second, B third	18
B first, A second, C third	12
B first, C second, A third	0
C first, A second, B third	1
C first, B second, A third	10

Under plurality with elimination rules, the new preference table leads to elimination of C in the first round. The new preference table for the run-off race between A and B is as follows.

Ranking	Votes
A first, B second	$0 + 18 + 1 = 19$
B first, A second	$12 + 0 + 10 = 22$

But the winner now is alternative B. ◆

If we look at the solution of Example 9.7 with the votes side by side, we can see that the only change is that four voters who first ranked C first, A second, and B third have improved the ranking of the winner A with A first, C second, and B third. Despite having even more support in the second election, A loses to B.

Ranking	First Election	Second Election
A first, B second, C third	0	0
A first, C second, B third	14	18
B first, A second, C third	12	12
B first, C second, A third	0	0
C first, A second, B third	5	1
C first, B second, A third	10	10

THE IRRELEVANT ALTERNATIVES CRITERION

The last criterion, the irrelevant alternatives criterion, concerns the effect of removing (or introducing) a candidate that has no chance of winning.

Irrelevant Alternatives Criterion

If an alternative is selected in an election and, in a reelection, one or more of the alternatives is removed, then the previous winner should be selected in the reelection.

EXAMPLE 9.8 Give an example of a preference table for five voters confronted with three candidates for which the plurality with elimination method violates the irrelevant alternatives criterion.

SOLUTION We name the candidates A, B, and C. We need to create a preference table in which one of the candidates, say A, wins the election by the plurality with elimination method, but there is another candidate, say B, that wins if C is eliminated before the election.

We assign first place votes so that candidate B will be eliminated in the first round of voting.

A	1	1			
B			1		
C				1	1

For the voters who do not have A as their first choice, we assign second and third place votes so that candidate A benefits from the elimination of B and candidate B benefits from the elimination of C.

A	1	1	**2**	**3**	**3**
B			1	**2**	**2**
C			**3**	1	1

Finally, we fill in the second and third place choices of the first and second voters. Since A will not be eliminated under any of the possible scenarios, these assignments can be arbitrary.

A	1	1	2	3	3
B	**3**	**3**	1	2	2
C	**2**	**2**	3	1	1

Now we see if it works. On the first round of voting the first place votes are distributed as follows:

A 2 votes
B 1 vote
C 2 votes

so candidate B is eliminated.
When B is eliminated, the rankings of the other candidates, A and C, move up appropriately, giving the new preference table

A	1	1	1	2	2
C	2	2	2	1	1

On the second round of voting we have

A 3 votes
C 2 votes

so candidate A is elected.

But what happens if the losing candidate C drops out of the race before the first round of voting? The elimination of C means the voters who ranked C first and B second, now rank B first. The new preference table is as follows:

A	1	1	2	2	2
B	2	2	1	1	1

When voting is held under this preference table, candidate B wins with three votes to candidate A's two votes. ◆

We have seen that each of the voting methods we introduced in Section 9.1 can be made to violate some reasonable criterion. You might ask why we do not present a better method, one that satisfies all the criteria all of the time. The reason we do not present such a "perfect" voting method is that *none exists!* The Arrow Impossibility Theorem states that even if all the voters assign preferences to all the alternatives, there is no voting method that will always satisfy the majority, head-to-head, monotonicity, and irrelevant alternatives criteria.

The following tables indicate which criteria are satisfied by the various voting methods.

Plurality Method	
majority criterion	always satisfied
head-to-head criterion	sometimes not satisfied
monotonicity criterion	always satisfied
irrelevant alternatives criterion	sometimes not satisfied

Borda Count Method	
majority criterion	sometimes not satisfied
head-to-head criterion	sometimes not satisfied
monotonicity criterion	sometimes not satisfied
irrelevant alternatives criterion	sometimes not satisfied

Plurality with Elimination Method	
majority criterion	always satisfied
head-to-head criterion	sometimes not satisfied
monotonicity criterion	sometimes not satisfied
irrelevant alternatives criterion	sometimes not satisfied

Pairwise Comparison Method	
majority criterion	always satisfied
head-to-head criterion	always satisfied
monotonicity criterion	sometimes not satisfied
irrelevant alternatives criterion	sometimes not satisfied

The pairwise comparison method satisfies the majority criterion, and it was designed to satisfy the head-to-head criterion. The pairwise comparison method does not satisfy the monotonicity criterion or the irrelevant alternative criterion, but still it may look like the best method we have seen. In fact, however, there is another problem with the pairwise comparison method: it often fails to produce a winner. For example, with the preferences in the following table, A is preferred to B by an 8 to 4 margin, B is preferred to C by an 8 to 4 margin, and C is preferred to A by an 8 to 4 margin. So A is preferred to B, B is preferred to C, and C is preferred to A; a totally useless set of conclusions.

Ranking	Votes
A first, B second, C third	4
A first, C second, B third	0
B first, A second, C third	0
B first, C second, A third	4
C first, A second, B third	4
C first, B second, A third	0

INITIAL PROBLEM SOLUTION

The Compromise of 1850 averted civil war in the United States for 10 years. This compromise began as a group of eight resolutions presented to the Senate on January 29, 1850 by Henry Clay of Kentucky. Six months of speeches, bargaining, and amendments culminated in the defeat of Clay's measure on July 31, 1850. Yet shortly thereafter, essentially the same proposals were shepherded to passage by Stephen Douglas of Illinois. How is this possible?

SOLUTION Within any set of voting rules, if a sufficiently large group of voters is presented with more than two alternatives, almost anything can happen. At the beginning of 1850 there were 30 states in the Union, and thus 60 Senators, so there were plenty of voters. There were many issues under consideration in Clay's eight resolutions, so despite the fact that each bill must pass by a majority, the parliamentary maneuvering that is possible in packaging and amending proposals has a decisive effect. As combined by Clay, the measures could not pass. But Douglas was able to build a majority for each piece of the legislation.

PROBLEM SET 9.2

1. Show that the Borda count method violates the majority criterion for the following preference table of nine voters with three choices:

A	1	3	1	3	1	1	1	3	3
B	3	2	3	1	3	2	3	2	2
C	2	1	2	2	2	3	2	1	1

2. Show that the Borda count method violates the majority criterion for the following preference table of eleven voters with four choices:

A	2	3	3	4	2	3	4	3	1	4	3
B	1	2	4	3	1	1	1	1	4	1	4
C	3	1	1	1	3	2	3	2	2	2	1
D	4	4	2	2	4	4	2	4	3	3	2

3. Give an example of a preference table for five voters with three choices in which the Borda count method violates the majority criterion.

4. Give an example of a preference table for nine voters with four choices in which the Borda count method violates the majority criterion, and the winner has no first place votes. (Hint: the candidate with a majority should be given the worst votes possible if there's a choice.)

5. Show that the plurality method violates the head-to-head criterion for the following preference table of nine voters with three choices:

A	1	3	1	3	1	1	2	3	3
B	3	2	3	1	3	2	1	2	2
C	2	1	2	2	2	3	3	1	1

6. Show that the plurality method violates the head-to-head criterion for the following preference table of eleven voters with four choices:

A	2	3	1	4	2	3	4	3	1	1	4
B	1	2	4	3	1	1	1	1	4	3	3
C	3	1	3	2	3	2	3	2	2	2	1
D	4	4	2	1	4	4	2	4	3	4	2

Solutions to Problems 7 and 8 illustrate that the plurality method does not satisfy the head-to-head criterion.

7. Give an example of a preference table for 10 voters confronted with three alternatives, A, B, C, for which alternative A wins by a plurality, but alternative B is preferred to alternative A in a head-to-head comparison and alternative B is also preferred to alternative C in a head-to-head comparison. (Hint: give alternative A four of the first place votes, the smallest possible number to win by a plurality.)

8. Give an example of a preference table for thirteen voters confronted with three alternatives, A, B, C, for which alternative B wins by a plurality, but alternative C is preferred to alternative B in a head-to-head comparison and alternative C is also preferred to alternative A in a head-to-head comparison. (Hint: give alternative B five of the first place votes, the smallest possible number to win by a plurality.)

9. Show that the Borda count method violates the head-to-head criterion for the following preference table of nine voters with three choices:

A	1	2	1	3	1	1	2	3	3
B	3	3	3	1	3	3	1	2	2
C	2	1	2	2	2	2	3	1	1

(a) Find the winner if the pairwise comparison method is used.

(b) Find the winner using the Borda count method.

10. Give an example of a preference table for seven voters with three choices in which the Borda count method does not satisfy the head-to-head criterion.

11. Show that the plurality with elimination method violates the head-to-head criterion for the 51 voters with four choices whose ballots are grouped as follows, with the number above each ballot indicating the number who voted this way:

18	10	9	7	4	3
A	C	C	B	D	D
D	B	D	D	B	C
C	D	B	C	A	A
B	A	A	A	C	B

(a) Find the winner if the pairwise comparison method is used.

(b) Find the winner using the plurality with elimination method.

12. Give an example of a preference table for 25 voters with three choices in which the plurality with elimination method does not satisfy the head-to-head criterion. (Hint: there are six possible types of ballots. Write these down and assign numbers of votes totaling 25 to get the desired results. It may take some trial and error.)

Solutions to problems 13 and 14 illustrate that the plurality method does not satisfy the irrelevant alternatives criterion.

13. Give an example of a preference table for seven voters confronted with three alternatives, A, B, C, for which alternative A wins by a plurality, but when alternative C is eliminated, alternative B wins by a plurality. (Hint: give alternative A three first place votes, the smallest possible number to win by a plurality.)

14. Give an example of a preference table for 10 voters confronted with three alternatives, A, B, C, for which alternative A wins by a plurality, but when alternative C is eliminated, alternative B wins by a plurality. (Hint: give alternative A four first place votes, the smallest possible number to win by a plurality.)

Solutions to problems 15 and 16 illustrate that the Borda count method does not satisfy the majority criterion and thus also does not satisfy the head-to-head criterion.

15. Give an example of a preference table for five voters confronted with three alternatives, A, B, C, for which alternative A wins by the Borda count method, but alternative B has a majority of the first place votes.

16. Give an example of a preference table for seven voters confronted with three alternatives, A, B, C, for which alternative A wins by the Borda count method, but alternative B has a majority of the first place votes.

Solutions to problems 17 and 18 illustrate that the Borda count method does not satisfy the irrelevant alternatives criterion.

17. Give an example of a preference table for five voters confronted with three alternatives, A, B, C, for which alternative A wins by the Borda count method, alternative B comes second, and alternative C is third, but when alternative C is eliminated, the Borda count method applied to the resulting new preference table gives the victory to alternative B.

18. Give an example of a preference table for seven voters confronted with three alternatives, A, B, C, for which alternative A wins by the Borda count method, alternative B comes second, and alternative C is third, but when alternative C is eliminated, the Borda count method applied to the resulting new preference table gives the victory to alternative B.

Solutions to problems 19 and 20 illustrate that the plurality with elimination method does not satisfy the head-to-head criterion.

19. Give an example of a preference table for five voters confronted with four alternatives, A, B, C, D, for which alternative A wins by the plurality with elimination method, but for which alternative D is favored when compared head-to-head with each of the other alternatives. (Hint: give both alternatives A and B two first place votes, and alternative C one first place vote, but give alternative D all the second place votes.)

20. Give an example of a preference table for seven voters confronted with four alternatives, A, B, C, D, for which alternative A wins by the plurality with elimination method, but for which alternative D is favored when compared head-to-head with each of the other alternatives. (Hint: give both alternatives A and B three first place votes, and alternative C one first place vote, but give alternative D all the second place votes.)

Solutions to problems 21 and 22 show that the plurality with elimination method does not satisfy the monotonicity criterion.

21. Give an example of two preference tables for 10 voters confronted with five alternatives, A, B, C, D, and E. The first preference table should have alternative A winning by the plurality with elimination method. In the second table every voter should either give alternative A the same ranking or a higher ranking, and yet alternative E should be the winner by the plurality with elimination method. (Hint: in the first table give alternative A four first place votes, alternative B three first place votes, alternative C two first place votes, and alternative D one first place vote. Alternative E should be given no first place votes, but all the second place votes.)

22. Give an example of two preference tables for 15 voters confronted with six alternatives, A, B, C, D, E, and F. The first preference table should have alternative A winning by the plurality with elimination method. In the second table every voter should either give alternative A the same ranking or a higher ranking, and yet alternative F should be the winner by the plurality with elimination method. (Hint: in the first table give alternative A five first place votes, alternative B four first place votes, alternative C three first place votes, alternative D two first place votes, and alternative E one first place vote. Alternative F should be given no first place votes, but all the second place votes.)

Solutions to problems 23 and 24 illustrate that the plurality with elimination method does not satisfy the irrelevant alternatives criterion.

23. Give an example of a preference table for nine voters confronted with three alternatives, A, B, C, for which alternative A wins by the plurality with elimination method, but if alternative B is eliminated the resulting new preference table gives alternative C a majority of the first place votes. (Hint: give alternative A four first place votes, alternative B three first place votes, and alternative C two first place votes.)

24. Give an example of a preference table for 12 voters confronted with three alternatives, A, B, C, for which alternative A wins by the plurality with elimination method, but if alternative B is eliminated the resulting new preference table gives alternative C a majority of the first place votes. (Hint: give alternative A five first place votes, alternative B four first place votes, and alternative C three first place votes.)

Solutions to problems 25 and 26 illustrate that the pairwise comparison method does not satisfy the irrelevant alternatives criterion.

25. Give an example of a preference table for five voters confronted with five alternatives, A, B, C, D, and E, for which alternative A wins by the pairwise comparison method, but when alternatives B, C, and D are eliminated, alternative E is preferred to alternative A. (Hint: for this problem, it may be easier to arrange the alternatives on each ballot, instead of assigning numbers in a table. Put E just above A on three ballots. Put A far above E on two ballots—A at the top, E at the bottom. The alternatives B, C, and D are straw men for A to beat earning points for the pairwise comparison method.)

26. Give an example of a preference table for seven voters confronted with five alternatives, A, B, C, D, and E, for which alternative A wins by the pairwise comparison method, but when alternatives B, C, and D are eliminated, alternative E is preferred to alternative A. (Hint: for this problem, it may be easier to arrange the alternatives on each ballot, instead of assigning numbers in a table. Put E just above A on four ballots. Put A far above E on three ballots—A at the top, E at the bottom. The alternatives B, C, and D are straw men for A to beat earning points for the pairwise comparison method.)

EXTENDED PROBLEMS

In addition to the shortcomings that have been discussed in this section, several of the voting methods can be manipulated through insincere voting or other means. Insincere voting (not voting in accordance with your true preferences) may be practiced to gain long-run strategic advantage, such as during a sequence of run-offs.

27. Nine voters confronted with three alternatives have the following preference table summary. A majority is needed to win.

Ranking	Votes
A over B over C	4
B over A over C	3
C over B over A	2

Since A has the most votes, but not a majority, a run-off is held between B and C, with the winner running against A.

(a) Determine the winner of the run-off and the election following the run-off. Show the vote totals.

(b) If voters who support A as their first choice feel strongly enough, how could they benefit from voting insincerely? What is the mininum numbers of voters who would need to vote this way to ensure the election of A?

28. Consider twenty voters confronted with three alternatives, A, B, C, and use the plurality with elimination method.

(a) If the following table summarizes the voters' preferences, who will win the election?

Ranking	Votes
A over B over C	9
C over A over B	6
B over C over A	5

(b) If the election is held in two stages, how can supporters of A use insincere voting to their advantage?

29. Suppose there are 15 voters with three choices, and the ballots are grouped as follows, with the number above each ballot indicating the number who voted this way:

2	4	3	2	1	3
A	A	B	B	C	C
B	C	A	C	A	B
C	B	C	A	B	A

The voters use a sequential run-off election in which they first select between two of the alternatives, and then the winner runs against the remaining alternative. Assuming all voters vote sincerely, find (if possible) an order for the run-offs so each of the alternatives could end up the winner.

30. In this section, we focused on four criteria for a good voting system and four methods of voting. In terms of satisfying the criteria, which voting method is the strongest? What is the biggest weakness for that method? Explain your answers.

31. Explain why the plurality method must satisfy the monotonicity criteria regardless of the number of voters or alternatives.

32. Explain why any method that violates the majority criterion must also violate the head-to-head criterion.

9.3 APPORTIONMENT METHODS

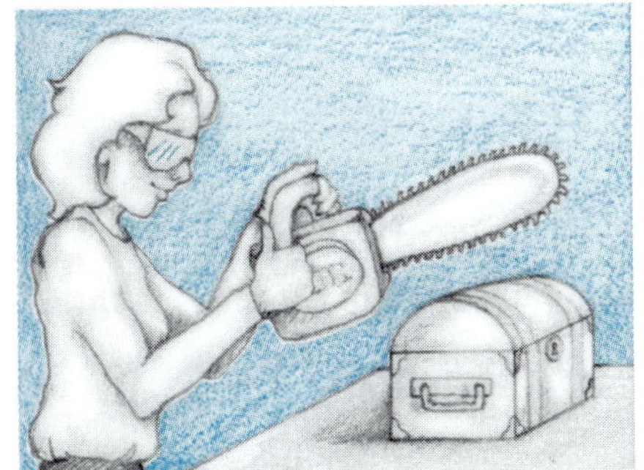

You, your brother, and your sister are the heirs to your grandfather's belongings, except for the house and other real estate. As your grandfather's favorite, you receive 48% of the belongings, while your sister receives 30%, and your brother receives 22%. Most of the items have been sold through an estate sale, but none of you wants to part with the 80 gold coins from a sunken Spanish galleon. These were your grandfather's most valued treasures, and all of you want your fair share of the coins. How can you decide, on a rational basis, how many coins each of you gets? What method will maximize your share?

In our society, most people want their fair share, and they want to be treated fairly, honestly, and with respect. They want their needs to receive equitable attention and their opinions to receive fair consideration. This is the sentiment behind the statement, "No taxation without representation." The cornerstone of a representative social democracy begins with determining how the seats in a decision-making body are to be allocated. How is the philosophy of "one person–one vote" implemented?

For the states in a national legislature, the counties in a state, the colleges in a university, or the heirs to an estate, the basic problem is the same: how do you divide, on an equitable basis, those things that are not individually divisible?

APPORTIONMENT

According to a dictionary, the verb "apportion" means "assign to as a due portion; allot; to divide into shares which may not be equal." The apportionment problem arises when one or more of the due portions has a fractional part, but what is being apportioned cannot be divided into fractions—for example, when assigning the makeup of seats in the House of Representatives to the states. The exact details of how those fractional parts are rounded to whole numbers matters a great deal; for example, members of Congress who might see their districts combined with others are very concerned. Even average citizens do not want to see their home state shortchanged in a reapportionment. Mathematically speaking, the **apportionment problem** is to determine a method for rounding a collection of numbers, some of which may be fractions, so that the sum of the numbers is unchanged.

In Section 2, of Article I, the Constitution of the United States required that "Representatives and direct taxes shall be apportioned among the several states which may be included within this Union, according to their respective numbers, which shall be determined by adding to the whole number of free persons, including those bound to service for a term of years, and excluding Indians not taxed, three-fifths of all other persons. The actual enumeration shall be made within three years after the first meeting of the Congress of the United States, and within every subsequent term of ten years, in such manner as they shall by law direct. The number of representatives shall not exceed one for every thirty thousand, but each state shall have at least one representative; . . ." (The three-fifths of all other persons is the Constitution's oblique way of referring to the slaves; this was superseded by Section 2 of Amendment XIV.) The first Congress had to determine a method of apportionment, and the question could be reconsidered every ten years. The various methods we will look at were proposed by famous statesmen and used for the purpose of apportioning the seats in the House of Representatives. These methods are named for their authors: Alexander Hamilton, Thomas Jefferson, and Daniel Webster.

HISTORY

Alexander Hamilton was the aide-de-camp to George Washington. He was the first secretary of the Treasury and a political adversary of Thomas Jefferson. In 1804, he was killed in a dual with Aaron Burr who served as vice president under Jefferson.

Hamilton's Method

Hamilton's method of apportionment proceeds in three steps.

1. Suppose the total population is P and the number of seats to be apportioned is M. The ratio $D = \frac{P}{M}$ gives the number of persons per seat over all. This number is called the **standard divisor.** If the seats could be divided into fractions, then we would want to assign
$$Q = \frac{\text{state's population}}{\text{standard divisor}}$$
to each state. That number, Q, is called the state's **standard quota.**
2. Round each state's standard quota Q downward to a whole number. Each state will get at least that many seats, but must have at least one seat.
3. If there are seats left over, and there will be unless all the standard quotas were whole numbers, then allocate those seats one at a time to the states ordered by the size of the fractional part of the standard quota, beginning with the state with the largest fractional part.

EXAMPLE 9.9 Let us consider the problem of apportioning seats in the legislature of the Republic of Freedonia. Suppose the population is 10,000,000, there are 200 seats in the legislature, and there are five states with the following populations:

State	A	B	C	D	E
Population (thousands)	1320	1515	4935	1118	1112

Find the standard divisor and standard quotas, and apportion the seats according to Hamilton's method.

SOLUTION The standard divisor is found by dividing the total population by the number of seats in the legislature. We find

$$\text{standard divisor} = \frac{10{,}000{,}000}{200} = 50{,}000.$$

The standard quotas are found by dividing the population of each state by the standard divisor of 50,000, or equivalently, we can divide each number in the population table by 50, since those numbers give the populations in thousands. For example, the standard quota of state A is

$$\frac{1320}{50} = 26.4.$$

Computing the standard quotas of the other states in the same way, we can fill in the following table.

State	A	B	C	D	E
Population (thousands)	1320	1515	4935	1118	1112
Standard Quota	26.4	30.3	98.7	22.36	22.24
Integer Part	26	30	98	22	22
Fractional Part	0.4	0.3	0.7	0.36	0.24

If we total the integer parts of the standard quotas, we get 198, so under the rules of Hamilton's method, two additional seats must be handed out. Since the standard quotas of A and C, 26.4 and 98.7 respectively, have the largest fractional parts, A and C each get one additional seat. The apportionment of the legislature in Freedonia is as follows:

State	A	B	C	D	E
Population (thousands)	1320	1515	4935	1118	1112
Standard Quota	26.4	30.3	98.7	22.36	22.24
Hamilton Apportionment	27	30	99	22	22

◆

Notice that when Hamilton's method is applied every apportionment is either the whole number just below any fractional standard quota, or is the whole number just above any fractional standard quota. This is a nice feature of Hamilton's method, since it would seem unfair to round up or down past the nearest whole number. Any method which has the property of always assigning either the whole number just above, or the whole number just below the standard quota is said to satisfy the **quota rule.**

Hamilton's method was approved by Congress to be the first method of apportionment used following the 1790 census, but it was vetoed by President Washington. The veto was sustained, and eventually a method proposed by Thomas Jefferson, which we describe next, was adopted. In the 1850s Hamilton's method was resurrected by Congressman Vinton of Ohio, and the method was used from 1852 until 1900. However, the apportionment of 1872 was done incorrectly.

HISTORY

Thomas Jefferson, the third president of the United States, drafted the Declaration of Independence and pushed through the Louisiana Purchase. He was knowledgeable in mathematics, and made practical use of calculus throughout his life.

Jefferson's Method

In Hamilton's method some states have their representation rounded up from the standard quota, and some states have their representation rounded down. Clearly, the states whose representation is rounded up obtain an advantage under Hamilton's system. **Jefferson's method** changes the standard divisor to obtain modified quotas that can all be rounded down to the desired total number of representatives. There are only two steps in Jefferson's method, but the first may require some effort.

1. Suppose the number of seats to be apportioned is M. Find a number, d, called the divisor, such that if all the numbers

$$mQ = \frac{\text{state's population}}{d},$$

called **modified quotas**, are *rounded down*, then the sum is M.
2. Assign to each state the integer part of its modified quota.

EXAMPLE 9.10 As in Example 9.9, we consider the problem of apportioning seats in the legislature of the Republic of Freedonia. Recall the population is 10,000,000, there are 200 seats in the legislature, and there are five states with the following populations:

State	A	B	C	D	E
Population (thousands)	1320	1515	4935	1118	1112

Apportion the seats according to Jefferson's method.

SOLUTION In the solution in Example 9.9, the standard divisor was found to equal 50,000. Since all the modified quotas are to be rounded down in Jefferson's method, we will need modified quotas that are slightly larger than the standard quotas (except in the miraculous circumstance that the standard quotas turn out to be whole numbers). Each quota is the quotient of the population and the divisor, so to make the quotients larger, we make the divisor smaller.

There is generally more than one choice of divisor which will work and there is no nice formula to find one, so it is usually most reasonable to find one by the guess and test method.

Suppose we try 49,500 as a modified divisor. To compute each modified quota we divide the state's population by the modified divisor. For example, the modified quota of state A is

$$\frac{1{,}320{,}000}{49{,}500} = \frac{1320}{49.5} \approx 26.67.$$

Computing the other modified quotas in the same way, we get a new table of quota values to consider.

State	A	B	C	D	E
Population (thousands)	1320	1515	4935	1118	1112
Standard Quota	26.4	30.3	98.7	22.36	22.24
Modified Quota	26.67	30.61	99.70	22.59	22.46
Integer Part	26	30	99	22	22

The total of the integer parts of the modified quotas is 199 which is still one short of the goal of 200. So we need to try a modified divisor that is smaller yet. A guess of 49,000 for the divisor will work, as we will now confirm. Again, we compute each modified quota by dividing the state's population by the modified divisor. For example, the modified quota of state A is

$$\frac{1{,}320{,}000}{49{,}000} = \frac{1320}{49} \approx 26.94.$$

Computing the other modified quotas the same way, we get the following table.

State	A	B	C	D	E
Population (thousands)	1320	1515	4935	1118	1112
Standard Quota	26.4	30.3	98.7	22.36	22.24
Modified Quota	26.94	30.92	100.71	22.82	22.69
Integer Part	26	30	100	22	22

The total of the integer parts of the modified quotas is 200 as desired. If the total had been too small we would have changed to a smaller divisor. If the total had been too large we would have made the divisor larger (but not larger than 49,500 which we already tried). ◆

As mentioned at the end of the discussion of Hamilton's method, a desirable property that an apportionment system ought to have is that the number of seats apportioned to a state should either be the integer obtained by rounding up the standard quota or the integer obtained by rounding down the standard quota. This requirement is called the quota rule.

Since Hamilton's method either rounds up or down from the standard quota, that method must necessarily satisfy the quota rule. Example 9.10 shows that Jefferson's method violates the quota rule, because state D is given 100 seats when its standard quota is 98.7. Jefferson's method was used to apportion the House of Representatives until 1840, but has not been used for that purpose since that time.

As you can tell, the quota rule is more of a suggestion than a rule.

HISTORY

Daniel Webster was a native of New Hampshire and was educated at Dartmouth College. Webster was a successful lawyer, famed orator, and member of the Senate for many years.

Webster's Method

In 1830, Webster suggested his apportionment method as a compromise between Hamilton's method and Jefferson's method. This method is called **Webster's method.**

1. Suppose the number of seats to be apportioned is M. Find a number, d, called the divisor, such that if all the numbers

$$mQ = \frac{\text{state's population}}{d},$$

called modified quotas, are *rounded off to the nearest integer*, then the sum is M.

2. Assign to each state the integer nearest its modified quota.

EXAMPLE 9.11 As in Examples 9.9 and 9.10, we consider the problem of apportioning seats in the legislature of the Republic of Freedonia. Recall the population is 10,000,000, there are 200 seats in the legislature, and there are five states with populations listed in the table.

Apportion the seats using Webster's method.

State	A	B	C	D	E
Population (thousands)	1320	1515	4935	1118	1112

SOLUTION In the solution to Example 9.9, the standard divisor was found to equal 50,000. In applying Webster's method, it is no longer automatically known whether one should increase or decrease the value of the standard divisor. Generally speaking one should try to move toward the apportionment that Hamilton's method would give. In this instance, we saw in Example 9.9 that Hamilton's method gives an extra seat to state A, which has a standard quota of 26.4. Standard rounding of 26.4 would round down, not up, so we use a slightly smaller divisor to modify that quota above 26.5. Again, there is no nice formula to find a modified divisor that works. Guess and test is the best approach.

We try a modified divisor of 49,700. To compute each modified quota, divide the state's population by the modified divisor. For example, the modified quota of state A is

$$\frac{1{,}320{,}000}{49{,}700} = \frac{1320}{49.7} \approx 26.56.$$

Computing the other modified quotas in the same way, we fill in the following table.

TIDBIT

The state of Massachusetts sued to have the 1991 House of Representatives apportionment changed, claiming that there had been a systematic under-count in the census. Further, the state claimed that the Huntington-Hill method that was used was unconstitutional because the Webster method more accurately reflected the principle of "one person–one vote."

State	A	B	C	D	E
Population (thousands)	1320	1515	4935	1118	1112
Standard Quota	26.4	30.3	98.7	22.36	22.24
Modified Quota	26.56	30.48	99.30	22.49	22.37
Rounded Quota	27	30	99	22	22

We have given the modified quotas in the table to two decimal places. The total of the rounded modified quotas is 200, as desired. If the total of the rounded off modified quotas had not been 200, then we would have needed to try another modified divisor. If the total is too small we use a smaller divisor, and if the total is too large we use a larger divisor.

The apportionment given here by Webster's method agrees with that given by Hamilton's method, but this is not always the case. ◆

Webster's method was used during the 1840s and again from 1900 until 1940. In 1941, the Webster method was modified to give us the Huntington-Hill method that is still in use at this time. (See Extended Problems in Section 9.4).

INITIAL PROBLEM SOLUTION

You, your brother, and your sister are the heirs to your grandfather's belongings, except for the house and other real estate. As your grandfather's favorite, you receive 48% of the belongings, while your sister receives 30%, and your brother receives 22%. Most of the items have been sold through an estate sale, but none of you wants to part with the 80 gold coins from a sunken Spanish galleon. These were your grandfather's most valued treasures, and all of you want your fair share of the coins. How can you decide, on a rational basis, how many coins each of you gets? What method will maximize your share?

SOLUTION This fair division, or apportionment problem, can be solved in many standard ways. If Hamilton's method is used, you receive 38 coins; your sister, 24; and your brother, 18. However, if Jefferson's method is used, you receive 39 coins; your sister, 24; and your brother, 17. (Note: you should verify these amounts.)

PROBLEM SET 9.3

Use the formats introduced in Examples 9.9 through 9.11 when working apportionment problems, and show all calculations to two decimal places.

In problems 1 through 4, three friends have pooled their resources to bid on 20 bottles of vintage red wine at a wine auction. They decide to divide the bottles based on the amount each contributed, using a standard apportionment method. The amounts they contributed to getting the wine are:

Jaron	$295
Mikkel	$205
Robert	$390

1. Apportion the bottles using Hamilton's method.
2. Apportion the bottles using Jefferson's method.

3. Apportion the bottles using Webster's method.
4. Suppose Mikkel had contributed an additional $20. How would the bottles be apportioned using Hamilton's method?

In problems 5 through 8, a small country with four states has 50 seats in the legislature. The populations of the states are as follows:

State	Population (1000's)
Gorge	275
Mane	767
Organ	465
Taxes	383

5. **(a)** What is the standard divisor?
   (b) What is each state's standard quota?
   (c) Apportion the legislature using Hamilton's method.
6. Apportion the legislature using Jefferson's method.
7. Suppose the size of the legislature is increased to 60 seats. Apportion the legislature using Jefferson's method.
   (a) Find an appropriate modified divisor.
   (b) Find each state's modified quota.
   (c) Find the apportionment.
8. If there are 60 seats in the legislature,
   (a) find the standard divisor.
   (b) find each state's standard quota.
   (c) apportion the legislature using Hamilton's method.

Problems 9 through 12 refer to the following data.

The faculty senate at Dartvard University decided to reorganize in 1993. The guidelines included the following principles:

1. There will be 30 seats in the new senate.
2. Representation will be based on enrollment.
3. A standard method will be used to apportion the senate seats.
4. The senate will be reapportioned every three years based on fall term enrollment.

Enrollment in the five colleges of the university in fall of 1993 were:

Arts and Letters	2540
Sciences	3580
Engineering	1410
Social Science	1830
Human Performance	1050

For each problem, give the standard and modified quotas used, and the final apportionment.

9. Apportion the faculty senate using Jefferson's method.
10. Apportion the faculty senate using Hamilton's method.

Suppose the fall term 1996 enrollment figures for the university are:

Arts and Letters	2930
Sciences	3320
Engineering	1290
Social Science	2140
Human Performance	1180

11. Apportion the faculty senate using Hamilton's method.
12. Apportion the faculty senate using Jefferson's method.

Problems 13 through 18 refer to the following data.

The mythical and mystical country of Mathematica is divided into four provinces: Algebrion, Geometria, Analystia, and Stochastica. There are 314 seats in the legislature and the population of 3,141,593 is scattered among the provinces with the following populations:

Province	Population (thousands)
Algebrion	892
Geometria	424
Analystia	664
Stochastica	1162

13. Find the standard divisor and standard quotas, and apportion the seats according to Hamilton's method.
14. Apportion the seats according to Jefferson's method.
15. Apportion the seats according to Webster's method.

In a move to increase participation in representation among the population, a new national law was passed in Mathematica increasing the number of seats in the legislature to 400.

16. Apportion the 400 seats according to Hamilton's method.
17. Apportion the 400 seats according to Jefferson's method.
18. Apportion the 400 seats according to Webster's method.

Problems 19 through 22 refer to the following data.

The School of Languages at Algebrion State University schedules classes for the coming term based on preregistrations. The school plans to offer 20 sections in the native language (Algebra). Preregistration figures are as follows:

Beginning Algebra	130
Intermediate Algebra	282
Advanced Algebra	188

19. How many sections of each class should be offered using the Hamilton method of apportionment?

20. How many sections of each class should be offered using the Jefferson method of apportionment?

21. How many sections of each class should be offered using the Webster method of apportionment?

Since the original fractional parts were very close, the dean of the school decided to offer another two sections in the native language so that all the section numbers could be rounded up. She thought this would appease the departments that had gotten less than their quota. However, when word of the additional resources was made known, one department insisted that ALL resources (including the new ones) should be apportioned by standard policies. They took their case to the academic senate, which overturned the administrative decision (this is a fairy tale, remember).

22. How are the 22 sections apportioned using each of the following methods?
(a) Hamilton
(b) Jefferson
(c) Webster

Problems 23 through 26 refer to the following data.

Under the constitution of the Republic of Freedonia (Example 9.9), the legislature is to be reapportioned every ten years, based on the census. The most recent census figures (1995) are as follows:

State	A	B	C	D	E
Population (thousands)	1592	1596	5462	1323	1087

23. Reapportion the Freedonia legislature using Hamilton's method.

24 Reapportion the Freedonia legislature using Jefferson's method.

25. Reapportion the Freedonia legislature using Webster's method.

If Freedonia wanted to preserve the ratio of one vote for every 50,000 citizens, the size of the legislature would be increased to 220 seats.

26. How would a 220-seat legislature be apportioned using the 1995 census and each of the following methods?
(a) Hamilton
(b) Jefferson
(c) Webster

Problems 27 through 30 refer to the following data.

The Gotham City Police Department has six precincts in the city and 180 officers. The officers are apportioned to the precincts based on the number of crimes reported in the precinct the previous year, which is given in the following table.

Precinct	A	B	C	D	E	F
Reports	456	835	227	526	338	446

27. Apportion the officers using Jefferson's method.

28. Apportion the officers using Hamilton's method.

29. Apportion the officers using Webster's method.

30. Suppose a state crime bill provides 10 new officers to the city. If Jefferson's method is used to apportion the officers, how are the ten new officers assigned if:
(a) a new apportionment is done using all 190 officers? Compare the new apportionment to the old one and assign the new officers to adjust the numbers.
(b) the new officers are apportioned as a separate force.

EXTENDED PROBLEMS

While many different apportionment methods have been devised and advocated, only four (Hamilton, Jefferson, Webster, and Huntington-Hill) have been implemented with the U.S. House of Representatives. One method was advocated by John Quincy Adams, the sixth president of the United States, and bears his name. In the **Adams method,** a divisor is chosen so that all quotas can be rounded upward, in contrast to the **Jefferson method** in which all quotas are rounded downward.

31. Use Adams' method to apportion the legislature of the Republic of Freedonia in Example 9.9.

32. Use Adams' method to apportion the legislature of the country of Mathematica in problem 13.

33. Use Adams' method to assign the officers to Gotham City precincts (problem 27).

34. Discuss the following statement, and use results from examples or problems to illustrate your points. "The Jefferson method favors large states, while the Adams method favors small states."

Use the following data for problems 35 through 37.

1790 United States Census

Connecticut	236,841
Delaware	55,540
Georgia	70,835
Kentucky	68,705
Maryland	278,514
Massachusetts	475,327
New Hampshire	141,822
New Jersey	179,570
New York	331,589
North Carolina	353,523
Pennsylvania	432,879
Rhode Island	68,446
South Carolina	206,236
Vermont	85,533
Virginia	630,560

35. In 1793, at the direction of President George Washington, Thomas Jefferson apportioned the U.S. House of Representatives. Jefferson used the census for 1790 and a divisor of 33,000, resulting in a total of 105 seats. How many seats were allocated to each of the states under the Jefferson method with this divisor?

36. If the Adams method had been used to apportion the House of Representatives in 1793 using a divisor of 33,000, how many seats would have been allocated to each state?

37. If the Adams method had been used in 1793 to apportion a total of 105 seats, what divisor would be needed (answers may vary slightly), and how many seats would have been allocated to each state?

38. An alternate method for finding the standard quotas is to divide each state's population by the total population and then multiply by the number of seats to be apportioned. In other words, each state receives the same percentage of seats as its percentage of the population.

(a) Set up an algebraic expression for this calculation, and show that it is equivalent to the standard method for finding the standard quotas.

(b) Does one of these methods have an advantage over the other, either philosophically or computationally, now or in the past?

39. Another alternate method for finding the standard quotas is dividing the number of seats by the total population and then multiplying by the number of people in each state.

(a) Write a philosophical rationale for this alternate method of finding standard quotas, such as was done in problem 38.

(b) Set up an algebraic expression for this calculation, and show that it is equivalent to the standard method for finding the standard quotas.

9.4 FLAWS OF THE APPORTIONMENT METHODS

The school district where you live receives a grant to buy 25 computers. The superintendent decides to apportion these computers among the schools by using Hamilton's method. The school in your neighborhood is to receive six computers. When the purchase is made, however, a price decrease allows the purchase of 26 computers rather than 25. On hearing this, the neighborhood school's principal reportedly says, "That's good news for the district, but it means our school will only get five computers." What is the principal talking about?! Will not every school still get the same number, but one school get an additional computer?

Just as there are many good voting systems, but none that are free of flaws, so it is with methods of apportionment. As we did with voting systems, we will consider several fair and reasonable properties that an apportionment method should have, and then see examples where these properties are not satisfied. To emphasize the importance of the problem, the U. S. House of Representatives has been apportioned approximately twenty times, and a number of serious difficulties have arisen. At different times, certain states have believed they were not getting their fair share of the seats in the House. Not only is equal representation an issue, but many policies and actions of the government are based on the number of representatives for each state. In several instances, the Supreme Court has had to make the final decision on the issue.

Problems can arise in several different situations, including:

- a reapportionment based on population changes,
- a change in the total number of seats, or
- the addition of one or more new states.

Much of the difficulty concerns the way we deal with quotas.

THE QUOTA RULE

There are two general types of method for apportionment: quota methods and divisor methods. A **quota method** is any method for which each state's apportionment is either the rounded up or rounded down standard quota, and a **divisor method** is any method that requires the use of a divisor other than the standard divisor.

Recall that it is often considered desirable to have each state's quota be the whole number just below or just above the state's standard quota; this is called the **quota rule.** It is automatic that a quota method must satisfy the quota rule. Interestingly, *no divisor method can always satisfy the quota rule.* Since the quota rule seems desirable, you might wonder why the method used to apportion the seats in the House of Representatives is a divisor rule (essentially Webster's method). As we will see, there are serious problems with quota rules, as well.

THE ALABAMA PARADOX

When Congress considered the apportionment of the House of Representatives for the 1880s, two possible sizes for the House were under serious consideration: 299 members or 300 members. Hamilton's method was the one in use at the time. It was discovered that adding one more seat to the House of Representatives in order to have 300 seats would actually decrease the number of seats for Alabama. This was the first time this paradoxical behavior was observed in Congressional apportionment, therefore it is referred to as the **Alabama paradox.**

When the number of seats is increased, the standard quota must also increase. So for a state to lose a representative (under a quota method) the decrease must be due to changing from rounding up to rounding down. The seat in question must go to some other state, so that other state's increase must be due to changing from rounding down to rounding up.

EXAMPLE 9.12 Show that the Alabama paradox arises under Hamilton's method if the number of seats in the legislature is increased from 99 to 100 in a country with population of 100,000 having four states with the following populations.

State	A	B	C	D
Population	40,650	38,650	10,400	10,300

SOLUTION With 99 seats in the legislature the standard divisor is

$$\frac{100{,}000}{99} \approx 1010.10.$$

The standard quota of each state is the state's population divided by the standard divisor. For example, the standard quota for state A is

$$\frac{40{,}650}{1010.10} \approx 40.2435.$$

Computing the standard quotas for other states in the same way, we can fill in the following table.

State	A	B	C	D
Population (thousands)	40,650	38,650	10,400	10,300
Standard Quota	40.2435	38.2635	10.296	10.197
Integer Part	40	38	10	10
Fractional Part	0.2435	0.2635	0.296	0.197

The integer parts are seen to add to 98, so there is one seat left over. Hamilton's method gives the leftover seat to the state with the largest fractional part, that is, to state C. Thus the apportionment is as follows:

State	A	B	C	D
Population (thousands)	40,650	38,650	10,400	10,300
Hamilton Apportionment	40	38	11	10

With 100 seats in the legislature the standard divisor is

$$\frac{100{,}000}{100} = 1000.$$

A new divisor will change the standard quotas and affect the apportionment. Now, for example, the standard quota of state A becomes

$$\frac{40{,}650}{1000} = 40.650.$$

We compute the standard quotas for all states in the same way.

State	A	B	C	D
Population (thousands)	40,650	38,650	10,400	10,300
Standard Quota	40.65	38.65	10.4	10.3
Integer Part	40	38	10	10
Fractional Part	0.65	0.65	0.4	0.3

The integer parts are unchanged so they again add to 98. But that means there are now two seats to be assigned on the basis of fractional parts. This time the extra seats go to states A and B because their quotas have the largest fractional parts. The new apportionment resulting from the addition of these seats is as follows.

State	A	B	C	D
Population (thousands)	40,650	38,650	10,400	10,300
Hamilton Apportionment	41	39	10	10

Thus state C loses a representative. In this situation, the bigger states benefit at the expense of a smaller state. The issue of power and representation among the states has been an important concern since the founding of the United States. ◆

POPULATION PARADOX

It is possible to construct a quota method that avoids the Alabama paradox, but there are other paradoxical situations that can arise instead. The next paradox, called the **population paradox,** involves two states with growing populations. When the legislature is reapportioned based on the new census, there is a transfer of a seat between the two states, but paradoxically the faster growing state is the one that loses the seat.

EXAMPLE 9.13 Suppose there is a country with three states having populations as given in the table. Show that if there are 100 seats in the legislature, then the population paradox occurs when Hamilton's method is used.

State	A	B	C
Old Population	9555	19,545	70,900
New Population	9651	19,740	70,900

SOLUTION The population of C did not change. We compute the rate of increase of the population of A and of B. The percentage increase in population of A is

$$\frac{9651 - 9555}{9555} = \frac{96}{9555} = 1.005\%,$$

and the percentage increase in the population of state B is

$$\frac{19{,}740 - 19{,}545}{19{,}545} = \frac{195}{19{,}545} = 0.998\%.$$

Both states are growing slowly, but state A is growing faster than state B.

Next we compute the standard quotas with the old population figures.

The old total population is 100,000 giving a standard divisor of

$$\frac{100{,}000}{100} = 1000.$$

The standard quotas are obtained by dividing the population of each state by the standard divisor.

State	A	B	C
Old Population	9555	19,545	70,900
Standard Quota	9.555	19.545	70.900
Integer Part	9	19	70
Fractional Part	0.555	0.545	0.9

The integer parts account for 98 seats. Hamilton's method gives leftover seats to the states with the largest fractional parts. In this case, the two remaining seats go to states A and C. The apportionment is as follows.

State	A	B	C
Old Population	9555	19,545	70,900
Hamilton Apportionment	10	19	71

Finally, we determine the apportionment using the new population figures. The new total population is 100,291 giving a standard divisor of

$$\frac{100{,}291}{100} = 1002.91.$$

We compute the standard quotas by dividing each state's population by the standard divisor of 1002.91.

State	A	B	C
New Population	9651	19,740	70,900
Standard Quota	9.623	19.683	70.694
Integer Part	9	19	70
Fractional Part	0.623	0.683	0.694

Again the integer parts account for 98 seats. Still using Hamilton's method, we assign the leftover seats to the states with the largest fractional parts. This time the two remaining seats go to states B and C.

State	A	B	C
New Population	9651	19,740	70,900
Hamilton Apportionment	9	20	71

We see that state A has lost a seat to state B, even though the population of state A grew more rapidly than the population of state B. ◆

NEW STATES PARADOX

Our last paradox was discovered after Utah was admitted to the Union in 1907 (when Webster's method was in use). If a new state is added, then new seats must be added to the legislature. How many seats? One answer that seems reasonable is to add as many seats as the integer part of what would be the new state's standard quota. For example, if there was a total population of 1,000,000 before the new state joined the country and there were 100 seats in the legislature, then the standard divisor is

$$\frac{1{,}000{,}000}{100} = 10{,}000.$$

If a new state with a population of 42,000 joins the country, then the new state would have a standard quota of

$$\frac{42{,}000}{10{,}000} = 4.2$$

and would be entitled to 4 representatives in the legislature. The size of the legislature would be increased by 4, bringing the total to 104 representatives.

The **new states paradox** occurs when a recalculation of the apportionment results in a change of the apportionment of some of the other states, not the new state.

EXAMPLE 9.14 Suppose there are only two states and that there are 100 representatives. The populations are indicated in the table, as is the apportionment of representatives in the Congress using Hamilton's method.

State	A	B
Population	9450	90,550
Hamilton Apportionment	9	91

Show that if a third state with a population of 10,400 is added to the union, and 10 new representatives are added to the legislature, state B will lose one seat to state A.

SOLUTION When the new state joins, the total population becomes 110,400 and the legislature is increased to 110 representatives. The standard divisor is

$$\frac{110{,}400}{110} \approx 1003.6364.$$

We compute the standard quotas by dividing each state's population by the standard divisor 1003.6364. We obtain the following data.

State	A	B	C
New Population	9450	90,550	10,400
Standard Quota	9.416	90.222	10.362
Integer Part	9	90	10
Fractional Part	0.416	0.222	0.362

The integer parts add to 109, so there is one seat leftover. Hamilton's method assigns the leftover seat to the state with the largest fractional part, that is, to state A. In the new apportionment that follows, we see that state B loses a seat in the legislature to state A.

State	A	B	C
Population	9450	90,550	10,400
Hamilton Apportionment	10	90	10

◆

We have discussed various paradoxes that apportionment methods can produce. Unfortunately, as mathematicians Michel L. Balinski and H. Peyton Young proved (**Balinski and Young's theorem**), there is no apportionment method that satisfies the quota rule and always avoids the Alabama, population, and new states paradoxes. You can have some of the good features such as obeying the quota rule and avoiding the paradoxes, but you cannot have *all* the good features / no matter what method is used. Thus there is no perfect apportionment method. The choice of an apportionment method is ultimately a political decision. Perhaps it is a disappointing realization that the democratic ideal of "one person-one vote" can never be perfectly achieved, but we must take the truth as we find it.

The school district where you live receives a grant to buy 25 computers. The superintendent decides to apportion these computers among the schools by using Hamilton's method. The school in your neighborhood is to receive six computers. When the purchase is made, however, a price decrease allows the purchase of 26 computers rather than 25. On hearing this, the neighborhood school's principal reportedly says, "That's good news for the district, but it means our school will only get five computers." What is the principal talking about?! Will not every school still get the same number, but one school get an additional computer?

SOLUTION The principal may sound a little crazy, but he may very well be correct. Losing a computer in this way would be an example of the Alabama paradox, which can happen under Hamilton's method of apportionment.

PROBLEM SET 9.4

1. A country with three states has 24 seats in the national assembly. The populations of the states are:

Medina	530,000
Alvare	990,000
Loranne	2,240,000

Show that the Alabama paradox arises under Hamilton's method when the national assembly is increased from 24 seats to 25.

2. In problem 1, does the Alabama paradox arise under Hamilton's method if the number of seats increases from 25 to 26? Justify your answer.

3. **(a)** Apportion the 24 seats of the national assembly of the country in problem 1 using Jefferson's method.

(b) Does the Alabama paradox arise under Jefferson's method when the number of seats is increased to 25? Justify your answer.

4. In problem 3, does the Alabama paradox arise under Jefferson's method if the number of seats increases from 25 to 26? Justify your answer.

5. Suppose that in 10 years time the states in problem 1 have the following populations.

Medina	680,000
Alvare	1,250,000
Loranne	2,570,000

Show that the population paradox arises under Hamilton's method with 24 seats.

6. In problem 5, does the population paradox arise under Hamilton's method with 25 seats? Justify your answer.

7. A small country with three states has 50 seats in the legislature apportioned using Hamilton's method. The populations are:

State A	99,000
State B	487,000
State C	214,000

(a) Verify that the standard divisor is 16,000, and find the apportionment for each of the states.

(b) Suppose a new state with a population of 116,000 is added to the country. Using the standard divisor of 16,000, the new state receives seven new seats in the legislature (116/16 = 7.25). Show that the new states paradox arises when the legislature is reapportioned with 57 seats.

8. In problem 7, does the new states paradox arise if Webster's method is used?

9. In problem 7, does the new states paradox arise if Jefferson's method is used? Remember that with Jefferson's method all quotas are rounded down and you may need a modified divisor. If so, this is the divisor that is to be used in assigning new seats to the new state.

The quota rule requires that an apportionment method used should always assign the whole number that is directly above or directly below the standard quota.

10. Explain why Hamilton's method will never violate the quota rule.

11. Explain why if Jefferson's method is used and the quota rule is violated, it must be a violation of the upper quota.

12. Explain why if Adams' method (see Section 9.3, Extended Problems) is used, any violation of the quota rule must be a violation of the lower quota.

Use the following data for problems 13 through 15.

Consider the country of Mathematica from the problem set in section 9.3. The populations of the four provinces were given as follows:

Province	Population (thousands)
Algebrion	892
Geometria	424
Analystia	664
Stochastica	1162

13. Use Hamilton's method and calculate the apportionment for legislatures having 314, 315, and 316 seats. Does the Alabama paradox occur? Which states benefit or lose?

14. Use Jefferson's method and calculate the apportionment for legislatures having 314, 315, and 316 seats. Does the Alabama paradox occur?

15. Use Webster's method and calculate the apportionment for legislatures having 314, 315, and 316 seats. Does the Alabama paradox occur?

16. Give an example of a country with three states and a population of 5,000,000 in which the Alabama paradox occurs when the number of seats in the legislature is increased from 100 to 101.

Use the following data for problems 17 through 19.

The Republic of Freedonia from Example 9.9 had the following population and apportionment data:

State	A	B	C	D	E
Population (thousands)	1320	1515	4935	1118	1112
Seats (200)	27	30	99	22	22

17. Use the Hamilton method and calculate the apportionment for legislatures having 201 and 202 seats. Does the Alabama paradox occur? Which states benefit or lose?

18. Use the Webster method and calculate the apportionment for legislatures having 201 and 202 seats. Does the Alabama paradox occur?

19. Use the Jefferson method and calculate the apportionment for legislatures having 201 and 202 seats. Does the Alabama paradox occur?

20. Give an example of a country with four states and a population of 5,000,000 in which the Alabama paradox occurs when the number of seats in the legislature is increased from 100 to 102.

Use the following data in problems 21 through 23.

Suppose new population figures for Freedonia are as follows:

State	A	B	C	D	E
Population (thousands)	1370	1565	5035	1218	1212

21. Use the Hamilton method and calculate the apportionment of the 200 seats in the legislature. Does the population paradox occur?

22. Use the Jefferson method and calculate the apportionment of the 200 seats in the legislature. Does the population paradox occur?

23. Use the Webster method and calculate the apportionment of the 200 seats in the legislature. Does the population paradox occur?

Use the following data with problems 24 through 26.

Suppose the country of Mathematica in problem 13 admits one of the new territories, Computvia, to full provincial status with "equal" representation in the legislature. Computvia is a rapidly growing frontier area with a population of 243,000. Since the legislature of 314 seats is based roughly on one seat for each 10,000 population, a law is passed to increase the number of seats by 24, to a total of 338 seats.

24. Use the Hamilton method and calculate the apportionment of the 338 seats in the legislature. Does the new states paradox occur?

25. Use the Jefferson method and calculate the apportionment of the orignal 314 seats and the 338 seats in the new legislature. Does the new states paradox occur?

26. Use the Webster method and calculate the apportionment of the 338 seats in the legislature. Does the new states paradox occur?

? EXTENDED PROBLEMS

The method currently being used to apportion the U. S. House of Representatives is the Huntington-Hill method. It is a variation of the Webster method and differs from Webster in that the decision to round the modified quotas is based on whether the number is less than or more than the geometric mean of the two whole numbers immediately before and after it rather than whether the fractional part is less than or greater than 0.5.

The geometric mean of two numbers is the square root of their product. For example, the geometric mean of 4 and 9 is 6, since $4 \times 9 = 36$, and the square root of 36 is 6.

As an example of the way quotas are rounded in the Huntington-Hill method, suppose that the quota in question is 4.475. The whole numbers involved are 4 and 5. The geometric mean of 4 and 5 is $\sqrt{20} \approx 4.4721$. Since 4.475 is greater than 4.4721, the quota is rounded up to 5.

If the quota in question was 5.475, then the geometric mean is $\sqrt{5 \times 6} \approx 5.4772$. In this case, the quota is rounded down to 5.

27. Use the Huntington-Hill method to find the apportionment for each state in a small country with four states. The quotas for each state are as follows:

State	A	B	C	D
Quota	10.47	3.47	5.47	7.59

28. Use the Huntington-Hill method to find the apportionment for each state in a country with three states and 40 seats is the legislature. The populations are:

State	A	B	C
Population	581,500	846,700	1,022,600

29. **(a)** Explain why the fractional part of the geometric mean of two consecutive whole numbers will always be less than 0.5.

(b) Explain why the fractional part of the geometric mean of two consecutive whole numbers will always be greater than 0.41.

30. A paradox of the Huntington-Hill method is that two states can have quotas that differ by more than 1 yet have the same apportionment. Give an example of a pair of quotas that show this.

Chapter Nine Problem

A person stops at the Pies-R-Us restaurant late at night for some dessert. The waiter tells the customer that only Cherry and Blueberry pie are available. The customer chooses Cherry. The waiter goes back to the kitchen and then returns to inform the customer that he had been mistaken; there is only Blueberry and Apple. The customer chooses Blueberry. The waiter goes back to the kitchen and surprisingly returns with a piece of Cherry pie. He explains to the customer that the last Blueberrry pie had just been taken, but there was a Cherry pie after all. However, since the customer preferred Cherry to Blueberry and had preferred Blueberry to Apple, the waiter knew that he should bring a piece of Cherry pie. "Oh no!" said the customer, "I would rather have Apple than Cherry. Please bring the piece of the Apple pie." What was wrong with the waiter's reasoning?

Solution

Strategy: Solve a Similar Problem

The waiter's reasoning is that if C is preferred to B, and B is preferred to A, then C must be preferred to A. This property is called transitivity and does not necessarily hold. The reason is that a person cannot use a single number to describe all the qualities of a pie. You have already seen cases where an election produces no winner in a pairwise comparison system. This is a similar example.

Suppose the customer believes there are three criteria for judging a pie: taste, freshness, and portion size. He rates the quality of each type of pie on each of these criteria. All three types of pie are very good in each category, but some are closer to excellent than others. When choosing between two types of pie, he chooses the one that rates higher on at least two of the three criteria. The customer's relative rating of each type of pie and each criteria is given in the table.

	Apple	Berry	Cherry
Taste	8	1	6
Freshness	3	5	7
Portion size	4	9	2

The numbers in the table above give a score where 1 means very good and 9 means excellent. Thus the Apple is very close to excellent in taste and somewhat closer to good than excellent for freshness and portion size.

Using these ratings, the customer would choose Berry over Apple because it is superior in freshness and portion size, he would choose Cherry over Berry because it is superior in taste and freshness, while he would choose Apple over Cherry because it is superior in taste and portion size.

Chapter Nine Review

Key Ideas and Questions

1. How do the various methods of voting take into account voters' second preferences? Which methods put the most weight on first preference? 424, 426, 428
2. What are the desired properties of a voting method? 435, 437, 438, 441
   Which voting methods have each of these properties?
   Explain what the Arrow impossibility theorem tells us. 442, 443
3. How do quota methods differ from divisor methods? 457
4. What are desirable properties of an apportionment method? 457, 459, 461
   How does Hamilton's method fail to satisfy each of these desirable properties? 457, 459, 461
   Do you think that Webster's method is more fair than Jefferson's? Explain. 450, 452

Vocabulary/Notation

Following is a list of key vocabulary, notation, and ideas for this chapter. Mentally review each of these items, write down the meaning of each term, and use it in a sentence. Then refer to the pages listed by number, and restudy any material that you are unsure of before solving the Chapter Nine Review Problems.

Section 9.1

Plurality Method 424
Borda Count Method 424
Plurality with Elimination Method 426
Run-off Election 426
Preference Table 426
Pairwise Comparison Method 428

Section 9.2

Majority Criterion 435
Head-to-Head Criterion 437
Monotonicity Criterion 438
Irrelevant Alternatives Criterion 441
Arrow Impossibility Theorem 442

Section 9.3

Apportionment Problem 448
Hamilton's Method of Apportionment 448
Standard Divisor 448
Standard Quota 448
Jefferson's Method 450
Modified Quota 450
Webster's Method 452

Section 9.4

Quota Method 457
Divisor Method 457
Quota Rule 457
Alabama Paradox 457
Population Paradox 459
New States Paradox 461
Balinski and Young's Theorem 462

Chapter Nine Review Problems

Use the following information in problems 1 through 3.

Suppose that Anne, Brad and Claire are running for class groundskeeper. The table below tells the number of first, second, and third place votes cast for each.

	First	Second	Third
Anne	6	2	4
Brad	4	7	1
Claire	2	3	7

1. Who is the winner using the plurality method?

2. Who is the winner using the Borda count method?

3. Is it possible to decide who would win the plurality with elimination method from the information above? Explain.

Use the following information in problems 4 through 8.

Below is a complete preference table for an election.

Alice	1	1	1	1	1	3	3	2	2	3	3
Bob	2	2	2	2	2	1	1	3	3	1	1
Claire	3	3	3	3	3	2	2	1	1	2	2

4. Who is the winner using the Borda count method?

5. Who wins if we use the pairwise comparison method?

6. Who wins if we use the plurality with elimination method?

7. Is it possible to change the votes in such a way that the following two things happen?

(a) The relative standing of the candidate who is the winner under the plurality with elimination method does not change on any ballot.

(b) That candidate is no longer the winner.

8. Is it possible to construct a preference table so that the same candidate has a majority but still loses the Borda count? If so, provide an example.

Use the following information in problems 9 through 16.

Suppose there are four states and there are 40 seats in the legislature. Suppose that the population of each state is given in the table below. Populations are in thousands.

State	A	B	C	D
Population	275	392	611	724

9. Find the standard divisor.

10. Find the standard quotas for each state.

11. How many seats will each state get using the Hamilton method of apportionment?

12. What is a satisfactory divisor for Jefferson's method?

13. How many representatives does each state get using Jefferson's method?

14. What is a satisfactory divisor for Webster's method?

15. How many representatives does each state get using Webster's method?

16. Suppose that another apportionment method is used in which state A gets 6 representatives, state B gets 6, state C gets 12, and state D gets 16 representatives. Is this a quota method? Why or why not?

PART 3

MATHEMATICS *in* THE WORLD

CHAPTER 10

GEOMETRY

MOUNTAIN PARADISE IN THE SOUTH PACIFIC?

The mountain island of Halo Siva in the south Pacific is a volcanic island that is almost perfectly formed in the shape of a cone. The volcano rises 13,000 feet above sea level creating a circular land mass that is nearly 100 miles long and more than 50 miles wide. As you approach Halo Siva by either air or sea, this magnificent island occupying 10 million cubic miles of land area seems to be a mountain rising from the sea. Its sandy white beaches give way to lush tropical jungles that rise ever more steeply to the volcano's top. The coral reefs are home to a dazzling array of colorful fish, and the trees seem to be filled with songbirds. Situated in the southern hemisphere, away from shipping lanes and other inhabited islands, Halo Siva is an unspoiled tropical paradise.

CHAPTER GOALS

1. Determine whether a regular or semi-regular tiling of the plane is possible.
2. Describe rigid motions and symmetries.
3. Write an equation for and graph a conic section.

This story about a mythical island is adapted from an airline travel magazine. It creates an image of an intriguing, and alluring, place. If you've ever dreamed about visiting an unspoiled tropical paradise, this could be it. But can an island described like this even exist, or is it simply a Fantasy Island? According to the story, the shape of the island is almost a perfect cone and the land mass is circular. First of all, if the mountain is a perfect cone with a vertical axis, then the shape of the island's shoreline should be perfectly circular. However, it is roughly 100 miles long by 50 miles wide—somewhat long and narrow; not a circle, but an oval. Another part of the article that does not make sense concerns the way land area is discussed. Area is measured in square miles; volume in cubic miles.

In this chapter, you will study a few special topics in geometry. The first topic deals with the possibilities of covering surfaces with geometric shapes in which there are no overlapping edges and no gaps of any type. Also, you will study the properties of curves known as conic sections: the circles, parabolas, ellipses, and hyperbolas that can be formed by intersecting a cone with a plane (as the water intersects the volcano in the story above). Today, we know only slightly more about the mathematical properties of this important family of curves than the Greeks discovered more than 2000 years ago. However, the list of important applications of these curves continues to grow, ranging from the paths of projectiles and planets to the development of projectors and receivers of light and sound.

THE HUMAN SIDE OF MATHEMATICS

Marjorie Rice

Marjorie Rice (1923–) was a housewife and mother of five children who had no mathematics education beyond that required when she graduated from high school in 1939. Although she had retained an interest in mathematics, she was an unlikely choice for someone who would find an answer to a question that had concerned mathematicians and others for a great many years: when is it possible to tile the plane with a given polygon? To put this into a meaningful context, suppose you have a tile (such as a ceramic floor tile) that has straight edges. When is it possible to cover a very large floor with tiles of the same size and shape so that there are no gaps or overlaps?

This question, and examples of some tiles that could be used to tile the plane, were published in *Scientific American* magazine in Martin Gardner's popular column "Mathematical Games and Diversions." The question actually focused on the special case, "Which pentagons will form a tiling of the plane?" It was thought that all such pentagons might have been discovered. When Rice read Gardner's article, she set out to discover if other tilings with pentagons might exist. She developed her own symbolic system for deciding if the pentagons would fit together flush and without overlaps. With her new method, Marjorie Rice discovered many new tilings by pentagons. Her mathematical work was at a level usually found only at the best research universities. She did this with no reward or compensation in mind, but for the pure joy of discovery.

As a high school student, Marjorie Rice had been advised not to take courses in mathematics and science; they were considered unnecessary for her. Instead, she was encouraged to take typing and shorthand as a means of earning a livelihood. Her interest in mathematics and science was rekindled when her children were taking courses in high school, and her talent was such as to take her to the highest level.

Hypatia

Hypatia (370?–415) was one of the first female mathematicians to be mentioned in the history of mathematics. That we know of few others is the fault of the historians rather than a lack of women doing first-rate mathematics—years earlier the Pythagorean society was run by Pythagoras' wife after his death. Hypatia was the daughter of a mathematician, Theon of Alexandria, who is chiefly known for his editions of Euclid's *Elements.* Alexandria had been founded by Alexander the Great in 331 B.C. It became an administrative, commercial, and cultural center under the Romans. Its library of 700,000 volumes was the greatest of ancient times, and the university was the center of scientific thought.

Hypatia became a famous lecturer in philosophy and mathematics at the university in Alexandria, although it is not known if she held an official teaching position. She was also said to have dressed in a tattered cloak and held public discussions in the center of the city, as was the custom of philosophers of that time. Hypatia is best known in mathematics for her work on conic sections, the curves known as circles, ellipses, parabolas, and hyperbolas.

When Hypatia was born, the Roman Empire was in decline, and its influence was waning in the eastern Mediterranean. There was widespread political and religious unrest. The university became the target of hostile protests by Christians in Alexandria because of the pagan Greek culture it represented. There were periodic outbreaks of violence, and during one of these incidents Hypatia was stoned to death by a mob of Christian fanatics.

10.1 TILINGS

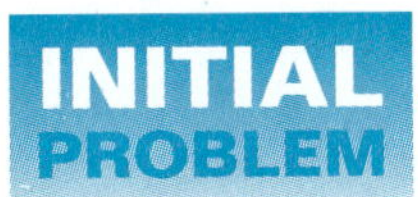

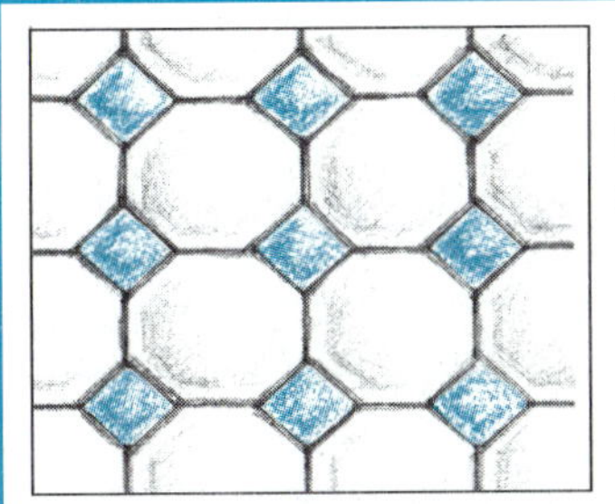

A ceramic tile wall composed of two different shaped tiles is shown. Explain why these two types of tiles fit together.

Patterns that were made by using tiles originated in early civilization. Stone walls and floors are most likely some of the first uses. As cultures became more sophisticated, tilings became works of art as well as serving a pragmatic purpose. Figure 10.1 illustrates tilings from around the world.

France – 12th century

Spain – 15th century

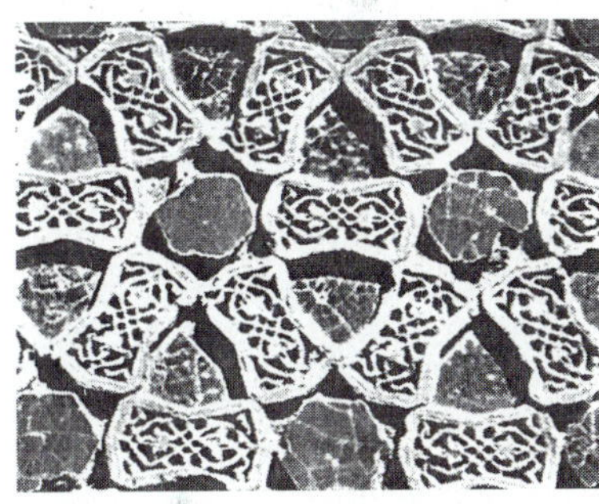

Central Asia – 12th century

Roman church – 13th century

Escher (Netherlands) – 20th century

FIGURE 10.1

These examples give us a glimpse of the beauty and variety that tilings offer. Mathematicians, true to their mission, analyze tiling patterns and have provided many useful classifications. One of the most complete references on this subject is the advanced book *Tilings and Patterns* by Grunbaum and Shephard (1987). In this section, tiling patterns will be classified in some of the more elementary cases.

POLYGONS

To a mathematician a tiling is a special collection of polygonal regions. Figure 10.2 is a tiling made up of rectangles.

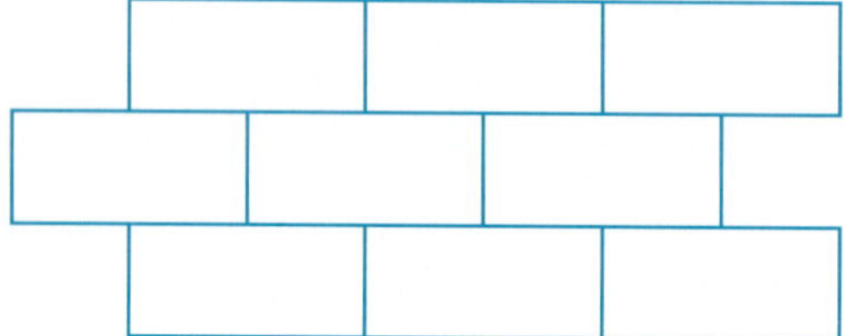

FIGURE 10.2

In general, a **polygon** is a figure consisting of line segments lying in a plane that can be traced so that the starting and ending points are the same and the path never crosses itself or is retraced (Figure 10.3).

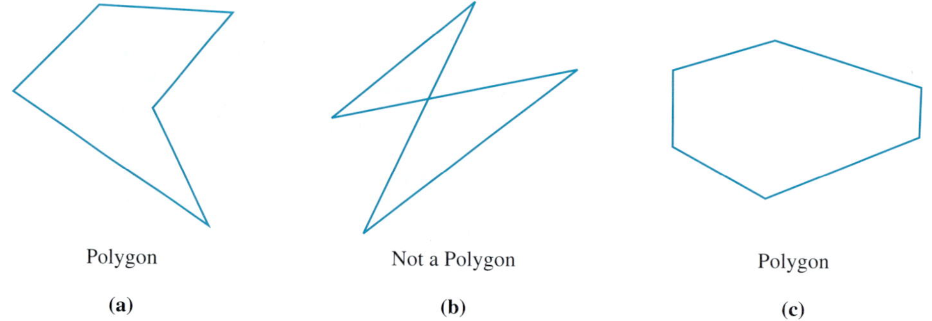

FIGURE 10.3

The line segments of a polygon are called its **sides** and the endpoints of the sides are called **vertices** (singular is **vertex**). A polygon with n sides, hence also n angles, is called an ***n*-gon**. (When n is small, there are more familiar names, such as triangle for a 3-gon, quadrilateral for a 4-gon, and so on.) A **polygonal region** is a polygon together with the portion of the plane (an idealized extended "flat" surface) that is enclosed by the polygon (Figure 10.4).

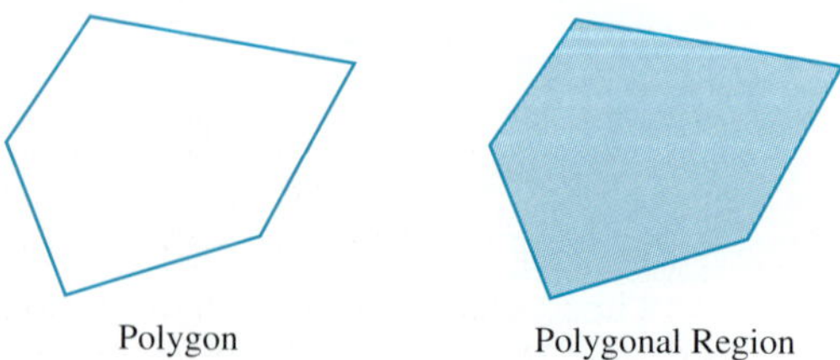

FIGURE 10.4

Polygonal regions form a **tiling** or **tessellation,** if

(1) the entire plane is covered without gaps, and
(2) no two polygonal regions overlap; that is, the only points common to polygonal regions are points on their common sides.

Figure 10.5 shows some partial tilings of the plane.

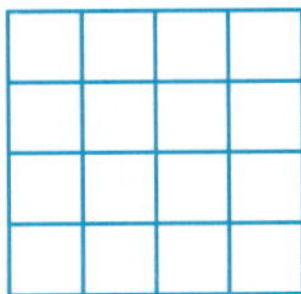

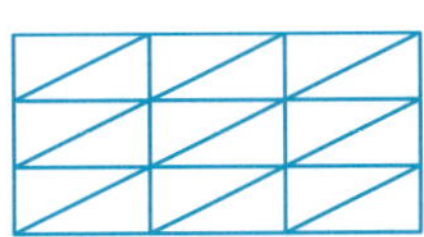

FIGURE 10.5

A tiling composed of triangles may be used to suggest a relationship among the angle measures of a triangle. Figure 10.6(a) shows a triangular region whose angle measures are a, b, and c. Identical copies of this region can be arranged to tile the plane [Figure 10.6(b)].

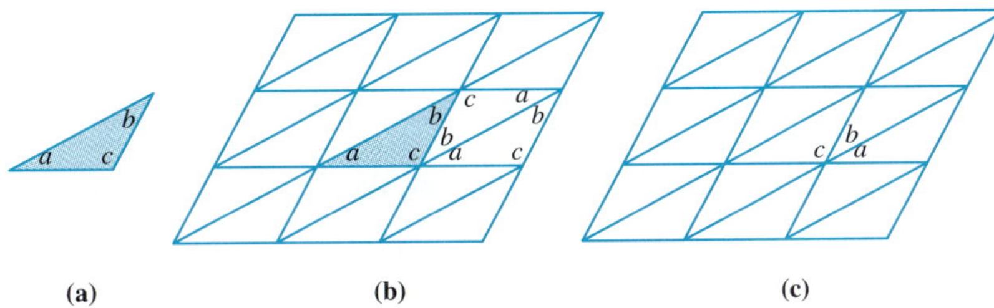

FIGURE 10.6

Notice that the sum of the angle measures in each of the triangular regions is $a + b + c$. However, because of the arrangement of these triangles, the angles c, b, and a in Figure 10.6(c) form a straight line. Hence $c + b + a = 180°$. This relationship is true for any triangle.

Angle Measures in a Triangle

The sum of the measures of the angles in a triangle is 180°.

The angles in a polygon are called its **vertex angles.** For example, in Figure 10.8 (a), the vertex angles in the pentagon are $\angle V$, $\angle W$, $\angle X$, $\angle Y$, and $\angle Z$; the symbol "$\angle$" indicates an angle. Line segments joining nonadjacent vertices in a polygon are called **diagonals.** In Figure 10.8(b), $\overline{WZ}$ and $\overline{WY}$ are two of the diagonals of $VWXYZ$. Note that line segments are indicated by listing the endpoints with a line drawn above.

TIDBIT

If the surface of a sphere is considered a "plane," then finding the sum of the angles in a triangle is a different matter. A triangle with two vertices on the equator and the third at the north pole will have at least two 90° angles. Thus the sum of its angle measures will exceed 180°. For example, the sum of the angles of the spherical triangle in Figure 10.7 is 270°.

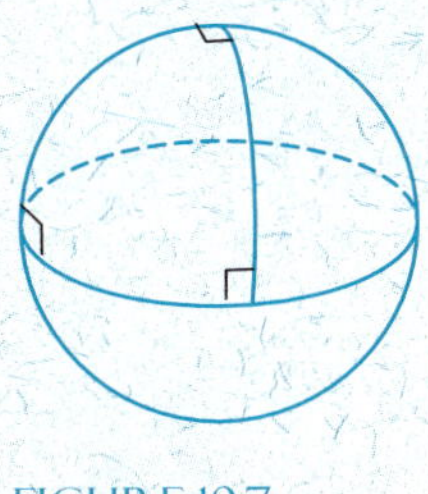

FIGURE 10.7

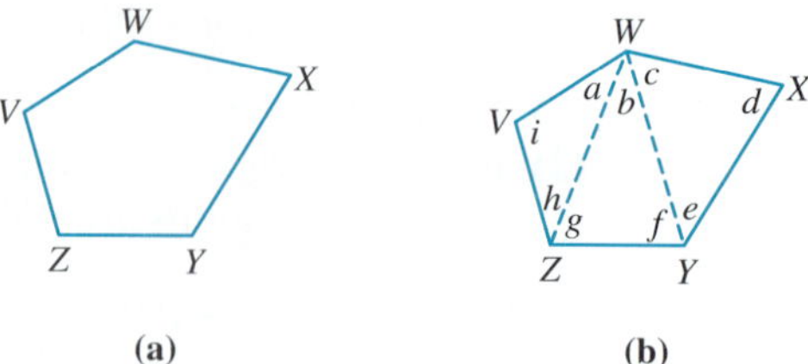

FIGURE 10.8

The sum of the measures of the vertex angles of a polygon can be found by subdividing the polygon into triangles using diagonals. For example, in Figure 10.8(b) we have $a + h + i = 180°$, $b + g + f = 180°$, and $c + d + e = 180°$. The sum of all the vertex angles in pentagon $VWXYZ$ is found by adding the angles in the triangles. That is, $a + b + c + d + e + f + g + h + i = 3(180°) = 540°$. This technique can be generalized to find the sum of the measures of the vertex angles in any polygon. In the pentagon, with five sides, three triangles were formed by drawing diagonals from one vertex. Analogously, in a polygon with n sides, $n - 2$ triangles will be formed. This leads to the next result.

Sum of the Angle Measures in a Polygon

The sum of the measures of the vertex angles in a polygon with n sides is $(n - 2)180°$.

EXAMPLE 10.1 Find the sum of the measures of the vertex angles of a 6-gon.

SOLUTION Substituting 6 for n in the formula: $(6 - 2)180° = 4(180°) = 720°$. ◆

Regular Polygons

Regular polygons are polygons in which all sides have the same length and all vertex angles have the same measure (polygons that are not regular are called **irregular polygons**). Squares and equilateral triangles are examples of regular polygons. Three other regular polygons are shown in Figure 10.9.

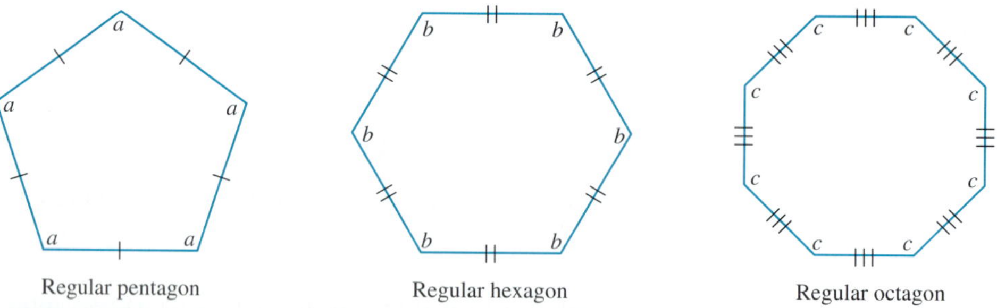

FIGURE 10.9

In a regular n-gon, there are n angles. Since the sum of all the vertex angles of an n-gon is $(n - 2)180$, the measure of any one of the vertex angles in a regular n-gon must be $\frac{(n-2)180}{n}$.

Vertex Angle Measure in a Regular Polygon

The measure of a vertex angle in a regular n-gon is $\frac{(n-2)180°}{n}$.

EXAMPLE 10.2 Find the measure of any vertex angle in a regular 6-gon.

SOLUTION Using the formula preceding this example we have

$$\frac{(6-2)180°}{6} = 4(30°) = 120°.$$

◆

Using this result, we can calculate the measure of a vertex angle in any regular polygon. Table 10.1 contains a list of several regular n-gons together with the measure of their vertex angles.

TABLE 10.1

n-gon	n	Measure of a vertex angle in a regular n-gon
Triangle	3	$(3-2)180°/3 = 60°$
Quadrilateral	4	$(4-2)180°/4 = 90°$
Pentagon	5	$(5-2)180°/5 = 108°$
Hexagon	6	$(6-2)180°/6 = 120°$
Heptagon	7	$(7-2)180°/7 = 128\frac{4}{7}°$
Octagon	8	$(8-2)180°/8 = 135°$
Nonagon	9	$(9-2)180°/9 = 140°$
Decagon	10	$(10-2)180°/10 = 144°$

TIDBIT

One of nature's sweetest tilings is found in a honeycomb. Bees actually build their honey chambers in the shape of cylinders, which then settle into tubes whose cross section is a tiling of regular hexagons.

Notice that as the number of sides in the regular polygon increases, the measure of each of the polygon's vertex angles also increases.

REGULAR AND SEMIREGULAR TILINGS

A **regular tiling** is a tiling composed of regular polygons in which all the polygons are the same size and shape. Figure 10.10 shows three regular tilings.

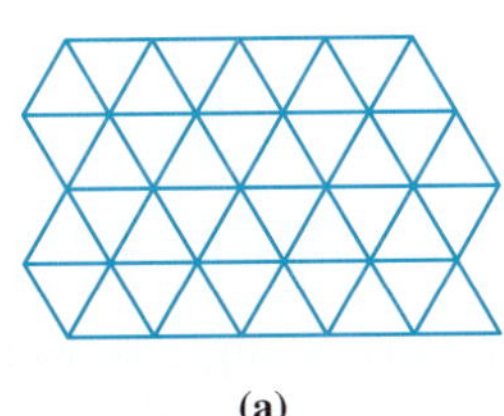

(a)

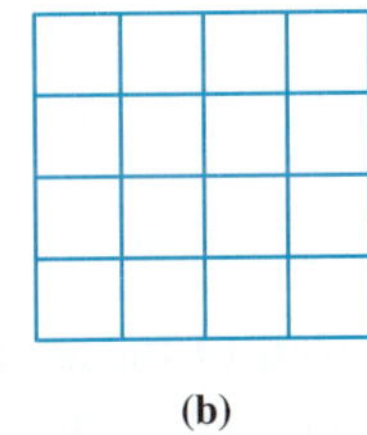

(b)

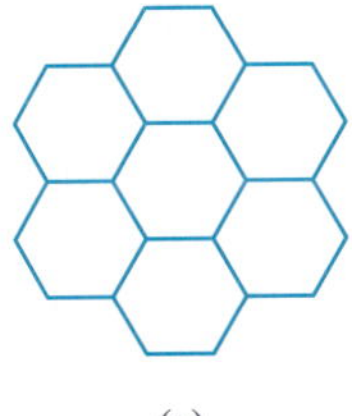

(c)

Regular Tilings

FIGURE 10.10

Notice that the polygonal regions in the tilings in Figures 10.10 have entire sides in common. Such tilings are called **edge-to-edge tilings.** Tilings do not need to be edge-to-edge (see Figure 10.2).

EXAMPLE 10.3 Construct a regular tiling with triangles that is not an edge-to-edge tiling.

SOLUTION Figure 10.11(a) shows a regular tiling using triangles that is edge-to-edge.

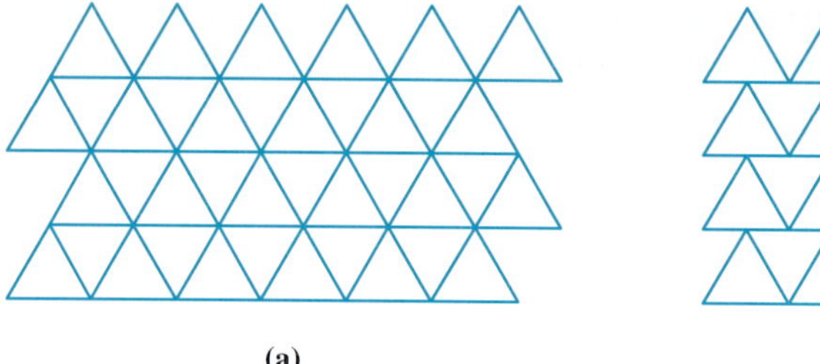

FIGURE 10.11

By sliding alternating rows to the right or the left so that the top vertices of the triangles are in the middle of a side, we obtain a new regular tiling as shown in Figure 10.11(b) that is not edge-to-edge. ◆

FIGURE 10.12

A typical question we might ask is "How many other regular tilings exist?" To determine this, we first observe that for every edge-to-edge tiling, the vertex angles of the tiles must meet at a point (Figure 10.12). This point is a vertex for each of the polygons in the tiling. In the case of a regular tiling, these angles must all have the same measure. For example, in the tiling with equilateral triangles in Figure 10.10(a), six 60° angles are formed at each vertex (Figure 10.13(a)), in the square tiling in Figure 10.10(b), four 90° angles are formed at each vertex (Figure 10.13(b)), and in the hexagonal tiling in Figure 10.10(c), three 120° angles are formed at each vertex (Figure 10.13(c)).

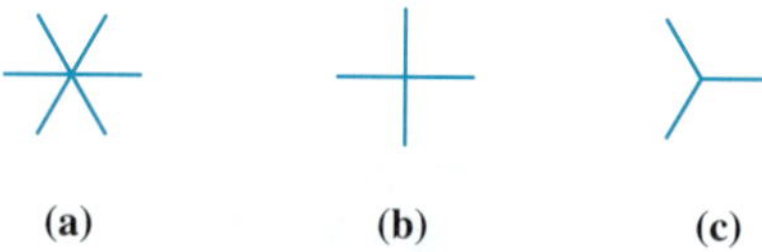

FIGURE 10.13

In each of these cases, the vertex angle is a factor of 360°. It follows that any regular tiling must have this property. By Table 10.1, the vertex angle measure in a regular 5-gon is 108°. Since 108 is not a factor of 360, regular 5-gons cannot form a regular tiling. (Three 5-gons surrounding one vertex total 3(108°) = 324°, which is less than 360°, and four 5-gons will overlap.)

Next we check all regular n-gons, for $n > 6$. Since there are infinitely many of these n-gons to check, we clearly cannot simply check them one at a time. However, we observe that for regular n-gons

(1) there must be at least three vertex angles at each point (two equal vertex angles would both be 180°; in that case, there is no polygon) and
(2) the vertex angle measures for n-gons, where $n > 6$, must all exceed 120° (Table 10.1).

Putting (1) and (2) together, the sum of the vertex angle measures in any n-gon tiling for $n > 6$ must exceed 360°, which will cause an overlap, and is impossible. This shows that the only regular (edge-to-edge) tilings are those in Figure 10.10.

Regular Tilings

The only regular tilings of the plane are those consisting of triangles, squares, or hexagons.

Thus far we have been concerned only with tilings involving the same regular polygons. However, there are many tilings that can be made with combinations of different regular polygonal regions. Figure 10.14 shows eight edge-to-edge tilings that are combinations of regular polygonal regions.

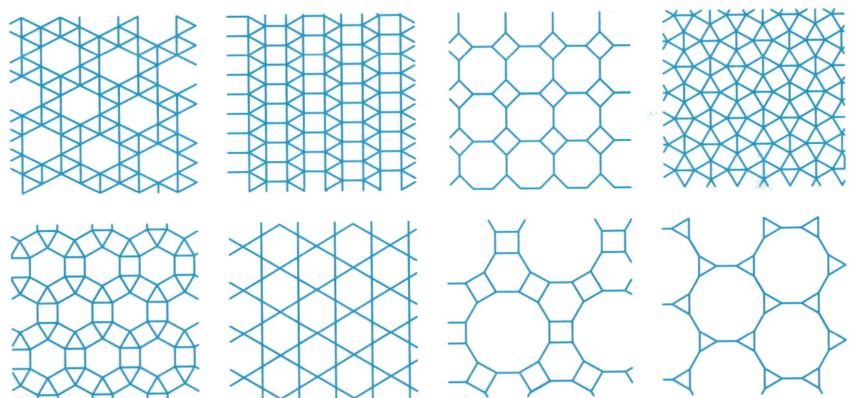

FIGURE 10.14

When we consider tilings using combinations of regular polygons, the possibilities greatly increase. We will limit our attention to edge-to-edge tilings. In an edge-to-edge tiling, several polygons will meet at each vertex. As we saw in studying regular tilings, much can be learned from understanding the way the polygons meet at the vertices. A **vertex figure** of a tiling is the polygon formed when line segments join consecutive midpoints of the sides of the polygons sharing that vertex. Figure 10.15 illustrates the vertex figures for the regular triangle tiling (a regular hexagon), for the regular square tiling (a square), and the regular hexagon tiling (an equilateral triangle).

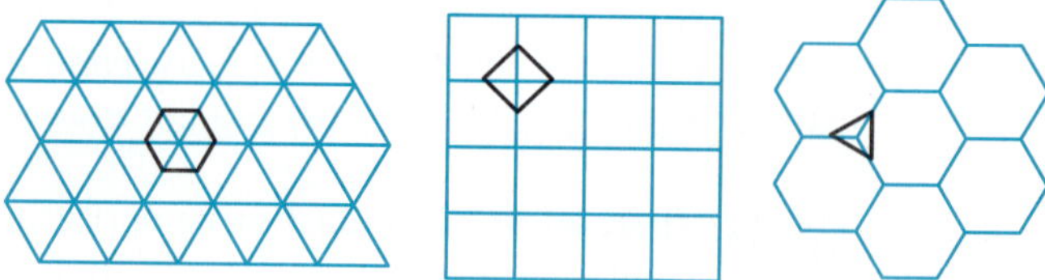

FIGURE 10.15

A **semiregular tiling** is an edge-to-edge tiling by two or more regular polygons such that their vertex figures are the same size and shape.

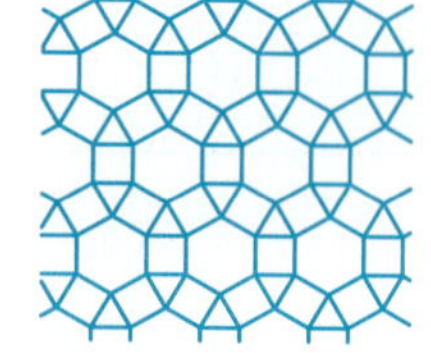

FIGURE 10.16

EXAMPLE 10.4 Show that the tiling in Figure 10.16 is semiregular.

SOLUTION In Figure 10.17 we have constructed a vertex figure at one vertex. Since all vertices contain identical square–equilateral triangle–square–regular hexagon configurations in the same order, each vertex figure will have to be the same size and shape as that in Figure 10.17. ◆

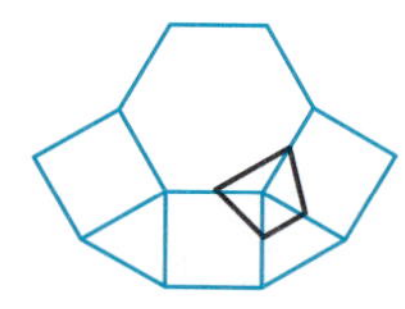

FIGURE 10.17

EXAMPLE 10.5 Show that a semiregular tiling of the plane using only triangles and squares must have three triangles and two squares meeting at every vertex.

SOLUTION The sum of the angles of the triangles and squares meeting at any vertex must add to 360°. We must have at least one square and one triangle in the desired tiling. The vertex angles of an equilateral triangle are 60°, and the vertex angles of a square are 90°. If four squares meet at a vertex, then the angles of those squares have a sum of 360°, leaving no room for a triangle. So there are at most three squares meeting at each vertex. We will go through the possibilities in a systematic way to see what combinations are possible. Note that once the total of the angles at a vertex exceeds 360°, adding a triangle only makes the angle total even larger; and once the total of the angles is less than 360°, removing a triangle makes the total even smaller.

Squares	**Triangles**	**Total of the Angles**
3	2	$3 \times 90 + 2 \times 60 = 390 > 360$
3	1	$3 \times 90 + 1 \times 60 = 330 < 360$
2	4	$2 \times 90 + 4 \times 60 = 420 > 360$
2	**3**	$\mathbf{2 \times 90 + 3 \times 60 = 360}$
2	2	$2 \times 90 + 2 \times 60 = 300 < 360$
1	5	$1 \times 90 + 5 \times 60 = 390 > 360$
1	4	$1 \times 90 + 4 \times 60 = 330 < 360$

The only combination with the correct angle total is two squares and three equilateral triangles meeting at each vertex. ◆

Although discovering and classifying all the semiregular tilings may appear to be a formidable task, it only requires arguments regarding angle measures and some work with fractions. Although we will leave this classification for the problem set, the results of the classification are given next.

> **Semiregular Tilings**
>
> The only semiregular tilings are the eight tilings pictured in Figure 10.14.

MISCELLANEOUS TILINGS

Up to this point we have classified tilings involving only regular polygonal regions. However, there are other tilings that involve many varied shapes, such as those originally seen in Figure 10.1. Next we discuss tilings made up of irregular figures of the same size and shape.

3-gons (triangles): Since the sum of the angles in a triangle is 180°, any triangle can form a tiling by forming infinite strips as shown in Figure 10.18. The fact that the angle sum is 180° is what allows us to bring all three angles together at a vertex with two edges falling on a line.

FIGURE 10.18

4-gons (quadrilaterals): The sum of the angles in a quadrilateral is 360°, thus the four angles of a quadrilateral will fit around a point as in Figure 10.19.

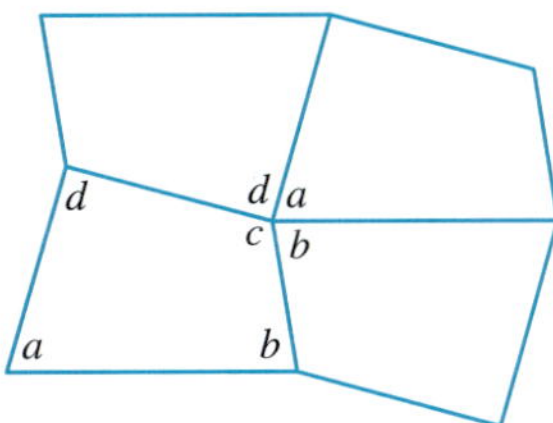

FIGURE 10.19

The tiling started in Figure 10.19 is extended in Figure 10.20.

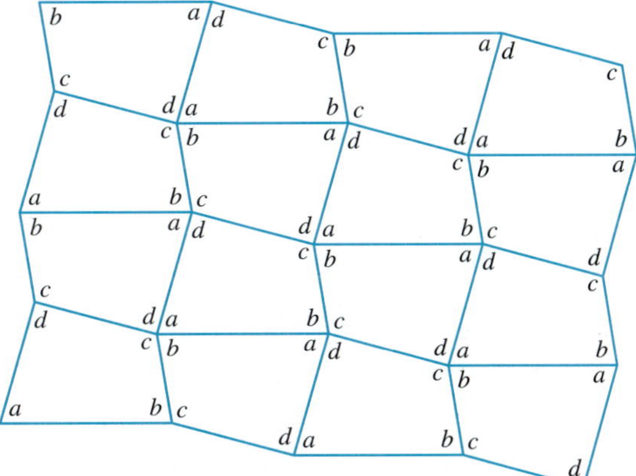

FIGURE 10.20

Triangles, as well as the quadrilaterals in Figure 10.20, are examples of convex polygons. A polygon region is called **convex** if, for any two points in the region, the line segment having the two points as endpoints is also in the region. Otherwise, it is a **concave polygonal region** (Figure 10.21).

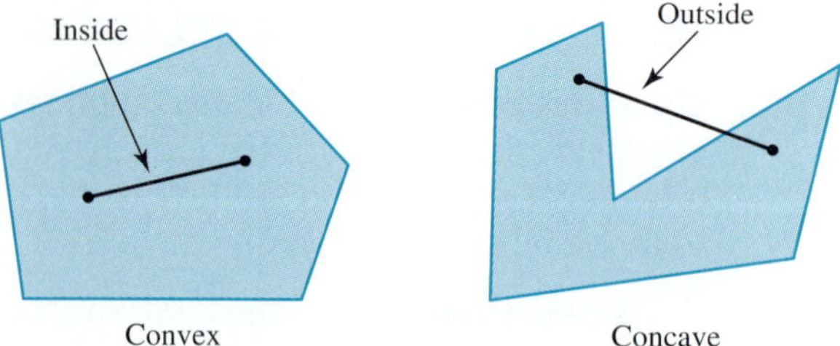

FIGURE 10.21

We have seen that all convex quadrilaterals tile the plane. Figure 10.22 shows the start of a tiling by a concave quadrilateral.

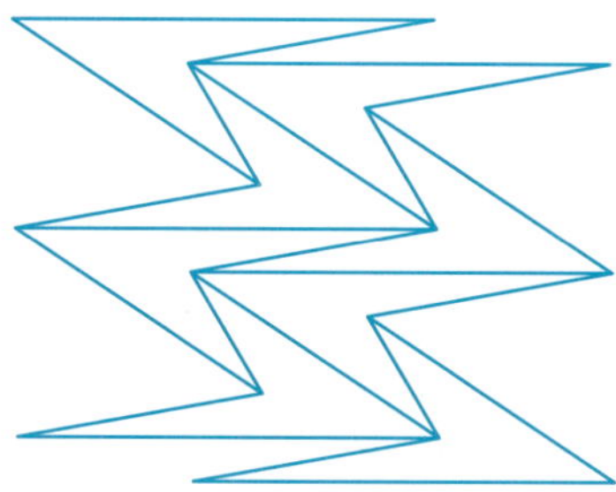

FIGURE 10.22

5-gons (pentagons): Earlier we saw that regular pentagons do not form a tiling. However, polygonal regions shaped like a baseball plate do tile the plane (Figure 10.23).

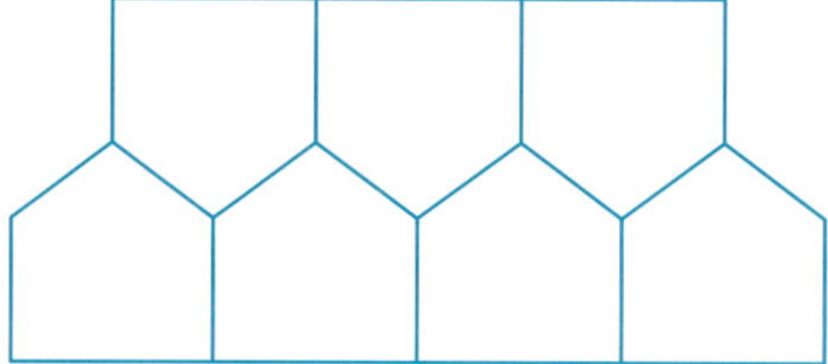

FIGURE 10.23

The tiling in Figure 10.23 is just one of at least 14 general types with irregular pentagons that tile the plane. It is unknown if there are any more.

6-gons (hexagons): We have seen that regular hexagons produce a regular tiling of the plane (Figure 10.10). By extending a pair of opposite sides of a regular

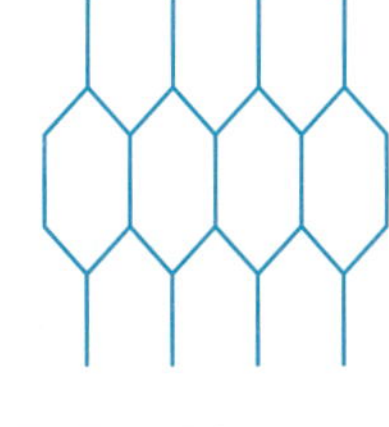

FIGURE 10.24

hexagon, we see that such irregular hexagons can also tile the plane (Figure 10.24). It has been shown that there are exactly three types of convex hexagons that tile the plane. These three types are illustrated in the problem set.

n-gons for $n \geq 7$: This situation is perhaps the most interesting of all, for it was proved by K. Reinhardt, in 1927, that no convex polygon with more than six sides could tile the plane!

These results are summarized in Table 10.2.

TABLE 10.2 Tilings by Convex Irregular Polygonal Regions of the same Size and Shape

Number of Sides	Number of Possible Tilings
3	All are possible
4	All are possible
5	Unknown, but ≥14
6	3
7 and more	None are possible

We have discussed tilings using regular polygonal regions (Figure 10.10), a mixture of the same regular polygonal regions (Figure 10.14), and irregular polygonal regions of the same size and shape (Figures 10.18, 10.20, 10.22, 10.23, and 10.24).

We close this subsection with a tiling involving a mixture of polygonal shapes. Consider the tiling in Figure 10.25.

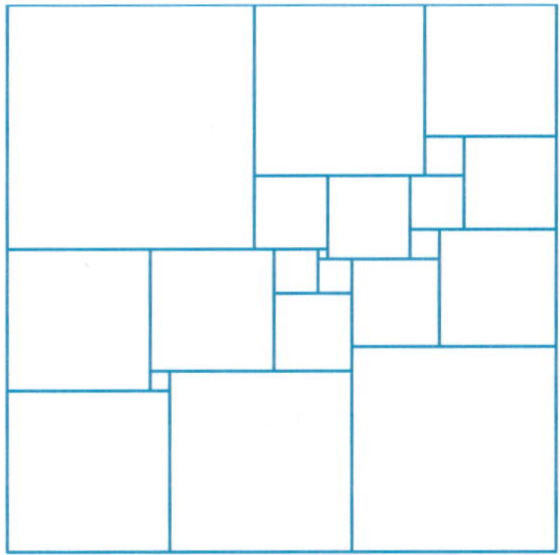

FIGURE 10.25

Here, a square is tiled by smaller squares of different sizes. Since the larger square will tile the plane, this shows that the entire plane can be tiled using such a combination of smaller squares. There is also a rectangle that is composed of squares of different sizes that can tile the plane. These two combinations of squares will be studied in the problem set.

THE PYTHAGOREAN THEOREM

The Pythagorean theorem is perhaps the most famous of all the theorems in mathematics. Although most remember it from algebra as $a^2 + b^2 = c^2$, this is only part of the statement of the theorem. In fact, the theorem is really one about geometry since it deals with right triangles. It is thought that the theorem may have been first discovered by observing a tiling such as the one in Figure 10.26(a).

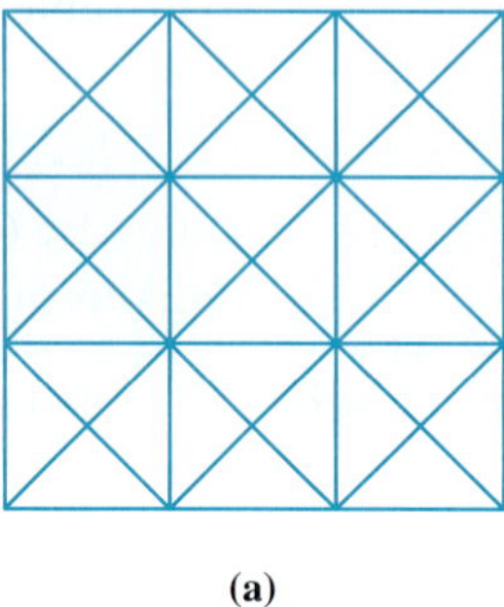

(a)

(b)

FIGURE 10.26

If we focus on the shaded triangle in Figure 10.26(b), the shorter two sides of the triangle can be seen to be the sides of squares, each composed of four of the smallest triangles. The longer side, the hypotenuse, is one side of a square composed of eight of the smallest triangles. Thus "the sum of the squares on the sides of the (right) triangle is equal to the square on the hypotenuse." Our example is a special case of the Pythagorean theorem when the right triangle has two sides of the same length.

Happily, the general case of the Pythagorean theorem can also be verified using tiles. A right triangular region with sides of length a and b and hypotenuse of length c is shown in Figure 10.27(a).

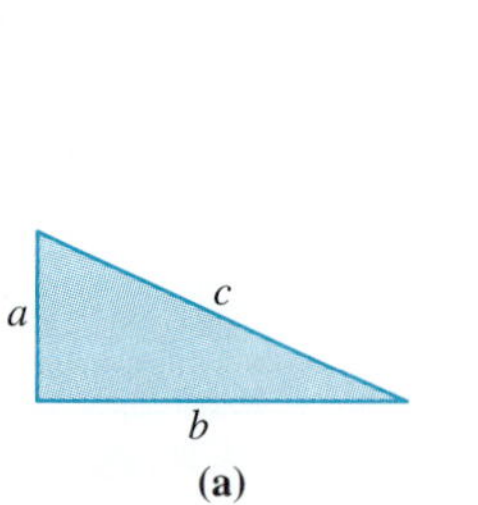

(a)

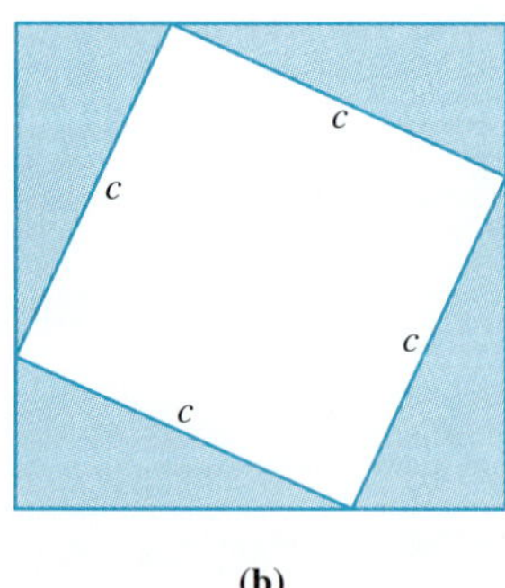

(b)

FIGURE 10.27

In Figure 10.27(b), four triangles identical to the one in Figure 10.27(a) are arranged into a square "donut," where the hole is a square with length c on each side (verify that the angles of the hole are 90°). Then, in Figure 10.28, a series of moves repositions the triangles leaving two squares, one whose sides have length a and one whose sides have length b.

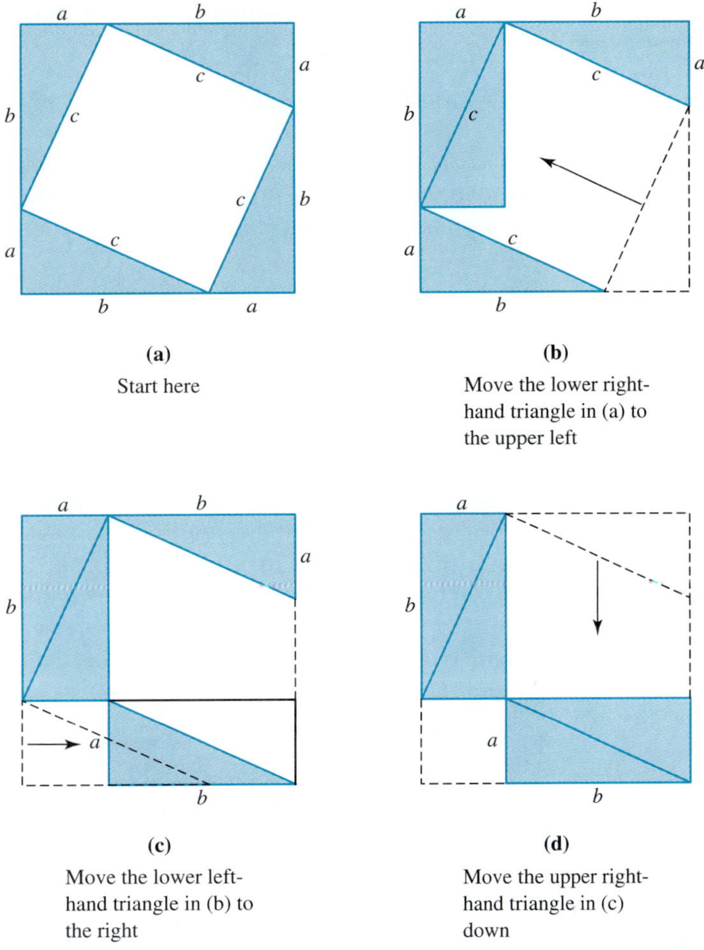

(a)
Start here

(b)
Move the lower right-hand triangle in (a) to the upper left

(c)
Move the lower left-hand triangle in (b) to the right

(d)
Move the upper right-hand triangle in (c) down

FIGURE 10.28

The two unshaded squares in Figure 10.28(d), namely the one whose sides have length a and the one whose sides have length b, take up exactly the same space as the original square whose sides had length c. Thus we have the following result.

The Pythagorean Theorem

In a right triangle, the sum of the areas of the squares on the sides of the triangle is equal to the area of the square on the hypotenuse.

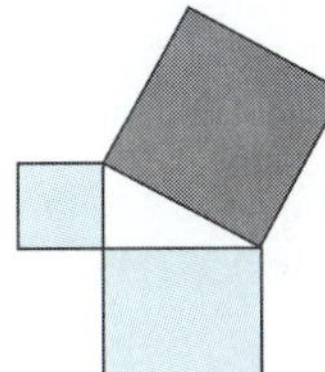

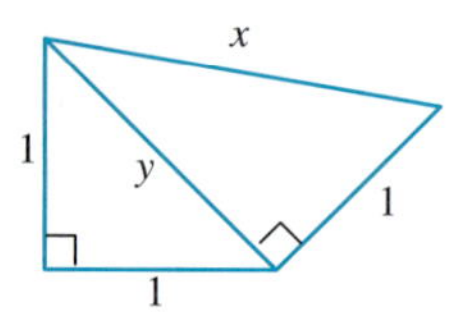

FIGURE 10.29

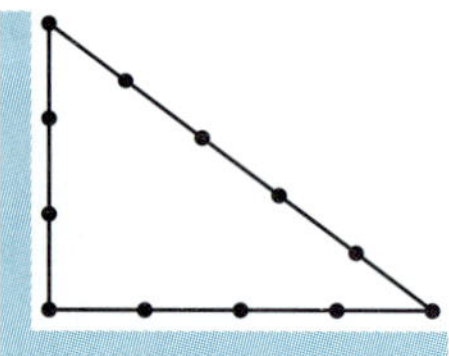
FIGURE 10.30

EXAMPLE 10.6 Use the Pythagorean theorem to find the length x in Figure 10.29.

SOLUTION First we apply the Pythagorean theorem to the isosceles right triangle with side length 1 and hypotenuse y to find that

$$y^2 = 1^2 + 1^2 = 2.$$

Then we apply the Pythagorean theorem to the right triangle with sides 1 and y and with hypotenuse x to find that

$$x^2 = 1^2 + y^2 = 1 + 2 = 3.$$

Therefore, we have $x = \sqrt{3}$. ◆

In Chapter 1 it was shown that the converse of an if-then statement may or may not be true. In the case of the Pythagorean theorem, its converse is true although a proof of this is beyond the scope of this book. The converse of this theorem states that if a triangle has sides of lengths a, b, and c, and $a^2 + b^2 = c^2$, then the triangle is a right triangle. One interesting application of this converse is a method used in ancient civilizations such as the Babylonians or Egyptians to make certain that two walls meeting in a corner actually meet at a right angle (Figure 10.30). The method involves using a string loop with knots tied at regular intervals. If this loop is stretched into a triangle whose sides are of length 3, 4, and 5, it forms a right triangle since $3^2 + 4^2 = 5^2$. This method is still in use in many countries around the world.

A ceramic tile wall composed of two different shaped tiles is shown. Explain why these two types of tiles fit together.

SOLUTION The angle measure in a square is 90° and in a regular octagon is 135° (Table 10.1). Each vertex contains a square and two regular octagons, and $90° + 2(135°) = 360°$. Thus the three tiles will fit at each vertex and, as long as their sides are the same length, will tile a wall. This is one of the eight semiregular tilings.

PROBLEM SET 10.1

1. Find the sum of the measures of the vertex angles of the following polygons.
   (a) an dodecagon (12-gon)
   (b) a decagon (10-gon)
   (c) a 15-gon
2. Find the sum of the measures of the vertex angles of the following polygons.
   (a) an icosagon (20-gon)
   (b) a nonagon (9-gon)
   (c) a 13-gon
3. Find the sum of the measures of the vertex angles of the following polygons.
   (a) a 16-gon
   (b) a 24-gon

4. Find the sum of the measures of the vertex angles of the following polygons.
   (a) an 18-gon **(b)** a 30-gon
5. Find the measure of each vertex angle in a regular nonagon (9-gon).
6. Find the measure of each vertex angle of a regular dodecagon (12-gon).
7. Find the measure of each vertex angle of a regular icosagon (20-gon).
8. Find the measure of each vertex angle in a regular 15-gon.

Another way to approach the question of the sum of the vertex angles of an n-gon is to think of beginning at one vertex and traveling around the perimeter of the polygon. When you come to the next vertex, you must make a turn of a certain number of degrees. If you extend the edge you just traveled, you will note that there are two angles formed with respect to the vertex, the side you extended, and the next side. One of these is the vertex angle, and the other we will call the exterior angle. The vertex angle and exterior angle add up to 180°. Since there are n vertices, the sum of measures of the vertex angles and exterior angles combined is $n(180°)$. As you travel around the perimeter of the polygon, you must turn 360°. Therefore, the sum of the measures of the vertex angles is found to be $n(180°) - 360° = n(180°) - 2(180°) = (n - 2)180°$.

In problems 9 through 12, we make use of the fact that the sum of the measures of the exterior angles of a polygon, as defined above, is 360°.

9. If the measure of each vertex angle in a regular polygon is 144°, how many sides does the polygon have? (Hint: what is the measure of the exterior angle?)
10. If the measure of each vertex angle in a regular polygon is 160°, how many sides does the polygon have? (Hint: what is the measure of the exterior angle?)
11. If the measure of each exterior angle of a regular polygon is 24°, how many sides does the polygon have?
12. If the measure of each exterior angle of a regular polygon is 30°, how many sides does the polygon have?
13. In Example 10.3, a regular tiling by triangles that is not edge-to-edge is constructed by sliding alternating rows to the right.
    (a) If arbitrary rows were shifted, would the result still be a regular tiling?
    (b) If the top vertex of each triangle were not in the middle of a side, would the result still be a regular tiling?
    (c) Explain your reasoning for parts (a) and (b).
14. Referring to Figure 10.11(a) in Example 10.3, slide alternating diagonal rows of triangles along a straight line so that the vertex of each triangle is in the middle of a side of another triangle.
    (a) Do you get a regular tiling that is not edge-to-edge?
    (b) Is this a different tiling than the one in Figure 10.11(b)? Explain your answer. (Hint: rotate the paper with the tiling for a full 360°. What do you notice?)
15. Figure 10.10(b) shows a regular tiling by squares. As in Example 10.3, construct a regular tiling by squares that is not edge-to-edge.
16. Referring to Figure 10.10(c), is it possible to construct a regular tiling by hexagons that is not edge-to-edge by sliding figures as in Example 10.3? Please explain.

Problems 17 through 24 refer to edge-to-edge tilings.

17. Explain why a regular tiling cannot be composed of regular 7-gons.
18. Explain why a regular tiling cannot be composed of regular octagons.
19. Show that a tiling cannot be composed of alternating regular pentagons and equilateral triangles around a vertex.
20. Show that a tiling cannot be composed of regular hexagons and squares.
21. Determine whether or not a tiling can be made up of two differently sized equilateral triangles. If so, provide a sketch. If not, provide a explanation.
22. Determine whether or not a tiling can be made up of two different size squares. If so, provide a sketch. If not, provide a explanation.
23. Determine whether or not a tiling can be made up of equilateral triangles and regular octagons. If so, provide a sketch. If not, provide an explanation.
24. Determine whether or not a tiling can be made up of squares and regular pentagons. If so, provide a sketch. If not, provide an explanation.
25. Sketch the vertex figure of the two middle tilings in the bottom row in Figure 10.14.
26. Sketch the vertex figure of the two middle tilings in the top row in Figure 10.14.
27. Determine whether or not the tiling in the following figure is semiregular. Give a reason for your answer.

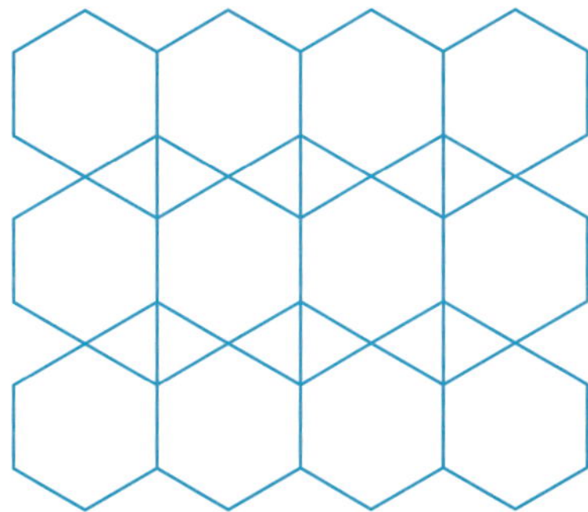

28. Sketch at least three tilings whose vertex figures are triangles.

29. Determine if each of the given irregular pentagons tile the plane. (Hint: trace and cut out several copies of each shape.)

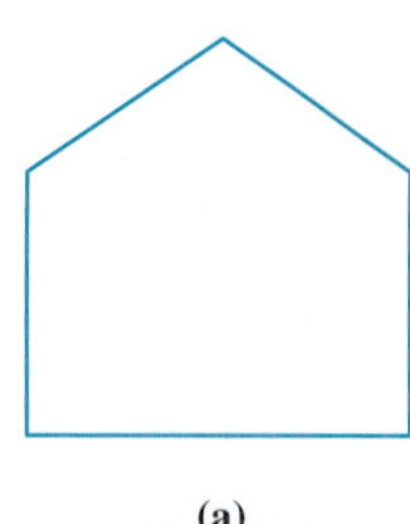

(a)

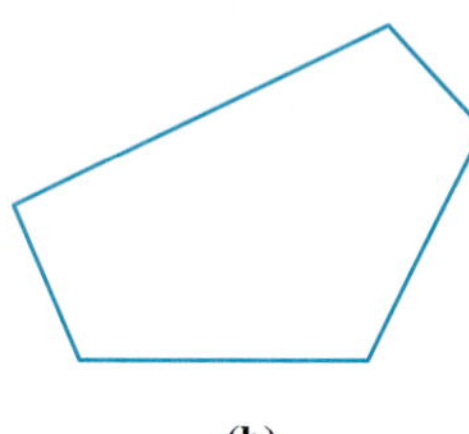

(b)

30. Determine if each of the following concave polygons tile the plane. (Hint: trace and cut out the shapes.)

(a)

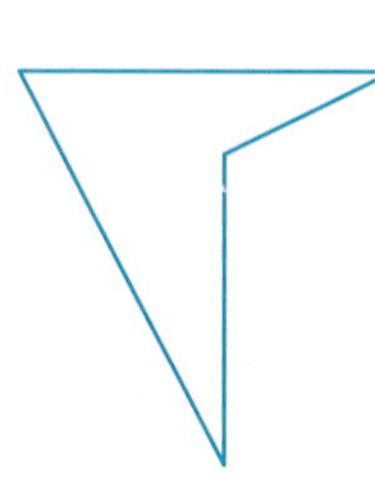

(b)

31. Use the Pythagorean theorem to find the length of the hypotenuse of a right triangle that has sides of lengths given.

(a) 5, 12
(b) 6, 8
(c) 15, 19

32. Use the Pythagorean theorem to find the length of a side of a right triangle if the hypotenuse and second side have the following lengths.

(a) hypotenuse: 20; side: 12
(b) hypotenuse: 2; side: 1
(c) hypotenuse: $\sqrt{113}$; side: 7

33. Use the converse of the Pythagorean theorem to determine whether or not a triangle with the given side lengths is a right triangle.

(a) 10, 24, 26
(b) $\sqrt{2}, \sqrt{3}, \sqrt{5}$
(c) 6, 8, 12

34. Use the converse of the Pythagorean theorem to determine whether or not a triangle with the given side lengths is a right triangle.

(a) 10, 20, 30
(b) $\sqrt{7}, \sqrt{8}, \sqrt{56}$
(c) $1, 5, \sqrt{26}$

35. Find the missing lengths in the figure. Round your answers to the nearest tenth.

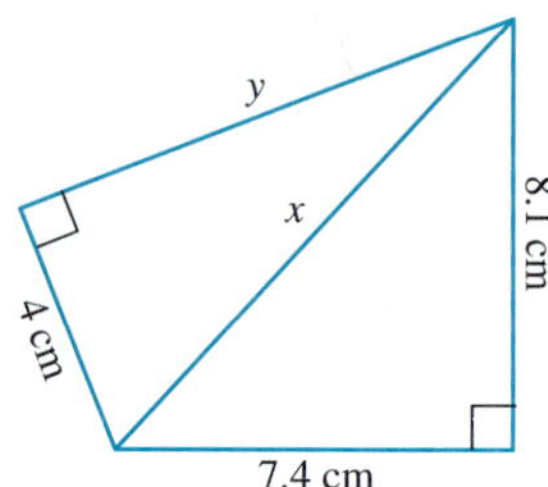

36. Find the missing lengths in the figure. Round your answers to the nearest tenth.

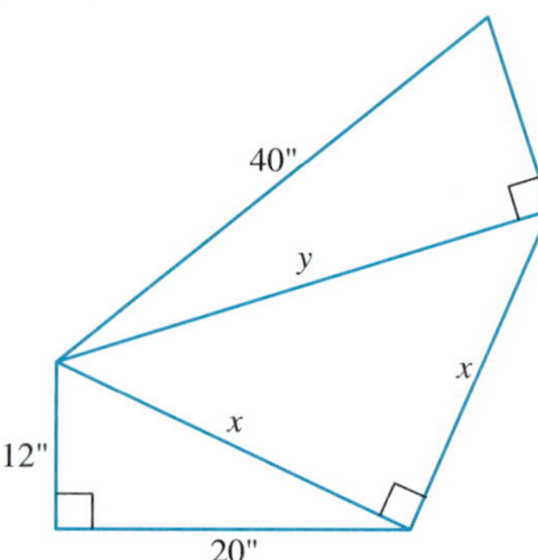

37. A baseball diamond is a square that measures 90 feet on a side. How far must the catcher throw the ball from home plate to second base to pick off a runner?

38. A 90-foot tall antenna on the flat roof of a building is to be secured with four cables. Each cable runs from the top of the antenna to a spot on the roof 30 feet from the base of the antenna. How much cable is needed? (Hint: draw a sketch.)

39. A 16-foot ladder will be used to paint a house. If the foot of the ladder must be placed at least 4 feet away from the house to avoid flowers and shrubs, what is the highest point on the house that the top of the ladder will reach? Round your answer to the nearest tenth of a foot.

40. If the diagonals of a square are 40 feet long, what is the length of each side? Round your answer to the nearest tenth of a foot.

41. The figure below is a rectangle composed of smaller squares. The numbers represent the side lengths of four of the squares. Find the side lengths of the other squares.

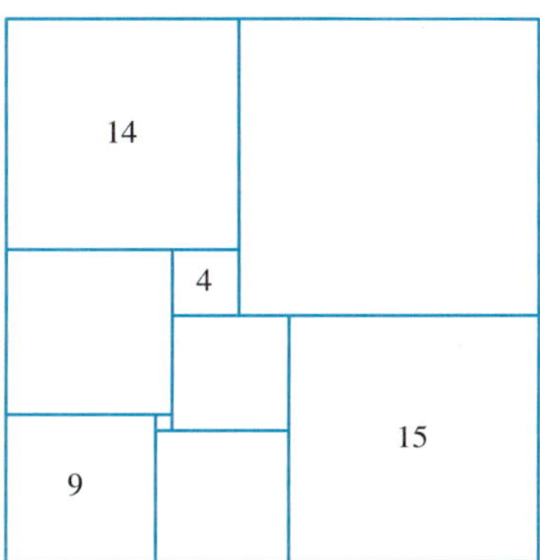

42. The figure below is a square composed of smaller squares. The numbers represent the side lengths of four of the squares. Find the side lengths of the other squares.

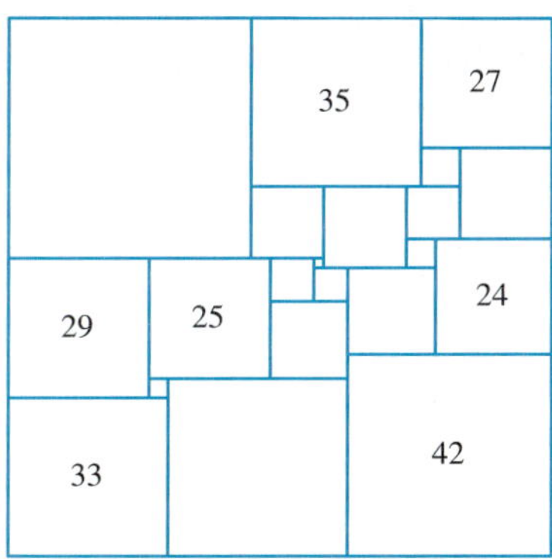

43. Draw two copies of the hexagon shown.

(a) Divide one hexagon into three identical parts so that each part is a rhombus (a quadrilateral having all sides the same length).

(b) Divide the second hexagon into six identical kites (a quadrilateral with two pairs of sides of equal length) which are non-overlapping.

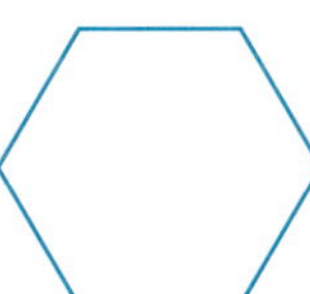

44. Draw two copies of the hexagon shown.

(a) Divide one hexagon into four identical trapezoids (quadrilaterals with one pair of parallel sides).

(b) Divide the second hexagon into eight identical polygons.

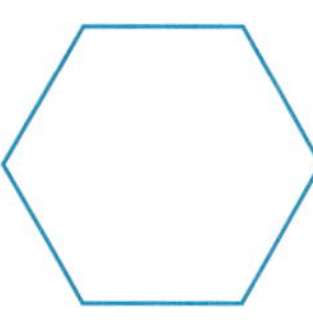

45. The following "tiling by regular polygons" was found in a coloring book. Why is it a fake?

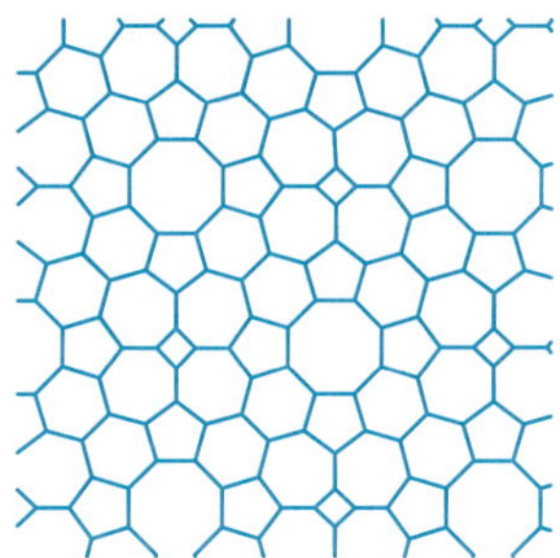

EXTENDED PROBLEMS

The following problems lead to a characterization of the semiregular tilings.

46. It was shown that a vertex angle of a regular n-gon has measure $\frac{(n-2)180}{n}$ degrees. If there are three regular polygons completely surrounding the vertex of a tiling, then

$$\frac{(a-2)180}{a}+\frac{(b-2)180}{b}+\frac{(c-2)180}{c}=360$$

where the three polygons have a, b, and c sides. Simplify this equation to arrive at the equation

$$\frac{1}{a}+\frac{1}{b}+\frac{1}{c}=\frac{1}{2}.$$

47. Problem 46 gives an equation that whole numbers a, b, and c must satisfy if a regular a-gon, a regular b-gon, and a regular c-gon completely surround a point.

(a) Let $a = 3$. Find all possible whole-number values of b and c that satisfy the last equation in problem 46.

(b) Repeat part (a) with $a = 4$.

(c) Repeat part (a) with $a = 5$.

(d) Repeat part (a) with $a = 6$.

(e) The above parts should provide all possible arrangements of regular polygons that completely surround a point. How many did you find?

48. The following triples give the number of sides for all possible arrangements of three regular polygons that *may* surround a point to form a tiling.

(3, 7, 42) (3, 8, 24) (3, 9, 18) (3, 10, 15) (3, 12, 12) (4, 5, 20) (4, 6, 12) (4, 8, 8) (5, 5, 10) (6, 6, 6)

The (6, 6, 6) arrangement yields a regular tiling. Also, Figure 10.14 shows that (3, 12, 12), (4, 6, 12), and (4, 8, 8) can be extended to form a semiregular tiling. Consider the (5, 5, 10) arrangement and the following figure.

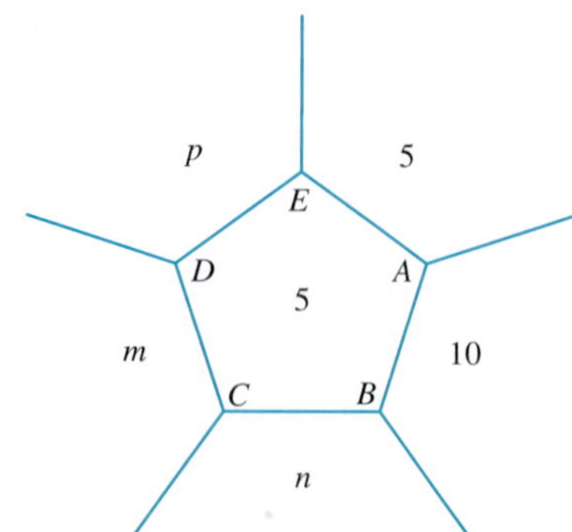

(a) Point A is surrounded by (5, 5, 10). If point B is surrounded similarly, what is n?

(b) If point C is surrounded similarly, what is m?

(c) If point D is surrounded similarly, what is p?

(d) What is the arrangement around E? This shows that (5, 5, 10) cannot be extended to a semiregular tiling.

(e) Show, in general, that this argument illustrates that the rest of the arrangements given in the list at the beginning of this problem cannot be extended to semiregular tilings.

49. When four polygons, an a-gon, a b-gon, a c-gon, and a d-gon, surround a point, it can be shown that the following equation is satisfied:

$$\frac{1}{a}+\frac{1}{b}+\frac{1}{c}+\frac{1}{d}=1.$$

(a) Find four combinations of whole numbers that satisfy this equation.

(b) One of these arrangements gives a regular tiling. Which one is it?

(c) The remaining three combinations can each surround a vertex in two different ways. Of those six arrangements, four cannot be extended to a semiregular tiling. Which are they?

(d) The remaining two can be extended to a semiregular tiling. Which are they?

50. (a) When five regular polygons surround a point, they satisfy the following equation.

$$\frac{1}{a}+\frac{1}{b}+\frac{1}{c}+\frac{1}{d}+\frac{1}{e}=\frac{3}{2}.$$

Find the two combinations of whole numbers that satisfy this equation.

(b) These solutions yield three different arrangements of polygons which can be extended to semiregular tilings. Illustrate those patterns.

51. (a) When six regular polygons surround a point, they satisfy the following equation:

$$\frac{1}{a}+\frac{1}{b}+\frac{1}{c}+\frac{1}{d}+\frac{1}{e}+\frac{1}{f}=2$$

Find the one combination of whole numbers that satisfies this equation.

(b) Can more than six regular polygons surround a point? Why or why not?

52. The following proof of the Pythagorean theorem, due to the Hindu mathematician Bhaskara (1114–1185), uses squares and right triangles. It is said that he simply presented the following picture and said "Behold!"

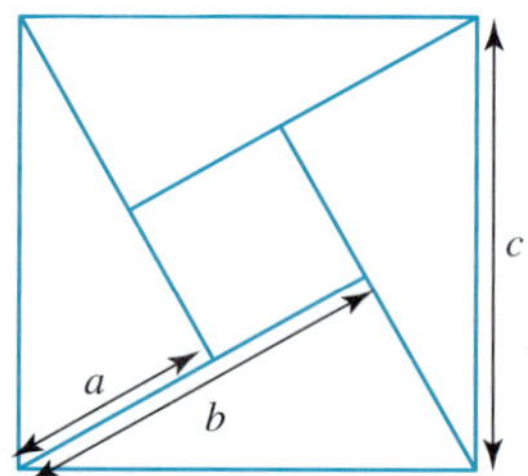

Use algebra and the areas of the squares and right triangles to verify the Pythagorean theorem.

10.2 SYMMETRY, MOTIONS, AND ESCHER PATTERNS

Art and pottery of most cultures can be described according to their symmetries. Describe the symmetries of the figure to the left.

In the previous section, tilings were classified according to the way that polygonal regions could be arranged around a point. Another way to view tilings and other patterns in art is through their symmetries. An excellent resource for advanced study on this topic is the book *Symmetries of Culture* by Washburn and Crowe. One fascinating aspect of this author team is that Washburn is an anthropologist and Crowe is a mathematician. Having seen each other's works, they decided to form a team to classify decorative patterns in various cultures.

STRIP PATTERNS AND SYMMETRY

Figure 10.31 displays a **strip** or **one-dimensional pattern** from the book by Washburn and Crowe.

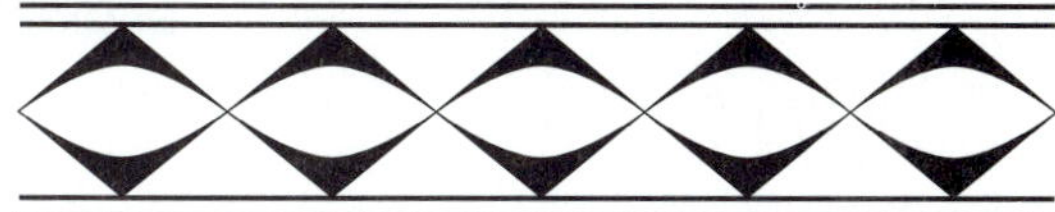

FIGURE 10.31

Informally, a figure has a **symmetry** if it can be moved so that the resulting figure looks identical to the original figure. If the pattern in Figure 10.31 is flipped or reflected across the vertical line [Figure 10.32(a) or the horizontal line [Figure 10.32(b)], the pattern will look the same after the reflection as it did initially.

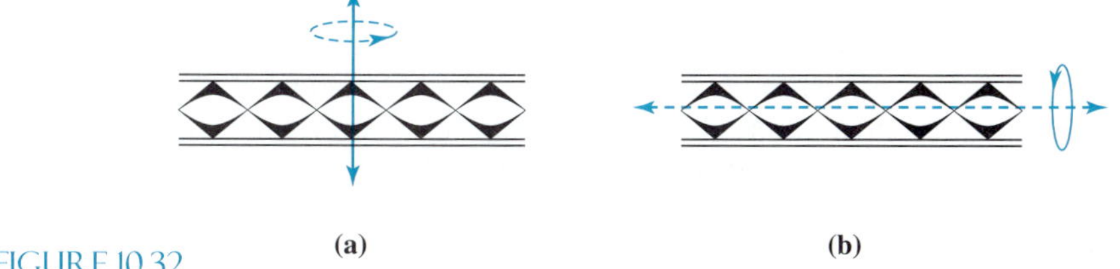

FIGURE 10.32

Thus we say that this pattern has **reflection symmetry.**

If the strip is turned or rotated a half-turn, or 180°, around a given point as shown inserted in Figure 10.33, the same pattern will be produced. Thus this pattern has **rotation symmetry.** (Note: if the only rotation symmetry that a pattern has requires a full, or 360°, turn, then we say that it has *no* rotation symmetry.)

FIGURE 10.33

Next, imagine that the strip pattern in Figure 10.31 extends indefinitely in two directions or wraps around an object such as a piece of pottery. The three dots at either end of the pattern indicate that it continues (Figure 10.34). If the pattern is slid or translated to the right as indicated by the dashed arrow, the same pattern will result. We say that this pattern has a **translation symmetry.**

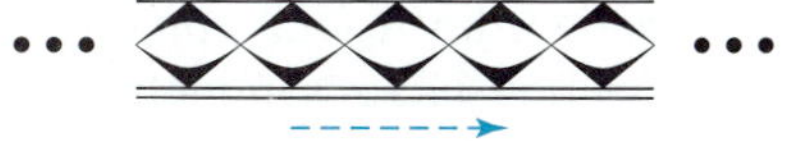

FIGURE 10.34

EXAMPLE 10.7 Find the symmetries in the patterns in Figure 10.35.

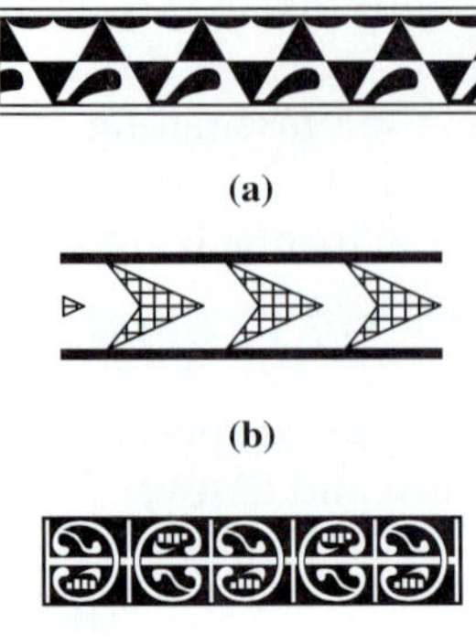

(a)

(b)

(c)

FIGURE 10.35

SOLUTION The pattern in Figure 10.35(a) has translation symmetry, but no reflection or rotation symmetry.

The pattern in Figure 10.35(b) has translation symmetry and reflection symmetry in the horizontal midline, but it has no rotation symmetry.

The pattern in Figure 10.35(c) has translation and rotation symmetries, but no reflection symmetry due to the alternating details. ◆

Finally, more complex patterns can be made by combining reflections, rotations, and translations that are performed one after the other. For example, in Figure 10.36(a) we have a basic figure. The figure is translated in (b), with copies of the figure left in place. In (c), the image following the translation is rotated with a copy left in place. If we look at Figure 10.35(c), we see that it can be made with any two adjacent rectangles in the same line using the translation and rotation as described.

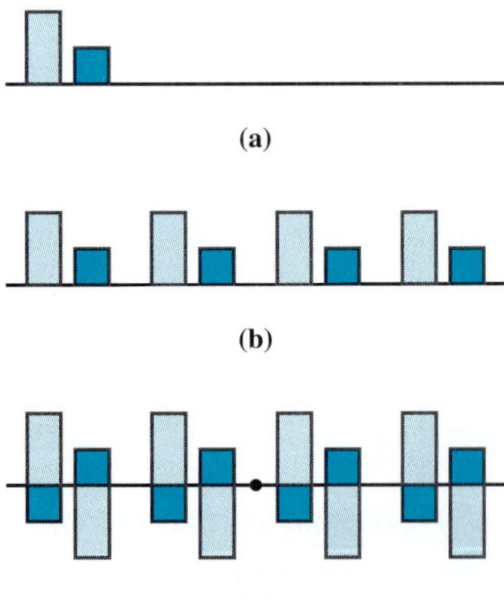

(a)

(b)

(c)

FIGURE 10.36

It is often assumed that any one-dimensional strip pattern will have translation symmetry, and this is commonplace for strip patterns in most cultures. However, we can create a horizontal *pattern* (an orderly sequence of repeated shapes) that does not have translation symmetry. As one example, suppose circles are placed to the left and the right from a starting circle in such a way that the distance between adjacent circles is increased one millimeter each time. The pattern could not have translation symmetry even though it has both reflection and rotation symmetry (Figure 10.37).

FIGURE 10.37

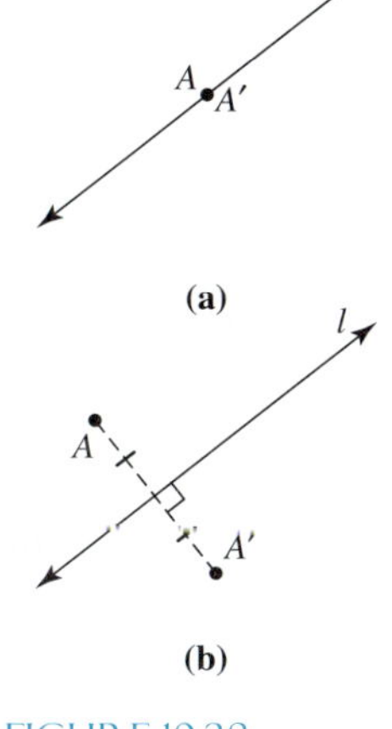

FIGURE 10.38

RIGID MOTIONS

Any combination of reflections in lines, rotations around a point, and translations is called a **rigid motion** or **isometry** (which means *same measure*). The study of reflections, rotations, translations, and combinations thereof is often called **motion geometry.**

The terms *reflections, translations,* and *rotations,* which have been used informally thus far, can be defined more formally in precise mathematical terms. A **reflection with respect to line** l is defined by describing the location of the image of each point of the plane as follows (A' represents the *image of* A with respect to the motion):

(i) If A is a point on l, then $A = A'$ (that is, a point on the line of reflection is its own image) [Figure 10.38(a)].

(ii) If A is not on l, then l is the perpendicular bisector of $\overline{AA'}$ [Figure 10.38(b)]. A and A' are the same distance away from l.

(Note: reflections are often thought of as mirror images with respect to the line of reflection.)

Next we consider the effect of the definition of a reflection when we apply it to a triangle instead of a single point.

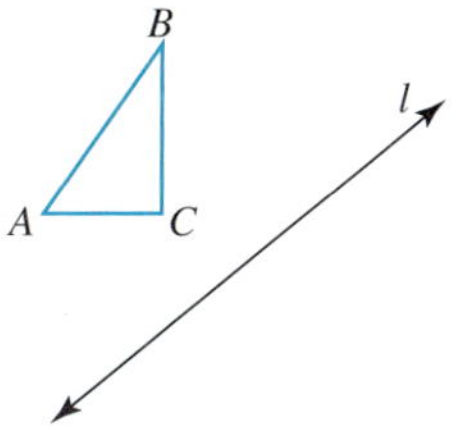

FIGURE 10.39

EXAMPLE 10.8 Describe the image of ΔABC under the reflection with respect to line l as shown in Figure 10.39.

SOLUTION Line l will be the perpendicular bisector of $\overline{AA'}$. So we construct a line through A that is perpendicular to l, then find A' on the other side of l so that $A'P = AP$ [Figure 10.40(a)]. Then we find points B' and C' in the same way. The respective images A', B', C' of A, B, and C are shown in Figure 10.40(b). Notice that $\Delta A'B'C'$ and ΔABC are the same size and shape, and that the orientation of ΔABC is clockwise (when vertices are read A-B-C), whereas the orientation of $\Delta A'B'C'$ is the opposite, namely counterclockwise.

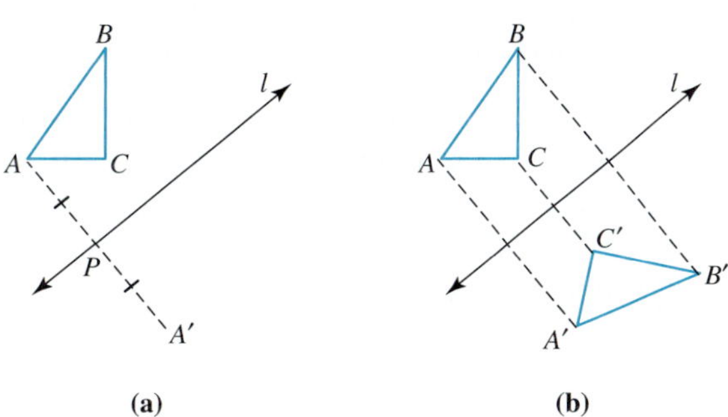

FIGURE 10.40

◆

Figure 10.41 shows a two-dimensional pattern where two lines of reflection have been inserted.

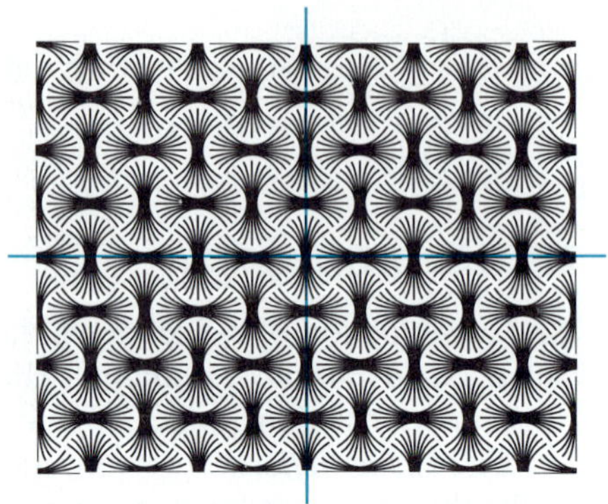

FIGURE 10.41

The rigid motion called a translation can be visualized as a puck sliding along the ice. Mathematically we represent a translation by a *directed* line segment. That is, a line segment where one end of the segment is the beginning point, and the other end (designated by an arrowhead) is the ending point. Such a directed line segment is called a **vector.** Associated with each vector is a length (the length of the line segment) and a direction (the measure of the angle the vector makes with the positive x-axis) [Figure 10.42(a)]. Now imagine moving every point in a plane the same distance and in the same direction as indicated by the vector v [Figure 10.42(b)]. Think of A as a puck.

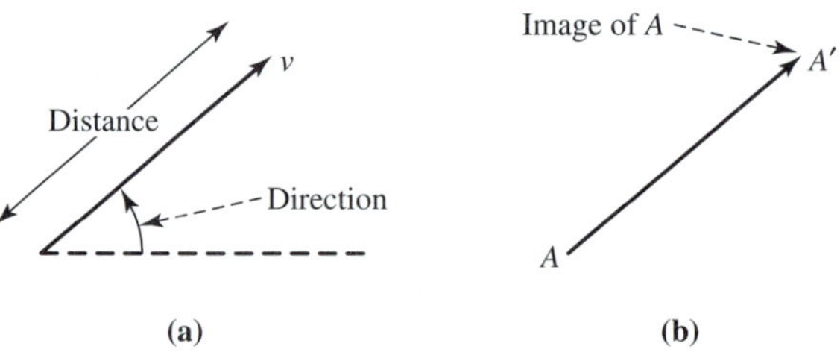

FIGURE 10.42

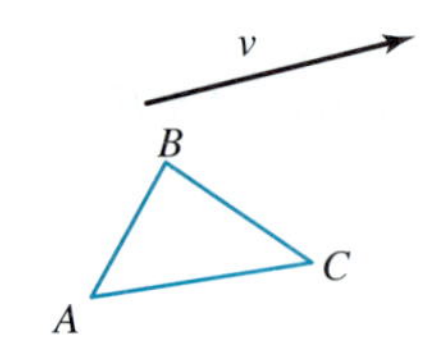

FIGURE 10.43

A vector is often denoted as v or as $\overrightarrow{AA'}$, where A is the initial point of the arrow and A' is the tip of the arrowhead as shown in Figure 10.42(b). A **translation** is defined by describing the location of the image of each point of the plane as follows.

> A vector, v, assigns to every point A in a plane, a point A' which is determined by the length and direction of v [Figure 10.42(b)].

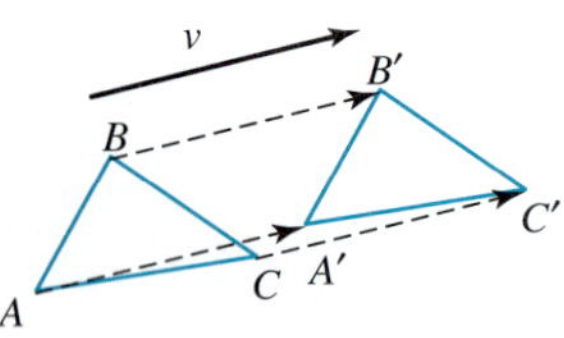

FIGURE 10.44

EXAMPLE 10.9 Describe the image of ΔABC under the translation determined by the vector v in Figure 10.43.

SOLUTION Since the vector v takes all points the same distance and direction, ΔABC is assigned to $\Delta A'B'C'$ as shown in Figure 10.44. Notice that vectors $\overrightarrow{AA'}$, $\overrightarrow{BB'}$, and $\overrightarrow{CC'}$ all have the same length and direction as the vector v. ◆

Figure 10.45 shows a two-dimensional pattern having translation symmetry in both the horizontal and vertical directions assuming *the pattern extends infinitely in all directions.*

FIGURE 10.45

The rigid motion called a rotation involves turning a figure clockwise or counterclockwise. A rotation is determined by a point, O, and a directed angle. A **directed angle** is an angle where one side is identified as the initial side, and its second side is the terminal side. An angle can be directed either clockwise or counterclockwise. In Figure 10.46(a), A is rotated counterclockwise 60° around O to A'. Here, $\overline{OA}$ is the initial side of $\angle AOA'$ and $\overline{OA'}$ is the terminal side. Angles that are directed counterclockwise are assigned a positive number. Thus we say that the measure of directed angle $\angle AOA'$ is 60°. In Figure 10.46(b) B is rotated clockwise 90° around O to B', and so directed angle $\angle BOB'$ has measure $-90°$, where the negative sign indicates its clockwise rotation. The point O is called the **center** of the rotation in each case.

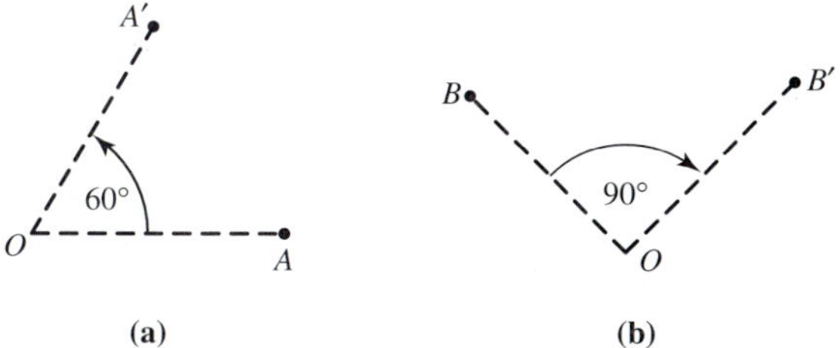

FIGURE 10.46

A **rotation** is defined by describing the location of the image of each point of the plane as follows:

> The image of a point X under the rotation determined by the directed angle $\angle AOB$ in Figure 10.47(a) is the point X' where
> (i) $OX = OX'$ and
> (ii) $\angle XOX' = \angle AOB$ as *directed* angles. [Figure 10.47(b)].

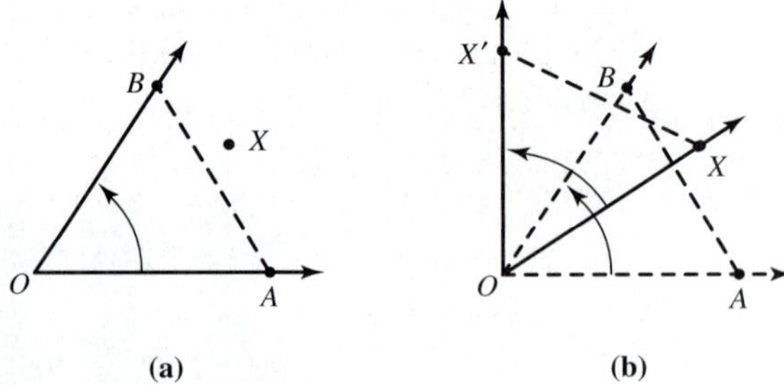

FIGURE 10.47

Once a rotation is defined by its center and its directed angle (whose vertex is the center of rotation), the image of every point in the plane is determined. That is, we can find the image A' of any point A in the plane.

Next we consider the effect a rotation has on a triangle.

EXAMPLE 10.10 Describe the image of ΔABC under the rotation with center O and directed angle $\angle XOX'$ as shown in Figure 10.48.

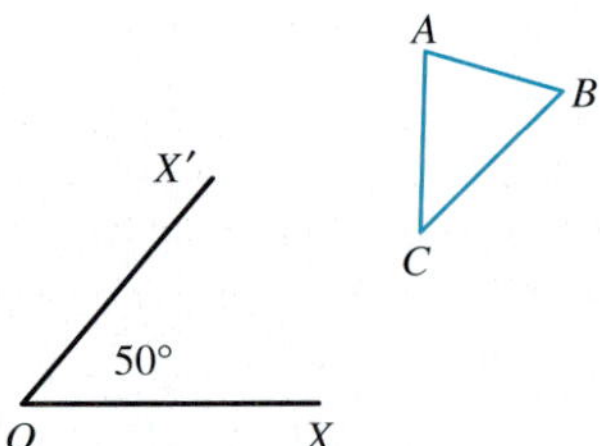

FIGURE 10.48

SOLUTION The respective images A', B', C' are shown in Figure 10.49. Notice that $OA = OA'$, $OB = OB'$, and $OC = OC'$. Also, $\angle AOA' = \angle BOB' = \angle COC' = 50°$.

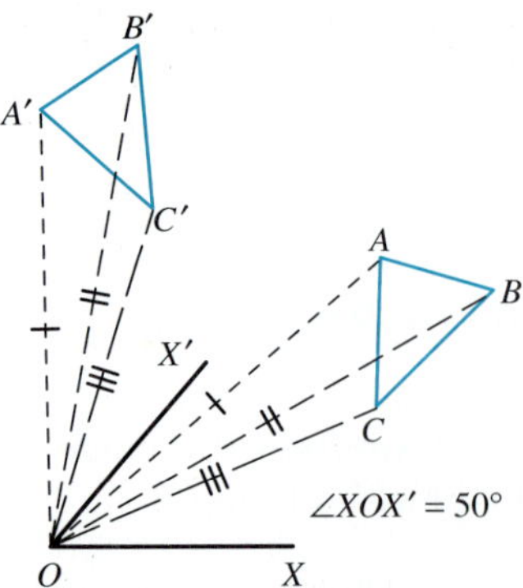

FIGURE 10.49

◆

FIGURE 10.50

Figure 10.50 shows a two-dimensional pattern having rotation symmetries of 120° and 240° around several different centers.

The center, O, of a rotation always corresponds to itself, thus we call it a fixed point. In general, a point A is called a **fixed point** under a transformation if A and its image, A', are the same point. Thus the center of any rotation is a fixed point. In a reflection in a line, the fixed points are the points of the line of reflection. Translations have no fixed points.

Another common rigid motion that is a combination of two rigid motions is motivated by footprints in the sand as pictured in Figure 10.51.

FIGURE 10.51

It is impossible to translate or rotate the left foot to the right foot because the feet have a different orientation. Also, no reflection line can be found to reflect one foot onto the other. But, as Figure 10.52 shows, the right foot can be obtained as the image of the left foot when it is transformed by a reflection that is followed by a translation.

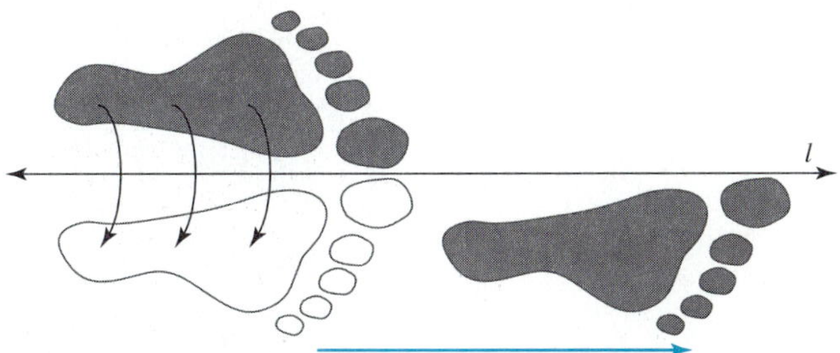

FIGURE 10.52

Notice that the left foot could also have been transformed to the right foot by first translating along l, then reflecting with respect to line l. The rigid motion pictured in Figure 10.52 is called a **glide reflection,** which means a reflection combined with a translation. It is usually assumed that the translation is in a direction parallel to the line of reflection. However, it can be shown that this assumption is not necessary. A thorough treatment of this topic can be found in *Symmetries of Culture* by Washburn and Crowe.

Using the rigid motions, it can be shown that there are only seven basic one-dimensional repeated patterns. The seven types of patterns are shown in Table 10.3 where the triangles are place holders for more complicated figures, and color or other factors are not considered.

TABLE 10.3 The Seven One-Dimensional Patterns

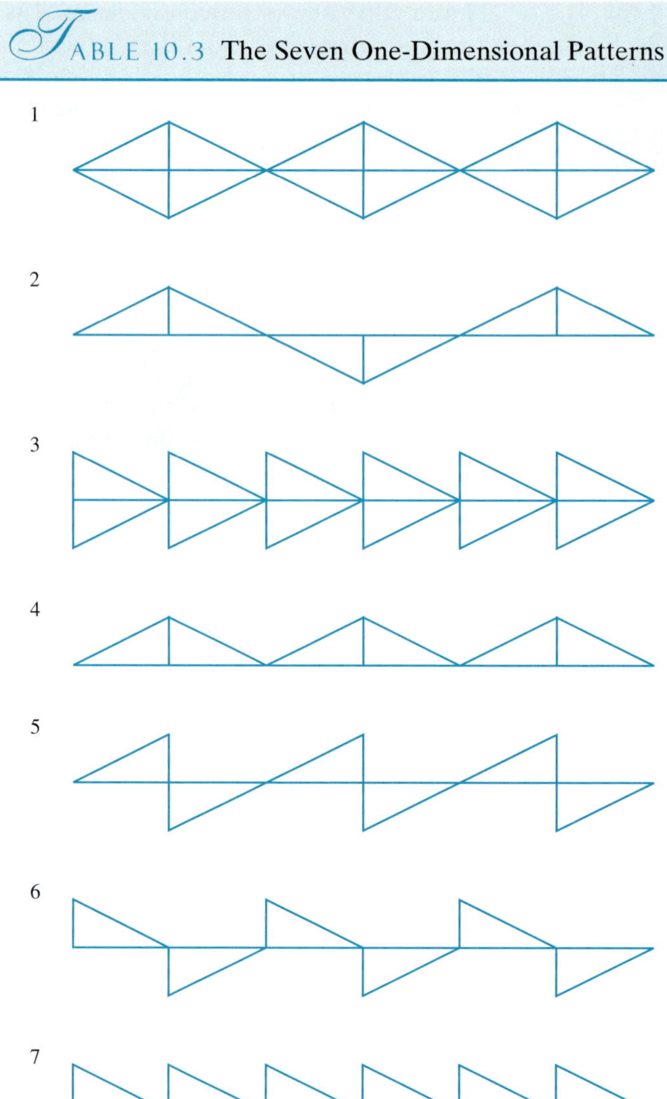

Some of the strip patterns that appear earlier in this section fit into this classification as follows:

Figure	Pattern
Figure 10.31	1
Figure 10.35(a)	7
Figure 10.35(b)	3
Figure 10.35(c)	5

In the case of two-dimensional patterns, there are exactly seventeen possible types of such patterns.

Patterns and symmetries have played an important role in the art and culture of many civilizations, in particular when they are incorporated in tilings.

Geometric forms have been an important part of our discussion about tilings, and we have talked about tiling the plane and using Euclidean geometry. But if the surface to be covered is not a plane, but a sphere or other form, new types of tilings can be created. Finally, if the shapes used for tilings are not polygons, but some other appropriate figures that cover the surface with no gaps or overlaps, even more interesting tilings can be created. During the 1930s, the graphic artist M. C. Escher began exploring new concepts in geometry and applying them to his art with stunning and beautiful results.

ESCHER PATTERNS

Maurits Escher (1898–1972) was born in the Netherlands. Although his school experience was largely a negative one, he looked forward with enthusiasm to his two hours of art each week. His father urged him into architecture to take advantage of his artistic ability. However, that endeavor did not last long. It became apparent that Escher's talent lay more in the area of decorative arts than in architecture, so he began the formal study of art when he was in his twenties.

Escher's works are varied, and much of it is based on mathematics. The *Circle Limit III* is based on a hyperbolic tiling where the fish seem to swim to infinity [Figure 10.53(a)]. In spherical geometry (see Figure 10.7) the sum of the angles in a triangle is more than 180°; on a surface in what is known as hyperbolic geometry, the sum is less than 180°.

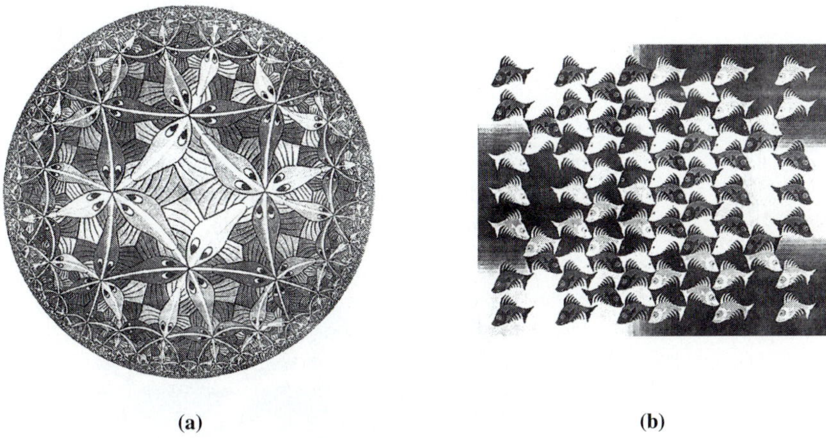

(a) (b)

FIGURE 10.53

Fish is one of Escher's many ever-changing pictures. Here arched fish evolve, appear, and disappear across the drawing [Figure 10.53(b)].

Our next goal is to see how Escher used rigid motions to prepare some of his patterns. First, we begin with a square [Figure 10.54(a)].

Next we cut a piece from the upper left corner [Figure 10.54(b)] and translate that piece to the bottom of the square [Figure 10.54(c)]. Repeat this process by reflecting the left side to the right side across a vertical line through the middle of

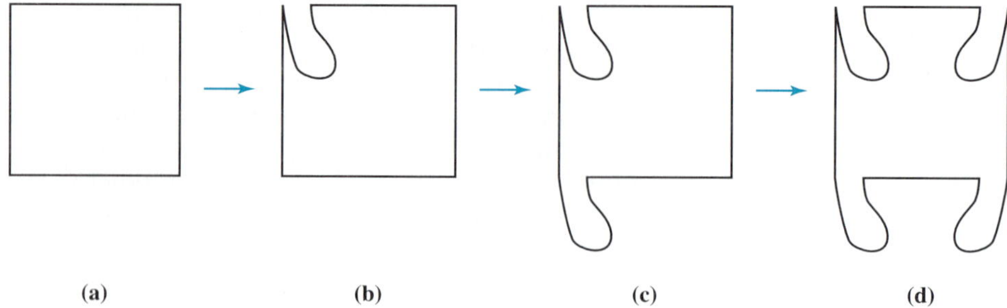

FIGURE 10.54

the square [Figure 10.54(d)]. If this process is applied to a grid of squares, and eyes and whiskers are added, "Voila!," a collection of kittens is born (Figure 10.55).

FIGURE 10.55

This tiling of cats has both vertical and horizontal translation symmetry and a vertical line reflection symmetry. Two other works of Escher shown next illustrate his use of rigid motions (Figure 10.56).

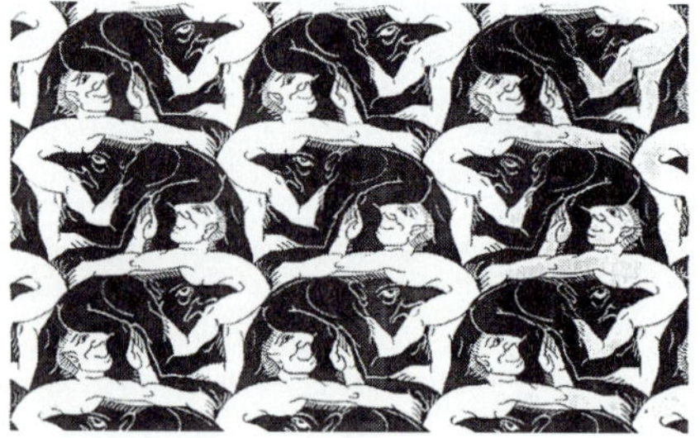

Translation symmetry

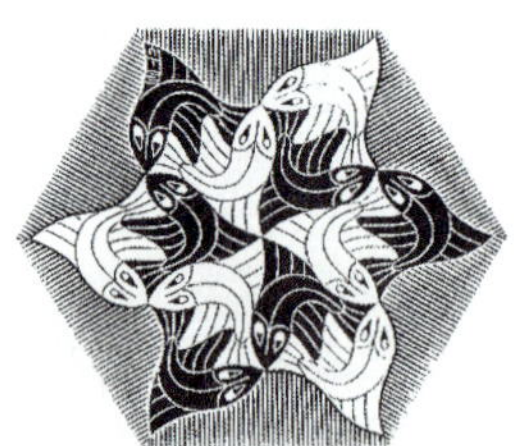

Rotation symmetry

FIGURE 10.56

Methods for constructing patterns having rotation symmetry will be given in the problem set.

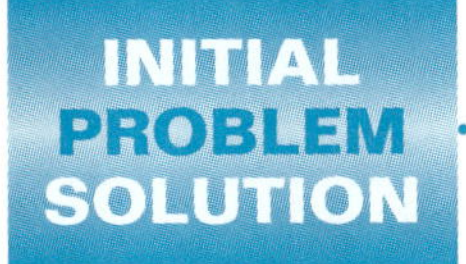

Art and pottery of most cultures can be described according to their symmetries. Describe the symmetries of the figure below.

SOLUTION The figure has a 180° rotation symmetry, but no reflection or translation symmetries.

PROBLEM SET 10.2

1. Find several symmetries in the following wall paper designs. Describe them clearly, or show an illustration.

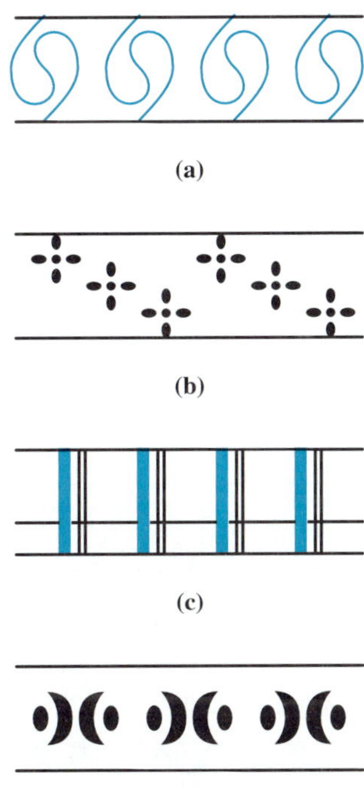

(a)

(b)

(c)

(d)

2. Find several symmetries in the following wall paper designs. Describe them clearly, or show an illustration.

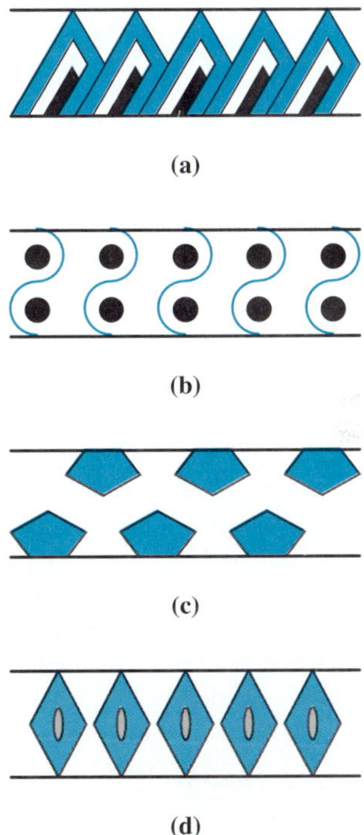

(a)

(b)

(c)

(d)

In problems 3 through 6, find four lines for reflection symmetry that produce the same pattern as the original. Each pattern continues indefinitely in all directions.

3.

4.

5.

6.

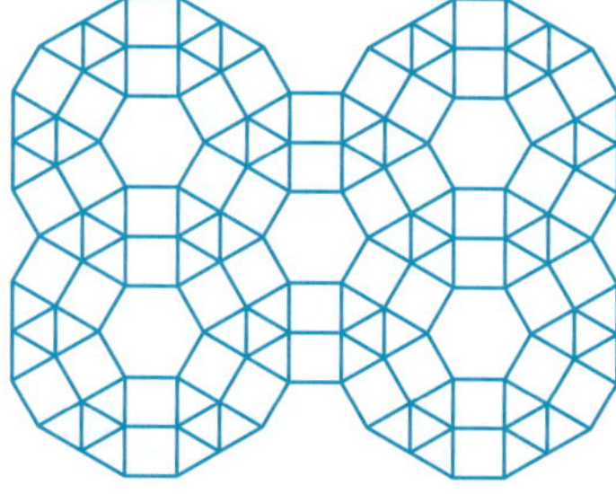

In problems 7 through 10, find four rotations that produce the same pattern as the original. Identify the centers of rotation as A, B, C, D and list the possible angles of each rotation. Each pattern continues indefinitely in all directions.

7.

8.

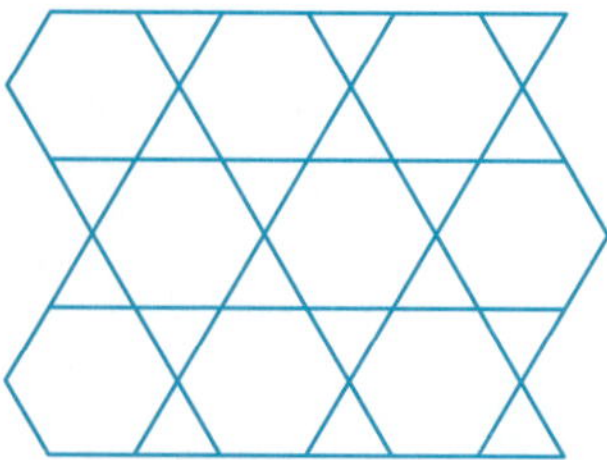

9.

10.

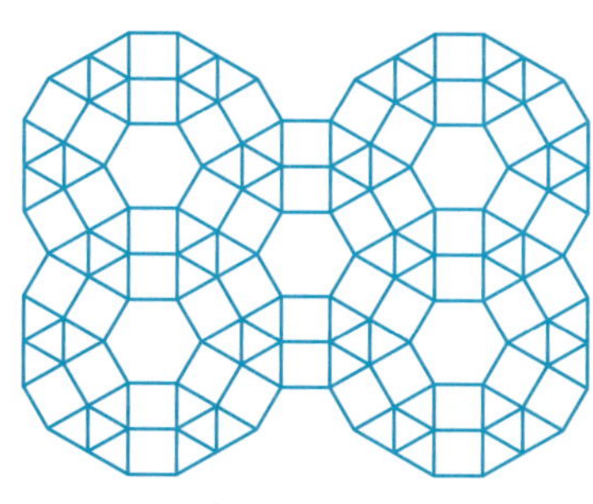

In problems 11 through 14, find four translations that produce the same pattern as the original. Identify the distance and direction with a vector that shows the movement of a point in the pattern. Each pattern continues indefinitely in all directions.

11.

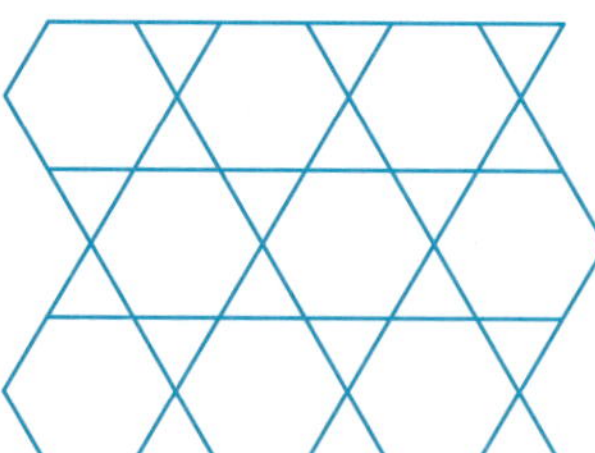

12.

13.

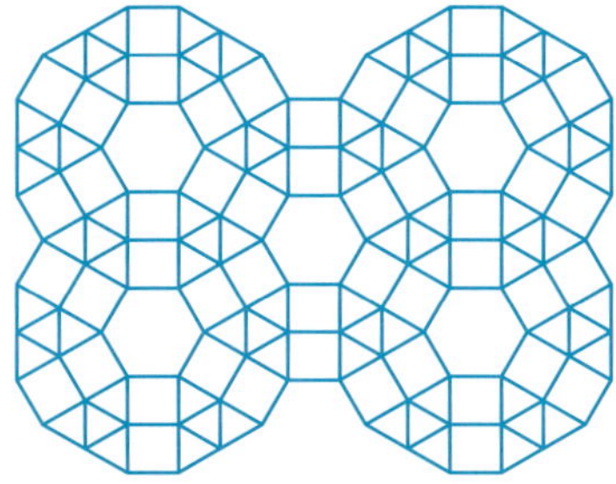

14.

In problems 15 to 18, identify the types of symmetry present in each of the patterns. Make a free hand sketch for each type of symmetry, and indicate appropriate lines, distances, centers, and angles as in previous problems. The complete pattern is shown.

15.

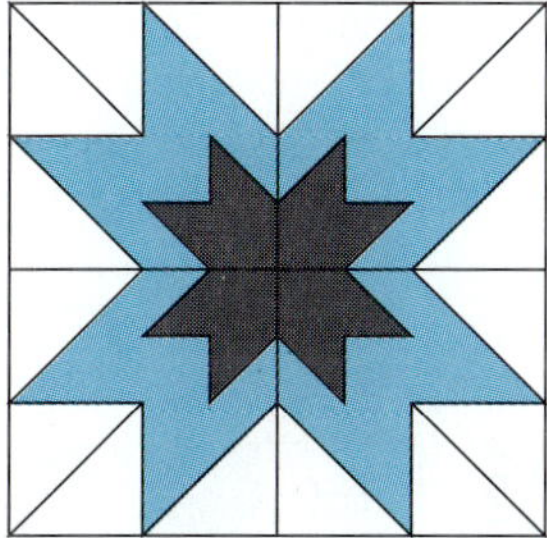

16.

17.

18.

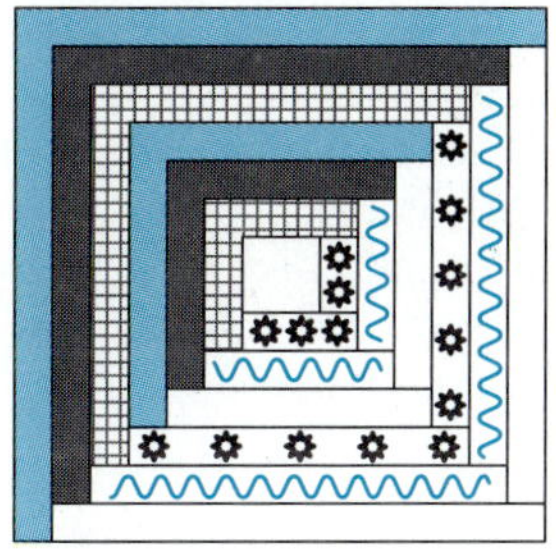

19. **(a)** Does the rectangle shown have reflection symmetry? If so, how many axes of symmetry does it have?

(b) Does the rectangle have rotation symmetry? If so, how many different rotation symmetries does it have?

20. **(a)** Does the rhombus shown have reflection symmetry? If so, how many axes of symmetry does it have?

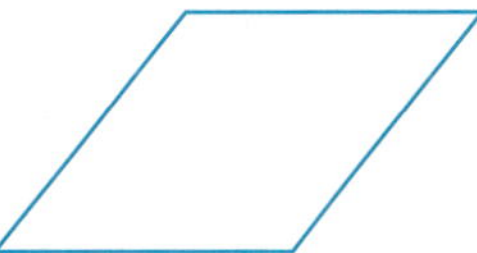

(b) Does the rhombus have rotation symmetry? If so, how many different rotation symmetries does it have?

21. **(a)** Does the isosceles trapezoid (the two nonparallel sides have the same length) shown have reflection symmetry? If so, how many axes of symmetry does it have?

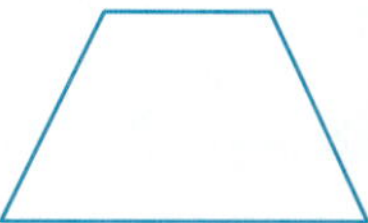

(b) Does the isosceles trapezoid have rotation symmetry? If so, how many different rotation symmetries does it have?

22. **(a)** Does the right isosceles triangle shown have reflection symmetry? If so, how many axes of symmetry does it have?

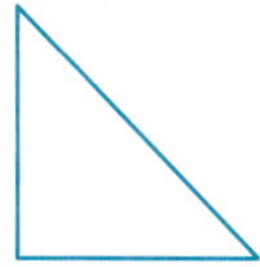

(b) Does the right isosceles triangle have rotation symmetry? If so, how many different rotation symmetries does it have?

23. **(a)** Draw the lines of symmetry in the regular n-gons in (i)–(iii). How many does each have?

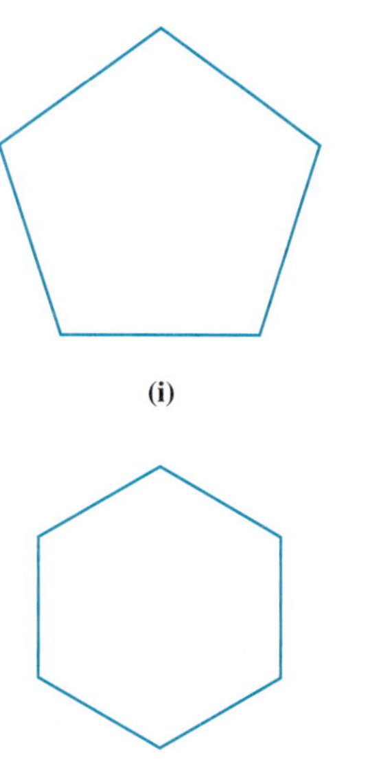

(i)

(ii)

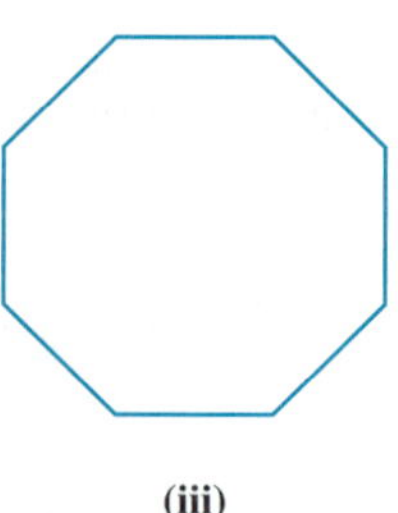

(iii)

(b) How many lines of symmetry does a regular n-gon have?

24. **(a)** Trace the regular n-gons in problem 23 to find all of the rotation symmetries for a regular pentagon, a regular hexagon, and a regular octagon.

(b) How many rotation symmetries does a regular n-gon have?

25. Bingo is played on a 5-by-5 grid of squares in which certain squares must be covered in order to win, usually five squares in a row vertically, horizontally, or diagonally.

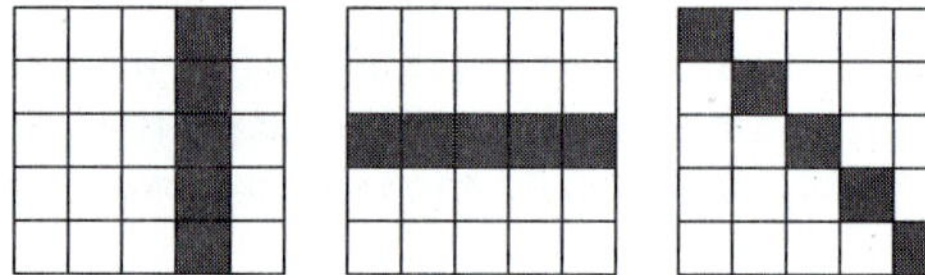

To make the games more interesting, other patterns on the grid are often chosen to be winners. Several examples are shown below. For each one, tell whether the pattern has reflection symmetry, rotation symmetry, or both.

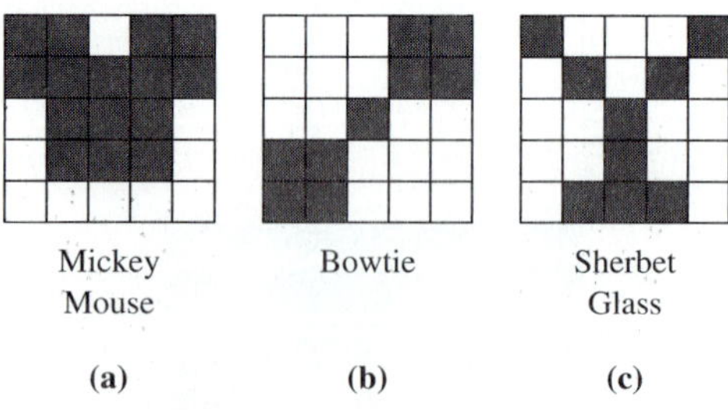

Mickey Mouse (a) | Bowtie (b) | Sherbet Glass (c)

26. For each one, tell whether the pattern has reflection symmetry, rotation symmetry, or both.

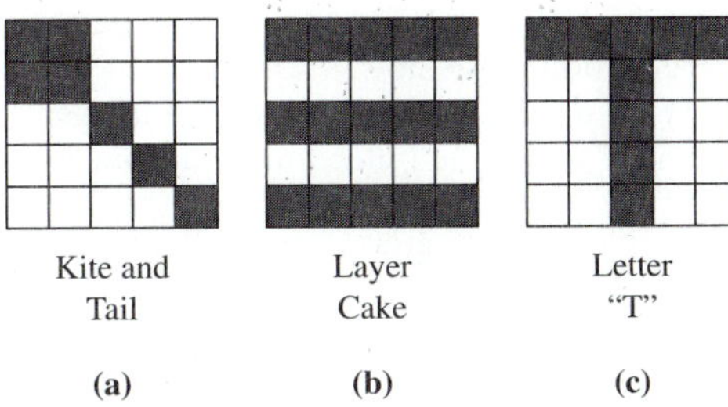

Kite and Tail (a) Layer Cake (b) Letter "T" (c)

27. Find a reflection and translation, if possible, that combine as a glide reflection that produces the original pattern.

28. Find a reflection and translation, if possible, that combine as a glide reflection that produces the original pattern.

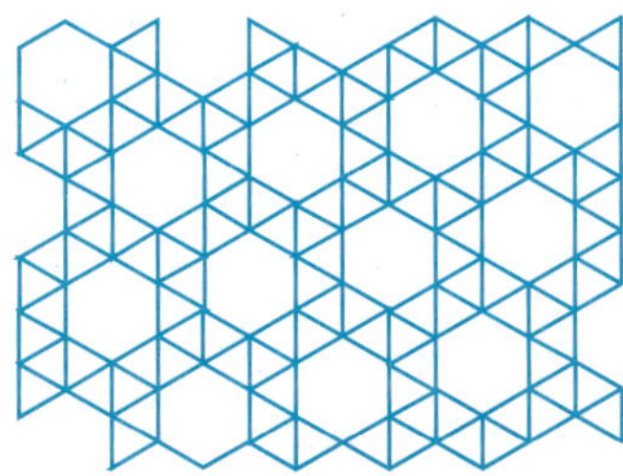

29. Find at least four translations in different directions that map the tiling onto itself. Indicate the starting and ending points of the translation.

30. Find at least four translations in different directions that map the tiling onto itself. Indicate the starting and ending points of the translation.

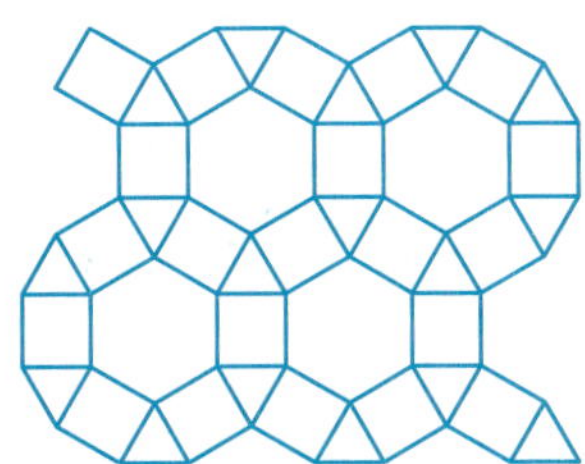

31. Find the 90° counterclockwise rotation of point P around point O. Add points to the graph as needed.

P O

(a)

P O

(b)

P O

(c)

32. Find the rotation of $\overline{AB}$ around point O with respect to the given angles.

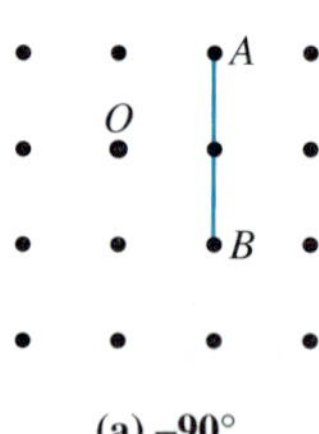

(a) –90°

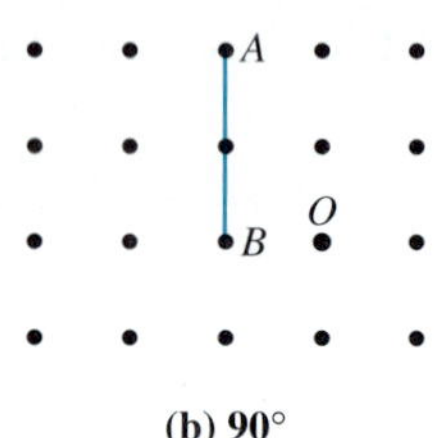

(b) 90°

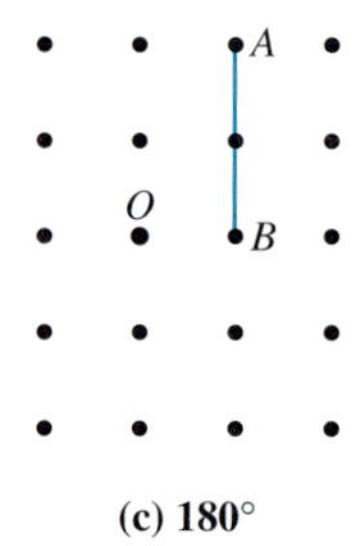

(c) 180°

33. **(a)** Graph ΔABC with A at (2, 1), B at (3, −5), and C at (6, 3). Then graph its image under the reflection in the y-axis.

(b) What are the coordinates of the images of A, B, and C. under the reflection in the y-axis?

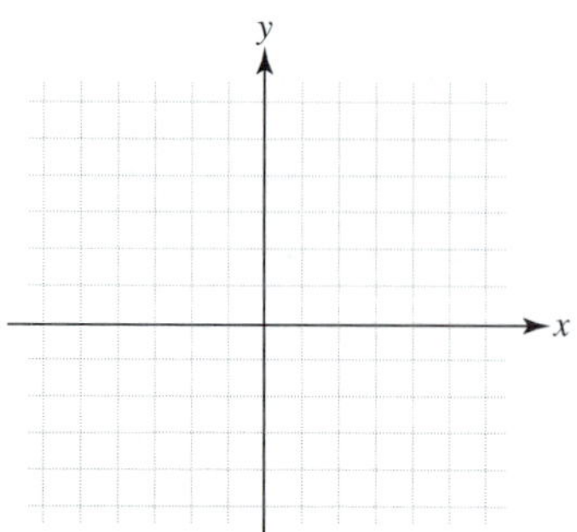

(c) If a point P has coordinates (a, b), what are the coordinates of its image under the reflection in the y-axis?

34. **(a)** Graph ΔABC with A at (3, 1), B at (4, 3), and C at (5, −2). Then graph its image under the reflection in the line as shown.

(b) What are the coordinates of the images of A, B, and C under the reflection in this line?

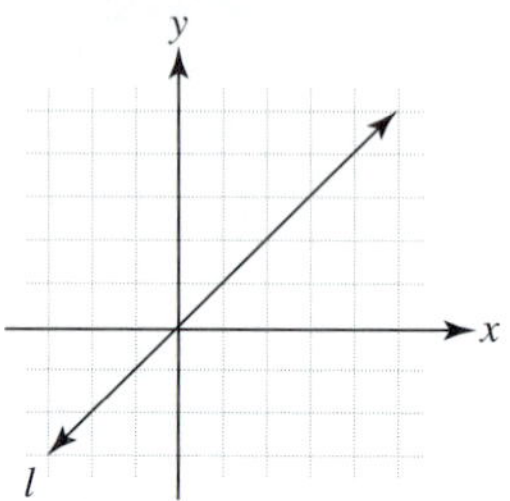

(c) If a point P has coordinates (a, b), what are the coordinates of its image under this reflection?

EXTENDED PROBLEMS

Escher-type tilings with irregular shapes can be produced by using the guidelines in the following problems.

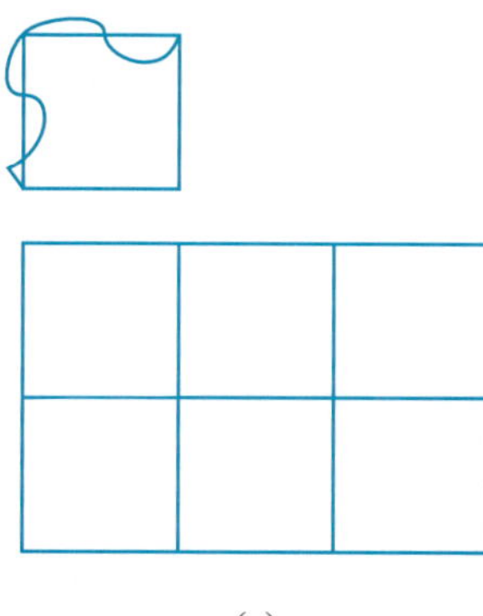

(a)

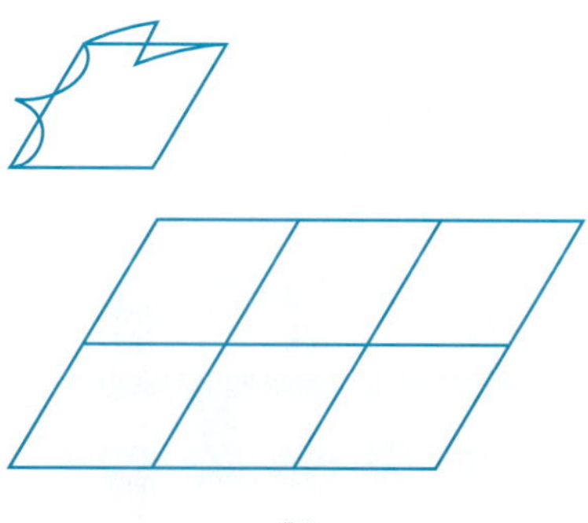

(b)

36. Translate each of the curves to the opposite sides of the parallelogram to create a shape that tiles. Use a grid to show that the shape will tile the plane.

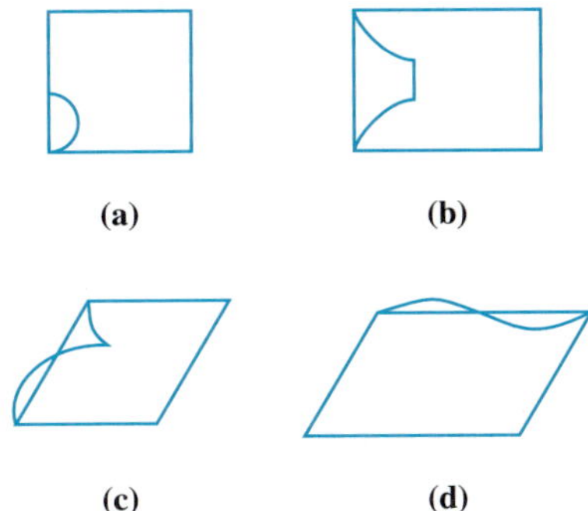

Rotations can also be used to make Escher-type drawings. The following figure shows a triangle that has been altered by rotation to produce an Escher-type pattern.

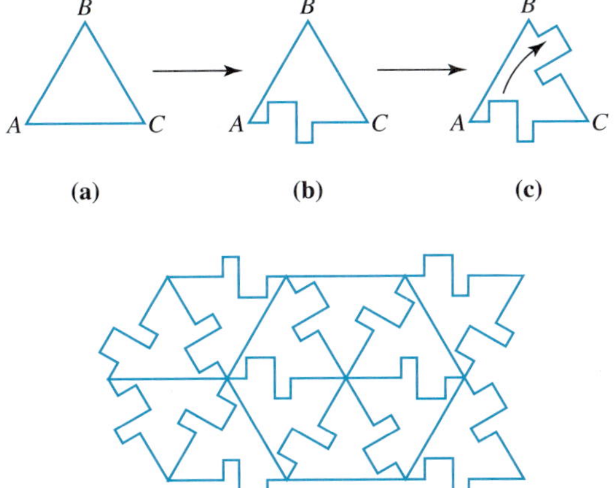

Side $\overline{AC}$ of ΔABC in (a) is altered arbitrarily in (b), provided that points A and C are not moved. Then, using point C as the center of a rotation, altered $\overline{AC}$ is rotated so that A is rotated to B in (c). The result is an alteration of $\overline{BC}$. The shape will tile the plane as shown.

37. Using the figure above, alter $\overline{AB}$ in a different way, rotate to $\overline{AC}$, using A as center of rotation. Verify that the shape will tile the plane.

38. **(a)** Alter $\overline{AB}$, $\overline{BC}$, and $\overline{CD}$.

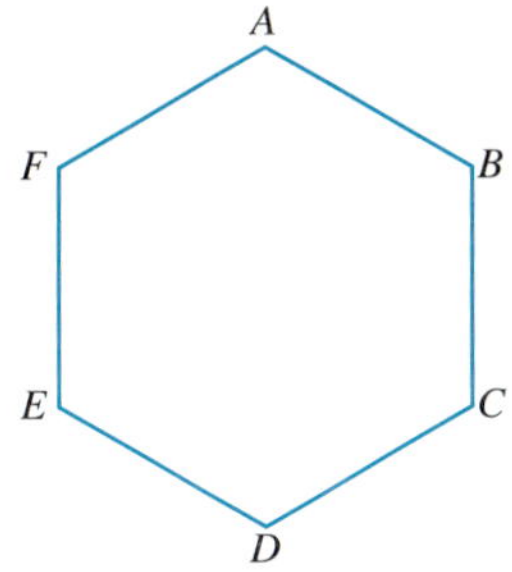

(b) Translate the changes to the opposite sides ($\overline{AB}$ to $\overline{ED}$, etc.)

(c) Use the grid to verify that the shape will tile the plane.

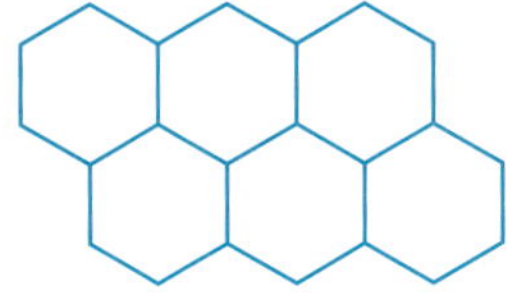

10.3 CONIC SECTIONS: PARABOLAS

A solar reflector is to be designed to collect sun rays and reflect them into a single point to create a concentrated stream of heat. What is a good design for such a reflector?

The first two sections of this chapter focused on figures in the plane, with an emphasis on patterns and tilings. Tilings played a major role in the cultures of many early societies, dating from the time of the Greek civilization. However, ancient tilings and those from the 12th century (as intricate and beautiful as they may be) are quite different from those of M. C. Escher. Like tilings, conic sections also have a long history dating back to the Greeks. Unlike tilings, however, little has been added to our knowledge of conic sections in the past 2000 years except for their representations in algebraic form and their applications related to advances in science and technology. Our understanding of the paths of projectiles, the orbits of planets, and the receivers and projectors of light and sound have been greatly enhanced by knowledge of the conic sections.

THE CONIC SECTIONS

An ice cream cone is a portion of a geometric figure that has been intensely studied throughout history both for its abstract properties and its many applications in science. The common mathematical definition of a **cone** is the surface formed as follows: consider a circle with center C on a plane and a point, V, above the plane so that $\overline{VC}$ is perpendicular to the plane [Figure 10.57(a)].

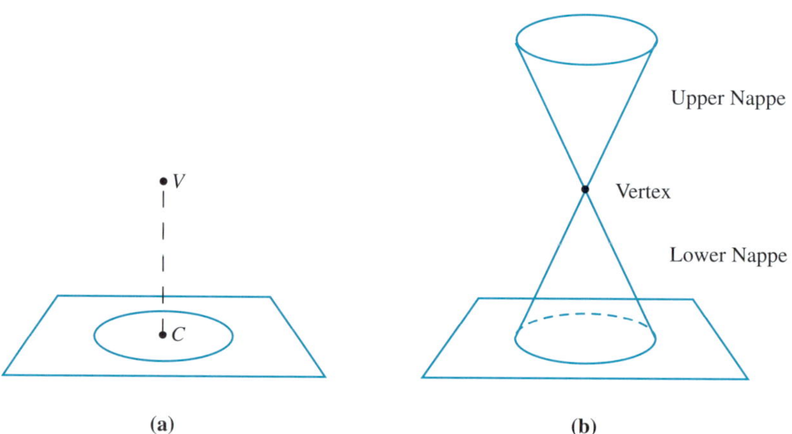

FIGURE 10.57

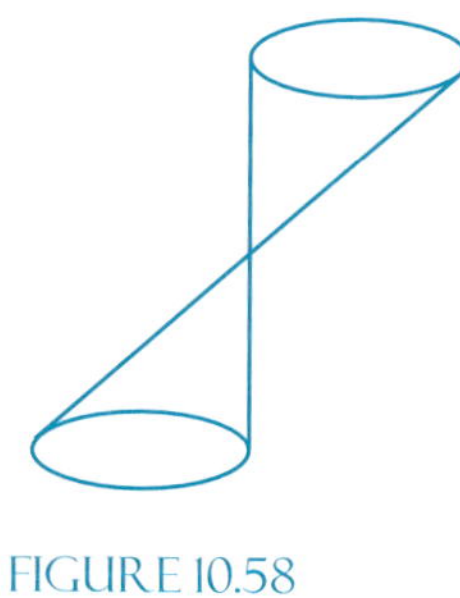

FIGURE 10.58

The figure formed by all lines passing through V and the circle is called a **right circular cone** [Figure 10.57(b)]. Point V is called the **vertex of the cone** and the two parts above V and below V are called the **nappes of the cone** (only a portion of the cone is shown—the two nappes are infinite in extent since they are composed of lines. If $\overline{VC}$ is not perpendicular to the plane, then the figure generated is called an **oblique circular cone** (Figure 10.58).

Slicing a cone with a plane produces several types of curves in the plane, called **conic sections** (Figure 10.59), which we will examine, together with some of their important applications. A **parabola** is formed when the intersecting plane is parallel to the side of the cone [Figure 10.59(a)], an **ellipse** when the intersecting plane intersects only one nappe of the cone but is not parallel to the side of the cone [Figure 10.59(b)], and a **hyperbola** when the intersecting plane intersects both nappes of the cone [Figure 10.59(d)]. A **circle,** which is a special ellipse, is formed in the case when the intersecting plane is parallel to the original plane [Figure 10.59(c)]. This section will study the parabola and its many important applications, and the next section will study the ellipse and hyperbola.

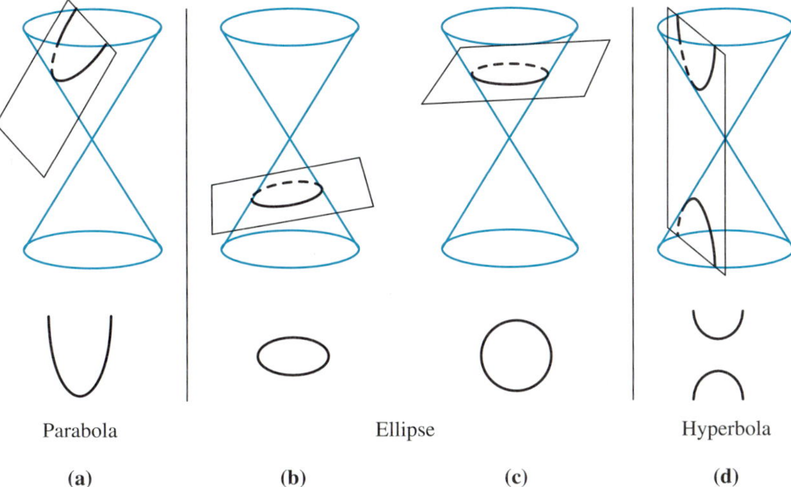

FIGURE 10.59

THE PARABOLA

In addition to being a particular type of slice of a cone, a parabola is defined as follows.

Parabola

A **parabola** is a figure in a plane determined by a fixed line and a fixed point, not on the line, as follows: a point is on the parabola if it is the same distance from the fixed line and the fixed point.

The line and the point that determine the parabola are called its **directrix** and **focus,** respectively. Figure 10.60 illustrates this definition.

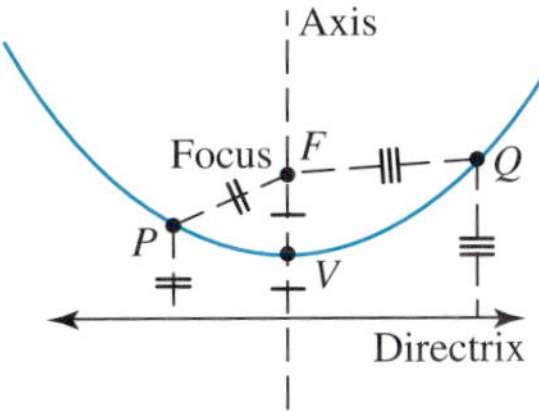

FIGURE 10.60

Notice that each of the three points P, Q, and V shown in Figure 10.60 is equidistant from the focus F and the directrix, and thus is on the parabola.

Another special line associated with a parabola is the **axis,** which is the line through the focus, F, perpendicular to the directrix. The axis is the line of symmetry of the parabola. Another special point is the **vertex,** V, which is the intersection of the parabola and the axis. The vertex is the point of the parabola closest to the focus. By the definition of a parabola, the vertex is midway between the focus and the directrix.

Applications of the Parabola

One of the most fascinating properties of the parabola is that all rays parallel to the axis that "hit" the parabola from within "bounce off" the parabola and "hit" the focus (Figure 10.61).

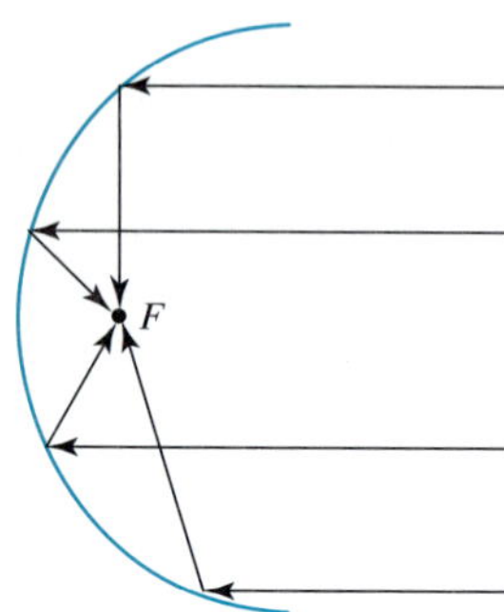

FIGURE 10.61

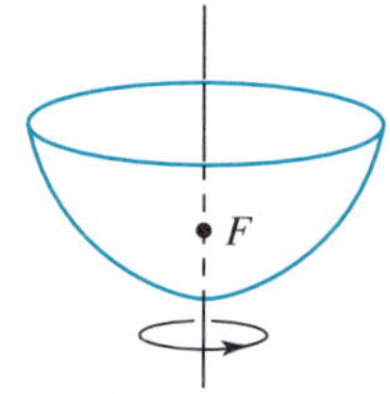

FIGURE 10.62

This property is very useful in three dimensions. If a parabola is rotated around its axis, the three-dimensional shell formed is called a **paraboloid** (Figure 10.62). By its definition, a paraboloid has the property that every plane cross section that contains the axis of the paraboloid is a parabola. Moreover, the focus (vertex) of the paraboloid is the same as the focus (vertex) of each of these parabolas. A television dish antenna that is a portion of a paraboloid is used to collect signals from satellites. The signals hit the dish in parallel rays and are, in turn, reflected to a receiver that is placed at the focus of the paraboloid (Figure 10.63).

TIDBIT

Parabolic reflectors are used to make solar furnaces. One located in France can create temperatures as high as 5400°F. Such a system was used by the villain in the James Bond movie *The Man with the Golden Gun.*

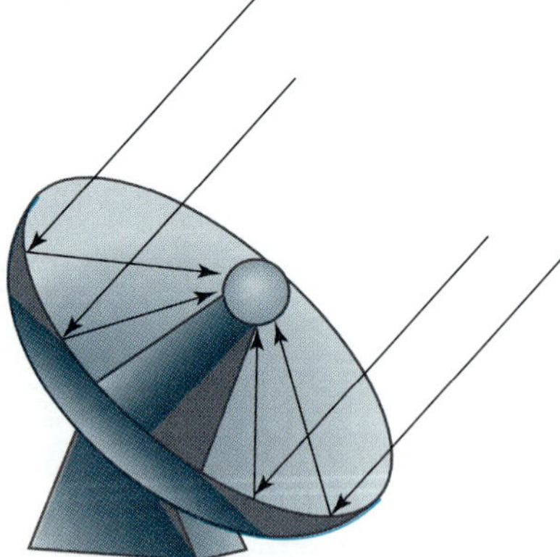

FIGURE 10.63

Conversely, flashlights, headlights, and so on are constructed using portions of paraboloids as reflectors. The light source, the bulb, is placed at the focus (Figure 10.64).

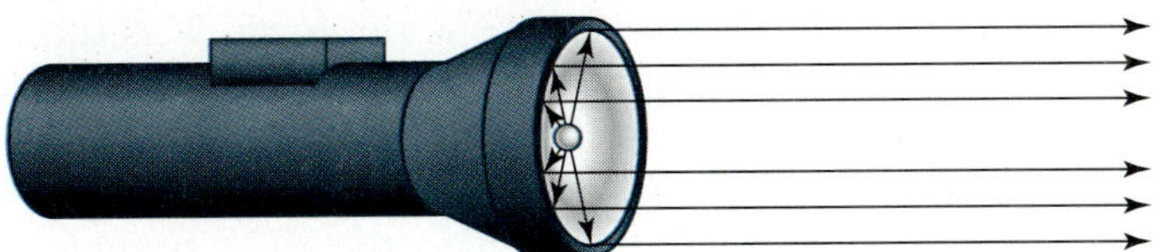

FIGURE 10.64

In this way, the light source is bounced off the reflector producing parallel light rays that form a cylinder of light. In practice, the entire bulb can not be placed exactly at the focus, and the beam will have some dispersion.

Portions of a pair of parabolas can be used to form a whispering gallery as shown in Figure 10.65.

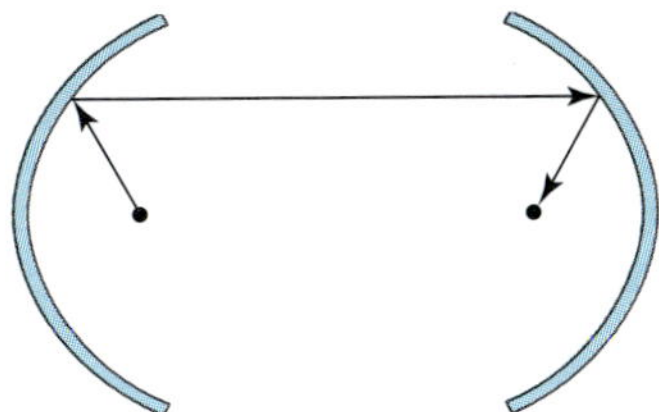

FIGURE 10.65

A person standing at the focus on the left whispers into the parabola, and the sound is reflected across to the parabola on the right. The sound is then reflected to the focus of the second parabola, where another person stands. Ellipses are also used to construct whispering galleries, as will be shown in the next section.

Equation of a Parabola

Conic sections can be described algebraically in the form of equations. Returning to the two-dimensional case, we now derive the equation of a parabola whose axis is the y-axis and whose vertex is on the x-axis (thus, the vertex is at the origin and the focus is on the y-axis) (Figure 10.66).

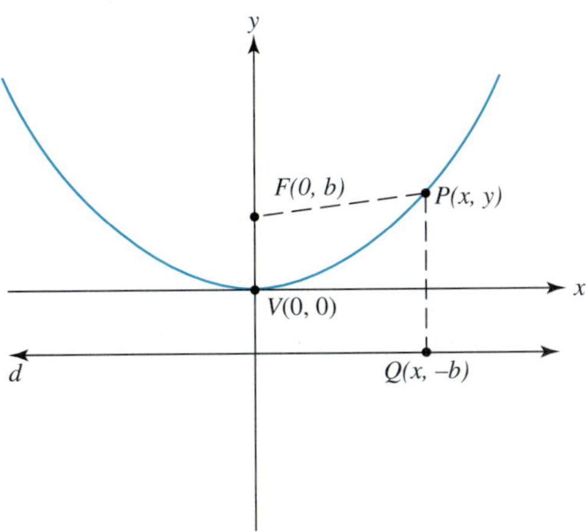

FIGURE 10.66

In Figure 10.66, F is the focus, V is the vertex, d is the directrix, and P represents a point on the parabola. Point Q is located on d, so that PQ is the distance from P to the directrix (that is, $\overline{PQ}$ is perpendicular to d). By definition of a parabola, $FP = PQ$. The distance from F to P is found by applying the Pythagorean theorem to the right triangle ΔFPR shown in Figure 10.67 (not to scale). Since the sides of the right triangle are horizontal and vertical lines, their lengths can be found as the difference in the horizontal and vertical coordinates, respectively.

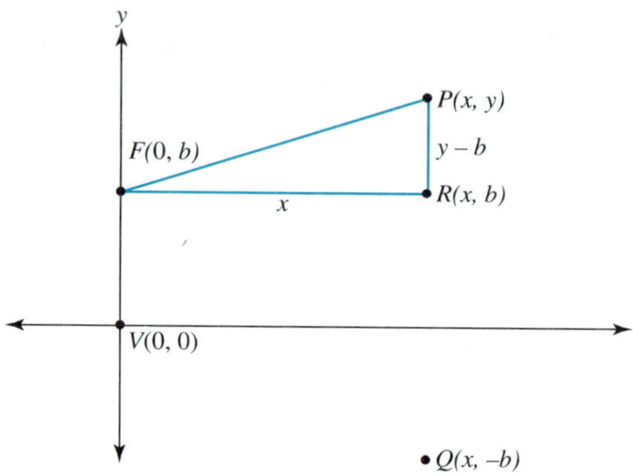

FIGURE 10.67

Since $\overline{PQ}$ is a vertical line, we have $PQ = y - (-b) = y + b$. Using the Pythagorean theorem, we see that $(FP)^2 = x^2 + (y - b)^2$. To put these facts together, we note that if $FP = PQ$, then

$$(FP)^2 = (PQ)^2.$$

Substituting, we have

$$x^2 + (y - b)^2 = (y + b)^2.$$

Simplifying the terms, we get

$$x^2 + y^2 - 2by + b^2 = y^2 + 2by + b^2, \text{ or}$$

$$x^2 = 4by, \text{ or } y = \frac{1}{4b}x^2$$

The derivation shows that certain parabolas can be expressed in the form $y = Ax^2$. Conversely, any equation of the form $y = Ax^2$ represents a parabola whose axis is the y-axis and whose focus is the point $(0, \frac{1}{4A})$. If the vertex of a parabola is at the origin and the focus is $(b, 0)$ on the x-axis, then the equation of the parabola is $x = \frac{1}{4b}y^2$.

More generally, the graphs of parabolas can assume any orientation and size on a coordinate plane. However, the derivation of their equations is more complicated and will be omitted here. For the sake of simplicity, we will confine our study to those parabolas whose axes are parallel to the y-axis.

Equation of a Parabola

The equation of a parabola whose axis is the y-axis, whose focus is $F(0, b)$, and whose vertex is $V(0, 0)$ is

$$y = \frac{1}{4b}x^2.$$

and conversely.

In general, the equation of a parabola with a vertical or horizontal axis is one of these two forms:

$$y = Ax^2 + Bx + c \quad \text{or} \quad x = Ay^2 + By + C,$$

where A, B, and C are constants, and $A \neq 0$.

EXAMPLE 10.11 Sketch the graph of the equation $y = \frac{1}{4}x^2$. Describe the figure.

SOLUTION From the information in the preceding box concerning the equation of a parabola, we know the graph is a parabola with focus $F(0, 1)$ and vertex $V(0, 0)$. To graph the equation we plot points on graph paper by substituting various values of x into the equation $y = \frac{1}{4}x^2$ (Figure 10.68).

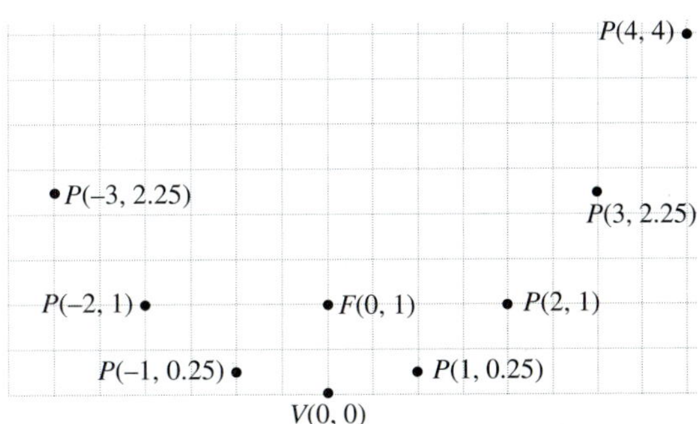

FIGURE 10.68

If many points are plotted, the shape of a nice looking parabola will be seen. A graphing calculator will produce this graph for you. If you are doing the graph by hand, you must connect the plotted points together as best you can in order to approximate the parabolic shape. One trick that can help near the vertex is to use an arc of a circle with a radius twice the distance from the focus to the vertex (Figure 10.69).

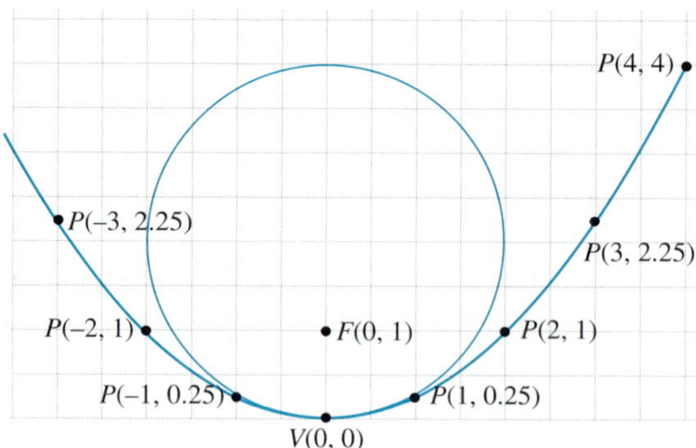

FIGURE 10.69

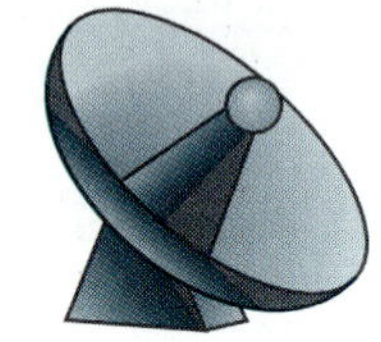

FIGURE 10.70

EXAMPLE 10.12 A folding dish antenna is to be constructed for a portable solar powered telephone (Figure 10.70). The center of the dish will be the vertex of a paraboloid which, when unfolded, will have a diameter of 2 feet and a depth at its center of 4 inches. Where should the receiver be located in the dish?

SOLUTION Figure 10.71 shows a cross section of the dish with its vertex at the origin and its axis the y-axis.

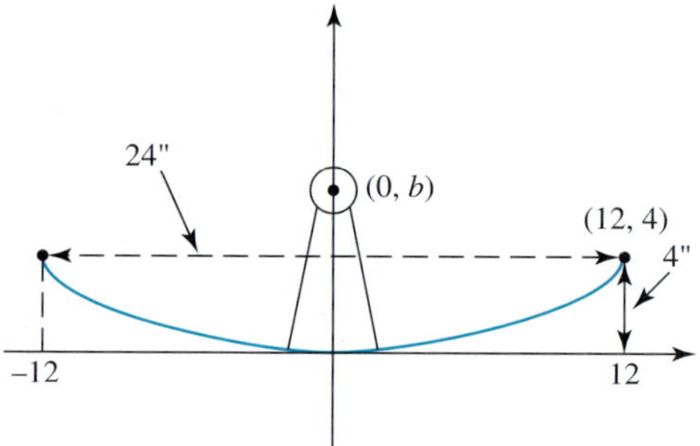

FIGURE 10.71

The receiver should be placed at the focus of the parabola because the signal will be concentrated at that point. This parabola has an equation of the form $y = \frac{1}{4b}x^2$ where $(0, b)$ is its focus, the exact location of the receiver. Since the dish is 24 inches across and 4 inches deep (we convert all units to inches), the point (12, 4) must be on the parabola. Substituting (12, 4) into the equation $y = \frac{1}{4b}x^2$, we have $4 = \frac{1}{4b} \times 12^2$. This simplifies to $4b = 36$, so $b = 9$. Thus the receiver should be located 9 inches above the vertex of the paraboloid. ◆

TIDBIT

Although sagging telephone lines, clothes lines, and electrical power lines appear to be in the shape of a parabola, they take the shape of a curve named a catenary.

In addition to the reflecting properties of parabolic shapes, the path of any free-falling object follows a parabola (provided we also agree that a straight line is a special type of parabola). The next example uses this fact.

EXAMPLE 10.13 A daring young woman is to be shot from a cannon inside a circular circus tent. The tent has a diameter of 150 feet, a height of 60 feet at its highest point, and a height of 30 feet around its perimeter. If the position of the woman above the ground at any time t is given by the equation

$$y = 16t^2 + 64t + 12 \text{ (in feet), } (t \text{ is the time in seconds})$$

will she fly through the roof of the tent or land safely inside? (Figure 10.72)

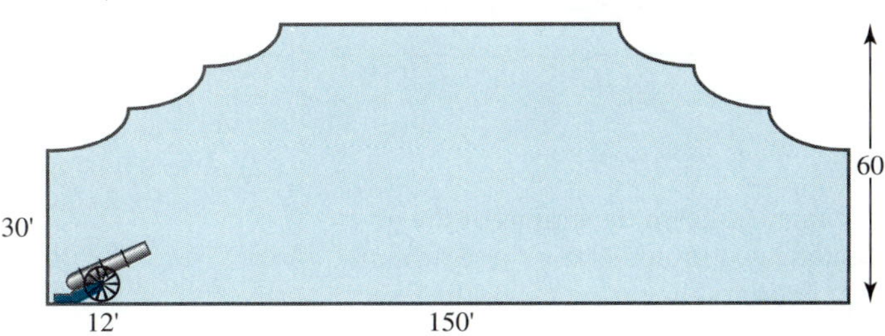

FIGURE 10.72

SOLUTION In the beginning, when $t = 0$, the woman is in the cannon, hence is 12 feet off the ground. Notice that y also equals 12 in the equation when $t = 0$ in the equation $y = -16t^2 + 64t + 12$. We simplify the problem by establishing a pair of coordinate axes with t, time, the unit on the horizontal axis and the woman's position from the ground, y, on the vertical axis (Figure 10.73). We graph the parabola that shows her distance above the ground during the flight.

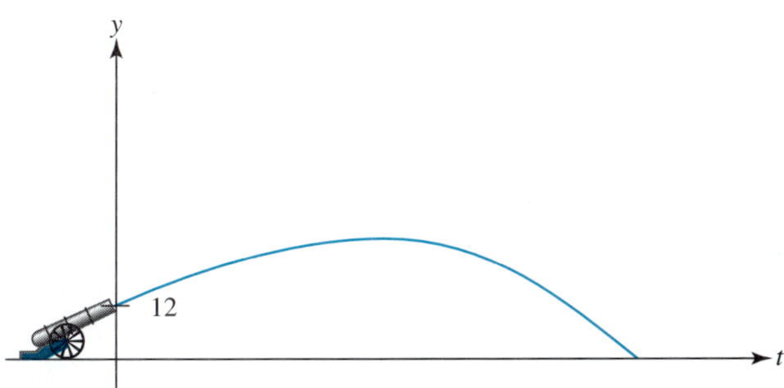

FIGURE 10.73

Next, we calculate four additional pairs of values for t and y and list them in Table 10.3.

TABLE 10.3

t	y
0	12
1	60
2	76
3	60
4	12

$t = 1$: $y = -16 \times 1^2 + 64 \times 1 + 12 = 60$

$t = 2$: $y = -16 \times 2^2 + 64 \times 2 + 12 = 76$

$t = 3$: $y = -16 \times 3^2 + 64 \times 3 + 12 = 60$

$t = 4$: $y = -16 \times 4^2 + 64 \times 4 + 12 = 12$

A parabola is symmetric with respect to its axis. Since the woman is at 12 feet when $t = 0$ and $t = 4$, the highest point on her flight should occur halfway between 0 and 4, namely $t = 2$. However, the tent is only 60 feet high and the cannoneer can reach a height of 76 feet, and she will hit the top of the tent. ◆

INITIAL PROBLEM SOLUTION

A solar reflector is to be designed to collect sun rays and reflect them into a single point to create a concentrated stream of heat. What is a good design for such a reflector?

SOLUTION The reflector should be in the shape of a paraboloid with the collection point at its focus.

PROBLEM SET 10.3

In problems 1 through 7, sketch the graphs of the given parabolas. If possible, you should graph these on standard graph paper. Poor graphs can be misleading and interfere with understanding.

1. $y = x^2$
2. $y = 2x^2$
3. $y = 3x^2$
4. $y = \frac{1}{2}x^2$
5. $y = -2x^2$
6. $y = -x^2$
7. $y = -\frac{1}{4}x^2$
8. The equations in problems 1 through 7 all had the form $y = Ax^2$. Explain the effect that the value of A (including negatives) has on the graph of the parabola. How does the graph change for different values of A?

The general equation for a parabola with axis parallel to the y-axis is

$$y = Ax^2 + Bx + C$$

In problems 9 through 20, the effect of the values of B and C on the graph will be considered. Sketch the graphs of the given parabolas.

9. $y = x^2 + 2x$
10. $y = x^2 - 2x$
11. $y = x^2 + 5x$
12. $y = x^2 + 3x$
13. $y = x^2 - 4x$
14. Explain the effect that the value of B (including negatives) has on the graph of the parabola for $y = x^2 + Bx$.
15. $y = x^2 - 2$
16. $y = x^2 + 2$
17. $y = x^2 + 3$
18. $y = x^2 - 4$
19. $y = x^2 - 5$
20. Explain the effect that the value of C (including negatives) has on the graph of the parabola for $y = x^2 + C$.
21. What is the equation of a parabola with its vertex at the origin and (0, 2) as its focus?
22. What is the equation of a parabola with its vertex at the origin and (0, −3) as its focus?
23. What is the equation of a parabola with its vertex at (0, 2) and (0, 4) as its focus? (Hint: draw a sketch.)
24. What is the equation of a parabola with its vertex at (0, 2) and (0, 2) as its focus? (Hint: draw a sketch.)

The general equation for a parabola with axis parallel to the x-axis is

$$x = Ay^2 + By + C.$$

In problems 25 through 30, sketch the graphs of the given parabolas.

25. $x = \frac{1}{4}y^2$
26. $x = 2y^2$
27. $x = y^2 + 2y$
28. $x = -y^2 + y$
29. $x = y^2 + 3$
30. $x = y^2 - 2$
31. Find the coordinates of the focus of a parabola whose axis is the y-axis, whose vertex is the origin, and whose equation is
(a) $y = .25x^2$ **(b)** $y = 3x^2$ **(c)** $y = -4x^2$
32. Find the coordinates of the focus of a parabola whose axis is the x-axis, whose vertex is the origin, and whose equation is
(a) $x = .25y^2$ **(b)** $x = -3y^2$ **(c)** $x = -y^2$
33. A television dish antenna is in the shape of a paraboloid. The receiver is located at the focus and is four feet above the vertex. Find an equation of a cross section of the dish, assuming that the vertex is at the origin.
34. The bulb in the headlight of a car is at the focus of a parabolic reflector behind it. The bulb is 1.5 inches from the vertex of the reflector. Find the equation of a cross section of the reflector. The reflector is six inches wide. How deep is it?
35. A satellite dish has a diameter of three feet and a depth at its center of six inches. The center of the dish is the vertex of a paraboloid. Where should the receiver be located in the dish?
36. The reflector in the head of a flashlight has a diameter of three inches and a depth at its center of one inch. The center of the reflector is the vertex of a paraboloid. Where should the center of the bulb be located in the head of the flashlight?
37. A man and woman are standing in a whispering gallery that is constructed with two congruent parabolas. The man is standing at one focus, exactly two

meters to the right of the vertex of the parabola on the left, and whispers into it.

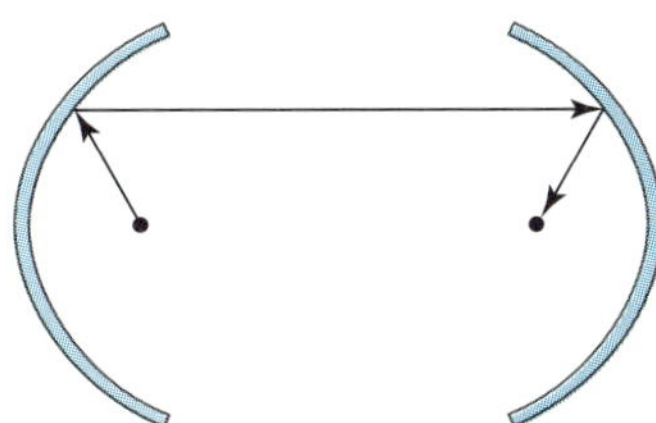

Where should the woman stand, relative to the parabola on the right, in order to hear the man's whisper?

38. Two parabolas with foci located on the coordinate axes share the same vertex and are congruent. The equation of one of the parabolas is $y = x^2$. What is the equation for the other parabola? List all possibilities.

39. A bullet is fired into the air and follows a parabolic path. The height of the bullet, in feet, can be described by the equation

$$y = -16t^2 + 96t + 10,$$

where t represents time, in seconds.

(a) What is the height of the bullet when the gun is fired?

(b) What is the maximum height attained by the bullet?

(c) When will the bullet strike the ground (to the nearest tenth of a second)?

40. A basketball player is shooting a free throw attempt. The ball followed a parabolic path described by the equation

$$y = -5t^2 + 10t + 8,$$

where y is the height of the ball, in feet, after t seconds.

(a) Find the maximum height of the ball.

(b) After how many seconds was the maximum height attained?

41. The simplest suspension bridges for short spans consist of a roadway connected to cables. The cables pass over towers, and are anchored at the ends of the bridge. The span of the bridge is the distance between the towers, and the sag is the distance between the highest and lowest points on the cable. The true shape of the cable is called a catenary, but it can be approximated with a parabola. Suppose the lowest point on the cable attached to a suspension bridge is 20 feet above the roadway, the span is 600 feet, and the towers are 110 feet high. How high is the cable above the roadway 100 feet from the center of the span? (Hint: set up a coordinate system with the lowest point on the cable being at the origin. Assume the shape of the hanging cable is a parabola. What are the coordinates for the tops of the towers?)

EXTENDED PROBLEMS

42. Form an approximation to a parabola by folding (wax) paper as follows:
Given focus, F, and directrix, d, and any point Q on d, fold Q onto F and make a crease. Repeat for many points on d. The shape formed by the intersecting crease lines approximates a parabola.

43. Form the shape of a parabola using a pencil and paper as follows:
Make an angle and place several equally spaced marks on each side of the angle. Then connect the point nearest the vertex on one side with the point farthest from the vertex on the other. Then connect the next point on the first side to the second farthest point from the vertex on the other. The following figure shows the result when six points are chosen on each side.

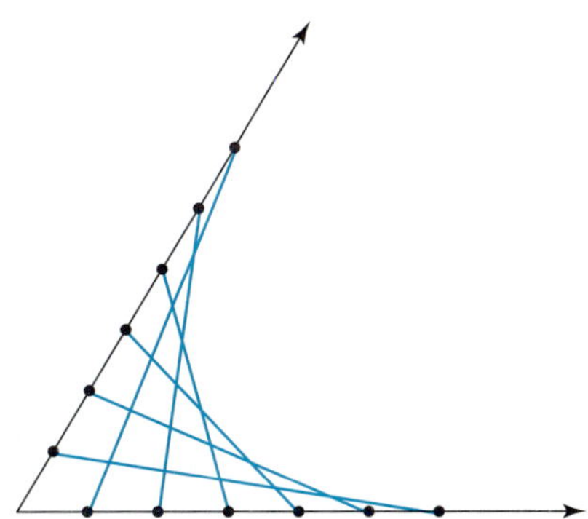

The parabolas we have been studying have had their vertices at the origin. Problems 44 through 47 deal with the situation where the vertices may be any point, but where the axes must still be horizontal or vertical.

44. Derive the following: the equation of a parabola with vertex at (h, k) and focus at $(h, k + a)$ is $(x - h)^2 = 4a(y - k)$. Refer to Figures 10.66 and 10.67, change the coordinates appropriately, and follow the steps in the derivation of the equation. (Note: it opens up if $a > 0$ and down if $a < 0$.)

45. Using the result in problem 44, find the equation of the following parabolas.

(a) Vertex: $(-2, 3)$, Focus: $(-2, 6)$

(b) Vertex: $(6, -2)$, Focus: $(6, -5)$

46. Derive the following: the equation of a parabola with vertex at (h, k) and focus at $(h + a, k)$ is $(y - k)^2 = 4a(x - h)$. (Note: it opens to the right if $a > 0$ and to the left if a < 0.)

47. Using the result in problem 46, find the equation of the following parabolas.

(a) Vertex: $(2, 3)$, Focus: $(6, 3)$

(b) Vertex: $(6, -2)$, Focus: $(-1, -2)$

10.4 CONIC SECTIONS: ELLIPSES AND HYPERBOLAS

A 60 foot by 48 foot whispering gallery is to be constructed. Where should the two people stand to maximize their chances of hearing each other?

To the Greeks, as well as the philosophers and wise men of many cultures, the circle epitomized perfection. Seeking perfection in the world and universe, Greek astronomers developed a model of the universe that placed the earth at the center, with the sun, moon, and planets having circular orbits around the earth. The stars (which appeared to the Greeks to be fixed in position relative to each other) were placed on perfect crystalline spheres surrounding the earth. As we now know, the planets follow elliptical orbits about the sun and the stars are spreading outward in the universe.

In this section, we study the ellipse (and as a special case, the circle) as well as the hyperbola. Once studied primarily for their mathematical and aesthetic qualities, the curves have found renewed importance in modern applications.

THE ELLIPSE

As mentioned in the previous section, an ellipse is formed when a plane intersects one nappe of a cone. A better working definition follows.

The Ellipse

An **ellipse** is a figure in a plane determined by two fixed points as follows: a point is on the ellipse if the sum of its distances from the two fixed points in the plane is a constant.

Figure 10.74 illustrates this definition.

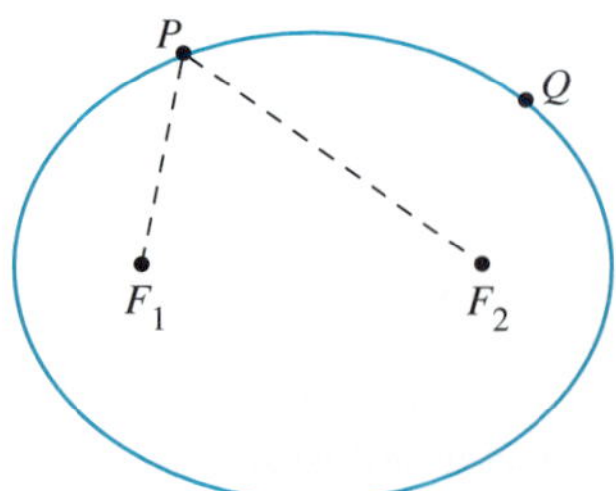

FIGURE 10.74

Point Q is on the ellipse containing P in Figure 10.74 if and only if $F_1Q + QF_2 = F_1P + PF_2$. The fixed points in the definition, F_1 and F_2, are called **foci** (singular is **focus**) **of the ellipse.** The **center of an ellipse** is the midpoint of the segment whose endpoints are the foci. Figure 10.75 shows how the thickness of an ellipse varies depending on the distance between the foci, assuming that the distance defining the ellipse is fixed.

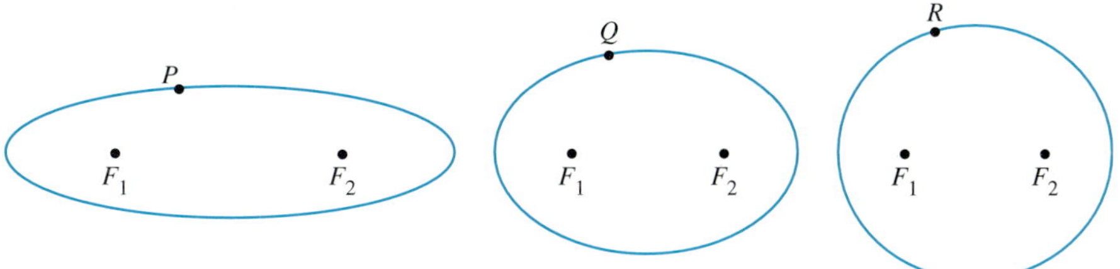

FIGURE 10.75

(Note: to verify that the distance that defines the ellipse is the same for all three ellipses, pick an arbitrary point on each ellipse and find the sum of the distances from the foci. In particular, in Figure 10.75, $F_1P = PF_2 = F_1Q + QF_2 = F_1R + RF_2$. When the two foci coincide the figure is a circle.)

One way to sketch an ellipse is to place a loop of string around two points (the foci) and a pencil (as shown in the figure), and then move the pencil completely around the two points (very carefully), keeping the string taut at all times (Figure 10.76).

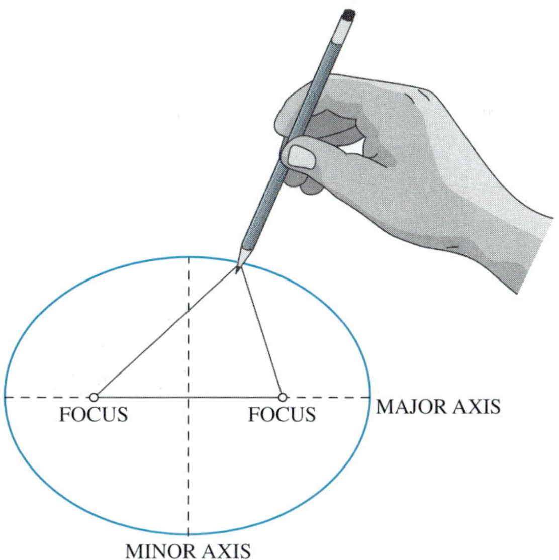

FIGURE 10.76

The line segment that contains the foci and has its endpoints on the ellipse is called the **major axis of the ellipse.** The segment that is the perpendicular bisector of the major axis and has its endpoints on the ellipse is called the **minor axis of the ellipse** (Figure 10.76). Each axis is a line of symmetry for the ellipse.

TIDBIT

The British Spitfire, used in dogfights in World War II, was noted for its superb maneuverability, part of which was due to its semielliptical shapes on the back of the wings and the front of the tail.

Another way to construct an ellipse by folding a circle is given in the problem set. The shape of an ellipse can be seen by filling a glass with a liquid and then tilting the glass (Figure 10.77).

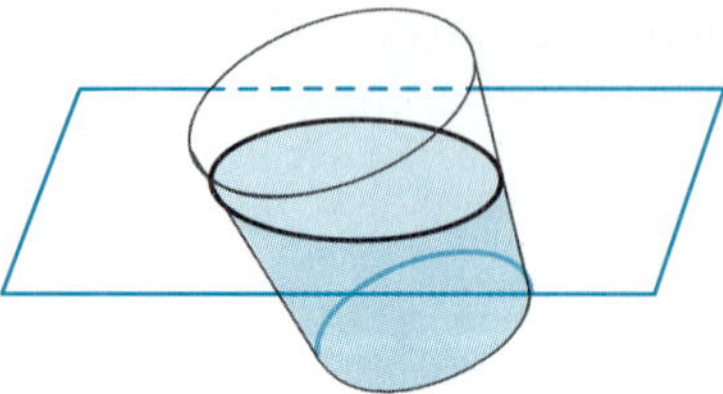

FIGURE 10.77

Applications of the Ellipse

Elliptical shapes appear in many situations in our physical world. In the early 17th century, Kepler observed that the planets moved in elliptical orbits around the sun with the sun at one focus, in contrast to the belief that the sun and planets moved around the earth in circular orbits. Seventy years later, Sir Isaac Newton developed the laws of universal gravitation and showed why this must be true.

Half the length of the major axis of the elliptical orbit of a planet is called the planet's **mean distance** from the sun. The ratio of the distance of the sun from the center of the planet's orbit to the planet's mean distance from the sun is called the **eccentricity** of the planet's orbit, that is,

$$\text{eccentricity} = \frac{\text{distance from the sun to the center of the orbit}}{\text{mean distance from the sun}}$$

The geometry of a planetary orbit is illustrated in Figure 10.78.

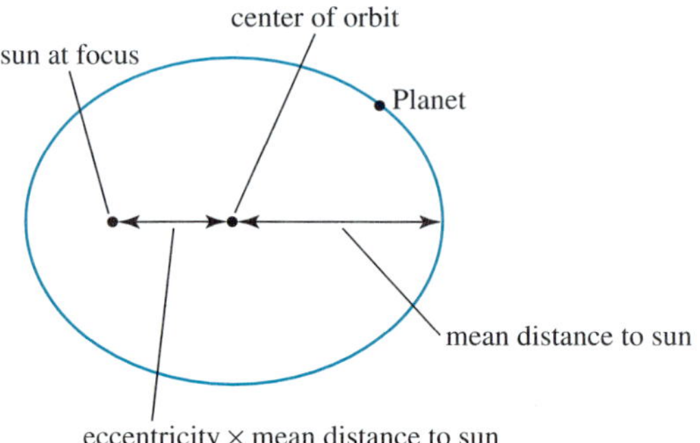

FIGURE 10.78

Newton also reasoned that comets would have elliptical orbits. British astronomer Edmund Halley (1656–1742) used this information to predict when a certain comet, now called Halley's Comet, would reappear. He found correctly that it reappears about every 77 years, the last time being 1986.

There are also many down-to-earth, even mundane, applications of the ellipse. For example, a pool table introduced in 1964 was the shape of an ellipse with a hole at one focus and a spot at the other. Its most interesting feature was that a ball shot from the spot (without any spin) would automatically go into the hole (Figure 10.79).

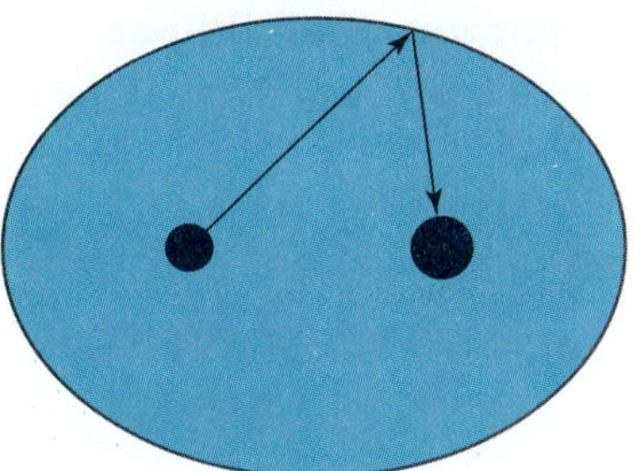

FIGURE 10.79

Sprocket wheels for racing bicycles have also been designed in elliptical shapes to maximize the power of the rider when pedaling. In most designs, the crank arms were in line with the major axis (Figure 10.80).

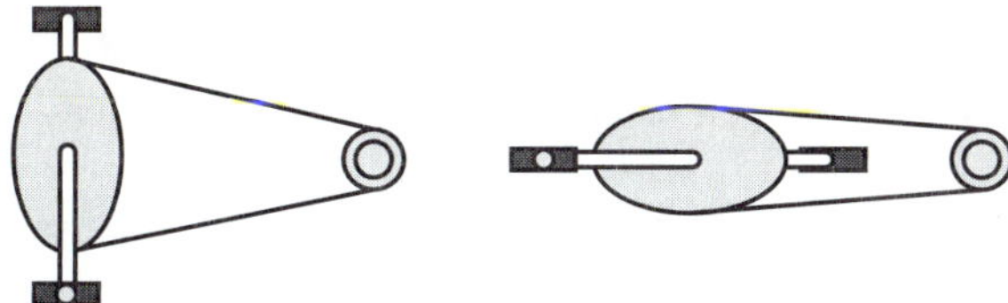

FIGURE 10.80

The elliptical shape of the sprocket compensates for the fact that a cyclist does not have as much leverage on the crank when the pedal is at the top. (Note: these sprocket wheels are generally not available any more.)

Another fascinating use of the elliptical shape is in the construction of whispering galleries. Such rooms are constructed so that when a person whispers at one focus, the sound can be heard clearly at the other focus, even if there is an obstruction between the two parties (Figure 10.81).

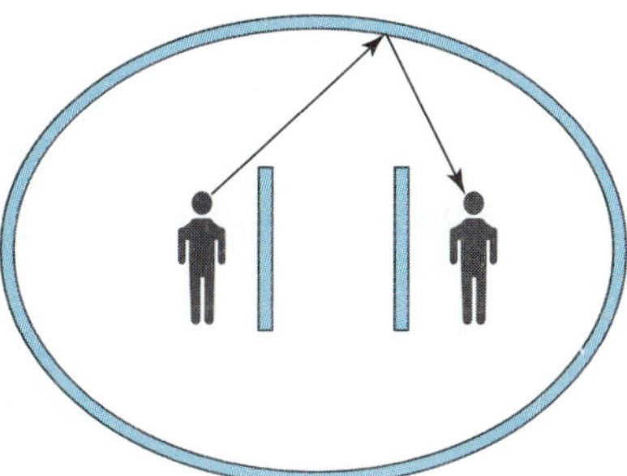

FIGURE 10.81

In the whispering gallery in the Taj Mahal in India, it is said that a groom of a honeymooning couple would whisper "To the memory of my undying love" and be heard by his bride standing more than 50 feet away even though there was a line-of-sight sound barrier between him and his bride.

The Equation of an Ellipse

Suppose that P is any point of the ellipse as shown in Figure 10.82.

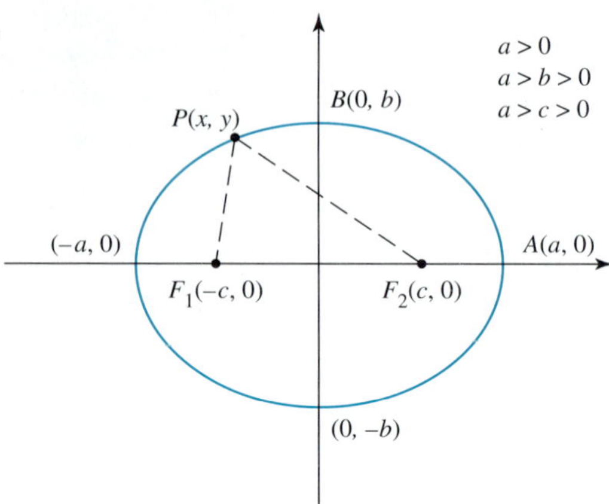

FIGURE 10.82

For convenience, the center of the ellipse has been placed at the origin, and the foci have been placed on the x-axis. The derivation of the equation of this ellipse is somewhat complex, but it is carefully outlined here and the completion of the steps is left to the problem set. However, an interesting relationship among a, b, and c can be found as follows: let A represent $(a, 0)$ and B represent $(0, b)$. Since A and B are points on the ellipse, the definition of the ellipse implies that

$$F_1A + AF_2 = F_1B + BF_2.$$

Using the Pythagorean theorem on the right-hand side, we have

$$(a + c) + (a - c) = \sqrt{c^2 + b^2} + \sqrt{(-c)^2 + b^2}$$

This simplifies to

$$2a = 2\sqrt{c^2 + b^2}, \text{ or } a = \sqrt{c^2 + b^2}.$$

Squaring both sides, we have $a^2 = c^2 + b^2$ or $c^2 = a^2 - b^2$. (Note: if the foci were on the vertical axis, by symmetry, the latter equation would become $c^2 = b^2 - a^2$.

The relationship among a, b, and c plays an important role in the following result whose derivation is developed in the problem set.

THEOREM

Equation of an Ellipse

An ellipse whose center is the origin and whose foci are at $(c, 0)$ and $(-c, 0)$ on the x-axis is the set of all points satisfying the equation

$$\frac{x^2}{a^2} + \frac{y^2}{b^2} = 1,$$

where $(a, 0)$, $(-a, 0)$, $(0, b)$, and $(0, -b)$ are points of the ellipse, $a > b$, and

$$c^2 = a^2 - b^2.$$

EXAMPLE 10.14 Find the equation of an ellipse centered at the origin with major axis of length 6, minor axis of length 4, and foci on the x-axis.

SOLUTION Because the foci are on the x-axis and the major axis of the ellipse has length 6, the points $(-3, 0)$ and $(3, 0)$ must be on the ellipse [the distance from $(-3, 0)$ to $(3, 0)$ is 6]. Also, because the minor axis has length 4, the points $(0, -2)$ and $(0, 2)$ must be on the ellipse. We conclude that $a = 3$, $b = 2$ and the equation is

$$\frac{x^2}{9} + \frac{y^2}{4} = 1.$$ ◆

In Figure 10.83 the foci of the ellipse are on the y-axis.

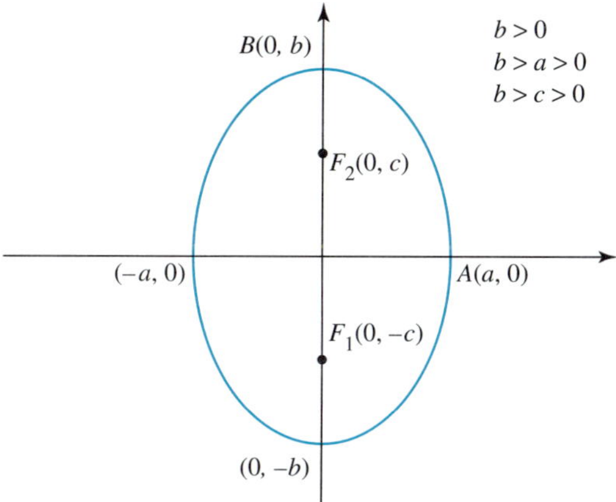

FIGURE 10.83

Equation of an Ellipse

An ellipse whose center is the origin and whose foci are at $(0, c)$ and $(0, -c)$ on the y-axis is the set of all points satisfying the equation

$$\frac{x^2}{a^2} + \frac{y^2}{b^2} = 1,$$

where $(a, 0)$, $(-a, 0)$, $(0, b)$, and $(0, -b)$ are points of the ellipse, $b > a$, and

$$c^2 = b^2 - a^2.$$

EXAMPLE 10.15 Find the foci and sketch the graph of the ellipse with the equation $36x^2 + 16y^2 = 576$.

SOLUTION If both sides of the equation are divided by 576 to make the right-hand side equal to 1, the equation will be in the same form as the definition. This gives us

$$\frac{36x^2}{576} + \frac{16y^2}{576} = \frac{576}{576}, \text{ or}$$

$$\frac{x^2}{16} + \frac{y^2}{36} = 1.$$

Here, $a = \sqrt{16} = 4$ and $b = \sqrt{36} = 6$. Next, we graph the ellipse.

If $a = 4$ and $b = 6$, the ellipse contains the points

$$(4, 0), (-4, 0), (0, 6), \text{ and } (0, -6).$$

Since b is greater than a, the major axis is vertical. Finally, since $c^2 = 36 - 16$, we have $c = \sqrt{20} = 2\sqrt{5}$ (Figure 10.84).

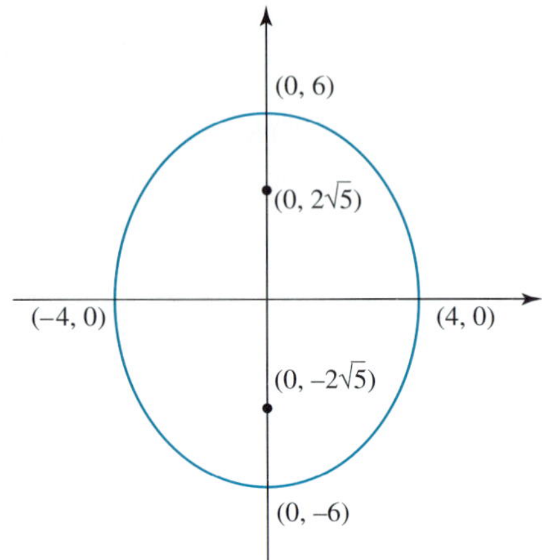

FIGURE 10.84

◆

EXAMPLE 10.16 Suppose you want to build an elliptical pool table that will be five feet by three feet. How can you draw an outline of the table and plan the layout before starting construction?

SOLUTION First draw a sketch of the ellipse having its center on the origin of a coordinate system and its axes on the x- and y-axes (Figure 10.85).

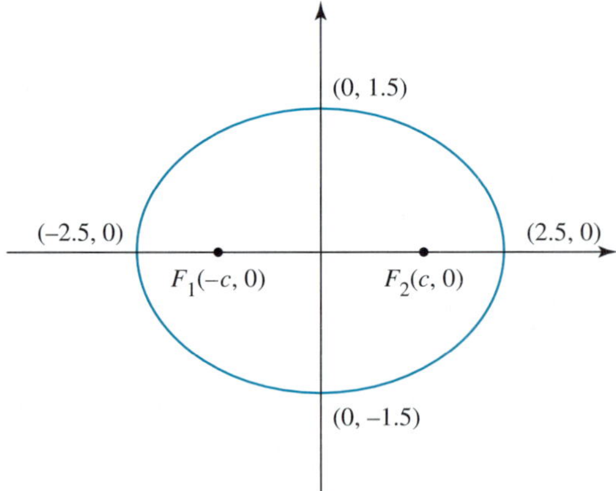

FIGURE 10.85

Here $a = \frac{5}{2} = 2.5$ and $b = \frac{3}{2} = 1.5$. Thus $c^2 = 2.5^2 - 1.5^2 = 6.25 - 2.25 = 4$, or $c = \sqrt{4} = 2$. To draw the outline of the table, we can use the method shown in Figure 10.76. The loop of string we use must have a length of $2a + 2c = 2(2.5) + 2 = 9$ feet. ◆

THE HYPERBOLA

The hyperbola, although composed of two disjoint parts of infinite extent, has a definition similar to that of the ellipse.

DEFINITION

Hyperbola

A **hyperbola** is a figure in a plane determined by two fixed points as follows: a point is on the hyperbola if the (positive) difference of its distances from the two fixed points is a constant.

Figure 10.86 illustrates this definition.

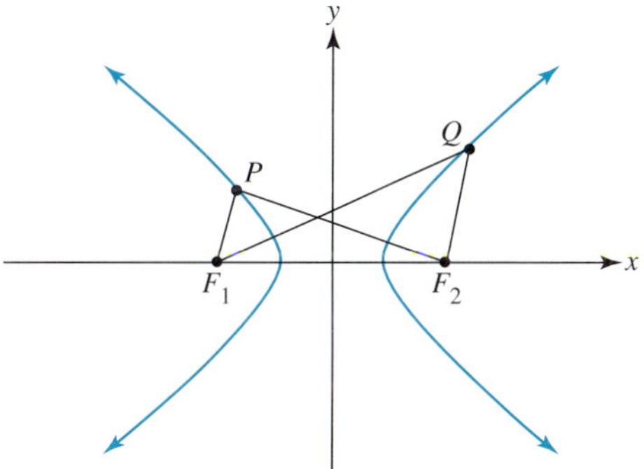

FIGURE 10.86

Point Q is on the hyperbola containing P in Figure 10.86 if and only if $F_1Q - QF_2 = F_2P - PF_1$. The fixed points F_1 and F_2 are called the **foci of the hyperbola** and the line determined by the foci is the **axis of the hyperbola.**

The three-dimensional counterpart of a hyperbola, a hyperboloid, is formed when a hyperbola is revolved around an axis. For example, the hyperbola in Figure 10.86 can be revolved around the y-axis to produce a hyperboloid.

Another way to produce a hyperboloid is illustrated below. Figure 10.87(a) displays a cylinder formed by connecting two circular disks with strings. When one of the disks is rotated less than 180°, the surface takes the shape of a portion of a hyperboloid [Figure 10.87(b)]. When one of the disks is rotated 180°, a portion of a cone is formed [Figure 10.87(c)].

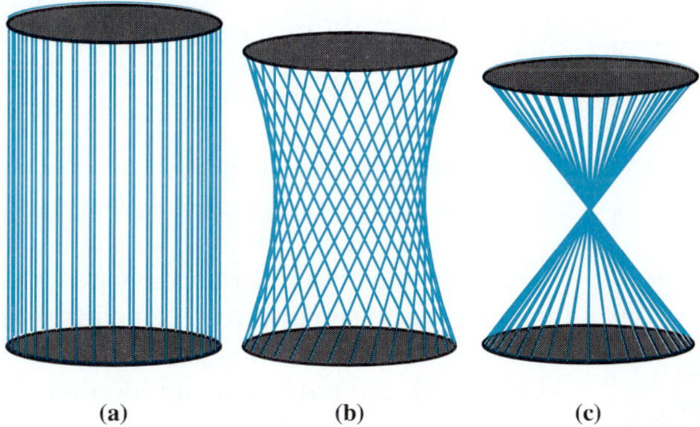

FIGURE 10.87

Applications of a Hyperbola

Hyperbolas occur in a variety of situations. When a lamp having a shade casts a shadow on a wall, the lighted areas form a hyperbola (Figure 10.88).

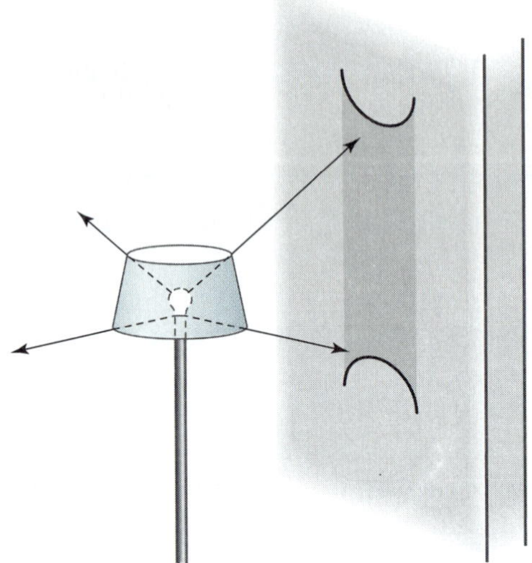

FIGURE 10.88

The lampshade in Figure 10.88 produces parts of two different hyperbolas depending on the size of the circles that form the top and bottom of the shade as well as the location of the lightbulb.

A supersonic plane leaves a conical shock wave which intersects flat terrain in one part of a hyperbola (Figure 10.89).

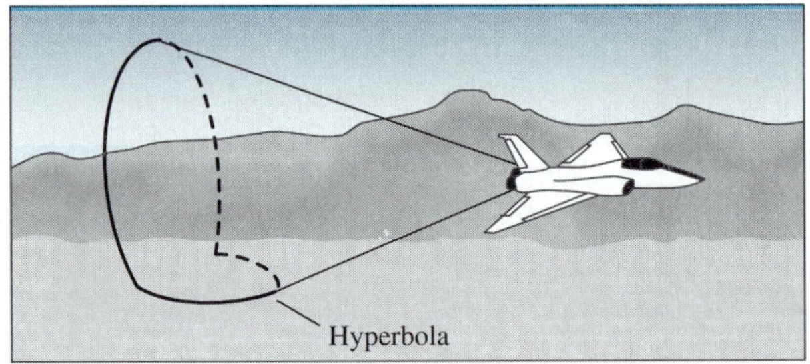

FIGURE 10.89

FIGURE 10.90

Cooling towers for nuclear reactors are approximately in the shape of a portion of a hyperboloid (Figure 10.90).

Perhaps one of the most useful applications of hyperbolas is LORAN (LOng RAnge Navigation), a radio-assisted navigational system. To determine a ship's location, radio signals are received from two overlapping pairs of transmitters, say F, G and F, H. Measuring the difference in reception times for each of the pairs F, G and F, H, the navigator can identify two hyperbolas having these pairs of points as foci. When these hyperbolas are plotted, their intersection is the location, S, of the ship (Figure 10.91).

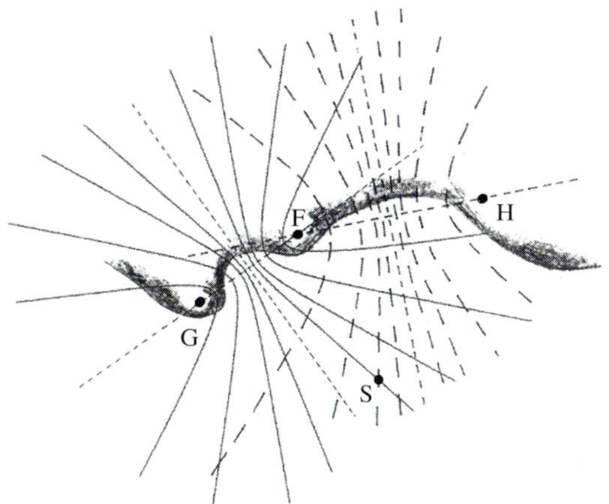

FIGURE 10.91

The Equation of a Hyperbola

The definition of a hyperbola can be used to derive an equation for a hyperbola that is similar to that of the ellipse. The derivation of this equation will be left for the problem set.

In Figure 10.92 the foci of the hyperbola are the points F_1 and F_2 on the x-axis, and P_1 and P_2 are the two points of the hyperbola on the x-axis.

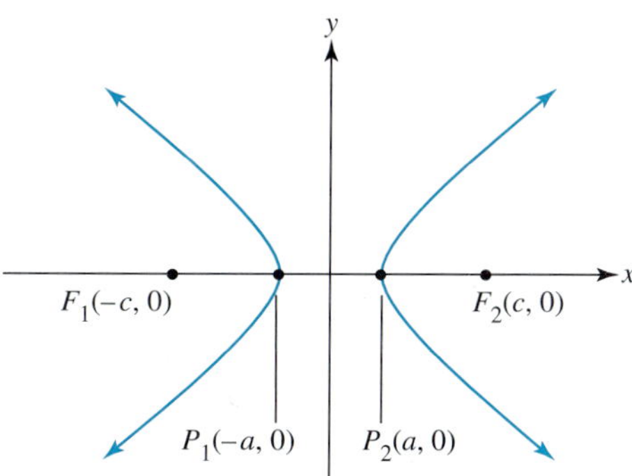

FIGURE 10.92

Equation of a Hyperbola

A hyperbola whose center is the origin and whose foci are at $(c, 0)$ and $(-c, 0)$ on the x-axis is the set of all points satisfying the equation

$$\frac{x^2}{a^2} - \frac{y^2}{b^2} = 1,$$

where $(a, 0)$ and $(-a, 0)$ are points of the hyperbola and

$$c^2 = a^2 + b^2.$$

The **center of the hyperbola** is the midpoint of the line segment whose endpoints are the foci of the hyperbola.

EXAMPLE 10.17 Find the equation of a hyperbola centered at the origin with foci 10 units apart on the x-axis and containing the point (4, 0).

SOLUTION Since the foci are centered on the x-axis with distance between them 10 units, the foci must be $(-5, 0)$ and $(5, 0)$. Thus $c = 5$. Since the point (4, 0) is on the hyperbola, we have $a = 4$. We know that

$$c^2 = a^2 + b^2,$$

so

$$b^2 = c^2 - a^2 = 5^2 - 4^2 = 25 - 16 = 9.$$

The equation of the hyperbola must be

$$\frac{x^2}{16} - \frac{y^2}{9} = 1.$$

◆

In Figure 10.93 the foci of the hyperbola are the points F_1 and F_2 on the y-axis, and P_1 and P_2 are the two points of the hyperbola on the y-axis.

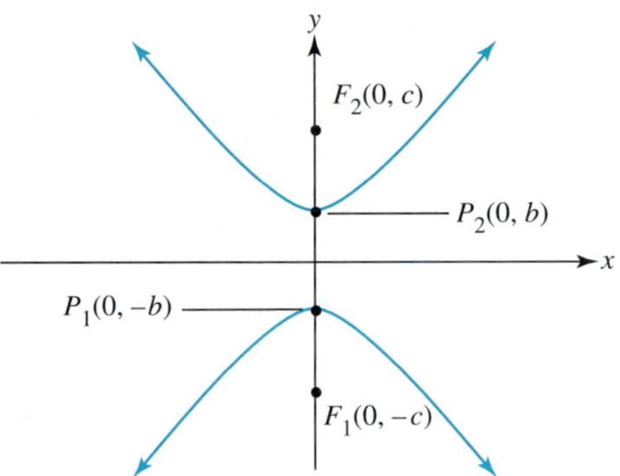

FIGURE 10.93

Equation of a Hyperbola

A hyperbola whose center is the origin and whose foci are at $(0, c)$ and $(0, -c)$ on the y-axis is the set of all points satisfying the equation

$$\frac{y^2}{b^2} - \frac{x^2}{a^2} = 1,$$

where $(0, b)$, and $(0, -b)$ are points of the hyperbola and

$$c^2 = a^2 + b^2.$$

EXAMPLE 10.18 Find the foci and sketch the graph of the equation $16y^2 - 9x^2 = 144$.

SOLUTION If both sides of the equation are divided by 144 to make the right-hand side equal to 1, we have

$$\frac{16y^2}{144} - \frac{9x^2}{144} = \frac{144}{144}, \text{ or}$$

$$\frac{y^2}{9} - \frac{x^2}{16} = 1.$$

Here $b = \sqrt{9} = 3$ so the points $(0, -3)$ and $(0, 3)$ are on the hyperbola. Also $a^2 = 16$, so

$$c^2 = a^2 + b^2 = 16 + 9 = 25, \text{ or}$$

$$c = \sqrt{25} = 5.$$

Thus the foci of the hyperbola are at $(0, -5)$ and $(0, 5)$.

The graph is sketched in Figure 10.94.

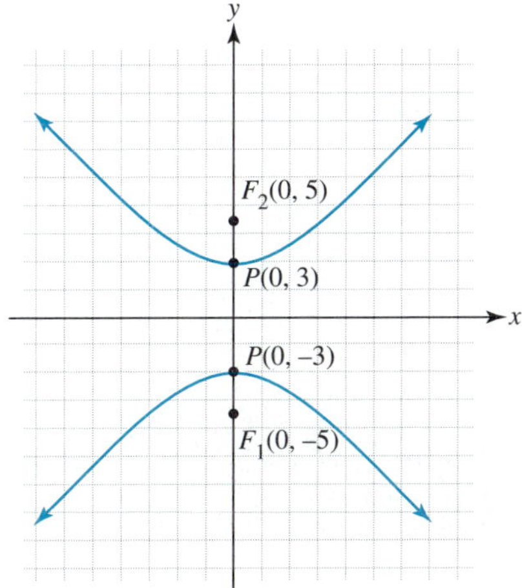

FIGURE 10.94

◆

INITIAL PROBLEM SOLUTION

A 60 foot by 48 foot whispering gallery is to be constructed. Where should the two people stand to maximize their chances of hearing each other?

SOLUTION The whispering gallery should be an ellipse with the listeners standing at the foci, F_1 and F_2 (Figure 10.95).

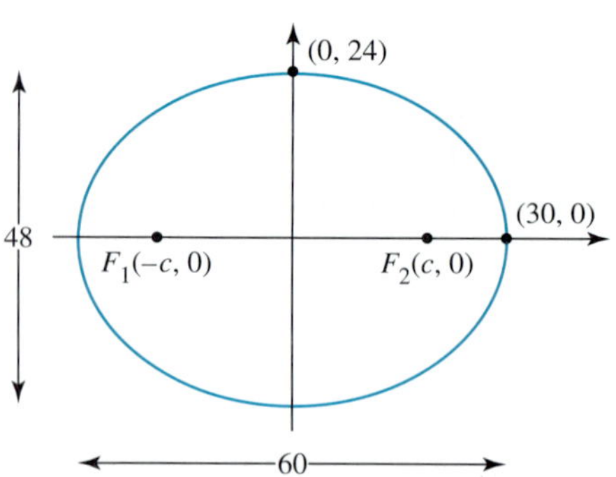

FIGURE 10.95

The points (30, 0) and (0, 24) are on the ellipse [referring to Figure 10.95; these are points $(a, 0)$ and $(0, b)$]. The foci are $F_1 = (-c, 0)$ and $F_2 = (c, 0)$, where $a^2 = b^2 + c^2$. This means that $c^2 = 30^2 - 24^2 = 900 - 576 = 324$. Thus $c = \sqrt{324} = 18$, so the listeners should stand at $(-18, 0)$, and $(18, 0)$.

PROBLEM SET 10.4

The equations in problems 1 through 9, represent conic sections centered at the origin. For each of these, do the following:

(a) Identify the curve.
(b) Identify the foci.
(c) Find the points where the curve intersects the coordinate axes.
(d) Sketch the graph of the curve and label the points from parts (b) and (c).

1. $\dfrac{x^2}{4} + \dfrac{y^2}{25} = 1$

2. $\dfrac{x^2}{16} + \dfrac{y^2}{9} = 1$

3. $\dfrac{x^2}{16} + \dfrac{y^2}{25} = 1$

4. $\dfrac{x^2}{36} + \dfrac{y^2}{20} = 1$

5. $\dfrac{x^2}{16} - \dfrac{y^2}{9} = 1$

6. $\dfrac{y^2}{16} - \dfrac{x^2}{16} = 1$

7. $\dfrac{y^2}{4} - \dfrac{x^2}{10} = 1$

8. $\dfrac{x^2}{36} - \dfrac{y^2}{25} = 1$

9. $\dfrac{x^2}{16} + \dfrac{y^2}{16} = 1$

10. The common equation for a circle is

$$x^2 + y^2 = r^2,$$

where r is the radius of the circle. Put this equation into the standard form for an ellipse and interpret the results with respect to major and minor axes and foci.

In problems 11 through 14, find the equation of the ellipse centered at the origin that fits the given description.

11. Major axis of length 20, minor axis of length 12, and the foci on the x-axis.

12. Major axis of length 14, minor axis of length 8, and the foci on the y-axis.

13. Major axis of length 8, minor axis of length 5, and the foci on the y-axis.

14. Major axis of length 100, minor axis of length 70, and foci on the x-axis.

In problems 15 through 18, find the equation of the hyperbola centered at the origin that fits the given description.

15. Foci on the y-axis 12 units apart, and containing the point $(0, -3)$.

16. Foci on the x-axis 9 units apart, and containing the point $(4, 0)$.

17. Containing the points $(5, 0)$ and $(-5, 0)$, with one focus at $(7, 0)$.

18. Containing the points $(0, -2)$ and $(0, 2)$, with one focus at $(0, -4)$.

19. Suppose an ellipse is centered at the origin with foci at $(4, 0)$ and $(-4, 0)$. Find the equation of the ellipse if there is a point (x, y) on the ellipse such that the distance to one focus is 6, and the distance to the other focus is 4. (Hint: draw a sketch and refer to the definition of an ellipse, Figure 10.74, and Figure 10.82.) How is the distance between the foci related to the distances from a point on an ellipse to the two foci?

20. Suppose a hyperbola is centered at the origin with foci at $(4, 0)$, and $(-4, 0)$. Find the equation of the hyperbola if there is a point (x, y) on the hyperbola such that the distance to one focus is 6, and the distance to the other focus is 4. (Hint: draw a sketch and refer to the definition of a hyperbola, Figure 10.86, and Figure 10.92. How is the distance between the foci related to the distances from a point on a hyperbola to the two foci?)

In problems 21 through 26, change the given equation to the standard form as given for the ellipse and hyperbola in the boxed equations. Then

(a) Identify the curve.
(b) Identify the foci.
(c) Find the points where the curve intersects the coordinate axes.
(d) Sketch the graph of the curve and label the points from parts (b) and (c).

21. $16x^2 - 20y^2 = 320$
22. $25y^2 - 64x^2 = 1600$
23. $25x^2 + 16y^2 = 400$
24. $24x^2 + 50y^2 = 1200$
25. $4x^2 + 16y^2 = 100$
(Hint: $(AB) \div C = A \div (C \div B)$)
26. $32x^2 - 128y^2 = 800$
(Hint: $(AB) \div C = A \div (C \div B)$).
27. A 30 m by 20 m elliptical whispering gallery is to be constructed. Where should two people be standing to maximize their chances of hearing each other whisper.
28. Suppose someone wants to build an elliptical pool table that measures 6 ft by 4 ft. Where should the hole be located?
29. The arch of a certain bridge makes a semielliptical shape. The major axis has a length of 18 m and the semi-minor axis has a length of 6 m. Find the height of the arch above the water at a point that is 3 m from one end of the arch.

30. A whispering gallery is constructed in the shape of a half an ellipsoid (a figure formed by revolving an ellipse around its major axis). If the length of the gallery is 60 m and the foci are located 10 m from each end, what is the height of the ceiling directly above each focus?
31. Thunder is heard by Hal and Bob, who are talking to each other by telephone, 8800 feet apart. Hal hears the thunder 4 seconds before Bob does. Sketch a graph of the locations where the lightning could have struck. Take the speed of sound to be 1100 feet per second. (Hint: suppose Hal is at (4400, 0) and Bob is at (−4400, 0) and think of the time in terms of distance.)
32. In Exercise 31, suppose Gary is also hooked into the conversation and is midway between Hal and Bob. If Gary hears the thunder one second after Hal does, determine where the lightning strike is in relation to the three persons involved.
33. During the earth's elliptical orbit around the sun (the sun is located at one of the foci), the earth's greatest distance from the sun is about 94.5 million miles and the shortest distance is about 91.5 million miles. Find the equation of the earth's elliptical orbit in the form $\frac{x^2}{a^2} + \frac{y^2}{b^2} = 1$. Assume that the center of the orbit is at the origin. (Hint: draw a sketch of the ellipse with the sun at one focus; the greatest distance is $a + c$ and the shortest distance is $a - c$.)
34. The same gravitational forces that Newton found to cause the planets to travel in elliptical orbits around the sun are responsible for the elliptical orbit of the moon around the earth. Find the equation of the moon's elliptical orbit (refer to problem 33). The greatest distance between the moon and the earth is approximately 252,000 miles, and the shortest distance is 222,000 miles.

The following definition should be used in problems 35 through 40.

The eccentricity of a conic section is defined as

$$\text{eccentricity} = \frac{\text{distance from center to focus}}{\text{distance from center to vertex}}$$

35. What is the eccentricity of the earth's orbit around the sun?
(Hint: refer to problem 33.)
36. What is the eccentricity of the moon's orbit around the earth?
(Hint: refer to problem 34.)
37. The eccentricity of an ellipse is a number between 0 and 1; or $0 < e < 1$. Explain why this is true.
38. Carefully explain why the eccentricity of a hyperbola will be a number greater than 1.
39. Explain what happens to the shape of an ellipse as the eccentricity gets close to 0. Include a sketch.
40. Explain what happens to the shape of an ellipse as the eccentricity gets close to 1. Include a sketch.

EXTENDED PROBLEMS

In problems 41 and 42, you are directed to create the figures of an ellipse and hyperbola by folding paper. In order to tell what you are doing more clearly you should do one of two things:

(1) Use wax paper, allowing you to see through more easily, or
(2) Use regular white paper and draw the circle and mark the point with a marking pen.

41. Create an ellipse by folding paper as follows: Draw a circle and select a point on the inside of the circle. Fold points from the circle onto the selected point and make a crease each time. Your figure should be similar to the picture below.

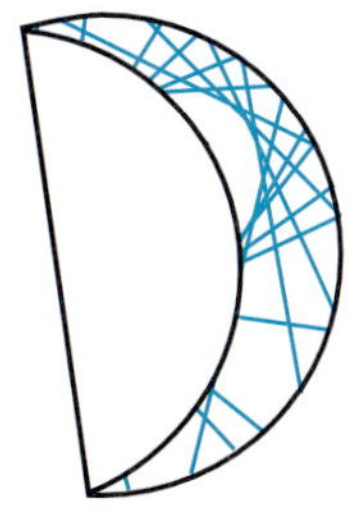

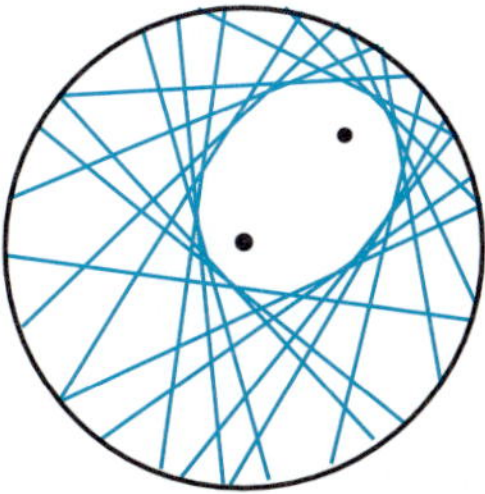

42. Create a hyperbola by folding paper in much the way as an ellipse is created in problem 41. This time however, the point is selected from outside the circle.

The equation of an ellipse with its center at (h, k) having a horizontal major axis of length $2a$ and a minor axis of length $2b$ is

$$\frac{(x-h)^2}{a^2} + \frac{(y-k)^2}{b^2} = 1.$$

The equation of an ellipse with its center at (h, k) having a vertical major axis of length $2a$ and a minor axis of length $2b$ is

$$\frac{(x-h)^2}{b^2} + \frac{(y-k)^2}{a^2} = 1.$$

43. Find the equation, and sketch the graph, of the ellipse whose center is $(3, 4)$ if the major axis is horizontal with a length of 12 and the minor axis has a length of 8.

44. Find the equation, and sketch the graph, of the ellipse whose center is $(-2, 5)$ if the major axis is vertical with a length of 18 and the minor axis has a length of 10.

45. Find the equation, and sketch the graph, of the ellipse whose center is $(2, -3)$ if the major axis is vertical with a length of 12 and the minor axis has a length of 8.

46. Find the equation, and sketch the graph, of the ellipse whose center is $(4, 1)$ if the major axis is horizontal with a length of 10 and the minor axis has a length of 6.

47. Draw a hyperbola using a pencil, a string, and a stick as follows: Let A and B be the foci. Place the stick so that A can serve as a center of rotation and attach a string at points B and C as shown in the following figure:

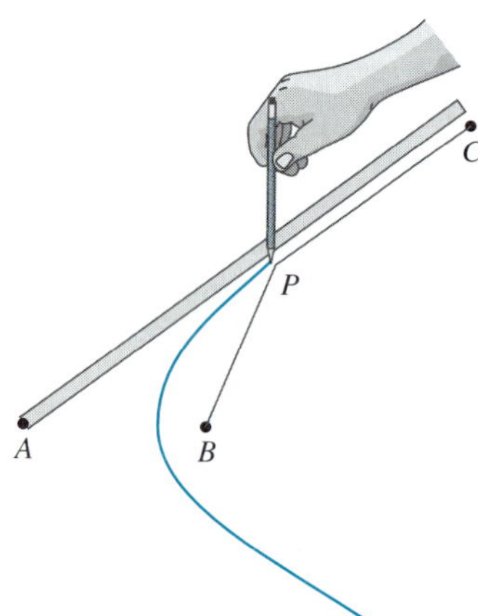

The pencil P keeps the string taut and against the stick as it is rotated counterclockwise. In this way, $BP + PC = S$, the length of the string. Also, $AP + PC = L$, the length of the stick. Thus $PC = S - BP$ as well as $PC = L - AP$. This means that $S - BP = L - AP$, which we can rearrange as $AP - BP = L - S$. That is, the difference $AP - BP$ must be a constant; hence a hyperbola is generated.

48. Following is the derivation of the equation of an ellipse whose center is the origin and whose axes are the x- and y-axes (refer to Figure 10.82). Fill in the missing steps.
An ellipse is the set of all points such that a point is on the ellipse if the sum of its distances from two fixed points is constant. Let $(c, 0)$ and $(-c, 0)$ be the coordinates of the two fixed points (the foci). Let the constant distance be $2a$. Then, the point (x, y) is on the ellipse if and only if the following holds

$$\sqrt{(x-(-c))^2 + (y-0)^2} + \sqrt{(x-c)^2 + (y-0)^2} = 2a.$$

This equation can be rewritten as

$$\sqrt{(x+c)^2 + y^2} + \sqrt{(x-c)^2 + y^2} = 2a.$$

Subtract one of the radical expressions from both sides. Then square both sides. You should be left with a radical on the right side. Isolate that radical on the right side, and then square both sides again. At this point, there should be no radical signs. By doing the necessary rearranging and canceling of terms, you should (after several additional steps) arrive at the equation

$$\frac{x^2}{a^2} + \frac{y^2}{b^2} = 1.$$

Note that, in this situation, $a^2 = b^2 + c^2$.

49. Following is the derivation of the equation of a hyperbola whose center is the origin and whose axes are the x- and y-axes (refer to Figure 10.92). Fill in the missing steps.
A hyperbola is the set of all points such that a point is on the hyperbola if the difference of its distances from two fixed points is constant. Let $(c, 0)$ and $(-c, 0)$ be the coordinates of the two fixed points (the foci). Let the constant distance be $2a$. Then, the point (x, y) is on the hyperbola if and only if the following holds.

$$\sqrt{(x-(-c))^2 + (y-0)^2} - \sqrt{(x-c)^2 + (y-0)^2} = 2a$$

This equation can be rewritten as

$$\sqrt{(x+c)^2 + y^2} - \sqrt{(x-c)^2 + y^2} = 2a.$$

Add the second radical to both sides, and then follow the instructions given for deriving the equation of an ellipse in the previous problem. You should arrive at the equation

$$\frac{x^2}{a^2} - \frac{y^2}{b^2} = 1.$$

Note that, in this situation, $c^2 = a^2 + b^2$.

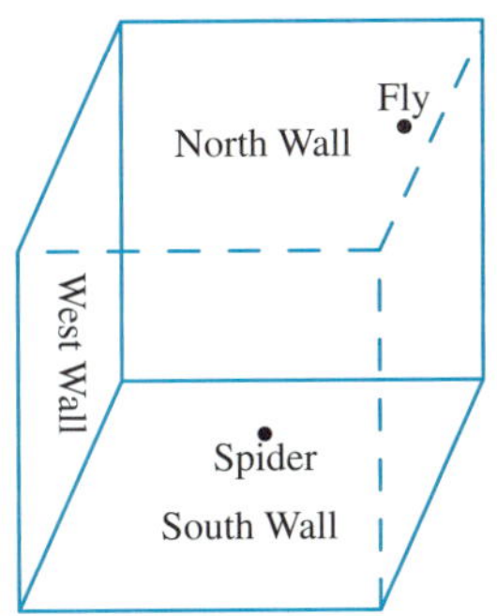

FIGURE 10.96

Chapter Ten Problem

A spider and a fly are in a room that is 10 feet by 10 feet and 10 feet high. The spider is on the south wall, 9 feet above the floor, and 4 feet east of the west wall. The fly is on the north wall, 4 feet above the floor, and 1 foot west of the east wall. The fly is asleep and the spider wishes to walk along the surface of the room to get to the fly in the shortest possible path. How long is this path?

Solution

Strategy: Make a model, draw a picture.

The natural way for the spider to go to the fly is to crawl toward the fly going up to the ceiling and then across to the north wall (Figure 10.97). Making a model of the cubical room shows us that there are other paths. The spider could crawl down to the floor and then across, or the spider could crawl across the east wall. Possibly there are other reasonable paths. The cube may be "unwrapped" in various ways. To find the distance the spider would travel across the ceiling, make a picture of the east wall, ceiling, and west wall unwrapped, lying flat on the paper. This is shown in Figure 10.97(a). The spider's path would be a straight line in this drawing, and it is the hypotenuse of a right triangle with base 5 feet and height 17 feet. This path is therefore $\sqrt{5^2 + 17^2} = \sqrt{25 + 289} = \sqrt{314}$. The spider's path along the floor [Figure 10.97(b)] has a length of $\sqrt{5^2 + 23^2} = \sqrt{25 + 529} = \sqrt{554}$, so the floor path is not the shortest. The path along the east wall has the same length as the path along the ceiling [Figure 10.97(c)]. Experimenting with other ways to unwrap the cube gives another path shown in Figure 10.97(d). This path cuts across part of the ceiling, dips down to the east wall, and curls around to the north wall. This is the shortest path, with length $\sqrt{12^2 + 12^2} = \sqrt{288} \approx 17$.

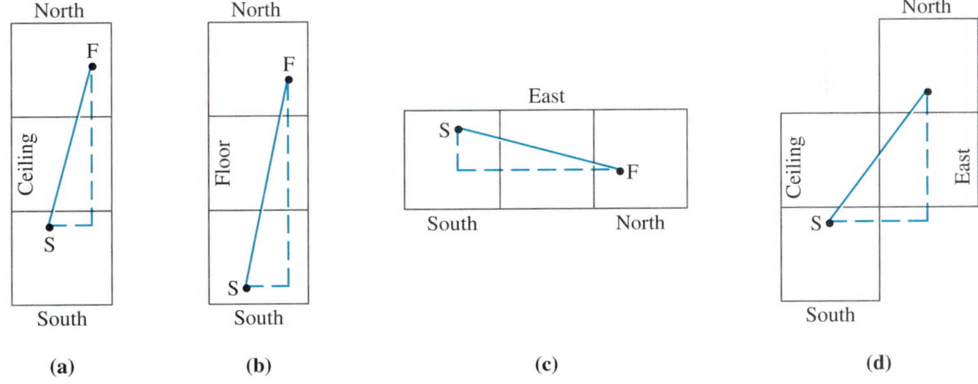

FIGURE 10.97

Chapter Ten Review

Key Ideas and Questions

The following questions review the main ideas of this chapter. Write your answers to the questions, and then refer to the pages listed to make certain that you have mastered these ideas.

1. Why are there no regular tilings for n-gons where n is greater than 6? 479
2. How do you test to see if a tiling of the plane is semiregular? 480
3. Suppose you have measured the sides of a triangle. What must be true for the triangle to be a right triangle? 486
4. Describe, in terms of rigid motions, the possible types of symmetry that a pattern may have. Give examples of each of these. 491
5. Describe the conic sections, both in terms of cross sections of cones and by their precise mathematics definitions. 508, 509, 518, 525
6. Suppose that a light source is put at the focus of a parabola. What happens to the rays of light? What if the light source is put at one of the foci of an ellipse? 510, 521

Vocabulary/Notation

Following is a list of the key vocabulary, notation, and ideas for this chapter. Mentally review each of these items; write down the meaning of each term and use it in a sentence. Then refer to the pages listed by number, and restudy any material you are unsure of before answering the questions in the Chapter Review.

Section 10.1

Polygon 474
Sides 474
Vertex/Vertices 474
n-gon 474
Polygonal Region 474
Tiling 475
Tessellation 475
Vertex Angle 475
Diagonals 475
Regular Polygon 476
Irregular Polygon 476
Regular Tiling 477
Edge-to-Edge Tiling 478
Vertex Figure 479
Semiregular Tiling 480
Convex Polygonal Region 482
Concave Polygonal Region 482
Pythagorean Theorem 483

Section 10.2

Strip Pattern/One dimensional Pattern 491
Symmetry 491
Reflection Symmetry 491
Rotation Symmetry 491
Translation Symmetry 491
Rigid Motion/Isometry 493
Motion Geometry 493
Reflection with respect to line l 493
Vector 494
Translation 494
Directed Angle 495
Center 495
Rotation 495
Fixed Point 497
Glide Reflection 497

Section 10.3

Cone 508
Right Circular Cone 508
Vertex of a Cone 508
Nappes of a Cone 508
Oblique Circular Cone 508
Conic Sections 508
Parabola 508, 509
Ellipse 508
Hyperbola 508
Circle 508
Directrix of a Parabola 509
Focus of a Parabola 509
Axis of a Parabola 509
Vertex of a Parabola 509
Paraboloid 510

Section 10.4

Ellipse 518
Foci/Focus of an Ellipse 519
Center of an Ellipse 519
Major Axis of the Ellipse 519
Minor Axis of the Ellipse 519
Mean Distance 520
Eccentricity 520
Hyperbola 525
Foci of the Hyperbola 525
Axis of the Hyperbola 525
Center of the Hyperbola 527

Chapter Ten Review Problems

1. What is the sum of the measures of the vertex angles of any 9-gon? What is the measure of a vertex angle in a regular 9-gon?
2. Why is it not possible for regular 9-gons to form a regular tiling of the plane?
3. Construct a tiling of the plane with equilateral triangles and squares. Why is this tiling semiregular?
4. A right triangle has side lengths 5 and 10. What is the length of the hypotenuse?
5. A flag pole is 40 feet high and casts a shadow which is 30 feet long. How far is the top of the flag pole from the top of the shadow?
6. Identify the types of symmetry in the following strip pattern.

7. How many lines of symmetry does the following regular pentagon have?

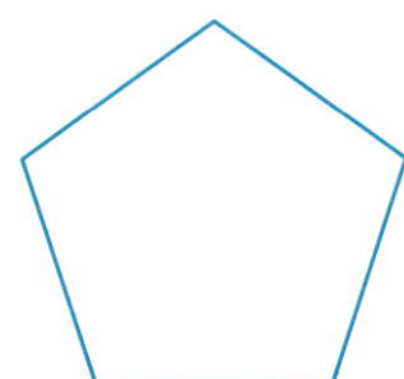

8. What rotational symmetry does the following picture possess?

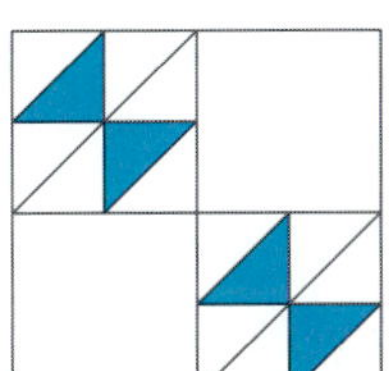

9. Find a glide reflection (a translation and a reflection) that takes the following pattern to itself.

10. Rotate point B 90° about point A.

11. Reflect the pentagon shown across the line AB.

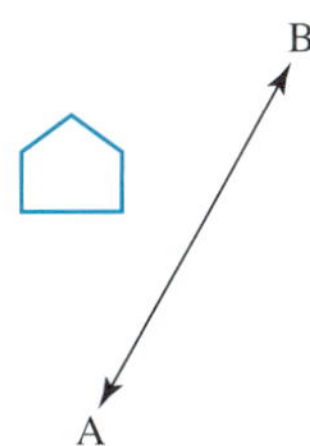

12. Sketch the graph of the equation $y = 4x^2$. What is the focus of this parabola? What is the vertex and axis?
13. A parabolic reflector is used as a solar energy collector. The reflector is 6 feet across and 2 feet deep. Where should the collector be placed so that it collects the most concentrated sunlight?
14. Find the foci and sketch the graph of the equation $25x^2 + 16y^2 = 400$.
15. Find the foci and sketch the graph of the equation $8y^2 - 12x^2 = 72$.
16. Find an equation of an ellipse with major axis 6 and minor axis 3.
17. Find the equation of a hyperbola whose vertices are $(-2, 0)$ and $(2, 0)$ with foci at $(-3, 0)$ and $(3, 0)$.

CHAPTER 11

GROWTH AND SCALING

50 FOOT WOMAN ABDUCTS COLLEGE PROFESSOR!

Scientists are baffled by reports of a woman more than 50 feet tall abducting a college professor in a small town. Shocked students who were on a biology field trip with the professor described the woman as tall and attractive, with a perfectly proportioned body. She was last seen entering the woods near Walden Lake after tearing down fences and wreaking havoc at the Frost farm. Footprints over seven feet long left a trail of smashed haystacks and ruined buildings.

Chapter Goals

1. Solve problems involving similar triangles.
2. Transform geometric objects into larger or smaller objects of similar shape through scaling.
3. Compute quantities such as perimeter, area, volume, and weight of scaled objects.
4. Model population growth, radioactive decay, and other quantities that change in proportion to their size.
5. Model physical objects and compute some related quantities.

You can often read stories this bizarre in supermarket tabloids. Creatures, especially humans, who have grown to gigantic proportions are part of the folklore in many cultures, or part of the popular culture. In some instances they can be our benefactors such as Paul Bunyan and his blue ox Babe. Two movies have been made called *The Attack of the 50 Foot Woman.* In the more recent movie, the 50 foot woman as played by Daryl Hannah looks like the 6-foot Daryl Hannah and every Daryl Hannah in between. Is such a thing possible? Could she have two very different sizes but the same shape?

Creatures that are large need large bones to support their weight. Elephant bones are very large in proportion to their bodies. The strength of a bone is generally dependent on a cross section of the bone, whereas the weight that must be carried is dependent on the volume. We will see in this chapter that as a creature grows, the volume grows more rapidly than the cross section of any part of the creature. If there were a 50 foot woman with the same proportions as a normal sized woman, her bones would break and she would collapse under her own weight. There is no possibility such a story is true; it was made up to entice people to buy the magazine.

This chapter will describe scale and growth effects in different contexts. An example of a growth problem involves the appearance of a new virus. How will the infected population grow? Will the infected population eventually level off, or will it continue to grow? You will be able to answer these types of questions after studying this chapter.

THE HUMAN SIDE OF MATHEMATICS

Emmy Noether

Emmy Noether (1882–1935) had a typical upbringing in a middle-class German Jewish household. She learned music, cooking, and how to run a house. German universities were just beginning to consider admitting women when she was eighteen. She was allowed to attend classes at the University of Erlangen and two years later was accepted into the doctoral program. After earning a Ph.D., Noether was ready to begin her career, except for the fact that women were not allowed to be professors. She therefore worked as a mathematician with no pay for the next eight years. Her fame grew and she was invited to the University of Gottingen, which was at that time the world's leading center for mathematics. She was not allowed to lecture under her own name—her lectures were announced under the name of the leader of the institute. Noether solved a basic problem on invariants (quantities that do not change under certain changes in scale) that was necessary for the development of Einstein's theory of relativity. Einstein, who had difficulty with mathematics, thanked her for her help when he was awarded the Nobel prize. When the Nazis came to power, Noether was forced to leave Germany. Emmy Noether was one of the best algebraists of all time, and a trailblazer for other women who would enter today's world of mathematics and science.

R. Buckminster Fuller

R. Buckminster Fuller (1895–1983) was an inventor and futurist philosopher. Fuller's family saved for years to allow him to go to Harvard, but he had difficulties being a student. He dropped out of Harvard after one year, having spent too much of his time and money in New York City, taking young Broadway actresses to dinner parties. Fuller never graduated from college, but educated himself while working at various industrial jobs and serving with the U.S. Navy in World War I. After several years, he decided to work full time on developing and marketing his own designs and inventions. His ideas were mainly ridiculed, and he lost the support of his investors; he became discouraged and even considered suicide. Unemployed and with no prospects, Fuller moved his family into a slum apartment and devoted the next two years to formulating his philosophical approach to technological innovation.

Bucky (as he came to be known worldwide) believed that human inventiveness had no limits, and technological progress could provide full and satisfying lives for everyone. His philosophy stressed "doing more with less," and many of his inventions were designed to eliminate barriers to mobility and reduce the dependence on limited resources and energy. Fuller saw the need to conserve energy and use it wisely at a time when it still seemed free to everyone else.

Fuller's best-known invention, the geodesic dome, finally brought him fame and fortune. Perfected in 1947, the geodesic dome encloses a greater volume with less material than any alternative structure, and has been considered the most significant structural innovation of the 20th century. Bucky also had a vision of energy efficient, floating cities that could travel across the globe. He designed cities whose buildings made a shell on the outside of a tetrahedron with sides two miles long. The volume of the buildings would be very small compared to the volume inside the shell. Because the inside air would be a degree or so higher than the outside temperature, the entire city would float like a balloon.

From 1959 until his death, Bucky was research professor of design sciences at Southern Illinois University, where he lived in a geodesic dome.

11.1 SCALING OF LENGTH AND AREA

INITIAL PROBLEM

You are thinking of building a rectangular shaped deck on the back of your house. At first, 8 feet by 10 feet seemed a nice size, but upon reflection you have decided on 12 feet by 15 feet as the correct size. Make a rough estimate of the factor by which the cost will be multiplied in going to the larger sized deck.

A common failing when dealing with objects that have the same shape but different sizes is to not realize that the relationships between the lengths, areas, and volumes of the two objects will be different. A modest change in length, for example, can lead to significant changes in volume.

In this chapter we will define and examine the concept of similar figures and the way in which length, area, and volume are affected by changes in size.

SIMILAR TRIANGLES

The fundamental concept needed to study the scaling of length, area, and volume is the notion of geometric similarity. The basic example is provided by similar triangles: Two triangles ΔABC and ΔDEF (Figure 11.1) are said to be **similar** if there is a correspondence of points $A \leftrightarrow D$, $B \leftrightarrow E$, and $C \leftrightarrow F$ such that corresponding angles are equal

$$(1)\ \angle A = \angle D, \qquad (2)\ \angle B = \angle E, \qquad (3)\ \angle C = \angle F,$$

and the ratios of lengths of corresponding sides are equal

$$(4)\ \frac{AB}{DE} = \frac{BC}{EF} = \frac{CA}{FD}.$$

HISTORY

The study of similarity of figures is part of the geometry studied by the Greeks and formalized in Euclid's *Elements* in 300 B.C.

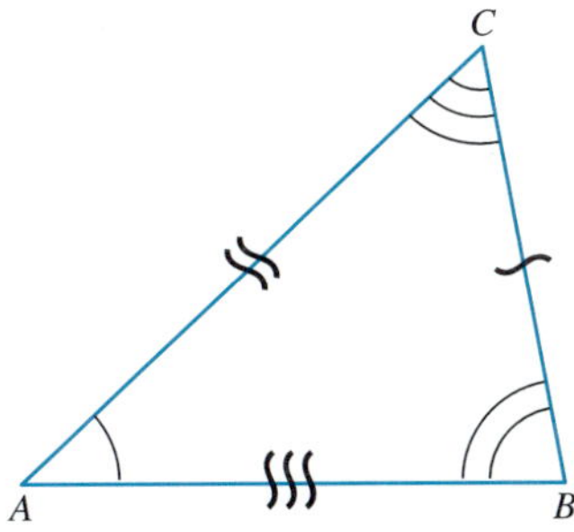

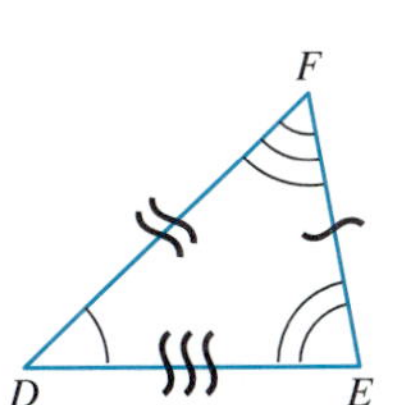

FIGURE 11.1

The arcs on the angles and the curved lines on the sides indicate that the corresponding angles have the same measure and the corresponding sides are proportional (in the same ratio).

It is conventional to use the notation $\overline{AB}$ to denote the line segment connecting A and B, and AB to denote the length of that line segment. It is also customary to write a pair of similar triangles so that the corresponding points are in the same order. The notation

$$\Delta ABC \sim \Delta DEF$$

means that the triangles are similar with $\angle A = \angle D$, $\angle B = \angle E$, $\angle C = \angle F$, and $\frac{AB}{DE} = \frac{BC}{EF} = \frac{AC}{DF}$.

If there are two pairs of angles equal in ΔABC and ΔDEF, say $\angle A = \angle D$ and $\angle B = \angle E$, then the third angles must be equal since the sum of the angles in a triangle is 180°. Also, as you can imagine, if the measures of the angles of a triangle are given, then its shape is determined, but not its size. Combining these two ideas leads to one of the fundamental properties of similar triangles.

PROPERTY

Angle-Angle Property of Similar Triangles

If two angles of one triangle are equal to two angles of another, then the triangles are similar.

EXAMPLE 11.1 Suppose $\Delta ABC \sim \Delta DEF$ with $AB = 5$, $BC = 8$, $AC = 11$, and $DF = 3$ (Figure 11.2). Find DE and EF.

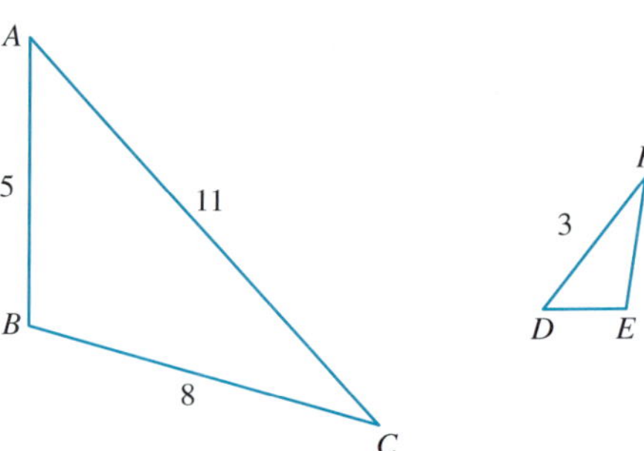

FIGURE 11.2

SOLUTION Since the triangles are similar under the correspondence $A \leftrightarrow D$, $B \leftrightarrow E$, $C \leftrightarrow F$, we know that $\frac{AB}{DE} = \frac{AC}{DF}$. Hence $\frac{5}{DE} = \frac{11}{3}$, so $DE = \frac{15}{11}$. Similarly, $\frac{BC}{EF} = \frac{AC}{DF}$, so $\frac{8}{EF} = \frac{11}{3}$, or $EF = \frac{24}{11}$. ◆

If we use r to represent the common ratio occurring in similar triangles ΔABC and ΔDEF, then we can write

$$\frac{AB}{DE} = r, \frac{BC}{EF} = r, \text{ and } \frac{CA}{FD} = r.$$

The value r is called the **scaling factor** of ΔABC with respect to ΔDEF. Once we know that the scaling factor of ΔABC with respect to ΔDEF is r, then we also know that the scaling factor of ΔDEF with respect to ΔABC is $\frac{1}{r}$. Notice that since $\frac{AB}{DE} = r$, we have $AB = r \times DE$. Similarly, $DE = \frac{1}{r} \times AB$.

EXAMPLE 11.2 Suppose $\Delta ABC \sim \Delta DEF$ with $AB = 5$, $BC = 8$, $AC = 11$, and $DF = 3$, as in Figure 11.2. What is the scaling factor of ΔABC with respect to ΔDEF?

SOLUTION To find the scaling factor, we must locate a pair of corresponding sides. Since the triangles are similar under the correspondence $A \leftrightarrow D$, $B \leftrightarrow E$, $C \leftrightarrow F$, $\overline{AC}$ corresponds to $\overline{DF}$. Thus the scaling factor of ΔABC with respect to ΔDEF is $\frac{AC}{DF} = \frac{11}{3}$. Notice that $\frac{11}{3} \times DF = \frac{11}{3} \times 3 = 11 = AC$. ◆

The scaling factor between the lengths of corresponding sides is also involved when we look at the perimeter and area of a triangle.

Suppose that ΔABC and ΔDEF are similar triangles and that the scaling factor of ΔABC with respect to ΔDEF is 3. Let us compare the perimeters of the two triangles.

$$\begin{aligned} \text{perimeter of } \Delta ABC &= AB + BC + CA \\ &= (3 \times DE) + (3 \times EF) + (3 \times FD) \\ &= 3 \times (DE + EF + FD) \\ &= 3 \times \text{perimeter of } \Delta DEF \end{aligned}$$

The perimeter of ΔABC can be obtained by multiplying the perimeter of ΔDEF by the scaling factor of ΔABC with respect to ΔDEF. The same discussion holds for any pair of similar triangles.

THEOREM

Scaling of Perimeter

If $\Delta ABC \sim \Delta DEF$ and r is the scaling factor of ΔABC with respect to ΔDEF, then

$$\text{perimeter of } \Delta ABC = r \times \text{perimeter of } \Delta DEF.$$

Now let's consider the areas of similar triangles ΔABC and ΔDEF with the scaling factor of ΔABC with respect to ΔDEF equal to 3. Since each side of ΔABC is three times the length of the corresponding side from ΔDEF, three triangles the size of ΔDEF can be constructed on each side of ΔABC (Figure 11.3).

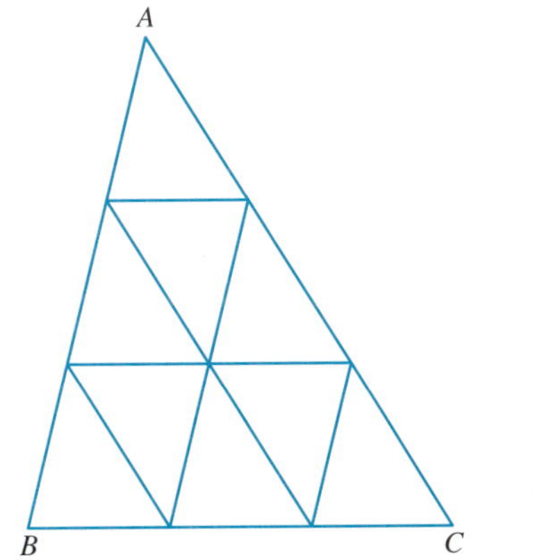

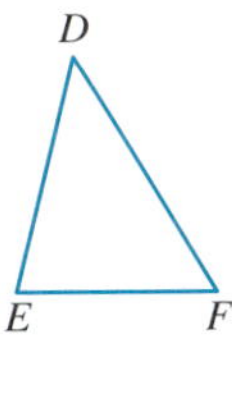

FIGURE 11.3

There are nine congruent copies (the same size and shape) of the smaller triangle that exactly fit together to form the larger triangle. So the area of ΔABC is obtained by multiplying the area of ΔDEF by 9, the square of the scaling factor of ΔABC with respect to ΔDEF. This observation has the following generalization.

THEOREM

Scaling of Area

If $\Delta ABC \sim \Delta DEF$ and r is the scaling factor of ΔABC with respect to ΔDEF, then

$$\text{area of } \Delta ABC = r^2 \times \text{area of } \Delta DEF.$$

EXAMPLE 11.3 The scale of the map of regional air routes in Figure 11.4 is one inch equals approximately 190 miles. Find the length of a flight from Vancouver to Calgary to Edmonton and back to Vancouver. What land area is circumscribed by such a flight?

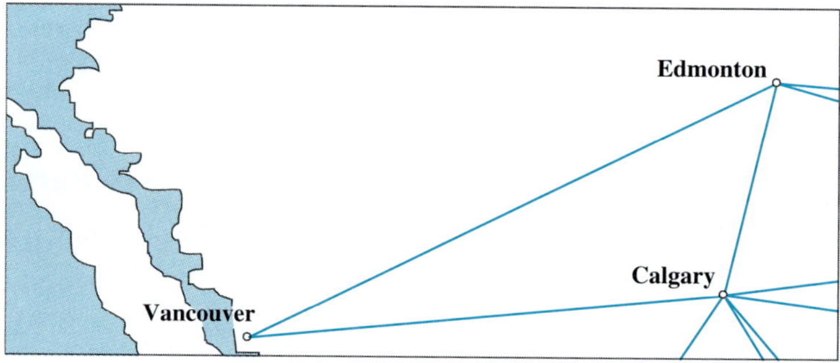

FIGURE 11.4

SOLUTION The proposed flight is represented on the map by a triangle like ΔVCE (Figure 11.5) with the distances measured to the nearest $\frac{1}{16}$ in.

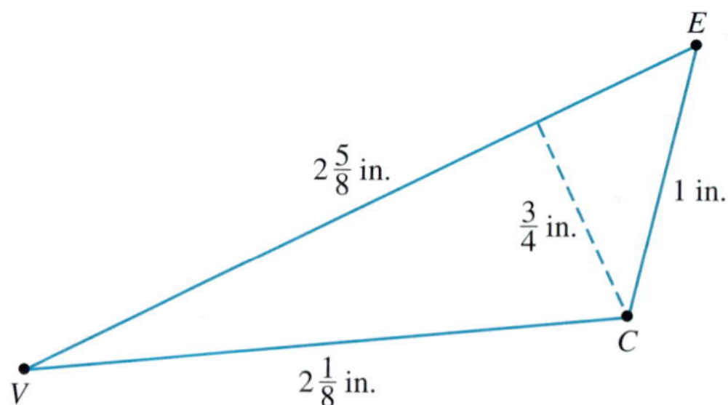

FIGURE 11.5

The perimeter of the triangle on the map is

$$2\frac{1}{8} + 2\frac{5}{8} + 1 = 5\frac{3}{4} = 5.75 \text{ inches.}$$

The formula that is used to find the area of a triangle is $A = \frac{1}{2}bh$, where b is the length of one side of the triangle, and h is the distance from that side to the opposite vertex. Thus the area of the triangle on the map is

$$A = \frac{1}{2}bh = \frac{1}{2} \times 2\frac{5}{8} \times \frac{3}{4} = 0.5 \times 2.625 \times 0.75 \approx 0.984 \text{ square inches.}$$

The scaling factor of the real-world triangle with vertices Vancouver, Calgary, and Edmonton with respect to the ΔDEF is

$$190 \frac{\text{miles}}{\text{inches}}.$$

So the length of the flight is

$$5.75 \text{ inches} \times 190 \frac{\text{miles}}{\text{inches}} = 5.75 \times 190 \text{ miles} \approx 1090 \text{ miles.}$$

The land area circumscribed varies as the square of the scaling factor, thus is

$$0.984 \text{ square inches} \times 190^2 \frac{\text{square miles}}{\text{square inches}} = 0.984 \times 36{,}100 \text{ square miles}$$
$$\approx 35{,}500 \text{ square miles.}$$

◆

SIMILITUDES

A triangle can be transformed into a similar triangle by applying a series of transformations. The transformations needed are translation, rotation about a point, reflection in a line, and a size transformation. The first three of these transformations preserve the size and shape of geometric figures; these were studied in Chapter 10. To visualize a size transformation, think of the plane as a photograph on your desk, perhaps a photograph of a target on a vertical post. Suppose the bullseye of the target is the point about which you want to expand the picture [Figure 11.6(a)]. To perform a size transformation you need to produce an enlargement of the picture, and put it on your desk with the bullseye still in the same place and with the post still vertical [Figure 11.6(b)].

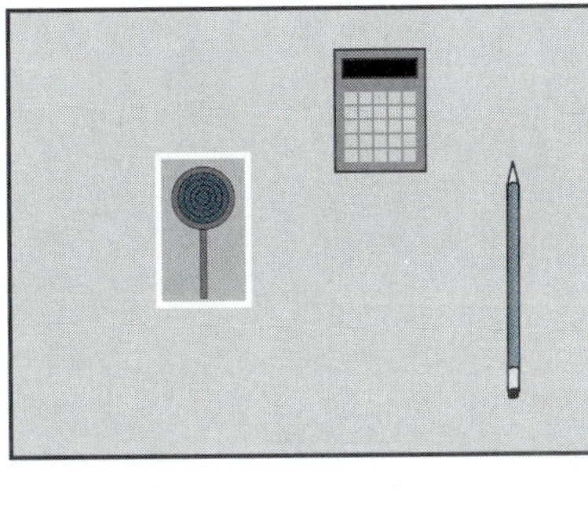

(a)

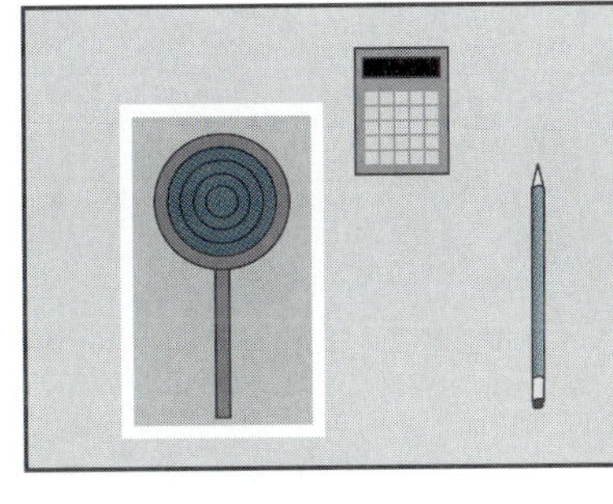

(b)

FIGURE 11.6

More formally, a **size transformation** is defined as follows:

For a fixed point C, called the **center,** and any nonnegative number k, called the **scaling factor,** any point P (other than C) corresponds to the point P' where P'

is on the ray from C through P with $\frac{CP'}{CP} = k$ (Figure 11.7). Equivalently, $CP' = k \times CP$.

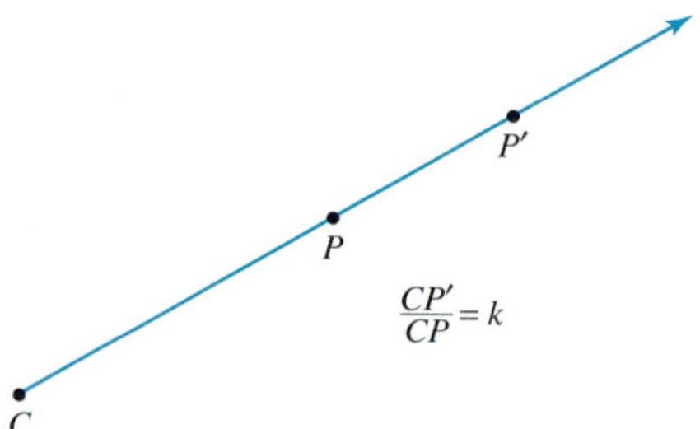

FIGURE 11.7

A size transformation is also known as a dilation, a magnification, or a dilatation. When $r > 1$, the result of a size transformation is a larger similar figure than the original; we will call such a size transformation an **expansion.** When $r < 1$, the result of a size transformation is a smaller similar figure than the original; we will call such a size transformation a **contraction.** The term size transformation is used to refer to either possibility.

To illustrate the connection between similar triangles and transformations, we will describe a sequence of the four basic transformations that transforms a given triangle into a similar triangle. In the figures, we will show the effects of the transformations on the triangle at each step of the sequence. In the figures we will also carry along a line segment $\overline{DY}$ that is not one of the sides of the triangle and a square, so you can see what happens to other points in the plane. The similar triangles will be ΔABC and ΔDEF, and we will show how to transform ΔDEF to ΔABC. For later use, we note that the scaling factor of ΔABC with respect to ΔDEF is 3.

The first transformation is to simply translate the plane so that D goes to A. (Figure 11.8).

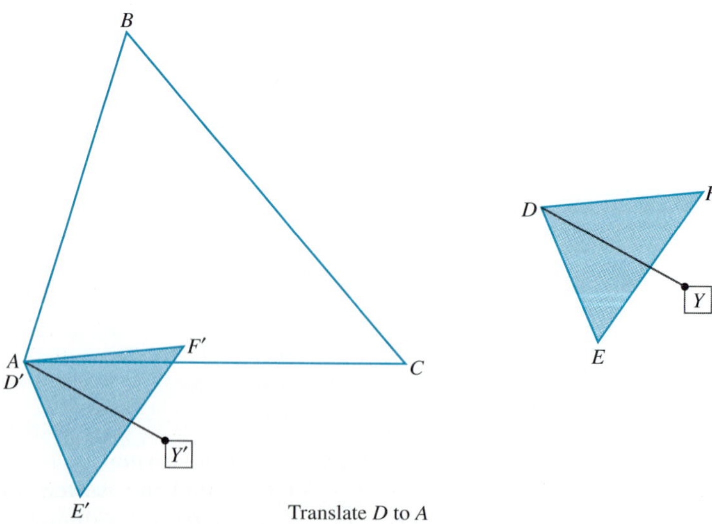

FIGURE 11.8

To keep track of the triangle ΔDEF before and after a transformation we call the transformed version of ΔDEF the image of ΔDEF, and we give the vertices of the transformed triangle slightly different new names. Thus in Figure 11.8, the image of ΔDEF after translation is $\Delta D'E'F'$.

The second transformation is to rotate the plane so that the side $\overline{D'F'}$ lies on $\overline{AC}$ (Figure 11.9).

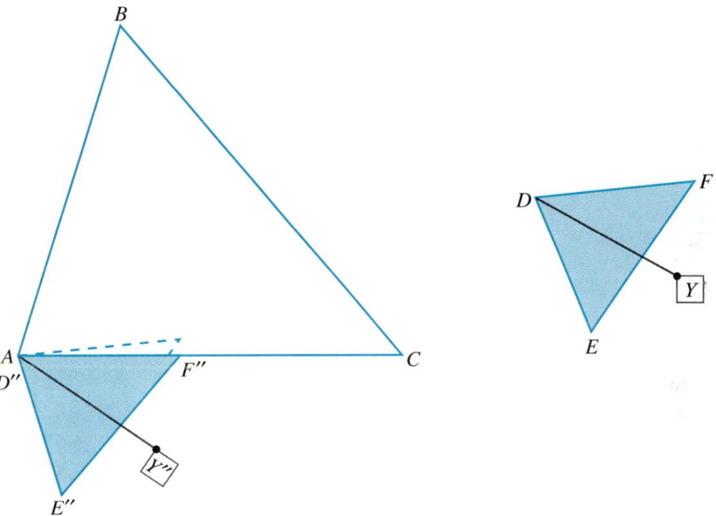

Translate D to A, then rotate $\Delta D'E'F'$ to $\Delta D''E''F''$

FIGURE 11.9

The image of $\Delta D'E'F'$ after rotation is $\Delta D''E''F''$ where $A = D' = D''$ is the point about which we have rotated.

The third transformation is the reflection of the plane through the line determined by $\overline{AC}$. This will put the image of E'' on the same side of the line as is B (Figure 11.10).

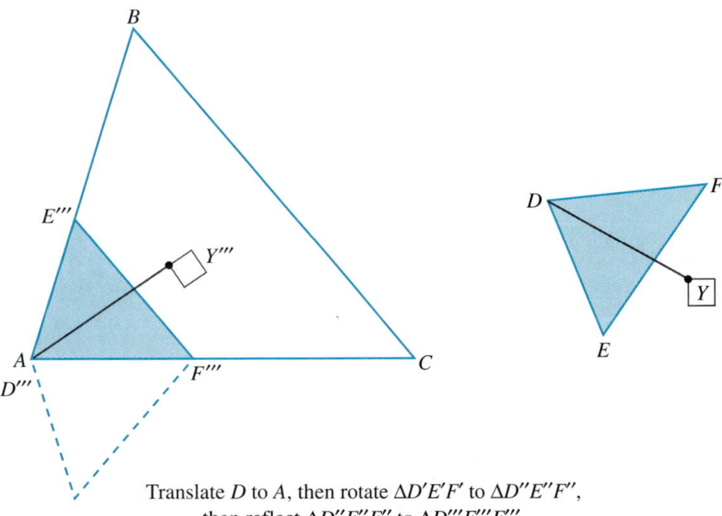

Translate D to A, then rotate $\Delta D'E'F'$ to $\Delta D''E''F''$, then reflect $\Delta D''E''F''$ to $\Delta D'''E'''F'''$.

FIGURE 11.10

The final transformation is the expansion about A which multiplies the lengths of all rays from A by the scaling factor of ΔABC with respect to ΔDEF (Figure 11.11).

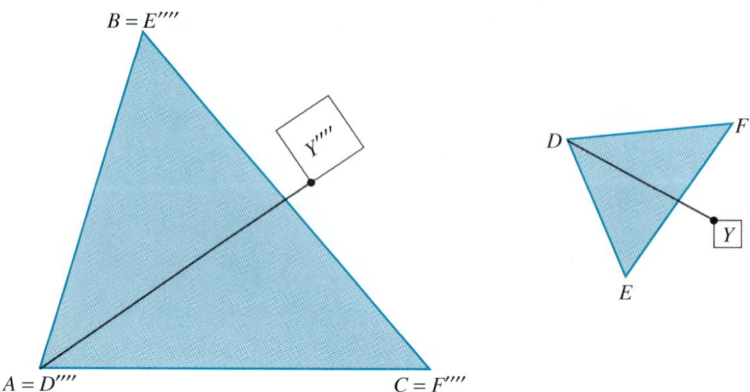

FIGURE 11.11

The example we just went through, which transformed ΔDEF to the similar triangle ΔABC, required all four types of basic transformations. For other pairs of similar triangles, some of the transformations may not be needed; for example if A and D coincide to begin with, then no translation is necessary.

Any combination of the four basic transformations, namely, translations, rotations, reflections, or size transformations, is called a **similitude.** When similitudes are applied to objects in the plane, shapes are preserved, but sizes may not be. If one similitude is followed by another similitude, then the combination is also a similitude, and the scaling factor of the combination is the product of the scaling factors.

EXAMPLE 11.4 Describe the similitude that transforms ΔABD to the triangle ΔADC, where these triangles are as shown in Figure 11.12.

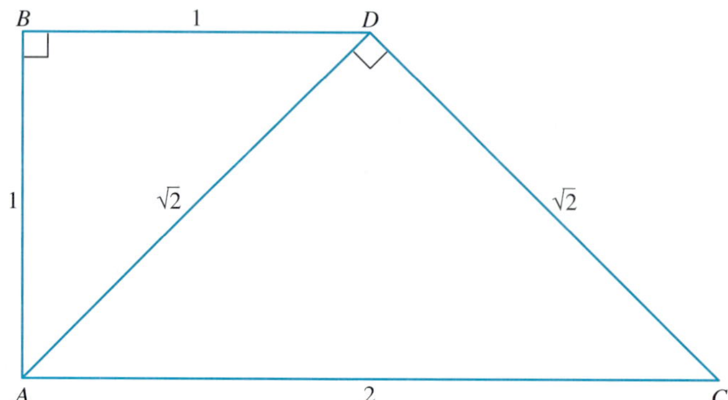

FIGURE 11.12

SOLUTION Observe that ΔABD and ΔADC are isosceles right triangles, hence their corresponding angles are congruent. Thus the triangles are similar. Since $\overline{AD}$ and $\overline{AB}$ are corresponding sides, $\frac{AD}{AB} = \frac{\sqrt{2}}{1} = \sqrt{2}$ is the scaling factor of ΔADC with respect to ΔABD. A clockwise rotation of 45° with center A brings the line determined by A and D into coincidence with the line determined by A and C (Figure 11.13).

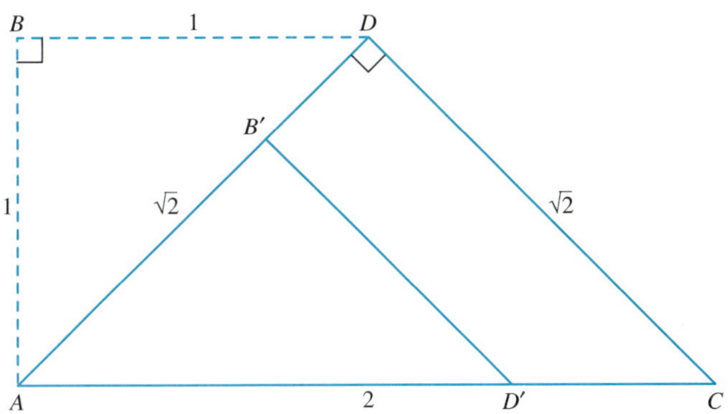

FIGURE 11.13

To keep A fixed while taking D' to C, we must dilate $\Delta AB'D'$ by a factor of $\sqrt{2}$, with center at A. This also takes B' to D (Figure 11.14).

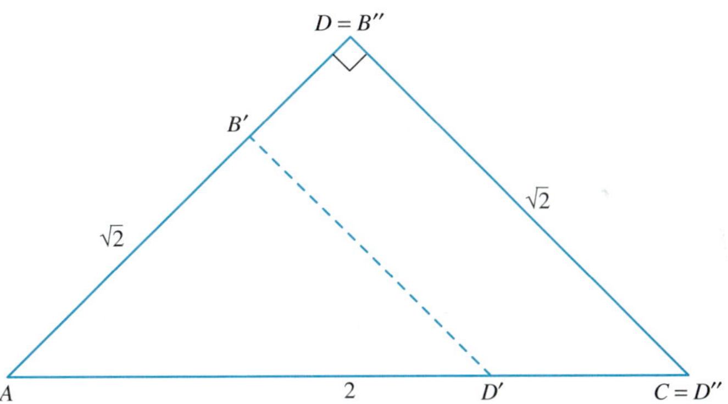

FIGURE 11.14

◆

INITIAL PROBLEM SOLUTION

You are thinking of building a rectangular shaped deck on the back of your house. At first 8 feet by 10 feet seemed a nice size, but upon reflection you have decided on 12 feet by 15 feet as the correct size. Make a rough estimate of the factor by which the cost will be multiplied in going to the larger sized deck.

SOLUTION A rectangle can be thought of as being made up of two triangles. (Just draw one of the diagonals.) Since the two triangles making up the larger deck are similar to the two triangles making up the smaller deck with a scaling factor of $\frac{15}{10} = \frac{12}{8} = 1.5$, the area is increased by a factor of $1.5^2 = 2.25$. The cost is very closely related to the area of the surface, so multiplying the cost by 2.25 provides a good estimate.

PROBLEM SET 11.1

1. Suppose ΔABC and ΔDEF are similar under the correspondence $A \leftrightarrow D$, $B \leftrightarrow E$, and $C \leftrightarrow F$ and that $AC = 4$, $BC = 9$, $AB = 12$, and $DE = 7$.

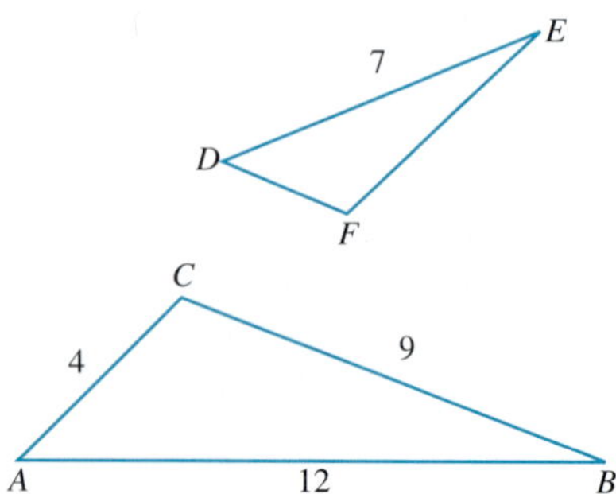

(a) What is the scaling factor of ΔABC with respect to ΔDEF?

(b) Find EF and DF.

2. Suppose ΔABC and ΔDEF are similar under the correspondence $A \leftrightarrow D$, $B \leftrightarrow E$, and $C \leftrightarrow F$ and that $AC = 10$, $BC = 8$, $AB = 6$, and $DE = 7$.

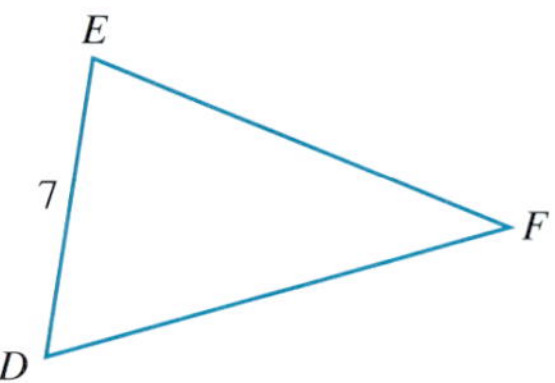

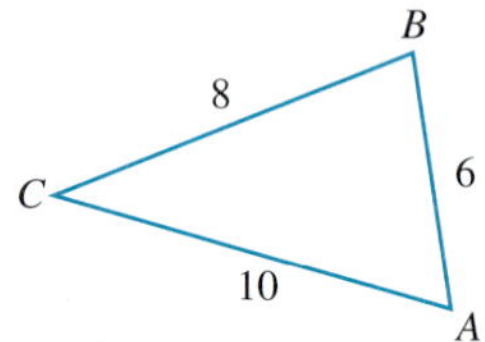

(a) What is the scaling factor of ΔABC with respect to ΔDEF?

(b) Find EF and DF.

In problems 3 through 10, include a properly labeled sketch with your work. The relative lengths of the sides in your sketches are not as important as having the sides properly identified. Two triangles that look similar will be sufficient for the purpose of these exercises.

3. Suppose ΔABC and ΔDEF are similar under the correspondence $A \leftrightarrow D$, $B \leftrightarrow E$, and $C \leftrightarrow F$ and that $AC = 5$, $BC = 8$, $AB = 7$, and $DE = 10$.

(a) What is the scaling factor of ΔABC with respect to ΔDEF?

(b) What is the perimeter of ΔDEF?

4. Suppose ΔABC and ΔDEF are similar under the correspondence $A \leftrightarrow D$, $B \leftrightarrow E$, and $C \leftrightarrow F$ and that $AC = 5$, $BC = 10$, $AB = 8$, and $EF = 6$.

(a) What is the scaling factor of ΔABC with respect to ΔDEF?

(b) What is the perimeter of ΔDEF?

5. Suppose ΔABC and ΔDEF are similar under the correspondence $A \leftrightarrow D$, $B \leftrightarrow E$, and $C \leftrightarrow F$ and that $AC = 5$, $BC = 8$, $AB = 7$, and $DF = 10$.

(a) What is the scaling factor of ΔABC with respect to ΔDEF?

(b) What is the perimeter of ΔDEF?

6. Suppose ΔABC and ΔDEF are similar under the correspondence $A \leftrightarrow D$, $B \leftrightarrow E$, and $C \leftrightarrow F$ and that $AC = 5$, $BC = 10$, $AB = 8$, and $DF = 6$.

(a) What is the scaling factor of ΔABC with respect to ΔDEF?

(b) What is the perimeter of ΔDEF?

7. Suppose ΔABC and ΔDEF are similar under the correspondence $A \leftrightarrow D$, $B \leftrightarrow E$, and $C \leftrightarrow F$ and that $AC = 5$, $BC = 8$, $AB = 7$, and perimeter of $\Delta DEF = 30$.

(a) What is the scaling factor of ΔABC with respect to ΔDEF?

(b) Find DF, EF, and DE.

8. Suppose ΔABC and ΔDEF are similar under the correspondence $A \leftrightarrow D$, $B \leftrightarrow E$, and $C \leftrightarrow F$ and that $AC = 6$, $BC = 12$, $AB = 8$, and perimeter of $\Delta DEF = 50$.

(a) What is the scaling factor of ΔABC with respect to ΔDEF?

(b) Find DF, EF, and DE.

(c) What is the perimeter of ΔDEF?

9. Suppose ΔABC and ΔDEF are similar under the correspondence $A \leftrightarrow D$, $B \leftrightarrow E$, and $C \leftrightarrow F$ and that $AC = 8$, $BC = 10$, $DF = 12$, and perimeter of $\Delta DEF = 36$. Find AB, EF, and DE.

10. Suppose ΔABC and ΔDEF are similar under the correspondence $A \leftrightarrow D$, $B \leftrightarrow E$, and $C \leftrightarrow F$ and that $AC = 12$, $BC = 9$, $DF = 8$, and perimeter of $\Delta DEF = 22$. Find AB, EF, and DE.

Use the following facts from geometry to answer the questions in problems 11 and 12.

(i) When the midpoints of two sides of a triangle are connected, the resulting line is parallel to the third side of the triangle.

(ii) When two parallel lines are transversed (crossed) by a third line, the corresponding angles formed are equal.

11. In ΔABC, D is the midpoint of $\overline{AB}$ and E is the midpoint of $\overline{AC}$, $AB = 8$, $BC = 16$, and $AC = 12$.

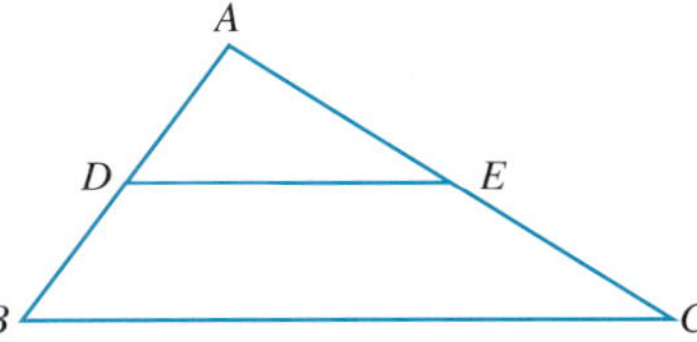

(a) Show $\Delta ADE \sim \Delta ABC$.
Find each of the following:
(b) DE
(c) perimeter of ΔADE
(d) Could you find the perimeter of ΔADE without first finding DE? Explain.

12. In ΔRST, suppose $RS = 9$, $RT = 12$, $ST = 15$, $RP = 3$, and $\overline{PQ}$ is parallel to $\overline{ST}$.

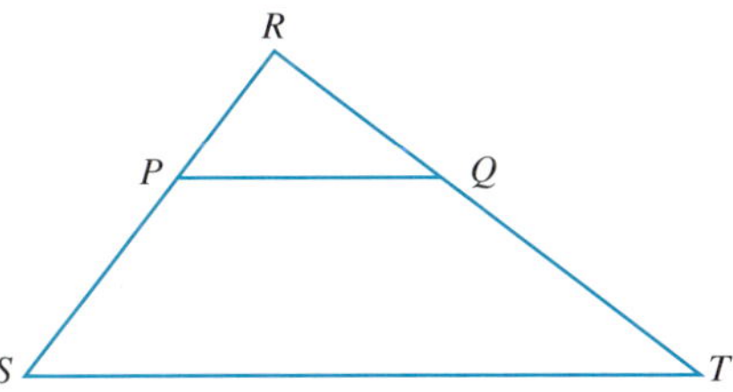

(a) Show $\Delta RPQ \sim \Delta RST$.
Find each of the following:
(b) PQ and RQ
(c) perimeter of ΔRPQ
(d) Could you find the perimeter of ΔRPQ without first finding PQ and RQ? Explain.

13. Suppose ΔABC (below) and ΔDEF are similar under the correspondence $A \leftrightarrow D$, $B \leftrightarrow E$, and $C \leftrightarrow F$; and that $DE = 12$.

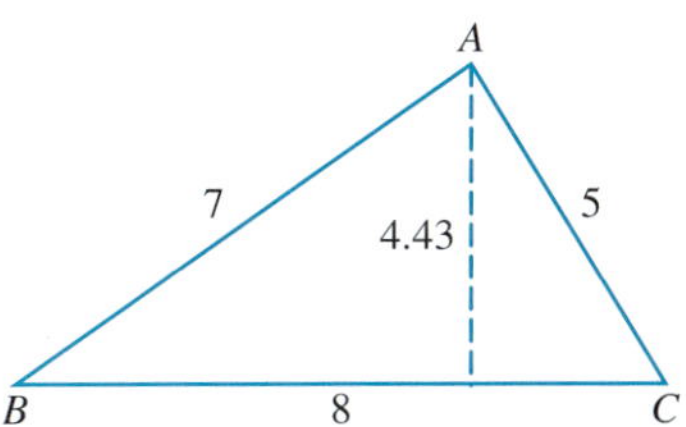

(a) What is the scaling factor of ΔABC with respect to ΔDEF?
(b) What is the area of ΔDEF?

14. Suppose ΔABC (below) and ΔDEF are similar under the correspondence $A \leftrightarrow D$, $B \leftrightarrow E$, and $C \leftrightarrow F$, and that $EF = 6$.

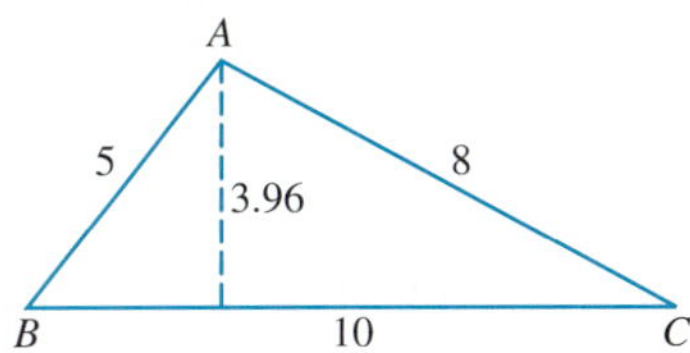

(a) What is the scaling factor of ΔABC with respect to ΔDEF?
(b) What is the area of ΔDEF?

15. Suppose ΔABC in problem 13 and ΔDEF are similar under the correspondence $A \leftrightarrow D$, $B \leftrightarrow E$, and $C \leftrightarrow F$, and that $DF = 15$.
(a) What is the scaling factor of ΔABC with respect to ΔDEF?
(b) What is the area of ΔDEF?

16. Suppose ΔABC in problem 14 and ΔDEF are similar under the correspondence $A \leftrightarrow D$, $B \leftrightarrow E$, and $C \leftrightarrow F$, and that $EF = 12$.
(a) What is the scaling factor of ΔABC with respect to ΔDEF?
(b) What is the area of ΔDEF?

17. Suppose ΔABC and ΔDEF are similar under the correspondence $A \leftrightarrow D$, $B \leftrightarrow E$, and $C \leftrightarrow F$, and that $AC = 5$, $BC = 10$, $AB = 12$, perimeter of $\Delta DEF = 54$, and area of $\Delta ABC = 26$. Find area of ΔDEF.

18. Suppose ΔABC and ΔDEF are similar under the correspondence $A \leftrightarrow D$, $B \leftrightarrow E$, and $C \leftrightarrow F$, and that $AC = 12$, $BC = 9$, $AB = 18$, perimeter of $\Delta DEF = 26$, and area of $\Delta ABC = 48$. Find area of ΔDEF.

19. Suppose you had originally budgeted \$150 for the decking material to build the 8 ft by 10 ft deck in the initial problem at the beginning of this section. How much should you budget for the 12 ft by 15 ft deck?

20. The carpeting for a 9 ft by 12 ft room is estimated to cost \$264. What would you estimate as the cost for the carpeting needed for a 12 ft by 16 ft room?

In problems 21 through 34, complete the sketch to show the result of the transformation that is described. Extend the dotted region if needed.

21. ΔABC is transformed by a translation in which D is the image of A in the given figure.

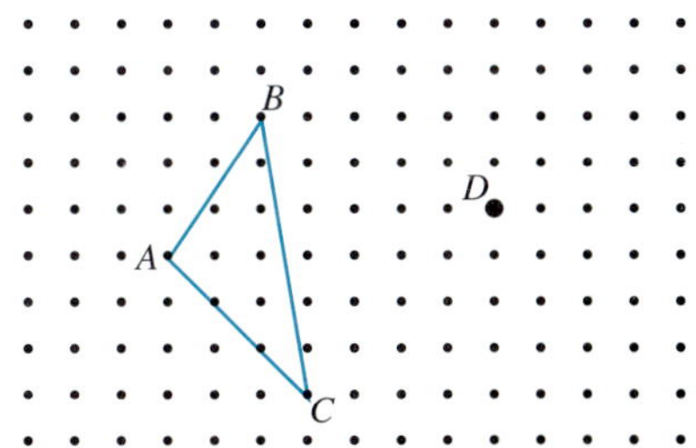

22. ΔABC is transformed by a translation in which D is the image of A in the given figure.

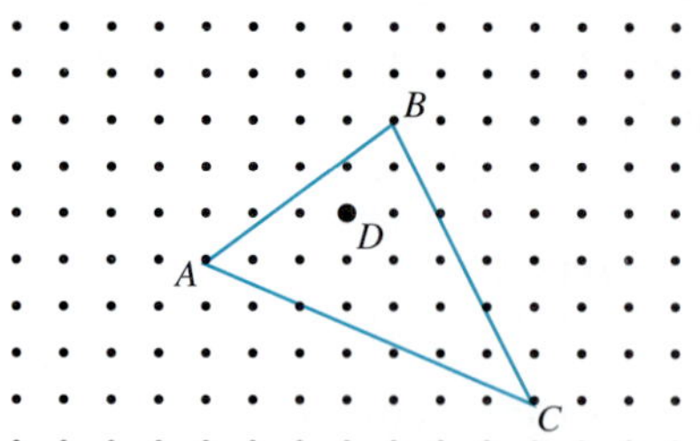

23. ΔABC is transformed by a reflection with respect to the line determined by B and C.

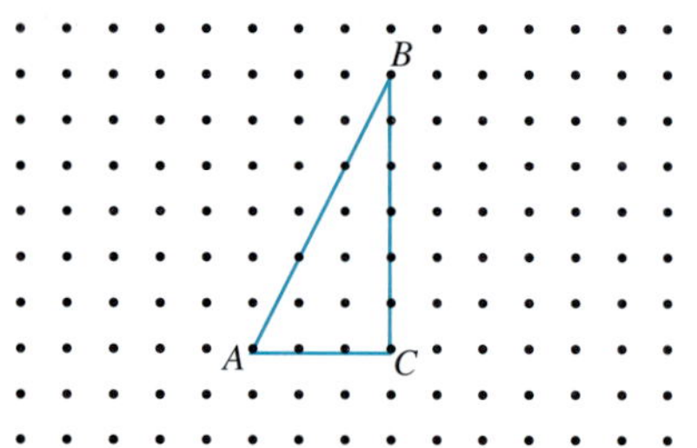

24. ΔABC is transformed by a reflection with respect to the line determined by A and B.

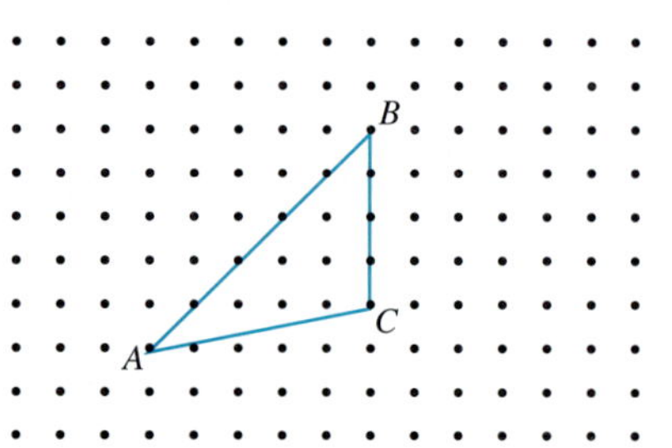

25. ΔABC is transformed by a dilation with center at B and a scaling factor of 2.

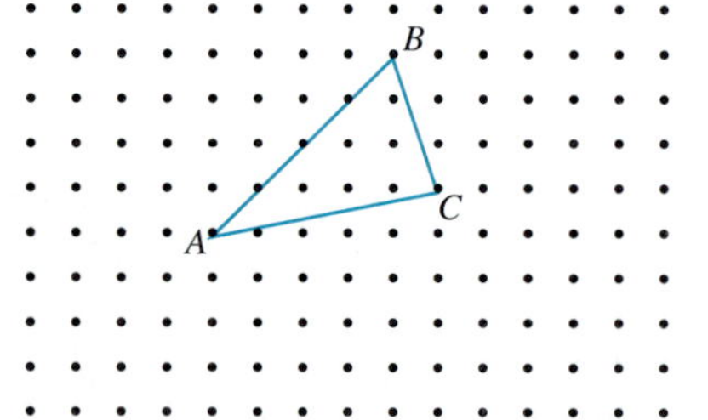

26. ΔABC is transformed by a contraction with center C and a scaling factor of 0.5.

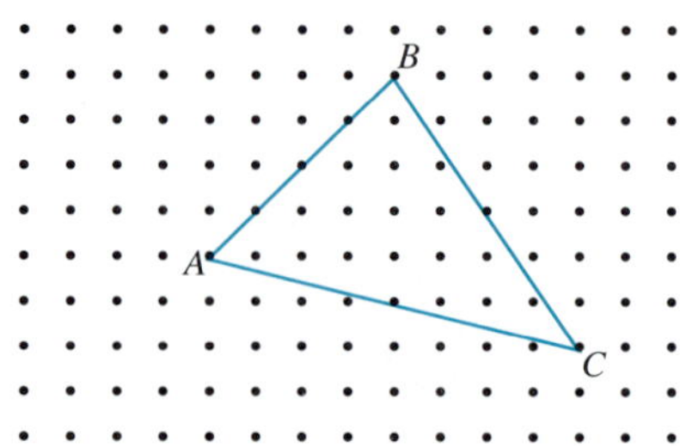

27. ΔABC is rotated so that $\overline{AC}$ coincides with the line determined by A and D.

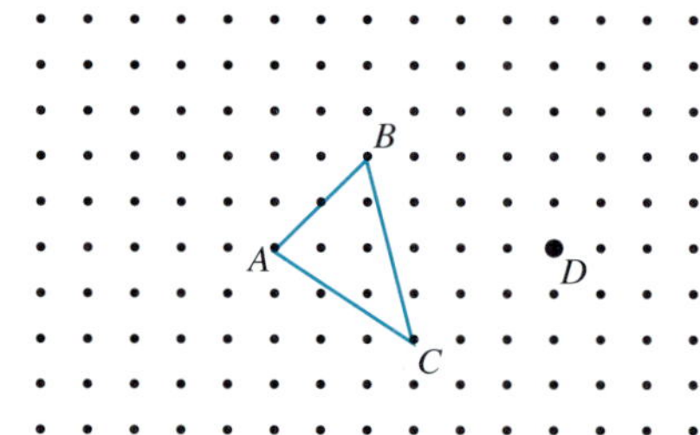

28. ΔABC is rotated so that $\overline{BC}$ coincides with the line determined by B and D.

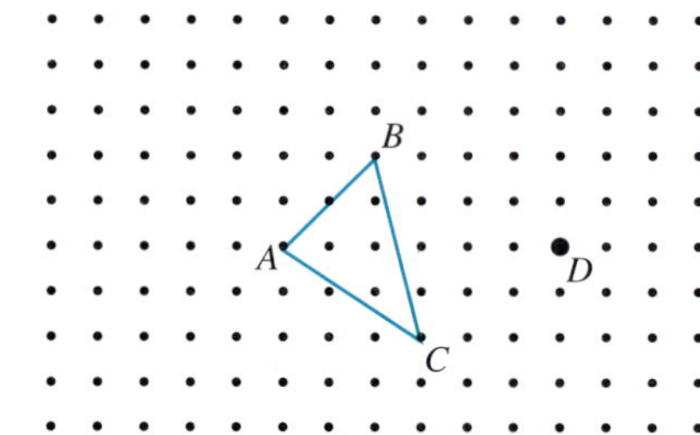

29. ΔABC is first translated so that A coincides with D and then reflected with respect to the line determined by the images of B and C.

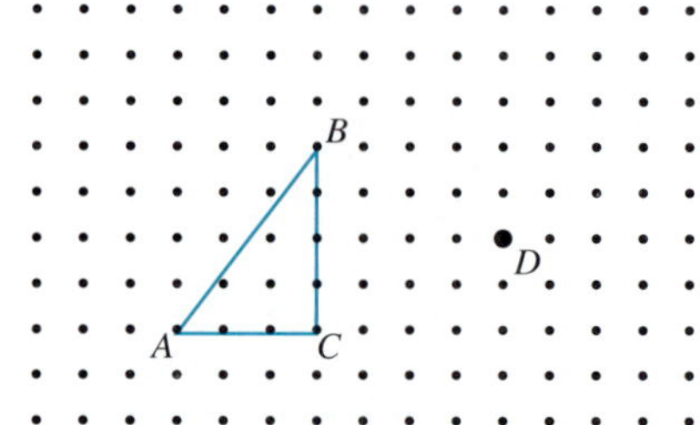

30. ΔABC is first translated so that A coincides with D and then reflected with respect to the line determined by the images of A and C.

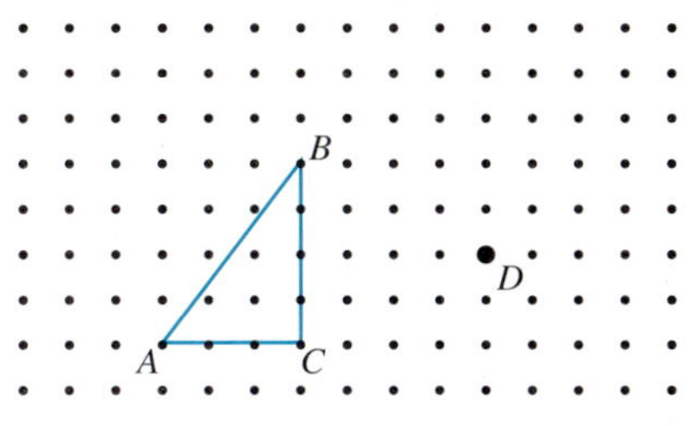

31. ΔABC is first translated so that C coincides with D and then rotated 90° in a clockwise direction about the image of A.

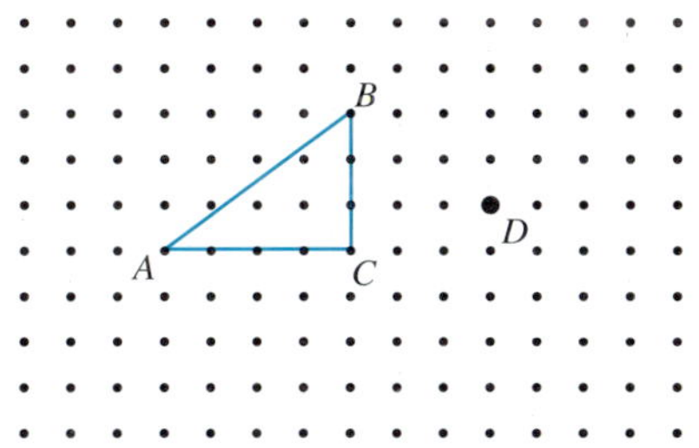

32. ΔABC is first translated so that A coincides with D and then rotated 90° in a counter-clockwise direction about the image of A.

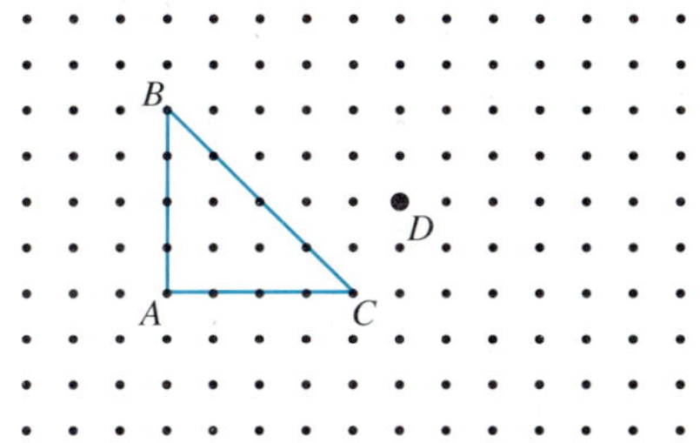

33. ΔABC is first translated 3 units to the left and 2 units upward. The image is then rotated 90° clockwise about the image of A.

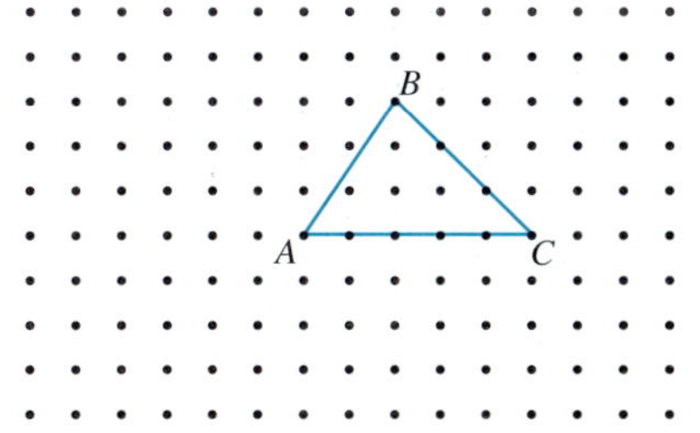

34. ΔABC is first translated 2 units to the left and 2 units downward. The image is then rotated 90° counter-clockwise about the image of B.

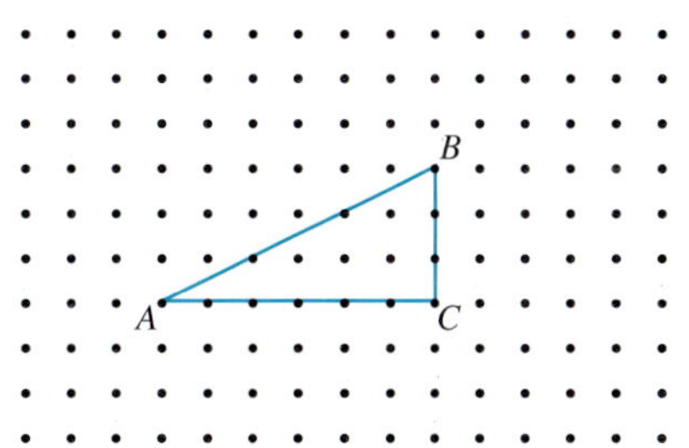

35. The Greek mathematician, Thales, is sometimes referred to as the father of Greek geometry. Among the accomplishments attributed to Thales was measuring the height of the Great Pyramid at Cheops. Although he couldn't measure the height directly, Thales could have accomplished this (or the measurement of any vertical object) as follows: In the proximity of the object he placed an upright stick and measured its length and the length of its shadow. The right triangle formed by the stick and its shadow are similar to the right triangle formed by any other nearby vertical object and its shadow. By measuring the length of the shadow (from a point directly below the highest point of the object to the tip of the shadow) and using the appropriate ratio, he could determine the height. Suppose that Thales used a stick that was 10 feet long (in the appropriate Greek units) and cast a shadow of 16 feet. If Thales's measurement of the shadow of the Great Pyramid (taken at the same time) was equivalent to 770 feet, what would he calculate as the height of the Great Pyramid?

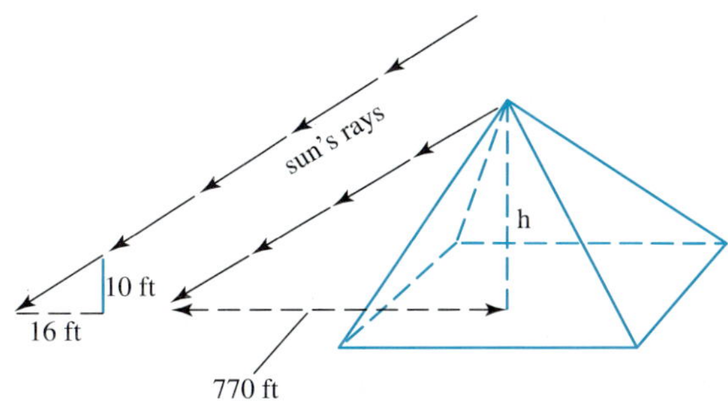

36. Suppose a six-foot post and a tree are casting shadows that are four feet and twenty feet long, respectively. What is the approximate height of the tree?

37. Use similar triangles to find the height of the tree in the figure.

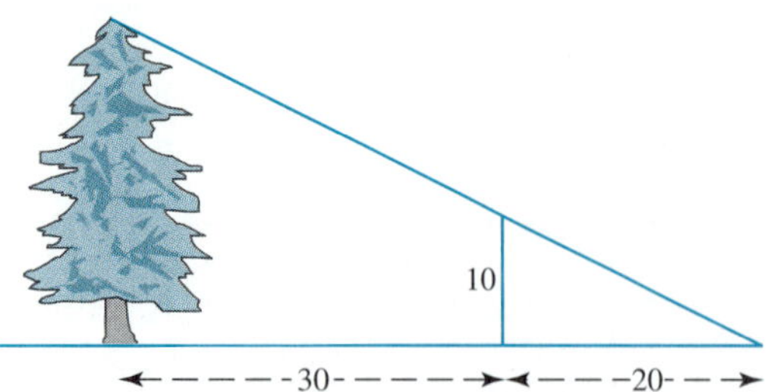

38. Use similar triangles to find the height of the building in the figure if the shadow of the building is 20 feet.

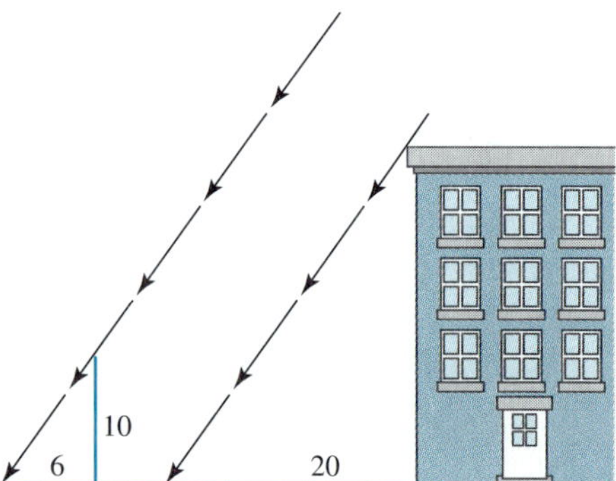

EXTENDED PROBLEMS

In problems 39 through 42, describe a similitude that maps ΔABC into ΔDEF (there will usually be more than one possible similitude). Be as specific as you can with regard to units involved in a translation, degrees and direction of rotation, reference lines for reflections, and scaling factors. These transformations may not all be present in a given situation.

39.

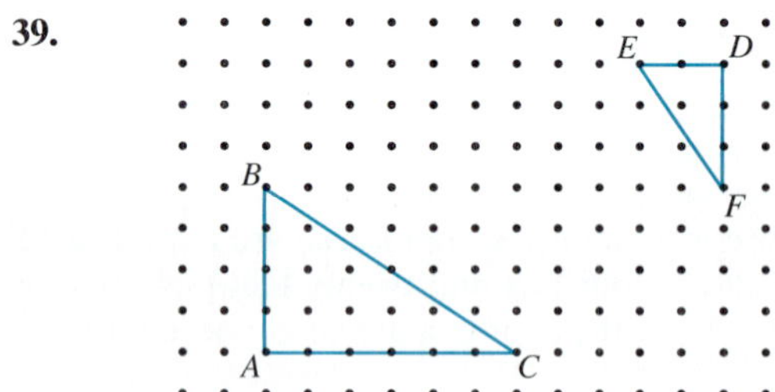

40.

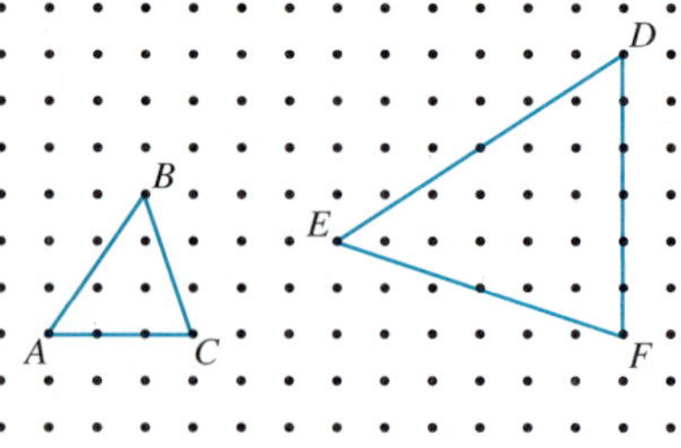

41.

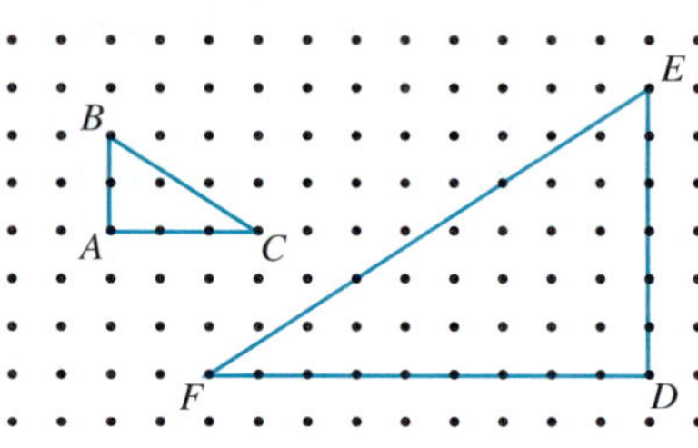

42.

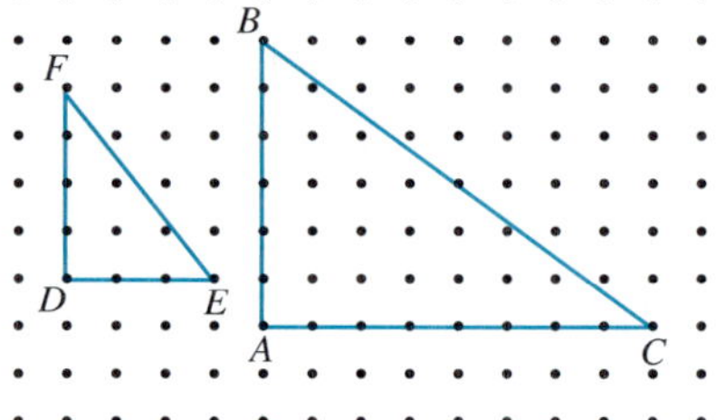

43. Thales, the Greek mathematician referred to in problem 35, won recognition and renown for his contributions to Greek mathematics and philosophy. He was said to have predicted a solar eclipse in 585 B.C. What were some of the other accomplishments of Thales? How did he use similarity as a tool?

44. Computer graphics programs allow the user to manipulate a figure drawn on the screen using transformations such as those described in this section. Most popular graphics programs have features such as *flip horizontal, flip vertical,* and *free rotate.* Many students have performed size transformations while using simpler drawing graphics programs to "size" the pictures and graphs they work with, and translations have become a standard feature referred to as "drag and drop." Experiment with a graphics program or read through the manual to see what features are included.

11.2 SIMILARITY AND SCALING

INITIAL PROBLEM

The largest and oldest of the Egyptian pyramids is the Great Pyramid of Cheops. It has a square base originally 755 feet on each side. The original height was 481 feet. The volume of the Great Pyramid was approximately 91,400,000 cubic feet, and the area of the four sides was approximately 923,300 square feet.

The Memphis Sports Arena is designed as a replica of the Great Pyramid with a scaling factor of $\frac{2}{3}$. What is the approximate volume and the area of the four sides?

In the last section we began our investigation of similar figures with a focus on similar triangles. Now we consider the implications of changes in the sizes of all geometric objects and the way in which area and volume are affected.

SIMILAR PLANE FIGURES

Suppose we have two plane figures P and Q that are not triangles. What should it mean to say these two figures are similar? We should want the figures to be the same shape, before we call them similar. Your eyes may tell you that two figures have the same shape, but we need a more mathematical definition. If P and Q are polygons having the same shape, then, like triangles, their corresponding angles have the same measure. However, if we look at the two rectangles in Figure 11.15, we notice that all the angles in both figures are right angles.

Equal angles, but not similar

FIGURE 11.15

Even though the angles in those rectangles are equal, they do not have the same shape, hence are not similar.

We might require that the ratios of corresponding sides be equal, since that also works for triangles. But both a square and a rhombus without right angles have sides of equal length so the ratios of corresponding sides will be equal (Figure 11.16).

Congruent sides, but not similar

FIGURE 11.16

Here again, though, the figures are not similar.

As it is in the case of triangles, two plane figures are similar if both attributes are present; namely, their corresponding angles are equal and their corresponding sides are proportional. However, we will use the transformation definition of similarity. Two plane figures P and Q are **similar** if there is a similitude that transforms one figure into the other. Remember that a similitude is a combination of four basic types of transformation: translation, rotation, reflection, and size transformation. You should be able to convince yourself that each of these four basic transformations preserves shape (Figure 11.17).

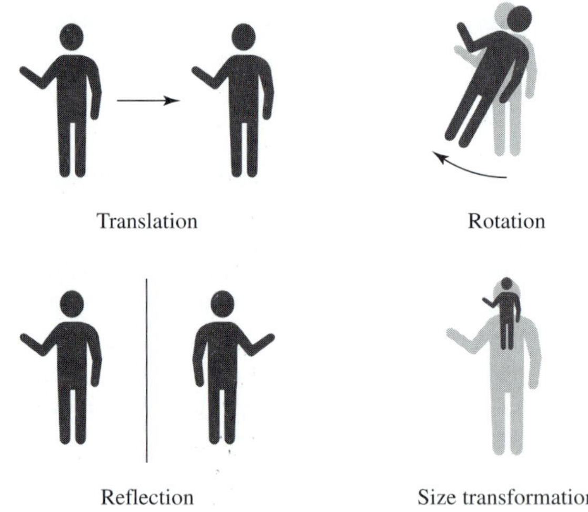

FIGURE 11.17

SCALING LENGTH AND AREA

We now want to consider the relationship between lengths and areas of similar plane figures. Figure 11.18 shows two similar one-dimensional plane figures P and Q.

The similitude that transforms figure P to figure Q requires a translation, a rotation, and an expansion. Note that the scaling factor of the expansion is 2. Notice that every line segment making up P gets twice as long when it is transformed into the corresponding line segment in Q; that is, the length of each line segment is multiplied by the scaling factor of the similitude. Thus the lengths in Q

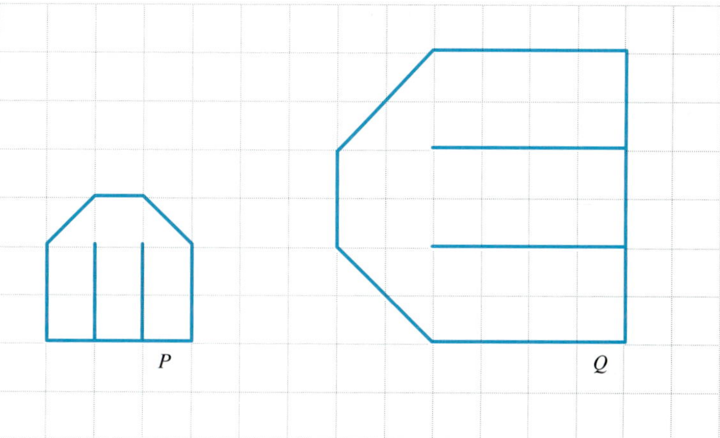

FIGURE 11.18

can be obtained from the lengths in P by multiplying by the scaling factor of the similitude that transforms figure P to figure Q. This illustrates the general rule that applies to all one-dimensional figures, not just those consisting of line segments.

THEOREM

Scaling of the Length of Plane Figures

If P and Q are one-dimensional plane figures, $P \sim Q$, and the similitude that transforms P into Q has scaling factor r, then

$$\text{length of } Q = r \times \text{length of } P.$$

Now we consider the area of the similar two-dimensional figures S and T shown in Figure 11.19.

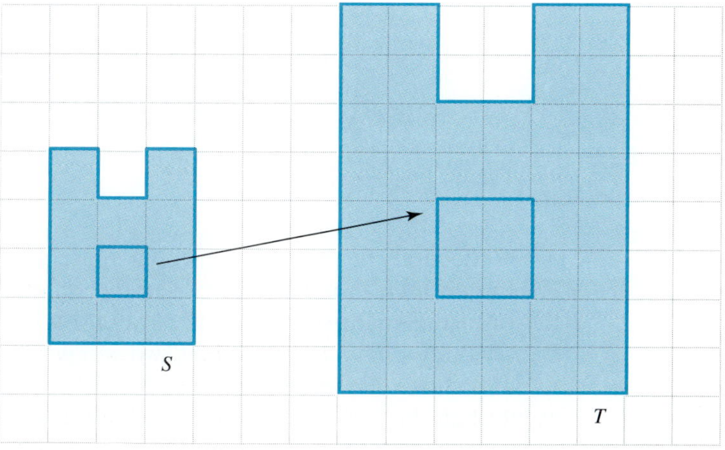

FIGURE 11.19

The similitude that transforms S into T again requires a translation and an expansion with scaling factor 2. Notice that any square in the grid that is inside S transforms into a square inside T that contains $2 \times 2 = 4$ squares of the grid. Therefore, the area of T will be 4 times the area of S. This illustrates the general rule that applies to all two-dimensional figures, not just those enclosed by line segments.

THEOREM

Scaling of the Area of Plane Regions

If S and T are two-dimensional plane regions, $S \sim T$, and the similitude that transforms S into T has scaling factor r, then

$$\text{area of } T = r^2 \times \text{area of } S.$$

SIMILAR THREE – DIMENSIONAL FIGURES

Objects in three-dimensional space are also called **similar (similarity in three dimensions)** if they have the same shape, but not necessarily the same size. If two objects in three-dimensional space are similar, then there is a similitude that transforms one into the other. As before, there are four basic types of transformations to consider: translations, rotations, reflections, and expansions or contractions. The translation of an object in three-dimensional space moves the object without rotating it, for example, a car traveling on a straight, level road. The rotation of an object in three-dimensional space occurs about an axis, not just about a point (Figure 11.20).

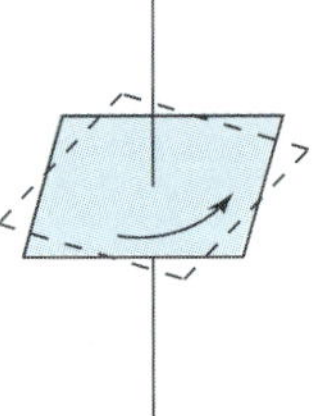

Rotation in Three Dimensional Space

FIGURE 11.20

TIDBIT

Besides having an extraordinary sense of aesthetic beauty, classical sculptors were remarkably adept at three-dimensional similarity. Michelangelo's famed sculpture of *David* is over 12 feet tall.

Reflections in three-dimensional space are through a plane, instead of a line. When you look in a mirror you see the reflection of objects through the plane of the mirror. Expansion can be illustrated by blowing up a balloon. If P and Q are similar three-dimensional objects, we will write $P \sim Q$.

If a similitude in three-dimensional space changes the length of any line segment, then the similitude changes the lengths of all line segments by multiplying by the same factor, which we call the scaling factor of the similitude. The effect of a similitude on the length of one-dimensional figures and the areas of two-dimensional figures is exactly as in the case of plane figures: lengths are multiplied by the scaling factor and areas are multiplied by the scaling factor squared.

SCALING VOLUME

Now we consider what happens to the volume of an object when it is transformed by a similitude. Once again the critical piece of information is the scaling factor of the expansion (or contraction) involved. In Figure 11.21 the cylinder on the left has been transformed into the cylinder on the right by a similitude with scaling factor 2.

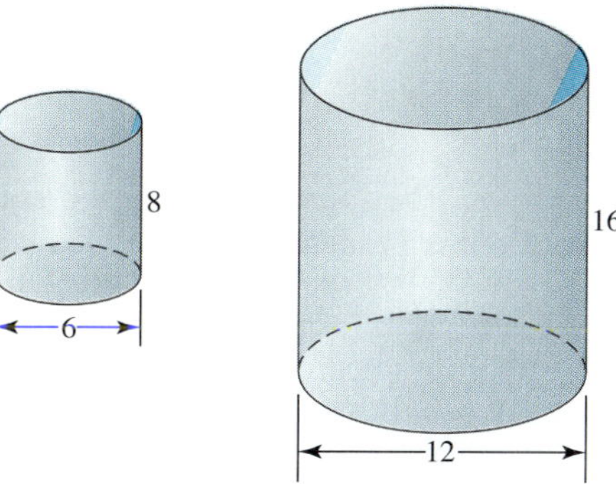

FIGURE 11.21

The formula for finding the volume of a cylinder is $V = \pi r^2 h$, where r is the radius of the base and h is the height. The radius of the cylinder on the left is 3 and its height is 8, so its volume is

$$V = \pi \times 3^2 \times 8 = 72\pi.$$

The radius of the cylinder on the right is $6 = 2 \times 3$ and its height is $16 = 2 \times 8$, so its volume is

$$\pi \times 6^2 \times 16 = \pi \times (2 \times 3)^2 \times (2 \times 8) = 2 \times 2 \times 2 \times \pi \times 3^2 \times 8 = 2^3 \times 72\pi.$$

The volume of the cylinder on the right equals the scaling factor cubed times the volume of the cylinder on the left. This illustrates the general rule that applies to all three-dimensional objects.

Scaling of the Volume of Three-Dimensional Objects

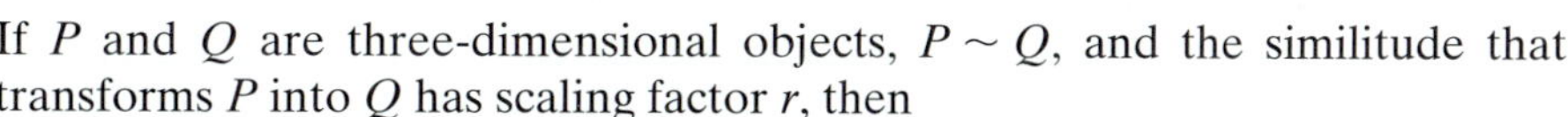
If P and Q are three-dimensional objects, $P \sim Q$, and the similitude that transforms P into Q has scaling factor r, then

$$\text{volume of } Q = r^3 \times \text{volume of } P.$$

EXAMPLE 11.5 Suppose a small souvenir football you own is about 5.5 inches long. You can measure the volume of this little football in your kitchen by immersing it in a completely full bowl of water to see how much water it pushes out of the bowl (Figure 11.22).

FIGURE 11.22

Suppose the souvenir football displaces 12.25 fluid ounces of water. If a regulation football is 11 inches long, what is its approximate volume?

SOLUTION Assuming the souvenir football is the same shape as the regulation football, we conclude that the souvenir football can be transformed into the regulation football by an expansion with scaling factor $\frac{11}{5.5} = 2$. Thus the regulation football must have a volume $2^3 (= 8)$ times as large as the volume of the souvenir football, or about 98 fluid ounces—a little over three quarts. ◆

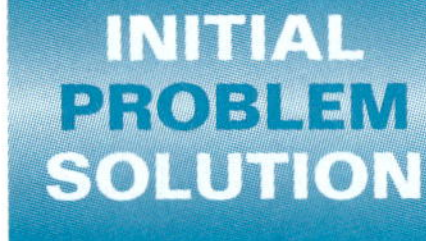

The largest and oldest of the Egyptian pyramids is the Great Pyramid of Cheops. It has a square base originally 755 feet on each side. The original height was 481 feet. The volume of the Great Pyramid was approximately 91,400,000 cubic feet and the area of the four sides was approximately 923,300 square feet.

The Memphis Sports Arena was designed as a replica of the Great Pyramid with a scaling factor of $\frac{2}{3}$. What is the approximate volume, and the area of the four sides?

SOLUTION If the linear dimension of an object is scaled by a factor of $\frac{2}{3}$, then the area is scaled by a factor of $(\frac{2}{3})^2$, and the volume is scaled by a factor of $(\frac{2}{3})^3$. This means that for the Memphis Sports Arena

$$\text{Volume} \approx \left(\frac{2}{3}\right)^3 \times 91{,}400{,}000 \approx 27{,}100{,}000 \text{ cubic feet, and}$$

$$\text{Area of sides} \approx \left(\frac{2}{3}\right)^2 \times 923{,}300 \approx 410{,}350 \text{ square feet.}$$

PROBLEM SET 11.2

Problems 1 through 6 refer to regular hexagons. A hexagon is a six-sided plane figure. The term regular means that all sides are the same length and all angles have the same measure. As a result, all regular hexagons are similar in the same way that all squares are similar.

1. If the length of the side of a regular hexagon is 2, then the area is $6\sqrt{3}$. What is the area of a regular hexagon with a side length of 4? (Hint: what is the scaling factor between the two objects?)
2. If the length of the side of a regular hexagon is 2, then the area is $6\sqrt{3}$. What is the area of a regular hexagon with a side length of 6?

3. If the length of the side of a regular hexagon is 2, then the area is $6\sqrt{2}$. What is the area of a regular hexagon with a side length of $\frac{4}{3}$?
4. What is the area of a regular hexagon that has a perimeter of 20? (Hint: what is the side length?)
5. The area of a regular hexagon with a side length of 2 is $6\sqrt{3} \approx 10.4$. What is the side length of a regular hexagon that has an area of 20?
6. If the area of one regular hexagon is twice that of another, what is the ratio of the perimeter of the first hexagon to the second?
7. Suppose a plane figure undergoes a size transformation in which the scaling factor is 3. How do the perimeter and area of the image of the plane figure compare to the original?
8. Suppose a plane figure undergoes a size transformation in which the scaling factor is $\frac{2}{3}$. How do the perimeter and area of the image of the plane figure compare to the original?
9. Suppose we have a three-dimensional region with a volume of 45 units. What would be the volume of a similar region if the scaling factor in going from the first to the second is 2.5?
10. Suppose we have a three-dimensional region with a volume of 80 units. What would be the volume of a similar region if the scaling factor in going from the first to the second is 0.5?

The surface areas of similar three-dimensional regions are related in the same manner as the areas of similar plane regions. That is, if P and Q are three-dimensional regions with $P \sim Q$ and the scaling factor going from P to Q is r, then

$$\text{surface area } (Q) = r^2 \times \text{surface area } (P)$$

11. The surface area and volume of a cube with side length s is given by $A = 6s^2$ and $V = s^3$. How would you express the surface area and volume of the cube with side length $2s$?
12. The surface area and volume of a cube with side length s is given by $A = 6s^2$ and $V = s^3$. How would you express the surface area and volume of the cube with side length $3s$?
13. Suppose a three-dimensional region has a surface area of 76 in^2 and a volume of 40 in^3. What would be the surface area and volume of a similar three-dimensional region if the scaling factor in going from the first to the second is 1.5?
14. Suppose a three-dimensional region has a surface area of 1300 in^2 and a volume of 3000 in^3. What would be the surface area and volume of a similar three-dimensional region if the scaling factor in going from the first to the second is 0.4?
15. Suppose the Memphis Sports Arena had been built as a scaled replica of the Great Pyramid of Cheops, with a scaling factor of $\frac{1}{2}$. If the volume of the Great Pyramid is approximately 91,400,000 cubic feet, what would the volume of the Memphis Sports Arena be?
16. The volume of the Great Pyramid of Cheops is approximately 91,400,000 cubic feet. What would be the volume of a scaled replica of the Great Pyramid if the scaling factor was 0.25?
17. What scaling factor would be needed to produce a scaled replica of the Great Pyramid of Cheops that had a volume of 1,000,000 cubic feet?
18. What scaling factor would be needed to produce a scaled replica of the Great Pyramid of Cheops that had twice the volume of the original?
19. One of the famous problems from Greek antiquity was the duplication of the cube. According to legend, the people of Delos were suffering from a plague and consulted the great oracle. The oracle told them that to rid themselves of the plague they must build a new altar to one of their gods that would be geometrically similar to an existing one, but double the volume.
    (a) How would the volume of the new altar compare to the old one if all the linear dimensions were doubled?
    (b) What scaling factor is needed to produce the new altar?
20. If an exact replica of Michelangelo's famous statue of *David* was made life-size from the same material, how would the volume and surface area of the replica compare (as a fraction or percentage) to the volume and surface area of the original? Assume David was 5 feet 6 inches tall, and the statue is 12 feet tall.

EXTENDED PROBLEMS

The following figure consists of concentric circles of radius $1, 2, \frac{1}{2}, 4, \frac{1}{4}, 8, \frac{1}{8}, \ldots$. This figure is self-similar because the similitude which dilates by a factor of 2 about the common center takes the figure to itself.

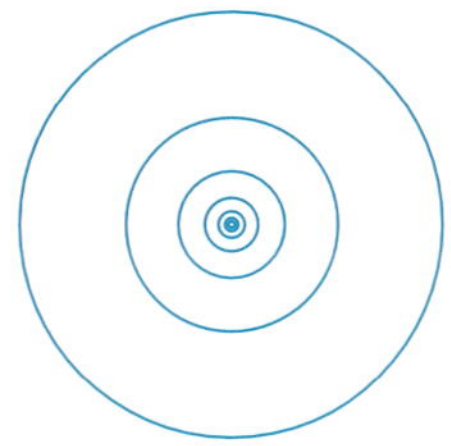

To construct such a self-similar figure begin with any pair of similar figures, say A and B. Let A_0 be the figure A. Let A_1 be the image of A_0 under the similitude that takes A to B, and let A_{-1} be the image of A_0 under the similitude that takes B to A. Let A_{n+1} be the image of A_n under the similitude that takes A to B, and let A_{-n-1} be the image of A_{-n} under the similitude that takes B to A. The self-similar figure is the union of $A_0, A_1, A_{-1}, A_2, A_{-2}, \ldots$. To construct the preceding figure, take A to be the circle of radius 1 and B to be the concentric circle of radius 2.

21. Construct the self-similar figure based on the following choice of A and B: the first figure A is a square and the second figure B is the square formed by connecting the midpoints of the edges of A.

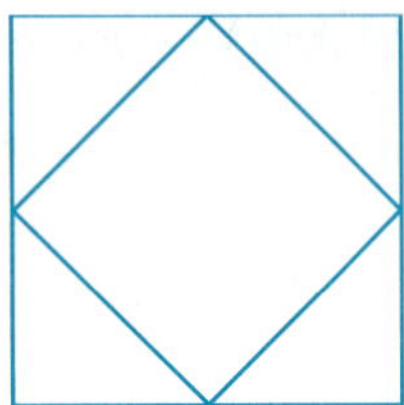

22. Construct the self-similar figure based on the following choice of A and B: the first figure A is an equilateral triangle and the second figure B is the equilateral triangle formed by connecting the midpoints of the edges of A.

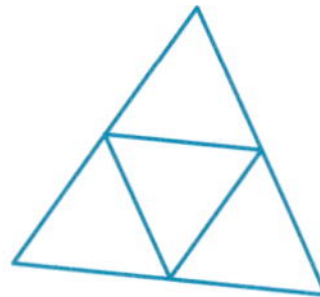

23. Construct a self-similar spiral figure based on extending the relationship between the following pair of figures: the first figure is the upper half of a circle of radius 1 and the second figure is the lower half of a circle of radius $\frac{3}{4}$, with center at distance $\frac{1}{4}$ from the center of the first and sharing an endpoint with the first.

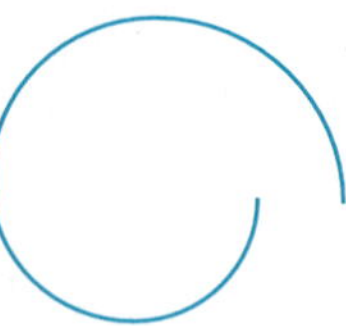

24. Construct a self-similar spiral figure based on extending the relationship between the following pair of figures: The first figure is a vertical line segment of length 1 and the second figure is a horizontal line segment of length $\frac{3}{4}$, sharing an endpoint with the first.

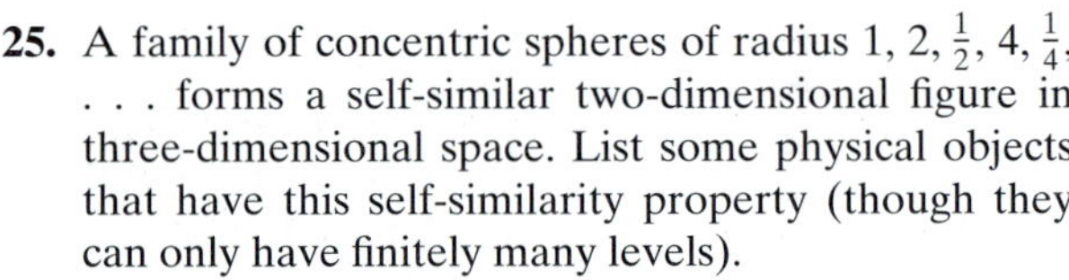

25. A family of concentric spheres of radius $1, 2, \frac{1}{2}, 4, \frac{1}{4}, \ldots$ forms a self-similar two-dimensional figure in three-dimensional space. List some physical objects that have this self-similarity property (though they can only have finitely many levels).

26. Do self-similar spiral surfaces exist? Theoretically? Physically?

11.3 APPLICATIONS TO POPULATION GROWTH AND RADIOACTIVE DECAY

Consider the following table of world population figures.

Year	Population (in millions)
1700	579
1750	689
1800	909
1850	1086
1900	1556
1950	2543
1994	5642
2000	6000 (est.)

Based on this data, can we make a reasonable prediction of the world population in the year 2050?

One of the major challenges facing the world is the balance between population and resources. Growth and scaling are important considerations in any situation where there is a limitation to resources, whether it is on a global, national, or local scale. The ability to accurately predict future population size is important for the planning and allocation of resources.

In this section we will introduce two approaches to population prediction. We will also consider one type of systematic decrease in the size of a population, known as decay.

POPULATION GROWTH

One of the first useful models of population growth was suggested by the English demographer and economist Thomas Malthus (1766–1834). Malthus made the assumptions that over one unit of time, usually a year, the number of births and the number of deaths in a population would be proportional to the population. Thus if the population is P, then during one year

$$\text{number of births} = b \times P \text{ and number of deaths} = d \times P,$$

where b and d are constants called the **birth rate** and the **death rate.** The net change in population over one year would be the number of births minus the number of deaths or

$$\text{net change} = b \times P - d \times P = (b - d) \times P = r \times P,$$

where $r = b - d$ is a new constant called the annual **growth rate.** The growth rate must be determined from data about the population and may change over time. It is expected to be different for different species. For example, the population growth rate is greater for small mammals than for large mammals.

Suppose we make Malthus's assumptions and also know the growth rate, r. If the population is P_0 at the beginning of the year, then after one year the population is $P_0 + rP_0 = (1 + r) \times P_0$. (This is only an approximation, because there is no reason to think $r \times P_0$ is a whole number even though the actual increase in population must be a whole number of people.) Suppose we use P_1 to represent the population after one year. If a second year passes, then the process is repeated with P_1 replacing P_0, so the population becomes

$$P_1 + r \times P_1 = (1 + r)P_1 = P_2.$$

Substituting $(1 + r)P_0$ for P_1, we have

$$\begin{aligned} P_2 = (1 + r)P_1 &= (1 + r)(1 + r)P_0 \\ &= (1 + r)^2 \times P_0. \end{aligned}$$

The pattern continues in this fashion: For each year that passes, the effect on the population is to multiply by another factor of $(1 + r)$.

Malthusian Population Growth

If the population is initially P_0, then after m years the population will be

$$(1 + r)^m \times P_0$$

where r is the growth rate.

EXAMPLE 11.6 Suppose a population grows by 5% each year. If the initial population is 20,000, what is the approximate population after 20 years?

SOLUTION The information that the population grows by 5% each year tells us that $r = 0.05$. Clearly, we are given $P_0 = 20{,}000$. Inserting this information into the formula for Malthusian Population Growth and using a calculator, we obtain

1.05 [x^y] 20 [×] 20,000 [=] [53065.9541]

So we conclude that the population is about 53,000 after 20 years. ◆

The Malthusian model for population is often quite good, especially in short to moderate time periods. But when applied to a long time period, the results are disturbing. For example, if we redo Example 11.6 with the time period of 20 years replaced by 300 years, we get the following result from the calculator

1.05 [x^y] 300 [×] 20,000 [=] [45479922572].

That is, the Malthusian model for population predicts that a small population of 20,000 and a constant growth rate of 5% will increase to a population of about 45 billion in 300 years time. This is about 8 times the entire world population (5.6 billion) in 1994.

As long as the growth rate r is positive, the Malthusian population model always leads to an enormous population estimate, no matter how small the original population may be (Figure 11.23). Even with low growth rates, the Malthusian model is one of gloom and doom. With a growth rate of 1%, the 1994 world population would be predicted to reach over 15 billion in 100 years. With a growth rate of 2%, the prediction is a staggering 40 billion in 100 years.

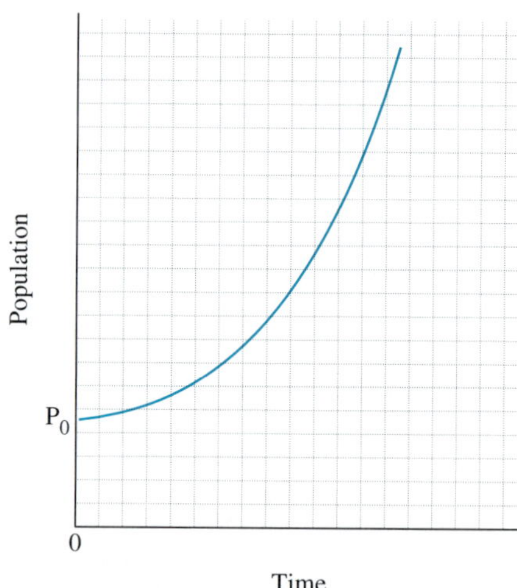

Malthusian Population Model

FIGURE 11.23

In the real world something must happen to stop this runaway population growth. Malthus felt that inevitably the brake on growth would be catastrophic: starvation, disease, or, in the case of humans, war. Human history provides many examples of such catastrophes, where certain populations have essentially disappeared. However, it is also possible for the growth rate to change, so a catastrophic end is not inevitable.

To apply the Malthusian population model, it is necessary to know the growth rate. Typically the growth rate is found from previous data recorded over a moderate number of years using the method that is presented next.

Suppose, for example, we know that the population changes from 2000 to 10,000 over a period of 16 years. Then, r can be found as follows: we consider 2000 to be the initial population, P_0, so the formula gives the population in 16 years as

$$P = (1 + r)^{16} \times 2000.$$

The population 16 years later is known to be 10,000, so we must have

$$10{,}000 = (1 + r)^{16} \times 2000.$$

Since r is the only unknown in the last equation, it is possible to solve for it as follows:

Divide both sides of the equation by 2000: $5 = (1 + r)^{16}$

Raise both sides to the $\frac{1}{16}$ power: $5^{\frac{1}{16}} = 1 + r$

We use a calculator to compute $5^{\frac{1}{16}}$: $5^{\frac{1}{16}} \approx 1.1058$

So $1.1058 \approx 1 + r$.

Finally, we subtract 1 from both sides: $r \approx 0.1058$ (or approximately 10.6%).

Tracing our steps backward, we see that $r = \left(\frac{10000}{2000}\right)^{\frac{1}{16}} - 1$.

This procedure leads to the following general formula.

Formula for the Growth Rate

If the population changes from P to Q in m years and a Malthusian model is assumed, then the growth rate r is given by

$$r = \left(\frac{Q}{P}\right)^{\frac{1}{m}} - 1.$$

EXAMPLE 11.7 Suppose the size of a population is initially 5000, and increases to 8000 in 10 years. What is the predicted population in 20 years?

SOLUTION First we need to find r, the growth rate that takes the population from 5000 to 8000 in 10 years. In the Formula for the Growth Rate, $Q = 8000$, $P = 5000$, and $m = 10$ (so $\frac{1}{m} = 0.1$). Using a calculator we find

8000 [÷] 5000 [=] [x^y] 0.1 [−] 1 [=] [0.048122389].

Thus $r = 0.048122$.

Then we calculate $(1 + r)^m \times P_0$ as in Example 11.6 where $P_0 = P = 5000$ and m is now 20 years.

1.048122 [x^y] 20 [×] 5000 [=] [12799.90487]

So the estimated population is 12,800. ◆

Another approach to the previous problem is to change the units of time. Instead of dealing in years, we could use decades (10 years). Then 20 years is two of these units. With this frame of reference, r is the growth rate per decade, and $8000 = (1 + r) \times 5000$. This means that $1 + r = 8000 \div 5000 = 1.6$. We then use the Malthusian Population Growth formula to find the new population after two (decade long) time periods:

$$(1.6)^2 \times 5000 = 12{,}800.$$

This calculation is exact. The slight difference from the previous result 12,799.9 is due to the unavoidable internal errors in the calculator's computations.

You can very often use the trick of changing the time unit to avoid finding the growth rate. The time unit to use is the length of time between the instants for which the population is known.

TIDBIT

One phenomenon similar to the Ponzi scheme occurs in a speculative bubble when the price of a commodity increases rapidly. New investors are rapidly attracted by the success of the early investors increasing the upward pressure on prices. The best example is the tulipomania which occurred in the Low Countries in the 1600s. The price of an individual tulip bulb rose as high as 2600 guilders at roughly the same time that 63 guilders purchased Manhattan Island from the Indians. The last people to buy tulips before the price collapse lost fortunes.

PONZI SCHEMES

Charles Ponzi was a confidence man who, in December, 1919, began what turned into a massive fraud which became known as the **Ponzi Scheme.** He claimed that great profits could be made on currency exchange rate differences by purchasing International Postal Reply coupons in a first country, exchanging them for stamps in a second country, selling the stamps in that second country, and finally converting the currency of the second country back to that of the first country. His slogan was "40% in 90 days." Ponzi actually made *no* purchases of International Postal Reply coupons. Instead he paid the old investors with the money received from new investors. In 1995, a similar scheme by an investment consultant caused a large number of nonprofit organizations in the United States to lose many millions of dollars.

In examining how Ponzi's scheme (40% in 90 days) works, we will use 90 days as our time unit. This time unit is called a quarter (of a year). For simplicity let us assume that each person invests the same amount of money in Ponzi's company. Let S_m be the number of persons who have an investment in Ponzi's company in the mth quarter. (Of course, S stands for "sucker.") For example, if 1000 people each invest \$1000 in, say, the tenth quarter, then at the end of that quarter, those investors need to get back their original investments of $\$1000 \times 1000 = \$1{,}000{,}000$ plus 40% interest, a total of \$1,400,000. Now, if the \$1,000,000 invested in the tenth quarter has not been spent, then 400 people need to each invest \$1000 in the eleventh quarter, to pay the \$400,000 interest. But, in all likelihood, the \$1,000,000 invested in the tenth quarter has been spent to pay back the ninth quarter investors. In that case 1,400 new investors are needed to pay off the tenth quarter investors. In general, to pay all S_m investors a profit of 40% in the mth quarter, Ponzi had to have

$$S_{m+1} \geq 1.4 \times S_m$$

investors in the $(m + 1)$st quarter. If $S_{m+1} = (1.4) \times S_m$, then this would be a case of Malthusian population growth with a growth rate of 0.4 and a time period of one quarter. Because of the inequality, the growth is even faster than the prediction of the Malthusian population model.

What actually happened, and usually does, is that as word of the success of the early investors spread, the number of new investors greatly exceeded the minimum number required to keep the scheme going. As the reservoir of potential new investors was used up, there came a time when the minimum growth in the number of suckers could not be sustained.

The final act involved the last investors losing all their money. Ordinary business failure involves bankruptcy, with either a reorganization to restore the company to financial stability, or an orderly selling off of assets and distribution of the remaining value of the company. But Ponzi's company had no assets such as factories to sell.

RADIOACTIVE DECAY

We know that the Malthusian population model is only an approximation, and that it can only be a good approximation over the limited time that resources allow a sustained growth rate. Radioactive materials lose particles in proportion to the amount of material present in a process called **radioactive decay.** Thus the behavior of radioactive materials is exactly modeled by the Malthusian scheme, but with a negative growth rate.

More precisely, if an amount A of a radioactive substance is present, then there is a constant d such that $d \times A$ of the substance decays in one time unit. The time unit that is convenient depends on how fast the particular substance decays, but we will use years for the present discussion. We will call d the **decay rate** of the substance. Unlike the growth rate for a population, which may change because of changing conditions, the decay rate is a constant that is associated with the atomic structure of the particular substance.

If we start with A_0 grams of the substance, then after one year there remains $A_0 - d \times A_0 = (1 - d) \times A_0$ grams of the substance. After another year there are $(1 - d)^2 \times A_0$ grams of the substance.

HISTORY

The theory that radioactive elements decay into other elements was proposed in 1903 by Ernest Rutherford and Frederick Soddy to explain the observed emission of particles from naturally occurring radioactive substances. Radioactivity was itself only discovered seven years before in 1896 by Antoine Henri Becquerel. The rays emitted from radioactive substances were originally called Becquerel rays.

Radioactive Decay

If a radioactive substance has an annual decay rate of d and there are initially A_0 units of the substance present, then after m years there will be

$$(1 - d)^m \times A_0$$

units of the radioactive substance present.

Notice that since $1 - d$ is strictly less than 1, as m increases the amount of the radioactive substance present gets smaller and smaller and, in fact, approaches zero. The decay rate d is hardly ever mentioned in practice. Instead because the amount present tends to zero, there must be a time after which there is half of the initial amount of the substance present. This is called the **half-life** of the substance. The decay rate can be found from the half-life and vice versa.

THEOREM

Decay Rate and Half–Life

If d is the annual decay rate of a substance and h is the half-life of the substance (in years), then

$$d = 1 - \left(\frac{1}{2}\right)^{\frac{1}{h}}.$$

TIDBIT

The body's metabolism of drugs works like radioactive decay. Among the data available to prescribing physicians is the half-life of particular drugs in the average person's body. There is a lot of variability from person to person; for example, the half-life of caffeine varies from three to seven hours.

EXAMPLE 11.8 Suppose the half-life of a particular radioactive substance is 25 years.

(a) What is the annual decay rate of the substance?
(b) If you have 128 grams of the substance in 1995, how much will remain in 2095?

SOLUTION

(a) To find the annual decay rate we can use the formula

$$d = 1 - \left(\frac{1}{2}\right)^{\frac{1}{25}} \approx 1 - 0.9727 = 0.0273.$$

The annual decay rate is approximately 2.73%.

(b) Now that the annual decay rate is known, we could use the formula to find the amount of the substance remaining after 100 years, but it is easier to reason as follows: After the first 25 years there will be $\frac{1}{2}$ of the original 128 grams left, that is, 64 grams. During the next 25 years, the 64 grams decays to 32 grams. During the third 25 years the 32 grams decays to 16 grams. In the final 25 years the 16 grams decays to 8 grams. ◆

Notice that in Example 11.8, the passing of each half-life had the effect of multiplying the amount of the substance by $\frac{1}{2}$. This gives us the following general rule.

Radioactive Decay and Half–Life

If initially there are A_0 units of a radioactive substance present, and if the half-life of the substance is h, then after time m (measured in the same time units as h) there will be

$$\left(\frac{1}{2}\right)^{\left(\frac{m}{h}\right)} \times A_0$$

units of the radioactive substance present.

Obtaining the half-life from the decay rate requires using one of the more advanced functions on your calculator and will not be covered.

Half-life information can be found on a typical periodic table of the chemical elements. For reference we provide Table 11.1, which contains information about selected radioactive isotopes.

TABLE 11.1

Element	Symbol	Atomic Weight	Half-Life
Carbon	C	14	5600 years
Plutonium	Pu	242	3.8×10^5 years
Plutomium	Pu	241	13 years
Plutonium	Pu	239	24,300 years
Radium	Ra	226	1620 years
Radon	Rn	222	3.82 days
Strontium	Sr	90	28 years
Strontium	Sr	89	51 days
Strontium	Sr	85	64 days
Uranium	U	238	4.5×10^9 years
Uranium	U	234	2.5×10^5 years
Uranium	U	235	7.1×10^8 years
Uranium	U	233	1.6×10^5 years

CARBON DATING

The radioactive isotope ^{14}C of carbon occurs naturally in the air, and the ratio of normal carbon ^{12}C to radioactive carbon ^{14}C has been constant for millions of years. All living things absorb carbon from the atmosphere and thus have the same ratio of normal carbon ^{12}C to radioactive ^{14}C as the atmosphere. Once the living thing dies, no more ^{14}C can enter, but what was already there decays. Using the knowledge that the half-life of ^{14}C is 5568 years, scientists can determine how much time has elapsed since the death of the sample. Carbon dating is included in the extended problems.

LOGISTIC POPULATION MODELS

Earlier in this chapter we looked at the population model suggested by the English economist Thomas Malthus. The Malthusian model predicts that any population will continue to grow, even to immense numbers, as long as there is a positive growth rate. Malthus himself argued that because populations eventually increase very rapidly and the means of sustenance (food, energy, and other resources) increase at a much slower rate, a population would eventually exceed the ability of the environment to sustain it.

In some situations, particularly over shorter time periods, the Malthusian model with its constant growth rate makes sense, but it is unreasonable in many others. Growth is inevitably constrained by factors such as food and space, and in many populations the growth rate declines in response to the diminishing available resources. Eventually, a population may reach a level where there are insufficient resources to support any increase; and either the population levels off, or the quality of life decreases, with fewer resources for each individual. For a given environment with limited resources (such as Earth, perhaps) there may be a maximum population size that can be sustained. This population size is called the **carrying capacity** of the environment.

One model for population growth that takes into account the carrying capacity of the environment is called the **logistic population model.** This model reduces the growth rate by a factor that reflects the relative size of the population in comparison to the carrying capacity. In using the model, we still need to have the growth rate constant r which would apply if there were no resource pressure restraining growth. For convenience, we make the assumption that our population grows over well-defined breeding seasons. This is often the case for animal populations.

Suppose the initial population is P_0, the population after one breeding season is P_1, the population after two breeding seasons is P_2, and so on. If there were no resource pressure, then after $m + 1$ seasons the population would be

$$P_{m+1} = (1 + r) \times P_m.$$

In order to include the effect of the carrying capacity, c, a new rule must be devised. One such rule is

$$P_{m+1} = (1 + r) \times P_m - \left(\frac{1 + r}{c}\right) \times P_m^{\,2}$$

which is called the **logistic law** (or **logistic equation**).

EXAMPLE 11.9 Suppose a population is governed by the logistic law, with $r = 0.5$ and $c = 6000$. Assume that the initial population size is 500. Compute the population after each of the next 15 breeding seasons, and plot the data on a graph.

SOLUTION Substituting for r and c in the logistic law we get

$$\begin{aligned} P_{m+1} &= (1 + r) \times P_m - \left(\frac{1 + r}{c}\right) \times P_m^{\,2} \\ &= 1.5 \times P_m - \left(\frac{1.5}{6000}\right) \times P_m^{\,2} \\ &= 1.5 \times P_m - 0.00025 \times P_m^{\,2}. \end{aligned}$$

From this we calculate

$$\begin{aligned} P_0 &= 500, \\ P_1 &= 1.5 \times 500 - 0.00025 \times 500^2 = 750 - 62.5 \approx 687, \\ P_2 &= 1.5 \times 687 - 0.00025 \times 687^2 \approx 913, \\ P_3 &= 1.5 \times 913 - 0.00025 \times 913^2 \approx 1161. \end{aligned}$$

Similarly, we find

$$\begin{array}{llll} P_4 \approx 1405, & P_5 \approx 1614, & P_6 \approx 1770, & P_7 \approx 1872, \\ P_8 \approx 1932, & P_9 \approx 1965, & P_{10} \approx 1982, & P_{11} \approx 1991, \\ P_{12} \approx 1995, & P_{13} \approx 1997, & P_{14} \approx 1998, & P_{15} \approx 1999. \end{array}$$

We plot this data, with the number of the breeding season on the horizontal axis and the population on the vertical axis (Figure 11.24).

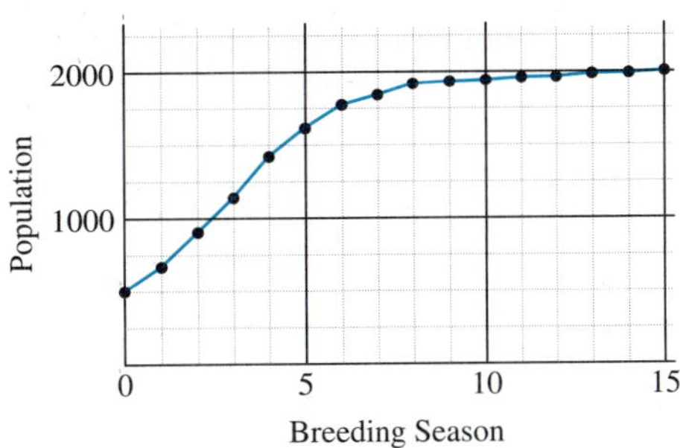

FIGURE 11.24

◆

Notice that the graph of the population in Figure 11.24 levels out near the value 2000. There is a good reason for this. Looking at the logistic law we see that if the population size ever takes on the value

$$P_m = \frac{r \times c}{1 + r}$$

then, substituting the expression for P_m in the logistic law, we find that

$$P_{m+1} = (1 + r) \times \frac{r \times c}{1 + r} - \left(\frac{1 + r}{c}\right) \times \left(\frac{r \times c}{1 + r}\right)^2.$$

Simplifying the last expression, we have

$$P_{m+1} = \frac{r \times c \times (1 + r)}{1 + r} - \frac{r \times r \times c}{1 + r}$$

$$= \frac{r \times c + r \times r \times c - r \times r \times c}{1 + r}, \text{ or}$$

$$P_{m+1} = \frac{r \times c}{1 + r} = P_m.$$

This means that the population stays the same. Using the values of $r = 0.5$ and $c = 6000$ from Example 11.9 we would have

$$\frac{r \times c}{1 + r} = \frac{0.5 \times 6000}{1.5} = 2000.$$

If the population is ever equal to 2000, then it will stay at 2000. This is a **stable population.** A population governed by the logistic law typically behaves in this manner.

Stable Population under the Logistic Law

If a population has a natural growth rate of r and the environment has a carrying capacity of c, then the stable population size is

$$\frac{r \times c}{1 + r}.$$

Other population levels (except 0) are not stable. For many choices of r and c, the population level behaves as illustrated in Figure 11.25.

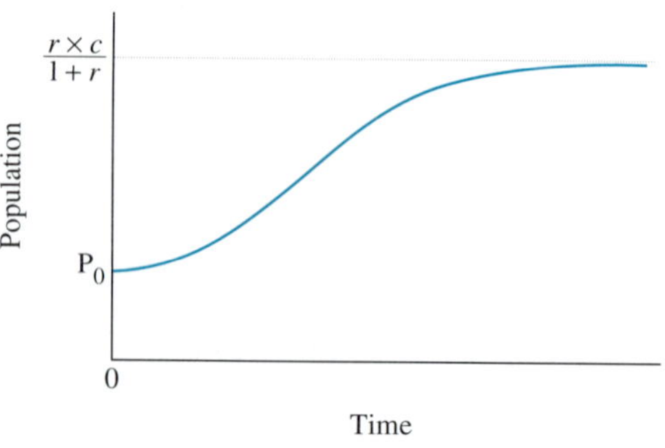

FIGURE 11.25

For some choices of r and c, the population may cycle through a sequence of high and low values, or the population may just seem to be a random sequence of values (it only seems random—in fact it is determined by the equation).

The study of the behavior of sizes of populations, **population dynamics,** is not merely an academic exercise. The survival of modern civilization may depend on human population dynamics. Many industries rely on harvesting living things, and the economics of those industries depend on the population dynamics of the living things being harvested. Generally such industries face regulation on their harvests, either for tax purposes or to protect the long-term viability of the industry. Case studies indicate that an industry will not necessarily maintain its own viability. For example, the passenger pigeon was harvested for local consumption and for marketing. These birds numbered in the billions and traveled in immense flocks consisting of millions of individuals. The flocks would nest in apparently unpredictable places. The advent of the telegraph and railroad allowed hunters to locate and reach the nesting sites rapidly. Each individual hunter sought to maximize his or her own profit by killing as many birds as possible. The birds were literally hunted to extinction. The last passenger pigeon died in captivity in 1914.

Now, almost everyone appreciates the need to harvest natural resources on a sustainable basis. Even the most rapacious laissez-faire capitalist has learned to say *sustainable.* The problem for our society is to adequately understand all the parameters that enter into the dynamics of a population, so that the best decisions can be made. Unfortunately, the scientific data currently available often does not give a clear answer, and some harvesters may still continue their current practices until the evidence is overwhelming. At that point, however, it may be too late to take actions that prevent the collapse of the population.

Consider the following table of world population figures.

Year	Population (in millions)
1700	579
1750	689
1800	909
1850	1086
1900	1556
1950	2543
1994	5642
2000	6000 (est.)

Based on this data, can we make a reliable prediction of the world population in the year 2050?

SOLUTION One possibility to check is whether the growth rates over the 50-year periods agree. Using the half-century as our time unit makes the growth rate simply the ratio between the population at the beginning and end of the half-century period. We find the following rates:

Time Period		Growth Rate
1700 to 1750	$689 \div 579 \approx 1.19$	19%
1750 to 1800	$909 \div 689 \approx 1.32$	32%
1800 to 1850	$1086 \div 909 \approx 1.19$	19%
1850 to 1900	$1556 \div 1086 \approx 1.43$	43%
1900 to 1950	$2543 \div 1556 \approx 1.63$	63%
1950 to 2000	$6000 \div 2543 \approx 2.36$	136%

Since the growth rates vary widely, the Malthusian model, which assumes a constant growth rate, does not fit well and would be only a very unreliable predictor. The logistic law also does not apply since the growth rate has actually increased while the population increased, which is contrary to what the logistic law would predict. Based on the data, we cannot make any reliable prediction of the world's population in the year 2050.

PROBLEM SET 11.3

In problems 1 through 6, use the Malthusian model for population growth.

1. Suppose a population grows at the rate of 3% each year. If the initial population is 50,000, what is the population after 15 years?
2. Suppose a population grows at the rate of 4% each year. If the initial population is 75,000, what is the population after 12 years?
3. In 1980, two cities had populations of 20,000 and 30,000. If their growth rates are 4% and 3%, respectively, how will their populations compare in the year 2000?
4. In 1990, the population of the United States was approximately 255 million and increasing at a rate of 0.7% per year. What is the anticipated size of the U.S. population in the year 2000?
5. If the growth rate of the United States population continues at 0.7%, what is the anticipated size of the population in the year 2020? (See problem 4)
6. **(a)** If the growth rate from 1990 to the year 2000 were 0.4%, what would be the anticipated population size in the year 2000? (See problem 4)
   (b) If the growth rate for the period were 1.0%, what would be the anticipated population size in the year 2000? (See Problem 4)

7. In 1626, Peter Minuit of the Dutch East India Company purchased the island of Manhattan for the equivalent of \$24 in trading goods. What would the value of these goods be in 1996 if their value had grown by a constant rate of 3% since that time? What would their value be if the rate of growth was 4%?
8. Suppose you purchased a home in 1993 for \$80,000. A review of real estate records for the past ten years indicates that the average value of houses in your area has increased by an average of 4.5% per year. If this rate of growth continues, how much should the house be worth in 2003?

In problems 9 through 14, use the Malthusian model for population growth.

9. The population of a given city grew from 20,000 to 33,000 during the past 10 years. What is the predicted population in 20 years? Use two methods to obtain your answer.
10. The population of a given city grew from 35,000 to 42,000 during the past 10 years. What is the predicted population in 20 years? Use two methods to obtain your answer.
11. During the past eight years, the population of Greenville grew from 28,000 to 35,500. What is the predicted population in 15 years?
12. Between 1985 and 1993, Braxton grew from a population of 62,400 to 70,800. If this rate of growth continues, what is the predicted population for 2010?
13. The current growth rate in the United States is approximately 0.7%. If this rate continues, how long will it be until the population of the United States doubles in size?
14. Kenya, with a population growth rate of 4%, had the highest growth rate of any country in the world in 1994. How long will it take Kenya's population to double at this rate?
15. The half-life of Strontium 90 is 28 years. What is the annual decay rate for Strontium 90?
16. The half-life of radium-226 is 1620 years. What is the annual decay rate for radium-226?
17. Plutonium-241 has a half-life of 13 years. If a sample of 100 grams was produced in 1950, how much of the sample remained in 1994?
18. Suppose the half-life for a radioactive substance is 400 years. How much will remain of an initial amount of 100 grams after 2000 years?
19. The half-life of sodium-22 is 2.6 years.
    (a) What is the annual decay rate for sodium-22?
    (b) If you have 200 grams of sodium-22 in 1995, how much will remain in 2050?
20. Suppose the half-life of a radioactive element is 73 days.
    (a) What is the annual decay rate for the substance?
    (b) If 200 grams of the substance was available on January 1, 1995, how much would remain on December 31, 1995?
21. Suppose the half-life of a radioactive element is 8.5 years.
    (a) What is the annual decay rate for the substance?
    (b) If you have 100 grams of the substance in 1995, how much will you have left in 2025?
22. The half-life of argon-41 is 1.8 hours.
    (a) What is the annual decay rate for the substance?
    (b) If 50 grams of the substance was available at 12:00 noon, how much would remain by 12:00 midnight?

In problems 23 through 26, assume that a population is governed by the logistic law. For each of the problems, do the following.

(a) Calculate and plot the population size after each of the next 10 breeding seasons.

(b) Discuss the behavior of the population size. Does it appear to behave smoothly, or is it erratic from year to year?

23. Carrying capacity: $c = 2000$
    Growth rate: $r = 0.5$
    Initial population is 1000.
24. Carrying capacity: $c = 2000$
    Growth rate: $r = 1.2$
    Initial population is 1000.
25. Carrying capacity: $c = 2000$
    Growth rate: $r = 2.2$
    Initial population is 1000.
26. Carrying capacity: $c = 2000$
    Growth rate: $r = 2.7$
    Initial population is 1000.

In problems 27 through 30, assume that a population is governed by the logistic law; its carrying capacity, growth rate, and initial population are given. For each problem, do the following.

(a) Make a table for the population size after each of the next 10 breeding seasons.

(b) Make a table for the population for the next 10 years based on the Malthusian model. These tables should be made side by side for easy comparisons.

(c) Plot the population figures for each model on a separate graph.

27. Carrying capacity: $c = 2000$
    Growth rate: $r = 0.6$
    Initial population is 1000.
28. Carrying capacity: $c = 2000$
    Growth rate: $r = 0.9$
    Initial population is 1000.
29. Carrying capacity: $c = 2000$
    Growth rate: $r = 0.35$
    Initial population is 1000.

30. Carrying capacity: $c = 2000$
Growth rate: $r = 2.5$
Initial population is 1000.

EXTENDED PROBLEMS

If a population exactly follows the logistic law, then the carrying capacity is given by

$$c = \frac{P_m{}^2 P_{m+2} - P_{m+1}{}^3}{P_m P_{m+2} - P_{m+1}{}^2}$$

and the growth rate is given by

$$r = \frac{P_m{}^2 P_{m+2} - P_{m+1}{}^3}{P_m P_{m+1}[P_m - P_{m+1}]} - 1,$$

where P_m, P_{m+1}, and P_{m+2} are the populations after three successive breeding seasons.

31. Suppose a population governed by a logistic law has size 1500, 2500, 3500 after three successive breeding seasons.
(a) What are the carrying capacity and growth rate?
(b) What is the population going to be after the next breeding season?
(c) Plot the population values over the next six breeding seasons.
(d) Does the population seem to behave smoothly or does it seem erratic?

32. Suppose a population governed by a logistic law has size 2000, 3000, 4000 after three successive breeding seasons.
(a) What are the carrying capacity and growth rate?
(b) What is the population going to be after the next breeding season?
(c) Plot the population values over the next six breeding seasons.
(d) Does the population seem to behave smoothly or does it seem erratic?

33. If you started a Ponzi scheme like the one in the text (40% in 90 days) with ten customers, how long would it take until every man, woman, and child in the United States would need to be a customer in order to keep the scheme going with new customers? Assume there are 260,000,000 people.

34. In 1995, many charitable institutions across the country lost many millions of dollars working with an individual management consultant. Research this or some other famous recent fraud cases. Were any fundamentally Ponzi schemes?

35. How does the budget deficit run by the United States compare with the interest on the national debt? At what point could one say the United States government is running a Ponzi scheme?

Carbon-14 dating is used by scientists to determine the age of artifacts. Cosmic ray bombardment of the atmosphere produces neutrons, which in turn react with nitrogen to produce radioactive carbon-14 (^{14}C). Radioactive ^{14}C enters all living tissues through carbon dioxide, which is first absorbed by plants. As long as a plant or animal is alive, ^{14}C is maintained at a constant level in the living organism. Once the organism dies, however, the ^{14}C decays at a constant rate, according to formula

$$A = A_0 \times \left(\tfrac{1}{2}\right)^{(m/h)},$$

where A_0 is the amount of carbon-14 originally available, m is the age of the substance in years, and h is the half-life of 5600 years. The percentage of ^{14}C present in a substance can be found using $\frac{A}{A_0} = \left(\frac{1}{2}\right)^{(m/h)}$.

36. If 500 milligrams of ^{14}C is present in a sample from a skull at the time of death, how many milligrams would be present in the skull after
(a) 5000 years?
(b) 25,000 years?
(c) 50,000 years?

37. What percentage of ^{14}C would be present in a bone that was
(a) 10,000 years old?
(b) 20,000 years old?
(c) 50,000 years old?

38. An archeologist discovers a burial site that he believes to be 8000 years old. Examination of bones from the site shows that 40% of the ^{14}C is still present. Is the archeologist correct? Justify your answer.

39. Charcoal from a suspected ancient campfire is tested and found to contain only 0.2% of the ^{14}C. Determine the age of the charcoal to the nearest 5000 years. (Hint: try "guess and test.")

11.4 SCALING PHYSICAL OBJECTS

Your small child has a fever for which you are going to administer some acetaminophen (e.g. Tylenol). How should the dosage be determined?

In previous sections we learned that if the shape of an object is kept the same while linear dimensions are increased by a scaling factor of r, then areas are increased by r^2 and volumes are increased by r^3. These relationships have important implications for physical objects, including living creatures.

WEIGHT

To determine the weight of a large object, it is often not practical to put the object on a scale. In such cases we can find (or estimate) the volume of the object and then multiply by the density of the material from which it is made. The **density** of a material is the ratio of the material's weight to its volume, or put another way, the density is the weight of a unit volume of the material. Table 11.2 lists the densities of some common materials.

TABLE 11.2

Material	Density in Pounds per Cubic Foot
Lead	710
Iron and Steel	500
Marble	170
Aluminum	170
Concrete	150
Water	62.4
Ice	57.4

Density of materials can be used to solve many applied problems.

EXAMPLE 11.10 Suppose a swimming pool is 5 feet deep, 50 feet long, and 25 feet wide. About how much does the water in the pool weigh?

SOLUTION The volume of the water is

$$5 \times 50 \times 25 = 6250 \text{ cubic feet.}$$

From Table 11.2 we see that each cubic foot of water weighs about 62.4 pounds, so the water in the pool weighs about

$$6250 \times 62.4 = 390{,}000 \text{ pounds} = 195 \text{ tons.}$$ ◆

We can also estimate the volume from the weight.

EXAMPLE 11.11 What is the volume of a typical human being?

SOLUTION An average weight for a full grown human is about 150 pounds. Since most people are slightly buoyant, the density of a human is about the same as that of water, or 62.4 pounds per cubic foot. Thus we estimate a volume of

$$\frac{150 \text{ lb}}{62.4 \text{ lb/ft}^3} \approx 2.4 \text{ ft}^3.$$

Small humans are about 100 pounds, big humans (football linemen) 250 pounds. So the range of volumes is from about 1.6 to 4.0 cubic feet. ◆

For examples and problems in this book, we will assume that the average human weighs 150 pounds, a small human weighs 100 pounds, and a large human weighs 250 pounds.

If the size of an object is increased by a scaling factor of r, then the volume is increased by a factor of r^3, so the weight of the object is also increased by a factor of r^3, provided the object is composed of the same material.

The Weight Rule

If the linear dimensions of a physical object are scaled by a factor of r and a new object is composed of the same material as the old, then

$$[\text{weight of the new}] = r^3 \times [\text{weight of the old}].$$

Next we use scaling factors to find weight.

EXAMPLE 11.12 Brobdingnagians described by Jonathan Swift in *Gulliver's Travels* were the same shape as we are, but twelve times as tall. The Lilliputians described in the same work were also the same shape as normal humans, but one twelfth as tall. How much did each weigh?

SOLUTION Taking 150 pounds as the average human weight, we see that typical Brobdingnagians were $12^3 \times 150 = 259{,}200$ pounds or about 130 tons, while the Lilliputians were $\left(\frac{1}{12}\right)^3 \times 150 \approx 0.0868$ pounds or about 1.4 ounces. ◆

PRESSURE

The **pressure** on a surface is the force on the surface divided by the area of the surface. Consider a piece of ice in the shape of a cube two inches on each edge. Further, suppose the cube weighs four ounces (that is a slight underestimate). If the cube is sitting on your kitchen counter, the weight is borne by the bottom face, which has a surface area of four square inches. The pressure on the bottom of the cube is the weight, $\frac{4}{16}$ pound, divided by the supporting area, four square inches. The common English unit for pressure is **pounds per square inch**, abbreviated **psi.** For this hypothetical cube of ice we have a pressure of about $\frac{4}{16} \div 4 = \frac{1}{16}$, or 0.0625 psi.

TIDBIT

A 100-pound woman in high heels exerts about 100 psi on the floor. A bull elephant weighs about six tons and has feet bigger than 12 inches in diameter, hence exerts about 27 psi. Thus the woman in high heel shoes puts greater pressure on the floor than the elephant does.

EXAMPLE 11.13 What is the pressure on the bottom of a block of ice in the shape of a cube two feet on each side?

SOLUTION We can think of the cube of ice two feet on each side as a scaled up version of a cube of ice two inches on each side. (Recall that the weight of the smaller cube was about $\frac{4}{16}$ pounds.) The scaling factor is 12 because there are 12 inches in a foot. Scaling up by a factor of 12 multiplies the length of each side by a scaling factor of 12. The volume is multiplied by a factor of $12^3 = 1728$. Since the weight is proportional to the volume, the weight is also multiplied by the same factor of 1728 (this is the weight rule above). The weight of the large ice cube is more than $1728 \times \frac{4}{16} = 432$ pounds.

Now consider the area of the bottom of the cube that is supporting the weight. Since the scaling factor is $r = 12$, the area of the bottom face is multiplied by a factor of 12^2. The pressure on the bottom is the ratio of weight to area. Since the weight is multiplied by the factor 12^3 and the area is multiplied by 12^2, the net effect on the pressure on the bottom is simply to multiply by the same 12 that scaled the linear dimension. The large ice cube exerts a pressure of about $12 \times 0.0625 = 0.75$ psi on the bottom. ◆

Example 11.13 shows us that although weight may increase dramatically when size is scaled up, the increase in pressure on the bottom is not as dramatic. Even if we are not absolutely certain of the precise value of the pressure exerted by the small cube, we know that increasing the linear dimensions by a scaling factor of 12 will increase the pressure by a factor of 12. If the cube of ice is now a giant, say 200 feet on each side (scaled up from Example 11.13 by a factor of 100) we know the weight will be multiplied by the factor of $100^3 = 1{,}000{,}000$ (the weight rule again), making a spectacular total weight in the hundreds of millions of pounds. The pressure on the bottom of the cube will also be multiplied by 100; that is, the pressure will be $100 \times 0.75 = 75$ psi. (That is not an unimaginable pressure. Your bicycle tires may take 50 psi.) However, 75 psi is enough to deform the ice and cause it to flow. Of course a cube of ice 200 feet on each side is not something you are going to make in your refrigerator or anywhere else, but 200 feet is a reasonable thickness for a small glacier. The ice on Greenland is as thick as 11,000 feet.

The Pressure Rule

If the linear dimensions of a physical object are increased by a scaling factor of r and the new object is composed of the same material as the old, then

$$[\text{pressure on the bottom of the new}] = r \times [\text{pressure on the bottom of the old}]$$

BOLTS

A bolt is a wire rod with a head at one end and threads at the other. Two plates can be fastened together by putting the bolt through holes in the plates and screwing a nut on the threaded end (Figure 11.26).

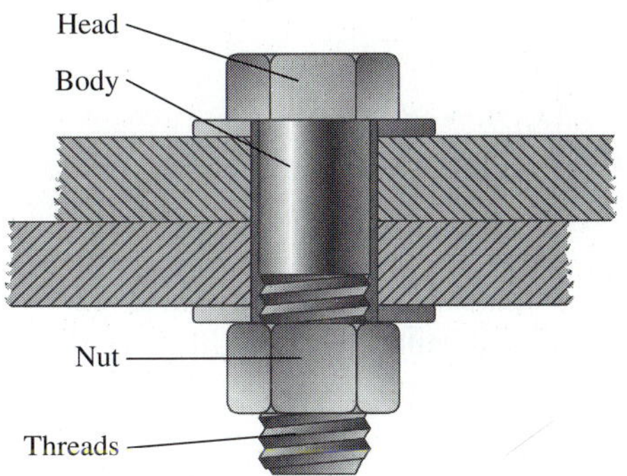

FIGURE 11.26

If the plates are pulled apart the bolt is stretched. The ability of the bolt to resist this stretching is called its **tensile strength.** The tensile strength of a bolt is measured as the maximum force the bolt can survive divided by the cross-sectional area of the bolt.

EXAMPLE 11.14 Suppose the diameter of a bolt is $\frac{1}{2}$ inch and the bolt has atensile strength of 64,000 psi. How much force can the bolt withstand?

SOLUTION The cross section of the bolt is a circle of diameter $\frac{1}{2}$ inch, or radius 0.25 inch. The area of the cross section is $\pi \times 0.25^2 \approx 0.19635$ square inches.

From our definition,

$$\text{tensile strength} = \text{maximum force} \div \text{cross section}.$$

Equivalently,

$$\text{maximum force} = \text{tensile strength} \times \text{cross section}.$$

Substituting our known values, we find

$$\text{maximum force} = 64{,}000 \, \frac{\text{pounds}}{\text{square in.}} \times 0.19635 \text{ square inches} \approx 12{,}566 \text{ pounds.}$$ ◆

From Example 11.14 we see that Figure 11.27 illustrates a safe and secure structure. The half-inch bolt can support 12,500 pounds.

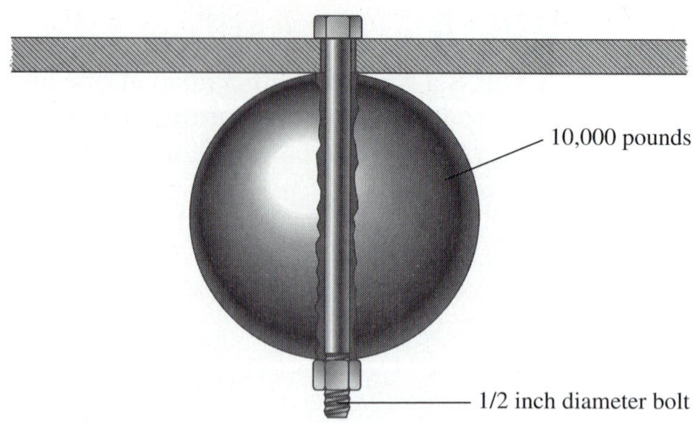

FIGURE 11.27

Now, what happens if we double the linear scale of Figure 11.27? By the weight rule, the weight becomes $2^3 \times 10{,}000 = 80{,}000$ pounds. The bolt is now one inch in diameter, so it has cross-sectional area of

$$\pi \times 0.5^2 \approx 0.785 \text{ square inches}$$

and can support

$$64{,}000 \times 0.785 = 50{,}240 \text{ pounds.}$$

The scaled up structure is going to break, because the weight increased more rapidly than the strength of the bolt holding it together.

If the diameter of a rod or wire is multiplied by a factor of r, then the area of the cross section is multiplied by r^2. If the same material is used to make the rod or wire, the strength of the rod or wire is also multiplied by r^2. If an entire structure is increased by a scaling factor of r, the weight of the components is increased by r^3. If the fasteners holding the structure together (which are essentially rods and wires) are to have the same relative ability to resist gravity, it will not suffice to increase the fasteners by a factor of r. Instead the fasteners should be scaled by $r^{\frac{3}{2}}$. This gives us the following rule of thumb.

THEOREM

Fastener Rule

If the linear dimensions of a structure are scaled by a factor of r, then the diameters of the fasteners holding the structure together should be scaled by $r^{\frac{3}{2}}$.

EXAMPLE 11.15 If the linear scale of the structure in Figure 11.27 is to be doubled, what size bolt should be used?

SOLUTION The weight of the object in Figure 11.27 will scale up by a factor of $2^3 = 8$, so the weight to be supported is 80,000 pounds. Since we are scaling by a factor of two, we should scale the bolt by a factor of

$$2^{\frac{3}{2}} \approx 2.83.$$

This means a bolt of diameter $0.5 \times 2.83 = 1.415$ inches or larger is needed. To double-check our calculations for this bolt,

$$\text{Maximum force} = 64{,}000 \text{ psi} \times \pi \times (0.708)^2 \approx 100{,}800 \text{ pounds.}$$ ◆

As important as the diameter of a bolt may be in determining the maximum force it can support, the quality of the steel plays an equally important role. Bolts have coded markings on the heads to distinguish those made with higher tensile strength steel. These markings are shown in Figure 11.28.

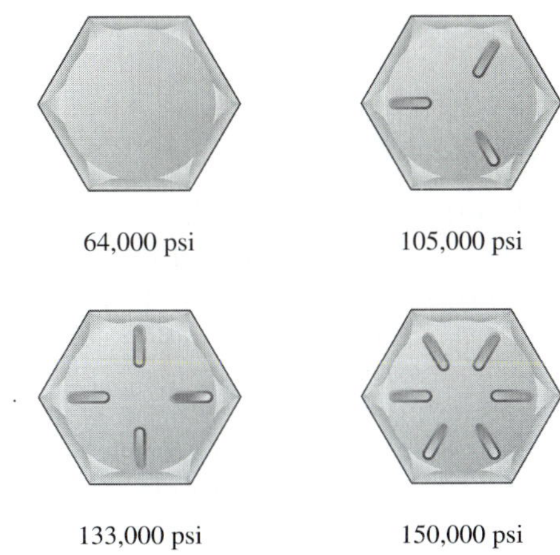

Head Markings of Steel Bolts
Minimum Tensile Strength

FIGURE 11.28

MODELING

Scaling of objects correctly is of great importance in design work. Engineers often do experiments on a small scale. Sometimes a scale model is the only sensible procedure, such as the model used by the Army Corps of Engineers to study the Mississippi river basin. Our discussion has indicated how to consider the effects of scaling on volume, weight, stress, and surface areas. Many of the situations modeled by engineers also involve the movement of various parts, which requires a more sophisticated analysis.

EXAMPLE 11.16 Scale models of automobiles are often made to $\frac{1}{4}$ inch scale meaning that a part one foot long on the real object is modeled by a part one-quarter inch long. How large should models of people be made if they are to be put alongside the model cars made to this scale?

SOLUTION Since people are generally between five feet and six feet tall, the model people should be between $5 \times \frac{1}{4} = 1.25$ inch and $6 \times \frac{1}{4} = 1.5$ inch tall. ◆

Given the dimensions of a real object and the model of that object, we can compute the scale involved, as in the next example.

EXAMPLE 11.17 The Suez Canal is about 103 miles long. A small tanker is about 500 feet long. If a model of the Suez Canal were 20 feet long, how long would the appropriate model of a small tanker be?

SOLUTION The Suez Canal is about $103 \times 5280 = 543{,}840$ feet long. A 20-foot long model would be scaled down from reality by a factor of

$$\frac{20}{543840} = \frac{1}{27192}.$$

Thus a model tanker would be only

$$500 \times \frac{1}{27192} = 0.0184 \text{ feet long, or about } \tfrac{1}{4} \text{ inch.}$$ ◆

MONSTERS

Returning to the Brobdingnagians discussed in an earlier example, we should note that the cross-sectional area of any bone in a Brobdingnagian's body increases by a factor of $12 \times 12 = 144$. Since the weight of the Brobdingnagian has increased by a factor of $12 \times 12 \times 12 = 1728$, the stress on the bones due to gravity has increased by a factor of $\frac{1728}{144} = 12$. Human bones frequently break in ordinary use, sometimes with hardly any movement involved. If there were any actual Brobdingnagians, they would probably be spending a lot of their lives in casts waiting for broken bones to heal. Large animals typically have relatively thicker bones than small animals because of the additional stress their weight puts on their bones.

EXAMPLE 11.18 The largest reliably recorded bird capable of flight is a buzzard that weighed almost 40 pounds. The smallest type of bird is the bee hummingbird of Cuba and the Isle of Pines weighing as little as 0.056 ounces. Do you think there could be any birds in Brobdingnagia capable of flight?

SOLUTION If the bee hummingbird is scaled up by a factor of 12, then its weight is scaled up by a factor of $12^3 = 1728$ to at least

$$0.056 \times 1728 = 96.768 \text{ ounces} \approx 6 \text{ pounds},$$

well under the weight of the largest birds known to be able to fly. Thus it is possible that there could be birds in Brobdingnagia capable of flight. ◆

INITIAL PROBLEM SOLUTION

Your small child has a fever for which you are going to administer some acetaminophen (e.g. Tylenol). How should the dosage be determined?

SOLUTION The acetaminophen will be distributed through the child's entire body by being dissolved in the blood. The crucial issue for effectiveness is the amount of drug per unit volume of blood, so ideally the dosage should be determined by the volume of blood in the child. Although it is impractical to measure the volume of blood in your child, you can find the approximate volume of the entire child by calculating from the weight, and since the blood is essentially a constant proportion of the volume of the child, the proper dosage should be based on the weight of the child.

PROBLEM SET 11.4

1. The weight of a 1-foot cube of steel is about 500 pounds. What is the weight of
(a) a 1-inch cube of steel?
(b) a 1-yard cube of steel?

2. The weight of a 1-foot cube of marble is about 170 pounds. What is the weight of
(a) a 1-inch cube of marble?
(b) a 1-yard cube of marble?

3. Suppose a block of ice measures 3 feet long, 1 foot wide, and 1.5 feet high. What is the weight of the block? A cubic foot of ice weighs about 57.4 pounds.

4. A marble slab is 5 feet long, 3 feet wide, and 6 inches deep. What is the weight of this slab of marble?

5. Suppose 10 gallons of water in a large container freezes to solid ice. What is the volume of the ice in cubic feet? (Hint: a gallon of water weighs 8.3 pounds.)

6. A large block of ice measuring 10 feet long, 5 feet wide, and 4 feet deep is placed in a tank and allowed to melt. What is the volume of the water?

7. What is the weight of a life-size statue of a typical human being (as defined in the text) if the statue is made of
(a) solid steel? (see problem 1)
(b) marble? (see problem 2)
(Hint: see Example 11.11.)

8. What is the weight of a half-scale statue of a typical human being (as defined in the text) if the statue is made of solid aluminum, given that aluminum weighs approximately 170 lbs per cubic foot? (Hint: see Example 11.11.)

9. Hakeem Olajuwon and Kenny Smith were teammates on the Houston Rockets 1995 NBA champion basketball team. Smith, one of the starting guards, was listed as being 6'4" tall and weighing 190 lbs. Olajuwon, the center, was listed at 7'0" and 255 lbs. If Smith were scaled up to Olajuwon's height, how much would he weigh?

10. At 7'0" tall and weighing 255 lbs, Hakeem Olajuwon is one of the largest basketball players in the NBA. What would be the height and weight of his counterpart in
(a) the Brobdingnagian Basketball Assoc.?
(b) The Lilliputian Basketball Assoc.?
(Hint: see Example 11.12.)

11. The radius of the Earth is approximately 3964 miles while that of the Moon is 1080 miles. If the Earth and Moon were made of the same materials, how would the masses of the two compare? (Note: we don't use the idea of weight because weight refers to the action of gravity on an object's mass. The mass of an object does not change with its position in space.) (Hint: compare the volumes. The volume of a sphere is given by $V = \frac{4}{3}\pi r^3$)

12. Re-do problem 11 as a scaling problem between the Earth and the Moon.

13. The average density of an object is the mass of the object divided by its volume. Earth has a radius of 3964 miles. Venus has a radius of about 3760 miles and a mass that is 81.5% that of Earth. Which planet has the greater average density, Earth or Venus? Justify your answer. (Note: because Earth and Venus are not made of all one material like a block of steel or granite, the average density is used rather than density.)

14. Suppose a building material is made from two substances, which we will refer to as A and B. The building material uses 2 parts of A, which has a density of 180 pounds per cubic foot, for each part of B, which has a density of 310 pounds per cubic foot. What is the average density of the building material? (See problem 13.)

15. What is the pressure in pounds per square inch of a 1-foot cube of steel sitting on a flat surface?

16. What is the pressure in pounds per square inch of a 1-foot cube of marble sitting on a flat surface?

For problems 17 through 20, use the following formula: the volume of a pyramid or a cone is given by $V = \frac{1}{3} \times (\text{Area of the base}) \times (\text{height})$ or $V = \frac{1}{3}Ah$.

17. What is the pressure on the base of a cone (circular base) that has a diameter of 10 feet and a height of 8 feet if the cone is made of solid marble? Marble weighs 170 lbs per cubic foot.

18. What is the pressure on the base of a pyramid (square base) that is 8 feet on each edge of the base and 10 feet high if the pyramid is made of solid steel? Steel weighs 500 lbs per cubic foot.

19. Suppose we have a cone (circular base) and a pyramid (square base) such that the diameter of the base of the cone is eight feet, the edge of the base of the pyramid is eight feet, and both are six feet high. If the cone and pyramid are made of the same solid material, how do the pressures on the bases of the two compare?

20. Suppose we have a cone and pyramid that are the same height and made of the same solid material. What would the diameter of the cone have to be if the pyramid has a base that is six feet on each side, and the pressures on the base of the cone and pyramid are the same?

For problems 21 and 22, use the following formula: The volume of a cylinder is $V = \pi r^2 h$.

21. What is the pressure (in pounds per square inch) at the base of a large wooden pole that is in the shape of a cylinder two feet in diameter and 40 feet high if the density of the wood is 60 lb/cu ft?

22. What happens to the pressure on the base of a solid cylinder if
(a) the height is doubled and the radius is halved?
(b) the radius is doubled and the height is halved?

Scale models of cars and small planes are often made to $\frac{1}{4}$ inch scale. This means that an object one foot long on the real object will be $\frac{1}{4}$ inch long on the model. This is also referred to as $\frac{1}{48}$th scale.

23. The 1984 Thunderbird had an overall length of 198" and a fuel tank capacity of 18 gallons. What would be the overall length and fuel tank capacity of an exact $\frac{1}{4}$ inch scale model?

24. The 1984 Thunderbird had a wheel base of 104" and luggage capacity of 14 cubic feet. What would be the wheelbase and luggage capacity of an exact $\frac{1}{4}$ inch scale model?

25. One of the most popular sizes (or gauges) for model trains is HO-gauge, where an exact scale of 1 to 87 is used. This means an object 87 feet long on a real train would be 1 foot long on an HO-gauge model.
(a) How does the length of a real locomotive compare to that of an HO-gauge model?
(b) How does the volume of a real locomotive compare to that of an HO-gauge model?
(c) How does the weight of a real locomotive compare to that of an HO-gauge model if the model is made of exactly the same material?

26. O-gauge trains are built to approximately $\frac{1}{4}$ inch scale. If a sixty foot locomotive were built in O-gauge scale, how long would it be, and how would the volume of the model compare to that of the real locomotive?

27. Suppose an architect's office builds an exact $\frac{1}{4}$ inch scale model of a residence that will have 3580 square feet of living space and a lap pool that is 25 meters long. How many square feet will the living space occupy in the model, and what will be the length of the model lap pool?

28. Doll houses and their furnishings are often built on a scale of one inch to one foot. Suppose a doll house were built as an exact replica of a real house.
(a) How would their square footage compare?
(b) How would their weights compare?

EXTENDED PROBLEMS

29. Write a short paper on the development of skyscrapers. What factors played important roles in the development? Has culture been a factor? How have improvements in building materials contributed to the building of ever taller buildings?

30. Write a chronology for the world's tallest buildings. Is there a limit to how tall buildings can be built? What are the current limitations?

31. Apartment buildings in ancient Rome were often built to heights of four or five stories. The collapse of such buildings was far too common. Have there been cases of catastrophic collapse of buildings in modern times—without external causes such as earthquakes?

32. What is the relationship between bone structure (size and width) and the size and weight of mammals? What would happen if you could scale a mouse up to the size of an elephant?

Chapter Eleven Problem

A new farmer decided to save time by storing his hay in one very large haystack. He did this, and his haystack amazingly caught fire and exploded! Explain what happened. You may assume these facts: haystacks look like half of a sphere in an hour, hay produces heat at a rate of 12 calories per cubic foot, heat escapes at a rate of 90 calories per square foot. (Note: in the metric system, a calorie is a unit of heat energy.)

SOLUTION

Strategy: Draw a picture, use a variable, and make a table.

Modeling the haystack as a half of a sphere with radius r, draw a picture of the haystack.

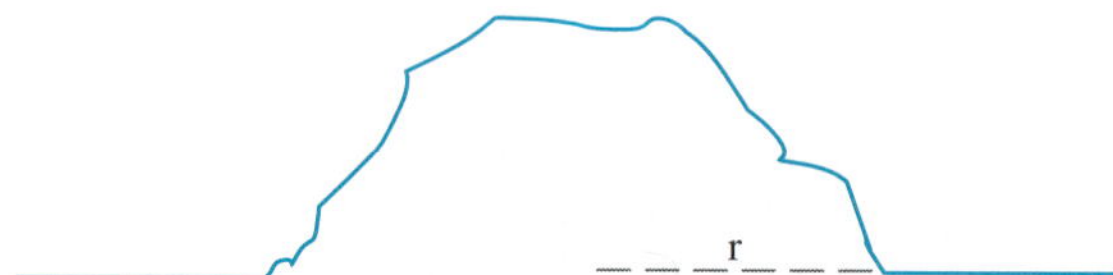

The formula for the volume of a sphere is $\frac{4}{3}\pi r^3$, and the formula for the surface area of a sphere is $4\pi r^2$. The volume and surface area of the haystack is half that.

The amount of heat produced in an hour is

$$12 \times \frac{1}{2} \times \frac{4}{3}\pi r^3 = 8\pi r^3 \approx 25.13r^3.$$

The amount of heat lost per hour across the surface area is

$$90 \times \frac{1}{2} \times 4\pi r^2 = 180\pi r^2 \approx 565.49r^2.$$

The total heat gain (or loss) for the haystack will be the difference $25.13r^3 - 565.49r^2$. We make a table of the heat gain/loss for various values of the radius r:

r	**heat gain**
10	−31,419
20	−25,156
30	169,569
40	703,536

This means that for haystacks of radius less than 20 feet, heat is lost at least as fast as it is created. But for large haystacks of radius 30 or 40 (60 or 80 feet across), heat is created faster than it may be dissipated. Thus a large haystack can become hot and explode.

Chapter Eleven Review

Key Ideas and Questions

The following questions review the main ideas of this chapter. Write your answers to the questions, and then refer to the pages listed to make certain that you have mastered these ideas.

1. Describe a similitude and scaling factor. 543, 540
   How do lengths, areas, and volumes transform when the scaling factor is r? 555–557
2. What are the fundamental similitudes? 546
3. How do you model populations that grow at a constant rate? 561
   What are limitations of this model? How does the logistic law make for a more realistic model? 568
4. How are the half-life and decay rate of a radioactive substance related? 566
5. What is the relationship between density, weight, and volume? 574, 575
   Why does the pressure exerted by similar objects scale like the scaling factor? 575, 576

Vocabulary/Notation

Following is a list of the key vocabulary, notation, and ideas for this chapter. Mentally review each of these items; write down the meaning of each term and use it in a sentence. Then refer to the pages listed by number, and restudy any material you are unsure of before answering the questions in the Chapter Eleven Review Problems.

Section 11.1

Similar Triangles 539
Scaling Factor 540
Scaling of Perimeter 541
Scaling of Area 542
Size Transformation 543
Center of a Size Transformation 543
Scaling Factor of a Size Transformation 543
Expansion 544
Contraction 544
Similitude 546

Section 11.2

Section 11.3

Section 11.4

Chapter Eleven Review Problems

1. Suppose that ΔABC and ΔDEF are similar under the correspondence A ´ D, B ´ E, C ´ F and that $AB = 5, DE = 15, EF = 21$, and $FD = 9$. Find the scaling factor. What are the side lengths of ΔABC? What are the perimeters of the two triangles?
2. Suppose that ΔABC and ΔDEF are similar under the correspondence A ´ D, B ´ E, C ´ F and that $AB = 6, BC = 10, CA = 12$ and the perimeter of ΔDEF is 40. What are the side lengths of ΔDEF?
3. Suppose that the length of the shadow of a tree is 20 feet and that the length of the shadow of a 6 foot tall person standing nearby is 2 feet. How tall is the tree?
4. Suppose that ΔABC and ΔDEF are similar under the correspondence A ´ D, B ´ E, C ´ F and that the area of ΔABC is 26 and the length of AB is 8. Suppose that the length of DE is 10. How large is the area of ΔDEF?
5. Suppose that ΔABC is translated so that the point B is moved to the point D. Sketch the result of this transformation.

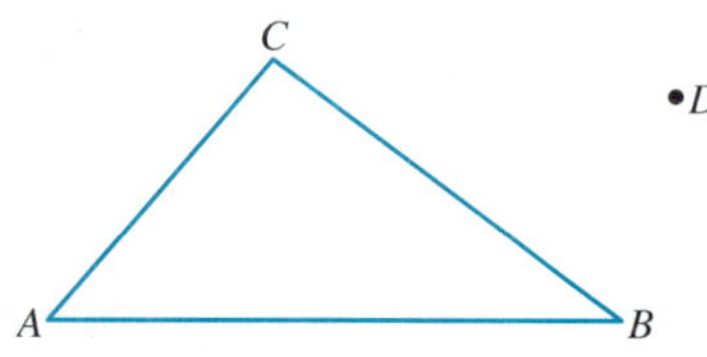

6. Suppose that ΔABC is reflected across the line containing $\overline{BC}$. Sketch the result of this transformation.

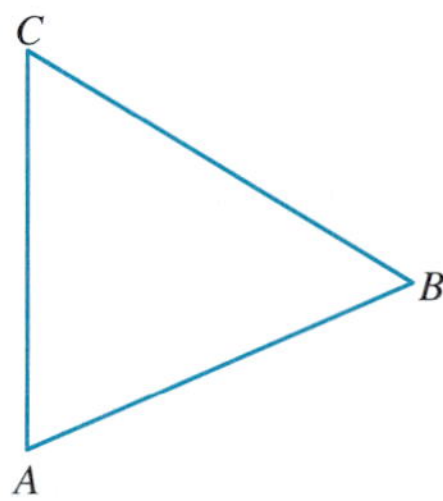

7. Suppose that ΔABC is rotated with center A so that the image of $\overline{AC}$ is on the line through A and D. Sketch the result of this transformation.

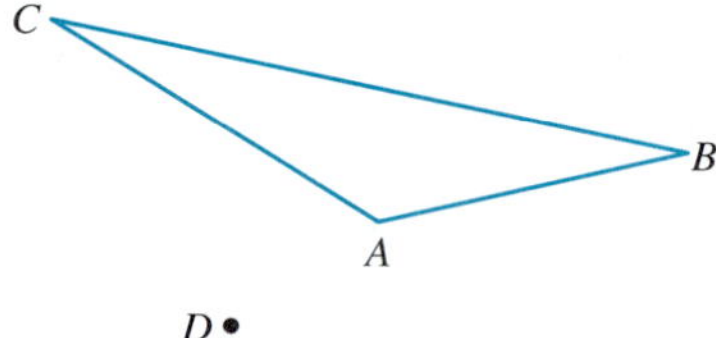

8. Suppose that ΔABC is first translated so that the image of A is at D and then reflected across the image of the line $\overline{BC}$. Sketch the result of this transformation.

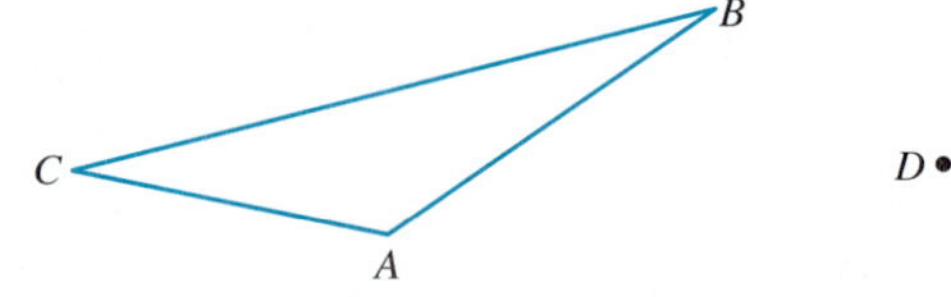

9. Describe a similitude that will transform ΔABC onto ΔDEF.

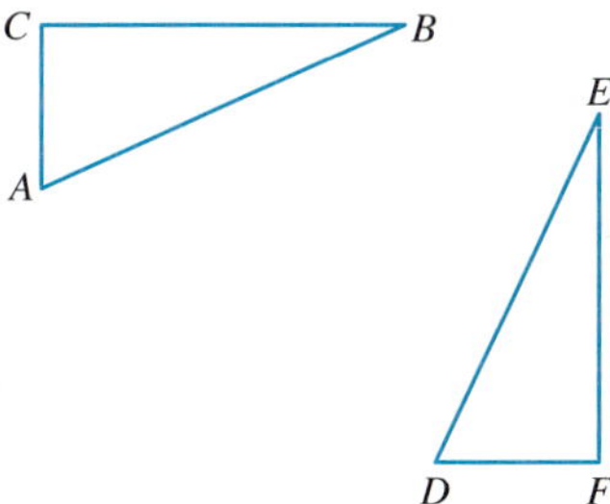

10. Suppose that a pentagon has perimeter 50 and area 172. If a similar figure has sides twice as large as the original pentagon, what are its perimeter and area?
11. An equilateral triangle with side length 2 has area 1.73. What is the area of an equilateral triangle with side length 4.6?
12. A volume of a sphere for storing natural gas is 113,000 cubic feet. Suppose that another storage sphere is made with a scaling factor of 3. What volume of natural gas would this storage sphere have?
13. The country of Contraria has a growth rate of 0.5% per year. Suppose that this is a small country with 160,000 citizens and that the growth rate stays constant. What will the population be in 10 years using the Malthusian model? In 50 years? In 100 years?
14. How long will it take for the population in Contraria in problem 13 to double?
15. The value of a stamp collection appreciates (grows) by 8% per year. If it is worth $20,000 now, what will it be worth in 10 years?
16. Suppose that a city has a constant growth rate. In a 5-year period it has grown from 41,000 people to 47,000 people. What will the population be in another 12 years?
17. A radioactive substance has a half-life of 2.5 years. What is the decay rate?
18. In problem 17, suppose that we have 28 grams of this substance. How much will be left in 2 years?
19. Suppose that a population has a constant growth rate of 4%. If the current population is 20,000, what will it be in a year?
20. A population is governed by the logistic law. The carrying capacity is 400,000 and the growth rate is 4%. If the population this year is 20,000, what will the population be in a year?
21. In problem 20, suppose that the population is 140,000. What will it be in a year?
22. The weight of a cubic foot of water is 62.4 pounds. What is the weight of a cube of water that has a side of 4 feet?
23. Suppose that a person weighs 140 pounds. What is this person's volume if we assume that the weight per cubic foot for humans is the same as for water?
24. Aluminum has a density of 170 pounds per cubic foot. Suppose that an aluminum block of side lengths 20 feet by 20 feet and height 2 feet is in a warehouse. How much does the block weigh? What is the pressure on the floor in pounds per square inch (psi)? Notice you will have to use the fact that one foot is 12 inches.
25. Suppose that the aluminum block in problem 24 is put on a pedestal that is one foot square. How much pressure is on the pedestal in pounds per square inch (psi)?
26. A scale model of a resort is made that is one inch to the foot. There are to be 120 rooms for overnight visitors, and the total living space will be 96,000 square feet. The swimming pool will hold 15,000 cubic feet. The model is to have a swimming pool, built to scale with real water. What will be the volume of the swimming pool in the model? What will be the living space in the model? How many rooms will the model have?

CHAPTER 12

RECURSION AND FRACTALS

MARINE BIOLOGIST COMPLETES 3000 MILE TREK ON OREGON COAST

In what may be the most ambitious study of the Oregon coast ever undertaken, Hal Walker, Ph.D., has just completed the first inventory of all accessible tide pools on the Oregon coast. Dr. Walker covered the entire coastline of Oregon by foot—from the California state line, near Brookings, to Astoria at the mouth of the Columbia River. In following the coastline, the scientist tried to stay within one yard of the water's edge except for impassable stretches and dangerous rocks. Bridges were used only when rivers and bays were too deep to be waded. To measure the distance covered, Dr. Walker used a pedometer, recording 2,928.5 miles by the trip's end. For the return trip from Astoria to Brookings, he used a small plane which hugged the coastline at low level. The scientist said the 317 mile air trip was the most beautiful he had ever taken. The support team from the university used a van to return to Brookings on Highway 101, a distance of 339 miles.

Chapter Goals

1. Use recursion to compute the terms of a sequence.
2. Compute and determine some surprising properties of iterations.
3. Compute sums of geometric series.
4. Construct geometric objects of infinite complexity and beauty by applications of a simple rule.

The primary mathematical question raised by the story is not "How many tide pools are there on the Oregon coast?", but "How far is it from Brookings to Astoria?" or "How do we measure distance or length?" As a practical application, the answer depends on the tools used and the way they are used. On a highway we can use a wheel as a measuring tool; we count the number of revolutions and multiply by the circumference of the wheel. The pedometer uses a stride for its measure, but since strides are measured along straight lines, the distance measured walking on the highway with a pedometer would not be exactly the same as using the revolving wheel.

Many objects from the natural world result from processes that are repeated: for example, the branching of a tree, the growth of bark on the trunk of the tree, the growth of moss on the bark, the patterns in the leaves, and so forth. Traditional geometry generally only gives us approximations for natural objects, and there have been several revolutions in mathematics that have produced new geometries that increased our ability to describe and understand the natural world.

In this chapter, we will consider algebraic and geometric situations in which the repeated application of simple rules leads to unexpected and sometimes complex behavior and forms. We will also introduce the new science of fractal geometry. In this new frame of reference for describing the natural world, the fundamental idea of dimension will change. Fractal geometry provides new tools to scientists and professionals from many disciplines to explore and understand the world in which we live. For those who are interested solely in the beauty of nature, you may never look at the world in quite the same way again.

THE HUMAN SIDE OF MATHEMATICS

Edward Lorenz

Edward Lorenz (1917–) was fascinated by the weather as a child, but his real love was mathematics and mathematical puzzles. Lorenz intended to study mathematics in college, but World War II intervened, and the Army Air Corps assigned him to weather forecasting. After the war, Lorenz completed college and joined the faculty at MIT, where he investigated the problem of long-range weather forecasting using mathematical models. In principle, forecasting the weather was thought to be like predicting the motion of planets, only more complicated, with more equations and variables. Since high speed computers were just becoming available, it seemed possible to Lorenz that long-range forecasting could become a reality. Lorenz developed a simplified model of the weather that could be run on his computer. With this model, which consisted of a set of interrelated equations, he could try different numbers that represented weather conditions and see what future weather patterns would arise. He made the surprising discovery that if two sets of weather conditions were only slightly different, then future weather patterns arising from them would only be similar for a short while. Over the long term, completely different weather patterns emerged. This phenomenon is now known as the *butterfly effect* after the title of a paper written by Lorenz—"Predictability: Does the Flap of a Butterfly's Wings in Brazil Set Off a Tornado in Texas?" Because we can never measure all the weather conditions exactly, there will always be a limit to how far in the future we can accurately forecast the weather. Lorenz's discovery was the birth of chaos theory.

Benoit Mandelbrot

Benoit Mandelbrot (1924–) brought the language and universality of fractals to the world's attention. "Clouds are not spheres, mountains are not cones, coastlines are not circles, and bark is not smooth, nor does lightning travel in a straight line." This is how Mandelbrot describes his inspiration for the geometry of fractals. Fractals are objects of infinite and intricate detail. A cloud can be modeled using fractals. If you look at a cloud you will first see the rough outline of the shape. This shape is not smooth, but composed of bumps of smaller and smaller sizes. Look at almost any natural shape, and you will see this kind of pattern within patterns within patterns.

Mandelbrot has worked on a variety of topics, including the distribution of large and small incomes in the economy. When an economist from Harvard invited him to give a lecture, Mandelbrot was surprised on his arrival to see what he thought were his findings already charted on the blackboard. The chart actually represented eight years of cotton prices the economist was studying. The data had produced variations that did not conform to any accepted assumptions about how such data behaved. Mandelbrot had begun to see the strange charted outline in other, very different places; and when he used IBM's computers on the cotton price data, he found they produced symmetry when examined from the point of view of scaling.

As he studied these problems, the idea of a new geometry emerged, and with it the concept of a fractal, an object based on self-similarity. When Mandelbrot turned his attention to the problem of "noise" in data transmission, he created a workable model for his new geometry. With the aid of computers and the new field of computer graphics this new geometry grabbed the attention of mathematicians and scientists everywhere, and a new means for understanding our world was created.

12.1 FIBONACCI NUMBERS, THE GOLDEN RATIO, AND RECURSION

INITIAL PROBLEM

The following expression is known as a continued fraction.

$$1 + \cfrac{1}{1 + \cfrac{1}{1 + \cfrac{1}{1 + \cfrac{1}{1 + \dots}}}}$$

Before the decimal system came into being, numbers like this (but ending after several levels) were used to approximate irrational numbers such as π. How can we find the decimal equivalent of this number?

Many forms and processes in nature involve the repetition of some geometric figure or computational rule. We begin by looking at a set of numbers that is generated by the repetition of a simple rule. This set is connected to nature in many surprising and fascinating ways.

FIBONACCI NUMBERS

One of the most famous sequence of numbers in the history of mathematics is the **Fibonacci sequence,**

$$1, \quad 1, \quad 2, \quad 3, \quad 5, \quad 8, \quad 13, \quad 21, \dots$$

in which each number after the first two is the sum of the preceding two numbers, for example, $8 + 13 = 21$. (A **sequence** $a_1, a_2\, a_3, \dots$ is any ordered set of numbers where a_n is the nth number.) Any number in the Fibonacci sequence is called a **Fibonacci number.** We will see that this sequence of numbers gives the answer to the following problem originally posed by Fibonacci:

If you begin with one pair of adult rabbits, how many pairs of rabbits will there be after one year if every month each pair of rabbits produces a new pair, and each new pair of rabbits begins to produce young two months after birth?

TIDBIT

Leonardo de Fibonacci (1170–1250) was one of the greatest mathematicians of the Middle Ages. He is credited with popularizing the Hindi–Arabic number system we use today. You may not have enjoyed learning long division, but try it with Roman numerals! In the Middle Ages, when Roman numeration was used, the ability to do division problems accurately was considered evidence of witchcraft.

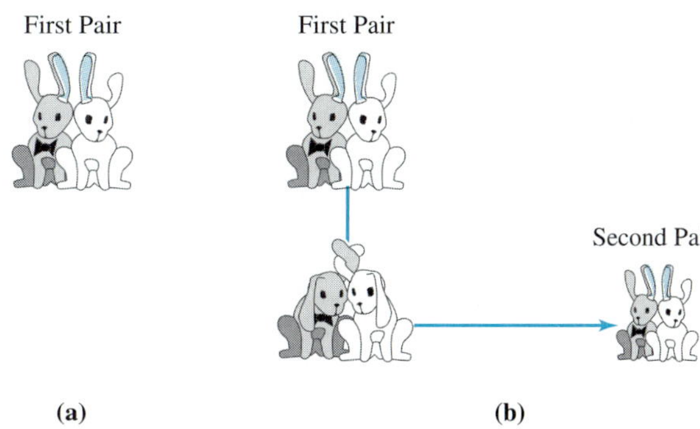

FIGURE 12.1

We can keep track of the pairs of rabbits in Fibonacci's problem using a few pictures. At the beginning of the year, January 1, we have one pair of rabbits [Figure 12.1(a)], and after another month, that is on February 1, they produce a pair of baby rabbits [Figure 12.1(b)].

March 1: the original pair of rabbits produce another pair of babies. The first pair of baby rabbits mature, but do not produce any babies yet (Figure 12.2).

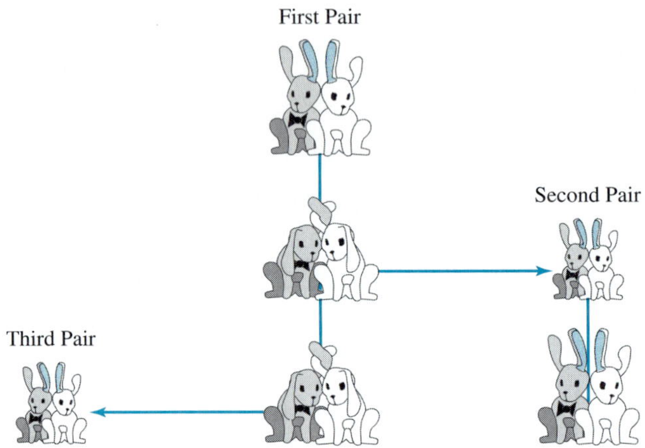

FIGURE 12.2

April 1: the original pair of rabbits again produce a pair of baby rabbits. The first pair of baby rabbits, that matured one month after birth, produce their first pair of baby rabbits. The second pair of baby rabbits mature, but do not yet produce any babies (Figure 12.3).

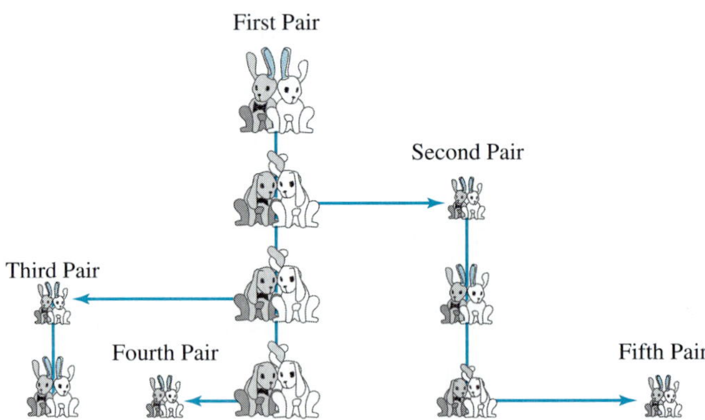

FIGURE 12.3

Each month each pair of adult rabbits produces a pair of babies. Each month the rabbits that begin as babies mature into adults. This is illustrated in Figure 12.4.

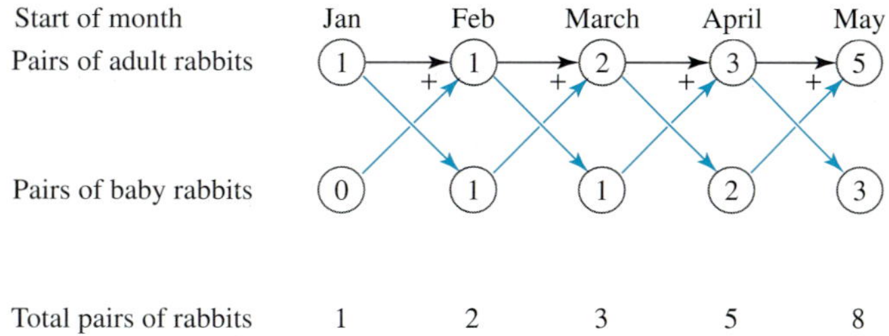

FIGURE 12.4

If we focus our attention on the number of pairs of adult rabbits, we see that the number can be obtained by adding together the number of adult rabbits in the previous two months (Figure 12.5).

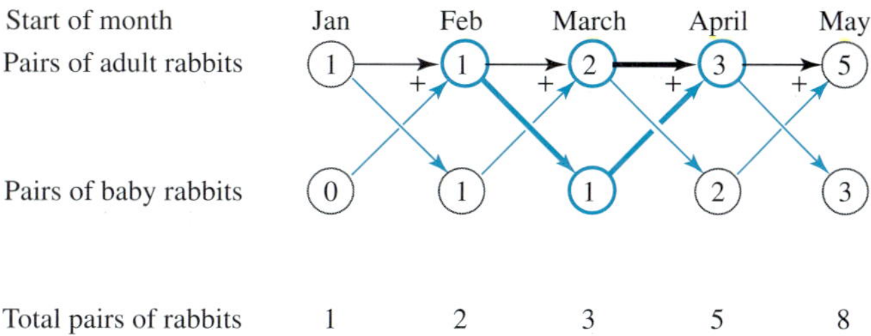

FIGURE 12.5

Notice that in Figure 12.4 there are actually three sequences in which each number is the sum of the previous two: the number of pairs of adult rabbits, the number of pairs of baby rabbits, and the total number of pairs of rabbits. These three sequences just start a little differently. We have

Month	**Jan**	**Feb**	**Mar**	**Apr**	**May**	**Jun**	**Jul**	**Aug**	**Sep**	**Oct**	**Nov**	**Dec**	**Jan**
Pairs of adults	1	1	2	3	5	8	13	21	34	55	89	144	233
Pairs of babies	0	1	1	2	3	5	8	13	21	34	55	89	144
Total pairs	1	2	3	5	8	13	21	34	55	89	144	233	377

So the answer to Fibonacci's question is that there will be 377 pairs of rabbits after one year.

Fibonacci numbers often occur in plants—in the number of petals in flowers and in the branching behavior along a trunk or stem. It is believed that the spiral nature of plant growth accounts for this. For example, the usual number of petals on various flowers are Fibonacci numbers as shown in Table 12.1.

TABLE 12.1

Petals	Plants
2	Enchanter's nightshade
3	Lily and iris
5	Wall lettuce and wild rose
8	Delphiniums and bloodroot
13	Ragwort and corn marigold
21	Aster and black-eyed Susan
34	Field daisies
55	African daisies
89	Michaelmas daisies

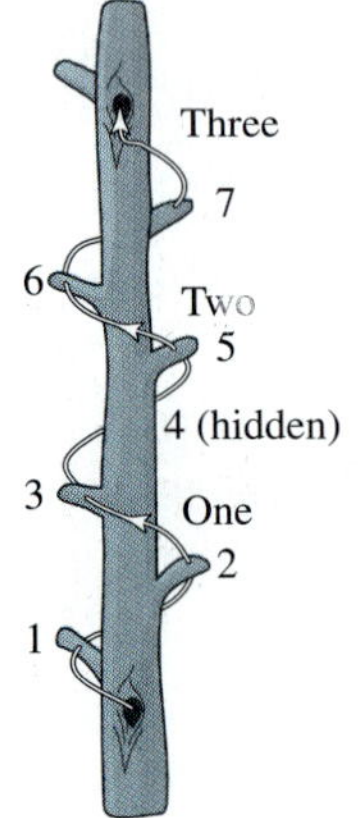

FIGURE 12.6

Figure 12.6 shows the spiral growth pattern of branches from a trunk. Notice in Figure 12.6, there are eight branches in three complete spirals. This is called a phyllotactic ratio of 3/8, that is, the number of spirals divided by the number of branches. Most phyllotactic ratios are ratios of Fibonacci numbers (Table 12.2).

TABLE 12.2

Ratio	Plants
$\frac{2}{3}$	Grasses and elm
$\frac{1}{3}$	Blackberry and beech
$\frac{2}{5}$	Apple, cherry, and plum
$\frac{3}{8}$	Pear and weeping willow
$\frac{5}{13}$	Pussy willow and almond

The Fibonacci numbers also occur in the growth patterns of sunflowers and pine cones (Figure 12.7).

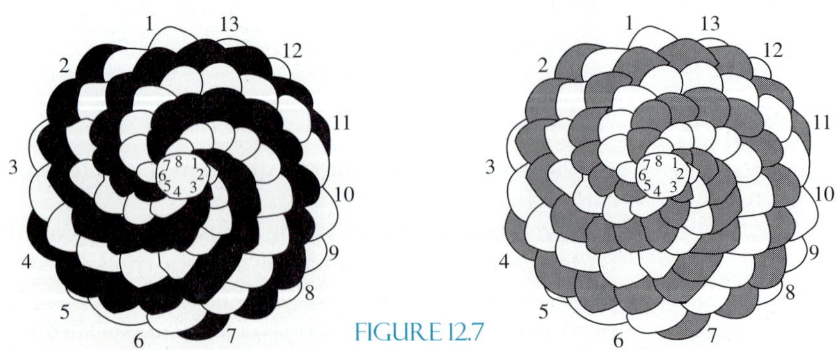

FIGURE 12.7

THE GOLDEN RATIO

Consider the ratios of pairs of successive Fibonacci numbers:

$$\frac{1}{1} = 1.0, \quad \frac{2}{1} = 2.0, \quad \frac{3}{2} = 1.5, \quad \frac{5}{3} = 1.66\ldots,$$

$$\frac{8}{5} = 1.6, \quad \frac{13}{8} = 1.625, \quad \frac{21}{13} = 1.61538\ldots, \ldots$$

Perhaps it is surprising, but the ratios of the Fibonacci numbers get closer and closer to one special number, as suggested in Figure 12.8.

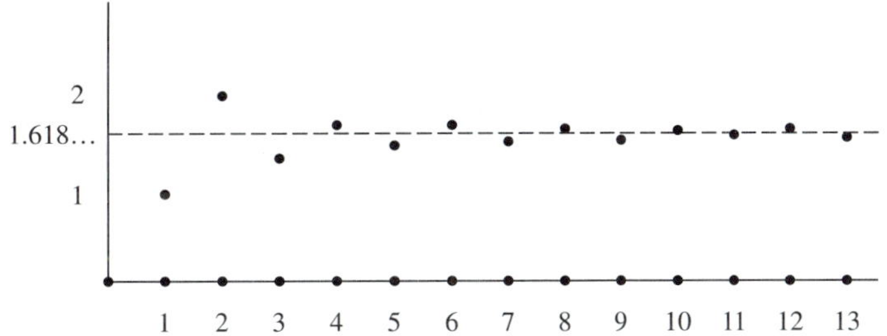

FIGURE 12.8

This number is called the **Golden Ratio** (also called the **Golden Mean**), which equals

$$\frac{1 + \sqrt{5}}{2} \approx 1.61803.$$

The Golden Ratio has figured prominently in mathematics, art, and architecture for more than two thousand years. This number gives the proportions of the Parthenon, the Temple of Athena built in Athens, Greece in the 5th century B.C. (Figure 12.9).

FIGURE 12.9

The Greeks were able to construct a "golden rectangle" using what is now called the Pythagorean Theorem [Figure 12.10(a)]. Interestingly, when the square is constructed on the left of the golden rectangle in Figure 12.10(a), the smaller rectangle to its right is also a golden rectangle. The Golden Ratio can also be seen in the art of Leonardo da Vinci [Figure 12.10(b)].

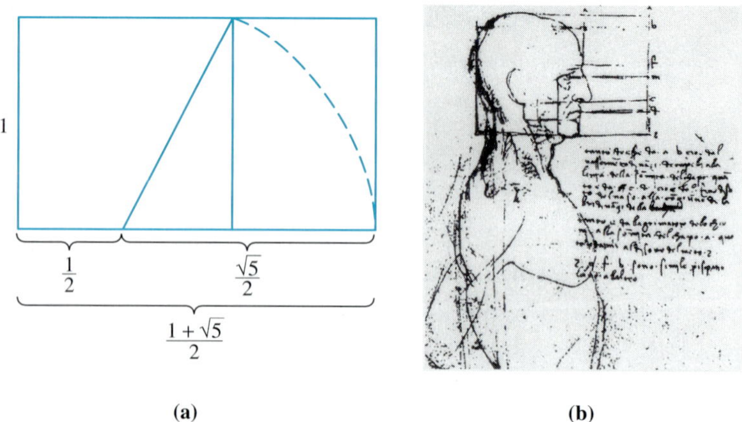

FIGURE 12.10

RECURSION

The Fibonacci sequence is just one example of a general class of sequences generated by a process called **recursion** where each number in the sequence is found using the previous numbers in the sequence. The rule used to determine any term is called the **recursion rule.** For the Fibonacci sequence, we begin with two 1's, and the recursion rule states that you add the preceding two numbers to get the next number. It helps to keep track of the position of the numbers in the sequence; in our notation, f_n refers to the nth term in the sequence. Thus for the Fibonacci sequence we set

$$f_1 = 1,\ f_2 = 1,\ f_3 = 2,\ f_4 = 3,\ f_5 = 5,\ f_6 = 8,\ f_7 = 13,\ f_8 = 21,\ \ldots$$

The recursion rule generating the Fibonacci sequence is written

$$f_n = f_{n-1} + f_{n-2}.$$

As another example, consider the sequence of numbers

$$1,\ 2,\ 6,\ 24,\ 120,\ 720,\ 5040,\ 40320,\ 362880,\ \ldots$$

Can you guess how each number is obtained from the preceding number? To determine the rule, we introduce some notation to keep track of the position of the numbers in the sequence by setting

$$a_1 = 1,\ a_2 = 2,\ a_3 = 6,\ a_4 = 24,\ a_5 = 120,\ \ldots$$

Now, it is easier to see that $a_2 = 2 \times a_1, a_3 = 3 \times a_2, a_4 = 4 \times a_3$, and so on. Thus the recursion rule for generating the sequence is

$$a_n = n \times a_{n-1}.$$

The complete information needed to define a sequence using recursion is

(a) the starting value(s) and
(b) a recursion rule.

We illustrate this in the next example.

EXAMPLE 12.1 From the starting value $t_1 = 1$ and the recursion rule

$$t_n = n + t_{n-1},$$

generate the first ten numbers in the sequence.

SOLUTION Starting from

$$t_1 = 1,$$

we compute t_2 as follows,

$$t_2 = 2 + t_1 = 2 + 1 = 3.$$

$t_1 = 1$, already known

$n = 2$

$n - 1 = 2 - 1 = 1.$

Now that t_2 is known, we can compute t_3 by

$$t_3 = 3 + t_2 = 3 + 3 = 6,$$

and then compute t_4 by

$$t_4 = 4 + t_3 = 4 + 6 = 10.$$

Continuing in this manner, we have

$$t_5 = 5 + 10 = 15,$$
$$t_6 = 6 + 15 = 21,$$
$$t_7 = 7 + 21 = 28,$$
$$t_8 = 8 + 28 = 36,$$
$$t_9 = 9 + 36 = 45,$$
$$t_{10} = 10 + 45 = 55.$$

The first ten numbers in the sequence are 1, 3, 6, 10, 15, 21, 28, 36, 45, 55. ◆

The numbers generated by the recursion in Example 12.1 are called the triangle numbers because they count the number of squares used in making successive triangular arrays of squares, as shown in the sequence of triangles in Figure 12.11.

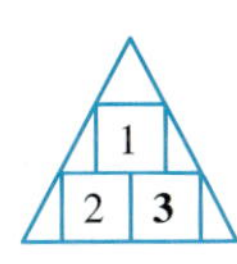

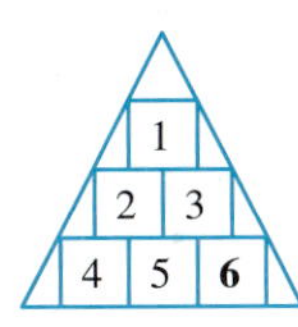

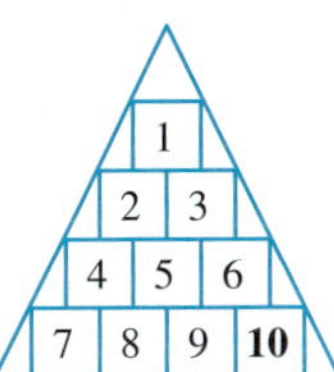

FIGURE 12.11

Recursions are typically used to solve problems in which you can see how to go from one case to the next as in Fibonacci's question about the pairs of rabbits.

A recursion is called an **iteration** if the same rule is applied each time. For example, the recursion defining the Fibonacci sequence is an iteration because if you are given two consecutive Fibonacci numbers, say 610 and 987, you can find the next Fibonacci number, namely, $610 + 987 = 1597$. The recursion defining the triangle numbers (Example 12.1) is not an iteration because finding each new triangle number involves adding a different value of n. If you are given a triangle number, say 465, you cannot find the next triangle number without additional information, namely, which triangle number 465 is. (By the way, 465 is the 30th triangle number.)

The simplest kind of iteration is one in which each term is obtained by applying the same rule to the preceding term. We call this a **simple iteration.** For example, we can define a simple iteration by starting with $a_1 = 1$ and using the recursion rule

$$a_n = \frac{1}{1 + a_{n-1}}.$$

This is an iteration because the same rule is applied to compute each new number, and it is a simple iteration because the rule uses only the immediately preceding number. Simple iterations are desirable because the computations can be carried out easily.

EXAMPLE 12.2 Compute the first ten numbers in the simple iteration defined by $a_1 = 1$ and

$$a_n = \frac{1}{1 + a_{n-1}}.$$

SOLUTION We are given $a_1 = 1$. We compute

$$a_2 = \frac{1}{1 + a_1} = \frac{1}{1 + 1} = \frac{1}{2}.$$

To continue the calculation, we use a calculator. After entering the current number, $a_2 = \frac{1}{2}$ (0.5), we add 1, giving us $\frac{3}{2}$ (1.5). Then we press the $\boxed{\frac{1}{x}}$ key, to obtain a_3, namely $\frac{2}{3} \approx 0.666667$. Then we add 1 and press the $\boxed{\frac{1}{x}}$ key, to obtain a_4, namely $\frac{3}{5} = 0.6$. Repeating these operations, we can generate the numbers:

$$a_3 \approx 0.666667, \quad a_4 \approx 0.6, \quad a_5 \approx 0.625, \quad a_6 \approx 0.615385,$$
$$a_7 \approx 0.619048, \quad a_8 \approx 0.617647, \quad a_9 \approx 0.618182, \quad a_{10} \approx 0.617978.$$

◆

You may recall that the Fibonacci ratios get closer and closer to the Golden Ratio, which is approximately 1.61803. Recursively generated sequences often have this property of approaching one particular number. In the case of a simple iteration you can determine whether this happens and find the number if it does. The next example illustrates this.

EXAMPLE 12.3 Compute the 10th, 20th, 30th and 40th terms of the sequence defined by the simple iteration with starting value $a_1 = 256$ and recursion rule

$$a_n = \sqrt{a_{n-1}}.$$

SOLUTION This sequence can be calculated quickly by entering 256 into a calculator, then repeatedly pushing the square root key. The first few terms are

$$a_2 = \sqrt{a_1} = \sqrt{256} = 16, \; a_3 = \sqrt{16} = 4, \; a_4 = \sqrt{4} = 2.$$

Repeating the operation as many times as needed, we find

$$a_{10} \approx 1.01088928605, \quad a_{20} \approx 1.00001057664,$$
$$a_{30} \approx 1.00000001032, \quad a_{40} \approx 1.00000000001.$$

The numbers in the sequence get closer and closer to 1. ◆

DEFINITION·······

Point Attractor

When the numbers in an iteration approach some specific value we call that value a **point attractor.**

The point attractor for the iteration in Example 12.3 does not depend on the starting value of the iteration. For example, if we begin with $b_1 = 155$ instead of $a_1 = 256$, and use the same recursion rule

$$b_n = \sqrt{b_{n-1}},$$

then

$$b_2 \approx 12.449899598,$$
$$b_5 \approx 1.3705526735,$$
$$b_{10} \approx 1.00989911496,$$
$$b_{20} \approx 1.00000961961.$$

So we see the b_n's are also approaching 1. For this particular iteration any positive starting value will generate a sequence that is attracted to 1. Starting with 0 gives a sequence of all zeros, and starting with a negative number is not allowed. (Why?)

DEFINITION·······

Stable Point Attractor

When the numbers in an iteration approach the same point attractor for all starting values in some interval with length greater than zero, then the point attractor is called **stable.**

The iteration defined by $a_1 = 1$ and the rule

$$a_n = \frac{1 - (a_{n-1})^2}{3}$$

has a stable point attractor that seems to be about 0.3. To find the equation satisfied by this point attractor, we let $x = a_n$ in the left-hand-side of the iteration and $x = a_{n-1}$ in the right hand side of the iteration. We obtain the equation

$$x = \frac{1 - x^2}{3}.$$

This equation can be rewritten as

$$3x = 1 - x^2 \qquad \text{or} \qquad x^2 + 3x - 1 = 0.$$

We solve this equation using the quadratic formula from algebra:

If $ax^2 + bx + c = 0$,

then

$$x = \frac{-b \pm \sqrt{b^2 - 4ac}}{2a}.$$

We conclude that

$$x = \frac{-3 + \sqrt{13}}{2} \approx 0.302776.$$

EXAMPLE 12.4 The iteration defined by $v_1 = 1$ and the rule

$$\sqrt{1 + v_{n-1}}$$

has a stable point attractor that seems to be about 1.6. Find the equation that must be satisfied by the stable point attractor. Solve the equation if possible.

SOLUTION We let $x = v_n$ in the left-hand side of the iteration and $x = v_{n-1}$ in the right-hand side of the iteration to obtain the equation

$$x = \sqrt{1 + x},$$

which the stable point attractor must satisfy. Squaring both sides of this equation, we get

$$x^2 = 1 + x \qquad \text{or} \qquad x^2 - x - 1 = 0.$$

From the quadratic formula, the possible solutions are

$$x = \frac{1 + \sqrt{5}}{2} \qquad \text{or} \qquad x = \frac{1 - \sqrt{5}}{2}.$$

Our stable point attractor is $\frac{1 + \sqrt{5}}{2} = 1.61803398\ldots$ which happens to be the Golden Mean. ◆

INITIAL PROBLEM SOLUTION

The following expression is known as a continued fraction.

$$1 + \cfrac{1}{1 + \cfrac{1}{1 + \cfrac{1}{1 + \cfrac{1}{1 + \cdots}}}}$$

Before the decimal system came into being, numbers like this (but ending after several "levels") were used to approximate irrational numbers such as π. How can we find the decimal equivalent of this number?

SOLUTION Continued fractions were a means by which earlier mathematicians were able to approximate numbers they knew could not be expressed as the ratio of two whole numbers; for example, $\sqrt{2}$ is such a number.

We can find the value of the continued fractions in several ways. First, we will use recursion to generate a sequence of similar fraction forms, which we will evaluate. Then we will use the problem solving strategy "look for a pattern." One way to define the recursion is as follows: we begin with the value $a_1 = 1 + 1$ and use the recursion rule $a_n = 1 + \frac{1}{a_{n-1}}$. We will generate the first four numbers of the sequence and simplify the results.

$$a_1 = 1 + 1 \qquad = 2$$

$$a_2 = 1 + \frac{1}{1+1} = 1 + \frac{1}{2} \qquad = \frac{3}{2}$$

$$a_3 = 1 + \frac{1}{1 + \frac{1}{1+1}} = 1 + \frac{1}{\frac{3}{2}} = 1 + \frac{2}{3} \qquad = \frac{5}{3}.$$

Note that the last denominator in a_3, $1 + \frac{1}{1+1}$, was previously evaluated as $a_2 = \frac{3}{2}$. We then made the substitution and use the fact that 1 divided by $\frac{3}{2}$ equals $\frac{2}{3}$. Continuing in this way, we find that

$$a_4 = 1 + \frac{1}{1 + \frac{1}{1 + \frac{1}{1+1}}} = 1 + \frac{1}{\frac{5}{3}} = 1 + \frac{3}{5} \qquad = \frac{8}{5}.$$

At this point, perhaps you recognize these values as the ratios of consecutive pairs of Fibonacci numbers. This sequence has a stable point attractor of approximately 1.61803.

We also could have defined the sequence with the following:

$$b_1 = 1; \; b_n = 1 + \frac{1}{b_{n-1}}.$$

If there is a stable point attractor, then it satisfies the equation $x = 1 + \frac{1}{x}$. This is equivalent to

$$x^2 = x + 1 \qquad \text{or} \qquad x^2 - x - 1 = 0.$$

This is the same equation solved in Example 12.4, and the value is the Golden Mean, $x \approx 1.61803$.

PROBLEM SET 12.1

The rule for finding the numbers in the Fibonacci sequence is

$$f_1 = 1,\ f_2 = 1;\ f_n = f_{n-1} + f_{n-2}$$

This means that you are given the initial values in the sequence and a rule to find the values that follow. In this case, each number is the sum of the two preceding it.

In problems 1 through 6, the initial values of a sequence will be given together with a rule for determining the values that follow in the sequence. You should be sure that you understand the rule by verifying that the numbers given satisfy the rule.

1. Initial values: 2 and 5
   Rule: Each of the next terms is the sum of the two preceding terms.

   2, 5, 7, . . .

   Find the next five terms of the sequence.
2. Initial values: 5 and −3
   Rule: Each of the next terms is the sum of the two preceding terms.

   5, −3, 2, . . .

   Find the next five terms of the sequence.
3. Initial values: 1, 2, and 4
   Rule: Each of the next terms is the sum of the three preceding terms.

   1, 2, 4, 7, . . .

   Find the next four terms of the sequence.

4. Initial values: 2 and -1
Rule: Each of the next terms is the product of the two preceding terms.

$$2, -1, -2, \ldots$$

Find the next five terms of the sequence.

5. Initial values: 4 and 3
Rule: Each of the next terms is found using the two preceding values by multiplying the first term by 2 and adding the next one.

$$4, 3, 11, \ldots$$

Find the next five terms of the sequence.

6. Initial values: 4 and 3
Rule: Each of the next terms is the quotient of the preceding terms.

$$4, 3, \tfrac{4}{3}, \ldots$$

Find the next five terms of the sequence.

In problems 7 through 9, the initial values for a sequence will be given. Use the recursion rule

$$a_n = a_{n-1} + a_{n-2},$$

and do the following:

(i) find the first 12 terms of the resulting sequence;
(ii) find the ratio of the twelfth term of the sequence divided by the eleventh term.

7. **(a)** $a_0 = 1, a_1 = 3$
(b) $a_0 = 3, a_1 = 1$
8. **(a)** $a_0 = -1, a_1 = 1$
(b) $a_0 = 1, a_1 = -1$
9. **(a)** $a_0 = 4, a_1 = 3$
(b) $a_0 = 1, a_1 = -5$
10. The following is a description of a mental experiment. You are not expected to actually carry it out.
(a) Select any two numbers (other than 0 and 0) as initial values.
(b) Use the recursion rule

$$a_n = a_{n-1} + a_{n-2}$$

to produce additional terms of a sequence.
(c) Produce the first 100 terms of the sequence.
(d) Divide the 100th term by the 99th term.
Based on the results from any of the problems 7 through 9, can you guess the value obtained in part (d)?

In problems 11 through 13, use the recursion rule

$$a_n = 2a_{n-1} - a_{n-2}$$

with the given initial values.

(i) Find the first 10 terms of the sequence.
(ii) Identify the pattern, if any, in the sequence.

11. **(a)** $a_0 = 0, a_1 = 2$
(b) $a_0 = 3, a_1 = 0$
12. **(a)** $a_0 = 0, a_1 = 3$
(b) $a_0 = 2, a_1 = 0$
13. **(a)** $a_0 = 1, a_1 = -5$
(b) $a_0 = -2, a_1 = 3$
14. Based on the results from any of the problems 11 through 13, what do you think would happen if the same recursion rule was used with the initial values $a_0 = 3$, $a_1 = 10$? Guess the next five terms without computing them by the rule, and then check your work with the rule.
15. Compute the next 10 numbers of the recursion given by

$$a_n = a_{n-1} + a_{n-2} + a_{n-3}$$

(a) $a_0 = 0, a_1 = 1, a_2 = 1$
(b) $a_0 = 1, a_1 = 0, a_2 = 1$
16. Compute the first 10 numbers of the recursion given by

$$a_n = a_{n-1} + a_{n-2} + a_{n-3}$$

(a) $a_0 = 1, a_1 = 3, a_2 = 1$
(b) $a_0 = 3, a_1 = -2, a_2 = 1$

In problems 17 through 20, a recursion rule and initial value(s) are given. Find the first 10 terms of the sequence. Round values to four decimal places when appropriate.

17. $b_1 = 1; b_n = 2b_{n-1} + n$
18. $t_1 = 1; t_n = 2n - t_{n-1}$
19. $b_1 = 0, b_2 = 1; b_n = n + b_{n-1} + b_{n-2}$
20. $v_1 = 1; v_n = \dfrac{n}{v_{n-1}}$

In problems 21 through 24, a simple iteration rule is given. Do the following.

(a) Use $a_1 = 1$ as the initial value.
(b) Find the first 10 terms of the iteration. Round values to four decimal places when appropriate.

21. $a_n = 2a_{n-1} + 1$
22. $a_n = 2 + a_{n-1}$
23. $a_n = \dfrac{2}{2 + a_{n-1}}$
24. $a_n = 2 + \dfrac{1}{a_{n-1}}$

For problems 25 through 30, a simple iteration rule and an initial value are given. Do the following.

(i) Find the next five values of the sequence.
(ii) Show there is a stable point attractor and find its value.

25. $u_n = 1 + 0.4u_{n-1}; u_1 = 1$
26. $u_n = 1 + 0.6u_{n-1}; u_1 = 3$
27. $u_n = 1 - 0.4u_{n-1}; u_1 = 1$
28. $u_n = 0 + 0.6u_{n-1}; u_1 = 3$
29. $u_n = 4 + 0.4u_{n-1}; u_1 = 1$
30. $u_n = 5 - 0.6u_{n-1}; u_1 = 6$

EXTENDED PROBLEMS

In problems 31 through 36, an initial value and a simple iteration rule are given. Do the following:

(a) Find the first 10 terms of the iteration. Round values to 4 decimal places.

(b) Does there appear to be a point attractor?

(c) If so, what equation must it satisfy? If you can solve the equation, do so.

31. $a_1 = 0; a_n = \sqrt{1 + 2a_{n-1}}$

32. $a_1 = 0; a_n = \sqrt{1 + 3a_{n-1}}$

33. $a_1 = 1; a_n = 2 + \dfrac{1}{3a_{n-1}}$

34. $a_1 = 1; a_n = 1 + \sqrt{a_{n-1}}$

35. $a_1 = 0; a_n = \dfrac{1}{1 + a_{n-1}}$

36. $a_1 = 1; a_n = 1 + \dfrac{1}{\sqrt{a_{n-1}}}$

37. $a_1 = 0; a_n = \sqrt[3]{2 + 3a_{n-1}}$

38. $a_1 = 0; a_n = \sqrt[3]{3 - 2a_{n-1}}$

In problems 39 through 41, use the following simple iteration to find the square root of a positive number c.

$$u_n = \frac{1}{2}\left(u_{n-1} + \frac{c}{u_{n-1}}\right)$$

Take a convenient initial value (a good guess) and iterate this function until the desired degree of accuracy is achieved.

39. **(a)** $c = 75$, nearest hundredth
(b) $c = 700$, nearest hundredth

40. **(a)** $c = 180$, nearest hundredth
(b) $c = 1050$, nearest hundredth

41. **(a)** $c = 350$, nearest thousandth
(b) $c = 2000$, nearest thousandth

42. Show that the stable point attractor, x, for the recursion

$$u_n = \frac{1}{2}\left(u_{n-1} + \frac{c}{u_{n-1}}\right)$$

satisfies the condition that $(x)^2 = c$.

43. The number e, which was used to find continuously compounded interest and is intimately involved in population growth, can be found as the stable point attractor for a recursion defined by:

$$e_1 = 2; e_n = \left(1 + \frac{1}{n}\right)^n$$

Because the sequence approaches the point attractor very slowly, find the following terms of the sequence:

$$e_1, e_{10}, e_{100}, e_{1000}, e_{10000}, e_{100000}$$

The value of e is 2.71828182845. . .
How does it compare to the point attractor?

12.2 GEOMETRIC RECURSION

A "superball" is dropped from a height of 20 feet onto a hard, smooth floor. The ball is made of a special material that enables it to rebound 98% of the distance it falls. That is, if it falls 100 feet, it will rebound 98 feet. What is the total vertical distance traveled by the ball as it bounces up and down until it comes to rest on the floor?

In Section 12.1 we applied recursion to generate sequences of numbers. We can also apply recursion to geometric objects and to natural and physical processes; this is often called feedback. As recursion is applied to a geometric object, the problems of finding length, area, and volume take on new meanings. Even the dimension of geometric objects can be looked at differently. The visual effects of these recursions are often attractive (even stunning) to look at. But more important, there is an opportunity to gain some new mathematical insight, which we can apply to our study and understanding of the world.

As a first example consider the following sequence of geometric figures:

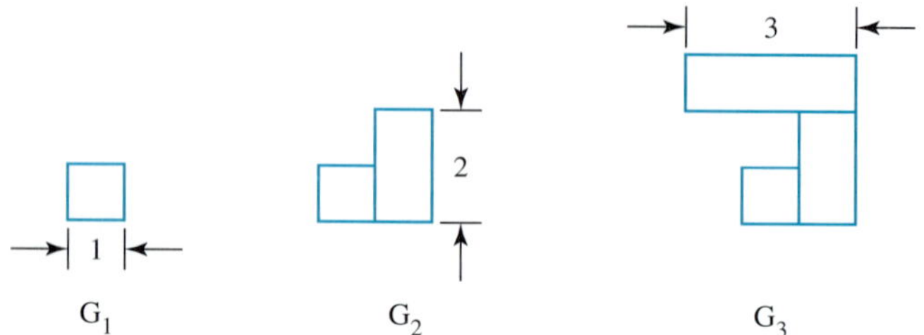

FIGURE 12.12

G_1 is a square of side 1. We obtain G_2 from G_1 by adding a rectangle of side lengths 1 and 2. In general, we obtain G_n from G_{n-1} by adding a rectangle of side lengths 1 and n, but the description of how to add that rectangle is more clearly explained by Figure 12.12 than by a description in words.

EXAMPLE 12.5 Construct the next three figures in the recursion begun in Figure 12.12.

SOLUTION The next three figures are shown in Figure 12.13.

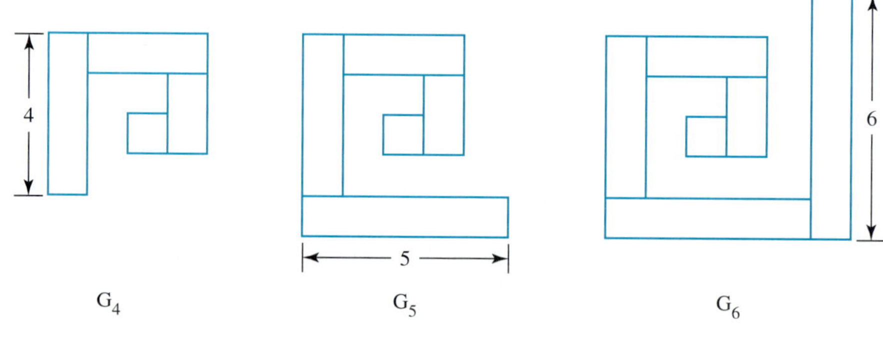

FIGURE 12.13

◆

You should also be able to imagine the infinitely large spiral using this recursive construction. Of course, we cannot draw the completed figure. This process of building a figure step by step by repeating some rule is called **geometric recursion.**

GEOMETRIC SERIES

Consider the geometric figure H_1 shown in Figure 12.14.

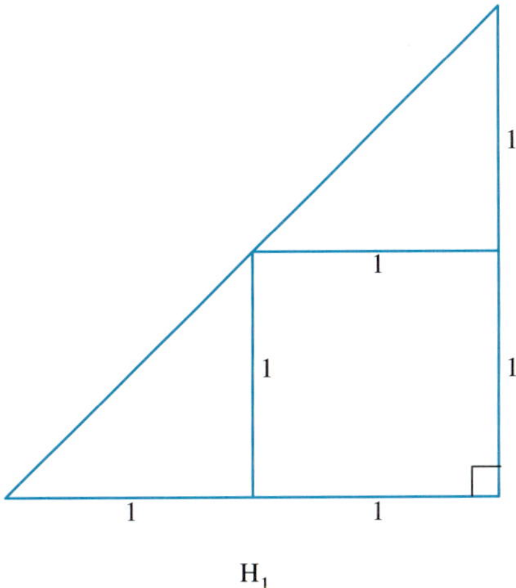

FIGURE 12.14

Notice that H_1 is a right triangle with equal sides of length 2 that has been subdivided into two similar triangles and a square. An entire copy of H_1 can be scaled by a factor of $\frac{1}{2}$ and placed in the top subtriangle. This will give us H_2 shown in Figure 12.15.

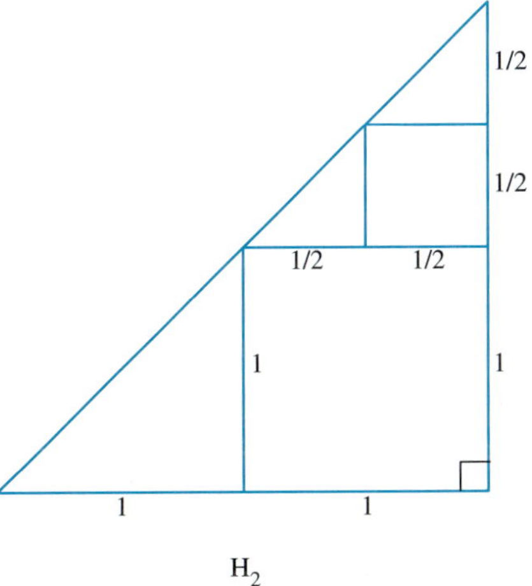

FIGURE 12.15

Once H_2 has been constructed, another copy of H_1, this time scaled by a factor of $\frac{1}{4}$, can be placed in the top-most triangle to give us a figure H_3. This defines a geometric recursion in which successive figures are more and more complicated in the top right corner. But we want to go one step farther and imagine the construction of the entire sequence of figures has been completed, giving us a figure of infinite complexity in the top right corner as indicated in Figure 12.16. This final figure plays the role that the point attractor did for numerical recursion: you cannot get there in a finite number of steps, but that is where the recursion is going.

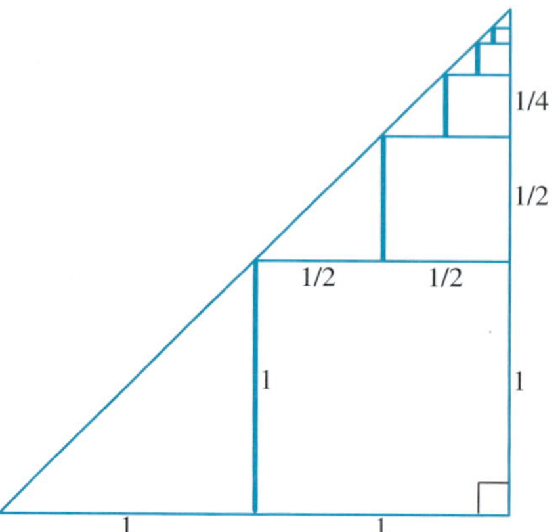

FIGURE 12.16

Figure 12.16 is more than just a curiosity, because if we look at the infinitely many vertical edges that have been highlighted, we see that their lengths must all add up to the length of the side of the figure, that is, to 2. But the lengths of those highlighted edges are known: they are $1, \frac{1}{2}, \frac{1}{4}, \frac{1}{8} \ldots$ We conclude that

$$1 + \frac{1}{2} + \frac{1}{4} + \frac{1}{8} + \ldots = 2.$$

The infinite sum on the left is called a geometric series.

DEFINITION

Geometric Series

A **geometric series** is any infinite sum where each summand is obtained from the last by multiplying by a fixed number called the **ratio.**

In the geometric series

$$1 + \frac{1}{2} + \frac{1}{4} + \frac{1}{8} + \ldots$$

the ratio is $\frac{1}{2}$.

Fortunately, there is a general formula for the sum of a geometric series, which is given next.

THEOREM

Sum of a Geometric Series

For any number r with $-1 < r < 1$,

$$A + (A \times r) + (A \times r^2) + (A \times r^3) + (A \times r^4) + \ldots = \frac{A}{1-r}.$$

EXAMPLE 12.6 Find the sum of the geometric series

$$1 - \frac{1}{2} + \frac{1}{4} - \frac{1}{8} + \ldots.$$

SOLUTION The ratio for this series is $-\frac{1}{2}$. Using the formula for the sum of a geometric series with $A = 1$ and $r = -\frac{1}{2}$, we have

$$1 + 1\left(-\frac{1}{2}\right) + 1\left(-\frac{1}{2}\right)^2 + 1\left(-\frac{1}{2}\right)^3 + \ldots = 1 - \frac{1}{2} + \frac{1}{4} - \frac{1}{8} + \ldots = \frac{1}{1-\left(-\frac{1}{2}\right)} =$$

$$\frac{1}{1+\frac{1}{2}} = \frac{1}{\frac{3}{2}} = \frac{2}{3}.$$ ◆

THE VON KOCH CURVE

The von Koch curve is a complex curve constructed in stages using the following special type of geometric recursion: the construction is based on a beginning curve called the **initiator,** and a second curve, called the **generator.** In constructing the next stage, each part of the curve that is similar to the initiator is replaced by a scaled copy of the generator.

For the von Koch curve, the initiator is a line segment and the generator is a line segment that has been divided into three equal sections, with the middle section replaced by the top of an equilateral triangle (Figure 12.17).

FIGURE 12.17

Note that if the initiator is one unit in length, then the generator consists of four line segments each $\frac{1}{3}$ unit in length. At each stage in the process smaller and smaller copies of the initiator are replaced by smaller and smaller copies of the generator. The basic form is repeated, over and over, producing figures with similar, but finer, details. The first four stages of the van Koch curve are shown in Figure 12.18.

The **von Koch curve** results when this process is carried out indefinitely. In the geometric recursions that we used to study the geometric series, the total length of

von Koch Curve with Four Stages

FIGURE 12.18

the various segments was finite. The behavior in the von Koch curve is completely different as the next example shows.

EXAMPLE 12.7 What is the sum of the lengths of the line segments at each stage in the development of the von Koch curve?

SOLUTION If the original line is one unit in length, then each of the four line segments produced in the first stage will be $\frac{1}{3}$ unit long, and they will have a combined length of $\frac{4}{3}$ units. In the next stage we replace each of the four line segments by a copy of the generator, so we have 16 line segments. Each of the copies of the generator is scaled down by a factor of $\frac{1}{3}$, so the line segments are $\frac{1}{9}$ units in length. The combined length is $\frac{16}{9}$ units. Continuing in this way, the number of line segments is quadrupled each time, but the new segments are only $\frac{1}{3}$ the previous length. We summarize these values in Table 12.3.

TABLE 12.3

Stage	Number of line segments	Length of each segment	Total Length
0	1	1	1
1	4	$\frac{1}{3}$	$\frac{4}{3}$
2	16	$\frac{1}{9}$	$\frac{16}{9}$
3	64	$\frac{1}{27}$	$\frac{64}{27}$
4	256	$\frac{1}{81}$	$\frac{256}{81}$
.	.	.	.
n	$4n$	$\left(\frac{1}{3}\right)^n$	$\left(\frac{4}{3}\right)^n$

At the 10th stage of the von Koch curve, the total length is about 18 units. At the 20th stage the total length is about 315 units. At the 30th stage the total length is about 5600 and increasing rapidly. By the 100th stage the total length is in the trillions. ◆

As we can see from these numbers, something strange is happening to the length. At each iteration, the distance, as the crow flies, from one endpoint of the von Koch curve to the other endpoint is one unit. However, the distance along the path, as the ant crawls, is increasing rapidly. The von Koch curve is infinitely jagged everywhere and has an infinite length.

THE SIERPINSKI GASKET AND CARPET

In 1916, the Polish mathematician Waclaw Sierpinski introduced two geometric figures that have come to be known as the Sierpinski Gasket and the Sierpinski Carpet. The basic concepts in their construction are comparable.

The starting figure for the Sierpinski Gasket is the triangle, S_1, shown in Figure 12.19(a). At each stage of the construction a collection of similar triangles is obtained by dividing each triangle from the previous stage into four congruent triangles and deleting the interior of the middle triangle (or, what is the same, keeping the three corner triangles). The second stage of the construction, S_2, is shown in Figure 12.19(b).

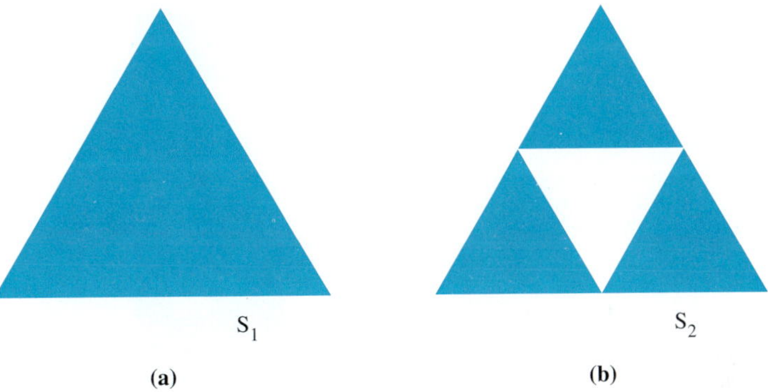

FIGURE 12.19

The next two stages of the construction are shown in Figure 12.20.

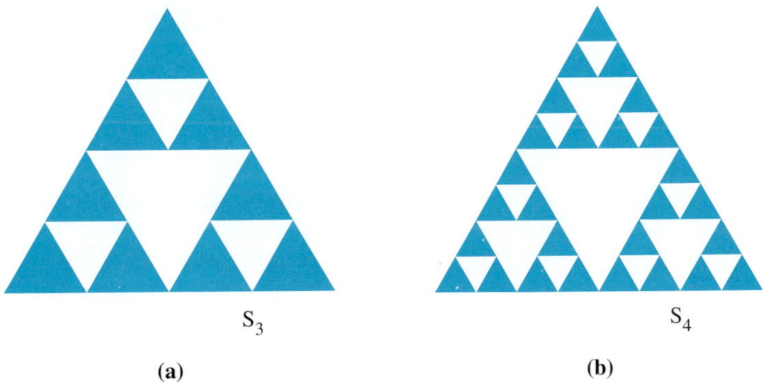

FIGURE 12.20

The **Sierpinski Gasket** is the figure left after this infinite process is completed.

The Sierpinski Carpet is constructed by beginning with a unit square. To obtain the second stage, divide the square into nine congruent squares and remove the interior of the middle square. At each successive stage, divide each of the remaining squares into nine congruent squares and remove the interior of the middle square. The **Sierpinski Carpet** is what is left after the infinite construction is completed.

To illustrate this, begin with a unit square, C_1, in the plane [Figure 12.21(a)]. Each side of the square is subdivided into three equal parts that subdivide the square into nine smaller squares of equal size. The interior of the middle square is then dropped out, leaving eight squares of side length $\frac{1}{3}$ forming the second stage C_2 [Figure 12.21(b)]. The area removed is the area of a square of side length $\frac{1}{3}$ or $\frac{1}{3} \times \frac{1}{3} = \frac{1}{3^2} = \frac{1}{9}$.

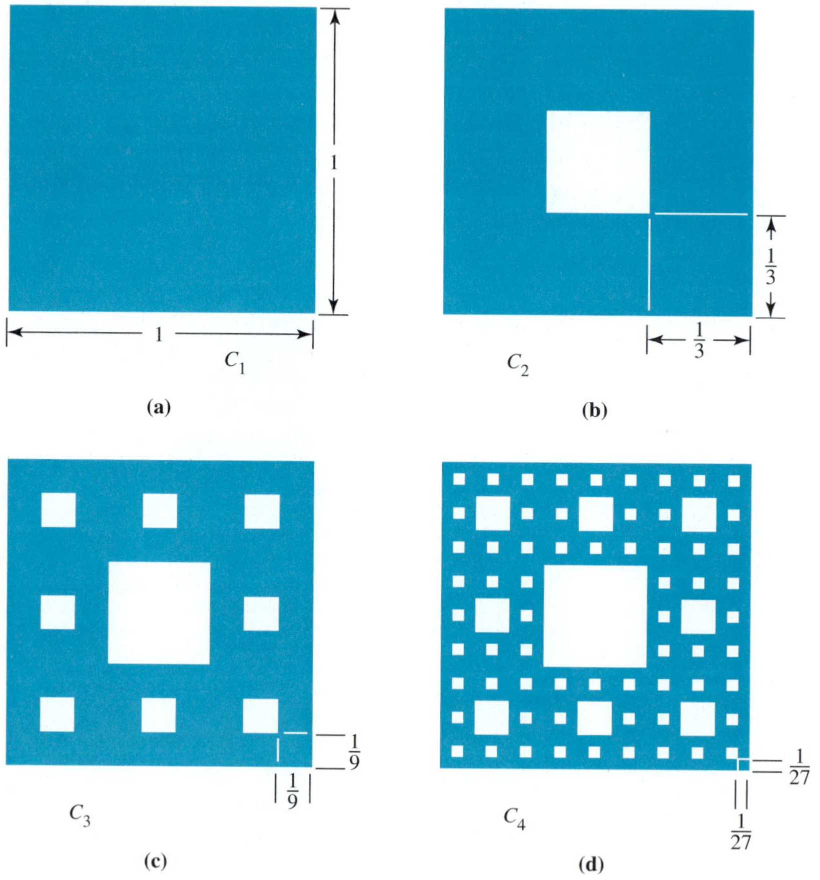

FIGURE 12.21 The first four stages in constructing the Sierpinski Carpet

To construct the third stage, each of the eight squares forming C_2 is subdivided into nine squares of side length $\frac{1}{3} \times \frac{1}{3} = \frac{1}{3^2}$, and the interior of the middle square is dropped out. That leaves $8 \times 8 = 8^2$ squares of side length $\frac{1}{3^2}$ forming C_3 [Figure 12.21(c)]. The area removed is eight times the area of a square of side length $\frac{1}{3^2}$, or $8 \times \frac{1}{3^2} \times \frac{1}{3^2} = \frac{8}{9^2}$.

To construct the fourth stage, each of the 8^2 squares forming C_3 is subdivided into nine squares of side length $\frac{1}{3} \times \frac{1}{3^2} = \frac{1}{3^3}$ and the interior of the middle square is dropped out. That leaves $8 \times 8^2 = 8^3$ squares of side length $\frac{1}{3^3}$ forming C_4 [Figure 12.21(d)]. The area removed is 8^2 times the area of a square of side length $\frac{1}{3^3}$ or $8^2 \times \frac{1}{3^3} \times \frac{1}{3^3} = \frac{8^2}{9^3}$.

The general pattern is that Cn is formed by 8^{n-1} squares of side length $\frac{1}{3^{n-1}}$. When C_{n+1} is constructed, 8^{n-1} middle squares of side length $\frac{1}{3} \times \frac{1}{3^{n-1}} = \frac{1}{3^n}$ are dropped out removing an area of

$$8^{n-1} \times \frac{1}{3^n} \times \frac{1}{3^n} = \frac{8^{n-1}}{9^n}.$$

The total area removed is thus

$$\frac{1}{9} + \frac{8}{9^2} + \frac{8^2}{9^3} + \dots$$

$$= \frac{1}{9} + \frac{1}{9}\left(\frac{8}{9}\right) + \frac{1}{9}\left(\frac{8^2}{9^2}\right) + \dots .$$

We see that this is a geometric series in which $A = \frac{1}{9}$ and $r = \frac{8}{9.}$ Using the formula for the sum of a geometric series, the total area removed is

$$\frac{\frac{1}{9}}{1 - \frac{8}{9}} = \frac{\frac{1}{9}}{\frac{1}{9}} = 1.$$

This somewhat surprising result means that all the area of the square is removed in the construction, but there is still something left since all the edges of the squares are in the final set! In fact, in the next section we will see that what is left is substantial enough to have dimension greater than 1.

THE CHAOS GAME

The construction of the Sierpinski Gasket followed a precise recursive rule. Each stage was completely determined by the preceding stage. Now we will give another construction of the Sierpinski Gasket, but this time the Gasket arises from a series of random choices. At the heart of the emerging field of **chaos theory** is the detection of regularity in a seemingly chaotic setting, and the following construction of the Sierpinski Gasket, called the **Chaos Game,** illustrates this.

We begin with the vertices of a triangle with the vertices labeled A, B, and C (Figure 12.22). Assign two numbers on a fair die to each vertex, say 1 and 2 to A, 3 and 4 to B, and 5 and 6 to C. Mark your initial position with a point on the segment from A to B (it can actually be anywhere on the plane). You are ready to begin. Roll the die and move your position halfway along an imaginary line from where it is now to the vertex corresponding to the roll of the die. Place a point at this location. Repeat this process over and over.

For example, if the first four rolls of the die come up 6, 2, 3, 5, mark the points as in Figure 12.22.

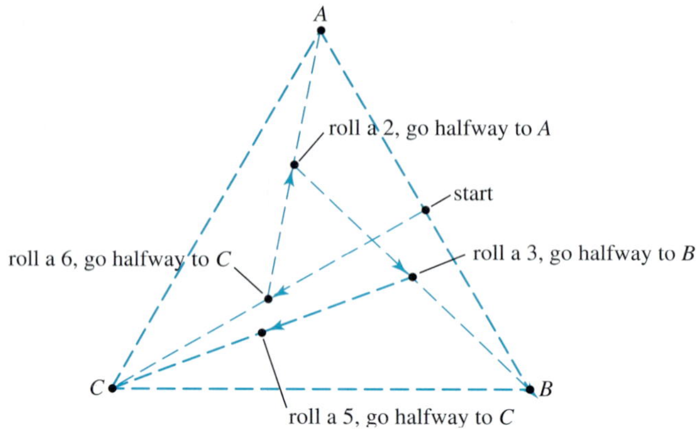

FIGURE 12.22

No pattern is discernible with just four rolls of the die. At 100 rolls a faint suspicion may be roused [Figure 12.23(a)], and at 1000 rolls of the die the Sierpinski Gasket begins to emerge [Figure 12.23(b)].

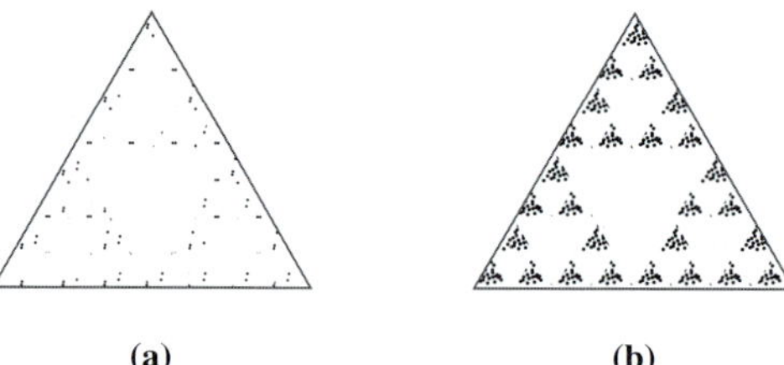

FIGURE 12.23

The best way to play the chaos game is to program a computer to do the work, using a random number generator to simulate the roll of the die.

INITIAL PROBLEM SOLUTION

A "superball" is dropped from a height of 20 feet onto a hard, smooth floor. The ball is made of a special material that enables it to rebound 98% of the distance it falls. That is, if it falls 100 feet, it will rebound 98 feet. What is the total vertical distance traveled by the ball as it bounces up and down until it comes to rest on the floor?

SOLUTION We can separate the total distance traveled into distance traveled downward and distance traveled upward and make a table showing what happens with each bounce.

Downward	**Upward**
20	0.98×20
0.98×20	$0.98 \times (0.98 \times 20) = (0.98)^2 \times 20$
$(0.98)^2 \times 20$	$0.98 \times (0.98)^2 \times 20 = (0.98)^3 \times 20$
$(0.98)^3 \times 20$	$0.98 \times (0.98)^3 \times 20 = (0.98)^4 \times 20$
etc.	etc.

The distance traveled downward is the sum of infinitely many terms:
Downward distance $= 20 + 0.98 \times 20 + (0.98)^2 \times 20 + (0.98)^3 \times 20 + \ldots$
This is a geometric series with $A = 20$ and $r = 0.98$. Since $r < 1$, the series converges to a sum. Using the formula for the sum of a geometric series, the distance downward is $\frac{A}{1-r} = \frac{20}{1-0.98} = \frac{20}{0.02} = 1000$ feet. Comparing the columns in the table, we see that the distance traveled upward would be 20 feet less, or 980 feet. Therefore, the total distance traveled by the superball is 1980 feet.

PROBLEM SET 12.2

In problems 1 through 8, write out the first five terms of an infinite geometric series with the given first term and common ratio. Express the terms as either fractions or decimals rounded to three places, whichever is appropriate.

1. $A = 20, r = \frac{1}{2}$

2. $A = 20, r = -\frac{1}{2}$

3. $A = 16, r = \frac{3}{2}$

4. $A = 100, r = \frac{3}{4}$

5. $A = 1, r = -0.6$

6. $A = 45, r = 0.9$

7. $A = 2000, r = 0.15$

8. $A = 25, r = 0.75$

In problems 9 through 12, determine the value of the ratio in the given infinite geometric series and write the next 2 terms of the series.

9. $10 + 8 + 6.4 + \ldots$

10. $200 + 120 + 72 + \ldots$

11. $80 - 40 + 20 \ldots$

12. $2 - 3 + 4.5 \ldots$

In problems 13 through 20, find the sum of the infinite geometric series. If you are unable to find the sum, explain why.

13. $4 + 2 + 1 + \ldots$

14. $20 + 10 + 5 + \ldots$

15. $8 - 4 + 2 - \ldots$

16. $20 - 12 + 7.2 - \ldots$

17. $4 + \frac{8}{3} + \frac{16}{9} + \ldots$

18. $6 - 4 + \frac{8}{3} - \ldots$

19. $1 + \frac{4}{3} + \frac{16}{9} + \ldots$

20. $81 + 108 + 144 + \ldots$

The Cantor Set is produced by a recursion that begins with a single line segment of a finite length; the line is divided into three equal lengths, and the middle third of the line is removed. This process is then repeated indefinitely. The first stage of the development of the set is shown.

21. Construct a table similar to Table 12.3 that shows the number of segments, the length of each segment, and the combined lengths of the first four stages of the Cantor Set.

22. Referring to problem 21, what is the rule for the sum of the lengths in the nth stage of the Cantor Set? What happens as the stages are continued indefinitely?

23. Construct the first three stages of a Cantor-like set in which you begin with a line segment, divide the segment into four equal parts, dropping the middle two. The first stage is shown; continue for two more stages.

24. Construct a table similar to Table 12.3 that shows the number of segments, the length of each segment, and the combined lengths of the first four stages of the set in problem 23.

In problems 25 and 27 you are to construct the next stage for curves similar to the von Koch curve. The initiator and generator are given for each curve.

25.

26. Construct a table similar to Table 12.3 that shows the number of segments, the length of each segment, and the combined lengths of the first four stages of the fractal curve in problem 25. The first two lines of the table are given; find the next three lines.

Stage	Number of segments	Length of segments	Total length
0	1	1	1
1	5	$\frac{1}{3}$	$\frac{5}{3}$

Find the general rule for the nth stage.

27.

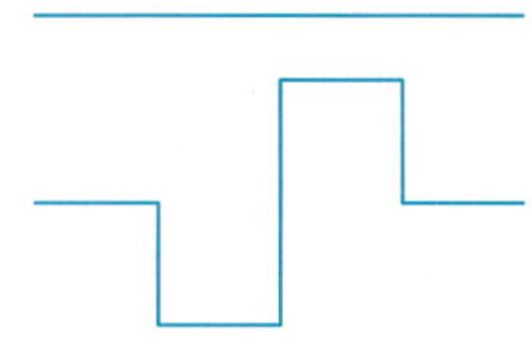

28. Construct a table similar to Table 12.3 that shows the number of segments, the length of each segment, and the combined lengths of the first four stages of the fractal curve in problem 27. The first two lines of the table are given; find the next three lines.

Stage	Number of segments	Length of segments	Total length
0	1	1	1
1	8	$\frac{1}{4}$	2

Find the general rule for the nth stage.

EXTENDED PROBLEMS

29. Construct a closed curve version of the von Koch curve (known as the von Koch snowflake or island) as follows:

(a) Begin with an equilateral triangle as the initiator. The generator for the snowflake is the same for the von Koch curve and is applied to each line segment of the snowflake.

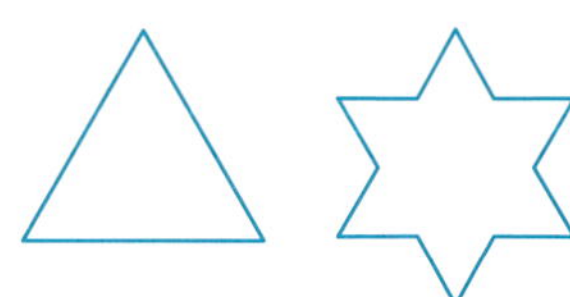

(b) In each subsequent stage, each line segment of the figure is replaced by a scaled down copy of the generator. Construct the next stage.

30. Construct a table similar to Table 12.3 that shows the number of segments, the length of each segment, and the combined lengths in the first four stages for the border of the von Koch snowflake. What is the general rule for the nth stage?

In Figures 12.19 and 12.20 the Sierpinski Gasket is constructed by removing portions of the figure at each stage. The relationship of the scaling of length and area is covered in section 11.2. Briefly, if two figures have the same shape, such as two squares or similar triangles, then the ratio of their areas is the square of the ratio of the lengths of corresponding sides. That is, if the length of the side of one square is $\frac{1}{3}$ that of the other, then its area is $\frac{1}{9}$ that of the first. If the lengths of sides in one triangle are $\frac{1}{2}$ those of another, then the area of the first triangle is $\frac{1}{4}$ that of the other. Use this relationship in the next several problems.

31. Using Example 12.7 as a model, calculate the area that remains at each stage in the construction of the Sierpinski Gasket. The initial triangle has an area of one square unit. The first two lines of a table are given.

Stage	Number of triangles	Area of triangles	Total Area
0	1	1	1
1	3	$\frac{1}{4}$	$\frac{3}{4}$

Find the general rule for the nth stage.

32. Referring to Example 12.7 and problem 31, find the area that is removed in the construction of the Sierpinski Gasket. Does your result agree with the result in problem 31? Explain.

33. While the Sierpinski Carpet and Gasket are formed by the subtraction of regions, the von Koch snowflake in problem 29 is formed by the addition of regions. To find the area of the von Koch snowflake, we will construct a table similar to the one for the Sierpinski Gasket, except we will concentrate on the area of the new triangles at each stage.

Stage	Number of new triangles	Area of each new triangle	Total new area
0	1	1	1
1	3	$\frac{1}{9}$	$\frac{1}{3}$

The total area is $1 + \frac{1}{3} = \frac{4}{3}$

Find the general rule for the total area for the nth stage. Find the area of the von Koch snowflake, if the area of the original triangle is one square unit.

34. Construct a closed curve version of the curve in problem 25.

(a) Begin with a square as the initiator. The generator is applied to each side of the square, with the bump on the outside of the square.

(b) In each subsequent stage, each line segment of the figure is replaced by a scaled down copy of the generator. Construct the next stage.

35. Construct a table similar to Table 12.3 that shows the number of segments, the length of each segment, and the combined lengths in the first four stages of the border of the figure in problem 34. What is the general rule for the *n*th stage?

36. Repeat problem 34 with the following changes.

(a) Use the same initiator and generator. However, this time reverse the direction of the generator by having the bump on the inside of the square.

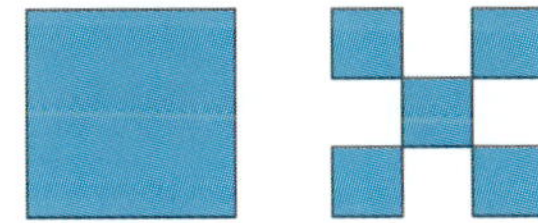

(b) Construct the next stage of this figure.

37. Construct a table similar to Table 12.3 that shows the number of segments, the length of each segment, and the combined lengths of the first four stages of curve in problem 36. What is the general rule for the *n*th stage?

38. Construct a closed curve version of the curve in problem 27.

(a) Begin with a square as the initiator. The generator is applied to each side of the square.

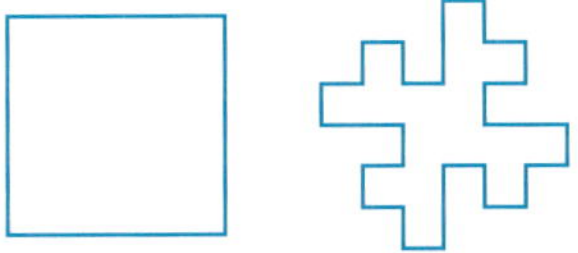

(b) In each subsequent stage, each line segment of the figure is replaced by a scaled down copy of the generator. Construct the next stage.

39. Construct a table similar to Table 12.3 that shows the number of segments, the length of each segment, and the combined lengths of the first four stages of the curve in problem 38. What is the general rule for the *n*th stage?

12.3 FRACTALS

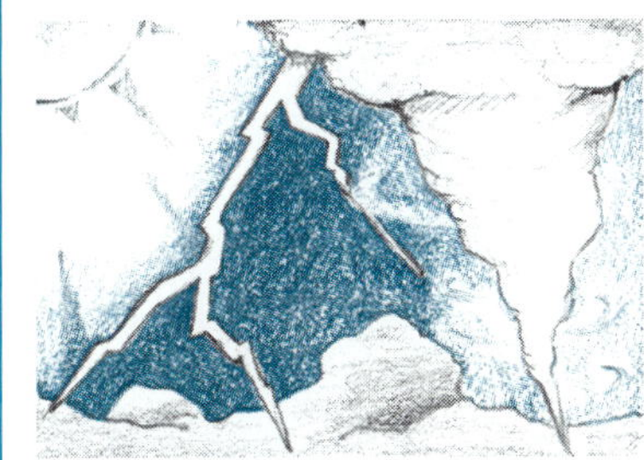

Will new mathematical techniques and discoveries in fractal geometry and chaos theory lead to more accurate and dependable models of phenomena such as weather?

In the last two sections we introduced the topics of recursion and investigated some unusual geometric forms. Although recursion is a topic that has been in the mathematical mainstream since the times of the earliest mathematicians, fractals are a fairly recent addition to the mathematical scene. As recently as 1900, the leading mathematicians referred to such shapes as mathematical monsters, because they were so unusual and difficult to comprehend. Now, personal computers let even school-age children experiment and discover new shapes and forms that are the result of what is commonly known as feedback in science. As we approach the year 2000, these former monsters may be the key to unlocking some of the mysteries of our physically turbulent world.

MATHEMATICS AS A MODEL FOR THE WORLD

The earliest well-developed mathematics that has been carried forward to the present is known as euclidean geometry—the classical geometry of the Greeks. Indeed, until the last decade, the geometry taught in high schools differed only slightly from that which was taught for the previous thousand years. The main characteristics of this geometry were its foundation on logic and its application to shapes and forms. Beginning even before the Greeks, geometry has been used to measure and describe the world in which we live. We came to believe that much of the world could be described in terms of the basic shapes of geometry.

Euclidean geometry deals with idealized shapes—straight lines, parabolas, circles, spheres, squares, cubes, etc.—and has proved itself to be very useful in many practical areas such as navigation, surveying, and architecture, to name but a few. Sometimes, however, the perfection of euclidean geometry has blinded people to the possibility of alternative models. Early astronomers attempted to explain the universe in terms of the perfect euclidean figure, the circle. The earliest models of the universe had the earth at the center, with the sun and other planets traveling around the earth in perfect circular paths, all surrounded by the stars imbedded in perfect crystalline spheres. Later, scientists developed more elaborate schemes to explain additional findings and observations (Figure 12.24).

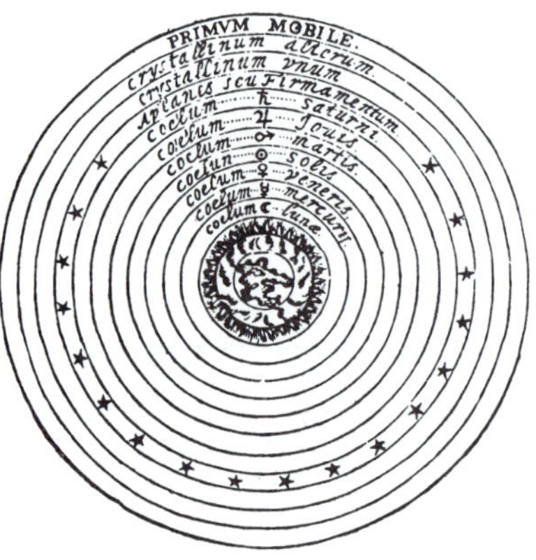

The world system of Ptolemy was centered on the earth. This version of the Ptolemaic system was included in a book published in 1647.

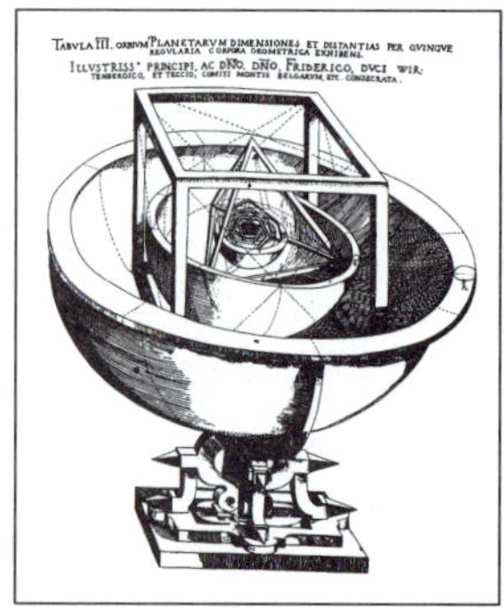

Kepler's Cosmic Mystery, the spheres of the six planets nested in the five perfect solids of Pythagorus and Plato.

FIGURE 12.24

It took a revolution in thought to conceive of the solar system having the sun at the center, with the earth and the other planets revolving around the sun.

Developments in other areas of mathematics together with increased powers of observation (the telescope and other tools) had a great impact on geometry and its role in describing the world. Of greatest importance was the development of coordinate geometry by René Descartes, and the application of algebra to geometry. Equations could describe the paths of moving objects and provide numerical models for natural phenomena such as the arc of a projectile or the orbit of a planet. Formulas provided answers regarding space and distance.

In the arts, the development of perspective geometry changed our point of view and provided new tools for artists for their depiction of the world. Figures

TIDBIT

"Now, as Mandelbrot points out . . . nature has played a trick on the mathematicians. The 19th century mathematicians may have been lacking in imagination, but nature was not. The same pathological structures that the mathematicians invented to break loose from 19th century naturalism turn out to be inherent in familiar objects all around us in nature."
Freeman Dyson, Physicist

12.25 and 12.26 show the change in realism afforded by the introduction of perspective geometry in the early fifteenth century.

FIGURE 12.25

FIGURE 12.26

In the 17th century, Isaac Newton and others developed the mathematics of calculus that led to the concept of the clockwork universe—a concept that everything fit into a mathematical framework. It was a system characterized by determinism.

Determinism refers to the philosophical position that if we have adequate information about the initial conditions surrounding an object, such as a planet, and an adequate model for its movement, then we can correctly predict how the object will behave in the future. The success experienced by Newton and others in predicting the reoccurrence of comets and the position of planets (even the existence of planets yet to be observed) reinforced these ideas.

The 20th century has seen an erosion of the solid positions held by determinism and classical geometry. The development of special and general relativity in physics was motivated, in part, by non-euclidean geometry, and the development of quantum mechanics has undermined any certainty that the world is deterministic.

The role for mathematics in describing nature and natural forms has also been questioned. Although shapes such as circles, spheres, triangles, and cones provide a framework for observing natural forms, most natural objects such as clouds, mountains, lightning, coastlines, and trees cannot be adequately described with these shapes. Classical geometry was incapable of dealing with these complex shapes, structures, and phenomena (Figure 12.27).

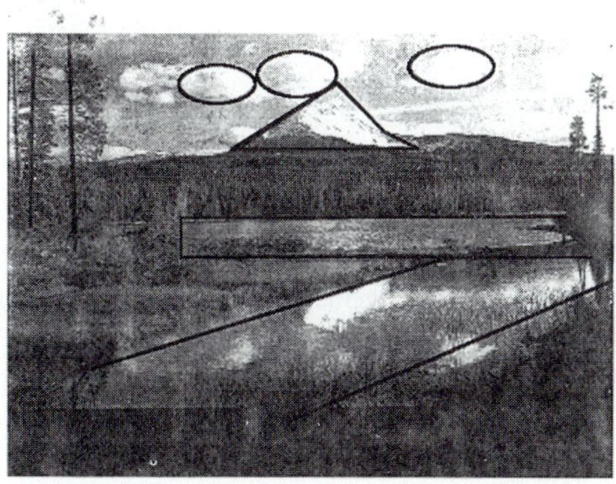

FIGURE 12.27 Two Models of Nature: The euclidean idealized world and the cartesian world of coordinates. In the euclidean world, the emphasis was on shape and form. In the cartesian world the emphasis was on location and distance.

In this section we will discuss some of the mathematical constructions that are being introduced to deal with the natural forms and phenomena that have been out of reach of classical geometry and deterministic models.

FRACTALS

The term **fractal** (derived from the Latin adjective *fracus*) was introduced by Benoit Mandelbrot (b. 1924) to refer to objects with a broken and irregular appearance. Mandelbrot's definition of a fractal is too technical for this book, but any set with a fractional dimension must be a fractal. It may seem impossible for any object to have a fractional dimension, but the following example may illustrate the idea. If an object is one-dimensional and you scale it up by a factor of 2, then the total length increases by a factor of 2^1, whereas if you scale it up by a factor of 3, the total length increases by a factor of 3^1. If an object is two-dimensional and you scale it up in all directions by a factor of 2, then the area is increased by a factor of 2^2. If a two-dimensional object is scaled up in all directions by a factor of 3, then the area is increased by a factor of 3^2. Thus, the **dimension** of the object corresponds to the exponent; for a fractal this is called its **fractal dimension.**

In this way of thinking, if an object is k-dimensional and is scaled by a factor of n, then n^k scaled copies of the object will be produced.

Consider the Sierpinski Gasket in Figure 12.28(a). The Gasket is scaled up by a factor of 2 in Figure 12.28(b).

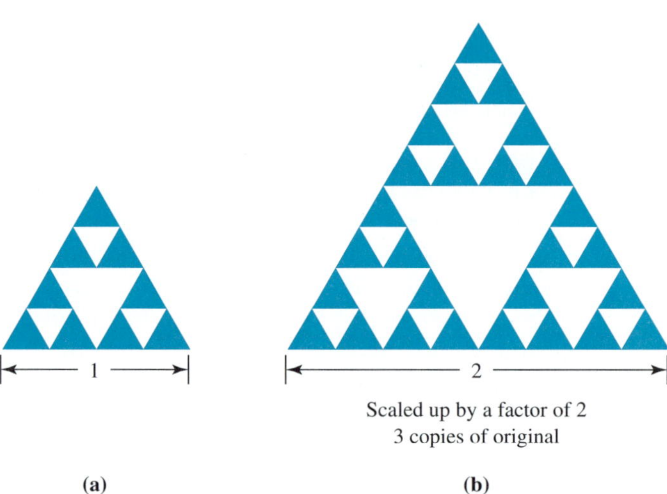

FIGURE 12.28

Notice that the bigger Gasket is not made from $2^1 = 2$ copies of the old Gasket, which would mean it was one-dimensional. Nor is it made from $2^2 = 4$ copies of the old Gasket, which would mean it was two-dimensional. Instead, the bigger Gasket contains three copies of the old Gasket.

Using our working definition of dimension, we are looking for a value k, such that

$$2^k = 3.$$

Since

$$2^{1.58} \approx 3$$

we conclude that the dimension of the Sierpinski Gasket is approximately 1.58.

The Sierpinski Gasket is a fractal. The Sierpinski Carpet and the von Koch curve are also fractals. Because they are constructed by recursions involving removing or replacing pieces that are similar, each of these figures are self-similar. This self-similarity is illustrated in Figure 12.29 for the von Koch curve.

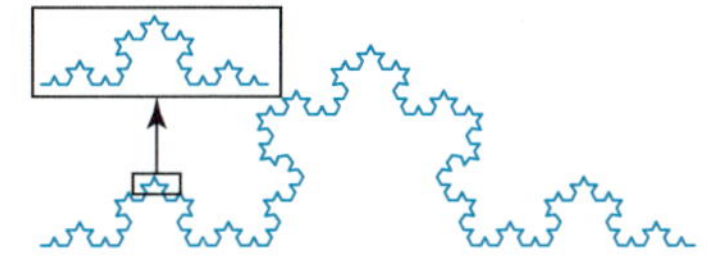

Self-similarity in the later stages of the von Koch curve

FIGURE 12.29

Self-similarity can often be used to estimate the dimension of a recursively constructed fractal as was done for the Sierpinski Gasket.

EXAMPLE 12.8 Use self-similarity to estimate the dimension of the von Koch curve.

SOLUTION In Figure 12.30 we have scaled up a von Koch curve by a factor of 3, but the bigger curve is four copies of the smaller curve.

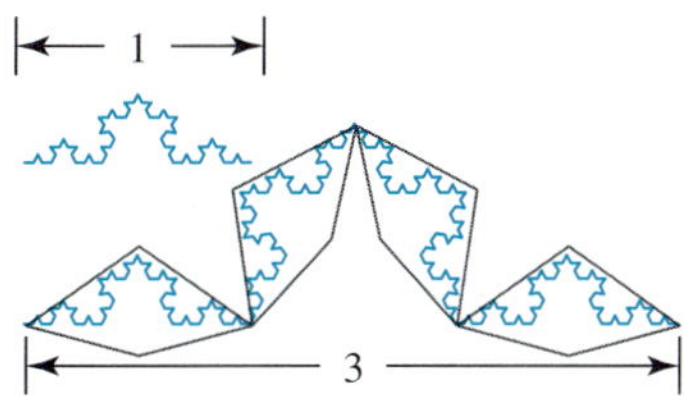

FIGURE 12.30

The dimension of the curve is the number x that satisfies the equation

$$3^x = 4.$$

Since $3^1 = 3 < 4$ and $3^2 = 9 > 4$, the value of x is between 1 and 2. We average our two guesses of 1 and 2, to get 1.5. With a calculator we find

$$3^{1.5} \approx 5.196152 > 4,$$

so 1.5 is also larger than x. We replace our high guess of 2 by our new high guess of 1.5, and again average our high and low guesses. The process can be continued until we agree that the results are close enough to accept.

Low Guess	**High Guess**	**Average**	3^{average}
1	2	1.5	5.196152 (high)
1	1.5	1.25	3.948222 (low)
1.25	1.5	1.375	4.529411 (high)
1.25	1.375	1.3125	4.228844 (high)
1.25	1.3125	1.28125	4.086125 (high)
1.25	1.28125	1.265625	4.016582

We will consider the bottom right value, 4.016582, to be close enough to 4, and give the dimension of the von Koch curve as approximately 1.27. ◆

When a fractal does not have self-similarity, it is much more difficult to compute the dimension. The precise definitions are again beyond the scope of this book.

FRACTALS IN NATURE

The fractals discussed above were first created to investigate mathematical questions of interest primarily to mathematicians. This theoretical work has been going on for a century and we expect it will continue for decades more, if not centuries, as scientists and mathematicians use fractal geometry to gain increased knowledge and insights in their disciplines. By contrast, the current popular interest in fractals was motivated by the beauty of the objects themselves and the surprising fact that fractals occur so frequently in nature.

There is even a natural way to assign a fractal dimension to many shapes and surfaces such as clouds, tree bark, and mountains. NASA photographs, for example, indicate that the surface of the earth has a fractal dimension of 2.1, compared with 2.4 for Mars.

TIDBIT

One English encyclopedia lists the coastline of Britain as about 4500 miles, while another lists it as 5000 miles

1. COASTLINES

A good example of a naturally occurring fractal is the coastline of Britain. You would probably expect that the coastline of Britain has some well-defined length. The length is usually computed by taking a map of Britain and measuring a path around the coast using dividers (or a compass or a ruler). To get a more precise estimate of the length of the coastline you use a bigger map and smaller dividers.

In 1967 Mandelbrot seriously investigated the question "How long is the coast of Britain?" and he discovered that the answer depends on the size of the map and the dividers used to do the measurement (Figure 12.31). It is possible to determine the fractal dimension from the way the estimate of the total length changes with the size of the map and dividers.

Length of Ruler	Approximate Coastline
200 miles	1600 miles
60 miles	2280 miles
33 miles	2550 miles
10 miles	5184 miles

FIGURE 12.31

If you look at a map, you will see that the coastline of Britain is jagged, and it is jagged at all scales, much like the von Koch curve. Such jaggedness at many scales is one of the indications that you have found a natural fractal. Finally, based on empirical work of L. F. Richardson, Mandelbrot was able to estimate the fractal dimension of the coast of Britain to be about 1.5.

TIDBIT

Big whorls have little whorls
Which feed on their velocity,
And little whorls have lesser whorls
And so on to viscosity.
—L. F. Richardson

TIDBIT

"Fractal geometry will make you see everything differently. . . . You risk the loss of your childhood vision of clouds, forests, galaxies, leaves, feathers, rocks, mountains, torrents of water, carpets, bricks, and much else besides."—Michael Barnsley, *Fractals Everywhere*, 1988

2. RIVER SYSTEMS

The tributaries of a river form a complex structure like a tree. Typically this tree structure is so complex that it is a fractal with a dimension between 1 and 2. The dimension of actual tributary systems can be estimated from aerial photographs. From the fractal dimension of the tributary system it is possible to estimate the size of fluctuations in the river. Knowing the fluctuations in river flow is crucial to building a bridge over the river, and this approach is used in third world countries where long-term records are often not available.

3. FOREST FIRES

The boundary between the burned and unburned areas in a forest following a fire is typically a fractal. The dimension is between 1 and 2 and can be estimated from aerial photographs. Computer simulations of forest fires also produce these fractal boundaries between the burned and unburned areas, and the fractal dimension taken from the aerial photographs provides information about the performance and accuracy of the computer models.

4. TURBULENCE IN FLUIDS

When the speeds in a fluid flow become sufficiently large, the flow becomes irregular with eddies on many different scales. While empirical data on the dimension of turbulence is lacking, understanding turbulence has always been of great interest in science and engineering. Mandelbrot's thesis that turbulence is fractal was part of the inspiration for the mathematical work of Vladimir Sheffer who has investigated the maximum possible dimension of the turbulent set (which lies in the four dimensions of space and time).

5. DATA TRANSMISSION ERRORS

Mandelbrot's first experience with fractals was in a project at IBM concerned with error correction in computer data transmission, where the data is transmitted as 0's and 1's. The type of noise that causes errors in such data transmission is random by nature, but was known to appear in clusters. This meant there might be long periods with no errors followed by clusters of errors. However, the more closely the clusters were examined, the more complex the patterns of errors became. Mandelbrot's surprising discovery was that the fractal dimension of the times at which errors occur is larger than 0. The practical consequence of this was that no matter what error correcting method is used, it will sometimes be overwhelmed. Thus perfect error correction was an unattainable goal.

6. MOUNTAINS

One of the features we observed in the artificial fractals was self-similarity. Self-similarity also occurs in nature and indicates that the object is a natural fractal. Once an object is recognized as a natural fractal, it can be simulated by a man-made fractal. Figure 12.32 shows natural fractals in mountains and their simulation.

7. PLANTS

Living things also sometimes exhibit self-similarity which then lends itself to imitation by man-made fractals (Figure 12.33).

Real mountain scapes

Fractal mountain scapes

FIGURE 12.32

Fractal Fern Fractal oak tree

FIGURE 12.33

TIDBIT

A study in the 1970s looked at the influence of turbulence caused by huge trucks meeting on the highways. The idea was that passing trucks might create a wind vortex that could be the seed for a tornado. The data collected showed a greater incidence of tornadoes on weekdays, when truck traffic was heavier, than on weekends.

8. CHAOS

At least since the time of Isaac Newton, scientists have been striving for a complete understanding of the way the world works. The basic method employed to analyze a system was to look at the behavior of its smallest parts using the principle that small changes will produce small effects (the principle of **linearity**), then put the small parts back together to see the overall behavior.

As humankind's technical skill increased, more precise measuring devices were perfected, and the ability to measure the small parts of a system increased steadily. The technical ability to perform mathematical calculations remained a stumbling block until the development of the electronic digital computer in the 1940s. With the aid of the computer, it appeared we could finally understand the most important systems in Nature.

Of course, one of the most important systems is the weather. In 1961, the meteorologist Edward Lorenz discovered that it is not always true that a small change will produce a small effect. In fact, in Lorenz's computer models of the weather small changes typically produced large effects, a phenomenon now known as **non-linearity.** Sometimes non-linearity is called the **butterfly effect** because a very small change, say a butterfly flapping its wings on one side of the earth, when interacting with other influences can eventually result in a dramatic effect, say a storm on the other side of the earth.

Computer models of the weather like Lorenz studied are not very complicated as models go, but the repeated recursions produced a very complicated behavior. When we studied recursions in Section 1, we introduced the idea of a point attractor. In chaotic recursions there may appear to be no overall pattern, but in fact there will be an attractor. The attractor, however, will be a fractal.

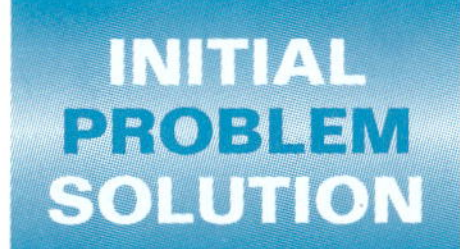

Will the new mathematical techniques and discoveries in fractal geometry and chaos theory lead to more accurate and dependable models of phenomena such as weather?

SOLUTION Work by mathematicians and scientists such as Mandelbrot and Lorenz has shown that many dynamical systems, including weather, display a sensitivity to initial conditions. This means that even slight changes in the numbers that are used in the model of the system can lead to dramatic changes in the results. Since all measurements are subject to some error, we can never get the initial conditions exact and can have little confidence in projections over any long period of time. The weather model used by Lorenz produced numbers in a three-dimensional space that have a complicated attractor. This attractor, known as the Lorenz attractor, has a fractal structure.

On the positive side, fractal geometry and chaos theory have already led to greater understanding of the physiological processes of growth and aging. Other scientists have successfully applied chaos theory to a wide range of topics such as insect populations, plant growth, chemical reaction, river turbulence, and many others, including the stock market.

PROBLEM SET 12.3

In problems 1 through 5, use self-similarity to estimate the dimension of the following fractals. All references are in Section 12.2.

1. Sierpinski Carpet (Figure 12.21)
2. The Cantor Set (problem 21)
3. The fractal curve described in problem 25.
4. The fractal curve described in problem 27.
5. The border of the von Koch snowflake (problem 29)

In problems 6 through 11 apply the concept of compass dimension illustrated with the problem of measuring the coastline of Britain to answer questions about borders of other geographic regions.

6. Which of the following states has a border with the higher compass dimension:
   (a) Florida or Alabama?
   (b) Nebraska or Illinois?
   (c) California or Nevada?
7. Name at least three states where the border has a compass dimension of 1.
   Hint: a straight line has a compass dimension of 1.
8. Using a map of the United States and a pair of dividers (a compass with two sharp points), find the approximate length of the border of the contiguous 48 states, as follows:
   (a) with the dividers set for 200 miles.
   (b) with the dividers set at 100 miles.
   (c) with the dividers set at 50 miles.
   Note: you should use a map that is at least 18 inches by 24 inches.

9. Repeat problem 8 for your home state, using a standard road map which is approximately 2 feet by 3 feet.
10. Using a state road map approximately 2 feet by 3 feet and a pair of dividers, which length will probably be estimated more accurately:
 (a) the border of Illinois, or
 (b) the border of Iowa
 Explain your reasoning.
11. Using a state road map approximately 2 feet by 3 feet and a pair of dividers, which length will probably be estimated more accurately:
 (a) the border of Texas, or
 (b) the border of Pennsylvania?
 Explain your reasoning.

EXTENDED PROBLEMS

In problems 12–14 do the following.

(a) Construct the first and second stages of a fractal figure that is described in the instructions.
(b) Make a table showing the number of segments, the length of each segment, and the combined lengths in the first four stages of the border of the figure. Find the general rule for the nth stage.
(c) Use self-similarity to estimate the dimension of the border of the fractal figure.

12. Construct a closed curve version of the fractal curve in problem 25 of Section 12.2.

 (a) Begin the figure with a square as the initiator. The generator is applied to each side of the square, with the "bump" on the outside of the square.
 (b) In each subsequent stages, each line segment of the figure is replaced by a scaled down copy of the generator.
13. Repeat problem 12 with the following changes: Use the same initiator and generator. However, this time reverse the direction of the generator by having the "bump" on the inside of the square.

14. Construct a closed curve version of the fractal curve in problem 27 in Section 12.2. Begin with a square as the initiator. The generator is applied to each side of the square.

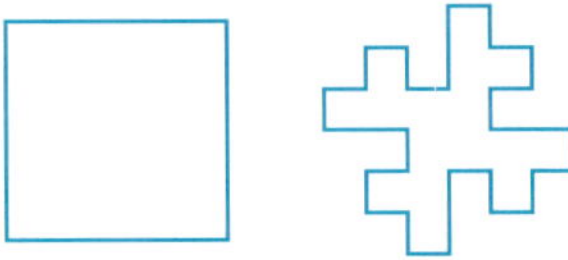

In problems 15-17 do the following:

(a) Consider the area of the region enclosed by the fractal curve that is described.
(b) Using problems 31 and 33 from Section 12.2 as examples, construct an appropriate table showing the area of the region at each stage in the development of the figure. Find the general rule for the total area for the nth stage.
(c) Use self-similarity to estimate the dimension of the area of this fractal region.

15. The closed region in problem 12.
16. The closed region in problem 13.
17. The closed region in problem 14.

Chapter Twelve Problem

Al Monbred is remodeling his family room. This room is 32 feet by 20 feet and one wall of 32 feet faces the outside. Al decides to build alcoves in the room. He will take the middle half of the outside wall, knock it out and build an alcove 8 feet outward. This alcove will be 16 feet long and have 8 feet in unused wall space on either side. Now Al cannot stop there; he really likes alcoves. There are two wall spaces he has not touched on either side. He decides to make alcoves there also. These will take up 4 feet of wall space for each of the two alcoves. Al wants them to project outward half as far as the larger alcove or 4 feet. Now Al has 3 alcoves. Why stop there? There are now 4 untouched lengths of wall each 2 feet long. He will build similar alcoves of depth 2 feet on these lengths, leaving 8 untouched lengths of wall where there will be smaller alcoves of depth 1 foot and so on. This process is taken to the limit. Now Al would like to carpet the room. Carpet costs \$18 per square yard and \$8 per foot for molding. What will this cost Al?

SOLUTION

Strategy: Draw a diagram and find a pattern

We begin by drawing a sketch showing the first stages of Al's house plan.

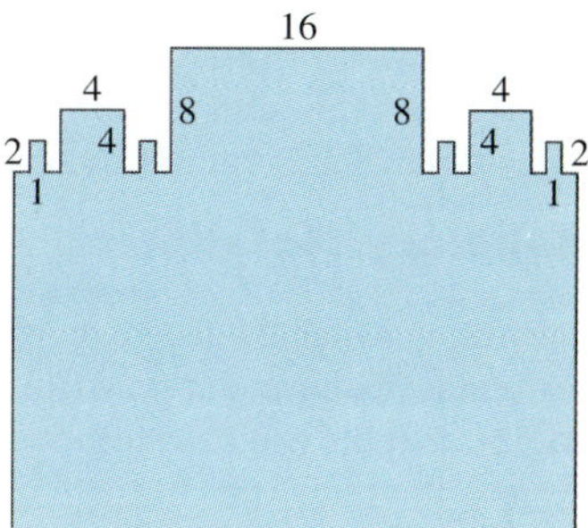

To find the cost of the carpeting, we must first calculate the area of the family room in square yards. We notice that the original room is 32 feet by 20 feet, so the original area is $32 \times 20 = 640$ square feet. To find the final area for the finished room, we need to add the areas of the alcoves. We organize the information in the form of a table.

Level	Number of Alcoves	Width	Depth	Area	Total Area	Total outward length
1	1	16	8	128	128	16
2	2	4	4	16	32	16
3	4	1	2	2	8	16

Notice that the total area for each level of alcoves is $\frac{1}{4}$ the previous area. In other words, the total area of the alcoves is a geometric series. The sum of the series is

$$128 + 128 \times \frac{1}{4} + 128 \times \left(\frac{1}{4}\right)^2 + \ldots = 128 \times \frac{1}{1 - \frac{1}{4}} = 128 \times \frac{1}{\frac{3}{4}} = 128 \times \frac{4}{3} = 170.7 \text{ sq. ft.}$$

Adding this amount to the original room area gives a total of $640 + 170.7 = 810.7$ square feet. There are 9 square feet in a square yard, so the area is $810.7 \div 9 = 90.07$ square yards. The cost per square yard of carpeting is \$18, so the carpet cost is $90.07 \times 18 \approx \$1621$. (The actual cost will be a bit higher.) To compute the cost for the molding, we look at the outside wall. Looking at the wall head on, we see there are still 32 feet visible. The added length is all in walls constructed outward to form the alcoves. Unfortunately, each level of alcoves adds 16 feet of wall length that needs molding. Taken to the limit, the total perimeter of the room is infinite. If Al decides on a minimum size of alcove that can be constructed, he can compute the molding costs.

Chapter Twelve Review

Key Ideas and Questions

The following questions review the main ideas of this chapter. Write your answers to the questions and then refer to the pages listed to make certain that you have mastered these ideas.

1. Describe a simple rule that defines a recursion and generates a sequence of numbers. 594 How does an iteration differ from a recursion? 595
2. Describe the nature of a stable point attractor. 597
3. How are fractal constructions able to give infinite detail, even though there is a simple recipe for this construction? 605 What does this suggest about fractals that occur in nature? 618–620
4. Describe what is meant by chaos and how a chaotic process can still produce order. 609

Vocabulary/Notation

Following is a list of the key vocabulary, notation, and ideas for this chapter. Mentally review each of these items; write down the meaning of each term and use it in a sentence. Then, refer to the pages listed by number and restudy any material you are unsure of before answering the questions in the Chapter Twelve Review Problems.

Section 12.1

Fibonacci Sequence 589	Recursion 594	Point attractor 597
Sequence 589	Recursion rule 594	Stable point attractor 597
Fibonacci number 589	Iteration 595	
Golden Ratio/Golden Mean 593	Simple iteration 596	

Section 12.2

Geometric recursion 602	Initiator 605	Sierpinski Carpet 608
Geometric series 604	Generator 605	Chaos Theory 609
Ratio 604	Von Koch curve 605	Chaos Game 609
Sum of a Geometric Series 605	Sierpinski Gasket 608	

Section 12.3

Fractal 616	Self–similarity 617	Non-linearity 621
Dimension 616	Natural fractals 618	Butterfly effect 621
Fractal dimension 616	Linearity 620	

Chapter Twelve Review Problems

1. Suppose we begin with 2 pairs of baby rabbits and 1 pair of adult rabbits. As in the example in the text, at each time unit a baby pair becomes an adult pair, and an adult pair produces a new baby pair. How many pairs will there be at times 1, 2, 3, through 10?
2. In problem 1, compute the successive ratios of the number of adult pairs. Do they appear to be converging to a number? If so, which number?
3. Write down the first seven terms of the sequence given by the recursion $t_1 = 1$ and $t_{n+1} = 2 + t_n$. What is the name for this set of numbers?
4. Consider the iteration $t_{n+1} = \frac{t_n}{2} + \frac{2}{t_n}$. Write down the first seven terms of the sequence beginning with the initial value of $t_1 = 1$. Do this again with the initial value $t_1 = 3$. Does there appear to be a stable point attractor?
5. Consider the iteration $t_{n+1} = 2 + \frac{1}{3}t_n$.
   (a) Write down the first seven terms of the sequence beginning with the initial value of $t_1 = 0$. Do this again with the initial value $t_1 = 6$. Does there appear to be a stable point attractor?
   (b) Find the equation this point attractor must satisfy, and solve the equation to find the value of the point attractor.
6. **(a)** What is the ratio in the geometric series $18 + 12 + 8 + \ldots$?
   (b) Diagram the geometric recursion to find the sum of this series.
   (c) Use the formula to find the sum of the series.

7. Consider the geometric recursion in which the initiator is a square. To generate the fractal, we divide the square into four equal squares and remove the upper right hand corner of the square. Draw stages 3 and 4 of this construction.

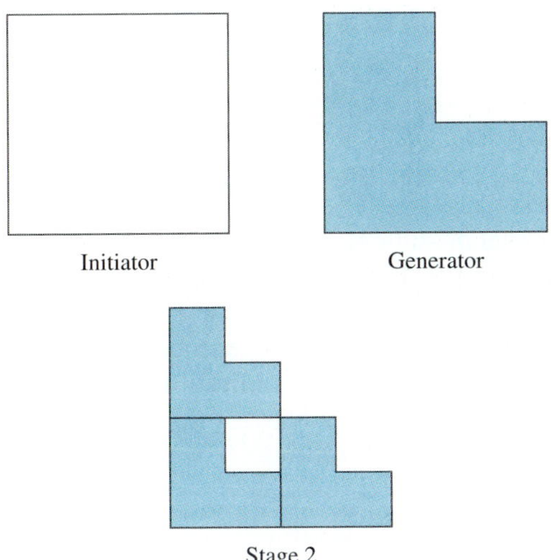

8. Play the chaos game for the case in which the three points are corners of a square. Draw several stages of this game. What do you think the fractal object generated will look like?

9. Find the dimension of the fractal obtained in problem 7.

10. Give several examples, not all taken from the text, of fractals occurring in nature.

TOPIC 1

SETS

Any collection of objects is called a **set,** and the objects are called **elements** or **members** of the set. In order to be useful, a set must be **well-defined;** that is, it must be clear whether any object belongs to the set or not. The United States is a well-defined set: Oregon belongs to the set, and Alberta does not. Sets can be defined in three basic ways: (1) a verbal description, (2) a **roster,** or listing, of the members, or (3) a description of characteristics of the elements using **set-builder notation.** For example, the verbal description "the set of all states in the United States that border the Pacific Ocean" can be presented in the other two ways as follows:

Roster: {Alaska, California, Hawaii, Oregon, Washington}.
Set-builder: $\{x \mid x$ is a U.S. state that borders the Pacific Ocean$\}$.

In both cases, a pair of braces ("{"and"}") are used to indicate the existence of a set. The set-builder statement is read: "The set of all x such that x is a U.S. state that borders the Pacific Ocean." Elements are separated by commas in the roster method.

Consider the verbal description: "The set of all even integers that are more than 7 and less than 13." It's natural to write the roster as {8, 10, 12}, and this is usually how it would be done. However, the order in which elements are written is not important. We could have written {12, 8, 10} for the set or any of the four other possible orderings of these three numbers. Also, elements are not double listed in sets, so the set {8, 8, 10, 12} should be written as {8, 10, 12}.

The set-builder notation also allows some flexibility. Compare the following descriptions:

$$\{x \mid x \text{ is an even integer and } 7 < x < 13\} \text{and}$$
$$\{y \mid y \text{ is an even integer and } 8 \le y \le 12\}.$$

Either one is satisfactory for describing the set {8, 10, 12}.

Often it is not convenient, or even possible, to list all the elements in a set using the roster method. If the set is too large to conveniently list all the elements, but we can establish the general nature of the elements, we can use an **ellipsis** (". . .") to indicate missing elements in the list. Thus the set of the first hundred even natural numbers can be given as {2, 4, 6, . . . , 200}. This approach can also be used for certain infinite sets; we can indicate the set of all even natural numbers as {2, 4, 6, . . .}.

Sets are usually denoted by capital letters such as A, B, W, and so on. The symbols "$\in$" and "$\notin$" are used to indicate that an object is or is not an element of a set, respectively. For example, if P represents the set of all U.S. states that border the Pacific, then Hawaii $\in P$ and Nevada $\notin P$.

The set without any elements is called the **empty set** or **null set** and is denoted by { } or ∅. The set of all U.S. states bordering Antarctica is an empty set, as is the set of all integers that are greater than 12 and less than 13. These two sets are equal; there is only one empty set.

EQUALITY AND EQUIVALENCE OF SETS

Two sets A and B are **equal,** written $A = B$ if and only if they have precisely the same elements. Two sets A and B can be shown to be equal if every element of A can be shown to be in B, and vice versa. If the two sets A and B are **not equal,** written $A \neq B$, we must be able to show that there is an element in one set that is not in the other. That is why there is only one empty set; there are no elements to check.

There is another way to compare sets that is often useful and forms the basis for counting. It uses the concept of a one-to-one (or 1-1) correspondence. A **one-to-one correspondence** between two sets A and B is a pairing of the elements of A with the elements of B so that each element of A is paired with exactly one element of B, and vice versa. If we can establish a one-to-one correspondence between sets A and B, then we say the sets are **equivalent,** written $\boldsymbol{A \sim B}$. There may be several one-to-one correspondences between the same two sets.

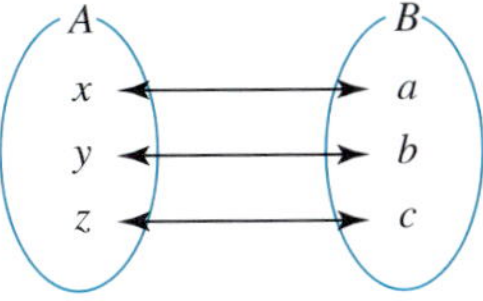

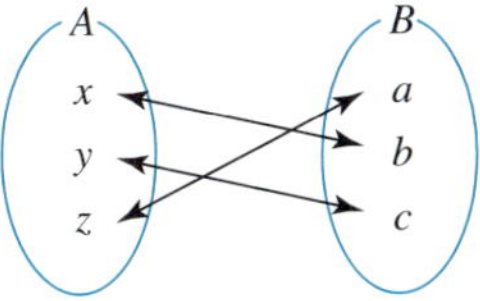

FIGURE T1.1

There are four other possible one-to-one correspondences between A and B. Notice that equal sets are always equivalent (since each element can be matched with itself), but equivalent sets are not necessarily equal. For example, $\{a, b\} \sim \{3, 7\}$ but $\{a, b\} \neq \{3, 7\}$. In the theory of numbers and counting, we say that a set has n elements if we can put it in a one-to-one correspondence with the set of the first n counting numbers, $\{1, 2, 3, \ldots, n\}$.

UNIVERSAL SET AND SUBSETS

The set that contains all the elements under consideration in a given discussion is called the **universal set,** and is usually denoted by U. Other letters are often used, however, to suggest the type of elements in question. We might, for example, use C to represent the universal set in a discussion of state capitals. For every problem, a universal set must be specified or implied, and it must remain fixed for that problem. However, when a new problem is begun, a new universal set can be specified.

To deal effectively with sets, it is important to first recognize and understand certain relationships among sets. Since every set being considered in a discussion consists of elements from the universal set, every set is, in a sense, part of the universal set. If one set A is part of another set B, we say A is a **subset** of B, written $\boldsymbol{A \subseteq B}$. If there is at least one element of A that is not an element of B, then A is not a subset of

B. If $A \subseteq B$, and B has an element that is not in A, then A is a **proper subset** of B, written $\boldsymbol{A \subset B}$. Notice that every set is a subset of itself ($A \subseteq A$), and the empty set is a proper subset of every nonempty set ($\varnothing \subset A$).

VENN DIAGRAMS

Although relationships among sets can be investigated and discussed in terms of the elements in the sets, it is often advantageous to use graphical representations of the sets instead. A helpful technique for examining the relationships among sets is the use of pictures called Venn diagrams. In a **Venn diagram,** the universal set is indicated by the inside region of a rectangle, and sets are represented by the inside regions of circles or other geometric forms.

In Figure T1.2(a), the sets A and B are shown as **disjoint sets** (sets having no common elements). Disjoint sets are also referred to as being mutually exclusive. In Figure T1.2(b), they are shown as overlapping sets; this is the standard way of drawing a Venn diagram with two sets unless we already know they are disjoint. In Figure T1.2(c), A is shown to be a subset of $B(A \subseteq B)$.

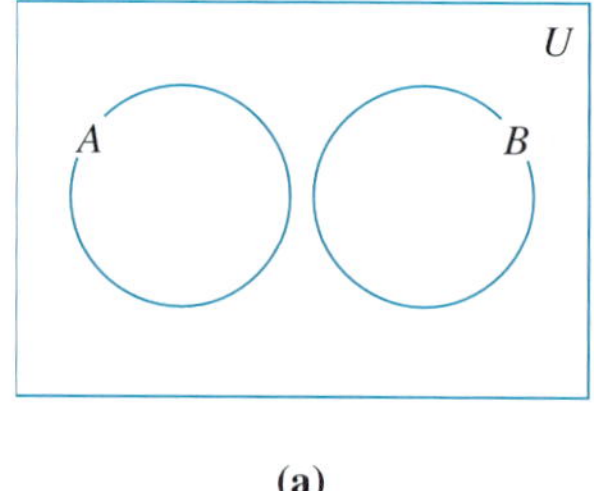

(a)

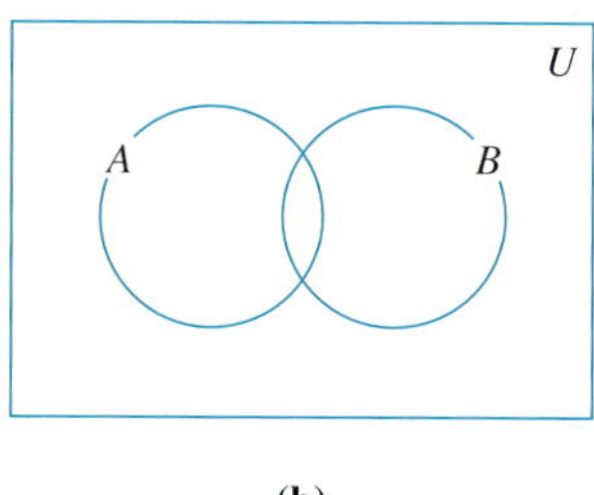

(b)

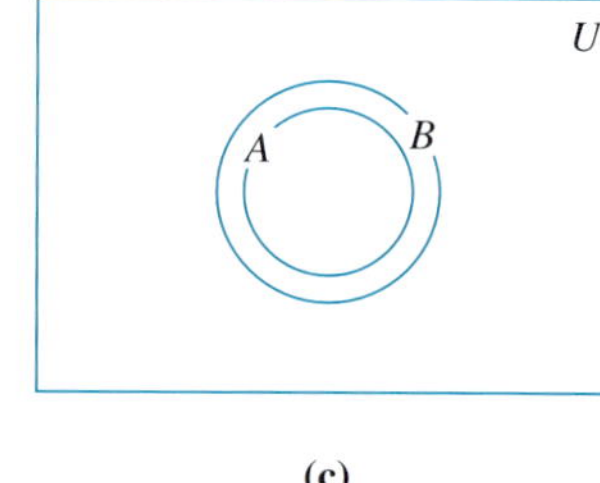

(c)

FIGURE T1.2

John Venn (1834–1923), a British logician, created these diagrams to illustrate principles of sets and symbolic logic. Venn applied logic to probability in order to reduce the confusion and difficulty arising in the use of language when dealing with descriptions and relationships in probability.

OPERATIONS ON SETS

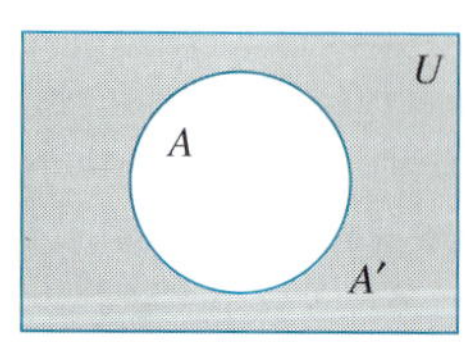

FIGURE T1.3

When a set A is well-defined, you are able to tell whether an object in the universal set belongs to the set A or not. In essence, there are two sets that are identified. The second set, called the **complement** of A (written A' and read "A prime"), consists of all elements of the universal set that are not in A. With set builder notation, we write $A' = \{x \mid x \in U \text{ and } x \notin A\}$. In many applications (in probability, for example), it is often more productive (and easier) to consider the elements in the complement of a set rather than the elements in the set itself. In Figure T1.3, the shaded region of the Venn diagram represents the complement of A. Notice that the complement of the empty set is the universal set and vice versa. It is also true that for any set A, we have $(A')' = A$.

Given any two sets A and B, we can form another set called the **union** of A and B (written $\boldsymbol{A \cup B}$) consisting of all elements of A *or* B *or both.* The shaded region in

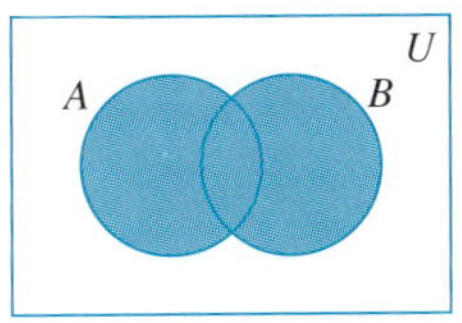

FIGURE T1.4

Figure T1.4 is $A \cup B$. Using set builder notation we write $A \cup B = \{x \mid x \in A$ or $x \in B\}$. The use of "or" in this context means that an element is in the union of two sets if it is in *at least one* of the sets. This use of *or* is customary in mathematics unless stated otherwise.

Another set that can be formed from two sets is called the intersection of the two sets. The **intersection** of A and B (written $\mathbf{A \cap B}$) consists of all elements that are common to both A and B. The shaded region in Figure T1.5 is $A \cap B$. In set-builder notation we write this as $A \cap B = \{x \mid x \in A \text{ and } x \in B\}$. Note that sets A and B are disjoint whenever $A \cap B = \varnothing$. Figure T1.5 also indicates that $A \cap B$ is a subset of A, B, and $A \cup B$. However, $A \cap B$ may be a proper subset of any of those three sets.

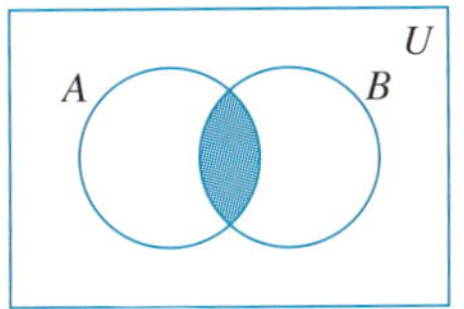

FIGURE T1.5

The **set difference** (or relative complement) of set B with respect to set A, written $\mathbf{A - B}$, is the set of elements in set A that are not in set B. Using set-builder notation we can write

$$A - B = \{x \mid x \in A \text{ and } x \notin B\}.$$

The shaded region in Figure T1.6 indicates the set $A - B$. You can think of $A - B$ as all elements remaining in A after any elements in B have been removed.

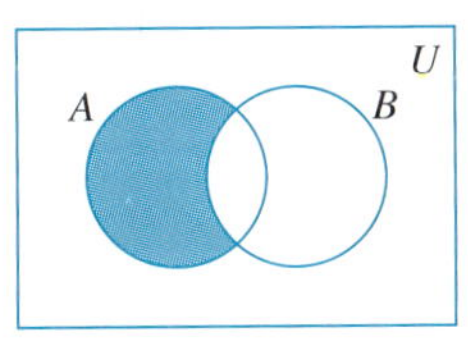

FIGURE T1.6

COMBINED OPERATIONS ON TWO SETS

A basic property in algebra is the distributive property: $a(b + c) = ab + ac$. Here we say that "multiplication is distributive over addition." There are similar, but different, properties with respect to sets and the operations we have defined on them that have important applications to other fields of mathematics, principally logic and probability. The British mathematician and logician Augustus De Morgan (1806–1871) applied algebraic operations to logic and helped put it on a sound mathematical basis. The following example illustrates one of the two properties for set operations known as De Morgan's laws.

Let $U = \{1, 2, 3, 4, 5\}$, $A = \{1, 2, 3\}$, and $B = \{1, 3, 4\}$. Find $(A \cup B)'$ and $A' \cap B'$. What do you notice?
To find $(A \cup B)'$, we first find $A \cup B$:

$$A \cup B = \{1, 2, 3\} \cup \{1, 3, 4\} = \{1, 2, 3, 4\}$$
$$(A \cup B)' = U - (A \cup B) = \{1, 2, 3, 4, 5\} - \{1, 2, 3, 4\} = \{5\}$$

Next, to find $A' \cap B'$, we first find A' and B'. As before, the complements are taken with respect to the universal set.

$$A' = \{4, 5\} \text{ and } B' = \{2, 5\}$$
$$A' \cap B' = \{4, 5\} \cap \{2, 5\} = \{5\}$$

By comparing the final results, we can see that $(A \cup B)' = A' \cap B'$ in this example.

While the above example isn't a proof, it does suggest validity of the statement $(A \cup B)' = A' \cap B'$. Another way to examine the validity of the statement would be by drawing Venn diagrams for both $(A \cup B)'$ and $A' \cap B'$ and comparing the results (see the problems). The following is a statement of De Morgan's laws on the relationships among set complement, union, and intersection.

De Morgan's Laws

For any set A and B, $(A \cup B)' = A' \cap B'$ and $(A \cap B)' = A' \cup B'$

In words, "The complement of the union is the intersection of the complements" and "The complement of the intersection is the union of the complements."

SETS AND COUNTING

When we refer to the number of elements in a set A, we will use the symbol $n(A)$. That is, if $A = \{a, e, i, o, u\}$, then $n(A) = 5$; if $B = \{2, 4, 6, \ldots, 200\}$ then $n(B) = 100$. Of particular interest is the number of elements in the union of two sets. Given $A = \{1, 3, 5, 7, 9\}$ and $B = \{2, 3, 5, 7\}$, find $n(A \cup B)$. We find $A \cup B = \{1, 2, 3, 5, 7, 9\}$ and $n(A \cup B) = 6$. Note that $n(A) = 5$ and $n(B) = 4$, so $n(A) + n(B) = 5 + 4 = 9$. Therefore, $n(A \cup B) \neq n(A) + n(B)$.
The last example suggests the following general result:

$$n(A \cup B) \leq n(A) + n(B).$$

The reason for the "less than or equal to" symbol rather than an "equals" sign is that A and B may have some elements in common, and duplication is ignored when forming the union of two sets. To find the number of elements in the union of two sets, we add the number of elements in each of the sets and then subtract the number of elements in the intersection of the sets so that the elements in the intersection are not counted twice. In our example, we can see that $n(A \cap B) = n(\{3, 5, 7\}) = 3$, so $n(A \cup B = 5 + 4 - 3 = 6$. In general, we have the following counting principle for sets.

The Number of Elements in the Union of Two Sets

For any two sets A and B, $n(A \cup B) = n(A) + n(B) - n(A \cap B)$.

This is an important result that is used in probability (which is defined in terms of the number of ways an event can happen) and other applications. We can also see the following special case of the preceding result.

The Number of Elements in the Union of Two Disjoint Sets

$n(A \cup B) = n(A) + n(B)$ if and only if $A \cap B = \emptyset$.

As a special case of this last statement, $n(U) = n(A \cup A') = n(A) + n(A')$. From this, we have either $n(A') = n(U) - n(A)$ or $n(A) = n(U) - n(A')$.

PROBLEM SET T1

1. Determine which of the following sets are well-defined:
(a) the set of the nine greatest baseball players of all time.
(b) the set of all recording artists who have sales over 10,000,000 copies.
(c) the set of all women who have played in the National Football League.
(d) the set of all prime numbers.

2. Determine which of the following sets are well-defined:
(a) the set of all elected politicians.
(b) the set of all honest politicians.
(c) the set of all movies produced and/or directed by George Lucas.
(d) the set of Steven Spielberg's three greatest movies.

3. List the members of the following sets with a roster.
(a) $\{x \mid x$ is the square of a negative integer greater than $-6\}$
(b) $\{S \mid S$ is a state bordering the Pacific Ocean$\}$
(c) $\{x \mid x$ is a prime number less than $30\}$
(d) $\{y \mid y$ is a multiple of 3 that is greater than $10\}$

4. Describe each of the following sets using set-builder notation.
(a) {a, e, u, o, i}
(b) {1, 8, 27, 64, 125}
(c) {Bush, Eisenhower, Ford, Nixon, Reagan}
(d) {m, a, t, h, e, i, c, s}

5. True or false?
(a) $6 \in \{3, 4, 5, 6, 7, 8\}$
(b) $6 \in \{x \mid x$ is a factor of $112\}$
(c) $6 \notin \{x \mid x$ is a prime number$\}$
(d) $6 \notin \{\ \}$

6. Let $A = \{1, 3, 7\}$ and $B = \{a, b, c\}$.
(a) How many one-to-one correspondences are there between A and B?
(b) Show three of the one-to-one correspondences between A and B.

7. True or false?
(a) $\{1, d, \$, 6, s\} = \{\$, 1, 6, d, s\}$
(b) $\{x \mid x$ is a prime number less than $12\} \sim \{x \mid x$ is a vowel in the word 'beautification'$\}$
(c) $\{x \mid x$ is a prime number less than $12\} \sim \{2, 4, 6, 8, 10\}$
(d) If A, B, and C are sets, with $A \sim B$ and $B \sim C$, then $A \sim C$.
(e) If A, B, and C are sets, with $A = B$ and $C = B$, then $A = C$.

8. True or false?
(a) $\{1, 3, 5, 7, 9\} \sim \{@, \#, \$, \%, \&\}$
(b) $\{x \mid x = 2^n$ for $n = 1, 2, 3\} = \{y \mid y$ is a factor of 16 less than $10\}$
(c) $\{x \mid x = 2^n$ for $n = 1, 2, 3\} = \{x \mid x = n^2$ for $n = 1, 2, 3\}$
(d) If A and B are two sets, with $A = B$, then $A \sim B$.
(e) If A and B are two sets, with $A \sim B$, then $A = B$.

9. For the following questions, the universal set is $U = \{2, 3, 6, 9, 10, 12, 15\}$.
(a) List the elements of A, if $A = \{x \mid x$ is an even integer$\}$
(b) List the elements of B, if $B = \{x \mid x$ is divisible by $3\}$
(c) List the elements of C, if $C = \{x \mid x$ is the square of an integer$\}$
(d) List the elements of D, if $D = \{x \mid x$ is divisible by 2 or $3\}$

10. List all the subsets of $S = \{a, 2, \#\}$.

11. List all the proper subsets of $A = \{a, b, c\}$.

12. List all the subsets of $Z = \{0\}$.

13. True or false?
(a) $\{\text{Ford, Bush}\} \subset \{x \mid x$ is a living ex-president$\}$ (Note: as of 1996)
(b) $3 \subset \{2, 3, 5, 7\}$
(c) $\{2, 5\} \subseteq \{1, 2, 3, 4, 5, 6\}$
(d) $\{2, 5\} \subset \{1, 2, 3, 4, 5, 6\}$
(e) $\varnothing \subseteq \{\ \}$

14. Use the most appropriate symbol $\in$, $\notin$, $\subset$, $\not\subset$, or $\subseteq$ in each of the following:
(a) 51 ___ $\{x \mid x$ is a prime number$\}$
(b) $\{2, 7\}$ ___ $\{1, 2, 3, 5, 7, 9\}$
(c) $\{5, 3, 2\}$ ___ $\{x \mid x$ is a factor of $100\}$
(d) $\{1, 2, 3\}$ ___ $\{3, 2, 1\}$
(e) 36 $\{x \mid x$ is the square of an integer$\}$

15. For parts (a) through (d) draw a Venn diagram like the following and shade it to represent the given set.

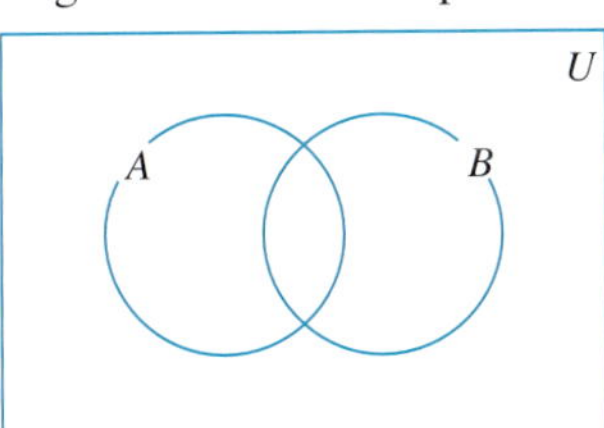

(a) A' **(b)** $(A \cup B)'$ **(c)** $A' \cup B$ **(d)** $A' \cap B$

16. For parts (a) through (c) draw a Venn diagram like the following and shade it to represent the given set.

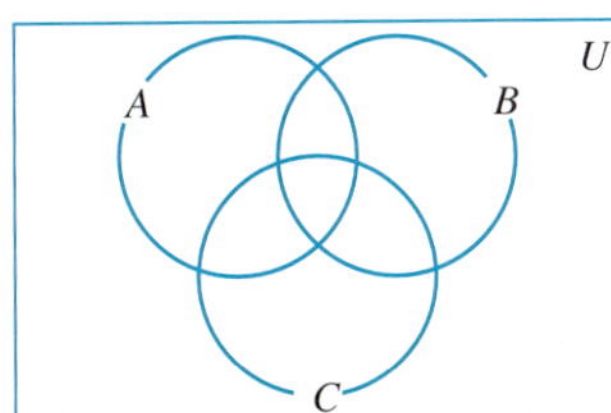

(a) $A \cap (B \cup C)$ **(b)** $A - (B \cap C)$ **(c)** $A \cup (B - C)$

17. Let $U = \{1, 2, 3, 4, 5, 6, 7\}$. Draw Venn diagrams that represent the given sets. Show all elements of U.
(a) $A = \{2, 3, 7\}$
(b) $A = \{3, 4, 5\}$, $B = \{1, 3, 5, 7\}$
(c) $A' = \{2, 3, 6\}$

18. Let $U = \{1, 2, 3, 4, 5, 6, 7\}$. Draw Venn diagrams that represent sets described as follows. Show all elements of U.
(a) $A = \{2, 3\}$, $B = \{2, 3, 6, 7\}$
(b) $A = \{2, 3, 5\}$, $B = \{1, 4, 6\}$
(c) $A \cup B = \{1, 3, 4, 6, 7\}$,
$B - A = \{3, 6\}$, $A \cap B = \{1\}$

TOPIC 2

CALCULATORS

Although a basic calculator can be bought for under \$10, there are features on \$15 to \$20 calculators that simplify many complicated calculations. For example, the exponentiation function $\boxed{x^y}$ is essential for many of the computations in this book. On the other hand, a calculator can be so complicated that it is hard to use, so you should not automatically think that more is better when it comes to calculator features. As a student, you should purchase your own personal calculator so that you will have it readily available for homework or tests. You should be as familiar as possible with the calculator's mode of operation. You will also be able to use it to solve problems or make calculations related to your daily life.

LOGIC

Calculators operate according to sets of rules and relationships called logic. There are three main types of logic available in calculators: Reverse Polish Notation, arithmetic, and algebraic.

REVERSE POLISH NOTATION

Reverse Polish Notation (RPN) is an excellent way to perform calculations and is found primarily on more sophisticated scientific calculators made by Hewlett-Packard. Our discussion will focus on the more common types used by the general student population. (Note: for ease of reading directions for keystrokes, we will write numerals without the usual squares around them to indicate that they are keys.)

ARITHMETIC LOGIC

In arithmetic logic, the calculator performs operations in the order they are entered. For example, if $3 \boxed{+} 4 \boxed{\times} 5 \boxed{=}$ is entered, the calculations are performed as follows: $(3 + 4) \times 5 = 7 \times 5 = 35$. That is, the operations are performed from left to right as they are entered.

ALGEBRAIC LOGIC

If your calculator has algebraic logic and $3 \boxed{+} 4 \boxed{\times} 5 \boxed{=}$ is entered, the result is different than the result when arithmetic logic is used. Here, the calculator evaluates expressions according to the usual mathematical convention for **order of operations,** listed in order of preference.

Within the innermost parentheses,

> *First:* Calculate exponentials.
> *Second:* Perform multiplications and divisions from left to right.
> *Third:* Perform additions and subtractions from left to right.

Repeat until all calculations have been performed.

For 3 [+] 4 [×] 5 [=], we get

$$3 + 4 \times 5 = 3 + 20 = 23$$

since multiplication takes precedence over addition.

If a calculator has keys for parentheses, they can be inserted to be sure that the desired operation is performed first. In a calculator that uses algebraic logic, the calculation $13 - 5 \times 4 \div 2 + 7$ will be performed in the following steps:

$$13 - 5 \times 4 \div 2 + 7 = 13 - 20 \div 2 + 7 = 13 - 10 + 7 = 3 + 7 = 10.$$

If you wish to calculate $13 - 5$ first, parentheses must be inserted.

$$(13 - 5) \times 4 \div 2 + 7 = 8 \times 4 \div 2 + 7 = 32 \div 2 + 7 = 16 + 7 = 23$$

For some makes of calculators (e.g., graphing calculators by Texas Instruments) the [=] key has been replaced by the [ENTER] key.

CALCULATOR FEATURES

During recent years, advances in calculator electronics and design have produced reasonably priced calculators (e.g., graphing calculators) that rival the capabilities of desktop computers from a decade before. A thorough discussion of calculator features and capabilities is not appropriate here, and you should study your user's manual for your calculator to become proficient in its use. However, we will discuss some of the common features that make a calculator more useful as a computational tool. Several keystroke sequences will be displayed to simulate the variety of calculator operating systems available.

PARENTHESES

As mentioned earlier when we were discussing algebraic logic, you must always be attentive to the order of operations when several operations are present. For example, the product $2 \times (3 + 4)$ can be found in two ways. First, by using commutativity, the following keystrokes will yield the correct answer:

3 [+] 4 [=] [×] 2 [=] [14]

Alternatively, the parentheses keys may be used as follows:

2 [×] [(] 3 [+] 4 [)] [=] [14].

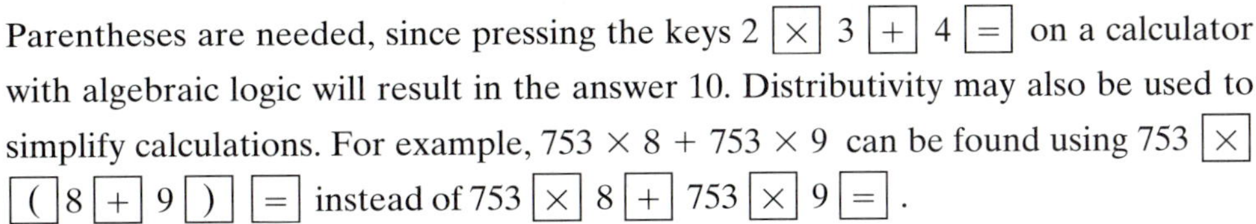

Parentheses are needed, since pressing the keys 2 [×] 3 [+] 4 [=] on a calculator with algebraic logic will result in the answer 10. Distributivity may also be used to simplify calculations. For example, $753 \times 8 + 753 \times 9$ can be found using 753 [×] [(] 8 [+] 9 [)] [=] instead of 753 [×] 8 [+] 753 [×] 9 [=].

SQUARING AND SQUARE ROOT FUNCTIONS

Scientific calculators can either square a number or take the square root of a number with a single keystroke. This will usually be done one of two ways, depending on the brand of calculator.

5^2 is either 5 [x^2] or 5 [x^2] [ENTER]

$\sqrt{5}$ is either 5 [$\sqrt{\ }$] or [$\sqrt{\ }$] 5 [ENTER]

RECIPROCAL

Scientific calculators can be used to find reciprocals using a single keystroke. The reciprocal key is usually either [$\frac{1}{x}$] or [x^{-1}]. The two ways that a reciprocal key is used are

5 [$\frac{1}{x}$], which produces [0.2] or

5 [x^{-1}] [ENTER], which also produces [0.2].

EXPONENTS

Many real-life problems (e.g., compound interest, exponential growth) require the use of exponents. On a basic calculator, 3^4 may be found using repreated multiplication:

$3^4 = 3 \times 3 \times 3 \times 3 =$ [81].

On scientific calculators the computation is simplified through the use of the exponentiation keys. There are two basic types, [y^x] (or [x^y]) and [∧], used as follows:

3 [y^x] 4, which produces [81] or

3 [∧] 4 [ENTER], which also produces [81].

MEMORY FUNCTIONS

Many calculators have memory functions designated by the keys M+, M−, MR, or STO, RCL, SUM. Your calculator's display will probably show an "M" to remind you that there is a nonzero number in the memory. Intermediate results or numbers that are going to be used several times in a sequence of calculations can be stored in memory and recalled as needed.

PROBLEM SET T2

Compute the following using your calculator.

1. $(12.3)(42.62)$
2. $45.98 + 10.05 + 423.52$
3. $432.77 - 29.76$
4. $(18.33)(2.4) - (7.86)(3.5)$
5. $1396.634 - 1234.105$
6. $3.075 - 4.672 + (12.6)(3.1)$
7. $\dfrac{71.28}{4.875}$
8. $\dfrac{231.75}{0.825}$
9. $\sqrt{612}$
10. $\sqrt{314.19}$
11. $(17.45)^2$
12. $(8.66)^4$
13. $(25.6)^{3.2}$
14. $(5)^{1.2}$

To calculate roots such as $\sqrt[5]{65}$, you must change from radical form to exponent form unless your calculator has special keys. Use the equivalent form $\sqrt[n]{x} = x^{1/n}$. To calculate $\sqrt[5]{65}$ on an algebraic calculator the keystrokes are:

65 x^y (1 ÷ 5) = or

65 x^y 5 $1/x$ =.

15. $\sqrt[4]{125}$
16. $\sqrt[5]{12.5 + 34.2}$
17. $\dfrac{(21.3)(12.75)}{(5.78)(3.45)}$
18. $\dfrac{65 + 27.2}{(15.7)(4.3)}$
19. $\dfrac{4.3^{25}\sqrt{52.7}}{1.8}$
20. $\dfrac{\sqrt[3]{245}}{12.7 - 4^{1.2}}$
21. $1000\left(1 + \dfrac{0.065}{12}\right)^{12\times 3}$
22. $750\left(1 + \dfrac{0.0925}{12}\right)^{12\times 5}$
23. $\left(1 + \dfrac{0.075}{12}\right)^{12} - 1$
24. $\left(1 + \dfrac{0.1125}{12}\right)^{12} - 1$
25. $\dfrac{2(14.95) + 7(21.50) + 15(19.69)}{2 + 7 + 15}$
26. $\dfrac{8(35.09) + 13(27.45) + 21(18.34)}{8 + 13 + 21}$
27. $\dfrac{300[(1 + 0.0075)^{60} - 1]}{0.0075}$
28. $\dfrac{5000(0.0125)}{(1 + 0.0125)^{60} - 1}$

TOPIC 3

MEASUREMENT AND THE METRIC SYSTEM

THE MEASUREMENT PROCESS

The measurement process is defined as follows.

DEFINITION

The Measurement Process

1. Select an object and an attribute of the object to measure, such as its length, area, volume, weight, or temperature.
2. Select an appropriate unit with which to measure the attribute.
3. Determine the number of units needed to measure the attribute.

TIDBIT

The science of weights and measures is called metrology.

There are two systems of standard units used in the United States. One is the English System, or customary system, which most of us learned in school and use daily, and the other is the *Systèm international d'unités* (SI), commonly called the metric system. This section presents the concepts and measurements central to the metric system. Since most students will be familiar with units for the customary system, these are simply listed in Table T3.5 at the end of this section with their equivalent metric system counterparts.

THE METRIC SYSTEM

The metric system arose by a distinctly different and more rational process than the English system. In 1790, the French Academy of Sciences, at the request of the French government, devised a system of measurement. The system incorporates all of the following features of an ideal system of units.

An Ideal System of Units

1. The fundamental unit can be accurately reproduced without reference to a prototype. (**Portability**)
2. There are simple (e.g., decimal) ratios among units of the same type. (**Convertibility**))
3. Different types of units (e.g., those for length, area, and volume) are defined in terms of each other, using simple relationships. (**Interrelatedness**)

LENGTH

The basic length unit was chosen to be one ten-millionth (1/10,000,000) of the distance from the North Pole to the equator measured along the meridian that passes through Paris. This gave a unit that was close to the yard, but also part of a decimal system. The name "meter" given to the new unit was taken from the Greek word *metron* meaning "to measure" (a meter is approximately 1.1 yards). Today the meter is actually defined as the length traveled by light in a vacuum during $\frac{1}{299\,792\,458}$ second.

The metric system is a decimal system of measurement in which multiples and fractions of the fundamental unit correspond to powers of ten. For example, one thousand meters is a **kilometer,** one tenth of a meter is a **decimeter,** one-hundredth of a meter is a **centimeter,** and one-thousandth of a meter is a **millimeter.** Table T3.1 shows some relationships among metric units of length.

TIDBIT

The units bigger than the basic unit have Greek root prefixes, while the units smaller than the basic unit have Latin root prefixes.

TABLE T3.1

Unit	Symbol	Fraction or Multiple of 1 Meter
1 millimeter	1 mm	0.001 m
1 centimeter	1 cm	0.01 m
1 decimeter	1 dm	0.1 m
1 meter	1 m	1 m
1 dekameter	1 dam	10 m
1 hectometer	1 hm	100 m
1 kilometer	1 km	1000 m

Notice the simple ratios among units of length in the metric system. From Table T3.1 we see that 1 **dekameter** (also spelled decameter) is equivalent to 10 meters, 1 **hectometer** is equivalent to 100 meters, and so on. Also 1 dekameter is equivalent to 100 decimeters, 1 kilometer is equivalent to 1,000,000 millimeters, and so on. You should verify these values. (For very large or very small quantities, the following prefixes are used: mega = million, giga = billion, micro = millionth, nano = billionth.)

AREA

In the metric system, the fundamental unit of area is the square meter. A square that is 1 meter long on each side (or is 1 meter square) has an area of **1 square meter,** written 1 m^2. Whereas areas are measured in square feet or square yards in the English system, they are measured in square meters in the metric system. A square meter is approximately 1.2 yd^2.

Smaller areas, such as the area of a piece of notebook paper or a photograph, are measured in square centimeters (cm^2). A **square centimeter** is the area of a square that is 1 centimeter long on each side. Figure T3.1 shows the relationship between a square centimeter and a square meter.

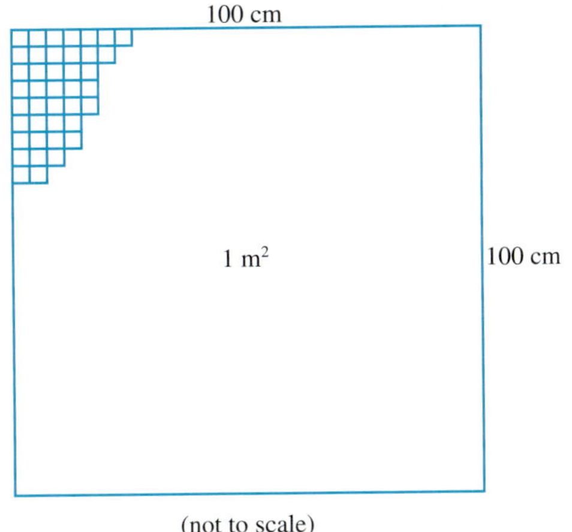

FIGURE T3.1

In Figure T3.1 we see part of an arrangement of square centimeters that will cover the square meter. There are 100 rows, each row having 100 square centimeters. Hence there are $100 \times 100 = 10{,}000$ square centimeters needed to cover the square meter. Thus $1\ m^2 = 10{,}000\ cm^2$. Very small areas, such as on a microscope slide, are measured using square millimeters. A **square millimeter** is the area of a square whose sides are each 1 millimeter long.

In the metric system the area of a square that is 10 m on each side is given the special name **are** (pronounced "air"). An are is approximately the area of the floor of a large two-car garage. It is a convenient unit for measuring the area of building lots. There are $100\ m^2$ in 1 are. An area equivalent to 100 ares is called a **hectare,** written 1 ha. Notice the use of the prefix "hect" (meaning 100). The hectare is useful for measuring areas of farms and ranches. As a rough conversion, 1 ha is about 2 1/2 acres. Finally, very large areas are measured in the metric system using square kilometers. Areas of cities or states, for example, are reported in square kilometers. One **square kilometer** is the area of a square that is 1 kilometer on each side.

As a rough conversion, $1\ km^2$ is about $\frac{3}{8}$ square mile. Table T3.2 gives the ratios among various units of area in the metric system. You should verify the entries in the table.

TABLE T3.2

Unit	Abbreviation	Fraction or Multiple of One Square Meter
square millimeter	mm^2	$0.000001\ m^2$
square centimeter	cm^2	$0.0001\ m^2$
square decimeter	dm^2	$0.01\ m^2$
square meter	m^2	$1\ m^2$
are (square dekameter)	a (dam^2)	$100\ m^2$
hectare (square hectometer)	ha (hm^2)	$10{,}000\ m^2$
square kilometer	km^2	$1{,}000{,}000\ m^2$

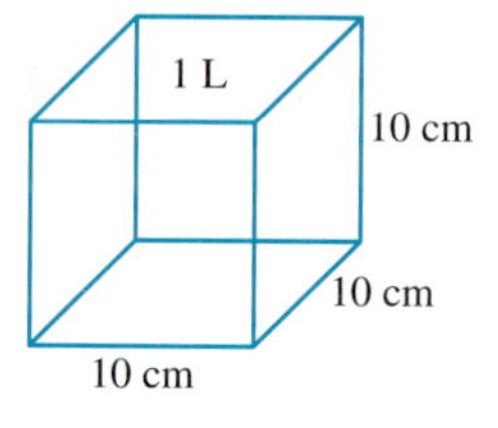

FIGURE T3.2

VOLUME

The fundamental unit of volume or capacity in the metric system is the liter. A **liter,** abbreviated L, is the volume of a cube that measures 10 cm on each edge (Figure T3.2). We can also say that a liter is 1 **cubic decimeter,** since the cube in Figure T3.2 measures 1 dm on each edge. Notice that the liter is defined with reference to the meter, which is the fundamental unit of length. The liter is slightly larger than a quart. Many soft-drink containers are now produced with capacities of 1 or 2 liters.

Imagine filling the liter cube in Figure T3.2 with smaller cubes, 1 centimeter on each edge. Figure T3.3 illustrates this.

TIDBIT

The medical profession tends to use the abbreviation "cc" for cubic centimeters. When dealing in small dosages by mouth (such as cough syrup), we finally find an accurate and easy conversion between the English and metric systems: 1 teaspoon equals 5 cc.

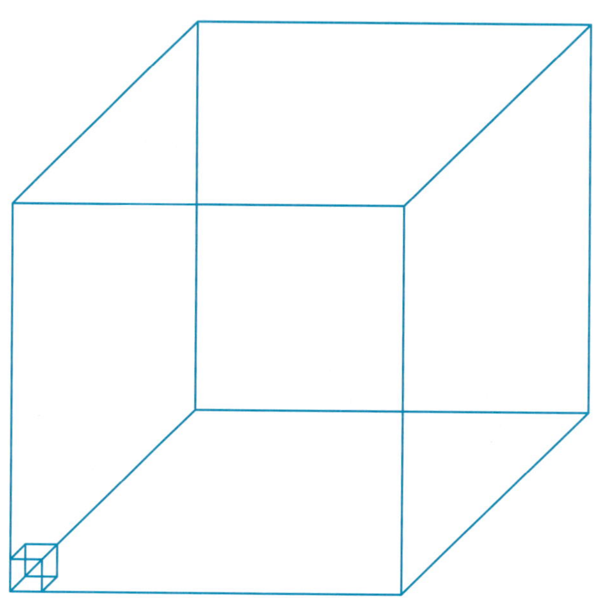

FIGURE T3.3

Each small cube has a volume of 1 **cubic centimeter** (1 cm^3). It will take a 10×10 array (hence 100) of the centimeter cubes to cover the bottom of the liter cube. Finally, it takes 10 layers, each with 100 centimeter cubes, to fill the liter cube to the top. Thus 1 liter is equivalent to 1000 cm^3. Thus we see that 1 **milliliter** is equivalent to 1 cubic centimeter since there are 1000 cm^3 in 1 liter. Small volumes in the metric system are measured in milliliters (cubic centimeters). Small containers of liquid, such as medicine or perfume, are frequently labeled in milliliters.

Large volumes in the metric system are measured using cubic meters. A **cubic meter** is the volume of a cube that measures 1 meter on each edge. Capacities of large containers such as water tanks, reservoirs, or swimming pools are measured using cubic meters. Table T3.3 gives the relationships among commonly used volume units in the metric system. In the metric system, capacity is usually recorded in liters, milliliters, and kiloliters.

TABLE T3.3

Unit	Abbreviation	Fraction or Multiple of One Liter
milliliter (cubic centimeter)	mL (cm^3)	0.001 L
liter (cubic decimeter)	L (dm^3)	1L
kiloliter (cubic meter)	kL (m^3)	1000 L

TIDBIT

The difference between weight and mass is that weight refers to the force exerted on an object by the earth's gravitational field, while mass refers to the force needed to accelerate the object. Since we are unlikely to find ourselves on another planet where the gravitational field is different from earth's, we will be casual and use "weight" and "mass" interchangeably.

MASS (OR WEIGHT)

In the metric system, the standard unit of mass is the kilogram. One **kilogram** is the mass of 1 liter of water in its most dense state. A kilogram is about 2.2 pounds in the English system. Notice that the kilogram is defined with reference to the liter, which in turn was defined relative to the meter. The metric units meter, liter, and kilogram are thus interrelated. An average size human weighs about 70 kilograms, with 20 kilograms, either way, being the rough size range (i.e. a 50 kilogram human is small and a 90 kilogram human is large).

From the information in Table T3.3, we can conclude that 1 milliliter of water weights $\frac{1}{1000}$ of a kilogram. This weight is called a **gram.** A paper clip weighs about 1 gram and a U.S. nickel weighs about 5 grams. In the metric system, grams are used for small weights such as ingredients in recipes or nutritional contents of various foods. Many packaged foods are labeled in both the English system (ounces, teaspoons, etc.) and in grams. One ounce in the English system is equivalent to about 28 grams. Large masses are measured using a **metric ton** (**tonne**) which is 1000 kilograms. We summarize the information in Table T3.3 with the definitions of the various metric weights in Table T3.4.

TABLE T3.4

Cube	Volume	Mass (Water)
1 m^3	1 kL	1 tonne
1 dm^3	1L	1 kg
1 cm^3	1mL	1 g

TEMPERATURE

In the metric system, temperature is measured in **degrees Celsius.** The Celsius scale is named after the Swedish astronomer, Anders Celsis, who devised it in 1742. This scale is commonly called **centigrade.** Two reference temperatures are used, the freezing point of water and the boiling point of water at sea level. These are defined to be, respectively, zero degrees Celsius (0°C) and 100 degrees Celsius (100°C). A metric thermometer is made by dividing the interval from freezing to boiling into 100 degrees. Figure T3.4 shows a comparison between readings on a Fahrenheit thermometer and the metric thermometer and some useful metric temperatures.

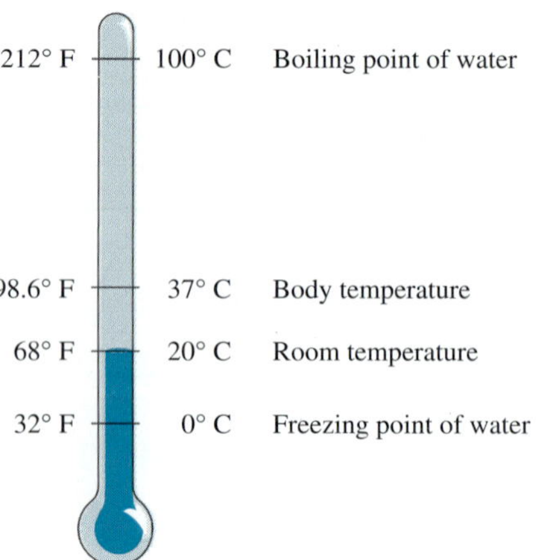

FIGURE T3.4

HISTORY

Gabriel Fahrenheit believed that the lowest temperature that could be reached was the temperature of a mixture of ice and salt, so he made that temperature 0°F. He used normal human body temperature (which is unfortunately quite variable, even in the same human) as his other naturally determined temperature, making it 100°F. The freezing and boiling points of water then turned out to be 32°F and 212°F.

The relationship between degrees Celsius and degrees Fahrenheit (used in the English system) is derived next. The basic facts are that the freezing point of water is 32 degrees Fahrenheit (32°F), and the boiling point of water is 212 degrees Fahrenheit (212°F). Thus the interval between the freezing and boiling points is measured by 180 degrees Fahrenheit, while it is measured by 100 degrees Celsius. So the size of 1 degree Celsius must be $\frac{180}{100} = \frac{9}{5} = 1.8$ degrees Fahrenheit. Suppose that C represents a Celsius temperature and F the equivalent Fahrenheit temperature. If C is a temperature above freezing, then the Fahrenheit temperature must be $1.8 \times C$ Fahrenheit degrees above freezing (since each Celsius degree is 1.8 Fahrenheit degrees). Thus the formula

$$1.8 \times C + 32 = F$$

can be used for converting temperatures from Celsius to Fahrenheit. (The same argument and formula also apply to temperatures below freezing.)

Next, we use algebra to solve $1.8 \times C + 32 = F$ for C in terms of F.

$$\begin{aligned} 1.8 \times C + 32 &= F \\ 1.8 \times C &= F - 32 \\ C &= \frac{F - 32}{1.8} \\ C &= \frac{5}{9} \times (F - 32) \end{aligned}$$

In summary, we have the following.

Fahrenheit/Celsius Conversion

$$F = 1.8C + 32 \text{ and } C = \frac{5}{9}(F - 32)$$

From the discussion above, we see that the metric system has all the features of an ideal system of units: portability, convertibility, and interrelatedness. These features make learning the metric system simpler than learning the English system of units. The metric system is the preferred system in science and commerce throughout the world. Moreover, there are only a handful of countries, including the United States, that use a system of units other than the metric system.

DIMENSIONAL ANALYSIS

When working with two (or more) systems of measurement, there are many circumstances requiring conversions among units. The procedure known as dimensional analysis can help simplify the conversion. In **dimensional analysis,** we use unit ratios that are equivalent to 1 and treat these ratios as fractions to convert one measurement to another. For example suppose that we wish to convert 17 feet to inches. We use the unit ratio 12 in/1 ft (which is 1) to perform the conversion.

$$17 \text{ ft} = 17 \cancel{\text{ft}} \times \frac{12 \text{ in.}}{1 \cancel{\text{ft}}} = 17 \times 12 \text{ in.} = 204 \text{ in.}$$

Hence a length of 17 feet is the same as 204 inches. Notice that the units in the fractions are treated as factors and can be "canceled" in the same way as when we work with common fractions. Dimensional analysis is especially useful if several conversions must be made. For example, the following equation shows how to determine the number of liters of water contained in a vase that holds 4286 grams of water when full. Since 1 mL of water weighs approximately 1 gm and 1L = 1000 mL, we have

$$4286 \text{ g} = 4286 \cancel{\text{g}} \times \frac{1 \cancel{\text{mL}}}{1 \cancel{\text{g}}} \times \frac{1 \text{ L}}{1000 \cancel{\text{mL}}} = \frac{4286}{1000} \text{ L} = 4.286 \text{ L}.$$

Consequently, the capacity of the vase is approximately 4.286 liters.

Table T3.5 is a brief table of more precise conversions between English and metric units.

TABLE T3.5

English Unit	Abbreviation	Multiply English by the ratio to convert to metric / Divide metric by the ratio to convert to English	Metric Unit	Abbreviation
Length		**Ratio**		
inch	in.	2.54	centimeter	cm
foot	ft	30.5	centimeter	cm
yard	yd	0.914	meter	m
mile	mi	1.61	kilometer	km
Area				
square inch	in^2	6.45	square centimeter	cm^2
square foot	ft^2	0.0929	square meter	m^2
square yard	yd^2	0.836	square meter	m^2
square mile	mi^2	2.59	square kilometer	km^2
acre	a	0.405	hectare	ha
Volume				
cubic inch	in^3	16.387	cubic centimeter	cm^3
cubic foot	ft^3	0.028317	cubic meter	m^3
cubic yard	yd^3	0.76455	cubic meter	m^3
cubic mile	m^3	4.16818	cubic kilometer	km^3
Capacity				
fluid ounce	fl oz	29.573	milliliter	mL
quart	qt	0.94635	liter	L
gallon	gal	3.7854	liter	L
Weight/Mass				
pound	lb	0.4545	kilogram	kg

For example, to convert 20 square inches to square centimeters, we calculate

$$20 \text{ in}^2 \times 6.45 \longrightarrow 129 \text{ cm}^2.$$

To convert 100 sq cm to square inches, we calculate

$$100 \text{ cm}^2 \div 6.45 \longrightarrow 15.5 \text{ in}^2$$

PROBLEM SET T3

1. Without referring to any table, match the prefix with its numerical equivalent.

(i)	deci	**(a)**	1000 times
(ii)	milli	**(b)**	0.01 times
(iii)	hecto	**(c)**	0.001 times
(iv)	kilo	**(d)**	100 times
(v)	centi	**(e)**	0.1 times
(vi)	deka	**(f)**	10 times

2. Arrange each of the following from largest to smallest.
   (a) 6.9 dm, 67 cm, 680 mm
   (b) 2.4 kg, 2500 g, 23,400 dg
   (c) 7.25 dam, 73.5 dm, 0.0715 km
3. Convert the following measurements to meters.
   (a) 654 cm **(b)** 32 hm **(c)** 600 mm
   (d) 35 km **(e)** 40 dm
4. Convert the following measurements to ares.
   (a) 10,000 cm^2 **(b)** 1.2 m^2
   (c) 1500 mm^2 **(d)** 500 dm^2
   (e) 1 dam^2

In problems 5 through 8, convert the given units to the units indicated.

5. **(a)** 2 m to mm
   (b) 4.25 m to cm
   (c) 850 cm to dm
   (d) 453 mm to dm
6. **(a)** 1.72 km to cm
   (b) 6540 dm to km
   (c) 2.75 hm to dm
   (d) 0.052 km to mm
7. **(a)** 1000 cm^2 to m^2
   (b) 1220 m^2 to cm^3
   (c) 2.7 km^2 to hm^2
   (d) 285 mm^2 to dm^2
8. **(a)** 100 cm^3 to mm^3
   (b) 37,500 mm^3 to m^3
   (c) 25 dam^3 to m^3
   (d) 5000 dm^3 to dam^3
9. Choose the most realistic measures for the following objects.
   (a) length of a paper clip
   28 mm or 28 cm or 28 m
   (b) height of a 12-year-old child
   148 mm or 148 cm or 148 m
   (c) volume of a pop bottle
   473 mL or 473cL or 473 L
   (d) volume of a bucket
   10 mL or 10 L or 10 hL
10. Convert the following Farenheit temperatures to degrees Celsius (to the nearest degree) and vice versa.
   (a) moderate oven (350°F)
   (b) a spring day (15°C)
   (c) ice skating weather (-20°C)
   (d) world's highest recorded temperature (136°F) at Azizia, Tripolitania, in northern Africa, Sept. 13, 1922.

In problems 11 through 14 convert the given units to the units indicated.

11. **(a)** 42.5 in. to cm
   (b) 60 yd^2 to m^2
   (c) 100 ft^3 to m^3
   (d) 17.5 gal to L
   (e) 8.3 lb to kg
12. (Hint: make a direct conversion from English to metric, and then convert within the metric system.)
   (a) 2.5 ft to cm
   (b) 20 ft^2 to cm^2
   (c) 1000 in^3 to m^3
   (d) 6 qt to mL
   (e) 64 oz to kg
13. **(a)** 200 cm to ft
   (b) 85 m^2 to ft^2
   (c) 100 m^3 to yd^3
   (d) 84 mL to fl oz
   (e) 64 kg to lb
14. (Hint: make a conversion within the metric system, and then convert directly to the English system.)
   (a) 2 m to in.
   (b) 4.2 m^2 to in^2
   (c) 10 m^3 to in^3
   (d) 28 L to fl oz
   (e) 5 kg to oz

TOPIC 4
SYSTEMS OF LINEAR EQUATIONS

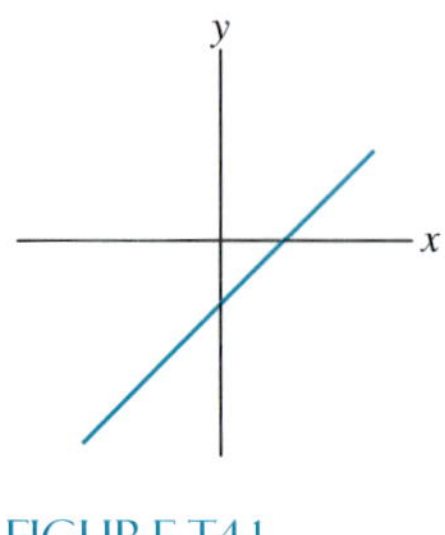

FIGURE T4.1

A **linear equation** is any equation we can write in the form $ax + by = c$, $ax + by + cz = d$, or a similar form with more variables, where x, y, z, etc. represent variables, and a, b, c, etc. represent fixed numbers. Such equations are called linear equations since the graph of an equation of the form $ax + by = c$ is a line. For example, the graph of the equation $x - y = 3$ (that is, all ordered pairs that make the equation true) is shown in Figure T4.1.

A **system of equations** is a set of two or more equations written in terms of the same variables. Solving a system of equations means finding all the values for the variables that satisfy each of the equations in the system. In the case of a system of equations of the form $ax + by = c$, one **solution** would be an ordered pair of numbers (x, y) that makes the equation true; the **solution set** for such an equation is the set of all such ordered pairs. In a system of linear equations having two variables, a solution, (x, y), would be a point that is on the graph of each equation in the system. Since each equation is represented by a line, a solution is a point where the lines intersect. The system

$$\begin{aligned} x - y &= 3 \\ 3x + y &= 5 \end{aligned}$$

is a linear system in two variables. The ordered pair $(2, -1)$ is a solution because it satisfies each equation in the system:

$$\begin{aligned} x - y &= 2 - (-1) = 2 + 1 = 3 \\ 3x + y &= 3(2) + (-1) = 6 - 1 = 5. \end{aligned}$$

As we will see, not every system has a solution, and some systems have more than one solution. For our example, we need to see if there are any more solutions. If we solve each equation for y, we will have

$$\begin{aligned} y &= x - 3 \\ y &= -3x + 5. \end{aligned}$$

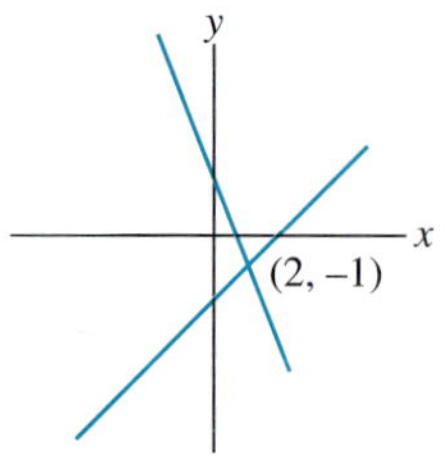

FIGURE T4.2

The graphs of these two equations are shown in Figure T4.2. Since the two lines can intersect in at most one point, the point of intersection, $(2, -1)$, is the only solution to the system.

A system of linear equations does not necessarily have a solution, and if it does, the solution may not be unique. This is easily demonstrated with a system of two equations having two variables since each equation has a line as its graph. Figure T4.3 shows the only three possible arrangements of pairs of lines.

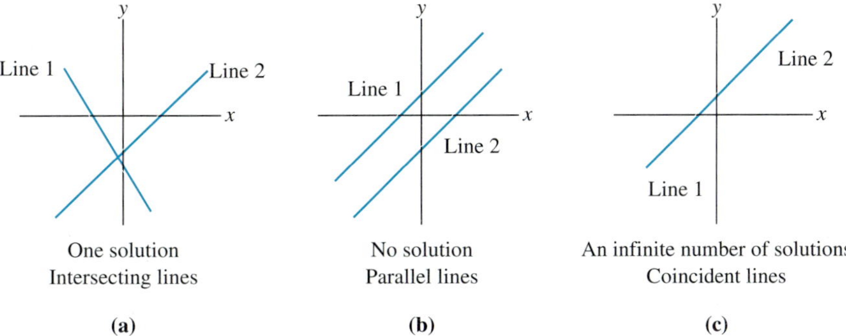

FIGURE T4.3

In Figure T4.3(a), the solution is the point of intersection of the lines. In Figure T4.3(b), there is no solution since there is no point common to both lines. In Figure T4.3(c), there are infinitely many solutions, namely all the points on the line (which represents both equations).

The graphs of two lines will be coincident if the equations are equivalent, that is, if one equation is a multiple of another. For example, the graphs of the equations

$$2x - 3y = 4 \qquad \text{and} \qquad 4x - 6y = 8$$

will be the same line.

SOLVING SYSTEMS OF LINEAR EQUATIONS

Next we will consider various techniques that can be used to find the solution sets of systems of linear equations.

GRAPHING

When working with a system of linear equations that has only two variables, it is often easy to graph the equations in the system. If the lines intersect, the point of intersection is the solution to the system. Graphing calculators can be used to solve a system of equations graphically. However, the disadvantage of the graphing method is that it may be impossible to identify the exact coordinates of the points of intersection if they are not integers. Thus we will use algebraic techniques to produce exact answers.

SUBSTITUTION

The simplest algebraic technique we can use is substitution. To apply this method, we solve one of the equations for one of the variables and then substitute for this variable in the other equations. To solve the system

$$\begin{aligned} 2x + 3y &= 11 \\ 3x - y &= 5 \end{aligned}$$

using substitution, we solve for one variable from either equation. Here we can solve for y in the second equation to obtain

$$\text{(i)} \quad y = 3x - 5.$$

This value of y, namely $3x - 5$, is then substituted for y in the first equation to obtain

$$\text{(ii)} \quad 2x + 3(3x - 5) = 11.$$

Next we solve for x in equation (ii).

$$\begin{aligned} 2x + 9x - 15 &= 11 \\ 11x &= 26 \\ x &= \frac{26}{11}. \end{aligned}$$

To find the complete solution, we now substitute this value of x in equation (i) to solve for y.

$$y = 3\left(\frac{26}{11}\right) - 5 = \frac{78}{11} - \frac{55}{11} = \frac{23}{11}$$

The solution set to the system is $\left(\frac{26}{11}, \frac{23}{11}\right)$, which can be verified in the original equations.

$$\text{Check: } 2\left(\frac{26}{11}\right) + 3\left(\frac{23}{11}\right) = \frac{52}{11} + \frac{69}{11} = \frac{121}{11} = 11 \quad \checkmark$$

$$3\left(\frac{26}{11}\right) - \frac{23}{11} = \frac{78}{11} - \frac{23}{11} = \frac{55}{11} = 5 \quad \checkmark$$

The substitution method can be applied to systems with three or more variables, but as the number of variables increases, the method becomes more unwieldy.

ELIMINATION

Another method for solving linear systems is the elimination method. In this method, you eliminate variables from the system one at a time until you have the value for one of the variables. The remaining values are then found through back substitution. The elimination method will also provide the necessary information regarding systems with no solutions or the presence of an infinite number of solutions. Because the elimination method is systematic, its application can be extended to any number of equations and variables.

The elimination method is based on two properties of equality you have previously used in algebra to work with equations.

Properties of Equations

If $A = B$, then $kA = kB$, where $k \neq 0$.

If $A = B$ and $C = D$, then $A + C = B + D$ and $A - C = B - D$.

In words, this means "If both sides of an equation are multiplied (or divided) by the same nonzero constant, the resulting equation is equivalent to the first," and "If equals are added to or subtracted from equals, then the results are equal."

The elimination method uses the two properties of equations to systematically eliminate variables from equations, hence reducing the system to an equivalent simpler one. Next we solve the previous system using the elimination method.

$$\begin{aligned} \text{E1: } & 2x + 3y = 11 \\ \text{E2: } & 3x - y = 5 \end{aligned}$$

We have labeled the equations E1 and E2 so that we may refer to them in the steps of the process.

By adding equations, we can eliminate a variable using the second property of equations, provided that we have equations in which the coefficients for one of the variables are opposites of each other. We will eliminate x from the second equation by multiplying the first equation by $-\frac{3}{2}$ and adding the resulting equation to the second equation. We obtain a new equation E3 that does not involve x.

$$\begin{aligned} -\frac{3}{2} \times \text{E1: } & -3x - \frac{9}{2}y = -\frac{33}{2} \\ \text{E2: } & 3x - y = 5 \\ \text{E3: } & -\frac{11}{2}y = -\frac{23}{2} \qquad \left(\text{sum of } -\frac{3}{2} \times \text{E1 and E2}\right) \end{aligned}$$

The original system of equations is equivalent to the new system

$$\begin{aligned} \text{E1: } & 2x + 3y = 11 \\ \text{E3: } & -\frac{11}{2}y = -\frac{23}{2}. \end{aligned}$$

But the fact that E3 does not involve x makes the new system easy to solve. We have $-11y = -23$ or $y = \frac{23}{11}$.

Now that we know the value of y, we can substitute that value into E1 to find x, a process called **back substitution.**

$$2x + 3 \times \frac{23}{11} = 11.$$

Simplifying, we have $2x = \frac{52}{11}$ or $x = \frac{26}{11}$. Thus the solution to the system is $\left(\frac{26}{11}, \frac{23}{11}\right)$ as we found earlier.

Thus far, we have been solving systems of linear equations where there is a single solution. However, as mentioned earlier, it may be that a system of equations has no solution or an infinite number of solutions. Next we will see how the elimination method deals with these two possibilities.

A SYSTEM WITH NO SOLUTION

Solve the following system using elimination.

$$\begin{aligned} \text{E1: } & 2x - y = 10 \\ \text{E2: } & 6x - 3y = 15 \end{aligned}$$

First, we set out to eliminate x from the second equation.

$$\begin{aligned} -3 \times \text{E1:}\ -6x + 3y &= -30 \\ \text{E2:}\ 6x - 3y &= 15 \\ \text{E3:}\ 0 &= -15 \qquad (\text{sum of } -3 \times \text{E1 and E2}) \end{aligned}$$

The original system is equivalent to the system

$$\begin{aligned} \text{E1:}\ 2x - y &= 10 \\ \text{E3:}\ 0 &= -15. \end{aligned}$$

Because the equation E3 can never be satisfied, the system has no solution. Thus there is no ordered pair for (x, y) that will satisfy both equations E1 and E2.

The elimination method of solution is a logical process based on the assumption that there is a solution to the system. When the process produces a contradiction, it means that the original assumption is wrong; that is, the system does not have a solution. A system that does not have a solution is called **inconsistent.**

A SYSTEM WITH INFINITELY MANY SOLUTIONS

Solve the following system using elimination.

$$\begin{aligned} \text{E1:}\ x - 2y &= -6 \\ \text{E2:}\ 5x - 10y &= -30 \end{aligned}$$

First we set out to eliminate x from the second equation.

$$\begin{aligned} -5 \times \text{E1:}\ -5x + 10y &= 30 \\ \text{E2:}\ 5x - 10y &= -30 \\ \text{E3:}\ 0 &= 0 \qquad (\text{sum of } -5 \times \text{E1 and E2}) \end{aligned}$$

The original system is equivalent to the system

$$\begin{aligned} \text{E1:}\ x - 2y &= -6 \\ \text{E3:}\ 0 &= 0. \end{aligned}$$

The equation $0 = 0$ is always true, and thus we may omit it from the system. The remaining equation $x - 2y = -6$ is satisfied by infinitely many pairs of numbers x and y. To find all the solutions to the system, let x have an arbitrary value. Then solving for y, we have

$$2y = x + 6 \quad \text{or} \quad y = \frac{x + 6}{2}.$$

Thus the solution set of the system may be written as

$$\left\{\left(x, \frac{x + 6}{2}\right) \middle| \ x \text{ is any real number}\right\}.$$

Several elements of this solution set can be listed by choosing values for x. If $x = 0$, then $y = 3$, so $(0, 3)$ is a solution. If $x = 1$, then $y = \frac{7}{2}$, so $\left(1, \frac{7}{2}\right)$ is a solution, and so on. These are referred to as particular solutions to the system.

The set of all solutions is called the **general solution** of a system of equations whereas individual solutions are referred to as **particular solutions.**

PROBLEM SET T4

In problems 1 and 2, determine whether the given values are or are not a solution to the system.

1. $(3, -1)$
$3x - 2y = 11$
$x + 3y = 2$

2. $(1, -2)$
$4x - y = 6$
$3x + 2y = -1$

Solve problems 3 through 8 using the substitution method.

3. $y = -2x + 3$
$3x + 2y = -17$

4. $5x - 2y = 18$
$x = 2y - 7$

5. $3x - 2y = 15$
$x + y = 10$

6. $4x - 3y = 13$
$7x - y = -7$

7. $6x + y = 7$
$5x - 2y = 3$

8. $2x + 2y = 15$
$3x - y = 10$

Solve problems 9 through 12 using the elimination method.

9. $4x + y = 1$
$3x - y = 6$

10. $3x - 2y = -13$
$x + 2y = 5$

11. $x - 2y = 12$
$2x + y = 4$

12. $3x + 2y = 7$
$x - 3y = 6$

In problems 13 through 16, determine whether the following systems have (i) one solution, (ii) no solutions, or (iii) infinitely many solutions. In case (iii), describe the solution set and find three particular solutions.

13. $6x + 3y = 12$
$4x + 2y = 8$

14. $4x - 4y = 20$
$3x - 3y = 12$

15. $2x + y = 12$
$-6x - 3y = -36$

16. $4x - y = 10$
$6x - 3y = 15$

T O P I C

PERIMETER AND AREA

PERIMETER

The **perimeter** of a closed plane figure is the length of the boundary that surrounds it. Formulas can be used to find perimeters for the following figures.

A **polygon** is a figure consisting of line segments in the plane that can be traced so that the starting and ending points are the same, and the path never crosses itself or is retraced.
A **triangle** is a polygon with three sides.
A **quadrilateral** is a polygon with four sides.
A **parallelogram** is a quadrilateral whose opposite sides are parallel.
A **rectangle** is a quadrilateral with four 90° angles.
A **rhombus** is a quadrilateral whose sides have the same length.
A **trapezoid** is a quadrilateral that has one pair of parallel sides.
A **square** is a quadrilateral whose sides have the same length and whose angles are all 90°.

DEFINITION

Perimeter of a Polygon

The **perimeter** of a polygon is the sum of the lengths of its sides.

This definition of perimeter leads to the following formulas for finding the perimeters of the quadrilaterals listed previously.

Perimeters of Common Quadrilaterals

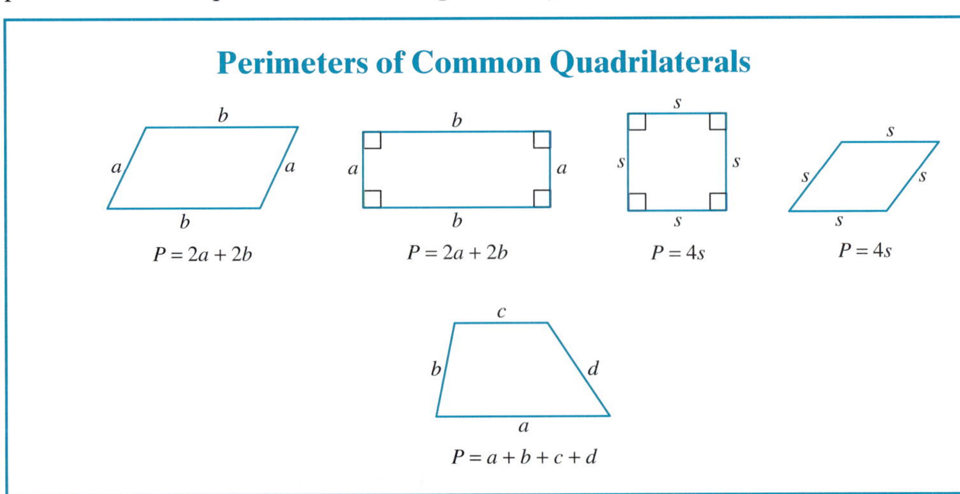

The perimeter of a circle, or the distance around the circle, is called its **circumference.** A **diameter** of a circle is any line segment containing the center of the circle and whose endpoints are points of the circle. A **radius** is any line segment whose endpoints are the center and a point of the circle. The words diameter and radius are also used to represent their respective lengths. If C represents the circumference and d represents the diameter of any circle, then $\frac{C}{d}$ is approximately 3.14, or $\frac{22}{7}$ as a fraction. The ratio $\frac{C}{d}$, which is represented by the Greek letter π (**pi**), is an irrational number; that is, it is a nonterminating, nonrepeating decimal.

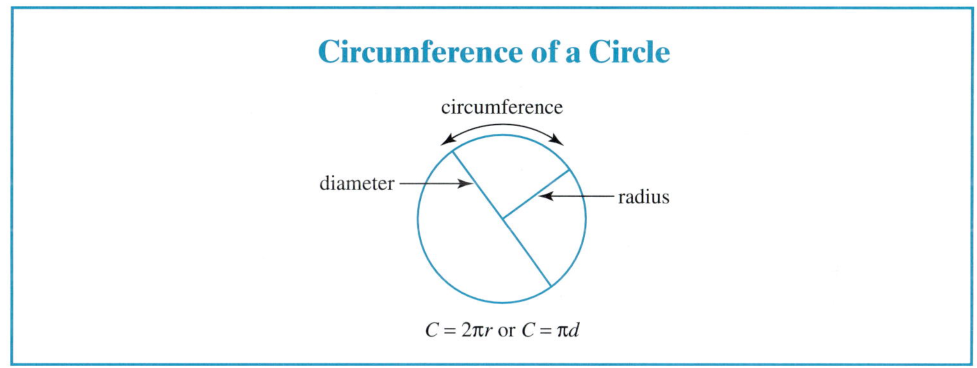

AREA

The **area** of a region in the plane is measured with respect to the number of small squares of standard size that are required to cover the region. For example, the region enclosed by the curve in Figure T5.1(a) can be estimated to be about six square units since it is covered by approximately six squares. If smaller squares are used to cover the figure, such as four per unit square as in Figure T5.1(b), then a better estimate of the area may be possible.

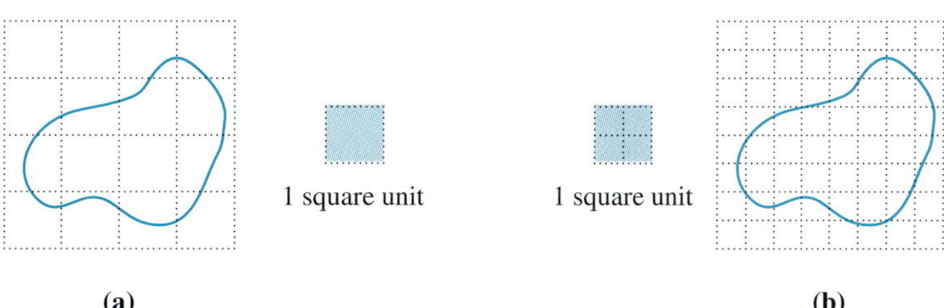

FIGURE T5.1

By making smaller and smaller squares, we can approximate the area more accurately. This suggests the following definition.

(a) For every region enclosed by a simple closed curve and unit square, there is a positive real number, called **area**, that gives the number of unit squares (and parts of unit squares) that exactly cover the region enclosed by the simple closed curve, and

(b) the area of a region enclosed by a simple closed curve is the sum of the areas of the smaller regions into which the region can be subdivided.

Suppose that a rectangle that measures 3 cm by 5 cm is covered by squares which are 1 cm by 1 cm (Figure T5.2).

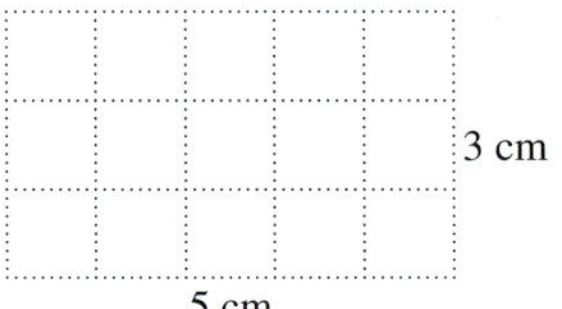

Three rows of 5 squares, or exactly 15 square centimeters, cover the rectangular region. We say that the area of the rectangular region is 15 square centimeters, written as 15 cm^2. Notice that $3 \times 5 = 15$. The length and width of any rectangle can be used to find its area.

The area of a rectangle is the product of its length and width.

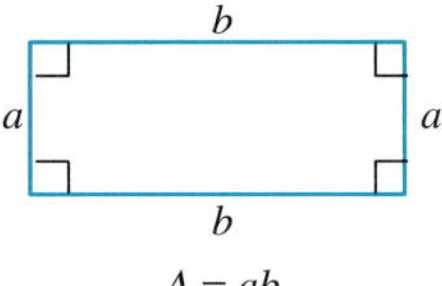

$A = ab$

A square is a special rectangle where the length and width are the same. So, in the case of a square with side of length s, its area is $s \times s$, or s^2.

The area of a square is the square of the length of any of its sides.

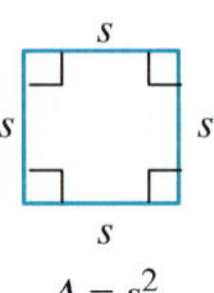

$A = s^2$

The area of a **right triangle,** that is, a triangle having a 90° angle, is one-half the area of a rectangle made up of two such triangles (Figure T5.3).

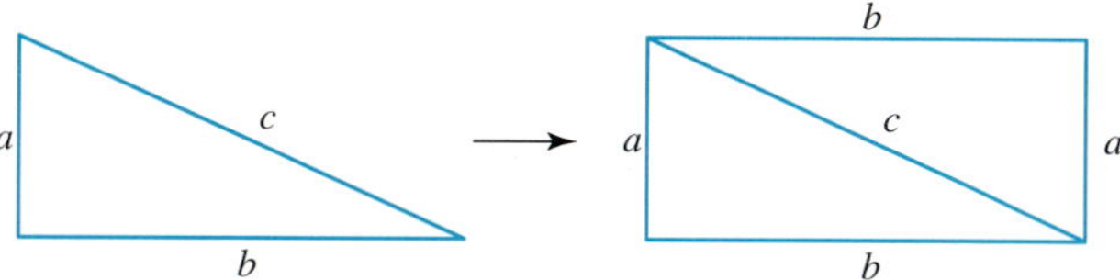

FIGURE T5.3

Area of a Right Triangle

The area of a right triangle is one-half the product of the lengths of its legs.

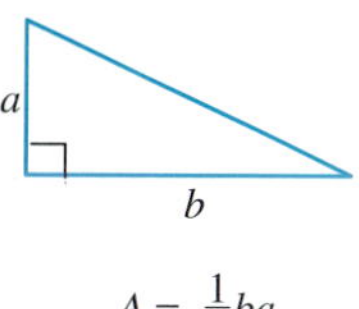

$$A = \frac{1}{2}ba$$

A line segment from a vertex of a triangle that is perpendicular to the opposite side of the triangle is called an **altitude** of the triangle. The length of an altitude is also referred to as the altitude or **height** to the particular side (Figure T5.4).

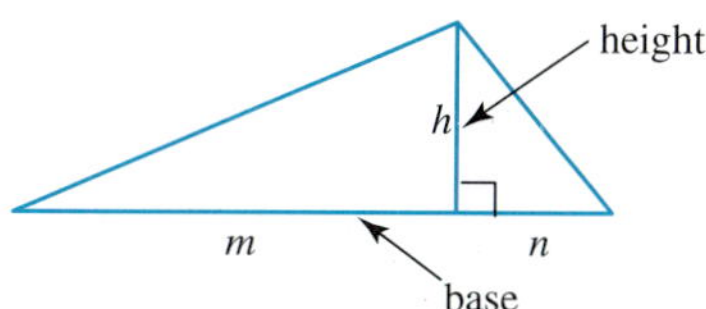

FIGURE T5.4

The side to which the altitude is drawn is called the **base** of the triangle associated with that altitude. The triangle in Figure T5.4 can be viewed as being made up of two right triangles—one with a base of m and an altitude of h, and the other with a base of n and an altitude of h. The area is the sum of the areas of the two smaller triangles, namely $\frac{1}{2}mh + \frac{1}{2}nh$ or $\frac{1}{2}(m + n)h$. That is, the area of the larger triangle is one-half the product of its base and the height.

Area of a Triangle

The area of a triangle is one-half the product of the lengths of one side and the height to that side.

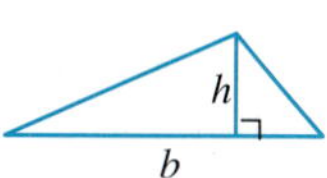

$$A = \frac{1}{2}bh$$

We can use the formula for finding the area of any triangle to find the areas of other polygons. Consider the parallelogram in Figure T5.5(a) whose base has length b and whose height, the distance between two parallel sides, is h.

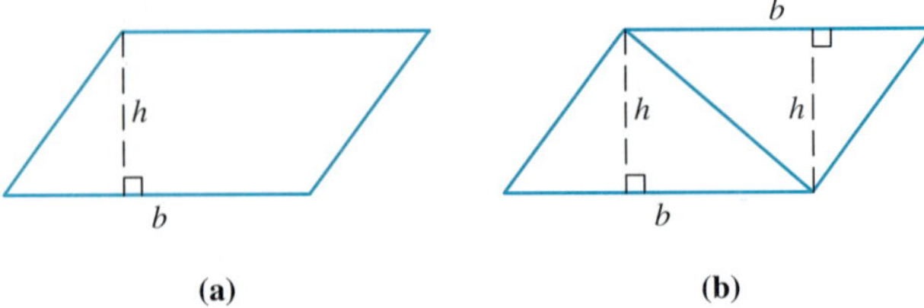

FIGURE T5.5

When a diagonal is drawn as shown in Figure T5.5(b), the parallelogram is divided into two triangles each having an area of $\frac{1}{2}bh$. Thus the area of the parallelogram is $2\left(\frac{1}{2}bh\right)$ or simply bh.

Area of a Parallelogram

The area of a parallelogram is equal to the product of the length of one of its sides and the height to that side.

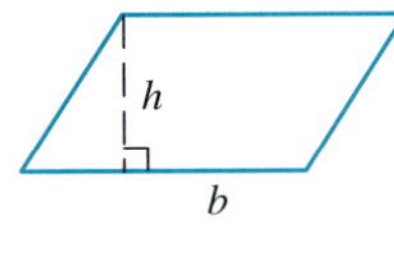

$A = bh$

The **height** of a trapezoid is the perpendicular distance between the two bases. The same technique used to derive the formula for the area of a parallelogram can be used to find a formula for the area of a trapezoid, namely, find the sum of the areas of the two triangles formed by the diagonal in Figure T5.6: $A = \frac{1}{2}ah + \frac{1}{2}bh = \frac{1}{2}(a + b)h$.

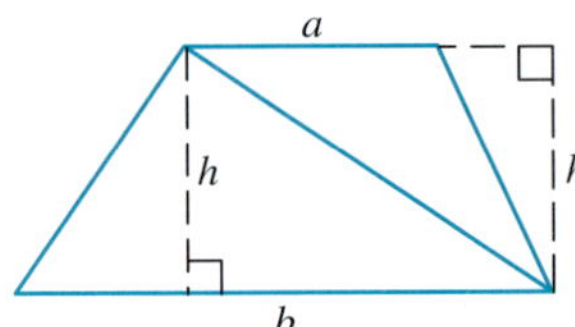

FIGURE T5.6

Area of a Trapezoid

The area of a trapezoid is equal to one-half the product of the sum of its bases and its height.

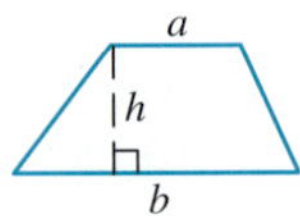

$A = \frac{1}{2}(a + b)h$

A **regular polygon** is a polygon whose sides have the same length and whose angles have the same measure. It can be shown that there is a point in the plane that is equidistant from all vertices of a regular polygon and from all the sides of the polygon. This point is called the **center** of the polygon. In Figure T5.7(a), point O is the center of the regular hexagon, and it is equidistant from all vertices. In Figure T5.7(b), we see that the center is equidistant from all the sides.

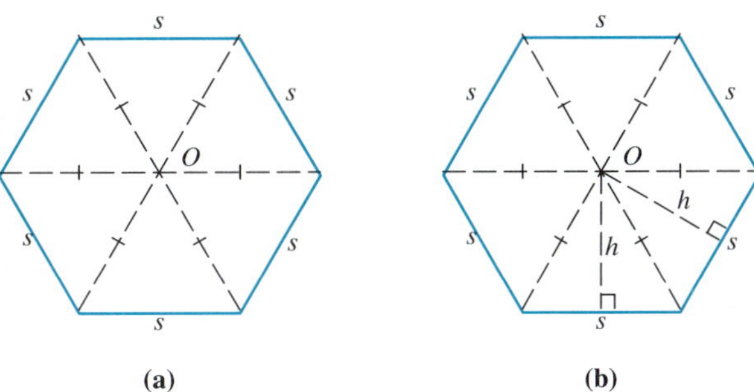

FIGURE T5.7

The area of any regular polygon can be found by conveniently subdividing the polygon into triangles as suggested by the regular hexagon in Figure T5.7. The area of each of the six triangles is $\frac{1}{2}sh$; where s is the length of a side of the polygon, and h is the perpendicular distance from the center to one of the sides [Figure T5.7(b)]. Thus the area of the hexagon is $6\left(\frac{1}{2}sh\right)$. This equation can be rewritten as $\frac{1}{2}h(6s)$, which is equivalent to $\frac{1}{2}hP$, where $P = 6s$ is the perimeter of the hexagon. This technique can be generalized to any regular polygon.

Area of a Regular Polygon

The area of a regular polygon is equal to one-half the product of the perimeter of the polygon and the perpendicular distance from its center to one of its sides.

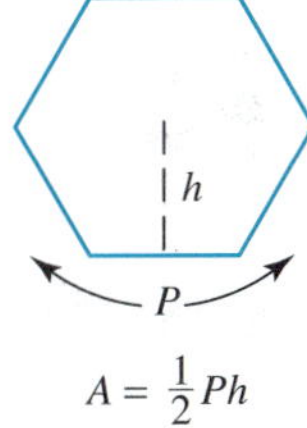

Now imagine a series of regular polygons where the number of sides increases to approach a circle (Figure T5.8).

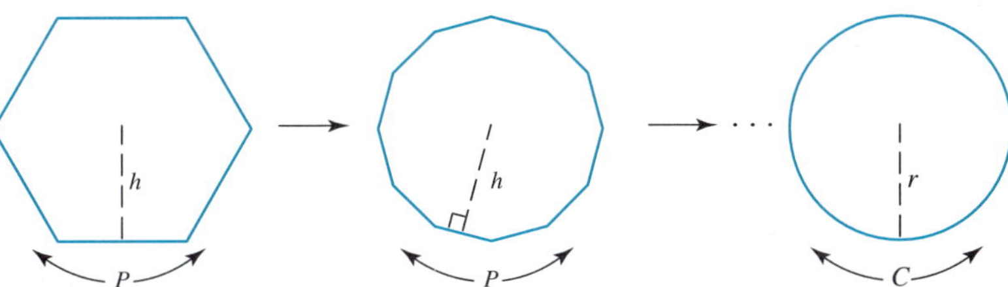

FIGURE T5.8

The height, h, in the formula for the area of a regular polygon becomes the radius, r, of the circle; and the perimeter, P, becomes the circumference, C, of the circle. By substituting r for h and C for P in the area formula for regular polygons, we can find the formula for the area of a circle. We will also use the formula for the circumference of the circle when we simplify the expression. That is,

$$A = \frac{1}{2}hP = \frac{1}{2}rC = \frac{1}{2}r(2\pi r) = \pi r^2.$$

Area of a Circle

The area of a circle with radius r is equal to πr^2.

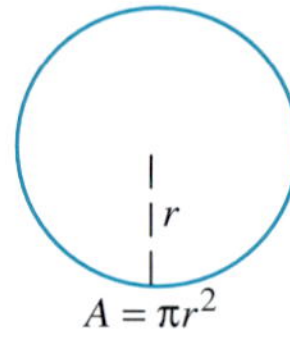

PROBLEM SET T5

1. Find the perimeter of each of the following.
   (a)

   (b)

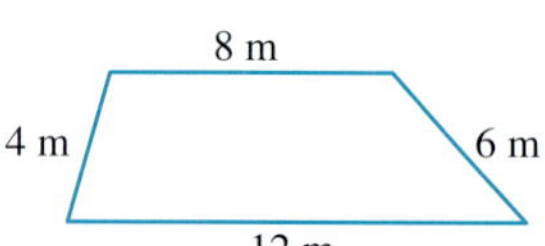

   (c)

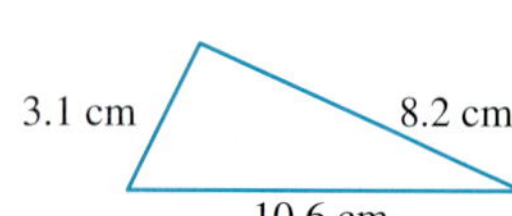

   (d)

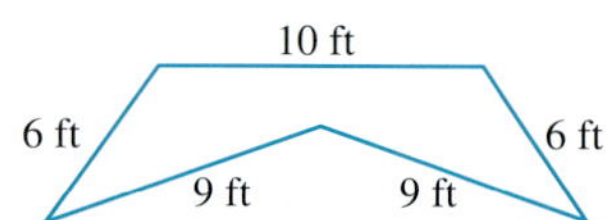

2. Find the perimeter of each of the following.
   (a) a square with sides of length 8.7 cm
   (b) a right triangle with legs of length 6 in. and 8 in.
   (c) an isosceles trapezoid with bases of length 4 ft and 10 ft and sides of length 5 ft.
   (d) a parallelogram with sides of length 71.4 in. and 36.9 in.
3. Find the perimeter of the following figure consisting of a rectangle topped by two congruent semicircles.

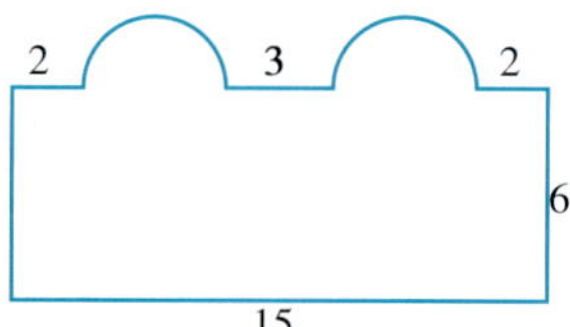

4. Find the perimeter of the following figure consisting of a square with sides of length 8 cm and 4 semicircles with radius 2 cm.

5. Find the circumference of a circle with radius 15 cm.

6. Find the diameter of a circle with a circumference of 66 m. (Use $\pi \approx \frac{22}{7}$.)

7. Find the area of each of the following figures.

(a)

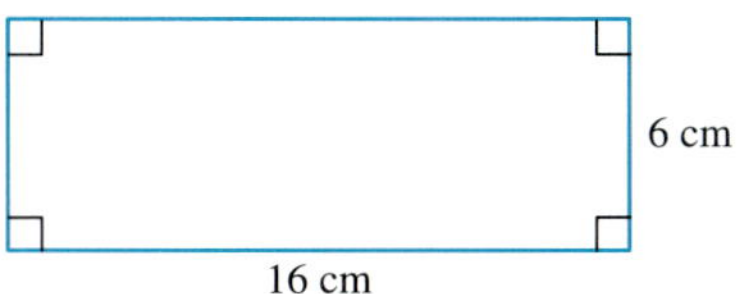

(b)

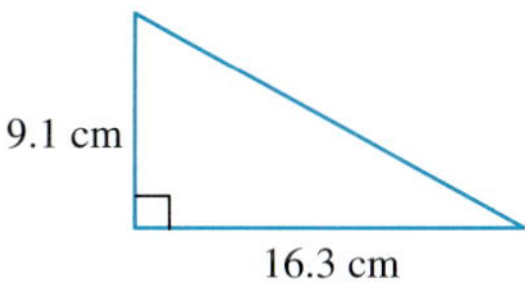

(c)

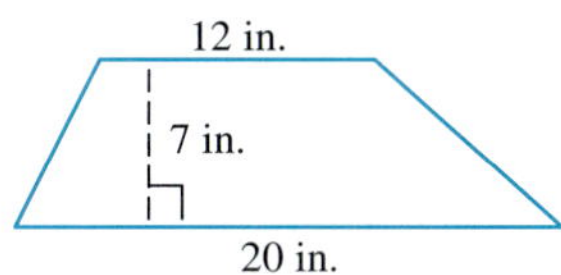

8. Find the area of each of the following figures.

(a)

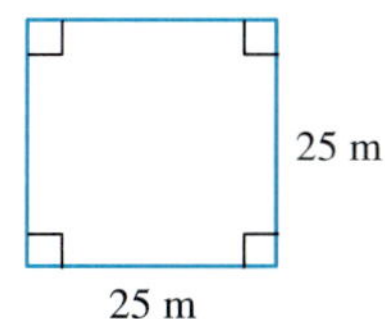

(b)

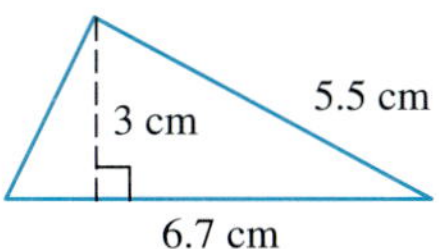

(c)

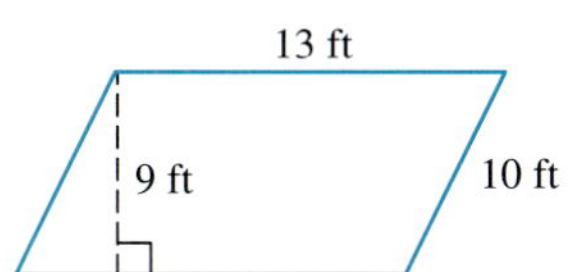

9. Find the area of the following figure consisting of a rectangle topped by two congruent semicircles.

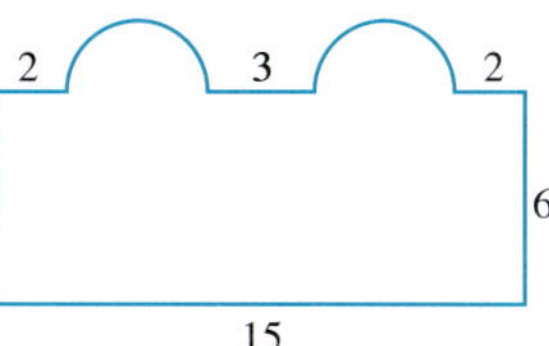

10. Find the area of the following figure consisting of a square with sides of length 12 in. and 4 semicircles with radius 3 in.

11. In a regular hexagon with sides of length 2, the distance from the center to the midpoint of any side is $\sqrt{3}$. Find the area of the hexagon.

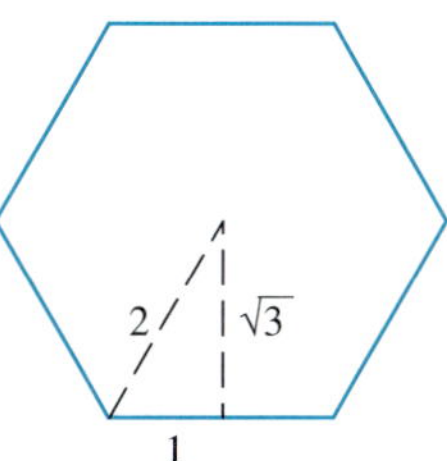

12. A regular octagon can be divided into eight congruent triangles, as shown. If the length of each side is 4 cm and the area of the octagon is 77.25 sq cm, find the distance from the center of the octagon to the midpoint of any side.

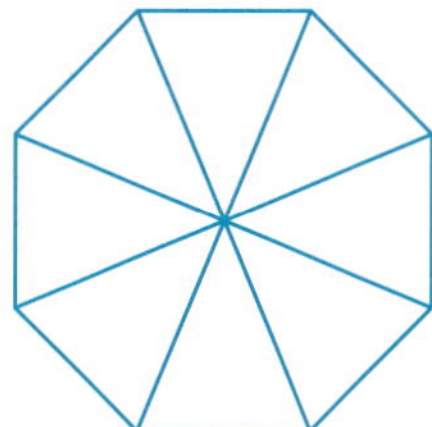

TOPIC 

SURFACE AREA AND VOLUME

PRISMS AND PYRAMIDS

PLANE THREE-DIMENSIONAL SHAPES

A **polyhedron** (plural is polyhedra) is a three-dimensional shape composed of polygonal regions, any two of which have at most a common side. In addition, it is an enclosed, connected finite portion of space without holes. Figure T6.1 shows three examples of polyhedra.

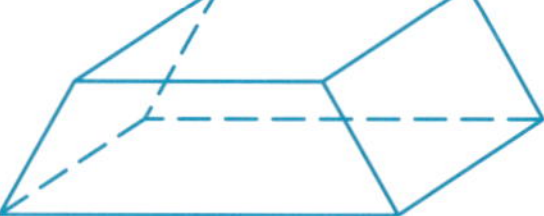

FIGURE T6.1

The polygonal regions of a polyhedron are called **faces,** the common line segments are called **edges,** and a point where edges meet is called a **vertex** (Figure T6.2).

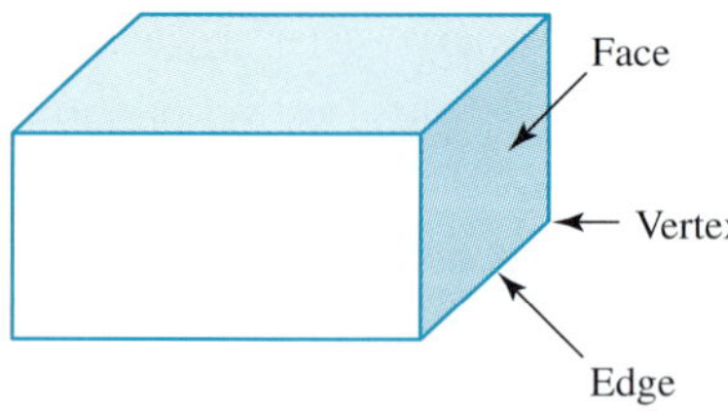

FIGURE T6.2

A **prism** is a polyhedron with two opposite faces, called **bases,** that are identical polygonal regions in parallel planes; the other faces are called **lateral faces.** Prisms are named by the shape of their bases (Figure T6.3).

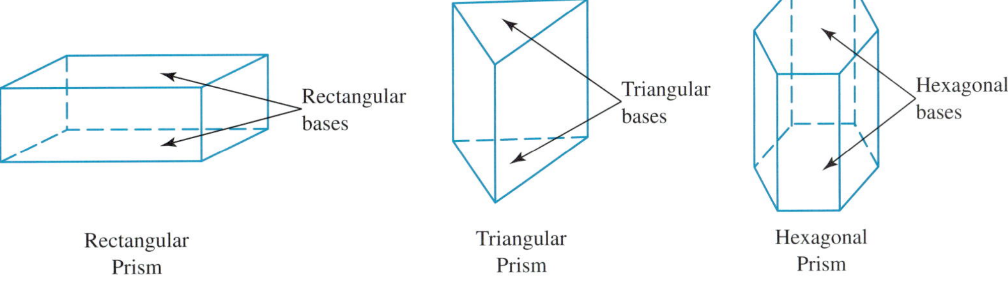

FIGURE T6.3

Prisms whose lateral faces are rectangular regions are called **right prisms;** otherwise they are called **oblique prisms** (Figure T6.4).

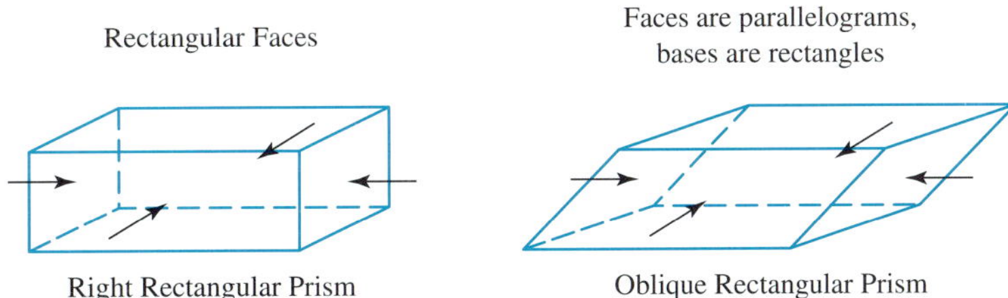

FIGURE T6.4

A **pyramid** is a polyhedron consisting of a polygonal region for its base and triangular regions as **lateral faces.** A pyramid is formed when a point that is not in the plane of the base is connected by line segments to the vertices of the base. This point is called the **apex.** The name of a pyramid is determined by the shape of its base. When the base of a pyramid is a regular polygon and the sides from the apex are all the same length, the pyramid is called a **right regular pyramid;** otherwise it is an **oblique regular pyramid.** The perpendicular distance from the apex to the base of a pyramid is called the **height** of the pyramid, and the **slant height** of a right regular pyramid is the height of any of its faces. (Figure T6.5).

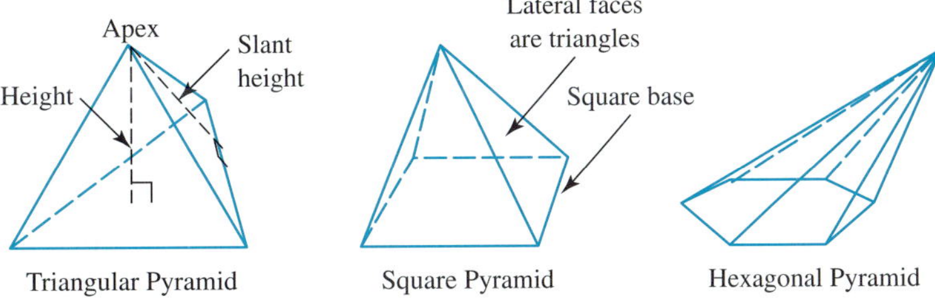

FIGURE T6.5

Surface Area

The **surface area of a polyhedron** is the sum of the areas of its bases and lateral faces. We use "*SA*" as an abbreviation for surface area. To find the surface area of the right rectangular prism in Figure T6.6, we find the area of each base is $8 \times 5 = 40 \text{ cm}^2$.

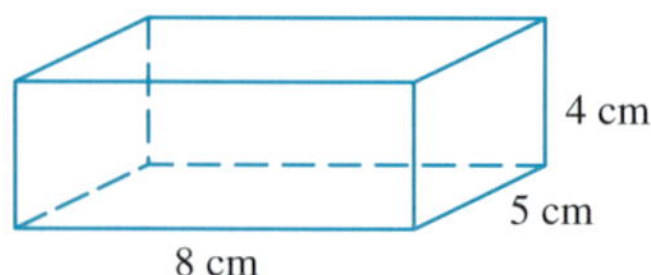

FIGURE T6.6

Two of the lateral faces have area $4 \times 5 = 20 \text{ cm}^2$, and two have area $8 \times 4 = 32 \text{ cm}^2$. So the surface area is

$$2 \times 40 + 2 \times 20 + 2 \times 32 = 2 \times (40 + 20 + 32) = 184 \text{ cm}^2.$$

The following method for finding the surface area for the prism above can be extended to all right prisms.

The area of the lateral faces can also be found by multiplying the perimeter of the base by the height of the prism:

$$(8 + 5 + 8 + 5) \times 4 = 104 \text{ cm}^2$$
$$SA = (\text{area of bases}) + (\text{area of faces}) = 80 + 104 = 184 \text{ cm}^2$$
$$SA = 2 \times (\text{area of base}) + (\text{perimeter of base}) \times \text{height}$$

Surface Area of a Right Prism

The surface area of a right prism is the sum of twice the area of its base plus the product of the perimeter of the base and the height of the prism.

$$SA = 2A + Ph$$

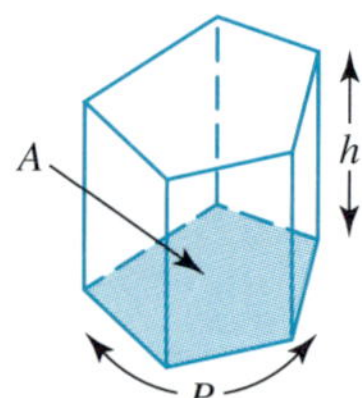

The surface area of a pyramid is the sum of the areas of its base and lateral faces (which are triangles).

Surface Area of a Right Regular Pyramid

The surface area of a right pyramid is the sum of the area of its base plus one-half the product of the perimeter of the base and the slant height.

$$SA = A + \frac{1}{2}Pl$$

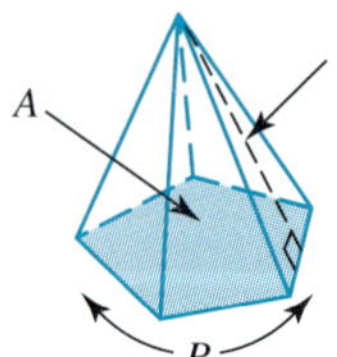

Volume

Volume is the three-dimensional analog of area. To find the volume of a shape, a unit cube must be designated. Then we must determine how many unit cubes completely fill a given three-dimensional shape (Figure T6.7).

FIGURE T6.7

Volume

(a) For every polyhedron and unit cube, there is a positive real number, called **volume,** that gives the number of unit cubes (and parts of unit cubes) that exactly fill the region enclosed by the polyhedron.

(b) The volume of the region enclosed by a polyhedron is the sum of the volumes of the smaller regions into which the region can be subdivided.

To find the volume of a right rectangular prism 3 cm by 4 cm by 2 cm, we use a unit cube of 1 cm by 1 cm by 1 cm and count the number of unit cubes that will fill the prism exactly (Figure T6.8).

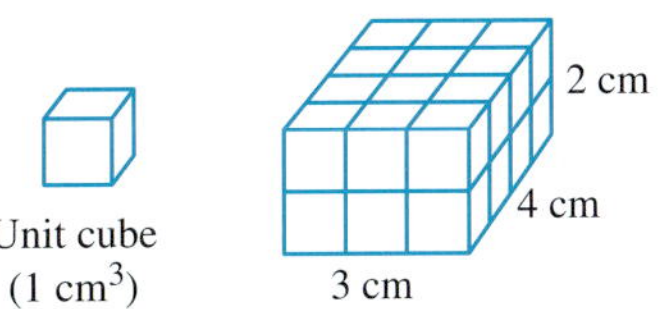

FIGURE T6.8

There are $3 \times 4 = 12$ cubes in each of 2 layers, or $3 \times 4 \times 2 = 24$ unit cubes in all. Thus the volume is 24 cm^3. This volume may be viewed as the product of the area of the base of the prism (3×4) and its height (2). This result holds true for all prisms, whether they are right prisms or oblique.

Volume of a Prism

The volume of a prism is the product of the area of its base and its height.

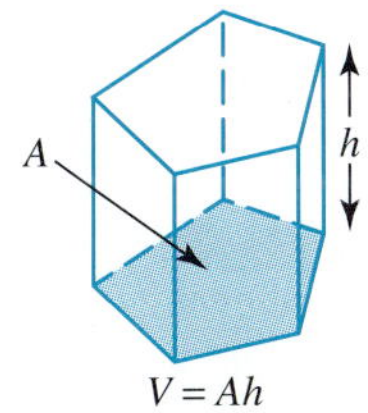

Formulas for a right rectangular prism and a cube are immediate consequences of this definition of volume. For the right rectangular prism, $V = lwh$ and for the cube, $A = s^3$.

Volume of a Right Rectangular Prism

The volume of a right rectangular prism is the product of its length, width, and height.

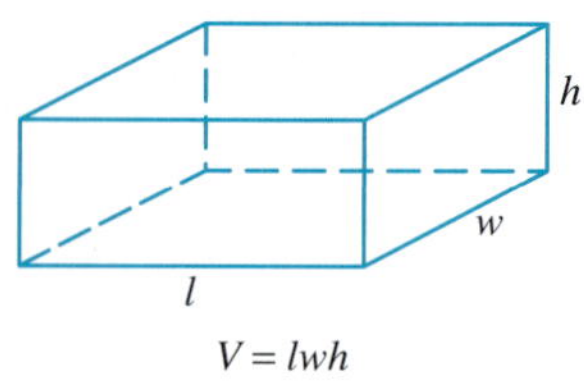

Volume of a Cube

The volume of a cube is the cube of the length of its side.

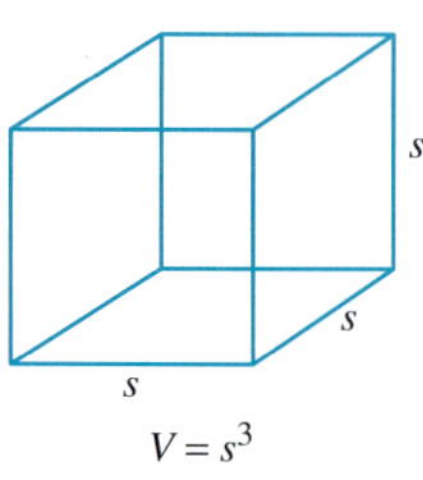

The volume of a pyramid may be inferred from Figure T6.9 where three pyramids identical to the one in Figure T6.9(a) are shown to be filling a right prism completely in Figure T6.9(b).

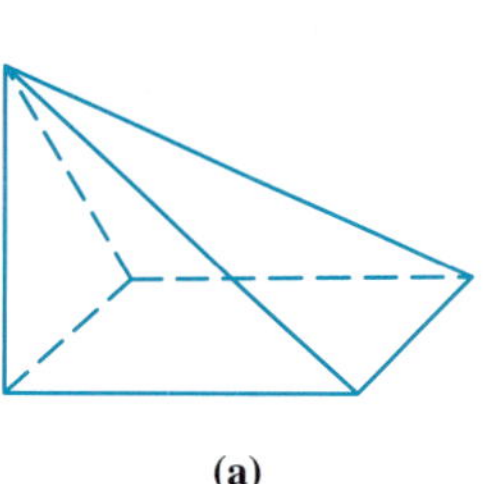

(a)

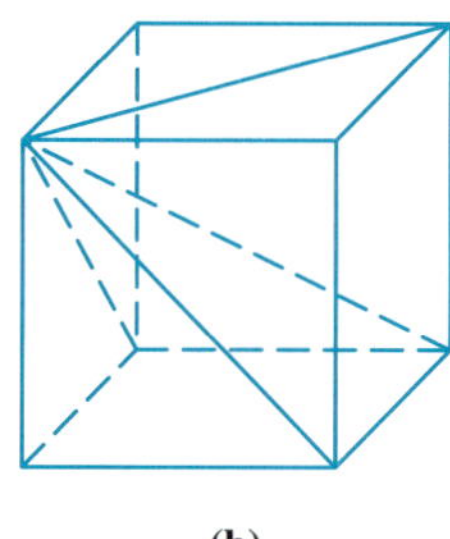

(b)

FIGURE T6.9

In general, the volume of a pyramid is one-third the volume of the corresponding prism determined by the base and height of the pyramid.

Volume of a Pyramid

The volume of a pyramid is one-third the product of the area of its base and its height.

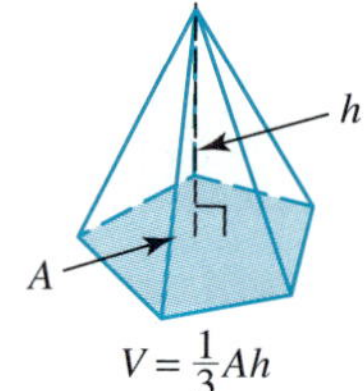

CYLINDERS, CONES, AND SPHERES

CURVED THREE-DIMENSIONAL SHAPES

A **circular cylinder** is the shape formed by two identical circular regions in parallel planes together with the surface formed by line segments joining corresponding points of the two circles (Figure T6.10).

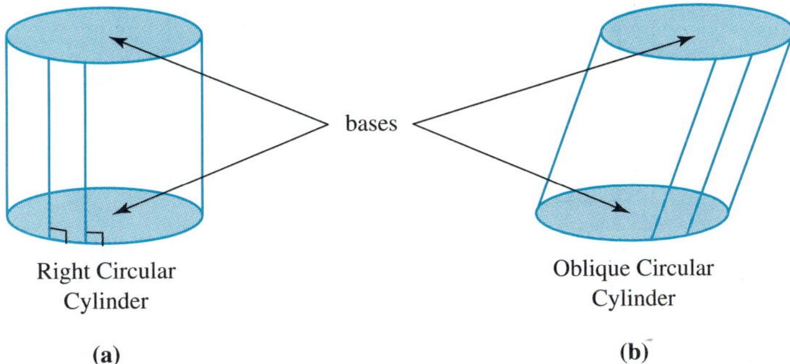

FIGURE T6.10

The two circular regions are called the **bases of the cylinder.** When the line segments joining corresponding points of the two bases are perpendicular to the planes containing the bases, the cylinder is called a **right circular cylinder** [Figure T6.10(a)]; otherwise it is an **oblique circular cylinder** [Figure T6.10(b)].

A **circular cone** is the shape formed by a circular region, called the base, together with the surface formed when the apex, a point not in the same plane as the circle, is joined by line segments to every point on the circle (Figure T6.11). Notice that the relationship of the cone to the cylinder is analogous to the relationship between a pyramid and a prism.

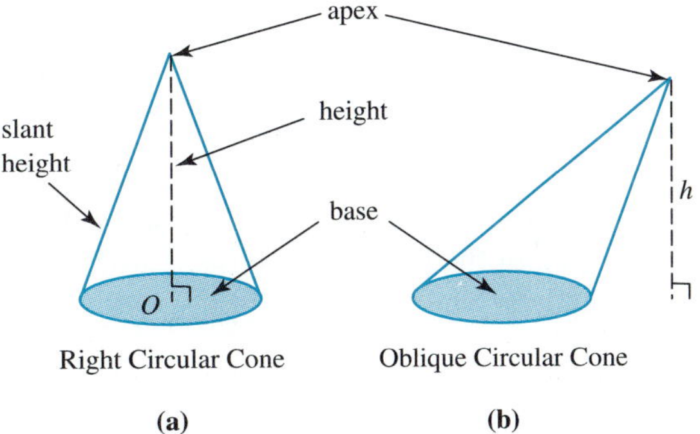

FIGURE T6.11

A **right circular cone** is a circular cone with its apex on the line that is perpendicular to the base at its center [Figure T6.11(a)]; otherwise circular cones are **oblique** [Figure T6.11(b)]. In the right circular cone, the perpendicular distance from the apex to the center of the base is called the **height of the cone.** For an oblique circular cone, the height is the perpendicular distance from the apex to the plane containing the base. For a right circular cone, the distance from the apex to a point on the edge of the base is called the **slant height.**

A **sphere** is the set of all points in space that are a fixed distance from a given point (the **center of the sphere**). A line segment whose endpoints are the center and a point on the sphere is called **a radius of the sphere;** the length of any radius is called **the radius of the sphere.** A line segment that contains the center of the sphere and whose endpoints are on the sphere is called **a diameter of the sphere;** the length of any diameter is called **the diameter of the sphere** (Figure T6.12).

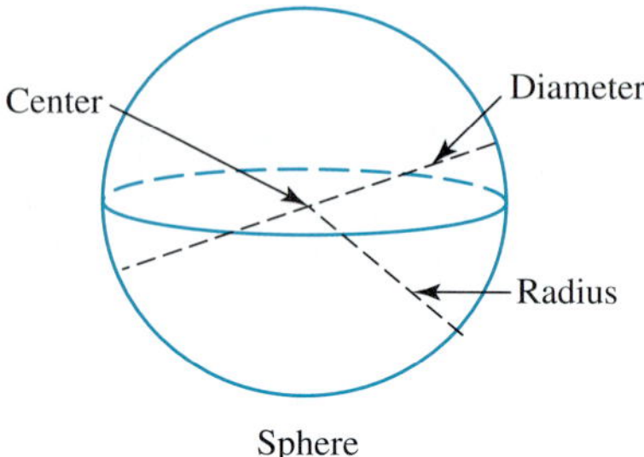

FIGURE T6.12

Surface Area

The surface area of a cylinder can be found by adapting the formula for the surface area of a prism as suggested by Figure T6.13.

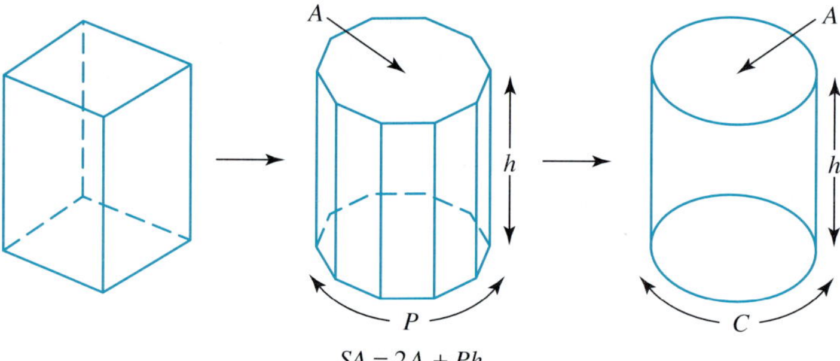

FIGURE T6.13

In the case of a right circular cylinder, whose base has radius r, A represents the area of the base, namely πr^2, and P represents the perimeter (C the circumference), namely $2\pi r$. This leads to the following.

Surface Area of a Right Circular Cylinder

The surface area of a right circular cylinder is twice the area of its base plus the circumference of its base times its height.

$$SA = 2A + Ch = 2\pi r^2 + 2\pi rh$$

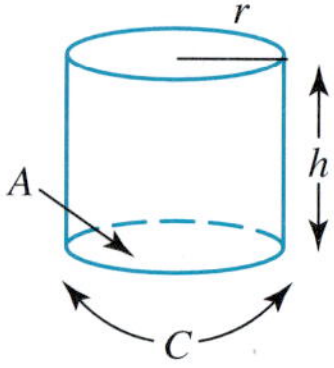

Using a method similar to the one we used to find the surface area of the cylinder, we can find the surface area of a cone by adapting the formula for the surface area of a pyramid as suggested in Figure T6.14.

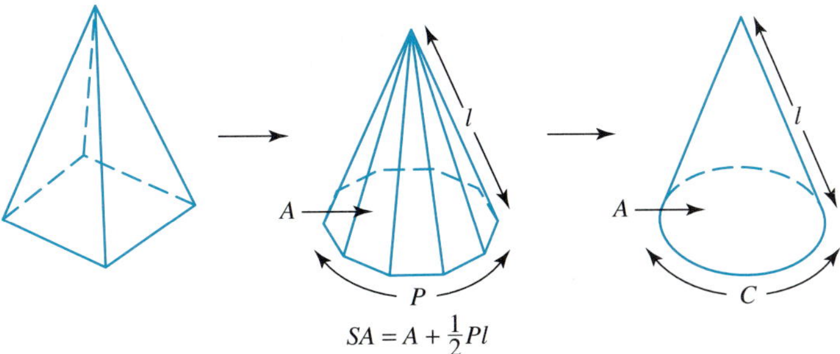

FIGURE T6.14

In the case of a right circular cone (where r is the radius of the base), SA represents the surface area, A represents the area of the base, namely πr^2, and C represents the circumference, $2\pi r$, of the base. Also, the slant height is represented by l. By substituting for A and P in the formula $A + \frac{1}{2}Pl$ we can arrive at another formulation for surface area of a right circular cone.

$$SA = A + \frac{1}{2}Cl = \pi r^2 + \frac{1}{2}(2\pi rl) = \pi r^2 + \pi rl = \pi r(r + l)$$

This discussion leads to our next result.

Surface Area of a Right Circular Cone

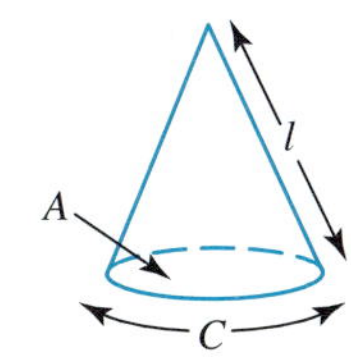

The surface area of a right circular cone is the sum of the area of its base and one-half the circumference of its base times its slant height.

$$SA = A + \frac{1}{2}Cl = \pi r^2 + \pi rl = \pi r(r + l).$$

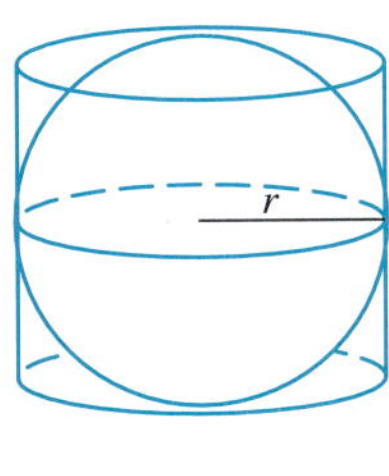

(a)

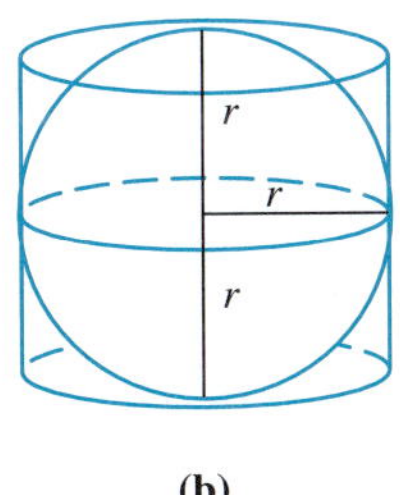

(b)

FIGURE T6.15

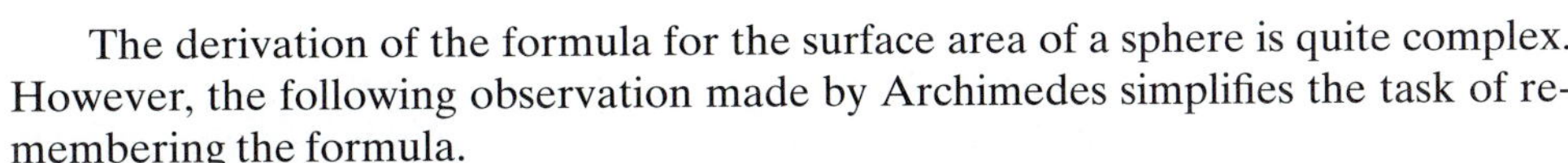

The derivation of the formula for the surface area of a sphere is quite complex. However, the following observation made by Archimedes simplifies the task of remembering the formula.

Consider the smallest right circular cylinder containing a sphere of radius r [Figure T6.15(a)]. Since the sphere has radius r, the base of the cylinder has radius r, and the height of the cylinder is $2r$ [Figure T6.15(b)]. Therefore, the surface area of the cylinder is as follows:

$$SA = 2A + Ch = 2\pi r^2 + 2\pi r(2r) = 2\pi r^2 + 4\pi r^2 = 6\pi r^2.$$

Archimedes's observation was that both the surface area and the volume of a sphere are two-thirds the respective surface area and volume of the smallest right circular cylinder containing the sphere. Thus the surface area of the sphere of radius r can be found using the previous equation as follows:

$$SA_{\text{Sphere}} = \frac{2}{3}SA_{\text{Cylinder}} = \frac{2}{3}(6\pi r^2) = 4\pi r^2.$$

Surface Area of a Sphere

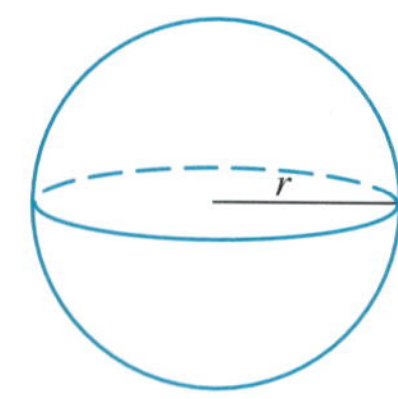

The surface area of a sphere with radius r is $4\pi r^2$.

Volume

The formulas for the volumes of cylinders and cones are obtained in a manner similar to that used to derive the formulas for their surface areas (Figure T6.16).

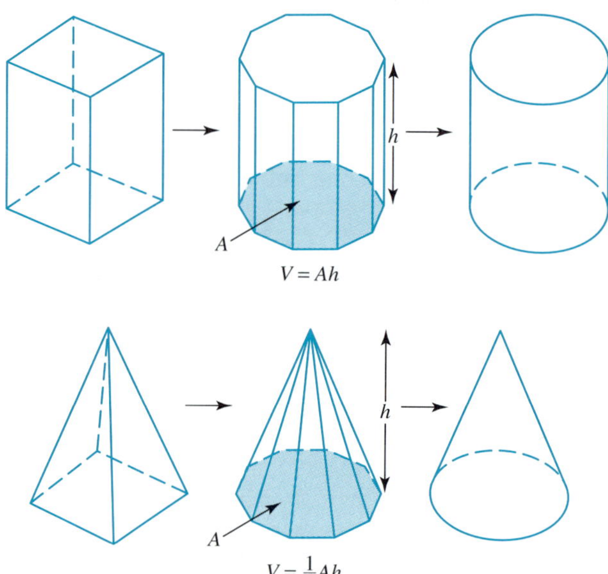

FIGURE T6.16

In the case of the circular cylinder with a base of radius r, A represents the area of the base, or πr^2. The height of the cylinder is represented by h. Figure T6.16 suggests the following.

Volume of a Right Circular Cylinder

The volume of a right circular cylinder is the product of the area of its base and its height.

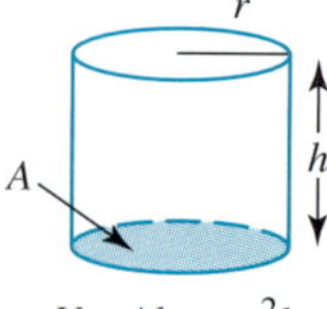

$V = Ah = \pi r^2 h$

Volume of a Right Circular Cone

The volume of a right circular cone is one-third the product of the area of its base and its height.

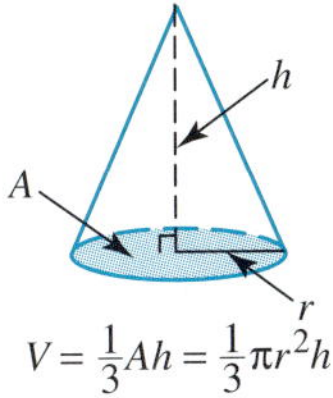

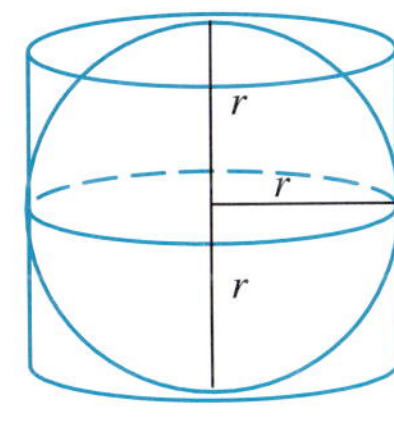

FIGURE T6.18

It was stated earlier that Archimedes observed that both the surface area and volume of a sphere are two-thirds of the respective surface area and volume of the smallest right circular cylinder containing the sphere (Figure T6.18). The sphere has radius r, so the base of the cylinder has radius r and the height of the cylinder is $2r$. Therefore, the volume of the cylinder is as follows:

$$V_{\text{Cylinder}} = Ah = \pi r^2(2r) = 2\pi r^3.$$

Thus the volume of the sphere of radius r can be found as follows.

$$V_{\text{Sphere}} = \frac{2}{3}V_{\text{Cylinder}} = \frac{2}{3}(2\pi r^3) = \frac{4}{3}\pi r^3.$$

We summarize this result in the following theorem.

Volume of a Sphere

The volume of a sphere is $\frac{4}{3}\pi$ times the cube of its radius.

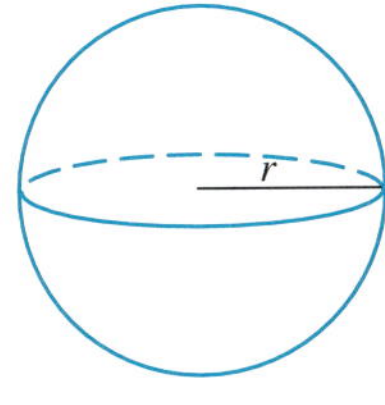

The following table summarizes the surface area and volume formulas we have studied. Connections formed by observing the similarities and differences in the various formulas will help you remember them more easily.

Geometric Shape	Surface Area	Volume
right prism	$SA = 2A + Ph$	$V = Ah$
right circular cylinder	$SA = 2A + Ch$	$V = Ah$
	$= 2\pi r(r + h)$	$= \pi r^2 h$
right regular pyramid	$SA = A + \frac{1}{2}Pl$	$V = \frac{1}{3}Ah$
right circular cone	$SA = A + \frac{1}{2}Cl$	$V = \frac{1}{3}Ah$
	$= \pi r(r + l)$	$= \frac{1}{3}\pi r^2 h$
sphere	$SA = 4\pi r^2$	$V = \frac{4}{3}\pi r^3$

PROBLEM SET T6

1. Find the surface area and volume of the following right rectangular prism.

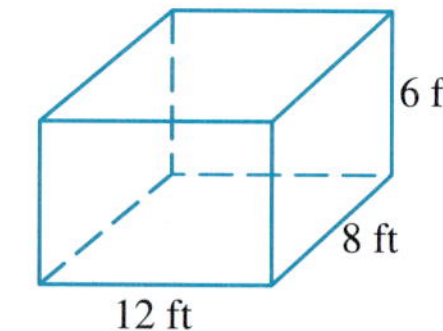

2. Find the surface area and volume of the following right triangular prism.

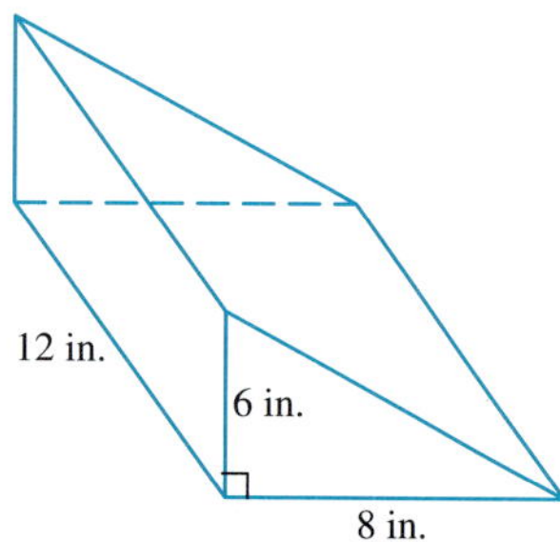

3. Find the surface area and volume of the following right pentagonal prism with base area 12 square inches.

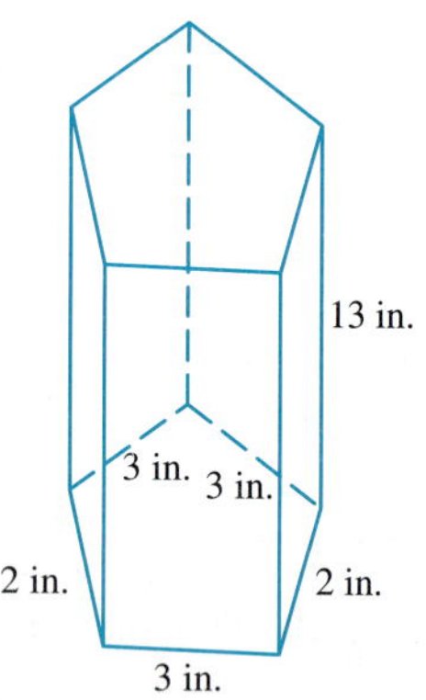

4. Find the surface area and volume of the following right hexagonal prism. Four sides of the base are 6 m and two are 7 m. (The sides are not to scale.) The area of the base is 100 m^2.

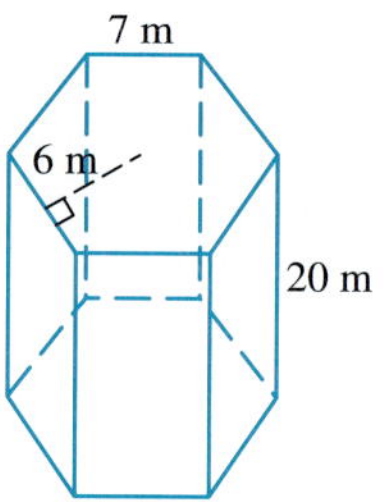

5. Find the surface area and volume of the following regular tetrahedron.

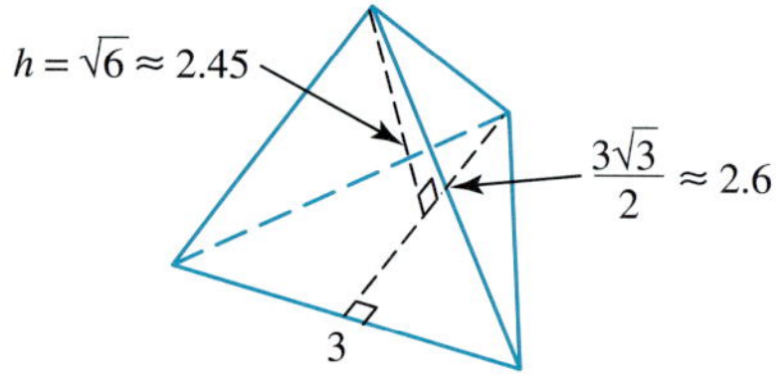

6. Find the surface area and volume of the following right rectangular pyramid with a height of 8.66 cm.

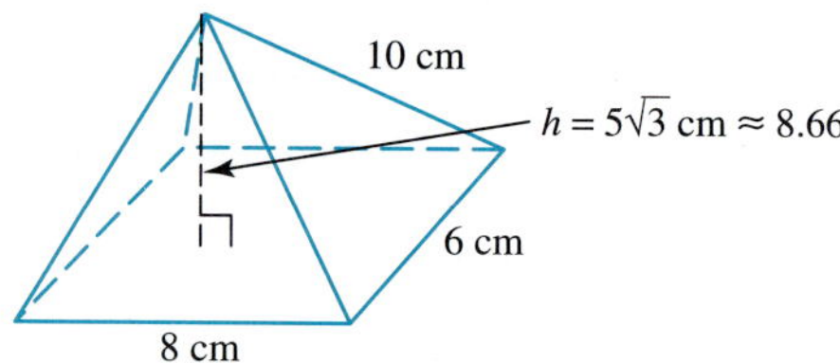

7. Find the volume of the following solid figure, rounding to the nearest cubic inch.

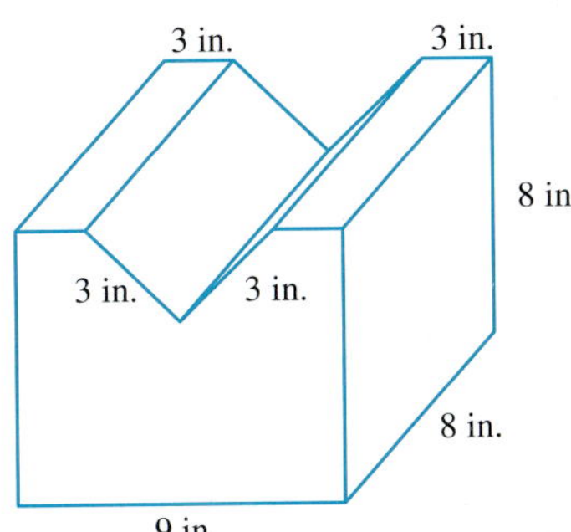

8. (a) How many square meters of tile are needed to tile the sides and bottom of the swimming pool illustrated?

(b) How much water does the pool hold?

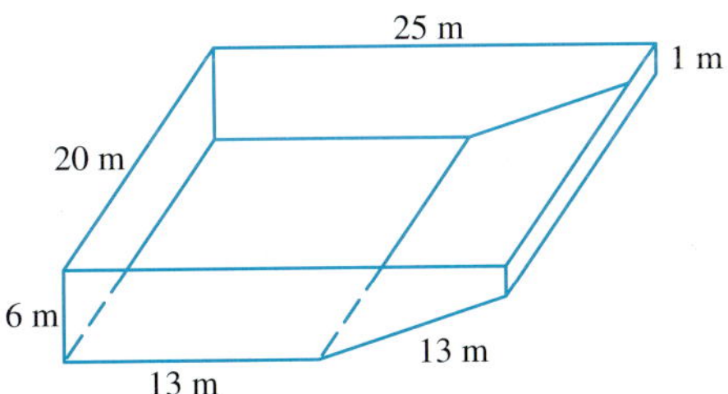

9. Find the surface area of each cone to the nearest whole unit.

(a)

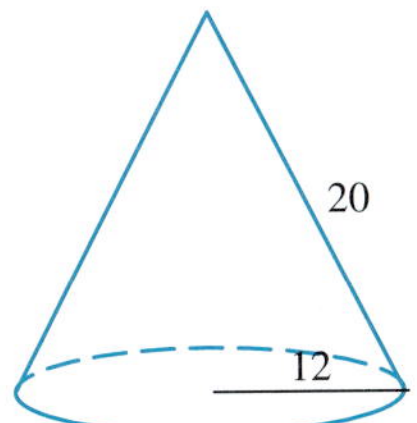

(b)

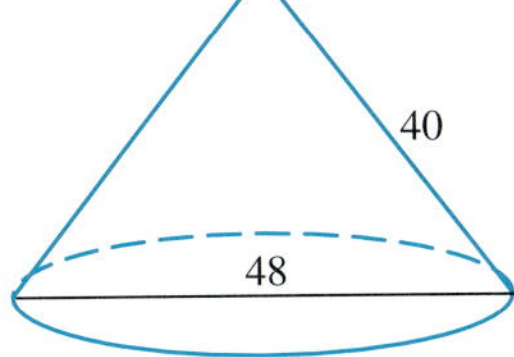

10. Find the surface area to the nearest whole unit.

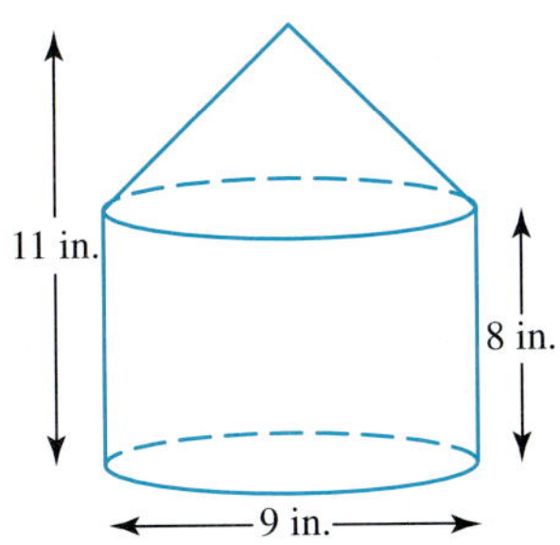

11. Find the total surface area of the hollow hemisphere shown. The shell has an inside radius and an outside radius.

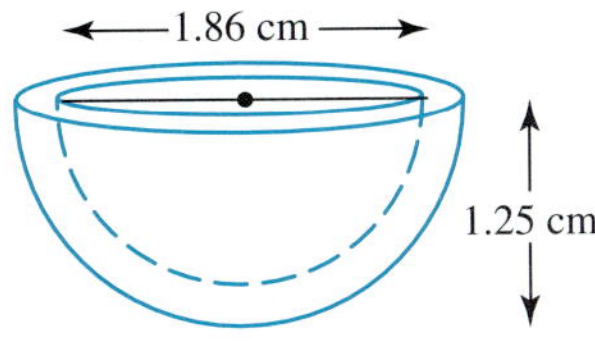

12. Find the total surface area of the hollow cylinder shown.

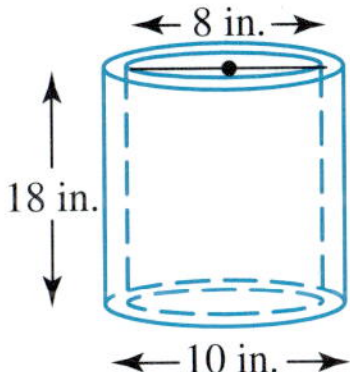

13. Find the surface area of the following figures.

(a) a sphere with radius 7 cm

(b) a cylinder with radius 7 cm and height 14 cm

(c) a cone with radius 7 cm and height 14 cm

14. Find the surface area of the following figures.

(a) a sphere with radius 4 in.

(b) a cylinder with radius 4 in. and height 8 in.

(c) a cone with radius 4 in. and height 8 in.

15. Find the surface area and volume of each can.

(a) coffee can
$r = 7.6$ cm
$h = 16.3$ cm

(b) soup can
$r = 3.3$ cm
$h = 10$ cm

16. Find the surface area and volume of each can.

(a) juice can
$r = 5.3$ cm
$h = 17.7$ cm

(b) shortening can
$r = 6.5$ cm
$h = 14.7$ cm

In problems 17 through 20 find the surface area and volume of the given shapes. Round your answers to the nearest whole square unit or cubic unit.

17. (a) a sphere with a radius of 6

(b) a sphere with a diameter of 24

18. (a) a sphere with a radius of 2.3

(b) a sphere with a diameter of 6.7

19. a right cone with a radius of 2 cm and height of 7.3 cm.

20. a right cylinder with a diameter of 14 ft and a height of 29 ft

APPENDIX RANDOM DIGITS

A Table of 10,000 Random Digits

	(1)	(2)	(3)	(4)	(5)	(6)	(7)	(8)	(9)	(10)
(1)	54046	02119	95330	54866	08524	94498	12197	66414	75829	34668
(2)	54271	09857	53749	28860	65198	27950	82096	90215	59648	06159
(3)	63506	24651	28863	26525	21482	66257	04988	44440	43015	49119
(4)	15835	01330	06920	54105	52336	34904	06887	01982	07432	69025
(5)	26679	73910	08177	20430	66350	65003	69071	25939	82676	96011
(6)	69827	23329	71606	57397	43175	14872	64902	31296	74987	00270
(7)	35931	89557	76436	42511	13366	69190	39565	97360	30176	76016
(8)	79227	28248	68837	64537	56899	52589	29645	36074	86866	57104
(9)	08877	81060	37465	97987	49601	94322	42005	47416	66812	80499
(10)	25858	17805	23451	71263	12157	05376	89692	70882	46914	46611
(11)	32761	50680	93832	11728	46925	81192	25590	70377	50302	88262
(12)	33984	91891	98083	93220	90187	28565	83485	93831	56991	60337
(13)	79276	92078	83915	18234	99601	14135	67616	57364	79477	36178
(14)	10632	28772	96439	79930	81729	24278	72601	96351	98720	24615
(15)	28967	32581	31511	04285	17582	59323	33194	44426	51345	20868
(16)	97142	46438	94544	73663	08300	70373	62145	59642	81052	45023
(17)	00280	83816	41501	47821	85616	14785	18620	60847	20251	85546
(18)	92193	52483	78474	71992	63523	31451	95817	61952	32923	11877
(19)	74398	55750	71812	33540	06225	28385	87825	64280	53160	69161
(20)	18462	69510	98533	32763	14365	97106	90149	80425	96982	95680
(21)	90809	44732	68370	04128	78355	84924	31949	10749	38516	94696
(22)	49170	56618	87990	62937	60382	45508	24096	13508	15138	30005
(23)	80447	49155	83385	79768	62396	27028	17587	50825	51250	34621
(24)	57225	94208	54036	15383	19452	91325	77174	82386	61545	07563
(25)	66137	24707	08957	14614	71472	45159	26807	65109	43992	47649
(26)	94785	75319	94402	75984	41696	39449	36773	53925	35961	36535
(27)	23405	88289	38692	98254	49899	20742	50268	03430	66928	58571
(28)	09340	80973	76280	78780	33911	62067	01446	05693	43945	72015
(29)	75910	08973	92632	50328	89542	79175	92392	60009	69986	08863
(30)	74752	99896	44973	64804	74747	34139	94061	44370	04777	43000
(31)	07465	86235	71492	99261	80512	11070	40568	46878	65712	21521
(32)	84918	41056	92630	08716	53663	19260	22894	88237	01109	91843
(33)	52765	52638	29146	40381	44421	43248	10410	16464	50802	94052
(34)	46705	84519	42397	02291	80035	75884	87792	01512	79932	96173
(35)	76466	56725	02121	78747	69405	62572	18322	06897	58842	91606
(36)	04531	57469	73681	36582	97566	64419	16028	24917	69178	35318
(37)	26992	84767	69128	23751	71064	96187	10674	27036	03397	71569
(38)	40766	09772	94284	18281	11526	40645	94912	79039	88461	80095
(39)	92802	52306	00716	62780	86989	34976	11549	61440	70600	33378
(40)	65075	07116	34649	36229	57326	63672	91447	09277	71012	48245
(41)	90782	05405	05767	48848	20514	76992	75398	37706	50958	17355
(42)	21639	24913	86301	85901	27913	22754	35872	98574	32208	85378
(43)	75015	48388	74630	50496	96517	07376	11824	47708	94897	82835
(44)	30176	40170	53779	99701	28413	89188	10471	47696	50027	33732
(45)	32325	70275	14816	66235	20564	73142	69907	57816	43463	01156
(46)	37031	91900	30574	66081	95892	59288	51756	81052	68777	09011
(47)	19986	75119	48021	22231	23373	90305	66258	26084	08282	67428
(48)	08921	84241	46943	88625	91536	28828	92375	31153	40040	51284
(49)	19673	09456	31734	47032	98717	86510	93621	99410	70013	36722
(50)	02983	06973	35923	17318	18954	49381	80462	25700	69538	87112

A Table of 10,000 Random Digits

	(1)	(2)	(3)	(4)	(5)	(6)	(7)	(8)	(9)	(10)
(51)	37058	12098	18640	93294	92975	54117	09236	00061	68291	63200
(52)	56879	80692	40347	20250	23590	22496	81878	15774	95842	74402
(53)	20216	43597	54055	83345	86752	73425	13223	27064	22236	97254
(54)	11448	01919	27862	14548	82615	16177	49891	00458	95769	25676
(55)	37187	16981	26345	26561	31185	23157	15303	42749	33004	15264
(56)	65035	43060	87341	23734	93558	42373	57531	56469	95450	98235
(57)	15921	87378	35576	05967	80782	71840	45301	73261	46487	34196
(58)	27607	77584	27571	37781	07598	94088	57131	53463	54019	79076
(59)	10562	64407	30695	96669	15457	05409	98091	93019	43961	77660
(60)	95661	23699	03099	85528	01977	59752	49330	77512	48136	91215
(61)	54284	64136	65389	63506	83105	82087	31294	60520	34471	88189
(62)	92624	28504	20690	49315	72756	21031	43020	22568	06337	22649
(63)	41662	93546	80253	13164	99232	35571	45597	22032	44562	27946
(64)	28708	78484	30665	97990	23412	69987	24062	40715	87743	50001
(65)	61354	87249	38380	28208	87319	45356	33472	45708	40085	88997
(66)	16326	41749	38954	91980	48269	26292	57497	23095	62661	30108
(67)	90854	85888	31102	29824	94268	54897	05561	65981	16166	91011
(68)	02943	98242	67002	49984	44023	07753	59236	69096	16990	55346
(69)	43020	02138	17589	06440	90184	43997	03186	29439	90539	37907
(70)	39632	19482	26640	42239	22821	27805	48396	09149	08943	48608
(71)	37718	36840	63815	51672	84777	30025	06589	85620	01955	62759
(72)	15190	55263	95256	20712	81457	75940	55985	67520	87083	76465
(73)	84438	15944	77048	71032	51905	48163	94674	23979	03380	97992
(74)	70115	07423	83180	22437	34994	74072	39483	02396	50581	24820
(75)	19946	74340	48535	50429	96338	46414	56891	11607	64558	34606
(76)	86118	99595	18465	49159	75503	61447	23898	14128	21885	09774
(77)	39726	00451	59442	39997	44513	39931	16658	59366	22145	82354
(78)	64514	27297	86966	76389	02698	13167	14572	37057	90934	46525
(79)	28124	80372	67664	91909	26342	98103	53203	29567	04743	60956
(80)	86120	70768	56364	98544	11804	23299	59111	36190	58033	31160
(81)	25128	81451	17921	73732	11386	21424	04583	75841	93904	99806
(82)	59937	07344	90219	40519	88933	87195	15121	67732	88433	01729
(83)	45142	03011	52111	24835	44372	09371	47486	13825	13834	94974
(84)	75661	56684	67764	11111	45642	98866	66630	01810	27488	77327
(85)	44227	47743	02095	12375	22627	87919	34751	21593	92333	00294
(86)	65690	31745	77309	22006	75256	61264	10965	47514	76735	32652
(87)	66313	49850	42705	17148	87190	37408	50516	32782	25893	30789
(88)	13554	96331	66167	06037	23461	69368	79943	23425	39299	71579
(89)	41265	66411	61565	71736	01893	20889	90219	74033	13311	90550
(90)	98591	64499	56413	91038	87101	23792	73611	84442	96369	24065
(91)	26444	90787	76186	29635	37585	91191	36129	62799	60295	74523
(92)	28804	27705	80581	48810	97747	07811	99154	76831	86648	09464
(93)	01423	19639	19837	96615	88709	46681	67316	27236	69139	63991
(94)	68684	76602	95877	44639	25844	06666	26578	96282	97548	32541
(95)	53092	24292	83326	58940	28094	77152	46692	42642	33689	31436
(96)	02050	03082	95709	28950	75174	35895	27978	24072	25241	14513
(97)	47420	76007	14245	19996	73566	45325	45892	94979	82583	35484
(98)	95123	94498	51347	35299	96937	12292	62267	05207	02047	53020
(99)	27501	57899	09601	85924	49875	68632	32170	47299	81933	16310
(100)	62635	52439	52903	25566	41685	13609	74806	84641	52844	25407

A Table of 10,000 Random Digits

	(1)	**(2)**	**(3)**	**(4)**	**(5)**	**(6)**	**(7)**	**(8)**	**(9)**	**(10)**
(101)	76336	60944	23973	79471	22981	55646	69503	65807	41500	16493
(102)	46622	87004	11871	93957	49770	91472	42107	10264	37946	94358
(103)	39497	58639	78329	58654	38966	35276	78308	72405	62333	84686
(104)	64346	41276	00335	60619	35156	81985	96181	97601	89852	32105
(105)	27489	27237	04277	05034	62710	13520	09810	23691	73735	28847
(106)	46073	28324	45660	13092	72370	05852	28168	04654	22549	62583
(107)	62852	35963	34228	40480	55330	73232	74232	56717	19496	13383
(108)	07327	30387	38950	29750	46502	30725	07607	32805	13539	07307
(109)	81686	43666	15825	47957	28679	33849	52765	76908	08753	59030
(110)	35743	56883	09112	14621	07237	22603	62995	46245	73436	89024
(111)	69987	87541	59423	24262	76190	68006	24656	89220	42519	34145
(112)	87427	83613	21764	43303	79505	14445	94733	58072	35188	31230
(113)	40967	95864	58906	95223	24569	41530	88661	75563	72906	53424
(114)	04504	52752	73395	13015	34800	64403	88072	74736	90739	64954
(115)	71557	04525	80775	66681	04885	54807	97426	84735	38018	44579
(116)	89091	98016	10608	26066	08597	22768	55424	98471	33822	29842
(117)	47243	56421	40785	00193	22204	27689	09100	99159	65913	45842
(118)	31750	53626	89308	92191	45504	22770	00711	83546	69993	47760
(119)	94597	23808	14344	31108	36627	56470	32783	93281	48727	41585
(120)	82882	96403	28099	43277	01136	26146	17332	75781	09534	59599
(121)	69496	81098	31226	74548	60457	13515	45331	80594	94410	55357
(122)	47111	16860	16556	66665	38527	49211	75681	53022	27336	03738
(123)	40802	99927	69595	87020	24185	39495	69367	16822	02449	60175
(124)	70188	49016	21827	84607	04884	69398	12620	36236	40206	59104
(125)	80579	06968	17499	36583	79415	94639	03892	18514	24423	33137
(126)	60056	75565	21336	26598	33623	67143	81426	74436	37726	60239
(127)	17486	78546	59956	90774	15736	09790	02010	75728	19058	74460
(128)	42042	62959	48768	04972	94706	12407	82134	11033	21722	86316
(129)	22431	00088	74335	15077	56174	45808	84716	69969	33449	72671
(130)	64135	44371	30865	63221	61786	75436	35250	31686	34983	46039
(131)	33071	60377	02427	11144	20902	83105	93215	56078	12317	70074
(132)	82828	31844	48838	92075	90037	23175	34338	87336	52995	98494
(133)	83939	06450	92821	99869	41360	00563	95854	36754	21789	85079
(134)	90487	35112	31688	28804	01371	53532	20886	26180	60976	99555
(135)	23718	37829	53809	62408	20962	05547	38943	83506	08600	82084
(136)	46020	88114	45538	32805	76440	04286	51410	79924	05306	48272
(137)	56137	56679	97805	32338	67164	66035	63867	54321	16879	68676
(138)	68504	99586	78448	49344	28871	75263	55415	18910	16570	52552
(139)	45473	83202	07313	90643	79745	62749	29193	77891	91835	24479
(140)	74829	73741	43693	18061	02237	84393	40095	73284	39440	57581
(141)	93880	56980	17961	73822	68452	63889	70479	69492	71051	24878
(142)	70798	49508	06664	96899	86879	82011	18380	52422	61822	05314
(143)	35220	80559	83496	25579	14731	09372	31654	25441	36833	76447
(144)	70351	79331	42311	63639	83826	94540	70491	61562	80925	96298
(145)	47421	07989	03905	02897	73418	12644	25372	42844	71721	31016
(146)	42069	00950	58165	16012	82142	34656	95671	78152	38586	08706
(147)	85886	34420	20299	12778	17488	50727	68119	44063	43086	53223
(148)	01137	91796	13369	89480	61693	67771	06095	43346	60019	61943
(149)	25875	21430	57845	98119	36757	72129	34042	88138	30684	95859
(150)	37601	54657	38086	34138	28513	88048	48569	67178	09775	54229

A Table of 10,000 Random Digits

	(1)	(2)	(3)	(4)	(5)	(6)	(7)	(8)	(9)	(10)
(151)	77138	52320	80734	53987	37793	18001	02355	37383	74790	97274
(152)	04014	12824	64924	65657	24891	62885	07122	13293	84612	66434
(153)	34062	32624	39204	03471	94476	77250	53420	18911	89360	70335
(154)	10981	66316	39424	83297	68150	97196	60577	06780	54075	34280
(155)	94541	51249	39568	26095	34326	97089	42190	13325	46655	41568
(156)	00395	05964	08591	20386	75836	77880	30581	79605	28822	42402
(157)	05329	72401	43440	87524	63722	27634	32652	45990	96840	39189
(158)	25493	29027	26728	17166	43517	18893	30409	33052	28683	65307
(159)	80985	74042	70284	46515	54761	31702	87760	06360	80195	95560
(160)	11655	53413	86057	34320	26499	54584	56036	55971	34398	63048
(161)	09134	12460	31092	87670	81484	85186	97415	83518	20541	28940
(162)	30167	26347	56102	43968	37783	00932	73000	49467	40468	67427
(163)	16816	77660	76966	10621	26513	74602	91234	75977	83967	25948
(164)	43188	86129	61975	82301	48668	57425	28113	61306	03370	28875
(165)	23068	60211	89642	11699	60278	29340	78834	58838	17592	17430
(166)	76891	37001	77157	71034	34935	59063	54695	29929	16156	04365
(167)	94218	84437	60587	96381	18002	60747	23781	02960	23705	46423
(168)	64074	18993	65291	73950	26480	38737	28522	70863	61716	22448
(169)	78927	42307	34451	78813	81814	17148	05762	34801	44815	19288
(170)	70704	37815	04392	19985	35559	57254	88382	61348	51749	58980
(171)	88321	15387	30036	14285	57252	46547	51798	70300	85793	29759
(172)	83999	36966	08282	47360	17642	13574	88578	04238	42555	99984
(173)	93625	81088	77747	80489	55257	95999	88313	99157	15114	65016
(174)	88903	66347	06140	53128	38427	50488	25253	30773	75443	75008
(175)	30555	52130	02013	78029	37620	56272	63767	81923	69112	16272
(176)	52894	32741	61380	91996	38785	99638	48878	44444	75006	48154
(177)	61949	78818	73429	44805	87840	64660	03067	38501	98402	97413
(178)	41239	27989	17483	01649	31649	14540	01493	88652	36308	63007
(179)	95782	67554	87880	23037	89643	45679	52355	84490	68879	52206
(180)	62468	69981	78325	31512	21739	33727	21688	08605	56494	71689
(181)	51724	44494	82102	71166	06989	99130	01561	32493	96104	58969
(182)	90875	72815	88470	66059	15380	93275	91756	29422	38695	16402
(183)	26175	77443	72881	56109	10047	65553	91723	82234	07961	78615
(184)	50872	16852	30848	15398	39573	71733	98101	81183	17734	19972
(185)	77488	74521	52941	23080	79857	57740	61945	70689	09832	14897
(186)	16030	57810	14295	54422	38555	26146	37321	29898	30596	39385
(187)	22360	71364	40763	95124	30922	88715	00038	79779	14609	76641
(188)	11718	14900	72827	28921	50110	06476	17625	60349	35179	75424
(189)	85131	61825	89023	39820	05017	15442	55793	30729	44891	59855
(190)	09763	85010	17293	41873	43113	10727	89485	03347	50378	65308
(191)	62902	36583	34079	24285	05802	77797	96582	38517	16820	98488
(192)	93627	01530	47174	34553	60288	28764	36087	84285	68422	97707
(193)	04496	97705	90738	89866	52821	40651	42205	03244	27006	94734
(194)	47489	95322	69082	96614	66668	05323	71168	54887	07726	68614
(195)	11736	92995	34274	40972	74304	30633	42268	67355	81871	54000
(196)	39830	78453	32858	53125	03335	67983	25539	31410	55779	77164
(197)	11758	72062	84486	26770	22150	45555	25664	34866	27206	18179
(198)	48315	07099	19414	15736	15131	37851	77880	06373	03647	90310
(199)	44981	20078	73645	99204	71879	28095	45255	58435	56500	39666
(200)	98294	91770	62550	68244	01986	36308	56546	54561	73700	76539

APPENDIX AMORTIZATION TABLE

Monthly Payment for Amortized Loan Of $1000

Interest rates: 2.00% to 9.75%
Terms: 1 year to 5 years

Rate	1 year	2 years	3 years	4 years	5 years
2.00	84.238867	42.540263	28.642579	21.695124	17.527760
2.25	84.352446	42.650241	28.751848	21.804406	17.637345
2.50	84.466111	42.760392	28.861376	21.914034	17.747362
2.75	84.579862	42.870716	28.971163	22.024008	17.857810
3.00	84.693699	42.981212	29.081210	22.134327	17.968691
3.25	84.807621	43.091881	29.191515	22.244991	18.080002
3.50	84.921630	43.202722	29.302080	22.356001	18.191745
3.75	85.035724	43.313736	29.412903	22.467356	18.303918
4.00	85.149904	43.424922	29.523985	22.579055	18.416522
4.25	85.264170	43.536281	29.635326	22.691098	18.529556
4.50	85.378522	43.647812	29.746924	22.803486	18.643019
4.75	85.492959	43.759515	29.858782	22.916218	18.756912
5.00	85.607482	43.871390	29.970897	23.029294	18.871234
5.25	85.722090	43.983437	30.083271	23.142713	18.985984
5.50	85.836785	44.095656	30.195902	23.256475	19.101162
5.75	85.951564	44.208047	30.308791	23.370581	19.216768
6.00	86.066430	44.320610	30.421937	23.485029	19.332802
6.25	86.181381	44.433345	30.535342	23.599820	19.449262
6.50	86.296417	44.546251	30.649003	23.714953	19.566148
6.75	86.411539	44.659329	30.762921	23.830428	19.683461
7.00	86.526746	44.772579	30.877097	23.946245	19.801199
7.25	86.642039	44.886000	30.991529	24.062403	19.919361
7.50	86.757417	44.999593	31.106218	24.178902	20.037949
7.75	86.872880	45.113356	31.221164	24.295742	20.156960
8.00	86.988429	45.227291	31.336365	24.412922	20.276394
8.25	87.104063	45.341398	31.451823	24.530443	20.396252
8.50	87.219782	45.455675	31.567537	24.648303	20.516531
8.75	87.335587	45.570123	31.683507	24.766503	20.637233
9.00	87.451477	45.684742	31.799733	24.885042	20.758355
9.25	87.567452	45.799532	31.916214	25.003920	20.879898
9.50	87.683512	45.914493	32.032950	25.123137	21.001861
9.75	87.799657	46.029624	32.149941	25.242691	21.124244

Monthly Payment for Amortized Loan of $1000

Interest rates: 2.00% to 9.75%
Terms: 6 years to 10 years

Rate	6 years	7 years	8 years	9 years	10 years
2.00	14.750442	12.767435	11.280872	10.125272	9.201345
2.25	14.860473	12.877995	11.392013	10.237027	9.313737
2.50	14.971023	12.989160	11.503843	10.349557	9.426990
2.75	15.082091	13.100928	11.616363	10.462862	9.541103
3.00	15.193676	13.213300	11.729572	10.576940	9.656074
3.25	15.305778	13.326274	11.843468	10.691792	9.771903
3.50	15.418397	13.439851	11.958052	10.807414	9.888587
3.75	15.531532	13.554028	12.073321	10.923807	10.006124
4.00	15.645183	13.668806	12.189275	11.040969	10.124514
4.25	15.759349	13.784184	12.305913	11.158898	10.243753
4.50	15.874030	13.900161	12.423234	11.277593	10.363841
4.75	15.989224	14.016737	12.541237	11.397052	10.484774
5.00	16.104933	14.133909	12.659920	11.517273	10.606552
5.25	16.221154	14.251678	12.779282	11.638256	10.729170
5.50	16.337887	14.370043	12.899322	11.759997	10.852628
5.75	16.455132	14.489002	13.020039	11.882496	10.976922
6.00	16.572888	14.608554	13.141430	12.005750	11.102050
6.25	16.691154	14.728700	13.263495	12.129757	11.228010
6.50	16.809930	14.849436	13.386233	12.254515	11.354798
6.75	16.929214	14.970764	13.509640	12.380022	11.482411
7.00	17.049006	15.092680	13.633717	12.506277	11.610848
7.25	17.169306	15.215184	13.758461	12.633275	11.740104
7.50	17.290112	15.338276	13.883871	12.761016	11.870177
7.75	17.411424	15.461953	14.009944	12.889497	12.001063
8.00	17.533241	15.586214	14.136679	13.018715	12.132759
8.25	17.655561	15.711059	14.264075	13.148668	12.265263
8.50	17.778385	15.836485	14.392129	13.279353	12.398569
8.75	17.901710	15.962492	14.520839	13.410767	12.532675
9.00	18.025537	16.089078	14.650203	13.542909	12.667577
9.25	18.149864	16.216242	14.780220	13.675774	12.803272
9.50	18.274691	16.343982	14.910887	13.809361	12.939756
9.75	18.400016	16.472296	15.042203	13.943666	13.077024

Monthly Payment for Amortized Loan of $1000

Interest rates: 2.00% to 9.75%
Terms: 15 years to 30 years

Rate	15 years	20 years	25 years	30 years
2.00	6.435087	5.058833	4.238543	3.696195
2.25	6.550848	5.178083	4.361307	3.822461
2.50	6.667892	5.299029	4.486167	3.951209
2.75	6.786216	5.421663	4.613109	4.082412
3.00	6.905816	5.545976	4.742113	4.216040
3.25	7.026688	5.671958	4.873162	4.352063
3.50	7.148825	5.799597	5.006236	4.490447
3.75	7.272224	5.928883	5.141312	4.631156
4.00	7.396879	6.059803	5.278368	4.774153
4.25	7.522784	6.192345	5.417381	4.919399
4.50	7.649933	6.326494	5.558325	5.066853
4.75	7.778319	6.462236	5.701174	5.216473
5.00	7.907936	6.599557	5.845900	5.368216
5.25	8.038777	6.738442	5.992477	5.522037
5.50	8.170835	6.878873	6.140875	5.677890
5.75	8.304101	7.020835	6.291064	5.835729
6.00	8.438568	7.164311	6.443014	5.995505
6.25	8.574229	7.309282	6.596694	6.157172
6.50	8.711074	7.455731	6.752072	6.320680
6.75	8.849095	7.603640	6.909115	6.485981
7.00	8.988283	7.752989	7.067792	6.653025
7.25	9.128629	7.903760	7.228069	6.821763
7.50	9.270124	8.055932	7.389912	6.992145
7.75	9.412758	8.209486	7.553288	7.164122
8.00	9.556521	8.364401	7.718162	7.337646
8.25	9.701404	8.520657	7.884501	7.512666
8.50	9.847396	8.678232	8.052271	7.689135
8.75	9.994487	8.837107	8.221436	7.867004
9.00	10.142666	8.997260	8.391964	8.046226
9.25	10.291923	9.158668	8.563818	8.226754
9.50	10.442247	9.321312	8.736967	8.408542
9.75	10.593627	9.485169	8.911374	8.591544

Monthly Payment for Amortized Loan of $1000

Interest rates: 10.00% to 18.75%
Terms: 1 year to 5 years

Rate	1 year	2 years	3 years	4 years	5 years
10.00	87.915887	46.144926	32.267187	25.362583	21.247045
10.25	88.032203	46.260399	32.384688	25.482813	21.370264
10.50	88.148603	46.376042	32.502444	25.603380	21.493900
10.75	88.265088	46.491855	32.620453	25.724283	21.617954
11.00	88.381659	46.607838	32.738717	25.845523	21.742423
11.25	88.498314	46.723992	32.857235	25.967098	21.867308
11.50	88.615054	46.840315	32.976006	26.089009	21.992607
11.75	88.731879	46.956809	33.095031	26.211255	22.118321
12.00	88.848789	47.073472	33.214310	26.333835	22.244448
12.25	88.965783	47.190305	33.333841	26.456750	22.370987
12.50	89.082863	47.307308	33.453626	26.579999	22.497938
12.75	89.200027	47.424481	33.573663	26.703581	22.625300
13.00	89.317276	47.541823	33.693952	26.827496	22.753073
13.25	89.434609	47.659334	33.814494	26.951743	22.881255
13.50	89.552027	47.777015	33.935287	27.076323	23.009846
13.75	89.669530	47.894864	34.056333	27.201234	23.138845
14.00	89.787118	48.012883	34.177630	27.326476	23.268251
14.25	89.904790	48.131071	34.299178	27.452050	23.398063
14.50	90.022546	48.249428	34.420977	27.577953	23.528281
14.75	90.140387	48.367954	34.543028	27.704186	23.658904
15.00	90.258312	48.486648	34.665329	27.830748	23.789930
15.25	90.376322	48.605511	34.787880	27.957639	23.921360
15.50	90.494416	48.724542	34.910681	28.084859	24.053191
15.75	90.612595	48.843742	35.033732	28.212406	24.185424
16.00	90.730858	48.963111	35.157033	28.340281	24.318057
16.25	90.849205	49.082647	35.280583	28.468482	24.451090
16.50	90.967637	49.202351	35.404383	28.597010	24.584521
16.75	91.086152	49.322224	35.528431	28.725864	24.718350
17.00	91.204752	49.442264	35.652728	28.855042	24.852576
17.25	91.323436	49.562472	35.777273	28.984546	24.987197
17.50	91.442204	49.682848	35.902066	29.114374	25.122214
17.75	91.561057	49.803391	36.027107	29.244525	25.257624
18.00	91.679993	49.924102	36.152396	29.375000	25.393427
18.25	91.799013	50.044980	36.277931	29.505797	25.529623
18.50	91.918118	50.166025	36.403714	29.636916	25.666209
18.75	92.037306	50.287238	36.529744	29.768357	25.803186

Monthly Payment for Amortized Loan of $1000

Interest rates: 10.00% to 18.75%
Terms: 6 years to 10 years

Rate	6 years	7 years	8 years	9 years	10 years
10.00	18.525838	16.601184	15.174164	14.078686	13.215074
10.25	18.652156	16.730644	15.306769	14.214419	13.353900
10.50	18.778970	16.860673	15.440016	14.350861	13.493500
10.75	18.906278	16.991271	15.573902	14.488010	13.633868
11.00	19.034079	17.122436	15.708426	14.625861	13.775001
11.25	19.162372	17.254167	15.843584	14.764412	13.916895
11.50	19.291156	17.386461	15.979374	14.903660	14.059544
11.75	19.420430	17.519317	16.115794	15.043601	14.202946
12.00	19.550193	17.652733	16.252841	15.184233	14.347095
12.25	19.680442	17.786707	16.390514	15.325550	14.491987
12.50	19.811179	17.921238	16.528809	15.467551	14.637617
12.75	19.942400	18.056324	16.667723	15.610231	14.783981
13.00	20.074105	18.191963	16.807255	15.753588	14.931074
13.25	20.206293	18.328153	16.947401	15.897616	15.078892
13.50	20.338962	18.464893	17.088160	16.042314	15.227429
13.75	20.472111	18.602179	17.229527	16.187677	15.376681
14.00	20.605739	18.740012	17.371501	16.333701	15.526644
14.25	20.739845	18.878387	17.514079	16.480383	15.677311
14.50	20.874427	19.017304	17.657257	16.627719	15.828679
14.75	21.009483	19.156761	17.801034	16.775705	15.980742
15.00	21.145013	19.296755	17.945405	16.924337	16.133496
15.25	21.281015	19.437284	18.090369	17.073612	16.286935
15.50	21.417488	19.578347	18.235923	17.223525	16.441054
15.75	21.554431	19.719941	18.382063	17.374073	16.595848
16.00	21.691841	19.862064	18.528786	17.525251	16.751312
16.25	21.829717	20.004714	18.676090	17.677055	16.907441
16.50	21.968059	20.147889	18.823971	17.829483	17.064230
16.75	22.106864	20.291587	18.972427	17.982528	17.221673
17.00	22.246131	20.435805	19.121454	18.136188	17.379765
17.25	22.385859	20.580541	19.271049	18.290458	17.538501
17.50	22.526046	20.725794	19.421210	18.445334	17.697876
17.75	22.666690	20.871560	19.571933	18.600812	17.857884
18.00	22.807791	21.017838	19.723214	18.756888	18.018520
18.25	22.949346	21.164625	19.875051	18.913557	18.179778
18.50	23.091354	21.311919	20.027441	19.070815	18.341654
18.75	23.233814	21.459718	20.180380	19.228659	18.504142

Monthly Payment for Amortized Loan of $1000

Interest rates: 10.00% to 18.75%
Terms: 15 years to 30 years

Rate	15 years	20 years	25 years	30 years
10.00	10.746051	9.650216	9.087007	8.775716
10.25	10.899509	9.816434	9.263833	8.961013
10.50	11.053989	9.983799	9.441817	9.147393
10.75	11.209480	10.152290	9.620927	9.334814
11.00	11.365969	10.321884	9.801131	9.523234
11.25	11.523446	10.492560	9.982395	9.712614
11.50	11.681898	10.664296	10.164690	9.902914
11.75	11.841314	10.837071	10.347982	10.094097
12.00	12.001681	11.010861	10.532241	10.286126
12.25	12.162987	11.185647	10.717438	10.478964
12.50	12.325221	11.361405	10.903541	10.672578
12.75	12.488370	11.538116	11.090523	10.866932
13.00	12.652422	11.715757	11.278353	11.061995
13.25	12.817364	11.894308	11.467004	11.257735
13.50	12.983185	12.073747	11.656449	11.454122
13.75	13.149873	12.254054	11.846660	11.651125
14.00	13.317414	12.435208	12.037610	11.848718
14.25	13.485797	12.617189	12.229275	12.046871
14.50	13.655009	12.799978	12.421629	12.245559
14.75	13.825038	12.983553	12.614647	12.444757
15.00	13.995871	13.167896	12.808306	12.644440
15.25	14.167497	13.352987	13.002582	12.844585
15.50	14.339903	13.538807	13.197452	13.045169
15.75	14.513078	13.725337	13.392895	13.246171
16.00	14.687007	13.912559	13.588889	13.447570
16.25	14.861681	14.100455	13.785413	13.649346
16.50	15.037086	14.289006	13.982446	13.851481
16.75	15.213211	14.478196	14.179971	14.053956
17.00	15.390043	14.668005	14.377966	14.256753
17.25	15.567571	14.858419	14.576414	14.459858
17.50	15.745782	15.049419	14.775297	14.663252
17.75	15.924666	15.240990	14.974598	14.866922
18.00	16.104210	15.433115	15.174299	15.070854
18.25	16.284403	15.625779	15.374386	15.275032
18.50	16.465234	15.818966	15.574842	15.479445
18.75	16.646690	16.012661	15.775651	15.684080

ANSWERS TO

ODD-NUMBERED PROBLEMS

PROBLEM SET 1.1

1. **(a)** not a statement **(b)** statement
(c) not a statement **(d)** statement

3. **(a)** statement **(b)** not a statement
(c) not a statement **(d)** statement

5. **(a)** "A kitten is not teachable." or "No kitten is teachable."
(b) "A kitten does not have whiskers." or "No kitten has whiskers."

7. **(a)** Phil did not buy a new car.
(b) The weather is not sunny.

9. **(a)** F **(b)** F
(c) T

11. **(a)** T **(b)** T
(c) T

13. **(a)** Roses are red and the sky is blue.
(b) Roses are red, and either the sky is blue or turtles are green.
(c) If the sky is blue, then roses are red and turtles are green.
(d) If turtles are not green and the sky is not blue, then roses are not red.

15. **(a)** $p \Rightarrow \sim q$ **(b)** $\sim q \Rightarrow \sim r$

17. **(a)** $q \Rightarrow \sim p$ **(b)** $q \wedge r$
(c) $r \Leftrightarrow (q \wedge p)$ **(d)** $p \vee q$

19. **(a)** $\sim p \Rightarrow q$ or $p \vee q$ **(b)** $\sim p \wedge \sim q$

21. **(a)** $p \Rightarrow q$ or $\sim q \Rightarrow \sim p$ **(b)** $\sim p \vee \sim q$

23. **(a)** T **(b)** F
(c) T **(d)** T

25. **(a)** F **(b)** F
(c) T **(d)** F

27. **(a)** T **(b)** F
(c) F **(d)** T

29. **(a)** Antecedent (hypothesis): The weather is good.
Consequent (conclusion): We will go to the game.
Converse: If we go to the game the weather will be good.
Inverse: If the weather is not good, we will not go to the game.
Contrapositive: If we do not go to the game, the weather will not be good.

(b) Antecedent (hypothesis): I do not go to the movie.
Consequent (conclusion): I will study my math.
Converse: If I study my math, I will not go to the movie.
Inverse: If I go to the movie, I will not study my math.
Contrapositive: If I do not study my math, I will go to the movie.

(c) Antecedent (hypothesis): I will get an A on the final.
Consequent (conclusion): I will get an A for the course.
Converse: If I get an A for the course, I will get an A on the final.
Inverse: If I do not get an A on the final, then I will not get an A for the course.
Contrapositive: If I do not get an A for the course, I will not get an A on the final.

31. **(a)** Antecedent (hypothesis): You do not have gasoline in the tank.
Consequent (conclusion): Your car will not start.
Converse: If your car will not start, then you do not have gasoline in the tank.
Inverse: If you have gasoline in the tank, your car will start.
Contrapositive: If your car will start, then you have gasoline in the tank.

(b) Antecedent (hypothesis): I can pass this class.
Consequent (conclusion): I will graduate.
Converse: If I graduate, I will pass this class.
Inverse: If I cannot pass this class, then I will not graduate.
Contrapositive: If I do not graduate, then I cannot pass this class.

33.

p	q	$(\sim p) \vee (\sim q)$	$(\sim p) \vee q$	$\sim(p \wedge q)$
T	T	F	T	F
T	F	T	F	T
F	T	T	T	T
F	F	T	T	T

$(\sim p) \vee (\sim q)$ and $\sim(p \wedge q)$ are logically equivalent.

35. (a)

p	q	$p \Rightarrow q$	$q \Rightarrow p$	$(p \Rightarrow q) \Rightarrow (q \Rightarrow p)$
T	T	T	T	T
T	F	F	T	T
F	T	T	F	F
F	F	T	T	T

Therefore $(p \Rightarrow q) \Rightarrow (q \Rightarrow p)$ is not always true.

(b)

p	q	$p \Rightarrow q$	$p \wedge (p \Rightarrow q)$	$[p \wedge (p \Rightarrow q)] \Rightarrow q$
T	T	T	T	T
T	F	F	F	T
F	T	T	F	T
F	F	T	F	T

Therefore $[p \wedge (p \Rightarrow q)] \Rightarrow q$ is always true.

37. (a) T **(b)** T
(c) Unknown **(d)** T
(e) F **(f)** Unknown

39. (a) T **(b)** F
(c) F

41. (a) It is not the case that both p and q occur is equivalent to either p does not occur or q does not occur.
It is not the case that either of p or q occur is equivalent to both p does not occur and q does not occur.
(b) It is not the case that the moon is dark and the night is cold has the same meaning as either the moon is not dark or the night is not cold.
It is not the case that either the moon is dark or the night is cold has the same meaning as the moon is not dark and the night is not cold.

PROBLEM SET 1.2

1. (a) Premises: If the room is warm, then I'll be uncomfortable. The room is warm.
Conclusion: I'll be uncomfortable.
(b) Premises: If the weather is bad, I'll go to the movies. It's raining heavily.
Conclusion: I'll go to the movies.
Unstated premise: Heavy rain is bad weather.

3. (a) Premises: If the weather is good, Barry will paint the house. Barry didn't paint the house.
Conclusion: The weather was not good.
(b) Premises: If you average at least 90% on the tests, you'll get an A for the term. You did not get an A for the term.
Conclusion: You did not average 90% on the tests.

5. $p \Rightarrow q$
q
$\therefore p$
Invalid, affirming the consequent.

7. $p \Rightarrow q$
p
$\therefore q$
Valid, modus ponens.

9. $p \Rightarrow q$
$q \Rightarrow r$
$\therefore p \Rightarrow r$
Valid, chain rule.

11. $p \Rightarrow q$
$\sim p$
$\therefore \sim q$
Invalid, denying the antecedent.

13. $p \Rightarrow q$
$\sim q$
$\therefore \sim p$
Valid, modus tollens.

15. $p \Rightarrow q$
$\sim p$
$\therefore \sim q$
Invalid, denying the antecedent.
Note: p = I can't go to the movie
$\sim p$ = I can go to the movie.

17. $p \Rightarrow q$
$\sim p$
$\therefore \sim q$
Invalid, denying the antecedent.

19. Chain rule and modus ponens.

21. Modus ponens.

23. Chain rule and modus ponens.

25. Disjunctive syllogism.

27.

$\sim p \Rightarrow \sim q$	which	$u \Rightarrow \sim p$
$\sim r \Rightarrow \sim s$	can be	$\sim p \Rightarrow \sim q$
$\sim q \Rightarrow \sim t$	arranged	$\sim q \Rightarrow \sim t$
$u \Rightarrow \sim p$	as	$\sim t \Rightarrow \sim r$
$\sim t \Rightarrow \sim r$		$\sim r \Rightarrow \sim s$

Conclusion: If a kitten has green eyes, then it will not play with a gorilla. Repeated application of the chain rule.

29. You did not do your assignments. Modus tollens.

31. Michelle finishes her assignment. Disjunctive syllogism.

33. Gunder is going to the technical university. Modus ponens.

35. Affirming the consequent.

37. Denying the antecedent.

39. Denying the antecedent.

41. If two lines are perpendicular to another line and if they are not parallel then the three lines form a triangle. The sum of the angles of a triangle is 180°. Two of the angles are 90° so the angle made by the other angle is 0°. Therefore, the two perpendicular lines are the same line and not part of a triangle. This is a contradiction so the lines must be parallel.

PROBLEM SET 1.3

1. **(a)** T **(b)** T
(c) F **(d)** F
(e) F

3. **(a)** T **(b)** F
(c) T **(d)** T
(e) F **(f)** T

5.

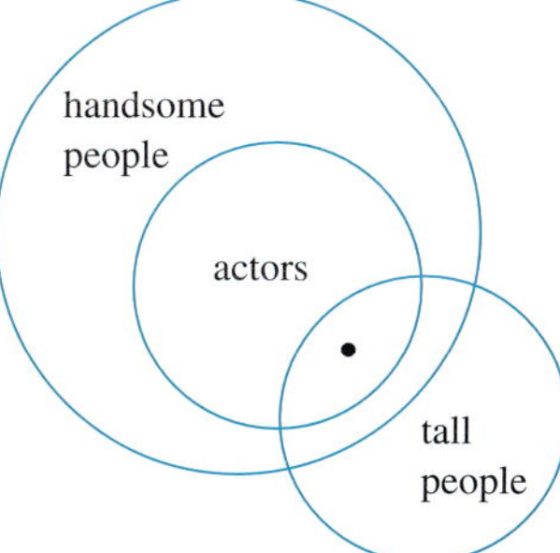

Valid

7.

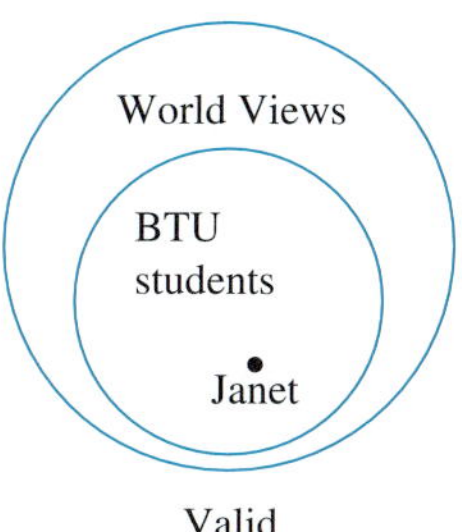

Valid

9.

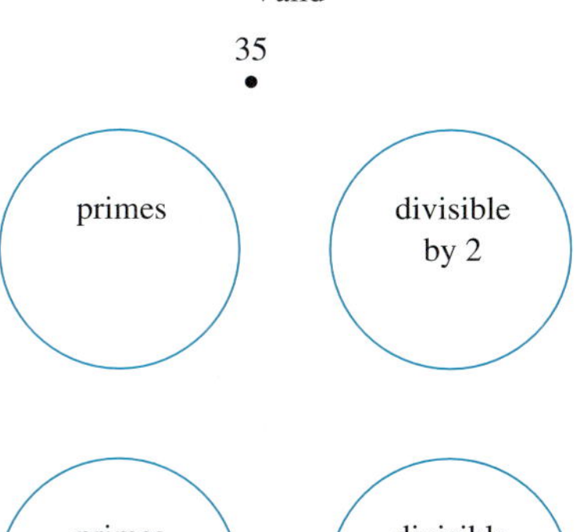

This is invalid. Both Euler diagrams above are consistent with the statements. One of them has 35 not a prime. Notice that these statements are not all true.

11.

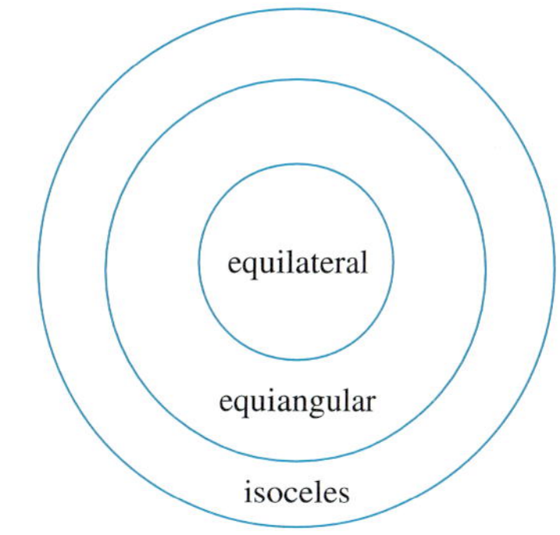

Valid

13.

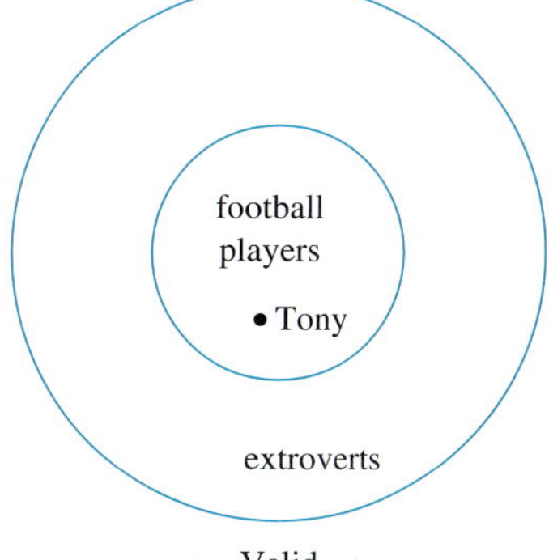

Valid

15.

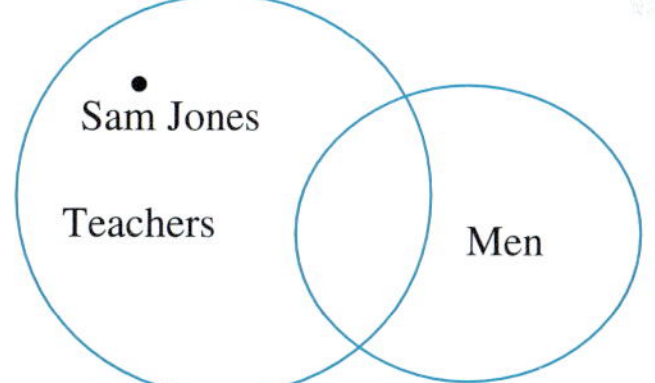

Invalid; Sam Jones may be a woman

17.

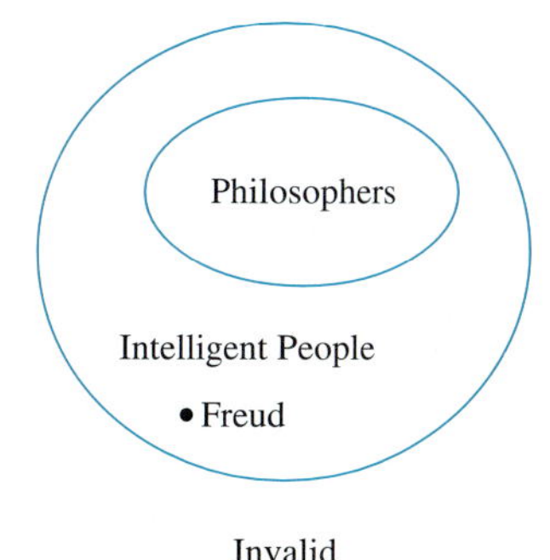

Invalid

19.

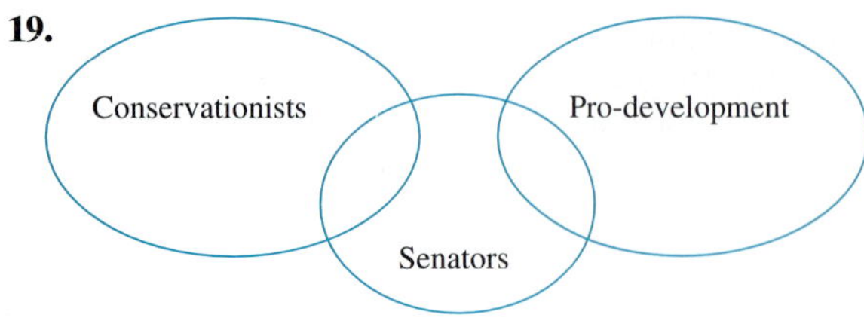

Valid

21.

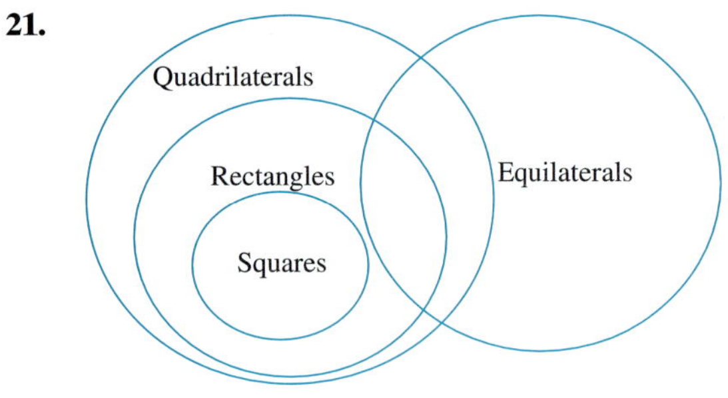

Invalid

23.

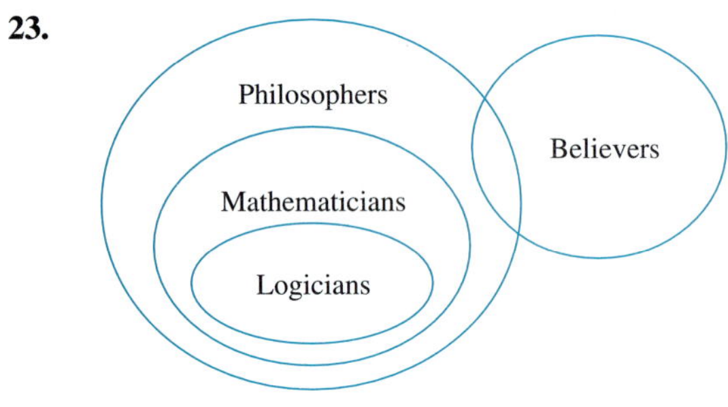

Invalid

25. Joe Montana was strong.
27. Some extraterrestrials are not carbon based life forms.
29. My roommate is not sane.
31. There are two possible Euler diagrams.

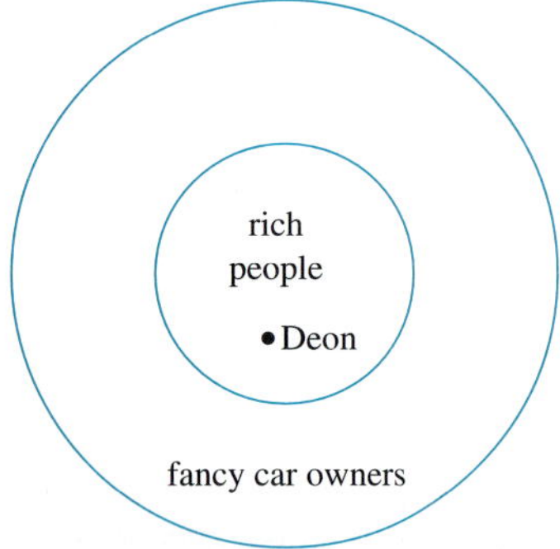

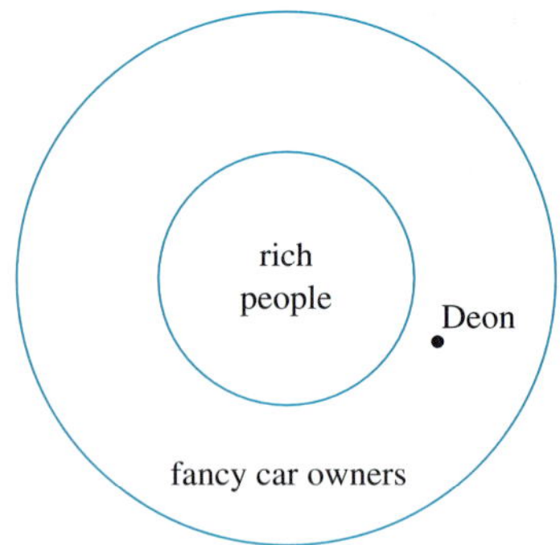

33. Some economists are not trustworthy.

35. Kareem must buy his own lock.

37. Jack is neither a composer nor a musician.

39. Tony is not big.

PROBLEM SET 1.4

1. 70; 252

3. **(a)** No. Actually no sum of four consecutive whole numbers has a factor of 4. The largest factor possible is 2.
   (b) The largest factor is 5.

5.

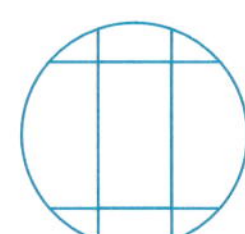

(a) There are other possibilities

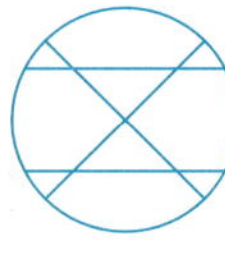

(b)

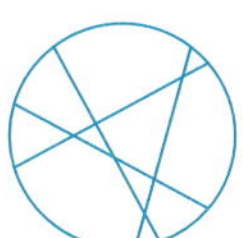

(c) The maximum is 11 pieces

7. There are at least 11 patterns.

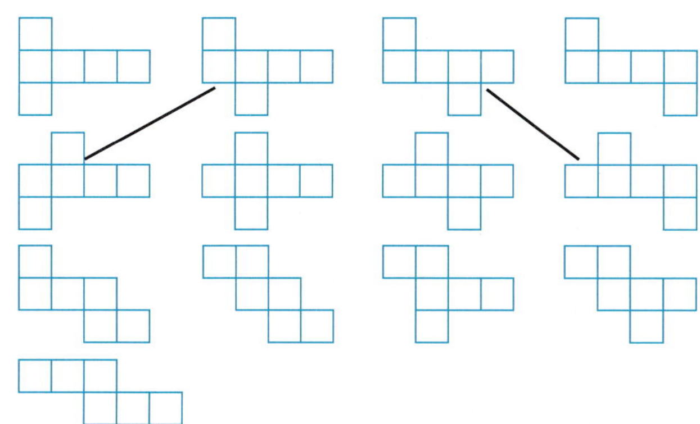

9. 8

11. $6 \div 6 + 6 + 6 = 13$
$6 + 6 \div 6 + 6 = 13$
$6 + 6 + 6 \div 6 = 13$

13.

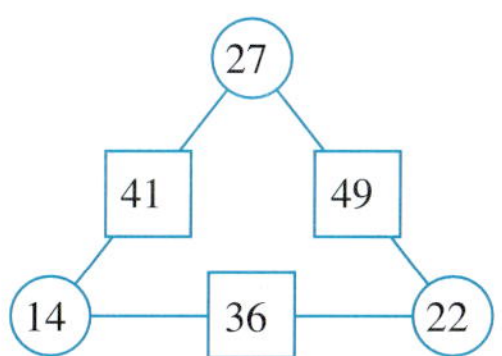

15. 15

17.

19. Bill

21. There are several possibilities, but 7, 8, and 9 are in the corners.

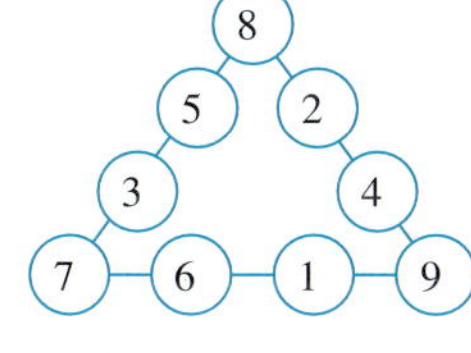

23.

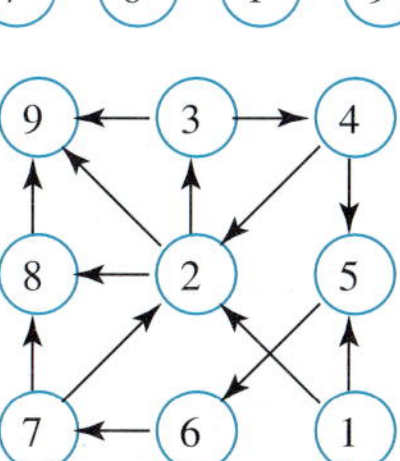

25. \$0.90 per hour

27.
$$\begin{array}{r} 9338 \\ 932 \\ \hline 10270 \end{array}$$

29. No. If 10 pockets have different numbers of bills in each, then the smallest number of bills is $0 + 1 + 2 + 3 + 4 + 5 + 6 + 7 + 8 + 9 = 45$. This is impossible since there are only 44 bills.

31. 35 years or 64 years

33. There are several possibilities. Two are

10	10	0		5	10	5
15		11	or	15		11
5	16	11		10	16	6

35. Let the 130 and 160 pounders go to the top. The 130 pounder goes down. Then the 210 pounder goes up. The 160 pounder goes down, gets the 130 pounder and they both ride up to the top.

37. $1234321 \times (1 + 2 + 3 + 4 + 3 + 2 + 1) = 4444^2 = 19{,}749{,}136$

39. **(a)** 26 **(b)** $\frac{5}{4}$
(c) 576 **(d)** 2347

41.

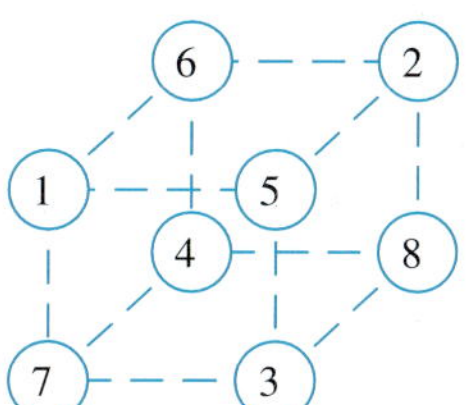

45. 13

47.

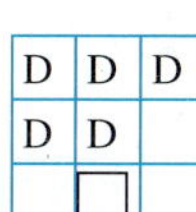

(a)

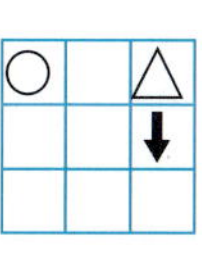

(b)

51. 1,000,000

53. At stage n there are $n^2 + (n - 1)^2$ squares. For example at stage 6 there are $6^2 + 5^2 = 61$ squares.

Extended Problems

55. $1^2 + 1^2 + 2^2 + 3^2 + \ldots + 144^2 = 144 \times 233 = 33{,}552$

57. $1 + 1 + 2 + 3 + 5 + \ldots + 144 = 377 - 1 = 376$

59. $1 + 3 + 8 + 21 + \ldots + 377 = 610 - 1 = 609$

61.

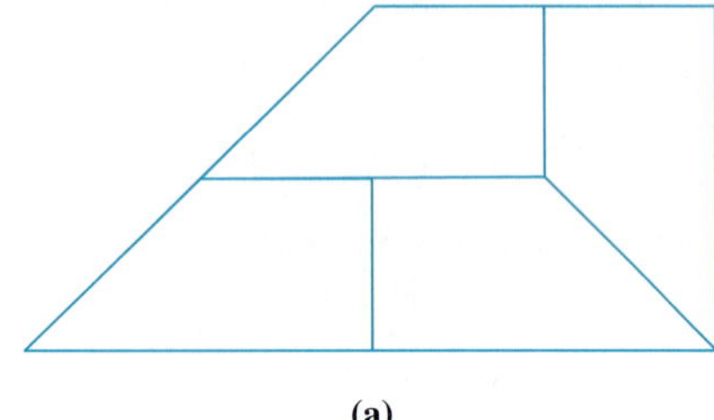

(a)

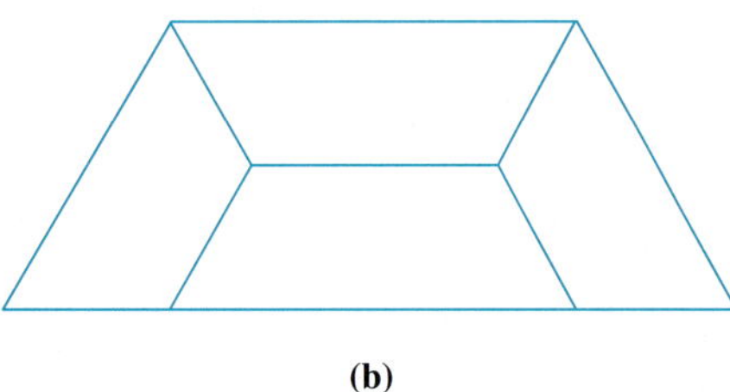

(b)

63. **(a)** 18
(b) 66
(c) $4^{n-1} + 2$

Chapter One Review Problems

1. **(a)** F **(b)** F
(c) T **(d)** F
(e) F **(f)** T

3. Converse: If the lawn gets wet, it will have rained.
Inverse: If it doesn't rain, the lawn will not get wet.
Contrapositive: The conditional and the contrapositive are true. If the lawn didn't get wet, then it didn't rain.

5. **(a)** Invalid; denying the antecedent
(b) Valid; disjunctive syllogism
(c) Invalid; denying the antecedent
(d) Valid; modus ponens
(e) Valid; chain rule

7.

p	q	$\sim p$	$p \Rightarrow q$	$\sim p \vee q$
T	T	F	T	T
T	F	F	F	F
F	T	T	T	T
F	F	T	T	T

9.

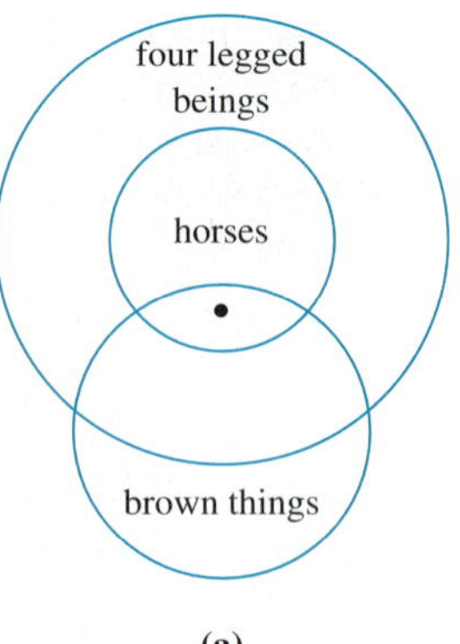

(a)

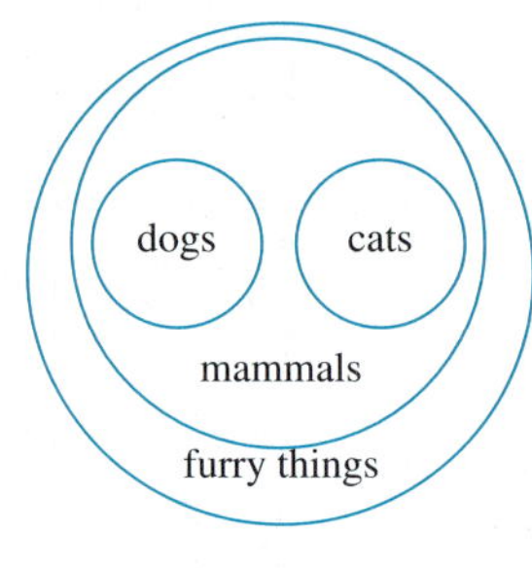

(b)

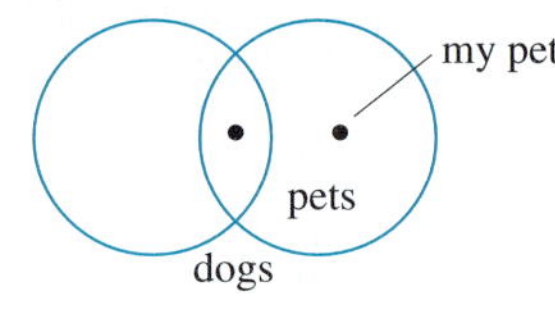

(c)

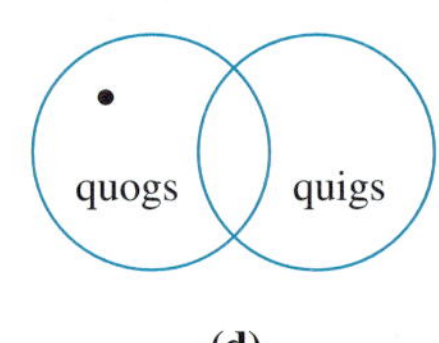

(d)

11. **(a)** T **(b)** F **(c)** F **(d)** F

13.

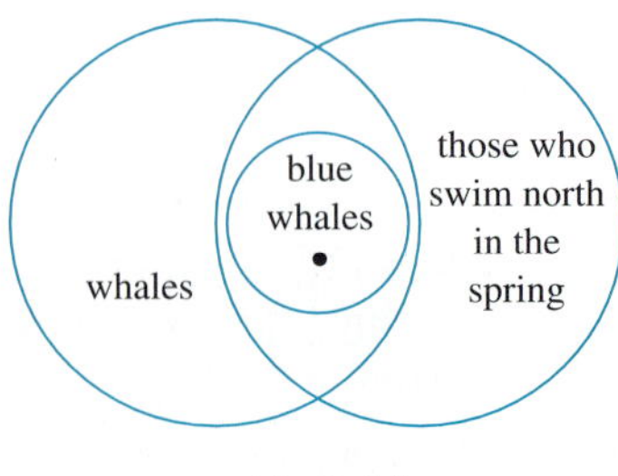

(a) Valid

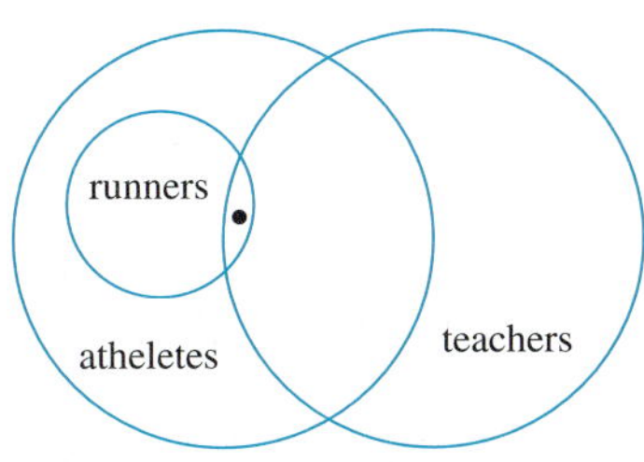

(b) Invalid

15. Your brother is 10. Use a variable.

17. There are 8 possible ways. Each may be obtained from the following way by reflections and/or rotations.

8	1	6
3	5	7
4	9	2

19. Two cuts: 5 pieces
Three cuts: 10 pieces

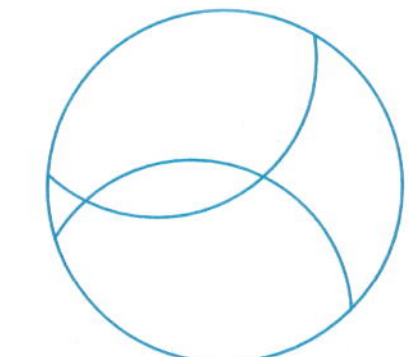

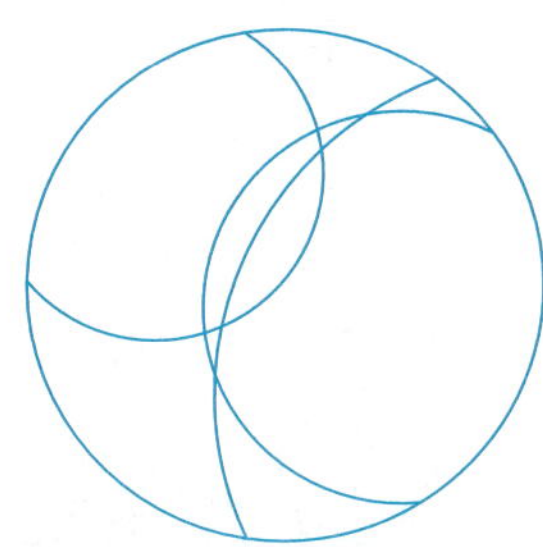

PROBLEM SET 2.1

1. **(a)** Ordinal
(b) Identification
(c) Cardinal
(d) Cardinal

3. **(a)** Identification/Ordinal
(b) Cardinal
(c) Ordinal
(d) Cardinal

5. **(a)** There are 45 days until the end of the semester.
(b) She won forty-fifth place in the decathlon.
(c) My address is 45 West Elm Street.

7. **(a)** The number of potato chips left in the bag is 0.
(b) 0 is not an ordinal number.
(c) The spy was so important, her code number was 0.

9. None of the above. A numeral is a way of writing a number. Cardinal, ordinal, and identification refer to the way a number is used.

11. **(a)** 42
(b) 404

13.

(a)

(b)

(c)

15. **(a)** 1991
(b) 976

17. **(a)** LXXVI
(b) CDXXXIV
(c) MCMXCIX

19.

(a)

(b)

21. **(a)** MCMXCIII
(b) The dealer would want everyone to know it was the 1993 model car.

23. **(a)** $4 \times 100 + 3 \times 10 + 7$
(b) $5 \times 1000 + 6 \times 100 + 0 \times 10 + 3$

25. **(a)** 105,842
(b) 22,060,300

27. **(a)** Three hundred forty-five thousand six hundred seventy-eight
(b) One hundred two million six hundred twenty thousand fifty-seven

29. **(a)** 7001
(b) 1020

31.

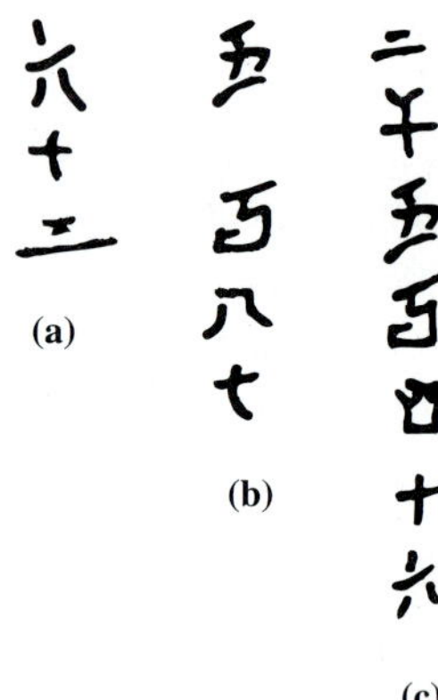

33. **(a)** 544
(b) 4207
(c) 992

35. **(a)** 124,797
(b) 8,724,640,224

37.

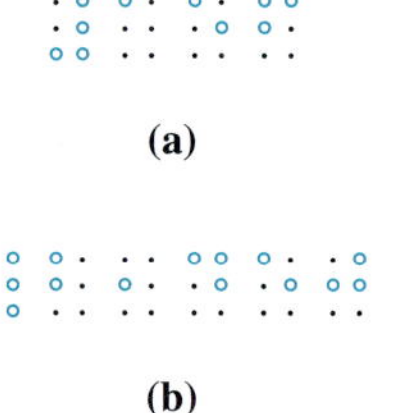

(a)

(b)

(c)

PROBLEM SET 2.2

1. **(a)** Associative property of addition
(b) Distributivity of multiplication over addition
(c) Associative property of multiplication
(d) Right distributivity of division over addition

3. **(a)** $(37 + 43) + 25 = 105$
(b) $47(15 + 85) = 4700$
(c) $(4 \times 25) \times 13 = 1300$
(d) $24(26 - 21) = 120$

5. 7, because $93 + 7 = 100$

7. **(a)** $38 \times 5 = 40 \times 5 - 2 \times 5 = 200 - 10 = 190$
(b) $483 \div 7 = 490 \div 7 - 7 \div 7 = 70 - 1 = 69$
(c) $(43 \times 6) + (7 \times 6) = (43 + 7) \times 6$
$= 50 \times 6 = 300$

9. **(a)** $288 + 102 = (288 + 2) + (102 - 2)$
$= 290 + 100 = 390$
(b) $310 + 199 = (301 - 1) + (199 + 1)$
$= 300 + 200 = 500$

11. **(a)** $43 - 17 = (43 + 3) - (17 + 3)$
$= 46 - 20 = 26$
(b) $132 - 96 = (132 + 4) - (96 + 4)$
$= 136 - 100 = 36$

13. **(a)** $18 \times 5 = (18 \div 2) \times (5 \times 2)$
$= 9 \times 10 = 90$
(b) $222 \times 5 = (222 \div 2) \times (5 \times 2)$
$= 111 \times 10 = 1110$

15. **(a)** $123 + 456 = 500 + 70 + 9 = 579$
(b) $467 - 134 = 300 + 30 + 3 = 333$

17. **(a)** 900
(b) 300

19. **(a)** 180,000
(b) 4000

21. **(a)** Range is from 600 to 800.
(b) Range is from 700 to 900.

23. **(a)** Range is from 6000 to 12,000.
(b) Range is from 60,000 to 120,000.

25. **(a)** 1230
(b) 680
(c) 23,790
(d) 10

27. **(a)** 770
(b) 550
(c) 10,120
(d) 670

29. **(a)** 4300
(b) 6600
(c) 2800
(d) 25,800

31. **(a)** $4776 \div 103 \approx 5000 \div 100 = 50$
(b) $46 \times 202 \approx 50 \times 200 = 10{,}000$

33. **(a)** $30 \times 23 = 690$
$30 \times 20 = 600$
$30 \times 25 = 750$
$31 \times 20 = 620$
(b) $35 \times 50 = 1750$
$30 \times 50 = 1500$
$35 \times 40 = 1400$
$40 \times 40 = 1600$

35. **(a)** 42,000 and 56,000. The true answer is 48,544.
(b) 10,000 and 40,000. The true answer is 20,449.

37. **(a)** $4 \times 350 = 1400$
(b) $60^3 = 216{,}000$

39. **(a)** $16 \times 21 = 8 \times 42 = 4 \times 84 = 2 \times 168 = 336$
(b) $4 \times 72 = 2 \times 144 = 288$

41. It would be cautious to overestimate, although a bit more expensive for the company, perhaps.

43. One should underestimate to better provide for contingencies.

45. **(a)** $1357 \times 90 = 122{,}130$
(b) $66 \times 6666 = 439{,}956$

(c) 789,345 × 6 = 4,736,070
(d) 12,345 × 67 = 827,115

PROBLEM SET 2.3

1. (a) -14 (b) 5 (c) -21
(d) -23 (e) 21 (f) 7

3. (a) -147 (b) -84 (c) 90
(d) 9 (e) -9 (f) -9

5. (a) Distributivity of multiplication over addition
(b) Associativity of addition
(c) Commutativity of multiplication
(d) Commutativity of addition

7. (a) $\frac{4}{5}$ (b) $\frac{5}{6}$ (c) $\frac{1}{3}$
(d) $\frac{3}{2}$ (e) $\frac{7}{3}$ (f) $\frac{9}{11}$

9. (a) Not equivalent
(b) Equivalent
(c) Not equivalent

11. (a) $\frac{3}{4}, \frac{18}{24}$
(b) $\frac{3}{15}, \frac{2}{10}$

13. (a) $\frac{11}{10}$ (b) $\frac{17}{18}$ (c) $\frac{4}{21}$
(d) $\frac{14}{15}$ (e) $\frac{9}{14}$ (f) $\frac{13}{168}$

15. (a) $\frac{-5}{6}$ (b) $\frac{4}{21}$ (c) $\frac{-2}{7}$
(d) $\frac{8}{5}$ (e) $\frac{-5}{6}$ (f) $\frac{14}{15}$

17. (a) Multiplicative commutativity, multiplicative identity
(b) Right distributivity of division over addition
(c) Multiplicative commutivity
(d) Multiplicative inverse

19. (a) 0 (b) $\frac{-m}{n}$

21. (a) 63° (b) 50° (c) 29° (d) $-16°$

23. It would drop about 60 degrees to a temperature of about -10°F.

25. $64; she was overdrawn once.

27. (a) 9 yard gain
(b) 2 point loss
(c) $200 gain

29. Texas

31. (a) $2\frac{1}{2}$ cups (b) $\frac{5}{8}$ cups (c) $2\frac{1}{12}$ cups

33. All numbers from top-down, left-to-right: -22, 2, -24, 4, -2, -22, -9, 13, -15, -7

35. There are 8 possible ways. Each may be obtained from the following way by reflections and/or rotations.

8	1	6
3	5	7
4	9	2

37. All of the properties for integer operations hold for this system.

39. All of the properties of integer operations hold. 12 corresponds to the additive identity. Additive inverse pairs are (1, 11), (2, 10), (3, 9), (4, 8), (5, 7), and (6, 6)

PROBLEM SET 2.4

1. (a) Three and forty-seven thousandths
(b) Three hundred and forty-seven hundredths
(c) Three hundred forty-seven thousandths

3. (a) Seventy five thousandths
(b) One hundred fifty and twenty-five hundredths
(c) Five and one hundred seventy-five ten thousandths

5. (a) 3.0105
(b) 327.27
(c) 0.000085

7. (a) 0.0305
(b) 0.004038
(c) 25.05287

9. (a) 413.46
(b) 413

11. (a) 28.875
(b) 87.70

13. (a) 0.68 (b) $\frac{3}{8}$ (c) $0.8\overline{3}$

15. (a) $2\frac{3}{8}$ (b) 0.5714 (c) $\frac{21}{25}$

17. (a) 23.346517
(b) 6.34277
(c) $412.3\overline{45}$

19. (a) 37,500
(b) 5652.9
(c) 0.00000045

21. (a) $\frac{324}{990} = \frac{18}{55}$
(b) $\frac{295}{9}$

23. \$7.93

25. **(a)** 28.6650 grams
(b) 28.6749 grams

27. **(a)** 328,000 Belgian francs
(b) 15,800,000 Italian lira
(c) 27,000,000 Brazilian cruzeiro

29. 6.21 miles

31. **(a)** 5.3 miles
(b) 6.9 miles

33. 0.647 should be changed to 0.657.

PROBLEM SET 2.5

1. **(a)** $\frac{1}{4}$, 0.25
(b) 425%, $4\frac{1}{4}$
(c) 40%, 0.4

3. **(a)** 37.5%, 0.375
(b) 1.12, $1\frac{3}{25}$
(c) 87.5%, $\frac{7}{8}$

5. Convert the fraction to a decimal and move the decimal two places to the right.

7. **(a)** 120 **(b)** $266\frac{2}{3}$ **(c)** 64

9. **(a)** 100 **(b)** 150 **(c)** 25%

11. **(a)** 40% increase
(b) $33\frac{1}{3}$% decrease
(c) 200% increase

13. **(a)** 104 **(b)** 25% **(c)** 300

15. **(a)** 156
(b) Decrease of 14%
(c) 56

17. \$2244

19. 236%

21. **(a)** \$2160 **(b)** 8%

23. 40%

25. \$23.70

27. \$2295

29. \$210

PROBLEM SET 2.6

1. **(a)** 2^{10} **(b)** 5^6 **(c)** b^9

3. **(a)** 8^4 **(b)** 8^3 **(c)** 30^2

5. **(a)** 5^6 **(b)** 3^8 **(c)** b^{18}

7. **(a)** 3^5 **(b)** 5^4 **(c)** b^4

9. **(a)** $\left(\frac{5}{8}\right)^3$ **(b)** $\left(\frac{4}{7}\right)^2$ **(c)** $\left(\frac{x}{y}\right)^5$

11. **(a)** 2.4025 **(b)** 295.84 **(c)** 0.000961

13. **(a)** 199,526.23 **(b)** 630,957.34 **(c)** 63,095,734

15. **(a)** 3802.0403 **(b)** 20,107.586 **(c)** 9.0658

17. **(a)** $\frac{1}{8}$ **(b)** $\frac{3}{25}$ **(c)** 16

19. $\frac{5}{4}$

21. -9

23. **(a)** 8.25×10^8 **(b)** 237,000
(c) 2.53×10^{-2} **(d)** 8.05×10^5

25. 2^{20}

27. 5^{-4}

29. $x = 10$

31. $x = 9$

33. 4.84×10^8

35. 1.2×10^{14}

37. 2.8×10^{-7}

39.

Planet	k
Mercury	1.0021
Venus	1.0008
Earth	1
Mars	1.0016
Jupiter	0.9988
Saturn	0.9991
Uranus	0.9987
Neptune	0.9971
Pluto	1.0002

Yes, k is approximately one in each case.

41. 333.85 years

43. Four times
$S' = 4\pi(2r)^2 = 4\pi 4r^2 = 4\,(4\pi r^2) = 4S$

45. 9.47 inches

PROBLEM SET 2.7

1. **(a)** 6 **(b)** 8 **(c)** -5

3. **(a)** 7.810 **(b)** 14.142 **(c)** -7.071

5. **(a)** $\sqrt[4]{16} = 2$ since $2^4 = 16$.
(b) $\sqrt[3]{-8} = -2$ since $(-2)^3 = -8$.
(c) $\sqrt[4]{-81}$ is undefined since $x^4 \geq 0$ for all x.

7. **(a)** 2 **(b)** 3 **(c)** -2

9. **(a)** $27^{1/2}$ **(b)** $15^{1/3}$ **(c)** $x^{3/4}$

11. **(a)** $45^{1/2}$ **(b)** $-64^{1/4}$ **(c)** $x^{3/2}$

13. **(a)** $\sqrt[3]{25}$ **(b)** $\sqrt[3]{100}$ **(c)** 4

15. **(a)** $2\sqrt{x}$
(b) $\sqrt{25y}$ or $5\sqrt{y}$
(c) $\sqrt{x^3}$

17. **(a)** 9 **(b)** 64 **(c)** $\frac{1}{32}$

19. **(a)** 1.442 **(b)** 8.550 **(c)** 181.109

21. **(a)** 3.162 **(b)** 3.121

23. **(a)** $b^{2/3}$ **(b)** $\frac{2}{3}\sqrt{x}$
(c) $3x^{10}$ **(d)** $27y^3$

25. **(a)** $\frac{4}{3}$ **(b)** $\frac{-1}{2}$ **(c)** 2

27. 2.821 inches

29. $2.66 \times 10^{11} = 266{,}000{,}000{,}000$ cubic miles

31. **(a)**

t	A
0	2000
1	2180
2	2376
3	2590
4	2823
5	3077
6	3354
7	3656
8	3985
9	4344
10	4735
11	5161
12	5625
13	6132
14	6683
15	7285
16	7941
17	8655
18	9434
19	10283
20	11209

(b) At $t = 9$; at $t = 17$

Chapter Two Review Problems

1. **(a)** Ordinal
(b) Cardinal
(c) Identification

3. Hindu-Arabic and Roman

5. MCMXIV

7. Nine billion eight hundred seventy-six million five hundred forty-three thousand two hundred ten
$9 \times 1{,}000{,}000{,}000 + 8 \times 100{,}000{,}000 + 7 \times 10{,}000{,}000 + 6 \times 1{,}000{,}000 + 5 \times 100{,}000 + 4 \times 10{,}000 + 3 \times 1{,}000 + 2 \times 100 + 1 \times 10.$

9. **(a)** 2200 **(b)** 7112

11. between 700 and 1000, 900, 870

13. **(a)** rational reciprocals
(b) rational number multiplication
(c) distributivity of multiplication over subtraction

15. $1\frac{4}{33}$

17. **(a)** $\frac{2}{5}$, 0.4 **(b)** 295%, $2\frac{19}{20}$
(c) 37.5%, 0.375

19. **(a)** 53.12 **(b)** 171.4%
(c) 523.81

21. \$11,433; \$9146

23. **(a)** 2.6744×10^{-5} **(b)** 1.8636×10^8

25. **(a)** 10.926 **(b)** 2.665

PROBLEM SET 3.1

1. **(a)**

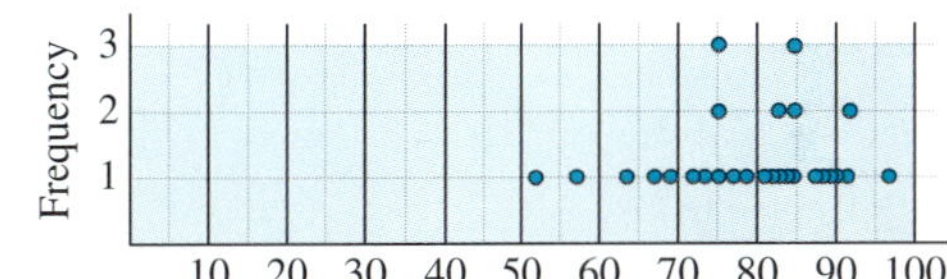

(b)

5	1 7
6	4 8
7	0 3 4 5 5 5 6 7
8	0 1 2 2 3 4 4 4 6 7 8
9	0 2 2 7

3. **(a)**

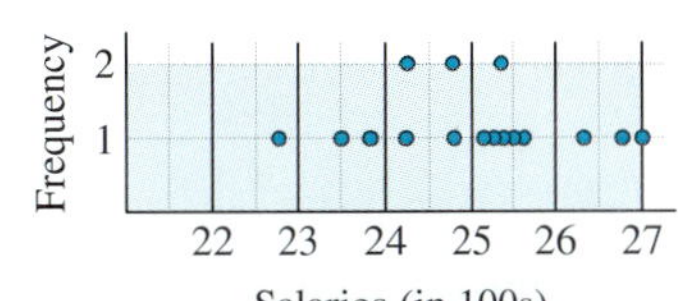

(b) **Salaries (in hundreds)**

22	8
23	5 8
24	2 2 8 8
25	1 2 4 4 5 6
26	3 8
27	0

5. Batting Champion Averages
1970–1994

Stem	Leaves
.31	8
.32	8 9
.33	2 3 3 3 6 7 9
.34	1 3 3
.35	0 7 9 9
.36	1 3 3 4 6 8
.37	
.38	8
.39	0

Most of the averages are between .332 and .368.

7.

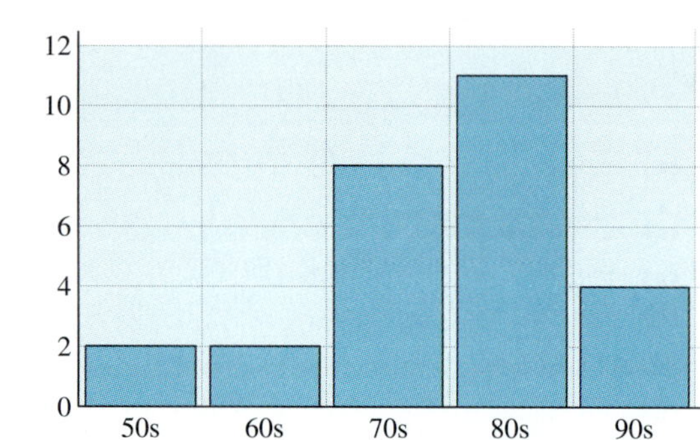

9.

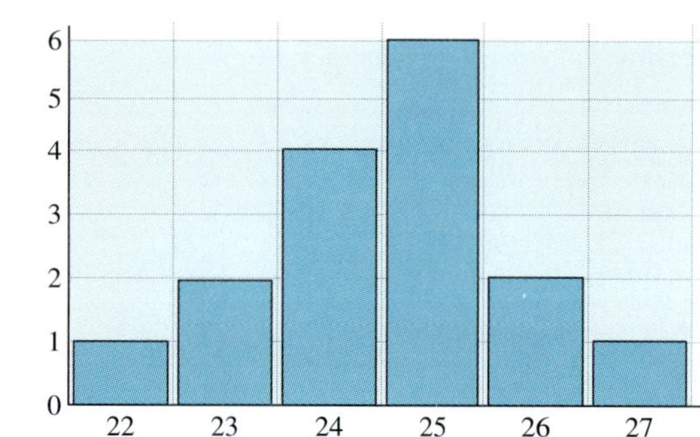

11. (a)

Stem	Leaves
8	4 5 7 7 8 8
9	0 1 3 3 3 3 4 6 6 6 6 7 8 9 9
10	0 0 0 0 1 1 2 3 4 4 5 6 7 8 9
11	0 0 0 3 3 4 4 4 6 6 8 9
12	0 0

(b)

Health Scores

15
10
5
0
80–89 90–99 100–109 110–119 120

13. (a)

Score	Frequency
3	2
4	5
5	6
6	4
7	4
8	5
9	3
10	1

(b) 30

15.

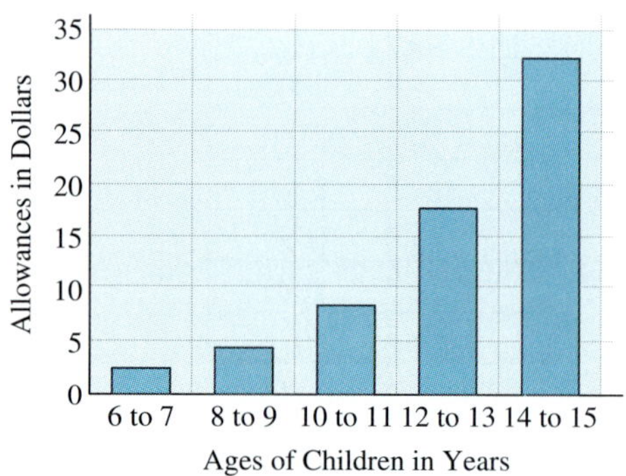

17.

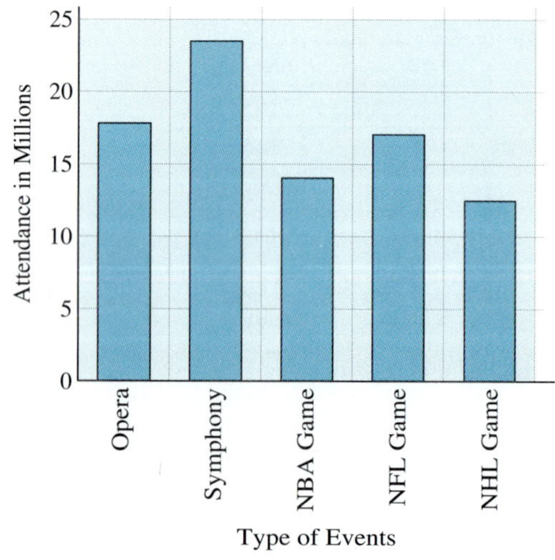

19.

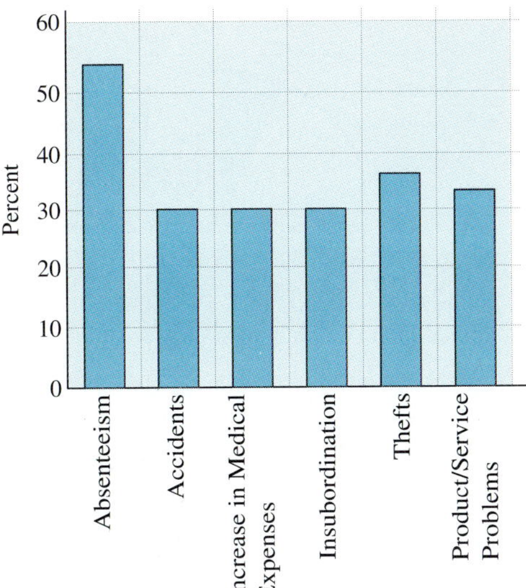

Types of Problems Caused by Drugs in the Workplace

21.

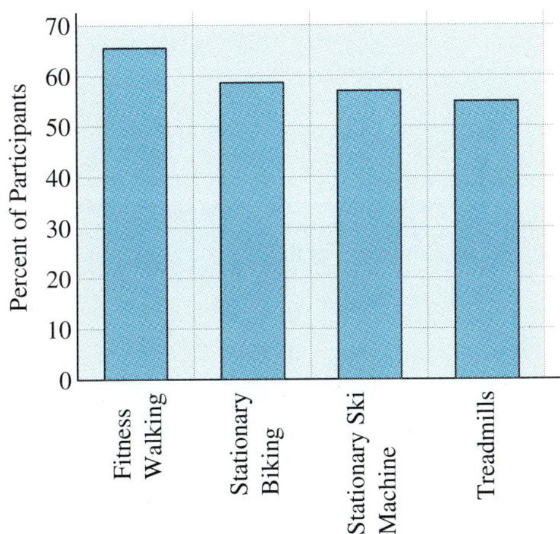

Types of Fitness Activities

23.

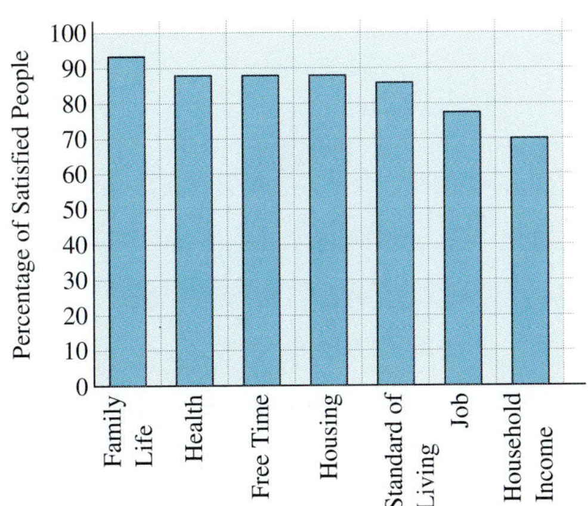

Portions of Life that People Were Satisfied With

25. Estimated populations used for answers:
1790—5 million, 1890—65 million, 1990—255 million

(b) 60 million
(c) 190 million
(d) 1200%
(e) 400%

27. Note: The values are plotted between the vertical lines.

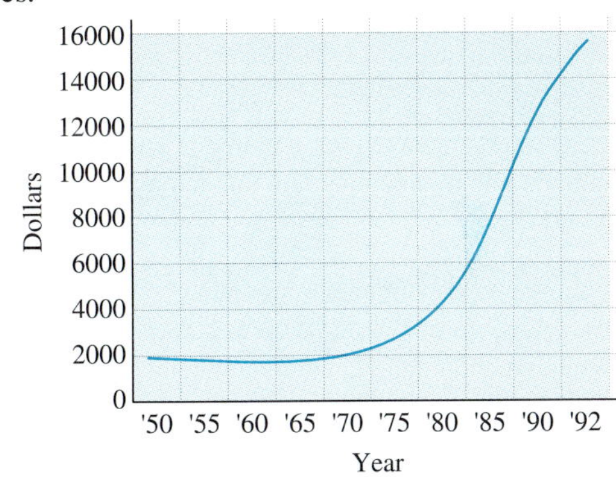

Per Capita Federal Debt

29. **(a)**

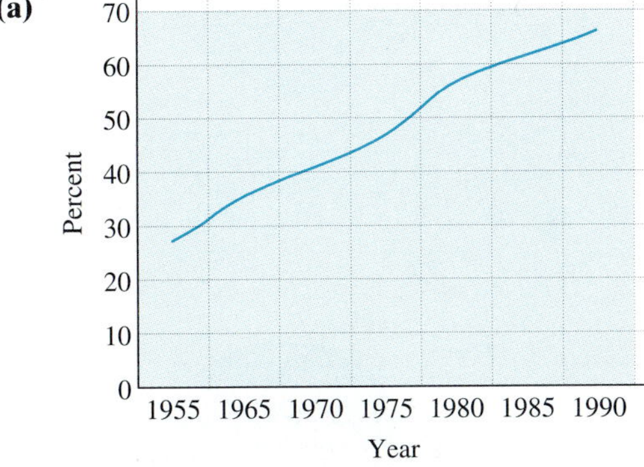

Mothers with Children Under 18 Participating in the Labor Force

(b) The percent increases from 1965 to 1975 all fall on the line connecting the percents for 1965 and 1975.

31.

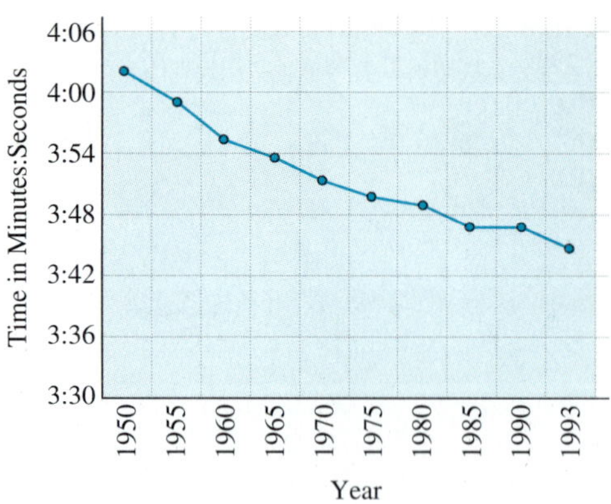

39.

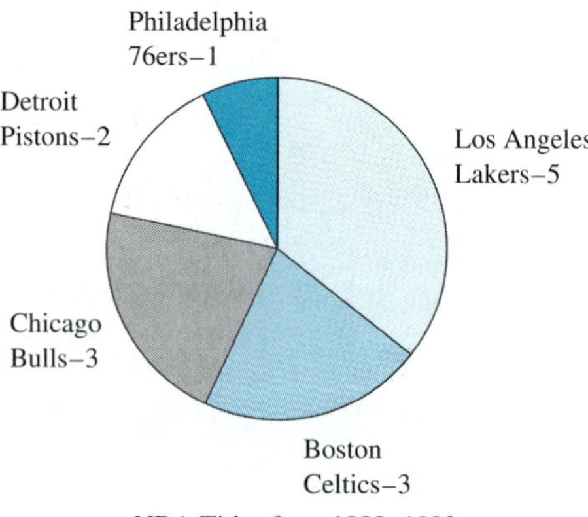

NBA Titles from 1980–1993

33.

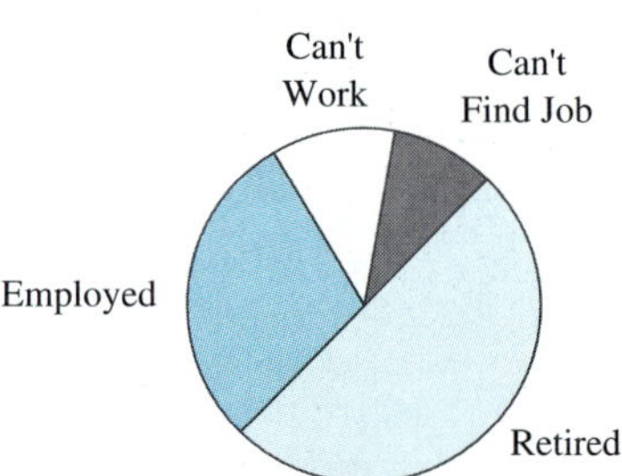

41.

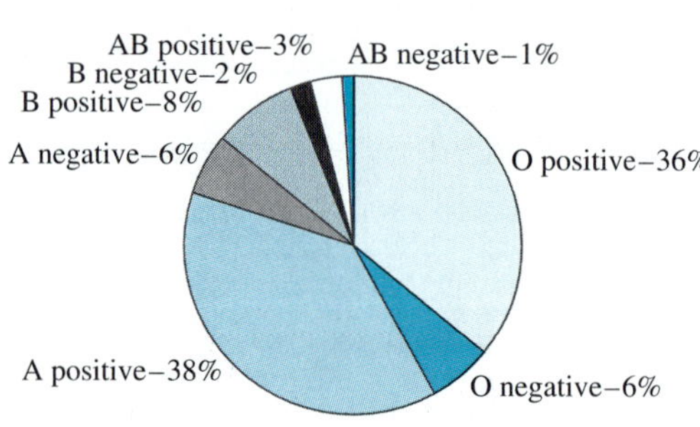

Percentage of Americans with Listed Bloodtypes

(a)

35.

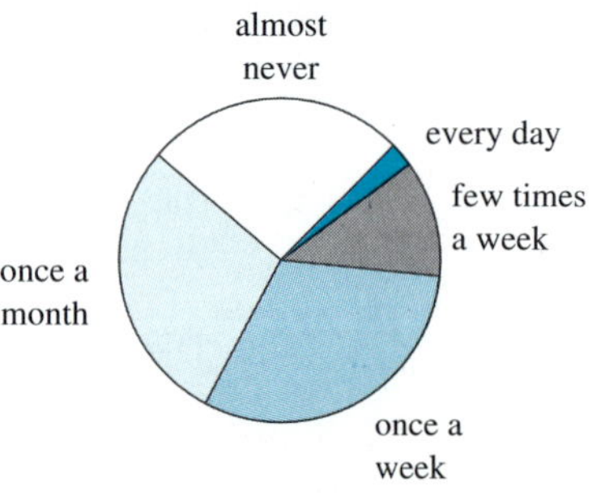

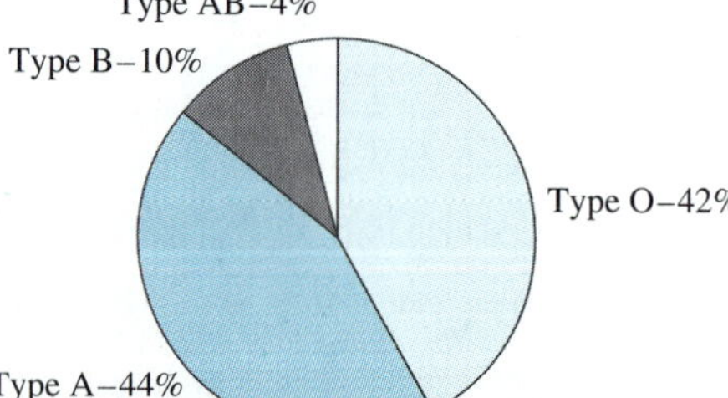

Relative Percentage of Americans Having the Listed Bloodtypes

(b)

37.

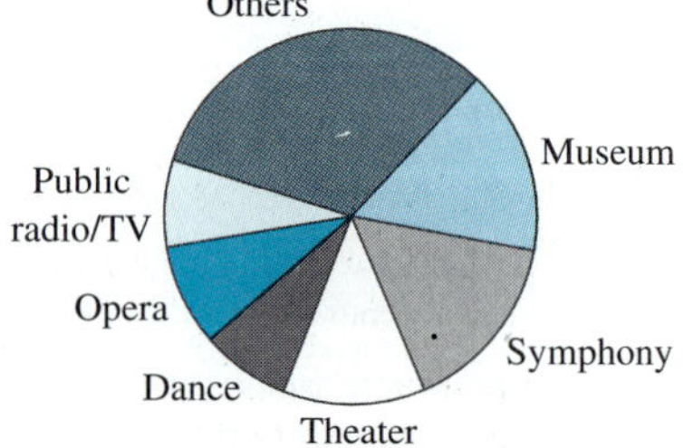

43.

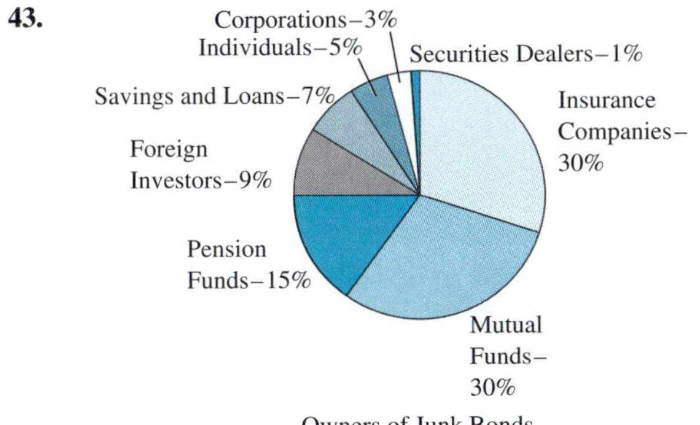

7.

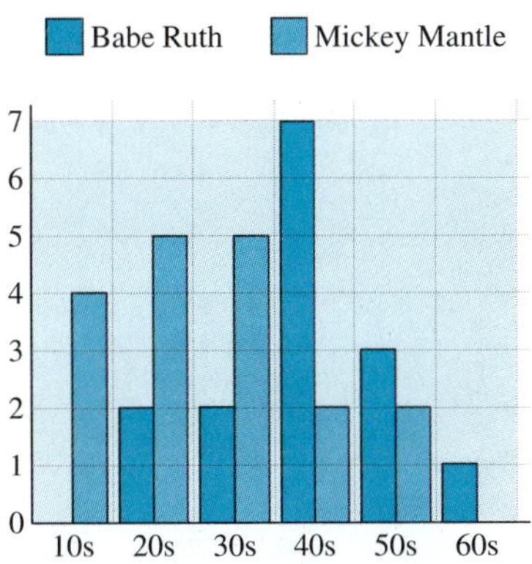

PROBLEM SET 3.2

1. Grades on Sociology Midterm

Class 1		Class 2
	5	4
643	6	6
76664433	7	033445678
766554440	8	00223344556688
5221	9	46

Both classes have most scores in the 70s and 80s.

3. Home Run Hitters

Babe Ruth		Mickey Mantle
	1	3589
52	2	12337
54	3	01457
9766611	4	02
944	5	24
0	6	

Babe consistently hit more home runs than Mickey.

5.

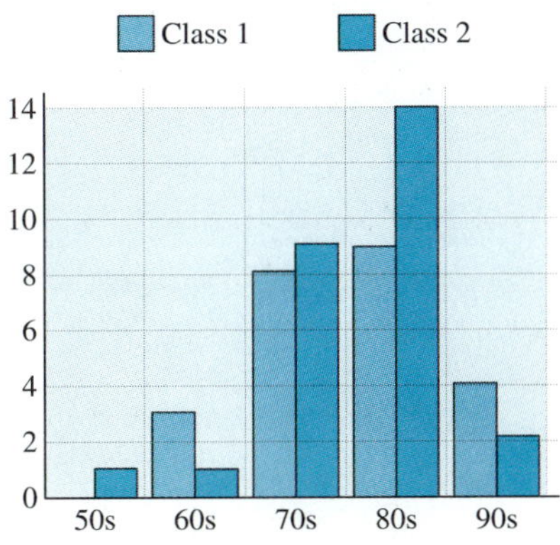

9.

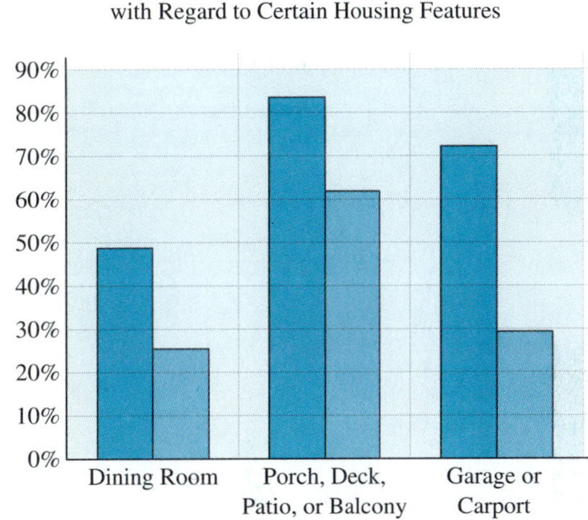

11.

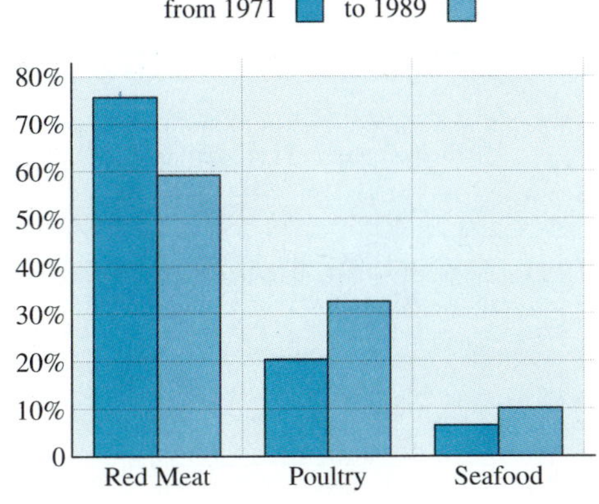

13.

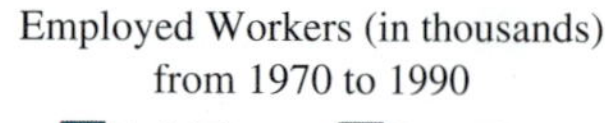

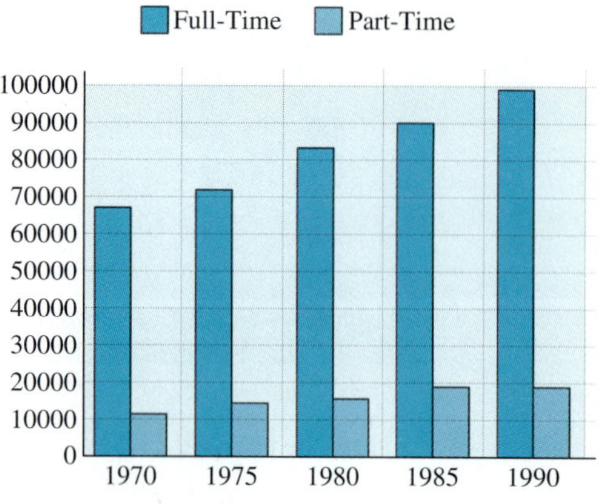

15.

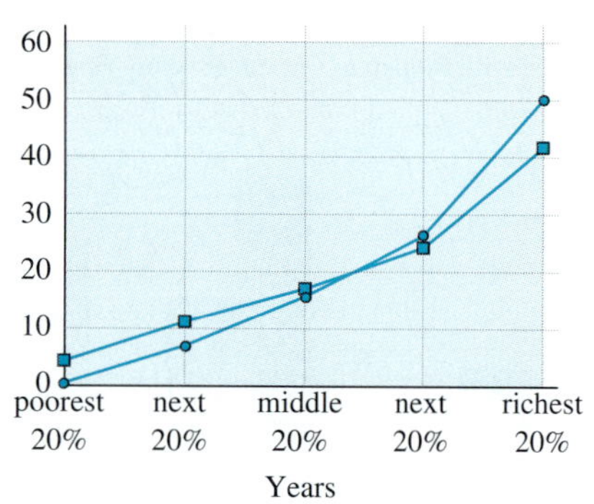

17.

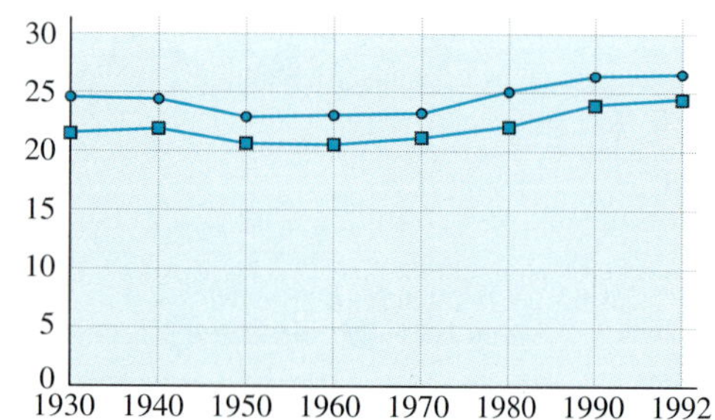

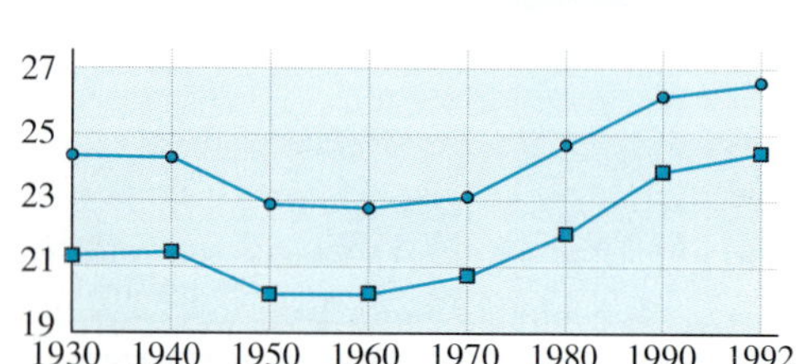

19.

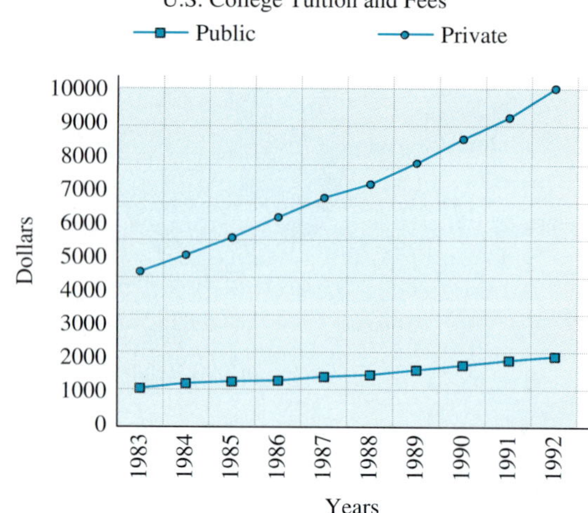

21. Fertilizer consumption is much higher in the industrial world. In North America it is down slightly while in Europe and the USSR it is up.

23. **(a)** 0.55 acres; 0.35 acres
(b) 13 pounds; 55 pounds
(c) The trend is a reflection of the fact that there is only a limited amount of arable land mass. To feed everyone agricultural production has been increased through use of fertilizers.

25.

Computers in Use

Others

Japan

Europe

U.S.

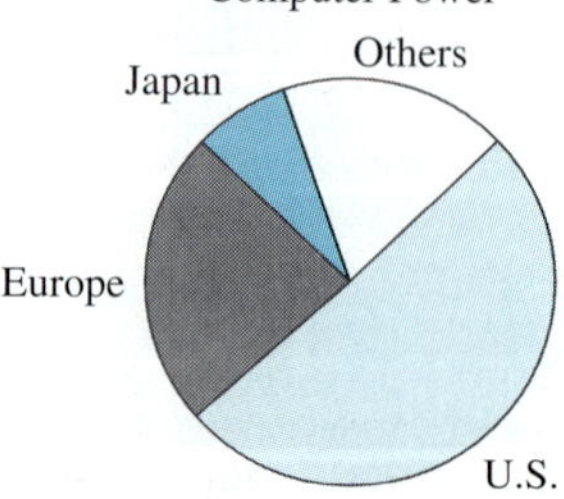

27.

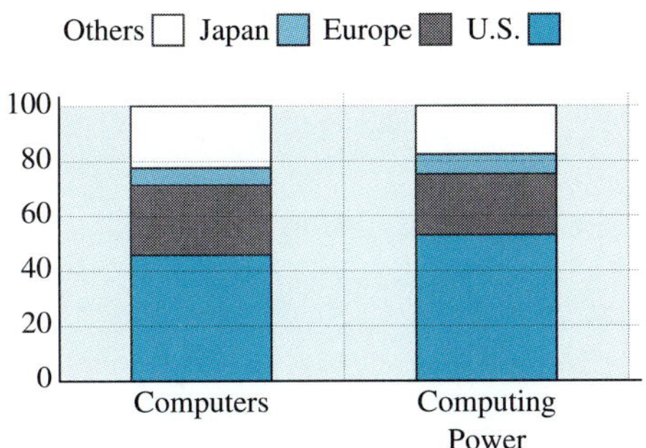

29. (a)

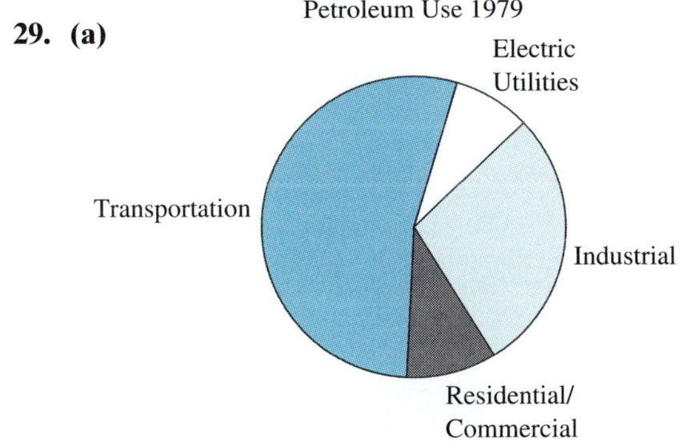

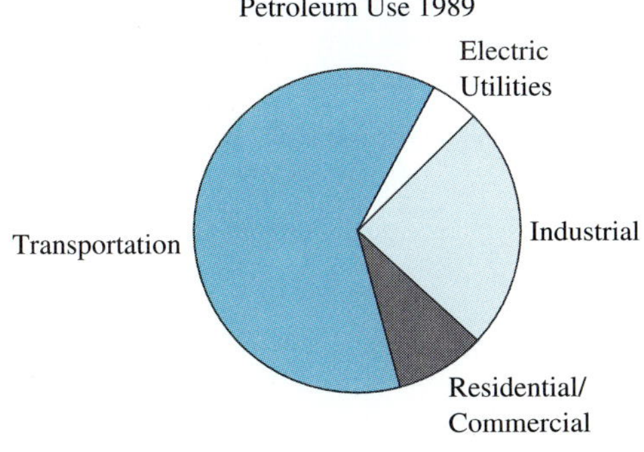

29. (b)

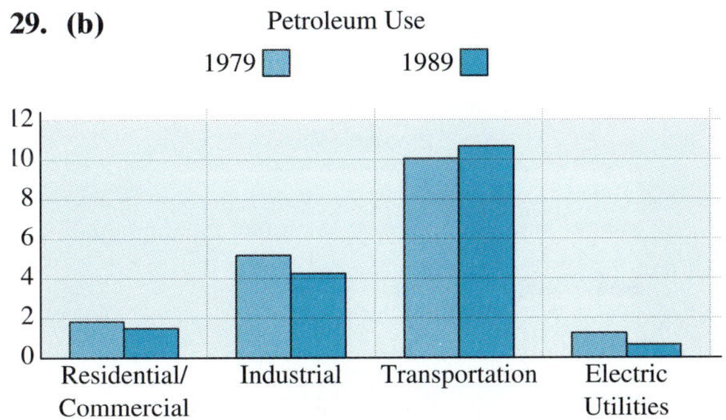

31. (a)

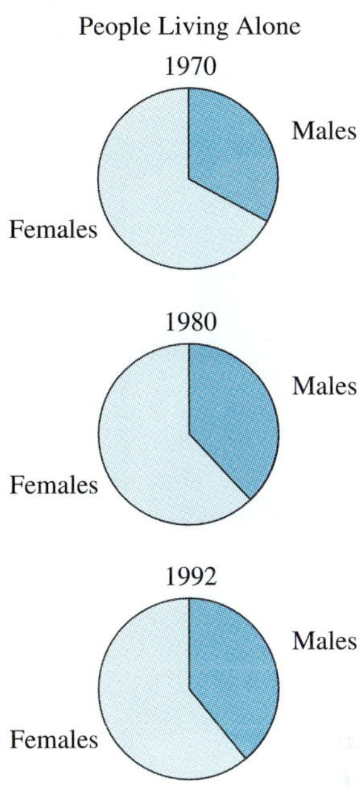

(b)

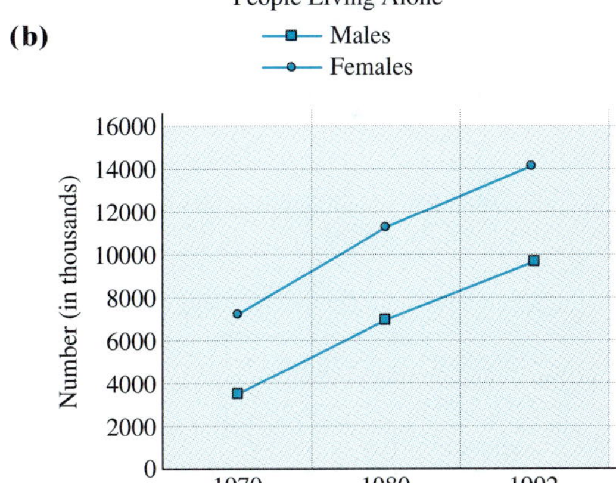

(c) The line graph shows trends more clearly whereas pie charts show the gender shift more clearly.

33. (a)

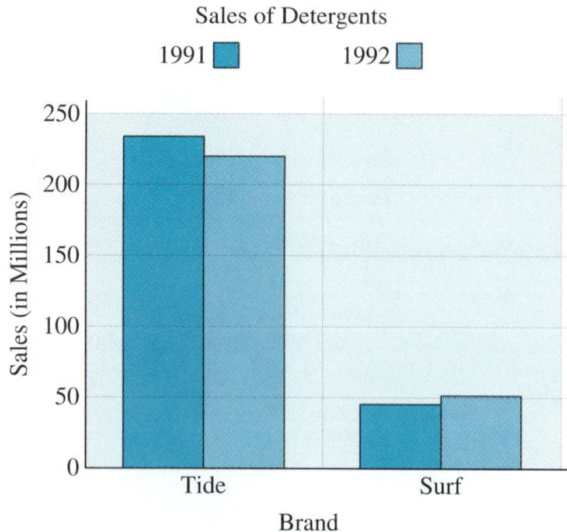

(b)

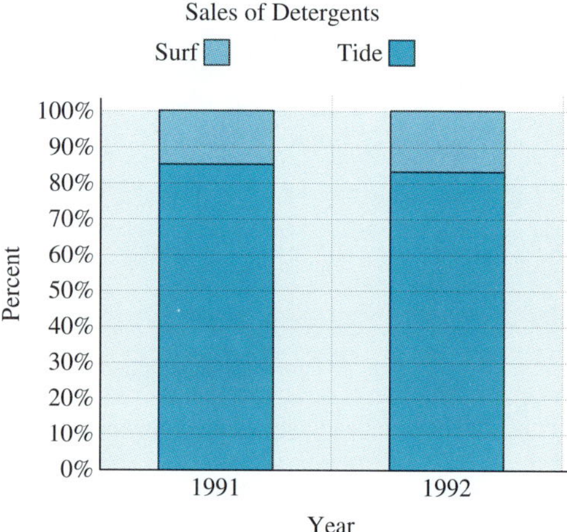

(c) The multiple bar graph allows for a clearer picture of the trends.

PROBLEM SET 3.3

1. (a)

Interval	Frequency
90–100	3
80–89	7
70–79	5
< 70	5

(b)

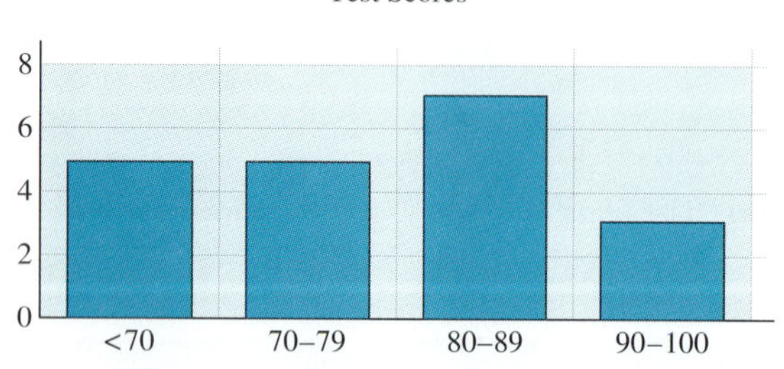

(c) The outliers are not shown.

3. (a)

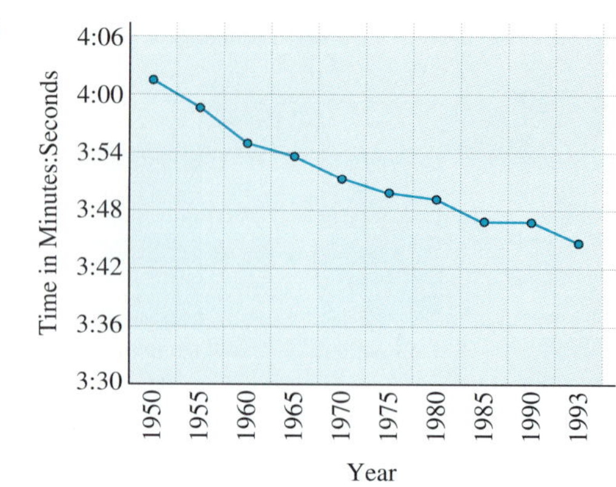

World Record Times For the Mile Run

(b) It makes the downward trend more apparent.

5. (a)

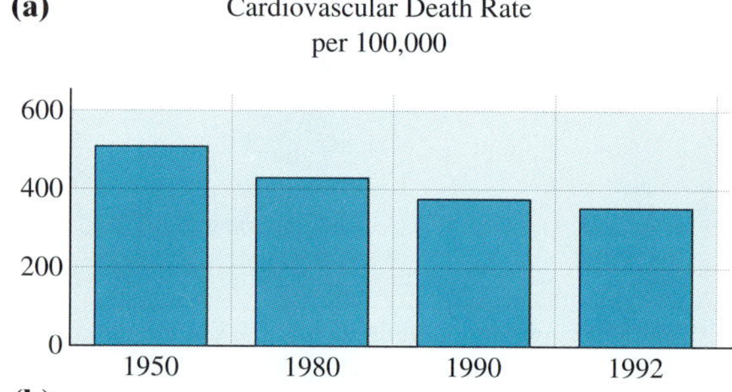

(b)

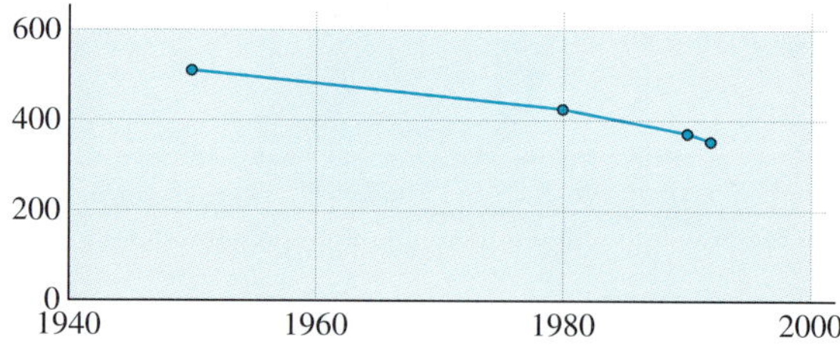

7.

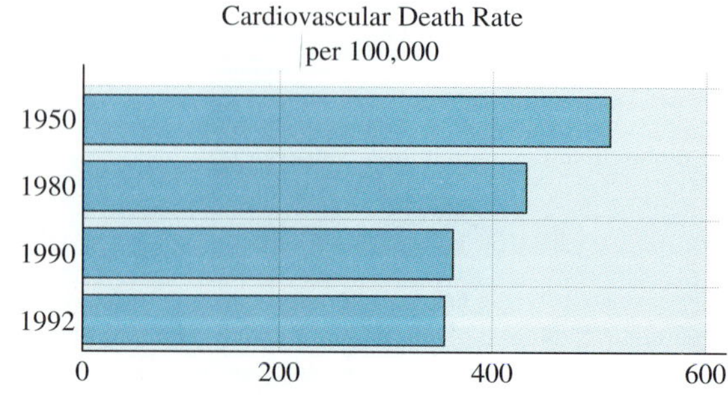

9.

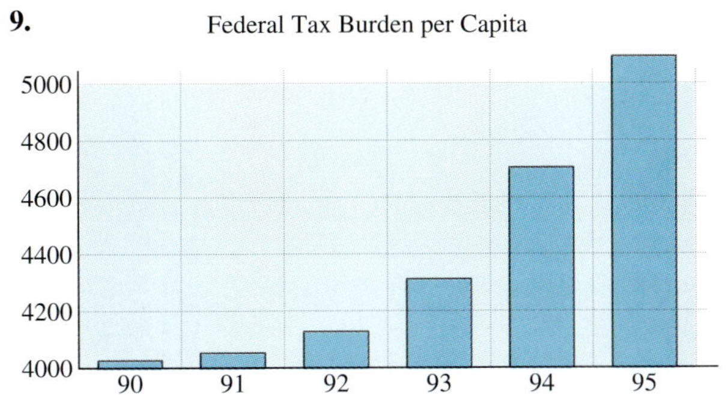

11.

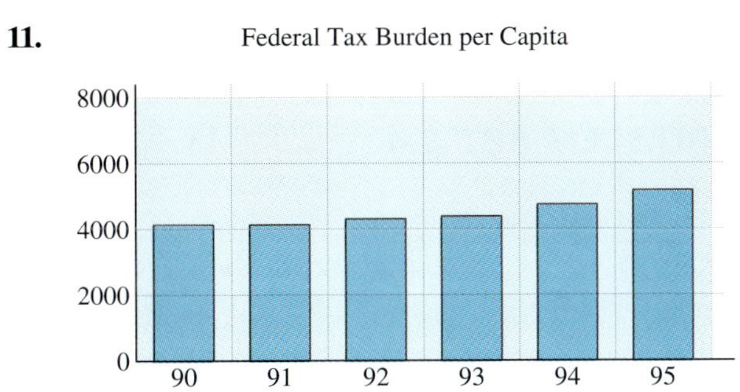

13.

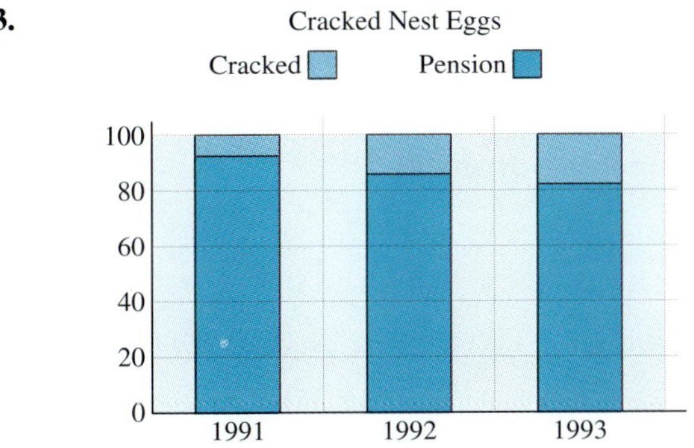

15.

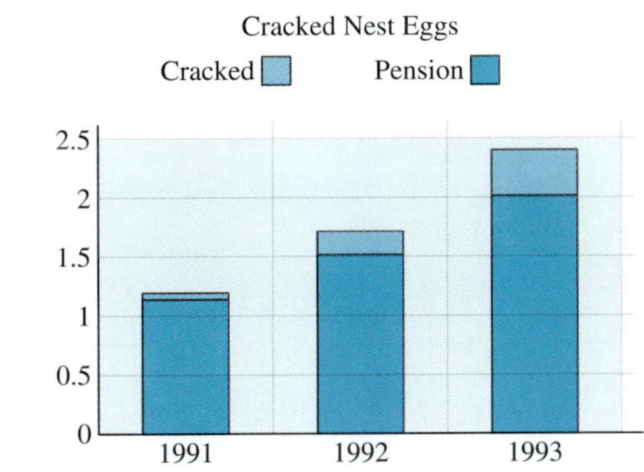

17.

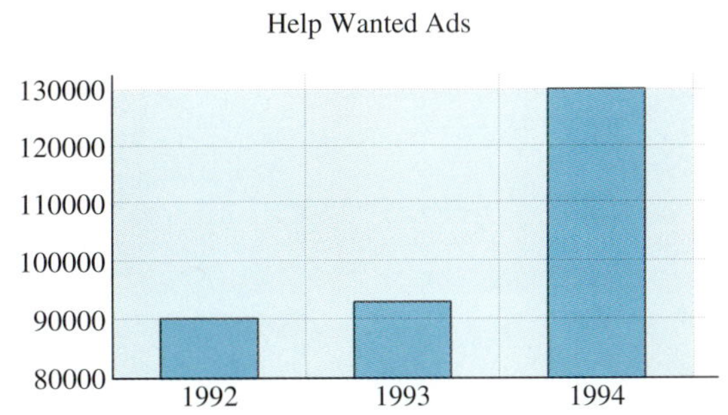

19.

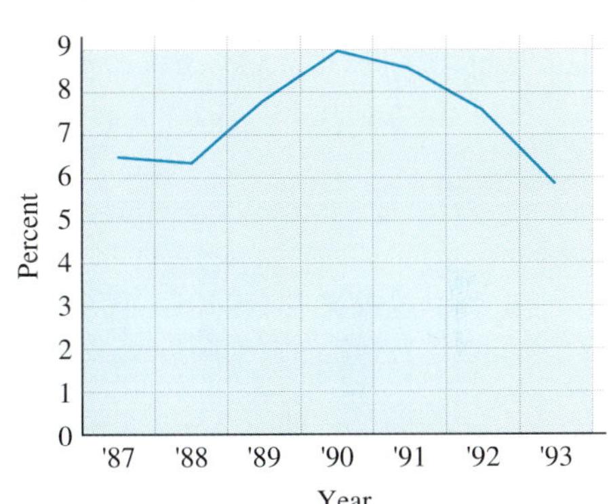

21.

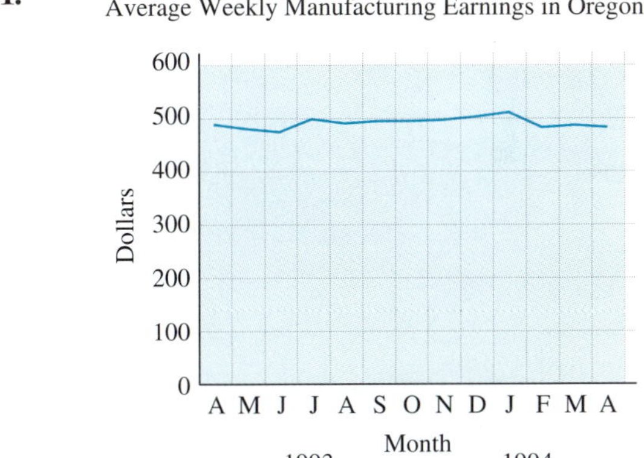

23. **(a)** $\sqrt{2}$ or about 1.4 in. Because the graphs are two-dimensional, their revenues vary as the square of their radii and 1^2: $\sqrt{2}^2 = 1{:}2 =$ 5,000,000:10,000,000.

(b) $\sqrt[3]{2}$ or about 1.3 in. Because the graphs are three-dimensional, their revenues vary as the cube of their radii and $1^3{:}\sqrt[3]{2}^3 = 1{:}2 =$ 5,000,000:10,000,000.

25.

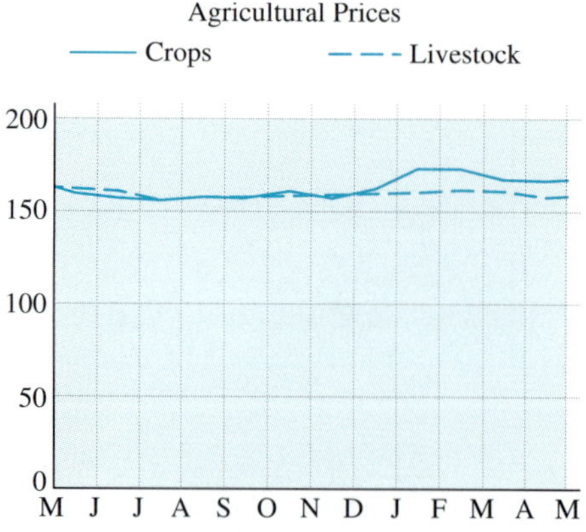

27.

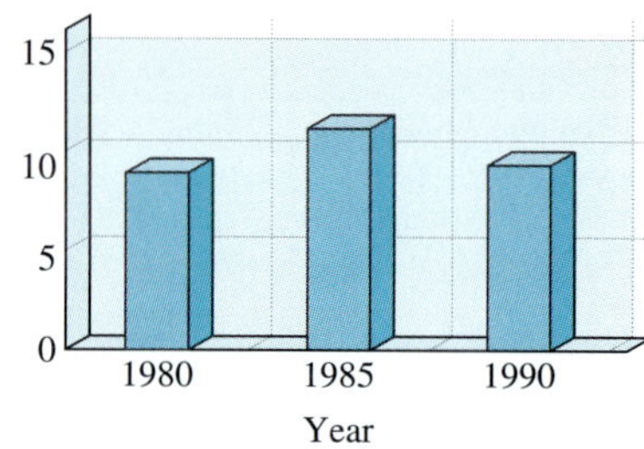

29.

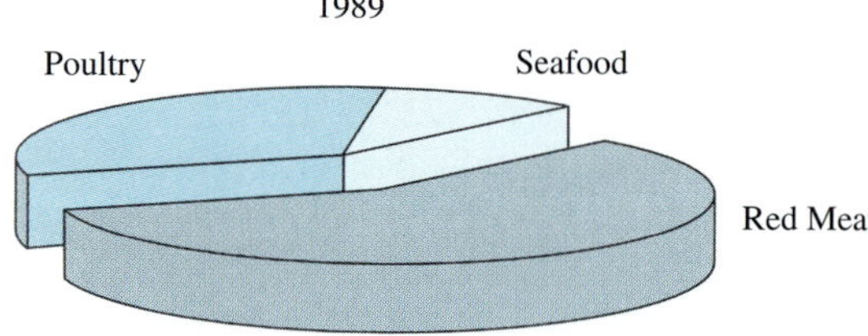

31.

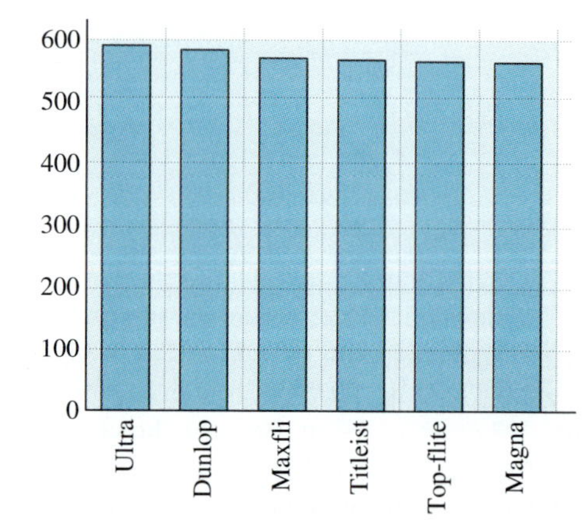

33.

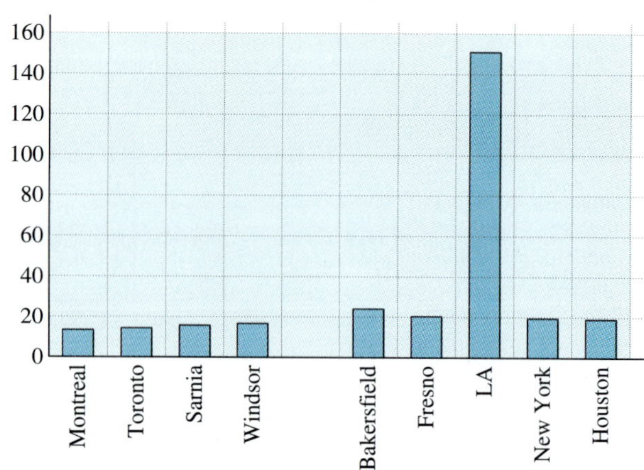

PROBLEM SET 3.4

1. 9.33

3. 10.25

5. 8

7. 15.625

9. A value in the set that is much larger than the others will pull the mean larger; similarly for a small value.

11. 26.6

13. 40.2

15. The answer to problem 13 is the sum of the answers to problems 11 and 12.

17. 79

19. {1, 3, 4, 5, 6, 8, 9, 10, 11, 12, 15}, 8, 4, 11, 7

21. {6, 9, 10, 12, 13, 14, 18, 21, 24, 26}, 13.5, 10, 21, 11

23. {2, 4, 5, 6, 8, 8, 9, 10, 10, 12, 15}, 8, 5, 10, 5

25. {1, 1, 2, 2, 3, 4, 4, 4, 5, 5, 6, 8, 9, 10}, 4, 2, 6, 4

27. (a) {73, 77.5, 79.5, 82.5, 95}

(b)

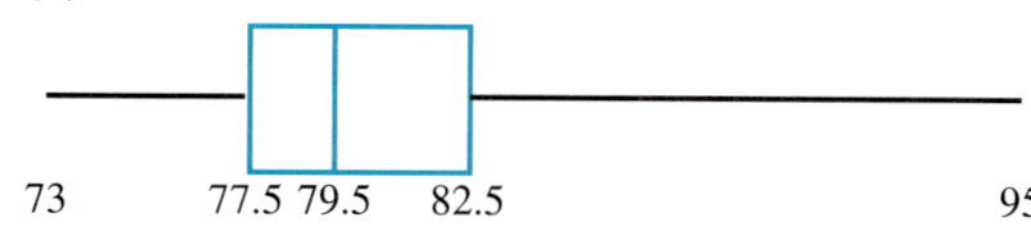

29. (a) 38.76, 41

(b)

6 27 41 51.5 60

31. **(a)** {20, 27, 34, 50, 86}
{5, 10, 18.5, 29, 33}

(b)

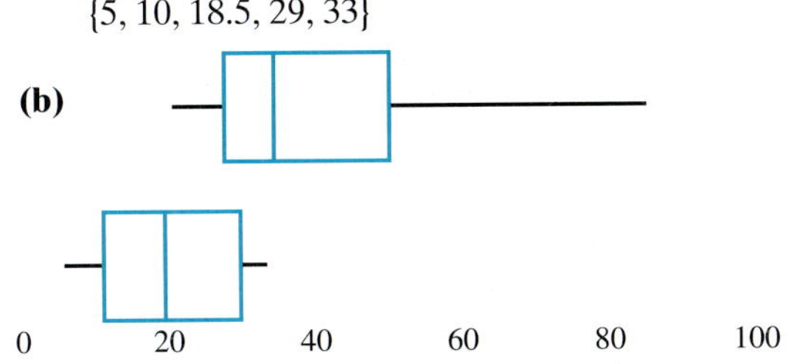

33. **(a)** 83rd percentile
(b) 17th percentile

35. **(a)** The score of 88 was better because it has a higher percentage than the 43 (86%)
(b) The score of 43 was better because it is a higher percentile

37. 18.72, 10.6, {0.4, 2.05, 10.6, 21.75, 96.4}

39.

0 20 40 60 80 100

41. **(a)** 5.81, 5.75, {4.3, 4.7, 5.75, 6.3, 9.2}
(b)

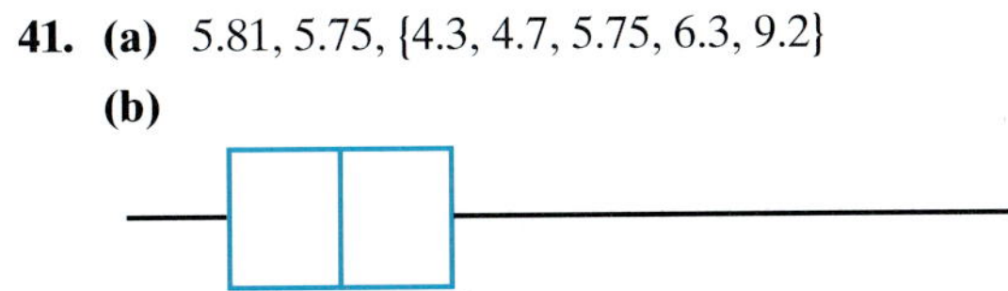

43. 95 is an outlier.

79.5
73 86 95
*
77.5 82.5

45. There are no outliers. The box-and-whiskers plot is unchanged.

6 27 41 51.5 60

47. **(a)** Any score above 21.75 + 29.55 = 51.3 is an outliner.
(b) No other outliers
(c)

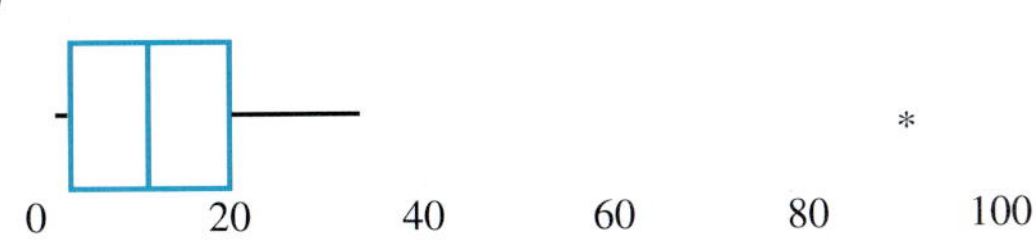

Chapter Three Review Problems

1. **(a)**

$28,500	$26,100	$27,100	$24,900
$26,900	$28,200	$27,700	$25,800
$27,800	$27,500	$26,500	$25,900
$26,500	$28,100	$26,700	$27,200

(b)

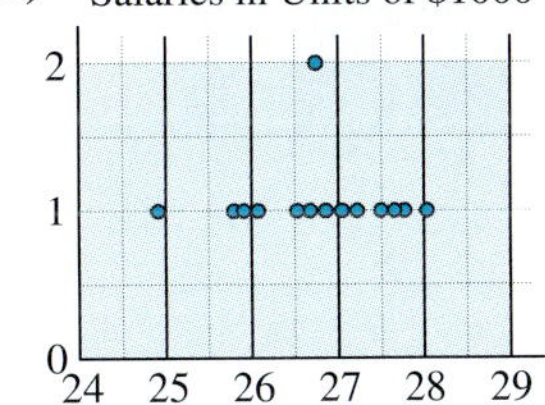

(c)

Stem	Leaves
28	1 2 5
27	1 2 5 7 8
26	1 5 5 7 9
25	8 9
24	9

3.

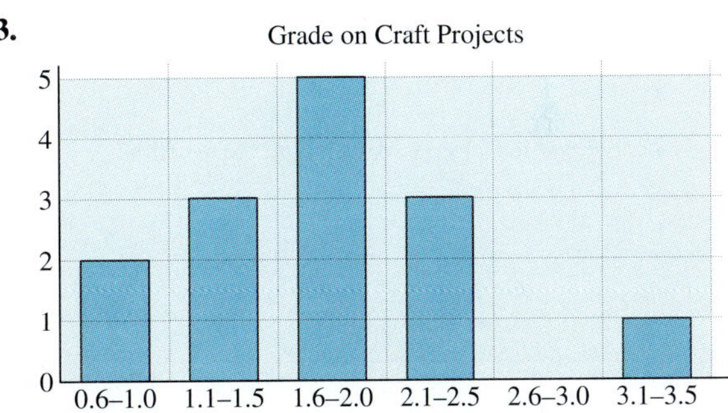

5. 1990

7. 40% increase; 14.3% decrease

9. The stem and leaf plot gives both complete information about the data and gives a graphical picture of it.

11. A multiple bar graph

13. (a)

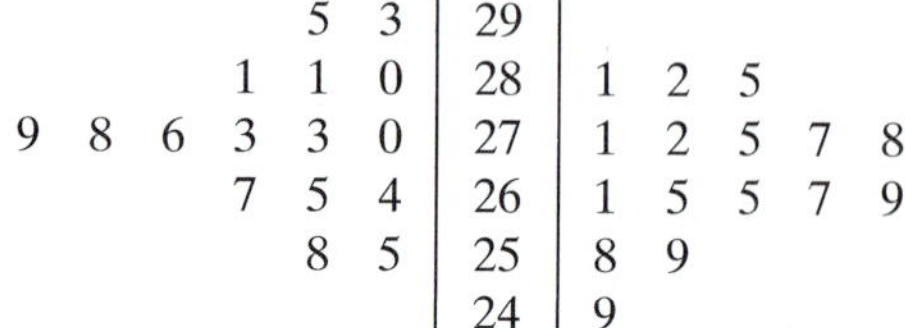

5 3	29	
1 1 0	28	1 2 5
9 8 6 3 3 0	27	1 2 5 7 8
7 5 4	26	1 5 5 7 9
8 5	25	8 9
	24	9

(b)

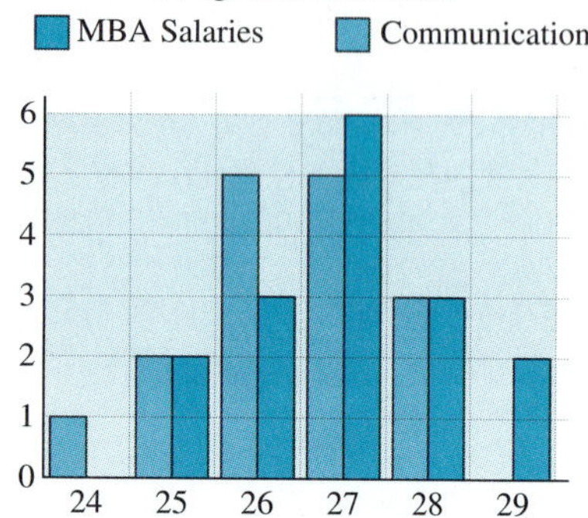

15. (a)

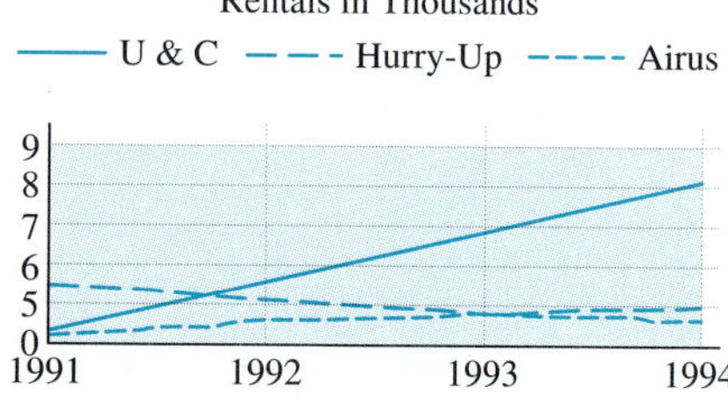

(b)

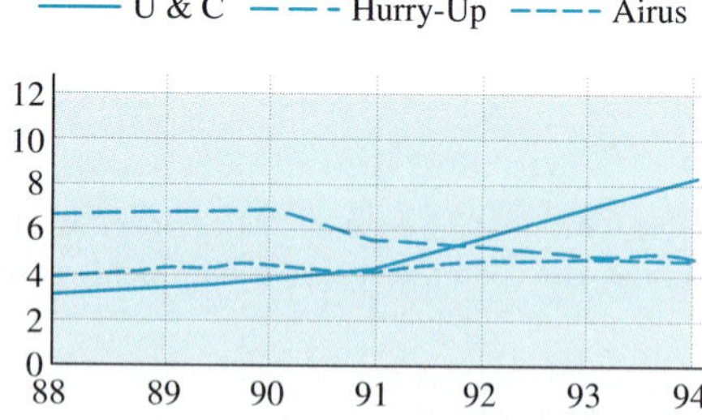

17.

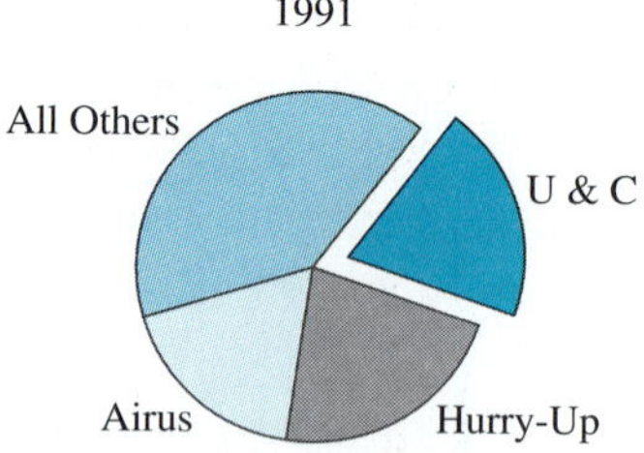

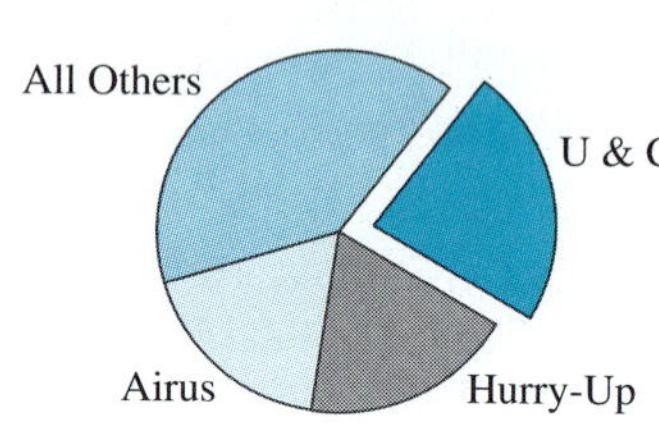

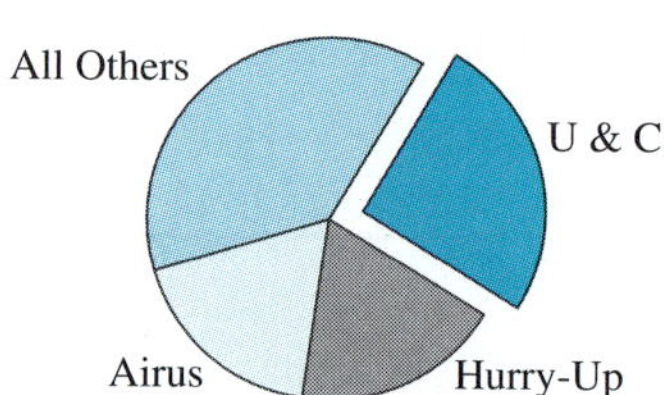

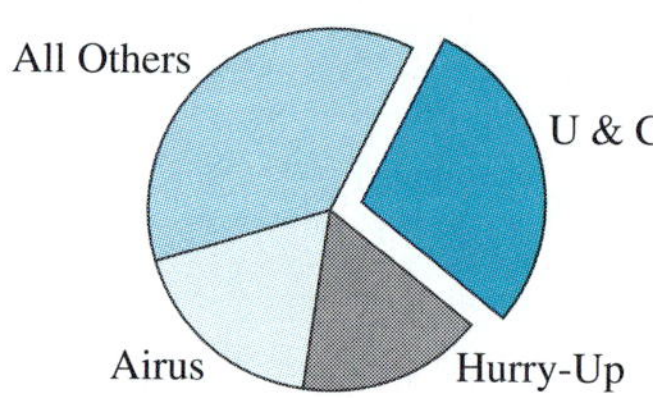

19. mean = 7.31, median = 7; {2, 4.5, 7, 9.5, 14}; 38th percentile

PROBLEM SET 4.1

1. {−4, 0, 5, 2, −3}

3. 10.8, 3.3

5. **(a)** 8, 10, 3.16
(b) 2.5, 22.58, 4.75
(c) 4.5, 0.75, 0.87

7. 4.64, 1.70, 1.30

9. Original data set: 8.8, 3.92
Modified data set: 13.8, 3.92

11. The new mean is $26.4 = 3 \times 8.8$. The new standard deviation is $11.76 = 3 \times 3.92$.

13. $\{-1.57, -0.67, 0.22, 0.89, 1.12\}$

15. $\{1.5, -0.5, -1.5, 0.5, 0\}$ The mean of z-scores is 0; the standard deviation of the z-scores is 1.

17. **(a)** Mean 15.3, standard deviation 1.23.

(b)

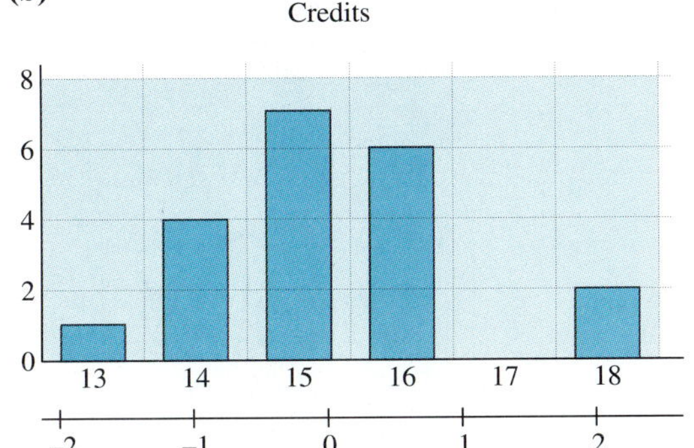

19. 15.85%, 2.5%, 32%

21. 2.35%, 97.5%, 5%

23. 16%, 68%, 160

25. 50%, 84%

27. 2.5%, about $2150

29. 17.5, 93.75, 9.68

31. 52.78, 106.17, 10.30

PROBLEM SET 4.2

1. Population = set of lightbulbs manufactured.
Sample = package of 8 chosen.

3. Population = set of full-time students enrolled at the university.
Sample = set of 100 students chosen to be interviewed.

5. Local dentists may be more likely to use a local product than dentists across the county.

7. Customers are more likely to prefer the lemon-lime so that they will be on television and to please the people making the commercial.

9. Population = the set of fish in the lake.
Sample = the 500 fish that are caught and are examined for tags. Bias results from the fact that some of the tagged fish may be caught or die before the sample is taken and the fish might not re-distribute throughout the lake.

11. Population = set of all doctors.
Sample = the set of 20 doctors chosen.
Bias results from the fact that they will commission studies until they get the result they want.

13. 11, 16, 28, 18, 32

15. 121, 066, 146, 060, 025

17. Dan, Amy, Tyler, Patty

19. Y, G, R, S, U, J

21. **(a)** Tom, Jane, Fred, Allen, Dan.
John, Matt, Fred, Jamie, Allen.
Patty, Allen, Dan, Fred, Jane.
Mary, Tyler, Fred, John, Patty.
(b) Margaret, Bill, Amy, and Chris were not in any sample.

23. **(a)** Arnold, Molly, Glenda, Oliver, Chris.
Natalie, Bob, Arnold, Victor, Ursula.
Glenda, Raul, Sandra, Ursula, Jason.
Kelly, Arnold, Lester, Victor, Wesley.
(b) 40%, 20%, 40%, 0%.

25. 19, 23, 28, 29, 55, 57, 60, 65, 72, 73, 76 are the numbers of the Vipers to be chosen.

27. Chris and Glenda

29. (column 6, line 101) 7, 11, 19, 29, 33.
Connecticut, Hawaii, Maine, New Hampshire, North Carolina.
28,110 23,354 18,895 22,659 18,702.

31.

Georgia	11
Illinois	20
Kansas	4
Kentucky	6
New Mexico	3
North Carolina	12
North Dakota	1
Texas	30
Washington	9

PROBLEM SET 4.3

1. population proportion = 0.05
sample proportion = 0.083

3. population proportion = 0.455
sample proportion = 0.5

5. **(a)** 0.6

(b) $\frac{1}{3}, \frac{1}{3}, \frac{1}{3}, \frac{2}{3}, \frac{2}{3}, \frac{2}{3}, \frac{2}{3}, \frac{2}{3}, \frac{2}{3}, 1$

(c)

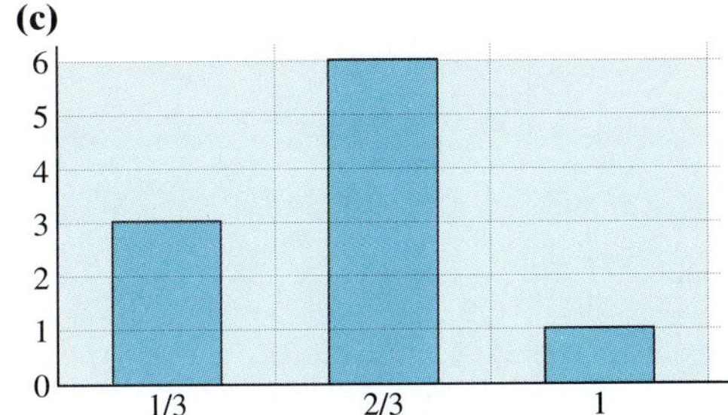

(d) The mean of sample proportions is

$$\frac{3 \times \frac{1}{3} + 6 \times \frac{2}{3} + 1}{10} = \frac{1 + 4 + 1}{10} = 0.6$$

7. population proportion = 0.05
mean of sample proportion = 0.05
standard deviation of sample proportion = 0.028.

9. population proportion = 0.455
mean of sample proportion = 0.455
standard deviation of sample proportion = 0.035.

11. 0.0203

13. 0.0189

15. 0.0223

17. 0.0333

19. **(a)** 48.5% and 2.4%
(b) 43.7% to 53.3%

21. **(a)** 61.4% and 2.9%
(b) 55.6% to 67.2%

23. 79.3% to 85.9%

25. 37.4% to 50.2%

27. **(a)** 0.566 to 0.662
(b) 0.527 to 0.701

29. **(a)** 0.800 to 0.852
(b) 0.785 to 0.867.

31. $46,500 to $163,500

PROBLEM SET 4.4

1. **(i)**

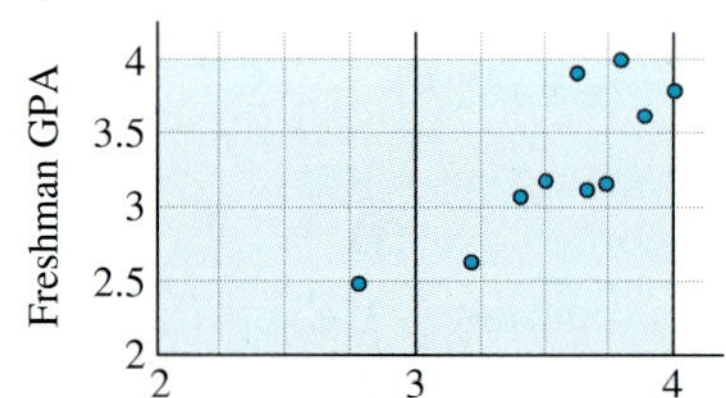

(ii) a strong positive correlation
(iii) no outliers

3. **(i)**

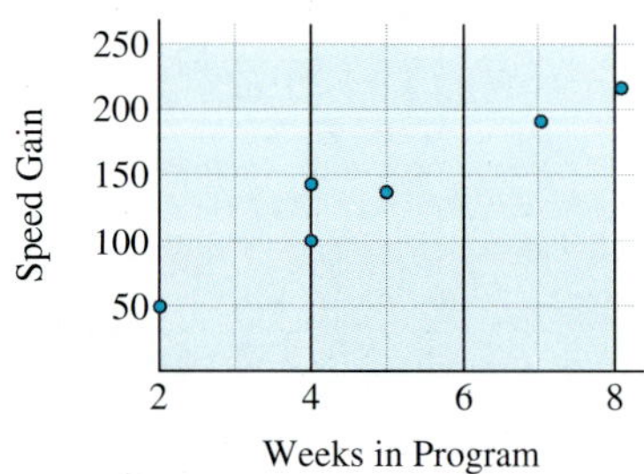

(ii) a strong positive correlation
(iii) no outliers

5.

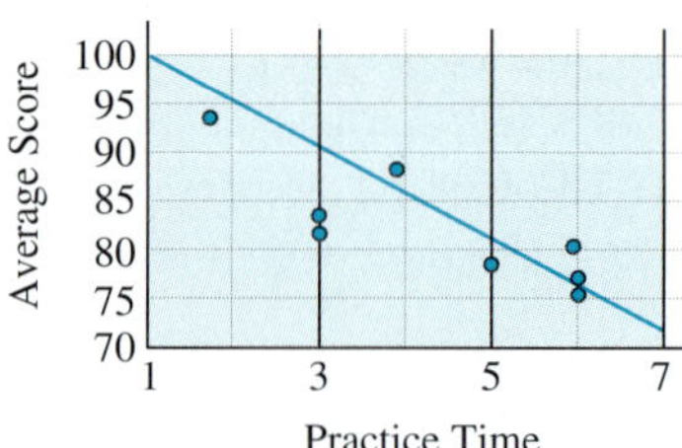

7.

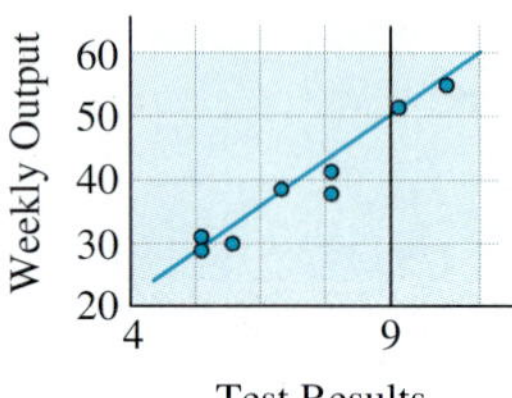

9. 42 per 1000 live births

11. **(a)** 80 or 81
(b) 85 to 87

13. **(a)** **(b)**

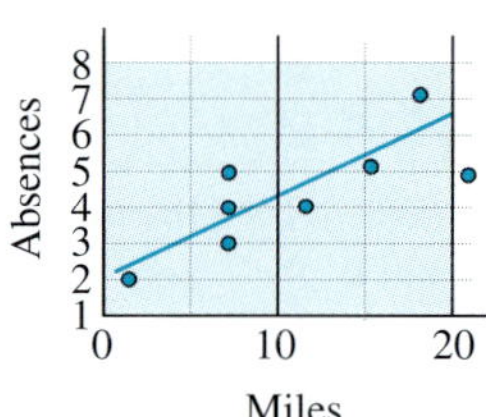

(c) 6

15. **(a)** **(b)**

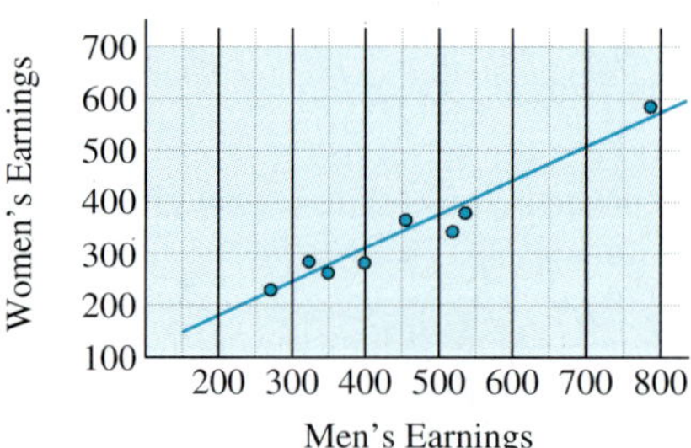

(c) 350

17. strong positive correlation

19. no correlation

21. (a)

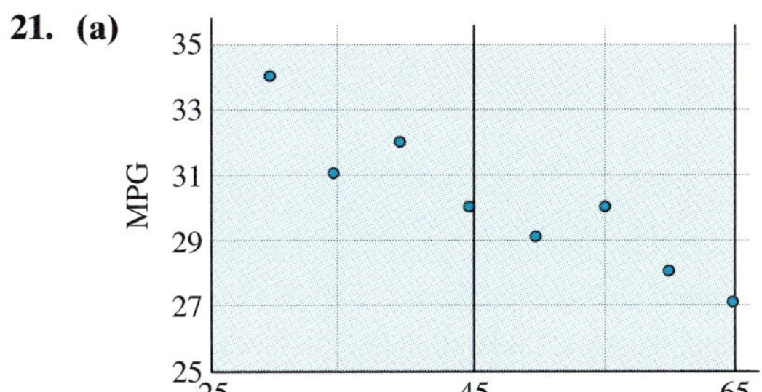

(b) strong negative correlation

23. (a)

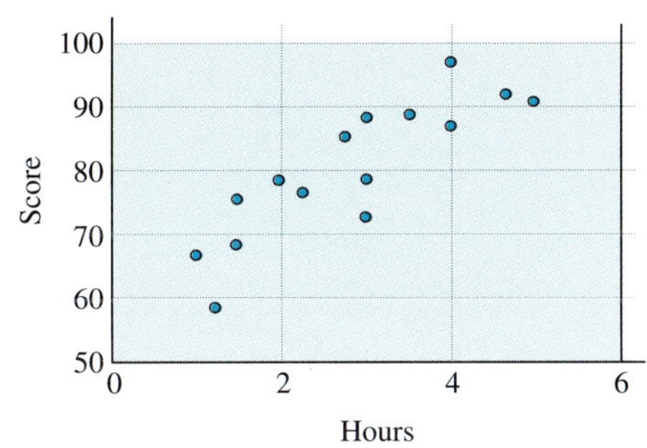

(b) strong positive correlation

25. $y = 1.18x - 0.87$

27. $y = -2.74x + 96$

29. 0.83

31. -0.739

Chapter Four Review Problems

1. 4, 11.6, 3.41, $\{-1.17, -0.59, -0.29, 0.29, 1.76\}$

3. 2.5%

5. Population: voters in town.
Sample: people who were surveyed.
Sources of bias: people near the high school are likely to be high school students and their friends who may be more likely to be rollerbladers.

7. Label the people, 0, 1, 2, 3, 4. Choose a starting point in a random number table. Record the first three numbers that are distinct and one of 0, 1, 2, 3, 4. These numbers tell which people to put in the sample.

9. 3 people. (This sample used column 5, line 109, first two numbers from the random number Table 4.1). The next sample used column 4, line 117, the first two numbers. In this sample there were two people with heart disease. One would expect these numbers to be often different because of sample variability.

11. 60

13. 12.5% to 37.5%

15. (a) Scatterplot 2
(b) Scatterplots 4 and 5
(c) Scatterplot 6

PROBLEM SET 5.1

1. (iii)

3. (a) {H, T}
(b) {A, B, C, D, E, F,}
(c) {1, 2, 3, 4}

5. (a) {HHHH, HHHT, HHTH, HTHH, THHH, HHTT, HTHT, HTTH, TTHH, THTH, THHT, HTTT, THTT, TTHT, TTTH, TTTT}
(b) {HHHH, HHHT, HHTH, HTHH, HHTT, HTHT, HTTH, HTTT}
(c) {HHHT, HHTH, HTHH, THHH}
(d) The entire sample space
(e) {HHTH, HHTT, THTH, THTT}

7.

	H	T
4	(H,4)	(T,4)
3	(H,3)	(T,3)
2	(H,2)	(T,2)
1	(H,1)	(T,1)

9. (a) $\frac{1}{5}$ **(b)** $\frac{28}{60} = \frac{7}{15}$ **(c)** $\frac{31}{60}$

11. (a)

outcome	0	1	2	3
Probability	$\frac{1}{8}$	$\frac{3}{8}$	$\frac{3}{8}$	$\frac{1}{8}$

(b) $\frac{7}{8}$

13. (a) $\frac{6}{36} = \frac{1}{6}$ **(b)** $\frac{9}{36} = \frac{1}{4}$
(c) $\frac{21}{36} = \frac{7}{12}$ **(d)** 0

15. $\frac{3}{8}$

17. (a) {2, 3, 4, 5, 6, 7, 8}
(b) $\frac{1}{4}, \frac{1}{2}, \frac{13}{16}$

19. (a) $\frac{2}{6} = \frac{1}{3}$ **(b)** $\frac{4}{6} = \frac{2}{3}$
(c) $\frac{4}{6} = \frac{2}{3}$ **(d)** $\frac{2}{6} = \frac{1}{3}$

21. **(a)** $P(A) = \frac{1}{4}, P(B) = \frac{1}{4}, P(A \cap B) = \frac{1}{16},$

$P(A \cup B) = \frac{7}{16}$

(b) $\frac{7}{16} = \frac{1}{4} + \frac{1}{4} - \frac{1}{16}$

23. **(a)**

(1, 1) (2, 1) (3, 1) (4, 1)
(1, 2) (2, 2) (3, 2) (4, 2)
(1, 3) (2, 3) (3, 3) (4, 3)
(1, 4) (2, 4) (3, 4) (4, 4)

(b) $P(A) = \frac{1}{4}, P(B) = \frac{1}{2}, P(A \cap B) = \frac{1}{8},$

$P(A \cup B) = \frac{5}{8}$

(c) $\frac{5}{8} = \frac{1}{4} + \frac{1}{2} - \frac{1}{8}$

25. **(a)** $P(A) = \frac{5}{12}, P(\bar{A}) = \frac{7}{12}$

(b) $\frac{7}{12} = 1 - \frac{5}{12}$

27. **(a)** 100 miles **(b)** 20 miles

(c) $P(A) = \frac{20}{100} = \frac{1}{5}$

29. $\frac{15}{24} = \frac{5}{8}$

PROBLEM SET 5.2

1.

H
T

3.

0
1
2
3
4
5
6
7
8
9

5.

2
6
9
12
13

7.

H — H, T
T — H, T

9.

m — *m*, *f*
f — *m*, *f*

11.

1 — 1, 2, 3, 4
2 — 1, 2, 3, 4
3 — 1, 2, 3, 4
4 — 1, 2, 3, 4

13.

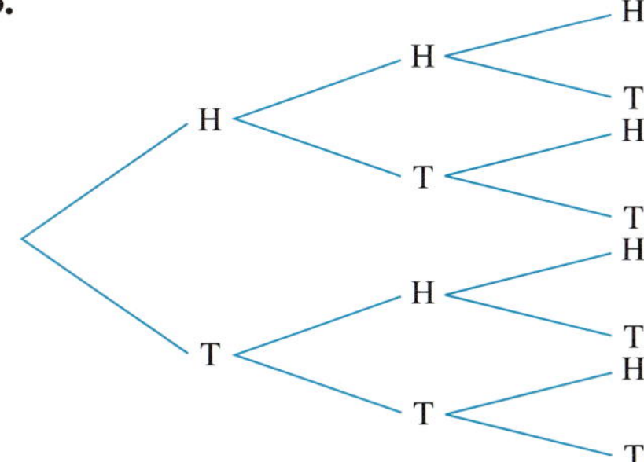

15.

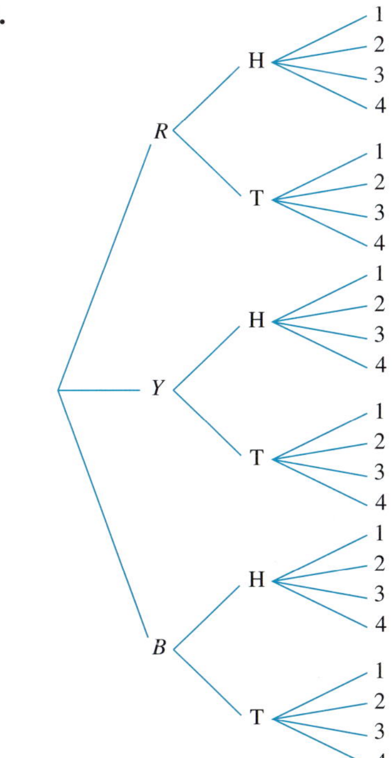

17.(a) − (d)

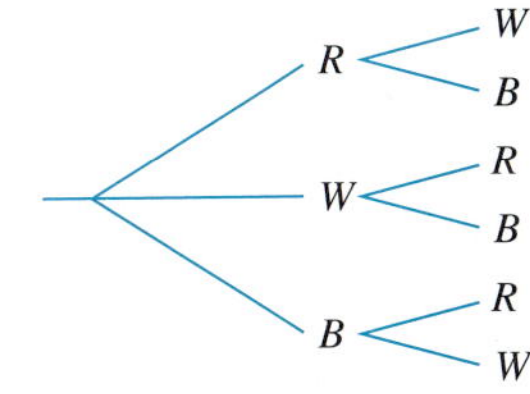

(e) 6

19.

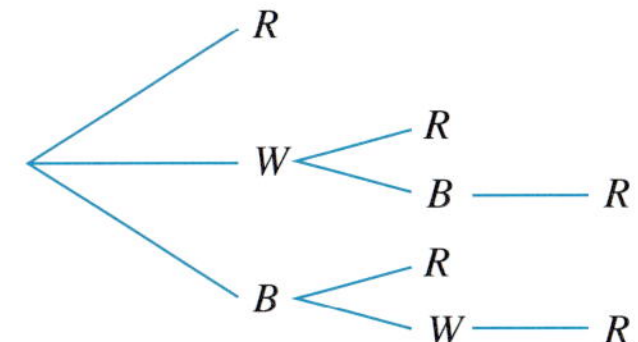

21. (a) 2 **(b)** 6 **(c)** 6 **(d)** 72

23. (a)

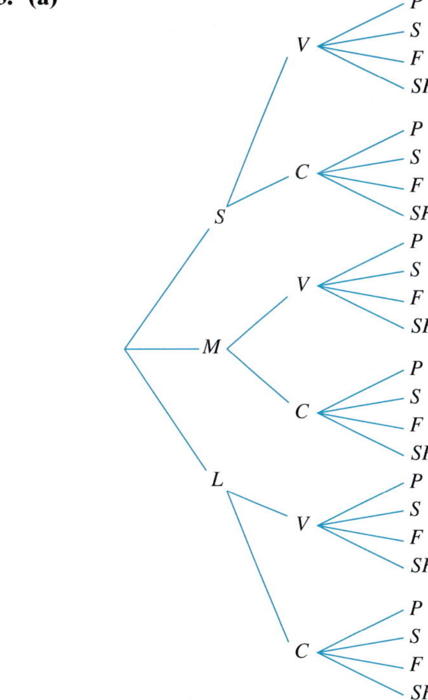

(b) $3 \times 2 \times 4 = 24$

25. $\frac{7}{47}$

27. (a) 6 **(b)** $\frac{6}{16} = \frac{3}{8}$

29. (a) $\frac{52 \times 51}{2} = 1326$ **(b)** $\frac{13 \times 12}{2} = 78$

(c) $\frac{78}{1326} = \frac{39}{663} = \frac{13}{221} = \frac{1}{17}$

31. (a)

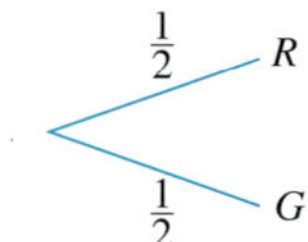

(b)

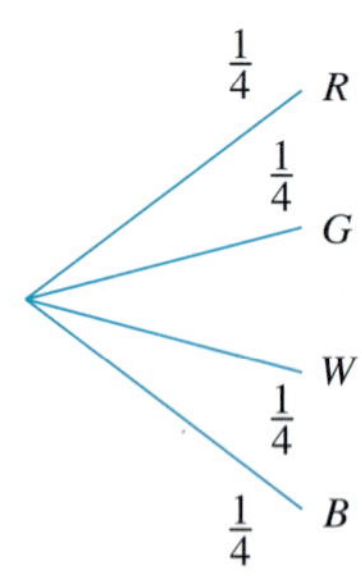

(c)

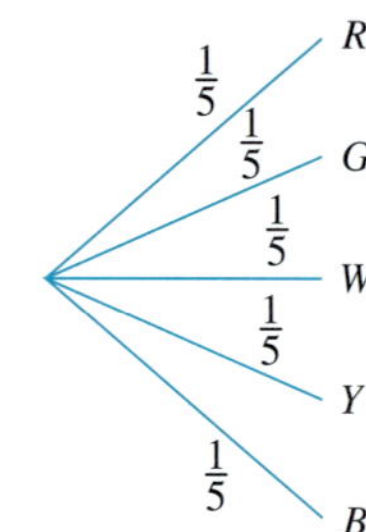

33. (a)

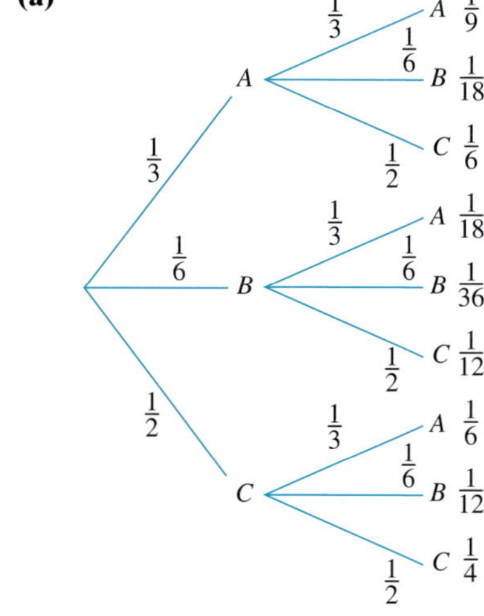

(b) $\frac{1}{36}$ **(c)** $\frac{1}{6}$

35.

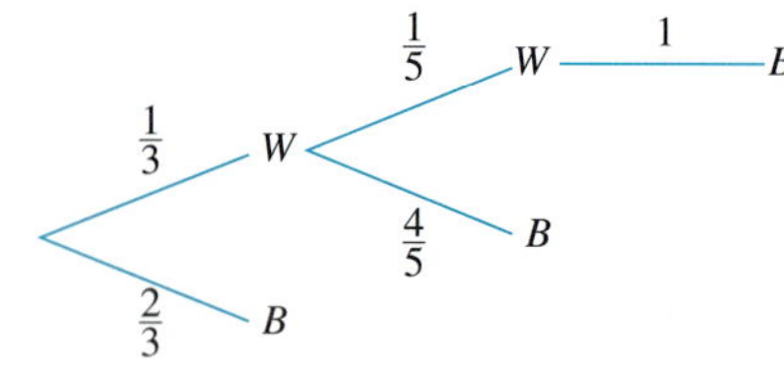

(a) $\frac{2}{3}$ **(b)** $\frac{4}{15}$ **(c)** $\frac{1}{15}$

37. $\frac{11}{21}$ if you can keep track of which key you've used and $\frac{24}{49}$ if you can't

39. (a) $\frac{1}{4}$ **(b)** $\frac{7}{18}$ **(c)** $\frac{4}{9}$

41. $\frac{1}{7059052} = 1.42 \times 10^{-7}$

43. $\frac{6 \times 38}{7059052} = 3.23 \times 10^{-5}$

Problem Set 5.3

1. $\{(2, 1), (2, 2), (2, 3), (2, 4), (2, 5), (2, 6), (4, 1), (4, 2), (4, 3), (4, 4), (4, 5), (4, 6), (6, 1), (6, 2), (6, 3), (6, 4), (6, 5), (6, 6)\}$

3. (a) $\frac{11}{36}$ **(b)** $\frac{5}{36}$ **(c)** $\frac{2}{5}$

5. (a) $\frac{8}{15}$ **(b)** $\frac{2}{5}$ **(c)** $\frac{1}{2}$ **(d)** $\frac{3}{8}$

7. (a) The probability that a drug dealer committed aggravated assault
(b) The probability that someone who committed aggravated assault is a drug dealer
(c) The probability that a drug dealer has not committed aggravated assault
(d) The probability that someone who is not a drug dealer did not commit aggravated assault.

9. (a) $\frac{3}{5}$ **(b)** $\frac{11}{12}$ **(c)** $\frac{31}{60}$
(d) 1 **(e)** $\frac{31}{55}$ **(f)** $\frac{31}{36}$

11. (a) $\frac{1}{6}$ **(b)** $\frac{1}{3}$ **(c)** 0

13. (a) $\frac{14}{30} = \frac{7}{15}$ **(b)** $\frac{87}{186} = \frac{29}{62}$
(c) $\frac{163}{205}$ (This assumes one must have a college degree to go to graduate school.)

15. (a) 0.045
(b) 0.005
(c) 0.68

17. $\frac{8}{15}$

19. $P(A) = \frac{1}{6}, P(B) = \frac{1}{6}, P(C) = \frac{1}{6}$

(a) $P(A \cap B) = \frac{1}{36} = P(A) \times P(B)$, independent

(b) $P(A \cap C) = \frac{1}{36}$, independent

(c) $P(B \cap C) = 0$, not independent

21.

value	0	1	2	3	4
probability	$\frac{1}{16}$	$\frac{4}{16}$	$\frac{6}{16}$	$\frac{4}{16}$	$\frac{1}{16}$

23. $0 \times \frac{1}{16} + 1 \times \frac{4}{16} + 2 \times \frac{6}{16} + 3 \times \frac{4}{16} + 4 \times \frac{1}{16}$

$= \frac{32}{16} = 2$

25. $4\frac{1}{3}$

27. 0.84

29. (a) \$0.85
(b) \$1.35
(c) \$1150

31. \$36.39

33. (a) \$2420
(b) \$1440
(c) Male drivers should be charged a higher premium since the average claim is substantially higher.

Chapter Five Review Problems

1. {10000, 20, 0}
$\frac{1}{100000}, \frac{1}{1000}, \frac{101}{100000}$

3. $\{PN, PD, PQ, ND, NQ, DQ\}$

(a) $A = \{PN, PD\}, P(A) = \frac{1}{3}$

(b) $B = \{PQ, NQ, DQ\}, P(B) = \frac{1}{2}$

(c) $C = \{PD, ND, DQ\}, P(C) = \frac{1}{2}$

(d) A and B

(e) $\frac{5}{6}, \frac{5}{6}$

5. $\frac{1}{12}$

7. $\frac{1}{5}$

9.

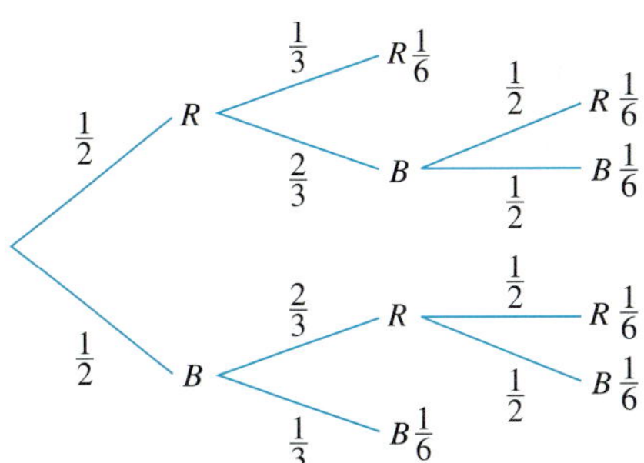

$\frac{1}{2}, \frac{1}{2}, \frac{1}{3}, \frac{1}{3}$

11.

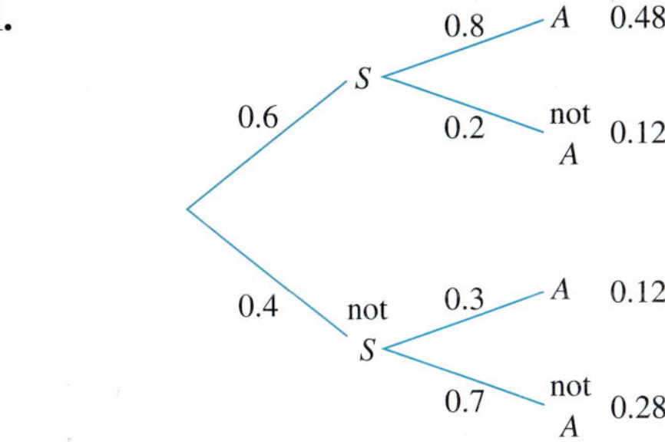

0.6, 0.8

13. $10\frac{1}{4}$ cents, $20\frac{1}{2}$ cents

PROBLEM SET 6.1

1. (a) \$126 (b) \$240 (c) \$864.50

3. (a) \$144 (b) \$420 (c) \$154.38

5. \$33.12

7. \$87.87

9. (a) \$84 (b) \$178.13

11. (a) \$427.70 (b) \$1170

13. \$2300

15. \$3975

17. \$3068.54

19. 3 months–\$1545
6 months–\$1591.35
9 months–\$1639.09
1 year–\$1688.26

21. (iii) 4.9 compounded monthly

23. 8.27%

25. 4%

27. 43.75%

29. \$2970.59

31. 8.3%

33. 6 years

35. 11.13 years

37. $12,609.67

39. **(a)** $625,000
(b) $416,667
(c) $250,000

41. Approximately 8.45%

47. 8 years. $2500 would become $5122.30 in 8 years.

49. 16 years. After 16 years $2500 would become $5129.17.

PROBLEM SET 6.2

1. **(a)** $2.70 **(b)** $5.39

3. **(a)** $3.99 **(b)** $3.72

5. $4.27, $324.77

7. $2.03, $149.88

9. $1.51, $137.43

11. **(a)** $120.45, $1.32
(b) $181.31

13. **(a)** $244.06, $2.99
(b) $287.94

15. $36.56

17. $28.92

19. $31.77

21. $3500

23. $795

25. **(a)** 20.3% **(b)** 11.1%
(c) 18.0% **(d)** 14.7%

27. $168

29. **(a)** 11.5% **(b)** 10%

31. **(a)** 14.5% **(b)** 16.5%

33. APR is approximately 22.5%

35. Add-on rate is 14%
APR is approximately 25.2%

37. Add-on rate is 16%
APR is approximately 28.8%

39. Add-on rate is 10%
APR is approximately 18%

41. 38%

PROBLEM SET 6.3

1. $1974.17

3. $12,754.29

5.

Payment	Interest	Net Payment	Balance
$111.23	$50	$61.23	$4938.77
$111.23	$49.39	$61.84	$4876.93
$111.23	$48.77	$62.46	$4814.47

7.

Payment	Interest	Net Payment	Balance
$123.78	$6.25	$117.53	$482.47
$123.78	$5.03	$118.75	$363.72
$123.78	$3.79	$119.99	$243.73
$123.78	$2.54	$121.24	$122.49
$123.77	$1.28	$122.49	$0

The last payment is $123.77

9. $132.16

11. Write $18,000 as the sum of the largest amounts that can be found in the table, and add the corresponding payments for a total of $382.47.

13. $400.40

15. $189.87

17. $264.30

19. $20.76

21. $242.77

23. $258.18

25. $179.75

27. $120.58

29. $283.27

31. $55,572.47

33. Using a factor of 10.44 we get an approximate answer of $57,500

35. $12,330

37. 9%

39. 11.5%

45. $41,369.98

PROBLEM SET 6.4

1. The house should cost $105,000 or less.
The total monthly payments should be $729 or less.

3. The house should cost $130,650 or less.
The total monthly payments should be $907 or less.

5. The house should cost $159,000 or less.
The total monthly payments should be $1104 or less.

7. $1387 (high), $913 (low)

9. $1900

11. $2300

13. $115,000

15. $90,000, $30,000

17. \$155,000
19. \$740.61
21. \$792.47
23. \$965.16
25. \$729
27. \$1291.99
29. \$1107.06

PROBLEM SET 6.5

1. \$1831.76
3. \$1556.11
5. \$34,649.70
7. \$1511.16
9. \$34,822.94
11. \$5581.60
13. \$23,223.51
15. \$5374.07
17. \$303.08
19. \$121.23
21. \$257.56
23. \$48,573
25. \$51,060
27. \$225,200
29. \$11,145.99
31. \$13,776.79
33. **(a)** \$56,407.15
(b) \$174

Chapter Six Review Problems

1. \$208
3. \$884.32, \$234.32
5. 9.38%
7. \$224.40, \$3.87
9. about 31%
11. \$220.40
13. \$153,000
15. mortgage \$726.43
taxes and insurance \$358.33
total monthly payment \$1084.76

Yes, the low estimate of what they can afford is \$1062; the high estimate is \$1615.

17. \$1806.59
19. \$127.800

PROBLEM SET 7.1

1. **(a)** 3 by 3 **(b)** -2

3. **(a)** 3 by 4 **(b)** 1

5. $\begin{bmatrix} 2 & 4 \\ -3 & -1 \end{bmatrix}$

7. $\begin{bmatrix} 3 & 4 & -2 \\ -3 & -1 & 2 \end{bmatrix}$

9. $\begin{bmatrix} 1 & 2 & 3 \\ 4 & 5 & -4 \\ -3 & -2 & -1 \end{bmatrix}$

11. $\begin{bmatrix} 1 & -6 & 2 & -5 \\ 3 & -4 & 4 & -3 \\ 5 & -2 & 6 & -1 \end{bmatrix}$

13.

		Patrick: One Finger	Patrick: Two Fingers
Patti	One Finger	10	−10
Patti	Two Fingers	−10	10

15.

		Rose: Ace	Rose: King
Carmen	Ace	1	2
Carmen	King	−2	−3

17.

		Pitcher: Fastball	Pitcher: Curveball
Batter	Fastball	.450	.200
Batter	Curveball	.240	.400

19.

		IRS: Audit	IRS: Don't Audit
Taxpayer	Cheat	−3000	1500
Taxpayer	Don't Cheat	−200	0

21.

1st Player Strategy	2nd Player Strategy	1st Player Payoff	2nd Player Payoff
I	I	1	−1
I	II	−2	2
II	I	−1	1
II	II	2	−2

23.

1st Player Strategy	2nd Player Strategy	1st Player Payoff	2nd Player Payoff
I	I	0	0
I	II	2	−2
I	III	−1	1
II	I	2	−2
II	II	1	−1
II	III	−2	2
III	I	−1	1
III	II	−2	2
III	III	2	−2

25.

1st Player Strategy	2nd Player Strategy	1st Player Payoff	2nd Player Payoff
I	I	0	0
I	II	−2	2
I	III	0	0
I	IV	−2	2
II	I	−1	1
II	II	1	−1
II	III	−1	1
II	IV	1	−1
III	I	2	−2
III	II	0	0
III	III	2	−2
III	IV	0	0

27. **(a)** II **(b)** II

29. **(a)** I **(b)** I

31. (a)

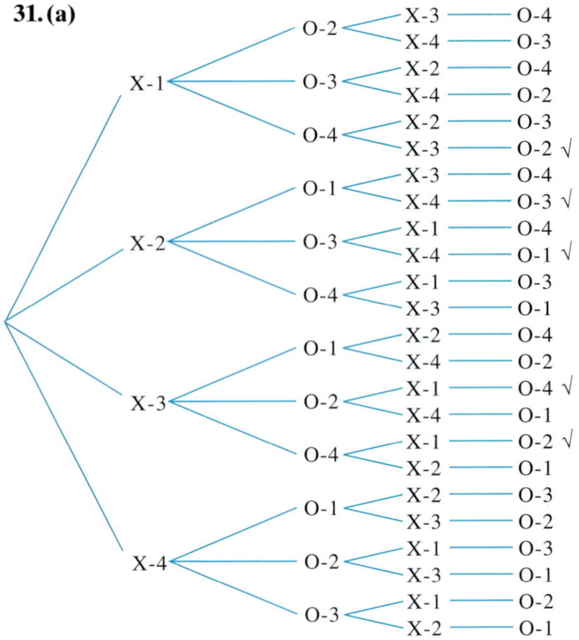

(b) Checked branches are to be pruned. Player X should choose either box 2 or box 3 first. There will always be a winning box to choose after O plays.

(c) Yes

PROBLEM SET 7.2

1. **(a)** Most conservative strategies:
1st player: 2nd strategy
2nd player: 1st strategy

(b) Not determined

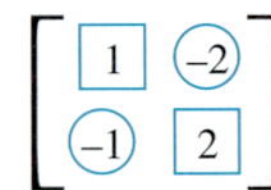

3. **(a)** Most conservative strategies:
1st player: 1st strategy
2nd player: 1st strategy

(b) Determined **(c)** 1

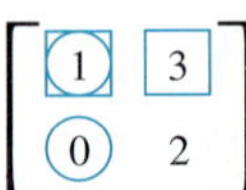

5. **(a)** Most conservative strategies:
1st player: 1st strategy
2nd player: 3rd strategy

(b) Determined

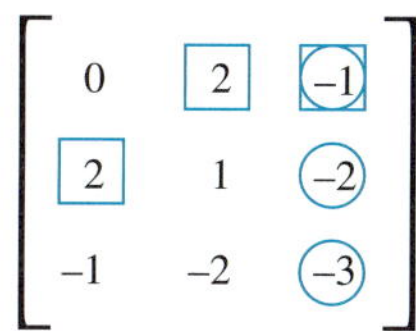

(c) -1

7. (a) Most conservative strategies:
1st player: 3rd strategy
2nd player: 1st strategy

(b) Determined

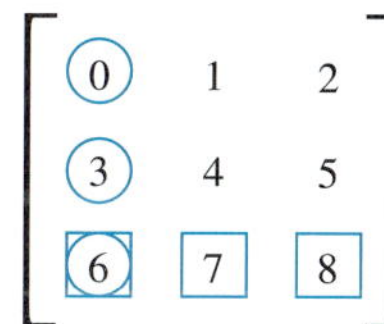

(c) 6

9. (a) Most conservative strategies:
1st player: 3rd strategy
2nd player: 2nd strategy

(b) Not Determined

$$\begin{bmatrix} 0 & -2 & 0 & \textcircled{-3} \\ \textcircled{-2} & \boxed{1} & -1 & \boxed{2} \\ \boxed{2} & \textcircled{0} & \boxed{2} & 1 \end{bmatrix}$$

11. (a)

$$\begin{bmatrix} 1 & -1 \\ -1 & 1 \end{bmatrix}$$

(b) All strategies are most conservative

(c) Not determined

13. (a)

$$\begin{bmatrix} .450 & .200 \\ .240 & .400 \end{bmatrix}$$

(b) The batter should prepare for a curve; the pitcher should throw a curve ball.

(c) Not determined

15. (a)

$$\begin{bmatrix} -2 & -1 \\ -5 & 3 \end{bmatrix}$$

(b) Jane should take the umbrella; it should rain.

(c) Determined

(d) -2

17. (a)

$$\begin{bmatrix} 1 & -5 \\ 3 & 1 \end{bmatrix}$$

(b) A choose 3, B choose 3 **(c)** Determined **(d)** 1

19. (a)

$$\begin{bmatrix} 0.5 & 0.4 \\ 0.65 & 0.5 \end{bmatrix}$$

(b) The most conservative strategy for each is locating at the mall.

(c) The game is determined.

(d) They will split the business.

21. Column 2 dominates 1 and 3
The value of the game is -2

23. Row 2 dominates row 3. Then, column 3 dominates column 2.

$$\begin{bmatrix} 4 & -2 \\ 0 & 1 \end{bmatrix}$$

25. Column 1 dominates all other columns.
The value of the game is 0.

27. No reduction is possible

29. (a)

Weather	Location	Payoff
Good	Stadium	285,000
Good	Coliseum	140,000
Bad	Stadium	195,000
Bad	Coliseum	140,000
Very bad	Stadium	−65,000
Very bad	Coliseum	140,000

b.

		Weather		
		Good weather	Bad weather	Very bad weather
Promoter	Stadium	285,000	195,000	−65,000
	Coliseum	140,000	140,000	140,000

(c) The most aggressive strategy is to use the stadium.

(d) The most conservative strategy is to use the coliseum.

37. (a) Strategy I for Japanese: go north.
Strategy II for Japanese: go south.
Strategy I for Kennedy: look north.
Strategy II for Kennedy: look south.

		Japanese Northern	Japanese Southern
Kennedy	Northern	2	2
	Southern	1	3

(b) 2

PROBLEM SET 7.3

1. $\frac{1}{4}$

3. 0

5. $-\frac{4}{9}$

7. First player: $\left(\frac{1}{2}, \frac{1}{2}\right)$

Second player: $\left(\frac{2}{3}, \frac{1}{3}\right)$

9. First player: $\left(\frac{1}{2}, \frac{1}{2}\right)$

Second player: $\left(\frac{3}{4}, \frac{1}{4}\right)$

11. Second strategy; the average payoff is 1

13. First strategy; the average payoff is $\frac{1}{2}$

15. First strategy; the average payoff is $\frac{1}{2}$

17. First player: $\left(\frac{1}{2}, \frac{1}{2}\right)$;

Second player: $\left(\frac{3}{8}, \frac{5}{8}, 0\right)$

19. First player: $\left(\frac{5}{8}, \frac{3}{8}\right)$;

Second player: $\left(0, \frac{3}{8}, \frac{5}{8}, 0\right)$

21. First player: $\left(0, \frac{2}{5}, \frac{3}{5}\right)$;

Second player: $\left(\frac{1}{2}, \frac{1}{2}, 0\right)$

23. The batter should prepare for a fastball 39% of the time, the pitcher should throw 49% fastballs.

25. Charlane should put 71% of her savings into stocks and 29% of her savings into money market funds, and none in bonds.

27. Action Faction should put its entire budget into TV advertising. Value of game is 1.25. This represents an increase of 1.25 million dollars in sales.

29. The Democrats should only campaign in the rural area. Value of game is 3.6. This represents a gain of 36,000 votes.

31. Action Faction should use 37.5% of its advertising for newspaper ads and 62.5% for television ads. This will result in with a payoff of 0.5 million dollars in increased sales. Mega Sports should evenly divide its advertising between newspapers and television, and will have 0.5 million dollars in decreased sales.

33. $r = \frac{5}{10}, s = \frac{3}{10}, 1 - r - s = \frac{2}{10}$

$p = \frac{1}{3}, q = \frac{1}{3}, 1 - p - q = \frac{1}{3}$

35. (a)

		Pitcher fastball	Pitcher curveball	Pitcher screwball
Batter	fastball	.400	.200	.100
	curveball	.220	.400	.160
	screwball	.120	.280	.380

(b) The pitcher should throw 44.6% fastballs, 3.2% curveballs, and 52.2% screwballs. The batter should be prepared for 40.5% fastballs, 26.6% curveballs, and 32.9% screwballs.

Chapter Seven Review Problems

1. 5

3. The first strategy is the most conservative for both players.

5.

		Bob	
		tight	loose
Alice	pitch	-1	4
	kick	5	-10

7. \$224.40, \$3.87

9. about 31%

PROBLEM SET 8.1

1. **(a)** No **(b)** No **(c)** Yes

3. **(a)** Yes **(b)** Yes **(c)** No

5. **(a)** No **(b)** Yes **(c)** No

7. **(a)** Yes **(b)** No **(c)** Yes

9.

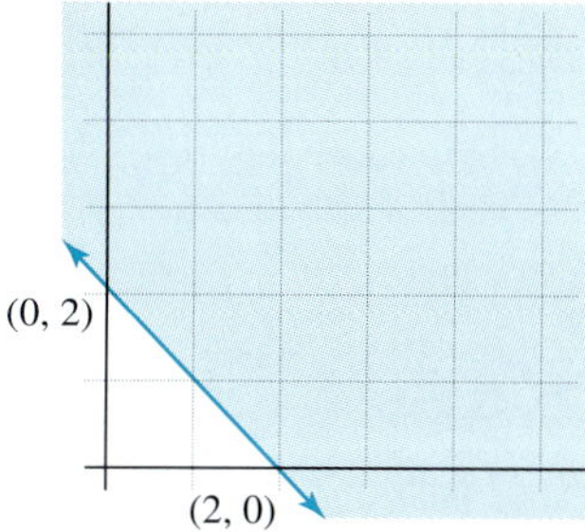

11.

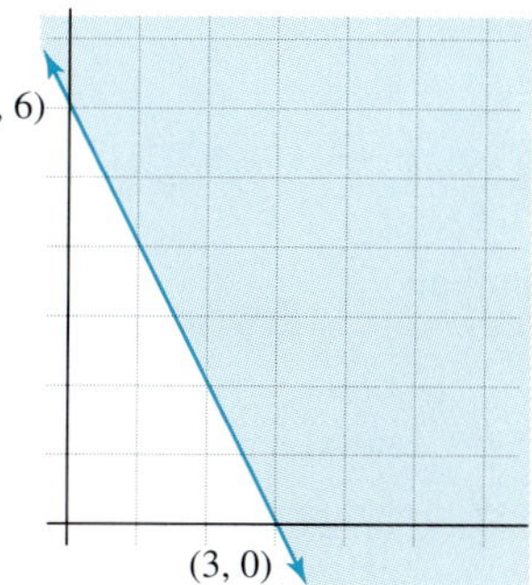

13.

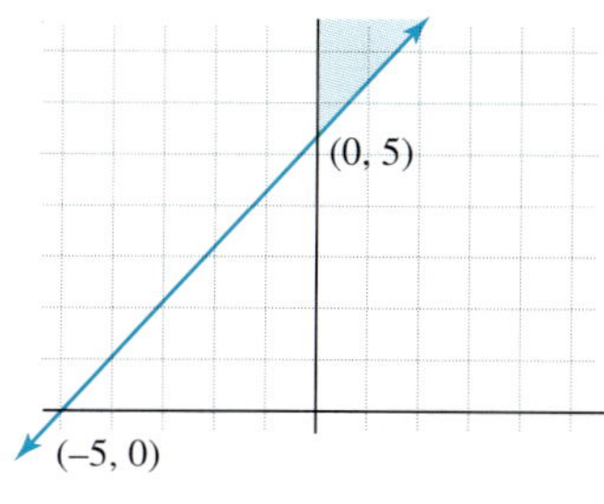

15.

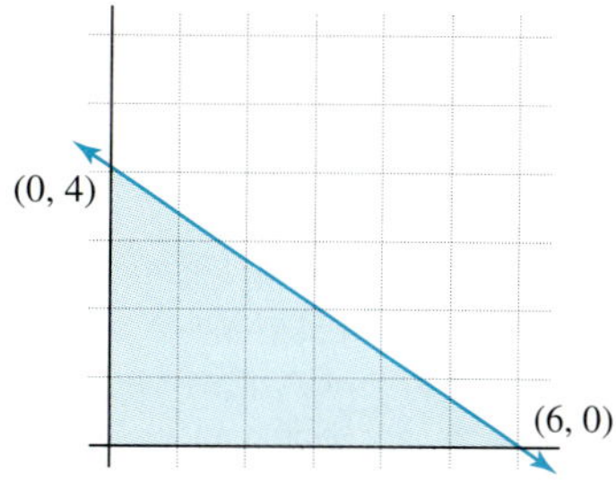

17. III

19. IV

21. II

23. IV

25.

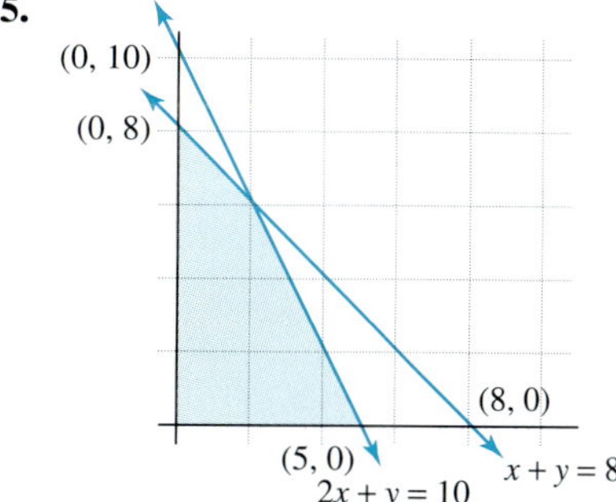

27.

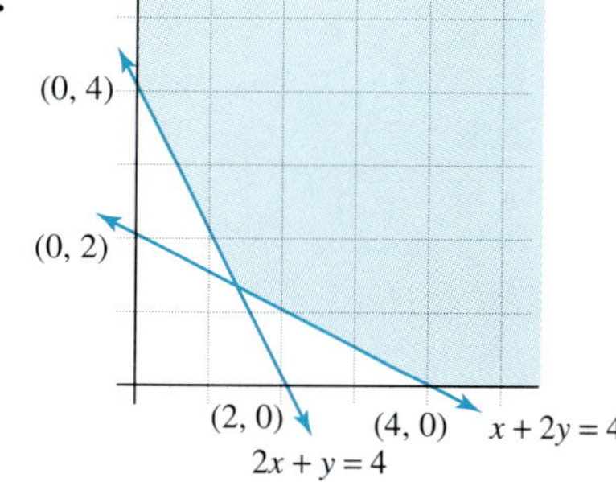

29.

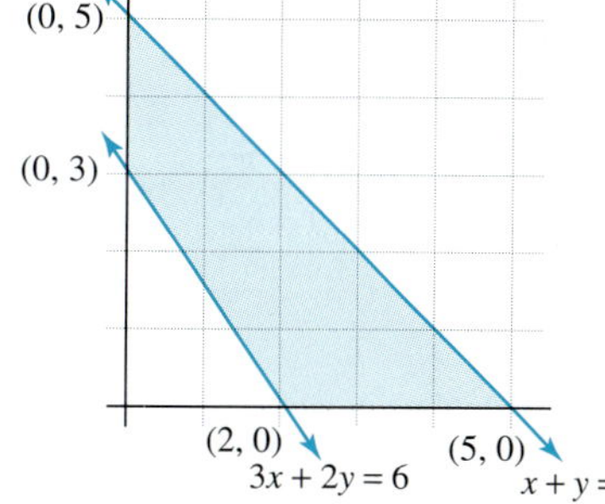

31.

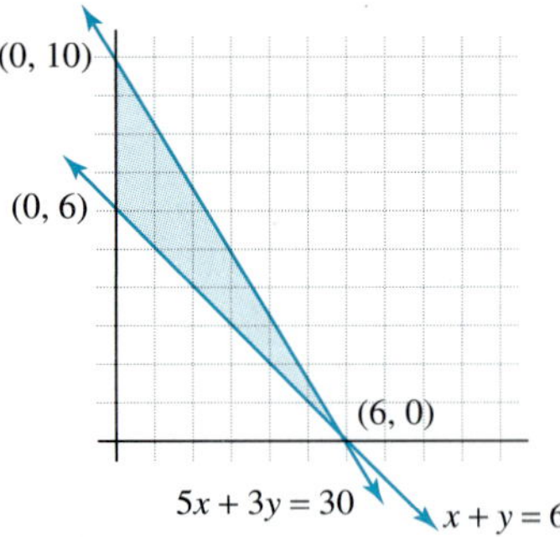

33.

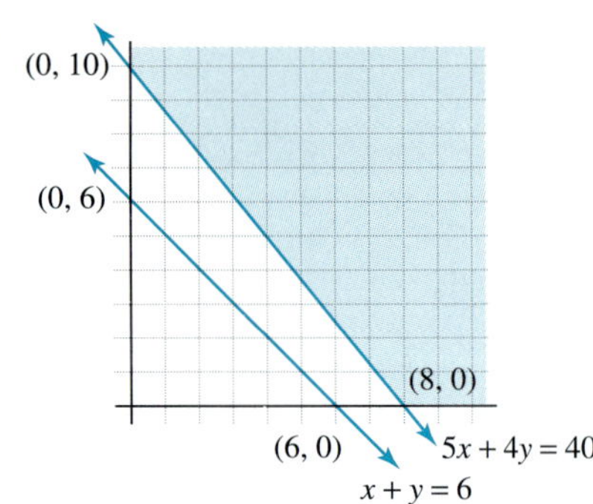

35.

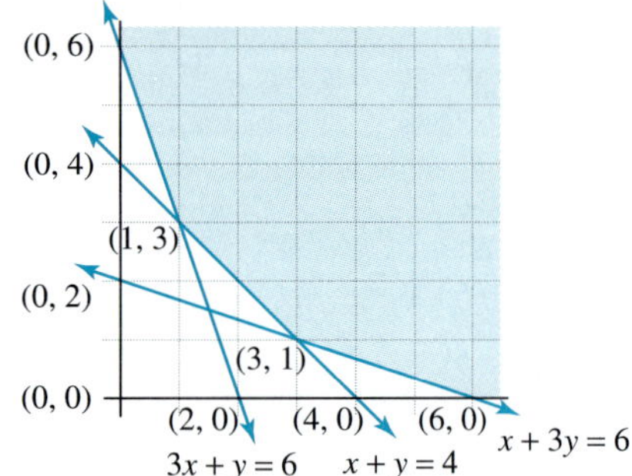

37.

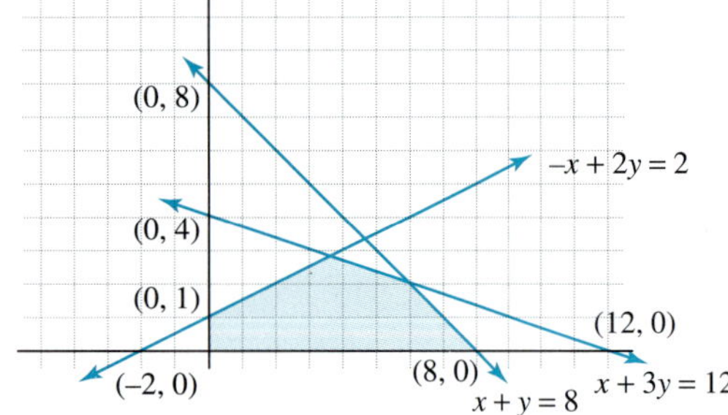

39. b = number of bookcases
t = number of tables
$40b + 68t \leq 800$
$b \geq 0, t \geq 0$
b and t are whole numbers

41. c = number of cord type drills
d = number of cordless drills
$2c + 3d \leq 600$
$c \geq 0, d \geq 0$

43. f = numbers of four-person tests
t = numbers of two-person tests
$100f + 60t \leq 9000$
$f \geq 0, t \geq 0$

45. b = number of bass
t = number of trout
$2b + 5t \leq 800$ ("A" food)
$4b + 2t \leq 800$ ("B" food)
$b \geq 0, t \geq 0$
b and t are whole numbers

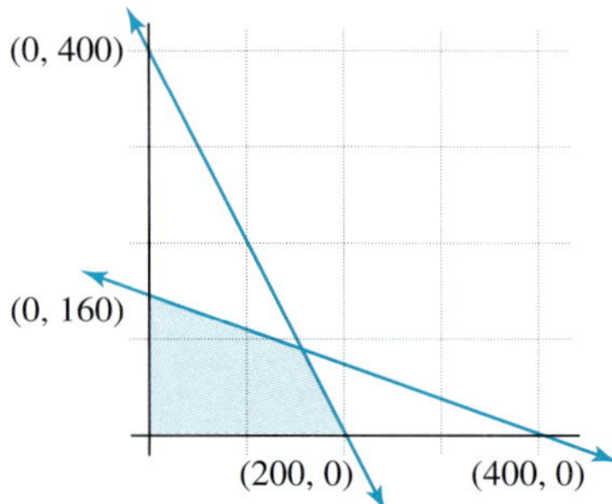

47. x = number of Type 1
y = number of Type 2
$0 < x + y \leq 10$
$100x + 150y \geq 1200$
$x \geq 0, y \geq 0$
x and y are whole numbers

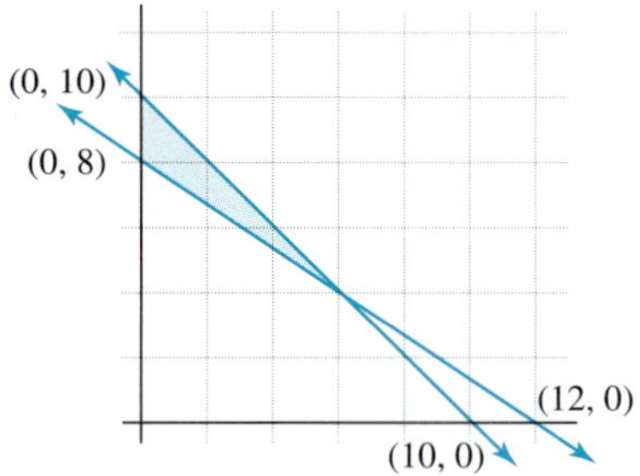

49. m = number of motel rooms
h = number of hotel rooms
$3m + 2h \geq 400$
$20(3m) + 40(2h) \leq 12{,}000$
$m \geq 0, h \geq 0$
m and h are whole numbers

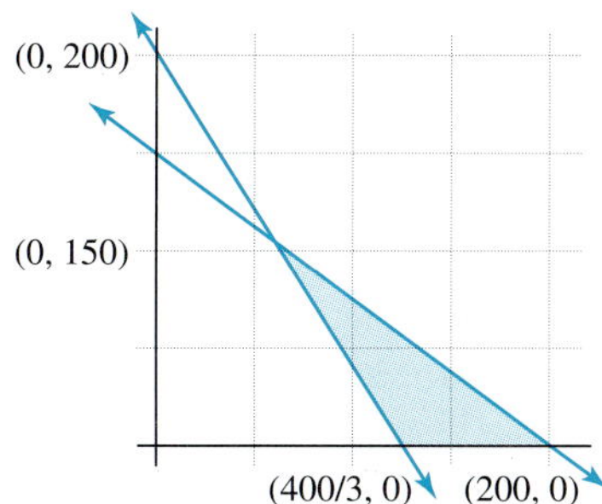

PROBLEM SET 8.2

1. **(a)** Yes; 3,7 **(b)** No
   (c) No **(d)** Yes; 2,1
3. **(a)** No **(b)** Yes; −1,2
   (c) Yes; 10,6 **(d)** No

5. 

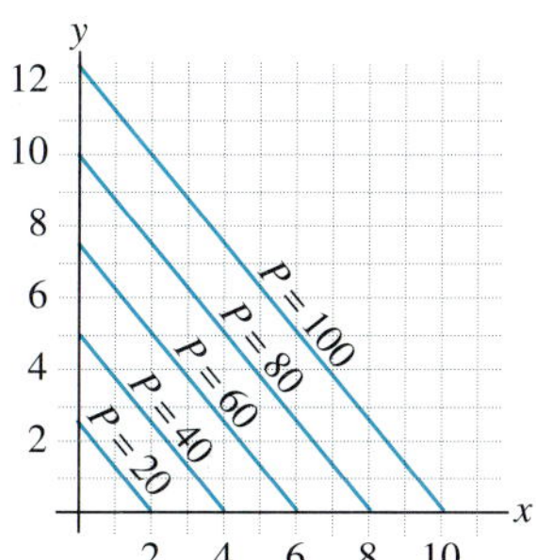

7.

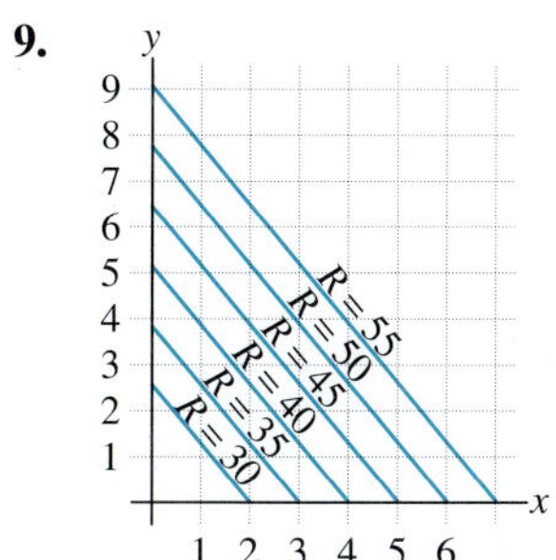

9.

11.

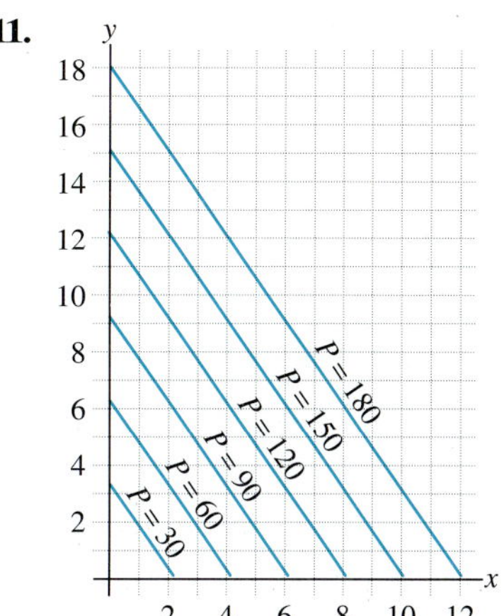

13.

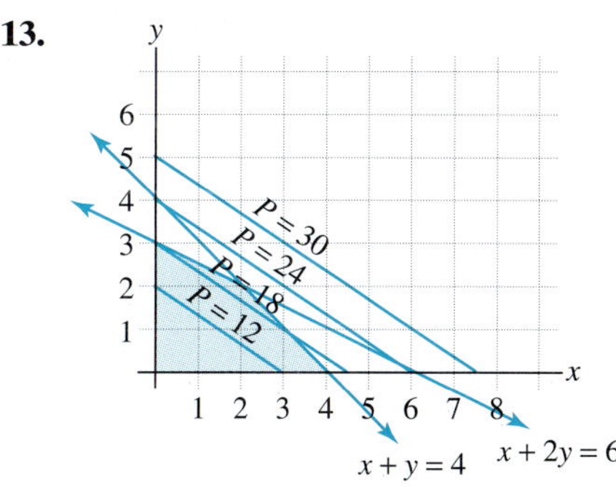

$P = 20$ is the estimated maximum.

15.

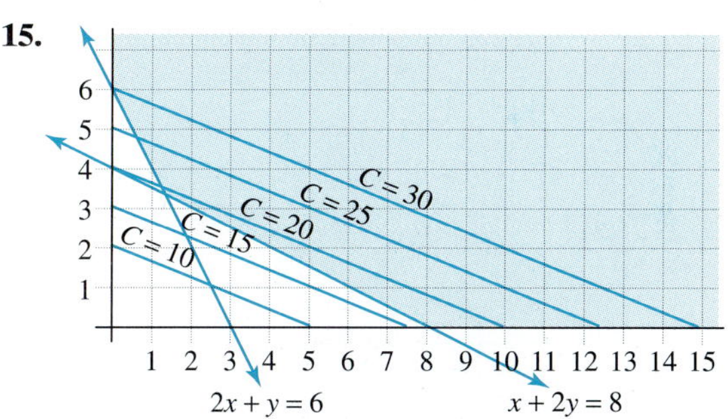

$C = 16$ is the estimated minimum.

17. (0,0), (0,4), (6,1), (7,0)
19. (0,0), (0,3), (4,2), (6,1), (7,0)
21. maximum: $P = 24$; minimum: $P = 0$
23. No maximum; minimum: $M = 17$
25. $F = 20$
27. $C = 36$
29. $P = 45$
31. c = number of cord type drills
    d = number of cordless drills
    $2c + 3d \leq 600$
    $c + d \leq 250$
    $c \geq 0, d \geq 0$
    Maximize $P = 45c + 60d$

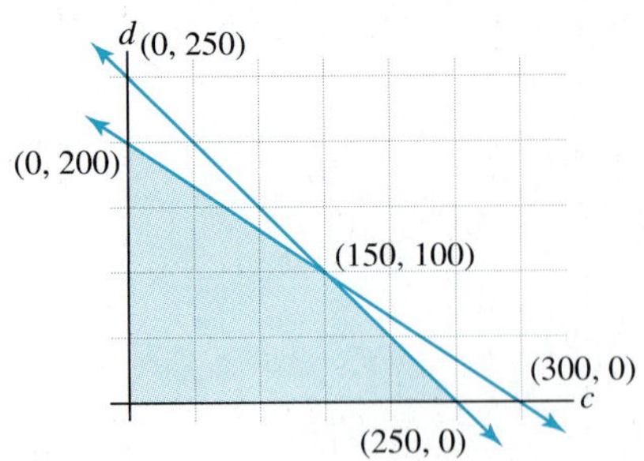

Points	Function Values
(0, 0)	\$0
(250, 0)	\$11,250
(0, 200)	\$12,000
(150, 100)	\$12,750

The maximum revenue is \$12,750 when 150 of the cord type and 100 of the cordless drills are produced.

33. b = number of bass
t = number of trout
$2b + 5t \le 800$
$4b + 2t \le 800$
$b \ge 0, t \ge 0$
Maximize $P = b + t$

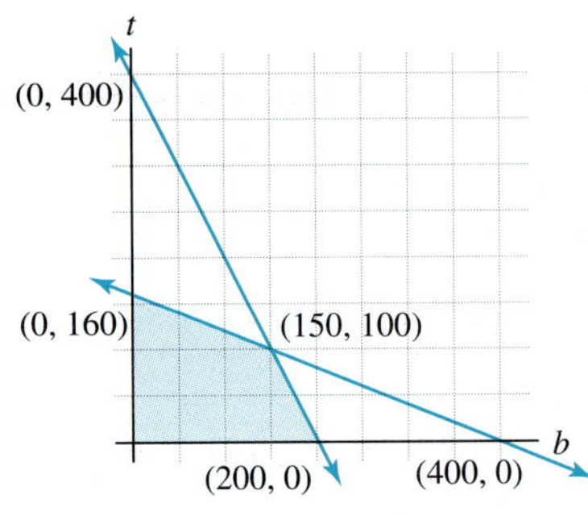

Points	Function Values
(0, 0)	0
(200, 0)	200
(0, 160)	160
(150, 100)	250

The maximum number of fish that can be supported is 250, 150 bass and 100 trout.

35. x = number of Type 1
y = number of Type 2
$x + y \le 10$
$100x + 150y \ge 1200$
$x \ge 0, y \ge 0$
x and y are whole numbers
Minimize $P = 9000x + 15000y$

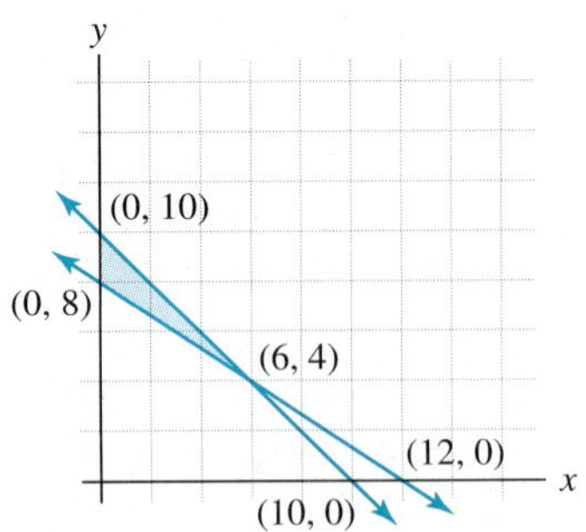

Points	Function Values
(0, 10)	\$150,000
(6, 4)	\$114,000
(0, 8)	\$120,000

The minimum cost is \$114,000 obtained using 6 Type 1 and 4 Type 2 aircraft.

37. m = number of motel rooms
h = number of hotel rooms
$3m + 2h \ge 400$
$20(3m) + 40(2h)$ or $60m + 80h \le 12{,}000$
$m \ge 0, h \ge 0$
m and h are whole numbers
Minimize $P = 135m + 120h$

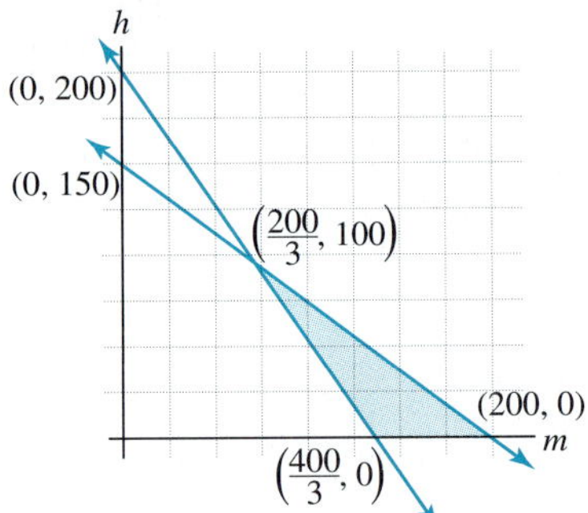

Points	Function Values
$\left(\frac{400}{3}, 0\right)$	\$18,000
(200, 0)	\$27,000
$\left(\frac{200}{3}, 100\right)$	\$21,000

Room costs are minimized at \$18,090 if 134 motel rooms are used.

39. **(a)** $b > 2a$ **(b)** $3b > a > \frac{1}{2}b$
(c) $a > 3b$ **(d)** $2a = b$
(e) $a = 3b$

41. d = number of day-time ads
p = number of prime-time ads

n = number of late-night ads
Constraints: $1200d + 2250p + 1600n \leq 30{,}000$
$d + p + n \leq 15$
$d + p + n \geq 10$
Minimum value constraints:
$d \geq 0, p \geq 0, n \geq 0$
Objective function:
Maximize $A = 12000d + 21000p + 16000n$

43. (a)

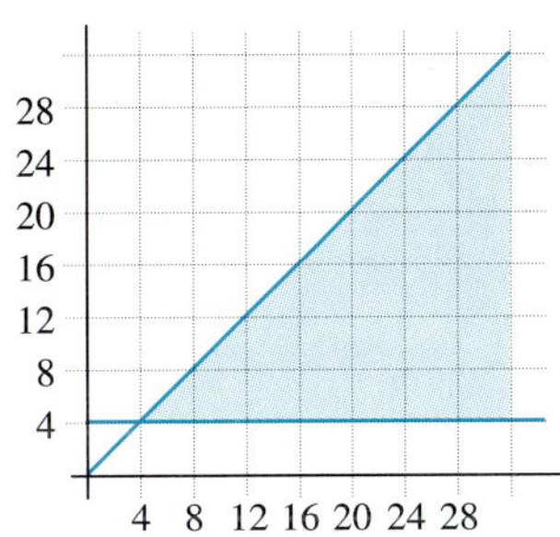

(b) $x = 13, y = 12$, and $x = 20, y = 6$ are the only pairs of whole numbers satisfying the constraints. Choosing 20 games per division rival and 6 for inter-divisional play gives the divisions more meaning.

PROBLEM SET 8.3

1.

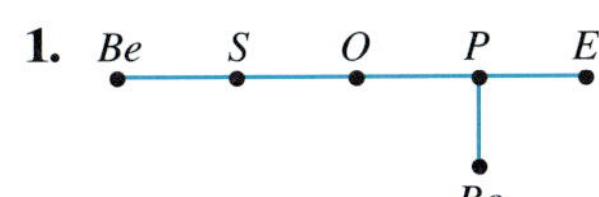

3.

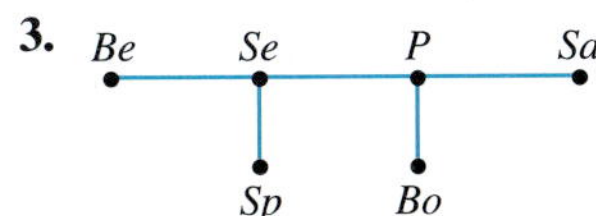

5. (a) A, B, C, D

(b) $(A, B), (A, D), (B, C), (B, D)$

7. (a) R, S, T, U

(b) $(R, S), (R, T), (R, U), (S, U), (T, U)$

9. Salem Portland Olympia Seattle

Boise Spokane

11. (a) T, U, K, J, R

(b) (T, U) − 2 edges, $(T, R), (R, K)$, $(R, J), (J, K), (U, K)$

13. (a) A, B, C, D, E

(b) $(A, E), (A, C), (C, D)$ − 2 edges, $(B, D), (B, B)$ (self-loop), (B, E)

15. (a) $A, B, C, D; AB, AC, AD, BC, BD$

(b) $A, B, C, D; AB, AC, AD, BC, BD$

17. They have the same vertices and the same edges, thus are essentially the same. They are different only in the arrangement of the vertices

19. (a) Two possible paths are

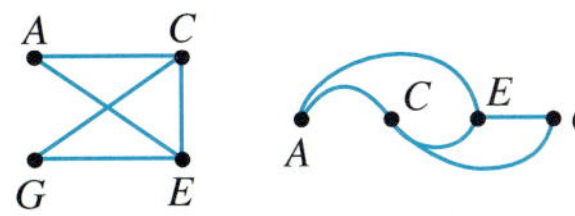

(b) Two possible paths are

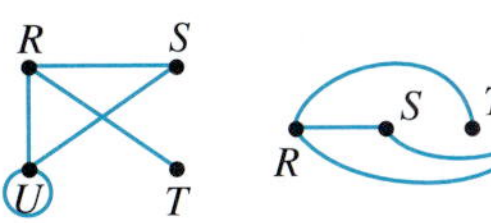

21. $T - 3, U - 3, K - 3, J - 2, R - 3$

23. $A - 2, E - 2, B - 4, C - 3, D - 3$

25. $A - 3, B - 4, C - 1, D - 2, E - 4$
Sum − 14, Edges − 7
The sum of the degrees is twice the number of edges.

27. $A - 5, B - 2, C - 3, D - 2, E - 3,$
$F - 2, G - 5, H - 2$
Sum − 24, Edges − 12
The sum of the degrees is twice the number of edges.

29. CD, DE, DF are bridges
Bridge CD

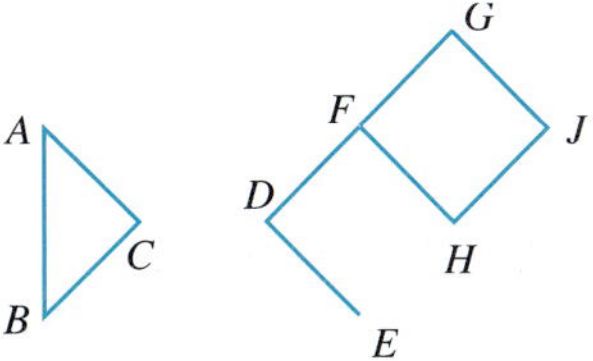

Bridge DF

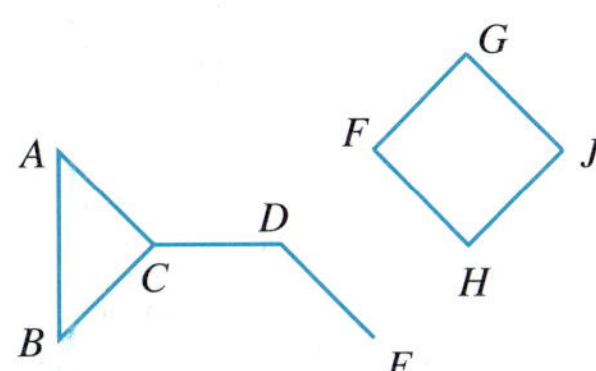

Bridge DE

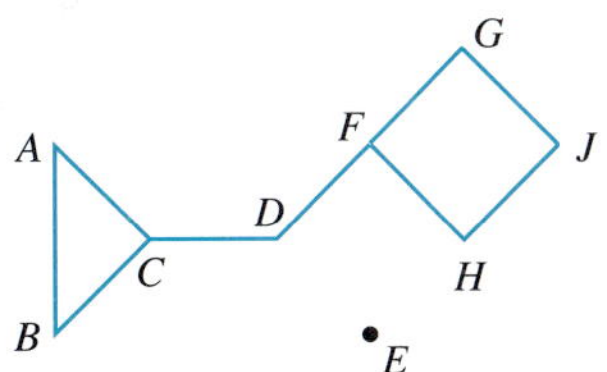

31. (a) Path **(b)** Not a path

(c) Circuit **(d)** Not a path

(e) Circuit; but not an Euler circuit.

33. **(a)** Path **(b)** Not a path
(c) Euler circuit **(d)** Euler circuit

35. U and R are the only two vertices of odd degree, thus this graph has an Euler path, but not an Euler circuit

37. $F, E, B, A, C, D, A, B, D, F$

39.

Vertex	Degree
R	5
S	4
T	2
U	3
Y	4

Since there are exactly two vertices of odd degree, there is at least one Euler path. One such path follows:

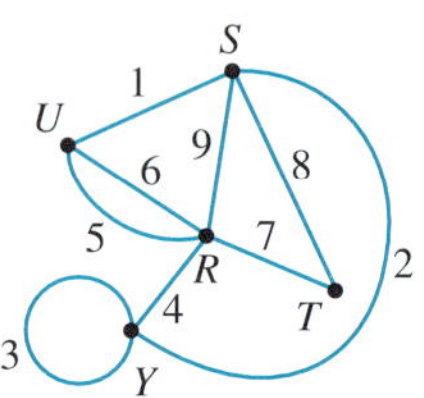

41.

Vertex	Degree
H	5
J	3
K	4
L	4
M	4
N	4

Since there are exactly two vertices of odd degree, there is at least one Euler path. One such path follows:

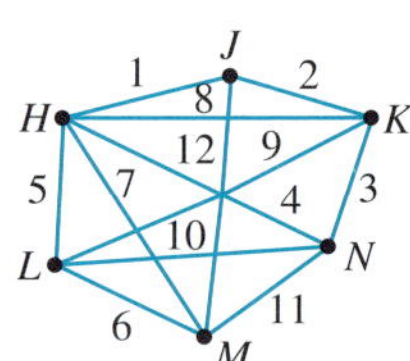

43.

Vertex	Degree
H	2
K	3
L	2
M	1
N	2
O	4
P	2
R	2
S	4

Since there are exactly two vertices of odd degree, there is at least one Euler path. One such path follows:

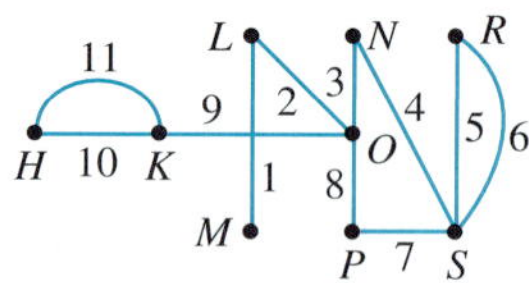

45. There are exactly two vertices of odd degree (B and C)
(a) 0 **(b)** 16
(c) 16 **(d)** 0

47. Let P = Portland, A = Albany, N = Pendelton, B = Burns, and O = Ontario.

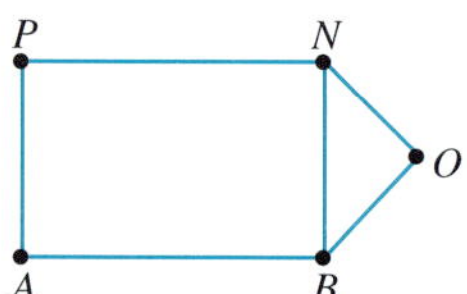

(a) NO, OB, BA, AP, PN, NB

(b) Cities N and B have an odd degree (3), thus there is an Euler path from N to B.

49. There is no Euler path starting at Portland since it has an even degree, thus the courier will have to travel some road twice.

51.

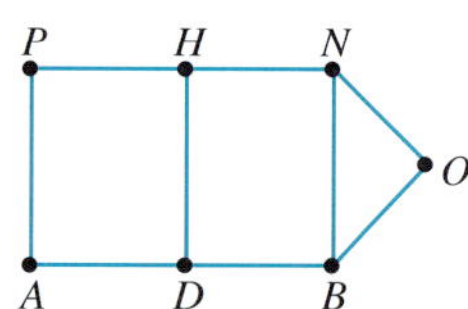

Since D, H, N, and B are of odd degree, there is no Euler path for this network.

53. Hollings Ave. since this would provide exactly two vertices with odd degree, so there would be an Euler path for the cruisers.

55. After we first leave a vertex of odd degree there are an even number of edges remaining, thus every time we come back there is at least one edge leading out.

57. Yes, add a bridge from the left bank to the lower island and a bridge from the right bank to the upper island.

59. Yes. One possibility:

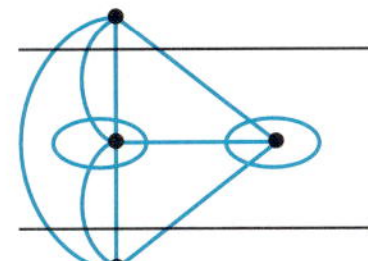

61. No. There are still two vertices of odd degree. There will be an Euler path but no Euler circuit

PROBLEM SET 8.4

1. (a)

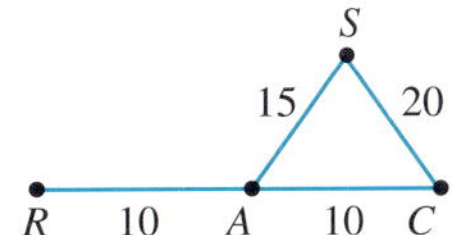

(b)

R 10 A 10 C

3.

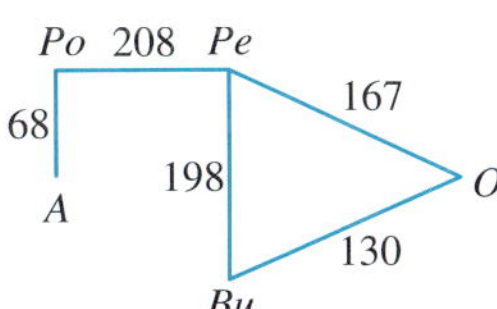

5. Notice that this is a valid graph but is no realistic geographically.

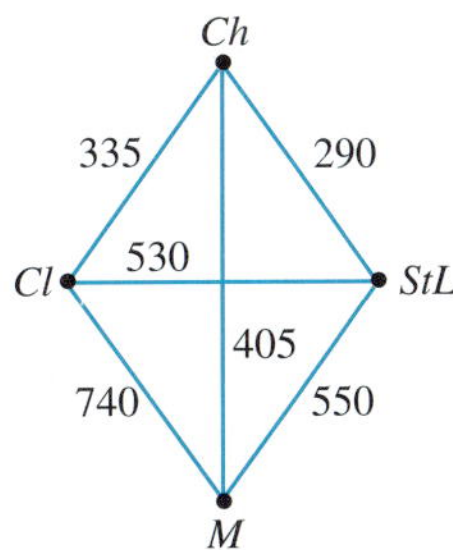

7. Notice that this is a valid graph but is not realistic geographically.

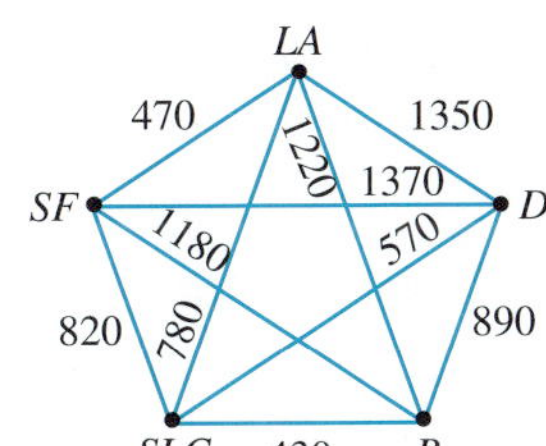

9. (a) Tree **(b)** Not a tree
(c) Not a tree

11.

13.

15. (i) 2
(ii) The top edge cannot be removed.
(iii) 9

17. (i) 1
(ii) The two edges sloping downward from left to right cannot be removed.
(iii) 4

19.

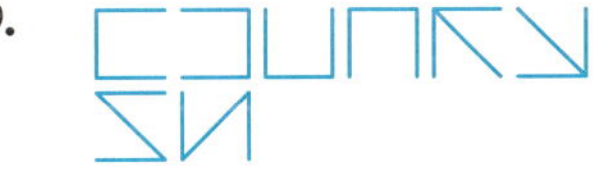

21.

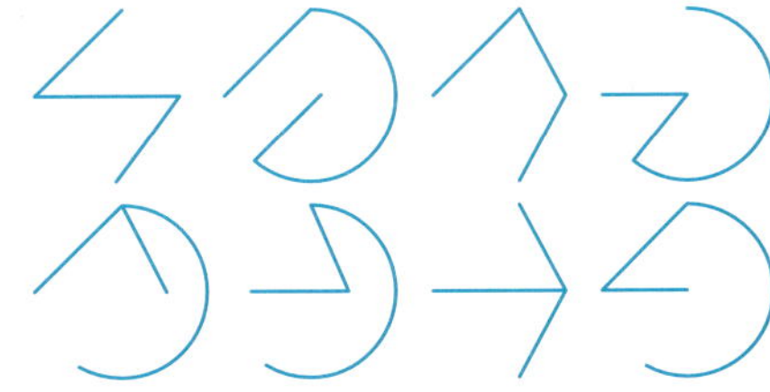

23.

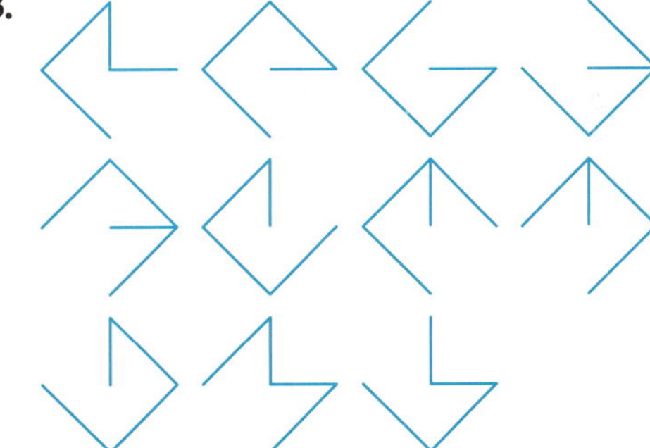

25.

27. **(a)** 30, 35, 40 **(b)** 7, 8, 9, 10

29. 5, 5, 5, 5, 6, 6, 8, 8, 10

31. 16, 15, 13, 12

33. 10, 10, 10, 8, 8, 8, 8, 6, 6

35. Yes. In the example preceeding this problem, the highway inspector's route is an Euler path (each edge is used once, and only once) but not a Hamiltonian path (three cities are visited twice each).

37. C must be visited more than once if all other vertices are visited and the path is a circuit.

39. **(a)** ABDECFA is one such Hamiltonian circuit.

(b) ABCDEFJIHGA is one such Hamiltonian circuit.

41. Nearest Neighbor Algorithm

AEBCDA

$9 + 11 + 12 + 11 + 10 = 53$

BEADCB

$11 + 9 + 10 + 11 + 12 = 53$

CDAEBC

$11 + 10 + 9 + 11 + 12 = 53$

DAEBCD

$10 + 9 + 11 + 12 + 11 = 53$

EADCBE

$9 + 10 + 11 + 12 + 11 = 53$

43. Nearest Neighbor Algorithm

ABFCDEA:

$7 + 10 + 13 + 5 + 21 + 14 = 70$

BAEFCDB:

$7 + 14 + 16 + 13 + 5 + 23 = 78$

CDABFEC:

$5 + 18 + 7 + 10 + 16 + 26 = 82$

DCFBAED:

$5 + 13 + 10 + 7 + 14 + 21 = 70$

EABFCDE:

$14 + 7 + 10 + 13 + 5 + 21 = 70$

FBAEDCF:

$10 + 7 + 14 + 21 + 5 + 13 = 70$

45. *ABCDENMLKJIHGQRSTPOFA*

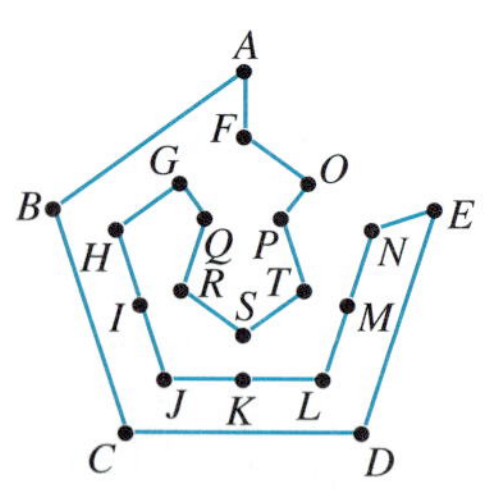

Chapter Eight Review Problems

1.

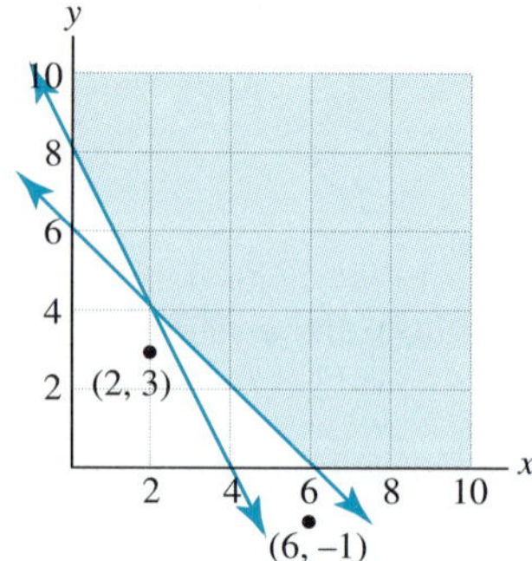

No, No

3. Maximum of $P = 2$ occurs at $(1, 0)$

Minimum of $P = -2$ occurs at $(0, 2)$

5.

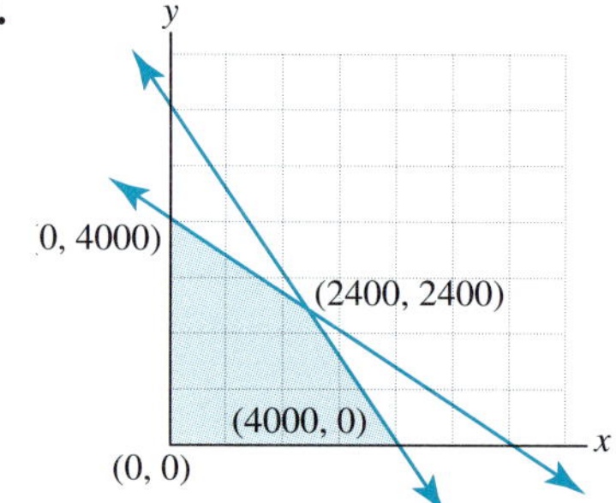

7. $P = \$0.60x + \$0.80y$ where x = amount of creamy and y = amount of chunky. Buy 2400 pounds of each type of peanut. Make 2400 pounds of each type of peanut butter. The maximum profit is $3360.

9. Yes. Each vertex has degree 3. There is no Euler path since there are more than two vertices of odd degree.

11. We can present the path on the graph in which every bridge has to be traveled twice as equivalent to an Euler path on the graph below. Each edge on the original graph has been replaced by two edges. Since every vertex is of even degree there is such a path.

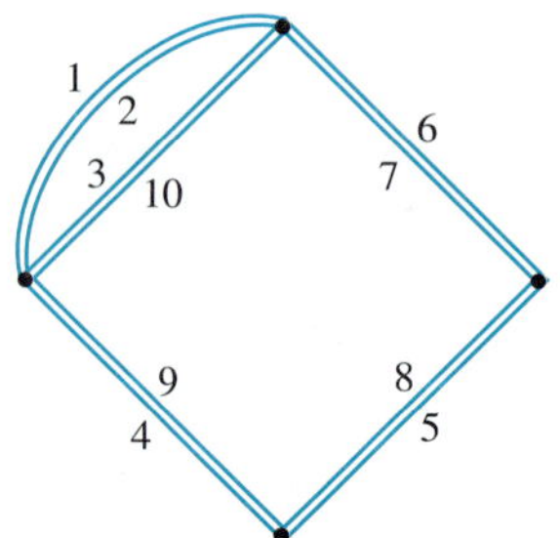

13.

15. AB, BC, BD

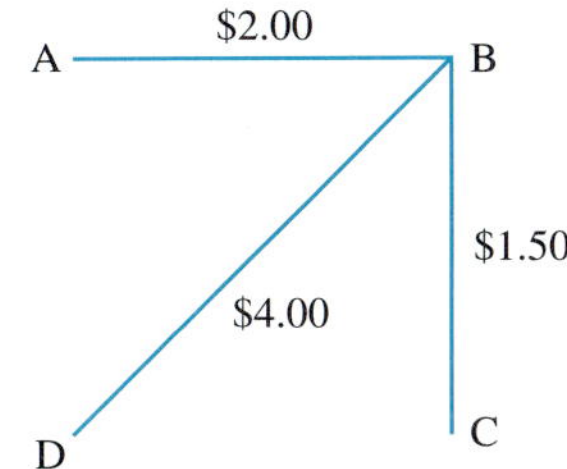

PROBLEM SET 9.1

1. Morrita

3. Davis Ave.

5. Yamada

7. 9th Street

9. Yamada

11. Beca Blvd.

13. Ann

15. Yellowstone

17. Albuquerque

19. Shawna

21. Charming

23. Carmen

25. Vikings, Spartans, Raiders, Titans

27. A

29. B and C tie.

31. C

33. C

35. Mike Griffey

37. Billy Bonds

39. Joe Aaron

41. Billy Bonds

43. Malcolm Adams and Angela Darden

45. **(a)** C
(b) B
(c) A

PROBLEM SET 9.2

1. A majority prefer A; however in the Borda count we have the totals: A–19, B–15, C–20; so C wins the Borda count.

3. One example:

A	1	1	1	3	3
B	2	2	2	1	1
C	3	3	3	2	2

5. *A* wins the plurality method but loses to *B* head-to-head, 5 to 4.

7. One example:

A	1	1	1	1	3	3	3	3	3	3
B	2	2	2	2	1	1	1	2	2	2
C	3	3	3	3	2	2	2	1	1	1

9. **(a)** A **(b)** C

11. **(a)** D **(b)** C

13. Alternative A wins by plurality.
A–3 first place votes
B–2 first place votes
C–2 first place votes

A	1	1	1	3	3	3	3
B	2	2	2	1	1	2	2
C	3	3	3	2	2	1	1

When alternative C is eliminated the table becomes

A	1	1	1	2	2	2	2
B	2	2	2	1	1	1	1.

Alternative B now has a majority of four to three.

15.

A	2	2	2	1	1
B	1	1	1	3	3
C	3	3	3	2	2

Alternative A receives 12 points under Borda count rules, alternative B receives 11 points, and alternative C receives 7 points. Alternative B has a majority of first place votes.

17.

A	2	2	2	1	1
B	1	1	1	3	3
C	3	3	3	2	2

Alternative A receives 12 points under Borda count rules, alternative B receives 11 points, and alternative C receives 7 points. When alternative C is eliminated the preference table becomes

A	2	2	2	1	1
B	1	1	1	2	2.

Alternative A receives 7 points under the Borda count rules, but alternative B receives 8 points, so alternative B is now the winner.

19.

A	1	1	4	4	3
B	3	3	1	1	4
C	4	4	3	3	1
D	2	2	2	2	2

In plurality with elimination, D is eliminated first, then C. A wins, 3 to 2. In head-to-head comparisons, D is the winner.

D beats A 3 to 2

D beats B 3 to 2

D beats C 4 to 1

21. Suppose the 17 voters have the following preferences:

6 voters prefer A to C to B
5 voters prefer B to A to C
4 voters prefer C to B to A
2 voters prefer C to A to B

Using the plurality with elimination method, B has the fewest first place votes and is eliminated. A is preferred to C by an 11 to 6 margin. A is the winner.

Suppose that the two voters who preferred C to A to B change their ballots to show that they prefer A to C to B. Clearly, this benefits A and no one else. The preference tables can now be grouped as follows:

8 voters prefer A to C to B
5 voters prefer B to A to C
4 voters prefer C to B to A

Using the plurality with elimination method, C has the fewest first place votes and is eliminated. B is preferred to A by a 9 to 8 margin. B is the winner.

23. Original table: alternative A wins using plurality with elimination method.

A	1	1	1	1	3	3	3	2	2
B	2	2	2	2	1	1	1	3	3
C	3	3	3	3	2	2	2	1	1

Table after alternative B is removed: Alternative C wins by a majority.

A	1	1	1	1	2	2	2	2	2
C	2	2	2	2	1	1	1	1	1

25.

Choices:	Ballot	Ballot	Ballot	Ballot	Ballot
first	A	A	B	C	D
second	B	C	E	E	E
third	C	D	A	A	A
fourth	D	B	C	D	B
fifth	E	E	D	B	C

Pairwise comparison:

A beats B, C, D for 3 points

B beats C and E for 2 points

C beats D and E for 2 points

D beats B and E for 2 points

E beats only A for 1 point.

27. **(a)** In the run-off between B and C, B receives 7 votes and C receives 2 votes, so B wins the run-off. In the final election between A and B, A receives 4 votes and B receives 5 votes, so B is the overall winner.

(b) If a minimum of 3 of the voters who prefer A over B over C were to vote (insincerely) for C over B in the run-off between B and C, then C would win the run-off. If C wins the run-off, then in the final, A receives 7 votes and C receives 2 votes, making A the overall winner.

29. If A, B first, then C wins

If B, C first, then A wins

If A, C first, then B wins

PROBLEM SET 9.3

1. Jaron–7 bottles
Mikkel–4 bottles
Robert–9 bottles

3. Jaron–6 bottles
Mikkel–5 bottles
Robert–9 bottles

5. **(a)** 37.8
(b) 7.28, 20.29, 12.30, 10.13
(c) 7, 20, 13, 10

7. **(a)** 30.6 **(b)** 8.99, 25.07, 15.20, 12.52
(c) 8, 25, 15, 12

9. Standard divisor: 347
Modified divisor: 320
7.94, 11.19, 4.41, 5.72, 3.28, 7, 11, 4, 5, 3

11. Standard divisor: 362
8.09, 9.17, 3.56, 5.91, 3.26, 8, 9, 4, 6, 3

13. Standard divisor = 10.00637

State	Alg	Geo	Ana	Sto
Pop (1000s)	892	424	664	1162
Std Quota	89.14	42.37	66.36	116.13
Integer Part	89	42	66	116
Fraction Part	0.14	0.37	0.36	0.13
Hamilton Apportion	89	43	66	116

15. Standard divisor = 10.00637
Try d = 9.98. (Other values are possible.)

State	Alg	Geo	Ana	Sto
Pop (1000s)	892	424	664	1162
Mod Quota	89.38	42.48	66.53	116.43
Rounded Quota	89	42	67	116

17. Standard divisor = 7.855
Try d = 7.82. (Other values are possible.)

State	Alg	Geo	Ana	Sto
Pop (1000s)	892	424	664	1162
Mod Quota	114.07	54.22	84.91	148.59
Rounded Quota	114	54	84	148

19. Standard divisor = 30

Course	BA	IA	AA
Pop	130	282	188
Std Quota	4.33	9.4	6.27
Integer Part	4	9	6
Fraction Part	0.33	0.4	0.27
Hamilton Appor	4	10	6

21. Standard divisor = 30
Try d = 29.6. (Other values are possible.)

Course	BA	IA	AA
Pop	130	282	188
Mod Quota	4.39	9.53	6.35
Rounded Quota	4	10	6

23. Standard divisor = 55.3
29, 29, 99, 24, 19

25. Try d = 55.5.
29, 29, 98, 24, 20

27. Try d = 15.4.
29, 54, 14, 34, 21, 28

29. Try d = 15.7.
29, 53, 14, 34, 22, 28

31. Try d = 50,600.
27, 30, 98, 23, 22

33. Try d = 16.
29, 53, 15, 33, 22, 28

35. Divisor = 33,000

State	Mod. Quota	Integer
CT	7.18	7
DE	1.68	1
GA	2.15	2
KY	2.08	2
MD	8.44	8
MA	14.40	14
NH	4.30	4
NJ	5.44	5
NY	10.05	10
NC	10.71	10
PA	13.12	13
RI	2.07	2
SC	6.25	6
VT	2.59	2
VA	19.11	19
		Total 105

37. Try d = 36,100.

State	Mod. Quota	Integer
CT	6.56	7
DE	1.54	2
GA	1.96	2
KY	1.90	2
MD	7.72	8
MA	13.17	14
NH	3.93	4
NJ	4.97	5
NY	9.19	10
NC	9.79	10
PA	11.99	12
RI	1.90	2
SC	5.71	6
VT	2.37	3
VA	17.47	18
		Total 105

PROBLEM SET 9.4

1. The 24 seats apportion as 4, 6, 14 (st. divisor = 156.67).
The 25 seats apportion as 3, 7, 15 (st. divisor = 150.40).

Thus Medina loses a seat and the Alabama paradox occurs.

3. 3, 6, 15 (divisor 145)
3, 7, 15 (divisor 141)
The Alabama paradox does not occur.

5. Original apportionment: 4, 6, 14 using Hamilton's method ($d = 156.67$).
New apportionment: 3, 7, 14 using Hamilton's method ($d = 187.5$).
Yes. Medina (with an increase of 28%) loses a seat, while Alvare (with an increase of 26%) gains a seat.

7. **(a)** $d = 16$, 6, 31, 13
(b) $d = 16.070$; 6, 30, 14, 7
State B loses a seat, state C gains a seat so the new states paradox occurs.

9. Original apportionment: 6, 31, 13 using Jefferson's method ($d = 15.5$).
New apportionment: 6, 31, 13, 7 using Jefferson's method ($d = 15.5$).
No, the new states paradox does not occur.

11. The Jefferson method uses a slightly lower divisor than the standard divisor and rounds down. Thus the modified quotients are higher than the standard quotients so each state's quota is at least as large as the integer part of the standard quota.

13. Standard divisor for 314 = 10,006
Standard divisor for 315 = 9,975
Standard divisor for 316 = 9,943

State	Alg	Geo	Ana	Sto
Pop (1000s)	892	424	664	1162
Rounded Quotas				
for 314	89	43	66	116
for 315	89	43	67	116
for 316	90	42	67	117

Yes, the Alabama paradox occurs. From 315 to 316, Geometria loses and both Algebrion and Stochastics gain.

15. For 314 use $d = 9980$.
For 315 use $d = 9976$.
For 316 use $d = 9970$.

State	Alg	Geo	Ana	Sto
Pop (1000s)	892	424	664	1162
Rounded Quotas				
for 314	89	42	67	116
for 315	89	43	67	116
for 316	89	43	67	117

No

17. Standard divisor for 201 = 49,751
Standard divisor for 202 = 49,505

State	A	B	C	D	E
Pop (1000s)	1320	1515	4935	1118	1112
Std. Quota For 201:	26.53	30.45	99.19	22.47	22.35
Hamilton Appor	27	30	99	23	22

D gains, thus the others lose some influence.
For 202:

Std Quota	26.66	30.60	99.69	22.58	22.46
Hamilton Appor	27	31	100	22	22

B and C gain, thus the others lose some influence, particularly D. The Alabama paradox occurs affecting D.

19. For 201, try 48,887.

State	A	B	C	D	E
Jefferson Appor	27	30	100	22	22

For 202, try 48,870.

State	A	B	C	D	E
Jefferson Appor	27	31	100	22	22

No Alabama paradox here.

21. Standard Divisor = 52,000

Pop (1000s)	1370	1565	5035	1218	1212
Std. Quota	26.35	30.10	96.83	23.42	23.31
Hamilton Appor	26	30	97	24	23
Growth $\left(\frac{\text{New}}{\text{Old}}\right)$	1.04	1.03	1.02	1.0894	1.0899

A grew faster than B, but lost a vote. D and E grew about the same amount, but D gained one on E.

Population paradox occurs.

23. Try $d = 51,800$.

Pop (1000s)	1370	1565	5035	1218	1212
Mod Quota	26.44	30.21	97.20	23.51	23.40
Webster Appor	26	30	97	24	23

Growth $\left(\frac{\text{New}}{\text{Old}}\right)$	1.04	1.03	1.02	1.0894	1.0899

A grew faster than B, but lost a vote. D and E grew about the same amount, but D gained one on E.

Population paradox occurs.

25. Try $d = 9930$.

State	Al	Ge	Aa	St	Co
Pop (1000s)	892	424	664	1162	243
Rounded Quota	89	42	66	117	24

No new states paradox because the same number of seats per state is apportioned as before Computiva was added.

27. 10, 4, 5, 8

29. **(a)** The following inequalities are equivalent and the last one is obvious.

$$\sqrt{n(n+1)} < n + \frac{1}{2}$$
$$n(n+1) < \left(n + \frac{1}{2}\right)^2$$
$$n(n+1) < n^2 + n + \frac{1}{4}$$
$$n^2 + n < n^2 + n + \frac{1}{4}$$
$$0 < \frac{1}{4}$$

(b)
$$n + 0.44 < \sqrt{n(n+1)}$$
$$(n + 0.41)^2 < n(n+1)$$
$$n^2 + 0.82n + 0.1681 < n(n+1)$$
$$0.1681 < 0.18n$$
$$\frac{0.1681}{0.18} < n$$
$$0.93 < n$$

Chapter Nine Review Problems

1. Anne

3. No. We do not know who Claire's voters will support. If Brad gets both votes, there will be a tie.

5. No winner

7. Switch the last two votes to Claire–1, Bob–2, Alice–3.

9. 50.05

11. 6, 8, 12, 14

13. 5, 8, 12, 15

15. 6, 8, 12, 14

PROBLEM SET 10.1

1. **(a)** 1800° **(b)** 1440° **(c)** 2340°

3. **(a)** 2520° **(b)** 3960°

5. 140°

7. 162°

9. 10

11. 15

13. **(a)** Yes **(b)** Yes **(c)** A regular tiling does not have to be edge to edge.

15. One possibility is the following:

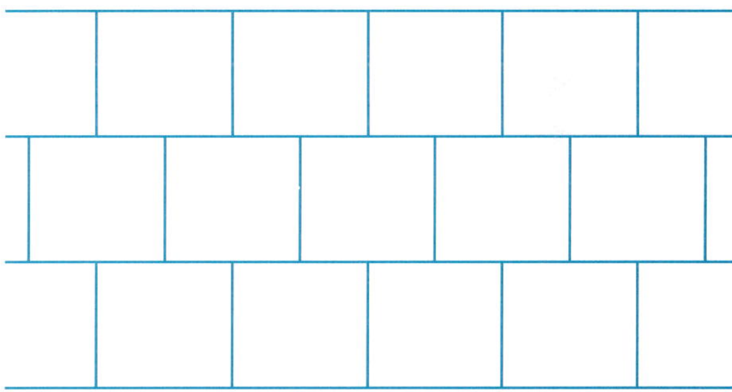

17. The vertex angle measure of a regular 7-gon does not divide into 360° evenly.

19. There are 108° in a vertex angle of a regular pentagon. There are 60° in a vertex angle of an equilateral triangle. There is no way to add multiples of these angles to sum to 360°.

21.

23. No, there are no combinations of multiples of the vertex angles (60° and 135°) that add up to 360°.

25.

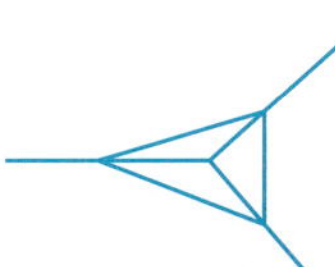

27. No, the vertex figures are different.

29. **(a)** Yes **(b)** No

31. **(a)** 13 **(b)** 10
(c) $\sqrt{586} \approx 24.2$

33. **(a)** Yes **(b)** Yes
(c) No

35. $x \approx 11.0$, $y \approx 10.2$

37. approximately 127.3 feet

39. 15.5 feet

41. The small square in the figure below has a side length of 1.

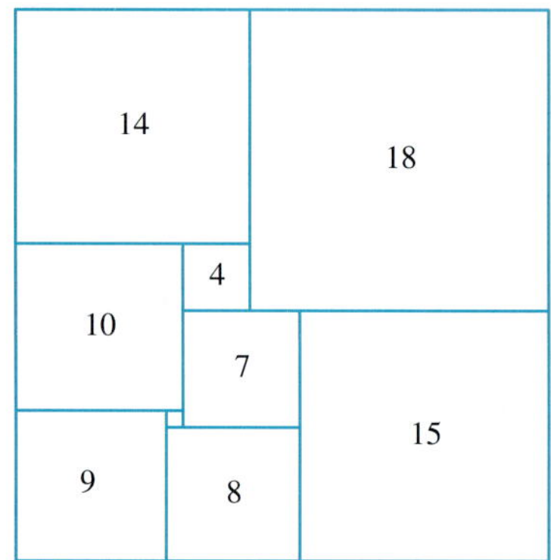

43. **(a)**

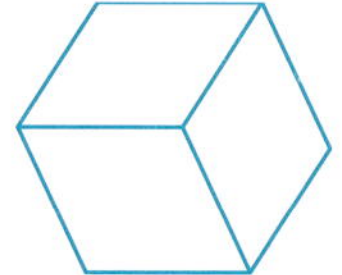

(b)

45. It shows a vertex surrounded by a regular 5-gon, a regular 6-gon, and a regular 7-gon. The sum of a vertex angle measure of each is $108° + 120° + 128\frac{4}{7}°$, which is *not* 360°. At another vertex, there are a regular 5-gon, 6-gon, and 8-gon. The sum of a vertex angle measure of each is $108° + 120° + 135°$; again, this is *not* 360°.

PROBLEM SET 10.2

1. **(a)** translation, rotation.
(b) translation, rotation.
(c) translation
(d) translation, reflection in the horizontal midline, reflections in vertical lines between two dots or two consecutive parentheses [(•|•) or •)|(•]., rotation about a point between two adjacent dots or two consecutive parentheses.

3. Any line that goes through a white or a black "propeller" through the middle of an arrow. There are three sets of such lines.

5. Any line containing the longer diagonal of a rhombus or the longer diagonal of a kite
There are six sets of such lines.

7. Any point at the center of the white or black "propeller": 120°, 240°

9. **(i)** Any point where four black or four white kites touch may be rotated 90°, 180°, or 270°.
(ii) Any point where two black and two white kites touch may be rotated 180°.

11. The vectors indicate four of the many possible translations.

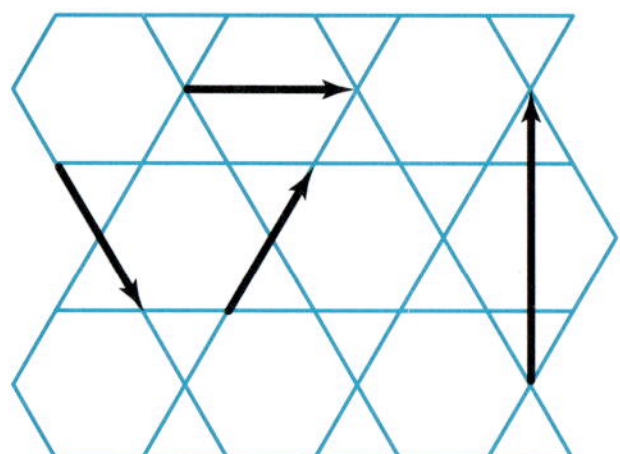

13. The vectors indicate four of the many possible translations.

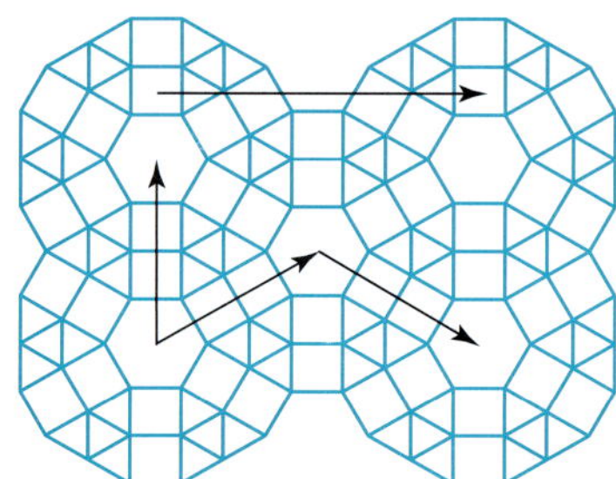

15. Reflections across horizontal, vertical and diagonal lines; rotations by 90°, 180°, or 270° about the center.

17. Reflections across diagonal lines; rotations by 180° about the center.

19. **(a)** Yes, 2
(b) Yes, one of 180° about the center

21. **(a)** Yes. One with the vertical midline.
(b) No

23. **(a)** (i) 5

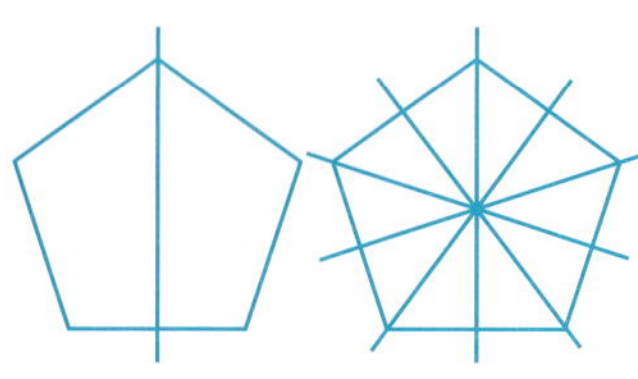

(ii) 6

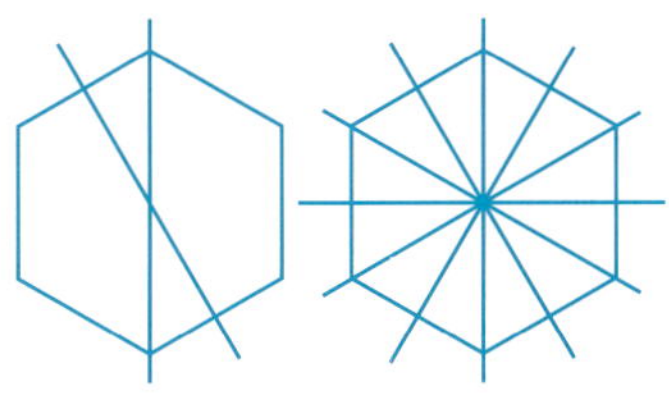

(iii) 8

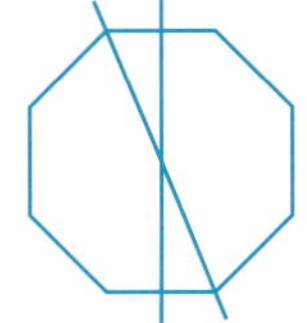

(b) n

25. **(a)** reflection
(b) both
(c) reflection

27. A reflection in the vertical line followed by a translation indicated by the arrow on the right.

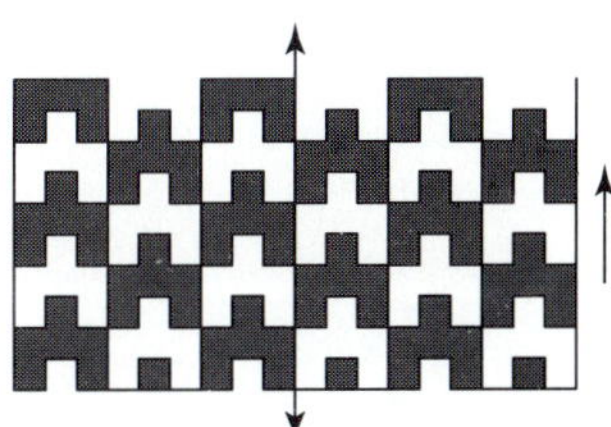

29.

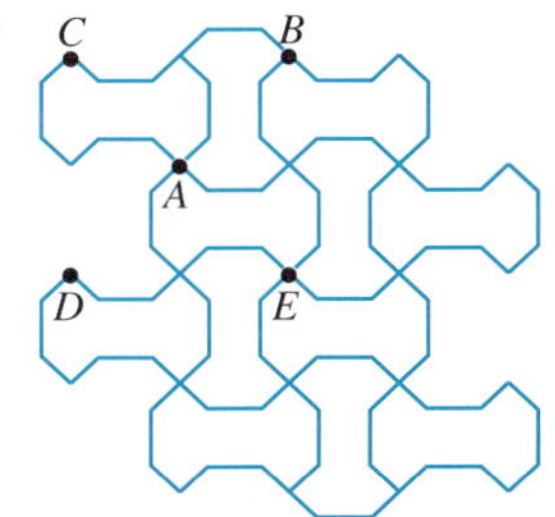

Translations from D to C, from C to B, from A to E and from B to E. Other translations are possible.

31. **(a)**

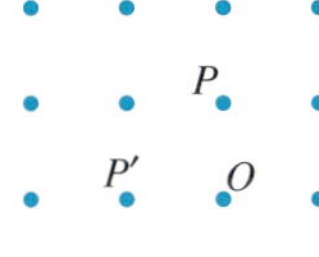

(b) P O P′

(c) P′ P O

33. **(a)**

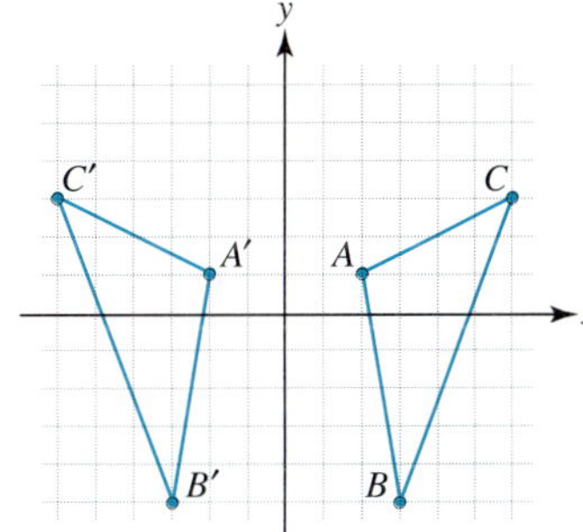

(b) $(-2, 1), (-3, -5), (-6, 3)$
(c) $(-a, b)$

35. **(a)**

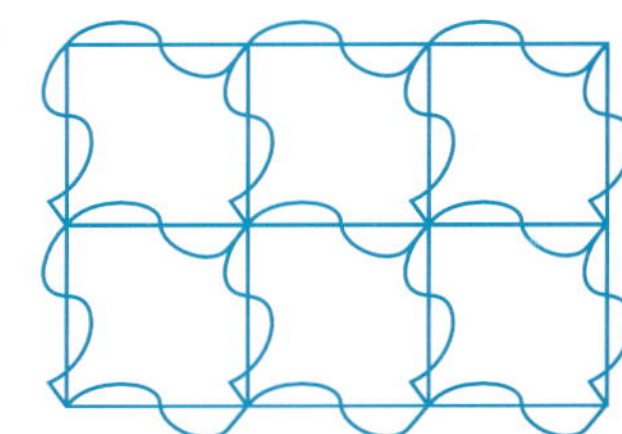

(b)

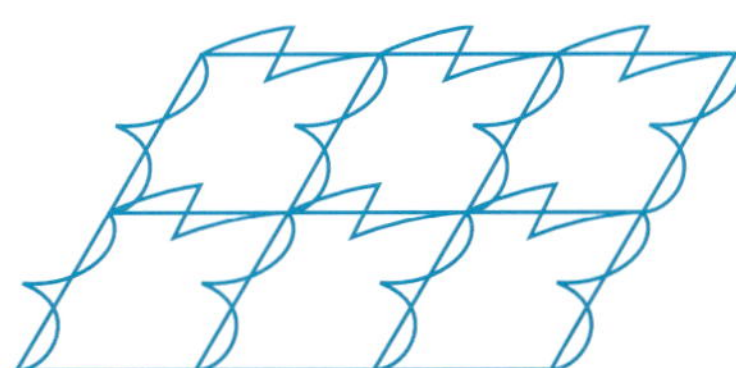

37. Other Escher-type patterns are possible.

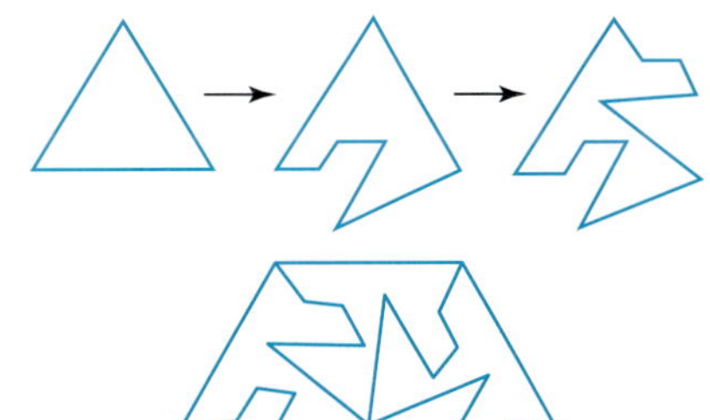

Other tessellations are possible.

PROBLEM SET 10.3

1.

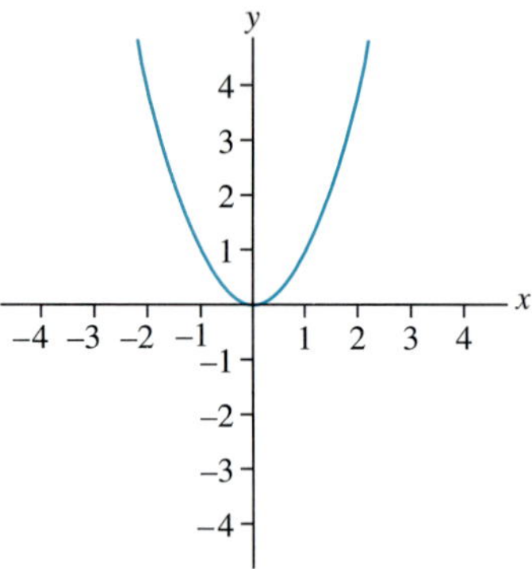

3.

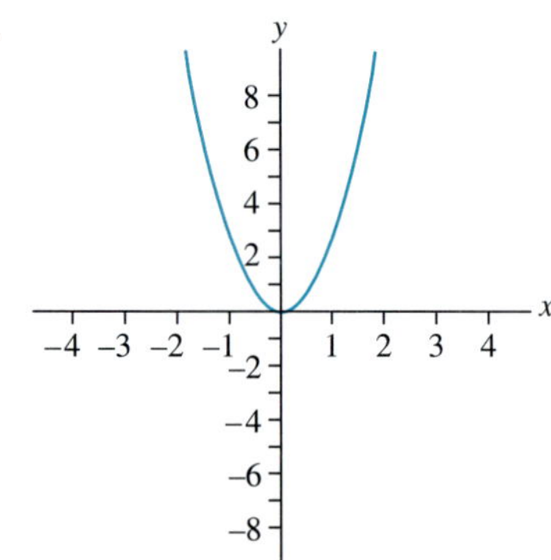

5.

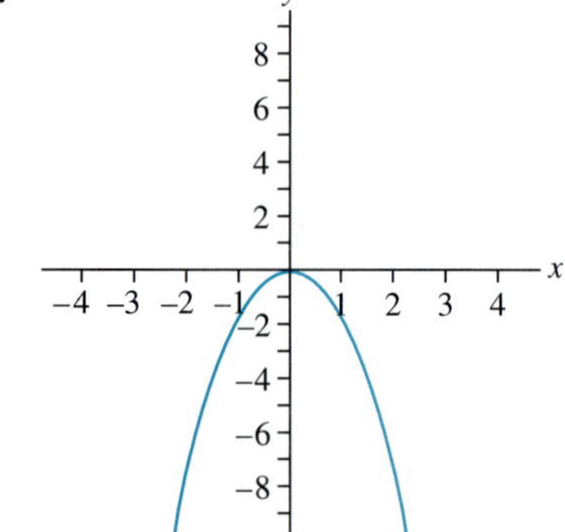

7.

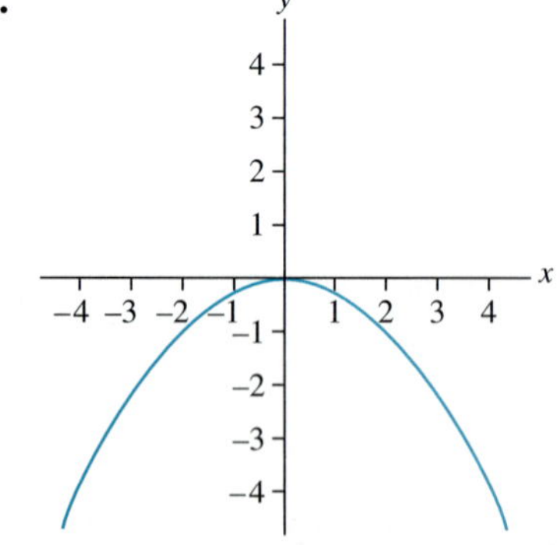

9.

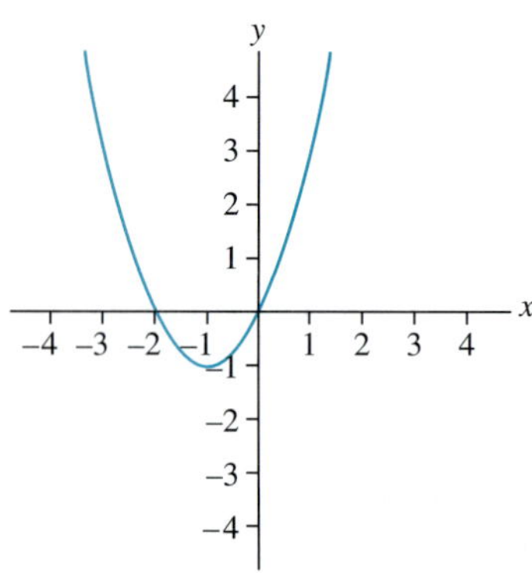

11.

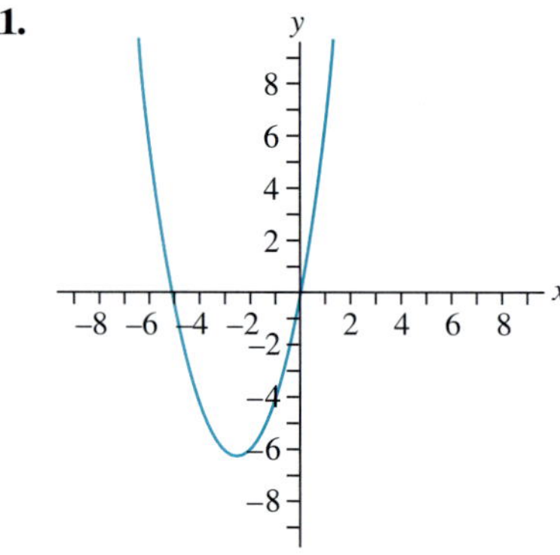

13.

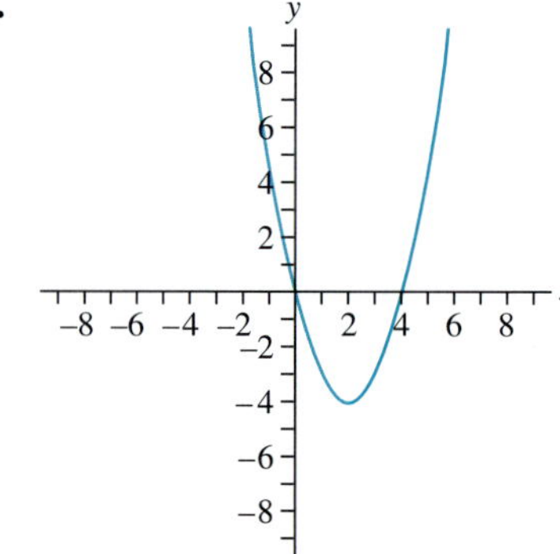

15.

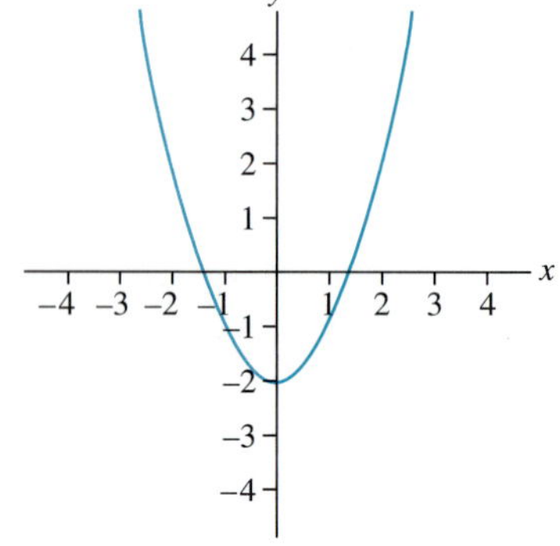

17.

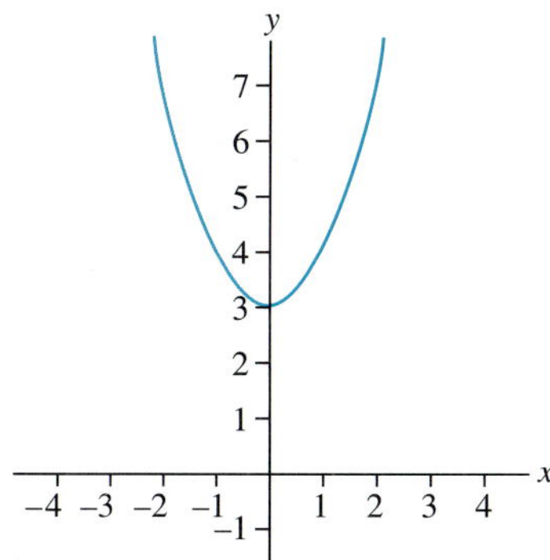

19.

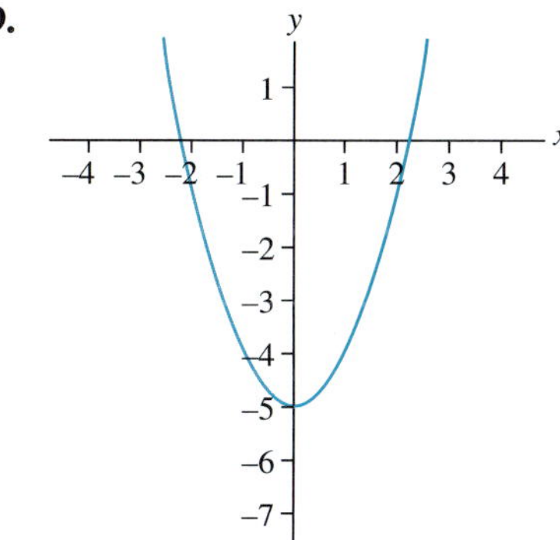

21. $y = \frac{1}{8}x^2$

23. $y = \frac{1}{8}x^2 + 2$

25.

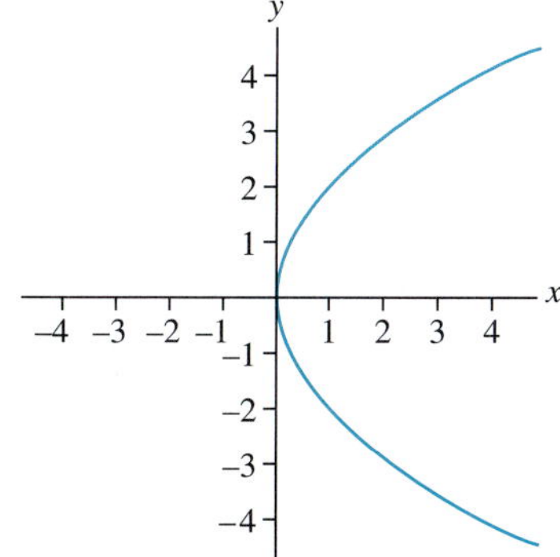

27.

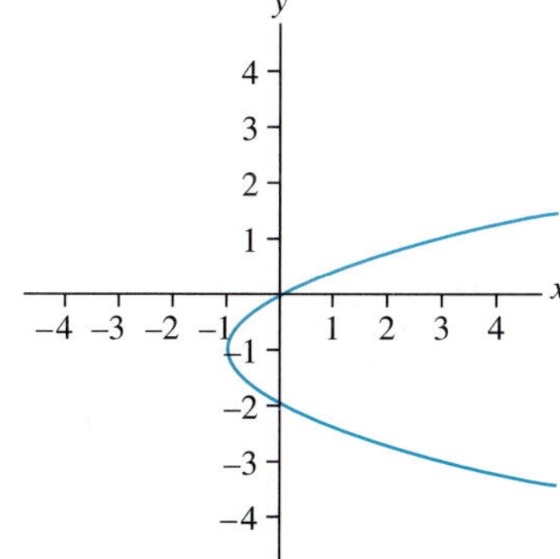

29.

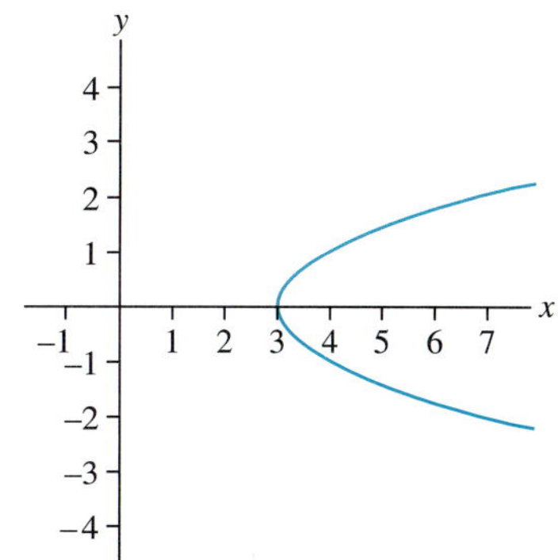

31. **(a)** $(0, 1)$ **(b)** $(0, \frac{1}{12})$
(c) $(0, -\frac{1}{16})$

33. $y = \frac{1}{16}x^2$

35. In the center $1\frac{1}{8}$ ft above the vertex.

37. Two meters to the left of the vertex of the parabola on the right

39. **(a)** 10 ft. **(b)** 154 ft.
(c) approximately 6.1 sec.

41. 30 feet

45. **(a)** $(x + 2)^2 = 12(y - 3)$
or $12y = x^2 + 4x + 40$
$y = \frac{1}{12}x^2 + \frac{1}{3}x + \frac{10}{3}$
(b) $(x - 6)^2 = -12(y + 2)$
or $12y = -x^2 + 12x - 60$
$y = -\frac{1}{12}x^2 + x - 5$

47. **(a)** $(y - 3)^2 = 16(x - 2)$
or $16x = y^2 - 6y + 41$
(b) $(y + 2)^2 = -28(x - 6)$
or $-28x = y^2 + 4y - 164$

PROBLEM SET 10.4

1. **(a)** ellipse
(b) $(0, -\sqrt{21}), (0, \sqrt{21})$
(c) $(2, 0), (-2, 0), (0, 5), (0, -5)$

(d)

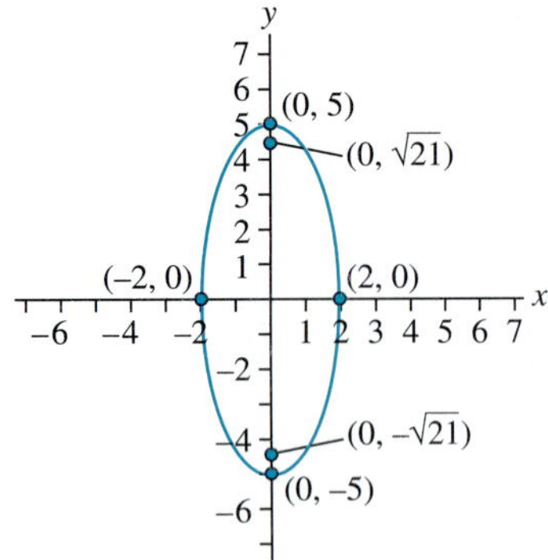

3. (a) ellipse **(b)** $(0,\ -3), (0, 3)$

(c) $(4, 0), (-4, 0)(0, 5), (0,\ -5)$

(d)

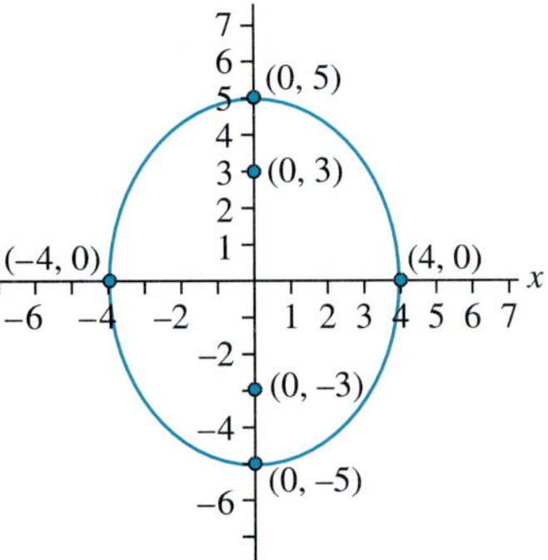

5. (a) hyperbola **(b)** $(\ -5, 0), (5, 0)$

(c) $(4, 0), (\ -4, 0)$

(d)

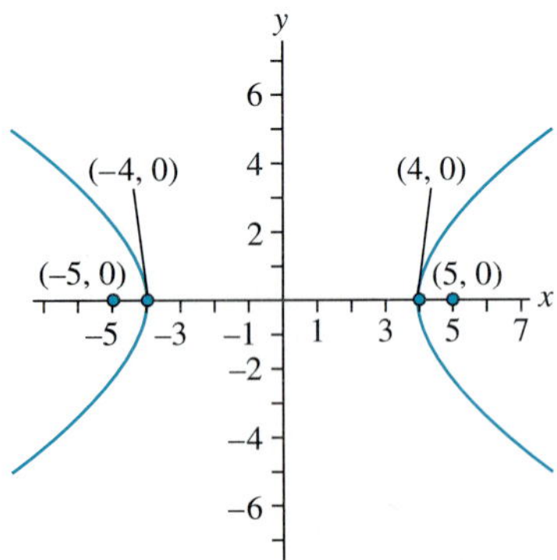

7. (a) hyperbola **(b)** $(0, -\sqrt{14}), (0, \sqrt{14})$

(c) $(0, 2), (0,\ -2)$

(d)

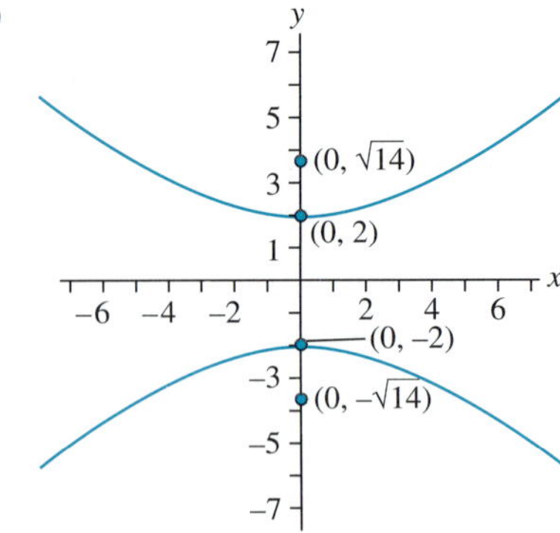

9. (a) ellipse (circle) **(b)** $(0, 0)$

(c) $(4, 0), (\ -4, 0), (0, 4), (0,\ -4)$

(d)

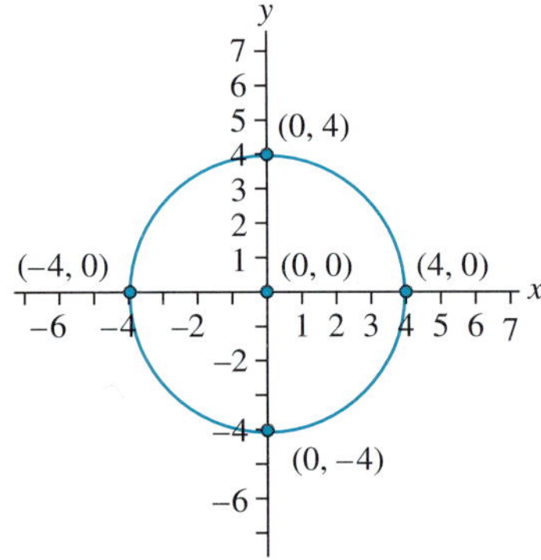

11. $\dfrac{x^2}{100} + \dfrac{y^2}{36} = 1$

13. $\dfrac{4x^2}{25} + \dfrac{y^2}{16} = 1$

15. $\dfrac{y^2}{9} - \dfrac{x^2}{27} = 1$

17. $\dfrac{x^2}{25} - \dfrac{y^2}{24} = 1$

19. $\dfrac{x^2}{25} + \dfrac{y^2}{9} = 1$

21. $\dfrac{x^2}{20} - \dfrac{y^2}{16} = 1$

(a) hyperbola **(b)** $(6, 0), (\ -6, 0)$

(c) $(\sqrt{20}, 0), (-\sqrt{20}, 0)$

(d)

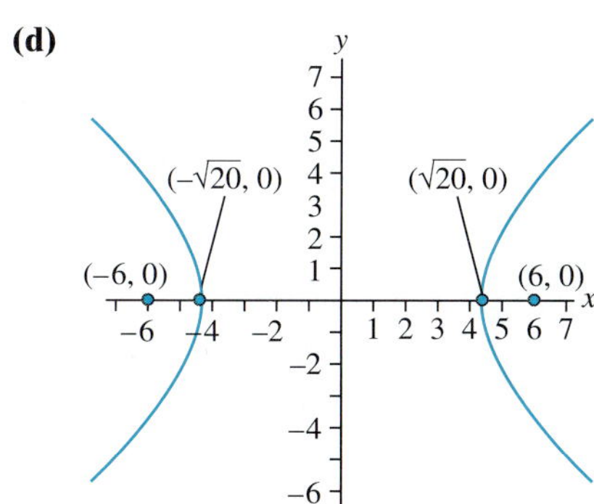

23. $\frac{x^2}{16} + \frac{y^2}{25} = 1$

(i) ellipse

(ii) (0, −3), (0, 3)

(iii) (4, 0), (−4, 0), (0, 5), (0, −5)

(iv)

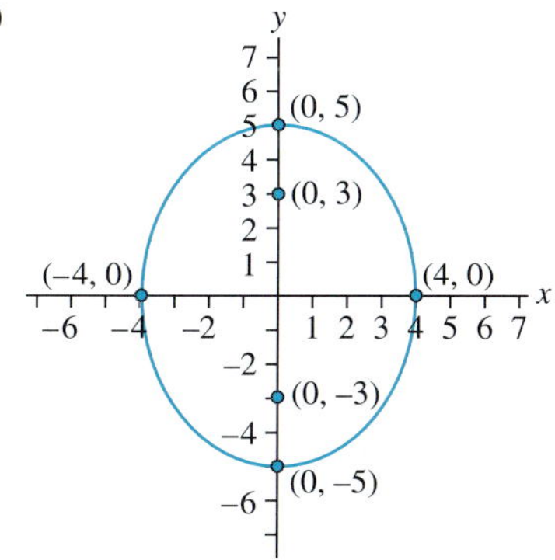

25. $\frac{x^2}{25} + \frac{4y^2}{25} = 1$

(i) ellipse

(ii) $\left(\frac{5}{2}\sqrt{3}, 0\right), \left(-\frac{5}{2}\sqrt{3}, 0\right)$

(iii) $(5, 0), (-5, 0), \left(0, \frac{5}{2}\right), \left(0, -\frac{5}{2}\right)$

(iv)

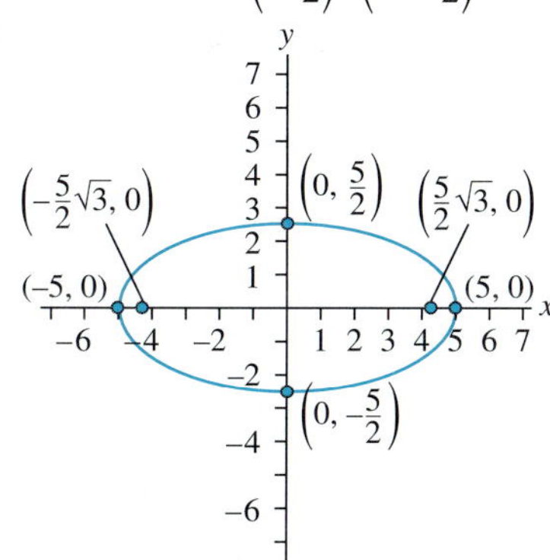

27. They should stand on a line through the vertices; about 3.82 meters to the left of the right vertex and about 3.82 meters to the right of the left vertex.

29. $\sqrt{20} \approx 4.47$m

31. The lightning could have struck anywhere along part of the hyperbola (closest to Hal) having an equation of

$$\frac{x^2}{4(1100)^2} - \frac{y^2}{12(1100)^2} = 1$$

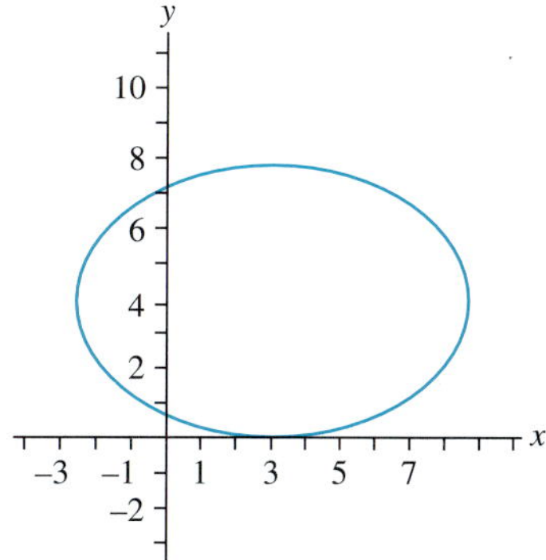

33. $\frac{x^2}{8.649 \times 10^{15}} + \frac{y^2}{8.64675 \times 10^{15}} = 1$

35. 0.016

37. The focus is closer to the origin than the vertex so $0 < c < a$ so $e = \frac{c}{a} < 1$.

39. The ellipse approaches a circle. The eccentricity is 0 if and only if $a = b$.

43. $\frac{(x-3)^2}{36} + \frac{(y-4)^2}{16} = 1$

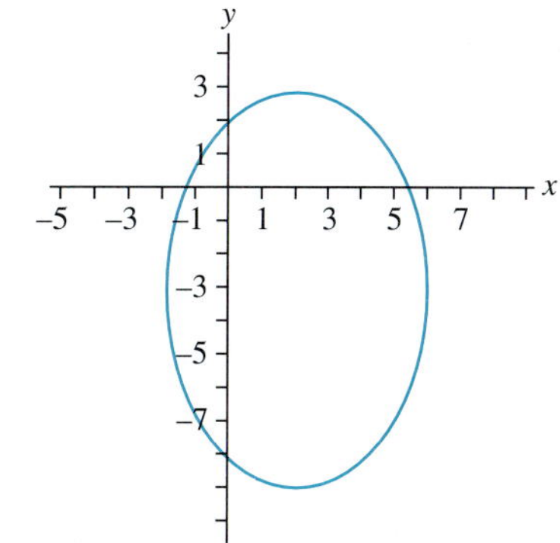

45. $\frac{(x-2)^2}{16} + \frac{(y+3)^2}{36} = 1$

49. $\sqrt{(x+c)^2+y^2} - \sqrt{(x-c)^2+y^2} = 2a$
$\sqrt{(x+c)^2+y^2} = 2a + \sqrt{(x-c)^2+y^2}$

Squaring both sides yields

$(x+c)^2 + y^2 =$
$4a^2 + 4a\sqrt{(x-c)^2+y^2} + (x-c)^2 + y^2$

Next, we isolate the radical

$(x+c)^2 - (x-c)^2 - 4a^2 = 4a\sqrt{(x-c)^2+y^2}$

which simplifies to

$4cx - 4a^2 = 4a\sqrt{(x-c)^2+y^2}$

and dividing both sides by 4.

$cx - a^2 = a\sqrt{(x-c)^2+y^2}$

Squaring both sides yields

$c^2x^2 - 2a^2cx + a^4 = a^2[(x-c)^2 + y^2]$

Simplifying the right hand side yields

$c^2x^2 - 2a^2cx + a^4 = a^2x^2 - 2a^2xc + a^2c^2 + a^2y^2$

which can be arranged as

$c^2x^2 - a^2x^2 - a^2y^2 = a^2c^2 - a^4$
$(c^2 - a^2)x^2 - a^2y^2 = a^2(c^2 - a^2)$

In a hyperbola, $c^2 - a^2 = b^2$.

Substitution gives us

$b^2x^2 - a^2y^2 = a^2b^2$

Dividing both sides by a^2b^2 we obtain

$$\frac{x^2}{a^2} - \frac{y^2}{b^2} = 1$$

Chapter 10 Review Problems

1. 1260°, 140°

3. The tiling is semiregular because because all the vertex figures are the same.

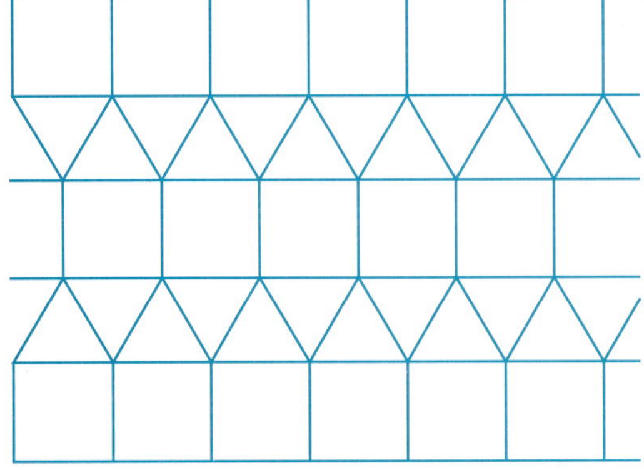

5. 50 feet

7. 5

9. Reflect about a vertical axis and translate one unit upward.

11.

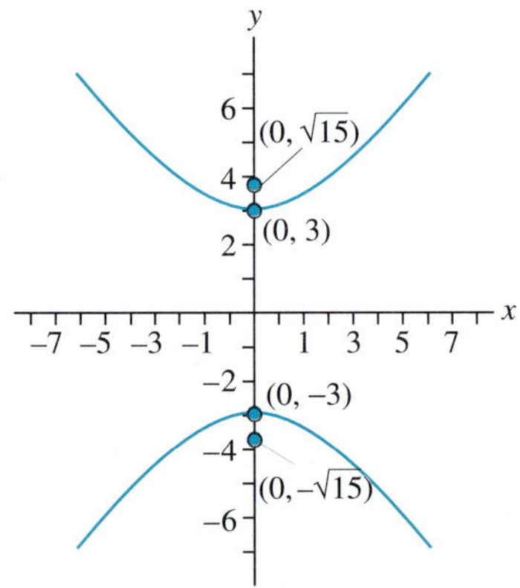

13. $\frac{3}{4}$ ft. (or 9 inches) above the vertex

15. $(0, \sqrt{15}), (0, -\sqrt{15})$

17. $\dfrac{x^2}{4} - \dfrac{y^2}{5} = 1$

PROBLEM SET 11.1

1. **(a)** $\dfrac{12}{7}$

(b) $EF = \dfrac{21}{4}, DF = \dfrac{7}{3}$

3. **(a)** $\dfrac{7}{10}$

(b) $DF = \frac{50}{7}$ and $EF = \frac{80}{7}$

Perimeter $= \frac{200}{7} = 28.57$

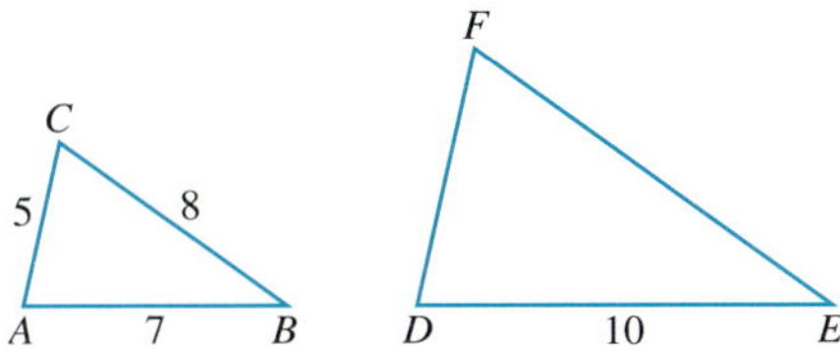

5. (a) 2

(b) Perimeter of $\triangle DEF = 40$

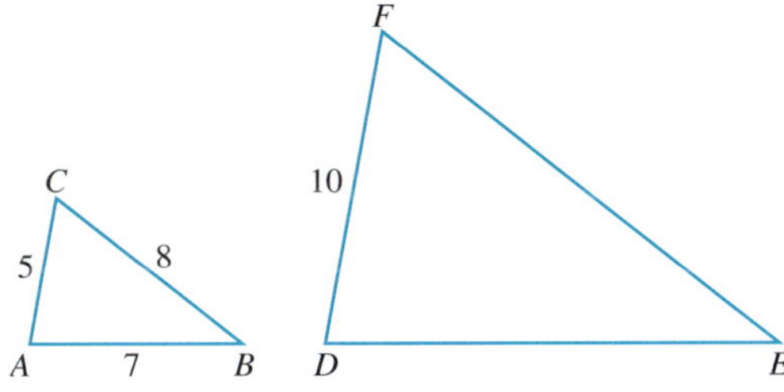

7. (a) $\frac{2}{3}$

(b) $DF = 7.5$, $EF = 12$, $DE = 10.5$

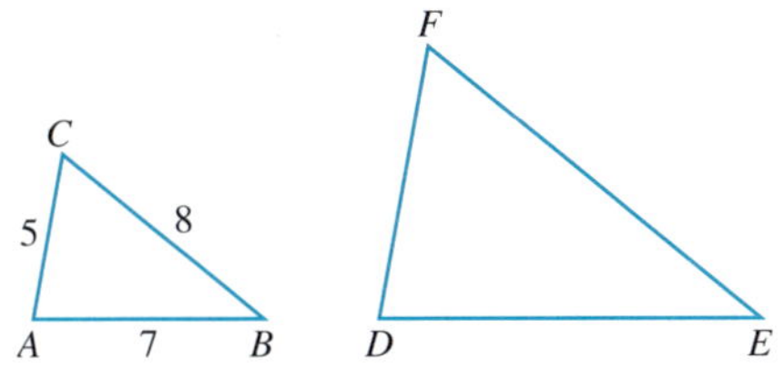

9. $AB = 6$, $EF = 15$, $DE = 9$

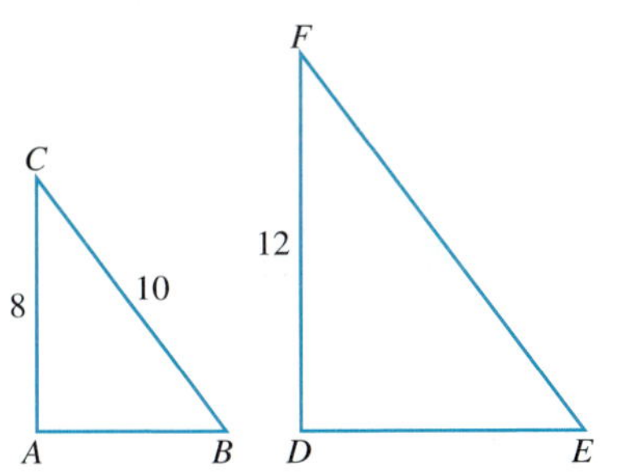

11. (a) Because $\overline{DE}$ is parallel to $\overline{BC}$, $\angle ADE = \angle ABC$ and $\angle AED = \angle ACB$. Triangles with equal angles are similar.

(b) 8

(c) 18

(d) Since the triangles are similar and AD is $\frac{1}{2}$ of AB, the scaling factor is $\frac{1}{2}$. The perimeter of $\triangle ADE$ is 36, so the perimeter of $\triangle ADE$ is $\frac{1}{2}(36) = 18$.

13. (a) Scaling factor is $\frac{7}{12}$

(b) Area $= 52.075$

15. (a) Scaling factor is $\frac{1}{3}$

(b) Area $= 159.48$

17. 104

19. \$337.50

21.

23.

25.

27.

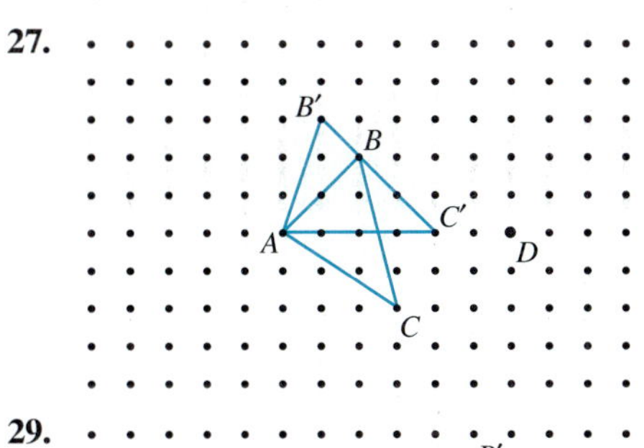

29.

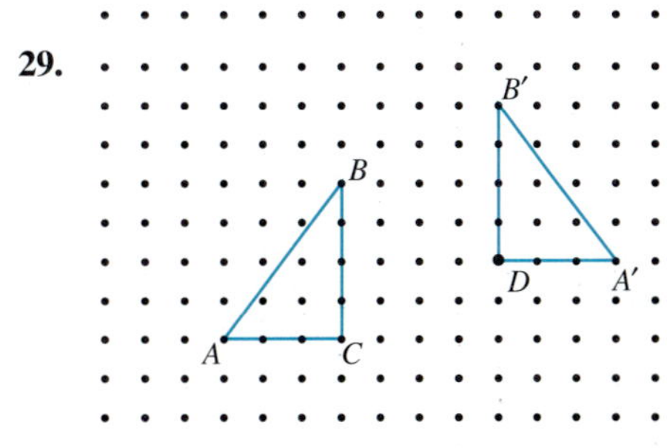

31.

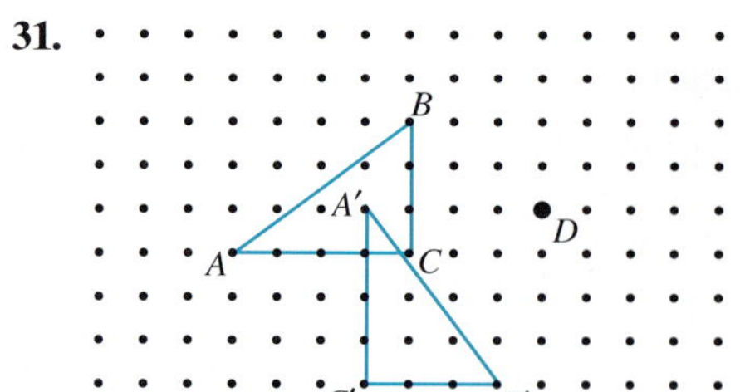

33.

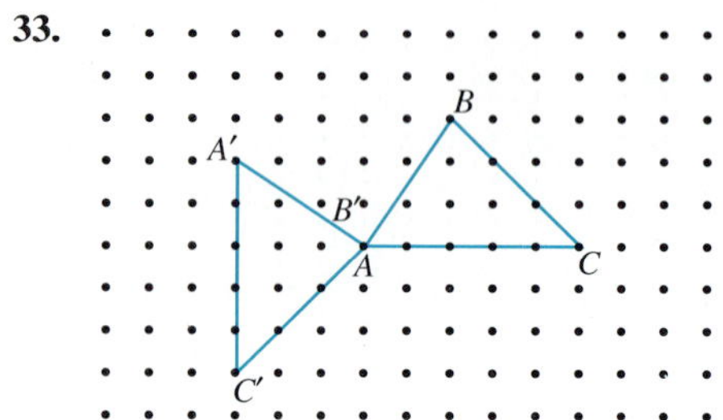

35. 481.25 feet

37. 25

39. **(i)** translate ΔABC 11 units right and 7 units up so that A coincides with D
(ii) reflect across AB
(iii) rotate 90° counterclockwise about A
(iv) shrink toward A with a factor of $\frac{1}{2}$

41. **(i)** reflect across AB
(ii) translate A to D; 11 units right and 3 units down
(iii) dilate by a factor of 3 away from A

PROBLEM SET 11.2

1. $24\sqrt{3}$

3. $\frac{8}{3}\sqrt{3}$

5. 2.77

7. The perimeter is 3 times larger and the area is 9 times larger.

9. 703.125

11. surface area $= 24s^2$, volume $= 8s^3$

13. surface area $= 171$ in^2, volume $= 135$ in^3

15. 11,425,000 ft^3

17. 0.222

19. **(a)** It would be 8 times as large.
(b) $\sqrt[3]{2}$ or about 1.26

21.

The figure keeps expanding outward and shrinking inward.

23.

25. ferns, some trees, many shells

PROBLEM SET 11.3

1. 77,898

3. 43,822 54,183
The second city is 1.24 the size of the first.

5. 314 million

7. At a 3% growth rate, the value is about $1,350,000. At 4%, about $48,000,000.

9. about 90,000

11. about 55,500

13. about 100 years

15. 2.4%

17. 9.57 grams

19. **(a)** 0.234 or 23.4%
(b) about 0.000086 grams

21. **(a)** 0.078 or 7.8%
(b) about 8.7 grams

23. (a)

Season	Population
0	1000
1	750
2	703
3	684
4	675
5	671
6	669
7	668
8	667
9	667
10	667

(b) Population stabilizes at 667.

25. (a)

Season	Population
0	1000
1	1600
2	1024
3	1599
4	1026
5	1599
6	1026
7	1599
8	1026
9	1599
10	1026

(b) Population cycles between the values 1599 and 1026.

27. (a) and (b)

Season	Population (Logistic)	Population (Malthusian)
0	1000	1000
1	800	1600
2	768	2560
3	757	4096
4	753	6554
5	751	10486
6	750	16777
7	750	26844
8	750	42950
9	750	68719
10	750	109951

(c)

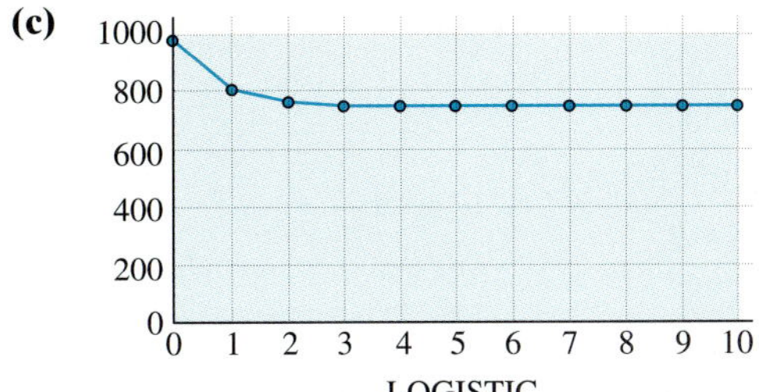

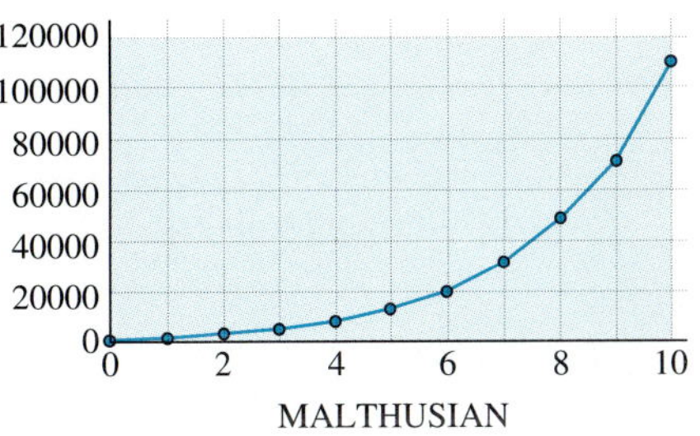

29. (a) and (b)

Season	Population (Logistic)	Population (Malthusian)
0	1000	1000
1	675	1350
2	604	1823
3	569	2460
4	550	3322
5	538	4484
6	531	6053
7	527	8127
8	524	11032
9	522	14894
10	521	20107

(c)

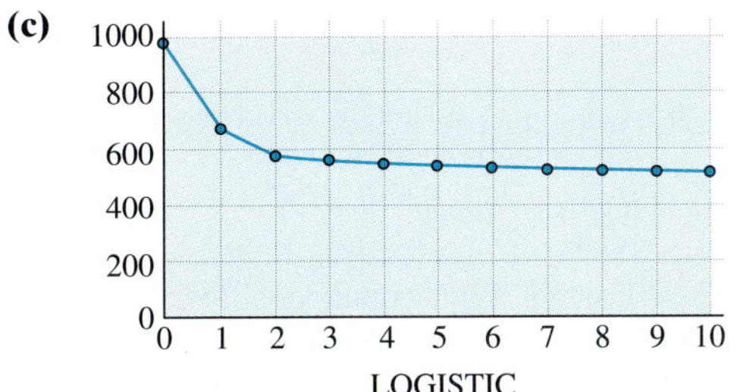

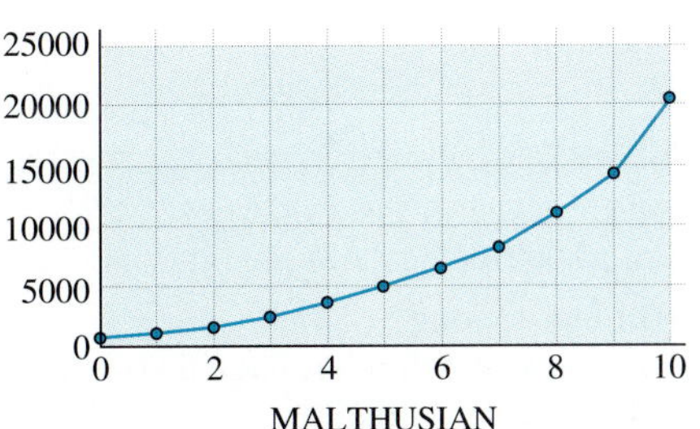

31. **(a)** $c = 7750, r = 1.06667$
(b) 3967
(c) 3500

3967
4002
4000
4000
4000
4000
4000

33. 51 iterations, 12.75 years

37. **(a)** 29%
(b) 8.4%
(c) 0.2%

39. About 30,000 years.

PROBLEM SET 11.4

1. **(a)** 0.289 lb or 4.63 oz.
(b) 13,500 lb

3. 258.3 lb

5. 1.45 cu ft

7. **(a)** 1200 lb
(b) 408 lb

9. 257 lb, or about the same as Olajuwon

11. The mass of Earth is about 49 times the mass of the moon.

13. Earth has a higher average density. The scaling factor for volume is $(\frac{3760}{3964})^3$ or 0.853. The volume of Venus is 85.3% of Earth's, but the mass is only 81.5% of Earth's. With equal densities, the ratio of the masses would also have this value.

15. 3.47 lb/sq in.

17. $453\frac{1}{3}$ lb/ sq ft

19. They are the same.

21. 16.7 psi

23. 4.125 in., 0.0001628 gallons

25. **(a)** 87 times as long
(b) $87^3 = 658{,}503$ times as large a volume.
(c) 658,503 times as heavy

27. 1.554 sq. ft of living space, the pool is 0.521 m or 1.709 ft long

CHAPTER ELEVEN REVIEW PROBLEMS

1. The scaling factor is $\frac{1}{3}$.

The sides of $\triangle ABC$ are 5, 7, 3.

The perimeter of $\triangle DEF = 45$, and the perimeter of $\triangle ABC = 15$

3. 60 ft

5.

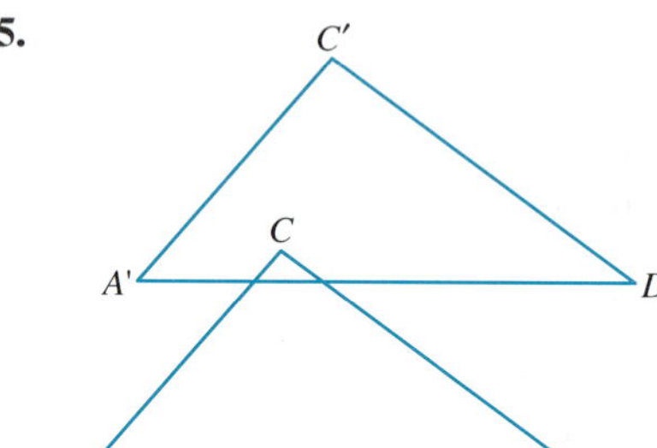

7.

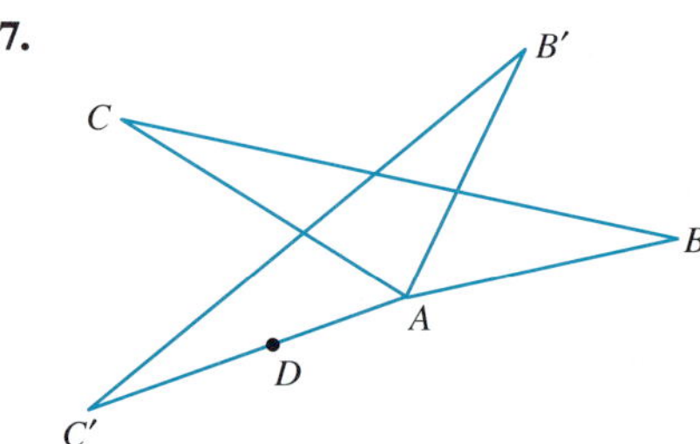

9. **(i)** reflect across AC
(ii) rotate about C 90° clockwise
(iii) translate C to F

Note: the order of the steps may vary.

11. 9.15

13. 10 years: 168,182
50 years: 205,316
100 years: 263,467

15. $43,178.50

17. 24.2%

19. 20,800

21. 96,640

23. 2.24 cubic feet

25. 944 psi

PROBLEM SET 12.1

1. 2, 5, 7, 12, 19, 31, 50, 81

3. 1, 2, 4, 7, 13, 24, 44, 81

5. 4, 3, 11, 17, 39, 73, 151, 297

7. **(a)** (i) 1, 3, 4, 7, 11, 18, 29, 47, 76, 123, 199, 322
(ii) 1.6180905

(b) (i) 3, 1, 4, 5, 9, 14, 23, 37, 60, 97, 157, 254
(ii) 1.6178344

9. **(a)** (i) 4, 3, 7, 10, 17, 27, 44, 71, 115, 186, 301, 487
(ii) 1.6179402
(b) (i) 1, −5, −4, −9, −13, −22, −35, −57, −92, −149, −241, −390
(ii) 1.6182573

11. **(a)** (i) 0, 2, 4, 6, 8, 10, 12, 14, 16, 18
(ii) (add 2 each time)
(b) (i) 3, 0, −3, −6, −9, −12, −15, −18, −21, −24
(ii) (add −3 each time)

13. **(a)** 1, −5, −11, −17, −23, −29, −35, −41, −47, −53 (add −6 each time)
(b) −2, 3, 8, 13, 18, 23, 28, 33, 38, 43 (add 5 each time)

15. **(a)** 0, 1, 1, 2, 4, 7, 13, 24, 44, 81, 149, 274, 504
(b) 1, 0, 1, 2, 3, 6, 11, 20, 37, 68, 125, 230, 423

17. 1, 4, 11, 26, 57, 120, 247, 502, 1013, 2036

19. 0, 1, 4, 9, 18, 33, 58, 99, 166, 275

21. 1, 3, 7, 15, 31, 63, 127, 255, 511, 1023

23. 1, 0.6667, 0.75, 0.7273, 0.7333, 0.7317, 0.7321, 0.7320, 0.7321, 0.7320

25. **(a)** 1, 1.4, 1.56, 1.624, 1.6496, 1.6598
(b) Stable point attractor is 1.6667.

27. **(a)** 1, 0.6, 0.76, 0.696, 0.7216, 0.7114
(b) Stable point attractor is 0.7143.

29. **(a)** 1, 4.4, 5.76, 6.304, 6.5216, 6.6086
(b) Stable point attractor is 6.6667.

31. (i) 0, 1, 1.7321, 2.1128, 2.2860, 2.3605, 2.3919, 2.4049, 2.4104, 2.4126
(ii) Yes
(iii) $x = \sqrt{1 + 2x}$, $x = 1 + \sqrt{2}$

33. (i) 1, 2.3333, 2.1429,2.1556, 2.1546, 2.1547, 2.1547, 2.1547, 2.1547, 2.1547,
(ii) Yes
(iii) $x = 2 + \dfrac{1}{3x}$, $x = 1 + \dfrac{2}{3}\sqrt{3}$

35. (i) 0, 1, 0.5, 0.6667, 0.6, 0.625, 0.6154, 0.6190, 0.6176, 0.6182
(ii) Yes
(iii) $x = \dfrac{1}{1 + x}$, $x = \dfrac{-1 + \sqrt{5}}{2}$

37. (i) 0, 1.2599, 1.7946, 1.9473, 1.9867, 1.9967, 1.9992, 1.9998, 1.9999, 1.99999
(ii) Yes
(iii) $x = \sqrt[3]{2 + 3x}$, $x^3 - 3x - 2 = 0$

39. **(a)** 8, 8.69, 8.66, 8.66
(b) 25, 26.5, 26.46, 26.46

41. **(a)** 18, 18.722, 18.708, 18.708
(b) 40, 45, 44.722, 44.721

43. $e_1 = 2$, $e_{10} = 2.5937$, $e_{100} = 2.7048$, $e_{1000} = 2.7169$, $e_{10000} = 2.7181$, $e_{100000} = 2.7182682$, $e = 2.7182818$

PROBLEM SET 12.2

1. $20 + 10 + 5 + \dfrac{5}{2} + \dfrac{5}{4} + \ldots$

3. $16 + 24 + 36 + 54 + 81 + \ldots$

5. $1 - 0.6 + 0.36 - 0.216 + 0.130 + \ldots$

7. $2000 + 300 + 45 + 6.75 + 1.013 + \ldots$

9. $r = 0.8$, $10 + 8 + 6.4 + 5.12 + 4.096$

11. $r = -0.5$, $80 - 40 + 20 - 10 + 5$

13. 8

15. $5\frac{1}{3}$

17. 12

19. $r = \dfrac{4}{3} > 1$ so sum does not exist.

21.

Stage	Number of Segments	Length of Segments	Total Length
0	1	1	1
1	2	$\frac{1}{3}$	$\frac{2}{3}$
2	4	$\frac{1}{9}$	$\frac{4}{9}$
3	8	$\frac{1}{27}$	$\frac{8}{27}$
4	16	$\frac{1}{81}$	$\frac{16}{81}$

23.

25.

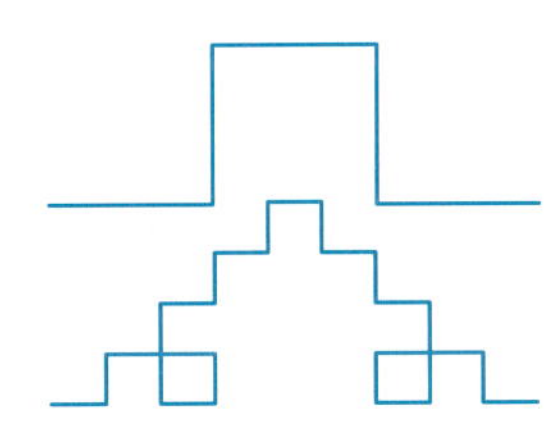

27.

29. **(a)**

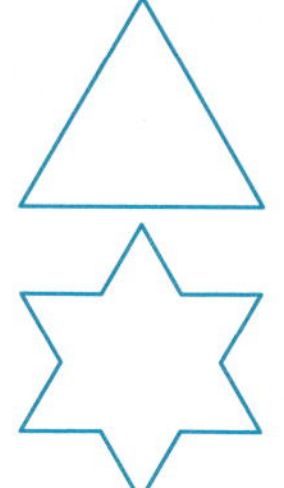

(b)

31.

Stage	Number of Triangles	Area of Each	Total Area
0	1	1	1
1	3	$\frac{1}{4}$	$\frac{3}{4}$
2	9	$\frac{1}{16}$	$\frac{9}{16}$
n	3^n	$\frac{1}{4^n}$	$\frac{3^n}{4^n}$

33. Note: When the length of the side is scaled down by a factor of r, the area is scaled down by a factor of r^2

Stage	Number of New Triangles	Area of Each New Triangle	Added Area	Total Area
0	1	1	1	1
1	3	$\frac{1}{9}$	$\frac{3}{9}$	$\frac{12}{9}$
2	12	$\frac{1}{81}$	$\frac{12}{81}$	$\frac{120}{81}$
3	48	$\frac{1}{729}$	$\frac{48}{729}$	$\frac{1128}{729}$
n	$3 \times 4^{n-1}$	$\frac{1}{3^{2n}}$	$\frac{4^{n-1}}{3^{2n-1}}$	

At stage n, for n at least 2, there are $3 \times 4^{n-1}$ triangles added, each with an area of $\frac{1}{3^{2n}}$

The total area can be expressed as follows:

$1 + \frac{3}{9} + \frac{12}{81} + \frac{48}{729} + \ldots$

This is 1 followed by a geometric series that has a common ratio of $\frac{4}{9}$.

That is, $1 + \frac{3}{9} + \frac{12}{81} + \frac{48}{729} + \ldots$

$= 1 + \left[\frac{3}{9} + \frac{12}{81} + \frac{48}{729} + \ldots\right]$.

The total area of the snowflake is $1 + \dfrac{\frac{3}{9}}{1 - \frac{4}{9}}$

$= 1 + \dfrac{\frac{3}{9}}{\frac{5}{9}} = 1 + \frac{3}{5} = \frac{8}{5} = 1.6$

35.

Stage	Number of Line Segments	Length of Each Segment	Total Length
0	4	1	4
1	20	$\frac{1}{3}$	$\frac{20}{3}$
2	100	$\left(\frac{1}{3}\right)^2$	$\frac{100}{9}$
3	500	$\left(\frac{1}{3}\right)^3$	$\frac{500}{27}$
n	4×5^n	$\left(\frac{1}{3}\right)^n$	$\frac{4 \times 5^n}{3^n}$

As n increases, the length of the border increases without limit.

37.

Stage	Number of Line Segments	Length of Each Segment	Total Length
0	4	1	4
1	20	$\frac{1}{3}$	$\frac{20}{3}$
2	100	$\left(\frac{1}{3}\right)^2$	$\frac{100}{9}$
3	500	$\left(\frac{1}{3}\right)^4$	$\frac{500}{27}$
n	4×5^n	$\left(\frac{1}{3}\right)^{2n-2}$	$\frac{4 \times 5^n}{3^{2n-2}}$

Although the curve seems quite different than the one in problem 35, the length of the border at each stage is the same. As n increases, the length of the border increases without limit.

39.

Stage	Number of Line Segments	Length of Each Segment	Total Length
0	4	1	4
1	32	$\frac{1}{4}$	8
2	256	$\left(\frac{1}{4}\right)^2$	16
3	2048	$\left(\frac{1}{4}\right)^3$	32
n	4×8^n	$\left(\frac{1}{4}\right)^n$	4×2^n

As n increases, the length of the border increases without limit.

PROBLEM SET 12.3

1. 1.893

3. 1.465

5. 1.262

7. Colorado, New Mexico, Utah, and Wyoming

9. Answers may vary.

11. Texas

13. (a)

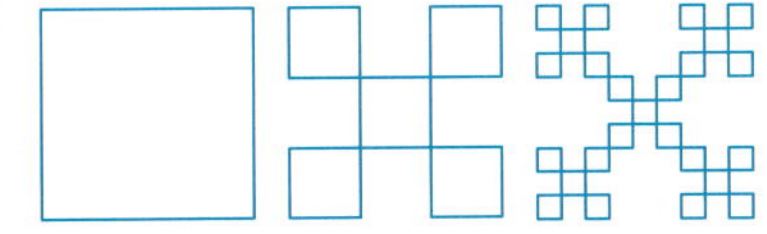

(b)

Stage	Number of Segments	Length of Segments	Total Length
0	4	1	4
1	20	$\frac{1}{4}$	5
2	100	$\frac{1}{16}$	$\frac{100}{16}$
3	500	$\frac{1}{64}$	$\frac{500}{64}$
4	2500	$\frac{1}{256}$	$\frac{2500}{256}$

General rule: Length $= 4\left(\frac{5}{4}\right)^n$

(c) $k = 1.465$.

15. (a)

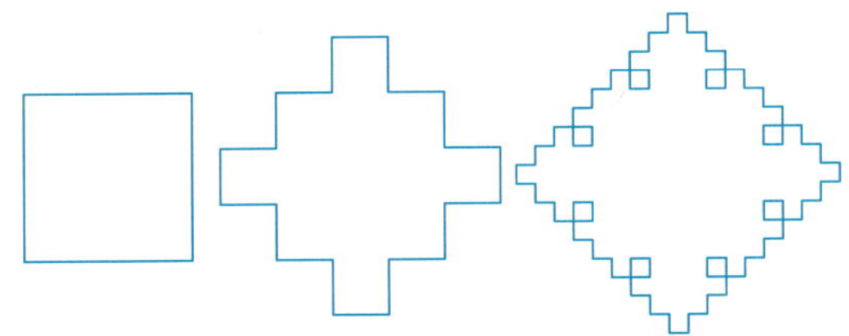

(b)

Stage	Number of new squares	Area of each new square	Total new area
0	1	1	1
1	4	$\frac{1}{9}$	$\frac{4}{9}$
2	20	$\frac{1}{81}$	$\frac{20}{81}$
3	100	$\frac{1}{729}$	$\frac{100}{729}$
4	500	$\frac{1}{6561}$	$\frac{500}{6561}$

General rule for nth stage:

Area $= 1 + \frac{4}{9} + \frac{4}{9}\left(\frac{5}{9}\right) + \frac{4}{9}\left(\frac{5}{9}\right)^2 + \ldots + \frac{4}{9}\left(\frac{5}{9}\right)^{n-1}$

(c) $k = 2.335$.

17. (a)

(b) If the area of the initiator is 1, the area of the nth stage is also 1. No new area is added. Rather, the existing area is re-arranged.

(c) $k = 2$.

Chapter Twelve Review Problems

1. 3, 4, 7, 11, 18, 29, 47, 76, 123, 199

3. 1, 3, 5, 7, 9, 11, 13; the odd numbers

5. (a) 0, 2, 2.6667, 2.8889, 2.9630, 2.9877
   (b) $x = 2 + \frac{1}{3}x$, $x = 3$

7.

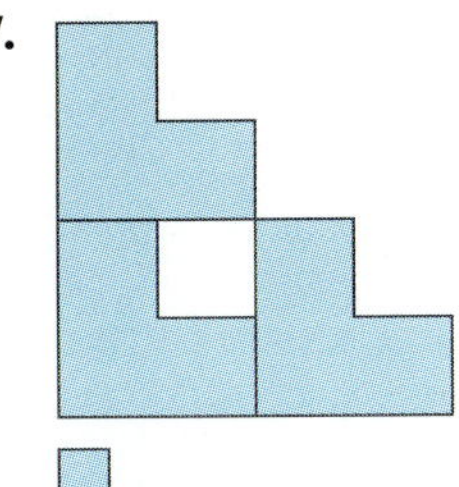

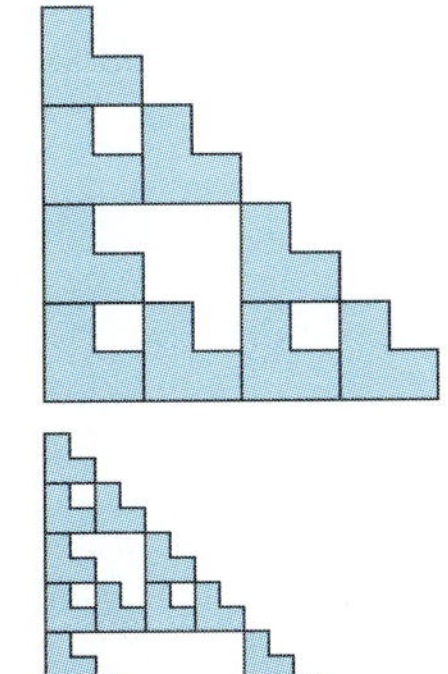

9. 1.585

TOPIC PROBLEM SETS

T1

1. (a) not well-defined
   (b) well-defined
   (c) well-defined
   (d) well-defined

3. (a) {1, 4, 9, 16, 25}
   (b) {California, Oregon, Washington, Alaska, Hawaii}
   (c) {2, 3, 5, 7, 11, 13, 17, 19, 23, 29}
   (d) {12, 15, 18, 21, . . . }

5. (a) T
   (b) F
   (c) T
   (d) T

7. **(a)** T
(b) T
(c) T
(d) T
(e) T

9. **(a)** $A = \{2, 6, 10, 12\}$
(b) $B = \{3, 6, 9, 12, 15\}$
(c) $C = \{9\}$
(d) $D = U = \{2, 3, 6, 9, 10, 12, 15\}$

11. Ø, {a}, {b}, {c}, {a, b}, {a, c}, {b, c}, {a, b, c}

13. **(a)** T (as of 1995)
(b) F
(c) T
(d) T
(e) T

15. **(a)**

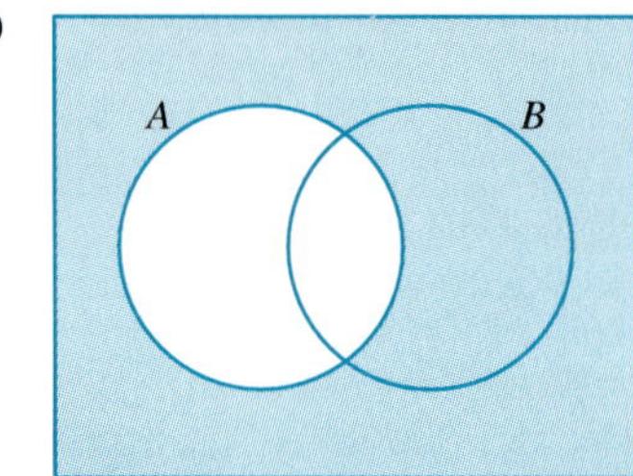

(b)

(c)

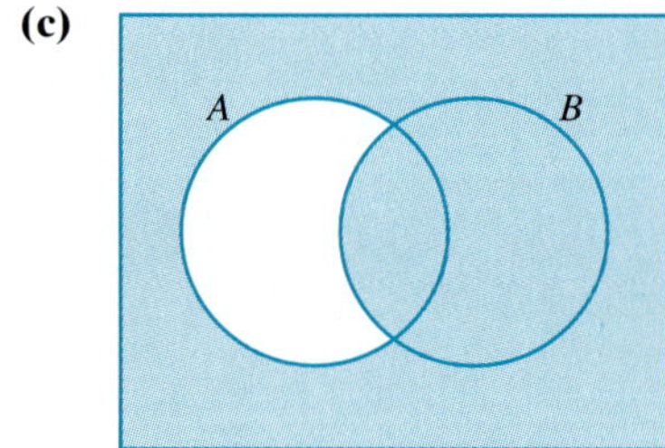

(d)

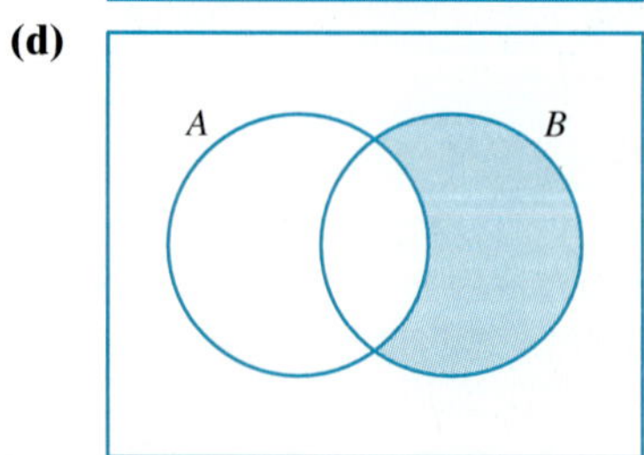

17. **(a)**

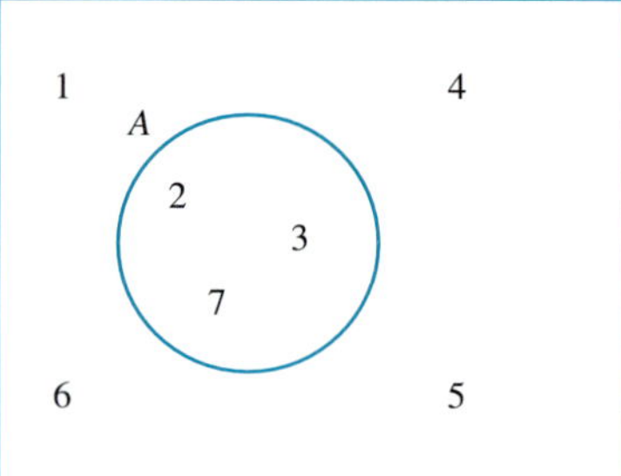

(b)

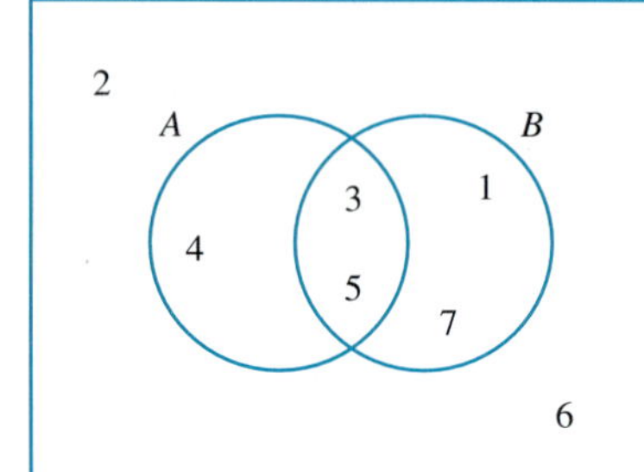

(c)

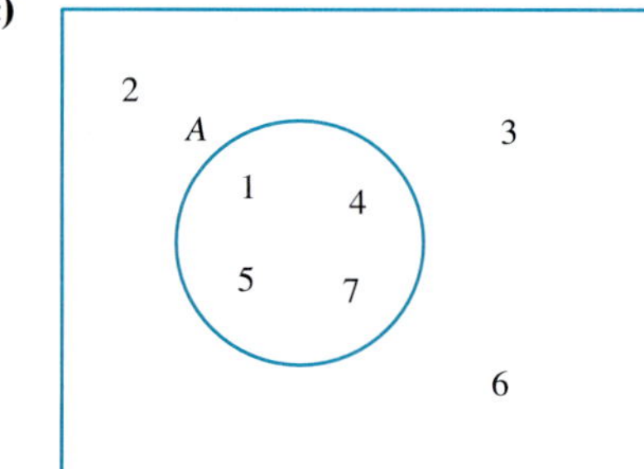

T2

1. 524.226
3. 403.01
5. 162.592
7. 14.6215
9. 2.4739
11. 304.5025
13. 32,089.8649
15. 3.3437
17. 13.6189
19. 154.6336
21. 1214.6716
23. 0.07763
25. 19.8229
27. 22,627.241

T3

1. (i)–(e)
(ii)–(c)
(iii)–(d)
(iv)–(a)
(v)–(b)
(vi)–(f)

3. **(a)** 6.54 m
(b) 3200 m
(c) 0.6 m
(d) 35,000 m
(e) 4m

5. **(a)** 2000 mm
(b) 425 cm
(c) 85 dm
(d) 4.53 dm

7. **(a)** 0.1 sq m
(b) 12,200,000 cm^2
(c) 270 hm^2
(d) 0.0285 dm^2

9. **(a)** 28 mm
(b) 148 cm
(c) 473 mL
(d) 10 L

11. **(a)** 107.95 cm
(b) 50.16 m^2
(c) 2.8317 m^3
(d) 66.2445 L
(e) 3.772 kg

13. **(a)** 6.557 ft
(b) 914.96 ft^2
(c) 130.8 yd^3
(d) 2.84 fl oz
(e) 140.814 lb

T4

1. not a solution

3. $(23, -43)$

5. $(7, 3)$

7. $(1, 1)$

9. $(1, -3)$

11. $(4, -4)$

13. infinitely many solutions
$(-1, 6)$, $(0, 4)$, and $(1, 2)$ are particular solutions

15. infinitely many solutions
$(0, 10)$, $(2, 8)$, and $(4, 4)$ are particular solutions

T5

1. **(a)** 30.6 in.
(b) 30 m
(c) 21.9 cm
(d) 40 ft

3. 46.57 units

5. 94.25 cm

7. **(a)** 96 cm^2
(b) 74.165 cm^2
(c) 112 in^2

9. 102.566 square units

11. $6\sqrt{3}$

T6

1. 432 ft^2, 576 ft^3

3. 193 in^2, 156 in^3

5. 15.59 square units, 3.18 cubic units

7. 544.8 in^3

9. **(a)** 1206 square units
(b) 4825 square units

11. 17.44 cm^2

13. **(a)** 615.8 cm^2
(b) 923.6 cm^2
(c) 498.2 cm^2

15. **(a)** 1141.3 cm^2, 2957.8 cm^3
(b) 275.8 cm^2, 342.1 cm^3

17. **(a)** 452 square units, 905 cubic units
(b) 1810 square units, 7238 cubic units

19. 59 cm^2; 31 cm^3

Photograph Credits

Chapter 1, page 4, top, *George Boole,* Mary Evans Picture Library; bottom, *Lewis Carroll,* Hulton Deutsch Collection Limited.

Chapter 2, page 56, *Gerbert of Aurillac (Pope Sylvester II),* Mary Evans Picture Library.

Chapter 3, page 122, top, *John Playfair,* The Granger Collection; bottom, *John W. Tukey,* Plasma Physics Laboratory, Princeton University.

Chapter 4, page 188, top, *William Sealy Gosset,* The Granger Collection; bottom, *Ronald A. Fischer,* courtesy of Waukesha College.

Chapter 5, page 236, top, *David Blackwell,* courtesy of George M. Bergman, University of California, Berkeley; bottom, *Marilyn vos Servant,* Jarvik Research Incorporated.

Chapter 6, page 278, top, *Charles Dow,* courtesy of Dow Jones & Company, Inc.; bottom, *Ralph Nelson Elliot,* courtesy of Elliot Wave International.

Chapter 7, page 322, top, *John von Neumann,* Bettmann Archives, New York; bottom, *Bertrand Arthur William Russell,* Hulton Deutsch Collection Limited.

Chapter 8, page 362, top, *George Dantzig,* Stanford University, Department of Operations Research, bottom, *Ronald L. Graham,* courtesy of AT & T.

Chapter 9, page 422, top, *The Marquis de Condorcet,* The Granger Collection; bottom, *Kenneth J. Arrow,* courtesy of Stanford University.

Chapter 10, page 472, top, *Marjorie Rice,* courtesy of Marjorie Rice; bottom, *Hypatia,* Bettmann Archives, New York.

Chapter 11, page 538, top, *Emmy Noether,* courtesy of Bryn Mawr College Photo Archives; bottom, *R. Buckminster Fuller,* courtesy of the Buckminster Fuller Institute.

Chapter 12, page 588, top, *Edward Lorenz,* courtesy of the Scripps Institution of Oceanography; bottom, *Benoit Mandelbrot,* courtesy of the IBM, Thomas J. Watson Research Center.

INDEX

C

D

N

O

P

Q

R

S

T